AF735463

SECOND EDITION

International Pharmaceutical Product Registration

edited by

Anthony C. Cartwright
Pharmaceutical Regulatory Consultant
Harpenden, U.K.

Brian R. Matthews
Alcon Laboratories
Hemel Hempstead, U.K.

informa
healthcare

New York London

DRUGS AND THE PHARMACEUTICAL SCIENCES
A Series of Textbooks and Monographs

Executive Editor

James Swarbrick
PharmaceuTech, Inc.
Pinehurst, North Carolina

Advisory Board

Larry L. Augsburger
University of Maryland
Baltimore, Maryland

Jennifer B. Dressman
University of Frankfurt Institute of Pharmaceutical Technology
Frankfurt, Germany

Anthony J. Hickey
University of North Carolina School of Pharmacy
Chapel Hill, North Carolina

Ajaz Hussain
Sandoz
Princeton, New Jersey

Joseph W. Polli
GlaxoSmithKline
Research Triangle Park
North Carolina

Stephen G. Schulman
University of Florida
Gainesville, Florida

Yuichi Sugiyama
University of Tokyo, Tokyo, Japan

Geoffrey T. Tucker
University of Sheffield
Royal Hallamshire Hospital
Sheffield, United Kingdom

Harry G. Brittain
Center for Pharmaceutical Physics
Milford, New Jersey

Robert Gurny
Universite de Geneve
Geneve, Switzerland

Jeffrey A. Hughes
University of Florida College of Pharmacy
Gainesville, Florida

Vincent H. L. Lee
US FDA Center for Drug Evaluation and Research
Los Angeles, California

Kinam Park
Purdue University
West Lafayette, Indiana

Jerome P. Skelly
Alexandria, Virginia

Elizabeth M. Topp
University of Kansas
Lawrence, Kansas

Peter York
University of Bradford School of Pharmacy
Bradford, United Kingdom

For information on volumes 1–151 in the *Drugs and Pharmaceutical Science* Series, Please visit www.informahealthcare.com

152. Preclinical Drug Development, *edited by Mark C. Rogge and David R. Taft*
153. Pharmaceutical Stress Testing: Predicting Drug Degradation, *edited by Steven W. Baertschi*
154. Handbook of Pharmaceutical Granulation Technology: Second Edition, *edited by Dilip M. Parikh*
155. Percutaneous Absorption: Drugs–Cosmetics–Mechanisms–Methodology, Fourth Edition, *edited by Robert L. Bronaugh and Howard I. Maibach*
156. Pharmacogenomics: Second Edition, *edited by Werner Kalow, Urs A. Meyer and Rachel F. Tyndale*
157. Pharmaceutical Process Scale-Up, Second Edition, *edited by Michael Levin*
158. Microencapsulation: Methods and Industrial Applications, Second Edition, *edited by Simon Benita*
159. Nanoparticle Technology for Drug Delivery, *edited by Ram B. Gupta and Uday B. Kompella*
160. Spectroscopy of Pharmaceutical Solids, *edited by Harry G. Brittain*
161. Dose Optimization in Drug Development, *edited by Rajesh Krishna*
162. Herbal Supplements-Drug Interactions: Scientific and Regulatory Perspectives, *edited by Y. W. Francis Lam, Shiew-Mei Huang, and Stephen D. Hall*
163. Pharmaceutical Photostability and Stabilization Technology, *edited by Joseph T. Piechocki and Karl Thoma*
164. Environmental Monitoring for Cleanrooms and Controlled Environments, *edited by Anne Marie Dixon*
165. Pharmaceutical Product Development: In Vitro-In Vivo Correlation, *edited by Dakshina Murthy Chilukuri, Gangadhar Sunkara, and David Young*
166. Nanoparticulate Drug Delivery Systems, *edited by Deepak Thassu, Michel Deleers, and Yashwant Pathak*
167. Endotoxins: Pyrogens, LAL Testing and Depyrogenation, Third Edition, *edited by Kevin L. Williams*
168. Good Laboratory Practice Regulations, Fourth Edition, *edited by Anne Sandy Weinberg*
169. Good Manufacturing Practices for Pharmaceuticals, Sixth Edition, *edited by Joseph D. Nally*
170. Oral-Lipid Based Formulations: Enhancing the Bioavailability of Poorly Water-soluble Drugs, *edited by David J. Hauss*
171. Handbook of Bioequivalence Testing, *edited by Sarfaraz K. Niazi*
172. Advanced Drug Formulation Design to Optimize Therapeutic Outcomes, *edited by Robert O. Williams III, David R. Taft, and Jason T. McConville*
173. Clean-in-Place for Biopharmaceutical Processes, *edited by Dale A. Seiberling*
174. Filtration and Purification in the Biopharmaceutical Industry, Second Edition, *edited by Maik W. Jornitz and Theodore H. Meltzer*
175. Protein Formulation and Delivery, Second Edition, *edited by Eugene J. McNally and Jayne E. Hastedt*
176. Aqueous Polymeric Coatings for Pharmaceutical Dosage Forms, Third Edition, *edited by James McGinity and Linda A. Felton*
177. Dermal Absorption and Toxicity Assessment, Second Edition, *edited by Michael S. Roberts and Kenneth A. Walters*
178. Preformulation Solid Dosage Form Development, *edited by Moji C. Adeyeye and Harry G. Brittain*
179. Drug-Drug Interactions, Second Edition, *edited by A. David Rodrigues*
180. Generic Drug Product Development: Bioequivalence Issues, *edited by Isadore Kanfer and Leon Shargel*
181. Pharmaceutical Pre-Approval Inspections: A Guide to Regulatory Success, Second Edition, *edited by Martin D. Hynes III*
182. Pharmaceutical Project Management, Second Edition, *edited by Anthony Kennedy*
183. Modified Release Drug Delivery Technology, Second Edition, Volume 1, *edited by Michael J. Rathbone, Jonathan Hadgraft, Michael S. Roberts, and Majella E. Lane*

184. Modified-Release Drug Delivery Technology, Second Edition, Volume 2, *edited by Michael J. Rathbone, Jonathan Hadgraft, Michael S. Roberts, and Majella E. Lane*
185. The Pharmaceutical Regulatory Process, Second Edition, *edited by Ira R. Berry and Robert P. Martin*
186. Handbook of Drug Metabolism, Second Edition, *edited by Paul G. Pearson and Larry C. Wienkers*
187. Preclinical Drug Development, Second Edition, *edited by Mark Rogge and David R. Taft*
188. Modern Pharmaceutics, Fifth Edition, Volume 1: Basic Principles and Systems, *edited by Alexander T. Florence and Jürgen Siepmann*
189. Modern Pharmaceutics, Fifth Edition, Volume 2: Applications and Advances, *edited by Alexander T. Florence and Jürgen Siepmann*
190. New Drug Approval Process, Fifth Edition, *edited by Richard A. Guarino*
191. Drug Delivery Nanoparticulate Formulation and Characterization, *edited by Yashwant Pathak and Deepak Thassu*
192. Polymorphism of Pharmaceutical Solids, Second Edition, *edited by Harry G. Brittain*
193. Oral Drug Absorption: Prediction and Assessment, Second Edition, *edited by Jennifer J. Dressman, Hans Lennernas, and Christos Reppas*
194. Biodrug Delivery Systems: Fundamentals, Applications, and Clinical Development, *edited by Mariko Morista and Kinam Park*
195. Pharmaceutical Process Engineering, Second Edition, *edited by Anthony J. Hickey and David Ganderton*
196. Handbook of Drug Screening, Second Edition, *edited by Ramakrishna Seethala and Litao Zhang*
197. Pharmaceutical Powder Compaction Technology, Second Edition, *edited by Metin Celik*
198. Handbook of Pharmaceutical Granulation Technology, *Dilip M. Parikh*
199. Pharmaceutical Preformulation and Formulation, Second Edition: A Practical Guide from Candidate Drug Selection to Commercial Dosage Form, *edited by Mark Gibson*
200. International Pharmaceutical Product Registration, Second Edition, *edited by Anthony C. Cartwright and Brian R. Matthews*

Informa Healthcare USA, Inc.
52 Vanderbilt Avenue
New York, NY 10017

© 2009 by Informa Healthcare USA, Inc.
Informa Healthcare is an Informa business

No claim to original U.S. Government works
Printed in the United Kingdom

Reprinted 2012

International Standard Book Number-10: 1-4200-8176-4 (Hardcover)
International Standard Book Number-13: 978-1-4200-8176-3 (Hardcover)

This book contains information obtained from authentic and highly regarded sources. Reprinted material is quoted with permission, and sources are indicated. A wide variety of references are listed. Reasonable efforts have been made to publish reliable data and information, but the author and the publisher cannot assume responsibility for the validity of all materials or for the consequence of their use.

No part of this book may be reprinted, reproduced, transmitted, or utilized in any form by any electronic, mechanical, or other means, now known or hereafter invented, including photocopying, microfilming, and recording, or in any information storage or retrieval system, without written permission from the publishers.

For permission to photocopy or use material electronically from this work, please access www.copyright.com (http://www.copyright.com/) or contact the Copyright Clearance Center, Inc. (CCC) 222 Rosewood Drive, Danvers, MA 01923, 978-750-8400. CCC is a not-for-profit organization that provides licenses and registration for a variety of users. For organizations that have been granted a photocopy license by the CCC, a separate system of payment has been arranged.

Trademark Notice: Product or corporate names may be trademarks or registered trademarks, and are used only for identification and explanation without intent to infringe.

Library of Congress Cataloging-in-Publication Data

International pharmaceutical product registration / edited by Anthony C. Cartwright, Brian R. Matthews. – 2nd ed.
p. ; cm. – (Drugs and the pharmaceutical sciences, v. 200)
Includes bibliographical references and index.
ISBN-13: 978-1-4200-8176-3 (hardcover : alk. paper)
ISBN-10: 1-4200-8176-4 (hardcover : alk. paper) 1. Pharmaceutical technology–Standards. 2. Drugs–Testing–Standards. 3. Drugs–Law and legislation. I. Cartwright, A. C (Anthony C.), 1940- II. Matthews, Brian (Brian R.) III. Series: Drugs and the pharmaceutical sciences, v. 200.
[DNLM: 1. Technology, Pharmaceutical–standards. 2. Clinical Trials as Topic–standards. 3. Legislation, Drug. 4. Quality Control. W1 DR893B v.200 2009 / QV 778 I603 2009]
RS189.I62 2009
615′.19–dc22

2009015392

For Corporate Sales and Reprint Permission call 212-520-2700 or write to: Sales Department, 52 Vanderbilt Avenue, 7th floor, New York, NY 10017.

Visit the Informa Web site at
www.informa.com

and the Informa Healthcare Web site at
www.informahealthcare.com

Preface

THE OBJECTIVES AND SCOPE OF THE NEW EDITION

The International Conference on Harmonisation of Technical Requirements for Registration of Pharmaceuticals for Human Use (ICH) Common Technical Document (CTD) provides for a harmonized structure and format for new pharmaceutical product registration applications. The CTD format has been adopted in the three ICH regions of the United States, the European Union, and Japan and is increasingly used elsewhere in the world. The chapters in the new second edition of this book cover all the major technical and clinical topics that are included under the main headings in the Common Technical Document—the Module 2 summaries and overviews, Module 3 Quality, Module 4 Nonclinical Safety, and Module 5 Clinical Efficacy. One chapter deals with the particular national and region specific administrative and prescribing information in Module 1 in terms of application forms, health professional, and patient information, etc.; other chapters deal with the electronic CTD and environmental risk assessment.

This book, therefore, provides a comprehensive overview of all the scientific areas involved in the development and registration of new drug and other pharmaceutical products intended for registration internationally. The book is written to meet the needs of both regulatory affairs staff involved in writing, assembling, compiling, publishing, and submitting registration files, and the professional staffs in the "data provider" teams who design the studies and provide the technical and scientific reports and documentation—the pharmaceutical R&D staff, chemical manufacturing, pharmaceutical manufacturing, analytical development, packaging development, QC, clinical trials staff, medical and scientific affairs, medical affairs, and project management.

We have also included one or two chapters dealing with "cutting edge" issues where the regulatory policy is not yet fully agreed. Thus, we have included chapters on "Modeling and Medicinal Product R&D," and a copy of an Agence française de sécurité sanitaire des produits de santé (Afssaps) Working Group on Nonclinical Innovation document entitled "Recommendations for Toxicological Evaluation of Nanoparticle Medicinal Products."

Medicinal products cover such an enormous range—from in vivo diagnostic agents to radiopharmaceuticals to specialist biological and biotechnological products—so that to cover in detail the needs of all types of product would have required writing an encyclopedia rather than a single book. Inevitably, the editors have had to omit some aspects in relation to these specialist products.

PROGRESS IN INTERNATIONAL HARMONIZATION

Since the first edition of this book, there have been enormous changes in the technical and scientific requirements for the dossier, as the work of the ICH has continued to devise and approve new guidelines in the categories of Quality, Safety, Efficacy, and Multidisciplinary. The CTD harmonized structure and modular format for new medicinal product registration files that were adopted in San Diego, is now the obligatory format in the European Union (EU), Japan, Canada, Switzerland, and Australia. It is the recommended format in the United States. As mentioned in Chapter 1 of this book, derivatives of the CTD are becoming widely accepted in other regions of the world.

THE CTD—A COMMON FORMAT, NOT A HARMONIZED CONTENT FOR SUBMISSIONS

Enormous efforts have been expended by the staffs of the regulatory agencies and the pharmaceutical industry in the work of the ICH, and this has achieved a remarkable degree of harmonization in many scientific and technical areas of the dossier. Despite this there are still national differences in the content of submissions not only in Module 1, the administrative and prescribing information, but also in other areas of the dossier. These arise from differences in regulatory practice and procedures, different practices of medicine and pharmacy, and differences in access to diagnostic and therapeutic procedures. We are, however, still far from a genuinely global single registration dossier. The chapter authors have tried to capture some of these differences wherever feasible to help those planning to market their products in different regions of the world.

INCREASED COOPERATION BETWEEN AGENCIES BASED ON ICH

Mutual Recognition Agreements between agencies in relation to Good Manufacturing Practice (GMP) are in operation between the EU and Canada, Australia, New Zealand, Switzerland, and Japan. Arrangements exist between many countries (including the ICH members) for exchange of pharmacovigilance and defect information.

The confidentiality arrangements between the EU and the FDA now allow for exchange of information on legal and regulatory issues, scientific advice, orphan drug designation, inspection reports, marketing authorization procedures, and postmarketing surveillance. In September 2004, the European Medicines Agency (EMEA) and the Food and Drugs Administration (FDA) instituted a pilot program of parallel scientific advice meetings for sponsors to obtain advice on scientific issues during the development phase of new medicinal products. Orphan indication products and pediatric products have been targeted in particular.

Under the aegis of the Transatlantic Economic Council, the EU and FDA have reviewed ways in which cooperation can reduce administration and enable them to share best regulatory practices. This has been followed by an EMEA announcement of proposals for a EU–US–Australian pilot program to make more efficient use of global GMP inspection resources.

Bilateral and trilateral collaboration has increased in 2008. Health Canada has agreed to exchange information with the European Commission and EMEA about the authorization and safety of drugs. Canada and Australia have started their parallel review project for biologicals (originally launched in 2006).

All of this has only been possible based on the prior work that has been done in harmonization of regulatory requirements and in the development of the CTD format of the dossier in ICH, which are discussed in this book.

SPREADING THE ICH MESSAGE—THE ICH GLOBAL COOPERATION GROUP

The ICH-affiliated and other developed countries, which have adopted the CTD format (the United States, EU, Japan, Canada, Switzerland, and Australia), comprise in total approximately 15% of the current (2008) world population of 6650 million people. The ICH Global Cooperation Group (GCG) was formed on March 11, 1999 as a subcommittee of the ICH Steering Committee. Its purpose is to make information available on ICH, ICH activities, and any ICH guideline to a wider group of countries.

Regional Harmonization Initiatives

A number of regional harmonization initiatives (RHIs) have been set up where a geographic grouping of countries harmonizes technical and scientific requirements and in some cases the format of submissions for member countries. These groups have been invited to designate permanent representatives to the GCG. They currently comprise:

- **Asia-Pacific Economic Cooperation** (APEC): 21 countries in the Asia-Pacific region.
- **Association of Southeast Asian Nations** (ASEAN): Brunei Darussalam, Cambodia, Indonesia, Laos, Malaysia, Myanmar, Philippines, Singapore, Thailand and Vietnam.
- **Gulf Cooperation Countries** (GCC): Saudi Arabia, Kuwait, United Arab Emirates, Oman, Bahrain, Qatar and Yemen.
- **Pan American Network on Drug Regulatory Harmonisation** (PANDRH): Argentina, Barbados, Bolivia, Brazil, Chile, Colombia, Costa Rica, Cuba, Guatemala, Jamaica, Mexico, Panama, Trinidad and Tobago and Venezuela.

- **South African Development Community** (SADC): Angola, Botswana, Democratic Republic of Congo, Lesotho, Malawi, Mauritius, Madagascar, Namibia, Seychelles, South Africa, Swaziland, Tanzania, Zambia and Zimbabwe.

The regional groups review the applicability of the ICH guidelines in their own specific countries. Particular topics of interest include the ICH stability guideline, GMP guidances, requirements for bioavailability and bioequivalence studies, clinical trials, export/import of medicines, traditional medicines, and market surveillance. Many of these topics are covered in this book.

DRUG SAFETY AND RISK MANAGEMENT

In 2008, both the EMEA and the FDA placed an increasing emphasis on drug safety and risk management. The FDA launched its Safety First initiative to strengthen its internal policies to improve drug safety review.

In Europe, the EMEA consulted on proposals to strengthen its pharmacovigilance system, and in December 2008, brought forward a new "pharma package" of draft legislation for consideration. This includes adding a new legal requirement for a risk management system for each new medicinal product. Adverse drug reporting and periodic safety update reports (PSURs) would be made more proportionate to the risks, and all data would be added to the Eudravigilance database. A new Pharmacovigilance Risk Assessment Advisory Committee would be set up to assess pharmacovigilance data.

Current requirements for pharmacovigilance and postmarketing are summarized in chapters in this book.

OUTSOURCING OF R&D, CLINICAL TRIALS, ETC.

There are currently single-digit percentage increases in drug sales in the United States and Europe, as the rate of new drug approval (particularly in the United States) slows down and there is increasing competition from generics. Companies are shedding staff and outsourcing R&D, manufacture, etc., increasingly to Asia and China. Again, the existence of internationally accepted guidelines (such as those discussed in this book) enables work to be outsourced and carried out to acceptable agreed standards.

In addition, many pharma companies are integrating clinical trials outside the ICH regions into global clinical development programs. These trials conducted according to ICH Good Clinical Practice and reported in accordance with ICH guidelines (as discussed in this book) will have access to well-trained physicians and a large number of treatment-naïve patients at significantly lower cost. Such global trials will enable clinical data to be obtained on patients with different ethnic backgrounds and will help facilitate a coordinated registration and reduce the "drug lag" in delayed registration outside the ICH regions.

New Chemical Entity or New Biotechnological Products

It is evident that a huge effort has been put into the harmonization discussions and negotiations between the major regulatory agencies and the pharmaceutical trade associations in the United States, Japan, and Europe. This book reviews the key international requirements for registering a New Active Substance product containing a New Chemical Entity (NCE) or new biological/biotechnological entity and to try to explain how these requirements could be met by defined programs of work. The book is intended to cover all the major scientific and technical topics in such a registration file.

Where harmonization has already taken place (either under the auspices of ICH or elsewhere), the authors of the individual chapters of this book have reviewed progress and have suggested the main directions that this is likely to take. Programs of drug discovery and development for new drug products usually take place over a 10- to 12-year period, and thus studies started now will often not be the subject of a submission to the authorities for some years. An astute research laboratory or a development manager is one who identifies and anticipates the trends in development of scientific and technical thinking and how these may eventually be translated into legal rules and guidelines for registration of medicinal products so that the company's submission can be seen to be "state of the art." We feel that this book will be of some assistance in this process.

Where harmonization has not taken place, there are often either no clearly defined requirements or even divergent requirements. In these cases, the authors have been asked to address the following questions in their chapters:

- What is the current state of science and technology in this area?
- In which direction is scientific development taking in terms of determining practical test requirements?
- How can the current scientific and technical regulatory requirements for a product intended for all major markets (and particularly the EU, the United States, and Japan) be best satisfied by a program of experimental work?

Generic Products, Biosimilars, etc.

The CTD format is applicable to all types of medicinal products, not just New Chemical Entities and biotechnological products. Many of the technical requirements in Modules 1, 2, 3, and 5 will also apply to generic chemical and biotechnological products ("biosimilars") and even Over-the-Counter (OTC) products. Thus, many of the relevant chapters also cover these technical requirements.

CHOICE OF AUTHORS

The editors have chosen authors based on their wide expertise and knowledge of the scientific and technical requirements involved in drug regulation. They have been chosen mainly from experts from the national authorities involved in the harmonization process, from international companies whose daily task is to provide the data for filing world-wide submissions, and from the consultants who advise such companies. The editors would like to express their gratitude to all the authors who have given so generously of their time and expertise to provide this text as well as to the organizations, for which they work, which have graciously allowed them to do so.

DISCLAIMER

The contents of this book represent the views of its authors. They do not necessarily reflect the policies or opinions of the national competent authorities (regulatory agencies), the health departments, advisory or regulatory committees, the pharmaceutical companies, the Contact Research Organizations, consultancy companies, academic institutions, hospitals, etc., for whom the authors work.

Anthony C. Cartwright
Brian J. Matthews

Acknowledgements

The recruitment of suitable authors for the large number of individual chapters in this second edition of the book relied heavily on the friends and colleagues of the editors. However, the range of subjects covered in the book is very wide and the editors wish in particular to thank Dr. David Snodin for his help in recruiting nonclinical authors and Professor Jean-Marc Husson for his help in recommending clinical authors. The willingness of the authors to take time out of their busy professional lives for writing chapters of this book is gratefully acknowledged.

Many of the authors work for official national competent authorities or as experts associated with these authorities. Others work for pharmaceutical companies, contract companies, academic institutions, and consultancies. The permission of these authorities, companies, and consultancies for their approval for individual authors to write and publish the texts of their chapters is acknowledged.

In some chapters the authors have used registered and other trademarks when mentioning particular products. The rights of the owners of those trademarks are hereby acknowledged. The mention of a particular product or service is not intended to represent an endorsement for the product or to necessarily imply fitness for a particular use.

The editors wish to thank Carolyn Honour, Sherri Niziolek, and the staff at Informa Healthcare USA, Inc. for their support, encouragement, and forbearance during production of this book.

Contents

Part 5: Clinical Studies

Part 6: Special Products and Modeling

Contributors

Paul Baldrick Scientific and Regulatory Consulting, North Yorkshire, U.K.

John Baldwin Kelvin Associates, Campton, U.K.

Jane Barrett MedicoLegal Investigations Ltd, Cambridge, U.K.

Rolf Bass Charité-Universitätsmedizin Berlin, Berlin, Germany

Gerd Bode Lecturer and Consultant, Goettingen, Germany

François-Henri Boissel NovaPharma, Lyon, France

Jean-Pierre Boissel UMR5558, Institute for Theoretical Medicine, RTH Laennec School of Medicine, Claude Bernard University, Lyon, France

Laura Brown Laura Brown Training, Chalfont St Peter, Buckinghamshire, U.K.

Anthony C. Cartwright Pharmaceutical Regulatory Consultant, Harpenden, U.K.

Jean-Roger Claude and Members of Afssaps Working Party Toxicology Department, Afssaps, Saint Denis, France

Roger Croswell Regulatory Consulting Services, Ludlow, Vermont, U.S.A.

Michel Cucherat Consultant, Lyon, France

Pina D'Angelo Cetero Research, Cary, North Carolina, U.S.A.

Henk J. de Jong Senior Advisor Pharmaceutical Sciences to Servier International Research Institute, Courbevoie, France

Murray P. Ducharme Cetero Research, Cary, North Carolina, U.S.A. and Faculté de Pharmacie, Université de Montréal, Montréal, Québec, Canada

Brian Edwards NDA Regulatory Science Ltd, Leatherhead, Surrey, U.K.

Chantal Le Floch Science Union, Courbevoie, France

François Gueyffier UMR5558, Institute for Theoretical Medicine, Grange-Blanche School of Medicine, Claude Bernard University, Lyon, France

Manfred G. Haase NDA Regulatory Science Ltd, Leatherhead, Surrey, U.K.

H. Guenter Hennings Regulatory Consultant, Bad Kreuznach, Germany

Danuta Herzyk Merck Research Laboratories, West Point, Pennsylvania, U.S.A.

Jean-Marc Husson European Diploma in Pharmaceutical Medicine, Lyon, France

David Jefferys Eisai Europe Ltd, Hertfordshire, U.K.

Roeline Jochemsen Servier International Research Institute, Courbevoie, France

Peter Kasper Federal Institute for Drugs and Medical Devices, Bonn, Germany

Richard Kay Independent Statistical Consultant, rkstatistics, Great Longstone, Bakewell, U.K.

Richard Lee Consultant, Hitchin, Hertfordshire, U.K.

Barry S. Levine Levine & Associates, LLC, Chicago, Illinois, U.S.A.

Patricia Maillère I.R.I.S., Courbevoie, France

Andrew Makin LAB Research Denmark (Formerly Scantox), Hestehavevej, Lille Skensved, Denmark

Brian R. Matthews Alcon Laboratories, Hemel Hempstead, U.K.

James W. McBlane Licensing Division, Medicines and Healthcare Products Regulatory Agency, London, U.K.

Iain J. McGilveray McGilveray Pharmacon Inc, Ottawa, Ontario, Canada

J. Michael Morris Irish Medicines Board, Dublin, Ireland

Don J. Nichols Pfizer Global Research and Development, Sandwich, Kent, U.K.

Patrice Nony Lyon's Public Hospital, UMR5558, Institute for Theoretical Medicine, Lyon, France

Klaus Olejniczak Federal Institute for Drugs and Medical Devices, Bonn, Germany

David J. Snodin PAREXEL Drug Development Consulting, Middlesex, U.K.

Mary Teeling NDA Regulatory Science Ltd, Leatherhead, Surrey, U.K.

Markku Toivonen NDA Regulatory Science Ltd, Leatherhead, Surrey, U.K.

Jean-Marie Vetel* Former Head of Geriatric Department, Le Mans Hospital, Sarthe, France

Don K. Walker Pfizer Global Research and Development, Sandwich, Kent, U.K.

Richard J. Weaver Servier R&D, Slough, Buckinghamshire, U.K.

Geoffrey Williams Roche Products Ltd., Welwyn Garden City, U.K.

Corinne Seng Yue Faculté de Pharmaciè, Université de Montréal, Montréal, Québec, Canada

*Currently President of the Association of French Gerontologists.

1 The Common Technical Document—A Global Format for Registration Files

Anthony C. Cartwright
Pharmaceutical Regulatory Consultant, Harpenden, U.K.

INTERNATIONAL HARMONIZATION

Harmonization of regulatory requirements for registering new medicinal products began in the 1970s as the European Community (now the European Union, EU) moved toward developing a single market for pharmaceutical products. European guidelines were developed by the Committee for Proprietary Medicinal Products (CPMP) and its scientific working parties. At the same time, there were bilateral discussions between Europe and the United States (U.S.A.) on the one hand and Europe and Japan on the other on the possibilities of harmonizing data requirements for registration files for pharmaceutical medicinal products. In 1989, the pharmaceutical industry proposed that a joint industry-regulatory authority initiative be set up to harmonize requirements for safety, quality, and efficacy. The International Conference on Harmonisation of Technical Requirements for Registration of Pharmaceuticals for Human Use (ICH) was set up in 1990. Its objective is to ensure that good quality, safe and effective medicines are developed and registered in the most cost-effective and efficient way.

The "players" in ICH are the regulatory agencies and the research-based industry in the three ICH regions of Europe, Japan, and the United States. They comprise the Food and Drug Administration (FDA) and the Pharmaceutical Research and Manufacturers of America (PhRMA) for the United States, the European Commission/European Union and the European Federation of Pharmaceutical Industries and Associations (EFPIA) for Europe, and the Ministry of Health, Labour and Welfare (MHLW) and the Japan Pharmaceutical Manufacturers Association (JPMA) for Japan. There are observers at meetings of the ICH from the World Health Organization (WHO), the European Free Trade Area (EFTA) (currently represented by Swissmedic), and Health Canada.

The ICH is managed by its Steering Committee, which meets twice yearly. Topics for harmonization agreed by the Steering Committee are considered by Expert Working Groups (EWGs), which meet in parallel with the Steering Committee to discuss scientific and technical aspects. The EWGs recommend texts of guidelines to the Steering Committee for issue for consultation and then for adoption.

WORK ON THE COMMON TECHNICAL DOCUMENT

The first topics considered for harmonization are related to safety, quality, and efficacy. The topics chosen were those where there were significant regional differences between the regulatory requirements, which added to the costs for developing of new drug products. After this first period of harmonization of key regulatory requirements, the ICH turned its attention to the format for documentation in a registration file. The objective was to remove redundancy and duplication so that as far as possible a single set of data could be provided to demonstrate safety, quality, and efficacy. The aim was also to reduce the delays and costs involved in converting registration files between the different national formats.

Adoption and Implementation of the CTD

The final text of the CTD was issued in November 2000 by the ICH Steering Committee and this has now been revised and updated (1–4). The CTD was implemented as an optional format in the European Union, the United States, Japan, Canada, and Switzerland. It became mandatory from July 2003 in the European Union, Japan, Canada, and Switzerland. It is the recommended format in the United States as ICH documents have always been considered as guidance by the

FDA, and the September 2000 Good Guidances Practice Final Rule requires that the CTD not be mandatory.

The CTD format has also been adopted in Australia by the Therapeutic Goods Administration (TGA) and it became the mandatory format after June 30, 2004. The New Zealand Medicines and Medical Devices Safety Authority (Medsafe) announced that the CTD format became mandatory for all intermediate-risk and high-risk new medicine applications from September 1, 2006.

MODULAR STRUCTURE OF THE ICH CTD (1)

The CTD has a highly modular structure

- Module 1: Administrative Information (different for each ICH region or country)
- Module 2: Summaries
- Module 3: Quality
- Module 4: Nonclinical Study Reports
- Module 5: Clinical Study Reports

The organizational structure is shown diagrammatically (Fig. 1), and the high-level structure is summarized below.

Module 2: Summaries

Module 2 contains seven high-level summaries of quality, nonclinical safety, and clinical safety and efficacy. It comprises the following:

Module 2.1: CTD Table of Contents
Module 2.2: CTD Introduction

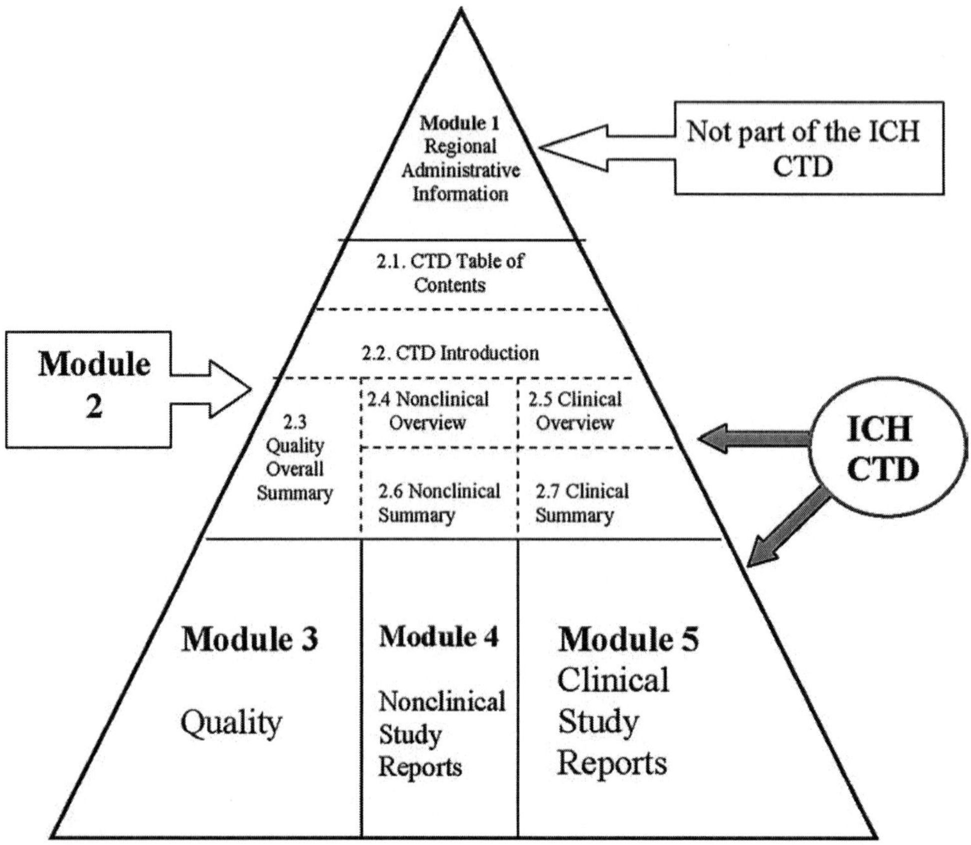

Figure 1 Organization of the CTD.

Module 2.3: Quality Overall Summary
Module 2.4: Nonclinical Overview
Module 2.5: Clinical Overview
Module 2.6: Nonclinical Summaries
Module 2.7: Clinical Summaries

Module 3: Quality
3.1 Table of Contents of Module 3
3.2 Body of Data
- 3.2.S Drug Substance
- 3.2.P Drug Product
- 3.2.R Regional Information

3.3 Literature References

Module 4: Nonclinical Study Reports
4.1 Table of Contents of Module 4
4.2 Study Reports
4.3 Literature References

Module 5: Clinical Study Reports
5.1 Table of Contents of Module 5
5.2 Tabular Listing of all Clinical Studies
5.3 Clinical Study Reports
5.4 Literature References

DETAILED MODULAR STRUCTURE OF THE CTD

Module 2: CTD Summaries
2.1 Table of Contents
2.2 CTD Introduction
2.3 Quality Overall Summary
- 2.3.S Drug Substance
  - 2.3.S.1 General Information
  - 2.3.S.2 Manufacture
  - 2.3.S.3 Characterization
  - 2.3.S.4 Control of Drug Substance
  - 2.3.S.5 Reference Standards or Materials
  - 2.3.S.6 Container Closure System
  - 2.3.S.7 Stability
- 2.3.P Drug Product
  - 2.3.P.1 Description and Composition of the Drug Product
  - 2.3.P.2 Pharmaceutical Development
  - 2.3.P.3 Manufacture
  - 2.3.P.4 Control of Excipients
  - 2.3.P.5 Control of Drug Product
  - 2.3.P.6 Reference Standards or Materials
  - 2.3.P.7 Container Closure System
  - 2.3.P.8 Stability
- 2.3.A. Appendices
  - 2.3.A.1 Facilities and Equipment
  - 2.3.A.2 Adventitious Agents Safety Evaluation
  - 2.3.A.1.3 Excipients
- 2.3.R Regional Information

2.4 Nonclinical Overview
- 2.4.1 Overview of the Nonclinical Testing Strategy
- 2.4.2 Pharmacology

2.4.3 Pharmacokinetics
2.4.4 Toxicology
2.4.5 Integrated Overview and Conclusions
2.4.6 List of Literature References
2.5 Clinical Overview
2.5.1 Product Development Rationale
2.5.2 Overview of Biopharmaceutics
2.5.3 Overview of Clinical Pharmacology
2.5.4 Overview of Efficacy
2.5.5 Overview of Safety
2.5.6 Benefits and Risks Conclusions
2.5.7 Literature References
2.6 Nonclinical Written and Tabulated Summaries
2.6.1 Introduction
2.6.2 Pharmacology Written Summary
2.6.3 Pharmacology Tabulated Summary
2.6.4 Pharmacokinetics Written Summary
2.6.5 Pharmacokinetics Tabulated Summary
2.6.6. Toxicology Written Summary
2.6.7 Toxicology Tabulated Summary

Module 3: Quality
3.1 Table of Contents of Module 3
3.2 Body of Data
3.2.S Drug Substance
3.2.S.1 General Information
3.2.S.2 Manufacture
3.2.S.3 Characterization
3.2.S.4 Control of Drug Substance
3.2.S.5 Reference Standards or Materials
3.2.S.6 Container Closure Systems
3.2.S.7 Stability
3.2.P Drug Product
3.2.P.1 Description and Composition of the Drug Product
3.2.P.2 Pharmaceutical Development
3.2.P.3 Manufacture
3.2.P.4 Control of Excipients
3.2.P.5 Control of Drug Product
3.2.P.6 Reference Standards or Materials
3.2.P.7 Container Closure System
3.2.P.8 Stability
3.2.A Appendices
3.2.A.1 Facilities and Equipment
3.2.A.2 Adventitious Agents Safety Evaluation
3.2.A.3 Novel Excipients
3.2.R Regional Information
3.3 Literature References

Module 4: Nonclinical Study Reports
4.1 Table of Contents of Module 4
4.2 Study Reports
4.2.1 Pharmacology
4.2.1.1 Primary Pharmacodynamics
4.2.1.2 Secondary Pharmacodynamics
4.2.1.3 Safety Pharmacology
4.2.1.4 Pharmacodynamic Drug Interactions

4.2.2 Pharmacokinetics
4.2.2.1 Analytical Methods and Validation Reports
4.2.2.2 Absorption
4.2.2.3 Distribution
4.2.2.4 Metabolism
4.2.2.5 Excretion
4.2.3 Toxicology
4.2.3.1 Single-Dose Toxicity
4.2.3.2 Repeat-Dose Toxicity
4.2.3.4 Carcinogenicity
4.2.3.5 Reproductive and Development Toxicity
4.2.3.6 Local Tolerance
4.2.3.7 Other Toxicity Studies
4.3 Literature References

Module 5: Clinical Study Reports

5.1 Table of Contents of Module 5
5.2 Tabular Listing of All Clinical Studies
5.3 Clinical Study Reports
5.3.1 Reports of Biopharmaceutic Studies
5.3.2 Reports of Studies Pertinent to Pharmacokinetics Using Human Biomaterials
5.3.3 Reports of Human Pharmacokinetics Studies
5.3.4 Reports of Human Pharmacodynamics Studies
5.3.5 Reports of Efficacy and Safety Studies
5.3.6 Reports of Post marketing Experience
5.3.7 Case Report Forms and Individual Patient Listings
5.4 Literature References

DEFINING THE GRANULARITY FOR A NEW CTD REGISTRATION FILE

The modular format defined above is further subdivided into individual documents that form the basis of the complete registration dossier (file). For example, Module 5.3.1 Reports of Biopharmaceutic Studies is further divided into the following:

5.3.1.1 Bioavailability (BA) Study Reports
5.3.1.2 Comparative BA and Bioequivalence (BE) Study Reports
5.3.13. In Vitro–In Vivo Correlation Study Reports
5.3.14 Reports of Bioanalytical and Analytical Methods for Human Studies

The question of which individual modular documents should be written and included in a new registration file is addressed in the ICH M4 guidance *Organisation of the Common Technical Document for the Registration of Pharmaceuticals for Human Use* (1). This defines the granularity of the registration file. The term *granularity* refers to the fineness with which the sections of the CTD data fields are subdivided, that is, the number of individual documents in any particular modular section of the file. In our example, if there were several BE studies in a particular registration file, each should be reported separately and each report should normally be as multiple individual documents consisting of a synopsis, a main body of the study report, and appropriate appendices.

Granularity of the Module 2 Quality Overall Summary and Module 3 Quality

There are options of granularity for the Quality Overall Summary (QOS) to suit different levels of complexity of products. The company can choose the level at which the QOS is managed. The three options are shown (Fig. 2). The greater the degree of granularity adopted, the easier it is to make subsequent changes to the documentation. For example, if the company wishes to change an analytical test method for the drug substance, the whole of 2.3.S would need to be resubmitted if this is all one document, whereas only 2.3.S.4 Control of Drug Substance would need to be resubmitted if a registration file with a greater level of granularity had been used.

Option 1	Option 2	Option 3
Module 2.1 Table of Contents	Module 2.1 Table of Contents	Module 2.1 Table of Contents
Module 2.2 CTD Introduction	Module 2.2 CTD Introduction	Module 2.2 CTD Introduction
Module 2.3 Body of Data	2.3.S Drug Substance	2.3.S.1 General Information
		2.3.S.2 Manufacture
		2.3.S.3 Characterization
		2.3.S4 Control of Drug Substance
		2.3.S.5 Reference Standards
		2.3.S.6 Container Closure
		2.3.S.7 Stability
	2.3.P Drug Product	2.3.P.1 Description & Composition
		2.3.P.2 Pharmaceutical Development
		2.3.P.3 Manufacture
		2.3.P.4 Control of Excipients
		2.3.P.5 Control of Drug Product
		2.3.P.6 Reference Standards
		2.3.P.7 Container Closure
		2.3.P.8 Stability
	2.3A Appendices	2.3.A.1 Facilities & Equipment
		2.3.A.2 Adventitious Agents Safety Evaluation
		2.3.A.3 Excipients
	2.3.R Regional Information	2.3.R Regional Information

Notes:

The Table of Contents is only needed for a paper CTD; it is not needed for an eCTD

Figure 2 Options for granularity in the Module 2.3 QOS.

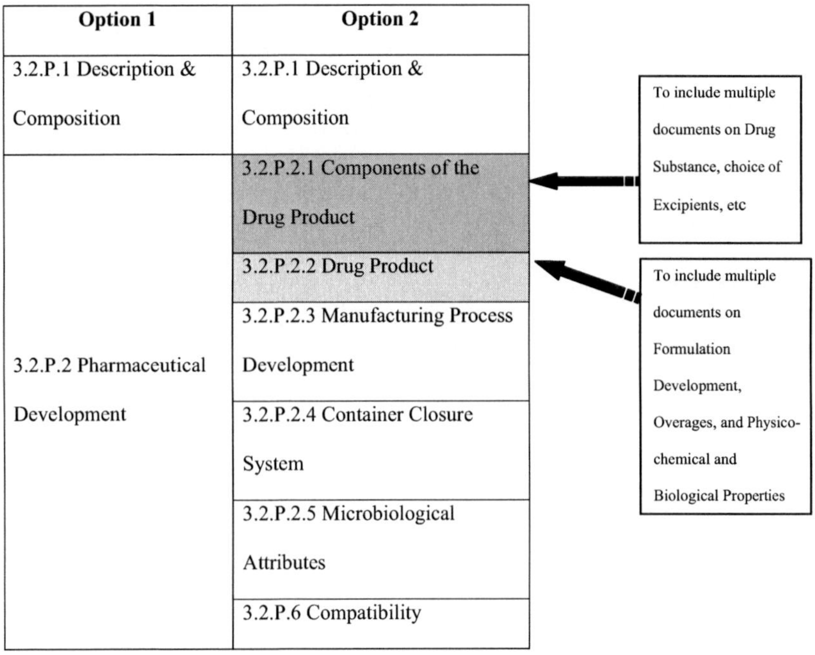

Figure 3 Options for granularity in the Module 3.2 quality file.

There are more limited options for granularity for Module 3 and again these are shown in the ICH Guideline. A major option is shown in Figure 3 for 3.2.P.2 Pharmaceutical Development.

CATEGORIES OF PHARMACEUTICAL PRODUCT THAT CAN BE SUBMITTED AS A CTD REGISTRATION FILE

Most categories of product can be submitted in all of the ICH regions as a CTD registration file. This includes the following:

- New drug substance (NCE) products
- New biological/biopharmaceutical products
- Radiopharmaceuticals
- Phytopharmaceuticals (herbal medicines)

In the case of the United States, CTD registration files may also be submitted for generic and OTC products and this format would be obligatory in the European Union for these products.

TABLES OF CONTENTS

The Module 1 TOC in a paper CTD filing should list out all individual documents included in the registration dossier.

The Module 2 to 5 TOCs should list all of the documents included in these Modules. A TOC can also be included for a particular section that contains multiple documents to enable the reviewer to follow the chronology and the subject matter. Long complex documents (such as Nonclinical or Clinical Study Reports) should include a TOC that lists all of the major sections.

PRESENTATION OF THE CTD

Individual documents included in a CTD registration dossier should be clearly presented so that the agency reviewers can find quickly the information they need. Font sizes for text and tables should be easily legible. Times New Roman, Courier, or Ariel, 12-point font can be used for narrative text. Each page of a document should be numbered. For a paper document, use of appropriate Headers and Footers in a uniform style (containing the name of the product, type of registration file, section of the application, etc.) is helpful in navigating through the file. In a paper file, each separate document is tagged so that it can be easily located.

Acronyms and abbreviations should be clearly defined in a document included in each module for ease of reference by the reviewer.

DIFFERENCES IN CTD CONTENT BETWEEN THE ICH REGIONS

As mentioned in the Preface, the CTD is a harmonized format for registration files; however, the content is not yet completely harmonized. There are still national or regional differences in the content of submissions—not only in Module 1 but also in other parts of the dossier. These arise from differences in regulatory practice and procedures, differences in practices of medicine and pharmacy, and differences in access to diagnostic and therapeutic procedures. What are the major differences in content and how can companies cope with them to make filings in the major developed world markets?

Module 1 Differences

Although the ICH CTD refers to Module 1 as comprising "Regional Administrative Information" (such as the application form, labeling, text of prescriber, and patient information), in practice there are other differences and some of key ones are summarized in Table 1 (5–9).

It is usual for companies to prepare a common Core Data Sheet for information to health practitioners that could then be used to prepare draft Prescribing Information and Patient

Table 1 Summary of Some Key National or Regional Differences in Module 1

Nation/region	Key differences
European Union	• 1.4 Information about the Experts (who sign the Module 2 Summaries) • 1.5 Specific Requirements for Different Kinds of Application (summaries to support generics, hybrid applications, bibliographic applications, extended market exclusivity applications, conditional marketing authorizations, etc.) • 1.6 Environmental Risk Assessment • 1.7 Information relating to Orphan Market Exclusivity • 1.8 Information relating to Pharmacovigilance • 1.9 Information relating to Clinical Trials
United States	• 1.3.5 Patent and exclusivity information • 1.9 Pediatric administrative information • 1.16 Risk management plans
Japan	• Patent status • Background of origin, discovery, and development • List of related products • Data for review of designation as poisons, deleterious substances, etc. • Draft of basic protocol for postmarketing surveillance
Canada	• 1.2.4 Patent information • 1.4 Health Canada Summaries 1.4.1 Certified Product Information Document (CPID) 1.4.2 Comprehensive Summary of BE • 1.5 Environmental Assessment Statement
Australia	• 1.4.1 Information about the Experts (who sign the Module 2 Summaries) • 1.5 Specific requirements about different types of application (literature based, orphan drug products, genetically modified organisms, comarketed medicines, etc.) • 1.6 Drug and Plasma Master Files and Ph Eur Certificates of Suitability • GMP clearance letters • Summary Biopharmaceutic Studies • Pediatric Development Program • Environmental Risk for non-GMOs containing medicines • Antibiotic resistance data

CPID: This is a condensed summary of current and specific chemistry and manufacturing information attested by the manufacturer and sponsor. There is no requirement for Expert signatures for the Module 2 Summaries in registration filings in the United States, Japan, and Canada.

Information for the United States, the Summary of Product Characteristics (SmPC), and Patient Information Leaflet for the European Union.

Module 2 Differences

Although the formal ICH requirements for the Module 2 Summaries are identical in all ICH and other countries, there are likely to be national or regional differences particularly in relation to the contents of the Module 3: Quality and Module 5: Clinical Studies modular files, and these will be reflected in these high-level summaries. For example, where a pharmacopoeial drug is the subject of a DMF, there will be just a reference to the DMF in Module 2 of a U.S. or Japanese dossier, whereas in the European Union there will be a summary of the Open Part of the European Union Active Substance Master File (DMF).

Module 3 Differences

Some of the key national or regional differences in content of Module 3 in relation to the drug substance and drug product are summarized in Table 2 and Table 3.

In addition, there are the 3.2.R Regional Differences. Examples quoted in the ICH CTD are as follows:

- Executed Batch Records for Drug Substance and Drug Product (the United States only): Copies of records with equipment specified, manufacturing conditions, packaging records, and batch reconciliation information (theoretical yield, actual yield, and packaged yield)
- Method validation package for drug substance and drug product (the United States only)
- Comparability protocols (the United States only)
- Process validation scheme (the European Union only), including a process validation protocol where validation studies on the manufacturing process for the drug product are not complete
- Medical device used in combination with the drug product (the European Union only)

In addition, highly abbreviated documentation (a "lite" document) on the manufacture of the drug substance in 3.2.S.2.2 Description of the Manufacturing Process and on 3.2.P.3.3 Description of the Manufacturing Process and Process Controls for the drug product may need to be supplied in countries where regulatory agencies do not always respect the confidentiality of data.

There may also be differences in marketing needs for different countries. Thus, blister or foil packs are usually the packaging material of choice for tablets or capsules in the European Union whereas in the United States high-density polyethylene (HDPE) bottles are much more commonly used. In such a case, 3.2.P.7 Container Closure will include different information and

Table 2 Summary of Some Key National or Regional Differences in Module 3.2.S

Nation/region	Key differences
European Union	• 3.2.S Drug substance data may be submitted as an EU 2-part DMF (Open Part to be reproduced in 3.2.S) or as a reference to a Ph. Eur Certificate of Suitability (for Ph Eur monograph substances) • 3.2.S.7 Stability: Storage requirements to be stated in accord with CHMP guideline
United States	• 3.2.S Reference may be made in the dossier to DMF information supplied directly by the drug substance manufacturer to FDA • 3.2.S.7 Stability: Storage requirements to be stated in accord with FDA labeling requirements
Japan	• 3.2.S Reference may be made in the dossier to DMF information supplied directly by the drug substance manufacturer
Canada	• 3.2.S Reference may be made in the dossier to DMF information supplied directly by the drug substance manufacturer
Australia	• 3.2.S Drug substance data may be submitted as a 2-part DMF (Open Part to be reproduced in 3.2.S) or as a reference to a Ph. Eur Certificate of Suitability (for Ph Eur monograph substances)

Table 3 Summary of Some Key National or Regional Differences in Module 3.2.P

Nation/region	Key differences
European Union	• 3.2.P.1 Description and Composition: Colors to be on the European Union permitted list. Excipients to be designated as conforming to Ph Eur or a European national pharmacopoeia where there is a monograph. • 3.2.P.4 Excipients: To conform to Ph Eur/European national pharmacopoeia if described in a monograph. • 3.2.P.5 Control of Drug Product: Assay limits to be ± 5% unless justified. A different manufacturing and shelf-life specification may be required. Products to conform to general monographs of the Ph Eur. • 3.2.P.7 Container Closure System: Name of manufacturer(s) not required unless product is critical (e.g., parenteral). • 3.2.P.8 Stability: Storage requirement to be in accord with CHMP guideline.
United States	• 3.2.P Reference may be made in the dossier to DMF information supplied directly to FDA by excipient and container/closure manufacturers. • 3.2.P.1 Description and Composition: Colors to be on FDA permitted list. Excipients to be designated as conforming to USP/NF where there is a monograph. • 3.2.P.4 Excipients: To conform to USP/NF if described in a monograph. • 3.2.P.5 Control of Drug Product: Assay limits allowed to be up to ± 10%. A single regulatory (shelf-life) specification is allowed. • 3.2.P.7 Container Closure System: Name of manufacturer(s) required. • 3.2.P.8 Stability: Storage requirement to be in accord with FDA requirements for wording.
Japan	• 3.2.P.1 Description and Composition: Colors to be on Japanese permitted list. • 3.2.P.4 Excipients: To conform to monographs of the Japanese Pharmacopoeia or Japanese Pharmaceutical Excipients.
Canada	• 3.2.P.7 Container/Closure: Reference may be made to a DMF from the supplier. • 3.2.P.8 Stability: Storage conditions to refer to Health Canada requirements (e.g., storage at controlled room temperature).
Australia	• 3.2.P.1 Description and Composition: Colors to be on Australian permitted list for colors in oral products • 3.2.P.8 Stability: Storage conditions to refer to TGA list of acceptable storage conditions (e.g., Store below 30°C).

also the primary stability data in 3.2.P.8 Stability for the European Union will be in blister or foil packs and in the United States it will be in HDPE bottles.

Module 4 Differences

There are usually no major differences in this module.

Module 5 Differences

Some of the key differences are summarized in Table 4.

Managing the Differences

If companies wish to file simultaneously in a number of the major developed world markets, the chemical, pharmaceutical, nonclinical, and clinical development program needs to be designed to meet all of the individual market regulatory needs. For example, additional "bridging" pharmacokinetic or clinical trials may be needed to support a foreign registration file in Japan. Most of the documentation in Modules 2 to 5 for a major new drug registration file can be identical, but where there are national differences in requirements (e.g., differences in 3.2.P.5.1 Drug Product Specification in terms of assay limits for the European Union and U.S. markets), it is usually more efficient to prepare the two versions of the document at the same time.

The Module 2.3 Quality Overall Summary and the Module 2.5 Clinical Overview could be prepared as identical "core documents" but they should then be reviewed by in-country staff and customized as necessary to meet any different technical or regulatory requirements of the different agencies.

Table 4 Summary of Some Key National or Regional Differences in Module 5: Clinical Studies

Nation/region	Key differences
Europe	• Where required BE studies for generic products need to use a European batch of reference product • Clinical trials should normally comply with CHMP Efficacy guidances where these exist • Clinical trials of new drug products versus European authorized "gold standard treatment" are important as well as placebo studies
United States	• Clinical trials should normally comply with FDA regulatory guidances where these exist • FDA Integrated Summaries of Safety and Efficacy (ISS/ISE) to be included in 5.3.5.3 Reports of Analyses from More than One Study (these are normally required in addition to the Clinical Overview and Clinical Summary in 2.5 and 2.7)
Japan	• "Bridging" pharmacokinetic and clinical studies may be needed to allow foreign data to be extrapolated to the Japanese population if the clinical studies performed outside Japan
Other Non-ICH countries	• "Bridging pharmacokinetic and clinical studies may be needed to allow foreign data to be extrapolated to the national population if the clinical studies are performed outside these countries

ADOPTION OF THE CTD FORMAT OUTSIDE THE ICH REGION

The CTD format is being adopted with local modifications as needed by other national regulatory agencies and regional groupings of agencies. The Association of Southeast Asian Nations (ASEAN) comprises Brunei Darussalam, Cambodia, Indonesia, Laos, Malaysia, Myanmar, Philippines, Singapore, Thailand, and Vietnam. These countries have a combined population of over 550 million. They have published the ASEAN CTD (10). The ICH format is allowed for NCE and Biological products, but compliance is needed with ASEAN technical requirements [e.g., the ASEAN guideline on Stability Study of Drug Product (10)] and the ASEAN CTD (ACTD) administrative requirements. The ACTD will be implemented across the whole region by January 1, 2009.

Organization of the ACTD

Part I: Table of Contents, Administrative Information, and Prescribing Information

Section A: Introduction
Section B: Overall ASEAN CTD Table of Contents
Section C: Documents required for registration (e.g., application forms, labeling, Product Data Sheet, prescribing information)

Part II: Quality Document

Section A: Table of Contents
Section B: Quality Overall Summary
Section C: Body of Data
  S: Drug Substance
  P: Drug Product
Section D: Key Literature References

Part III: Nonclinical Document

Section A: Table of Contents
Section B: Nonclinical Overview
Section C: Nonclinical Written and Tabulated Summaries
1. Table of Contents
2. Pharmacology
3. Pharmacokinetics
4. Toxicology

Section D: Nonclinical Study Reports

1. Table of Contents
2. Pharmacology
3. Pharmacokinetics
4. Toxicology

Section E: List of Key Literature References

Part IV: Clinical Document

Section A: Table of Contents

Section B: Clinical Overview

Section C: Clinical Summary

1. Summary of Biopharmaceutics and Associated Analytical Methods
2. Summary of Clinical Pharmacology Studies
3. Summary of Clinical Efficacy
4. Summary of Clinical Safety
5. Synopses of Individual Studies

Section D: Tabular Listing of All Clinical Studies

Section E: Clinical Study Reports

Section F: List of Key Literature References

For ASEAN member countries, the Nonclinical and Clinical Study Reports may not be required for NCE, Biotechnological Products, and other Major Variation Products if the Original Products are already registered and approved for marketing authorization in Reference Countries. Specific Study Reports may be requested by the national agencies if they feel it to be necessary. For generic products, Parts III and IV are not required and Part II Quality Documentation has an extra section in the Drug Product entitled P9 Product Interchangeability, where BA and BE studies are to be included to meet the requirements of the ASEAN guideline on Bioquivalence Study (10).

RECOGNITION OF PRODUCT REGISTRATIONS BY OTHER COUNTRIES

The ICH guidelines now provide internationally accepted standards for safety, quality, and efficacy against which pharmaceutical companies can develop medicinal products for submission of registration files to the regulatory agencies in the European Union, the United States, Japan, and elsewhere. The guidelines also provide standards for the agencies to review the registration files.

Once a product has been approved, pharmaceutical companies can use the registration approvals from well-recognized regulatory authorities to extend the acceptance of these products to other countries. The WHO Certification Scheme (11) allows Member States to support the export of pharmaceutical products via the issue of certificates for the products they have approved via an attestation that

- the specific product is authorized to be placed on the market;
- the manufacturing plant in which it is produced is subject to inspection and the manufacturer complies with GMP; and
- all submitted product information, including labeling, is currently authorized in the country.

Each individual country defines its own further requirements for pharmaceutical product registration in addition to the WHO Certificates, although these are now being increasingly harmonized via regional groupings such as Asia Pacific Economic Cooperation (APEC), the Gulf Cooperation Countries (GCC) the Pan American Network on Drug Regulatory Harmonisation (PANDRH), and the South African Development Community (SADC) (see Preface).

REFERENCES

1. Organisation of the Common Technical Document for the Registration of Pharmaceuticals for Human Use. M4. ICH Harmonised Tripartite Guideline. http://www.ich.org/LOB/media/MEDIA554.pdf. Accessed March 19, 2008.

2. The Common Technical Document for the Registration of Pharmaceuticals for Human Use: Quality—M4Q(R1). ICH Harmonised Tripartite Guideline. http://www.ich.org/LOB/media/MEDIA556.pdf. Accessed March 19, 2008.
3. The Common Technical Document for the Registration of Pharmaceuticals for Human Use: Safety—M4S(R2). ICH Harmonised Tripartite Guideline. http://www.ich.org/LOB/media/MEDIA559.pdf. Accessed March 19, 2008.
4. The Common Technical Document for the Registration of Pharmaceuticals for Human Use: Efficacy—M4E(R1). ICH Harmonised Tripartite Guideline. http://www.ich.org/LOB/media/MEDIA561.pdf. Accessed March 19, 2008.
5. Volume 2B Notice to Applicants, Medicinal products for human use. Presentation and format of the dossier, Common Technical Document (CTD), Edition June 2006. http://ec.europa.eu/enterprise/pharmaceuticals/eudralex/vol-2/b/ctd_06–2006.pdf. Accessed March 19, 2008.
6. Information in English on Japan Regulatory Affairs. English Information Task Force, Japan Pharmaceutical Manufacturers Association, Pharmaceutical Administration and Regulations in Japan, March 2007. http://www.nihs.g.jp/mhlw/jouhou/yakuji/yakuji-e0703.pdf. Accessed March 19, 2008.
7. FDA CTD Comprehensive Table of Contents and Hierarchy, version 1.2, July 6, 2005. http://www.fda.gov/cder/regulatory/ersr/5640CTOC-v1.2.pdf. Accessed March 19, 2008.
8. Module 1 Administrative Information and Prescribing Information for Australia, Notice to Applicants CTD—Module 1, TGA Edition, September 1, 2007. http://www.tga.gov.au/docs/pdf/euguide/tgamod1.pdf. Accessed March 20, 2008.
9. Draft Guidance for Industry, Preparation of New Drug Submissions in the CTD Format, Health Products and Food Branch Guidance Document, Draft June 25, 2003. http://www.hc-sc.gc.ca/dhp-mps/alt_formats/hpfb-dgpsa/pdf/prodpharma/ctdnds_ctdpdn_e.pdf. Accessed March 20, 2008.
10. The ASEAN Common Technical Dossier (ACTD) for the Registration of Pharmaceuticals for Human Use. http://www.hsa.gov.sg/publish/hsaportal/en/health_products_regulatory/western_medicine/guidelines.html. Accessed March 20, 2008.
11. Guidelines on the Implementation of the WHO Certification Scheme on the Quality of Pharmaceutical Products Moving in International Commerce, 2008. http://www.who.int/medicines/areas/quality_safety/regulation_legislation/certification/guidelines/en/index.html. Accessed March 20, 2008.

2 The Electronic Common Technical Document

Geoffrey Williams
Roche Products Ltd., Welwyn Garden City, U.K.

Electronic submissions have been part of the regulatory landscape for over 15 years since the first formats were developed in the early 1990s. The electronic Common Technical Document (eCTD) is an International Conference on Harmonisation (ICH) created and maintained format for electronic submissions that aims to support the regional and international submission requirements of the Common Technical Document (CTD) in the three ICH regions [European Union (EU), United States (U.S.A.), and Japan] and beyond.

ELECTRONIC SUBMISSIONS: A BRIEF INTRODUCTION

Since the desktop PC became a common part of the office working environment of the pharmaceutical industry and regulatory agencies, there have been efforts to use the technology to improve the efficiency of the review and approval process through the creation and use of electronic submissions. The fact that the data and documents that form part of the regulatory submission were being created using electronic systems led to efforts to find a way to submit these in an electronic format rather than printing them to paper and losing some of the opportunities to explore making review procedures more efficient.

The simplest definition of an electronic submission is any set of files submitted to aid the review of a dossier. The objective in submitting the content of a submission electronically is to take advantage of the technology to

- provide a more efficient navigation around the content of the dossier,
- provide content that can be searched and queried,
- provide content that can be more easily copied into assessment reports and other documents,
- reduce the amount of paper submitted, and
- allow the creation of different views of the submitted data.

Not all of the simplest electronic submissions will meet all of these objectives but all meet at least one of them. For example, the requests made by many agencies to receive copies of key summary and labeling documents in an electronic format are usually to aid in the review, preparing comments, and creation of assessment reports. The requests by some agencies to receive data files are usually to allow the reprocessing and analysis of the data and also to reduce the amount of paper received.

The Food and Drug Administration (FDA) in the United States started one of the first electronic submission programs when they began the CANDA (Computer-Aided New Drug Application) project in the early 1990s. Applicants could work with the agency to deliver a complete new drug application (NDA) submission in an electronic format to aid in the review of the dossier. Many companies who were making submissions in both the United States and Europe started to explore the way in which they could reuse this technology in Europe and European agencies started accepting electronic submissions in 1993 and 1994. Often, these early electronic submissions consisted not only of the set of submission files but also of the hardware and software necessary for the agency to be able to use and review the submission. The other main feature of these submissions was that the way in which the data was presented, although conforming to the dossier standards of the region they would vary because of the different software applications used by the applicants in their creation.

The late 1990s saw the introduction of a number of standardized ways of presenting a dossier in an electronic format. The electronic NDA (eNDA) in the United States grew from initial efforts to standardize the presentation of the data submitted in items 11 and 12 of the NDA and resulted in a published standard employing the adobe portable document file (PDF) and SAS formats, for documents and datasets respectively, for the entire content of the dossier. The FDA went further to develop the supporting processes to allow the electronic-only submission of the dossier without the need for any paper. In Europe, the German agency developed DAMOS (Drug Application Methodology with Optical Storage) and the French agency developed SEDAMM (Soumission Électronique des Dossiers d'Autorisation de Mise sur le Marché), both of which were full specifications for the way in which the complete dossier should be presented; DAMOS building on images in tagged image file (TIF) format and SEDAMM on standard graphical markup language (SGML).

The growing use of electronic submissions on a global basis demonstrated that the concept and the objectives behind their use were all valid. However, for global companies there remained issues: that the increased standardization of the electronic format was still taking place on a country or regional basis and that global companies could not get the greatest benefit of content reuse on an international basis. Against this background, the ICH approved the adoption of a topic to look at the creation of an electronic submission standard.

THE eCTD

The ICH Steering Committee approved the adoption of the global electronic submission standard as a project under the M2 (Electronic Standards for the Transmission of Regulatory Information, ESTRI) group. There was only one small problem—without a global standard for the presentation of regulatory content there would be a significant issue in developing an electronic submission specification. The M4 topic to create the CTD was approved to create the structure for the submission of regulatory information, and as this was developed, M2 was able to start the work to define the electronic specification.

The eCTD Business Case and Design Considerat ions

The main objectives of the ICH eCTD specification are the same as those of other electronic submission formats, to provide a global specification for the transport of a submission meeting the CTD structure from an applicant to the regulatory agency. The eCTD submission provides the specification to meet the basic requirements for the presentation of documents and the means to navigate within and between them in the dossier structure.

In addition to providing the specification to meet the structural and navigation requirements, the eCTD has sought to provide a standardized solution to the problem of relating submissions to each other. In the paper submission world, it has long been a desire of regulatory agencies to be able to see how each subsequent dossier relates to the previous ones. In some cases, it has been a requirement to submit the changed pages on different color paper so that they could be identified in an updated volume on the shelf. The possibilities offered by electronic submission technologies have long been seen as providing a more efficient way to resolve this problem. The FDA started a project known as the cumulative Table of Contents (c-TOC) in about 1998 to create an electronic submission solution to this problem. When the ICH adopted the eCTD project, the experience from c-TOC was brought to the project and it was decided that the specification should include a means to create and manage the relationships between documents and dossiers over the life of the drug product. Within the eCTD topic, this is known as life-cycle management.

Key to the resolution of the life-cycle management requirements is the ability to be able to submit submission content only once and to provide a means to refer to this from later submissions, without the need to resubmit the content. From this initial requirement, the specification must also be able to identify that when content is submitted and what its relationship is to previously submitted data. This will be described in the following sections.

Another major part of the ICH eCTD business case would be the ability to design a specification that could also be extended to cover the regional aspects of the CTD. It is understood that the ICH CTD could only describe content that would be common to all three of the ICH regions, but that each region would have regional requirements that would need to be defined locally. The eCTD specification would need to accommodate the global and regional needs and attempt to do so in a way that would be compatible so that processes and the applications for creating the eCTD could be broadly similar between regions.

The ICH M2 group also wanted to learn from and reuse other parts of the existing regional and national specifications. Therefore, individual parts of these would be copied and reused wherever they could be as the new specification was developed.

Lastly, wherever possible, the eCTD specification should employ open standards and avoid using proprietary formats. There was a concern that if the standard was to be employed for submissions that would need to be readable and reusable over the potentially long lives of drug products, then one of the best ways to ensure this was by using formats less susceptible to a vendors developmental whims.

The Development of the eCTD Specification

The development of the eCTD in the ICH M2 group began in 1999 with the first Step 2 document being released in June 2001. Step 4 was reached in September 2002 with the specification, Version 3, posted to the ICH Web site in October 2002. This was followed by a minor update to the specification, Version 3.2, in February 2004. In July 2008, Version 3.2.2 (1) was released, which contained only minor wording clarifications. The ICH Study Tagging File specification was released in November 2004 and updated with minor wording clarifications as Version 2.6.1 (2) in July 2008.

The ICH also works on questions and change requests raised by users of the specification. At each ICH Meeting, the latest change requests are considered and new clarifications on the specification are usually issued as official Questions and Answers, the latest version (as of December 2008) being Version 1.16 (3). As the occasion arises, the answers are integrated into the eCTD specification, as happened with Version 3.2.2.

THE COMPONENTS OF THE eCTD

Applications, Activities, and Sequences

An eCTD is usually organized with the following hierarchy. The terms are generic and mapped to terms on a local basis.

Application

The highest level of organization is usually referred to as the application. In the United States, this maps to the investigational new drug (IND) or NDA application with the top level folder named with the procedure number (4). In the European Union, the highest level is less clearly defined and depends on the review procedure for the product. In the Centralised Procedure, the top level is the procedure number (e.g., EMEA/H/C/1234) but can also be the INN or invented name if the procedure number is not known (5), but in the other procedures, it is more often the trade name of the product.

The specified naming of the top-level application folder in the United States does offer the opportunity to refer to content in another application, something that is not possible in the European Union where the top-level folder name is not specified. The ability to refer between applications in the United States is especially useful for the NDA when reusing content originally submitted in the IND phase.

Activities

The concept of a regulatory activity is not described specifically in the specifications but has emerged through the way that the United States and European Union have implemented the eCTD. It recognizes that the individual submissions are related to specific activities (an original application, a supplement or variation, etc.) and that sequences within an activity have a particular relationship. Moreover, the product life cycle will often have parallel activities taking place and that sequences in different activities have no particular relationship.

In both the United States and the European Union, sequences in the same activity are related using the related-sequence attribute in the administrative data section of the Module 1 (see later). In Japan, the eCTD life cycle is managed at the level of the activity.

Sequences

A single eCTD is often referred to as a Sequence because each submission is identified with a four-digit sequence number. The eCTD sequence is a self-contained submission, containing all of the necessary information to place the submission in context with other earlier sequences and the required files to demonstrate that the sequence is valid and can be viewed and consumed by the receiver of the sequence.

In general, sequences are submitted in sequential order but it is accepted that the sequences received by a single agency could have gaps or be received out of order (6). This is something that is most likely to happen in the European Union Mutual Recognition and Decentralised Procedures where country-specific sequences will not be submitted to all of the agencies in the procedure (7).

eCTD Components

The eCTD consists of four main component parts

- Files
- Folders
- XML (eXtensible Markup Language) backbone files
- Utility files to manage the XML backbone files

The first two components, files and folders, are the common elements that the eCTD shares with other earlier electronic submission formats. The defining feature of the eCTD is the use of XML to manage the metadata about the documents and dossier that allows the relationships over the product life cycle to be managed.

Files

The ICH eCTD specification gives a lot of detail about the acceptable files that can be included in the ICH eCTD (8). The files are often referred to as leaf files, a reference to the XML element that is used to refer to the file which is called a leaf element. The reason is that the analogy of a tree is used to describe the eCTD with the CTD structure forming the trunk and branches on to which the leaf documents are attached. In general, the regional specifications use the same leaf specifications as the ICH eCTD.

File Formats

The ICH specification identifies the acceptable file formats for the eCTD. These are PDF for narrative data, XML for structured data, and four file formats acceptable for graphics (JPEG, PNG, SVG, or GIF). It was noted in the design considerations that proprietary file formats would be avoided where possible. However, the wide use of PDF in business and beyond had established it as a de facto standard and Adobe has since made the latest version of the format an ISO standard.

In specifying PDF as the main file format for narrative data, the eCTD specification copies much of the detail about the requirements from the original FDA eNDA standard. This includes the requirements for the page size, text font size, page orientation, margins and the navigational aspects from the bookmarking, and hyperlinking of the files. In general, applicants are advised to provide bookmarks to match the TOC of the document, although it has been clarified that bookmarks are not an absolute requirement if the document does not contain any internal structure. Hyperlinks are an essential part of the navigation within a single file and to other files in the same sequence or other sequences. The requirement is for hyperlinks to be displayed as blue text or with a blue bounding box around black text.

The eCTD recommends the use of PDF v1.4 as the only acceptable PDF version in all of the ICH regions. This recommendation has arisen due to problems with the use of other PDF versions in Japan where problems with the conversion from the source document and display of Kanji and western characters have been observed.

The use of alternate file formats for graphics is recommended only when the applicant feels that appropriate quality of reproduction cannot be achieved within the PDF format.

The regional specifications tend to use PDF as their predominant format. The U.S. specification allows for the use of other formats to support the submission of datasets and both the

U.S. and the European Union specifications allow for the submission of labeling documents in an XML format.

File Naming

File names in the eCTD should use only lower-case letters, the digits 0–9, and hyphens. All other characters are disallowed. The file name should not exceed 64 characters and can have a three- or four-character file extension.

There has been much discussion about the file names given in the eCTD specification, but it has been clarified that the file names are only "highly recommended" and none of them are mandatory. This is because the general expectation is that the user of the eCTD will navigate to the files using the information contained in and displayed from the XML backbone. The metadata in the XML backbone allows for a much richer description of the individual content files and does not rely on this information being derived from the file name. Put the other way around, it is not expected that the eCTD user will have to navigate through the file folder structure to locate a document, so the file name does not need to be very descriptive, and it merely needs to be something that can be identified from within the XML backbone.

In reviewing the recommended file names, it should be noted that some file names are repeated (e.g., there is an introduction.pdf in Module 2.2, Module 2.3, and Module 2.6.1). Also, there are several parts of the CTD structure where the applicant is expected to use one's own study number or other identifier in place of the generic name given in the specification (e.g., replacing analytical-procedure-1.pdf in Module 3.2.S.4.1 with an applicant identifier).

File Size

In general, individual files should not exceed 100 MB in size. This limit is to maintain acceptable performance in accessing and working in the files. It is usual that this limit is approached only if the source document contains a large number of scanned pages or high-quality graphic images. When this limit is exceeded, the applicant is advised to split the document into smaller parts, but this is one exception to the general requirement that individual files be numbered starting at page 1 as the second and subsequent parts should continue pagination from the previous part.

File Granularity

The subject of granularity (how much information or data should be included in a single file) is addressed in an ICH-approved annex to the CTD guidance document (9). This document describes what constitutes a document in the paper CTD in each section of each Module and how it should be identified and paginated. In the eCTD, the definition of a file follows the CTD definition of a document in the paper submission. It should be noted that there is an optional granularity in a number of the sections of the CTD/eCTD. The general rule for the eCTD is that the granularity should not be changed over the life of the product as this can lead to problems in managing the life-cycle operations. Therefore, some thought must be given to the future management of a product when setting the initial granularity in many sections of the eCTD.

Repeated Content and Data Reuse

As has been previously noted, the intent behind the eCTD is that applicants submit data only once in the life cycle of the product represented by the eCTD. The leaf concept allows this because, among other things, it is a pointer to content (10). Therefore, if an individual document is used in more than one section of the dossier, such as a combined ADME (Absorption, Distribution, Metabolism, Excretion) study that should appear in the separate subsections of Module 4.2.2, then the content can be submitted once and separate leaf elements used to point at the content.

This concept can be carried further with the reuse of content permissible between sequences in the same eCTD life cycle and, in some cases, between applications. For example, a pivotal clinical study used to support the main indication in the original submission can be referred to from later sequences if the same report is then used to support safety findings for new indications. Other examples of this reuse could be analytical methods reused for new product formulations, toxicology studies referred to later in the product life to support safety follow-ups, the reuse of basic literature references over the life cycle, and many other examples that can arise.

Sections with No Data

The initial implementation of the CTD stated that sections with no data should contain documents stating "not applicable." In the eCTD, this would have brought problems with the management of document life cycle as including simple leaf documents to state "not applicable" would have meant that there were documents with a life cycle to maintain, just to say nothing was required. The CTD and eCTD were subsequently revised to make it clear that, where there was not scientific or regulatory content in a section, then there was no need to supply a document. This means that the eCTD only contains documents that include data.

One slight change to this has been the need in some parts of the dossier to justify why there is no data. This has led some applicants to prevent regulatory questions by supplying documents in sections such as Module 3.2.P.4.5 to state that no excipients of animal and human origin are included.

Folders

The folder structure of the eCTD is a convenient set of folders to help the applicant manage the many leaf files that make up the complete eCTD submission (11). The eCTD for a new chemical entity original application can contain several thousand files, particularly in the United States where the datasets and case report forms (CRFs) for each study must also be submitted. Managing this number of files is assisted by using a number of folders to split up the task.

The folder structure specified follows the basic structure of the CTD. There are folders for each main module and subsections of the modules as identified in the CTD specification. This allows a user to navigate the structure and identify files in this way should the need arise. However, it is still expected that the main way of navigating the eCTD will be the XML backbone.

Folder Naming

The folder naming follows the same rules as that of the file naming. This means that there is a restricted set of characters available and certain characters are disallowed.

The folder structure and naming is not mandatory, only highly recommended. Indeed, this flexibility is required to allow the management of folder naming to avoid ending up with overly long folder paths to the individual leaf files. It is recommended that the path length does not exceed 230 characters but this should be seen as an absolute maximum and applicants should generally try to keep the path as short as possible.

Sections Without Data

As noted previously where a section has no data then there is no need to submit a file. As a result, there is no need to provide a folder structure where there is no content at the lowest level. The general rule is that a folder must contain another folder or a file.

The XML Backbone

The XML backbone is the defining difference between the eCTD and other electronic submission formats. The XML backbone is used to manage the metadata about the documents and dossier that allows the relationships over the product life cycle to be managed.

In any one eCTD, there are actually two XML backbones. The ICH specification defines the makeup of the file index.xml that is used to manage the information for Modules 2 to 5. The second backbone is called xx-regional.xml (where xx is a two-letter code to define the region, e.g., the United States or European Union) and manages the information for the Module 1. In addition, the Study Tagging Files (STF) are akin to mini backbone files to manage the information for each study in Modules 4 and 5.

Metadata is often referred to as "information about information." It is not information that is part of the reviewable content of the submission but it is used to better understand, organize, or view the content. It is the metadata and ability to manage it over the life cycle of the eCTD that is invaluable in the eCTD format.

The metadata is used in a number of different ways in the eCTD and this is described in the following sections. Within the eCTD, the metadata is technically used in the XML as either elements or attributes of the elements, but within the discussion, this distinction is not detailed and the term attribute is used interchangeably.

Within the current eCTD specification there is no easy way to change the metadata, so it is vital to ensure that metadata is set correctly in the initial submission.

Structural Metadata

The first use of metadata is to help define the structure of the eCTD submission. In particular, the metadata is used to identify the repeated sections of the CTD structure. As is known, the CTD specifies that certain sections of the dossier structure are repeated to allow a single submission to contain multiple drug substances, drug products, excipient sections, and indications. Rather than build submission-specific structures with the identifiers for the repeated sections hardcoded into the backbone, the metadata allows an attribute to be defined as variable that the applicant sets the value for.

The structural metadata are used in the following sections:

- Drug substance (Modules 2.3.S, 3.2.S, and 3.2.A) (12)
    - Substance name
    - Substance manufacturer
- Drug product (Modules 2.3.P, 3.2.P, and 3.2.A) (13)
    - Product name
    - Dosage form
    - Product manufacturer
- Excipient (Module 3.2.P.4 and 3.2.A)
    - Excipient name
- Indication (Module 2.7.3 and 5.3.5) (14)
    - Indication

Defining the structure of the submission is a topic in itself, but will not be covered in this chapter, and key to setting this is the way that the drug substance and product metadata are set. As it is not easily possible to change these settings during the life of the eCTD, it is vital that this is set appropriately at the start.

Leaf Metadata

The leaf metadata defines important information about the leaf (15), its link to the content files, its integrity, and its relationship to previously submitted leafs. This is another difference between the eCTD and earlier electronic submission formats where many of the metadata items described were forced into the individual file names of the content files. This section describes some of the key metadata values.

Leaf title

The leaf title allows the applicant to provide a detailed descriptor of the content without any restrictions on the length of the text or the characters used. The title is then used as a display value by all of the eCTD tools.

File location

The leaf attribute that points to the content file is called xlink:href. It is this attribute that allows the content to be submitted once but leafs in other parts of the same sequence or in another sequence to point to the content.

ID

The ID attribute is a unique identifier for the leaf within the XML backbone file. It is used as a means to easily and unambiguously identify a specific leaf.

Operation and modified file

The operation attribute is the key metadata value to define the relationship of one leaf to a previously submitted leaf. The attribute for each leaf can have one of four values: new, replace, append, or delete (16).

The value "new" implies that the associated content file has no relationship to previously submitted and is applied to all content being submitted for the first time. The value "replace" implies that the previously submitted content file is now invalid for some reason and needs to be replaced. The value "append" is used when there is additional data provided to supplement to the already submitted content file but that the originally submitted data is still valid in its entirety. Lastly, the value "delete" is used when previously submitted data should no longer be used. The use of "delete" does not lead to content being removed physically—it is an instruction for viewing tools to not display the associated data.

Whenever the operation attribute has the values replace, append, or delete, then the modified-file attribute identifies the content to which the operation applies. It does so by identifying the sequence, backbone file, and ID attribute of the leaf the operation affects.

Checksum and checksum type

The checksum attribute is a means of identifying whether a file has been changed during the transmission or storage of the submission. The checksum is a value that is calculated using a widely available algorithm (defined by the checksum type). The algorithm uses physical characteristics of the file to generate the checksum value. If the physical characteristics of the file are changed (e.g., by adding a hyperlink, deleting content, a corruption from the transfer of the file), then the checksum value calculated will be different.

In the eCTD, the algorithm used is called MD5. The checksums values are calculated for each content file and recorded as an attribute of the leaf. The checksum of the index.xml file is also calculated and submitted, but this cannot be saved in the backbone (adding the value in would change the index.xml file and, thus, the checksum value), so it is recorded in a text file submitted along with the submission. The agency receiving the eCTD will calculate the checksums for all of the files they receive and then compare them with the values submitted by the applicant. Any discrepancies will then be challenged to ensure that the correct content has been received.

Submission Metadata

The regional eCTD specifications have defined metadata to help in the automatic processing and management of the submission. Although this metadata is defined regionally, it has similarities between the regions. The regions define values to identify the product [typically the international non-proprietary name (INN) and trade name but might also include the chemical name or company identification numbers], information about the applicant company, information about the submission type [in the United States, this includes the NDA supplement type while in the European Union there is a range of submission types to cover all of the submission types to support Marketing Authorisation Applications (MAAs)], and the related sequence that is used to define the regulatory activity. In the European Union, there are additional metadata for the country and agency and the review procedure.

Additionally, the structure of the submission metadata in the European Union allows for the application to include the local information for a number of different countries within a single eCTD sequence. Clearly, the country and agency identifiers will be different, but it is also likely that the applicant name and even the trade name could differ between countries, so having a means to separately manage this information is vital.

Study Metadata

The ICH Study Tagging File specification (2) is used primarily in the United States and helps reviewers identify and associate content with particular studies. The metadata is used by the review tools to automatically process and present the content in a standard way to the reviewer.

The study metadata includes the study identifier and title along with some specific further metadata values for particular study types (species of animal, study duration, route of administration for some preclinical studies, and type of control for some clinical studies). The individual content files have a specific study tag value that identifies the type of content the file contains. This allows the reviewer to easily identify and compare data received in a number of different studies.

Other European Union Regional Metadata

The European Union Module 1 specification (17) also includes other structural metadata that is used to help organize and present the content of the regional Module. The management of the different review procedures requires that some content be identified as being associated with a particular country or language, particularly where a single submission sequence can contain content for a number of different countries and their local languages. The specification also includes a metadata value to identify the type of labeling document. This is used to help organize the display of the content.

Utility Files

The use of XML technology in the eCTD brings with it the need for some utility files to support the backbone files.

Document Type Definition (DTD)

The DTD is the rules' file for the XML backbone. The DTD defines the elements and attributes that make up a technically valid backbone file. For example, it defines the headings in the CTD structure, their order, and whether they are repeatable. The DTD defines the metadata elements and attributes of each leaf.

Each valid European Union, U.S., and Canadian eCTD submission will have at least two DTDs, the ICH DTD and the regional Module 1 DTD. In the European Union, the DTD is supported by two MOD files as a more modular approach to the management of the regional backbone has been taken. In addition, if the U.S. submission has any studies in Modules 4 or 5, then the eCTD will also contain the STF DTD.

In Japan, the Module 1 is defined by a Schema. This file performs the same role as the DTD for the ICH and regional backbone files but is based on a more modern XML technology.

Stylesheets

All of the eCTD specifications include a simple means to view the submission using an Internet browser. This is a limited functionality means of viewing the submission as the stylesheets do not allow the viewing of the life-cycle relationships in the way that more advanced eCTD viewing tools allow.

Applicants can develop their own stylesheets for eCTD viewing using standard XML tools. However, it is not recommended to submit these as they are not used by the agencies and, depending on how they have been developed, they may upset the functioning of the agency review tools.

eCTD LIFE-CYCLE MANAGEMENT

The concept of eCTD life-cycle management has been touched on in a number of preceding sections. The ability to establish and manage the relationships between documents and dossiers is the biggest functional difference between the eCTD and previous electronic submission formats.

Document Life-cycle Management

The relationships between individual documents are established primarily by the operation and modified-file attributes of the leaf elements in the eCTD. The four allowed values of the operation attribute are reasonably self-explanatory but experience has shown that the use of the operation attribute is dependent crucially on the business processes and document granularity adopted by the applicant.

A finer document granularity allows management of the content in smaller chunks and this may be advantageous when submitting supplements or variations as the scale of the change can be controlled more tightly. However, the finer granularity gives the applicant a bigger document management task and also means that there may be more external hyperlinks between files, something that is more difficult to create and manage.

Dossier Life-cycle Management

The concepts of dossier life-cycle management are probably not as well developed in the eCTD specifications as those of the document life cycle. The concept is managed mainly by the definition of the application and the use of the related-sequence metadata value to show the regulatory activity.

An element of dossier life-cycle management also comes from the decisions taken over the structure of the submission, particularly in Module 3. This will then define the way in which the applicant manages future supplements and variations within the overall life cycle.

eCTD TOOLS

The original eCTD design requirements suggested that the eCTD should be something that could be created without recourse to sophisticated tools and experience has shown that single eCTD submissions can be created using basic text editor tools. However, the complexity of managing document and dossier life cycles within the regulatory submission timelines means that some tools are almost certainly required to assist in the creation, validation, viewing, and reviewing of the eCTD.

eCTD creation tools come with a variety of functionalities. The simpler tools require the user to create the content files separately and will create the XML backbone and folder structure. More complex tools might introduce functionality to integrate with an electronic document management system, to manage the creation of the PDF files from source files and to better manage the document relationships over the life of the eCTD.

Validation of the eCTD is a key requirement and tools can assist in this task greatly. The DTD defines the rules to create a technically valid eCTD and a parsing XML editor can check that these rules are being followed. However, there are many additional business rules, such as the checking of the MD5 checksums, that also apply for an eCTD to be valid and tools are required to assist in these checks.

Viewing of a single eCTD sequence can be achieved using the stylesheets that accompany the eCTD specifications. However, to display the full life-cycle relationships between documents and dossiers, a tool is required. Reviewing tools will add functionality to annotate the content and assist in the creation of assessment reports at the agencies.

eCTD IMPLEMENTATION STATUS

Only four regions/countries have fully implemented the eCTD since the ICH specification was first published in 2002 by publishing a regional specification to accompany the ICH one. These are Canada, the European Union, Japan, and the United States.

Status in the European Union

The European Union was the first region to publish its draft Module 1 specification in March 2002 with an initial statement to accept submissions in the eCTD format from June 2003. However, it was not until July 2004 that Version 1.0 of the European Union Module 1 specification was finally published.

Since then, there have been three further updates to the European Union Module 1 specification. The driver for all of these changes has been changes to the European Union regulation requiring new sections in the Module 1. At the same time, the opportunity has been used to review the technical specification of the Module 1 and to implement other updates. Version 1.1 was released in January 2006, Version 1.2.1 in October 2006, and Version 1.3 (17) in May 2008.

In February 2005, the European Union Heads of Medicines Agencies resolved that by the end of 2009 all the European Union Member States would be able to accept submissions electronically without paper and that the format would be the eCTD (18). Since then, the individual National Competent Authorities (NCAs) and the European Medicines Agency (EMEA) have worked to implement the eCTD. Individually, the Medicines Evaluation Board (MEB) in the Netherlands was the first NCA to publish guidance about electronic submissions and the acceptability of the eCTD, but they have been followed by at least 10 other NCAs who have also published guidance. The EMEA published a Statement of Intent about the acceptability of the eCTD in the *Centralised Procedure* in January 2008 making the procedure fully electronic from

January 1, 2009 (19) and making the eCTD the mandatory electronic submission format from January 1, 2010 (20).

Despite the early adoption of the standard, the numbers of eCTDs have been relatively low. Gathering data has been difficult due to the problems of getting figures from all of the European Union Member States, but the best estimate is that about 5000 eCTD sequences have been submitted. The EMEA Statement of Intent has led to eCTD numbers in the Centralised Procedure rising with 90% of products registered via this process having some submissions made using the format, although the overall percentage of eCTD is somewhere about 30% of all submissions. In the other procedures, the numbers are a lot smaller with a maximum estimate of 3% of submissions being made in the eCTD format. Work continues in Europe to increase the numbers of eCTDs through the transition to electronic working and then to the eCTD.

Status in the United States

The initial draft guidance for Module 1 and the use of the eCTD in Modules 2–5 were published in August 2003, but full implementation waited until the publishing of the ICH STF specification in November 2004 (21). The STF specification describes how to organize study information and is vital in allowing the eCTD to be used in the United States during the IND phase as well as the NDA phase. At the end of 2007, the FDA announced that the eCTD would become the only acceptable format for electronic submissions from January 1, 2008, although a waiver process was put in place so that the older eNDA could still be used under certain conditions (22).

As of the end of October 2008, the FDA had received just over 44,000 eCTD sequences associated with just over 4000 applications. These numbers show the overall success of the eCTD implementation in the United States. However it is worth noting that an analysis by the FDA of all of their 2007 submissions (23) showed that of around 167,000 submissions, 11% were electronic only and 8% were mixed paper and electronic, meaning that 81% of submissions were still in paper only. Further analysis showed that the percentage of eCTD submissions for the various NDA types (original applications and various supplement types) varied between 24% and 71%. The greatest percentage of paper submissions remains in the IND where paper accounts for over 90% of all submissions.

Status in Other Regions

The implementation of the eCTD in Japan has been slightly different from that in other regions (24). As has been noted, the eCTD life cycle is based around a regulatory activity, with a new eCTD life cycle being started if, for example, a new indication is submitted for a product which was previously submitted in eCTD format. The Module 1 in Japan is based on a Schema, rather than a DTD, which is a more modern XML technology. The backbone is also submitted with all of the previously submitted leaf elements included along with the new leafs. As of September 2008, the Pharmaceuticals and Medical Devices Agency (PMDA) had received 92 eCTD applications, with each application containing a maximum of five eCTD sequences.

The other region to implement the eCTD has been Canada. The Module 1 specification was published in October 2005 (25) and an implementation project has been under way since 2006. The eCTD is used for MAAs in a similar way to both the United States and the European Union. At the end of May 2008, Health Canada had received just over 900 sequences associated with 195 different applications.

FUTURE DEVELOPMENTS FOR THE eCTD

The ICH eCTD specification has been around in pretty much the same form since Version 3.0 was published in 2002 (the update to Version 3.2 in February 2004 was a fairly minor change to the specification). The challenge for the ICH M2 group has been to support the eCTD specification's adoption by only making changes absolutely necessary to support the business needs, but not making so many changes that industry and the eCTD tool vendors are unable to implement because of the frequent changes.

In 2007, the ICH Steering Committee approved the ICH M2 group to begin the collection of the business requirements for the development of the Next Major Version of the eCTD. In October 2008, the ICH Steering Committee approved the development of the Next Major Version of the eCTD through a process involving Standards Development Organisations (SDOs). SDOs are accredited organizations that develop standards to meet the requirements of individual national or international standards bodies. Some of the better-known SDOs in the health-care industry are the International Standards Organisation (ISO), the European Standards Organisation (CEN), and Health Level 7 (HL7), an SDO accredited by the American National Standards Institute (ANSI). The processes used by these individual organizations are required to be followed by the mandates of some of the individual regions within the ICH if the resulting standard is to have a legal status and basis in the region.

The intent for the ICH is to produce an ISO, CEN, and HL7 standard that will meet the requirements of all of the ICH regions. This development process will take into account the existing electronic submission formats, including the eCTD and the HL7 Regulated Product Submission (RPS) format. The exact process that will bring together the eCTD and RPS formats has yet to be determined but the work has started and is likely to result in a new standard, at the earliest, in 2011 or 2012.

Work in adopting the eCTD in other regions and countries is also continuing, though slowly. The only country known to be working on developing its own Module 1 specification is Switzerland, which intends to be ready to accept eCTDs from the end of 2009. It is hoped that the development of the eCTD through a process that involves international standards organizations may improve the chances of adoption of the standard outside the three ICH regions.

Although other countries are working on electronic submission formats, some of which borrow ideas from the eCTD, such as the Russian electronic submission format (26) that creates a text-based backbone file with MD5 checksums for the files referenced in the submission. These national formats are aimed at meeting many of the business objectives outlined at the beginning of this chapter. The hope is that as the eCTD becomes more widely adopted national formats will be withdrawn and that the eCTD can become a truly global standard.

REFERENCES

1. http://estri.ich.org/eCTD/eCTD_Specification_v3_2_2.pdf. Accessed January 2009.
2. http://estri.ich.org/STF/STFV2-6-1.pdf. Accessed January 2009.
3. http://estri.ich.org/eCTD/eCTDQAV1_16.xls. Accessed January 2009.
4. http://www.fda.gov/cder/regulatory/ersr/validation_1.0.pdf. Accessed January 2009. Item 1519, page 16.
5. http://www.emea.europa.eu/pdfs/human/regaffair/59688107en.pdf. Accessed January 2009. Page 21.
6. http://estri.ich.org/eCTD/eCTD_Specification_v3_2_2.pdf. Accessed January 2009. Page 6–1.
7. http://www.hma.eu/uploads/media/BPG_eCTD_in_MRP_DCP_Rev0.pdf. Accessed January 2009.
8. http://estri.ich.org/eCTD/eCTD_Specification_v3_2_2.pdf. Accessed January 2009. Appendix 7.
9. http://www.ich.org/LOB/media/MEDIA554.pdf. Accessed January 2009.
10. http://estri.ich.org/eCTD/eCTD_Specification_v3_2_2.pdf. Accessed January 2009. Page 6–5.
11. http://estri.ich.org/eCTD/eCTD_Specification_v3_2_2.pdf. Accessed January 2009. Appendix 3.
12. http://estri.ich.org/eCTD/eCTD_Specification_v3_2_2.pdf. Accessed January 2009. Table 6–10.
13. http://estri.ich.org/eCTD/eCTD_Specification_v3_2_2.pdf. Accessed January 2009. Table 6–11.
14. http://estri.ich.org/eCTD/eCTD_Specification_v3_2_2.pdf. Accessed January 2009. Table 6–9.
15. http://estri.ich.org/eCTD/eCTD_Specification_v3_2_2.pdf. Accessed January 2009. Table 6–8.
16. http://estri.ich.org/eCTD/eCTD_Specification_v3_2_2.pdf. Accessed January 2009. Page 6–3.
17. http://ec.europa.eu/enterprise/pharmaceuticals/eudralex/vol-2/b/update_200805/eu_m1_specification_v13.zip. Accessed January 2009.
18. http://www.hma.eu/uploads/media/23.feb.2005.pdf. Accessed January 2009.
19. http://www.emea.europa.eu/pdfs/human/regaffair/56336607en.pdf. Accessed January 2009.
20. http://www.emea.europa.eu/pdfs/human/regaffair/57245908en.pdf. Accessed January 2009.
21. http://www.fda.gov/cder/regulatory/ersr/ectd.htm. Accessed January 2009.
22. http://www.fda.gov/cder/regulatory/ersr/waiver.htm. Accessed January 2009.
23. Gensinger G. DIA 7th Annual Electronic Submissions Conference, San Diego, CA, Nov 6–7, 2008.
24. http://www.pmda.go.jp/ich/m4.htm. Accessed January 2009.
25. http://www.hc-sc.gc.ca/dhp-mps/prodpharma/applic-demande/guide-ld/ctd/ectd/index-eng.php. Accessed January 2009.
26. http://www.regmed.ru/downloads/examples/fgu2008/e%D0%C7%CD%C4.doc. Accessed January 2009.

3 CTD Module 1—Administrative Information

Roger Croswell
Regulatory Consulting Services, Ludlow, Vermont, U.S.A.

Anthony C. Cartwright
Pharmaceutical Regulatory Consultant, Harpenden, U.K.

MODULE 1: ADMINISTRATIVE INFORMATION

The ICH Harmonised Tripartite Guideline M4(R3) on Organisation of the Common Technical Document (1) indicates that the Common Technical Document (CTD) Module 1 should contain administrative documents specific to each region, for example, forms, labeling, etc., and that the format be specified by the relevant regulatory authorities. Therefore, Module 1 is the one module in the CTD that is not common among the various regions. This chapter reviews the contents of Module 1 in each of the three ICH regions in the following order--Europe, Japan, and the United States. The format and contents of Module 1 for other countries is available from Web sites of the national competent authorities such as that for the Therapeutic Goods Administration (TGA) in Australia (2).

EUROPE

The Module 1 requirements (Administrative and Prescribing Information) for the 27 European Union countries are defined in the current edition of Volume 2B of the *Notice to Applicants, Medicinal products for human use, Presentation and format of the dossier, Common Technical Document (CTD)* (3). This format is used for national marketing authorisation (MA) applications in each of the European Union countries, and for applications via any of the European procedures (mutual recognition, decentralized and centralized). A centralized procedure is made via the European Medicines Agency (EMEA). The same format can be used for applications in the European Economic Area (EEA) countries—Iceland, Norway, and Liechtenstein.

The major sections in the European Module 1 are shown in Table 1. Additional information will be needed for national, mutual recognition, and decentralized applications, and the requirements for each country are detailed in Chapter 7 of Volume 2 A of the *Notice to Applicants*, and also on the European Commission/Directorate General Enterprise Web site (4).

Cover Letter

A cover letter to the application is invariably included. This can have a "Notes to Reviewer" attachment, which provides further information to help the reviewers in the agency/ies such as presentation of the volumes, hyperlinking, etc.

Comprehensive Table of Contents

The Table of Contents (TOC) should include a list of all of the individual documents in Modules 1, 2, 3, 4, and 5 included in the particular MA application. The choice of individual documents reflects the granularity of the dossier. Granularity can be defined as the extent to which a system contains separate components (like granules). The more components are contained in a system, the greater the granularity. The ICH M4 Guidance document *Organisation of the CTD for the Registration of Pharmaceuticals for Human Use* (1) suggests the options in terms of granularity--levels of documentation.

Application Form

The European application form (5) is used for applications submitted to the EMEA under the centralized procedure or for applications to the European Union and EEA countries under national, mutual recognition or decentralized procedures.

Table 1 Major Sections Included in Module 1 of a European Union CTD (3)

Sections	Descriptions
1.0	Cover Letter
1.1	Comprehensive TOC
1.2	Application Form
1.3	Product Information
1.3.1	Summary of Product Characteristics, Labeling, and Package
1.3.2	Leaflet
1.3.3	Mock-up
1.3.4	Specimen
1.3.5	Consultation with Target Patient Groups
1.3.6	Product Information already approved in the European Union Member States Braille
1.4	Information about the Experts
1.4.1	Quality
1.4.2	Nonclinical
1.4.3	Clinical
1.5	Specific Requirements for Different Types of Applications
1.5.1	Information for Bibliographic Applications
1.5.2	Information for Generic, "Hybrid," or Bio-similar Applications
1.5.3	(Extended) Data/Market Exclusivity
1.5.4	Exceptional Circumstances
1.5.5	Conditional Marketing Authorisation
1.6	Environmental Risk Assessment
1.6.1	Non-GMO
1.6.2	GMO
1.7	Information relating to Orphan Market Exclusivity
1.7.1	Similarity
1.7.2	Market Exclusivity
1.8	Information relating to Pharmacovigilance
1.8.1	Pharmacovigilance System
1.8.2	Risk-Management System
1.9	Information relating to Clinical Trials
1.10	Information relating to Pediatrics
	Responses to Questions
	Additional Data

Product Information

A European MA application must include proposals for the Summary of Product Characteristics (SPC--data sheet document addressed to European health-care professionals), immediate container and secondary container labeling, and the Package Leaflet (PL)/Patient Information Leaflet (PIL) intended for use by the patient or carer. There is a mandatory format for templates for the product information (6). The templates need to be used in conjunction with relevant guidelines on the SPC and the PL/PIL included in Volume 2C of the *Notice to Applicants* (7). By the end of a European registration procedure, the applicant has to provide a complete set of SPC/labeling/PL texts in all of the relevant European languages.

"Mock-up" of the Outer and Immediate Packaging

A "mock-up" of the outer and immediate packaging is required to be included. A mock-up is defined as a copy of the flat artwork (sometimes called a "maquette") as a replica of the outer and immediate packaging, in full color.

Specimen

A specimen is an actual sample of the printed outer and immediate packaging and PL. The European Union countries or the EMEA may request such a specimen to check compliance with the legal labeling requirements.

Consultation with Target Patient Groups

There is a legal requirement (Articles 59 and 61 of Directive 2001/83/EC) that the PL (Package Information Leaflet or PIL) reflects the results of consultations with target patient groups so that it is legible, clear, and comprehensible. There is a guideline on readability testing (8) and a separate guideline on consultation with target patient groups for the PL (9). The European Federation of Pharmaceutical Industry Associations (EFPIA) has also issued some valuable general recommendations for readability user testing (10). A number of contract companies now offer a service in PIL design, layout, and user testing using panels of volunteers. They will produce a report with the results of this testing, which is included in this section of Module 1.

Braille

Braille consists of arrangements of raised dots within defined "cells," which make up letters, numbers, and punctuation intended for blind and partially sighted users. The name and strength of the product in Braille format needs to appear on the packaging (11). Since there are differences in Braille in different countries, the national requirements for Braille coding should be followed. A Web site is being constructed by the European Blind Union and the Royal National Institute for Blind People. A European Standard is under development, which will include recommendations and requirements for Braille for pharmaceutical labeling (prEN 15823 Packaging—Braille on packaging for medicinal products). A separate Comite European de Normalisation (CEN) document is available (12), which offers advice on the provision of product information for blind patients (CEN/TR 15753:2008 Packaging. PLs for medicinal products. Braille and other formats for visually impaired people). In addition, blind and partially sighted patient groups can request large print, CD-ROM, or audiocassette versions of the PIL.

Information About the Experts

Article 12 of Directive 2001/83/EC requires that experts must provide detailed summaries of the documents and information in Modules 3 (Quality), 4 (Nonclinical), and 5 (Clinical). Each of these summaries must be signed by one individual expert. Module 1.4 requires a formal signed declaration by each expert and a copy of brief *Curriculum Vitae* detailing their background, training, and experience.

Specific Information for Different Types of Applications

This section includes requirements for specific information for different kinds of marketing authorizations such as bibliographic, generic, bio-similar applications, and also for claims for additional market exclusivity for new indications or changes in legal classification [e.g., from prescription to over-the-counter (OTC) sale]. It also includes the requirements to support requests for authorization in exceptional circumstances (where the applicant is unable to provide comprehensive data on the efficacy and safety of the product) or for requests for conditional marketing authorizations (products such as those for serious debilitating disease or life-threatening conditions or where the product is to be used in emergency situations, or for orphan indications).

Information for Bibliographic Applications

A five-page summary is required of the grounds and evidence from the scientific literature to show that the constituents of the medicinal product have a well-established use with an acceptable level of safety and efficacy.

Information for Generic, "Hybrid," or Bio-similar applications

A five-page summary of the evidence is required to show that the product qualifies as a generic of a reference product, a so-called "hybrid" of a reference product (product with a different pharmaceutical form, strength, indications, or route of administration to the reference product), or a similar biological medicinal product to a reference product. The summary includes details of bioavailability or bioequivalence studies for generic or hybrid applications, and studies showing comparability to the reference product for the bio-similar application.

Exceptional Circumstances
Where companies consider that they are unable to provide comprehensive data on the efficacy and safety of the product, the justification needs to be included in this section of Module 1. This will include (13):

- The claim that comprehensive nonclinical or clinical data cannot be provided,
- A listing of the data that cannot be provided,
- Justification of the grounds for approval under exceptional circumstances, and
- Proposals for detailed information on the specific procedures to be followed (safety procedures, further program of studies, statements in the SPC and PL/PIL, etc.).

Conditional Marketing Authorization
This section of Module 1 needs to include the justification for the application for a conditional MA (14). The product needs to fall into one or more of the categories—treatment for seriously debilitating disease or life-threatening disease, a product to be used in emergency situations in response to public health threats, or an orphan product. The applicant should also summarize the proposals for ongoing studies, new studies, or collection of pharmacovigilance data.

Environmental Risk Assessment
The potential risks posed by the medicinal product to the environment arising from its use, storage, and disposal need to be assessed in this section. The requirements are considered in detail in chapter 4 of this book.

Information Relating to Orphan Drug Market Exclusivity
Article 18.1 of EC Regulation 141/2000 requires the European Union Member States not to accept another application for a similar orphan drug product for 10 years once it has been granted authorization. However, there are a number of derogations ("get-out" clauses) and if a second applicant wishes to pursue an application for a similar product within the period of exclusivity, the justification is included in this section (15).

Information Relating to Pharmacovigilance
In this section, the company's pharmacovigilance system and risk management system need to be described. The requirements for an EU Risk Management Plan (EU-RMP) are outlined in Volume 9A of *The Rules Governing Medicinal Products in the European Union—Guidelines on Pharmacovigilance for Medicinal Products for Human Use* (16).

Information Relating to Clinical Trials
In this section, a statement is needed to confirm that all clinical trials carried out outside the European Union have met the ethical requirements of the European Union Clinical Trial Directive 2001/20/EC.

Information Relating to Pediatrics
This section includes copies of documents in relation to the pediatric use of the product, such as an EMEA waiver for the product or class of products and any Pediatric Investigation Plan decision(s) together with copies of compliance reports (17).

Responses to Questions
This section is for responses to questions at Day 121 in the European Union Centralised Procedure or Day 60 in the Mutual Recognition Procedure. New or updated documents or reports are included in Module 3 for quality documents, Module 4 for nonclinical, and Module 5 for clinical.

Additional Data
For national, mutual recognition or decentralized applications additional national-specific data may be needed, such as particular translations or officially verified translations of particular documents, specific numbers of copies of particular sections of the application, mock-ups,

Table 2 12 Major Sections Included in Module 1 of a Japanese CTD

Sections	Descriptions
1.2.1	NDA Application Form
1.2.2	Certificates
1.2.3	Patent Status Information
1.2.4	Origin, Background of the Discovery, and R&D History
1.2.5	Conditions of Use in Foreign Countries
1.2.6	List of Other Pharmaceuticals with Similar Pharmacological Effect(s) and Indication(s)
1.2.7	Draft Package Insert (Labeling)
1.2.8	Documentation of Nonproprietary Name
1.2.9	Format for Designation of Poisonous/Deleterious Pharmaceutical Ingredients
1.2.10	Draft Protocol for Principle Postmarketing Surveillance
1.2.11	List of Attached Documents
1.2.12	Others

Source: From Ref. 20.

samples of drug substance and dosage form, etc. These are mainly specified in Chapter 7 of Volume 2A of the Notice to Applicants (4).

JAPAN

On June 21, 2001, The Japanese Ministry of Health, Labor and Welfare (MHLW) issued "Notification: ELD-No 899" (18), which indicated that a Japanese NDA (J-NDA) for new prescription drugs should be organized in accordance with ICH CTD Guidances (19). This guideline refers to several annexes, which should be followed for NDA dossiers of prescription drugs. Annex 2 of this document provides "Guidance on Administrative Information and Prescribing Information" for Module 1 and briefly describes the 12 major section of Module 1 for a Japanese CTD [Table 2 (20)]. Each of the items in Module 1 of a Japanese CTD is discussed in further detail below. Since this Module and Module 2 of a Japanese CTD are to be submitted in the Japanese language, there is very limited electronic information available in English that discusses these requirements.

Japanese CTD Section 1.1—Table of Contents

In addition to a traditional TOC for Module 1, this section in a Japanese CTD also includes a Summary Tabulated Format, which shows the overall organization of information within Modules 2 to 5 of the CTD.

Japanese CTD Section 1.2.1—NDA Application Form

A copy of the NDA Application Form based on Article 17 and 27 of the Pharmaceutical Affairs Law Enforcement Regulations (MHLW Ordinance No. 1, 1961) is included in this section of a Japanese CTD. The application form provides information on the drug substance and drug product, has a size similar to a Quality Overall Summary in the European Union and the United States, must be completed in Japanese, and is considered to be legally binding. Any change to this information requires a variation (21).

Japanese CTD Section 1.2.2—Certificates

This section includes copies of documents relating to any codevelopment/comarketing contracts or other agreements that may exist with third parties (e. g., supply agreements, letters of cross reference to Drug Master Files, etc.). Additionally, documentation of compliance with Good Laboratory Practices (GLPs) and Good Clinical Practices (GCPs) is also included in this section.

For example, with reference to GLPs, if a nonclinical GLP study is conducted in Japan, the GLP Compliance Statement, signed by the study director, should include a certification such as the following: "this study was conducted in compliance with the U.S. Food and Drug Administration (FDA) Good Laboratory Standards for Nonclinical Laboratory Studies (21 CFR part 58, October 5, 1987) and the Japanese Good Laboratory Practice Standards, Ordinance 21 of PAL (Pharmaceutical Affairs Bureau, MHLW, March 26, 1997)."

Japanese CTD Section 1.2.3—Patent Information

The status of any patents that pertain to the drug that is the subject of this application is provided in this section.

Japanese CTD Section 1.2.4—Origin, Background of the Discovery, and R&D History

The information to be included in this section of a Japanese CTD is defined in Article 18–3, Paragraph 1, Item 1 of MHLW Ordinance No. 1. It should provide a review of the entire development history of the product including information on the drug substance and drug product, the types of studies conducted (including a tabulated history of Quality testing, Pharmacology studies, ADME studies, toxicology studies, and clinical studies), and how they confirmed the safety and efficacy of the product and a discussion of the specific benefits of the product. Section 1.2.4 concludes with a brief discussion including the rationale for any other indications or dosage forms that may be under development. It is generally limited to 10 pages or less and may be omitted from Module 1 if it is included within the CTD Summary Module 2.

Japanese CTD Section 1.2.5—Conditions of Use in Foreign Countries

This section of a Japanese CTD provides information on the regulatory status of the product in other countries with special focus on those countries where it has been approved. If a country has decided not to approve the product, this should be noted with the reason for nonapproval. This section also includes a listing of countries where an MA application has been submitted but no final regulatory action has yet occurred. The information should include the following: name of country and product identification (i.e., trade name, dosage form, strength, indication, and dose). Copies of the labeling (Summary Product Characteristics or SPC, Package Insert) with a summary in Japanese should also be submitted from several countries where the product has been approved. If a Company Core Data Sheet (CCDS) is available, it should also be submitted in this section of the CTD. If approval in one or more countries is based on acceptance of approval in another country (e. g., the European Union Centralized Procedure, Mutual Recognition Procedure), this should be noted.

Japanese CTD Section 1.2.6—List of Other Pharmaceuticals with Similar Pharmacological Effects(s) and Indication(s)

In this section, the applicant is to include a tabulated list of other pharmaceutical products available in Japan, which have a similar pharmacological effect or indication or which may be structurally related to the chemical under discussion or those which have been used as comparators in clinical studies included in the application. The tabular listing should include the following information: product name (generic and trade), company name, approval date(s), regulatory status, chemical structure(s), product identification (i.e., dosage form, strength, indication, and dose), and copies of the product labeling. If one of more of these products were used as comparators in the clinical studies, this should be noted in the tabular listing.

Japanese CTD Section 1.2.7—Draft Package Insert

A draft package insert (labeling) is to be included in this section with emphasis on indications, dose, contraindications, precautions, and warnings. Justification for each of the proposed statements should be made including references to the specific clinical or preclinical data from Module 2 (CTD Summary) and the technical Modules 4 (Safety) and 5 (Efficacy) of the CTD.

Japanese CTD Section 1.2.8—Documentation of Nonproprietary Name

The official nonproprietary name in Japan is called the Japanese Accepted Name (JAN). The sponsor should apply for a JAN prior to submitting a Japanese NDA (J-NDA). The average approval time for a JAN is about six months. A copy of the JAN Approval Certificate and related documents is included in this section of the application form.

Japanese CTD Section 1.2.9—Format for Designation of Poisonous/Deleterious Pharmaceutical Ingredients

On March 18, 1998, the MHLW issued a notification regarding designation of Poisonous/Deleterious Substances. This notification specified the format of information to be summarized on this topic. This summary is to include information on the single- and repeated-dose toxicology studies as well as adverse drug events observed during clinical studies. The Japan specific classification/designation of the drug is based on a review of this summary information. The requirements for control of distribution and storage of the drug depend on the type of designation granted.

Japanese CTD Section 1.2.10—Draft Protocol for Principle Postmarketing Surveillance

As part of the approval of a J-NDA, the Pharmaceuticals and Medical Device Agency (PMDA) can require a continuing evaluation of safety and efficacy postapproval by requesting Phase IV or postmarketing surveillance studies within the first 6 months after launch in varying patient populations such as children, the elderly, pregnant or breastfeeding women, and patients with renal or hepatic dysfunction. The PMDA may also request the development of a longer-term Post Marketing Surveillance Plan. The reexamination period ranges from 4 to 10 years and is dependent on the type of product that is the subject of the application. Draft protocols or plans for the longer-term surveillance program are included in this section of the Japanese CTD.

Japanese CTD Section 1.2.11—List of Attached Documents

This section contains a list of references that describes all the research reports (e. g., quality reports/documents, preclinical pharmacology and toxicology study reports, clinical study reports) and other literature references which the applicant has referenced and included in support of Modules 3 (Quality), 4 (Safety), and 5 (Efficacy).

Japanese CTD Section 1.2.12—Others

If the application is for a new indication, dose, or dosage form of a product previously approved in Japan, this section includes copies of the NDA approval letter for the previous product indicating its current approval status and the Evaluation Report and Module 2 Summary of the approved product.

THE UNITED STATES

In the United States, the Food and Drug Administration (FDA) has issued a document entitled "Comprehensive Table of Contents Headings and Hierarchy" (22). This document lists headings for the 16 specific sections [Table 3 (20)], which are included in Module 1 of a U.S. CTD application. The FDA has issued this guidance to facilitate proper placement of all submission components within Module 1 of the U.S. CTD backbone. Not all sections

Table 3 16 Major Sections Included in Module 1 of a U.S. CTD

Sections	Descriptions
1.1	Forms
1.2	Cover Letters
1.3	Administrative Information
1.4	References
1.5	Application Status
1.6	Meetings
1.7	Fast Track
1.8	Special Protocol Assessment
1.9	Pediatric Administrative Information
1.10	Dispute Resolution
1.11	Information Amendment
1.12	Other Correspondence
1.13	Annual Report
1.14	Labeling
1.15	Promotional Material
1.16	Risk Management Plans

Source: From Ref. 24.

Table 4 Subsections of U.S. CTD Module 1.1

Subsections	Descriptions	Use
1.1.1	Application Form: FDA Form 1571	For submission of an Initial IND and all subsequent submissions
1.1.2	Application Form: FDA Form 356 h	For submission of an Initial NDA and all subsequent submissions (pre- and post-NDA approval)
1.1.3	User Fee Cover Sheet: FDA Form 3397	For submission of any NDA requiring a User Fee
1.1.4	Annual Report Transmittal: FDA Form 2252	For submission of the NDA Annual Report
1.1.5	Advertisements and Promotional Labeling Transmittal: FDA Form 2253	For submission of advertisements and promotional labeling
1.1.6	Transmittal of Labels and Circulars: FDA Form 2567	For submission of labeling (only applicable to biologic products)

Source: From Ref. 24.

described below are included in an initial investigational new drug (IND) or an initial new drug application (NDA) submission. Rather, the structure detailed by FDA is designed to cover essentially any submission that may be necessary during the entire life-cycle management of a product. For example, in Section 1.1.2, an FDA Form 356h is included with the NDA and every subsequent submission to that NDA; whereas in Subsection 1.5.1, the Withdrawal Request would be used only if a sponsor desired to withdraw a previously submitted application; or in Subsection 1.12.1, pre-IND Correspondence is used only in support of a pre-IND meeting.

A detailed review of each of the 16 major Sections of a U.S. CTD follows.

U.S. CTD Section 1.1—Forms (23)

Every MA submission to the FDA needs to include the applicable official FDA form. Accordingly, the FDA document includes six subsections, one for each specific form. These are described in Table 4 (24). The forms are self-explanatory and most include detailed instructions for their completion.

U.S. CTD Section 1.2—Cover Letter

The FDA has indicated that the use of a Cover Letter with electronic submissions is optional (25). If a cover letter is used, it should include the following information:

1. A description of the submission including appropriate regulatory information (e.g., if the submission is an Information Amendment, indicate the type of Information Amendment, the Serial Number of the submission, etc.);
2. A description of the approximate size of the submission (e.g., two gigabytes), the format used for Digital Linear Tapes (DLT), and the type and number of electronic media used (e.g., three CD-ROMs), if applicable;
3. A statement that the submission is virus free with a description of the software (i.e., name, version and company) used to check the files for viruses; and
4. The names and contact information of the regulatory and technical point of contacts for the submission.

U.S. CTD Section 1.3—Administrative Information

Section 1.3 has the five major subsections shown in Table 5 (24). Each of these is reviewed below in more detail.

Subsection 1.3.1—Contact/Sponsor/Applicant Information

Subsection 1.3.1 has five further subdivisions and provides a location to submit contact information, changes in Sponsorship or Ownership of an Application, and for information regarding any of the sponsor's obligations that may have been transferred to a third party. Two of these

Table 5 Subsections of U.S. CTD Module 1.3

Subsections	Descriptions
1.3.1	Contact/Sponsor/Applicant Information
1.3.1.1	Change of Address of Corporate Name
1.3.1.2	Change in Contact/Agent
1.3.1.3	Change in Sponsor
1.3.1.4	Transfer of Obligations
1.3.1.5	Change in Ownership of an Application
1.3.2	Field Copy Certification
1.3.3	Debarment Certification
1.3.4	Financial Certification and Disclosure
1.3.5	Patent and Exclusivity
1.3.5.1	Patent Information
1.3.5.2	Patent Certification
1.3.5.3	Exclusivity Request

Source: From Ref. 24.

subdivisions 1.3.1.2--Change in Contact/Agent and 1.3.1.4--Transfer of Obligations require additional discussion.

In reference to a U.S. Agent, the November 27, 2001 U.S. Federal Register (FR) (26) included a final regulation, effective February 11, 2002, which requires all foreign drug establishments (registrants) to register with the FDA and identify a U.S. Agent (27). The U.S. Agent may be an individual, firm, or company physically located in the United States. Each registrant can designate only one U.S. Agent. The agent assists the FDA in communications with the registrant, responds to questions concerning the registrant's products, and assists FDA in scheduling inspections of the registrant's foreign establishment (27).

Any changes in the U.S. Agent's name, address, or phone number must be reported to FDA within 10 business days of the change.

Some or all sponsor obligations under an IND may be transferred to a third party [e.g., a Contract Research Organization (CRO)]. Title 21 of the U.S. Code of Federal Regulations at Section 312.52 (21 CFR 312.52) (28) requires that the transfer be described in writing and that each obligation being assumed by the CRO is described. If *all* obligations are being transferred, a general statement to that effect is acceptable. Any obligation not included will be deemed not to have been transferred. It is important to also note that a CRO that assumes any obligation of a sponsor shall comply with the specific regulations applicable to the obligation being transferred and that the CRO is subject to the same regulatory action as a sponsor for failure to comply. This transfer of obligations is submitted in Subsection 1.3.1.4 of a U.S. CTD.

Subsection 1.3.2—Field Copy Certification

If you are filing a paper submission, the FDA requires (29) the initial NDA submission to include a statement certifying that a Field Copy of the application has been provided to the applicant's home FDA district office. The Field Copy contains the Chemistry, Manufacturing and Controls (CMC) technical section, a copy of the FDA Form 356 h, a copy of the application summary, and a certification that the field copy is a true copy of the CMC technical section contained in the archival and review copies of the application. The field copy is not required for an electronic NDA.

Subsection 1.3.3—Debarment Certification

Section 306(k)(1) of the Federal Food, Drug and Cosmetic Act (FD&C Act) (30) states that "Any application for approval of a drug product shall include (1) a certification that the applicant did not and will not use in any capacity the services of any person debarred ... in connection with such application, and (2) if such application is an abbreviated drug application [ANDA], a list of all convictions, described in subsections (a) and (b) which occurred within the previous five years, of the applicant and affiliated persons responsible for the development or submission of such application." This certification is referred to as the Debarment Certification (31). It is important to note that use of conditional or qualifying language when making this certification, such as "To the best of my knowledge...," is not acceptable. Information about individuals with convictions is discussed under Section 1.12.

Subsection 1.3.4—Financial Certification and Disclosure

On February 2, 1998, FDA published a final rule (32) effective on February 2, 1999, which requires anyone who submits an MA for any drug, biological product, or device to submit certain information concerning the compensation to, and financial interests of, any clinical investigator conducting clinical studies covered by the rule. This requirement is applicable to any clinical study submitted in a marketing application that the applicant or FDA relies on to establish that the product is effective, and any study in which a single investigator makes a significant contribution to the demonstration of safety. These studies are referred to as "covered clinical studies."[1] This rule requires applicants to either certify to the absence of certain financial interests of clinical investigators or to disclose those financial interests. The FDA may refuse to file (RTF) the application if the applicant does not include the required certification or disclosure, or does not certify that it was not possible to obtain the information. This rule was amended on December 31, 1998 (33), to reduce the need to gather certain financial information for studies completed before February 2, 1999. On March 20, 2001, FDA published a guidance (34) to provide clarification in interpreting and complying with these regulations.

Form FDA 3454 (35) is used to certify the absence of any financial interest. If a certification is made, it must document the following: that no financial arrangements have been made with an investigator where study outcome could affect compensation, that the investigator has no proprietary interest in the tested product, that the investigator does not have a significant equity interest in the sponsor of the covered study, and that the investigator has not received significant payments of other sorts. If there are financial disclosures of specified financial arrangements, these are to be made using Form FDA 3455 (36) with the details of such financial interest and a description of any steps taken to minimize the potential for bias are to be appended to the form.

Subsection 1.3.5—Patent and Exclusivity

The Patent and Exclusivity section of a U.S. CTD consists of three main subdivisions: 1.3.5.1 Patent Information, 1.3.5.2 Patent Certification, and 1.3.5.3 Exclusivity Request.

Patents and "exclusivity" work in a similar fashion but are distinctly different from one another. Patents are granted by the U.S. Patent and Trademark Office (PTO) during the development of a drug and can encompass a wide range of claims. "Exclusivity" provides exclusive marketing rights and is granted by the FDA upon approval of a drug. Exclusivity may run concurrently with a patent (37) and was designed to promote a balance between new drug innovation and generic drug competition.

Subdivision 1.3.5.1—Patent Information

All new drug applications (NDAs) (505(b)(1)[2] and 505(b)(2)[3] applications) and abbreviated new drug applications (ANDAs) (505(j)[4] applications) and amendments or supplements thereto are required to contain the patent information as described in 21 CFR 314.53 (38). Information on

[1] In general, these do not include Phase I tolerance studies or pharmacokinetic studies, most clinical pharmacology studies (unless they are critical to an efficacy determination), large open safety studies conducted at multiple sites, treatment protocols, and parallel track protocols. An applicant may consult with FDA as to which clinical studies constitute "covered clinical studies" for purposes of complying with financial disclosure requirements.

[2] A 505(b)(1) New Drug Application is an application that contains full reports of investigations of safety and effectiveness.

[3] A 505(b)(2) New Drug Application is an application that contains full reports of investigations of safety and effectiveness but where at least some of the information required for approval comes from studies not conducted by or for the applicant and for which the applicant has not obtained a right of reference.

[4] A 505(j) Abbreviated New Drug Application is an application that contains information to show that the proposed product is identical in active ingredient, dosage form, strength, route of administration, labeling, quality, performance characteristics, and intended use to a previously approved product.

Table 6 Types of Patent Certifications Required by 21 CFR 314.50

21 CFR citations	Title of certifications	Certifications
21 CFR 314.50(i)(1)(i)(A)(1)	Paragraph I Certification	That no patent information has been submitted to FDA.
21 CFR 314.50(i)(1)(i)(A)(2)	Paragraph II Certification	That the patent has expired.
21 CFR 314.50(i)(1)(i)(A)(3)	Paragraph III Certification	The date on which the patent will expire.
21 CFR 314.50(i)(1)(i)(A)(4)	Paragraph IV Certification	That the patent is invalid, unenforceable, or will not be infringed by the manufacture, use, or sale of the drug product for which the application is submitted.

patents claiming the active ingredient, the composition or formulation, or the method of use are to be reported in the application within this section of the CTD using Form FDA 3542a (submitted at time of initial submission) or using Form FDA 3542 (submitted within 30 days following approval of an application or if the patent information changes or new applicable patents are issued). Copies of all FDA forms, including the patent information forms, are available from the FDA Web site (39).

Subdivision 1.3.5.2—Patent Certification

All 505(b)(2) NDAs and 505(j) ANDAs are required to contain a Patent Certification. A patent certification must be made with respect to each relevant patent that claims the listed drug or any other drugs on which the investigations that are relied on by the applicant for approval of the application were conducted, or that claims a use for such drug. For each such patent, the applicant must provide the patent number and certify, in its opinion and to the best of its knowledge, to one of the four different certifications described in 21 CFR 314.50(i)(1)(i)(A)(1)--(4) (40) [Table 6 (24)].

In addition to including this certification in the application, at the time the applicant receives an acknowledgment letter from FDA stating that its application has been filed, the applicant must also notify the owner of the patent and the holder of any approved 505(b) application that is the subject of the certification. After providing these notifications, the applicant must amend its application to include a statement certifying that the required notifications have been provided and that the notification met the content requirement.

Subdivision 1.3.5.3--Exclusivity Request

As noted previously, "exclusivity" provides exclusive marketing rights and is granted by the FDA upon approval of a drug. The length and type of protection varies depending on the type of exclusivity that is granted (see Table 7). Exclusivity prevents either submission or approval of a 505(b)(2) or 505(j) application. Marketing exclusivity does not affect submission or approval of a 505(b)(1) NDA.

If applicants believe that they are entitled to the exclusivity provisions of 21 CFR 314.108 (37), they must submit a request for the desired period of exclusivity in the application. The contents of such a request are described in 21 CFR 314.50(j) (40). This request is submitted in Subsection 1.3.5.3 of a U.S. CTD.

U.S. CTD Section 1.4—References

This section of the CTD contains four different subsections. The Letter of Authorization included in Subsection 1.4.1 is where applicants place any letters that owners of Drug Master Files (DMFs) or other applications that the applicant is referencing have issued to the applicant permitting FDA to refer to the information concerned in the context of this new application. The list of persons authorized to reference a document or another application (e.g., a DMF) is included in Subsection 1.4.3--List of Authorized Persons to Incorporate by Reference.

If an applicant is relying on someone else's data and has obtained the right of reference to that data, this should be noted in Subsection 1.4.2—Statement of Right of Reference. If the applicant is submitting a 505(b)(2) NDA, where at least some of the information required for

Table 7 Types of Exclusivity and Periods of Protection

Type of exclusivity	Period of exclusivity	What is the protection
Orphan Drug	7 years	During the period of exclusivity, the FDA will not approve a 505(b)(1), 505(b)(2), or 505(j) for the same active moiety for the same indication.
New Chemical Entity (NCE)	5 years	During the period of exclusivity, no person may submit a 505(b)(2) NDA or 505(j) Abbreviated New Drug Application for a drug product that contains the same active moiety as in the new chemical entity with the exception that a 505(b)(2) or 505(j) application may be submitted after four (4) years if the application contains a Paragraph IV Patent Certification. In either case the FDA will not approve the application until the exclusivity has expired.
Other[a]	3 years	During the period of exclusivity, FDA will not approve a 505(b)(2) New Drug Application or 505(j) Abbreviated New Drug Application or a supplemental application thereto.
Pediatric Exclusivity	6 months	The 6-month period of exclusivity attaches to any other existing market exclusivity and remaining patent life for each drug product containing the active moiety and where the applicant submitting the studies is the holder of an approved NDA. Exclusivity attaching to remaining patent life is NOT patent extension, rather it extends the period during which FDA may not approve a 505(b)(2) or 505(j) application.

[a]Covers those drug products which contain an active moiety that has been previously approved in another application under Section 505(b) of the FD&C Act and which contained reports of new clinical investigations (other than bioavailability studies) conducted or sponsored by the applicant that were essential to approval of the application.
Source: From Ref. 41.

approval comes from studies not conducted by or for the applicant and for which the applicant has not obtained a right of reference, the applicant should include a statement about not having obtained the right of reference in Subsection 1.4.2.

In Subsection 1.4.4—Cross Reference to Other Applications, the applicant should list all applications or other documents (e.g., DMFs) that are being referenced in their new application.

U.S. CTD Section 1.5—Application Status

This subsection consists of seven subsections [Table 8 (41)] all of which relate to specific requests regarding the regulatory status of an application. For example, if sponsors want to request that FDA inactivate an IND, they would submit such a request and include it in Subsection 1.5.2.

U.S. CTD Section 1.6—Meetings

This section of the U.S. CTD and its three subsections provide locations for placing meeting requests (Subsection 1.6.1), background materials submitted in support of meetings (Subsection 1.6.2), and any other correspondence that might be available regarding those meetings (Subsection 1.6.3).

Table 8 Subsections of U.S. CTD Module 1.5

Subsections	Descriptions
1.5.1	Withdrawal Request
1.5.2	Inactivation Request
1.5.3	Reactivation Request
1.5.4	Reinstatement Request
1.5.5	Withdrawal of an Unapproved NDA
1.5.6	Withdrawal of a Listed Drug
1.5.7	Request for Withdrawal of Application Approval

Source: From Ref. 24.

U.S. CTD Section 1.7—Fast Track

The U.S. FDA has established the Fast Track program to facilitate the development and expedite the review of new drugs that are intended to treat serious or life-threatening conditions and that demonstrate the potential to address unmet medical needs. Detailed information on this program including a discussion of the criteria for qualifying for a fast-track drug development and the process for fast-track designation can be found in an FDA Guidance (42). According to the guidance, a sponsor can request fast-track designation anytime from the initial submission of the IND and up to just prior to receiving marketing approval. One of the key benefits of this designation is the opportunity for a greater and more frequent interaction with the FDA with the goal of insuring that the evidence necessary to support marketing approval will be developed and presented in a format that will enable an efficient review. If your product receives Fast-Track designation, you may be eligible for

1. priority Review of your MA application (currently this means FDA first action within six months rather than the standard 10 months);
2. piecemeal submission of your MA application (this is referred to as a "rolling review" and must be requested by the sponsor); and
3. accelerated approval (43) where FDA may grant marketing approval on the basis of adequate and well-controlled clinical trials demonstrating an effect on a surrogate endpoint rather than a defined clinical endpoint.

If a product is approved under these regulations, it may be subject to the further requirement to verify its clinical benefit, especially where there is uncertainty as to the relation of the surrogate endpoint to clinical benefit. The three subsections of Section 1.7 of a U.S. CTD include 1.7.1--Fast Track Designation Request, 1.7.2--Fast Track Designation Withdrawal Request, and 1.7.3--Rolling Review Request.

U.S. CTD Section 1.8—Special Protocol Assessment Request

Section 505(b)(4)(B) of the Federal Food Drug and Cosmetic Act was added following the implementation of Food and Drug Administration Modernization Act (FDAMA) of 1997 (44). This amendment requires that FDA meet with sponsors, providing certain conditions are met, in order to reach agreement on the adequacy of certain proposed studies associated with a New Drug or Biologic MA application. The guidance recognizes three kinds of study protocols that are eligible for review under the Special Protocol Assessment (SPA) provisions. These include

1. clinical study protocols for Phase III trials whose data will form the primary basis for an efficacy claim (submitted in CTD Subsection 1.8.1),
2. animal carcinogenicity study protocols (submitted in CTD Subsection 1.8.2), and
3. final product stability protocols (submitted in CTD Subsection 1.8.3).

Specific format and data requirements are described in the guidance. Of particular, note that FDA will not conduct an SPA for a study that is already in progress.

Table 9 Subsections of U.S. CTD Section 1.9

Subsections	Descriptions
1.9.1	Request for Waiver of Pediatric Studies
1.9.2	Request for Deferral of Pediatric Studies
1.9.3	Request for Pediatric Exclusivity Determination
1.9.4	Proposed Pediatric Study Request and Amendments
1.9.5	Proposal for Written Agreement
1.9.6	Other Correspondence Regarding Pediatric Exclusivity or Study Plans

Source: From Ref. 24.

U.S. CTD Section 1.9—Pediatric Administrative Information

In 2003, the U.S. Congress passed the Pediatric Research Equity Act (PREA) (45), which required most applications for drugs and biologics to include pediatric information, unless it was waived or deferred, on the use of pharmaceutical products when such studies are needed to ensure the safe and effective use of the products in children. With the passage of the Food and Drug Administration Amendments Act (FDAAA) of 2007 (46), Congress reauthorized and amended PREA through 2012. The amendments focused on improving FDA and applicant accountability for the agreed upon pediatric assessments. A draft guidance for Industry was issued in September, 2005, providing details on how to comply with PREA (47). A revised draft guidance incorporating the provisions of the 2007 FDAAA was not available at the time of writing this chapter.

A companion law providing voluntary incentives for generating pediatric use information was enacted by Congress in January, 2001. It was entitled the Best Pharmaceuticals for Children Act (BPCA) (48). This Act provided an additional six months of patent protection and different periods of nonpatent market exclusivity to applicants who conducted acceptable pediatric studies and for which the applicant has received a Pediatric Written Request (PWR). These incentives were scheduled to sunset on October 1, 2007. Title V of the 2007 FDAAA reauthorized and amended BPCA through 2012. Among the changes introduced by the 2007 FDAAA were a broadening of the definition of "pediatric studies" to include preclinical studies and a narrowing of the time frame for the sponsor to qualify for pediatric exclusivity.

Submissions to the FDA regarding the provisions of PREA and BPCA are submitted in the appropriate subsections of U.S. CTD Module 1.9 as shown in Table 9 (24).

U.S. CTD Section 1.10—Dispute Resolution

In February, 2000, the FDA issued a Guidance for Industry entitled "Formal Dispute Resolution: Appeals Above the Division Level" (49). The stated purpose of this document is to provide guidance for industry on procedures adopted by the Center for Drug Evaluation and Research (CDER) and the Center for Biologics Evaluation and Research (CBER) for resolving scientific and procedural disputes that cannot be resolved at the Division level. The regulations covering dispute resolution at the IND stage can be found at 21 CFR 312.48 (50) and those applicable to the NDA stage are located in 21 CFR 314.103 (51). Requests for dispute resolution are submitted in U.S. CTD Subsection 1.10.1. Any other correspondence relating to dispute resolution is included in U.S. CTD Subsection 1.10.2.

U.S. CTD Section 1.11—Information Amendment

In the United States, changes to an active IND or an NDA pending approval are submitted as amendments to the respective application. These amendments are usually included within the CTD Module to which they apply--for example, Module 2 Summary or in the individual technical sections of the CTD (i.e., Modules 3, 4, or 5). Occasionally, an amendment is of such a general nature that it is more appropriate to include it in the U.S. CTD Module 1. The FDA expects that if amendments need to be submitted to an application, which are not specifically covered by Modules 2 to 5, they should be placed in the U.S. CTD Section 1.11 as follows: Quality Information Amendments (Subsection 1.11.1), Safety Information Amendments (Subsection 1.11.2), and Efficacy Information Amendments (Subsection 1.11.3).

U.S. CTD Section 1.12—Other Correspondence

Section 1.12 of the U.S. CTD contains 16 subsections where various forms of correspondence are to be submitted during the life-cycle management of an IND and NDA. These 16 subsections are described below.

Subsection 1.12.1—Pre-IND Correspondence

A company and the FDA may agree to meet before the submission of an IND to discuss certain new or unique issues to a project. This meeting is called a pre-IND meeting. Copies of information relating to the pre-IND meeting are submitted in Subsection 1.12.1.

Subsection 1.12.2 and 1.12.3—Request to Charge and Notification for Charging Under a Treatment IND

The U.S. regulations as described below provide for certain instances where a sponsor may charge for investigational drugs. Subsections 1.12.2 and 1.12.3 provide a location for placing information relating to requesting and notifying FDA of a sponsor's intent to charge for such drugs.

"Charging for an investigational drug in a clinical trial under an IND is not permitted without the prior written approval of FDA. In requesting such approval, the sponsor shall provide a full written explanation of why charging is necessary in order for the sponsor to undertake or continue the clinical trial, e.g., why distribution of the drug to test subjects should not be considered part of the normal cost of doing business." [21 CFR 312.7(d)(1)] (52). This request to charge is submitted in U.S. CTD Subsection 1.12.2.

"A sponsor or investigator may charge for an investigational drug for a treatment use under a treatment protocol or treatment IND provided: (i) There is adequate enrollment in the ongoing clinical investigations under the authorized IND; (ii) charging does not constitute commercial marketing of a new drug for which a marketing application has not been approved; (iii) the drug is not being commercially promoted or advertised; and (iv) the sponsor of the drug is actively pursuing marketing approval with due diligence. FDA must be notified in writing in advance of commencing any such charges, in an information amendment submitted under Sec. 312.31. Authorization for charging goes into effect automatically 30 days after receipt by FDA of the information amendment, unless the sponsor is notified to the contrary." [21 CFR 312.7(d)(2)]) (52). This notification is submitted in the U.S. CTD Subsection 1.12.3.

Subsection 1.12.4—Request for Comments and Advice

If sponsors desire to obtain feedback from FDA about a given issue, they may submit a letter containing a description of the issue and related questions. This request, if not specifically identified in another section of the CTD, is filed in Subsection 1.12.4.

Subsection 1.12.5--Request for a Waiver

The U.S. regulations also provide that a sponsor may request a waiver from any of the applicable provisions in the regulations. For example, under the IND regulations, at 21 CFR 312.10(a) (53) it is stated: "A sponsor may request FDA to waive applicable requirement under this part. A waiver request may be submitted either in an IND or in an information amendment to an IND." This request, if not specifically covered by another subsection (e.g., 1.9.1--request for waiver of pediatric studies, or 1.12.15—request for waiver of in vivo bioavailability studies), is included in Subsection 1.12.5 of the U.S. CTD. A similar provision exists for NDAs (54) and ANDAs (55).

Subsections 1.12.6, 1.12.7, and 1.12.8—Informed Consent for Research

The U.S. informed consent regulations are described in 21 CFR 312 Part 50 (56). Subsections 1.12.6 through 1.12.8 of a U.S. CTD provide locations for submitting information relating to exceptions from informed consent for emergency research. For example, there are cases where it may be appropriate to request an exception to informed consent in emergency research where the human subjects are in a life-threatening situation, available treatments are unproven or unsatisfactory, the collection of valid scientific evidence is necessary to determine the safety and effectiveness of particular interventions, and the intervention under investigation must be administered before consent from the subjects' legally authorized representatives is feasible. The basis for an exemption is included in Subsection 1.12.6.

Public disclosure to the communities in which the clinical investigation will be conducted and from which the subjects will be drawn, prior to initiation and following completion of the clinical investigation, is also required by these regulations. This disclosure is submitted in Subsection 1.12.7 of the CTD.

Subsection 1.12.8 is used for any other correspondence regarding exceptions from informed consent for research.

Subsection 1.12.9—Notification of Discontinuation of Clinical Trial

Occasionally, during the course of a clinical investigation, a sponsor may discontinue the investigation. For example, as described in 21 CFR 312.56(d) (57), if a sponsor determines that its investigational drug presents an unreasonable and significant risk to subjects, the sponsor shall discontinue the investigations that present the risk, notify FDA and all institutional review boards, and all investigators who have at any time participated in the investigation of the discontinuance. The sponsor shall also assure the disposition of all stocks of the drug outstanding and furnish FDA with a full report of the sponsor's actions. This notification and related correspondence is submitted in Subsection 1.12.9, Notification of Discontinuation of Clinical Trial.

Subsections 1.12.10, 1.12.11, and 1.12.12—Special Statements for ANDAs

Subsections 1.12.10, 1.12.11, and 1.12.12 are all used to hold certain required pieces of information for a generic or ANDA.

Section 306(k)(2) of the FD&C Act (30) requires that an ANDA include a list of any individual, responsible for the development or submission of any component of an ANDA, who has been convicted of a crime within the previous five years, which could be the basis for a future debarment action. This listing is provided in Subsection 1.12.10.

21 CFR 314.94 (58) describes the content and format of an ANDA. Subsection (a)(3) of this regulation requires that the applicant list the basis for the ANDA. This information is required to contain the NDA number, the name of the Reference Listed Drug (RLD), and a statement about market exclusivity, if any, which exists for the RLD. The RLD is the drug product identified by the FDA in the *Orange Book* (59) as the product upon which applicants rely in seeking the FDA's approval of its ANDA and is the reference standard against which the bioequivalence of their product is compared. The RLD may also be the drug which a 505(b)(2) applicant relies in part on the FDA's previous findings of safety or efficacy. This information is submitted in CTD Subsection 1.12.11.

The information to be included in Subsection 1.12.12, Comparison of Generic Drug and Reference Listed Drug, is required by 21 CFR 314.94 subsections (a)(4) to (6). This includes statements that the conditions of use, the active ingredient, the route of administration, and the strength of the proposed product are identical to those of the RLD. If the proposed product differs in route of administration, dosage form, or strength, an ANDA can only be submitted if the FDA has approved a petition submitted under 21 CFR 314.93 (60) for these changes. The FDA will normally approve such a petition unless they determine that clinical investigations other than BE studies must be conducted to evaluate the safety and efficacy of the product that differs from the RLD.

Subsection 1.12.13—Request for Waiver of In Vivo Studies

If the applicant intends to request a waiver of in vivo studies in accordance with 21 CFR 312.10 (53) (for INDs) or 21 CFR 314.90 (54) or 21 CFR 314.126 (61) (for NDAs), then the waiver request, with its justification, is to be submitted in CTD Subsection 1.12.13. The FDA may delete this section in the future and place this information in CTD Subsection 1.12.5.

Subsection 1.12.14—Environmental Analysis

The regulations both for an IND (62) and for an NDA (40) require that the sponsor submit either a claim for a categorical exclusion (21 CFR 25.31) (63) for submitting an environmental analysis or a complete environmental analysis study (21 CFR 25.40) (64). INDs are routinely subject to categorical exclusions. An NDA can be categorically excluded from the requirements for a full environmental analysis if the approval of the NDA will not increase the use of the active moiety or if the use of the active moiety is increased but the estimated concentration at the point of entry into the aquatic environment is less than 1 ppb (part per billion). Documentation to support a categorical exclusion or the complete environmental analysis study is placed in CTD Subsection 1.12.14.

Subsection 1.12.15—Request for Waiver of In Vivo Bioavailability Studies

21 CFR 320.22 (65) describes criteria that, if met, may support a waiver of the requirement for the submission of evidence measuring the in vivo bioavailability or demonstrating the in vivo bioequivalence of the drug product that is the subject of the application. This waiver request, with its justification, is submitted in CTD Subsection 1.12.15.

Subsection 1.12.16--Field Alert Reports

21 CFR 314.81 (66) in the NDA regulations defines the requirements for various postmarketing reports, which must be submitted to FDA. Field Alert Reports are described in 21 CFR 314.81(b)(1) and are to be made for label mix-ups or in cases where the distributed drug product has undergone a chemical, physical, or other change, which causes it to fail to meet its specifications. These field alert reports are submitted to FDA in CTD Subsection 1.12.16.

U.S. CTD Section 1.13—Annual Report

The U.S. FDA regulations provide that Annual Reports be submitted each year for active INDs (21 CFR 312.33) (67) and approved NDAs [21 CFR 314.81(b)(2)] (66). Subsection 1.13 of the U.S. CTD hierarchy includes 14 locations where data for these reports are to be included in CTD submission format [Table 10 (24)]. It should be noted that in one case, Subsection 1.13.9, General Investigational Plan, is also used for the General Investigational Plan component of an original IND.

U.S. CTD Section 1.14—Labeling

In the United States, the labeling section of the CTD is composed of 5 major subsections and 13 further subdivisions as described below.

Subsection 1.14.1—Draft Labeling

The content of labeling is specified in 21 CFR 201.56 (68). The term "labeling" is used in a different way in the United States and European Union; in the European Union it mainly refers

Table 10 Subsections of U.S. CTD Section 1.13

Subsections	Descriptions	Use and 21 CFR citations
1.13.1	Summary for Nonclinical Studies	• IND Annual Report [312.33(b)(6)] • NDA Annual Report [314.81(b)(2)(i)]
1.13.2	Summary of Clinical Pharmacology Information	• IND Annual Report [312.33(b)(5)] • NDA Annual Report [314.81(b)(2)(i)]
1.13.3	Summary of Safety Information	• IND Annual Report [312.33(b)(1)–(4)] • NDA Annual Report [314.81(b)(2)(i)]
1.13.4	Summary of Labeling Changes	• NDA Annual Report [314.81(b)(2)(i)]
1.13.5	Summary of Manufacturing Changes	• IND Annual Report [312.33(b)(7)] • NDA Annual Report [314.81(b)(2)(i)]
1.13.6	Summary of Microbiological Changes	• IND Annual Report [312.33(b)(7)] • NDA Annual Report [314.81(b)(2)(i)]
1.13.7	Summary of Other Significant New Information	• NDA Annual Report [314.81(b)(2)(i)]
1.13.8	Individual Study Information	• IND Annual Report [312.33(a)]
1.13.9	General Investigational Plan	• Original IND [312.23(a)(3)(iv)] • IND Annual Report [312.33(c)]
1.13.10	Foreign Marketing History	• IND Annual Report [312.33(f)]
1.13.11	Distribution Data	• NDA Annual Report [314.81(b)(2)(ii)]
1.13.12	Status of Postmarketing Study Commitments	• NDA Annual Report [314.81(b)(2)(vii)]
1.13.13	Status of Other Postmarketing Studies	• NDA Annual Report [314.81(b)(2)(viii)]
1.13.14	Log of Outstanding Regulatory Business	• IND Annual Report [312.33(g)] • NDA Annual Report [314.81(b)(2)(ix)]

to container and carton labeling. In the United States, it includes text, tables, as well as figures and is commonly referred to as the package insert. Effective October 31, 2005, the text of all package insert submissions to FDA needs to be submitted electronically in Structured Product Labelling (SPL) format. SPL is a document markup standard based on XML, which was approved by Health Level Seven (69) (HL7) and adopted by the FDA. HL7 is one of several American National Standards Institute (ANSI-) accredited Standards Developing Organizations (SDOs) operating in the health-care arena. In April 2005, the FDA issued a guidance document (70), which describes in detail the SPL requirements. This guidance also indicates that "Copies of the formatted label and labeling and specimens of enclosures required elsewhere in the regulations (e.g., 21 CFR 314.50(e)(2)(ii)), including carton and container labels, must still be submitted either electronically in Portable Document Format (PDF) or on paper." Draft labeling text and Final labeling text is submitted in SPL format in Subsections 1.14.1.3 and 1.14.2.3, respectively.

The regulations for NDAs at 314.50(c)(2)(i) (40) also require that an Annotated Package Insert be included in the submission. An annotated package insert (PI) is the labeling text with each statement in the labeling referenced to the location of supporting statements in the NDA summary and the respective NDA technical sections. This annotated PI is placed in CTD Subsection 1.14.1.2.

A label comprehension study is a study designed to evaluate proposed OTC product labeling with the principal goal of determining if a typical consumer can understand the label to insure proper utilization of the OTC product. The results of such studies are submitted in CTD Subsection 1.14.1.4.

Labeling History, CTD Subsection 1.14.1.5, is where an applicant places a historical listing of changes made to previously approved labeling and the basis for each such change. For example, in the NDA Annual Report [21 CFR 314.81(b)(2)(iii)(c)] (66), the applicant is required to submit a summary of any changes in labeling that have been made since the last report. These are to be listed by date in the order in which they were implemented. If there were no changes, the applicant must include a statement of that fact. This listing from the NDA Annual Report is to be included in this CTD Subsection.

Subsection 1.14.2—Final Labeling

At the time of approval of an MA, the FDA will request 12 copies of all final printed labeling (including PI, cartons and containers labeling) and the final labeling text (in SPL format). This labeling is included in Subsection 1.14.2.

Subsection 1.14.3—Listed Drug Labeling

You will remember our earlier discussion about the RLD. This is the drug which is referenced as the basis for an ANDA or may be the drug upon which an applicant relies on FDA's previous findings of safety and efficacy for a 505(b)(2) NDA.

The regulations for an ANDA [21 CFR 314.94(8)((iii) and (iv)] (58) require that a statement be made concerning the applicant's proposed labeling. The statement needs to indicate it is the same as the RLD or that it is different in the particulars noted in the annotated labeling side-by-side comparison of the approved label for the RLD and the proposed label for the product subject to the MA application. This side-by-side comparison, with any changes annotated, is submitted in CTD Subsection 1.14.3.1. The approved labeling text for the RLD is included in Subsection 1.14.3.2. Subsection 1.14.3.3 will be deleted by FDA in the future as it is an inadvertent duplication of 1.14.3.2.

Subsection 1.14.4—Investigational Drug Labeling

This subsection applies to labeling used in support of IND clinical studies. Clinical trial supplies for use in clinical studies need to be properly labeled. Additionally, the IND regulations [21 CFR 312.23(a)(5)] (62) require that an Investigational Drug Brochure (i.e. investigational labeling or product information) be included in the IND and provided to the Institutional Review Board and to each investigator. Copies of the labeling for the clinical trial supplies are placed in Subsection 1.14.4.1 and a copy of the Investigational Drug Brochure is to be included in Subsection 1.14.4.2.

Subsection 1.14.5—Foreign Labeling

If the product, which is the subject of the MA, has been approved in countries outside the United States, the FDA requests that the applicant submit English language copies of the labeling approved for that product. These copies are included in Subsection 1.14.5.

U.S. CTD Section 1.15—Promotional Material

The U.S. Regulations [21 CFR 314.81(b)(3)(i)] (66) require that samples of mailing pieces and any other labeling or advertisements developed for promoting the drug product be submitted to the FDA at the time of initial dissemination or publication. Samples of these materials are included in Section 1.15 of the U.S. CTD.

U.S. CTD Section 1.16—Risk Management Plans

In March, 2005, FDA issued a guidance document entitled "Development and Use of Risk Minimization Action Plans" (71) (RiskMAPs). This guidance indicates that the FDA views risk management as an iterative process encompassing the assessment of risks and benefits, the minimization of risks, and the maximization of benefits. Two years later, in September, 2007, President Bush signed into law the FDAAA of 2007 (72). Title IX of FDAAA provided certain mechanisms to enhance the FDA's authority regarding postmarket safety. One of these provisions gave FDA authority to require Risk Evaluation and Mitigation Strategies (REMS) for new drugs and biologics if FDA determined that such was necessary to ensure the benefits outweigh the risks of the drug. RiskMAPs and REMS, if required by FDA, are placed in this section of the U.S. CTD.

ACKNOWLEDGMENTS

The authors wish to acknowledge the support and valuable contributions of Ms. Claudia Pluess Deputy Director/Regulatory Affairs Director Asia Pacific, Mundipharma Laboratories GmbH, St. Alban Rheinweg 74, 4020 Basel, Switzerland in preparation of the discussion about Japan.

REFERENCES

1. Organisation of the Common Technical Document for the Registration of Pharmaceuticals for Human Use M4, January 2004, ICH. http://www.ich.org/LOB/media/MEDIA554.pdf. Accessed October 2008.
2. Module 1 Administrative Information and Prescribing Information For Australia, Notice to Applicants, CTD-Module 1, TGA Edition September 2007, Therapeutic Goods Administration. http://www.tga.gov.au/docs/pdf/euguide/tgamod1.pdf. Accessed October 2008.
3. Volume 2B of the Notice to Applicants, Medicinal products for human use, Presentation and format of the dossier, Common Technical Document, European Commission, 2008. http://ec.europa.eu/enterprise/pharmaceuticals/eudralex/vol-2/b/update_200805/ctd_05_2008.pdf. Accessed October 2008.
4. Volume 2A of the Notice to Applicants, Medicinal products for human use, Chapter 7: General Information, European Commission. http://ec.europa.eu/enterprise/pharmaceuticals/eudralex/vol-2/a/vol2 a_chap7_rev_2008_07.pdf. Accessed October 2008.
5. Notice to Applicants, Medicinal Products for Human Use, Volume 2B, Module 1: Administrative information Application Form, May 2008, European Commission. http://ec.europa.eu/enterprise/pharmaceuticals/eudralex/vol-2/b/update_200805/applicformrevised_rev9.pdf. Accessed October 2008.
6. Annotated QRD Product Information template for Centralised Procedure, EMEA. http://www.emea.europa.eu/htms/human/qrd/docs/Hannotatedtemplate.pdf (Accessed October 2008) and CMD(h) Annotated QRD Template for MR/DC Procedures, CMD(h), 2006 http://www.hma.eu/uploads/media/QRD_annotated_template_CMDh.pdf. Accessed October 2008.
7. Notice to Applicants, Volume 2C: Regulatory guidelines, European Commission. http://ec.europa.eu/enterprise/pharmaceuticals/eudralex/vol2_en.htm. Accessed October 2008.
8. A guideline on the readability of the label and package leaflet of medicinal products for human use, Pharmaceutical Committee, September 1998, European Commission. http://ec.europa.eu/enterprise/pharmaceuticals/eudralex/vol-2/c/gl981002.pdf. Accessed October 2008.
9. Guidance concerning consultations with target patient groups for the package leaflet, May 2006, European Commission. http://ec.europa.eu/enterprise/pharmaceuticals/eudralex/vol-2/c/user_consultation_200605.pdf. Accessed November 2008.

10. EFPIA General Recommendations for Readability User Testing of Package Leaflets for Medicinal Products for Human Use Submitted or Approved under the European Centralised Procedure, revised version March 2003. http://www.efpia.eu/content/default.asp?PageID=559&DocID= 2332. Accessed November 2008.
11. Guidance concerning the Braille requirements for labelling and the package leaflet, ENTR/F2 D(2005), European Commission. http://ec.europa.eu/enterprise/pharmaceuticals/pharmacos/docs/doc2005/04_05/braille_text20050411.pdf. Accessed October 2008.
12. CEN/TR 15753:2008, Packaging. Packaging Leaflets for medicinal products. Braille and other formats for visually impaired people, ISBN 978 0 580 60795 0, August 2008, British Standards Institute Group.
13. CHMP Guideline on procedures for the granting of a marketing authorisation under exceptional circumstances, pursuant to Article 14(80 of Regulation (EC) No. 726/2004. December 2005, EMEA. http://www.emea.europa.eu/pdfs/human/euleg/35798105en.pdf. Accessed October 2008.
14. CHMP Guideline on the scientific application and the practical arrangements necessary to implement Commission Regulation No. 507/2006 on the conditional marketing authorisation for medicinal products for human use falling within the scope of Regulation No. 726/2004, December 2006, EMEA. http://www.emea.europa.eu/pdfs/human/regaffair/50995106en.pdf. Accessed October 2006.
15. Communication from the Commission, Guideline on aspects of the application of Article 8(1) and (3) of Regulation (EC) No 141/2000: Assessing similarity of medicinal products versus authorised orphan medicinal products benefiting from market exclusivity and applying derogations from that market exclusivity, C(2008) 4077 final, September 2008, European Commission. http://ec.europa.eu/enterprise/pharmaceuticals/orphanmp/doc/C_2008_4077.pdf. Accessed October 2008.
16. The Rules Governing Medicinal Products in the European Union, Volume 9A: Guidelines on Pharmacovigilance for Medicinal Products for Human Use, March 2007, European Commission. http://ec.europa.eu/enterprise/pharmaceuticals/eudralex/vol-9/pdf/vol9_2007-07_upd07.pdf. Accessed October 2008.
17. Communication from the Commission—Guideline on the format and content of applications for agreement or modification of a paediatric investigation plan and requests for waivers or deferrals and concerning the operation of the compliance check and on criteria for assessing significant studies, 24.9.2008 C 243/01, Official Journal of the European Union. http://www.emea.europa.eu/pdfs/human/paediatrics/Guideline_2008_C243_01.pdf. Accessed October 2008.
18. Notification ELD No. 899 of 21 June 2001, On Organization of Application Dossier Appended to New Pharmaceutical Application (NDA) for Approval, Director of Evaluation and Licensing Division, Pharmaceutical and Medical Safety Bureau, Ministry of Health, Labour and Welfare. http://123.55.252.122:8080/dbase.aboutchem.com/ich/pdfdoc/899e-front.pdf?MT=1211535723. Accessed June 2009.
19. http://www.ich.org/cache/compo/276-254-1.html. Accessed June 2009.
20. http://www.elecdoc.hu/japanmodule1spec.pdf. Accessed June 2009.
21. http://www.pmda.go.jp/english/services/reviews/file/guideline application.pdf. Accessed August 2008.
22. FDA Comprehensive Table of Contents Headings and Hierarchy" Version 1.2, July 6, 2005. http://www.fda.gov/cder/regulatory/ersr/5640CTOC-v1.2.pdf. Accessed July 2008.
23. http://www.fda.gov/AboutFDA/ReportsManualsForms/Forms/default.htm. Accessed June 2009.
24. FDA Comprehensive Table of Contents Headings and Hierarchy" Version 1.2, July 6, 2005. http://www.fda.gov/cder/regulatory/ersr/5640CTOC-v1.2.pdf. Accessed November, 2008.
25. http://www.fda.gov/downloads/Drugs/GuidanceComplianceRegulatoryInformation/Guidances/ucm072349.pdf. Accessed June 2009.
26. http://frwebgate.access.gpo.gov/cgi-bin/getdoc.cgi?dbname=2001_register&docid=01-29393-filed.pdf. Accessed June 2009.
27. http://edocket.access.gpo.gov/cfr_2008/aprqtr/pdf/21cfr207.40.pdf. Accessed June 2009.
28. http://edocket.access.gpo.gov/cfr 2008/aprqtr/pdf/21cfr312.52.pdf. Accessed June 2009.
29. See 21 CFR 314.50(d)(1)(v) at http://edocket.access.gpo.gov/cfr_2008/aprqtr/pdf/21cfr314.50.pdf. Accessed June 2009.
30. http://www.fda.gov/RegulatoryInformation/Legislation/FederalFoodDrugandCosmeticActFDC Act/FDCActChapterIIIProhibitedActsandPenalties/ucm107326.htm. Accessed June 2009.
31. FDA Draft Guidance for Industry: Submitting Debarment Certification Statements, September, 1998 at http://www.fda.gov/downloads/Drugs/GuidanceComplianceRegulatoryInformation/Guidances/ucm080584.pdf. Accessed June 2009. A copy of FDA's debarment list may be found at http://www.fda.gov/ICECI/EnforcementActions/FDADebarmentList/default.htm. Accessed June 2009.
32. http://frwebgate.access.gpo.gov/cgi-bin/getdoc.cgi?IPaddress=frwais.access.gpo.gov&dbname=1998_register&docid=98-2407-filed.pdf. Accessed June 2009.
33. http://frwebgate.access.gpo.gov/cgi-bin/getdoc.cgi?dbname=1998_register&docid=98-34546-filed.pdf. Accessed June 2009.

34. http://www.fda.gov/RegulatoryInformation/Guidances/ucm126832htm. Accessed June 2009.
35. http://www.fda.gov/downloads/AboutFDA/ReportsManualsForms/Forms/OtherForms/UCM048304.pdf. Accessed June 2009.
36. http://www.fda.gov/downloads/AboutFDA/ReportsManualsForms/Forms/UCM048310.pdf. Accessed June 2009.
37. http://edocket.access.gpo.gov/cfr_2008/aprqtr/pdf/21cfr314.108.pdf. Accessed June 2009.
38. http://edocket.access.gpo.gov/cfr_2008/aprqtr/pdf/21cfr314.53.pdf. Accessed June 2009.
39. http://www.fda.gov/AboutFDA/ReportsManualsForms/Forms/default.htm. Accessed June 2009.
40. http://edocket.access.gpo.gov/cfr_2008/aprqtr/pdf/21cfr314.50.pdf. Accessed June 2009.
41. http://www.fda.gov/RegulatoryInformation/Legislation/FederalFoodDrugandCosmeticActFDCAct/FDCActChapterVDrugsandDevices/ucm108125.htm. Accessed June 2009.
42. http://www.fda.gov/downloads/Drugs/GuidanceComplianceRegulatoryInformation/Guidances/UCM079736.pdf. Accessed June 2009.
43. http://edocket.access.gpo.gov/cfr_2008/aprqtr/pdf/21cfr314.510.pdf. Accessed June 2009.
44. See Food and Drug Modernization Act at http://www.fda.gov/RegulatoryInformation/Legislation/FederalFoodDrugandCosmeticActFDCAct/SignificantAmendmentstotheFDCAct/FDAMA/default.htm. Accessed June 2009. See also FDA Guidance on Special Protocol Assessments at http://www.fda.gov/downloads/Drugs/GuidanceComplianceRegulatoryInformation/Guidances/ucm080571.pdf. Accessed June 2009.
45. http://www.fda.gov/downloads/Drugs/DevelopmentApprovalProcess/DevelopmentResources/UCM077853.pdf. Accessed June 2009.
46. http://www.fda.gov/RegulatoryInformation/Legislation/FederalFoodDrugandCosmeticActFDCAct/SignificantAmendmentstotheFDCAct/FoodandDrugAdministrationAmendmentsActof2007/default.htm. Accessed June 2009.
47. http://www.fda.gov/downloads/Drugs/GuidanceComplianceRegulatoryInformation/Guidances/ucm079756.pdf. Accessed June 2009.
48. http://www.fda.gov/downloads/Drugs/DevelopmentApprovalProcess/DevelopmentResources/UCM049874.pdf. Accessed June 2009.
49. http://www.fda.gov/downloads/Drugs/GuidanceComplianceRegulatoryInformation/Guidances/ucm079743.pdf. Accessed June 2009.
50. http://www.fda.gov/downloads/Drugs/GuidanceComplianceRegulatoryInformation/Guidances/ucm079743.pdf. Accessed June 2009.
51. http://edocket.access.gpo.gov/cfr_2008/aprqtr/pdf/21cfr314.103.pdf. Accessed June 2009.
52. http://edocket.access.gpo.gov/cfr_2008/aprqtr/pdf/21cfr312.7.pdf. Accessed June 2009.
53. http://edocket.access.gpo.gov/cfr_2008/aprqtr/pdf/21cfr312.10.pdf. Accessed June 2009.
54. http://edocket.access.gpo.gov/cfr_2008/aprqtr/pdf/21cfr314.90.pdf. Accessed June 2009.
55. http://edocket.access.gpo.gov/cfr_2008/aprqtr/pdf/21cfr314.99.pdf. Accessed June 2009.
56. http://www.access.gpo.gov/nara/cfr/waisidx_08/21cfr50_08.html. Accessed June 2009.
57. http://edocket.access.gpo.gov/cfr_2008/aprqtr/pdf/21cfr312.56.pdf. Accessed June 2009.
58. http://edocket.access.gpo.gov/cfr_2008/aprqtr/pdf/21cfr314.94.pdf. Accessed June 2009.
59. http://www.fda.gov/Drugs/InformationOnDrugs/ucm129662.htm. Accessed June 2009.
60. http://edocket.access.gpo.gov/cfr_2008/aprqtr/pdf/21cfr314.93.pdf. Accessed June 2009.
61. http://edocket.access.gpo.gov/cfr_2008/aprqtr/pdf/21cfr314.126.pdf. Accessed June 2009.
62. http://edocket.access.gpo.gov/cfr_2008/aprqtr/pdf/21cfr312.23.pdf. Accessed June 2009.
63. http://edocket.access.gpo.gov/cfr_2008/aprqtr/pdf/21cfr25.31.pdf. Accessed June 2009.
64. http://edocket.access.gpo.gov/cfr_2008/aprqtr/pdf/21cfr25.40.pdf. Accessed June 2009.
65. http://edocket.access.gpo.gov/cfr_2008/aprqtr/pdf/21cfr320.22.pdf. Accessed June 2009.
66. http://edocket.access.gpo.gov/cfr_2008/aprqtr/pdf/21cfr314.81.pdf. Accessed June 2009.
67. http://edocket.access.gpo.gov/cfr_2008/aprqtr/pdf/21cfr312.33.pdf. Accessed June 2009.
68. http://edocket.access.gpo.gov/cfr_2008/aprqtr/pdf/21cfr201.56.pdf. Accessed June 2009.
69. http://www.hl7.org/. Accessed June 2009
70. http://www.fda.gov/Drugs/DevelopmentApprovalProcess/FormsSubmissionRequirements/ElectronicSubmissions/ucm114043.htm. Accessed June 2009.
71. http://www.fda.gov/downloads/Drugs/GuidanceComplianceRegulatoryInformation/Guidances/ucm071616.pdf. Accessed June 2009.
72. http://www.fda.gov/RegulatoryInformation/Legislation/FederalFoodDrugandCosmeticActFDCAct/SignificantAmendmentstotheFDCAct/FoodandDrugAdministrationAmendmentsActof2007/default.htm. Accessed June 2009.

4 Environmental Risk Assessment

David J. Snodin
PAREXEL Drug Development Consulting, Middlesex, U.K.

INTRODUCTION

Trace quantities of human pharmaceuticals or their metabolites have been identified in many countries in various aquatic environments including sewage treatment plant (STP) effluents, surface waters, seawaters, groundwater, and some drinking waters (1). Of particular concern is the potential for contamination of fresh water, which is a highly finite resource. As most of the water on our planet is salty, only 3% is fresh and only around a third of this is accessible for human use, the remainder being frozen in glaciers and permanent snow (2). Advances in environmental analysis have enabled the detection of increasingly low levels of drug residues and evidence of their widespread distribution, particularly in treated wastewater and surface water (1), could be a cause for concern mainly in respect of possible effects on aquatic organisms (3). On the other hand, drug residues may well be present at concentrations significantly below those that cause adverse effects in aquatic environments (4,5). Consequently, regulatory approaches in dealing with pharmaceuticals in the environment have assumed that, unless there is evidence to the contrary for a particular therapeutic class, adverse effects are unlikely unless a threshold concentration is exceeded, and that environmental testing is required only when this threshold is crossed.

ENTRY OF HUMAN PHARMACEUTICALS INTO THE ENVIRONMENT

Entry pathways of human pharmaceuticals into the environment can include manufacturing process effluents, patient use, disposal of unused or expired drug products or from accidental spills, especially during storage/transportation. The principal route is through patient use, and regulatory guidance on environmental risk assessment is focused on this aspect. Most drugs are metabolized to a greater or lesser extent during their passage through the patient's body, and drug-related material is excreted via urine and feces in the form of parent compound, metabolites, and conjugates. In some cases, there may be at least partial mineralization of the drug to carbon dioxide, which will be excreted in expired air. Volatile drug substances such as anesthetic gases will, in most cases, be excreted by exhalation directly into the atmosphere.

SEWAGE TREATMENT PROCESSES

Overview

Sewage treatment normally involves four main stages (6):

- *Preliminary*: Screening of large solids and grit removal by flow attenuation; discharge to coastal and estuarine waters.
- *Primary*: Settlement of suspended solids; discharge to "less-sensitive areas," for example, to the sea surrounding isolated small islands.
- *Secondary*: Biological treatment: (*i*) activated sludge process using aerated agitated liquor in which microorganisms convert organic matter to a carbon-rich sludge and (*ii*) filter beds in which sewage is trickled over coarse aggregates coated with bacteria. Discharge to fresh, estuarine, and coastal waters.
- *Tertiary*: Further treatment may be used to remove remaining organic matter (depending on the nature of the effluent and the local discharge conditions) and/or disinfect the treated water. Typical processes include phosphorus or nitrate reduction and disinfection by UV or filter membranes. Discharge to "sensitive areas."

In England and Wales, over 10 billion liters of sewage are produced every day (equivalent to approximately 200 l/person/day). Effluent is discharged to fresh, ground, or coastal water. Recovered sludge is applied to agricultural land (62%), incinerated (19%), used for land reclamation (11%), or used for composting or landfill (8%) (7). Raw sewage sludge is normally further processed before disposal by dewatering (using centrifugation, for example) and/or by anaerobic digestion.

Activated Sludge

The success of the (aerated) activated sludge process is dependent upon establishing a mixed community of microorganisms that will remove and consume organic waste material, which will aggregate and adhere (bioflocculation), and which will settle in such a manner as to produce a concentrated sludge (return activated sludge) for recycling. In the ideal system, filamentous organisms grow within a floc, with few filaments protruding into the surrounding bulk solution. In such a system, there is no interference with the compaction and settling rates of the activated sludge prior to its recycling. Bacteria, fungi, protozoa, and rotifers account for most of the biological mass of activated sludge. In addition, some metazoa, such as nematode worms, may be present. The predominant species of microorganism depends on environmental conditions, process design, the mode of plant operation, and the characteristics of the secondary influent wastewater. Microorganisms present in the greatest numbers are bacteria: some bacteria are strict aerobes whereas others are anaerobes, but the majority of activated sludge bacteria are facultative that are able to live in either the presence or absence of oxygen, an important factor in the survival of activated sludge when dissolved oxygen concentrations are low. Heterotrophic (carbonaceous matter dependent) predominate over autotrophic bacteria and are able to convert most of the organic compounds in sewage to carbon dioxide and water. Important genera of heterotrophic bacteria include *Achromobacter, Alcaligenes, Arthrobacter, Citromonas, Flavobacterium, Pseudomonas,* and *Zoogloea* (8).

Removal of Pharmaceuticals by Sewage Treatment Processes

In most developed countries, excreta containing drug-related material will enter the domestic/municipal sewer systems and eventually reach an STP. The principal concerns over pharmaceuticals relate to their potential impact on the treatment process, that is, harmful effects on sewage sludge organisms, and to the possibility of drug-related material reaching the wider environment, especially if it possesses pharmacological activity. In fact, pharmaceuticals in wastewater reaching an STP are likely to undergo a number of chemical and biological changes during the various stages of treatment. Any remaining (active) pharmaceutical residues will be retained on sludge sediment or discharged as an aqueous effluent. As shown in Figure 1, nonretained drug-related material (the majority of which is likely to be in the form of pharmacologically inactive metabolites and degradation products) will enter the surface water (normally rivers and estuaries) and soil compartments.

The principal mechanisms of removal of active pharmaceutical ingredients (APIs) appear to be biological oxidation and sorption onto sludge. Adsorption is dependent on hydrophobic and electrostatic interactions of the API (or metabolite) with organic particulates (particularly microorganisms). Acidic APIs including acetylsalicylic acid, the profen nonsteroidal anti-inflammatory drugs (NSAIDs) such as ibuprofen, ketoprofen, and fenoprofen, and other NSAIDs such as naproxen, diclofenac, and indomethacin, are poorly adsorbed onto sludge because they are present in the anionic form at neutral pH (with pK_a values normally around 4–5). In general, sorption of acidic APIs is unlikely to be of major importance in their elimination during sewage treatment and on entry to surface waters, only minor amounts having been detected in sludge and sediments. On the other hand, APIs that are lipophilic, basic, or are present as zwitterions can be adsorbed to a significant extent as has been shown for fluoroquinolone antibiotics (1).

A number of researchers have determined concentrations of pharmaceuticals at various stages of the sewage treatment process, and some of the more recent publications are summarized in Table 1. For many APIs, no residues could be detected in crude sewage, presumably largely due to human metabolism, and for those reaching the STP, removal efficiencies were normally $\geq$80%. APIs that were difficult to remove included the anticonvulsant carbamazepine, the iodinated contrast agent iopromide, and the NSAID diclofenac. Removal efficiency seems to be related much more strongly to chemical structure than to pharmacological class.

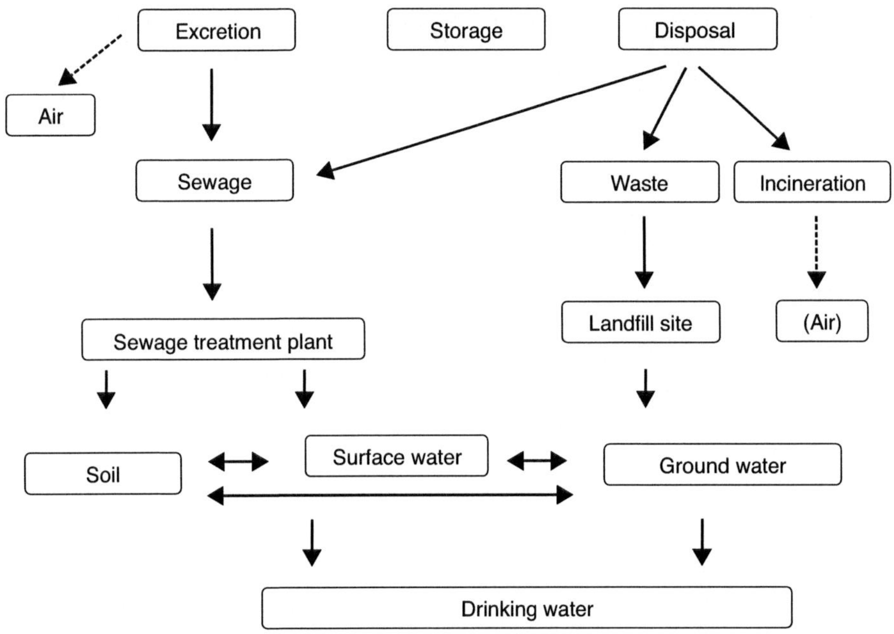

Figure 1 Entry paths into the environment following use, storage, and disposal of human pharmaceuticals (30).

Supplementary ozonation followed by activated carbon treatment was effective in removing virtually all residues of 26 different APIs (16).

Studies on elimination rates by STPs are based on measurements of influent and effluent concentrations, which are subject to multiple variables including STP construction and treatment technology, hydraulic retention time, season, and efficiency of the treatment processes. Hence, removal rates can vary considerably: for example, reported elimination rates for diclofenac range from 17% to 100% (1). One noticeable feature of most studies is that data are restricted generally to acidic or neutral APIs, which are not significantly adsorbed onto sludge, and the primary mechanism of removal is (aerobic) biotransformation/mineralization by sludge organisms.

ENVIRONMENTAL FATE

Although significant biodegradation of many APIs can occur during passage through the STP, other processes have been shown to be important (Table 2). In surface waters abiotic transformation is often more important than biodegradation. Whereas hydrolysis generally plays an insignificant role, photodegradation can be the main removal process as is the case for diclofenac, for example (1). The efficiency of photodegradation depends on the photochemical properties of the API, the strength of the solar radiation (i.e., latitude and season), and on the presence of natural photosensitizing constituents (such as nitrates and humic acid) that can generate hydroxyl radicals and singlet oxidation by the action of sunlight (18).

Biodegradation in soil is a further major post-STP elimination process, the main entry point being the application of sewage sludge to agricultural land. It is generally assumed that, when bioavailable, APIs can often transform more readily in soil than in water owing to the presence of a wide diversity of microorganisms including fungi, bacteria, and actinomycetes (17). Complete mineralization of virtually all pharmaceuticals is likely to occur in soil, although the rates of degradation may be quite slow in some cases, leading to the possibility of leaching of low concentrations into water courses. Examples of recent soil degradation reports are shown in Table 3.

Research and regulatory activities on pharmaceuticals in the environment are focused almost exclusively on the aqueous and terrestrial compartments. Information on the environmental impact of volatile APIs such as anesthetic gases is generally lacking, which is surprising given the major concerns over CFC (chloroflurocarbon) propellants some years ago, leading in

Table 1 Removal of Specific Pharmaceuticals by Sewage Treatment Processes

Number of APIs investigated	Summary of findings	Country, date (reference)
5	Aspirin, clofibric acid, fluoxetine, cyclophosphamide not detected in crude sewage or sewage effluent. Ibuprofen detected in all crude sewage; removal by sewage treatment normally 80–100%.	U.K., 2003 (9)
8	Concentrations of eight pharmaceuticals (carbamazepine, diazepam, diclofenac, ibuprofen, naproxen, roxithromycin, sulfamethoxazole, and iopromide) have been determined in municipal Sewage Treatment Plants (STPs) in Galicia, NW Spain. Aerobic activated sludge treatment reduced concentrations of all compounds detected, between 35% and 75%, with the exception of iopromide, which remained in the aqueous phase. The overall removal efficiencies within the STP were 40–65% for the anti-inflammatories and 60% for sulfamethoxazole.	Spain, 2004 (10)
4	Two nonsteroidal anti-inflammatory drugs (NSAIDs), ibuprofen, and acetaminophen, a β-blocker atenolol, and an antidepressant fluoxetine were investigated. The removal efficiency for a 6-hr batch activated sludge treatment was over 90% for the two NSAIDs whereas that for atenolol was as low as 10%; both agreed with the results obtained in conventional studies, which showed the concentration in the influent and effluent of sewage treatment plants. The removal efficiency for fluoxetine was also over 90%, but it was sorbed by sludge and not biodegraded.	Japan, 2005 (11)
7	The removal of seven pharmaceuticals in the biological units of various full-scale municipal wastewater treatment plants was studied. The observed removal of pharmaceuticals was mainly due to biological transformation and varied from insignificant (<10%, carbamazepine) to >90% (ibuprofen). However, no quantitative relationship between structure and activity was apparent and it was concluded that for compounds showing a sorption coefficient (K_d) of below 300 l/kg, sorption onto secondary sludge is not relevant and their transformation can consequently be assessed simply by comparing influent and effluent concentrations.	Switzerland, 2005 (12)
6	Elimination of six acidic pharmaceuticals (clofibric acid, diclofenac, ibuprofen, ketoprofen, mefenamic acid, and naproxen) was investigated in a real wastewater treatment plant (WWTP) using an activated sludge system and membrane bioreactors (MBRs). The MBRs exhibited greater elimination rates than did the real plant. The sludges in the MBRs had large specific sorption capacities mainly due to their large specific surface areas. Despite the sorption capacities of sludges, the main mechanism of elimination of the pharmaceuticals in the investigated processes was found to be biodegradation. Biodegradation of diclofenac, which has been believed to be refractory to biodegradation, seemed to occur very slowly.	Japan, 2007 (13)
>12	Several pharmaceutical products (analgesics and anti-inflammatory drugs, lipid regulators, antibiotics, etc.) were monitored during treatment of wastewater in a laboratory-scale membrane bioreactor (MBR). The results were compared with removal in a conventional activated-sludge (CAS) process in a wastewater-treatment facility. Pharmaceuticals were, in general, removed to a greater extent by the MBR integrated system than during the CAS process. For most of the compounds investigated, the performance of MBR treatment was better (removal rates >80%) and effluent concentrations of, e.g., diclofenac, ketoprofen, ranitidine, gemfibrozil, bezafibrate, pravastatin, and ofloxacin were steadier than for the conventional system. Occasionally, removal efficiency was very similar, and high, for both treatments (e.g., for ibuprofen, naproxen, acetaminophen, paroxetine, and hydrochlorothiazide). The antiepileptic drug carbamazepine was the most persistent pharmaceutical and it passed through both the MBR and CAS systems untransformed.	Spain, 2007 (14)

Table 1 Removal of Specific Pharmaceuticals by Sewage Treatment Processes (*Continued*)

Number of APIs investigated	Summary of findings	Country, date (reference)
12	In wastewater treatment processes, membrane bioreactors (MBR) showed limited target compound removal, but were effective at eliminating hormones and some pharmaceuticals (e.g., acetaminophen, ibuprofen, and caffeine). Membrane filtration processes by using reverse osmosis (RO) and nanofiltration (NF) showed excellent removal (>95%) for all target analytes.	South Korea, 2007 (15)
66	26 compounds out of 66 were detected in the influent ranging ng/l to mg/l order. Particularly, diisopyramide, sulpiride, and dipyridamole that have been rarely detected before in the WWTP (waste water treatment plant), occurred at concentration levels of more than 100 ng/L. The total concentration of the individual pharmaceuticals in the influent was efficiently reduced by 80% during the biological treatment. Ozonation followed by biological activated carbon treatment efficiently reduced all the residual pharmaceuticals below their quantification limits.	Japan, 2008 (16)

Table 2 Degradation/Removal Processes for Pharmaceuticals in STPs, Surface Water, and Soil (17)

Process	Process subtype	Location[a]	Comments
Abiotic	Aqueous photodegradation (partial or complete)	**Surface waters** Soil	For compounds that absorb in the UV-visible range (290–800 nm) may be rapid particularly in the presence of photosensitizers. If absorption occurs at <250 nm, photodegradation is unlikely except in the presence of photosensitizers
	Hydrolysis	Trickling filters Activated sludge Surface waters	Most pharmaceutical compounds are resistant to hydrolysis
	Adsorption	**Trickling filters** **Activated sludge** **Surface water** **Soils**	Most APIs are of low aqueous solubility; basic, lipophilic, and zwitterionic APIs in particular adsorb strongly to organic matter in sewage sludge, river sediment, and soil. Bound compounds are unavailable for microbial biodegradation, hydrolysis, and photolysis
Biotic	Aerobic biodegradation (partial or complete biotransformation/mineralization)	**Trickling filters** **Activated sludge** Surface waters **Soils**	Most drug substances are resistant to partial or complete mineralization in STPs. The diversity of microorganisms in soil often enables mineralization of drug substances that are resistant to STP biodegradation
	Anaerobic biodegradation (partial or complete biotransformation/mineralization)	**Anaerobic sludge digesters** **Soils**	Anaerobic–aerobic cycles of treatment may enhance biotransformation, especially when combined with biodegradation

[a] Principal location of process in **bold**.

Table 3 Biodegradation of Pharmaceuticals in Soils

Test system	APIs evaluated	Summary of findings	Reference
Field-scale study involving two agricultural soils	Chlortetracycline and tylosin	The dissipation of chlortetracycline (CTC) and tylosin A (TYL A) was assessed as a function of time using a sandy loam soil (field A) and a sandy soil (field B). Colony-forming units and TYL A were detected throughout the entire sampling period, with respective starting soil concentrations of 30 and 50 μg/kg soil declining to 1 and 5 μg/kg soil, on day 155. $t_{1/2}$ for CTC was 25 days (20–34) and 34 days (28–42) in fields A and B, respectively, and $t_{1/2}$ for TYL A was 67 days (54–86) and 49 days (40–64) in fields A and B, respectively. The level of both CTC- and TYL-resistant bacteria was affected in the soil by amendment of manure, but declined during the study to the same level as observed at the beginning.	21
Soil irrigated with reclaimed water	19 APIs	Several of the selected pharmaceuticals increased in total soil concentration at one or more of the sites. The four most commonly detected pharmaceuticals were erythromycin, carbamazepine, fluoxetine, and diphenhydramine. Typical concentrations of the individual pharmaceuticals observed were low (0.02–15 μg/kg dry soil). The existence of subsurface maximum concentrations and detectable concentrations at the lowest sampled soil depth might indicate interactions of soil components with pharmaceuticals during leaching through the vadose zone. Nevertheless, the present study demonstrates that reclaimed-water irrigation results in soil pharmaceutical concentrations that vary through the irrigation season and that some compounds persist for months after irrigation.	22
Field study on biosolids slurry	9 APIs	Biosolids slurry was applied at a commercial rate using either subsurface injection or broadcast application. Atenolol, carbamazepine, cotinine, gemfibrozil, naproxen, ibuprofen, acetaminophen, sulfamethoxazole, and triclosan were studied. In runoff from the injected plots, concentrations of the model compounds were generally below the limits of quantitation. In contrast, in the broadcast application treatment, the concentrations of atenolol, carbamazepine, cotinine, gemfibrozil, naproxen, sulfamethoxazole, and triclosan on the day following application ranged from 70 to 1477 ng/L in runoff and generally declined thereafter with first order kinetics. For ibuprofen and acetaminophen, concentrations in runoff first decreased and then increased, suggesting that these drugs were initially chemically or physically sequestered in the biosolids slurry, and subsequently released in the soil. Carbamazepine and triclosan were detected at low concentrations in a runoff event 266 days after broadcast application. Overall, this study showed that injection of biosolids slurry below the soil surface could effectively eliminates surface runoff of APIs.	23

Table 3 Biodegradation of Pharmaceuticals in Soils (*Continued*)

Test system	APIs evaluated	Summary of findings	Reference
Leaching behavior in different soils	6 APIs	Leaching potential was low for diazepam, ibuprofen, ivermectin, and carbamazepine, and so groundwater contamination with these substances seems unlikely. Clofibric acid and iopromide were very mobile under the experimental conditions. Lack of concordance with literature reports was found for carbamazepine, because it occurs very often in groundwater, which might be explained by the fact that the leaching tests were performed with soil, whereas groundwater contamination originates mainly from river sediments and subsoil from receiving waters.	24
Fate and mobility in two soil types	6 APIs	The sorption and mobility of six pharmaceuticals (carbamazepine, propranolol, diclofenac sodium, clofibric acid, sulfamethoxazole, and ofloxacin) were investigated in two soil types with different organic carbon and clay content, and in bacterial biomass (aerobic and anaerobic). Ofloxacin was particularly strongly adsorbed (except of the case of using anaerobic biomass for the solid matrix), while clofibric acid was found to be weakly adsorbed. Major differences in the mobility of the six pharmaceuticals were observed and correlated with the adsorption/desorption properties of the compounds.	25
Laboratory tests using two different soils	Ibuprofen and diazepam	Laboratory tests on the degradation of the 14 C-labeled model substances ibuprofen and diazepam were performed in two different soils. The kinetic behavior of both pharmaceuticals in soil is characterized by the participation of microbial processes in the formation of nonextractable residues and by low degradation rates out of the nonextractable fractions. An accumulation of those nonextractable residues may occur in soils following multiple applications.	26

Europe to the adoption of HFAs (hydrofluoroalkanes) such as HFA-134a (19) that are not damaging to the ozone layer. Anesthetic nitrous oxide, for example, has the potential to affect the ozone layer (20).

ENVIRONMENTAL RISK ASSESSMENT

European Union/European Economic Area

Regulatory Background

The European Union (EU) Directive 93/39/EEC (27) amending Directive 65/65/EEC stipulated that the applicant should provide "indications of any potential risks presented by the medicinal product to the environment." Subsequently, the requirement for submission of an Environmental Risk Assessment (ERA) became part of the "Notice to Applicants" (28). However, at that time not even a draft guideline had been published (although a group had been working on ERA guidance since 1990). This, in combination with US Food and Drug Administration's (FDA's) relaxation of the requirements for environmental assessments in 1998 (see below) produced an inconsistent approach by different member states, some requiring an ERA and others not. A draft guideline/discussion paper for an ERA on non-GMO (Genetically Modified Organism)-containing drugs was published by the European Medicines Evaluation Agency (EMEA) in 2001, but drew significant criticism particularly from the Scientific Committee on Toxicity, Ecotoxicity, and the Environment (CSTEE) (29). The focus of the CSTEE concerns was that the threshold surface water concentration of 0.01 μg/l that triggered the need for environmental testing was not scientifically based. A significantly revised draft guideline was released by the EMEA (European Medicines Agency) in 2003 for a further round of consultation and the final guideline was agreed in mid-2006, coming into operation on 1st December 2006 (30). The requirement for an assessment of the potential environmental impact of non-GMO medicinal products follows from the provisions of Article 8(3) of Directive 2001/83/EC; in addition, Annex 1 to this directive, updated by Directive 2003/63/EC (31), provides guidance on the location of an ERA in a Marketing Authorisation Application (MAA) (Module 1.6) and on the need for the ERA to be prepared by an appropriately qualified expert whose *curriculum vitae* should be included in the application. Provisions in Directive 2001/83/EC (as amended) relate to risks to the environment arising from use, storage, and disposal of medicinal products, and not risks arising from their synthesis or manufacture. However, it is explicitly stated that the environmental impact "should not constitute a criterion for refusal of a marketing authorization."

According to the EU ERA guideline (30), an ERA report is required to accompany most types of applications including those made under centralized, mutual recognition, decentralized or national procedures. For type II variations, line-extension and generic product applications, an ERA is required if there is a potential increase in environmental exposure. An ERA is not required for type I variations or where the active ingredient is unlikely to pose a significant risk to the environment (e.g., vitamins, electrolytes, amino acids, peptides, proteins, carbohydrates, lipids, vaccines, and herbal products). As protein drug substances are excluded from the scope of the guidance, biological and biotechnology derived drugs would normally not require an ERA. Notwithstanding the scope of the ERA guideline, because there is no explicit derogation in Directive 2001/83/EC for the various ERA exclusions that are permitted, many EU agencies may refuse to validate a MAA without an ERA report in Module 1.6. This is much more an administrative rather than a scientific issue: administrators want to tick the ERA box during validation without having to assess whether any claimed exemption is justified. Consequently, even for biotechnology drugs and for generic products, it is advisable to prepare an ERA report (including the expert's *curriculum vitae*) describing briefly the drug substance and any claimed exemption from the need for a scientific assessment. It may also be prudent to comment on the drug product particularly in terms of the environmental acceptability of excipients and packaging materials. Any label advice on disposal of the drug product should also be included. If a medicinal product requires a disposable device for its delivery, it would be appropriate to take this into consideration in any ERA report.

Scientific Assessment

The EU guideline (30) describes the assessment process as a "stepwise, phased procedure"; in fact three stages are involved: prescreening, screening, and extended, as shown in Table 4. In the guideline, frequent reference is made to Part II of the EU Technical Guidance Document (TGD), which provides comprehensive guidance on the environmental assessment of chemicals (32). The Part II TGD on environmental risk assessment, which runs to 328 pages, is often highly complex and far from user-friendly; moreover, the definitions of some parameters are not completely aligned between the TGD and the EU ERA guideline, and some expert guidance on "custom and practice" simplifying assumptions may be required in order to make a complete environmental assessment. General approaches in environmental risk assessment are shown in Figure 2 (from the TGD), and a synopsis of the main phases and decision points involved in the EU environmental risk assessment process is shown in Figure 3.

Phase I Prescreening

This initial phase involves determination of

- octanol–water partition coefficient (K_{OW}) of the API,
- market penetration factor (F_{pen}), and
- predicted environmental concentration (PEC) in surface water.

It is advisable to determine K_{OW} as early as possible in order to assess whether a Persistence, Bioaccumulation, Toxicity (PBT) assessment is triggered. The thinking behind this concept, which was developed by the Convention for the Protection of the Marine Environment of the North-East Atlantic (OSPAR), is that, if persistent compounds that may also be

Table 4 Key Parts of Stepwise, Phased EU Assessment Procedure (27)

Stage in regulatory evaluation	Stage in risk assessment	Objective	Method	Test/data requirements
Phase I	Prescreening	Estimation of exposure	Action limit	Consumption data, log K_{OW}
Phase II tier A	Screening	Initial prediction of risk	Risk assessment	Base set aquatic toxicity and fate
Phase II tier B	Extended	Substance and compartment-specific refinement and risk assessment	Risk assessment	Extended data set on emission, fate, and effects

bioaccumulative and toxic are clearly labeled as such, it should be easier to reduce releases to the marine environment. The OSPAR Convention is the current legislative basis for international cooperation on environmental protection in the North-East Atlantic. The OSPAR Commission is made up of representatives of the Governments of the 15 signatory nations (Belgium, Denmark, Finland, France, Germany, Iceland, Ireland, the Netherlands, Norway, Portugal, Spain, Sweden,

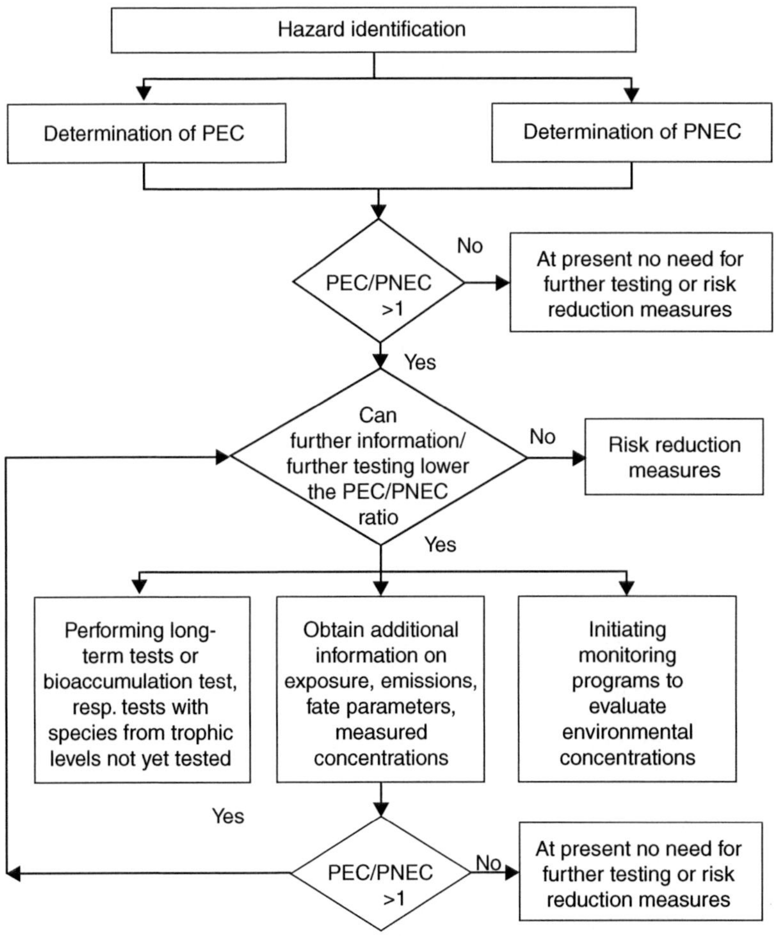

Figure 2 General procedure for environmental risk assessment (32).

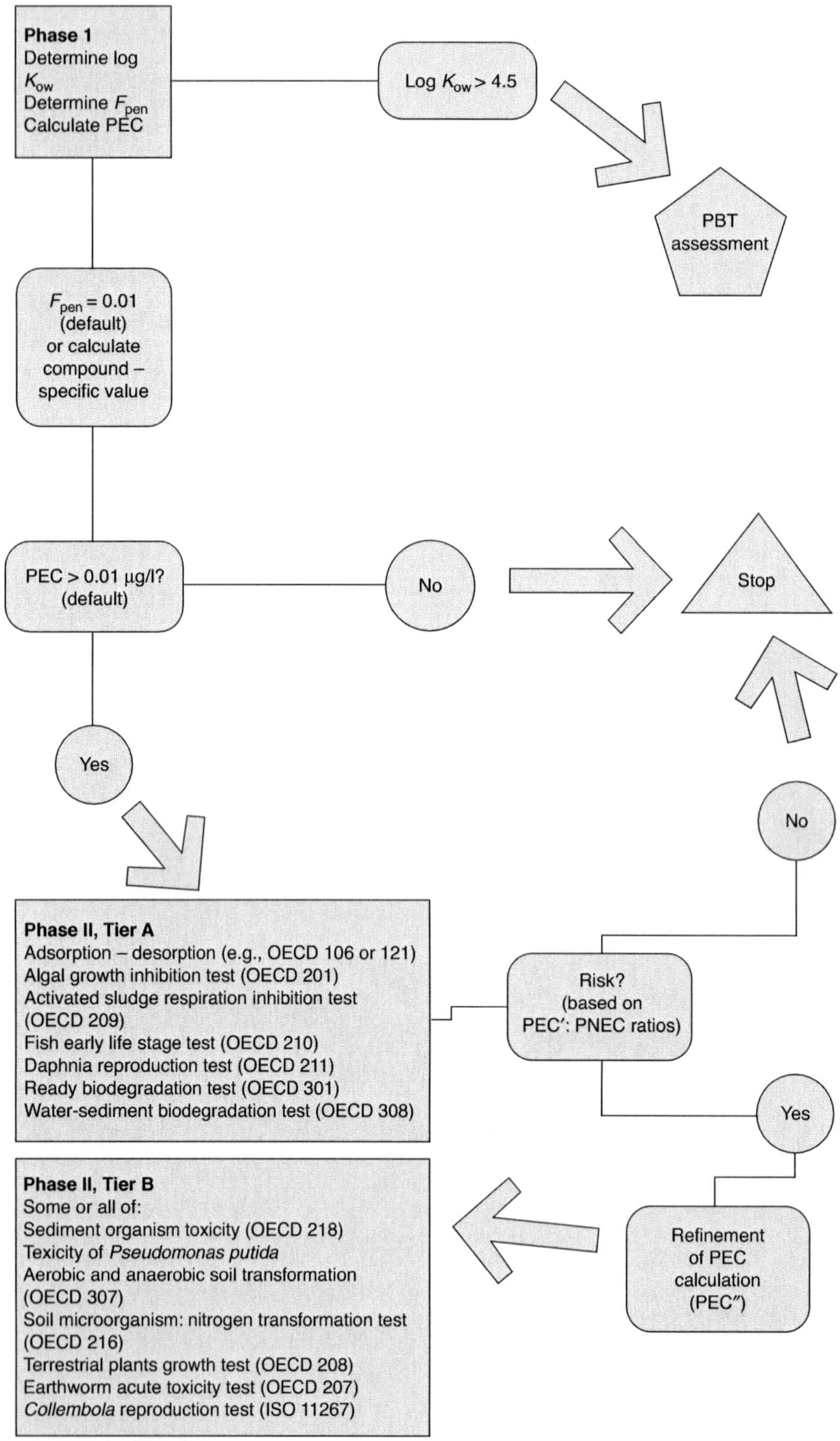

Figure 3 Stepwise EU Environmental Risk Assessment.

Table 5 Criteria for Identification of PBT and vPvB Substances (29)

Criterion	PBT criteria	vPvB criteria
P	Half-life >60 days in marine water >40 days in freshwater[a] or half-life >180 days in marine sediment or >120 days in freshwater sediment[a]	Half-life >60 days in marine or freshwater or >180 days in marine or freshwater sediment
B	BCF >2000	BCF >5000
T	Chronic NOEC <0.01 mg/l in CMR or endocrine disrupting effects	Not applicable

[a] For the purposes of marine environmental risk assessment half-life data in freshwater and freshwater sediment can be overruled by data obtained under marine conditions.

Abbreviations: BCF, bioconcentration factor; NOEC, no-observed-effect concentration; CMR, Carcinogenic, Mutagenic, and Reproductive toxicity.

the United Kingdom of Great Britain and Northern Ireland, Luxembourg, and Switzerland) and delegates from the European Commission, representing the European Union. Drug substances with log $K_{OW} > 4.5$ should be screened in a stepwise procedure for persistence, bioaccumulation, and toxicity. Although a number of PBT criteria have been suggested by various groups, those specified in the TGD should be applied. The specific PBT criteria are shown in Table 5; guidance on PBT testing strategies is provided in the TGD.

The most important part of the Phase I assessment is calculation of the predicted environmental concentration (PEC) of the API in surface water. The exposure estimation should be based only on drug substance irrespective of its route of administration, pharmaceutical form, metabolism, and excretion. This seems to be an unreasonable provision for drugs that are completely or nearly completely metabolized before excretion in that a phase IIA program of testing may be mandated on a drug substance for which there will be minimal environmental exposure. Certain substances such as highly lipophilic APIs and potential endocrine disruptors may need to be evaluated in Phase IIA/B irrespective of the quantity released into the environment. Initial calculation of the PEC relies on a number of assumptions that are summarized in Table 6.

PEC (mg/l) is then calculated from the expression

$$\mathrm{PEC}_{\mathrm{SURFACE\,WATER}} = \frac{\mathrm{DOSE}_{\mathrm{ai}} \times F_{\mathrm{pen}}}{\mathrm{WASTEW}_{\mathrm{inhab}} \times \mathrm{DILUTION}}$$

Table 6 Assumptions and Default Values for Initial Calculation of PEC in Phase I

Parameter	Symbol	Value	Unit
Maximum daily dose per patient (highest recommended dose)	$\mathrm{DOSE}_{\mathrm{ai}}$	Depends on drug/indication	mg/inhab/day
Fraction of market penetration[a]	F_{pen}	Default = 0.01	–
Amount of wastewater per inhabitant per day	$\mathrm{WASTEW}_{\mathrm{inhab}}$	200	l/inhab/day
Dilution factor	DILUTION	10	–
Predicted environmental concentration[b]	$\mathrm{PEC}_{\mathrm{SURFACEWATER}}$	Depends on drug/indication	mg/l

[a] The default value of 0.01 may be used or a drug/indication-specific value may be calculated based on published epidemiological and other data.
[b] Assumes that the predicted drug intake is evenly distributed over the year and throughout the geographical area, that the sewage system is the main route of entry of the drug substance into surface water, that there is no biodegradation or retention of the drug substance in the STP, and that metabolism in the patient is not taken into account.

The normal action limit triggering a Phase IIA assessment is 0.01 μg/l (0.01 ppb), which, based on the parameter values shown in Table 6, equates to a $DOSE_{ai}$ value of >2.0 mg/patient/day when the default F_{pen} of 0.01 is employed.

$$DOSE_{ai} = \frac{200 \times 10 \times 0.00001}{F_{pen}} = 2.0\,\text{mg/patient/day for default } F_{pen} \text{ of } 0.01$$

Because few pharmaceutical products are used at maximum doses of ≤2 mg/day, many drugs intended to treat relatively minor indications (leading to a modest level of environmental exposure) could be "caught" by the use of the standard F_{pen} of 0.01. The latter is based on consumption data for 800 licensed drug substances available on the German market in 2001 (30). The 95 percentile consumption equated approximately to a 1% market penetration factor. Major criticisms can be made of this approach when applied to new drugs in that

- Consumption of well-established licensed drugs is likely to be much higher than that for new drugs (particularly those used in niche indications).
- There is no transparency (and potential bias) over the selection of the 800 drugs (out of the 2700 drugs marketed in Germany at the time) used in the calculation.
- Calculation of F_{pen} for a new drug based on projected consumption data (which was recommended in the 2001 draft ERA guideline) will in virtually all cases result in a (much) lower value (see further discussion below).

For drugs that are used intermittently (such as antibiotics), an assumption of constant daily use will produce an overestimate of the PEC. Overall, the standard PEC calculation seems more targeted towards "capturing" as many drugs as possible for subsequent Phase IIA evaluation in order to build up a database on effects assessment rather than attempting to provide the best estimate of environmental exposure. Some regulators with a special interest in pharmaceuticals in the environment (33) have expressed frustration over the fact PEC values exceeding 0.01 μg/l are rarely reported if projected consumption data are used to calculate F_{pen}—a situation leading to the absence of effects assessment and risk characterization that "does not advance science or improve the knowledge of environmental risks possibly posed by pharmaceutical ingredients."

Determination of drug-specific F_{pen} values is often possible using epidemiological and other data, although it appears to be becoming increasingly difficult to obtain acceptance of such estimates by European regulators who seem to be focused on the perceived need to maximize the amount of environmental data made available to the competent authorities. The guidance on calculating F_{pen} for the Phase I prescreening assessment states: "The Applicant may use the default value or refine the F_{pen} by providing reasonably justified market penetration data, for example, based on published epidemiological data." Deciding on what constitutes "published epidemiological data" in the context of providing a reasonably justified market penetration estimate can often be a point of contention between the applicant and the assessor; some assessors may accept estimates that are based on existing patterns of use for drug categories within a therapeutic area whereas others may insist that the total number of treatable patients within an indication should form the basis of an F_{pen} calculation. Table 7 provides an example of an F_{pen} calculation for a new anticonvulsant with the intended indication "adjunctive treatment in adults with partial onset seizures, with or without a secondary generalization." A recent EU economic analysis provides an estimated prevalence of epilepsy in Europe in 2004 of 4.3 to 7.8 per 1000 (34). A German study indicates that 58% of epilepsies are focal epilepsies, such as partial onset (35). A U.K. report, NICE Implementation Uptake Report (36), indicates that gabapentin was the most prescribed newer antiepilepsy drug (AED) in England representing 13% of the total number of AED prescriptions. In this same report, 32.5% of the surveyed patients were on monotherapy and 28% were prescribed ≥2 AEDs. Assuming that the population of the EEA is 500 million, 495 million for EU27 (37), and 5 million for Norway, Iceland, and Liechtenstein combined, an estimate of the maximum number of patients treated with the new AED can be made, leading to an F_{pen} of 0.000165. Because the epilepsy indication described above is almost identical to that of Keppra (levetiracetam), a similar F_{pen} should apply. According to the EPAR (European Public Assessment Report) for Keppra (38), the PEC is 0.3 μg/l and the maximum

Table 7 Indication-Specific Assessment of F_{pen} (Adjunctive Therapy for Partial-Onset Epilepsy)

Parameter	Worst-case estimate	Population/patient numbers	Reference
EEA population		500,000,000	31
Epilepsy prevalence	7.8 per 1000	3,900,000	32
Partial onset seizure	58%	2,262,000	33
Taking medication	60.5%	1,368,510	34
Patients on multiple AEDs	46.3%	633,360	34
Market share after ≥10 years	13%	82,337	34
F_{pen}	0.000165 (82,337/5×10^8)		N/A

daily dose 3000 mg, and hence the F_{pen} can be calculated as 0.0002. Consequently, the most conservative F_{pen} estimate of 0.0002, based on a regulatory precedent, should be acceptable to EU regulators. This example includes an assumption that the market share of the new AED will not exceed that of the current best-selling drug for a similar indication, and so there could be some potential for disagreement during regulatory assessment as to whether the F_{pen} calculation complies in full with the criterion of using only "published epidemiological data." However, given that action limit of 0.01 μg/l is exceeded in this case and a Phase IIA assessment is required, arguments over the method of calculation are somewhat academic because F_{pen} can be refined later on in the EU assessment process. On the other hand, a thorough justification complying strictly with the criteria in the EU guideline would be required to support an initial PEC < 0.01 μg/l involving a drug-specific F_{pen} < 0.01.

Phase IIA Program

The purpose of the Phase IIA program is to provide information on environmental fate and toxicity obtained from a "base set" of tests. There are four tests relating to assessment of environmental fate and three chronic toxicity tests in algae, daphnia, and fish. The test methodology is already available, having been developed for industrial chemicals, biocides, and pesticides, and in general employs Organization for Economic Co-operation and Development (OECD) guidelines (39), tests developed by the International Organization for Standardization (ISO) being applicable in some cases (40). Most literature information on the effects of pharmaceuticals on aquatic organisms is based on the results of acute toxicity tests, but the EU guidance was changed quite late in the consultation process to require long-term tests because it is believed that this best reflects the environmental exposure of aquatic organisms to pharmaceuticals. Pharmaceuticals are sometimes described as being "pseudopersistent" in the environment, that is, although APIs are subject to a variety of abiotic and biotic degradation processes, additional human waste containing undegraded compounds is constantly entering STPs and surface water. The test material for these tests should be the API itself unless otherwise justified, for example, in the case of a pro-drug, where the active moiety would need to be tested (if feasible).

The Phase IIA tests, which should comply with Good Laboratory Practice (GLP) requirements, are as follows (shown also in Fig. 3):

- Adsorption–desorption [e.g., OECD 121 (41)]
- Ready biodegradation [OECD 301 (42)]
- Water-sediment study [OECD 308 (43)]—only if API is not readily biodegradable.
- Activated sludge respiration inhibition test [OECD 209 (44)]
- Algal growth inhibition test [OECD 201 (45)]
- Daphnia reproduction test [OECD 211 (46)]
- Fish early life stage test [OECD 210 (47)].

A ready biodegradability test should be conducted to assess the likely fate of the API in the STP. If the API is shown not to be readily biodegradable (which occurs in the majority of cases), its fate should be further investigated in a water-sediment study (OECD 308). The sorption characteristics of the API in sewage sludge are likely to be determined by the adsorption coefficient (K_{OC}), defined as the ratio between the concentration of the substance in sewage sludge's organic carbon and the concentration of substance in the aqueous phase at adsorption equilibrium. It is assumed that a substance with a high K_{OC} value will most probably be retained in the STP and may reach the terrestrial compartment through application of sewage sludge to agricultural land. Several tests are available for determination of adsorption–desorption characteristics, although OECD 121 (41) would possibly be the normal choice because this is the simplest to perform (using HPLC). A test [OECD 209 (44)] is required to evaluate the potential for toxic effects on activated sludge organisms in respect of their respiration (rate of oxygen uptake).

Long-term tests on species from three trophic levels are specified in the EU guidance to determine the predicted no-effect concentration of the API in surface water. The recommended tests would normally involve toxicity assessments in blue-green algae (e.g., *Pseudokirchneriella subcapita*) and daphnia, and an early life stage test in fish (e.g., fathead minnow, *Pimephales promelas*). The guidance indicates the need for using blue-green algae (Cyanophyta) when testing for effects of antimicrobials because these species are more sensitive than green algae, and so use of *P. subcapita* should be generally acceptable. Short descriptions of the methods employed in Phase IIA fate and toxicity tests are shown in Table 8.

The next step is to calculate predicted no-effect concentrations (PNECs) based on the no-observed-effect concentration (NOEC) values determined in the phase IIA program. PNECs are calculated by dividing the appropriate NOECs by an assessment factor (AF; i.e., safety factor) for which the default value is 10. The AF is considered to account for interspecies variation in sensitivity, intraspecies variation, and extrapolation of laboratory data to the field situation. The following PNECs can be calculated:

- $PNEC_{WATER}$: based on lowest NOEC from base set of long-term chronic toxicity tests.
- $PNEC_{MICROORGANISM}$: based on NOEC from activated sludge respiration inhibition test.
- $PNEC_{GROUNDWATER}$: based on NOEC in daphnia toxicity study.

Assessment Criteria for Phase IIA Fate and Effects Analysis

It is possible to refine the initially calculated value for $PEC_{SURFACEWATER}$ at this stage, but only in respect of taking into account information on the sales forecast of the product. This kind of refinement was explicitly described in the 2005 draft guideline, but is mentioned only fleetingly in the finalized guideline; not allowing for realistic market penetration at this stage of the assessment could produce a significant overestimate of environmental exposure and lead to unnecessary further testing. The impact of other factors such as human metabolism and partitioning into sewage sludge are allowed as part of the Phase IIB assessment. Consequently, in the example shown above concerning a new AED, a value for $F_{pen} \leq 0.000165$ rather than 0.0002 could be used. Moreover, the *average* rather than the maximum daily dose in the specific patient population could be employed to calculate the refined PEC. In subsequent text, PEC′ is used to denote this first refinement of PEC.

A summary of the Phase IIA assessment criteria is shown in Table 9. The standard approach to assessment of groundwater does not apply for APIs with $K_{OC} > 10000$ l/kg (because adsorption to sewage sludge and/or soil is likely) or for APIs that are readily biodegradable or those that have a DT_{50} (biodegradation half-life) of <3 days. $PEC'_{GROUNDWATER}$ can be estimated from the expression:

$$PEC'_{GROUNDWATER} = 0.25 \times PEC'_{SURFACEWATER}$$

If the ratio of $PEC'_{GROUNDWATER}:PNEC_{GROUNDWATER}$ is <1, there is no need to proceed to a further groundwater assessment in Phase IIB.

Other criteria for deciding on the need for further testing are as follows:

- If $PEC'_{SURFACEWATER}:PNEC_{WATER}$ is <1, then it is concluded that the API and/or metabolites are unlikely to pose a risk to aquatic organisms and no further testing in the aquatic compartment is required. If this ratio is >1, further evaluation, preferably on the fate of the API and/or metabolites in the aquatic compartment, is needed in tier B.
- If $PEC'_{GROUNDWATER}:PNEC_{GROUNDWATER}$ is >1, a further fate assessment is required, as above.

Table 8 Short Descriptions of Phase IIA Test Methods

Test	Test guideline	Brief test description
Adsorption–desorption	OECD 121	The API is injected onto an appropriate HPLC column simulating the relationship between retention time and adsorption coefficient of the API on the organic part of a soil. No quantitative analysis is required and only the determination of retention time is necessary.
Ready biodegradation	OECD 301	Measurement of oxygen (biochemical oxygen demand; BOD) uptake uses a suitable respirometer. Oxygen uptake is recorded every 112 min over the 28-day test period. COD (chemical oxygen demand) is determined by standard techniques and pH and temperature are monitored and adjusted as necessary. Sodium benzoate is normally employed as a positive control.
Water-sediment study	OECD 308	Humidified air is drawn through replicate test vessels containing high and low organic matter sediments and river water for periods up to 99 days. Radiolabeled API is added to each test vessel at a suitable concentration (determined from a preliminary study). Vessels are sacrificed at day 0 and periodically thereafter. Any radiolabeled CO_2 produced is trapped and measured. Various analyses are performed including radioactivity in the overlying water, sediment, and glassware. Radio-TLC and MS analyses are undertaken on the overlying water and sediment extracts.
Activated sludge respiration inhibition test	OECD 209	The respiration rate of each culture is measured after 3 hr and compared with the mean respiration rate of two control cultures. Rate of oxygen uptake is measured and the EC_{50} value is estimated using appropriate computer software.
Algal growth inhibition test	OECD 201	Replicate 100 ml cultures of *P. subcapitata* are incubated at 22 ± 2°C, using cool white lighting and orbital shaking at 160 rpm. Control medium and test cultures are randomly employed in the incubator. Algal cell particle densities of the inoculum and test cultures are determined by electronic particle counting. After 24, 48, and 72 hr samples are removed from each test and blank vessel. The appropriate blank particle count is subtracted from that of the test culture to obtain the cell particle density. Microscopic observations are made on samples taken from a single replicate of the control and each test substance concentration.
Daphnia reproduction test	OECD 211	Ten replicates per exposure concentration are normally employed. Initially, one randomly selected daphnid (P_0 generation) is added to each test vessel; thereafter, observations are made daily from day 0 for the presence of live and dead offspring (F_1 generation). The test material should be stable in the test medium and so test solutions are prepared on day 0 and may require frequent renewal (possibly daily) throughout the study. The normal test duration is 21 days. At the end of the test the length (apex of helmet to base of spine) of each surviving P_0 daphnid is determined.

(*Continued*)

Table 8 Short Descriptions of Phase IIA Test Methods (*Continued*)

Test	Test guideline	Brief test description
Fish early life stage test	OECD 210	Exposure of fish embryos (normally 15 embryos per replicate test tank (60 per concentration), giving an approximate loading of 3.2 eggs/embryos/l of test solution, begins at day 0. Live and dead egg/embryo numbers are recorded daily, dead eggs/embryos being discarded. With hatching at least 90% complete, or 24 hr after the first hatch, larvae are released into the appropriate replicate test tank. When the hatch is complete, the number of live and dead larvae in each replicate is recorded. The percentage hatch per replicate and per concentration is then calculated. Exposure is terminated at 28 day posthatch when the surviving larvae are counted and measured for standard length and dry weight.

- If $PEC'_{SURFACEWATER}$:$PNEC_{MICROORGANISM}$ is >0.1, a fate assessment in tier B is required as above.
- If $K_{OW} > 1000$, transfer of the API into aquatic organisms is likely and determination of a bioconcentration factor should be considered in tier B.
- If $K_{OC} > 10000\,l/kg$, the API is likely to bind to sewage sludge in the STP and an assessment of the potential effect in the terrestrial compartment should be conducted unless the substance is readily biodegradable.
- If the API is not readily biodegradable and the results of the water-sediment study indicate a significant shift of the drug substance (or radioactivity) to the sediment (i.e., if more than 10% of the substance at any time point on or after 14 days is bound to sediment in the OECD 308 test), then effects on sediment-dwelling organisms should be investigated in tier B. This situation occurs quite frequently because binding to sediment is likely to occur when the API is basic or zwitterionic in character, although it may not be particularly lipophilic.

Further Refinement of PEC for Phase IIB Assessment

Before proceeding to tier B studies, further refinement of the PEC is recommended based on the extent of human metabolism and removal in the STP. The second refinement of the PEC is calculated as follows:

$$PEC''_{SURFACEWATER} = \frac{Elocal_{water} \times F_{stpwater}}{WASTEW_{inhab} \times CAPACITY_{stp} \times FACTOR \times DILUTION}$$

$$\text{where,}\ Elocal_{water} = DOSE_{ai} \times F_{excreta} \times F_{pen} \times CAPACITY_{stp}$$

Table 9 Phase IIA Assessment Criteria

Phase II tier a results	Action
$PEC'_{SURFACEWATER}$:$PNEC_{WATER} < 1$	No additional aquatic testing required
$PEC'_{SURFACEWATER}$:$PNEC_{WATER} > 1$	Phase IIA tier B fate assessment required
$PEC'_{GROUNDWATER}$:$PNEC_{GROUNDWATER} > 1$	Phase IIA tier B fate assessment required
$PEC'_{SURFACEWATER}$:$PNEC_{MICROORGANISM} > 0.1$	Phase IIA tier B fate assessment required
$K_{OW} > 1000$	Phase IIA tier B bioaccumulation study required
$K_{OC} > 10000$ l/kg	Phase IIA tier B terrestrial compartment effects assessment required (unless API is readily biodegradable)
Shift to sediment in OECD 308 > 10% over 14 days	Unless API is readily biodegradable, phase IIA tier B test required on sediment dwelling organism (e.g., *Chironomus* spp)

Table 10 Parameters and Defaults for $PEC''_{SURFACEWATER}$ Calculation in Tier IIB[a]

Parameter	Symbol	Value	Unit
Input			
Wastewater per inhabitant per day	$WASTEW_{inhab}$	200	l/inh/day
Capacity of STP	$CAPACITY_{stp}$	10,000	inh
Fraction of emission reaching surface water	$F_{stp\ water}$	Case-specific	N/A
Dilution factor	DILUTION	10	N/A
Factor accounting for adsorption to suspended matter	FACTOR	Case-specific	N/A
Local emission of API in wastewater	$Elocal_{water}$	Case-specific	mg/day
Output			
Local surfacewater concentration	$PEC''_{SURFACEWATER}$	Case-specific	mg/l

[a] Guidance on calculation of parameters is provided in the TGD.

Parameters and defaults used in calculating $PEC''_{SURFACEWATER}$ for the tier B assessment are shown in Table 10.

$F_{stpwater}$ represents the fraction of API passing through the STP and is mediated by biodegradation and adsorption onto sewage sludge particulate matter. The parameter can be estimated in a variety of ways including STP simulation studies [OECD 303 (48)], direct measurement of adsorption onto sludge [OPPTS 835.1110 (49)], or use of the "SimpleTreat" model described in the TGD (highly conservative). The SimpleTreat approach requires a value of log K_{OW} and for log H (Henry's law constant)—which can be obtained by using a predictive system such as HENRYWIN (50). In terms of the "FACTOR" in the denominator of the equation for determination of the second refinement of PEC, which is intended to correct for adsorption to suspended matter in surface water, neither the ERA guideline nor the TGD is particularly helpful. Removal of API from surface water via adsorption to suspended (organic) matter will depend mainly on the extent of binding and the concentration of suspended matter. In most cases, the simplest and most conservative approach is to assume a value of 1 for the "FACTOR."

Adsorption is described by the distribution coefficient (K_D):

$$K_D(\mathrm{l/kg}) = \frac{\mathrm{SorbedConcentration(mg/kg)}}{\mathrm{DissolvedConcentration(mg/l)}}$$

Values for K_D can be obtained from the OECD 308 study, and the K_D obtained by using high organic-matter sediment would represent the worst case in terms of binding of drug in surface water to sediment. Thus, the concentration of drug in surface water sediment ($PEC_{SEDIMENT}$; units mg/kg) can be estimated by multiplying either $PEC'_{SURFACEWATER}$ or $PEC''_{SURFACEWATER}$ by K_D. A similar approach to estimate the sequestering by sludge of drug substance in the STP would be inappropriate because the proportion of sediment to liquid is much lower in the STP than in the OECD 308 system.

Extended Effects Analysis—Water Sediment Effects

According to the ERA guideline, determination of the effects of the API on a sediment dwelling organism such as *Hyalella*, *Lumbriculus*, or *Chironomus* spp. is required if the results of the OECD 308 study show a significant shift of the drug substance (radioactivity) to sediment. Various protocols are available for such studies [e.g., OECD 218 (51)] and involve adding radiolabeled drug to a water-sediment system.

It may be possible to make a preliminary evaluation of the potential for effects on sediment-dwelling organisms because there is evidence suggesting that such organisms generally have similar or slightly higher tolerance to the toxic effects of chemicals compared to *Daphnia magna*. A report by Fiedler et al. (52) indicates that toxicity to *Chironomus* was similar to toxicity to *Daphnia* for six compounds (cadmium chloride, trinitrotoluene, 3,4-dichloroaniline, 2,4-dichlorophenol, pentachlorophenol, and tributyltin chloride). The information is presented in more detail in a

Ph.D. thesis (53) by Gildemeister who concludes that "acute toxicity data of eight chemicals in *D. magna* significantly correlate with data from *L. variegatus* and *C. riparius* ($p < 0.05$)." However, he draws attention to the small data set and the variation in sensitivity of the organisms. Highly lipophilic compounds such as 3,4-dichloroaniline and DDT tended to be the most toxic to both *Chironimus* and *Lumbriculus*. Compounds with log K_{OW} up to 3 showed LC_{50} values in the 10 to 100 μM range, whereas highly lipophilic compounds had values in the 0.01 to 0.1 μM range.

A $PNEC_{SEDIMENT\ DWELLER}$ can be determined from an experimental NOEC in a sediment-dwelling organism by application of an appropriate assessment factor (AF); an AF of 100 is recommended in OECD 218. If $PEC_{SEDIMENT}$ is calculated from $PEC'_{SURFACEWATER}$ or $PEC''_{SURFACEWATER}$ as shown above, an estimate of the $PEC_{SEDIMENT}$:$PNEC_{SEDIMENT\ DWELLER}$ ratio can be obtained, which should be <1.

Extended Effects Analysis—Microorganisms

If a risk of toxicity to sewage microorganisms is detected in tier A then a further evaluation of such effects is required, particularly in respect of performing standardized tests in single microbial species such as *Pseudomonas putida* [TGD (32)].

Terrestrial Environmental Fate and Effects Analysis

If K_{OC} for the API is >10,000 l/kg, unless the compound is readily biodegradable, its concentration in the terrestrial compartment should be estimated. The SimpleTreat model—a somewhat simplistic model that considers the volatility (Henry's constant, *H*), the partition coefficient, adsorption coefficient, and biodegradation (54) in the European Union System for the Evaluation of Substances [EUSES (55)] (http://ecb.jrc.it) can be used to calculate PEC_{SLUDGE} and PEC_{SOIL}. SimpleTreat model outputs are also tabulated in the TGD (32). A base set of tests investigating biodegradation in soil, toxicity to soil invertebrates, and acute effects on terrestrial plants and microorganisms should be undertaken (Fig. 3). Interpretation of the results should follow relevant guidance in the TGD.

The Environmental Risk Assessment Report

An environmental risk assessment (ERA) report should be included in Module 1.6 of the MAA when appropriate (see section regulatory background, above). Key parts of the ERA report include:

- An estimate of the potential environmental exposure (probably using PEC as a surrogate) with a statement of underlying assumptions.
- An assessment of possible risks to the environment in respect of the use of the drug product and a presentation and evaluation of data in support of the risk evaluation.
- An evaluation of any precautionary and safety measures to be taken regarding the environmental release from use in patients, disposal of unused products, or disposal of waste materials derived from such products.
- Proposals for labeling in respect of the Summary of Product Characteristics (SmPC) and Patient Information Leaflet (PIL) if any precautionary measures are deemed necessary for reducing risks to the environment in respect of administration to patients and disposal of waste products.

The expert preparing the ERA report should provide a signature and a *curriculum vitae* and the relationship to the applicant should be stated.

Comments on EU Environmental Risk Assessment Guidance

One of the key characteristics of the EU guidance, causing much discussion, is the rather low threshold (2 mg/day) triggering a Phase IIA assessment. Assuming chronic administration, this dose would amount to 730 mg/patient/year. If a (default) market penetration factor of 0.01 is also assumed, the annual environmental burden would be 7.3 kg/million of population, equivalent to a threshold of 438 kg/year for the UK and 3650 kg/year for all 30 EEA countries. By way of comparison, in England, the total volumes of drug substance for the top 20 most prescribed drugs in 2007 ranged from 53 kg (levothyroxine) to 769,000 kg (paracetamol) (56). It should be relatively simple (not relying on epidemiological data and so more robust) to switch from the F_{pen} approach to one involving marketing estimates for the first 5 or 10 years following authorization. Furthermore, the conventional model for estimating PECs (57) also takes into account the removal rate of the drug due to loss by adsorption to sludge particles, by biodegradation, and by abiotic processes such as hydrolysis (but not human metabolism):

$$\mathrm{PEC}_{\mathrm{SURFACEWATER}_{\mathrm{conventional}}} = \frac{A \times (1 - R/100)}{365 \times P \times V \times D}$$

A is the predicted amount of API in kilogram used per year in the relevant geographical area, R is the removal rate due to biotic and abiotic processes, P is the number of inhabitants of the geographical area, V is the volume of wastewater per capita per day (default of 200 l in ERA guideline), and D is the dilution of wastewater by surface water flow (average factor 10). PECs have been estimated for 11 drugs marketed in Norway comparing the conventional with the F_{pen} (ERA guideline approach) and the results are shown in Table 11. Perhaps the most striking feature is that the F_{pen} approach significantly overestimates PECs for the five antibiotics, emphasizing the need to modify the $\mathrm{DOSE_{ai}}$ for drugs used only intermittently. The higher conventionally estimated PECs for drugs such as metaprolol and paracetamol that are used extensively are not unexpected.

Overall, the evidence suggests that the current F_{pen} approach is likely to overestimate PECs for many drugs, particularly those with low market penetration factors and those that are used intermittently. The alternative equivalent approach of comparing marketing estimates with the annual consumption figure (7.3 kg/million of population per year) implied by the EMEA criteria appears to offer a better alternative. It seems feasible, however, that there may be attempts to reduce the threshold PEC that triggers a phase IIA assessment. A value of 0.4 ng/l has been advocated (58) for the aquatic compartment (i.e., 25 times lower than the current threshold concentration), which would be equivalent to 292 g API per million of population (17.5 kg/year for the United Kingdom and 146 kg/year for the EEA).

In terms of the technical guidance, based on an evaluation of the biodegradation of three representative compounds in the water-sediment system (OECD 308), several recommendations for changes in the EU guideline have been made (59). These include using only aerobic conditions as the default, moving OECD 308 to tier B, and replacing it in tier A with an inherent biodegradation or a die-away test.

USA

Regulatory Background

In the mid-1990s, as part of President Clinton's "Reinventing Regulation" program, a review was commissioned on environmental assessment of pharmaceutical products. Up to that time, the National Environmental Policy Act (NEPA) required all federal agencies, including the FDA, to assess the environmental impact of their actions that "may significantly affect the quality of

Table 11 Estimation of PECs According to the EMEA Guideline and the Conventional Approach (40)

API	$\mathrm{DOSE_{ai}}$ (mg)	Log K_{OW}	$\mathrm{PEC_{EMEA}}$ (μg/l)	$\mathrm{PEC_{conventional}}$ (μg/l)
Cefuroxime	9000	-	45	0.029
Ciprofloxacin	1500	0.4	7.5	0.27
Cyclophosphamide	500	0.63	2.5	0.00026
Diclofenac	150	4.51	0.75	0.48
Ethinyl oestradiol	0.035	4.15	0.00018	0.00057
Ibuprofen	2400	3.97	12	8.0
Metoprolol	200	1.88	1.0	1.7
Paracetamol	3000	0.46	15	42
Sulfamethoxazole	2400	0.89	12	0.067
Tetracycline	1000	−1.19	5.0	0.31
Trimethoprim	480	0.91	2.4	0.16

the human environment." The FDA's environmental impact regulations were established in 1977. A drug could not be approved unless the manufacturer had submitted an acceptable Environmental Assessment (EA). On the basis of FDA analysis of the EA, the Agency could either issue a "finding of no significant impact" (FONSI), or decide that a full environmental impact statement (EIS) must be prepared. Preparation of an EA was usually quite expensive yet, in virtually every case, FDA issued a FONSI.

Up to the mid-1990s, the pharmaceutical industry submitted approximately 50 to 60 full EAs and about 50 abbreviated EAs to the Center for Drug Evaluation and Research (CDER) per year. Pharmaceutical firms also sent 20 to 25 EAs annually to the Center for Biologics Evaluation and Research (CBER), some in abbreviated form. It could take up to six months to review an EA, obtain additional information from the firm to correct any deficiencies, and issue a FONSI. And yet, in the preceding few years, the FDA had identified only one product, Taxol, as presenting any potentially significant environmental concerns, which were related to harvesting of Pacific yew trees, an endangered species. In Taxol's case, CDER incorporated by reference the EIS prepared by the U.S. Forest Service to address the resource question; the manufacturing process and use were addressed through the routine EA and were found to have no significant impact.

As a consequence of these findings, the FDA proposed a significant amendment to EA and EIS requirements. The FDA's intention was to reduce the number of EAs required to be submitted by industry and, consequently, the number of FONSIs prepared by the Agency under NEPA by increasing the number of categorical exclusions based upon little or no impact on the environment from the use of the drug. Based upon its experience gained up to that time in reviewing environmental assessments, FDA believed that nearly all product approvals would qualify for categorical exclusion. For example, it was considered that virtually all drug approvals should result in only minute releases of the drug into the environment as a result of human use and such releases would not be environmentally significant. It was believed that these changes would substantially benefit industry and improve regulatory efficiency without having any adverse impact on public health or the environment. Industry was expected to save from $40,000 to $150,000 on each EA (60).

All applications or petitions requesting FDA action must be accompanied by either an EA or a claim of categorical exclusion. Failure to submit one or the other is a sufficient ground for refusing to file or approve the application (61,62). An EA that is adequate for filing is one that addresses the relevant environmental issues. An EA adequate for approval is one that contains sufficient information to enable the Agency to determine whether the proposed action may significantly impact the quality of the human environment. Various categorical exclusions for human drugs and biologics are allowed (61,62). However, FDA will require an EA if "extraordinary circumstances" indicate that the specific proposed action may significantly affect the quality of the environment.

The permitted categorical exclusions listed in Part 25 of Title 21 of the Code of Federal Regulations (CFR) (63) are as follows (including citations of specific subparts of 21 CFR 25):

1. New Drug Applications (NDAs), Abbreviated New Drug Applicatons (ANDAs), applications for marketing approval of a biologic product, and supplements to such applications if the FDA's approval of the application does not increase the use of the active moiety [21 CFR 25.31(a)].
2. NDAs, ANDAs, and supplements to such applications if the FDA's approval of the application increases the use of the active moiety, but the estimated concentration of the substance at the point of entry into the aquatic environment will be below 1 part per billion (ppb) [21 CFR 25.31(b)].
3. NDAs, ANDAs, applications for marketing approval of a biologic product, and supplements to such applications for substances that occur naturally in the environment when the approval of the application does not alter significantly the concentration or distribution of the substance, its metabolites, or degradation products in the environment [21 CFR 25.31(c)].
4. Applications for Investigational New Drugs (INDs) [21 CFR 25.31(e)].
5. Applications for marketing approval of a biologic product for transfusable human blood or blood components and plasma [21 CFR 25.31(j)].

Expected Introduction Concentration (EIC)

EIC is defined in the FDA's guidance document on environmental assessment (61) as "the concentration, based on fifth-year marketing estimates, of the active moiety (drug substance/API) that can enter the environment due to human use. Depletion mechanisms that occur prior to introduction into the environment and human metabolism may be considered in the calculation." The EIC at the point of entry into the aquatic environment is calculated (in ppb) from

$$\text{EIC} = [A \times B \times C \times D]$$

where A = kilogram per year produced for direct use as active moiety (maximum production/year in a five-year cycle based on marketing estimates); B = (liters/day)$^{-1}$ entering the POTW (Publicly Owned Treatment Works), estimated at 1.214×10^{11} l; C = year/365; $D = 10^9$ μg/kg (conversion factor). Substituting the default parameters, EIC (in ppb) can be calculated as follows:

$$\text{EIC} = \frac{A \times 10^9}{1.214 \times 365 \times 10^{11}} = \frac{A}{44{,}311}$$

Hence, an estimated amount of ≥44,311 kg of API needs to be manufactured per year in order to equal or exceed the threshold of 1 ppm for EIC-aquatic. Categorical exclusions are not normally granted if EIC > 1 ppb, although there are some exceptions.

In calculating EIC-aquatic, information on one or more various depletion mechanisms can be applied in individual cases:

- Human metabolism.
- Hydrolysis/photolysis and biodegradation during passage through the POTW.
- Partitioning of the API into sewage sludge.

When evaluating depletion by metabolism, only metabolites that are inactive or less pharmacologically active than the active moiety can be considered to contribute to depletion. Metabolites of unknown pharmacological activity are assumed to have similar activity to the API.

Assuming a US population of 300 million, the threshold amount of API calculated above is equivalent to 148 kg/million of population, 20 times higher than the threshold calculated for the EU. In addition, making an allowance for metabolism, dilution, and other depletion mechanisms is likely in virtually all cases to create a significantly higher discrepancy in the EU/US environmental assessment estimated aquatic concentration thresholds.

Scientific Assessment (61)

The potential environmental effect of a released substance needs to be assessed if no categorical exclusion is deemed to apply using the criteria listed above (most importantly EIC-aquatic > 1 ppb). Physicochemical parameters such as water solubility, dissociation constant, octanol–water partition coefficient (K_{OW}), Henry's law constant or vapor pressure, and adsorption–desorption properties to biosolids (if log $K_{OW} > 3$) can be used to determine if the compound is most likely to amass predominantly in aquatic, terrestrial, and/or atmospheric environments.

The FDA guidance distinguishes between EIC and EEC (Expected Environmental Concentration). The latter is the concentration of the active moiety after consideration of depletion factors such as dilution, degradation, sorption, and/or biodegradation. In the majority of cases, EEC-aquatic would be expected to be significantly less than EIC due mainly to dilution, a factor of 10 being the normal assumption for calculation of the EEC-aquatic.

Irrespective of the EIC, if a rapid and complete depletion mechanism is identified then no testing to determine environmental effects needs to be performed except for a microbial inhibition test to assess the potential for disruption of waste treatment processes. Criteria in respect of $t_{1/2}$ for rapid depletion are as follows: hydrolysis (pH 5–9) ≤ 24 hours; aerobic biodegradation ≤ 8 hours; soil biodegradation ≤ 5 days.

From this point the environmental assessment follows a systematic, tiered approach as shown in Figure 4. If no rapid depletion mechanism is identified and log $K_{OW} \leq 3.5$, a tier 1 acute toxicity test with one aquatic species is recommended. If the EC_{50} or LC_{50} from this tier 1 acute test divided by the maximum expected environmental concentration (MEEC = EIC or EEC, whichever is greater) is ≥1000, no further testing is required. Overall, three tiers of testing are described with assessment factors of 1000, 100, and 10 being used in tiers 1 to 3, respectively. Tier 1 testing merely involves an acute test in one aquatic species as indicated above. Tier 2

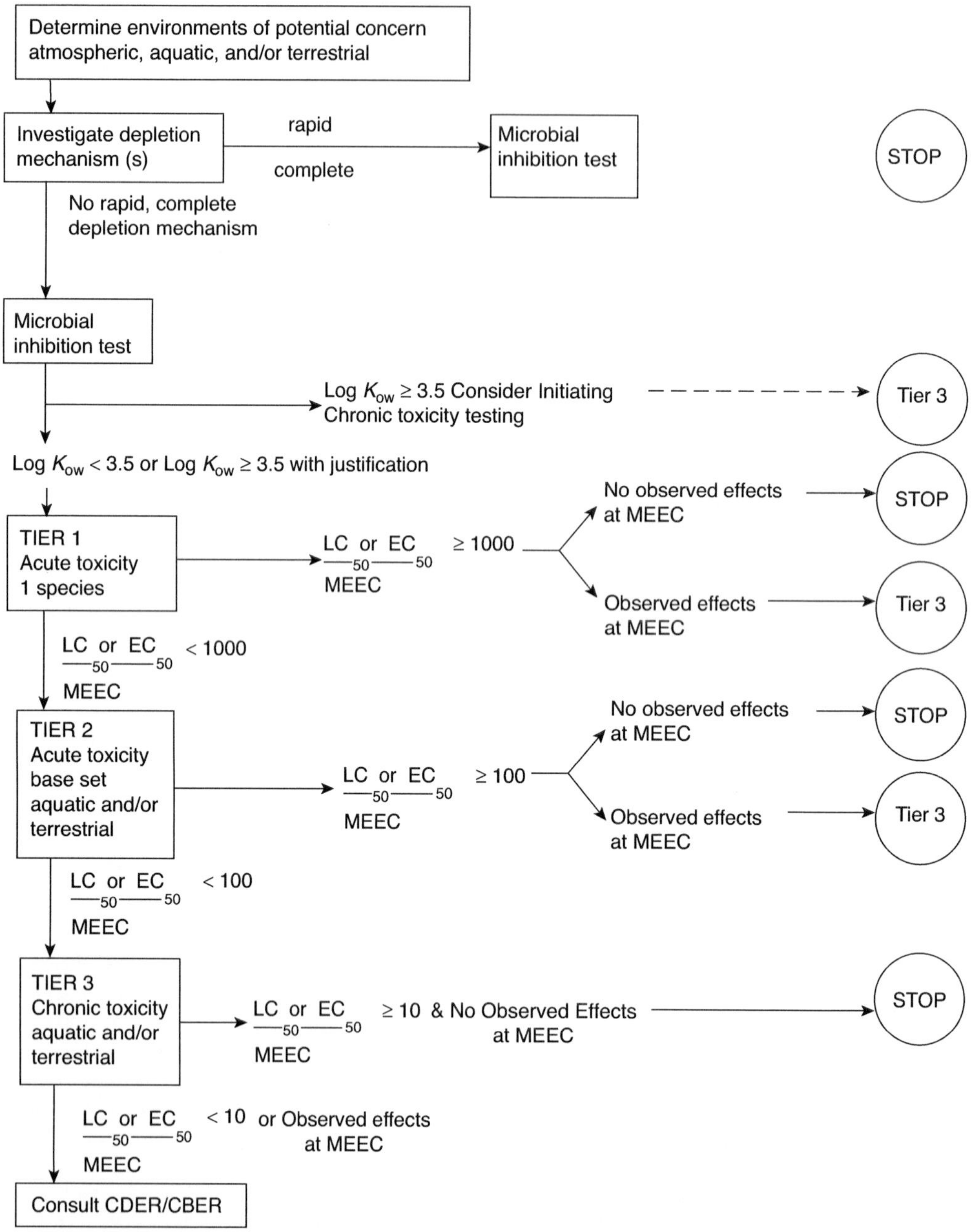

Note: MEEC = EEC or EIC whichever is greater

Figure 4 US environmental assessment: tiered approach to fate and effects testing (61).

involves acute ecotoxicity testing on a base set of aquatic and/or terrestrial organisms (fish acute toxicity, aquatic invertebrate acute toxicity, an algal species bioassay, plant early growth tests, earthworm toxicity tests, and soil microbial toxicity tests). The earthworm test is indicated only if the substance binds tightly to soil. Tier 3 involves chronic toxicity testing in aquatic and/or terrestrial organisms.

Comments on US Environmental Assessment

Although the main focus of the FDA guidance relates to use of an API, there are specific provisions for disposal that may come into play if no categorical exclusion applies (62). However, FDA does not request manufacturing and disposal information routinely in an EA (61). Manufacturing is controlled by the FDA's Good Manufacturing Practice (GMP) requirements (64) and effluent discharges from manufacturing facilities are regulated by the Environmental Protection Agency (EPA) (62).

Comparing the FDA's system with that in the EU, the former seems to be designed to minimize the need to undertake extensive fate and effects testing, while the opposite appears to be the case for the EU. Moreover, even when an evaluation is actually required (i.e., in the absence of a categorical exclusion) there is a gradual escalation of study requirements starting with an acute toxicity test in one species, with evaluation criteria based on EC_{50} or LC_{50} values. This contrasts sharply with the EU base set of tests involving various fate evaluations and chronic toxicity studies in three species, assessments being based on NOEC values. A similar more graduated testing scheme was recommended by the CSTEE (29) in its comments on the draft EU ERA guideline, but no such modifications were made to the EU guideline.

Relatively few commentators in the United States have expressed major concerns over the direct impact of pharmaceuticals on the environment, although the potential for pharmaceuticals to be present (in trace amounts) in drinking water has been extensively publicized (65–68). However, safety assessments in both the United States (4) and Europe (69) have concluded that the potential intakes of human pharmaceuticals via drinking water are so exceedingly low that pharmacological effects in consumers are most unlikely.

DISCUSSION

Only the United States and EU have so far put in place legal requirements with accompanying technical guidance for the environmental risk assessment of human pharmaceuticals. Given the marked differences between these two existing systems, it must be difficult for countries such as Canada and Japan, reported to be developing ERA guidance (70), to decide exactly where to position their own systems in the spectrum of possibilities. In Canada, the New Substances Notification Regulations, 1999 (71), are already in place and apply to Food and Drugs Act substances such as human pharmaceuticals. Supply/production thresholds of 100, 1000, and 10,000 kg/year are defined for all new substances with an escalating set of environmental testing requirements as each threshold is reached. Substances supplied at <100 kg/year do not require notification. Notified new substances at 100 to 1000 kg/year require no a priori environmental testing, whereas if the amount exceeds 1000 kg/year then data from an acute toxicity study in one aquatic species and from a ready biodegradation study are required (72). In 2001, Health Canada/Environment Canada announced their intention to develop "environmental assessment regulations," which would apply to human pharmaceuticals (73), but little progress seemed to have been made by 2006 (74) and no firm detailed proposals have yet emerged. Assuming a Canadian population of 33 million, API supply thresholds of 241 and 4884 kg/year can be calculated on the basis of the EU and U.S. criteria.

Perhaps the country with the most advanced classification system (FASS) for the environmental impact of human pharmaceuticals is Sweden, the main participants being the Swedish Medical Products Agency (MPA), the Swedish Association of Pharmaceutical Industry (LIF), the National Corporation of Swedish Pharmacies (Apoteket AB), the Swedish Association of Local Authorities and Regions (SKL), and Stockholm County Council (75). The system is voluntary and relies on an assessment of information submitted by pharmaceutical companies (76). In 2004, the MPA released a report in which 27 prioritized APIs were assessed for their impact on the environment and 15/27 (Table 12) were judged (on the basis of somewhat incomplete data

Table 12 Potential Environmental Risks of 15 Prioritized APIs (57)

API	Structure	Environmental risk
Atenolol	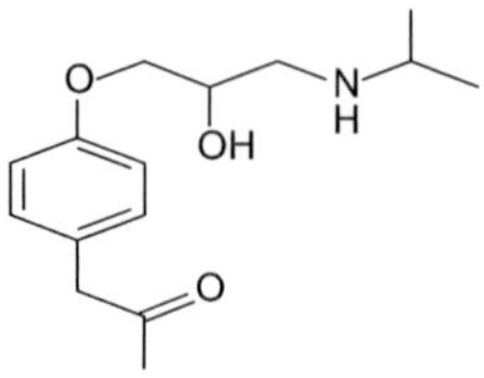	A
Dextropropoxyphene	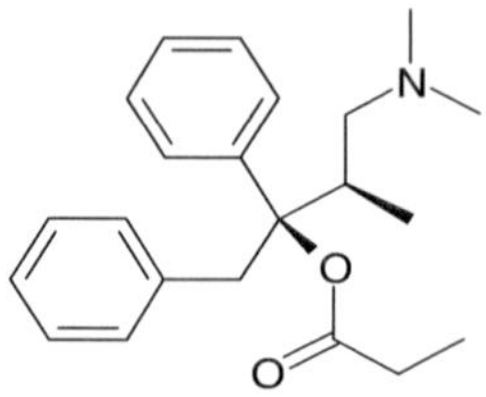	T
Diazepam	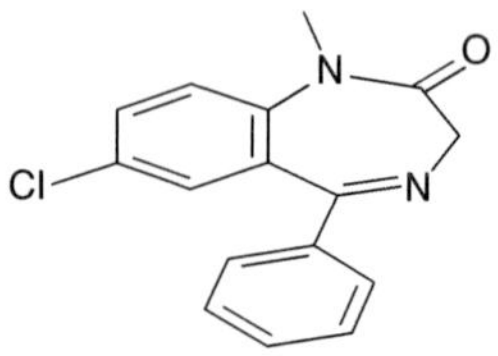	T
Diclofenac	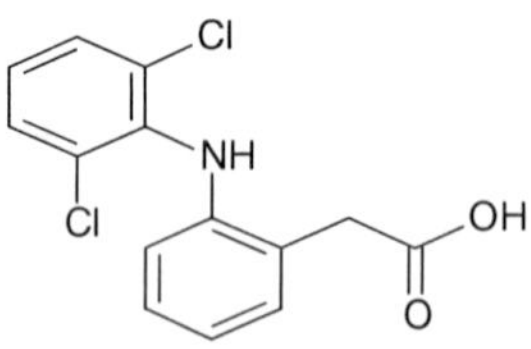	B
Ethinyl oestradiol	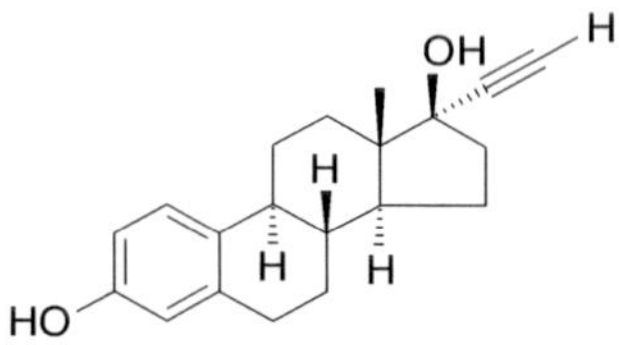	A
Ketoprofen	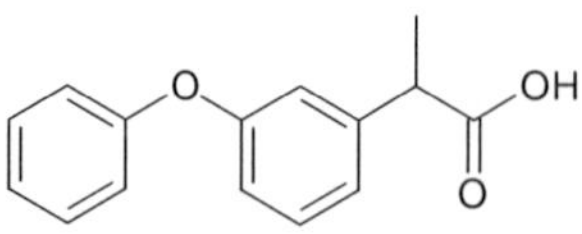	A
Norethisterone	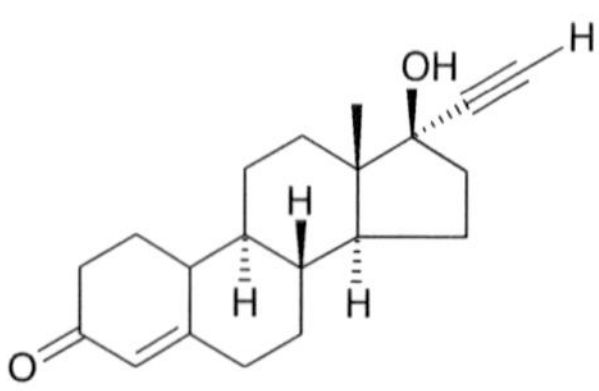	A

Table 12 Potential Environmental Risks of 15 Prioritized APIs (57) (*Continued*)

API	Structure	Environmental risk
Oxazepam	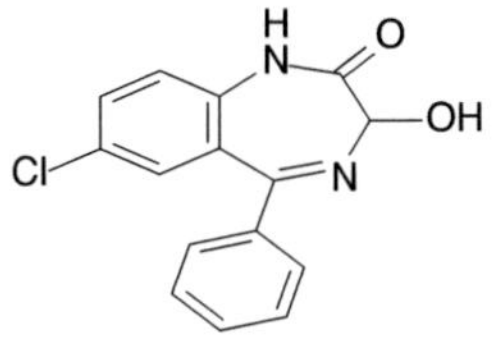	A
Oxytetracycline	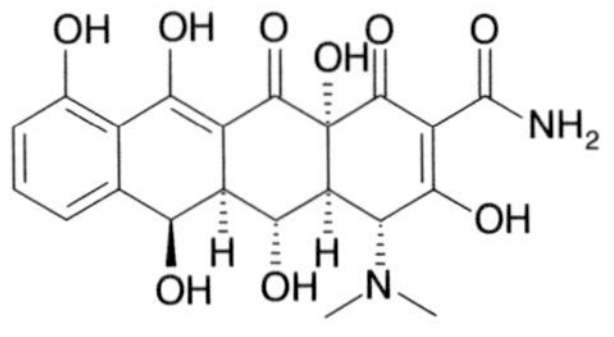	M
Paracetamol	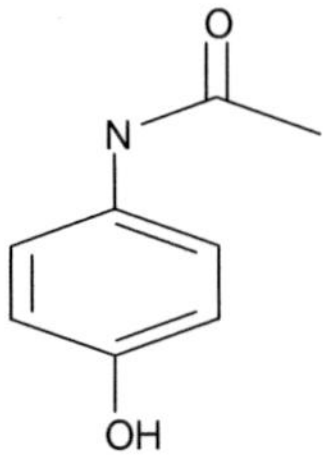	A
Ranitidine	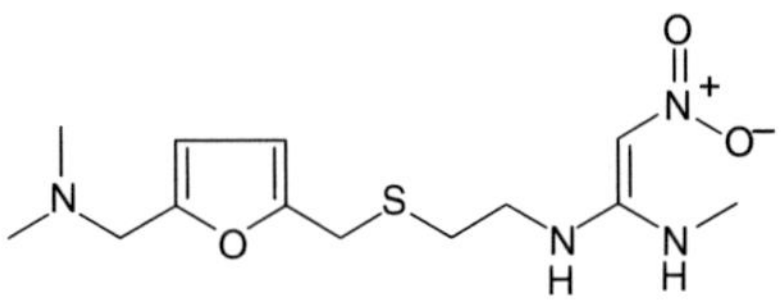	T
Simvastatin	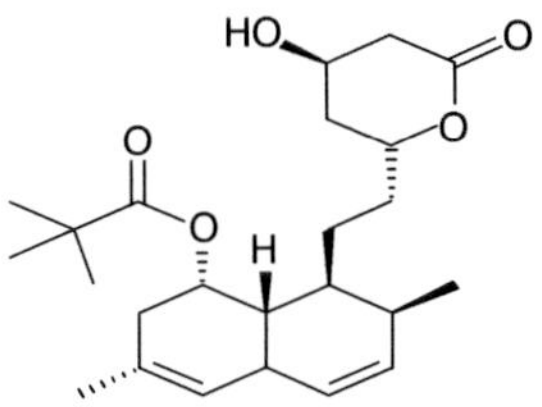	A, B
Tetracycline	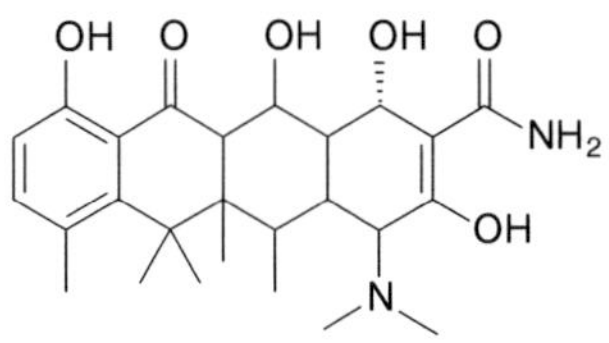	M

(*Continued*)

Table 12 Potential Environmental Risks of 15 Prioritized APIs (57) (*Continued*)

API	Structure	Environmental risk
Oestradiol	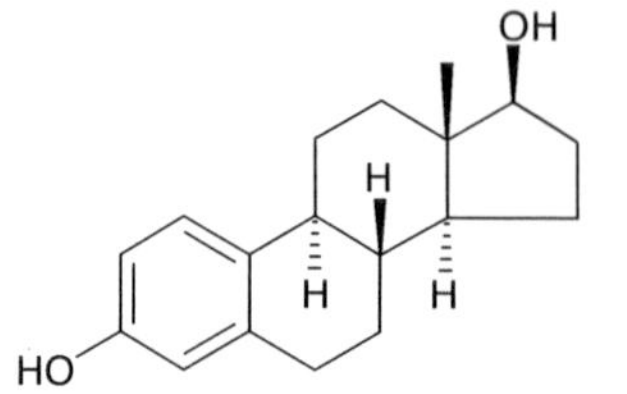	A
Oestriol	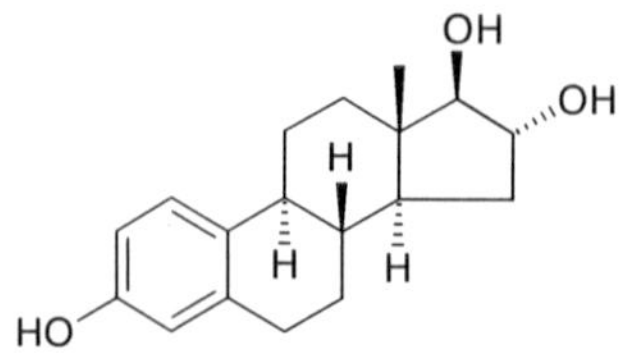	A

Abbreviations: A, aquatic; T, terrestrial; B, bioaccumulation; M, microbial.

sets) to possess the potential to cause adverse effects (77). Excipients in cosmetics and pharmaceutical products were also assessed and several preservatives were found to be potentially harmful to the environment; however, an MPA assessment published in 2006 concluded that the environmental risks posed by excipients used in pharmaceutical products are likely to be negligible (78).

Dissatisfaction with various elements of the current environmental testing paradigm has been reported in a number of publications (79–85): perceived deficiencies include by-rote testing to standard protocols based on methods developed for industrial chemicals; lack of utilization of mammalian pharmacological and toxicological information; absence of understanding of physiological function and pharmacological targets in aquatic species; lack of biomarker development, and little use of case-specific testing regimens. Recommendations for assessing the ecological risk of pharmaceuticals are summarized by Ankley et al. (84) as follow:

- Use similar effects assessment approaches for veterinary and human drugs.
- Reassess the basis of using production volumes or PEC values as testing triggers.
- Evaluate persistence as a function of chemical stability and input relative to hydrodynamic regime.
- Allow mode of action (MOA) data to guide species and endpoint selection for testing.
- Support the development and use of MOA-specific biomarkers for both diagnostic and prospective applications.
- Build flexibility into the regulatory process to allow a priori knowledge to guide testing.

In contrast to industrial chemicals, pharmaceuticals are intended to exert specific pharmacodynamic effects and their impact on aquatic species may well reflect the mammalian pharmacotoxicological profile, as is the case for endocrine-disrupting drugs such as ethinyl oestradiol (86). Despite the logic of such case-specific pharmacologically based approaches, with opportunities to make testing programs much more targeted and efficient, the current regulatory paradigm seems unlikely to change in the short or medium term. Possibly after 5 or 10 years' experience with the current EU guideline, regulators may be willing to open up the debate on alternative testing strategies. By that time it is expected that many of the ongoing environmental research programs, such as ERAPharm (87), will have generated sufficient data to support alternative testing strategies.

A variety of risk management strategies have been suggested for dealing with the potential impacts of human pharmaceuticals on the environment (88) including labeling, restricting use to hospitals, and responsible disposal (as already adopted by the EU). More fundamental approaches include substitution or phase-out of existing environmentally suspect drugs and incorporation of environmental considerations into molecular design of new drugs (89,90). This latter type of proposal, which tends to arise exclusively from environmental scientists who may not have a full awareness of the exquisitely balanced structure–activity relationships that often have to be negotiated in drug development, raises many technical, ethical, and commercial considerations. There may well be conflicting factors when designing a drug that is readily degraded in the environment and which also optimizes human safety and efficacy. A readily metabolized/degraded drug may have low human bioavailability or may produce (toxic) reactive metabolites. Moreover, in the current EU ERA guidance, there is virtually no incentive to develop readily metabolized or biodegraded drugs because the base set of tests needs to be undertaken on the parent API irrespective of the extent of human metabolism or evidence of rapid depletion following excretion. The FDA approach to factoring in human metabolism at the beginning of the environmental assessment process represents a readily transferable set of criteria that could be adopted by EU regulators to provide some encouragement to industry regarding the development of "green pharmacy" drugs.

Other global trends may in due course impact on environmental risk assessment. Regulatory caution over assessment of environmental impact can be driven by the concern that almost any new drug might eventually become a massive seller. Although it is certainly true that many well-established drugs in the "essential formulary" such as ibuprofen, paracetamol, etc., have a significant potential environmental footprint, such drugs will most likely continue to be used for many years and it may be inappropriate to assume that new drugs will come along to displace them. A major trend in drug development is to abandon the "blockbuster" model and to focus on a larger number of "niche" pharmaceuticals; so the future of environmental risk assessment may be characterized by the evaluation of smaller environmental burdens to more new drugs used in specialized indications. A further consideration is sustainability given that many experts believe that we have passed or are close to passing peak oil supply (91). STPs and drinking water plants can be high users of energy and so there may well be a gradual move towards more sustainable wastewater treatment practices including wind- and solar-powered facilities and use of composting toilets, all of which could impact significantly on the underlying concepts of environmental risk assessment for human pharmaceuticals.

REFERENCES

1. Fent K, Weston AA, Caminada D. Ecotoxicology of human pharmaceuticals. Aquat Toxicol 2006; 76:122–159.
2. The Global Education Project, Earth, Fresh Water: http://www.theglobaleducationproject.org/earth/fresh-water.php.
3. Jones OA, Voulvoulis N, Lester JN. Potential ecological and human health risks associated with the presence of pharmaceutically active compounds in the aquatic environment. Crit Rev Toxicol 2004; 34(4):335–350.
4. Schwab BW, Hayes EP, Fiori JM, et al. Human pharmaceuticals in US surface waters: A human health risk assessment. Regul Toxicol Pharmacol 2005; 42(3):296–312.
5. Christensen FM. Pharmaceuticals in the environment—A human risk? Regul Toxicol Pharmacol 1998; 28(3):212–221.
6. Sewage treatment in the UK, DEFRA, 2002. http://www.defra.gov.uk/environment/water/quality/uwwtd/report02/pdf/uwwtreport2.pdf.
7. Energy and Sewage, Parliamentary Office of Science and Technology Postnote, April 2007, No. 282. http://www.parliament.uk/documents/upload/postpn282.pdf.
8. Jenkins D, Richard MG, Daigger GT. Manual on the Causes and Control of Activated Sludge Bulking and Foaming, 2nd ed. Boca Raton: Lewis Publishers, 1993.
9. Kanda R, Griffin P, James HA, et al. Pharmaceutical and personal care products in sewage treatment works. J Environ Monit 2003; 5(5):823–830.
10. Carballa M, Omil F, Lema JM, et al. Behavior of pharmaceuticals, cosmetics and hormones in a sewage treatment plant. Water Res 2004; 38(12):2918–2926.
11. Yamamoto H, Hayashi A, Nakamura Y, et al. Fate and partitioning of selected pharmaceuticals in aquatic environment. Environ Sci 2005; 12(6):347–358.

12. Joss A, Keller E, Alder AC, et al. Removal of pharmaceuticals and fragrances in biological wastewater treatment. Water Res 2005; 39(14):3139–3152.
13. Kimura K, Hara H, Watanabe Y. Elimination of selected acidic pharmaceuticals from municipal wastewater by an activated sludge system and membrane bioreactors. Environ Sci Technol 2007; 41(10):3708–3714.
14. Radjenovic J, Petrovic M, Barceló D. Analysis of pharmaceuticals in wastewater and removal using a membrane bioreactor. Anal Bioanal Chem 2007; 387(4):1365–1377.
15. Kim SD, Cho J, Kim IS, et al. Occurrence and removal of pharmaceuticals and endocrine disruptors in South Korean surface, drinking, and waste waters. Water Res 2007; 41(5):1013–1021.
16. Okuda T, Kobayashi Y, Nagao R, et al. Removal efficiency of 66 pharmaceuticals during wastewater treatment process in Japan. Water Sci Technol 2008; 57(1):65–71.
17. Velagaleti R. Behaviour of pharmaceutical drugs (human and animal health) in the environment. Drug Inf J 1997; 31:715–722.
18. Andreozzi R, Raffaele M, Nicklas P. Pharmaceuticals in STP effluents and their solar photodegradation in aquatic environment. Chemosphere. 2003; 50(10):1319–1330.
19. Kunka R, Andrews S, Pimazzoni M, et al. From hydrofluoroalkane pressurized metered dose inhalers (pMDIs) and comparability with chlorofluorocarbon pMDIs. Respir Med 2000; 94(Suppl. B):S10–S16.
20. Rhoderick GC, Dorko WD. Standards development of global warming gas species: Methane, nitrous oxide, trichlorofluoromethane, and dichlorodifluoromethane. Environ Sci Technol 2004; 38(9):2685–2692.
21. Halling-Sørensen B, Jacobsen AM, Jensen J, et al. Dissipation and effects of chlortetracycline and tylosin in two agricultural soils: A field-scale study in southern Denmark. Environ Toxicol Chem 2005; 24(4):802–810.
22. Kinney CA, Furlong ET, Werner SL, et al. Presence and distribution of wastewater-derived pharmaceuticals in soil irrigated with reclaimed water. Environ Toxicol Chem 2006; 25(2):317–326.
23. Topp E, Monteiro SC, Beck A, et al. Runoff of pharmaceuticals and personal care products following application of biosolids to an agricultural field. Sci Total Environ 2008; 396(1):52–59.
24. Oppel J, Broll G, Löffler D, et al. Leaching behaviour of pharmaceuticals in soil-testing-systems: A part of an environmental risk assessment for groundwater protection. Sci Total Environ 2004; 328(1–3):265–273.
25. Drillia P, Stamatelatou K, Lyberatos G. Fate and mobility of pharmaceuticals in solid matrices. Chemosphere 2005; 60(8):1034–1044.
26. Richter R, Kullmer C, Kreuzig R. Metabolic fate modeling of selected human pharmaceuticals in soils. Water Sci Technol 2004; 49(3):213–221.
27. European Union Directive 93/39/EEC. Council Directive 93/39/EEC of 14 June 1993 amending Directives 65/65/EEC, 75/318/EEC and 75/319/EEC in respect of medicinal products. http://www.ikev.org/docs/eu/393L0039.htm.
28. Notice to Applicants. Volume 2B. Presentation and Content of the Dossier. http://ec.europa.eu/enterprise/pharmaceuticals/eudralex/vol2_en.htm.
29. CSTEE: Opinion on draft CPMP discussion paper on environmental risk assessment of medicinal products for human use. http://ec.europa.eu/health/ph_risk/committees/sct/documents/out111_en.pdf.
30. Guideline on the environmental risk assessment of medicinal products for human use. http://www.emea.europa.eu/pdfs/human/swp/444700en.pdf.
31. Directive 2003/63/EC (amending Directive 2001/83/EC). http://ec.europa.eu/enterprise/pharmaceuticals/eudralex/vol-1/dir_2001_83_cons/dir2001_83_cons_20081230_en.pdf.
32. European Chemicals Bureau (2003), Technical Guidance Document, Part II, Chapter 3—Environmental Risk Assessment. http://ecb.jrc.ec.europa.eu/documents/TECHNICAL_GUIDANCE_DOCUMENT/EDITION_2/tgdpart2_2ed.pdf.
33. Carlsson C, Johansson AK, Alvan G, et al. Are pharmaceuticals potent environmental pollutants? Part I: Environmental risk assessments of selected active pharmaceutical ingredients. Sci Total Environ 2006; 364(1–3):67–87.
34. Pugliatti M, Beghi E, Forsgren L, et al. Estimating the cost of epilepsy in Europe: A review with economic modeling. Epilepsia 2007; 48(12):2224–2233.
35. Freitag CM, May TW, Pfäfflin M, et al. Incidence of epilepsies and epileptic syndromes in children and adolescents: A population-based prospective study in Germany. Epilepsia 2001; 42(8):979–985.
36. NICE (National Institute for Clinical Excellence) Implementation Uptake Report. Epilepsy—Newer Drugs. http://www.nice.org.uk/niceMedia/pdf/TA7679NICEImplUptakeEpilepsy.pdf.

37. European Commission, Eurostat. http://epp.eurostat.ec.europa.eu/tgm/table.do?tab = table&language = en&pcode = tps00001&tableSelection = 1&footnotes = yes&labeling = labels&plugin = 1.
38. EPAR for Keppra: http://www.emea.europa.eu/humandocs/PDFs/EPAR/keppra/198600en6.pdf.
39. OECD Guidelines for the Testing of Chemicals. Section 1: Physical Chemical Properties. http://puck.sourceoecd.org/vl = 182668/cl = 30/nw = 1/rpsv/cw/vhosts/oecdjournals/1607310x/v1n1/ contp1–1.htm; Section 2: Effects on Biotic Systems. http://puck.sourceoecd.org/vl = 3513667/cl = 15/nw = 1/rpsv/cw/vhosts/oecdjournals/1607310x/v1 n2/contp1–1.htm. Section 3: Degradation and Accumulation. http://puck.sourceoecd.org/vl = 3513667/cl = 15/nw = 1/rpsv/cw/vhosts/oecdjournals/1607310x/v1 n3/contp1–1.htm.
40. ISO Standards. Environmental Testing of Chemicals. http://www.iso.org/iso/iso_catalogue.htm.
41. OECD Guidelines for the Testing of Chemicals. Test No. 121: Estimation of the Adsorption Coefficient (K_{OC}) on Soil and on Sewage Sludge Using High Performance Liquid Chromatography (HPLC). http://puck.sourceoecd.org/vl = 3513667/cl = 15/nw = 1/rpsv/ij/oecdjournals/1607310x/v1n1/s21/p1.
42. OECD Guidelines for the Testing of Chemicals Test No. 301: Ready Biodegradability. http://puck.sourceoecd.org/vl = 3513667/cl = 15/nw = 1/rpsv/ij/oecdjournals/1607310x/v1 n3/s2/p1.
43. OECD Guidelines for the Testing of Chemicals. Test No. 308: Aerobic and Anaerobic Transformation in Aquatic Sediment Systems. http://puck.sourceoecd.org/vl = 3513667/cl = 15/nw = 1/rpsv/ij/oecdjournals/1607310x/v1 n3/s11/p1.
44. OECD Guidelines for the Testing of Chemicals. Test No. 209: Activated Sludge, Respiration Inhibition Test. http://puck.sourceoecd.org/vl = 3513667/cl = 15/nw = 1/rpsv/ij/oecdjournals/1607310x/v1 n2/s10/p1.
45. OECD Guidelines for the Testing of Chemicals. Test No. 201: Alga, Growth Inhibition Test. http://puck.sourceoecd.org/vl = 3513667/cl = 15/nw = 1/rpsv/ij/oecdjournals/1607310x/v1 n2/s2/p1.
46. OECD Guidelines for the Testing of Chemicals. Test No. 211: *Daphnia magna* Reproduction Test. http://puck.sourceoecd.org/vl = 3513667/cl = 15/nw = 1/rpsv/ij/oecdjournals/1607310x/v1 n2/s12/p1.
47. OECD Guidelines for the Testing of Chemicals. Test No. 210: Fish, Early-Life Stage Toxicity Test. http://puck.sourceoecd.org/vl = 3513667/cl = 15/nw = 1/rpsv/ij/oecdjournals/1607310x/v1 n2/s11/p1.
48. OECD Test Guideline 303. http://puck.sourceoecd.org/vl = 3513667/cl = 15/nw = 1/rpsv/ij/ oecdjournals/1607310x/v1n3/s6/p1
49. OPPTS 835.1110. Activated Sludge Sorption Isotherm. Office of Prevention, Pesticides and Toxic Substances (OPPTS), United States Environmental Protection Agency. http://fedbbs.access.gpo.gov/library/epa_835/835–1110.pdf.
50. US Environmental Protection Agency. Exposure Assessment Tools and Models (including HENRYWIN). http://www.epa.gov/oppt/exposure/pubs/episuitedl.htm.
51. OECD Guidelines for the Testing of Chemicals. Test No. 218: Sediment-Water Chironomid Toxicity Using Spiked Sediment: http://puck.sourceoecd.org/vl = 3513667/cl = 15/nw = 1/rpsv/ij/ oecdjournals/1607310x/v1 n2/s19/p1.
52. Fiedler S, Gildemeister T, Nagel R. Comparing aquatic and sediment toxicity data of the sediment organisms *Vallisneria Americana, Hyalella azteca, Lumbriculus variegatus* and *Chironomus riparius*. Poster. SETAC, 2003. http://www.cefic-lri.org/uploads/events-2003-session/SETAC2003_comparison_toxicity_post.pdf.
53. Gildemeister T. Comparative toxicity of eight model substances to the sediment dwelling invertebrates *Lubriculus variegatus* and *Chironmus riparius*. Ph.D. Thesis. University of Dresden, 2006. http://hsss.slub-dresden.de/documents/1178202259866–9719/1178202259866–9719.pdf.
54. Risk Assessment of Chemicals. http://www.denehurst. co.uk/id28.html.
55. European Union System for the Evaluation of Substances (EUSES). http://ecb.jrc.ec.europa.eu
56. Richman C, Castensson S. Impact of waste pharmaceuticals: An environmental hazard or "greenwash"? Pharm J 2008; 280(7494):335–336, 341–342.
57. Grung M, Källqvist T, Sakshaug S, et al. Environmental assessment of Norwegian priority pharmaceuticals based on the EMEA guideline. Ecotoxicol Environ Saf 2008; 71(2):328–340.
58. Montforts M. The trigger values in the environmental risk assessment for (veterinary) medicines in the European Union: A critical appraisal. http://rivm.openrepository.com/rivm/bitstream/10029/7379/1/601500002.pdf.
59. Ericson JF. An evaluation of the OECD 308 water/sediment systems for investigating the biodegradation of pharmaceuticals. Environ Sci Technol 2007; 41(16):5803–5811.
60. Environmental Assessments for Human Drugs. http://govinfo.library.unt.edu/npr/library/reinvent (drug&medical%20devicereg).htm.

61. Environmental Assessment of Human Drug and Biologics Applications. Food and Drug Administration. CDER/CBER. http://www.fda.gov/downloads/Drugs/GuidanceComplianceRegulatoryInformation/Guidances/ucm070561.pdf.
62. Velagaleti R, Burns PK, Gill M, et al. Impact of current good manufacturing practices and emission regulations and guidances on the discharge of pharmaceutical chemicals into the environment from manufacturing, use, and disposal. Environ Health Perspect 2002; 110(3):213–220.
63. Code of Federal Regulations. Title 21—Food and Drugs. Part 25—Environmental Impact Considerations. https://www.accessdata.fda.gov/scripts/cdrh/cfdocs/cfCFR/CFRSearch.cfm?CFRPart = 25.
64. Code of Federal Regulations. Title 21—Food and Drugs. Part 210. Current Good Manufacturing Practice in Manufacturing, Processing, Packing, or Holding Of Drugs; General. https://www.accessdata.fda.gov/scripts/cdrh/cfdocs/cfCFR/CFRSearch.cfm?CFRPart = 210. Part 211. Good Manufacturing Practice for Finished Pharmaceuticals. https://www.accessdata.fda.gov/scripts/cdrh/cfdocs/cfCFR/CFRSearch.cfm?CFRPart = 211.
65. Heberer T, Reddersen K, Mechlinski A. From municipal sewage to drinking water: Fate and removal of pharmaceutical residues in the aquatic environment in urban areas. Water Sci Technol 2002; 46(3):81–88.
66. Drewes JE, Heberer T, Reddersen K. Fate of Pharmaceuticals During Indirect Potable Reuse. Water Sci Technol 2002; 46(3):73–80.
67. Stackelberg PE, Furlong ET, Meyer MT, et al. Persistence of pharmaceutical compounds and other organic wastewater contaminants in a conventional drinking-water-treatment plant. Sci Total Environ 2004; 329(1–3):99–113.
68. Gibs J, Stackelberg PE, Furlong ET, et al. Persistence of pharmaceuticals and other organic compounds in chlorinated drinking water as a function of time. Sci Total Environ 2007; 373(1):240–249.
69. Webb S, Ternes T, Gibert M, et al. Indirect human exposure to pharmaceuticals via drinking water. Toxicol Lett 2003; 142(3):157–167.
70. Van Noorloos B. Risk assessment of pharmaceuticals in the environment,2007. http://www.notox.nl/download.php?file = cHVibGljYXRpb25fZG9jL3B1YmxpY2F0aW9uX2RvY18xMi5wZGZGaWVsRFNlcFBhcHBsaWNhdGlvbi9wZGY = .
71. New Substances Notification Regulations. Canada. http://www.ec. gc. ca/Substances/nsb/eng/home_e.shtml.
72. Stringer G. Environmental Assessment of *Food and Drugs Act* Substances, 2007. www.enviroadvisory.com/2007presentations/Stringer. ppt.
73. Notice of intent to develop environmental assessment regulations for products regulated under the Food and Drugs Act, 2001. http://www.hc-sc.gc.ca/ewh-semt/contaminants/person/impact/200109_noi-adi-eng.php.
74. The development of environmental assessment regulations for new substances contained in products regulated under the *Food and Drugs Act*, 2006. http://dsp-psd.pwgsc. gc. ca/Collection/H164–11-2006E.pdf.
75. Environmentally Classified Pharmaceuticals. http://www. sll.se/upload/Milj%C3%B6/A4.1.klassific.pdf.
76. FASS. Environmental Classification of Pharmaceuticals. http://www.fass.se/LIF/lakarbok/lakemedelhalsa_artikel.jsp?articleID = 76074.
77. Environmental effects of cosmetics, pharmaceuticals and hygiene products. Medical Products Agency, 2004. http://www.noharm.org/details.cfm?ID = 1026&type = document.
78. Carlsson C, Johansson AK, Alvan G, et al. Are pharmaceuticals potent environmental pollutants? Part II: Environmental risk assessments of selected pharmaceutical excipients. Sci Total Environ 2006; 364(1–3):88–95.
79. Länge R, Dietrich D. Environmental risk assessment of pharmaceutical drug substances—Conceptual considerations. Toxicol Lett 2002; 131(1–2):97–104.
80. Länge R, Steger-Hartmann T, Schweinfurth H. The environmental risk assessment of human pharmaceuticals in the overall EU regulatory affairs process. Regul Toxicol Pharmacol 2006; 45(3):223–228.
81. Seiler JP. Pharmacodynamic activity of drugs and ecotoxicology—Can the two be connected? Toxicol Lett 2002; 131(1–2):105–115.
82. Dorne JL, Skinner L, Frampton GK, et al. Human and environmental risk assessment of pharmaceuticals: Differences, similarities, lessons from toxicology. Anal Bioanal Chem 2007; 387(4):1259–1268.
83. Eason C, O'Halloran K. Biomarkers in toxicology versus ecological risk assessment. Toxicology 2002; 181–182: 517–521.
84. Ankley GT, Brooks BW, Huggett DB, et al. Repeating history: Pharmaceuticals in the environment. Environ Sci Technol 2007; 41(24):8211–8217.

85. Williams M, Saison CL, Williams DB, et al. Can aquatic distribution of human pharmaceuticals be related to pharmacological data? Chemosphere 2006; 65(11):2253–2259.
86. Sumpter JP. Xenoendorine disrupters—Environmental impacts. Toxicol Lett 1998; 102–103: 337–342.
87. Knacker T, Duis K, Ternes T, et al. The EU-project ERAPharm. Incentives for the further development of guidance documents? Environ Sci Pollut Res Int 2005; 12(2):62–65.
88. Williams RT. Human health pharmaceuticals in the environment: An introductory perspective. In Science for Assessing the Impacts on Human Pharmaceuticals on Aquatic Ecosystems (ed. Williams R.). SETAC Press, Pensacola, FL 2005:1–46.
89. Adler NE, Koschorreck J, Rechenberg B. Environmental impact assessment and control of pharmaceuticals: The role of environmental agencies. Water Sci Technol 2008; 57(1):91–97.
90. Dieter HH, Mückter H. Assessment of so called organic trace compounds in drinking water from the regulatory, health and aesthetic-quality points of view, with special consideration given to pharmaceuticals. Bundesgesundheitsblatt Gesundheitsforschung Gesundheitsschutz 2007; 50(3):322–331.
91. Aleklett K. International Energy Agency accepts Peak Oil. Uppsala Hydrocarbon Depletion Study Group, 2004. http://www.peakoil.net/uhdsg/weo2004/TheUppsalaCode.html.

5 Quality Overall Summary

Anthony C. Cartwright
Pharmaceutical Regulatory Consultant, Harpenden, U.K.

2.3 QUALITY OVERALL SUMMARY

The Quality Overall Summary (QOS) is a summary that follows the format of the data included in Module 3. It does not include any information, data, etc., that has not been included in Module 3 or elsewhere in the Common Technical Document (CTD).

The major parts of this section of the CTD are as follows (1):

2.3. Introduction
2.3.S Drug Substance
2.3.P Drug Product
2.3.A Appendices
2.3.R Regional Information

PURPOSE OF THE QOS

The QOS provides summary information from each section of Module 3 to give the Quality/Chemistry Manufacture and Controls reviewer an overview of Module 3. It should emphasize critical key parameters of the product. Where the development of the product did not follow ICH, relevant regional, or national guidelines or pharmacopoeial requirements, the QOS should explain and justify the approach followed. The objective of providing a well-written QOS is so that it can then be used by the quality reviewer in the regulatory agency to facilitate their review and help them produce their own Assessment Report on the product.

INTERFACE ISSUES

Key issues should be reviewed and cross-reference made to other Modules as appropriate, in particular for what may be termed "interface issues" (issues which need to integrate information from the Module 4 Nonclinical and Module 5 Clinical). Interface issues will include references to the bioavailability and bioequivalence data in Module 5, the different compositions of clinical trial batches of drug product used in trials cited in Module 5, and the nonclinical data for biological qualification of impurities in Module 4. Cross-referencing may be either by citing the volume, document and document page number in a paper CTD application, or by hyperlinking in an eCTD filing.

INTRODUCTION

The introduction should include the proprietary name of the drug product, nonproprietary or common name of the drug substance, the company name, a list of the dosage forms, the route of administration, and a summary of the proposed indications.

ORGANIZATION AND STRUCTURE OF THE QOS

Length of the QOS

The ICH Guideline on the CTD Quality (M4Q(R1) (2) provides detailed guidance on the contents of the QOS. It indicates that the QOS should normally not exceed 40 pages of text, excluding tables and figures. For biotechnology products, the QOS should not normally exceed 80 pages, excluding tables and figures. In both cases if the product is more complex (e.g., a generic product with several different manufacturing sources of the drug substance each with its own DMF, or a drug-device combination), or manufactured using a more complex process, the QOS can be longer.

Granularity

Granularity can be defined as the extent to which a system contains separate components (like granules). The more components are contained in a system, the greater the granularity. The ICH M4 guidance document *Organisation of the CTD for the Registration of Pharmaceuticals for Human Use* (1) suggests three options in terms of granularity for the QOS.

Option 1

A single 2.3 QOS document comprising the Introduction, Drug Substance, Drug Product, Appendices, and Regional Information.

Option 2

Separate documents as follows:

- Introduction
- 2.3.S Drug Substance
- 2.3.P Drug Product
- 2.3.A Appendices
- 2.3.R Regional Information

Option 3

Separate documents as follows:

- Introduction
- 2.3.S.1 General Information
- 2.3.S.2 Manufacture
- 2.3.S.3 Characterization
- 2.3.S.4 Control of Drug Substance
- 2.3.S.5 Reference Standards or Materials
- 2.3.S.6 Container Closure System
- 2.3.S.7 Stability
- 2.3.P.1 Description and Composition of the Drug Product
- 2.3.P.2 Pharmaceutical Development
- 2.3.P.3 Manufacture
- 2.3.P.4 Control of Excipients
- 2.3.P.5 Control of Drug Product
- 2.3.P.6 Reference Standards or Materials
- 2.3.P.7 Container Closure System
- 2.3.P.8 Stability
- 2.3.A.1 Facilities and Equipment (biotechnology products)
- 2.3.A.2 Adventitious Agents Safety Evaluation
- 2.3.A.3 Excipients
- 2.3.R Regional Information

The applicant company can choose the option with the most appropriate granularity for a particular registration file. The more complex the product in general terms, the greater the granularity that will be needed.

Variations, Changes, Updates, and Granularity

Applications for changes requiring approval from a regulatory agency will usually need to be supported with a replacement document or documents that include the aspect of the product which it is proposed to change. If Option 1 is chosen, the entire QOS document must be resubmitted in support of the change. If a greater the degree of granularity of the QOS is chosen (e.g., Option 3), the less documentation will need to be submitted (and reviewed by the agency). Thus the option with the greater degree of granularity is best for ease of making subsequent changes to the product.

Table 1 Tables and Figures That Can Be Imported from Module 3 (2)

Sections of QOS	Table or figure that can be imported	Module 3 file from which imported
2.3.S.2 Manufacture	Flow diagram of drug substance manufacture	3.2.S.2.2
2.3.S.3 Characterization	Tabulated summary of impurities and their structure	3.2.S.3.2
2.3.S.4 Control of Drug Substance	• Specification	3.2.S.4.1
	• Tabulated summary of batch analyses	3.2.S.4.4
2.3.S.7 Stability	Summary of stability results	3.2.S.7.3
2.3.P.1 Description and Composition	Composition	3.2.P.1
2.3.P.2 Pharmaceutical Development	Tabulated summary of the composition of the formulations used in clinical trials, with dissolution profiles where relevant	3.2.P.2.2.1
2.3.P.3 Manufacture	Flow diagram of manufacture	3.2.P.3.3
2.3.P.5 Control of Drug Product	• Specification	3.2.P.5.1
	• Tabulated and graphical summary of batch analyses	3.2.P.5.4
2.3.P.8	Tabulated and graphical summary of stability results	3.2.P.8.3
2.3.A.2 Adventitious Agent Safety Evaluation	Tabulated summary of reduction factors for viral clearance	3.2.A.2

WRITING THE QOS

Tables and Figures in the QOS

The ICH M4Q(R1) Guideline (2) indicates which individual documents will be needed in the case of Option 3 in terms of granularity and summarizes what information should be included in each document. Many of the tables and figures can be imported directly from Module 3 as summarized in Table 1. Large sections of the QOS can thus be drafted by writing them in parallel with the production of the individual documents for Module 3.

Major Issues to Be Discussed in the QOS

The ICH M4Q(R1) Guideline (2) summarizes for each modular section of the drug substance and drug product, which are the major issues that need to be included in the QOS. As an example, the critical information in 2.3.S.2 Manufacture includes the following (for a chemical synthetic drug substance):

- Information on the manufacturer (name, address, and function of each separate site involved in manufacture and release of the active substance).
- Brief description of the manufacturing process and the controls that are intended to result in the routine and consistent production of a drug substance of appropriate quality.
- A discussion of the selection and justification of critical manufacturing steps, process controls, and acceptance criteria. Critical process intermediates need to be highlighted.
- A description of the process validation and evaluation.
- A brief summary of the major manufacturing changes made through development and conclusions as to the assessment of product consistency. The QOS should refer to the nonclinical and clinical studies that used batches affected by these manufacturing changes.

Who Should Write the QOS?

There are a number of options for writing the QOS. For the European Union and for Australia, the marketing authorization application has to include a signed declaration in Module 1 confirming that the QOS has been drawn up and signed by a suitably qualified expert. The expert has to provide a summary Curriculum Vitae (CV). Thus for a QOS for the European Union or Australia, the expert has to either write the QOS or at the very least review, amend, and agree the text drafted by others. For a U.S. or Japanese registration filing, there is no requirement for an expert signature so that the QOS can be drafted by an individual or a team of the relevant company professional staff. However, caution should be applied during the review of a draft QOS since if all critical elements are deleted in a (vain) attempt to hide issues, the result will be a bland document, which will be much less useful to the agency reviewer and ultimately to the applicant.

EUROPEAN MARKETING AUTHORIZATION APPLICATIONS

Marketing authorisation applications (MAAs) in Europe can use several different procedures

- The centralized procedure [where an application is made to the European Medicines Agency (EMEA)],
- The decentralized procedure (where an application is made to a number of countries simultaneously) and one country (Rapporteur) makes the first review of the MAA and coordinates the review in other countries, and
- The mutual recognition procedure where an application is filed and approved first in one country (Rapporteur) and then submitted in other European countries.

In the centralized procedure, one CHMP member acts as the Rapporteur and is responsible for the assessment of the MAA in his own country, and another acts as the Co-Rapporteur and is responsible for the assessment in his country. In the centralized procedure, other countries do not normally assess the MAA and will only receive a copy of Modules 1 and 2 (the summaries including the QOS) and will then mainly rely on the assessment carried out by the Rapporteur and Co-Rapporteur.

In the decentralized and mutual recognition procedures, many European Union countries will mainly rely on the assessment of the Rapporteur.

Thus in Europe the Module 2 summaries are vital since they may be the principal source of information on the scientific dossier that most countries receive. Therefore, the quality of the Module 2 summary is crucial and this should be a self-standing document, capable of being read and understood without reference to the Module 3 dossier.

Signature for the QOS in Module 1

Module 1.4 in the European Union requires a signed declaration by a suitably qualified expert using a standard form. The expert is required to state that they have drawn up the QOS and that it is in accord with the technical requirements of the Annex to Directive 2001/83/EC as amended (3). The Annex to this Directive (4) sets out the technical requirements for a marketing authorisation application in terms of the quality and nonclinical and clinical files, and the expert is stating that, in his professional opinion, the MAA conforms to those requirements. The expert annexes a short (one page) summary CV to the declaration.

Who Should Be the European Union Quality Expert?

Article 12 of Directive 2001/83/EC requires that the detailed summaries (such as the QOS) are drawn up "by experts with the necessary technical and professional qualifications." These are no longer defined in European regulation. However, common sense indicates that an expert signing the QOS, who is expected to cover all topics in the Quality dossier, will need a suitable pharmaceutical professional training and broad experience of pharmaceutical Research and Development (R&D). They do not need to be a European national.

Assessment Reports in the European Union

The initial Assessment Reports written by the national agency Rapporteur (and Co-Rapporteur) team in a European assessment procedure (centralized, decentralized, and mutual recognition) have to conform with the various EMEA and Coordination Group for Mutual Recognition and Decentralised Procedures—Human (CMD(h)) Guideline documents (5–10).

The *Day 80 Critical Assessment Report—Quality* (6) and the *Day 80 Critical Assessment Report (Generic medicinal product)—Quality* (9) guidelines identify the key elements in relation to the drug substance and drug product on which the Quality reviewer is expected to summarize and comment in each section of Module 3. If the QOS has been written appropriately by the company, all of the relevant information should be in it and much of the text from the QOS can

Table 2 Example of Critical Quality Criteria in Relation to the Drug Substance from the EMEA Day 80 Critical Assessment Report—Quality (6)

Sections of Module 3	Assessment criteria
3.2.S.4 Control of Drug Substance	Tabulated specification and analytical methods Validation of analytical procedures—state if in accordance with ICH or not, and mention any deviation. Are the methods adequate to control the substance on a routine basis? Batch analysis results ($n = ?$); do these confirm consistency and uniformity of the product? Do they indicate that the process is under control? Is the applicant's proposed justification of the specification adequate or not, bearing in mind the intended use of the drug substance in the product?

then be cut and pasted to form the Assessment Report—Quality. Thus, when writing the QOS, the expert should have in mind the criteria defined in the Day 80 Critical Assessment Report—Quality guideline, and try to address each item fully so that there are no major unresolved issues.

Examples of assessment criteria in the Day 80 Critical Assessment Report—Quality guideline are summarized in Table 2 and 3.

Table 3 Example of Critical Quality Criteria In Relation to the Drug Product from the EMEA Day 80 Critical Assessment Report—Quality (6)

Section of Module 3	Sections of Module 3.2.P.2	Assessment criteria
3.2.P.2 Pharmaceutical Development	3.2.P.2.1.1 Components of the Drug Product—Drug Substance	Has the company identified those physicochemical properties of the drug substance that are of relevance for product performance? Have these properties been adequately specified? On what basis have the limits been justified? Where potentially key parameters are not controlled, is the justification for their omission acceptable?
	3.2.P.2.1.2 Components of the Product—Excipients	Use of materials of animal or human origin—have these been justified? Have important, novel, or unusual excipients been identified regarding their impact on product performance? The applicant's choice and function of key excipients should be mentioned, e.g., those modifying release or disposition of the drug substance. Assessors should also refer to 3.2.A.3 Novel Excipients where a more detailed evaluation of the excipient may be given.

AUSTRALIAN REGISTRATION FILINGS

Module 1 of a CTD application to the Australian Therapeutic Goods Administration (TGA) is different from Module 1 of the European MA as it has features applicable to Australian needs. In Module 1.4, it includes a section for signatures of the experts similar to that required in the European MA (11). However, there is a special form for any local (Australian) experts. This includes the following statement:

I, the undersigned, declare that I have

- the suitable technical or professional qualifications to act in this capacity (for more information, refer to the enclosed curriculum vitae).

- fully examined the data provided by the applicant and have provided references to the literature to support statements made that are not supported by the applicant's original data. This report represents an objective assessment of the nature and extent of the data.
- provided a report based on my independent assessment of the data provided.
- based my recommendations, regarding suitability for registration, on the data provided herewith. I have considered the attached data and have recommended as to suitability for registration of the intended dose forms and presentations according to the proposed product information document.

I further declare that this expert report represents my own view.

Further, I declare the following to be the full extent of the professional relationship between myself and the applicant.

FDA REGISTRATION FILINGS (NDAs, ANDAs)

The FDA Office of Generic Drugs (OGD) has developed the question-based review (QbR) for evaluation of abbreviated new drug applications (ANDAs) (12–15). This is an assessment system based on what FDA feels are the critical pharmaceutical quality attributes. The QbR approach was publicized in 2005 through workshops, webcasts, meetings, and industry talks. In 2006, there were three workshops on how to prepare an effective QOS that would address the critical quality attributes and meet the FDA expectations in terms of evidence of quality by design. Companies were encouraged to submit a Quality Overall Summary that answered specific QbR review questions with every ANDA submission. FDA provided a document entitled "*Questions to be completed by ANDA Sponsors for the preparation of a QbR-Quality Overall Summary*" (15). Table 4 and 5 illustrate some QbR questions in relation to the drug substance and drug product.

The FDA Web site provides two very useful examples of model QOSs—for an extended release capsule (16) and an immediate release tablet (17).

The Office of New Drug Chemistry (ONDC) of FDA has not issued any detailed formal guidance on the required content of the QOS but their guidance on Risk-Based Pharmaceutical Quality Assessment System (18) does emphasize that "a comprehensive QOS can facilitate the development of a 'Big Picture' assessment protocol" by the Pharmaceutical Assessment Lead (PAL). It goes on to mention that if properly documented, the QOS can serve as a written summary of the new drug application (NDA), thus obviating the need for the ONDC staff to prepare their own summary.

JAPANESE REGISTRATION FILINGS

Before the international harmoniztion of the format of NDAs in the ICH CTD, much of the assessment of a Japanese NDA file was carried out on summaries of the data—the "Gaiyo." Since the adoption of the ICH CTD format in Japan, the Module 2 summaries now form the

Table 4 Examples of FDA QbR Questions to Be Completed by ANDA Sponsors—Drug Substance (15)

Sections of QOS	QbR questions
2.3.S.2 Manufacture	Who manufactures the drug substance?
	How do the manufacturing processes and controls ensure consistent production of drug substance?
2.3.S.3 Characterization	How was the drug substance structure elucidated and characterized?
	How were potential impurities identified and characterized?
2.3.S.4 Control of Drug Substance	What is the drug substance specification? Does it include all critical drug substance attributes that affect the manufacturing and quality of the drug product?
	For each test in the specification, is the analytical method(s) suitable for its intended use and, if necessary, validated? What is the justification for the acceptance criteria?
2.3.S.5 Reference Standards	How were the reference standards certified?
2.3.S.7S.7 Stability	What drug substance stability studies support the retest or expiration date and storage conditions for the drug substance?

Table 5 Examples of FDA QbR Questions to Be Completed by ANDA Sponsors—Drug Product (15)

Sections of QOS	QbR questions
2.3.P.1 Description & Composition	What are the functions and composition of the final product? What is the function of each excipient? Does any excipient exceed the Inactive Ingredients Database (IIG) limit for this route of administration? Do the differences between this formulation and the reference-listed drug (RLD) present potential concerns with respect to therapeutic equivalence?
2.3.P.2.1.1 Drug Substance	Which properties or physical chemical characteristics of the drug substance affect drug product development, manufacture, or performance?
2.3.P.2.1.2 Excipients	What evidence supports compatibility between the excipients and the drug substance?
2.3.P.2.2 Drug Product	What attributes should the drug product possess? How was the drug product designed to have these attributes? Were alternative formulations or mechanisms investigated? How were the excipients and their grades selected? How was the final formulation optimized?
2.3.P.2.3 Manufacturing Process Development	Why was the manufacturing process selected for this drug product? How are the manufacturing steps related to the drug product quality? How were the critical process parameters identified, monitored, and controlled? What is the scale-up experience with the unit operations in this process?

Gaiyo. These summaries must be extensive and complete stand-alone documents for them to be used by the Japanese reviewers.

The concept of "design space" for critical manufacturing steps has been the subject of discussion within the Japanese National Institute of Health Sciences (NIHS) as well as in the European Union and the United States. The group developing this concept within NIHS has produced a draft mock QOS in English for a tablet formulation—Sakura Tablets (19). The mock QOS comprises approximately 50 pages, but this summarizes only Modules 3.2.P.1 to 3.2.P.5 of the Quality file. This draft will perhaps be of particular interest to those concerned with possible future Japanese requirements in terms of design and evaluation of the drug product manufacturing process.

CANADIAN REGISTRATION FILINGS

Health Canada has provided a useful template (20) for the QOS for synthetic or semisynthetic chemical entity products (New Drug Submissions and Abbreviated New Drug Submissions), which should be used for filings in Canada. This format is not obligatory, and the guidance makes it clear that other approaches can be used. The template includes a number of tables, which are very useful ways of summarizing the relevant quality information, and these could be used in the QOS for Canadian registration filings and also for submissions in other countries.

ASEAN REGISTRATION FILINGS

The ASEAN CTD (21) is based on the ICH CTD and is structured as follows:

Part 1: Administrative information
Part II: Quality
Part III: Nonclinical
Part IV: Clinical

Structure of the ASEAN Quality File

The structure of Part II: Quality (22) is as follows:

Section A: Table of Contents
Section B: Quality Overall Summary

Section C: Body of Data
1. Drug Substance
2. Drug Product

Section D: Key Literature References

The structure of the QOS is again based on the ICH CTD. The ASEAN Part II Guideline includes a table of the ICH and other headings and a note of their applicability for NCE products, biotechnology products, major variations, minor variations, and generic products. One major difference is a new section P9 Product Interchangeability Equivalence Evidence. This consists of comparative in vitro dissolution data and in vivo bioequivalence data and is required for major variations and generic products.

Singapore (a member of ASEAN) has some of its own regulatory guidances and these include two useful examples of model QOSs—one for New Drug Applications and Generic Drug Applications (Chemicals) and one for a New Drug Application (Biological). The model QOS (Chemical) is for Paracetamol 100-mg tablets (23), and the model QOS (Biological) is for Rabinab—Humanized Anti-rabies immunoglobulin (24).

WRITING THE QOS FOR A GLOBAL REGISTRATION FILING

Each national or regional authority needs the quality information in the QOS to be constructed in the light of both ICH and any national or regional guidelines and to take into account also pharmacopoeial monograph requirements for that dosage form or particular pharmaceutical form. In addition, there may be differences in the presentation of products to be marketed in different regions—for example, use of blister packs for capsules and tablets in Europe, use of High Density Polyethylene (HDPE) bottles for the same products in the United States, need for more protective packaging for Zone IVb countries such as the ASEAN region. However, it is normally possible for the technical R&D team responsible to produce a "core QOS" as a master document, which can then be customized to meet national or regional needs or to follow a particular format (in the case of Health Canada's QOS-CE template). The customization will need to be carried out by or with input from the relevant expert in the case of European or Australian filings so that the expert can sign the declaration in Module 1.

EUROPEAN PHARMACOPOEIA CERTIFICATION AND THE QOS

The European Directorate for the Quality of Medicines and Healthcare (EDQM) gives detailed guidance on the content of the EDQM Certification applications in the document "*Content of the Dossier for Chemical Purity and Microbiological Quality*" (25). Applications for a European Pharmacopoeia Certificate (CEP) for a pharmacopoeial drug substance (or excipient) have to include a Quality Overall Summary, and the EDQM Web site includes a template for the QOS (26). This has to be written by an expert, and a copy of the CV of the expert appended.

Most of the information in the EDQM template is identical to that in the ICH QOS Module 2.3.S Drug Substance. However, it does include some extra information such as the history of the commercialization of the substance. The declarations appended to the application form have to be summarized

- Manufacture of the substance in accordance with ICH Q7A GMP rules (for an API).
- Commitment of the manufacturer to keep the proposed authorization holder informed of any changes to the documentation.
- Willingness of the manufacturer to be inspected.
- Use/nonuse of materials of human or animal origin in the manufacturing process.

A key element in the EDQM application is the need to identify new impurities not mentioned in the impurity section of the pharmacopoeial monograph and to mention the levels found, the Limit of Detection (LOD) and Limit of Quantitation (LOQ) of the analytical method used to measure them. The suitability of the monograph to control potential impurities is to be discussed. There is a specific section on possible genotoxic impurities that refers to the requirements of the Committee for Human Medicinal Products (CHMP) Safety and Quality Working Parties Guideline (27). There is also a special tabular format for reporting solvents, reagents, and catalysts and this is shown in Table 6.

Table 6 EDQM Tabular Format for Reporting Residual Solvents/Reagents/Catalyst(s) (26)

Solvent/reagent/ catalyst	Used in step x/y	Applicant's limits	ICH class/ limit	Levels (ppm)	LOD of the method	LOQ of the method

EUROPEAN ACTIVE SUBSTANCE MASTER FILES AND THE QOS

The European Active Substance Master File (ASMF), sometimes known as the European Drug Master File (EDMF), consists of two parts—the Applicant's Part (AP) provided to the MA applicant and to the authorities, and the Restricted Part (RP) provided to the authorities only. The EMEA Guideline on ASMF procedure (28) defines the content of each. The AP contains information that is nonconfidential to the MA applicant. The RP contains the remaining information, such as detailed information on the individual stages of the manufacturing method and quality control during the manufacture of the drug substance.

A separate QOS has to be provided for both the AP and the RP. In the case of the QOS for the RP, this is written by a suitable person at the drug substance manufacturer and supplied with a signed declaration and CV of the signatory similar to that in Module 1.4 of a European CTD Marketing Authorisation (MA) application.

In the case of the QOS for the AP, this could be written by either the drug substance manufacturer or the MA applicant. Whoever writes the QOS on the AP, it is included together with the remainder of the QOS on the drug product, the appendices, and the regional information. It is the responsibility of the drug product manufacturer to ensure that suitable information is included in the relevant sections of 2.3.S in the QOS for the AP as it is a single expert who has to sign off on all sections of the QOS in the declaration in Module 1.4.

REFERENCES

1. Organisation of the Common Technical Document for the Registration of Pharmaceuticals for Human Use, M4, January 2004, ICH. http://www.ich.org/LOB/media/MEDIA554.pdf. Accessed September 18, 2008.
2. The Common Technical Document for the Registration of Pharmaceuticals for Human Use: Quality—M4Q(R1), Quality Overall Summary of Module 2, Module 3: Quality, September 2002, ICH. http://www.ich.org/LOB/media/MEDIA556.pdf. Accessed September 18, 2008.
3. Volume 2B Notice to Applicants, Medicinal Products for Human Use, Presentation and format of the dossier, Common Technical Document, Edition May 2008, Module 1.4 Information on the Experts, 1.4.1 Quality, European Commission. http://ec.europa.eu/enterprise/pharmaceuticals/eudralex/vol-2/b/update_200805/ctd_05–2008.pdf. Accessed September 18, 2008.
4. Commission Directive 2003/63/EC amending Directive 2001/83/EC, Official Journal of the European Union, L159/46, 27 June 2003. http://ec.europa.eu/enterprise/pharmaceuticals/eudralex/vol-1/dir_2003_63/dir_2003_63_en.pdf. Accessed September 18, 2008.
5. Guidance Document Day 80 Critical Assessment Report, Overview and List of Questions, D80-Overview-Rev3–2007, EMEA. http://www.emea.europa.eu/pdfs/human/chmptemplates/CHMP-D80-AR-Guidance/D80AR-Overview-Guidance.pdf. Accessed September 18, 2008.
6. Guidance Document Day 80 Critical Assessment Report, Quality, D80 Quality-Rev2–2006, EMEA. http://www.emea.europa.eu/pdfs/human/chmptemplates/CHMP-D80-AR-Guidance/D80AR-Quality-Guidance.pdf. Accessed September 18, 2008.
7. Guidance Document Day 80 Critical Assessment Report, Quality (Template), D80-Quality-Rev2–2006, EMEA. http://www.emea.europa.eu/pdfs/human/chmptemplates/CHMP-D80-AR-Guidance/D80AR-Quality-Template.pdf. Accessed September 18, 2008.
8. Guidance Document Day 80 Critical Assessment Report (Generic medicinal product), Quality, D80-AR-QUAL-Generics-2006, EMEA. http://www.emea.europa.eu/human/chmptemplates/D80_AR_Generics_Quality_Guidance.pdf. Accessed September 18, 2008.

9. Guidance Document Day 80 Critical Assessment Report (Generic medicinal product), Quality (Template), D80-AR-GEN-QUAL-2006, EMEA. http://www.emea.europa.eu/human/chmptemplates/D80_AR_Generics_Quality_Template.pdf. Accessed September 18, 2008.
10. Co-ordination Group for Mutual Recognition and Decentralised Procedures—Human Guideline on the Assessment Report for Mutual Recognition and Decentralised Procedures, Rev.3, September 2006. http://www.hma.eu/uploads/media/actd-guideline.pdf. Accessed September 18, 2008.
11. Australian Government, Department of Health and Ageing, Therapeutic Goods Administration, Notice to Applicants, CTD-Module 1, TGA Edition September 2007, Module 1: Administrative Information and Prescribing Information for Australia, Module 1.4.1 Information about the experts. http://www.tga.gov.au/docs/pdf/euguide/tgamod1.pdf. Accessed September 19, 2008.
12. Question-Based Review for CMC Evaluation of ANDAs, 2007, FDA. http://www.fda.gov/cder/ogd/QbR.htm. Accessed September 18, 2008.
13. Office of Generic Drugs White Paper, Question-Based Review (QbR) for Generic Drugs: An Enhanced Pharmaceutical Quality Assessment System, August 2005, FDA. http://www.fda.gov/cder/ogd/QbR_white_paper.htm. Accessed September 18, 2008.
14. OGD Update on Question-based Review Submissions, August 2007, FDA. http://www.fda.gov/cder/ogd/QbR/QBR-submissions_upd.htm. Accessed September 18, 2008.
15. Office of Generic Drugs QbR Quality Overall Summary Outline, Questions to be completed by ANDA Sponsors for the preparation of a QbR-Quality Overall Summary, June 2007, FDA. http://www.fda.gov/cder/ogd/QbR_Summary_outline.htm. Accessed September 18, 2008.
16. Example Quality Overall Summary, Controlled Release Capsules, FDA. http://www.fda.gov/cder/ogd/QbR/OGD_Model_QOS_ER_Capsule.pdf. Accessed September 19, 2008.
17. Example Quality Overall Summary, Immediate Release Tablet, FDA. http://www.fda.gov/cder/ogd/QbR/OGD_Model_QOS_IR_Tablet.pdf. Accessed September 19, 2008.
18. ONDC's New Risk-Based Pharmaceutical Quality Assessment System, 2004, FDA. http://www.fda.gov/cder/gmp/gmp2004/ondc_reorg.htm. Accessed September 18, 2008.
19. English Mock QOS, P2, 2.3 Quality Overall Summary, Sakura Tablet, National Institute of Health Science, June 2008. http://www.nihs.go.jp/drug/section3/MockP2(E).pdf. Accessed September 19, 2008.
20. Health Canada Quality Overall Summary—Chemical Entities (New Drug Submissions/Abbreviated New Drug Submissions) (QOS-CE (NDS/ANDS)), April 1, 2004. http://www.hc-sc.gc.ca/dhp-mps/prodpharma/applic-demande/templates-modeles/qoscenotice_sgqecavis-eng.php. Accessed September 19, 2008.
21. Singapore Government Health Sciences Authority, Singapore Drug Registration Information and Guidelines, The ASEAN Common Technical Document Dossier (ACTD) for the Registration of Pharmaceuticals for Human Use, Organisation of the Dossier. http://www.hsa.gov.sg/publish/etc/medialib/hsa_library/health_products_regulation/western_medicines/files_guidelines.Par.22449.File.dat/ACTD_OrganizationofDossier.pdf. Accessed September 19, 2008.
22. Singapore Government Health Sciences Authority, Singapore Drug Registration Information and Guidelines, The ASEAN Common Technical Document Dossier (ACTD) for the Registration of Pharmaceuticals for Human Use, Part II: Quality. http://www.hsa.gov.sg/publish/etc/medialib/hsa_library/health_products_regulation/western_medicines/files_guidelines.Par.28201.File.dat/ACTD_PartIIQuality_Apr05.pdf. Accessed September 19, 2008.
23. Singapore Government Health Sciences Authority, Singapore Drug Registration Information and Guidelines, Guidance on Medicinal Product Registration in Singapore—Singapore Quality Overall Summary for Chemical Drugs. http://www.hsa.gov.sg/publish/etc/medialib/hsa_library/health_products_regulation/western_medicines/files_guidelines.Par.51871.File.dat/07_SingaporeQOSforChemicalDrugs_sample.pdf. Accessed September 19, 2008.
24. Singapore Government Health Sciences Authority, Singapore Drug Registration Information and Guidelines, Guidance on Medicinal Product Registration in Singapore—Singapore Quality Overall Summary for Biologics. http://www.hsa.gov.sg/publish/etc/medialib/hsa_library/health_products_regulation/western_medicines/files_guidelines.Par.70975.File.dat/09_SingaporeQOSBiologics_sample.pdf. Accessed September 19, 2008.
25. European Directorate for the Quality of Medicines and HealthCare, Certification of Suitability of Monographs of the European Pharmacopoeia, Content of the dossier for chemical purity and microbiological quality, May 2007, Strasbourg. http://www.edqm.eu/medias/fichiers/Content_of_the_Dossier_for_Chemical_Purity_Microbiological_Quality.pdf. Accessed September 19, 2008.
26. European Directorate for the Quality of Medicines and HealthCare, Certification of Suitability of Monographs of the European Pharmacopoeia, Template for Quality Overall Summary to be submitted for Certification applications. http://www.edqm.eu/medias/fichiers/Template_Quality_Overall_Summary_Word_format.doc. Accessed September 19, 2008.

27. Committee for Medicinal Products for Human Use (CHMP), Guideline on The Limits of Genotoxic Impurities, EMEA/CHMP/QWP/251344/2006, January 2007, EMEA. http://www.emea.europa.eu/pdfs/human/swp/519902en.pdf. Accessed September 19, 2008.
28. Committee for Medicinal Products for Human Use (CHMP), Committee for Medicinal Products for Veterinary Use (CVMP), Guideline on Active Substance Master File Procedure, CPMP/QWP/227/02 Rev 2 Consultation, August 2006, EMEA. http://www.emea.europa.eu/pdfs/human/qwp/022702en.pdf. Accessed September 19, 2008.

6 The Nonclinical Overview and Summary

Richard Lee
Consultant, Hitchin, Hertfordshire, U.K.

INTRODUCTION

The very title of this chapter poses an immediate semantic question as to the meaning of the term "non-clinical." It is a misnomer in the sense that "clinical" by definition implies factual observation, and much of the information discussed in "nonclinical" assessment is based on observing how animals react to the administration of potential human medicinal compounds. Formerly, the term "preclinical" was used, but this, too, has similar etymological problems in that the timing of the studies in the development program is not always prior to first studies in humans.

If, however, "clinical" is taken, erroneously, to mean "referring to human treatment" then "nonclinical" can be seen as acceptable for the purpose of medicines regulation.

Thus, the first task of this chapter is to define what is intended herein by the term "nonclinical." It is the information generated from in vitro studies (including data from isolated human cell/tissue experiments) and from studies using animal models to predict the likely human pharmacology, pharmacokinetics, metabolism, and toxicology of a substance administered by one or more dose routes and as different formulations.

Some of the information gleaned from in vitro assay systems and animal models can be superseded by data from human clinical trials, that is, pharmacology (primary and safety) pharmacokinetics and relatively short-term toxicology, plus there is reassurance in knowing that the nonclinical data can be predictive. However, confirmation of some nonclinical information cannot ethically be sought in humans, for example, genotoxicity, effects on embryo/fetus, and long-term toxicity, including carcinogenic potential.

There is thus a need to rely on some nonclinical safety data throughout the period of use of a compound as a human medicine.

The function of the Nonclinical Overview (NCO) and its attendant summaries is to provide a concise but fully informative documentation of these types of study, including an assessment and interpretation of the data for judging the probability of effective and safe use of an active substance in humans.

The various aspects of how a Nonclinical Overview and Summary can best achieve its objective are the subject of the remainder of this chapter.

HISTORY OF THE NONCLINICAL OVERVIEW

The Nonclinical Overview has, as one of its ancestors, the Preclinical Expert Report adopted by the expert advisors of the European Regulatory Agencies when they came together as the Committee on Proprietary Medicinal Products (CPMP). Its origin lies in the twin Directives 75/318/EEC and 75/319/EEC issued on 20 May 1975.

Directive 75/319 Article 2 indicated that the Marketing Authorization Application should provide an assessment of the pharmacotoxicological data and Annex 1 of Directive 75/318 refers to such assessment as an "expert report."

Prior to this, the launching of a pharmaceutical product in France had required the sign-off by appropriate experts, one of whom would deal with the information from animal studies and other "preclinical data." The sign-off was usually accompanied by an opinion about the likely safety of the medicinal product—obviously always positive, in terms of benefit versus risk for the patient, and giving few details or reasons as to why this conclusion was reached. The Expert was the expert and trusted to be truthful.

This concept, of an opinion provided on behalf of the applicant, was taken on board by the European regulators, but expanded to provide a summary of the preclinical information with comments on its relevance (or otherwise) to safety in human use of the chemical (and later biological) compound, as a medicine in defined indication(s).

This contrasted at the time with the executive summary preferred by the Food and Drug Administration (FDA) in the United States, whose attitude could be summarized as being that the regulatory assessors held the prerogative of interpretation and that the data should be presented in a summarized format, but without comment by the NDA applicant.

Thus, when the concept of a "global dossier" was put forward at the 3rd International Conference on Harmonization (Yokohama, 1995), the requirements of the FDA clashed with those of the CPMP in respect of the type of commentary on the pre or nonclinical data that would be acceptable. (The Japanese position seemed to be aligned with that of the United States in not wishing the commercial applicant to preempt the regulatory assessment.)

This seeming impasse was, however, rapidly bypassed by the proposal to settle for a Common Technical Document (CTD) rather than an all embracing Global Dossier, which would need much time and legal adjustment to achieve. Within that CTD, the problem of whose opinion was to be accepted was resolved by permitting the submission of an Overview, which would allow interpretation of results but remain anonymous for those regulatory authorities who did not wish to consider the opinion of a named expert.

Hence, Overviews (both clinical and nonclinical) were divided into two sections, with the signature of the expert separated onto one page, which would be included in Module 1 (the unharmonized part of the global dossier or CTD as it is now commonly called).

The Summary of the data and interpretation was consigned to Module 2 of the CTD and was thus available to all regulatory assessors, because Modules 2 to 5 of the CTD were agreed as harmonious within the United States, Japan, and Europe.

From this point on, the nonclinical element of Module 2 of the CTD has evolved to include not only the Overview but also separate summaries of pharmacology, pharmacokinetics, and toxicology—both in Written and Tabular formats.

For the European Regulatory Assessment today, there is reference to an EC Regulation (1), a Directive (2), and the published Notice to Applicants, all of which may be found in The Rules Governing Medicinal Products, particularly Volume 2B, issued by the European Commission (Directorate General DG III). The NCO is a required item for all marketing applications.

For the United States, a relevant document is "Guidance for Industry M4S: The CTD—Safety" (3) and in Japan the Gaiyo (synopsis) should be prepared (4). Both mention the NCO as an essential element in the CTD required to be submitted in a marketing application.

At this point it may be worth recalling the remarks of Kurt Briner in the Opening Plenary Session of the 3rd ICH Conference in Yokohama: "The global dossier has progressed a long way. But is this really progress? There is a danger that the exercise has resulted in requirements that, although harmonized, are considerably more demanding than even the sum of (*previous*) nonharmonized requirements, to the extent that industry's capacity to invest in innovation may be affected." (5)

Indeed, today a great deal of care and resource must be applied to Module 2 (nonclinical) to ensure that the three elements of Overview, Written Summary, and Tabular Summary tell the same story and give the same interpretation, especially if the authors of each part are not the same.

OBJECTIVES OF THE NONCLINICAL OVERVIEW

The aim of an NCO should be to assess the significance of the in vitro and animal model data for judging the likely safety issues relevant to the use of the active substance as a human medicine. The specific aspects (pharmacology, kinetics, and metabolism plus the various types of toxicity study) will need to be succinctly described and then analyzed as far as knowledge permits for any action in the models that could translate into activity within patients that would have a safety element and impinge on the benefit to such an extent as to warrant warnings to the physician and hence to the patient.

Thus, the effects in the nonclinical models may be usefully categorized as

A. meaningful at the concentration/exposure anticipated as required to generate the benefit in patients or

B. biologically significant at concentration/exposure higher than is probable in the majority of patients (a relatively low-safety margin in previous terminology) or
C. of biological significance only at concentration/exposure much in excess of that likely to be encountered in patient use or
D. specific to the test model by virtue of the species differences between that model and humans.

Such categorization alerts the regulatory assessor to the degree of interrogation of the nonclinical data necessary for an effective consideration of the information presented in the marketing authorization application.

The conclusion of an Overview will necessarily have the objective of convincing a regulatory agency assessor that the nonclinical information is supportive of the submission requesting a marketing authorization (because anything less than supportive would be self-defeating in its aim). However, the arguments put forward to support the submission should be soundly and scientifically based and all possible safety concerns should be addressed and justified in terms of the benefit:risk calculation.

Aside from this main objective, the NCO is the vehicle to provide assurance to a regulatory authority assessor that the tests/assays/experiments in the nonclinical studies have, where necessary, been conducted according to approved guidance on standards such as Good Laboratory Practice and that the reports have been audited to ensure that a trail from conclusions back to raw data can be followed if required.

The provision of certain basic information about the product under assessment is usefully included in the Introduction to the NCO; this is seen as helpful to the specialist Regulatory Assessor, who may not have convenient access to the whole dossier. Such information could include:

A. The claimed indications for human medicinal use.
B. The posology—enabling a comparison of human exposure in terms of dose with the same parameter values in the animal models used.
C. The impurity profile of the finished product and how this compares with the material used in toxicity testing—to give an assessment of comparative exposure in clinical use and during the nonclinical safety testing.

Another recommended aim of the NCO is to indicate how closely the nonclinical development program of studies matched relevant guidance (ICH/CHMP/FDA/MHW or other) on what would be expected for the type of application under consideration and to justify deviation from that expectation. Similarly, where there is guidance on the design of such studies, the adherence to the recommended protocols should be assessed and variations explained and justified.

Finally, and perhaps most importantly, the NCO should indicate and justify the nonclinical testing strategy followed by the applicant to try to achieve the desired level of support for the type of application being considered (6).

Thus, an application relating to a new active substance would presumably have had a development strategy that included all the recommended nonclinical study types in ICH guidance, plus any others that had scientific justification unique to the active ingredient and its mode of action. In other words, the developing company had reviewed the nonclinical safety package as anticipated by regulatory assessors based on ICH and local guidelines, and selected from it the elements considered to be relevant and added any other tests that would occur to any intellectual scientist. Such a selective nonclinical development programme should be devised for all new medicinal products; there is no such thing as a "one-size-fits-all" set of studies.

Obvious examples of situations where the expected full program of nonclinical safety studies may be modified are the omission of carcinogenicity studies in rodents when the medicine is only to be administered for short periods (some antibiotics) or the omission of reproduction toxicity studies when the contraindications for the medicine include pregnant women and women of child-bearing potential (some psoriasis medication).

For products that are biological or produced by biotechnology, the nonclinical safety program may need to be modified to comply with scientific principles. Thus, the testing of biological substances in conventional bacterial mutation assays may not be sensible, because the large molecules involved may not reach their target site within cells, and the presence of biological impurities or breakdown products could invalidate the assay by providing the bacteria with histidine, for example.

For generic products, the thought process may be simpler but the development program still requires inquisition as to the necessity for studies to support, for example, a different impurity profile; special consideration will also be needed for the use of a new excipient. The role of the NCO is to explain, with justification, the chosen method of supporting the generic application, even if it is only to state that there is complete reliance on originator's data. (*Nota bene*: It has become an accepted exercise to update the originator's information with more recent data from peer-reviewed, published work and to review this for safety significance in the NCO.)

In other circumstances, the NCO may be a short statement to confirm that the nonclinical data have been assessed previously and, presumably, approved by a regulatory agency—this could be the case for abridged applications and line extensions where relatively simple changes to an existing product are the subject of the submission.

A modified version of an NCO may be useful at an earlier stage of the development of a human medicine, when an assessment of the nonclinical data is required to support applications to conduct clinical trials. The principle is the same—to provide a succinct summary of the available pharmacology, pharmacokinetic, and toxicology data and supplement that with discussion of the significance of the data, leading to a set of conclusions that should prove helpful to the regulatory assessors and the clinicians involved in the trials.

FORMAT AND POSITION OF THE NCO

The overall structure of the NCO is similarly prescribed in the guidance issued by the three regulatory members of ICH—that is, FDA for the United States, CHMP for the EU, and the Ministry of Health and Welfare in Japan. This provides consistency in so far as the main section headings are concerned so that the NCO has five sections—Introduction (1), section on Pharmacology (2), section on Pharmacokinetics, including metabolism (3), section on Toxicology (4), and an Integrated Summary and Conclusions (5).

It is recommended that the way in which each of these sections is constructed follows a consistent format suitable for electronic submission; this applies to the whole of the application dossier and is the subject of another chapter in this book.

Within this formalized structure, the use of subsections is left to the author. These may be identified simply by title, although some authors favor a numerical system so that, for example, 3.2.3.2.1 could be the subsection title for Pharmacokinetics (3) metabolism (3.2) in dogs (3.2.3) secondary reactions (3.2.3.2) conjugation by glucuronide (3.2.3.2.1) by comparison with 3.2.2.2.2, which would indicate, by analogy, the subsection on secondary metabolism in rats—conjugation with glycine.

The difficulty that some regulatory authorities have with this type of notation is that it becomes cumbersome and does not necessarily mean the same data set for different marketing authorization applications—so perhaps an underlined or italicized heading *Pharmacokinetics—secondary (conjugative) metabolism in dogs* would suffice for the example given.

A contents page for the NCO is helpful, but not demanded by the format guidance.

Any Literature Cited should be added after the Integrated Summary and Conclusions. The format of these references should follow internationally accepted standards as used, for example, in *Chemical Abstracts* or in *Entrez PubMed* on the Internet.

Reference made to the Tabulated Summary that accompanies the NCO should give Table Number and Study or Report Number as listed. This Study/Report Number should be a unique identifier, which can be easily found in any Written Summary that also is a part of the NCO package.

The purpose of having a prescribed format is consistency of presentation across the areas of the world covered by ICH, allowing easy exchange between regulatory authorities, particularly within Europe when various types of application require different member states (MS) to perform different functions such as Rapporteur or Corapporteur or Reference MS or Concerned MS.

In the age of electronic submissions and data exchange, it is worth reiterating that a consistent format is crucial and should include recommended mechanisms for referencing between the NCO and other modules in the CTD part of the application.

A recommendation from the regulatory guidance documentation is that an NCO should be about 30 pages maximum in length. This is generally taken to mean the text with included tables and figures rather than the total length including literature references, etc. Written Summaries are separate documents, as are the Tabulated Summaries, for the purposes of page counting.

For a complex set of nonclinical data supporting a novel therapeutic substance, a dispensation of 20% should present no difficulties in acceptance by regulatory agencies.

The position of the NCO is in Module 2 (section 2.4) of the CTD.

DETAILED CONSIDERATION OF THE DATA

There are many guidance documents to take into account—some are ICH sponsored and apply to applications for marketing made in the United States, Japan, and the European Community; the list for the last-named regulatory region is attached as an Appendix to this chapter (7).

For a New Active Substance

As indicated in the section on format (above), the scientific data should be assessed in sufficient detail to make a justification of actions taken in relation to those data and in the order Pharmacology, Pharmacokinetics, and Toxicology.

Each of these is usually divided into subsections and these are considered to be as follows:

Primary Pharmacology

The NCO should outline the results of pharmacology studies performed to test the concept of the mode of action and indicate the strength of the argument that these results, from in vitro or animal model systems, could be justifiably used to propose that a similar set of effects would occur in humans.

Examples:

- Action at a receptor site in an isolated pharmacological system should be correlated with the same mode of action in vivo in the species used for testing both proof of concept (including humans) and for examining safety.
- If different receptor sites or more than one receptor site operate(s) in the nonhuman systems, then the implications for the relevance of data generated in the nonclinical models should be fully examined and justification provided in the NCO for the use of the selected species or test systems.
- Inhibition (rarely stimulation) at an enzyme site as mode of action should be demonstrated as preserved across species (including humans) so that consequences of the effects on the enzyme may be validly compared. The NCO should provide reassurance on this issue.
- In chemotherapeutic action, the NCO should present the evidence that the foreign organism (bacteria, virus, neoplasm, or parasite) is attacked within the test systems, in case the model may not be valid because of the presence or absence of sickness engendered by the invading organism, which could in turn affect the distribution of the therapeutic agent to the site of action.

Secondary Pharmacology (including Safety Pharmacology)

Pharmacological activity, at sites unrelated to those involved in the desired mode of action, may be innocuous or only occur at such high concentrations as to make it irrelevant to safety. The presence of this type of activity in animal models should be discussed in the NCO with the aim of concluding that it is not pertinent to the consideration of the active substance as a human medicine. If, however, it cannot be concluded that the activity is irrelevant, then the action comes under the definition of "safety pharmacology" and its implications for human medicinal use must be discussed fully with the objective of alerting clinicians to the possibilities of adverse reaction—ideally confirmed or dismissed in clinical trial data analysis so that the appropriate wording for the Summary of Product Characteristics (SmPC) can be devised.

In particular, there is guidance (ICH) on the aspects of major organ pharmacology in animal models that are held to be significant for human use of a medicine and there is specific guidance on the possible cardiac effects, including ECG changes such as the prolongation of the QT interval (7). Any such changes should be assessed in the NCO for likely importance in clinical trial observations and in pharmacovigilance postmarketing.

For biological/biotechnology products, the testing of pharmacology has previously been seen as superfluous because the high-molecular-weight substances were frequently replacement agents such as hormones. More recently large molecules that can penetrate to intracellular reaction sites, as well as act extracellularly, have been developed and illustrate the need for an understanding of their mode of action and their potential for other pharmacological action—especially on vital organ systems. Thus, the NCO has a role to play in explaining and justifying the strategy for nonclinical safety testing—as in for example, botulinum toxin or complex mycin–based chemotherapy.

By the time of Marketing Authorization Application (MAA), the animal model pharmacology data will probably have been superseded by information from humans in clinical trials. The NCO can usefully point out when this has occurred and to what extent the nonclinical pharmacology has become of historical interest only.

Pharmacokinetics and Biotransformation (Metabolism)

The first task of the NCO is to provide the historical rationale for the selection of dose for the Phase I clinical trials when these are aimed at determining the systemic (usually plasma) pharmacokinetics of a new active substance in humans. This rationale is based on the data from animal models, taking into account the likely difference in elimination rate from the systemic circulation between animals (particularly rodents) and humans. Linking pharmacodynamics (PD) with pharmacokinetics (PK), that is, investigating the possible effect/time/concentration relationship and selecting the model that best describes this in mammals can provide a suitable explanation and justification for the dosing protocol in the first human studies.

It is reassuring to a regulatory assessor to have a reasoned trail of progression from the animal PD through the associated PK to the human dose(s) and hence to the human pharmacodynamics (PD)/PK relationship that will be used to set the posology in the SmPC and to justify the selection of patient dose, because most regulatory authorities like to feel that they have authorized a near optimal patient dose regimen, rather than one plucked from the air, which happens to give acceptable efficacy and safety.

The second major function of the NCO, in this section, is to assess the degree to which the metabolic pathways acting on the active substance are preserved across the species used to study the pharmacology and toxicology.

Thus, if the primary biotransformation is by oxidation in rats and dogs (or other nonclinical species of relevance) followed by a conjugation and largely biliary excretion of the metabolite(s), then it follows that this should be the major transformation mechanism and route of elimination in humans in order that the patient is not exposed to any metabolite structure that was not present in significant concentration in the animal model systemic circulation. If this is not the case, then the NCO needs to justify the strategy pursued in further nonclinical investigation to study the effects of such human metabolites in the animal models. It may be possible to synthesize the human metabolites and administer them to the models, but this does not take account of the very local effects that could occur at the point of transformation, especially if there are unstable intermediates involved in the metabolic process.

Similarly, hydrolytic breakdown, deamination, hydration, or other possible primary biotransformations of molecules should follow the same pattern in humans and in the animal models used for safety testing. Consequential conjugation is important for reducing exposure to primary metabolites (the nature of the conjugant is less critical, unless its supply is easily exhausted) and may also contribute to further systemic exposure if it is hydrolyzed in the GI tract, after biliary excretion, and reabsorbed.

The provision of a Figure showing the detailed chemical structure of the molecules being discussed is a great kindness to a Regulatory Assessor and up-to-date NCOs contain one.

The NCO should, therefore, take into account all these possibilities and construct a rational explanation that justifies the use of the active substance as a human medicine by comparing the various exposures to metabolites in the animal models and humans.

The final kinetic task of the NCO is to validate the toxicology studies by analyzing the pharmacokinetic data obtained from blood samples taken at various intervals from the animals in those studies. As with the link between PD and PK, it is reassuring for those who are asked to approve medicines to know the relationship between the range of human systemic exposure and the same exposure in animal models [toxicokinetics (TK)] where the onset of toxicity and its change with increasing dose are established.

The bridge here is the concentration/time profile of the active substance (and/or its metabolites) in blood (plasma) samples, which allows a correlation of several systemic exposure values with the absence, or degree, of adverse reaction occurring in the animal models used for toxicity testing.

The maximum profile associated with an absence of observable adverse effect can then be compared to the profile in humans considered to be the upper end of the range engendered by administering the recommended dose. Any favorable difference between them is termed a safety margin based on TK.

To expand this statement, analysis of safety margins would be most useful when it is possible to assay the concentration of active substance and any active metabolite in the vicinity of a receptor, enzyme, or other binding site that is directly linked to an adverse reaction. This is not always feasible, so generally the circulating plasma concentration is substituted with the hope and assumption that equilibrium between this and the binding site will be rapid and render the data valid.

From the plasma concentration/time profile, duration of dosing (TK profile) and the associated toxic response of the animal model, it may be reasonable to deduce one or more of the following conclusions:

1. There is an identified dose level and duration of repeated administration that produces no significant adverse reaction in the model and has a known TK profile.
2. There is a dose level and duration of repeated administration with TK profile that is associated with a defined toxic response from the model and identifies target organ(s) but is by nature not so concerning as to preclude further development.
3. There is a dose level and duration of repeated administration (with TK profile) that is not lethal but suggests unacceptable toxicity if translated to humans.

If the data allow all three conclusions, and the responses are graded, then a judgment can be made on the likely effects in humans (based on their plasma concentration/time profile) at the recommended posology.

Ideally only conclusion 1 is established—at a level that permits the claim of an unequivocally acceptable safety margin based on a known relationship between dose and systemic exposure as described by the kinetics.

Overall the NCO serves the purpose of linking PD, PK, TK, and toxicity in the relevant animal models to the PD, PK, and safety in humans.

Toxicology

Having established from the systemic exposure (PK, TK) data at what blood/plasma concentration profile (and hence dose) the toxicity is likely to be found, it is necessary to give a brief description of the signs and symptoms in the NCO and then proceed to interpretation and prediction of the significance of these findings to the safety of a healthy volunteer or patient in a clinical trial.

For all types of toxicity testing, it is the function of the NCO to consider the appropriateness of

- the animal models used (were they most sensitive, most resembling humans?),
- the number of animals per group (was it suitable for statistical analysis?),
- the dose route (did it correspond to that proposed for administration to patients?),

- the range of dose levels (how relevant to the likely human dose regimen?), and
- the duration of the studies (suitability for supporting the clinical trials?)

The general toxic effects of an active substance in a rodent and a nonrodent species should be assessed for possible significance in human medicinal use. Often these occur at higher exposures than will result from the human dose regimen or will only manifest themselves in the animals after a long period of sequential daily dosing. There are many possible scenarios: the NCO should address the specific concerns relating to the particular active substance, bearing in mind that there will be a human database of adverse effects for comparison. Any toxic reaction that is apparently confined to an animal model should be thoroughly assessed.

The function of the NCO is to indicate what animal data are superseded by human safety data—usually, these will be from the shorter period studies initially, depending on the length of the clinical trials. As the human database increases in terms of individual patient years of treatment, so more and more of the animal general toxicity information becomes historical and of very limited relevance to the continued use of the active as a human medicine.

Not so the data on genotoxicity: it is most unlikely that any studies on mutation or clastogenicity will be undertaken in living humans, thus the in vitro and in vivo animal model information remains relevant for a long period until human medicinal use has reached a longevity that indicates there is insignificant risk to the generation taking the medicine and the subsequent offspring.

Thus, at the time of MAA submission it will be essential for the NCO to discuss and justify the genotoxicity assays submitted in the dossier in terms of the type of studies conducted and the significance of the results for human safety. The ICH notes for guidance suggest three first-line studies, with follow-up assays if any of these three produce equivocal positive reactions. Deviation from this guidance needs to be justified and explained as to why the alternative strategy was considered to be as good or better.

Similarly, the nonclinical data on carcinogenic potential for the active substance will not be superseded by human information for several years and must therefore be carefully assessed by the NCO at the time of MAA submission, when the long-term exposure of patients is unlikely to be sufficient to warrant ignoring the animal data.

The NCO needs to give an opinion on any significant increase in tumor incidence when comparing treated and control rodents—this opinion could cover such circumstances as

- Tumor is rodent specific and has no meaning for human treatment.
- Tumor incidence is driven by increased stimulation of cell neoplasia resulting only from very high exposures, so there is an exposure threshold (which is, or is not, reached or exceeded by the human recommended dose schedule).
- Rodent tumor is hepatic in origin and likely the result of the metabolic pathway, which includes reactive intermediates. This route of biotransformation does, or does not, occur in humans.

These examples are given to illustrate the type of assessment that the NCO should undertake—there are many others that require an equally specific thought process.

Toxicity to the reproductive processes in mammals is another topic that requires careful consideration, because it will be some years into the life of a pharmaceutical treatment before anything certain will be established about the activity of the medicine in pregnant patients. Thus, there is reliance on the animal model data for a relatively long period after the MAA has been approved.

The SmPC for any product containing a new active substance will give advice to clinicians on the advisability, or otherwise, of prescribing the treatment for pregnant women and this will be based on the evidence from the animal model studies. Thus, the comments made in the NCO need to be compatible with the SmPC advice and frequently claim that the nonclinical safety assessment is reflected in the warning of section 4.6 of the SmPC and that some detail justifying the warning may be found in section 5.3.

When considering the Local Tolerance of an active substance, the NCO should indicate whether this is adequately covered by the section on general toxicity—which it will be if the route of administration to the selected toxicity test species is the same as that to be recommended for human medication.

If other routes of dosing were used in toxicity testing, the local effects at the site(s) of contact should be assessed, although the only relevance to human use would be a mistake in method of administration. Specifically, the intravenous route of administration is liable to error in that the dose may leave the needle perivenously. The likely adverse effect of this may have been included in the animal toxicity testing and should be interpreted by the NCO, particularly for European applications where some Member States require an assessment of this possible misdosing in patients.

Relatively recently, and partly in response to the greater frequency of marketing applications for "biologic" medicines, the topic of immunotoxicity has become of interest to regulatory assessors. There are two aspects to the possible interaction between medication and the immune system.

First, the new active substance may be immunogenic in its own right and the NCO should consider the possibility that antibodies will be formed that could react with later doses and neutralize its effect—thereby causing a safety problem through lack of efficacy.

An extension of this is when the active combines with a native substance (usually protein) to form an immunogen. The antibodies to this could then not only neutralize the active/protein complex, but also sequester the native protein with serious adverse consequences to the proper functioning of the body tissues.

Such reactions are difficult to determine in animal models, because their immune systems will very probably differ in response from those of humans. Where human proteins (or modified forms thereof) are the potential new medicine, administration of these to rodent and nonrodent models will almost certainly engender an immune reaction, masking any autoimmune response possibilities.

Second, the administered active may be immunotoxic *sensu stricto*, that is, will impair the immune system's capability to respond to challenges. This phenomenon can be tested in animal models by quantitatively comparing the model's ability to deal with the known effects of an established immunogen.

As with all regulatory documentation the final section recommended for consideration under the heading Toxicology is "Other results." These will be product specific, but the NCO should take the familiar line of assessing the data as evidence for any implications on safety for human use.

For the convenience of the regulatory assessors, the salient points from each of the above sections should be pulled together in a short Integrated Summary with Conclusions drawn from the assessment of these sections.

For Europe (as opposed to the United States and Japan), this Integrated Summary can usefully contain information about the type of application for which the NCO is intended—examples being a novel active substance submitted into the Centralized Procedure via the EMEA or a generic active substance by using the Decentralized Procedure. In the latter case, the Reference Medicinal Product should be identified, and in each case the Articles of the European Directives that cover the situation can be quoted so that the regulatory assessor for nonclinical data is informed without the need to reference other parts of the marketing authorization (or clinical trial) application.

In particular, this Integrated Summary and Conclusions section of the NCO is helpful to those regulatory agencies who are not directly involved in the assessment of the application, but do wish to have a synopsis of the nonclinical part of the dossier.

For an Established Active Substance

Established active substances are usually those that have been used as human medicines for a continuous period to the point in time when patent protection and data protection no longer apply. At this time, the developers of generic products can apply for MAAs with dossiers that consist of limited information on nonclinical issues.

Nevertheless, an NCO is required, if only to cite the product originator's data as relevant and reliable (based on the situation that the original product is still available in the market with no serious objections to its place there).

By the time a new active becomes available to the generic market, the human database will probably be sufficient to support a Marketing Authorization Application (MAA) without the need to refer in the NCO to the nonclinical data, other than to state that it exists and is available to Regulatory Agencies under the terms of the appropriate Directive or Official Journal approval.

It has become a custom of late to supplement this simple statement with a literature search, and assessment of the data revealed by that search, for any evidence that the existing wording of the SmPC should be modified in its safety assessment of the product.

The construction of the NCO for generic products should be the same as for new actives, starting with an Introduction to give relevant background information—particularly if the impurity profile is novel or known to differ significantly from that of the originator's product or if the excipients have changed.

Such changes may be dealt with in a simple statement or may require separate sections in the NCO for fuller assessment of the safety implications.

The pharmacology/pharmacokinetics (PK)/toxicology format can then be used to consider the information revealed by the published scientific literature and the whole should be rounded off with an Integrated Summary and Conclusions, providing a final section that can be incorporated into an agency assessor's report if there is agreement on the issues.

AUTHORSHIP OF THE NCO

For the purposes of presenting to the US FDA and the Japanese MHW an assessment that represents the opinions of the applicant company, there is no author in the sense of "someone who is required to take responsibility for the NCO."

For European MA applications, the situation is different in that the regulatory authorities require a sign-off page (in Module 1) with the signature and *Curriculum Vitae* of the person taking responsibility on behalf of the applicant for the content of the NCO.

Irrespective of whether the author is identified or not, the writing of the NCO may be done by one or many people, although obvious differences in style of composing the various sections should be avoided.

In larger pharmaceutical companies, a house-style is often adopted for all documentation that involves scientific assessment of data. The NCO is often written by people who are intimately knowledgeable about the development of the putative product and the discussions that have occurred within the project team throughout that development.

Notwithstanding the commitment and sagacity of such authors, there is a case for a critical review of the NCO by someone who has no prior exposure to the project, if only because this is the situation of the regulatory authority assessor, who needs to be informed as rapidly and succinctly as possible about the major safety issues arising from the nonclinical data. This reviewer could also be from within the applicant company, but independent consultants can be a good source of unbiased opinion as to the presentation of the NCO and can provide an insurance that all the available data have been considered, not just those results which support the ingrained company view.

For generic companies, which seldom employ specialist pharmacologists or toxicologists, there is a role for independent consultants in the writing of the NCO, because they are usually well placed to scan the scientific literature and have the necessary skill to judge the revealed information for its relevance to the generic product under development.

In every case, it is important to ensure that the signatory has access to all the data, and all the discussions about the data, as he/she may find themselves being asked to answer questions arising from the regulatory agency assessment.

There is thus no room for insufficiently informed famous names in the Module 1 part of the NCO construction process.

Nothing in the guidance prevents an outside expert from writing an NCO, but it would be prudent to ensure that the expert is fully conversant with the safety issues that arose during the development of the medicine not merely presented with the so-called facts from the nonclinical studies.

STYLE OF WRITING FOR THE NCO

The impression taken from reading many NCOs is that the style corresponds to that of contributions to scientific journals using the past tense and third person. One personal anathema is the use of the phrase "rats were dosed with x mg of compound y." It is not the rats that were dosed but the compound—hence the construction "x mg of compound y was dosed to rats" is surely not too much to wish for?

Thus for the sake of those pedantic assessors working in Regulatory Agencies, the NCO should be grammatically correct lest it cause such frustration as to distort the impartiality of the scientific assessment.

For some years before the CTD became accepted in Europe, Japan, and the United States, the use of the word drug was actively discouraged in European documentation, having unfortunate association with illicit substances. However, since the CTD, the American use of the word has prevailed in many situations, notably as "drug substance" in the Pharmaceutical part of a dossier. Common usage seems to demand its acceptance again.

Where appropriate, the use of illustrative Tables and Figures can be a help to the understanding of the reasoning given for specific safety issues.

The layout can be made more user friendly by the judicious use of headings, italics, and different type faces to emphasize points of discussion.

The mechanism for referencing cited literature may be simple with only a number in parenthesis indicating the position in the list at the end of the Overview, or may use first author name and date (e.g., Smith et al., 1999) to link with an alphabetical listing in the section of the Overview containing the referenced publications.

THE NCO IN RELATION TO OTHER SUMMARIES AND OVERVIEWS

There will be need for the NCO to refer to the Chemical and Pharmaceutical data in order to consider the degree of exposure of animal models in toxicity tests to any impurities of significance. The batches of active substance tested in each of the toxicity tests or assays will need to be recorded and the assumed (from 100% bioavailability) systemic exposure for the animal species, and for patients on the recommended maximum dose, calculated for consideration in the NCO.

Thus reference to the Clinical Overview is required to discover what the recommended maximum daily dose is for patients. It is to be hoped that (at least at the highest dose/concentration) the toxicity test systems will have been exposed to significantly more impurity than will occur in correct patient use and it generally falls to the NCO to make this claim or to explain why there is a deviation from the ideal.

The NCO should also refer to information on any new/unusual excipients so that the toxicity testing of these may be assessed for appropriateness of the study designs and the consequences of the results of those studies.

Whatever the outcome of the analyses of the data in the NCO, its conclusions must be consistent with those of the other Summary and Overview documents, and any recommendations for wording of the SmPC should not differ between the NCO and the Clinical Overview. This is particularly relevant to the results from studies on animal reproduction, which are often quoted in Section 5.3 of the SmPC as support for what is recommended in Section 4.6 of the SmPC.

Further reference to the Clinical Overview is also appropriate when considering which (if any) of the nonclinical information has been made redundant by human data that supersedes it, and a statement to this effect is useful, for example, the mechanism of action derived from primary pharmacology studies in rats and dogs has been confirmed in humans and the results of the animal studies are now considered of historic interest, but are second in importance to the human data.

EPILOGUE

There are no formal conclusions to be drawn from the suggestions given in the preceding sections of this chapter. The essence of a Nonclinical Overview is that it should convey the thought processes of the company in presenting the development strategy for the pharmaceutical agent in question and provide an assessment of the data generated to satisfy that strategy.

The result should be a clear statement of the significance of any of the nonclinical information at the time-point in the development, for example, prior to a first clinical trial or at the time of marketing authorization application (MAA).

The Overview should be suited to its purpose in support of the various possible types of MAA and there is, therefore, no one set of contents.

The acceptance of the ICH concept of a Common Technical Document has made the Nonclinical Overview a familiar document to regulatory assessors in the United States and Japan (and by extension to many other countries' authorities), but for Europe it is essentially the former "Expert Report" for Pharmacology/Toxicology, the construction of which was the subject of a pamphlet written in 1993 (8).

The Integrated Summary and Conclusions at the end of an NCO may be the only part of the document of interest to regulatory agencies, which are relying on the assessment by another agency, so it should be a clear but succinct statement of the nonclinical issues and the way in which the applicant sees them as being resolved.

REFERENCES

1. Council Regulation (EEC) No. 2309/93 of 22 July 1993 laying down Community procedures for the authorization and supervision of medicinal products for human and veterinary use and establishing a European Agency for the Evaluation of Medicinal Products *OJ No L 214 of 24. 8. 1993.*
2. Directive 2001/83/EC of the European Parliament and of the Council of 6 November 2001.
3. Guidance for Industry M4S: The CTD—Safety U.S. Department of Health and Human Services Food and Drug Administration Center for Drug Evaluation and Research (CDER) Center for Biologics Evaluation and Research (CBER) August 2001.
4. Japanese Ministry of Health, Labor and Welfare. Gaiyo (Application Summary).
5. D'Arcy PF, Harron DWG. Proceedings of the Third International Conference on Harmonisation (Yokohama 1995). Belfast, U.K.: Queen's University, 1996.
6. Olejniczak K, Günzel P, Bass R. Preclinical testing strategies. Drug Inf J 2001; 35:321–336.
7. EudraLex—Volume 3. Scientific guidelines for medicinal products for human use, Commission of the European Communities, http://www.emea.europa.eu/htms/human/humanguidelines/nonclinical.htm. Accessed September 1, 2008.
8. Matthews BR, Lee RM, Eisen SM. How to Write an Expert Report, 2nd ed. London, U.K.: Rostrum (Phoenix International), 1998. ISBN 1 873292 13 9.

ANNEX

Nonclinical Guidelines

Title	C	D	A	O	Reference number	Publication date	Effective date	Other remarks
Pharmacology								
The nonclinical Evaluation of the potential for delayed Ventricular Repolarization (QT Interval Prolongation) by Human Pharmaceuticals (ICH S7B)			◆		CPMP/ICH/423/02	May 2005	Nov 2005	
Safety pharmacology studies for human pharmaceuticals (ICH S7A)			◆		CPMP/ICH/539/00	Nov 2000		Under revision
Non-Clinical Safety Studies For The Conduct Of Human Clinical Trials For Pharmaceuticals (ICH M3[M]) modification			◆		CPMP/ICH/286/95	Nov 2000	Nov 2000	
Points to Consider on the assessment of the potential for QT interval prolongation by noncardiovascular medicinal products					CPMP/SWP/986/96	Dec 1997		Replaced by ICH S7A and taken off EMEA website on 18/05/06
Pharmacokinetics								
Toxicokinetics: the assessment of systemic exposure in toxicity studies (ICH S3A)			◆		CPMP/ICH/384/95	Nov 1994	Jun 1995	
Pharmacokinetics: Guidance for repeated dose tissue distribution studies (ICH S3B)			◆		CPMP/ICH/385/95	Nov 1994	Jun 1995	
Pharmacokinetics and metabolic studies in the safety evaluation of new medicinal products in animals			◆		3BS11A		Apr 1994	
Toxicology								
Single-dose toxicity								
Single-dose toxicity			◆		3BS1A	Apr 2001		
Repeat-dose toxicity								
Repeated-dose toxicity			◆		CHMP/SWP/488313/07	Release for consultation Feb 2008		Deadline for comments May 2008

(*Continued*)

ANNEX (*Continued*)

Nonclinical Guidelines

Title	C	D	A	O	Reference number	Publication date	Effective date	Other remarks
Repeated-dose toxicity			♦		CPMP/SWP/1042/99	Jul 2000	Oct 2000	
Duration of chronic toxicity testing in animals (rodent and nonrodent toxicity testing) (ICHS4A)			♦		CPMP/ICH/300/95	Nov 1998	May 1999	
Genotoxicity								
Guidance on Genotoxicity Testing and Data Interpretation for Pharmaceuticals intended for Human use (ICH S 2 (R1))		♦			CHMP/ICH/126642/08	Release for consultation Mar 2008		Deadline for comments May 2008
Limits of genotoxic impurities			♦		CPMP/SWP/5199/02 CHMP/QWP/251344/2006	Jun 2006	Jan 2007	Questions and Answers on the guideline
Reflection Paper on the assessment of the Genotoxic Potential of Antisense Oligodeoxynucleotides			♦		CHMP/SWP/199726/04	Jan 2005		
Position Paper on the genotoxic and carcinogenic potential of phenolphthalein			♦		CPMP/818/97	Dec 1997		
Genotoxicity: a standard battery for genotoxicity testing of pharmaceuticals (ICHS2B)			♦		CPMP/ICH/174/95	Sept 1997	Mar 1998	
Specific aspects of regulatory genotoxicity tests for pharmaceuticals (ICHS2A)			♦		CPMP/ICH/141/95	Sept 1995	Apr 1996	
Carcinogenicity								
Carcinogenicity Evaluation of Medicinal Products for the Treatment of HIV Infection		♦	♦		EMEA/194898/2006	Jan 2008	Jul 2008	
CHMP SWP Conclusions and recommendations on the use of genetically modified animal models for carcinogenicity assessment			♦		CPMP/SWP/2592/02 Rev 1	Jun 2004		
Carcinogenic potential			♦		CPMP/SWP/2877 /00	Jul 2002	Jan 2003	
Points to consider on the Nonclinical assessment of the carcinogenic potential of human insulin analogues			♦		CPMP/SWP/372/01	Nov 2001		

Carcinogenicity: testing for carcinogenicity of pharmaceuticals (ICH SIB)			♦	CPMP/ICH/299/95	Sep 1997	Mar 1998	
Need for carcinogenicity studies of pharmaceuticals (ICH SIA)			♦	CPMP/ICH/140/95	Dec 1995	Jul 1996	
Dose selection for carcinogenicity studies of pharmaceuticals (ICH S1C (R2))			♦	CPMP/ICH/383/95	Apr 2008	Oct 2008	
Reproductive and developmental toxicity							
Risk Assessment of Medicinal Products on Human Reproduction and Lactation: From Data to Labeling		♦		EMEA/CHMP/203927/05	Release for consultation Mar 2006		Deadline for comments Sep 2006
Need for Nonclinical Testing in Juvenile Animals on Human Pharmaceuticals for Pediatric Indications		♦	♦	CHMP/SWP/169215/05	Mar 2008	Aug 2008	
Points to consider on the Need for assessment of reproduction toxicity of human insulin analogues			♦	CPMP/SWP/2600/01	Mar 2002	Mar 2002	
Reproductive toxicology: Detection of toxicity to reproduction for medicinal products including toxicity to male fertility (ICH S5A)			♦	CPMP/ICH/386/95 (ICH S5A)	Sep 1993	Mar 1994	
Local Tolerance							
Nonclinical local tolerance testing of medicinal products			♦	CPMP/SWP/2145/00	Feb 2001	Feb 2001	
Other toxicity							
Nonclinical guideline on drug-induced hepatotoxicity		♦		CHMP/SWP/150115/06	Release for consultation Jan 2008		Deadline for comments Aug 2008
Nonclinical Investigation of the Dependence Potential of Medicinal Products			♦	CHMP/SWP/94227/04	Mar 2006	Sep 2006	
Immunotoxicity studies for Human Pharmaceuticals (ICH S8)			♦	CHMP/ICH/167235/04	Oct 2005	May 2006	
Need for revision of the Note for Guidance on photosafety testing (CPMP/SWP/398/01)	♦			CHMP/SWP/534549/07	Release for consultation Jan 2008		Deadline for comments May 2008
Photosafety testing			♦	CPMP/SWP/398/01	Jun 2002	Dec 2002	

(*Continued*)

ANNEX (*Continued*)

Nonclinical Guidelines

Title	C	D	A	O	Reference number	Publication date	Effective date	Other remarks
Background to the CPMP Position Paper on possible preclinical studies to investigate addiction and dependence/withdrawal related to the use of selective serotonin uptake inhibitors (SSRIs)			♦		CPMP/2278/00	Dec 2000		
Background to the CPMP Position Paper on selective serotonin uptake inhibitors (SSRIs) and dependency/withdrawal reactions			♦		EMEA/CPMP/2775/99	Apr 2000		
Replacement of animal studies by in vitro models			♦		CPMP/SWP/728/95	Feb 1997		
In Vitro investigation of mitochondrial toxicity of anti-HIV nucleoside reverse transcriptase inhibitors	♦	♦			CHMP/SWP/8212/07	Release for consultation Dec 2007		Deadline for comments Jul 2008
General guidelines								
Strategies to identify and mitigate risks for first-in-human clinical trials with investigational medicinal products		♦	♦	♦	CHMP/SWP/28367/07	Jul 2007	Sep 2007	
Non-clinical studies required before first clinical use of gene therapy medicinal products		♦			CHMP/GTWP/125459/2006	Release for consultation Mar 2007		Deadline for comments Nov 2007
Scientific Requirements for the Environmental Risk Assessment of Gene Therapy Medicinal Products	♦	♦			CHMP/GTWP/125491/06	Release for consultation Feb 2007		Deadline for comments Aug 2007
Specification Limits for Residues of Metal Catalysts		♦ ♦	♦		CPMP/SWP/QWP/4446/00	Feb 2008	Sep 2008	
Environmental Risk Assessment of Medicinal Products for Human Use			♦		CPMP/SWP/4447/00	Jun 2006	Dec 2006	
Development of a CHMP Guideline on the Non-Clinical Requirements to Support Early Phase I Clinical Trials with Pharmaceutical Compounds	♦				EMEA/CHMP/SWP/91850/06	Release for consultation Mar 2006		Deadline for comments Jun 2006

Annex Guideline on Similar Biological Medicinal Products containing Biotechnology-Derived Proteins as Active Substance: Nonlinical and Clinical Issues – Guidance on Similar Medicinal Products containing Recombinant Erythropoietins			♦	EMEA/CHMP/94526/05	Mar 2006	July 2006	
Quality, Preclinical, and Clinical aspects of Gene Transfer Medicinal Products—Annex on Nonclinical testing for Inadvertent Germline transmission of Gene Transfer Vectors		♦		EMEA/273974/05	Release for consultation Nov 2005		Deadline for comments May 2006
Nonclinical Documentation for Mixed Marketing Authorization Applications			♦	CPMP/SWP/799/95	Oct 2005	Apr 2006	
Nonclinical Development of Fixed Combinations of Medicinal Products		♦	♦	CHMP/SWP/258498/05	Mar 2008	Aug 2008	
Evaluation of Control Samples for Nonclinical Safety Studies: Checking for Contamination with the Test Substance			♦	CPMP/SWP/1094/04	Mar 2005	Sep 2005	
Environmental risk assessment for human medicinal products containing or consisting of GMOs			♦	3BR1A		Jan 1995	(under revision)
Development of a CHMP Guideline on nonclinical testing for inadvertent germline transmission of Gene Transfer Vectors	♦			CPMP/SWP/110180/04	Nov 2004		
Position Paper on the nonclinical safety studies to support clinical trials with a single micro dose			♦	CPMP/SWP/2599/02	Jun 2004	Jun 2004	
Dossier Structure and Content for Pandemic Influenza Vaccine Marketing Authorization Application			♦	CPMP/VEG/4717/03	Mar 2004	Apr 2004	
EMEA Public Statement on Thiomersal in Vaccines for Human Use—Recent Evidence Supports Safety of Thiomersal-containing Vaccines			♦	CPMP/VEG/1194/04 v02[a]	Mar 2004		
Comparability of medicinal products containing biotechnology-derived proteins as active substance—Annex on nonclinical and clinical issues			♦	CPMP /3097/02[a]	Dec 2003	Jun 2004	

(Continued)

ANNEX (*Continued*)

Nonclinical Guidelines

Title	C	D	A	O	Reference number	Publication date	Effective date	Other remarks
Impurities in new drug products (Revision)			◆		CPMP/ICH/2738/99	Feb 2003	Aug 2003	
Maintenance of Note for guidance on impurities: Residual solvents			◆		CPMP/ICH/1940/00 corr	Nov 2002	Mar 2003	
Impurities testing: impurities in new drug substances (Revision)			◆		CPMP/ICH/2737/99	Feb 2002	Mar 2002	
Limitations to the use of ethylene oxide in the manufacture of medicinal products			◆		CPMP/QWP/159/01	Mar 2001	Apr 2001	
Pre-clinical evaluation of anticancer medicinal products			◆		CPMP/SWP/997/96	Jul 1998	Jan 1999	
Preclinical pharmacological and toxicological testing of vaccines			◆		CPMP/SWP/465/95	Dec 1997	Jun 1998	
Preclinical safety evaluation of biotechnology-derived pharmaceuticals (ICH S6)			◆		CPMP/ICH/302/95	Sep 1997	Mar 1998	
Investigation of chiral active substances			◆		3CC29A	Jun 1997	Jan 1998	
Impurities: Residual solvents			◆		CPMP/ICH/283/95	Sep 1995	Mar 1996	
Herbal medicinal products								
Assessment of genotoxicity of herbal substances/preparations	◆	◆	◆	◆	EMEA/HMPC/107079/07	Jun 2008	Dec 2008	
Nonclinical Documentation for Herbal Medicinal Products in Applications for Marketing Authorization (Bibliographical and Mixed Applications) and in Applications for Simplified Registration			◆	◆	EMEA/HMPC/32116/05	Sep 2006	Jul 2006	

[a]new guidelines.
3 July 2008.
Source: From Ref. 7.
Abbreviations: C, concept paper; D, draft guideline; A, adopted guideline; O, overview of comments.

7 The Clinical Overview and Summary

Chantal Le Floch
Science Union, Courbevoie, France

Patricia Maillère
I.R.I.S., Courbevoie, France

In the framework of the International Conference on Harmonization of Technical Requirements for Registration of Pharmaceuticals for Human Use (ICH), the involved parties [European Union (EU), Japan, and the United States (U.S.A.)] have established common reporting standards and template formats. The compilation of these reports forms the basis of the Common Technical Document (CTD) (1–3) (Fig. 1), which has become an internationally agreed format for the preparation of a well-structured presentation for applications to be submitted to regulatory authorities in the European Union, Japan, and the United States. As a result, the applicants have to provide a variety of documents that comply with the ICH guidelines to ensure a rapid and efficient review of their submission. The ICH guidelines provide the framework for the modules that are required as part of the submission of the CTD format marketing authorisation (MA) application.

The CTD contains five standard modules. Module 2 of the CTD is the repository for the Clinical Overview and Summary (Fig. 1). In this module, the Clinical Overview is located in Section 2.5, the Clinical Summary of Efficacy in Section 2.7.3 and the Summary of Clinical Safety in Section 2.7.4 (Table 1, p. 109) (4).

The Clinical Overview should provide a **critical analysis** of relevant clinical efficacy and safety information presented in the different parts of the CTD (Clinical Summaries, individual study reports and other relevant reports) together with any other relevant information (e.g., animal data that may have clinical implications) whereas the Clinical Summary of Efficacy and the Clinical Summary of Safety provide a **detailed factual summarization** of the clinical information presented in the individual clinical study reports, and other relevant reports.

The CTD defines some very specific criteria to be followed regarding numbering and the use of heading for Clinical Overview and Clinical Summary of Efficacy and Clinical Summary of Safety (Tables 2–4). It also states clearly into which sections or subsections the applicant is required to enter the relevant information (Tables 2–4). The CTD has a back-to-front approach to the presentation of the clinical data. The most detailed information on the reviews of individual study results is found primarily in Module 5 of the CTD while summary information is included earlier in the Clinical Summaries and in the Clinical Overview (Fig. 1).

The Clinical Overview and Summaries are intended primarily for use by regulatory agencies in the review of the clinical section of an MA application. In preparing an MA, the applicant should work with the thought that a well-written objective summary of the clinical findings in the CTD Clinical Summary of Efficacy (2.7.3) and in the Clinical Summary of Safety (2.7.4) and a credible critical analysis of the clinical data in the CTD Clinical Overview (2.5) can potentially become the foundation of the public assessment report.

FORMAT AND CONTENT OF CLINICAL OVERVIEW AND SUMMARY

Clinical Overview

The Clinical Overview is a key document of the CTD that supports the product information. It should provide a critical analysis of the clinical efficacy and safety data. This analysis must be accurate and clear, stating only those conclusions that can be supported by the data, and

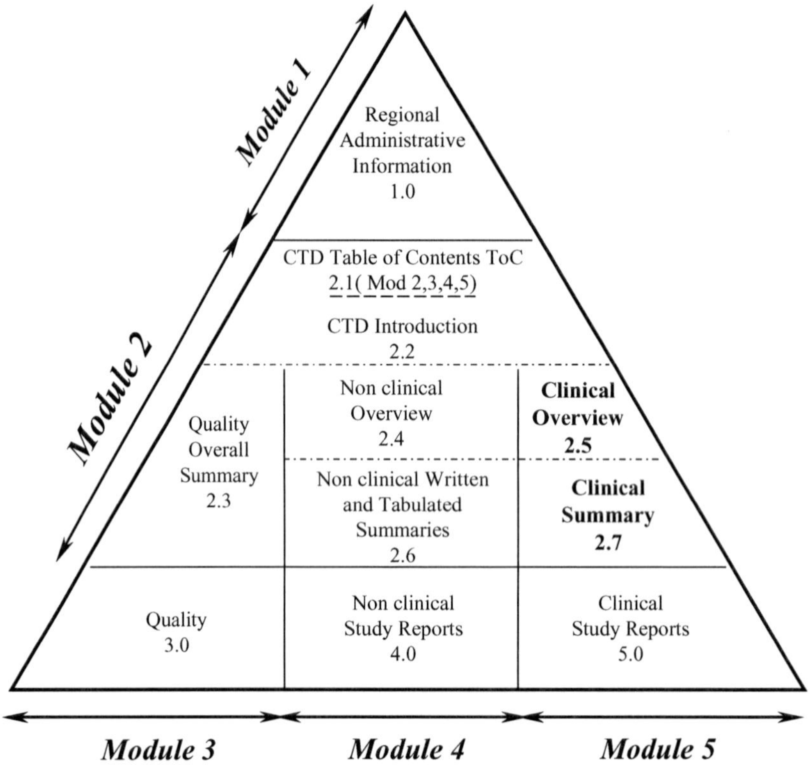

Figure 1 Schematic representation of the five modules in the common technical document (CTD). Drawn from international conference on harmonization of technical requirements for registration of pharmaceuticals for human use (ICH) M4E guideline on CTD.

should be without bias and free of promotional material. It should (according to the CTD guideline) (2)

- present the strengths and limitations of the development program and study results,
- analyze the benefits and risks of the medicinal product in its intended use, and
- describe how the study results support the prescribing information.

For this purpose, the Clinical Overview should (2)

- describe and explain the clinical development overall approach including critical study design decisions;
- assess the quality of the design and performance of the studies;
- include a statement regarding GCP compliance;
- provide a brief overview of the clinical findings, including important limitations (e.g., lack of comparisons with relevant active comparator, absence of information on subpopulations, or on use in combination therapy);
- explore unresolved issues, explain why they should not be considered as barriers to approval, and describe plans to resolve them;
- provide an evaluation of benefits and risks based on the conclusions of the relevant studies; and
- explain how the clinical data supports the information included in the labeling.

The Clinical Overview should generally be a short document (about 30 pages). The use of concise tables, extensive cross-references, and hypertext linking to the related pieces of information throughout the CTD (e.g., Clinical Summary or individual study reports) is encouraged for brevity and to assist the reviewer.

Table 1 CTD Module 2—Table of Contents

2 COMMON TECHNICAL DOCUMENT SUMMARIES
2.1 CTD TABLE OF CONTENTS
2.2 CTD INTRODUCTION
2.3 QUALITY OVERALL SUMMARY
2.4 NONCLINICAL OVERVIEW
2.5 CLINICAL OVERVIEW
2.5.1 Product Development Rationale
2.5.2 Overview of Biopharmaceutics
2.5.3 Overview of Clinical Pharmacology
2.5.4 Overview of Efficacy
2.5.5 Overview of Safety
2.5.6 Benefits and Risks Conclusions
2.5.7 Literature References
2.6 NONCLINICAL WRITTEN AND TABULATED SUMMARY
2.7 CLINICAL SUMMARY
2.7.1 Summary of Biopharmaceutic Studies and Associated Analytical Methods
2.7.2 Summary of Clinical Pharmacology Studies
2.7.3 Summary of Clinical Efficacy
2.7.4 Summary of Clinical Safety
2.7.5 Literature References
2.7.6 Synopses of Individual Studies

The table of contents of the Clinical Overview is as follows (2):

2.5.1 Product Development Rationale
2.5.2 Overview of Biopharmaceutics
2.5.3 Overview of Clinical Pharmacology
2.5.4 Overview of Efficacy
2.5.5 Overview of Safety
2.5.6 Benefits and Risks Conclusions
2.5.7 Literature References

Section 2.5.1 is a discussion of the rationale for the development of the product. It should provide the reviewer with information such as the pharmacological class of the product, targeted indication, scientific background supporting the investigations performed, description of the clinical development program (including a table of all relevant clinical studies), concordance or lack of concordance with current standard research approaches (e.g., Good Clinical Practice, the Declaration of Helsinki, local regulations in each participating country), and reference to regulatory guidance and official advice documents. In case of regulatory advice (e.g., Scientific Advice, end of Phase II meeting), the applicant should discuss how this advice was implemented. Extensive hypertext linking and cross-referencing to regulatory advice documents in Module 5 will facilitate the review process.

Section 2.5.2 is an overview of the biopharmaceutics. It should provide the reviewer with information such as analytical methods, pharmacokinetic analysis of experimental data, dosage form(s) and in vitro dissolution profiles, and data on absorption and bioavailability. Relevant bioavailability issues that may have potential impact on the formulation to be marketed (e.g., influence of food on exposure) should be addressed.

Section 2.5.3 is an overview of the clinical pharmacology. It should provide a critical analysis of the pharmacokinetic parameters (absorption, distribution, metabolism, and elimination) according to intrinsic (e.g., renal or hepatic impairment) and extrinsic factors (e.g., concomitant drugs or food), pharmacodynamic, and related in vitro data. The analysis should explain why and how the data support the conclusions drawn and included in the proposed labeling.

The efficacy Section 2.5.4 should provide a critical analysis of the efficacy data that considers all relevant data, **whether positive or negative**. It should also explain why and how the data support the proposed indication and prescribing information. Finally, it should demonstrate

Table 2 Outline's Contents of the Clinical Overview Sections (2)

2.5.1 Product Development Rationale

- Identify the pharmacological class of the active ingredient(s)
- Describe the intended indication
- Summarize the scientific background that supports the investigations of the medicinal product
- Briefly describe the clinical development program including ongoing and planned clinical studies (and include table of all relevant clinical studies)
- Explain concordance or lack of concordance with current standard research approaches (e.g., Good Clinical Practice, the Declaration of Helsinki, local regulations in each participating country) with reference to pertinent published literature
- Include a reference to regulatory guidance and official advice documents

2.5.2 Overview of Biopharmaceutics

- Describe analytical methods
- Describe pharmacokinetic (PK) analyses of experimental data (including population PK approach)
- Describe dosage form(s), in vitro dissolution profiles and in vitro–in vivo correlation (if applicable)
- Present in vitro and in vivo data on absorption and bioavailability data
- Present any relevant bioavailability issues that may have clinical implications of the to-be-marketed formulation(s) (e.g., influence of food on exposure)

2.5.3 Overview of Clinical Pharmacology

- Address PK in different populations (healthy subjects, special populations); intrinsic factors (e.g., age, sex, race, renal and hepatic impairment), extrinsic factors (e.g., smoking, concomitant drugs); absorption; distribution, metabolites; excretion; time-dependent changes; stereochemistry issues; PK interactions with other products or other substances
- Address pharmacodynamic (PD), e.g., mechanism of action, PD profile, PK–PD relationships; clinical interactions, and possible genetic differences in response
- Address immunogenicity studies, clinical microbiology studies, or other drug classes–specific PD studies

2.5.4 Overview of Efficacy

- Present critical analysis of all relevant efficacy data, whether positive or negative
- Identify the studies considered relevant for the efficacy evaluation
- Identify the impact of prematurely terminated studies
- Address the following issues:
  - Differences between the studied population(s) and those expected to receive the product.
  - The choice of endpoints (use of surrogates to be validated, validation of scales to be discussed, and clinical relevance to be discussed)
  - Statistical methods and any issues that could affect the interpretation of the study results
  - Justification of the noninferiority margin (ICH E10)(5), demonstration of the assay sensitivity
  - Similarities and differences in results among studies or in different patient subgroups within studies
  - Observed relationships between efficacy and dose for each indication, in both the overall population and in the different patient subgroups (ICH E4)(6)
  - Efficacy findings pertinent to the maintenance of long-term efficacy
  - Efficacy in special populations

2.5.5 Overview of Safety

- Consider relevant animal toxicology and product quality information
- Discuss limitations of the safety database and the potential implications on the safety of the product in the marketplace
- Address common and nonserious adverse events focusing on events of higher frequency compared to placebo/active control
- Discuss serious adverse events with absolute numbers and frequency, including deaths
- Note similarities and differences in results among studies and their potential effects on the interpretation of the safety data
- Discuss relation of adverse events to dose, dose regimen, and treatment duration
- Discuss long-term safety (E1a)(7)
- Identify methods to prevent, mitigate, or manage adverse events
- Discuss safety in special subgroups (e.g., the elderly, those with liver insufficiency, those with renal insufficiency, the pediatric population, in pregnancy and lactation) and in special situations (drug abuse, drug interaction)

(*Continued*)

Table 2 Outline's Contents of the Clinical Overview Sections (2) (*Continued*)

- Discuss reactions due to overdose; the potential for dependence, rebound phenomena and abuse or lack of data on these issues
- Summarize all relevant postmarketing experience available to the applicant (published or unpublished) with any new or different safety issues identified and any regulatory actions related to safety

2.5.6 Benefits and Risks Conclusions
- Integrate all of the conclusions of the previous sections
- Provide an overall benefits and risks assessment relating to the use of the medicinal product in clinical practice
- Discuss the implications of any deviations from regulatory advice or guidelines and any important limitations of the available data
- Clarify the expected place of the product in the armamentarium of treatment
- Identify the efficacy of the product for each proposed indication
- Present significant safety findings and any measures that may enhance safety
- Explain dose–response and dose–toxicity relationships and optimal administration regimens
- Discuss efficacy and safety in subpopulations
- Present data in children of different age groups, if applicable, and any plans for a development program in children
- Give any risks to the patient of known and potential interactions
- Discuss any potential effect of the product that might affect ability to drive or operate heavy machinery

2.5.7 Literature References
- Include list of references

that the efficacy data are robust and that the magnitude of the effect is clinically relevant (with direct comparison or comparison with historical data).

In the safety Section 2.5.5, beyond the critical analysis of the safety data, the applicant should assess which adverse event can be related possibly with the use of the product and therefore be considered an adverse reaction. The applicant should describe the criteria used to identify adverse reactions and justify how the table of adverse reactions included in the labeling was generated. This assessment should result from an examination of the pooled data to reveal consistent differences from control across studies, and evidence dose response.

The data on the safety in special groups (e.g., the elderly, those with liver insufficiency, those with renal insufficiency, the pediatric population and use in pregnancy and lactation) and in special situations (overdose, drug abuse) map directly to the label and will be the source for the Summary of Product Characteristics.

In the benefit–risk conclusions 2.5.6, the applicant should provide arguments to answer the question *"why should the product be marketed with this product information?"* with respect to the data included in the dossier, the other products already marketed, and the balance of risks to benefit. The benefit–risk evaluation should show how the efficacy and safety findings support the proposed dose and target indication. It should also assess how prescribing information and other approaches (risk minimization measures) will optimize benefits and manage risks.

An outline of the Clinical Overview sections' content is provided in Table 2 (p. 110).

Clinical Summaries

The Clinical Summary of Efficacy and the Clinical Summary of Safety should provide a global factual view of the development program. They should include a factual narrative and tabulated study results to support conclusions and critical issues identified in the Clinical Overview (2). The results should be presented

– on individual study basis and
– across studies
  - comparison and analysis of results of individual studies
  - data combined from various studies
  - cross-study analysis
  - pooled analysis

Table 3 Outline's Contents of the Summary of Clinical Efficacy Sections (2)

2.7.3.1 Background and Overview of Clinical Efficacy
- Include a brief overview of the design of the controlled studies that were conducted to evaluate efficacy (dose–response studies, comparative studies, long-term studies, and studies in specific populations)
- Discuss critical features of study design, e.g., randomization, blinding, choices of control treatment, choice of patient population, unusual design features such as crossover or randomized withdrawal designs, use of run-in periods, other methods of "enrichment," study endpoints, study duration, and prespecified plans for analysis of the study results

2.7.3.2 Summary of Results of Individual Studies
- Include tabular listing of all relevant efficacy studies
- Narrative descriptions (similar to an abstract for a journal article) for important studies including description of critical design features and critical results
- Narratives of bridging studies [i.e., certain studies intended to evaluate the ability to extrapolate certain types of foreign clinical data (including PK and PD data) to the new region (see ICH E5)] (8).

2.7.3.3 Comparison and Analyses of Results Across Studies
- Summarize all available data that characterize the efficacy of the drug (use comparison of results of individual studies, and analysis of data combined from various studies)
- Cross-reference to important evidence from Section 2.7.2 "Summary of Clinical Pharmacology studies" (data that support the dosage and administration section of the labeling)

2.7.3.3.1 Study Populations
- Describe demographic and other baseline characteristics of patients across studies
- Describe the characteristics of the disease
- Include any prior treatment in the study subjects
- List inclusion/exclusion criteria
- Summarize differences in baseline characteristics of the study populations in different studies or groups of studies
- Identify the population expected to receive the drug when it is marketed
- Describe the number of patients who dropped out of the studies and reasons for their discontinuation

2.7.3.3.2 Comparison of Efficacy Results of All Studies
- Summarize results of any bridging studies
- Summarize and compare the results of all efficacy studies (positive or negative)
- Compare results across studies with a focus on prespecified primary endpoints and change over time if appropriate
- Discuss important differences in outcome between studies
- Clarify if meta-analysis of the clinical studies is a predefined protocol or is a post hoc exercise

2.7.3.3.3 Comparison of Results in Subpopulations
- Compare efficacy in all relevant subpopulations with analyses across multiple studies to evaluate effects of major demographic factors (age, sex, and race) and of other predefined or relevant intrinsic and extrinsic factors (e.g., disease severity, concomitant drugs, alcohol, tobacco, and body weight) on efficacy
- Analyze efficacy in the pediatric population (if applicable)

2.7.3.4 Analysis of Clinical Information Relevant to Dosing Recommendations
- Summarize and analyze all data (reference can be made to nonclinical studies) that pertain to the dose–response or blood level–response relationships of effectiveness (including dose–blood level relationships) supporting dose selection and choice of dose interval
- Describe any identified deviations from relatively simple dose–response or blood–level response relationships due to nonlinearity of pharmacokinetics, delayed effects, tolerance, enzyme induction, etc.
- Discuss any evidence of differences in dose–response relationships that result from a patient's age, sex, race, disease, or other factors

2.7.3.5 Persistence of Efficacy and/or Tolerance Effects
- Summarize available information on persistence of efficacy over time (e.g., number of patients, length of exposure)
- Note any evidence of tolerance (loss of therapeutic effects over time)
- Examine any apparent relationships between dose changes over time and long-term efficacy
- Consider the effect of premature discontinuation of therapy or switching to other therapies for long-term efficacy trials

2.7.3.6 Appendix

Table 4 Outline's Contents of the Summary of Clinical Safety (11)

2.7.4.1 Exposure to the Drug

2.7.4.1.1 Overall Safety Evaluation Plan and Narratives of Safety Studies

- Include an introduction
- Describe the safety evaluation plan
- Describe the studies to be included
- Cross-reference the table of studies in the appendix
- Identify primary, secondary, or supporting studies
- Describe how the studies were grouped and why

2.7.4.1.2 Overall Extent of Exposure

- Include all subjects who received at least one dose of study medication
- Describe cumulative information
- Describe drug concentration data

2.7.4.1.3 Demographic and Other Characteristics of Study Population

- Summarize the demographic characteristics of the population exposed to the product
- Provide separate tables if the relative exposure of demographic groups in the controlled trials differed from overall exposure

2.7.4.2 Adverse Events

2.7.4.2.1 Analysis of Adverse Events

2.7.4.2.1.1 Common Adverse Events

- Present frequency of adverse events (AEs) in text and tables
- Place longer tables (more than one page) in the appendix
- Discuss differences between dose groups, indications, or previously submitted data, as appropriate
- Be consistent with presentation in individual study reports without reiterating data exactly

2.7.4.2.1.2 Deaths

- List all deaths that occurred during the studies in an appendix table (or in text if small numbers)
- Include deaths that occurred shortly after treatment termination (e.g., within 30 days or as specified in the study protocol) as well as other deaths that occurred later but may have resulted from a process that began during the studies

2.7.4.2.1.3 Other Serious Adverse Events

- Define serious adverse event (SAEs)
- Present number of subjects with SAEs. Include SAEs that occurred after drug discontinuation
- Present analyses or assessments of SAEs across studies
- Examine SAEs for frequency over time, particularly for drugs that may be used chronically

2.7.4.2.1.4 Other Significant Adverse Events

- Present AEs that resulted in the discontinuation or interruption of study medication
- Discuss reasons for premature treatment discontinuations
- Compare rates of discontinuations across studies and also rates of discontinuations for placebo and/or active treatment

2.7.4.2.1.5 Analysis of Adverse Events by Organ System or Syndrome

- Assess the causality and risk factors for events such as deaths, SAEs, and other significant events
- Assess the risk associated with class-specific AEs

2.7.4.2.2 Narratives

- Create links to the narratives in each of the individual study reports

2.7.4.3 Clinical Laboratory Evaluations

- Provide a brief overview of the major changes in laboratory values across clinical studies
- Describe changes in patterns of laboratory tests with study drug use
- Provide normal laboratory ranges for each analysis
- Present laboratory data for hematology, clinical chemistry, urinalysis, and other data

2.7.4.4 Vital Signs, Physical Findings, and Other Observations Related to Safety

- Present cross-study observations and comparisons of vital signs, weight, and other data related to safety
- Present any dose–response or drug concentration–response relationship to individual variables (e.g., disease demographics, concomitant therapy)
- Give particular attention to changes considered to be AEs

(*Continued*)

Table 4 Outline's Contents of the Summary of Clinical Safety (11) (*Continued*)

- Give particular attention to studies that were designed to evaluate specific safety issues (e.g., studies of QT interval prolongation)
- Present antibody and immunogenicity data (if applicable)
- Present relationship between safety and PK

2.7.4.5 Safety in Special Groups and Situations

2.7.4.5.1 Intrinsic Factors

- Summarize analyses of demographics or other intrinsic factors (e.g., age, sex, race, renal and hepatic impairment)

2.7.4.5.2 Extrinsic Factors

- Summarize analyses based on extrinsic factors (e.g., smoking, concomitant drugs)

2.7.4.5.3 Drug Interactions

- Summarize the impact of potential drug–drug or drug–food interactions on safety

2.7.4.5.4 Use in Pregnancy and Lactation

- Summarize all known safety information on use during pregnancy or breast-feeding

2.7.4.5.5 Overdose

- Summarize available clinical information relevant to overdose

2.7.4.5.6 Drug Abuse

- Summarize relevant studies and information regarding the dependence potential in humans

2.7.4.5.7 Withdrawal and Rebound

- Summarize analysis of withdrawal and rebound effects

2.7.4.5.8 Effects on Ability to Drive or Operate Machinery or Impairment of Mental Ability

- Summarize safety data related to impairment in any factor that would diminish the ability to drive a vehicle or operate machinery or that would impair mental ability

2.7.4.6 Postmarketing Data

- Summarize all relevant postmarketing data available to the Applicant (published and unpublished)
- Include PSUR(s) in Module 5, if applicable
- Describe postmarketing findings in subgroups if any

2.7.4.7 Appendix

The size of Clinical Summaries will vary according to the extent of product development and the regulatory needs (generally 50–400 pages without tables).

Clinical Summary of Efficacy

This document should include an accurate analysis and factual comparison of the adequate and well-controlled studies that the applicant believes demonstrate the effectiveness of the new drug in the proposed indication(s). The applicant should also either demonstrate the consistency of the effectiveness findings across studies or explain any inconsistencies.

The table of contents of the Clinical Summary of Efficacy is as follows (2):

2.7.3.1 Background and Overview of Clinical Efficacy
2.7.3.2 Summary of Results of Individual Studies
2.7.3.3 Comparison and Analyses of Results Across Studies
2.7.3.4 Analysis of Clinical Information Relevant to Dosing Recommendations
2.7.3.5 Persistence of Efficacy and/or Tolerance Effects
2.7.3.6 Appendix

Section 2.7.3.1 is an overview of the clinical efficacy data. It should provide a brief overview of the design of the efficacy-controlled studies. It can refer to nonclinical data and clinical pharmacological data. Critical study design features (e.g., randomization, statistical approaches, study duration) of the controlled studies supporting effectiveness should be discussed and compared. Particular attention should be paid to any limitations of the efficacy studies (e.g., lack of placebo-controlled studies). If an effectiveness claim is based on a surrogate endpoint not used previously by the Regulatory Authorities as a basis for approval the basis for choice of the endpoint should be discussed and its validity as a predictive of clinical outcome should be supported.

Section 2.7.3.2 is a summary of results of individual studies. It should include a tabular listing of data relevant to drug efficacy together with a short study narrative description.

Section 2.7.3.3 is a comparison and an analysis of results across studies. These comparisons and analyses should use appropriate methods for comparing studies of similar design, controlling for design differences, weighting by sample size, and examining by common covariates or stratifications. Comparisons of results across studies should focus on a prespecified endpoint. Confidence intervals for treatment effects should be given to facilitate the interpretation of point estimates. If differences are shown between placebo and test drugs in the change from baseline, the baseline values and the magnitude of effect in all treatment groups should generally be presented in the table or in the text accompanying a figure. If the objective of an active control trial was to show equivalence or noninferiority, the difference in the ratio of outcomes between treatments should be given with the confidence interval. The results should be evaluated by using the predefined criteria for defining equivalence or noninferiority and the rationale for the criteria and support for the determination that the studies have assay sensitivity should be provided (ICH E10) (5). This section should generally use two kinds of analyses: comparison of results of individual studies and analysis of data combined from various studies. If meta-analysis (Integrated Summary of Efficacy) is performed generally, a detailed description of the methodology and of the results should be submitted in a separate report in Module 5. It should be clearly stated in the Integrated Summary of Efficacy if the analysis was performed according to predefined protocol or if it was a post hoc analysis. Finally, the results of individual studies or overview of efficacy in specific populations (e.g., elderly, pediatric) should be summarized in this section. Well-constructed tables and graphs that summarize the key effectiveness findings and highlight their consistency are very useful. Extensive hypertext linking and cross-referencing to listings, data sets, and other supportive data in Module 5, including separate report of the Integrated Summary of Efficacy (if applicable), will facilitate the review process.

Section 2.7.3.4 should provide an integrated summary and analysis of all data (e.g., data from individual dose–response studies) that pertain to the dose–response relationship of effectiveness. Any differences in dose–response relationships that result from intrinsic factors (e.g., age, race) or other factors should be described.

Section 2.7.3.5 should provide an assessment of treatment response over time (i.e., persistence and tolerance or loss of therapeutic effects over time). Therapeutic effects of a treatment can decline over time because of tolerability issues or from the development of drug resistance or tolerance.

An outline of the content of the Clinical Summary of efficacy sections is provided in Table 3 (p. 112).

Clinical Summary of Safety

This document should include a summary of data relevant to safety integrating the results of individual clinical study reports as well as the integrated analyses of safety.

The table of contents is as follows (2):

2.7.4.1 Exposure to the Drug
2.7.4.2. Adverse Events
2.7.4.3 Clinical Laboratory Evaluations
2.7.4.4 Vital Signs, Physical Findings, and other Observations Related to safety
2.7.4.5 Safety in Special Groups and Situations
2.7.4.6 Postmarketing Data
2.7.4.7 Appendix

Section 2.7.4.1 should start with an overall safety evaluation plan describing how studies were grouped together with short narrative study descriptions including exposure of study subjects.A summary should also be provided in this section of the overall extent of drug exposure from all phases of the clinical study development program as well as demographic characteristics and other characteristics (e.g., severity of the disease, concomitant medications).

Section 2.7.4.2 should include an assessment of the common adverse events associated with the use of the product. All adverse events occurring or worsening after the first administration of the study drug should be summarized in tables listing each event, the number of subjects in whom the event occurred, and the frequency of occurrence in different treatment groups. Because most Phase III trials generally are not powered for hypothesis testing on the individual safety endpoints such as incidence rates for individual adverse events or laboratory abnormalities, the assessment of safety relies on exploration and estimations of event rates in both the individual studies and in the pooled data from all clinical trials. The method used to code adverse event terms for the purpose of pooling safety data should be clearly described. The methods used (e.g., a checklist) to collect adverse event data should also be provided. The tables of common adverse events, pooling adverse event data from the Phase II and III studies, should take into consideration study design (e.g., placebo controlled vs. active controlled), studies of a similar duration of exposure, and the dosage regimens tested in each study. It should be noted that the pool of all controlled studies is generally welcomed by the reviewer as it allows for comparison of the incidence rates in the drug-treated group with a control group (9).

If treatment groups have taken drug for different lengths of time (e.g., combining a six-month study with a six-week study, or if subjects are followed until death), it is recommended to present adverse event rate analysis by patient-year or patient-month of exposure to facilitate the interpretation of the results. Another important objective of this section is the pooled data analyses on adverse event dose dependency and adverse events time dependency in order to better characterize the safety profile of the drug.

The applicant can facilitate the work of the reviewer by identifying the patients who have presented serious adverse events, potentially serious adverse events, and adverse events leading to premature discontinuation of the study drug of the kind that can affect the approval decision or severely limit the use of the drug. In this respect, the applicant should use individual patient listings for deaths, pooled results by treatment group of mortality data, serious adverse event listings, and dropouts listings. Further, the applicant should provide navigational tools to guide the reviewer to relevant individual patients data. Listings should have hyperlinks to patient narratives and case report forms (if applicable). The criteria that were used for including deaths, serious adverse events, and dropouts due to adverse events in the data base should also be clearly described as well as the methods used to analyze overall mortality, cause-specific mortality, the overall incidence rate of serious adverse events, analysis of the relationship of serious adverse events to study drug, and the incidence of dropouts due to adverse events (10).

The rare and medically important adverse events should also be identified. This aspect is more difficult and depends on the size of the safety database. The applicant should also discuss in the Clinical Summary of Safety the adverse reactions potentially associated with the therapeutic class of the product.

Section 2.7.4.3 should describe major changes in laboratory values (e.g., biochemistry and hematology) across the clinical studies. For each analysis, a comparison of the treatment and control groups should be made. For each parameter, the description must be done with the general trend and with the number of subjects with abnormal and potentially clinically significant values. Any dose–response or relationship to intrinsic (e.g., age) or extrinsic factors (e.g., concomitant treatment) should be identified.

Section 2.7.4.4 should describe major changes in vital signs (e.g., heart rate, blood pressure) and other observations related to safety (e.g., electrocardiogram). Any dose–response relationship to intrinsic (e.g., age) or extrinsic factors (e.g., concomitant treatment) should be identified.

Section 2.7.4.5 should summarize safety data in special groups according to intrinsic factors (e.g., age), extrinsic factors (e.g., concomitant treatment), drug interactions, and in special situations (e.g., drug abuse) in order to identify important demographic drug or other extrinsic and intrinsic factors that may predict the occurrence of adverse experiences.

Section 2.7.4.6 should include all relevant postmarketing data available (published and unpublished). The periodic safety update reports can be included in Module 5.

An outline of the content of Clinical Summary of Safety sections is provided in Table 4 (pp. 113–114).

Table 5 FDA and European Regulations Applicable to the Content of Clinical Overview and Summaries

CTD section	U.S. regulation	European regulation
2.5 Clinical Overview (~30 pages)		Volume 2B Notice to Applicants Medicinal products for Human Use—Presentation and format of the dossier Common Technical Document—Module 2—July 2003 (14)
2.5.4 Overview of Efficacy 2.5.5 Overview of Safety	Not a U.S. requirement, but recommended by ICH guidance	
2.7 Clinical Summary (~50–400 pages)		
2.7.3 Summary of Clinical Efficacy	Draft for guidance for Industry —Integrated Summary of Effectiveness (12)	
2.7.4 Summary of Clinical Safety	21 CFR 314.50 (c)(2)(viii) (13)	

COMPARISON OF CLINICAL DOCUMENTS AND SUMMARIES REQUIREMENTS

Table 5 lists the FDA and European regulations applicable to the content of Clinical Overview and Summaries.

The format and the content of the Clinical Overview and Summary are described in the ICH guidelines (2). The applicant should respect the ICH recommendations to ensure a rapid and efficient review of their submission. Moreover, they have to anticipate any specific requirements in the region in which the CTD-formatted MA application is going to be submitted in order to incorporate specific Regional expectations in the filing submitted.

REFERENCES

1. ICH Harmonised Tripartite Guideline. Organisation of the Common Technical Document for the Registration of Pharmaceuticals for Human Use—M4(R3)—Status: Step 5 September 2002.
2. ICH Harmonised Tripartite Guideline. Organisation of the Common Technical Document for the Registration of Pharmaceuticals for Human Use—M4E (R1): EFFICACY—Module 2: Clinical Overview and Clinical Summary/Module 5: Clinical Study Reports—The clinical section of the Application: Step 5 September 2002.
3. International Conference on Harmonization of Technical Requirements for Registration of Pharmaceuticals for Human Use—M2 EWG Electronic Common Technical Document Specification: July 2008.
4. Molzon JA. The International Conference of Harmonization common technical document—global submission format? Food Drug Law J 2005; 60:447–451.
5. ICH Harmonised Tripartite Guideline. Organisation of the Common Technical Document for the Registration of Pharmaceuticals for Human Use—E10 Choice of Control Group and Related Issues in Clinical Trials: Step 5 July 2000.
6. ICH Harmonised Tripartite Guideline. Organisation of the Common Technical Document for the Registration of Pharmaceuticals for Human Use—E4 Dose-Response Information to Support Drug Registration—Step 5 March 1994.
7. ICH Harmonised Tripartite Guideline. Organisation of the Common Technical Document for the Registration of Pharmaceuticals for Human Use—E1 The Extent of Population Exposure to Assess Clinical Safety for Drugs Intended for Long-Term Treatment of Non-Life-Threatening Condition—Step 5 October 1994.
8. ICH Harmonised Tripartite Guideline. Organisation of the Common Technical Document for the Registration of Pharmaceuticals for Human Use—E5(R1) Ethnic Factors in the Acceptability of Foreign Clinical Data—Step 5 February 1998.
9. Todic M. Dossier for marketing authorization in the European Union. Bosn J Basic Med Sci 2003; 3:56–60.
10. Vanderhoof M, Umen MJ, Schwartz DN. Preparing regulatory submissions with the FDA reviewer in mind: Use of the clinical review template and clinical safety reviewer guidance by sponsors. Drug Inf J 2008; 42:349–357.

11. AMWA (American Medical Writer Association) 65th Annual Conference September 29–October 1, 2005.
12. Draft Guidance for Industry—Integrated Summary of Effectiveness—August 2008.
13. Code of Federal Regulations—Title 21, Chapter 1 Food and Drug Administration, Department of Health and Human Services—Subchapter D: Drug for Human Use—Part 314: Application for FDA Approval to Market a New Drug—Subpart B: Applications §314.50 (c) (2) (viii). April 2008.
14. Notice to Applicants Volume 2B: Medicinal Products for Human Use—Presentation and Format of the dossier of Common Technical Document—Module 2—Edition July 2003.

8 Drug Substance—General Information

Anthony C. Cartwright
Pharmaceutical Regulatory Consultant, Harpenden, U.K.

WAYS IN WHICH TO SUBMIT DRUG SUBSTANCE INFORMATION

Where a major pharmaceutical company has developed its own new drug substance and will itself manufacture batches for inclusion in batches of new drug product for world-wide marketing after registration approvals, all of the documentation in Modules 3.2.S.1 to 3.2.S.7 can be included directly in the CTD Module 3.2.S documentation.

Where a manufacturer of a new drug has a co-marketing partner, it will often wish to keep some of the information on manufacture of the drug substance confidential, in which case a Drug Master File [termed an Active Substance Master File in the European Union (EU)] may be appropriate. A Drug Master File may also be appropriate for existing drug substances (whether described in a pharmacopoeial monograph or not) if the material is purchased from a separate supplier.

In Europe, information on a drug substance for which there is a published *European Pharmacopoeia* (Ph Eur) monograph can be submitted and approved separately via the European Pharmacopoeia Certificate of Suitability Scheme (CEP).

The format and contents of the United States (US), Canadian, Australian, European, and Japanese Drug Master Files, and the Certificate of Suitability to the Monograph of the European Pharmacopoeia (CEP) are discussed separately in chapter 9, "Drug Substance Manufacture."

INTRODUCTION

Module 3.2.S.1 of the CTD comprises the general information on the nomenclature, structure, and general properties of the drug substance.

3.2.S.1.1 NOMENCLATURE

Information on the nomenclature should be provided. 3.2.S.1.1 of the CTD gives a number of examples of type of nomenclature information that are required.

- Recommended International Nonproprietary Name (INN)
- Compendial name, for example, Ph. Eur, United States Pharmacopeia (USP), Japanese Pharmacopoeia (JP)
- National approved name, for example, United States Approved Name (USAN), Japanese Authorised Name (JAN)
- Company or laboratory code
- Systematic chemical name—International Union of Pure and Applied Chemistry (IUPAC)
- Other proprietary name(s)
- Other nonproprietary name(s)
- Chemical Abstracts Service (CAS) registry number (RN)

These are summarized in Table 1 with details on the organizations responsible for drug substance nomenclature and references as to how to apply for the various names.

Chemical Name

When a new unique chemical is developed as a candidate new drug substance, a chemical name should be allocated using IUPAC nomenclature principles. This would include the stereochemical assignment according to the Cahn–Ingold–Prelog priority rules ("CIP rules") for chiral drug substances using one-letter designators for stereocenters such as *R*, *S* for tetrahedral carbons (15). Commercial nomenclature software (AutoNom 2000, ChemDraw Ultra 11.0, OpenEye Lexichem, ChemInnovation Nomenclator 6.0, and ACD/Name 11.0) is available, which facilitates the generation of chemical names. Eller (16) has compared the scope

Table 1 Drug Substance Nomenclature, Responsible Bodies, and How to Apply

Name	Body responsible for name	How to apply for name [reference(s)]
Recommended international nonproprietary name (INN)	Quality Assurance and Safety: Medicines (QSM), Medicines Policy and Standards (PSM), World Health Organization, Geneva, Switzerland	1–6
Compendial name		
• Ph Eur	• EDQM—Council of Europe, 7 allée Kastner, CS 30026, F-67081 Strasbourg, France	7
• USP	• US Pharmacopeia, 21601 Twinbrook Parkway, Rockville, MA 20852, U.S.A.	8
• JP	• Society of the Japanese Pharmacopoeia, 2–1215, Shibuya, Shibuya-ku, Tokyo 150–0002, Japan	9
Chemical name	–	Use International Union of Pure and Applied Chemistry (IUPAC) nomenclature rules (10)
Company or laboratory code	Pharmaceutical company	Internal company name and/or code
Other nonproprietary name		
• USAN	• United States Approved Names Council, American Medical Association, Chicago, USA	11–13
• USAN	• National Institute of Health Sciences, Yoga, Japan	
Chemical abstracts service (CAS) Registry number	Chemical Abstracts Service, PO Box 3012, Columbus, Ohio 43210, USA	14

and limitations of AutoNom 2000 with previous versions of ChemDraw and ACD/Name. ACD/Name will generate systematic chemical names of two types—IUPAC names and index names according to the Chemical Abstracts Service (CAS) Index nomenclature rules. An American Chemical Society CAS Registry Number and name should be formally requested.

US Approved Name

In the United States, an Investigational New Drug (IND) application has to be filed before applying for a US Approved Name. The following items are needed in a USAN:

- Suggested name—one to three nonproprietary names or the INN if already approved
- Chemical name or description (according to IUPAC rules)
- CAS Registry Number and CA index name
- Structural formula
- Molecular formula
- Molecular weight
- Company code designation
- Trademark (including any foreign trademarks)
- Trivial names
- Manufacturer
- Principal therapeutic uses
- Pharmacological action
- Date clinical trials began
- IND number

International Nonproprietary Name

For a new drug substance to be used in a product to be marketed globally, a WHO International Nonproprietary Name (INN) is essential. The USAN Council works in conjunction with the WHO INN Committee and other national nomenclature groups to standardize drug substance nomenclature.

An application for an INN includes:

- Suggested name(s)—up to three suggested names using the recommended INN stems, which do not conflict with existing chemical names, nonproprietary names, or trademarks.
- Chemical name and description according to IUPAC rules.
- Graphic formula according to the WHO guideline.
- Molecular formula.
- Chemical Abstracts Service (CAS) registry number.
- Trademarks (known or contemplated) with names of countries where registered.
- Any other name or code (laboratory name, etc.).
- Principal therapeutic use and posology, with reprints to support the claimed therapeutic use.
- Pharmacological action, with reprints to support, as this may influence the stem selected for the compound.
- Date of clinical trial.
- Availability of suggested names (from a check on the Cumulative List of INNs).
- Permission to publish the CAS Registry Number.

The proposed INNs are reviewed by a WHO Expert Panel and then published for comment or objection. During a four-month period any person can make comments or lodge objections to a proposed INN.

Nonproprietary Stem Lists

Both the USAN Council and the WHO publish list of "stems" (5,6) for which chemical and pharmacological categories have been established. A new proposed USAN or INN should be based on the use of these stems. Examples of such stems and drug substance names devised using them are –lukast for leukotriene receptor antagonists (such as monte**lukast**), –azepam for anxiolytic agents (such as lor**azepam**), and –conazole for systemic antifungals (such as flu**conazole**).

Compendial Name

Where a registration application is made for a drug product containing an existing drug substance, there may be a pharmacopoeial (compendial) monograph for the drug substance in the Ph Eur, USP, British Pharmacopoeia (BP), JP, etc. In such a case, 3.2.S.1.1 should include the compendial monograph name for the substance.

3.2.S.1.2 STRUCTURE

The information on the structure can include:

- One or more drawings to show the overall chemical structure of the drug substance, including the stereoconfiguration. Chemical drawing software may be used conveniently for this.
- Molecular formula—one-line molecular formula constructed in accordance with accepted practice. For salts, the active moiety and the other species should be shown separately, for example, $C_{18}H_{26}N_2 \cdot H_2SO_4$ (dexamfetamine sulphate).
- Relative molecular mass (M_r): This should be calculated using the latest guidelines for atomic weights of the elements recommended by IUPAC (17). Where the drug substance is a salt or hydrate, the M_r of the therapeutically active moiety should be stated. In the example of dexamfetamine sulphate, the M_r of dexamfetamine sulphate is 368.50 and the M_r of dexamfetamine base is 270.42.

Active Entity Strength Declaration for the EU

It should be noted that it is a legal requirement in the EU (Annex 1 to Directive 2001/83/EC as amended) for medicinal products containing an active substance that is the subject of an application for marketing authorization in a European Member State for the first time for the quantitative statement of an active substance, which is a salt or a hydrate in the product to be systematically expressed in terms of the mass of active entity. This is known as "active entity" or "active moiety" strength declaration. Clinical trials of new drug products for registration in the EU should use convenient strengths expressed in terms of the active entity and not the salt or hydrate. This is not a legal requirement outside the EU (although it is usual practice for major pharmaceutical companies) and this has caused difficulties where the strength of a new drug product developed outside the EU was originally expressed in terms of milligram of a salt and has to be expressed differently for EU registration and marketing in terms of an inconvenient number of milligram of the active entity.

Protein Drugs

For protein drugs, the structural information should include:

- Schematic amino acid sequence indicating the glycosylation sites
- A general description of the molecule
- Number of amino acid residues
- Molecular mass

3.2.S.1.3 GENERAL PROPERTIES

In addition to a general description of the drug substance (appearance, color, physical state), a list of the physicochemical and other relevant properties should be provided, in particular those which are likely to affect pharmacological efficacy and toxicological safety. The particular physical form or forms that will be used routinely in the drug product manufacture should be identified (polymorph, isomer, solvate, hydrate). The properties could include (as appropriate):

- Melting point or boiling point
- Optical rotation
- Solubility profile in water and other solvents (including aqueous pH solubility profile)
- Solution pH
- log P (partition coefficient) in octanol/water
- pK_a value(s) [negative logarithm of the dissociation constant(s)]
- Permeability—based on the extent of absorption of the drug substance in man, or via an in vitro cell culture method capable of predicting drug absorption in man (18).

Protein Drugs

For protein drug substances, the following information may be needed in addition:

- Isoelectric point
- Extinction coefficient
- Any unique spectral characteristics

REFERENCES

1. General Principles for Guidance in Devising International Non-Proprietary Names for Pharmaceutical Substances, WHO, Geneva. http://www.who.int/medicines/services/inn/GeneralprinciplesEn.pdf. Accessed March 24, 2008.
2. Request for an International Nonproprietary Name (INN), WHO, Geneva, 2007. http://www.who.int/medicines/services/inn/innform_06–2007.pdf. Accessed March 24, 2008.
3. Guidelines on the Use of International Nonproprietary Names (INNs) for Pharmaceutical Substances, WHO, Geneva, 1997. http://whqlibdoc.who.int/hq/1997/WHO_PHARM_S_NOM_1570.pdf. Accessed March 24, 2008.
4. Annexes 1–8 to Guidelines on the Use of International Nonproprietary Names (INNs) for Pharmaceutical Substances, WHO, Geneva. http://who.int/medicinedocs/fr/d/Jh1806e.7. Accessed March 24, 2007.

5. The Use of Stems in the Selection of International Nonproprietary Names (INN) for Pharmaceutical Substances, WHO, Geneva, 2006. http://www.who.int/medicinedocs/collect/medicinedocs/index/s14147e/s14147.pdf. Accessed March 24, 2008.
6. Addendum 1 to "The Use of Stems in the Selection of International Nonproprietary Names (INN) for Pharmaceutical Substances," WHO, Geneva, 2007. http://www.who.int/medicinedocs/collect/medicinedocs/index/s14141e/s14141.pdf. Accessed March 24, 2008.
7. News of the European Directorate for the Quality of Medicines. EDQM, 2008. http://www.edqm.eu/site/page_628.php. Accessed March 24, 2008.
8. U.S. Pharmacopeia, 2008. http://www.usp.org/aboutUSP/contactUs.html. Accessed March 24, 2008.
9. The Society of Japanese Pharmacopoeia, November 1, 2007. http://www.sjp.jp/eng/01/01.htm. Accessed March 24, 2008.
10. Leigh JH, Favre HA, Metanomski WV. Principles of Chemical Nomenclature—A Guide to IUPAC Recommendations. Malden, MA: Blackwell Science Publications, 1998.
11. United States Adopted Names. American Medical Association, 2008. http://www.ama-assn.org/ama/pub/category/2956.html. Accessed March 24, 2008.
12. New USAN Application Policy Effective March 15, 2007, American Medical Association, 2007. http://www.ama-assn.org/ama/pub/category/13077.html. Accessed March 24, 2008.
13. Rules for coining names, American Medical Association, 2008. http://www.ama-assn.org/ama/pub/category/print/4769.html. Accessed March 24, 2008.
14. CAS Registry and CAS Registry Numbers. American Chemical Society, 2008. http://www.cas.org/expertise/cascontent/registry/resys.html. Accessed March 24, 2008.
15. Cahn RS, Ingold CK, Prelog V. Spezifikation der molekularen Chiralität. Angew Chem 1966; 78:413–447; Angew Chem Int Ed 1966; 5:385–415.
16. Eller GA. Improving the quality of published chemical names with nomenclature software. Molecules 2006; 11:915–928.
17. IUPAC Commission on Atomic Weights and Isotope Abundance—Atomic Weights of the Elements 2007. IUPAC. http://www.chem.qmul.ac.uk/inpac/AtWt/. Accessed March 25, 2008.
18. Guidance for Industry, Waiver of In Vivo Bioavailability and Bioequivalence Studies for Immediate-Release Solid Oral Dosage Forms Based on a Biopharmaceutics Classification System. CDER, FDA, August 2000. http://www.fda.gov/cder/guidance/3618fnl.pdf. Accessed March 25, 2008.

9 Drug Substance—Manufacture

Anthony C. Cartwright
Pharmaceutical Regulatory Consultant, Harpenden, U.K.

DRUG MASTER FILES AND CERTIFICATES OF SUITABILITY TO MONOGRAPHS OF THE EUROPEAN PHARMACOPOEIA

As already mentioned in chapter 8: Drug Substance—General Information, in cases where the full details of manufacture of the drug substance cannot be included in Module 3.2.S of the Common Technical Document (CTD) registration file, the regulatory agencies have devised other means of submitting and reviewing the documentation to protect valuable and commercially confidential "know-how" on the details of manufacturing process for the drug substance. Two key methods are the Drug Master File (called an Active Substance Master File in the European Union) and Certificates of Suitability to the Monograph of the European Pharmacopoeia (CEPs) for pharmacopoeial monograph drug substances. The CEP is used in the European Union, Australia, and Canada.

Drug Master Files and Active Substance Master Files

There are major differences between the Drug Master Files (DMFs) used in the United States, Japan, and elsewhere, and the Active Substance Master Files (ASMFs) used in the European Union. These are summarized in Table 1.

Drug Master Files (1)

In the United States, there are five types of DMF:

Type I: Manufacturing Site, Facilities, Operating Procedures, and Personnel
Type II: Drug Substance, Drug Substance Intermediate, Material Used in Preparation, Drug Product
Type III: Packaging Material
Type IV: Excipient, Colorant, Flavour, Essence, or Material Used in Their Preparation
Type V: FDA-Accepted Reference Information

In the context of drug substances, this chapter is only concerned with the U.S. Type II DMFs for Drug Substances, Drug Substance Intermediates, and Materials Used in Their Preparation. A DMF should be limited to a single drug intermediate, drug substance, drug product, or type of material used in their preparation. The DMF can include all of the relevant information detailed in Module 3.2.S on manufacture and controls of the drug intermediate or substance. It should detail the DMF Holder, the corporate headquarters of the DMF Holder, the street address of the manufacturing/processing facility, contact for correspondence, and the details of any agent. It needs to include a signed statement by the holder certifying that the DMF is current and the DMF Holder will comply with the statements included in it.

Canadian Product Master Files are classified into the following categories:

Type I: Active Drug Substance
Type II: Packaging Material
Type III: Colorants, Flavours, and other Additives
Type IV: Dosage Form

Product Master Files may be submitted in advance of a registration filing, but may only be accessed with the permission of the Master File Holder. The Canadian Type I Active Drug Substance Master File is submitted as two separate volumes—the "closed part" and the "open part." A copy of the open part is provided to the sponsor of the drug product registration submission, and this is included in their registration dossier. The guideline on Product Master

Table 1 Comparison of U.S./Japanese, etc. DMFs and ASMFs

Parameters	US drug master file	Active substance master file (the European Union)/Canadian Type I product master file
Structure of file	Single DMF file containing all information in 3.2.S	Two-part file—Applicant's Part (open part) and Restricted Part
Documentation to be included in Module 3.2.S of the CTD registration file	Reference to the number of the DMF submitted by the DMF Holder and letter of authority to refer to the file	All of the Applicants Part of the ASMF in Module 3.2.S, with letter of authority to refer to the ASMF in Module 1
When submitted to the agency	Can be submitted in advance of the CTD drug product registration file	Submitted at the same time as the CTD file
Information disclosed to the drug product registration file applicant	None	Applicants Part of the ASMF

Files is being updated. The content of the Active Drug Substance Master File is nearly identical to the European Union Active Substance Master File (see below).

DMFs in Japan relate to drug substances, and again need to include all of the relevant information detailed in Module 3.2.S. Drug substance DMFs are also supplied directly to the authorities in many other countries (e.g., India, Korea).

Active Substance Master Files (2)

In the European Union, an ASMF may be filed for a new or existing chemical active drug substance, a herbal substance, or a herbal preparation. The complete ASMF is submitted by the drug substance manufacturer at the same time as the registration dossier (Marketing Authorisation Application or MAA) with a covering letter citing the particular MAA (name of product, registration procedure being used, and the European Union Member States where the ASMF is being filed) and version numbers for the Applicant's Part and Restricted Part. The Marketing Authorisation (MA) Applicant has to include a Letter of Access to the ASMF from the ASMF Holder authorizing the competent authority/EMEA to refer to and review the ASMF in support of the MA submitted. The ASMF Holder has to commit formally in writing to ensure batch consistency and to inform both the MA Holder/Applicant and the competent authority/EMEA of any change to the ASMF.

An overview of the division of the ASMF contents into the Applicant's Part and Restricted Part is given in Table 2. The rationale behind the two-part European ASMF is that the drug product MA holders have to be given sufficient information by the drug substance manufacturer in the Applicant's Part so that they can be responsible and legally liable for the quality of the drug substance they are using to manufacture the drug product. If there are product defects or adverse drug reactions, it is the drug product MA holders who will be the point of contact for the authorities and it is they who may be sued by patients.

If an ASMF is filed, the MA applicant is required to reproduce the information in the Applicant's Part in 3.2.S Drug Substance of the CTD registration file. In addition, a summary of the information in the Applicant's Part should be included in the Module 2.3 Quality Overall Summary (see chap. 5).

The Australian Therapeutic Goods Administration applies the same ASMF requirements for its own Drug Master File system (3). However, it has its own specific requirements for DMFs for antibiotics produced wholly or in part by fermentation.

Certificates of Suitability to Monographs of the European Pharmacopoeia (4–6)

CEPs are accepted by the authorities in the European Union countries for national applications or applications to the EMEA, and their use is also being evaluated in Canada. Australia also accepts CEPs (3).

Table 2 Overview of the Active Substance Master File Contents

Modules in 3.2.S	Module number(s)	Name of modules	Applicant's Part (open part)	Restricted part
3.2.S.1 General Information	3.2.S.1.1–1.3	General Information	√	
3.2.S.2 Manufacture	3.2.S.2.1	Manufacturer(s)		
	3.2.S.2.2	Flow Chart of Manufacture	√	
	3.2.S.2.2	Details of Manufacture	a	√
	3.2.S.2.3	Control of Materials		√
	3.2.S.2.4	Control of Critical Steps and Intermediates		√
	3.2.S.2.5	Process Validation and Evaluation		√
	3.2.S.2.6	Manufacturing Process Development		√
3.2.S.3	3.2.S.3.1 and 3.2	Characterization	√	b
3.2.S.4	3.2.S.4.1–4.5	Control of Drug Substance	√	c
3.2.S.5		Reference Standards or Materials	√	
3.2.S.6		Container closure system	√	
3.2.S.7	3.2.S.7.1–7.3	Stability	√	

[a]Details of any sterilization procedure on the bulk drug should be provided to the MA Applicant in cases where there is no further sterilization of the final product.
[b]Information on possible impurities need not be supplied in the Applicants Part where the ASMF Holder justifies that there is no need to control these impurities in the final substance.
[c]Information on justification for control of particular impurities in the Specification can be omitted where the ASMF Holder justifies that there is no need to control those impurities in the final substance.

Applicants send the following documents to the Certification Division of the EDQM in Strasbourg for evaluation:

- A complete application form including the fee form
- A single copy of the CEP dossier in CTD format
- A single copy of the Module 2.3.S Quality Overall Summary in Microsoft Word format

An applicant can apply for a CEP for a sterile substance, and the additional information is defined in the document "Certificates of suitability for sterile active substances (PA/PH/Exp. CEP/T (06) 13" (7). A database of approved CEPs is published on the EDQM Web site.

Once approved, an EDQM CEP can be cited in a registration filing using its reference number. However, any additional tests used by the dosage form manufacturer (e.g., particle size of the drug substance) need to be documented, data provided to show that they are validated, the specification acceptance limits need to be justified, and batch data supplied.

United States Pharmacopeia Drug Substance Verification Program

The United States Pharmacopeia (USP) has started its own program for certification of drug substance suppliers (8). Participants in the program receive permission to use a special USP Certification Mark in conjunction with a certificate of analysis. The scheme certifies that the ingredient has been manufactured according to GMP (as defined in the Manual for the Programme). The program includes the following:

- Evaluation of the participants' quality systems through a GMP audit
- Review of manufacturing and quality control documents including review of characterization, stability, and release data for compliance with the USP monograph
- Laboratory testing of batches of the drug substance

- Grant of the Certification Mark
- Postverification surveillance testing
- Postverification audits
- Reporting by participants of changes to the manufacture or testing of drug substances bearing the Certification Mark

FDA is still, however, requiring separate submission of information on the drug substance in the form of the Drug Master File.

INTRODUCTION TO MODULE 3.2.S.2 DRUG SUBSTANCE MANUFACTURE

The description provided here represents the formal commitment by the applicant as to how the manufacture of the drug substance will be carried out and controlled. The contents are defined in The Common Technical Document for the Registration of Pharmaceuticals for Human Use: Quality—M4Q(R1) dated September 2002 (9).

3.2.S.2.1 MANUFACTURERS

The name, address, and role in the manufacturing process should be provided for each company involved. This will include contract manufacturers and test laboratories. The street address of each site should be given, and this should be the address where the operation is carried out and not the corporate headquarters (HQ) of the company. For U.S. filings, the drug establishment number should be given as defined in 21 CFR part 207 (10). Manufacturing facilities should be ready for GMP inspection when the registration dossier is submitted.

3.2.S.2.2 DESCRIPTION OF MANUFACTURE

A flow diagram and a complete description of the manufacturing process and process controls should be provided. Any manufacturing stages which are likely to have an impact on quality of the drug substance and which are critical should be identified and described in detail (see 3.2.S.2.4 Control of Critical Steps and Intermediates). If an alternative manufacturing process is proposed, the relevant information should be provided in each section of this module.

Flow Diagram

This should show the process of manufacture for a representative scale commercial batch beginning with the starting materials through to drug substance release testing. The flow diagrams should include (12)

- each manufacturing stage with a note of those that are critical. Steps can include chemical reactions, isolation (e.g., centrifugation to separate solid material), purification (e.g., recrystallization, chromatography), milling or micronizing, and release testing;
- chemical structures and names of starting materials, intermediates, reagents, and the active drug substance (including stereochemistry);
- an indication of whether intermediates are used in situ or separated before being used in the next stage of the manufacture;
- solvents and catalysts used in the manufacture;
- molecular formulae and molecular masses of starting materials, intermediates, and active drug substance; and
- weights of starting materials, intermediates, reagents, etc. used in the manufacture;
- operating conditions for each step (e.g., temperature, pressure, pH); and
- typical range of yields seen for each synthetic manufacturing stage.

The flow diagram given in the Applicant's Part of an ASMF for a generic drug is often given in less detail than recommended in the above, and typically includes only the chemical names, structures and molecular formulae of starting materials, intermediates, reagents, and the active drug substance with an indication of the scale of the manufacture (approximate batch size expressed as the yield in kilograms of the purified final drug substance).

Description of the Manufacturing Process

A detailed sequential procedural description should be provided for the manufacturing process. This is usually provided for each major stage of the proposed commercial scale manufacture.

For each stage in manufacture (Step 1, Step 2, etc.), the description would typically include the following:

- Structural formulae, chemical names and chemical structures, molecular formulae and relative molecular masses.
- Names of reagents, catalysts, solvents used.
- Conditions of the reaction (temperature etc.).
- List of the materials used in the manufacturing step with quantities used in a typical batch
- Details of the manufacturing procedure used in the step with reaction conditions (with limits where critical), holding times, separation details for any isolated intermediate, and any analytical testing done during the step. Maximum storage times for any isolated intermediate may need to be defined. Where an in-process control is carried out, the type of analytical procedure should be stated (e.g., high pressure liquid chromatography-HPLC).

A process flow diagram using commercial flow-charting software and allocating specific shapes to starting materials, reagents, solvents, unit operations, in-process controls, etc. may be useful in displaying the information for ease of understanding by the agency reviewer. Figure 1 is an example.

Process controls applied during production can include those on operating parameters (such as temperature, pH, time, and mixing conditions), environmental controls (temperature and humidity), analytical tests (such as residues of a reactant), and in-process testing (such as on an isolated intermediate or the unfinished crude bulk of the active substance).

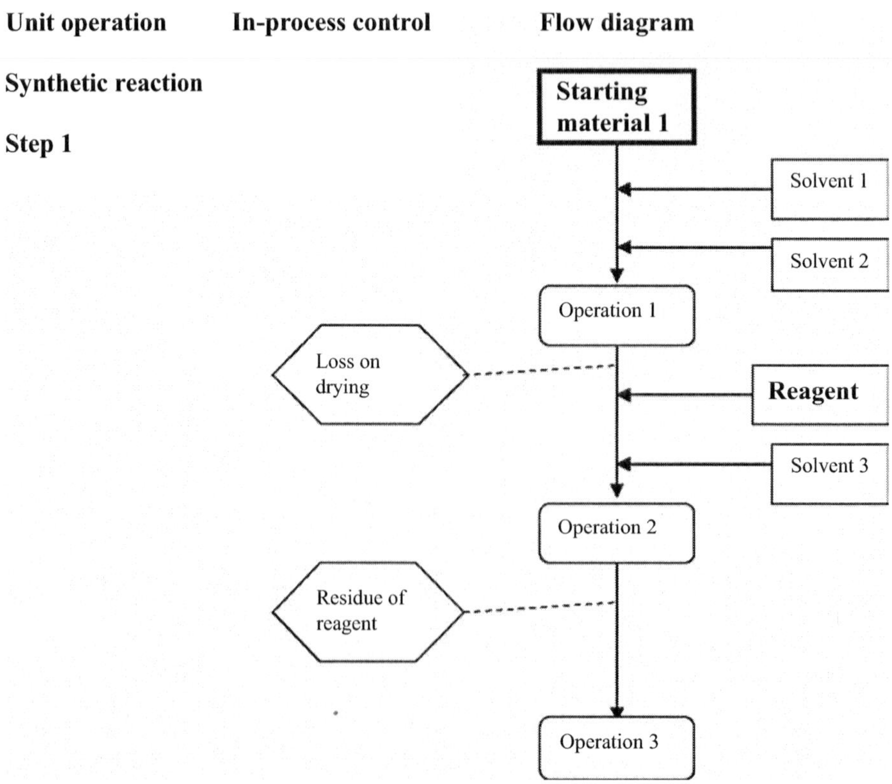

Figure 1 Process flow diagram for drug substance manufacturing process.

Alternative Processes

Alternative manufacturing processes should be described in the same detail as the primary process (with both a flow diagram and a detailed process description). Alternative processes are sometimes proposed where there are two or more different sites of manufacture, perhaps in different countries. If alternative steps or solvents are proposed, they should be justified, and sufficient tabulated batch data provided to show that the final quality of material (particularly in relation to the impurity profile) remains the same whichever process is used in manufacture.

Reprocessing and Reworking

Reprocessing (if proposed to be carried out on a frequent basis) should be mentioned in 3.2.S.2.2 Manufacture.

Reprocessing can be defined as subjecting an intermediate or drug substance that does not conform to its specification to a repetition of at least part of the manufacturing process, for example, by repeating a crystallization or some other manipulation (e.g., distillation, chromatographic separation, or milling). If reprocessing is a one-off event, the records of that batch should merely be retained and available for GMP audit or inspection, unless the reprocessing could affect adversely the physical or chemical properties of the drug substance.

Reworking can be defined as subjecting an intermediate or drug substance that does not conform to its specification to one or more manufacturing steps which are different to those described in the registration file. The ICH Q7 Guideline "Good Manufacturing Practice Guide for Active Pharmaceutical Ingredients" (11) recommends that an investigation into the reasons for nonconformance to specification is carried out. Reworking is generally a one-off event and may need to be the subject of a postapproval variation (the European Union) or prior approval supplement (the United States) or equivalent (other countries).

Recovered Solvents

The use of recovered solvents (e.g., from mother liquor or filtrates) should be described in 3.2.S.2.2, and the point at which they are to be used should be stated. Information should be provided as to whether any processing is done to improve the quality of the solvent. An appropriate specification for each recovered solvent should be included in 3.2.S.2.3 Control of Materials.

3.2.S.2.3 CONTROL OF MATERIALS

Information should be provided in this section on the materials used to manufacture and purify the drug substance—the starting materials, reagents, solvents, and other materials.

Designation of Starting Materials

Drug substance starting materials are defined in the ICH Q7 Guideline on "Good Manufacturing Practice Guide for Active Pharmaceutical Ingredients" (11) as a raw material, intermediate, or an active pharmaceutical ingredient (API) that is used in the production of an API and that is incorporated as a significant structural fragment into the structure of the drug substance. An API Starting Material can be

- an article of commerce,
- a material purchased from a supplier under a contract or commercial agreement, or
- produced in-house.

The company needs to designate and justify the substance chosen as the drug substance starting material. The name and street address of the supplier should be stated. From the point of the designated starting material, GMP has to be applied during the manufacture of the drug substance. The ICH Q7 Guideline recommends how starting materials enter into the manufacturing process for extraction, purification, fermentation products, and how GMP should be applied.

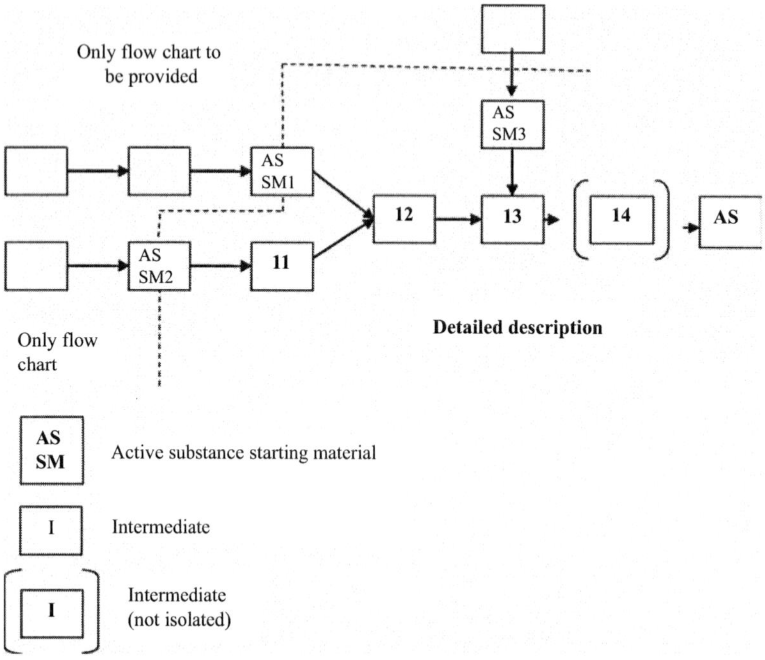

Figure 2 Schematic diagram of drug substance manufacture.

Since the manufacture of a starting material may often be purchased or contracted out to a supplier and because GMP applies downstream in the synthesis from this material, the correct designation of starting material is crucial. In the United States, it is usual to discuss the choice of starting materials with FDA during review of the IND (e.g., at an end-of-phase II meeting), and thus suitable commercial arrangements for sourcing the material can be put in place well in advance of the registration filing. This allows the timely manufacture of launch stock of the bulk drug substance.

For registration filings in the European Union, the 2003 Committee for Proprietary Medicinal Products (CPMP) "Guideline on the Chemistry of New Active Substances" (12) recommends that the description of the manufacturing process should include a justification for the choice of starting material. A flow chart should be included indicating the synthetic process prior to the introduction of the starting material, as it is felt that this may be useful in evaluating the suitability of the specification. A schematic diagram is suggested as in Figure 2. The European review process has the disadvantage that the acceptability of the choice of starting material is not certain until the MAA has been reviewed.

Some of the factors that the agencies will consider in evaluating the acceptability of the designation of a particular chemical as a starting material include the following:

- The commercial availability of the material for uses other than drug substance production.
- How many steps in the manufacturing process are between the starting material and the final drug substance—the more steps there are, the less is the risk of variation in quality of the starting material affecting the quality of the drug substance.
- Potential for carryover of impurities from the starting material.
- Complexity of structure—whether, for example, the starting material contains a number of features and functional groups that can give rise to isomers and analogues.

Information on the Starting Materials

The information in this section of the CTD will comprise

- a list of the proposed starting materials (with chemical name, Chemical Abstracts Service (CAS) Registry Number, chemical structure, relative molecular mass and relevant physical characteristics such as appearance, physical state, and melting point),

- a justification for the choice of starting materials (where needed),
- a flow diagram, and
- a specification for each starting material.

The starting material should have a specification which ensures that it is readily distinguishable from potential isomers and analogues which might carry through the synthesis. The specification should include identification test(s) for the starting material, a limit for individual impurities and for total impurities. Acceptance criteria for limit tests for organic solvents may also need to be considered. The specification should be based on the historical batch analyses of starting materials used in the successful manufacture of key development and production batches of the drug substance. Where the starting material is only a few steps away from the final drug substance, there is a greater possibility of carryover of impurities from the starting material to the drug substance, and control of impurities may be more critical. The detailed analytical methods should be provided as part of the specification.

No clear ICH or national guidance is given as to the need to provide analytical validation data for the tests included the starting material specifications, but this information seems to be increasingly requested by the major agencies.

Reagents and Solvents

A list of the reagents and solvents should be provided. A specification should be given for each material, which should include all of the acceptance criteria and refer to the analytical methods to be used. Commercial grades of xylene or toluene used as solvents may often contain benzene as an impurity, which needs to be limited; although residues of up to 890 ppm of toluene in a drug substance are acceptable as an ICH Class 2 solvent, residues of benzene are to be avoided (limit of 2 ppm) (13). Specifications for solvents used in the final stages of the synthesis require tighter specifications than those used earlier in the synthesis.

Other Materials Used in Manufacture

Details should also be given of any other materials used in the manufacture, such as diatomaceous earth used as a filter aid and activated charcoal used as a decolorizing agent.

3.2.S.2.4 CONTROL OF CRITICAL STEPS AND INTERMEDIATES

Critical Steps

A critical step could be defined as one where the controls used during production to monitor or adjust the process in that step must be maintained within predetermined limits to ensure that the final active drug substance meets it specification. Tests and acceptance criteria for all of the critical steps identified in 3.2.S.2.2 Manufacturing Process should be provided.

Examples of critical steps include (12) the following:

- Mixing of multiple components
- Phase change and phase separation steps
- Steps where control of temperature and pH are critical
- Intermediate steps that introduce an essential structural element into the molecule
- Final purification step(s).

Steps that affect the solid-state properties (crystal or polymorphic form, particle size distribution) will always be considered critical, particularly if the active drug substance is used in a solid dosage form or in a suspension, since these factors could affect dissolution and hence bioavailability.

When designing a synthetic manufacturing process, the range of values for process controls such as temperature, pressure, and time may derive from actual experimental studies to optimize the yield from a reaction or drive it to completion, but may also just reflect a range of typical values obtained practically when using a particular set of equipment etc.or even just the synthetic chemist's experience as to what is reasonable. Where a particular range of control conditions (such as temperature range) are found to be critical for yield or completion (so that a particular intermediate is fully converted), they may represent part of the "design space" for the reaction. Spiking studies (adding extra levels of an impurity) may sometimes be used to evaluate how well subsequent stages in the manufacture remove a particular impurity and this may show how critical is a particular step in the manufacture. A report on such studies could be important in demonstrating the removal of a particular toxic reagent.

Application of Process Analytical Technology to Drug Substance Manufacture

The process analytical technology (PAT) initiative was originally introduced by FDA as a collaborative effort with the pharmaceutical industry to introduce systems for design, analysis, and control of manufacturing processes based on timely measurements of critical quality attributes of materials and products. The PAT approach has been accepted by the European Union authorities. The application of PAT to drug substance crystallization processes has been reviewed by Yu et al. (14). Crystallization of drug substances that possess multiple polymorphic forms is a critical process step. The authors cite two case studies where in situ monitoring of polymorphic form by Raman spectroscopy was used to monitor and control the production of the desired polymorph. The examples quoted were for production of the desired polymorphic form of progesterone, and of MK-A a compound under development at Merck Research. Further examples were the monitoring of the production of the desired polymorphic form of trovafloxacin mesylate by near infrared (NIR) at Pfizer and real-time monitoring of the production of the stable β form of glycine by X-ray diffraction.

Intermediates

Specifications should be established for intermediates that are isolated and tested before being used in the further downstream process. The specification normally includes an appearance, assay, and limits for impurities (such as by-products of the reaction carried out in that step, unreacted intermediate from the previous step, etc.).

The FDA distinguishes between intermediates and "postsynthesis materials." A postsynthesis material is a material that appears after the final intermediate but before the final drug substance (such as a free base where the final drug is a salt). They also distinguish "unfinished" grades of the drug substance (such as the crude grade before final recrystallization, drying or milling). In both cases, the information could conveniently be included in this section of the drug substance CTD. The crude drug would nearly always merit its own specification.

In the European Union, it is recommended that the analytical methods for "key" intermediates should be suitably validated (according to the ICH requirements) (12).

3.2.S.2.5 PROCESS VALIDATION AND EVALUATION

The ICH M4Q(R1) Guideline on the Quality files (2.3.QOS and Module 3) merely states that process validation and evaluation is needed for aseptic processing and sterilization (9). The CPMP Guideline on The Chemistry of New Active Substances goes further and asks that steps that are identified in 3.2.S.4 as critical for the quality for the drug substance should be validated.

The ICH Q7 "Good Manufacturing Practice Guide for Active Pharmaceutical Ingredients" includes a section on validation and recommends that the critical parameters/attributes should normally be identified during the development stage or from historical data and the ranges necessary for critical process parameters (temperature, pH, pressure, time, etc.) for reproducible operation should be defined. The Q7 Guide suggests that for the process validation program, three successful production batches is a good guide, but that for complex API processes or API processes with prolonged completion times additional process runs may be needed. The critical process parameters should be controlled within the defined limits and monitored during the validation runs. The validation report should include all of the process parameter information as well as analytical information on intermediates and the finished product. The impurity profile of the final bulk drug should be within the proposed specification.

For sterile drug substances, validation data is needed for the processes used for sterilizing both the bulk drug and packaging components. The July 1994 internal FDA Guide to Inspections of Sterile Drug Substance Manufacturers (15) suggests that at least three consecutive successful validation runs are necessary before an aseptic process (such as aseptic crystallization of the sterile drug substance; aseptic isolation of the drug substance; aseptic drying, milling, and blending of the sterile substance; and aseptic packaging) can be considered to be validated. The results of environmental and personnel monitoring should be included in the validation report.

3.2.S.2.6 MANUFACTURING PROCESS DEVELOPMENT

A description of the manufacturing process through the various stages of drug development should be provided. A tabular comparison is useful to the reviewer. This should show the relationship between the manufacturing process used in producing nonclinical, clinical, scale-up, and final production scale batches (if available) and any associated changes in chemical or physical properties (e.g., particle size, polymorphic form). The use of batches of drug substance for major nonclinical and clinical studies with significant differences in particle size or impurities to the proposed final manufacturing process should be discussed in relation to the biological qualification of individual impurities and the justification for the proposed drug substance specification.

FUTURE HARMONIZATION ACTIVITY

The ICH Steering Committee released in April 2008 a Final Concept Paper (16) for a new guideline to be written—Q11: "Development and Manufacture of Drug Substances (chemical entities and biotechnological/biological entities)." The proposed timetable is for a draft document to be available and issued for consultation in the fourth quarter of 2009. The Concept Paper indicates that there are no general developed harmonized approaches to demonstrating process understanding, particularly regarding sources of variability in product quality. It indicates that there is currently little consensus or agreement across the industry and the regulatory bodies with respect to manufacturing process information and justification to be included in the dossier.

The Q11 Guideline will harmonize the scientific and technical principles relevant to the design, development, and manufacture of drug substances. It will include general principles and concepts for the following:

- Selection of materials and components for manufacture of the drug substance;
- Identification and justification of key and critical manufacturing steps, process controls, process parameters, choice of analytical procedures, and acceptance criteria;
- The capacity of the process to reduce or remove impurities and product-related substances;
- Evaluation of process robustness;
- Suitability of scaled-down models for evaluation and validation; and
- Identification and control of critical intermediates.

DEFICIENCIES IN APPLICATIONS

A review of the main deficiencies in CEP dossiers for chemical purity for 87 applications has been published by the EDQM (17). In relation to 3.2.S.2 Manufacture, the key deficiencies are summarized in Table 3. These are probably also typical of deficiencies found in DMFs and for information included directly in Module 3.2.S in registration files.

Table 3 Module 3.2.S.2 Major Deficiencies in CEP Chemical Purity Applications to the EDQM

Modules	Deficiencies	Percentages of dossiers where deficiency found
3.2.S.2.3 Starting Materials	Lack of information on starting materials—route of synthesis, absence or inadequacy of its impurity profile, and possible carryover of impurities to final substance	58.5
	Specifications for starting materials do not include suitable impurity limits	42
3.2.S.2.2 Description of Manufacture	Maximum/typical batch size and yields not stated	34
3.2.S.2.4 In-process Controls	Description of tests performed and analytical methods insufficiently detailed	24

CHIRAL ACTIVE SUBSTANCES

Over the last 20 years or more, the development of new drugs as racemates (equimolar mixtures of enantiomers) has markedly decreased as the authorities have required relevant preclinical and clinical data to be submitted on both enantiomers to assess their effects and fate in vivo. There are cases where both enantiomers are pharmacologically active (e.g., ibuprofen, bupivicaine, quinolone antibiotics), cases where one enantiomer is inactive (l-propranolol is a β-blocker, d-propranolol is not), and cases where toxicity is associated with one isomer (e.g., granulocytopenia is related to the d-isomer of levodopa, vomiting is caused by the d-isomer of levamisole). Guidance on the specific requirements for chiral drugs has been issued in the past by both the European Union (18) and the FDA (19).

In relation to the manufacture of the drug substance, the step where the stereogenic center or centers is formed needs to be described in detail, including the in-process controls, together with data showing that the desired stereochemical control has been achieved. Where a chiral starting material is used, it must be fully characterized in terms of its stereochemical purity using a validated stereospecific method. Where the preferred enantiomer is isolated by resolution (e.g., by diastereoisomer crystallization), this step must be described in detail with information on the in-process controls. In the cases where a single enantiomer is produced, the other enantiomer will be considered as an impurity.

REFERENCES

1. Guideline for Drug Master Files, CDER, FDA, September 1989. http://www.fda.gov/cder/guidance/dmf.htm. Accessed March 28, 2008.
2. Guideline on Active Substance Master File Procedure. Committee for Medicinal Products for Human Use. CPMP/QWP/227/02 Rev 2 Consultation. London April 27, 2005. http://www.emea.europa.eu/pdfs/human/qwp/022702en.pdf. Accessed March, 2009.
3. Australian Regulatory Guidelines for Prescription Medicines, June 2004. http://www.tga.gov.au/meds/argmap11.pdf. Accessed April 8, 2008.
4. New Certificate of Suitability Applications. http://www.edqm.eu/site/New-Applications-29.html. Accessed March 31, 2008.
5. Certification of suitability of Monographs of the European Pharmacopoeia, Content of the dossier for chemical purity and microbiological quality, PA/PH/CEP (04) 1 4R, EDQM, February 2007. http://www.edqm.eu/medias/fichiers/Content_of_the_Dossier_for_Chemical_Purity_Microbiological_Quality.pdf. Accessed March 31, 2008.
6. Certification of suitability of Monographs of the European Pharmacopoeia, Submission of electronic documentation for applications for certificates of suitability (CEPs), PA/PH/Exp. CEP/T (07) 48, EDQM, December 2007. http://www.edqm.eu/medias/fichiers/NEW_PAPHExp_CEPT_07_48_Submission_of_electronic_documentation_for_applications_for_CEPs.pdf. Accessed March 31, 2008.
7. Certification of suitability of Monographs of the European Pharmacopoeia, Certificates of suitability for sterile active substances, PA/PH/Exp. CEP/T (06) 13, EDQM, March 2006. http://www.edqm.eu/medias/fichier/Certificates_of_Suitability_for_Sterile_Active_Substances.pdf. Accessed March 31, 2008.
8. United States Pharmacopeia Drug Substance Verification Program, United States Pharmacopeial Convention, V.3.1, December 2007. http://www.usp.org/pdf/EN/USPVerified/pivpParticipantManual.pdf. Accessed April 8, 2008.
9. ICH Harmonised Guideline, The Common Technical Document for the Registration of Pharmaceuticals for Human Use: Quality—M4Q(R1), Quality Overall Summary of Module 2, Module 3: Quality, ICH, September 2002. http://www/ich.org/LOB/media/MEDIA556.pdf. Accessed April 7, 2008.
10. Electronic Code of Federal Regulations (e-CFR), Title 21—Food and Drugs, Chapter 1—Food and Drug Administration, Department of Health and Human Services, Chapter C—Drugs: General, Part 207—Registration of Producers of Drugs and Listing of Drugs in Commercial Distribution, 2008. http://ecfr.gpoaccess.gov/cgi/t/text-idx?c=ecfr&tpl=/ecfrbrowse/Title21/21cfr207_main_02.tpl. Accessed April 4, 2008.
11. ICH Q7 Good Manufacturing Practice Guide for Active Pharmaceutical Ingredients, ICH, November 2000. http://www.ich.org/LOB/media/MEDIA433.pdf. Accessed April 8, 2008.
12. CPMP Guideline on the Chemistry of New Active Substances, CPMP/QWP/130/96 rev. 1. EMEA, December 2003.
13. ICH Guideline, Impurities: Guideline for Residual Solvents, Q3C(R3), November 2005. http://www.ich.org/LOB/media/MEDIA423.pdf. Accessed April 4, 2008.

14. Yu LX, Lionberger RA, Raw AS, et al. Applications of process analytical technology to crystallisation processes. Adv Drug Deliv Rev 2004; 56(3):349–369.
15. FDA Guide to Inspections of Sterile Drug Substance Manufacturers, FDA, July 1994. http://www.fda.gov/ora/Inspect_ref/igs/subst.html. Accessed April 7, 2008.
16. ICH Final Concept Paper Q11: Development and Manufacture of Drug Substances (chemical entities and biotechnological/biological entities) dated April 11, 2008, ICH. http://www.ich.org/LOB/media/MEDIA4523.pdf. Accessed April 23, 2008.
17. Certification of suitability of Monographs of the European Pharmacopoeia. How can the content of applications for a certificate of suitability for chemical purity be improved? The TOP TEN deficiencies found in applications, EDQM, Strasbourg, December 2006. http://www.edqm.eu/medias/fichiers/Top_10_Deficiencies_found_in_Applications.pdf. Accessed April 7, 2008.
18. CPMP Guideline on Investigation of Chiral Active Substances, October 1993. http://www.emea.europa.eu/pdfs/human/qwp/3cc29aen.pdf. Accessed April 8, 2008.
19. FDA's Policy Statement for the Development of New Stereoisomeric Drugs, CDER, FDA, May 1992 updated July 2005. http://www.fda.gov/cder/guidance/stereo.htm. Accessed April 8, 2008.

10 Drug Substance—Characterization

Anthony C. Cartwright
Pharmaceutical Regulatory Consultant, Harpenden, U.K.

David J. Snodin
PAREXEL Drug Development Consulting, Middlesex, U.K.

3.2.S.3 CHARACTERIZATION

This section of the Common Technical Document (CTD) includes reports of the investigation and establishment of the chemical structure of the drug substance, its physicochemical characteristics, and information on potential and actual impurities, including degradation products (1).

3.2.S.3.1 ELUCIDATION OF STRUCTURE AND OTHER CHARACTERISTICS

Evidence of Chemical Structure

This section of the CTD provides reports of the evidence confirming the chemical structure of the drug substance, including its stereochemistry and potential for isomerism.

Choice of Batch of Drug Substance to Provide Evidence of Chemical Structure

The batch of drug substance used to provide the evidence of structure needs to originate from the synthetic process proposed for the marketed product in the registration dossier. Where several different synthetic routes were used during the drug development, it is essential to provide evidence of chemical structure for each synthetic route to provide assurance that pivotal nonclinical and clinical studies have been carried out with batches of material of the correct structure.

It is important to use a batch of drug substance that has been suitably purified for elucidation and proof of structure to eliminate possible interferences due to related substances. Part of a batch of the primary reference standard of the drug substance is typically used for these studies (see chap. 18, "Reference Standards or Materials").

A single report (with a Table of Contents) could include the methods, results, and interpretation of all the investigations by using physical and chemical techniques, which have been carried out to confirm the chemical structure.

Elemental Analysis

The elemental analysis carried out on the reference sample (for C, H, N, O, etc.) should be tabulated together with theoretical (calculated) values. Elemental analysis equipment for CHN and CHNS is based on the Pregl–Dumas method, where a sample is combusted in a pure oxygen atmosphere and the combustion gases measured. Oxygen is determined by using a modified Unterzaucher method, where the sample is pyrolyzed in a helium/hydrogen atmosphere and the reaction products converted to carbon monoxide. Modern automated analyzers typically have an analytical performance of $<0.5\%$ for both accuracy and precision. The values found should be within the expected experimental errors for the instrument.

Infrared Absorption

Legible copies of infrared (IR) spectra can be provided in the report by using either transmission IR or multiple internal reflection spectroscopy (MIR) instruments. MIR spectroscopy can utilize the pure drug substance and transmission IR will usually utilize a Nujol mull or a halide disc such as potassium bromide (KBr) or potassium chloride (KCl). Details of the IR instrument and spectrophotometric method used should be given. The principal bands in the spectrum should be tabulated and assignments given to key chemical features in the molecule.

Ultraviolet Absorption

The ultraviolet (UV) absorption spectrum of an aqueous or other suitable solution should be provided in the range 200 to 350 nm, giving details of the instrument and method. The molar absorptivity of any band(s) should be provided and assignments given for the chemical structure(s) causing this absorption.

Spectra should be provided of solutions of different pH, if appropriate, and the pH-dependent shifts in the UV spectrum should be discussed.

Nuclear Magnetic Resonance Spectra

Nuclear magnetic resonance (NMR) spectra are used to provide structural identification and normally a variety of experiments are used. Details of the NMR instrument(s), methods, and sample treatment should be given. Legible copies of the spectra should be provided with assignments for the peaks. For new organic chemical drug substances, a combination of ^{1}H and ^{13}C NMR spectra is usually provided. ^{1}H NMR may include one-dimensional and two-dimensional experiments such as Correlation Spectroscopy (COSY), Total Correlation Spectroscopy (TOCSY), Heteronuclear Single Quantum Correlation spectroscopy (HSQC), and Heteronuclear Multiple Bond Correlation spectroscopy (HMBC). ^{13}C NMR experiments may include Attached Proton Test (APT), Distortionless Enhancement by Polarization Transfer (DEPT), and Heteronuclear Correlation (HETCOR) spectroscopy.

Mass Spectrometry

Details of the mass spectrometry (MS) instrument and methods should be given including any sample treatment (such as conversion of a salt to free acid). Experiments may include electron impact and chemical ionization spectra. Legible copies of the mass spectra should be provided together with tabulated spectral peak assignments.

Synthetic Route

The synthetic route used may provide evidence of structure if well-known chemical reactions are used for the key steps in the synthesis. The synthesis used should be summarized, and literature references to reaction mechanisms should be provided, if necessary.

Evidence of Structure from Key Intermediates

It may be difficult to carry out single crystal X-ray analysis if it is not possible to grow a crystal of the drug substance. In this case, it may be necessary to use a single crystal from the immediate precursor intermediate, the free acid, or the free base of a salt. Similarly, it may have been necessary to convert the drug substance to the free acid from its salt to carry out the mass spectrometry. In these cases, evidence of structure can be presented from the data on the free acid, intermediate, etc.

Chemical Reactions That are Diagnostic of Structural Elements

Although most structural evidence will be deduced from interpretation of X-ray and spectroscopic data, it may be useful to include any information from chemical reactions (such as colorimetric reactions), which are diagnostic of structural elements in the drug substance molecule.

X-Ray Crystallography

Powder X-Ray Data

The powder X-ray diffraction (XRPD or XRD) pattern can be used to provide a "fingerprint" of the crystal morphology and polymorphic form. Details of the diffractometer and sample treatment should be provided. The diffraction pattern should be reproduced.

Single Crystal X-Ray Analysis

Details of the instrument, wavelength, etc., should be provided. A copy of the structure determined by the single crystal X-ray diffraction analysis should be provided.

Optical Rotation
The specific optical rotation should be measured for chiral drugs. Details of the polarimeter, solvent used, and wavelength used for the test (normally the D-line of sodium, $\lambda = 589.3$ nm) should be reported. If the drug is an equimolar racemate, the absence of optical rotation should be reported for batches because it confirms the racemic nature of the drug substance molecule.

Scientific Discussion of the Evidence of Structure
This section of a report in the dossier can provide an overview of the evidence from the spectroscopic and other investigations outlined above to provide the evidence of chemical structure. The configuration, conformation, and potential isomerism should be discussed. The review should include the stereochemical properties of the molecule, that is, geometrical isomerism, the number of chiral centers, and configuration at each center (2). Issues such as counterion stoichiometry should also be discussed.

Evidence of Structure for Generic Drug Substances
The evidence outlined above is mainly that needed for a new drug substance. For a generic drug substance, a simpler package of data to establish structure is usually sufficient. This may, for example, comprise IR, UV, and NMR data comparing spectra of a batch of the generic drug with those from a suitable pharmacopoeial reference standard of the drug substance.

Physicochemical Characteristics

Polymorphism
Polymorphism is characterized as the ability of a drug substance to exist in different crystalline forms, which differ in their internal solid-state structure and will possess different chemical and physical properties—thermodynamic, spectroscopic, and mechanical. Polymorphism may also encompass different solvation products (known as hydrates if the solvent is water) and amorphous forms as defined in the ICH Q6A Specifications guideline (3). The different properties of polymorphs can have a direct effect on drug substance processing, drug substance stability, drug product dissolution, bioavailability, and stability. A key regulatory concern is the effect of different forms on different solubilities and therefore on bioavailability. Historical examples include carbamazepine (4), where polymorphism contributed to the erratic dissolution and clinical failures of carbamazepine tablets and a major recall, and chloramphenicol palmitate (5). In 1998, ritonavir oral capsules became unavailable in the market due to the appearance of the lower energy (more stable) polymorphic form II, which slowed dissolution of the capsules and compromised the bioavailability of the product (6,7).

The polymorphic form can be affected not only by the choice of solvents for recrystallization during manufacture of the bulk drug substance but also by other crystallization parameters such as temperature, cooling, seeding, and ripening. Where polymorphism occurs, the optimum polymorphic form should be identified early in the drug development process and the same form should be used for the pivotal nonclinical and clinical studies. The drug product should possess the same stable polymorphic form throughout its shelf life.

A complete report on the investigations into possible polymorphism of a new drug or generic drug substance should be included in the CTD. This would include recrystallization from different solvents using a range of different techniques. Because the different crystalline states or forms may show different levels of thermodynamic stability, and an unstable form can melt at a different temperature to the thermodynamically more stable form, this property may be used to detect polymorphism. Thus hot-stage microscopy, differential scanning calorimetry (DSC), and thermogravimetric analysis (TGA) are amongst the techniques used. Hot-stage microscopy enables a sample of the drug substance to be subjected to a programmed temperature increase and any changes in appearance can then be examined in polarized light. Two types of DSC apparatus are available—those using power compensation to maintain a null temperature difference between sample and reference and those that apply a constant rate of heating and detect temperature differential as a difference in heat flow between sample and reference. DSC thermograms on a sample can show the appearance of more than one peak (endothermic or exothermic), which would indicate polymorphic behavior. X-ray powder diffraction (XRPD) is of particular value in fingerprinting crystal phases by their unique diffraction patterns. Different polymorphs may sometimes exhibit different IR spectra [including Near Infra Red (NIR) in the range 800–2500 nm] and this could be used as part of the identification test in the drug substance specification. Other spectroscopic techniques that might be used to investigate polymorphism include Raman spectroscopy and solid-state NMR.

For determining the potential for manufacture of different polymorphic forms during manufacture, it is important to check the possibility of phase conversions during processing steps such as drying, milling, micronization, spray drying, and compaction. Phase conversions may also occur during storage and under exposure to elevated temperature and relative humidity. The report should include results of these investigations.

The ICH Q6A guideline on Specifications includes two Decision Trees in relation to identification of polymorphism and the possible need for routine testing on batches of the drug substance to ensure that the correct polymorph has been manufactured. If there are significant effects on solubility, which is likely to affect product performance, for example, for a sparingly soluble drug for an oral dosage form, then a routine control (e.g., using XRPD) would be needed.

In addition to controls to ensure that the correct polymorphic form has been used, it may also be important to control levels of crystalline and amorphous materials in the drug substance. Byard et al. have shown that solid-state NMR is a complementary technique with respect to XRPD for studying crystallinity in a micronized drug substance (8).

Polymorphism Control and Generic Drug Substances

Adequate information on drug substance polymorphism is also important in drug development, manufacturing, and testing of a new generic product. This may come from the scientific literature, patents, or the pharmacopoeias. However, if this information is not readily available, studies similar to the above would be needed. If a manufacturer utilizes the polymorphic form used in the originator company's product or the most stable polymorphic form this may be sufficient.

It may not be essential to use the same polymorphic form of drug substance as used by the originator in the reference product (in the EU) or Reference Listed Drug (RLD) (in the United States). Raw and his colleagues from the FDA have given a number of examples where different polymorphic forms have been accepted in ANDAs in the United States (9). These examples include enalapril maleate (which exists as polymorphs I and II), ranitidine hydrochloride (which exists as two polymorphs I and II), terazosin hydrochloride (which exists as three anhydrous forms, a monohydrate, and a dihydrate), and cefuroxime axetil (which exists in amorphous and crystalline forms). Where the choice of polymorphic form does not affect the solubility significantly, for example, and the generic product has been shown to be bioequivalent to the originator product, using a different polymorphic form may be acceptable to the authorities.

Solubility

Solubility is usually determined by the equilibrium solubility method where a saturated solution is obtained by stirring an excess of the drug substance into the solvent for a prolonged period until equilibrium is achieved. The results should be expressed numerically (e.g., as mg/ml).

Aqueous Solubility

Aqueous solubility at 20°C should be determined in water and a range of solutions of different pH. The Biopharmaceutics Classification System (10) is based on measurement of three parameters—solubility, permeability, and dissolution. A drug substance is considered highly soluble when the highest dose strength of drug product is soluble in 250 ml or less of aqueous media over the pH range 1 to 7.5. Thus it is essential to measure the solubility at 37°C in pH 1.0, 4.5, and 7.5 solutions and to express the results in 250 ml aqueous solutions in terms of the highest dose strength product, which will be marketed.

Solubility in a Range of Solvents

The solubility should be determined in a range of solvents. They should include those used in the analytical methods for the drug substance and dosage form. Details of the method used should be stated.

Physical Characteristics
Results of physical properties should be described such as particle size distribution, melting point, and boiling point. A reference to the test method should be provided for each determination (e.g., a reference to a pharmacopoeial general test method).

pK_a and pH Values
The pK_a values of the drug substance can be determined based on the Henderson–Hasselbalch equation by UV spectrometry, titration or via solubility determinations. pK_a values can also be determined for acidic, basic, and multivalent compounds by pressure-assisted capillary electrophoresis (PACE) (11). Details of the method used should be given.

The pH of a stated strength of aqueous solution should be provided.

Partition Coefficient
The octanol–water partition coefficient (log P_{ow} value) is an indicator of the drug lipophilicity. Classical measurements use the "shake flask" method (12) for log P_{ow} values in the range 2 to 4 or potentiometry for ionizable substances. Newer automated methods such as microemulsion electrokinetic chromatography (MEEKC) (13) and reversed phase HPLC for log P_{ow} values in the range 0 to 6 (14) are also available. Details of the method used should be stated.

3.2.S.3.2 IMPURITIES

Classification of Impurities
The ICH Q3A(R2) Guideline on *Impurities in New Drug Substances* (15) classifies impurities as follows:

- organic impurities (process- and drug-related),
- inorganic impurities, and
- residual solvents.

Organic impurities can arise during the manufacturing process and/or storage of the drug substance. They may include:

- starting materials;
- by-products of the manufacture;
- intermediates;
- degradation products; and
- reagents, ligands, and catalysts.

Inorganic impurities can result from the manufacturing process. They may include:

- reagents, ligands, and catalysts;
- heavy metals or other residual metals;
- inorganic salts; and
- other materials (e.g., filter aids).

CTD Report on Impurities
The report on impurities in a CTD registration file can include the following sections:

- Discussion of potential and actual impurities that might be present in the drug substance from its manufacture.
- Summary of laboratory studies to detect impurities in the new drug substance, including results from batches manufactured during the drug development and (if possible) on commercial manufacturing scale, with details of analytical methods (including limits of detection and limits of quantitation).
- Studies carried out to characterize the structure of actual impurities in the drug substance, including whether actual samples of the impurities have been synthesized for evaluation and data on the structural confirmation of the synthesized impurities.
- Justification for selecting limits of actual impurities based on safety and toxicity data and the process capability of the manufacturing process.

Table 1 Suggested Format for Tabulated Summaries of Actual Impurities Detected in Batches (53)

		Use of batch and batch numbers		
Impurity	**Proposed limit**	**Batches used in nonclinical studies**	**Batches used in clinical studies**	**Commercial manufacturing process**
Organic impurities				
Total				
Inorganic impurities				
Residual solvents				

Discussion of Potential and Actual Impurities

According to the ICH Q3A Impurities guideline (15) and the CPMP guideline on *The Chemistry of New Active Substances* (16), this section should review potential and actual impurities likely to arise from the synthesis, purification, and storage of the drug substance. It is based on a review of the process of manufacture, impurities associated with starting materials and intermediates, by-products of the synthesis, and possible degradation products. It is often a subject of contention within pharmaceutical companies as to how extensive and speculative this discussion should be. However, the Q3A guideline states clearly that the discussion should be limited to impurities "that might reasonably be expected based on knowledge of the chemical reactions and conditions involved."

This discussion could include a review of each stage in the manufacture of the drug substance of potential and actual impurities in starting materials and intermediates and information on the levels of those impurities actually found (with an indication of the limit of quantitation of the methods used). Spiking studies are sometimes used to detect what levels of a particular impurity in a starting material or intermediate can be tolerated without significantly affecting the quality of the drug substance.

Summary of Laboratory Studies on Impurities

The impurity profiles of batches of the drug substance during drug development (nonclinical, clinical, and commercial manufacturing scale) can be reported in the form of a table (see Table 1 for a suggested format), which enables the reviewer to compare the impurity profile of nonclinical, clinical, and commercial manufacturing batches. A detailed table showing the uses to which the batches have been put should be included in 3.2.S.4.4 Batch Analyses, which gives information on the batches used for all of the nonclinical, clinical, biopharmaceutical, and stability batches. For each batch of a new drug substance, the report should include:

- batch number and size,
- date of manufacture,
- manufacturing site,
- manufacturing process, and
- reference to the analytical procedures used.

Batch analytical data is also provided on materials used in nonclinical tests in module 2.6.7.4. in the Nonclinical Tabulated Summary.

Characterization of the Structure of Impurities

Studies need to be undertaken to identify impurities if these are present at a level greater than the identification threshold in the ICH Q3A Impurities guideline table (see Table 2). These studies should be reported in the CTD marketing authorization application (MAA) file.

Table 2 ICH Q3A Thresholds for Reporting, Identifying, and Qualification of Impurities in the Drug Substance (15)

Maximum daily dose of drug substance per day	Reporting threshold	Identification threshold	Biological qualification threshold
≤2 g/day	0.05%[a,b]	0.10% or 1.0 mg/day intake—whichever is the lower[b]	0.15% or 1.0 mg/day intake—whichever is the lower[b]
>2 g/day	0.03%	0.05%	0.05%

[a]Higher reporting thresholds should be scientifically justified.
[b]Lower reporting thresholds can be appropriate if the impurity is unusually toxic.

Studies to identify manufacturing impurities can include chromatography in combination with other techniques—such as HPLC/MS, Gas Chromatography/Mass Spectroscopy (GC/MS), and HPLC-NMR. Bertram has published a review (17) of profiling and structural elucidation of impurities in multisource pharmaceutical raw materials, which advocates on-line enrichment methods such as Solid Phase Microextraction (SPME) and Programmable Temperature Vaporization (PTV) with GC and Solid Phase Extraction (SPE) with Liquid Chromatography (LC), which can be used for impurity profiling.

Where an impurity is found at a level above the identification threshold, then, if possible, it is important to either purchase or manufacture the impurity so that its retention time can be compared to that of the unknown impurity as part of its identification. Computer-assisted synthesis design software such as SYLVIA (18) and the Methods in Organic Synthesis database from the Royal Institute of Chemistry monthly bulletin (19) are often used to help manufacture small amounts of organic impurity compounds. Copies of chromatograms should be provided. Where an impurity is manufactured for use in impurity identification and subsequent analytical development and validation studies, information should be provided on its structural identification (elemental analysis, IR, MS, NMR, etc.). Copies of spectra should be provided together with interpretations. Any impurity samples intended for biological testing (e.g., Ames assays) should be of high purity because it is possible that novel impurities (not present in the drug substance) may be generated by the stand-alone synthesis of an impurity.

Module 3.2.S.7.1 Stability will include information on the stress stability studies where a batch of the drug substance is exposed to a variety of adverse stress storage conditions (light, acid, base, oxidation, temperature, and relative humidity) to force degradation. These studies are carried out to develop stability-specific assay methods to determine potential degradation products and to help develop analytical methods for the actual degradation products. Cross-reference to these studies should be given to show the identity of degradation products, which might be formed during manufacture and storage of the bulk drug substance.

It is useful to summarize the actual impurity data for the reviewer of the CTD MAA file with details of the structure, source of impurity, levels found, and the analytical method. Table 3 gives a suggested format (which is similar to those used by a number of companies).

Justification for Proposed Limits for Impurities

Impurity Specifications

According to the ICH Q3A(R2) Impurities guideline (15), the specification for organic impurities in the drug substance should be listed in the following way:

- each specified identified impurity,
- each specified unidentified impurity,
- any unspecified impurity with an acceptance criterion of not more than the identification threshold (in Attachment 1 of the guideline), and
- total impurities.

Table 3 Suggested Format for Summary Table of Impurities Found

Impurity name	Chemical structure	Source of impurity	Analytical method	Levels found

The "specified" impurities are those actually found in batches manufactured by the proposed commercial process. They are listed individually in the specification with specific acceptance criteria.

The ICH Q3A(R2) guideline allows some impurities to be included in the specification as "unidentified." The analytical procedures used to try to identify such impurities need to be summarized and the assumptions made as to how the level is controlled should be stated (for example, where the drug substance is used as a chromatographic standard and it is assumed that the response factor of the drug substance and the impurity to the analytical detection system are the same). Specified unidentified impurities need to be characterized in some suitable way such as "Unidentified Compound A." An example of this approach is in the *European Pharmacopoeia* (Ph Eur) monograph for nystatin. Nystatin is an antifungal substance produced from fermentation using strains of *Streptomyces noursei*. It consists of a mixture of tetraenes, the principal component being nystatin A1. The monograph specifies that it contains a minimum of 85% of nystatin A1; other compounds are limited to a maximum of 4.0% each.

Some care needs to be taken in pursuing this approach of allowing unidentified impurities in the specification because by their nature "unidentified" impurities do not have a completely defined chemical structure and it is therefore not possible to make an assessment of structural alerts for potential toxicity issues (including genotoxicity).

Residual solvents and inorganic impurities are included separately as part of the specification.

Setting Proposed Impurity Limits—How?

Limits for impurities in drug substances are based on two main considerations:

- *Safety aspects*: What levels of the impurity are acceptable?
- *Process capability*: To what levels of impurity is the manufacturing process capable of controlling during routine commercial scale manufacture?

The main consideration in setting the limits ought to be the safety consideration rather than process capability, particularly where there are limited batch data. However, this is often the subject of dispute between the company and the agency MAA reviewers. In the case of Class 2 residual solvents this has now been resolved in that in the EU it is sufficient if the specification limits are restricted to the safety limit in the ICH Q3C(R3) guideline (48).

The justification for the choice of particular impurities and their proposed limits in the specification in the MA should be based on an integrated pharmaceutical/toxicological assessment of the manufacturing process, analytical data, and the safety data. It should be unnecessary to evaluate the toxicological implications related to potential carry-over of a highly reactive reagent (e.g., an acid chloride used for ester formation) if the chemical processing conditions ensure that any residues will be removed or deactivated. Equally, the review of the justification by the health authorities should be a multidisciplinary integrated assessment by the chemical/pharmaceutical and nonclinical reviewers.

Impurities—Safety Considerations

Since the 1990 survey by Cartwright on the toxicology of impurities (20), there have been a number of reports of serious adverse reactions caused by impurities in drug substances. The eosinophilia-myalgia syndrome (EMS) outbreak that occurred in the United States in 1989 was caused by impurities in L-tryptophan produced by Showa Denko of Japan. A number of compounds present in the L-tryptophan have been identified as case-associated impurities. These include 3a-hydroxy-1,2,3,3a,8,8a-hexahydropyrrolo-[2,3-b]-indole-2-carboxylic acid (21), 1,1′-ethylenebistryptophan (22), 3-phenylamino-1-alanine (21), and 2-(3′-indolylmethyl)-indole, 2(2-hydroxyindoline)-tryptophan (21), and a number of other compounds (23). Some commercially available melatonin sources were also found subsequently to contain impurities associated with EMS (24). In 1995, convulsions after parenteral injections of pethidine were attributed to the impurity *N*-methyl-4-phenyl-1,2,3,6-tetrahydropyridine (MPTP) (25). This impurity metabolizes to the neurotoxic *N*-methyl-4-phenyl-pyridinium cation (MPP^+). In 2000, 66 deaths and hundreds of patients suffering from severe side effects were reported after administration of batches of gentamicin sulfate. Wienen et al. (26) have compared the impurity profiles of multiple sources of gentamicin. In 2008, over 149 Adverse Event reports were received by the FDA that included one or more allergic symptoms or symptoms of hypotension and death associated with heparin products (27). Heparin was imported by Baxter Healthcare from the Changzhou SPL plant in China. Analysis of samples of imported batches of heparin by FDA showed that from 2% to 50% of over-sulfated chondroitin sulfate (OSCS) was present as an impurity. Kishimoto et al. (28) found that OSCS activated directly the kinin-kallikrein pathway in human plasma, which can lead to generation of bradykinin, which is a potent vasoactive mediator. In addition, OSCS induced generation of C3a and C5a potent anaphylatoxins derived from complement proteins. In all, 10 countries reported the presence of contaminated heparin within their borders. These were Australia, Canada, China, France, Germany, Italy, Japan, The Netherlands, New Zealand, and the United States.

Biological Qualification of Impurities

The biological qualification of impurities is the process of acquiring and evaluating data that show the safety of an individual impurity or impurity profile. If an impurity is present in a new drug substance that has been fully tested in the major nonclinical and clinical studies, it would generally be considered to be qualified. Issues may be raised, however, if the impurity profiles of batches used for one or more of the major nonclinical studies are different and if a key study (e.g., reproductive toxicity) has been carried out with a batch of drug substance that lacks one of the typical impurities likely to be present in commercial batches: this impurity cannot be regarded as qualified in terms of that particular study.

The levels of impurity present in a new drug substance used in nonclinical studies can be used to justify the proposed impurity limit. The Q3B(R2) guideline indicates that a higher level of impurity can sometimes be justified based on actual amounts of impurity administered in relevant safety studies. Although the guideline does not define how this may be done the following equations can be considered. The animal studies used for these calculations would usually be the chronic studies in rodents (usually rat) and nonrodents (usually the dog), although data from rodent carcinogenicity studies (if conducted) would also be important.

$$\text{Safety factor} = \frac{\text{mg/kg NOAEL dose of drug in repeat} - \text{dose animal study}}{\text{Highest therapeutic mg/kg dose in man}}$$

$$\text{Highest safe \% impurity} = \text{\% impurity in drug in animal study} \times \text{Safety factor}$$

The NOAEL is the no-observed-adverse-effect level found in relevant repeat-dose toxicity studies. Use of data from a study showing atypical species-specific toxicity would be inappropriate. This calculation does of course make a number of assumptions about the pharmacokinetics and toxicokinetics of the drug substance and the impurity in the animal species and man, but it does give some useful indication of the likely safety margin of the impurity. Related substances that are close homologues of the drug substance containing no structural alerts for genotoxicity are unlikely to be of toxicological concern. Literature information may be available on some impurities from stand-alone toxicological evaluations. It may help to search for such information if a Chemical Abstracts Service (CAS) registry number of the particular impurity is available.

Toxic Impurities

The ICH Q3A Impurities guideline includes limits for reporting, identification, and qualification of impurities (see Table 2 of this chapter) but these do not apply to impurities that are expected to be unusually potent, producing toxic, or pharmacological effects at a level not more than the identification threshold. In these cases appropriate limits should be applied. Some examples are the *British Pharmacopoeia* monograph control of 2,3,7,8-tetrachlorodibenzo-*p*-dioxin (also known as TCDD, which is a carcinogen, teratogen, causes endocrine changes, immunological changes, and the skin condition known as chloracne) in hexachlorophene at a level not exceeding 2 ppb, the Ph Eur monograph control of 1-methyl-4-phenyl-1,2,3,4-tetrahydropyridine (also known as MPTP, which is neurotoxic) at a level of not more than 0.1 ppm in parenteral pethidine dosage forms and 10 ppm in oral products, the Ph. Eur monograph control of hydrazine (a probable carcinogen, neurotoxic, nephrotoxic compound) at a level of not more than 500 ppm in isoniazid and at a level of not more than 10 ppm in hydralazine hydrochloride (used in tablets and an injection).

Use of Commercial Software for Predicting Potential Toxicity

Once impurities have been identified there are a number of structure–activity relationship software tools that can provide a rapid validated assessment of their potential toxicity based on chemical structure. There are basically two categories of commercially available software systems for toxicity prediction (29):

- Knowledge-based (expert system) software that allows for the association of a particular fragment or substructure with some toxicological event.
- Quantitative Structure–Activity Relationship (QSAR)–based linking databases of toxicity with molecular descriptors such as molecular fragments.

Table 4 lists some commercially available toxicity prediction software programs. The software screening should also be applied to starting materials, intermediates, and potential degradation products, although it could be quite expensive to undertake multiple assessments for each drug substance unless the software package is available in-house. Use of predictive software is not foolproof and there is no substitute for expert toxicological assessment. In many cases, adverse effects may be apparent only at high doses because at lower doses effective defence mechanisms, of which there are many, may come into play. Moreover, it is strongly advisable to avoid the use of yes/no criteria and to evaluate a safe human dose whenever possible. The principal concern relates to genotoxicity, and it is relatively straightforward to identify structural alerts by visual inspection of impurity structures and by using free databases such as PubMed, PubChem, and TOXNET. If no structural alert is identified, then the impurity would be controlled by using conventional Q3A(R2) qualification criteria.

Process understanding is critical in deciding whether a particular impurity is likely to survive the conditions of manufacture and particularly the last two or three stages. In designing the drug substance manufacturing process, it is better to avoid the use of a toxic reagent if a less toxic one is readily available provided that the substitution does not impair the efficiency of the synthesis. However, such a change should not be necessary if it can be established clearly that residues of any toxic reagent are removed efficiently to safe levels in an intermediate or the drug substance.

Analytical Identification of Highly Toxic Impurities

It has been common practice to analyze impurities perceived to be highly toxic (such as alkylating agents, nitrosamines, and epoxides) by using suitable analytical methods with appropriate limits of detection. While GC with a flame ionization detector (FID) or HPLC with a UV detector are suitable methods for conventional impurities, those of greater concern that may need to be controlled at the parts per million level are likely to need high-sensitivity methods such as GC–MS or LC–MS (e.g., by single ion monitoring on a quadrupole analyzer).

Genotoxic Impurities

One major challenge is the issue of identification and quantitation of potentially genotoxic (and carcinogenic) impurities in drug substances. As van Aerts et al. have illustrated (30), only highly

Table 4 Commercially Available Software for Prediction of Toxicity

Name	Category	Main endpoints predicted	Supplier	Contact details
DEREK for windows	Knowledge-based (expert system)	Mutagenicity, carcinogenicity, skin sensitization, acute toxicity	Lhasa Ltd.	www.lhasalimited.org
HazardExpert pro	Knowledge-based	Oncogenicity, mutagenicity, teratogenicity, membrane irritation, immunotoxicity, neurotoxicity	CompuDrug	www.compudrug.com
OncoLogic (free program)	Knowledge-based	Carcinogenicity	US Environmental Protection Agency (EPA)	www.epa.gov/oppt/newchems/tools/oncologic.htm
CSGenoTox	QSAR based	Mutagenicity	ChemSilico	www.chemsilico.com
MCASE	QSAR based	Carcinogenicity, mutagenicity, teratogenicity, acute and chronic toxicity	Multicase Inc	www.multicase.com
MDL	QSAR based	Carcinogenicity, mutagenicity, reproductive effects, skin and eye irritation	Symyx Technologies	www.symyx.com
DS TOPKAT	QSAR based	Carcinogenicity, mutagenicity, development toxicity, acute and chronic toxicity	Accelrys Inc	www.accelrys.com
Tox boxes	QSAR based	Acute toxicity, genotoxicity	Pharma Algorithms Inc	www.pharma-algorithms.com
Leadscope toxicity	QSAR based	Carcinogenicity, genotoxicity, reproductive toxicity, irritation	LeadScope Inc	www.leadscope.com

potent mutagenic impurities are likely to be identified when the drug substance itself is tested in conventional mutagenicity tests such as those described in the ICH S2 Guidance on *Genotoxicity Testing and Data Interpretation* (31). For example, a conventional Ames assay would be expected to detect genotoxic activity due to methyl methanesulfonate (MMS) or due to an impurity with similar mutagenic potency only when present in the drug substance at levels of 0.33% or higher. Consequently, the use of a minimum concentration of 250 μg/plate of impurity is recommended in Ames assays on impurities (32).

The 2006 Committee for Human Medicinal Products (CHMP) Guideline on *The Limits of Genotoxic Impurities* (33), together with the August 2008 *Question and Answer* document (34) on the CHMP guideline provides recommendations on the toxicological assessment of genotoxic impurities and the determination of acceptable limits. As indicated above for screening of other toxic impurities, it is permitted to screen impurities for the presence of structural alerts or by using suitable software such as DEREK or MCASE. If a structural alert is identified, several options are available for clarifying whether there is a potential safety concern (35).

- Availability of genotoxicity data on the impurity in the literature.
- If genotoxic is the compound, a DNA-reactive genotoxin or a non-DNA reactive compound (clastogen)?
- If DNA-reactive has the compound been tested for carcinogenic potential and, if so, what is the carcinogenic potency (TD_{50}) value? Can a compound-specific VSD (virtual safe dose) be calculated? (see Table 5).

Table 5 Calculation of Virtual Safe Doses for Typical Reagents and Starting Materials Used in API Syntheses

Compound[a]	**Carcinogenic potency TD_{50} (mg/kg/day)**[b]	**VSD (μg/day)**[c]	**Nonsignificant risk level (NSRL) (μg/day)**[d]
Acrylonitrile	6.32 (mouse)	7.6	0.7
2-Amino-4-nitrophenol	839 (rat)	1007	–
4-Amino-2-nitrophenol	309 (rat)	371	–
1-Chloro-2-nitrobenzene	157 (mouse)	188	–
4-Chloro-*o*-phenylenediamine	214 (rat)	257	40
4-Chloro-*o*-toluidine HCl	25.8 (mouse)	31	3.3
p-Chloroanisidine HCl	7.62 (rat)	9.1	–
p-Cresidine	54.3 (rat)	65	5
2,4-Diaminotoluene	2.47 (rat)	3.0	0.2
1,2-Dichloropropane	276 (mouse)	331	9.7
N,N-Dimethylaniline	125 (rat)	150	–
Ethyl acrylate	119 (rat)	143	
Methyl methanesulfonate	31.8	38	7
5-Nitro-*o*-anisidine	63.9 (rat)	65	–
2-Nitro-*p*-phenylenediamine	614 (mouse)	737	–
Nitrobenzene	25.5 (rat)	31	–
Vinyl bromide	18.5 (rat)	22	(Vinyl chloride = 3)
2,4,5-Trimethylaniline	6.13 (mouse)	7.4	–

[a]All Ames-positive.
[b]From CPDB; only species tested or species with lower TD50.
[c]Based on a 60-kg patient using linear extrapolation of TD50 (i.e., dividing by 50,000).
[d] No significant risk level from California OEHHA assessment (46); based on 70 kg adult using various conservative extrapolation techniques.

- If DNA-reactive has a threshold been established (as is the case for ethyl methanesulfonate and also for other sulfonate esters)?

If no literature information is available, additional genotoxicity testing of the impurity should be considered typically using a bacterial reverse mutation assay (Ames test). If the results of this Ames test are negative, this is sufficient to conclude that the impurity is unlikely to be DNA-reactive. If the Ames test is positive, it identifies the impurity as a suspected DNA-reactive mutagen. If the Ames response is weak, additional (in vivo) testing may be appropriate in order to better characterize the genotoxicity profile. In the absence of data on the impurity itself from carcinogenicity studies or evidence for a threshold mechanism for genotoxicity (as is assumed for non-DNA reactive genotoxic compounds, i.e., clastogens), the implementation of an approach defined by the Threshold of Toxicological Concern (TTC) is proposed for DNA-reactive genotoxins. A TTC value of 1.5 μg/day of a genotoxic impurity is considered to be an acceptable risk for chronic exposure. From this threshold value, a permitted level in the drug substance can be calculated based on the expected daily dose. The TTC approach is based on the concept of "acceptable risk" based on a predicted ≤ 1 in 10^5 probability in a human lifetime of an increased cancer incidence from exposure to the impurity (32). A TTC value higher than 1.5 μg/day may be acceptable under certain conditions such as short-term exposure or for treatment of a life-threatening condition when life expectancy is less than 5 years. The "staged-TTC concept" is now accepted by the EU in which the acceptable dose of genotoxic impurity reduces gradually over 12 months from 120 μg for a single-dose exposure to the default 1.5 μg/day (34). However, the guideline recommends that it is necessary to reduce the level of a known or suspected mutagenic impurity to a level as low as reasonably practicable (ALARP), where it is a structure of very high concern, for example, N-nitroso, aflatoxin-like, and azoxy compounds. The CHMP limit of 1.5 μg/day is the subject of some dispute. It has been supported by Müller et al. (36) but criticized by Delaney (37) who suggests that the 1.5 μg/day limit is too conservative and that a higher default limit can be defended, that is, both in keeping with TTC principles and the best interests of patients. In addition, it has been proposed that the TTC in its present form is unreliable for a variety of reasons such as the lack of transparency in respect of the make-up of the 730-compound database from which it is derived, the likelihood of transcription errors in terms of TD_{50} values and the recent acceptance in the EU of a threshold dose of 25 mg/kg/day in mice for ethyl methanesulfonate (42,43). The threshold dose for ethyl methanesulfonate was determined in studies commissioned by Roche in order to resolve safety issues relating to contamination of the antiviral drug Viracept® (nelfinavir mesilate) with this impurity.

One class of drug substances where concern has been expressed regarding the possibility of genotoxic impurities is the mesilate salts. The Ph Eur monographs for all of these salts have a Production Statement requiring that the production method must be evaluated to determine the potential for formation of genotoxic alkyl mesilates "which are particularly likely to occur if the reaction medium contains lower alcohols." This concern seems to be almost entirely hypothetical and it is interesting to note that no regulatory body has proposed publicly a reaction mechanism explaining alkyl mesylate formation and justifying the concerns. Where necessary, the production method is validated to demonstrate that alkyl mesilates are not detectable in the final product. The Ph Eur monographs to which this applies include benzatropine mesilate, betahistine mesilate, bromocriptine mesilate, codergocrine mesilate, desferrioxamine mesilate, dihydroergocristine mesilate, dihydroergotamine mesilate, doxazosin mesilate, loprazolam mesilate, pefloxacin mesilate, pergolide mesilate, phentolamine mesilate, and prochlorperazine mesilate. More recently, however, several publications have cast doubt on whether standard synthetic methods are likely to produce alkyl mesilate/sulfonate impurities:

- Alkyl sulfonate formation occurs in an anhydrous binary alcohol-sulfonic acid system only when the pH is low enough ($pH < 0.5$) to ensure protonation of the alcohol moiety, because the ester formation reaction occurs by sulfonate anion displacement of the hydroxonium ion from a protonated alcohol (38–41). The same considerations apply to an isolated sulfonic acid salt of a base in the presence of an alcohol; highly acidic conditions will be necessary to generate any protonated alcohol and so if the pH is >3, no sulfonate ester will form (41,42). Sulfonic acid salts can be crystallized from ethanol or aqueous ethanol with no alkyl sulfonate formation (40).
- If either water or an equimolar amount of base (e.g., an amine) is present, no alkyl sulfonate formation will occur (39,40).

Even if any alkyl sulfonate were to be produced, it has now been established that a human PDE (permitted daily exposure) of 2 mg/kg/day (nearly 100,000 times the default TTC of 1.5 μg/day) can be determined based on the threshold dose of 25 mg/kg/day in mice mentioned above (43,44). This appears to be because low-*s* (Swain–Scott parameter) selective alkylating agents covalently bind only nitrogen atoms on DNA and the alkylated DNA is repaired with 100% fidelity (44,45). Therefore, other low-*s* alkylating agents such as methyl methanesulfonate, epichlorohydrin, dimethyl sulfate, and ethylene oxide are likely to behave similarly.

The EU guideline on genotoxic impurities is not particularly well understood in its present form and several of its concepts now appear outdated. Consequently, it is believed that an updating exercise should now be undertaken paying particular emphasis to validation of the TTC concept (if feasible), criteria for setting specification levels of genotoxic impurities, and integration of toxicological and chemical considerations (35).

Inorganic Impurities

The CHMP guideline on the *Specification Limits for Residues of Metal Catalysts* (47) recommends maximum acceptable levels of metal residues arising from the use of metals as catalysts or reagents in a drug substance synthesis. The guideline defines Permitted Daily Exposures (PDEs) for each of the metals evaluated. The PDE is the maximum acceptable exposure to residual metal levels on a chronic basis that is unlikely to produce any adverse health effects. Metals are grouped separately into three classes:

- *Class 1 metals*: Metals of significant safety concern, known or suspected human carcinogens, or possible causative agents of other significant toxicity.
- *Class 2 metals*: Metals with low safety concern, generally well tolerated at levels relevant to the guideline (e.g., trace metals needed for nutrition or present in food).

Table 6 Class Exposure and Concentration Limits for Individual Metals Used as Catalysts or in Reagents (47)

	Oral exposure		Parenteral exposure	
Class of metal	**PDE (μg/day)**	**Concentration (ppm)**	**PDE (μg/day)**	**Concentration (ppm)**
Class 1A: Pt, Pd	100	10	10	1
Class 1B: Ir, Rh, Ru, Os	100	10	1	
Class 1C: Mo, Ni, Cr, V	300	30	3	
Class 2: Cu, Mn	2500	250	250	25
Class 3: Fe, Zn	13,000	1300	1300	130

- *Class 3 metals*: Metals with minimal safety concerns with well-established safety profile and tolerated at doses exceeding those relevant to the guideline.

A general set of exposure limits from the guideline and based on PDEs is given in Table 6. Alternatively, an allowable concentration can be calculated from the PDE values in Table 5 by using the maximum daily dose (MDD) of the drug. The equation to be used is as follows:

$$\text{Concentration (ppm)} = \frac{\text{PDE}}{\text{MDD}}$$

The PDE is given in terms of μg/day and the MDD in g/day.

Residual Solvents

Residual solvents are organic volatile impurities that are used during the manufacture or purification of a drug substance.

Toxicological Classification of Solvents

The ICH Q3C(R3) *Impurities: Guideline for Residual Solvents* (48) classifies solvents into three classes:

- *Class 1 solvents*: Solvents to be avoided—known human carcinogens, strongly suspected human carcinogens, and environmental hazards, for example, benzene, carbon tetrachloride, and 1,1-dichloroethane.
- *Class 2 solvents*: Solvents to be limited—nongenotoxic animal carcinogens or possible causative agents of other irreversible toxicity such as neurotoxicity or teratogenicity, for example, dichloromethane, hexane, methylbutylketone, and toluene.
- *Class 3 solvents*: Solvent with low toxic potential where no health-limited exposure limit is needed, that is, ethanol, 1-butanol, 1-propanol, tetrahydrofuran.

Permitted Daily Exposure Values for Solvents

The guideline provides permitted daily exposures (PDEs) in parts per million for each solvent in Classes 1 and 2. For Class 2 solvents, the EMEA Inspectors-Quality Working Party has now clarified (49) that in the EU it is sufficient to restrict the specification in terms of concentration (ppm) to the limit in the ICH guideline and that there is no need for further tightening in line with actual batch data. For Class 3 solvents, the amount of any of these residual solvents of 50 mg or less per day (corresponding to 5000 ppm) is acceptable without further justification. Higher amounts of Class 3 solvents may be accepted if realistic in terms of manufacturing capability and Good Manufacturing Practice (GMP).

The PDEs are based on chronic exposure to these solvents. Where a drug is to be used for short or intermittent treatments, it is possible to argue a case for somewhat higher limits.

In general, solvents in Class 1 should not be used in the manufacture of drug substances, excipients, and drug products, because of their unacceptable toxicity. However, if it can be shown that their use is unavoidable to produce a drug product that constitutes a significant therapeutic advance, their levels should be restricted to the levels shown in the table in the guideline, for example, less than 2-ppm benzene. But what is a "significant therapeutic advance?" The 2006

CHMP *Guideline on the Procedure for Accelerated Assessment* (50) discusses the definition of what constitutes "a product of major public health interest from the point of view of therapeutic innovation." It indicates that the justification should present the arguments to support the claim that the medicinal product introduces new methods of therapy or improves on existing methods thereby addressing to a significant extent the greater unmet needs for maintaining and improving the health of the [European] Community. Some further clarification on the use of Class 1 and 2 solvents is given in the Annexes to the ICH Residual Solvents Guideline (51). This indicates that Class 1 solvents are allowed when used as starting materials early in the synthesis well before the key starting material. In this case, they should be controlled either in an intermediate or in the final active substance. Where a Class 2 solvent is used prior to the last stage in a synthesis, it may be exempted from routine control in the drug substance if it has been shown in a number of pilot or industrial scale batches of an intermediate or the finished substance that the levels are not more than 10% of the ICH limits for the particular solvent.

The ICH guideline offers the possibility of using the overall level of a particular Class 2 solvent in the drug product by assessing the levels in the drug substance and the excipients as long as the overall level in the drug product is less than the PDE value in the guideline table for Class 2 solvents. This might allow the level of the solvent in the drug substance to be higher than the PDE value.

The guideline lists a number of solvents for which no adequate toxicological data were found to support a PDE value. Examples of these solvents are isopropyl ether, methyl isopropyl ketone, and petroleum ether. If a manufacturer wishes to use one of these solvents (or any other solvent not listed in the guideline), a justification based on safety considerations would need to be supplied in the CTD MAA file for the proposed residue level in the drug substance. In many cases, sufficient toxicological information may be available in the literature to determine a NOAEL or LOAEL (lowest observed adverse effect level) and then a PDE can be calculated by using the assessment-factor approach described in ICH Q3C(R3).

Impurities—Assessing Process Capability

The batches used to assess process capability need to be a representative of the final proposed commercial manufacturing process as far as possible. Thus the batches reviewed could include pilot scale batches made by using equipment typical of the final manufacturing scale and also any manufacturing scale batches. Where two or more different manufacturing sites are proposed for the drug substance, batch data would need to be provided to support the capability of each site to manufacture according to the proposed specification. Where sufficient batch data are available to do a meaningful statistical analysis, the general rule of thumb in Decision Tree #1 in the ICH Q6A Specifications guideline (3) is that where there is no safety concern; the upper specification limit for impurities may be set at a maximum of 3σ (three standard deviations) above the mean value of the impurity determinations in representative batches.

Where there are more limited data on representative batches of drug substance, the setting of suitable acceptance limits poses more of a challenge for companies and the regulators. The ICH Q6A Specifications guideline discusses the situation where limited data are available at the time of filing and acknowledges that this can influence the process of setting acceptance criteria. It emphasizes that the basis for setting acceptance criteria should focus on safety and efficacy. The Q6A guideline suggests that initial acceptance criteria are set that can be reviewed as more information is collected. The initial acceptance criteria for impurities can utilize the safety margins calculated from the nonclinical studies and will then allow some "wriggle room" to allow for drug substance manufacturing variability postapproval. Where such an interim specification is proposed, the application should include a formal postapproval commitment statement with

- the rationale for setting an interim specification,
- Details of the number of future batches that will be tested before the acceptance criteria will be finalized, and
- the proposed mechanism for finalizing the specification acceptance limits for impurities.

Table 7 Module 3.2.S.3—Major Deficiencies in CEP Chemical Purity Applications to the EDQM (52)

Module	Deficiency	Percentage of dossiers where deficiency found
3.2.S.3.2 Impurities	Demonstration of the absence of particular reagents in the final substance (e.g., catalysts, alkylating agents).	45
	Demonstration that all solvents used during the process are removed or suitably limited in the final substance is not complete.	42
	Discussion on impurities is not satisfactory—all of the potential impurities should be listed with their origin (by-product or degradation). The suitability of the European Pharmacopoeia to detect and limit the related substances of a particular synthetic process should be demonstrated.	26

DEFICIENCIES IN APPLICATIONS

As mentioned in chapter 9, "Drug Substance Manufacture," a review of the main deficiencies in Certificates of Suitability of the European Pharmacopoeia (CEP) dossiers for chemical purity for 87 applications has been published by the EDQM (52). In relation to 3.2.S.3 Characterization the key deficiencies are summarized in Table 7. These are probably also typical of deficiencies found in DMFs and for information included directly in Module 3.2.S.3 in MAA registration files. It will be noted that a major deficiency is the lack of adequate investigation of carry-over of reagents, catalysts, solvents, etc.

Carry-over of starting materials, intermediates, reagents, catalysts, and solvents can usually be investigated conveniently by evaluating their levels in the next stage in a synthetic manufacturing process. It is only when they carry through to subsequent stages or are used in one of the late stages in the synthesis that it should be necessary to evaluate their levels in the batches of the final purified drug substance.

REFERENCES

1. ICH Guideline The Common Technical Document for the Registration of Pharmaceuticals for Human Use: Quality—M4Q(R1), Quality Overall Summary of Module 2, Module 3: Quality, 2002, ICH. http://www.ich.org/LOB/media/MEDIA556.pdf. Accessed November 19, 2008.
2. CHMP Note for Guidance on Investigation of Chiral Active Substances, III/3501/91, 1993, EMEA. http://www.emea.europa.eu/pdfs/qwp/3cc29aen.pdf. Accessed April 22, 2008.
3. ICH Q6A Guideline on Specifications: Test Procedures and Acceptance Criteria for New Drug Substances and New Drug Products: Chemical Substances, 2006, ICH. http://www.ich.org/LOB/media/MEDIA430.pdf. Accessed April 22, 2008.
4. Kobayashi Y, Ito S, Itai S, et al. Physicochemical properties and bioavailability of carbamazepine polymorphs and dihydrate. Int J Pharm 2000; 193:137–146.
5. Aguiar AJ, Krc J, Kinkel AW, et al. Effect of polymorphism on the absorption of chloramphenicol from chloramphenicol palmitate.J Pharm Sci 1967; 56:847–853.
6. Bauer J, Spanton S, Henry R, et al. Ritonavir: An extraordinary example of conformational polymorphism. Pharm Res 2001; 18(6):859–866.
7. Morissette SL, Soukasene S, Levinson D, et al. Elucidation of crystal form diversity of the HIV protease inhibitor ritonavir by high-throughput crystallization. Proc Natl Acad Sci U S A 2003; 100(5):2180–2184.
8. Byard SJ, Jackson SL, Smail A, et al. Studies on the crystallinity of a pharmaceutical development drug substance.J Pharm Sci 2005; 94(6):1321–1335.
9. Raw SA, Furness MS, Gill DS, et al. Regulatory considerations of pharmaceutical solid polymorphism in abbreviated new drug applications (ANDAs). Adv Drug Deliv Rev 2004; 56:397–414.
10. Guidance for Industry: Waiver of In Vivo Bioavailability and Bioequivalence Studies for Oral Immediate-Release Solid Oral Dosage Forms Based on a Biopharmaceutics Classification System. 2000, CDER, FDA. http://www.fda.gov/cder/guidance/3618fnl.pdf. Accessed April 22, 2008.
11. Poole SK, Patel S, Dehring K, et al. Determination of acid dissociation constants by capillary electrophoresis.J Chromatogr A 2004; 1037(1–2):445–454.

12. Test No. 107: Partition Coefficient (*n*-octanol/water): Shake Flask Method. OECD Guidelines for the Testing of Chemicals, Section 1: Physical-Chemical Properties. July 27, 1995, OECD.
13. Jia Z, Mei L, Lin F, et al. Screening of octanol–water partition coefficients for pharmaceuticals by pressure-assisted microemulsion electrokinetic chromatography. J Chromatogr A 2003; 1007(1–2):203–208.
14. Test No. 117: Partition Coefficient (*n*-octanol/water): HPLC Method. OECD Guidelines for the Testing of Chemicals, Section 1: Physical-Chemical Properties. November 23, 2004, OECD.
15. ICH Q3A(R2) Guideline on Impurities in New Drug Substances, 2006, ICH. http://www.ich.org/LOB/media/MEDIA422.pdf. Accessed April 22, 2008.
16. CPMP Guideline on The Chemistry of New Active Substances, CPMP/QWP/130/96, Rev 1, 2003, EMEA. http://www.emea.europa.eu/pdfs/human/qwp/013096en.pdf. Accessed April 22, 2008.
17. Bertram J. Multisource pharmaceutical raw materials—Method profiling and structural elucidation of impurities. Pharmeuropa 2003; 15(2):280–289.
18. SYLVIA. SciTegic Inc, San Diego, CA. http://www.molecular-networks.com/software/sylvia. Accessed April 22, 2008.
19. Methods in Organic Synthesis, Database of novel synthetic methodologies. Accelrys Software Inc. http://www.accelrys.com/products/accord/chemical-databases/methods-in-organic-synthesis. Accessed April 22, 2008.
20. Cartwright AC. Toxicology of impurities in organic synthetic drugs. Int Pharm J 1990; 4(4):146–150.
21. Williamson BL, Johnson KL, Tomlinson AJ, et al. On-line HPLC-tandem mass spectrometry structural characterisation of case-associated contaminants of L-tryptophan implicated with the onset of eosinophilia-myalgia syndrome. Toxicol Lett 1998; 99(2):139–150.
22. Adachi J, Asano M, Ueno Y. Tetrahydro-beta-carboline-3-carboxylic acids and contaminants of L-tryptophan. J Chromatogr A 2000; 881(1–2):501–515.
23. Müller B, Pacholski C, Simat T, et al. Synthesis and formation of an EMS correlated contaminant in biotechnologically manufactured L-tryptophan. Adv Exp Med Biol 1999; 467:481–486.
24. Laforce R, Rigozzi K, Paganetti M, et al. Aspects of melatonin manufacturing and requirements for a reliable active component. Biol Signals Recept 1999; 8(1–2):143–146.
25. Lessel J. MPTP—Neurotoxin and model substance in Parkinson research. Pharm Unserer Zeit 1994; 23(2):106–107.
26. Wienen F, Deubner R, Holzgrabe U. Composition and impurity profile of multisource raw material of gentamicin—A comparison. Pharmeuropa 2003; 15(2):273–279.
27. Information on Adverse Event Reports of Heparin, May 11, 2008, CDER, FDA. http://www.fda.gov/cder/drug/infopage/heparin/adverse_events.htm. Accessed November 5, 2008.
28. Kishimoto TK, Viswanathan K, Ganguly T, et al. Contaminated heparin associated with adverse clinical events and activation of the contact system. N Engl J Med 2008; 358(23):2459–2467.
29. Cronin M. Opportunities for Computer-Aided Prediction of Toxicity in Drug Discovery. Business Briefings Ltd a division of Ambient PLC, Islington, London: Pharmagenerics, 2003:50–53.
30. Aerts LAGJM van, Kasper P, Queckenberg O, et al. International workshop on qualification and control of impurities. Drug Inf J 2006; 40(2):155–163.
31. ICH Topic S2 (R1). Draft Guidance on Genotoxicity Testing and Data Interpretation for Pharmaceuticals Intended for Human Use, March 2008. Geneva, Switzerland: ICH.
32. Dobo KL, Greene N, Cyr MO, et al. The application of structure-based assessment to support safety and chemistry diligence to manage genotoxic impurities in active pharmaceutical ingredients during drug development. Regul Toxicol Pharmacol 2006; 44(3):282–293.
33. CHMP Guideline on The Limits of Genotoxic Compounds, EMEA/CHMP/QWP/251334/2006, June 2006, EMEA. http://www.emea.europa.eu/pdfs/human/swp/519902en.pdf. Accessed April 22, 2008.
34. CHMP Safety Working Party Question and Answers on the CHMP Guideline on the Limits of Genotoxic Impurities, EMEA/CHMP/SWP/431994/2007 Revision June 1, 2008, EMEA. http://www.emea.europa.eu/pdfs/human/swp/43199407en.pdf. Accessed November 6, 2008.
35. Snodin D. EU guideline on genotoxic impurities needs updating. RAJ Pharma 2008; 19:593–598.
36. Müller L, Mauthe RJ, Riley CM, et al. A rationale for determining, testing and controlling specific impurities in pharmaceuticals that possess potential for genotoxicity. Regul Pharmacol Toxicol 2006; 44:198–211.
37. Delaney E. An impact analysis of the application of the threshold of toxicological concern concept to pharmaceuticals. Regul Pharmacol Toxicol 2007; 49:107–124.
38. Lee DG, Demchuk KJ. A carbon-13 nuclear magnetic resonance study of the basicities of aliphatic alcohols. Can J Chem 1987; 65:1769–1774.
39. Snodin DJ. Residues of genotoxic alkyl mesylates in mesylate salt drug substances: Real or imaginary problems? Regul Toxicol Pharmacol 2006; 45(1):79–90.

40. Teasdale A, Eyley SC, Delaney E, et al. Mechanism and processing parameters affecting the formation of methyl methanesulfonate from methanol and methanesulfonic acid, an illustrative example for sulfonate ester impurity formation. Org Process Res Dev 2009; 13(3):429–433.
41. Elder D, Snodin D. Drug substances presented as sulfonic acid salts: Overview of utility, safety and regulation. J Pharm Pharmacol 2009; 61(3):269–278.
42. CHMP European Public Assessment Report for Viracept, 2007, EMEA. http://www.emea.europa.eu/humandocs/PDFs/EPAR/Viracept/Viracept-H-164-Z-109-AR.pdf. Assessed November 5, 2008.
43. Questions and Answers on the Follow-up to the Contamination of Viracept (*nelfinavir*) with Ethyl Mesilate, July 24, 2008, EMEA. www.emea.europa.eu/humandocs/PDFs/EPAR/Viracept/Q&A_Viracept_37580708en.pdf. Accessed November 2008.
44. Müller L, Gocke E, Lave T, et al. Elevated ethyl methanesulfonate (EMS) in nelfinavir mesylate (Viracept®, Roche): Animal studies confirm toxicity threshold and absence of risk to patients. Poster LBPE1167 XVII International AIDS Conference, August 3–8, 2008, Mexico City, Mexico.
45. Vogel EW, Nivard MJ. International commission for protection against environmental mutagens and carcinogens, the subtlety of alkylating agents in reactions with biological macromolecules. Mutat Res 1994; 305(1):13–32.
46. Proposition 65 Safe Harbor Levels. No Significant Risk Levels for Carcinogens and Maximum Dose Levels for Chemicals Causing Reproductive Toxicity, Office of Environmental Health Hazard Assessment, California Environmental Protection Agency, January 2008. www.oehha.org/prop65/pdf/Feb2008StatusReport.pdf. Accessed November 5, 2008.
47. CHMP Guideline on The Specification Limits for Residues of Metal Catalysts. EMEA/CHMP/SWP/4446/2000, February 2008, EMEA. http://www.emea.europa.eu/pdfs/human/swp/444600enfin.pdf. Accessed November 6, 2008.
48. ICH Q3C(R3) Guideline on Impurities: Guideline for Residual Solvents. 2005, ICH. http://www.ich.org/LOB/media/MEDIA423.pdf. Accessed April 22, 2008.
49. Questions and Answers, European Medicines Agency Inspections and Quality Working Party, EMEA. http://www.emea.europa.eu/Inspectors/qwp/q12.htm. Accessed November 6, 2008.
50. CHMP Guideline on the Procedure for Accelerated Assessment Pursuant to Article 14(9) of Regulation (EC) No. 726/2004. EMEA/419127/05 July 2006, EMEA. http://www.emea.europa.eu/pdfs/human/euleg/41912705en.pdf. Accessed April 22, 2008.
51. CHMP/CVMP Annexes to the ICH Guideline on Impurities: Residual Solvents, Annex I: Specifications for Class 1 and Class 2 Residual Solvents in Active Substances, Annex 2: Residues of Solvents used in the Manufacture of Finished Products, CPMP/QWP/450/03, July 2004, EMEA. http://www.emea.europa.eu/pdfs/human/qwp/045003en.pdf. Accessed November 6, 2008.
52. Certification of suitability of Monographs of the European Pharmacopoeia. How can the content of applications for a certificate of suitability for chemical purity be improved? The TOP TEN deficiencies found in applications, EDQM, Strasbourg, December 2006. http://www.edqm.eu/medias/fichiers/Top_10_Deficiencies_found_in_Applications.pdf. Accessed April 7, 2008.
53. Guidances for Industry, Preparation of the Quality Information for Drug Submissions in he CTD Format: Conventional Biotherapeutic Products, May 2004, Health Canada. http://www.hc-sc.gc.ca/dhp-mps/brgtherap/applic-demande/guides/qualit/prod/tech-doc-biologic/ctd_convbio-eng.php. Accessed November 19, 2008.

11 Control of Drug Substance

Anthony C. Cartwright
Pharmaceutical Regulatory Consultant, Harpenden, U.K.

3.2.S.4 CONTROL OF DRUG SUBSTANCE

This section of a CTD registration file includes the specification, details of analytical procedures, reports on the validation of the analytical procedures, batch analyses of relevant batches, and a justification of the proposed specification (choice of tests and the proposed acceptance criteria).

3.2.S.4.1 SPECIFICATION

The ICH Q6A Specifications Guideline (1) defines a specification as "a list of tests, references to analytical procedures, and appropriate acceptance criteria which are numerical limits, ranges or other criteria." Specifications are critical quality standards that are approved by the regulatory authorities on the basis of the evidence submitted by the applicant in the marketing authorisation application file. The specification lists all of the tests with which each batch of drug substance should comply. Where a drug substance is purchased from a supplier with a certificate of analysis (C of A), it should be clearly stated which tests will be carried out routinely only by the supplier and which will be carried out by or on behalf of the drug product manufacturer.

The specification included in the registration file should be for the quality of drug substance, which is to be used in the manufacture of the drug product. Thus if the drug substance is manufactured by the supplier and then subsequently milled or micronized by a contractor or the drug product manufacturer, the specification should be for the milled or micronized grade. The specification for the intermediate (unmilled) grade should be included in 3.2.S.2.4 Control of Intermediates.

Need for Signed and Dated Specifications

Some regulatory agencies (such as the Therapeutic Products Directorate of Health Canada) require the specification to be formally signed and dated by the Quality Control manager or designate (2). The specification should include a version number. Electronic signatures are accepted if certified in accord with an acceptable standard (3).

Tests to Be Included in a Drug Substance Specification

The ICH Q6A Guideline classifies tests and acceptance criteria in a specification into "universal tests" (applicable to all drug substances) and "specific tests" (to be considered on a case-by-case basis for drug substance and included where they have a significant impact on the quality of the drug substance). The key tests mentioned in the guideline are listed in Table 1.

A pharmacopoeial monograph for a drug substance usually contains the following key elements (4):

- Definition (percentage purity limits calculated with respect to the anhydrous substance)
- Characters (appearance, solubility, and melting or boiling point)
- Identification (Infrared-IR, ultraviolet-UV, melting point, boiling point, specific optical rotation, Thin Layer Chromatography-TLC, Gas Chromatography-GC, and chemical reactions)
- Tests (appearance of solution, pH, optical rotation, UV spectrophotometry, impurities, readily carbonizable substances, foreign anions or cations, heavy metals, loss on drying, Karl Fischer, sulphated ash, and residual solvents)
- Assay
- Storage and labeling statements

A company's proposed specification is usually presented in tabular form with tests, acceptance criteria, and a reference to the specific analytical procedure or procedures to be used. Where a manufacturer's or applicant's analytical procedure is used, it is usually referred to by

Table 1 List of Universal and Specific Tests for a Drug Substance

Universal tests	Specific tests
Description (physical state and color)	Physicochemical properties
Identification	Particle size
Assay	Polymorphic form
Impurities	Water content (loss on drying or Karl Fischer titration)
• Organic impurities	Microbial limits and endotoxin
• Inorganic impurities	
• Residual solvents	

its reference number; pharmacopoeial procedures can be cited by reference to the appropriate general test method. Pharmacopoeial procedures are generally preferred by the reviewers in the regulatory authorities to in-house procedures.

Internationally Harmonized General Test Methods

ICH Quality Topic Q4B is Evaluation and Recommendation of Pharmacopoeial Texts for Use in the ICH Regions (5). This describes the process of selecting and adopting pharmacopoeial texts for use as general test methods, which are interchangeable in the ICH regions, as discussed and agreed in the tripartite Pharmacopoeial Discussion Group (PDG) consisting of USP, Ph Eur, and the JP. ICH then issues topic-specific annexes with information on the texts. The first of these, the Q4B Annex 1 on Residue on Ignition/Sulphated Ash General Chapter, was finalized in November 2007 and has now been adopted in all regions (6).

Periodic or Skip-Lot Testing

A periodic or skip-lot test is one that is carried out on either preselected batches (e.g., every tenth batch) or at predetermined intervals (e.g., once a month) rather than on a batch-by-batch basis. A periodic test is one that is felt to be indicative of quality but where available information indicates that it need not be carried out routinely. Any test that is proposed as a periodic test should be identified as such in the specification. An example might be a test for polymorphic form of the drug substance where considerable evidence exists from a large number of production batches using a particular manufacturing procedure that only one particular polymorph is produced.

Periodic Test Failures

Any periodic test failure must be investigated in the same way as a failure of any other test and may result in a batch rejection. The investigation would need to include testing of earlier batches and the possible need for the test to be upgraded so as to be carried out routinely on each batch.

Key Drug Substance Tests

Description

Description includes the physical form (solid, gas, or liquid) and color. Crystalline or amorphous form is sometimes specified. The terminology "white" should be avoided as minor traces of colored impurities may affect color. "White to off-white" or "white or almost white" would be preferred. Colored substances may be described in terms of one of the major colors (e.g., brown) or using a compound term (e.g., blue-green).

Identification

Identification testing should show specificity for the drug substance against other compounds manufactured or purchased in-house. Specificity is also desirable against other compounds manufactured by a drug substance supplier. This is normally achieved by comparison of the properties of a batch sample (e.g., spectrum, chromatography, chemical reactions) with that of a reference standard. Where one procedure is not sufficiently specific, it can be complemented by another procedure. Where a drug substance is a salt such as a hydrochloride or sulphate, the identification should include a test for the counterion as well as the active moiety.

Spectrophotometric methods such as IR or NMR using a comparison with the spectrum of a reference standard or with a published reference spectra are usually considered as first choice method for nonionized organic substances. Some inorganic salts of organic drug substance such as sulphates and phosphates can be distinguished by IR. Where the test is insufficient because closely related substances have similar spectra, another complementary test may also be needed (e.g., chromatographic retention time, melting point).

Identification based on a chromatographic retention time is unacceptable on its own as not being sufficiently specific (1); however, a combination of tests such as High Pressure Liquid Chromatography/Diode Array Detector (HPLC/DAD), High Pressure Liquid Chromatography/Mass Spectrometry (HPLC/MS) or Gas Chromatography/Mass Spectrometry (GC/MS) will be accepted.

UV and visible absorption spectrophotometry are regarded as less specific unless the absorption curve shows several maxima and minima. An identification test will usually consist of identification of an absorption peak (λ_{max}) and measurement of the specific absorbance (the absorbance of a 10 g/l solution in a 1-cm cell), or the ratio of absorbances of two peaks.

Thin-layer chromatography needs the use of a reference substance. Selectivity can be improved by using chemical reactions with a spray reagent detection system.

Chemical reactions for a specific part of the molecule are often used in pharmacopoeial tests (usually to provide identification when instrumental methods are not readily available). These are used less often within the pharmaceutical industry apart from their use to identify inorganic salts of organic drug substances.

Assay

An assay is normally needed for drug substances except in the very rare case where its omission can be fully justified because all foreseeable impurities are identified and limited and other quantitative tests can be carried out with sufficient precision to define purity. An example is the Ph Eur monograph for clofibrate where impurities are controlled by a GC volatile-related substances test, a GC limit test for 4-chlorophenol, and where there are tight limits for the refractive index and relative density of clofibrate (which is a liquid).

The ICH Q6A Specifications Guideline recommends that the assay be specific and stability indicating. Although many of the pharmacopoeial monographs still use methods such as a UV assay or nonaqueous titration, these are less specific, and chromatographic methods such as liquid chromatography (LC) or gas chromatography are now used for new drug substances.

For well-defined salts (where the stoichiometry is reproducible), the assay of one of the ions, preferably the active moiety, is normally sufficient. Both active moiety and counterion must be determined when the stoichiometry is not consistent.

As is the case for pharmacopoeial drug substances, the assay limits for a new drug should be expressed in terms of the anhydrous drug substance. Where the drug substance is a hydrate, it will also be useful to determine the "as is" assay of a batch of drug substance so that this assay can be used to factorize the amount of drug substance added to a batch of drug product. Factorization is the adjustment of the amount of the drug substance added so that it ensures that the assay of the drug product is 100% label claim.

$$\text{Factorised Amount} = \text{Theoretical Batch Quantity} \times \frac{100}{\text{'As is' assay}}$$

Impurities

Organic, inorganic impurities and residual solvents all need to be controlled as defined in the ICH Q3A and Q3C Guidelines (7,8) and are discussed in detail in chapter 10 of this book. Organic impurities are measured by a variety of techniques, and often their analysis is combined with the assay for the parent drug molecule using a suitable chromatographic method such as LC or GC to separate impurities from the drug itself. The most common method for controlling

Table 2 Recommendations for Heavy Metal Limits for Drug Substances

Daily intake of drug substance and length of treatment	Heavy metals limit
> 0.5 g/day for less than 30 days	20 ppm
> 0.5 g/day for more than 30 days	10 ppm
< 0.5 g/day for more than 30 days	Parenteral use 10 ppm, other use 20 ppm
< 0.5 g/day for less than 30 days	No test needed

impurities is LC, although GC or capillary electrophoresis may be the preferred method in some instances—for example, where the analytes are volatile. Many pharmacopoeial drug substance monographs still employ TLC, but this is being phased out and replaced by LC or GC as far as practicable. Thus TLC should generally be avoided for new drug substances.

The drug substance can often be used as a standard to estimate the level of impurities by using the response factor (k), which is often determined as part of the analytical validation study. This is based on the relative peak areas of equal weights of the impurity and the drug substance itself under the conditions of the test, where the response of the drug substance is given as unity. Response factors of less than 0.2 or more than 5 are not generally used. The correction factor to be applied in the formula for calculating the level of impurity is then the reciprocal of the response factor (1/k).

Where an inorganic acid and base is used in the final steps of the manufacture of the drug substance, it may be useful to include a test for foreign anions or cations as an impurity (such as tests for chloride or sulphate).

Heavy Metals

It may be more appropriate to determine levels of inorganic impurities such as catalysts by specific methods such as atomic absorption spectrophotometry.

The pharmacopoeial test for heavy metals uses the production of dark-colored sulphides in comparison to a standard containing a known quantity of lead. Metals that can form sulphides include lead, copper, silver, mercury, cadmium, bismuth, gold, platinum, vanadium, arsenic, and molybdenum. The European Pharmacopoeia Technical Guide (4) provides recommendations for limits in relation to dose of the drug and duration of treatment and these are shown in Table 2. The Ph Eur has seven different methods for heavy metals, depending on the solubility characteristics of the drug substance, the limit of quantitation needed, etc. The method to be used should be specified.

Appearance of Solution

This test is almost always used where a drug substance will be used in the manufacture of a liquid product. It can comprise one or both of the following tests:

- Clarity and degree of opalescence of a solution (when compared to a reference preparation)
- Degree of coloration of a solution (used for substances that may contain colored impurities or degrade to form colored degradation products).

Physicochemical Properties

A range of physicochemical properties can be considered for inclusion in the specification. These include pH of an aqueous solution or suspension, melting point/boiling point, and refractive index. Pharmacopoeial general test methods are preferred by the regulatory agencies wherever possible. A pH test can allow the limitation of acidic or alkaline impurities and may also allow the stoichiometric composition of some salts to be verified.

Loss on Drying

Where a drug substance is a solvate or hydrate, it is usual to include both upper and lower limits for loss on drying. Otherwise an upper limit is usually applied. The pharmacopoeias have a number of standard methods for testing loss on drying; stable drugs are usually dried in an oven at 105°C to constant mass.

Moisture Content Using the Karl Fischer Method
Where the drug substance is a solvate, or considerable levels of solvent residues may be present, it may be more appropriate to determine water content by using one of the commercial Karl Fischer reagents (preferably pyridine-free). This is a more specific method.

Residue on Ignition/Sulphated Ash
This test is used as measure of the total amount of foreign cations present in an organic drug substance (6). The typical pharmacopoeial limit is 0.1%.

Polymorphic Form
Polymorphism is discussed in chapter 10. Where necessary (where the polymorphic form has an effect on stability or bioavailability), the solid state form should be specified and controlled using a suitable test such as X-ray powder diffraction, IR, solid state NMR, or thermal analysis (TGA or DSC).

Particle Size
For drug substances used in solid oral dosage forms or suspensions and in other formulations such as ophthalmic suspensions, particle size can have an effect on dissolution rate and hence bioavailability and can also affect patient acceptability of the formulation. It can also affect stability of the drug substance. For very potent drugs, the particle size can affect content uniformity of the dosage form if mixed in a dry powder. For dry powder inhalation products, the drug substance particle size has typically to be less than 5 μm using a suitable method with acceptance criteria set at multiple points across the size distribution (9). The ICH Q6A Specifications Guideline includes a Decision Tree on acceptance criteria for the need for a particle-size distribution and this has been adapted and updated as Figure 1.

Microbiological Testing
There may be a need to specify the total count of aerobic microorganisms, the total count of yeasts and moulds, and the absence in the sample tested of specific pathogens such as *Staphylococcus aureus*, *Escherichia coli*, *Salmonella*, etc. A Decision Tree is given in ICH Q6A Specifications Guideline, which addresses the properties of the drug substance, the possible need for sterility, whether the drug substance manufacturing process will involve steps that inherently reduce microorganisms (such as alcohol used for the final crystallization solvent), and the results of any batch tests.

Drug substances to be used in manufacture of parenteral and ophthalmic products may be produced sterile and thus each batch may need to be tested for endotoxins (e.g., using the Limulus Amoebocyte Lysate test) and sterility. Outside the United States, endotoxin controls for ingredients/products are not usual for ophthalmic products. It is more usual to buy in nonsterile actives and achieve sterility by a separate sterilization process or using sterile filtration in solution.

3.2.S.4.2 ANALYTICAL PROCEDURES

The analytical procedures for the drug substance should be provided. Within companies, it is often a matter of debate as to how much detail should be included in the procedures filed as part of the registration dossier. If full details are provided, then a relatively small change in the analytical procedure can require a variation (the European Union) or may need to be mentioned in the Annual Report (the United States). If the level of detail provided is similar to that in a pharmacopoeial monograph, this is acceptable to most authorities, but then the company may still need to maintain its own detailed procedure kept in parallel to the regulatory document, and this will be monitored as part of the regulatory compliance role within the company.

Where the analytical procedure used is the current version cited in an official compendium, then a reference to this is sufficient. However, it should be kept in mind that European reviewers generally do not have the USP or JP on their desks, and U.S. reviewers do not have the Ph Eur, so that copies of an official procedure should be provided if a "foreign" pharmacopoeia procedure is cited and should be provided in English if this is the language chosen for the dossier. Where

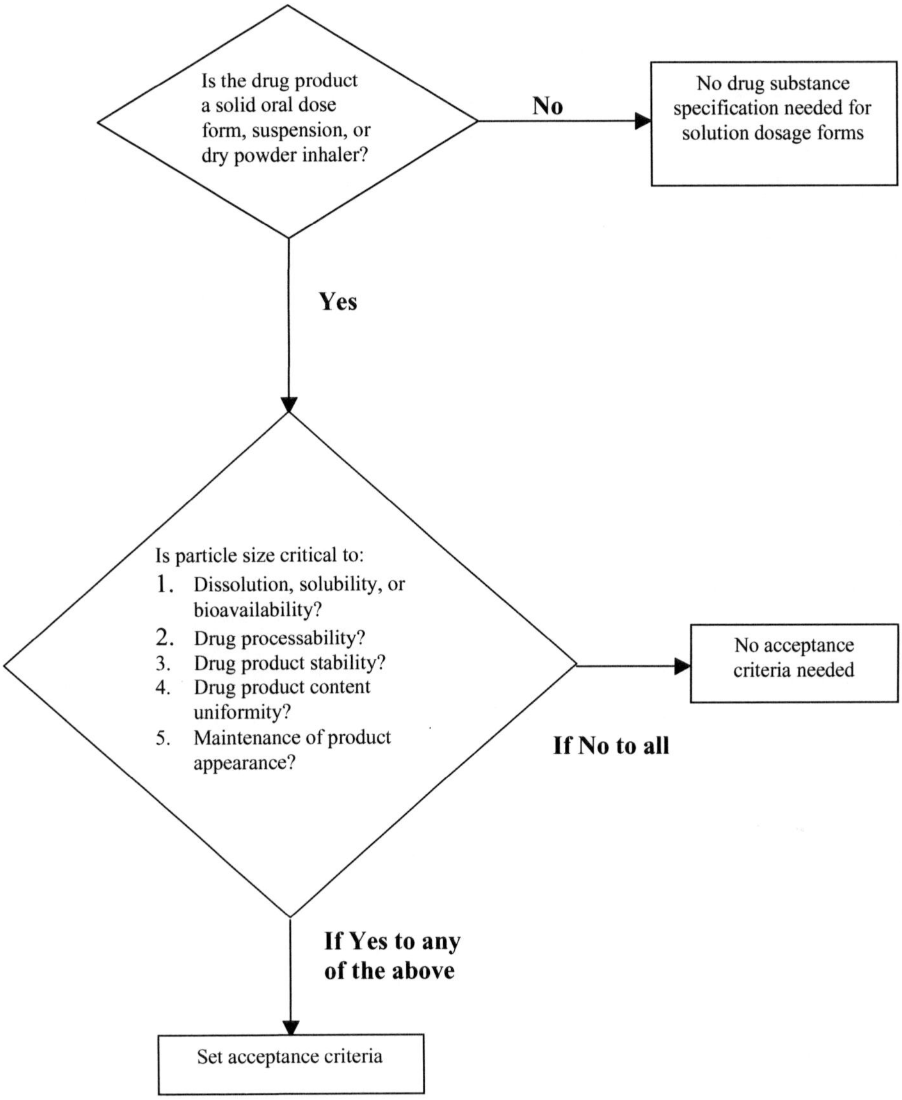

Figure 1 Decision Tree on the need for a drug substance particle-size distribution.

a number of variants of an official procedure are available (e.g., the seven different methods in the Ph Eur for Heavy Metals), the exact procedure used should be stated.

List of Information to Be Included in an Analytical Procedure

The FDA Draft Guidance on Analytical Procedures and Method Validation (10) includes a useful list of the information that should typically be included in a description of an analytical procedure:

A. **Principle:** Statement of the principle (e.g., separation based on isocratic reversed phase HPLC with UV detector).
B. **Sampling:** Details of samples selected, how used, and the number of replicate analyses per sample.
C. **Equipment and equipment parameters:** Listing of the equipment (instrument, detector, column type, and dimensions) and settings (flow rate, temperatures, run time, wavelength for detector, etc.).

D. **Reagents:** List of reagents and grades, with information where necessary on preparation of any in-house reagents.
E. **System suitability testing:** Details of the routine testing to ensure that the system is working properly at the time of analysis (e.g., for an HPLC procedure the tailing factor, relative retention, resolution between the peak and an impurity or the peak and the internal standard, relative standard deviation, capacity factor and number of theoretical plates).
F. **Preparation of standards:** Procedures for preparation of standard solutions (e.g., working standards, internal standards).
G. **Preparation of samples:** Preparation of samples for testing.
H. **Procedure:** Description of the analytical procedure.
I. **Calculations:** Mathematical formulas to be used, with explanation of symbols and numerical factors.
J. **Reporting of results:** Summary of format for results (e.g., percentage on a dried basis, parts per million).

3.2.S.4.3 VALIDATION OF ANALYTICAL PROCEDURES

The 1994 ICH Guideline on Validation of Analytical Procedures and the 1996 Guideline on Methodology were merged in November 2005 and these now comprise the ICH Q2(R1) Guideline (11). The FDA Draft Guidance on Analytical Procedures and Methods Validation is complementary to the ICH Guideline. The European Pharmacopoeia 2005 Technical Guide for the Elaboration of Monographs (4) also includes a substantial section on Analytical Validation with useful details as to the validation requirements for specific analytical methods.

The purpose of validation is to show that the analytical procedures are suitable for their intended purpose. In the European Union, the method validation data (on the drug substance, excipients, dosage form, and packaging materials) will be evaluated by the Module 3 Quality reviewer in the European national agency. The analytical methods may be assessed in laboratories associated with the assessment team.

In the United States, in addition to the evaluation by the FDA reviewing chemist, there is an additional requirement for a separate "methods validation package" of analytical procedures, validation reports, specifications, samples (of standards, drug substance, impurities, and degradation products), results, etc., which have to be submitted. This package enables the methods to be evaluated by one of the FDA laboratories to check the suitability of the analytical procedures.

The consideration of the suitability of the analytical procedure may be based on the ICH criteria. For example, if the reporting level for unknown impurities is 0.05%, the test procedure must be able to quantify impurities at this level with appropriate precision and accuracy.

The ICH Guideline applies to the four of the most common analytical procedures – identification tests, quantitative test for impurities, limit tests for impurities, and quantitative tests of the active moiety in samples of the drug substance. The European Pharmacopoeia Technical Guide includes information on validation of other tests such as optical rotation and noninstrumental limit tests. The FDA Draft Guidance on Analytical Procedures and Method Validation includes guidance on validation of particle-size test methods.

Typical validation parameters include accuracy, precision (repeatability and intermediate precision), specificity, limit of detection, limit of quantitation, linearity, and range.

The ICH validation requirements for identification, impurity testing, and the assay are set out in Table 3.

Information to Be Included in a Validation Report

Analytical validation studies are carried out according to a predetermined validation protocol, and the protocol is often submitted in the registration file as well as the validation report. However, this is not necessary: the protocol can be merely kept available on file if needed for a GMP inspection. The validation protocol will include the specified procedures for statistical analysis of the validation data, the amount of data to be collected, and the criteria to be used to determine the acceptability of the analytical procedure.

The report should include information on the validation characteristics recommended in the ICH Guideline (accuracy, precision, linearity, range, etc.) both as a summary and with all

Table 3 ICH Analytical Validation Requirements

Validation parameter	Identification	Impurity testing: Quantitative test	Impurity testing: Limit test	Assay
Accuracy	−	+	−	+
Precision				
• Repeatability	−	+	−	+
• Intermediate Precision	−	+[a]	−	+[a]
Specificity[b]	+	+	+	+
Limit of detection	−	−[c]	+	−
Limit of quantitation	−	+	−	−
Linearity	−	+	−	+
Range	−	+	−	+

[a] Where reproducibility testing (precision test between laboratories) has been carried out, intermediate precision is not needed.
[b] Lack of specificity of one analytical procedure could be compensated by another procedure.
[c] May be needed in some cases.
−, Signifies that this parameter not normally required.
+, Signifies that this parameter is normally required.

of the individual data in a tabulated format. As recommended in the FDA Draft Analytical Procedures and Methods Validation Guideline, the report should also include the following:

- Data showing the stability of analytical samples through the time required to carry out the analysis.
- Legible copies of chromatograms or instrumental output, including records for standard and sample.
- Copies of typical calculations using the raw data to show how the levels of individual and total impurities are calculated.
- Information from stress studies.
- Discussion on possible formation and control of polymorphs and enantiomers.
- Identification and characterization of each inorganic and organic impurity and residual solvent.
- A list of the known impurities with structures if possible.

Identification

As mentioned earlier, the identification test(s) should be able to discriminate between substances of closely related structures, which may be present at the company or may have been received from the supplier. This could be shown by applying the test to materials that are closely structurally related to the drug substance, and showing that a positive result is not obtained.

Impurity Testing

Specificity

Discrimination of the analyte in the presence of the drug substance itself and other impurities needs to be demonstrated. For chromatographic procedures, representative chromatograms should be included labeled with the individual components.

Critical separations can be shown from the resolution of the two closely eluting components (12):

$$R_s = \frac{1.18(t_{R2} - t_{R1})}{w_{h1} + w_{h2}}, \text{ where } t_{R2} > t_{R1}$$

where t_{R1} and t_{R2} are the retention times or distances along the baseline from the point of injection for the two adjacent peaks; w_{h1} and w_{h2} are the peak widths at half height.

A resolution greater than 1.5 corresponds to baseline separation.

The discrimination can be shown by spiking the drug substance with appropriate levels of potential or actual impurities and degradation products and demonstrating their separation in the analytical procedure. For a new drug substance or a generic version of an existing drug substance, the specificity is also shown using the information from the stress stability studies (thermal degradation, acidic and basic hydrolysis, oxidation, and photodegradation) as detailed in chapter 19. The stress studies should show that the individual peaks due to the active substance, impurities, and degradation products are all suitably resolved.

Range

The specified range for the validation studies on an impurity is that over which it has been shown to have an acceptable linearity, accuracy, and precision. For impurities, it should be from the reporting level (as specified in ICH Q3A, Q3B, and Q3C Guidelines) to 120% of the specification. For toxic impurities, the range should encompass a range around the specification limit (say 50–120% of the limit).

Linearity

The linearity over the required range can be shown by carrying out the analytical procedure on dilutions of a standard solution of the impurity. The results are statistically analyzed to show the correlation coefficient, y-intercept, slope of the regression line, and residual sum of squares. A copy of the plot of signal versus impurity analyte concentration should also be provided.

Accuracy

The accuracy is assessed by spiking the drug substance with known amounts of impurity, using a minimum of three concentrations with three replicates of the total analytical procedure. Accuracy is reported as the percentage recovery by the assay of known added amount of the impurity analyte in the sample.

Precision

The repeatability is determined by a test carried out by a single analyst on a single day using the proposed procedure on a minimum of either three concentrations with three replicates of the total analytical procedure or a minimum of six replicate determinations at the test concentration.

The intermediate precision is a test to evaluate the day-to-day, analyst-to-analyst, and equipment variability. A typical study includes days, analysts, columns, etc.

The validation report will include the standard deviation, relative standard deviation, and confidence interval for each type of precision.

Limit of Detection

The ICH Guideline suggests a number of ways to determine the detection limit for an impurity test in the following:

- Visual examination of results of tests on drug substance samples with known concentrations of impurity to detect the minimum level at which the impurity can be detected.
- Signal-noise ratio (for analytical procedures which show baseline noise) by comparing measured signals from samples with known low concentrations of the impurity with blank samples. A signal-noise ratio of 2:1 or 3:1 is acceptable.
- Based on the standard deviation of the response and slope.

$$\mathrm{DL} = \frac{3.3\,\sigma}{S}$$

where DL is the detection limit, σ is the standard deviation of the response, and S is the slope of the calibration curve.

The value of σ can be determined either by measuring the background response for a number of blank samples and then calculating the standard deviation, or by calculating the residual standard deviation of the calibration regression line, or from the standard deviation of y-intercepts from regression lines.

If DL is based on visual evaluation or based on signal to noise ratio, the ICH Guideline requires copies of the relevant chromatograms to be included in the validation report.

Limit of Quantitation

The ICH Guideline suggests a number of ways to determine the detection limit for an impurity test in the following:

- Visual examination of results of tests on drug substance samples with known concentrations of impurity to detect the minimum level at which the impurity can be quantified with acceptable accuracy and precision.
- Signal-noise ratio (for analytical procedures which show baseline noise) by comparing measured signals from samples with known low concentrations of the impurity with blank samples, and by establishing a minimum concentration at which the impurity can be quantified. A typical signal-noise ratio is 10:1.
- Based on the standard deviation of the response and slope.

$$QL = \frac{10\,\sigma}{S}$$

where QL is the detection limit, σ is the standard deviation of the response, and S is the slope of the calibration curve.

The value of σ can be determined either by measuring the background response for a number of blank samples and then calculating the standard deviation, or by calculating the residual standard deviation of the calibration regression line containing samples in the range of QL, or from the standard deviation of y-intercepts from regression lines.

Assay

Specificity

It is essential to show that the assay will discriminate between the analyte and other materials in the presence of impurities and degradation products. This can be done by spiking a pure drug substance sample with an appropriate level of impurities. For chromatographic procedures, representative chromatograms should be included in the registration package, which are labeled with the identity of the individual components to show the resolution of the drug substance from the other components. Peak purity tests such as diode array or mass spectrometry may also be useful, in addition, to show that the chromatographic peak for the analyte is attributable to only one component.

Range

The specified range for the determination of the linearity of the assay should be from 80% to 120% of the test concentration.

Linearity

The linearity over the required range can be shown by carrying out the analytical procedure on dilutions of a standard solution of the impurity. For the establishment of linearity, a minimum of five concentrations are recommended. The results are statistically analyzed (e.g., by least-squares regression) to show the correlation coefficient, y-intercept, slope of the regression line, and residual sum of squares. A copy of the plot of signal versus impurity analyte concentration should also be provided.

Accuracy

The ICH Q2 Guideline suggests several methods for determining accuracy in the following:

- Application of the assay to a batch of drug substance of known purity (such as a reference material);

- Comparison of the results of the proposed analytical procedure with those of another independent procedure with defined accuracy; or
- Inferring the accuracy from the results of precision, linearity, and specificity.

Precision

Repeatability is determined using either a minimum of nine determinations covering the specified range (three concentrations with three replicates of the complete assay procedure) or a minimum of six determinations at 100% of the test concentration.

The intermediate precision is a test to evaluate the day-to-day, analyst-to-analyst, and equipment variability. A typical study includes days, analysts, columns, etc.

The validation report will include the standard deviation, relative standard deviation, and confidence interval for each type of precision.

Optical Rotation

A specific optical rotation test is normally specified for a chiral drug—with a specific range to verify the optical purity of an enantiomer or a range of +0.10 to −0.10 degree for a racemate. Wherever possible, the contribution of impurities to the specific optical rotation should be reported. Samples of different age batches should be evaluated.

Validation criteria should include specificity and intermediate precision.

Particle-Size Testing

A number of different methods are used to characterize particle size. They can be divided into fractionation methods (sieving, cascade impactor, and sedimentation) and nonfractionation methods (microscopy, light scattering, and electrozone sensing). The instrument may need to be calibrated using an appropriate reference substance. The methods validation should include intermediate precision and robustness.

Robustness

During the development and optimization of the analytical method, its robustness should be evaluated. This will show the reliability of the analysis with regard to deliberate or accidental perturbations of the method parameters. Where variations in a particular parameter are shown to affect the measurement, the conditions need to be controlled.

The ICH Guideline gives the following examples of typical variations that need to be evaluated for an HPLC method to check their effect on the measurement:

- Variations in pH of the mobile phase
- Variations in mobile phase composition
- Different lots of column or different column suppliers
- Temperature
- Flow rate

There is a certain ambiguity in the guidelines as to how much of the experimental data on robustness should be included in the formal validation report. The ICH Q2 Guideline merely states that "it should be considered during the development phase" and the FDA Draft Guidance states that the data should be "discussed and/or submitted." A useful compromise would be to provide a summary of the major variations investigated with data (including relevant chromatograms) being provided on factors that significantly affected the analytical results. Since the evaluation of robustness is part of the analytical development, the system suitability test is usually devised from the results of the robustness studies.

Revalidation

When a company makes a substantive change in an analytical procedure for the drug substance, or where there is a significant change in the manufacture of the drug substance (with possible changes in the impurity profile), the changes may be such as to require revalidation of the analytical procedure. The degree of revalidation depends on the nature of the change. If the impurity profile might have changed, it will be necessary to show that the method is still specific for the active substance and known and unknown impurities. Where a different analytical procedure is used, the new procedure should be validated.

Life-Cycle Concept of Validation

The analytical laboratory needs to be able to guarantee that an analytical procedure will maintain its validated status. Monitoring and review of batch release results, stability test data, method technology transfers, and control charts enable this to be done. Ermer and Ploss (13) have shown how the analysis of such a database can be used for benchmarking and improvement in data quality and reliability.

3.2.S.4.4 BATCH ANALYSES

Tabulated batch analyses should be supplied to support the proposed specification. Batch analyses to be provided should include the following:

- Batches of material used in nonclinical and clinical studies (including bioavailability/bioequivalence studies)
- Batches from stability and scale-up validation batches (including the primary stability testing batches)
- Batches from routine production. A minimum of three recent batches manufactured on not less than 10% of commercial production scale are recommended to show that the proposed manufacturing procedure will routinely give material that falls within the proposed specification limits. If multiple manufacturing sites for the drug substance are planned, consideration should be given to supplying batch data from each site.

Where a historical batch record is supplied, which uses an earlier version of the specification, with either earlier limits or (more importantly) different tests, it is useful to include an explanatory note stating the impact of the change(s). It may also be useful to get the analytical laboratory to reanalyze a retention sample of the batch using the current proposed method and specification. The batch analysis should include the results of all tests, including those not currently part of the specification.

The batch results should include the following:

- Batch number and size
- Date of manufacture
- Site of manufacture
- Manufacturing process, where applicable (if more than one has been used during the development)
- Use of batch (bioavailability, stability)

Signed and Dated Certificates of Analysis

Most agencies require copies of C of As to be supplied in addition to a tabulated summary of batch data. These documents should be signed and dated. The Therapeutic Products Directorate of Health Canada requires C of As for batches used in pivotal clinical and bioavailability studies (2). FDA requires C of As for all drug substance batches used for nonclinical studies, clinical efficacy and safety studies, bioavailability, bioequivalence, and primary stability studies.

3.2.S.4.5 JUSTIFICATION OF SPECIFICATION

The regulatory submission should contain a justification for each analytical procedure and each associated acceptance criterion. The justification can refer to a number of different information sources as shown in Figure 2. In addition, the expected analytical and manufacturing variability should be considered.

Choice of Test

The choice of tests in the specification should be justified. Any test usually performed on a particular type of drug substance (e.g., in pharmacopoeial monographs for chemically related drug substances), which is omitted in the specification, should be justified. Where a test was included in the batch analyses but is now to be omitted, this should be justified.

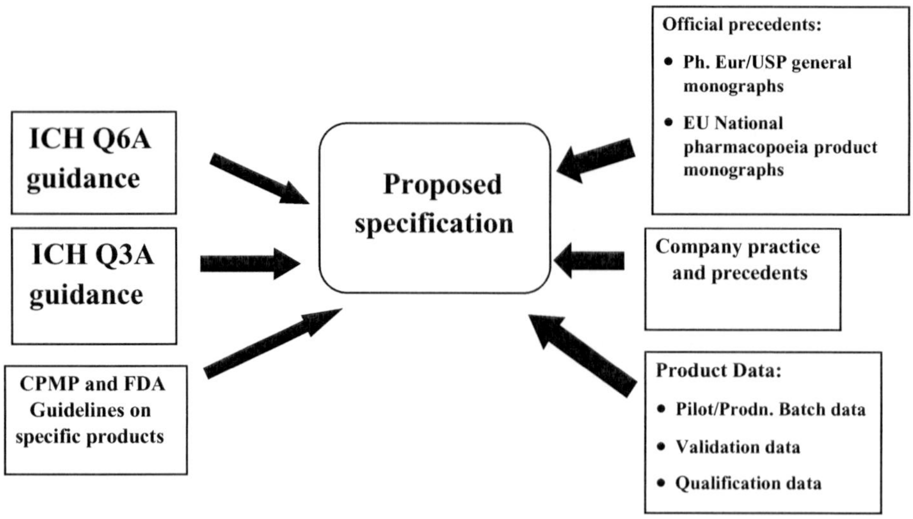

Figure 2 Reference material to be considered in the justification of the proposed drug substance.

Where a non-stability-indicating assay is proposed for routine use, this would need to be justified—explaining why such a procedure was not available or needed (e.g., for an inorganic salt).

The FDA allows a proposal to be made for a "sunset test protocol" where the results of tests on laboratory and pilot batches may indicate that a certain test is not needed routinely but where there is limited or no production batch experience. In such a case, a proposal can be made to run the test on a certain number of batches and then to omit it from the specification when an agreed number of batches have met the criteria. There would need to be a suitable mechanism for notifying FDA of the test results when the test is dropped from the specification.

Proposed Acceptance Criteria in the Specification

Justification should be provided for the acceptance criteria in the drug substance specification. Results from nonclinical tests, clinical studies, stability studies, and manufacturing studies should be taken into account. The proposed acceptance criteria can include a reasonable allowance for analytical and manufacturing variability. (A tolerance of 3σ might not be seen as reasonable.)

Setting specification limits for individual and total impurities needs to involve consideration of a number of factors. One key element is the biological qualification data for the impurities from the results of the nonclinical and clinical studies. The ICH Q6A Specifications Guideline states that "At the time of filing it is unlikely that sufficient data will be available to assess process consistency. Therefore it is considered inappropriate to establish acceptance criteria which tightly encompass the batch data at the time of filing." Despite this admonition, the regulatory agency reviewers often seem to just "eyeball" the limited batch data to see how closely the specification limits encompass the individual results, and if there is any disparity, ask the company to "tighten or justify" the limits. Thus, it is important to produce a well-argued case for the impurity limits. This case will include the biological qualification (safety) data as well as data on the manufacturing capability. The nonclinical data are very important in showing what levels of a particular impurity are acceptable, and the scientific case in the justification needs to involve both analysts and toxicologists. Where a multiple of the human dose was used in particular chronic feeding animal toxicity studies at the no adverse effect level, this gives a safety factor, which can be used to justify higher levels of impurity than were actually present in the drug substance batches used in the study. The ICH Q6A Specifications Guideline includes Decision Tree #1 on establishing an acceptance criterion for a specified impurity in a new drug substance. This proposes establishing the manufacturing and analytical process capability by determining the mean and upper confidence limit for the impurity (where the upper confidence limit is three times the standard deviation from batches used in the development, pilot, and scale-up studies). This is a good approach where very extensive clinical trials have been carried out involving production scale manufacture of many batches of the drug substance for use in manufacture of clinical drug product, but is more difficult where only a small number of relevant drug substance batches have been manufactured.

The issues relating to setting limits for toxic impurities have been extensively discussed in chapter 10.

CHIRAL DRUGS

The Q6A Specifications Guideline includes Decision Tree #5, which summarizes when chiral identity tests, impurity tests, and assays may be needed for new drug substances.

An enantioselective assay for the drug substance should normally be part of the specification—either using a chiral assay procedure or by the combination of an achiral assay with limits on enantiomeric impurity.

The identity test should be able to distinguish both enantiomers and the racemic mixture for a single enantiomer drug substance.

DEFICIENCIES IN APPLICATIONS

The EDQM survey in 2006 of deficiencies in Certification of Suitability of Monographs of the European Pharmacopoeia (CEP) identified a number of deficiencies that are probably typical of drug substance applications, particularly those for generic drug substances (14). These are summarized in Table 4.

The 2000 FDA Draft Analytical Procedures and Methods Validation Guidance provides a useful list of methods validation problems causing delays in review. These include the following:

- Failure to provide sufficient detail, or use of unacceptable analytical procedures. Examples include the following:
  - Use of arbitrary arithmetical corrections
  - Failure to provide system suitability test results
  - Differing content uniformity and assay analytical procedures without showing equivalence factors for defining corrections (requirement of USP)
- Failure to submit complete or legible data, for example:
  - Failure to label instrument output to indicate sample identity
  - Failure to label the axes
- Failure to supply suitable samples of impurities, degradation products, internal standards, or reference standards.

Table 4 Drug Substance Deficiencies in CEP Applications

Section of the dossier	Percentage of applications containing the deficiency	Deficiency
3.2.S.3.2	45	Insufficient demonstration of absence of particular reagents from the drug substance (particularly catalysts, alkylating agents, or other toxic reagents)
3.2.S.3.2	42	Incomplete demonstration of removal or limitation in the specification of solvents used in the synthesis
3.2.S.4	33	Limits for impurities not set according to ICH Q3 Guidelines
3.2.S.3.2	26	Unsatisfactory discussion on impurities
3.2.S.4.4	24	Incomplete or missing batch results on three recent batches. Results should be expressed as actual values instead of "complies" or "conforms"

REFERENCES

1. ICH Q6A Guideline Specifications: Test Procedures and Acceptance Criteria for New Drug Substances and New Drug Products: Chemical Substances. ICH, 1999. http://www.ich.org/LOB/media/MEDIA430.pdf. Accessed April 2, 2008.
2. Pharmaceutical Sciences—Questions and Answers. May 2007, Therapeutic Products Directorate, Health Canada. http://www.hc-sc.gc.ca/dhp-mps/prodpharma/applic-demande/guide-ld/chem/ps_qa_sp_qr-eng.php. Accessed July 5, 2008.
3. FDA Guidance for Industry, Part 11, Electronic Records; Electronic Signatures—Scope and Application. August 2003, Food and Drug Administration. http://www.fda.gov/cder/guidance/5667fnal.pdf. Accessed July 5, 2008.
4. European Pharmacopoeia Technical Guide for the Elaboration of Monographs, 4th ed, 2005, EDQM, Council of Europe, 67075 Strasbourg, France. http://www.edqm.eu/media/fichiers/technical_guide_english.pdf. Accessed July 5, 2008.
5. ICH Q4B Evaluation and Recommendations of Pharmacopoeial Texts for Use in the ICH Regions, 2007, ICH. http://www.ich.org/LOB/media/MEDIA3093.pdf. Accessed July 5, 2008.
6. ICH Q4B Annex 1: Evaluation and Recommendations of Pharmacopoeial Texts for Use in the ICH Regions. Residue on Ignition/Sulphated Ash General Chapter, 2007, ICH. http://www.ich.org/LOB/media/MEDIA3093.pdf. Accessed July 5, 2008.
7. ICH Q3A(R2) Guideline on Impurities in New Drug Substances, 2006, ICH. http://www.ich.org/LOB/media/MEDIA422.pdf. Accessed April 22, 2008.
8. ICH Q3C Guideline on Impurities: Guideline for Residual Solvents, 2005, ICH. http://www.ich.org/LOB/media/MEDIA423.pdf. Accessed April 22, 2008.
9. CHMP Guideline on the Pharmaceutical Quality of Inhalation and Nasal Products, EMEA/CHMP/QWP/49313/2005 Corr., EMEA, 2006. http://www.emea.europa.en/pdfs/human/qwp/493105en.pdf. Accessed July 5, 2008.
10. FDA Draft Guidance for Industry: Analytical Procedures and Methods Validation. August 2000, Food and Drug Administration. http://www.fda.gov/cder/guidance/2396dft.pdf. Accessed July 5, 2008.
11. ICH Q2(R1) Guideline on Validation of Analytical Procedures: Text and Methodology, ICH, 2005. http://www.ich.org/LOB/media/MEDIA417.pdf. Accessed July 5, 2008.
12. Chromatographic Separation Techniques. Ph. Eur method 2.2.46. European Pharmacopoeia, 6th ed, Strasbourg, 2007.
13. Ermer J, Ploss H.-J. Validation in pharmaceutical analysis Part II: central importance of precision to establish acceptance criteria and for verifying and improving the quality of analytical data. J Pharm Biomed Anal 2005; 37:859–870.
14. The TOP TEN deficiencies found in applications—How can the content of the applications for a certificate of suitability for chemical purity be improved? PA/PH/Exp. CEP/T (06) 35, December 2006, EDQM, Strasbourg. http://www.edqm.eu/medias/fichiers/Top_10_Deficiencies_found_in_Applicatoins.pdf. Accessed July 5, 2008.

12 Drug Substance and Drug Product—Container/Closure

Brian R. Matthews
Alcon Laboratories, Hemel Hempstead, U.K.

INTRODUCTION

Information on the container-closure system is required in a dossier for a marketing authorisation application (MAA) and the relevant information will be included in a number of places in the Common Technical Document (CTD) format (1). The primary placement of information on container-closure systems is in Module 3 at Section 3.2.S.6 (for the drug substance) and at Section 3.2.P.7 for the drug product (dosage form). In addition, there will be some information included in Module 2.3 (Quality Overall Summary) based on the information in Module 3. A certain amount of information will also need to be added to Module 1—perhaps in the application form but certainly in the product information (which may be intended for the prescriber or the patient or both).

However, container-closure systems will have to be taken into account in a number of other sections of the dossier—for example the sections on pharmaceutical development and on stability studies (including in-use stability studies) to be reported on the product in the file. One example of the importance of this information is that a justification may be needed for the choice of container-closure system, and this is often included in the pharmaceutical development section of the dossier (see chapter 14). A second example is that the primary stability data submitted in the file need to relate to the intended container-closure system to be used for storage of the drug substance and of the drug product when the product is in commercial scale production. Any other container-closure system may be considered to be at best supporting data. This is further discussed in chapter 19.

As with all aspects of MAAs, due account will also need to be taken of local requirements. In the European Union (EU) there are requirements for containers included in the *European Pharmacopoeia* (Ph Eur) (which includes a number of chapters on pharmaceutical containers, which include glass vials to various plastics) and these need to be taken into account in setting specifications. There is also guidance in a number of documents of indirect relevance to containers—for example the notes for guidance on development pharmaceutics and on stability testing. These and other documents will be discussed further below.

Once the container-closure system has been defined in an application it is usually necessary to seek approval for changes to this. In some cases any change to the shape or capacity of the container will be considered by the regulatory agencies to be an important change—especially for products such as injections, eye drops and those in aerosol containers.

There may be specific labelling requirements that apply due to container-closure system components—for example where latex is present.

Local requirements for child resistance for containers will need to be taken into account, as will any local requirements for ease of access to the product by elderly patients. These requirements might both apply to the same product in which case an explanation of the approach taken in design and testing of the container-closure system should be discussed.

Finally, it is noted that the types of container-closure systems in use and in development include a number of components that might, in their own right, be considered to be "medical devices." An example of this would be the prefilled syringe where the syringe component would otherwise be regulated as a medical device. Interested readers are pointed in the direction of chapter 45 for further discussion on this topic.

APPLICATION FORMS

European Union

The EU application form (2) requires information on the container-closure system in a number of places. For example, in Section 2.2 "Strength, pharmaceutical form, route of administration, container and pack sizes," the information is required in 2.2.3 Container, closure and administration device(s), which is required to include a description of material from which the container-closure system is constructed using terms from the current list of European Department for the Quality of Medicines and Healthcare (EDQM) standard terms (3). For each type of pack, it is necessary to also give information on package size(s).

Japan

The Japanese application allows for information on the container closure to be added under the column on manufacturing method (4).

United States

The U.S. Food and Drug Administration (FDA) application form (Form FDA 356h) does not require information on the container-closure system (5).

CTD MODULES 2 AND 3

European Union

It is not possible for a container manufacturer to submit a Drug Master File in the EU.

The European CTD Module 2 requires the following information to be submitted:

2.3.S.6 Container-Closure System: A brief description and discussion of the information, from 3.2.S.6.

2.3.P.7 Container-Closure System: A brief description and discussion of the information in 3.2.P.7.

For the CTD Module 3 the following information is to be submitted in Europe:

3.2.S.6 Container-Closure System: A description of the container-closure system(s) including the identity of materials of construction of each primary packaging component together with the relevant specifications. Drawings of the container-closure components should be included where appropriate. Nonpharmacopoeial methods of test should be validation and the description of the method and the validation data included where appropriate. Only a brief description is required for secondary packaging components that do not provide additional protection or serve some other purpose. Additional information will be required when the secondary packaging provides additional function. This part of the dossier should include a discussion of

- the suitability of the container-closure materials of construction and this needs to consider the choice of materials and their safety,
- whether the container-closure system offers adequate protection from moisture (loss or absorption) and light and
- the compatibility of the materials of construction with the drug substance (including sorption and leaching characteristics if necessary—e.g. for liquid or semiliquid products).

Much of the detailed discussion of these topics might be more appropriately included in development pharmaceutics section and the stability section of the dossier, with appropriate cross-references.

3.2.P.2.4 Container-Closure System: The suitability of the container closure system should be discussed, including

- the choice of materials,
- the adequacy of the protection offered against moisture (loss or absorption) and light,

- the compatibility of the materials of construction with the dosage form (and any other material that will come into contact with the container—e.g. diluents or other products added to the container),
- sorption and leaching characteristics,
- the safety of materials of construction and
- performance characteristics (such as reproducibility of the dose delivery from the device when presented as part of the drug product).

3.2.P.7 Container-Closure System: A description of the container-closure system(s) should be provided including

- the identity of materials of construction of each primary packaging component and
- specifications for each primary packaging component including description, identification and critical dimensions, with drawings where appropriate; where nonpharmacopoeial test methods are used, details of these should be included and adequate validation data submitted.

For functional secondary packaging components reasonably detailed information should be provided. For nonfunctional secondary packaging components only a brief description should be provided.

Suitability of the container-closure system should be discussed in 3.2.P.2 (Pharmaceutical development)–(see chap. 14).

The above information is derived from the *Notice to Applicants* (1) but, as has been pointed out in the Introduction to this chapter, there are other sources of information that have to be taken into account when putting together an MAA for submission in the regions. The EU is no exception to this.

The Ph Eur includes a number of relevant sections relating to container-closure systems. There may be requirements of relevance to a particular product in the general texts, general chapters, and general monographs of the Ph Eur. There may also be specific requirements in the monographs for specific materials or products in national pharmacopoeias of the Member States of the European Union—for example, in the *British Pharmacopoeia* (BP) (where there may be additional information of relevance to items included in general or specific monographs of relevance to the United Kingdom and other countries in which the BP is the official requirement) (6). Examples of requirements of this type include the texts on

- Glass Containers for Pharmaceutical Use,
- Plastic Containers and Closures,
- Rubber Closures for Containers for Aqueous Parenteral Preparations,
- Sterile Single-Use Plastic Syringes, and
- Materials Used for the Manufacture of Containers/Polyolefines.

In addition there are specific guidelines produced by the Committee for Human Medicinal Products (CHMP) of which the most directly relevant is the Guideline on Plastic Immediate Packaging Materials (7) of which more later. Other documents that have relevant information on containers include the following:

- The guideline on the use of ionizing radiation (where there is information on validation data required) (8).
- ICH Q8 Pharmaceutical Development (9) which discusses the data needed related to the container (e.g. the choice and rationale for selection of the container-closure system for the commercial product) including the intended use of the drug product and the suitability of the container-closure system for storage and transportation, the justification of the choice of materials for primary packaging; studies performed to demonstrate the integrity of the container and closure (particularly for sterile products), and possible interactions between the product and container or label. Where a dosing device is used the need to demonstrate that a reproducible and accurate dose of the product is delivered under testing conditions which simulate the use of the product is indicated. It might also be necessary to provide data on extractives from primary labels (e.g. from ink or adhesive) and this might be included in the pharmaceutical development or the stability data sections of the MAA.

- The ICH Q1AR2 Guideline on stability testing of new drug substances and products containing them (10) has the following requirements for container closure system:
  For the drug substance: "The stability studies should be conducted on the drug substance packaged in a container closure system that is the same as or simulates the packaging proposed for storage and distribution."
  For the drug product: "Stability testing should be conducted on the dosage form packaged in the container closure system proposed for marketing (including, as appropriate, any secondary packaging and container label). Any available studies carried out on the drug product outside its immediate container or in other packaging materials can form a useful part of the stress testing of the dosage form or can be considered as supporting information, respectively."
  There are specific additional requirements for semipermeable containers, particularly with liquid products such as injections, ear drops, and eye drops; and relaxations for products in impermeable containers. Similar requirements apply to new drug products containing known active ingredients (11) (see chap. 19).
- There are specific requirements relating to container-closure systems in the guideline on the pharmaceutical quality of inhalational and nasal products (12) which includes the following: "For non-compendial plastic and for rubber container closure components that are in contact with the formulation during storage (e.g. valves), a study should be conducted to determine the extractables profile. Details and justification of the study design (e.g. solvents used, temperature, storage time) and the results should be provided. It should be determined whether any of the extractables are also leachables present in the formulation at the end of the shelf life of the product or to the point equilibrium is reached, if sooner. The leachables profile should also be determined for compendial plastics and rubber container closure components."
- There are requirements to be addressed in the development pharmaceutics part of the dossier on the suitability of any measuring devices to be used to administer liquid medicines (and their graduation) (13).

As indicated above, there is a specific CHMP Guideline on plastics containers. This is quite detailed and deserves a separate discussion, and is of direct relevance to this chapter.

The CHMP Guideline covers plastic immediate packaging materials other than elastomers and natural and synthetic rubber (for which there may be pharmacopoeial or other requirements). It covers packaging materials, parts of the container, the closure or seal or of other parts of the container closure system(s) in direct contact with the active substance or medicinal product. The data requirements apply to new applications but any changes to container closure systems for approved products may need to take its recommendations into account.

Applicants will need to take the recommendations into account when preparing Sections 3.2.S.6, 3.2.P.2.4, and 3.2.P.7 of an application. The documents referred to above will also need to be taken into account. Due account should also be taken of the legislation in Europe concerning food contact uses of polymers in relevant cases.

The general requirements for containers described above continue to apply—that is general information on type and nature of the material, specifications of the plastic material, and results of extraction and interaction studies where necessary—but there is a requirement to explain the basis of selection of the specific polymers used. There might also be a requirement for toxicological documentation in some cases.

For the formulated product, the development pharmaceutics section may need to be specifically expanded to include extraction and interaction studies and, in some cases, these will give rise to a need for toxicological data. It may also need to include information on the photostability of the polymers and the extraction of the photodegradation products into the drug substance or the drug product may need to be examined, particularly if they might present a toxicological hazard. Consideration might also need to be given to the effect of the manufacturing processes—especially sterilization processes—on the polymer.

For all plastic materials that are used as immediate packaging material for actives substances or medicinal products the chemical name of the material and the chemical name(s) of any monomers used have to be stated. Additional information is required in all cases where the material in the container-closure system is not a solid, although even with solids additional information may be required unless it can be shown that the requirements of European food contact legislation are met. The guideline includes two decision trees which are useful in determining the data requirements for a specific product.

For polymers used with materials that are not solid the data requirements will include the qualitative composition of the plastic material (including additives such as antioxidants, stabilizers, plasticizers, lubricants, solvents and/or dyes) if the active substance is in a plastic immediate container that does not comply with the relevant Ph Eur requirements.

If the finished product is not a solid then the following requirements apply: For plastic materials used in packaging of nonsolid medicinal products and where the medicinal product is intended for inhalation, injection, or ophthalmic administration the name of the material supplier and the complete qualitative composition of the plastic material is to be included unless the material is described in the Ph Eur or in the pharmacopoeia of a Member State. This information is also needed where the monograph authorizes the use of several additives. The qualitative composition should also be provided for nonpharmacopoeial packaging materials used for medicinal products other than solids intended for oral and topical (except ophthalmic) administration unless the supplier can certify compliance with European foodstuff legislation.

It was indicated earlier that specifications are required for container-closure system components. This is true for plastics components but the requirements are defined more closely. Where compliance with a pharmacopoeia monograph is indicated this will need to be demonstrated. Where the material is of a nonpharmacopoeial grade or material an in-house specification will need to be provided. This should address the description of the material, its identification and its characteristic properties including mechanical and physical properties. Where the container is to be used for active ingredients or for products that are not solids (other than those intended for oral administration or for noncritical topical use—which does not include ophthalmic use) then the following additional data will be required: identification of the main additives, especially those which could migrate into the product such as antioxidants, plasticizers, catalysts, initiators, etc.; colouring matter; and the nature and amount of extractable substances based on the results of the extraction studies. A batch analysis certificate should be submitted to show that the material to be used complies with the in-house monograph.

It is indicated above that information might be required on extractives from the polymer in some circumstances. It would be reasonable to assume that this is intended to cover semiliquid and liquid contents and, in the case of formulated product, those intended for use by injection and ophthalmic and inhalational routes of administration. Such data are likely to be required when the polymer is neither described in the Ph Eur nor in the pharmacopoeia of a Member State nor approved for foodstuff packaging. For nonpharmacopoeial plastic material used for products intended for inhalation, injection or ophthalmic administration, extraction studies will be required whether or not the plastic has been approved for use in food packaging.

Extraction studies expose a sample of the plastic to a suitable solvent (preferably the placebo of the formulated product—in any case it should simulate the extractive properties of the product) under stress conditions to increase the rate of extraction.

Interaction studies would not normally be required where the contents of the container will be solid. In other cases comprehensive interaction studies (and migration studies where extractives have been found) will be required—looking for leachates and for sorption. Toxicological data will usually be required for extractable and leachable substances identified in the studies for products intended for administration by injection, ophthalmic and possibly inhalational routes of administration. Such data would not usually be required when the material to be contained is a solid and the plastic is described in the Ph Eur or in national pharmacopoeias of the Member States, or the plastic has been approved for food contact use.

Japan

It is possible for a container manufacturer to submit a Drug Master File in Japan.

Information is required in the Japanese Modules 2 and 3 relating to the container-closure system. As has been mentioned elsewhere in this book, the Japanese version of the Quality

Overall Summary is a much more detailed document than its counterpart in the EU or US. The information that needs to be included for container-closure systems is outlined below.

The information on the immediate container and inner bag or wrapper should state the composition even if a glass container is used. In the case of plastic containers it is sufficient to state the polymer's main ingredient and its physical properties such as colour but it is not necessary to state the additives present in the polymer. Nonetheless, it is required that plastic containers do not affect their contents and this should be addressed in the stability studies.

Plastic containers for aqueous injectable products should comply with the requirements of the Japanese Pharmacopoeia (JP). An English translation of the 15th edition of JP is available (14). There are sections relating to tests for glass containers for injections, test methods for plastic containers and a test for rubber closures for aqueous infusions, and for plastic containers for pharmaceutical products. Due account will also need to be taken of the relevant general notices, general monographs, general tests, processes and apparatus and specific monographs. A statement to the effect that the container-closure meets those requirements should be included in the manufacturing method column of the marketing application. In other cases the entry should state: "Filled into a synthetic resin containers with the attached specification." The specification and test methods for the container closure should then be attached together with drawings to show the container shape and size unless they are of the shape and size indicated in *Yakushin* Notification No. 832 of December 1992 (etc.) (15) when it is sufficient to state that the container shape is as defined in that document. For certain topical preparations (nasal, eye, and ear drops) it is usually sufficient to state the nature of the container material. For ophthalmic products (eye drops) there is an expectation that the container should meet the requirements of *Yakuhatsu* Notification No. 336 of March 1996 but the author was unsuccessful in tracing a copy of this document in English on the Internet. There are special requirements for aerosol containers (16).

United States

The US Drug Master File procedure can be used for container-closure components, but this will not be discussed further in this chapter other than to observe that the FDA guidance has a section on DMFs as they relate to container-closure systems (17, pages 37–39).

The US follows the usual ICH requirements relating to container-closure systems as amended by the following guidance documents:

- Container closure systems for packaging human drugs and biologics (17)
- Container closure systems for packaging human drugs and biologics Questions and answers (18)
- Draft guidance: Inhalational drug products packaged in semipermeable container closure systems (19)
- Container and closure system integrity testing in lieu of sterility testing as a component of the stability protocol for sterile products (20) (which is really more relevant to chapter 19)

In addition to the guidelines applicants should also take into account relevant requirements in the general notices, general chapters, general monographs, and specific monographs in the *United States Pharmacopeia* (USP). For solid dosage forms these requirements are largely for the design characteristics of the container; for injection products they also relate to materials of construction. See particularly the General Notices on "Preservation, packaging, storage and labeling" and the General Tests and Assays for Injections, Transfusion and infusion assemblies, Elastomeric closures and injections, Aerosols, Containers, Light transmission, Chemical resistance—glass containers; Biological tests—plastics and other polymers; Physicochemical tests—plastics; Containers for ophthalmics—plastics; Polyethylene containers, Polyethylene terephthalate bottles and polyethylene terephthalate G bottles, Single-dose containers and unit-dose containers for nonsterile solid and liquid dosage forms, Containers—permeation; and Ophthalmic ointments.

It should be kept in mind that under U.S. legislation the drug is considered to be adulterated if the container contains any poisonous or deleterious substance that may render the contents injurious to health. There are specific requirements for tamper-resistant closures in the

Code of Federal Regulations (21) and for child-resistant containers (22). Due account should also be taken of legal requirements relating to indirect food additives (23).

Aspects of U.S. requirements relating to the application of good manufacturing practices to the packaging of pharmaceuticals will not be discussed in this chapter.

The container-closure system is approved as part of the new drug application (NDA) or abbreviated new drug application (ANDA) procedure. This approval relates to a specific product and the container closure approved for use with one product may or may not be accepted for use with another product. Suitability for its intended use should be demonstrated in each NDA or ANDA citing a particular container-closure system. Suitability for use relates to protection from light or temperature, loss of solvent, exposure to reactive gases, absorption of water vapour, and exposure to microorgansisms and of compatibility with the contents so that no unacceptable changes are induced in the contents due to loss of potency through adsorption or absorption, pH shift, discoloration, leaching from the container matrix or changes to the brittleness of the container closure.

The container-closure safety will need to be addressed—for example, no harmful materials should leach or migrate from it. The amount of information required will depend on the details of the product—the dosage form and the route of administration. More detailed information is likely to be required for injectable, ophthalmic or inhalational products than for a tablet for oral administration. Liquid products are likely to have more information than solid products. This is likely to include specific extraction studies and a toxicological evaluation of the substances found in the extraction study. USP biological reactivity test data and results from USP elastomer tests are likely to be required.

The FDA guidance (17) indicates that the degree of concern related to the route of administration and the likelihood of interaction between the packaging components and the contents of the container are related and are summarized in Table 1. For solid oral products, data relating to the indirect food additive regulations may suffice. If the container closure has a functional feature it should be shown that this works properly. This could well require data on container-closure functional testing where the system is intended to help improve patient compliance or where it achieves drug delivery (such as a prefilled syringe, a transdermal delivery system, a dropper, a spray bottle or a metered dose aerosol inhaler). The specification and tests for such container-closure systems should be tailored to the specific product. The FDA guidance includes a detailed table of the types of data that should be considered according to route of administration and function provided by the container-closure system. (17, pages 12 and 13).

The FDA also requires information on the quality controls applied to container-closure system components. This will include information on the physical characteristics of the components (e.g. dimensional criteria, weight of components, performance characteristics such as metering valve function, syringe plunger function). There will also need to be information

Table 1 Examples of FDA Packaging Concerns for Drug Products

Degree of concern associated with the route of administration	Likelihood of packaging component–dosage form interaction		
	High	Medium	Low
Highest	Inhalational aerosols and solutions Injections	Sterile powders and powders for injection Inhalational powders	
High	Ophthalmic liquid products Transdermal ointments and patches Nasal aerosols and sprays		
Low	Topical liquid products Topical and lingual aerosols Oral liquid products	Topical powders Oral powders	Oral tablets Oral capsules (hard and soft)

Source: Container closure systems for packaging human drugs and biologics (16).

on the chemical composition of the components. It should be noted that a change in the composition of the components will be considered by the FDA to be a change in the specifications. Such changes will need to be investigated by the applicant—minor changes in composition might affect suitability for the intended use of the container closure component. Changes in manufacturing practices on the part of the supplier of these components could also affect suitability for use.

Appropriate information will be required for components that come into contact with the pharmaceutical product but which do not form part of the container-closure system. Examples include measuring spoons, dosing cups and syringes, and vaginal delivery tubes.

In most cases, a brief description of secondary packaging components will suffice unless these are intended to provide some additional protection for the drug product. Account might also need to be taken of the permeability of the primary container when choosing and setting specifications for the secondary packaging.

In the NDA or ANDA (or the equivalent application for a biologic) submission, the following information will be required for container-closure components (17, page 22):

- Identification of the component and name/address of the manufacturer with a physical description,
- Identification of the materials of construction by specific material designation and identification of any alternate materials and their source(s),
- Description of any operations such as washing, coating, sterilization, or depyrogenation undertaken by the applicant,
- Information on the suitability of the component for its intended use—complete chemical composition, qualification and characterization test methods, and results including relevant USP and other procedures (including gas transmission, solvent loss and container integrity), results form extraction, and toxicological testing or justified alternative approaches (such as USP biological reactivity or indirect food additive acceptability in appropriate cases), etc.
- Information on quality control (much of which relates to good manufacturing requirements compliance in the broader sense) and
- Stability data relating to packaging.

For injectable drug products there are likely to be more demanding data expectations. These are considered by the FDA to represent one of the highest risk groups of drug product. Ophthalmic products come only slightly lower on the FDA "concern" listing. There is considerable discussion of these products in the guideline (17, pages 23–27). Applications for these types of product will be expected to be accompanied by additional data on suitability of the container-closure components for their intended use, including (17, page 26)

- Protection from light exposure, reactive gases, moisture permeation, solvent loss, sterility maintenance or increased bioburden and seal integrity or leak testing.
- Safety data—chemical composition, elastomer component USP test results, glass component USP test results, chemical resistance of glass components, USP biological reactivity results for plastic components and coatings (e.g. of metal tubes), extraction data using the drug product if this is expected to differ from that of water and if the weight of extractives is greater than with water then the extraction profile, and for components that are sterilized extraction profile data obtained at 121°C using a suitable solvent.
- Compatibility data—coating integrity tests for coating on metal tubes, swelling effects on elastomeric components, USP containers physicochemical tests results for plastics, particulate matter and eye irritants for ophthalmic products, and appropriate supporting stability studies.

Liquid products and topical delivery systems are addressed in the FDA Guideline (17, pages 27–33). Solid oral dosage forms and powders for reconstitution are also addressed (17, pages 33–36).

The FDA has specific guidance for metered dose inhalers and dry powder inhalers (24) and for nasal spray and inhalational solution, suspension, and spray drug products (25), which have some relevant information on container-closure systems and to which interested readers are directed.

It should be noted that for ophthalmic products there is a recommended colour coding system for the caps of the containers developed by the American Association of Ophthalmology (26). This differs in certain respects from the list in the FDA guidance (17, page 27).

For inhalational drug products in semipermeable container closure systems the FDA has expressed concern following a study undertaken on the examination of the content of inhalational products in non-overwrapped unit dose low-density polyethylene containers, which showed that the majority of products examined contained chemical contaminants of various types derived from the primary and secondary container and labelling components. The relevant guidance document (19) identifies that it might not be possible to identify all potential chemical contaminants that might be present in the formulated product or to determine their toxicological profile. It is identified that while the clinical consequences of this contamination are uncertain many chemical contaminants are potential respiratory irritants. Because of the uncertainties associated with these issues the FDA has recommended the use of protective measures to reduce the access of environmental chemical contaminants to the product while manufacturers are also recommended to pay particular attention to factors such as adhesive lamination, heat-seal lamination, temperature conditions, etc. It has also been recommended that documentation be submitted in the application to address leaching of contaminants into the formulated product from the primary container, from any protective secondary container or other packaging components (such as the carton) to demonstrate that the issue has been adequately documented, quantified, and qualified. In addition, the potential for effects from paper labels should be addressed—contaminants could derive from adhesives, varnishes or lacquers, inks and other chemicals used in the manufacture of the paper. The FDA has recommended that the direct application of paper labels should not be made to semipermeable containers (but rather that alternative approaches such as direct embossing or debossing or other means be found to display the necessary information). It has also been recommended that the number of units within a protective overwrap should be minimized to reduce the possible exposure of unused units to environmental contaminants prior to use.

REFERENCES

1. ICH Common Technical Document—primary source used for this chapter is the EU Notice to Applicants, Volume 2B. http://ec.europa.eu/enterprise/pharmaceuticals/eudralex/vol-2/b/update_200805/ctd_05-2008.pdf. Accessed September 20, 2008.
2. http://ec.europa.eu/enterprise/pharmaceuticals/eudralex/vol-2/b/update_200805/applicformrevised_rev9.pdf. Accessed September 20, 2008.
3. List of Standard Terms, accessed by means of an electronic subscription. http://www.edqm. eu/site/-590.html. Accessed September 20, 2008.
4. Drug Approval and Licensing Procedures in Japan. Tokyo: Jiho: 595.
5. http://www.fda.gov/opacom/morechoices/fdaforms/356Hes.pdf. Accessed September 20, 2008.
6. Information accessed on the CD-rom version of the British Pharmacopoeia 2008, which was valid at the time of writing (September 2008).
7. http://www.emea.europa.eu/pdfs/human/qwp/435903en.pdf. Accessed September 20, 2008.
8. http://www.emea.europa.eu/pdfs/human/qwp/3aq4aen.pdf. Accessed September 20, 2008.
9. http://www.emea.europa.eu/pdfs/human/ich/16706804en.pdf. Accessed September 20, 2008.
10. http://www.emea.europa.eu/pdfs/human/ich/273699en.pdf. Accessed September 20, 2008.
11. http://www.emea.europa.eu/pdfs/human/qwp/012202en.pdf. Accessed September 20, 2008.
12. http://www.emea.europa.eu/pdfs/human/qwp/4931305en.pdf. Accessed September 20, 2008.
13. http://www.emea.europa.eu/Inspections/qwp/q14.htm. Accessed September 20, 2008.
14. http://jpdb.nihs.go.jp/jp15e/JP15.pdf. Accessed September 20, 2008.
15. This document could not be traced on the internet but information is included in Drug Approval and Licensing Prcedures in Japan 2006. Tokyo: Jiho: 204–205.
16. Drug Approval and Licensing Prcedures in Japan 2006 Tokyo: Jiho: 206–207.
17. http://www.fda.gov/CDER/GUIDANCE/1714fnl.pdf. Accessed September 20, 2008.
18. http://www.fda.gov/Cber/gdlns/cntanrq&a.pdf. Accessed September 20, 2008.
19. http://www.fda.gov/CDER/GUIDANCE/4168dft.pdf. Accessed September 20, 2008.
20. http://www.fda.gov/cber/gdlns/contain.pdf. Accessed September 20, 2008.
21. 21 CFR 211.132, http://www.accessdata.fda.gov/scripts/cdrh/cfdocs/cfcfr/CFRSearch.cfm?fr = 211.132. Accessed September 20, 2008.

22. 16 CFR 1700, http://www.access.gpo.gov/nara/cfr/waisidx_04/16cfr1700_04.html. Accessed September 20, 2008.
23. 21 CFR 174–186, starting at http://www.cfsan.fda.gov/~lrd/fcf174.html. Accessed September 20, 2008.
24. http://www.fda.gov/cder/guidance/2180dft.pdf. Accessed September 20, 2008.
25. http://www.fda.gov/CDER/GUIDANCE/4234fnl.pdf. Accessed September 20, 2008.
26. http://www.aao.org/about/policy/upload/Color_Codes_for_Topical_Ocular_Medications.pdf. Accessed September 20, 2008.

13 Drug Product—Description and Composition

Brian R. Matthews
Alcon Laboratories, Hemel Hempstead, U.K.

INTRODUCTION—IT IS ALL IN A NAME

When preparing a submission for a new drug or for an existing product (e.g. as a generic application or an abbreviated or abridged application) there will be a need to identify the product concerned throughout the text. This is not quite as simple as it sounds, since during a development project for a pharmaceutical product a number of names may have been applied. This could include laboratory codes (which could be referred to in pharmaceutical development reports as well as nonclinical and clinical study reports), proposed and adopted trade names, and proposed and adopted generic (nonproprietary) names. It is difficult to go back and amend the earlier documents once the name in which a submission is to be made has been finally agreed. Therefore, the applicant for a marketing authorization will need to consider how to address the nomenclature adopted in the final text. Often this will require the preparation of a glossary of the various names or identifiers used at various stages in the development of the product. However, it would be useful to use a standard format in the International Conference on Harmonization (ICH) Common Technical Document (CTD) Module 1, 2, 3, 4, and 5 texts (but not necessarily in attachments such as study reports).

In the CTD Module 1 document it will of course be necessary to follow the local requirements for nomenclature where the particular version of the CTD file is to be submitted. The application form requirements will be dependent on the country or region in which the application is to be submitted.

In the European Union (EU), the product name and pharmaceutical form will appear in the Declaration page and will also appear in a number of other places (for the product for which an authorization is sought or for a reference product in the case of a generic or hybrid product application, for example). It will also appear on the page where information is given on the name of the product for which an authorization is sought—this includes the invented name, the name of the active substance and the strength and pharmaceutical form using *Standard Terms* (see later). The route of administration and the container description are also included in the application form and will need to be stated using *Standard Terms*.

The name for a dosage form will consist of at least two parts. For a new drug product this will usually include a trade mark (invented name) and a dosage form. The trade name should not cause confusion with other products (when written or oral communications are taken into account). It should not be based closely on generic stems such as those used in international nonproprietary names (INNs) and, in the EU, there is a preferred sequence from which the name of active substances and excipients should be selected—INN, *European Pharmacopoeia* (Ph Eur) monograph title, national pharmacopoeia monograph title, common name, and then scientific name. In other jurisdictions similar requirements are applied except that the preference for the local or regional pharmacopoeia—*United States Pharmacopeia* (USP) or *Japanese Pharmacopoeia* (JP), for example—will be substituted. Additional information such as Chemical Abstracts Registry Numbers may also be included in appropriate places in the dossier—for example where information is included on the nomenclature of the active substance.

Where there could be confusion with the names of other products that have already been approved it is possible for regulatory agencies to decline to accept a proposed name—in the EU, the Invented Names Review Group at the European Medicines Agency (EMEA) refuses to allow a large percentage of the names proposed for use with new applications intended to be submitted to the Centralised Procedure, for example. Conflict with other products' names (when written or spoken) can be seen as a safety issue and can be grounds for refusal to grant a marketing authorization. Similar concern at potential conflicts in proposed trade names will be found in all regions and countries. In order to minimize the potential delays in the registration process it is important to undertake a worldwide search for potential conflicting trade and nonproprietary names (not limiting this to the intended field of application of the product) and to be prepared for objections by having alternate trade names available.

For a generic product (and for those cases where a trademark has not been finalized at the time of submission) the first part of the name of the dosage form will usually be the INN of the drug substance (which can be accessed by means publications from the WHO or a subscription to a WHO database) or a locally approved generic name (such as the British Approved Name (BAN)—which are in all but a very few cases identical with the INN), the United States Adopted Name (USAN)—which can be accessed at http://www.ama-assn.org/ama/pub/category/9615.html—or the Japanese Accepted Name (JAN)—a list of which can be seen at http://jpdb.nihs.go.jp/jan/scripts/name_list_eng.asp. The procedures associated with obtaining approval for nonproprietary names are discussed in more detail in chapter 8.

The second part of the title will be a descriptor for the type of dosage form that the product is presented as—for example film-coated tablets, eye drops (solution or suspension), intravenous infusion. There are local requirements for the format of the dosage form descriptor. In many cases, the name can be predicted by looking at national or regional pharmacopoeias if they include dosage forms—for example, the USP or the *British Pharmacopoeia* (BP) or the JP. In the EU there is a specific requirement that *Standard Terms* [developed and published by online subscription by the European Department for the Quality of Medicines (EDQM), a part of the Council of Europe rather than of the EU] should be used. This online database includes a wide range of dose form descriptors and their translations into many European languages. Many of the standard terms are quite long and complicated and can result in difficulties when designing the labelling for containers. Examples of these terms are as follows:

Eye drops, (solution *or* suspension)
Eye drops, powder, and solvent for (solution *or* suspension)
Powder and solvent for (solution *or* suspension) for injection
Powder for solution for infusion
Concentrate for solution for infusion
Concentrate for haemodialysis solution
Gastro-resistant capsule, (hard *or* soft)

In some cases a shortened version of the full title is available for labelling small containers.

In addition to providing mandatory names for dosage forms, the *Standard Terms* also include terms for routes of administration and containers. These should be used for application forms, labelling, summaries of product characteristics and patient information as well as in all parts of the CTD format application.

The information on the dosage form should also indicate whether it is intended for immediate or modified release in appropriate cases. There is some inconsistency over the use of terms such as "modified release," "slow release," "prolonged release" etc. There are some relevant *Standard Terms* to consider—"modified release" has been used in some cases such as modified-release granules, modified-release capsules, and modified-release tablets as has "prolonged release" in prolonged-release granules, prolonged-release capsules, prolonged-release eye drops, and prolonged-release tablets. There is also a *Standard Term* for one specific type of formulation—"gastro-resistant" (formerly enteric coated) capsules and tablets. There are some instructions on the use of these terms on the *Standard Terms* Web page: modified release should be used only where gastro-resistant and prolonged release are not appropriate, for example.

In the US, the FDA Centre for Drug Evaluation and Research has a number of Data Standards Manuals available on their Web pages (accessible via http://www.fda.gov:80/cder/dsm/index.htm). Of particular relevance to this topic are the manuals on

Ingredient names (http://www.fda.gov/cder/dsm/DRG/drg00401.htm),
Proprietary names (http://www.fda.gov/cder/dsm/DRG/drg00402.htm),
Potency (http://www.fda.gov/cder/dsm/DRG/drg00501.htm),

Dosage form (http://www.fda.gov/cder/dsm/DRG/drg00201.htm), and
Package type (http://www.fda.gov/cder/dsm/drg/Drg00907.htm).

USP monograph titles are widely applied.

For Japan it is understood that emphasis is given on the use of terms found in the JP.

Having determined how the name of the product is to be expressed, attention can now be paid to the description and composition of the pharmaceutical product.

DESCRIPTION AND COMPOSITION OF THE PHARMACEUTICAL PRODUCT

The composition of the pharmaceutical product will need to be declared in the CTD Module 1 (application form etc.) as well in Module 2 and Module 3. The format and content of the information should follow the requirements in the country or region in which the marketing authorization application is to be submitted. Both the active substance and the composition of the other components of the formulation will need to be declared in various places in the application dossier. The components other than the active substance may be described as "inactive ingredients" or as "excipients," but in this chapter, the term "excipients" is preferred (since not all excipients are free from some form of activity in the formulation—e.g. penetration enhancers, antioxidants, preservatives are not inactive).

APPLICATION FORMS

European Union

For the EU the following information on the composition of the product for which an authorization is sought is required in Module 1 (1.2 Application Form Section 2.6.1):

- Active substance(s)
  - Quantity/Unit/Reference or monograph standard
- Excipients
  - Quantity/Unit/Reference or monograph standard

The active substance(s) should be declared as, and the quantities stated in terms of, the active moiety accompanied by its salt or hydrate form as appropriate. Any overages of the active substances or the excipients should be stated below the formulation columns in the application form. Compatible information will also be required in Module 1 Section 1.3. 1 product information—consisting of the Summary of Product Characteristics, labelling and package leaflet. (1), (2).

The quantity should be expressed in terms of the amount of the active ingredient or excipient per dosage unit or volume of liquid.

The monograph references should be (in preference order): Ph Eur, national pharmacopoeia, or other pharmacopoeia. If there is no pharmacopoeial specification then the manufacturer's in-house specification can be cited.

It is an added complication that appropriate information (including the name of the product and of the active ingredient) has to be included on product labels in Braille in the EU.

United States

In the United States (US), FDA form 356h (available online as a document that can be filled out on the Web page (3)) has to accompany any application for a new drug application (NDA) or abbreviated new drug application (ANDA). This requires the "established name" of the product using a "proper name" or USP/USAN name as well as the trade name (if any) to be stated. Details of the excipient content of the product are not required in the application form, but the dosage form will need to be stated.

Japan

The application form is Form 22–1, as referred to in Article 38 of the Pharmaceutical Affairs Law Enforcement Regulations (4). The names used should be those used in the current edition of the JP. Other names may also be used but the JP monograph title should always be included.

Reference may also be included to terms used in the *Japan Pharmaceutical Codex, Japan Pharmaceutical Excipients, Japan Specifications and Standards for Food Additives*, and the former *Japan Standards for Cosmetic Ingredients*. Active ingredients have to be expressed quantitatively according to the dosage form. Excipients may also need to be stated quantitatively. Less complete information can be included in justified cases for excipients present in vehicles, pH regulators, ointment bases, suppository bases, solvents used in injections and other preparations, bases used in oral liquid products in capsules, lubricants and ingredients used for sugar-coating tablets.

COMMON TECHNICAL DOCUMENT

Module 2.3 Quality Overall Summary—2.3.P.1 Description and Composition of the Drug Product

Module 3.2 Body of Data—3.2.P.1 Description and Composition of the Drug Product

While there are separate entries for these two sections of the CTD, the information in Module 2.3.P.1 is derived from that in Module 3.2.P.1 so that the two sections can be conveniently considered in one place. (5)

The formal requirements for Module 3.2.P.1 are that the information summarized in Table 1 should be included in the dossier. The same information should be used in Module 2.3.P.1.

The function of the excipients present in the formulation should be identified using appropriate descriptive terms. These might include stabilizer, surfactant, antifoaming agent, solubilizer, plastcizer, lubricant, buffer, pH adjustment, adsorbent, binding agent, suspending

Table 1 Information to Be Included in Module 3.2.P.1 of the CTD

Description of the dosage form	Qualitative description including colour, shape, and size. Use appropriate descriptors from pharmacopoeias or other official sources (e.g. *Standard Terms* in the EU).
Composition	List all components (active ingredients and excipients) of the dosage form and their amounts per unit including overages (if any) and overfill (if any). Statement of the function of the components. Reference to their quality standards, preferably pharmacopoeial or other official specifications or in their absence the manufacturer's in-house specification. Acceptable quality standards can include appropriate food colour specifications, cosmetic ingredient specifications, as appropriate and available. Where possible, references may also be included to drug master files (but not for the EU for excipients). The composition of all drug product components should be listed, including the printing inks, capsule shells, tablet coating materials, flavours, and colours. In some regulatory jurisdictions it might be possible for the suppliers of compounded excipients to provide the information on the detailed composition of these materials to the regulatory in a drug master file or using some other mechanism. Additional information may be included to complete the description of the composition of the drug product. This might include the weight or volume of the dosage form, the inclusion of any tracers or markers, etc.
Description of any accompanying reconstituting media or diluents	Detailed information will need to be included in a separate Part P for each such material.
Type of container and closure used for the dosage form and any accompanying reconstitution medium	More detailed information will be included on the dosage form container and closure in Module 3.2.P.7.

Primary sources: ICH M4Q(R1): Quality Overall Summary of Module 2 Module 3: Quality.
ICH M4Q: CTD—Quality.
ICH M4Q(R1): Implementation Working Party: Questions and Answers.
ICH Q6A: Specifications: Test procedures and acceptance criteria for new drug substances and new drug products: Chemical substances. All documents available in English from the ICH Web site: www.ich.org. They may also be accessed from the FDA or the EMEA Web sites.

agent, coating agent, moisture-adjusting agent, filler, colour, flavouring agent, fragrance, tonicity adjusting agent, emulsifier, viscosity-adjusting agent, thickening agent, dispersing agent, propellant, disintegrating agent, antioxidant, preservative, compression aid and solvent, for example.

It should be noted that inclusion of materials derived from human or animal sources will give rise to additional data requirements.

The inclusion of a novel excipient (or one present in a novel concentration or for a novel use) will give rise to additional data requirements. In the US, it is possible to use the FDA's Inactive Ingredient database to identify whether an excipient has been used previously in the type of product subject to the application and to determine the amount of the excipient used. There is no global source of such information in the EU (6).

ACKNOWLEDGMENTS

I thank my colleagues at Alcon Laboratories for helpful comments during the preparation of this chapter, and in particular Yuichi Fukazawa, Yoji Hamano and Yvonne Lynch.

REFERENCES

1. Rules Governing Medicinal Products in the European Union Volume 2B Notice to Applicants, http://ec.europa.eu/enterprise/pharmaceuticals/eudralex/vol-2/b/update_200805/ctd_05-2008.pdf. Accessed August 29, 2008.
2. Guideline on Summary of Product Characteristics, http://ec.europa.eu/enterprise/pharmaceuticals/eudralex/vol-2/c/spcguidrev1–oct2005.pdf. Accessed August 29, 2008.
3. http://www.fda.gov/opacom/morechoices/fdaforms/356Hes.pdf. Accessed August 29, 2008.
4. An English translation of the Japanese application form can be found in Drug Approval and Licensing Procedures in Japan 2006. Tokyo: Jiho Inc, Japan at page 595.
5. The European version of the CTD guideline was used, reference [1].
6. Some information can be found in Matthews, BR. Pharmaceutical Excipients—A Manufacturer's Handbook. PDA, Bethesda, Maryland: DHI Publishing LLC, River Grove Illinois, 2005.

14 Drug Product—Pharmaceutical Development

J. Michael Morris
Irish Medicines Board, Dublin, Ireland

INTRODUCTION

Pharmaceutical development studies are conducted routinely during the development of a pharmaceutical product to better understand the properties and qualities that may need to be controlled in order to ensure consistency and quality in manufacture. As such these studies should be presented in the chemistry and pharmacy section of a marketing authorization dossier in order to convince the regulatory authorities that the applicant has a satisfactory knowledge and understanding of the physicochemical attributes of the product. Traditionally, in the European Union (EU), information on development pharmaceutics would have been expected to be presented at the beginning of the chemistry and pharmacy dossier after the description of the proposed formulation. Such information, amplified as appropriate in the pharmaceutical expert report, provided a very useful overview of the rationale for the formulation and the properties of the drug substance pertinent to the intended therapeutic use of the product. Accordingly, a note for guidance entitled "*Development pharmaceutics and process validation*" (1) was produced by the EU authorities to assist applicants for marketing authorization and to expand on the more general requirements laid down in European Community (EC) Directives (2) and Volume 2B of the "*Notice to Applicants*" (3).

While there were no direct parallel guidelines produced by the US Food and Drug Administration (FDA) or other internationally recognized regulatory authority, the basic principles found in the EC document were common to all authorities. Clearly, it is essential that the basic scientific principles upon which pharmaceutical formulations are developed are made clear in any application for marketing of a medicinal product.

Subsequently, this gap has been filled by the appearance in 2005 of the ICH guideline Q8 pharmaceutical development (4), which has undergone further expansion to address the implementation of modern technology. It is the intention of this chapter therefore to focus on these basic principles and to examine the different approaches for obtaining and presenting such information to the regulatory authorities.

Having considered the particular characteristics of a drug substance, and the proposed route of administration, a decision must be made about the type of formulation in which it is to be incorporated. These characteristics together with the properties of the different types of formulations whether they are liquid, solid, or semisolid, and depending upon their route of administration, will be considered in turn.

SCIENTIFIC PRINCIPLES

Preformulation Studies

The chemical characteristics of active drug substances and excipients are largely determined by the structure and spatial orientation of their individual molecules. Thus, clear predictions can usually be made of the chemical attributes of new molecular species from a study of closely related species. The knowledge of such properties is important in establishing the likely stability and compatibility of such substances when formulated into their intended dose forms and packaged in their appropriate containers.

In contrast, however, the physical properties of pharmaceutical substances are not always so readily predicted. Although the physical properties of a drug substance depend in turn on its chemical properties, for example, polymorphism and crystal shape, they are not always so readily predicted. Nevertheless, such physical properties may affect dramatically the quality and the performance of a medicinal product during manufacture, storage, transportation, and administration to a patient and therefore they may need to be controlled in the same way as chemical properties in the substance specifications.

The physicochemical properties of drug substances and of new excipients or those with significant impact on products therefore need to be reviewed before any attempt is made to formulate them into pharmaceutical dose forms. In fact thorough preformulation, studies should provide considerable assistance to pharmaceutical developers in producing successful formulations. Conducting of preformulation studies on new drug substances is not only a routine practice but it is also essential for good regulatory practice to include the conclusions from such studies in the registration dossier. These initial studies provide the basis for the developmental studies on the final formulations, and also of the successful choice of packaging materials in order to maintain the desired quality of the product during manufacture and throughout its proposed shelf life. In examining physicochemical properties of product components at the preformulation stage, some or all of the following characteristics may need to be taken into consideration, bearing in mind that many physicochemical properties of a given substance are themselves interdependent.

Solubility

The solubility of a new substance in water and a range of common solvents should be established as early as possible in the development of a product to provide help in approaches to formulation and also to help the successful design of the early clinical studies. Clearly, the absorption of a drug depends upon its solubility in the physiological medium. Parenteral and topical liquid preparations may rely on the drug being dissolved before administration to ensure appropriate biological activity. In the case of orally administered dose forms, those with poor aqueous solubility are likely to show low or variable bioavailability.

Where possible, a poorly water soluble drug in liquid preparations may be formulated using co-solvents, and useful leads for this can be obtained by the study of solubility characteristics in media other than water. Alternatively, solubilized systems or suspensions may be used depending upon the proposed route of administration. The solubility characteristics in water and common organic solvents can also be predictive of the drug's behavior as an analyte in analytical development, for example, assisting in the choice of solvent systems to be used with standard or reversed phase liquid chromatography columns.

Solubility properties depend in turn on other physical parameters such as temperature and dissolution medium pH and these need to be standardized in the measurement procedures in order to derive meaningful results. Solubility properties may also be dependant on other physical characteristics such as polymorphism, particle size, and moisture content.

pH and pK_a

The pH of a drug substance (in an aqueous environment) is clearly dependant upon its chemical structure and the presence of particular functional groups (such as –COOH or $-NH_2$), determining the acidic or basic characteristics of an organic molecule. For those dugs that ionize in water, it is useful to determine the pH of aqueous solutions of standard concentrations, for example, 0.1% (M/v) as an indication of concentrations likely to be present in liquid formulations. Strongly acidic or basic drugs or certain electrolyte solutions may have unacceptably low or high pH values under these circumstances, and therefore may require buffering to more physiologically acceptable pH values, for example, salicylic or acetylsalicylic acid buffered as salts. The pH may also contribute to the choice of analytical methods, for example, in the use of ion-pair liquid chromatographic techniques for highly charged species.

The pH and pK_a of an ionized species are particularly important in determining the absorption characteristics of weak electrolytes, especially when a drug is administered via the gastrointestinal tract. The pH partition theory (5) proposed that it is the more lipid-soluble unionized species that are more readily absorbed across a physiological membrane, such as from the gastrointestinal (GI) tract. Thus, for a weak acid, this will occur at low pH (below the pK_a) and the absorption of weak acids will be favored in the acidic conditions found in the stomach (pH 0.8–2.0), while weak bases can be expected to be better absorbed in the ileum (pH 6–7.5). These considerations may, however, need to be balanced against local (e.g., gastric) irritation produced by highly charged species or strong electrolytes. Absorption is additionally dependant upon a number of other factors, for example, the significantly greater surface area available for absorption in the small intestine. Nevertheless, the three parameters of solubility, pH, and pK_a remain inextricably linked and must be considered carefully when designing dose forms to deliver specific active substances.

Particle Size

The particle size of certain active substances is also an important physicochemical parameter that may need to be determined in the preformulation phase. Particle size may be an important determinant of solution rate, particularly for poorly soluble substances. The surface area of a solid is also inversely proportional to particle size, and therefore dissolution of sparingly soluble drugs may be enhanced by reduction of particle size, which leads to an increase in the surface area of the solid, for example, micronization of griseofulvin or spironolactone is necessary to give acceptable absorption following oral administration. Thus, for sparingly soluble substances, bioavailability may depend upon control of particle size for the solid drug or the drug in suspension.

Particle size may also need to be controlled for physical or mechanical reasons such as to aid suspendibility of a solid in a dispersion medium, to aid mixing and compressibility in a solid dose form, and also to reduce the risks of physical abrasion and discomfort in sensitive body tissues (for example, in eye drop suspensions). For semisolid preparations, rheological properties are also heavily dependant upon particle size.

The importance of particle size in determining these other physical parameters leads on to consideration of its importance in determining stability of a substance in a given dose form. The greater surface area of a substance of low particle size might result in a greater level of physicochemical changes induced on storage, adversely affecting quality. Smaller particles could also tend to show a greater level of particulate aggregation when present in suspension, resulting in physical instability on storage.

Evaluation of the particulate properties of a substance prior to incorporation in a proposed formulation should therefore lead to the specification of an appropriate particle size for routine control of the substance. This can be achieved either by comminution to an appropriate target size or better still by carefully controlled synthesis in particular at the precipitation or crystallization stage. Careful attention to selection of appropriate crystal properties and polymorphs is also highly desirable in order to achieve this end.

Various techniques are used routinely to determine particle size, but it should be remembered that the values obtained are often not interchangeable because a different parameter such as particle diameter, surface area, or density may actually be determined. It is therefore essential to link a particular particle size value to the actual technique being used. This applies in particular to indirect methods such as sieving, electro-resistance, and light scattering or densitometry methods. Wherever possible use of the standardized methods of the pharmacopoeias is to be recommended. Specifications should be set for the substances with reference to properly validated test procedures and appropriate acceptance criteria.

Polymorphism and Crystal Properties

Polymorphism means the existence of at least two different forms of a substance with different arrangements of its molecules in the solid state, thereby resulting in different physicochemical properties. Five or more different polymorphic forms of an active drug substance is not uncommon. These can be characterized using a range of analytical techniques such as infra red (IR) spectroscopy, thermal analysis, or X-ray diffraction, in which the different polymorphs behave differently. The stability of individual polymorphs will vary and are described empirically as unstable ($t_{1/2}$ less than one day), metastable ($t_{1/2}$ one day or greater), and stable. In general terms, the less stable polymorphs have a higher solubility. It is therefore important in such cases not only that the correct polymorph is chosen but also that solubility problems do not occur following reversion to the more stable polymorph.

Pseudopolymorphism is a similar phenomenon that occurs when different solvate forms exist following the use of different solvents in the manufacturing process and in particular following solvent removal. Such forms show similar variability in physicochemical properties to true polymorphs. The use of different purification or crystallization solvents in the preparation of an active substance can thus give rise to the occurrence of different polymorphs or pseudopolymorphs. Transformations can also occur on heating the more stable polymorph being associated with the lowest energy levels. For example, three polymorphs of sulfathiazole exist with transition temperatures of 103°C, 120°C, and 240°C, respectively.

Crystalline forms can also exist with different external crystal shapes or "crystal habit." These may again arise through the presence of trace amounts of solvents or other impurities in the active ingredient. Changes in crystal habit can once again affect other physical properties, for example, the different hardness profiles of anhydrous glucose and dried glucose monohydrate.

Moisture Content

Changes in moisture content of a drug substance can affect its physicochemical properties such as the crystal habit as indicated above, as well as contributing to chemical instability. Particle size may alter during a drying process, which can occur inadvertently due to excessive heating during processing, for example, during sieving or milling, which generate friction. Alternatively, a substance may take up moisture from the environment with resultant detrimental effects, both physical and chemical. Where preformulation studies conducted on active substances show the need to control their moisture content, this can usually be achieved in practice.

Excipients

Choice and Function

The function of each excipient in a formulated product as well as the scientific rationale behind their choice should be made clear. Where a decision to use a particular excipient, especially one which is key to the function of the formulation, has been made, information should be provided on the likely quantitative content of the excipient in the formulation. For some excipients, such as antimicrobial preservatives, the concentration range might be critical whereas for simple fillers it should not be.

The characteristics that can influence the performance of the medicinal product, for example, stability or bioavailability or the ability to assist in its routine manufacture, should be discussed relative to the function of each excipient. This should include substances used in the manufacture of the product, which might disappear during manufacture, such as water or solvents.

Compatibility

Compatibility between the drug substance and the excipients with which they are formulated should be addressed during pharmaceutical development. For example, the suitability of preservatives intended to be combined in dual or multicomponent preservative systems should be established—clearly, it is not appropriate to mix an anionic antimicrobial preservative with a cationic one. Care should be taken to choose excipient combinations to minimize such potential interactions.

Incompatibilities may also occur between different active substances that might be formulated together in fixed dose combinations products, in order to exclude the possibility of adverse interactions. The same principle extends also to drugs intended to be coadministered and also to any reconstitution or dilution solvents recommended in the accompanying product literature.

Dose Form Development

Liquid Dose Forms

Liquid preparations include solutions, emulsions or other colloidal preparations, and suspensions and may be intended for administration parenterally, orally, or topically, including administration into body cavities. Requirements for the quality specifications developed for such products will reflect the intended route of their administration but will be based largely upon their physical form. Solutions are usually aqueous but may be prepared by using nonaqueous solvents or cosolvents, where the capacity for local irritancy or toxicity is not a problem. Clearly, the solubility characteristics of the drug and other components of the formulation will dictate the concentrations used. For substances that are poorly soluble in aqueous solution, various alternatives are available including the use of nonaqueous solvents—for example, oils, addition of cosolvents, or solubilizing agents—as well as the formulation of such products as suspensions. The requirements for parenteral products, in particular those for intravenous or intrathecal administration, are most demanding. These require particle-free, sterile, apyrogenic aqueous solutions, or emulsions. There is a gradation to less stringent requirements for other products where the administration risks are proportionally less.

Physical parameters of the dose form itself will need to be examined during product development in order to determine ranges and tolerances, etc. Such parameters might include pH, particle size, and shape (for example, for colloidal or suspension formulations) and rheological properties as indicated above. The inclusion of excipients in a formulation, or indeed other active substances, may well influence the solubility of the desired species and this must be borne in mind during formulation design.

For liquid preparations formulated in multidose containers and intended for use on more than one occasion, it is important that the preparations are preserved adequately both chemically and microbiologically. Many drug substances are prone to degradation in the presence of oxygen, particularly in the presence of light and moisture and antioxidants are often required to be added to reduce such degradation. They tend to act by being preferentially oxidized themselves to harmless or inert species, for example, ascorbic acid, and therefore will be progressively used up during the shelf life of the product. While such antioxidants are often necessary in the unopened state, their use is of particular importance after first opening, when air ingress occurs. These aspects will need to be addressed as part of in-use stability testing.

Even more important, however, is the presence of antimicrobial preservatives that may be added to prevent problems arising from adventitious microbial contamination of a liquid formulation. It is a requirement that any liquid multidose preparation be adequately preserved to reduce the chance of inadvertent microbial contamination and overgrowth, which might result in a safety hazard or loss of efficacy, due to the presence of microorganisms, their degradation products, or even microbial-induced degradation of the drug substance or product. Many pharmaceutical products, particularly those based on aqueous liquids, provide a ready substrate for the growth of microorganisms, including those responsible for pathogenic or opportunistic infections in their target patients. At the development stage, studies should be undertaken to demonstrate the ability of the multidose formulation to withstand the possible exposure to microorganisms (4).

The incorporation of an antimicrobial preservative in such multidose preparations is not always essential because certain formulations may be self-preserving. Thus, products containing a high content of glycerol or other polyols are unlikely to require added preservatives, while those with high sugar content (70% sucrose) are unlikely to become contaminated with bacteria because only fungi are normally able to withstand the resultant high osmotic pressure. The value of preservatives in oily solutions is also debatable. However, in the absence of a self-preserving system, a suitable antimicrobial preservative is necessary. Careful attention should be given to the choice of the antimicrobial substance to provide an effective broad spectrum activity without potential incompatibilities with the drug substance or other components of the formulation. Safety considerations will also limit the choice of antimicrobial preservatives, because the most effective ones are usually also toxic. The choice will therefore be largely determined by the route of administration of the product, there being few safe nontoxic preservatives acceptable for use in parenteral or ophthalmic preparations. By contrast, the range of preservatives used in the cosmetics industry is much broader, given the lower risk due to topical application. Typically, benzalkonium chloride and para-hydroxybenzoates are used in medicinal products.

When considering antimicrobial preservatives, and other formulation components, due consideration must also be given to the way in which the product will be used, stored, and presented to the patient. Products requiring dissolution, reconstitution, or dilution before administration must also be examined to ensure that the preservative system remains effective under these conditions. For example, parenteral preparations consisting of lyophilized solids for reconstitution prior to administration will need to be examined throughout their proposed period of use after reconstitution. Effective preservation will need to be demonstrated throughout shelf life. As this type of formulation is often used for substances with poor stability in aqueous systems, the presence of a preservative may pose an added risk of degradation, and this must be investigated as part of the pharmaceutical development and stability studies. Preservative efficacy studies would normally be undertaken during the development stage of the product and repeated at the end of the proposed shelf life. Appropriately validated, preferably compendial, preservative efficacy test procedures should always be used.

In the case of biological preparations, which are frequently formulated in preserved systems, particular attention must be given to compatibility of the often inherently unstable macromolecule with the antimicrobial preservative. Biological products, particularly those presented as lyophilized powders for injection, may need to incorporate a range of stabilizers or other additives intended to safeguard the chemical, physicochemical, or macromolecular homogeneity of the product.

The uniformity of dose delivery may need to be established; particularly in the case of liquid preparations in unit dose presentations and this again should be addressed at the development stage rather than routinely on the finished product. In designing experiments to verify dosing uniformity, certain physical properties of the product should be borne in mind, for example, density and viscosity of the liquid.

For liquid suspensions, homogeneous samples must be examined from a well-shaken system to ensure meaningful results. However, the suspendibility of the preparation must be such that resuspension of a sedimented product can be readily achieved by the user with a reasonable degree of agitation so that the product usage is adequately reflected.

Semisolid Dose Forms

Semisolid products tend to share many of the properties of liquids and a broadly similar approach to development studies should therefore be followed. Such products are often intended for topical application and appropriate parameters should therefore be addressed at the development phase. The physical forms of these products tend to be solutions or dispersions (analogous to liquid suspensions) of an active substance in a vehicle. In the latter case, particle size may need particular attention to ensure correct dispersion properties as well as to guard against possible mechanical abrasion generated by large "gritty" particles. Rheological properties of semisolid preparations may also need to be studied carefully to ensure the correct flow properties for the particular application. Homogeneity of dispersion is of particular importance to ensure uniform dosing at the target site where potent active ingredients are being used, for example, in the case of topical fluorinated corticosteroids.

Many topical semisolids are required to be sterile when they are intended for use in critical areas, such as the eye or wounds or large areas of broken skin, especially in immunocompromised patients. In such cases, particular attention must be given to ensure that the particular formulation properties can be maintained during the proposed method of manufacture, including sterilization. Even nonsterile semisolids may require the inclusion of an antimicrobial preservative if intended for use on more than one occasion if they are aqueous based and are not self-preserving. As with liquids, such products should be subjected to appropriate preservative efficacy testing using methods adapted where necessary to suit the particular material, for example, in the presence of large amounts of oily material. The presence of oils may also make the inclusion of antioxidants necessary.

Solid Dose Forms

Most solid dose forms are intended ultimately for oral administration, although one can consider more specialized forms such as sterile implants, some of which will be considered in more detail in a subsequent section of this chapter. Finely divided powders may also be administered topically as dusting powders for skin and body cavities. Powders intended for reconstitution with solvents prior to administration of the resultant liquids have already been discussed. The major distinction is drawn in practice, however, between bulk and unit dose solid—the pharmaceutical development studies required for the latter group being considerably greater.

Solid dose forms are fundamentally more chemically stable and in general show less capacity for interaction, for example, with containers, and these aspects are usually addressed in the context of the stability program. Homogeneity of the formulation, however, does become a major consideration in solid dose forms, particularly where the active ingredient is present in lower concentration than the excipients. Powders in dry mixes tend to segregate because of their different surface properties, crystallinity, particle size, etc., and some form of mixing process is normally required to ensure even distribution of the active species. In the case of dusting powders, for example, the major component is usually talc or starch (chosen for lubricant or absorbent properties). Homogeneity of bulk product from the mixing process will need to be addressed at the development and/or the process validation stage. In practice, studies carried out at the development stage, can provide useful predictions of the validation protocols applied to the large-scale mixing process.

In the case of unit dose solid preparations, homogeneity becomes even more important because it is essential to demonstrate the uniformity of dosing both within a batch and between batches. Potency determination on a mixed sample of the batch alone will not indicate the possible distribution of content between individual dosage units. It is usual therefore that the finished product specification addresses such uniformity on a batch-by-batch basis. This should be supported by studies carried out during the development phase as well as in-process testing and/or process validation to reinforce the routine testing, particularly for highly potent substances present in low concentration in the formulation.

The type of excipients used will vary considerably with the intended therapeutic purpose. For example, pessaries or suppositories intended to deliver active drug quickly to the vagina or rectum, respectively, are usually based on low melting point fats or waxes in which the drug is dispersed. Sterile implants, on the other hand, may need to deliver the drug much more slowly and will therefore have a higher melting point. Some products such as estrogens or androgens may consist entirely of the active substance in a compressed form. Aerosolized powders for bronchial application and transdermal devices are highly specialized forms, which will be considered in more detail later in this chapter.

In practice, the majority of drugs consumed today are likely to be in the form of compressed tablets or capsules administered orally to maximize patient acceptability and compliance. These dosage forms usually consist of active drug substances presented with a range of excipients chosen for their specific properties to optimize the delivery of the drug via the GI tract. The function of all the excipients in the formulation should be clearly explained in the development pharmaceutics part of the dossier. Clearly, the primary objective is usually to ensure the maximum stability of the product in its packaging while achieving rapid (or in some cases controlled) delivery of the dose form to the target site in such a way that the availability of the drug for absorption is optimized.

For the solid unit dose form containing active drug substance intimately mixed with excipients, especially in the case of compressed tablets administered by the oral route, absorption of the drug from GI tract depends upon a number of factors. Apart from physiological properties, absorption requires disintegration of the dose form, release of the active ingredient, and dissolution in the liquid contents of the GI tract. Only after dissolution has taken place can the drug molecules cross the semipermeable barrier of the gut wall, except in the case of active or facilitated transport of certain molecular species.

Thus, one of the requirements of the development studies carried out on such dose forms is to monitor their performance after administration. Because in vivo bioavailability studies cannot be carried out on each batch, the intention is to correlate them with in vitro tests that can be performed routinely on the finished product and that link is made through the development testing. From studies carried out at this stage, the basis of, and the need for, routine batch testing is established.

The achievement of maximum solubility is important in optimizing pharmacokinetic behavior, particularly for orally administered substances, that are weakly basic or acidic. In order to deliver a drug substance in a form or species, which can be absorbed by the body, it must be first dissolved in the aqueous environment of the GI tract. Absorption will then be determined by the ability of the dissolved species to cross the wall of the GI tract. The biopharmaceutical classification system developed by Amidon and others (6) in conjunction with US FDA recognizes that there are two parameters influencing uptake of drugs from the GI tract and their relative bioavailabilities following oral administration. These properties are solubility and permeability, that is, to say the degree of the uptake through biological membranes such as the GI wall of the undissociated species are responsible for in vivo activity.

Disintegration testing is normally applied in accordance with pharmacopoeial practice, to each batch of finished product, or perhaps an intermediate stage such as uncoated cores, of tablets or other compressed dose forms to demonstrate the effective break up of the tablet (and thus confirm the performance of the disintegrating and deaggregating excipients) and the liberation of the active substance. In many cases, routine disintegration testing may be replaced by the more meaningful dissolution test and the choice needs to be justified at the development stage.

The dissolution test developed from a realization that the presence of active substance in a unit dose formulation alone was insufficient to ensure comparability of two similar products or even two batches of the same product. Because the compendial tests measure the actual amount of drug liberated from the dose form in an aqueous reservoir, the test reflects the properties of the active substance (solubility, etc.) as well as those of the formulation and is therefore considered to be a possible predictor of in vivo behavior.

However, in practice, the many factors that affect absorption of drugs from the GI tract include not only the physicochemical properties of the drug and dose form but also a constantly changing physiological environment of pH, enzyme secretions, motility, absorption surfaces and routes, etc. (7). Thus, it is clear that a complete in vivo–in vitro correlation (*ivivc*) may be difficult to achieve in practice. At the extremes, highly lipid soluble drugs may cross the GI membrane rapidly, while small water soluble drugs can pass freely through water-filled pores. In either case, the absorption rate achieves rapid equilibrium, and blood flow to the GI tract becomes the absorption rate limiting factor. The majority of orally active drugs fall in between these extremes and are subject to absorption limited by a range of influences other than blood flow. Although the dissolution test does not provide a simplified version of an in vivo study, its performance can lead to the provision of a range of useful data. It is therefore essential that the dissolution characteristics of such products be investigated in the development phase in order to avoid problems due to inadequate or variable drug release from batches used during the clinical testing phase. From these studies, a decision can be made as to the relevance of the dissolution test result to the likely in vivo behavior.

Irrespective of the degree of correlation obtained, a dissolution test can be incorporated into the finished product specification as a useful quality control parameter. Any difficulties in the dissolution behavior of a given drug can be identified and an appropriate dissolution test developed (8). Alternatively, omission of the dissolution test from the finished product specification should be justified. The routine dissolution testing of products containing poorly soluble drugs or slowly dissolving formulations may not be meaningful. However, for drugs whose physical properties, for example, particle size or polymorphism, affect absorption markedly or in the presence of excipients or coatings that alter absorption, the performance of the test can be very useful. In certain products even changes in brands of certain key excipients may affect bioavailability (9).

Development studies and the validation of the dissolution test should address the conditions of the test in order to provide justification of the eventual choice of test conditions used for routine testing. Properly validated apparatus and test conditions, for example, those laid down in the pharmacopoeias, should be utilized wherever possible, because these are likely to be required in the eventual marketing authorization dossier. Test media of fixed volume should be aqueous based and preferably buffered to physiological pH except where investigating the influence of different pH values on release rates, for example, in studying enteric-coated formulations. Care should be taken that the varying ionic strength of buffer solutions does not cause distortion of results. In this context, it is very important that aqueous-based media are chosen, and the use of organic solvents is resisted, thereby maximizing the discriminating power of the test procedure. For poorly soluble drugs, small amounts of surfactants such as sodium dodecyl sulfate may be added or flow through methods explored.

Manufacturing Processes and Their Development

The selection, development, and control of manufacturing processes for the drug product should also be addressed in the development phase. Process optimization in scale up from pilot to commercial scale should ensure that any critical processing parameters are identified and controlled. The suitability of the product components and equipment should be discussed. Process development studies should provide the basis for processing improvements, process validation, and ongoing process controls and their verification. The knowledge gained from these studies should be used to justify the finished product specifications and approaches to continue product improvements. Such studies should address physical, chemical, and microbiological attributes.

Overages

As a general rule, the use of overages of active substances in formulations is discouraged unless their use can be justified completely. The routine inclusion of overages by manufacturers of certain products without adequate justification is unlikely to be permitted. Similarly, the use of overages should not be an excuse for poor manufacturing, formulation, or analytical procedures. Where loss takes place during manufacture of the product, then the replacement might well be justified as a "manufacturing overage" because the amount present within the dose form at the time of release is likely to remain within the approved specification. However, large (for example 10%) overages used to compensate for poor product stability, in general, are not considered to be acceptable.

The presence in excess of inactive constituents in a formulation should also be explained and justified.

Container–Closure System

The choice, rationale, and function of the container closure system for the proposed market pack should be considered in the context of the nature of the product and its intended use. The container should ensure sufficient protection for the product during storage, transport, and distribution. Choice of materials for primary packaging should be considered carefully and the integrity of the packaging material should be studied.

The possibility of interactions with the container and/or closure (primary packaging material) in contact with liquid or semisolid preparations should also be addressed at the development phase. Any interactions that might have adverse effects on the product quality will need to be considered more fully in the design of formal stability studies. The capacity for potential interactions may lead to the selection of different container types to minimize the possibility of such interactions, particularly where nonaqueous solvents or cosolvents are presented in plastic packaging materials. Such decisions need to be made as early as possible in the development of a product to avoid wasting time.

The capacity for interaction with packaging materials may also need to be addressed in the formulation of semisolids, although this is less likely to be as critical as in the case for parenteral liquid preparations, which tend to show a greater capacity for extraction of plasticizers and other container compounds. Ophthalmic semisolids in plastic containers would be prime candidates for such studies, depending on the nature of their formulation.

Controlled Release Preparations

In certain preparations, the release of the active substance from the dose form has been altered compared with conventional release preparations in order to achieve some particular therapeutic goal. Such modifications may be carried out for a variety of purposes including prolonged effect, reduced frequency of dosing, reduction in toxic effects, or even a change in indications or an increase in activity. Special consideration needs to be given to the development of such preparations, and normally this should be reflected by the presentation of appropriate data in the development part of a registration dossier.

For such controlled release products, the rate of release of the active substance(s) from the dose form upon administration to the patient is critical for the intended usage of the product, and therefore needs to be carefully addressed in the development phase. The development studies should identify and explain the rationale behind the particular therapeutic objective and demonstrate how the particular physicochemical properties of the dose form have been modified and controlled in order to meet that therapeutic objective.

Solid Dose Oral Preparations

In practice, there are two major approaches for the development of oral solid dose preparations depending upon whether a delayed or prolonged release pattern is required. In the former case, a "lag time" occurs after administration before drug is released after which the release characteristics return to that of the conventional formulation. The classic example of such a product is the gastro-resistant (enteric-coated) preparation designed not to release the drug following oral administration until the product has passed through the acid environment of the stomach and reached the small intestine.

In the second case, the rate of release of the active substance from the formulation is altered in order to meet the therapeutic target of extending a therapeutic action. A wide range of mechanisms has been used in practice in order to achieve a slowing of the release of active substance from the dose form. Examples include embedding the drug substance in a solid matrix, coating of a single tablet core (nondisintegrating unit), or multiunit pellets filled into a hard gelatin capsule or compressed into a tablet. More sophisticated systems have used microspheres prepared from hydrophilic polymers such as polyvinyl alcohol (10) or polycarbonate (11) containing the drug.

Any specialized testing procedures developed to demonstrate appropriate physicochemical properties should be described. These would consist principally of in vitro dissolution tests designed carefully for the closest achievable correlation with in vivo performance characteristics. The test should therefore provide a clear discrimination between those batches of product with acceptable and unacceptable in vivo performance characteristics.

The choice of the conditions used during these tests is governed by the type of release system. Thus, a gastro-resistant preparation is tested by a procedure that demonstrates the resistance to an acidic medium followed by rapid dissolution at pH 6 to 8. In contrast, the prolonged release preparation will normally be tested at a range of pH values to determine a pH-dissolution profile (12). In addition, the effect of rotation speed of the dissolution apparatus, ionic strength, and/or surface tension of the medium may also need to be investigated in order to clearly discriminate batches with acceptable in vivo performance from those with unacceptable performance.

From investigation of the in vitro behavior of those batches used in the in vivo testing program, both clinical and BA/BE studies, meaningful specifications for routine batch control can be derived, taking into account the variabilities in the respective manufacturing processes and the assay procedures. The specifications may also need to be validated for variations in the physicochemical properties of the drug substance or the excipients, which determine bioavailability.

It is clear that an attempt at correlation of in vivo and in vitro studies should be attempted for prolonged release preparations in order that the specifications applied are meaningful. This relationship constitutes a major objective of the development studies for a controlled-release system (13) and will be dealt with in more detail in the various regulatory approaches described below.

Liquid Preparations

Liquid formulations for oral use have also been developed based on colloidal or particulate microspheres in an aqueous continuous phase. For example, a method of coating ion exchange beads impregnated with dextromethorphan with a semipermeable film allowed diffusion control of the rate of drug release from the beads. In this way the beads could be either administered in a gelatine capsule or suspended in a liquid vehicle (14). Such preparations could be expected to be investigated and controlled in a similar manner to multiparticulate solid dose forms.

A number of liquid pharmaceutical formulations have been investigated based on liposomes. These are microscopic vesicles composed of bi-layered phospholipid membranes enclosing water, which can be used to entrap drug substances. Liposomes used for drug delivery range in diameter from 25 nm to several micrometers and are usually suspended in an aqueous medium (15). There are two basic types of liposomes, unilamellar vesicles with a fluid core and multilamellar vesicles consisting of several lipid bilayers separated by fluid. As these liposomes are made up of the same phospholipid material as cell membranes, they are nontoxic and can be adsorbed on cell surfaces where they fuse or are endocytosed within the target cell. The liposomes can pass unchanged through the vascular system and deliver doses of medication to an appropriate diseased tissue (15).

Liposomes provide a flexible system for incorporation of drug molecules—water-soluble substances being incorporated into the aqueous interior of the vesicles, while lipid-soluble substances will dissolve in the lipid bilayer. The degree of uptake of drug molecules will vary with the size of the vesicle, the larger multilamellar vesicles having a much greater uptake capacity than smaller vesicles. However, the larger vesicles are more likely to be cleared by the reticuloendothelial system after administration and therefore may not reach the target site. Thus, a balance must be struck between drug uptake and targeting the dose in determining a suitable liposome diameter (16).

Although the therapeutic usage of such preparations has to date been limited, there are certain advantages in selectively delivering certain drugs to their target sites. More important, however, has been the formulation of parenteral liposomal preparations of certain highly toxic drug substances in an attempt to reduce their systemic toxicity, that is, amphotericin (15), chloroquine (17), and doxorubicin (18); and biological preparations have included products such as vaccine immunoadjuvants, toxoids, interleukins (16), and peptides (19). The use of such liposomal preparations also reduces dependence on large quantities of toxic solvents or lipids.

The development studies on the liposome formulation must address details of the vesicle size and composition as well as the usual physicochemical properties of the active substance and vehicle itself. Various techniques have been used to characterize liposomal properties including sizing by electron microscopy and other methods, entrapment and partitioning studies, and calorimetric investigations (20). The capacity for leaking or breakdown of the liposomal structure, leading to an increase in free drug over liposome-bound drug should also be addressed in the development phase as a prelude to stability studies. Residual solvents from the manufacturing process may need to be monitored during the development phase and presented in this section of the marketing authorization dossier.

Apart from parenteral usage, liposomal preparations have also been used for topical application on the skin or in the eye. The oral administration of blood coagulation Factor VIII entrapped in liposomes has also been reported to cause elevation of plasma levels of this Factor in hemophilia patients (15).

A range of other prolonged release liquid or solid dose forms has been developed intended for application to skin, eyes, the uterine cavity, etc. For these more specialized products, the development studies can only be examined on a case-by-case basis taking into account the drug, the therapeutic purpose, and the technical properties of the dose form following the general principles elaborated above.

Other Dose Forms

Apart from the general development study principles outlined above for liquid, semisolid, or solid dose forms, and the more specialized requirements for the controlled release preparations, certain other more specialized dose forms are worthy of mention.

Injections and Implants

Certain specialized injectable products intended to achieve a sustained therapeutic effect are based on the formulation of drug substance in suspension or in oily or other nonaqueous vehicles. Thus, dissolution of the drug in physiological fluid at the (intramuscular or subcutaneous) target site is necessary before absorption can take place. The suspension formulation (e.g., medroxyprogesterone acetate) will need to be examined carefully from the point of view of particle size and suspendibility of the solid particles (21).

In the case of oily, long-acting injection solutions such as the depot neuroleptic preparations, a partitioning of the drug from the oil into the aqueous phase is a prerequisite for therapeutic effect. Careful choice of solvent from the toxicity and solubility points of view must be made, while the required viscosity and density of the vehicle may also need to be addressed at the development stage. In both cases, reduced frequency of dosing can usefully be achieved, in excess of one month or longer.

An alternative to long-acting liquid injections is the use of the solid dose implant, the simplest example of which is the subcutaneous pellet of pure drug substance such as the hormonal preparations of estradiol or testosterone in the form of their poorly soluble esters. For these preparations, absorption will largely be governed by the physicochemical properties of the drug, although the physical properties of particle size and polymorphism may need to be addressed. All of these forms can be used to overcome compliance problems, although they cannot really be described as "controlled release" preparations in the sense of purpose-designed release characteristics.

More recent developments of depot preparations have utilized drug impregnated in a variety of biodegradable or nondegradable polymers, such products include intrauterine contraceptive devices containing an active drug substance such as estrogenic hormones or even copper. Mechanical properties and dimensional examination of these products, which border on categorization as medical devices, need to be carefully addressed. The disadvantage of the nondegradable type, such as the sponges commonly used in veterinary medicine, is the need to remove the spent "device" at the end of the treatment period.

The bioerodable devices offer a more complex approach that has applications in the field of cancer treatment. A biodegradable implant of goserelin (Zoladex, AstraZeneca plc) has been marketed for the treatment of hormone responsive prostate cancer (22). The capacity of the polymeric carriers of such drugs for degradation in vivo needs to be carefully examined at the development stage, resulting in the generation of suitable specifications and routine tests to ensure the consistent desired behavior during use.

The development of the pen-injection device to aid in the regular administration of parenteral preparations is a welcome move towards patient convenience and compliance. These devices have been devised for use with insulin, glucagons, and growth hormone—complex biological preparations that require regular, frequent, and long-term parenteral administration. The special features of these products lie in the devices—the immediate packaging is a conventional glass or polymeric cartridge. Many of the features of these devices lie outside the scope of medicines regulation, and assessment of their manufacture and construction will largely be dealt with by bodies responsible for regulation of medical devices, for example, notified bodies established under EU medical device legislation (23).

However, from a pharmaceutical point of view, the performance of these devices should be assessed in conjunction with the intended product. For example, the precision and accuracy of delivery of the intended dose should be assessed during the development phase of the pharmaceutical product. Such studies should take into account the density and viscosity of the preparation in question, for example, insulin zinc suspensions of different composition. Microbiological assessment of the ability of the design to withstand microbial challenge during usage is also necessary. This would include not only preservative efficacy testing but also some demonstration of the ability of the device to protect the contents against deliberate exposure to microorganisms, thereby challenging the device, for example, by exposure to microorganisms in the laboratory with a simulated "worst case" scenario.

Ocular Preparations

The development of more specialized eye preparations has largely focussed on the unit dose presentations of eye drops or eye ointments. As such preparations are intended for single dose administration, they tend to be more simple pharmaceutical formulations without the need for inclusion and therefore testing of preservatives. However, solid dose eye preparations of pilocarpine have been developed, which release drug over a long period of time from a biodegradable polymer system. Assessment of the release rates of drugs from such devices is therefore necessary.

A range of other drug substances have been studied as nondegradable implants intended for administration in to the posterior chamber of the eye, for example, preparations containing antiviral agents designed to leach drugs slowly over a long period of time.

Inhalation Therapy

Therapeutic agents can be formulated as inhalations to be administered via the nose or mouth in order to achieve a local effect in the respiratory tract. Nasal preparations of sprays or aerosols are normally presented as simple solutions or dry powders for local action in the nasal cavity and would therefore have few specialized pharmaceutical characteristics other than isotonicity and microbial quality for aqueous solutions. By contrast, however, many products administered through the oral cavity are intended for local delivery of drug to the lower respiratory tract.

Such an approach allows the use of a much lower dose of drug than by oral administration, particularly for those drugs that show poor absorption, GI metabolism/first pass effects, or indeed inactivation. Certain peptides may also be administered intranasally to avoid first pass metabolism.

Aerosols are technically colloidal preparations of liquid or solid particles dispersed in a gaseous phase. Particles in a pressurized gas stream flow in the direction of propulsion until deflected by bends in the airway, where the gas flow changes direction. The size of inhaled particles in the respiratory tract is critical for their effective deposition in the airways. Particles greater than 10 μm are filtered out in the upper passages, while small particles (<0.5 μm) float up and down under Brownian motion. Therefore the drug particle size distribution needs to be controlled within these extremes (0.5–5 μm) for effective deposition within the lower respiratory tract (24). Deposition and clearance of inspired drug, and therefore the formulation particle size, is a key variable in the design of aerosol inhalation devices.

Three basic devices are currently used in inhalation therapy, namely, the metered dose inhaler (MDI), dry powder inhaler (DPI), and nebulizers. These will be considered, in turn, in terms of similarities and differences and features of relevance to the development dossier emphasized.

The MDI contains drug present in solution or suspension in liquid propellant, which is ejected via a metering valve upon actuation of the device. The aerosolized droplets rapidly shrink on exposure to air following evaporation of the propellant, ultimately producing particles of the desired range—2 to 5 μm being the particle size range of the formulated drug. Solution formulations are the simplest type of MDI but suspension formulations may be more useful when solubility is limited although problems may arise with powder clogging of the metering valve. These latter require particle size control within the drug formulation. The concentration of drug in the liquid needs to be carefully tailored to the potency of the substance, and the actual amount of liquid delivered by the metering valve (typically 50–100 μl) in an actuation.

The performance of the metering valve needs to be assessed carefully in terms of accuracy and consistency and this would normally be undertaken during the development phase. The possibility of chemical interactions between the valve components and the formulation proposed for marketing should also be excluded or investigated in development and stability studies. Where particular adaptations of the inhaler design have been proposed, such as the use of a downstream spacer chamber to overcome coordination and compliance problems during use, these may need to be considered in determination of performance testing.

Environmental problems arising from the use of chlorofluorocarbons (CFC), their replacement with hydrofluroalkane (HFA) propellants, and user coordination problems have led to an upsurge of interest in propellant-free dry powder inhalers (DPI). Although DPIs have problems of complexity and inefficiency during use, their major advantage is the facility for breath-actuation and the absence of propellants. The range of devices available is large and growing, and the technical details are outside the scope of the pharmaceutical licensing procedure provided dose reproducibility can be demonstrated. Significant interactions with the product in a solid state are also much less likely to occur with DPIs than with MDIs.

Solid preparations for inhalation are dispensed in single or multidose containers depending upon the type of administration device. Single dose containers may be hard gelatin capsules or multicompartment blister packs. Multidose containers have special powder or solid block reservoirs from which the drug is dosed into an inhalation channel by an automatic, reproducible mechanism. Both systems use single or multiple active substances, with or without diluents such as lactose. For DPIs, the moisture content of the powders needs to be carefully monitored to ensure correct powder flow during actuation.

Nebulization systems rely upon the use of an external pump or source of pressurized gas to drive a compressed gas through a narrow tube, thereby creating an area of negative pressure that draws the liquid to be administered up the tube from a reservoir. When the liquid reaches the negative pressure area, it is broken up into droplets that are separated by baffles—the smaller drops being inhaled by the patient. Because the devices tend to be available separately and are relatively standardized, the major determinants of dosing are the physicochemical characteristics of the liquid preparations, such as density, viscosity, surface tension, etc, and these should be examined in the development phase. Products are generally prepared as sterile unit dose preparations to avoid the risk of microbial contamination. Nebulizers based on ultrasonic mechanisms are also used.

Performance of inhalation products is a key area of interest, which normally would be investigated at the development stage. The in vivo performance of these products will depend upon the formulation, the device efficiency, and the administration technique in question. The first two can be controlled as outlined and in vitro testing carried out for the purpose of comparison with in vivo data generated in bioavailability or other clinical performance studies. Applicants for marketing authorizations should demonstrate the ability of the product to deliver an adequate dose of active substance to the target site. To this end, the pharmacopoeial twin impinger test for deposition of the emitted dose, and modifications thereof, has been applied widely. Such tests are used in the development phase and should be also applied in routine control and in stability studies to monitor dose delivery at release and upon storage. Harmonization of standards for such tests in this specialized area is welcomed and is being achieved through mutual collaboration between the pharmacopoeias and industry, especially in the development of the next generation impacter testing device.

Transdermal Systems

The drive towards continuous release of an active ingredient with a reduced dosing interval for medium- to long-term treatment has focused on the use of topical application of drugs to intact skin in order to achieve systemic delivery of therapeutic levels. Although the skin permeability varies considerably at different body sites and between individuals, the transdermal devices or "patches" allow a slow release of active substance, which is lower than the absorption rate from the least permeable site. Thus, absorption becomes limited by the release of drug from the patch, and a constant systemic absorption rate can be achieved, which has the advantage of avoiding GI deactivation and allows rapid termination of treatment by removal of the patch.

Considerations for the drugs selected for transdermal application include physicochemical properties, potency, biocompatibility, and clinical need. To date nicotine, steroid hormones, antihypertensive, antianginal, analgesic, and anti-inflammatory drugs have been investigated (25). Physicochemical attributes of potential candidate drugs include molecular weight (<1000 Da) together with appropriate solubility, melting point, pK_a, and partition coefficients. Those properties which single out certain drug substances as suitable for inclusion in such devices should be discussed in the pharmaceutical development report.

To date, several types of mechanisms have been used to control release of active substances from transdermal devices including gradient reservoir, diffusion from a polymer matrix, microencapsulation in a polymer vehicle, or membrane control.

The drug reservoir is encased in an impermeable backing and held to the skin by means of an adhesive ring, leaving the drug/membrane surface in direct contact with the skin. Study of the devices in the development phase would include a consideration of the rate of release of the active substance in specially designed diffusion cells, which can be compared with the dissolution test apparatus for solid dose forms. In these cells, carefully stirred receptor compartments are maintained in a controlled environment and samples are removed at appropriate times. In this way a compartmental model (26) can be derived, which should be correlated with in vivo bioavailability data, as can be demonstrated for nicotine (27).

REGULATORY REQUIREMENTS

European Union Requirements

The requirements for data on development pharmaceutics in an EU marketing authorization dossier emanate from Directive 75/318/EEC, as codified in Directive 2001/83/EC (2) as amended, which calls for an explanation of the choice of composition, constituents, and containers and the intended function of the excipients in a finished medicinal product, "supported by scientific data on development pharmaceutics." Such studies are necessary to demonstrate that the type of dosage form selected and the formulation and container proposed are satisfactory for the intended use of the medicinal product specified in the application (2). The studies further aim to identify aspects of the formulation and the method of processing, which are crucial for batch reproducibility and which therefore must be controlled routinely (1).

The formal requirement of the EC Directive is expanded in Volume 2B of the "Notice to Applicants" (3), which lays down the format and content of applications for marketing authorizations, as accepted by all EU Member States. However, the "Notice to Applicants" lays down little additional guidance on pharmaceutical development and refers only to the recommendation that tests carried out during pharmaceutical development should be described in detail, citing a solid oral dose form dissolution test as an example (3). Formerly, there was useful advice presented for those responsible for preparing the "pharmaceutical expert report" as an intrinsic part of the EC application format. This focused on bioavailability of solid dose forms and content of key excipients and also linked in with information required in other parts of the dossier relating to chiral active ingredients. It is clear from the foregoing that little specific advice is given to applicants for marketing authorizations in either the mandatory or nonmandatory texts described. Although in the pharmaceutical report the expert gave an overview of pharmaceutical development studies, the provision of appropriate data in the body of the dossier was still considered to be essential. Given the format of an EC marketing authorization application (MAA), the presentation of data at the beginning of the pharmaceutical file was a very useful opportunity to present a clear overview of the rationale for the development of the formulation.

With this in mind an expanded note for guidance on the subject was prepared in 1988 as an aid to MAA applicants (1) and this was revised in 1997 (28). The guideline is divided into five main areas covering constituents, composition of liquid, semisolid and solid dose forms, and containers. As the first such specific guideline in this area, the framework has already been used above as means of introducing the scientific principles that lie behind pharmaceutical development studies. Nevertheless, the guideline can be described under these five main headings insofar as it provides an agreed format for use within the EU. The text has in fact been largely superseded with the adoption of the ICH text-Q8 in 2006 (4), although the principles above remain valid. The sections of the guideline areas follow:

Constituents: Active substances and excipients, properties, and compatibilities.
The guideline reminds the applicant that a new excipient proposed for medicinal usage for the first time, or for a new route of administration, needs to be described fully and its safety should be established by proper investigation where necessary. Excipients already approved in food use will require no further qualification provided they are being administered by the oral route—additional routes of administration may require additional studies. An EC guideline on quality of excipients was prepared in 1993 and was revised in 2008 (29).

Liquid dose forms: In the case of multidose liquid dose forms containing an antimicrobial preservative, the guideline requires the demonstration of antimicrobial preservative efficacy using the European Pharmacopoeia (Ph Eur) preservative efficacy test. The Ph Eur test allows two levels of compliance, but applicants should demonstrate compliance at the higher level unless the lower level can be justified in certain cases. These requirements also apply to *semisolid dose forms*, and the guideline stresses that the preservative test should address both antibacterial and antifungal activities. In contrast, the USP test is less stringent in its requirements and normally would not be accepted in the EU.

In the case of *solid dose forms*, dissolution testing, as an indicator of product performance, is of paramount importance and should always be carried out by official Ph Eur methodology. Where this is not practical, dissolution test equipment described in a national pharmacopoeia of a Member State or other pharmacopoeia can be used, or failing this an appropriate noncompendial method, provided that appropriate justification is provided. In effect the methodology of the test described in Ph Eur, USP, and JP is now harmonized and one of these methods can be used in all three territories. For conventional release, products investigations at the development phase will determine whether routine finished product testing is necessary. For prolonged release preparations, however, the dissolution test would be routinely required. Finally, the demonstration of product homogeneity, in addition to that required in the uniformity of content test of the Ph Eur, should be addressed to ensure even distribution of active substance with the excipients present.

The final section of the EC guideline deals with aspects of the container, which is dealt with in more detail elsewhere in this chapter. Nevertheless, the development pharmaceutics guideline addresses certain aspects of the container, some of which are described above in relation to liquid, semisolid, or finely powdered solid formulations, and the capacity for interaction with the container (28). Basically, such interactions are of two types—sorption and leaching. In the former case, sorption of the drug or a key excipient such as an antimicrobial preservative onto the container surface might take place, leading to loss in product potency or safety. These phenomena are known to occur with rubber closures, plastic containers and with certain administration sets during use. The converse relates to leaching of some component from the packaging material, especially in the case of plastic packaging materials, into the formulation during storage or processing of the product. Leaching could be of the container material itself or additives, substances from the closure, or even in certain cases labeling adhesive following migration through the container wall (28). EU has developed specific guidelines on plastic packaging materials, which addresses many of these points in more detail, for such increasingly common container materials, which might be expected to be less inert than glass containers (30).

Finally, where a dosage delivery device is included with the product, for example, metered dose aerosol, pen injection device, and liquid or powder measure, etc., the ability to deliver the desired dose reproducibly and accurately should be demonstrated for each specific product.

The specialized requirements for controlled release preparations have already been described in an earlier section. However, there is an EC guideline laying down requirements for assessing the quality requirements for oral solid dose prolonged release preparations (31). This EC guideline focuses on specific requirements in all parts of a MAA dossier for prolonged release preparations, which are defined as products in which the rate of release of active substance from the formulation after administration has been reduced for various therapeutic purposes. Such products must be distinguished from delayed release preparations in which the modification is in the site of release. Both product types are, however, covered by the general Ph Eur definition of "modified release," although only prolonged release preparations are covered by the EC guideline.

Terminology is critical in describing these types of products, since different definitions appear to be used in different territories to describe the same product behavior pattern, while certain terms can be unclear or even misleading. The therapeutic objectives and rationale for developing the formulation should be provided and the physicochemical and pharmacokinetic characteristics described. The manner and mechanism of release should also be explained.

As far as in vitro testing is concerned, a discriminatory dissolution test procedure is required as explained earlier. In the development phase, it is expected that a dissolution profile will be produced for each strength of each prolonged release preparation for which marketing authorization is sought. Where it is proposed that tablets may be halved for administration purposes then this should be fully justified. Clearly, in the case of coated tablets or other preparations for which maintenance of physical integrity prior to administration is required, in order to retain in vivo performance characteristics, tablet halving cannot be accepted. However, certain systems, for example, multipelleted products, can be halved without loss of prolonged release characteristics. For these, dissolution profiling of the half tablet is additionally required. Changes in the formulation made during the development phase will require repetition of the dissolution profile studies. From these profiles, a definitive specification for routine testing can be established. The content of any key excipient, which determines release rate of active substance, may need to be controlled additionally if performance cannot be guaranteed solely by dissolution behavior.

Attempts should be made to correlate the in vitro release characteristics with the in vivo bioavailability characteristics. Various techniques are proposed in the guideline in decreasing order of predictive power as to how such correlation can be achieved. However, the choice of method is left to the applicant who must justify the correlation obtained as part of the validation of the dissolution test procedure. It is also stressed that bioavailability testing should be carried out on production scale batches where possible because performance equivalence cannot always be maintained upon scale up. This link between the development phase and production phase is made via process validation for such products whose performance is critical, in particular, and indeed for all products. For this reason, EU has developed a guideline setting out process validation principles (32). Similarly, all test procedures, whether they relate to developmental or production tests, require validation, and a guideline addressing validation of analytical methods has been addressed within ICH Q2(R) (33). In this way, the principles of development pharmaceutics are inextricably linked with those of process and analytical validation.

US Requirements

Similar basic principles apply to pharmaceutical development, process, and methods validation in the United States as in the EU. However, although there is extensive guidance on methods validation, no specific FDA guidance on development pharmaceutics had been produced before the adoption of ICH Q8 (4). The general guidance on chemistry, manufacturing, and controls formerly applied to the manufacture of drug substances and drug products imply a similar approach to that undertaken by the EC note for guidance. These guidelines stressed the importance of physicochemical control of drug and product, insofar as this affects quality, safety, and efficacy considerations, but have effectively been replaced by the ICH Common Technical Document (34) and ICH Guideline Q8(4).

For the drug substance, physicochemical properties such as solubility profile, solid-state form, melting point, pH, specific rotation, and refractive index should be addressed. The solid-state properties, which might affect bioavailability, such as polymorphism, solvate and amorphous forms, and particle size, should be examined and carefully controlled. Similarly, inactive substances (excipients) should be controlled where appropriate.

For the drug product, further detail is spelled out in the guideline on manufacture and control of the drug product (35), intended for the submission of INDs, NDAs, and ANDAs. This guideline calls for a complete statement of the formulation with alternatives justified in terms of bioavailability. In the section on specifications, it is clear that these must be chosen carefully to reflect the desired characteristics, which will therefore need proper justification. This approach is therefore analogous to the situation in the EU.

Further detailed guidance on packaging materials (36) has also been produced by the FDA to supplement the data on drug substance and product. Throughout this document the themes of integrity and compatibility with product are stressed repeatedly and cross-reference is made to the requirements of the USP. Detailed requirements for containers of different materials for the various types of dose form are laid down in this document, from the most stringent (such as parenterals) to the least stringent. Similarly, the emphasis on closures is to ensure compatibility and effective sealing to maintain quality of the product. Aerosols are singled out for special mention when it is stressed that the valve mechanism forms the most critical part of the container–closure system and evidence to demonstrate accuracy and precision of the dose delivered together with maintenance of the correct particle size is sought accordingly. For dosage forms requiring unique component systems, special data concerning suitability for intended use might be requested and the examples of medicine droppers and intravenous administration sets are given. The major difference between the FDA and EU systems in this regard, however, is that in the former case no formal requirement for a "development package" is mentioned and there is no distinction made between routine testing and developmental work.

To verify more carefully the data supplied in support of a marketing authorization application (NDA or ANDA), the FDA introduced, in 1991, the concept of the preapproval inspections (36). In its Guide to Foreign Inspections, the FDA has highlighted the specific documentation and operations to be covered in such inspections (37). In addition to demonstration of GMP and GCP compliance, and provision of validation data, a Development Report is called for to provide documentation based in pilot or development batches to support the proposed manufacturing process.

For oral and topical dose forms, the development report should cover the formulation, equipment, manufacturing procedures, in-process test data, final dosage form test results, conclusions, and formal certification. In the case of sterile products, this information should be supplemented by data on certain physical factors (such as extrusion force or particulate controls), sensitivity studies to light or heavy metals, filter compatibility and integrity, oxygen sensitivity, container requirements, preservative efficacy testing, and heat sensitivity where appropriate (e.g., when autoclaving) (39).

It is clear that this development report covers much of the same ground as the corresponding EC guideline, although the FDA document (40) strays into areas covered in different ways under the EU system. Nevertheless, the FDA approach described reinforces the close connection between development pharmaceutics, process, and methods validation, and therefore ties the development laboratory more closely to the production and quality control facility from a regulatory perspective.

In the area of controlled release preparations again, the FDA has requirements that closely parallel those of the EU, although terminology may differ. Thus, the terms "controlled release" and "modified release" appear to be used almost interchangeably by FDA (40). USP and Ph Eur, official in their respective territories, are more or less in agreement in their usage of modified release to define alteration in place or time of release of active substance. As far as subdivisions of the terms, the USP and FDA again appear to agree on the concept of delayed release, in line with Ph Eur, and in fact USP has specific monographs for at least three such products, including aspirin and erythromycin. However, it is in the definition of "prolonged release" preparations (EU), where divergence appears because such products are referred to in USP as "extended release," while FDA do not appear to have focused on a particular term and refer to prolonged release, sustained release, or controlled release almost interchangeably (41). It is therefore important to have a clear understanding of different terminology when moving between the EU and US markets and hence the desire of the EU authorities to restrict the number of terms used.

As far as United States requirements for such products are concerned, there are a number of similarities to the EU in terms of the types of development studies that need to be undertaken. Thus, dissolution profiling at a range of pH values (1–7) should be carried out and the choice and volume of medium and apparatus are justified as outlined in USP General Chapter Drug Release <724> (42). Helpful advice is also given about which apparatus may be best suited for specific dose forms, for example, USP Apparatus 3 for "bead-type," and Apparatus 7 for nondisintegrating oral modified release forms (42).

Concerning in vitro–in vivo correlation this is defined in USP as "The establishment of a relationship between a biological property or a parameter derived from a biological property, produced by a dosage form and a physicochemical characteristic of the same dosage form." The levels of correlation achievable in descending order of quality are

Level A: 1:1 relationship between in vitro and in vivo dissolution data.
Level B: mean in vitro dissolution is compared to mean in vivo data.
Level C: mean in vitro correlates with single in vivo parameter.

Helpful advice is provided in the guideline as to how to attempt to develop such correlation (41).

Japanese Requirements

As is the case with the Ph Eur and USP, the Japanese Pharmacopoeia (JP) lays down certain standards for physical parameters for the substances described. These and additional parameters may need to be taken into consideration in the specification of active substances for incorporation into appropriate products. As with the US regulatory system no separate guideline for development pharmaceutics exists but the general guidelines for preparation of new drug approval applications (NDAs) have an implicit concern for physicochemical properties of new drug substances including solubility, hygroscopicity, melting point and thermal analysis, solution pH and pK_a, partition coefficient, polymorphism, and crystal properties. Advice previously issued by the MHLW to the Japanese Industry called for documents concerning the background to discovery and the development process including names and dates to justify the choice of specification parameters. Once again these have been superseded by ICH guidelines.

For the drug product, the rationale for the formulation design should be presented including reasons for selection of the formulation excipients and their concentrations and the selection of the manufacturing procedures. Where necessary the relationship between formulation design and bioavailability should be discussed (40). General test requirements for dose forms are laid down in the JP. In addition, the draft guideline (40) indicates for new and established substances a tabular presentation of the necessary testing protocol, although no differentiation between routine testing and developmental testing is given. Drug release tests are routinely required for solid dose forms while certain forms require particle size measurement, for example, powders and suspensions used as aerosols, ophthalmic, and parenteral preparations.

Compatibility studies of liquid preparations (injections, syrups, etc.) with diluents should be conducted and discussed in the stability section. Justification for omission of any of the tests described above should be made with any supporting data. Japanese requirements also exist for oral prolonged release dosage forms (41), which describe testing protocols to be followed. The results of such tests together with those of evaluation tests performed during the course of final formulation development should be presented as part of the NDA application.

A number of these guidelines have been superseded in Japan following the significant revision of the regulatory system and the adoption of the ICH common technical document in 2001. The Japanese summary document (Gaiyo) has been effectively replaced by a modified version of the Quality Overall Summary described in Module 2 of the CTD format (42).

Requirements in Other Territories

Of the three major marketing territories covered by ICH and described above, only the EU had formal guidance for development pharmaceutics packages and this in turn has been largely overtaken by the ICH guideline Q8 (4). As the requirements of the principal territories apply also in other parts on the world, for example, EC requirements are accepted in EEA, Switzerland, and other European countries, the United States and Canadian requirements are becoming harmonized and Canada also participates actively in ICH. There are few other major guidelines available on this subject. However, individual countries, for example, Australia may have some specific requirements in this area.

ICH Guidelines

The adoption of the ICH Guideline M4 (Common Technical Document, CTD) at Step 4 in November 2000 signified a major step forward in harmonization of medicines regulatory requirements across these three regions and elsewhere (Canada, Switzerland, Eastern, and Southern Europe, etc.) (34). Although primarily the CTD is concerned with the format not the content, there was inevitably tacit agreement on a number of content-related areas including the inclusion of a pharmaceutical development section following the EU model. This led in turn to the agreement on the need for a guideline to supplement the headings included in section P2 of the quality CTD.

The ICH Steering Committee agreed a concept paper for this topic in 2002, and Step 2 of Guideline Q8 was signed in November 2003. The final guideline "ICH Q8 Pharmaceutical Development" reached Step 4 in 2005 (4) and was adopted as a EU document shortly afterwards. The guideline should also be considered as part of the wider new quality by design initiative in conjunction with ICH guidelines Q9 Quality Risk Management (2006) (43) and Q10 Quality Systems (2008) (44). However, the basic part of the Q8 guideline describes the submission of knowledge gained from the application of scientific principles and quality risk management during pharmaceutical development of a product to ensure that all products are of optimal quality to suit the patients' need. Typically, these principles should be applied to support the original marketing authorization of the product and provide a platform on which to build on new knowledge gained postapproval during the product lifecycle. Thus, it will provide the reviewer or GMP inspector with a comprehensive understanding of the product, an opportunity for those applicants who are prepared to expend energy and resources to gain a more in-depth understanding of the product and its manufacturing processes and thereby enhance product quality, and to create a basis for greater regulatory flexibility.

Data aspects covered by the guideline include those aspects of drug substance, excipients, container closure systems, and manufacturing processes that are critical to product quality and which therefore need to be controlled to minimize product variability. In addition, there are two further sections dealing with microbiological attributes and compatibility, respectively, which are additional to the headings listed in the EU guideline (28) (which this document largely supersedes).

Microbiological attributes should be detailed in CTD section 3.2.P.2.5 and would include the rationale for performing or not performing microbial testing on nonsterile products as highlighted in Decision Tree #8 in the ICH guideline Q6A *Specifications: Test Procedures and Acceptance*

Criteria for New Drug Substances and Products: Chemical Substances (45). Selection and demonstration of effectiveness of antimicrobial preservatives or of self-preserving formulations should also be addressed. The integrity of container–closure systems in relation to prevention of microbial contamination should be demonstrated, while microbial challenge under test conditions that simulate actual patient use may also need to be considered.

The compatibility of the drug product with reconstitution diluents or solvents should be addressed where these are specified in product literature in order to support label warnings. Such studies should cover recommended in-use shelf lives under actual storage conditions at extremes of concentration. In addition, admixture or dilution of products prior to administration might also need to be addressed. The guideline is completed with a glossary.

The main part of the guideline Q8 (Part 1) provides details on the requirements for pharmaceutical development, which mirror those described above (4). The guideline should be seen as a two part initiative with all of the detail in Part 1 being relevant for products generally, while an enhanced product understanding, as described in Part 2, can lead to a "design space" in which greater control can be expected to be recognized by allowing freer movement within such a design space. The applicant can choose to conduct additional studies that can lead to an enhanced knowledge of product parameters over a range of material attributes, processing options, and process parameters. The resultant enhanced knowledge can facilitate risk-based regulatory decisions, process improvements with minimum regulatory restrictions, reduction in postapproval submissions, and the introduction of real-time quality control and product release.

Part 2 of the guideline has been developed as an annex to the guideline Q8 that supplements the general pharmaceutical development principles in the main document and describes the principles of quality by design (QbD) (46). It explains how tools such as process analytical technology (PAT) and design space can be put into practice. The document sets out to explain how the more empirical (minimal) approach to product development can be enhanced by more systematic approaches described as QbD, which includes results of studies using design of experiments, use of quality risk management (ICH Q9), and knowledge management (see ICH Q10) throughout the product lifecycle. The annex, although developed separately through the ICH process, was agreed and signed off at Step 4 in November 2008 and is in the process of regional implementation. However, the two texts have been merged into a single ICH document referred to as ICH Q8 (R1) and divided into the two parts referred to above (46).

The enhanced QbD approach described in Part 2 typically includes a systematic evaluation of formulation and manufacturing processes, including identifying through prior knowledge, experimentation and risk assessment, the material attributes, and process parameters that effect critical quality attributes (CQA) of the product. It is also important to determine the functional relationships linking material attributes and process parameters to the CQAs. Enhanced product and process understanding coupled with quality risk management can help to establish an appropriate control strategy that might include proposals for design space or real-time release. This more systematic approach facilitates continuous improvement and innovation throughout the life cycle.

The guideline goes on to expand in part 2 terms such as critical quality attributes for different types of product and their linkage to material attributes and process parameters. Design space is based on a relationship between ranges of material attributes and process parameters with their effect on product CQAs. Thus, once a valid design space has been developed, demonstrated, and approved, the manufacturer can work within the design space making adjustments that do not need prior approval because they are not regarded as changes. This brings about a greater flexibility to the manufacturer and lays down a framework for regulatory relief because of a greater assurance of product quality. Design space is not simply a combination of acceptable ranges for parameters, although proven acceptable ranges will result in product meeting appropriate quality when other variables remain constant. However, the defining of proven acceptable ranges together with determining "edge of failure" for process parameters can assist in defining a design space.

Three examples of design space are provided in Appendix 2 to the guideline. The first of these shows how dissolution behavior for a tablet can be controlled by parameters in a granulation operation either by using a nonlinear combination of ranges or by using a smaller range of granulation parameters to produce a linear combination in both cases, delivering satisfactory dissolution behavior defined as greater than 80%.

In the second example, a design space is determined for two CQAs, namely, tablet friability and dissolution, based on control of factors of a granulation operation that affect the dissolution rate (e.g., excipient attribute, water amount, granule size). Plotting these variables against friability and dissolution gives an overlapping region, which defines the design space.

In the third example, a design space is developed for a drying operation depending upon the path of temperature and/or pressure over time to result in a target for moisture content 1% to 2%. Above the upper limit of the design space is likely to result in thermal degradation, while operating below the lower limit can cause particle attrition. These examples relate primarily to solid oral dose forms but is intended in future developments to add additional appendices to the guideline dealing with other product types, as more experience is gained in this field.

The guideline also describes how a control strategy can be devised to ensure that products meet consistently the desired quality. The comprehensive development approach will generate product and process understanding and identify potential sources of variability, which can be controlled. This in turn allows for an adaptive approach where a step responsive to input materials can be supported with appropriate process controls to compensate for input variables, while maintaining product quality. Enhanced understanding of product performance can justify the use of alternative approaches to testing and may lead to real-time release testing. For example, combination of monitoring of tablet weight variation coupled with near infra red assay in process can result in improved quality assurance compared to traditional release testing of uniformity of content by compendial testing, in view of the larger numbers of samples tested. Combinations of end product testing and in process testing elements can be envisaged in a single product control strategy.

The guideline goes on to explain how and where to submit these kinds of elements in the marketing authorization application based on the common technical document. There is an additional glossary included in Part 2 to address the various new terminology described and to supplement the existing glossary in Part 1. A table in Appendix 1 compares the traditional empirical approach to pharmaceutical development with the enhanced QbD approach, and some additional illustrative examples of tools to assist the enhanced approach are provided in Appendix 2.

NEW DEVELOPMENTS

A vast amount of research effort continues in the field of formulation development aimed at providing new more carefully controlled and targeted drug release patterns to respond to continuing therapeutic challenges. From a regulatory perspective, it is essential that new developments are based on sound scientific principles in order to provide improvements in pharmaceutical dose forms, which can result in improvements in quality, safety, and efficacy, in accordance with obligations in legislation and in new ICH Q8/Q9/Q10 principles. The clearer and more carefully those such studies are planned and documented, with regulatory requirements in mind, the more quickly can such products receive approval.

In the past development, studies might not always have been based on rigorous science, and while those times are past, there has still been a conservative approach to adopt new technology, which is widely used in other sectors of industry. This reluctance to fully adopt process improvements, despite their clear advantages to product quality and process efficiency, reflect an innate conservatism amongst pharmaceutical industry and regulators alike.

New developments occurring in all formulation areas will continue to challenge not only formulation scientists but also regulatory scientists as new and more complex formulations are developed. The recognition of encouraging pharmaceutical innovation, as exemplified by the newer ICH guidelines, should be better captured in the desire for continuous improvement and innovation.

Parenteral Products

In the parenteral area, prefilled syringes and pen injection devices are now commonplace. They provide the advantages of convenience for users with integrated systems, which should reduce product exposure during administration and therefore the opportunities for microbial and particulate contamination. New injection systems are being introduced including implantable osmotic pumps and biofeedback devices where continuous adjustment of dose is made in response to a measured biological parameter (21). While the device components may be regulated separately in EU (23), the relevant authorities will require reassurance of the dose delivery, accuracy, and container integrity of such combined products.

Ophthalmic Products

Modern ophthalmic preparations are moving away from conventional multidose preparations in favor of single use unpreserved products, or more carefully designed multidose container products with reduced or minimal preservative content in order to reduce toxicity problems. Mucoadhesive agents or contact lenses impregnated with drug substances are being used to achieve slow release (47). Intraocular implants that release drug slowly over long periods of time are also being used to treat chronic infections in certain patient groups. Apart from safety and local tolerance issues, compatibility demonstration of such systems will be required. In addition new techniques capable of providing good in vitro measurement of drug release from such systems could usefully be developed preferably with some degree *of in vivo* correlation.

Transdermal Products

In the transdermal area, the first generation of transdermal patches relied on passive permeation of lipophilic drugs, such as nicotine and opiates, across the skin and therefore the rate of drug release controlled absorption. Work will undoubtedly continue to find ways of improving transdermal dosing of less potent and less lipophilic drugs, while minimizing absorption variability. Second generation systems are attempting to achieve active drug uptake from this route using permeability enhancers and electrical or iontophoretic assisted transdermal delivery (27). Again suitable test systems will be required to demonstrate appropriate release characteristics and consistency of function of such systems, while control of key processing parameters and good validation studies will be needed to ensure consistency in manufacture.

Inhalation Products

In the inhalation area, the usage of dry powder inhalers (DPIs) is on the increase, particularly because the move away from CFC propellants on environmental grounds. A variety of presentations of DPIs are used, in particular those based on blister-strip or other continuous feed single dose units. There is also a move towards DPIs based on pure drug without excipient, which eliminates excipient interference and therefore assists dosing accuracy and precision, because the drugs administered by this route are extremely potent and therefore are presented in low dose. The performance of the devices in conjunction with low doses of actives results in increased regulatory vigilance of such dosage forms. It is also encouraging that DPIs might present a useful approach for certain peptide and protein drug, where the absence of excipients would be an advantage. The EU has developed specific guidance dealing with preparations for inhalation (48) and specifically with CFC replacement aspects (49).

Gastrointestinal Tract

The buccal and rectal routes are both available for use by systemically acting drugs with the advantage of avoidance of first pass metabolism. In the oral cavity, use is being made of adherent materials for long-term local action. Polymer fibers impregnated with antibiotic have been utilized for long-term treatment of local infections (48). For systemic purposes, the buccal mucosa provides a good site for absorption for those products where GI degradation is likely and a number of peptides have been administered in this way. The rectal mucosa provides a similar absorptive surface but patient acceptability is clearly much lower. Likely formulations will need particular physicochemical characteristics and a good in vitro dissolution model system.

The majority of medicinal products with systemic actions will continue to be administered orally and therefore efforts will continue to be made to counter GI degradation and metabolism. Liquid-filled gelatin capsules are used to reduce homogeneity problems amongst powder preparations. Semisolid matrix formulations can be prepared with thermosetting or thixotropic properties provided their release characteristics are carefully monitored. Where heat is used in the manufacturing process, full investigation of the thermal stability of the product is required in the development package.

Increased targeting of drugs restricted to specific sites within the GI tract may become feasible based upon increased use of techniques such as γ-scintigraphy to monitor the movement of the products in vivo (49), while administration of insulin or other peptides, normally inactivated by this route, can be facilitated.

The challenge will be to find an appropriate modification of the dissolution test capable of demonstrating the localization of the drug at specific GI sites or control other test parameters to provide appropriate in vivo correlation.

Other Routes

Other sites will also be used for the delivery of protein drugs—the nasal route is already used, but also the buccal, pulmonary, or transdermal routes are under consideration. There will be an increased use of microencapsulation and liposome technology to target these drugs more carefully (50). Meanwhile the parenteral routes will remain the norm for biological products that are often orally inactive. Pen-type injection devices have been developed for such products as insulin and human growth hormone, and it is likely that the technology will be expanded to cover other products where routine injection is necessary.

Biological products are characterized by a number of features relating to an inherent variability in their production and analysis and their capacity for microbial and viral contamination that render them different from conventional chemically derived substances. Thus, it is necessary to examine not only their physicochemical properties but also complex tertiary and quaternary molecular structures of peptides or nucleic acids as well as glycosylation patterns. Maintenance of these appropriate characteristics throughout the manufacture, storage, and distribution of these products should also be demonstrated. Formulation development therefore must address such properties and the addition of various substances such as human albumin in order to stabilize these characteristics, for example, during a freeze drying or sterilization procedure, needs to be fully explained and justified in the development package.

As many biological products are heat labile, terminal heat sterilization is not an option and other approaches have to be used—these aspects should be addressed at the development phase. The high cost of these products often results in their presentation as multidose parenterals, which require the inclusion of antimicrobial preservatives, antioxidants, and buffers to maintain quality and potency during usage. Investigation of the shelf life of these products after reconstitution and opening needs to be addressed in both development and stability sections. Careful investigation of the suitability of the antimicrobial or other preservative should also be undertaken bearing in mind the particular properties of the biological substance in question.

Nanotechnology

The growing field of nanotechnology—the science of engineering materials at the nanoscale (1–100 nm)—is offering opportunities in many fields of medicine including pharmaceuticals and drug delivery systems. The EC Commission in 2004 issued a communication on a European Strategy for nanotechnology, which identified the need for "an integrated, safe, and responsible approach" to regulate health and environmental aspects of this new technology (51). Various materials and devices at the nanoscale are already in use and undergoing clinical testing (52); nanomedicine is already dealt with elsewhere (chap. 44). However, there are many development aspects to be considered in the usage of nanotechnology for such purposes such as masking taste, protection of components during processing and digestion, and enhancing permeability and bioavailability. Nanotechnology can enhance solubility and therefore improve the pharmacokinetics of poorly water-soluble drugs, increase specificity towards target cells, reduce immunogenicity, and enable a more controllable release of therapeutic compounds both individually and for combination therapy (52). Hence, it will become a critical tool in drug development and its usage will have a significant impact on the pharmaceutical development dossier of the future.

CONCLUSIONS

The continued development of new medicinal products of both biological and conventional origin and the search for new drug delivery systems will result in increased regulatory pressures on pharmaceutical companies. Regulatory authorities will need to be reassured that sufficient effort has been put into the development of effective formulations to ensure that these are translated into the industrial scale production of safe and effective medicinal products of acceptable quality. In this way, the attention of regulatory authorities will be focused not only on production and quality control, ensuring that the requirements of GMP are being met, but also on the research and development function to ensure compliance with best practice in the areas of formulation development, stability testing, and analytical research.

Regulatory authorities will in turn have to provide a lead to applicants for marketing authorizations in order that they can continue to comply with registration requirements. The link between research and production is thus reinforced in the marketing authorization dossier by the interdependence of development pharmaceutics with process and analytical methods validation, insofar as they affect pharmaceutical quality assurance. Thus, pharmaceutical process validation should be considered to be supported, on the one hand, by thorough development studies and on the other hand, by properly validated test procedures. This is apparent in the former dossier structure used in EU and the emergence of the development report in the United States as a useful tool for the preapproval inspection process. Both of these have given way in some measure to the introduction of the common ICH CTD (and e-CTD) format (34) and the new ICH quality guidelines, in particular Q8, as discussed above (4). Hence, the link between formulation development and production and quality control as part of the overall quality assurance process is maintained.

It is clear that the basic requirements of the three major markets are in tune with each other even when these similarities may not always be clearly stated. The new types of product and delivery system development will doubtless require an updated regulatory lead. Nevertheless, the basic requirements of regulatory authorities in this area remain applicable to likely developments in the foreseeable future.

REFERENCES

1. The Rules Governing Medicinal Products in the European Community, Volume III, Guidelines on the Quality, Safety and Efficacy of Medicinal Products for Human Use, Commission of the European Communities. http://ec.europa.eu/enterprise/pharmaceuticals/eudralex/vol3_en.htm. Accessed February 25, 2009.
2. The Rules Governing Medicinal Products in the European Community, Volume I. The Rules Governing Medicinal Products for Human Use, Commission of the European Communities. http://ec.europa.eu/enterprise/pharmaceutical/eudralex/vol1_en.htm. Accessed February 18, 2009.
3. The Rules Governing Medicinal Products in the European Community, Volume II, Notice to Applicants for Marketing Authorisation for Medicinal Products for Human Use in the European Community, Commission of the European Communities. http://ec.europa.eu/enterprise/pharmaceutical/eudralex/vol2_en.htm. Accessed February 18, 2009.
4. ICH Guideline Q8 Pharmaceutical Development, November 2005. http://www.ich.org/LOB/media/MEDIA4986.pdf. Accessed February 2009.
5. Shore PA, Brodie BB, Hogben CAM. J Pharmacol Exp Ther 1957; 119:361.
6. Amidon GL, Lennernas H, Shah VP, et al. A theoretical basis for a biopharmaceutics drug classification: The correlation of in vitro drug product dissolution and in vivo bioavailability. Pharm Res 1995; 12(3, March):413–420.
7. Banakar UV. Dissolution and bioavailability of drug delivery systems. Manufacturing Chemist 1987; 18–23.
8. Abdou HM. Dissolution, bioavailability & bioequivalence. In: Gennaro A, Migdalof B, Hassert GL, Medwick T, eds. Easton, PA: Mack Publishing Co., 1989.
9. Newton JM, Chow AK, Jeewa KB. The effect of excipient source on spherical granules made by extrusion/spheronisation. Pharm Technol Int 1992; October:52–58.
10. Thanoo BC, Sunny MC, Jayakrishnan A. Controlled release of oral drugs from cross linked polyvinyl alcohol microspheres. J Pharm Pharmacol 1993; 45:16–20.

11. Thanoo BC, Sunny MC, Jayakrishnan A. Oral sustained release drug delivery systems using polycarbonate microspheres capable of floating on the gastric fluid. J Pharm Pharmacol 1993; 45:21–24.
12. Davis SS, Hardy B, Wilson CG, eds. Drug Delivery to the Gastrointestinal Tract. Chichester, U.K.: Ellis Horwood, 1989.
13. Tabusso G. Regulatory aspects of development pharmaceutics (2). Regulatory Affairs J 1992; 3:1000–1003.
14. Lilienfield LB, Zapolski EJ. Controlled release dextromethorphan using advanced ion-exchange technology. Curr Ther Res 1983; 33:692–702.
15. Alpar O. Liposomes as drug carriers. Pharm J 1989; 263:254–255.
16. Gregoriadis G. Liposomes. Immunol Today 190; 11:89–97.
17. Titulaer HAC, Eling WMC, Crommelin DJA, et al. Parenteral controlled release of liposome encapsulated chloroquine in mice. J Pharm Pharmacol 1990; 42:529–532.
18. Cable C, Florence AT. Binding of doxorubicin to surfactant vesicles. J Pharm Pharmacol 1988; 37(Suppl.):30.
19. Weiner AL. Lipid-based vehicles for peptide and protein drugs, Part II: Manufacturing variables. Pharm Technol Int 1990; 46–49.
20. Oldfield E, Chapman D. FEBS Lett 1972; 23:285–297.
21. Livingstone C, Livingstone D. Injections and implants. Pharm J 1989; 262:313–314.
22. Livingstone C, Livingstone D. Injections and Implants. Pharm J 1988; 261:76–78.
23. Council Directive 93/42/EC of 14 June 1993, concerning medical devices, Official Journal No L169/1, 12/7/1993, as amended-consolidated text, 1993 0042-EN-11.10.2007. www.ec.europa.eu/enterprise/medical_devices /legislation_en.htm. Accessed February 2, 2009.
24. Gonda I, Byron PR. Perspectives on the biopharmacy of inhalation aerosols. Drug Dev Ind Pharmacy 1978; 4:243.
25. Berba J, Banakar U. Clinical efficacy of current transdermal drug delivery systems: A retrospective evaluation. Am Pharmacy1990; NS30:657–655.
26. Guy RH, Hadgraft J. J Controlled Release 1985; 1:177.
27. Bannon YB, Corish J, Corrigan OI, et al. Proceedings of the 16th International Conference on Controlled Release Bioactive Materials, Chicago, IL, 1989, 312.
28. ICH Guideline Q2(R1), Analytical Validation, Text and Methodology, incorporated from separate texts 2005. http://www.ich.org/LOB/media/MEDIA417.pdf. Accessed February 2009.
29. US Food and Drug Administration, Center for Drugs and Biologics, 'Guidelines for the format and content of chemistry, manufacturing and controls section of an application'. February 1987, Department of Health and Human Services. Replaced in July 2004 by Draft guideline "Drug Substance: Chemistry, Manufacturing and Controls Information, 1/7/2004 but this is now withdrawn and being revised, so the 1987 Guideline remains official.
30. US Food and Drug Administration, Center for Drugs and Biologics, 'Guidelines for submitting documentation for the manufacture of and controls for drug products' February 1987, Department of Health and Human Services. Replaced in January 2003 by Draft guideline "Drug Product: Chemistry, Manufacturing and Controls Information, 1/28/2004. Both documents were withdrawn in June 2006 and are being replaced by revised documents. http://www.fda.gov/cder/guidance/old029fn.pdf.
31. US Food and Drug Administration, Center for Drugs and Biologics, 'Guidelines for submitting documentation for packaging for human drugs and biologics' February 1987, Department of Health and Human Services. http://www.fda.gov/cder/guidelines/1714fnl.pdf. Accessed February 2009.
32. US Food and Drug Administration 'Requirements for pre-approval Inspections',October 1991, Department of Health and Human Services. Superseded February 2003
33. US Food and Drug Administration, 'Foreign Inspection Guide'. Office of Regional Operations International Programs and Technical Support Branch, 14 September 1992. http://www.fda.gov/cder/dmpq/compliance_guide.htm, Accessed February 2009.
34. Skelly JP, Barr W. Regulatory assessment in controlled drug delivery, 2nd ed. In: Robinson J, Lee VH, ed. New York: Marcel Dekker, 1992:294.
35. Dighe SV, Adams WP. Bioavailability and bioequivalence of oral controlled release products: A regulatory perspective in Pharmaceutics: Regulatory, industrial and academic perspectives. In: Welling PG, Tse FLS, eds. , New York: Marcel Dekker, 1988:307; Chap. 8.
36. USP. In vitro–in vivo correlation for extended release oral dosage forms. Pharm Forum 1988; 14(4):4160–4161.
37. USP <1090>, In vitro and in vivo evaluation of dosage forms.
38. USP General Chapter, Drug release <724>.
39. Japanese Guideline,(1992) Guidelines for preparation of Japanese new drug approval application, documents describing Category B data, 1992. http://www.nihs.go.jp/mhlw/jouhou/yakuji/yakuji-e0703.pdf.

40. Japanese Guideline, the Guiding Principles for Preparing the 'Document Summary (GAIYO)' of Applications for Approval of New Drugs. Director of New Drugs Division, Notification No. 21, 31 March 1992. http://www.nihs.go.jp/mhlw/jouhou/yakuji/yakuji-e0703.pdf.
41. Japanese Guidelines, Guidelines for the design and evaluation of oral prolonged release dosage forms (PAB/FERD Notification No. 5, 11 March, 1988. http://www.nihs.go.jp/drug/be-guide(e)/CR-new.pdf. Accessed February 2009.
42. ICH M4Q(R1) Guideline, The Common Technical Document, 2000, Quality. http://www.ich.org/LOB/media/MEDIA556.pdf. Accessed February 2009.
43. ICH Guideline Q9, Quality Risk Management, 2006. http://www.ich.org/LOB/media/MEDIA1957.pdf. Accessed February 2009.
44. ICH Guideline Q10, Pharmaceutical Quality Systems, 2008. http://www.ich.org/LOB/media/MEDIA3917.pdf. Accessed February 2009.
45. ICH Guideline, Q6A, Specifications, Test Procedures and acceptance criteria for new drug substances and products, 1998. http://www.ich.org/LOB/media/MEDIA430.pdf. Accessed February 2009.
46. ICH Guideline Q8(R1), Pharmaceutical Development, 2008. http://www.ich.org/LOB/media/MEDIA4986.pdf. Accessed February 2009.
47. Livingstone C, Livingstone D. Formulations for the eye, nose and ear. Pharm J 1989; 262:688–689.
48. Livingstone C, Livingstone D. Buccal and rectal routes. Pharm J 1989; 262:68–69.
49. Marriott C. Gastrointestinal transit of dosage forms. Pharm J 1989; 263:106–107.
50. Tissier B. Drug delivery of peptides. In: Bouchard P, Haour F, Franchimont P, Schatz B, eds. Recent Progress on GnRH and Gonadal Peptides. Paris, France: Elsevier, 1990:159–168.
51. Opinion on the Appropriateness of the Risk Assessment Methodology in Accordance with the Technical Guidance Documents for New and Existing Substances for Assessing the Risks of Nanomaterials, Scientific Committee on Emerging and Newly-Identified Health Risks, European Commission Health and Consumer Protection Directorate, 21/22 June 2007. http://ec.europa.eu/health/ph_risk/committees/04_scenihr/docs/scenihr_o_010.pdf. Accessed 21 November 2008.
52. Sanvicens N, Marco MP. Multifunctional nanoparticles—properties and prospects for their use in medicine. Trends Biotechnol 2008; 26(8):425–431.

15 Drug Product—Manufacture and Process Validation

Brian R. Matthews
Alcon Laboratories, Hemel Hempstead, U.K.

INTRODUCTION

Manufacture of the finished product (also known as the drug product) is one topic that will be required to be addressed for all applications for Marketing Authorization Applications (MAAs) whether the application concerned is for a new drug or is abridged (abbreviated). Thus the information will be required for all application types and procedures—New Drug Applications (NDA), Abbreviated New Drug Application (ANDA), full application, or abridged application. The information submitted to regulatory agencies on the method of manufacture of the pharmaceutical product will also need to be kept up to date and will therefore be subject to various types of updating process (which might include variations, supplements and annual reports as applicable in the jurisdiction where the product is licensed).

For the purpose of this chapter attention will be paid to the manufacturing information relating to small-molecule products. The data requirements for biological or biotechnology products will differ and for this the reader is referred to relevant guidance available from regulatory agencies or the International Conference on Harmonization (ICH).

One general issue that applies to all types of products is that they need to be manufactured under the applicable Good Pharmaceutical Manufacturing Practices (GMPs) [which may be called different things in different jurisdictions—e.g., Current Good Manufacturing Practices in the United States (US)]. The topic of GMPs is, however, outside the scope of this chapter and will not be discussed further here.

There is, however, one aspect of manufacturing practices that will need to be addressed in the MAA: process validation. This will be addressed in the current chapter.

INFORMATION REQUIRED AS PART OF THE COMMON TECHNICAL DOCUMENT FORMAT FOR MARKETING AUTHORIZATION APPLICATIONS

As the Common Technical Document is the defined format for all MAAs within the ICH regions—the European Union (EU), Japan, and the US—the placing of the information relating to manufacture of the finished product and process validation is defined in the ICH ICH M4 documents which can be accessed *via* the ICH web site (1).

The name of the manufacturing site and its address are usually required in the application form included in Module 1. Information on the method of manufacture and process validation will need to be included in the Quality Overall Summary (Module 2). It will also be required to be included in Module 3 under section 3.2.P.3 Manufacture, which includes

- 3.2.P.3.1 Manufacturer(s)
- 3.2.P.3.2 Batch formula
- 3.2.P.3.3 Description of manufacturing process and process controls
- 3.2.P.3.4 Controls of critical steps and intermediates
- 3.2.P.3.5 Process Validation

All of these will be expanded later in this chapter.

There are few guidelines from ICH on the information requirements for manufacture of the finished product or for data requirements for process validation but there is some text included in ICH M4QR1 (2) and the relevant question and answer document (3). The information in these two documents is summarized below.

3.2.P.3.1.1 Description and Composition of the Drug Product

A description of the drug product and its composition should be provided. The information provided should include, for example:

- A description of the dosage form.
- The composition—a list of all components of the dosage form and their amount on a per-unit basis (including overages, if any—justification for overages should be included in 3.2.P.2.2.2.), the function of the components and a reference to their quality standards (e.g. pharmacopoeial monographs or manufacturer's specifications). All drug product components should be listed here. The composition (e.g. components of the capsule shell, components of inks) should be included here. In some regions, the qualitative composition of proprietary components can be replaced with reference to appropriate Drug Master Files. When called for, additional information can be included to describe adequately the composition of the drug product—for example (*i*) total weight, volume, etc., of unit; (*ii*) tracers or markers; (*iii*) composition statement for (purchased) mixtures; and (*iv*) capsule shells.
- The description of accompanying reconstitution diluent(s). If the diluent is co-packaged with the drug product the information on the diluent should be placed in a separate Drug Product section. The compatibility of the drug product with reconstitution diluents should be discussed in 3.2.P.2.6 and
- The type of container and closure used for the dosage form and accompanying reconstitution diluent, if applicable.

3.2.P.2.3 Manufacturing Process Development

- The selection and optimization of the manufacturing process described in 3.2.P.3.3, in particular its critical aspects, should be explained.
- The method of sterilization should be explained and justified where necessary and
- Any differences between the manufacturing process(es) used to produce pivotal clinical batches and the process described in 3.2.P.3.3 that can influence the performance of the product should be discussed.

2.3.P.3 Manufacture

Information in the Quality Overall Summary should include:

- Information on the manufacturer.
- A brief description of the manufacturing process and the controls that are intended to result in the routine and consistent production of product of appropriate quality.
- A flow diagram, as provided under 3.2.P.3.3.
- A brief description of the process validation and/or evaluation, as described in 3.2.P.3.5.

The information in Module 3 for 3.2.P.3 Manufacture should include:

3.2.P.3.1 Manufacturer(s)

- The name, address, and responsibility of each manufacturer including contractors and each proposed production site or facility involved in manufacturing and testing should be provided.

3.2.P.3.2 Batch formula

- A batch formula should be provided that includes a list of all components of the dosage form to be used in the manufacturing process, their amounts on a per batch basis, including overages, and a reference to their quality standards—e.g. pharmacopoeial monograph or manufacturer's specification.

3.2.P.3.3 Description of manufacturing process and process controls

- A flow diagram should be presented giving the steps of the process and showing where materials enter the process.

- The critical steps and points at which process controls, intermediate tests or final product controls are conducted should be identified.
- A narrative description of the manufacturing process giving the sequence of steps undertaken and the scale of production should be provided, including packaging.
- Any novel processes or technologies and packaging operations that affect directly product quality should be described with a greater level of detail.
- Equipment should be identified at least by type (e.g. tumble blender, in-line homogenizer) and working capacity, where relevant.
- All steps in the process should have the appropriate process parameters identified such as time, temperature or pH. Associated numeric values can be presented as an expected range. Numeric ranges for critical steps should be justified in section 3.2.P.3.4. All process controls should be identified in this section. For critical controls additional information should be provided in 3.2.P.3.4.
- In certain cases, environmental conditions (e.g. low humidity for an effervescent product) should be stated.
- When called for justification of sterilization should be included in 3.2.P.2.3.
- Proposals for the reprocessing of materials should be justified. Any data to support this justification should be either referenced or filed in this section. If there are critical controls associated with the reprocessing operation then the critical controls should be included in 3.2.P.3.4. If validation information is needed it should be included in 3.2.P.3.5.
- Any over-fill should be identified and
- A statement regarding manipulation of ruminant-derived materials in the drug product manufacturing facility should be included in this section. If a potential for cross-contamination with adventitious agents exists additional information should be provided in 3.2.A.1 and 3.2.A.2.

3.2.P.3.4 Controls of critical steps and intermediates

Critical steps:

- Tests and acceptance criteria should be provided (with justification, including experimental data) performed at the critical steps, as identified in 3.2.P.3.3, of the manufacturing process to ensure that the process is controlled.

Intermediates:

- Information on the quality and control of intermediates isolated during the process should be provided.

Test(s) performed in lieu of a finished product test:

- The same information should be provided for an in-process material test performed in lieu of a finished product test as that submitted for a finished product test (analytical procedure, methods validation information).

3.2.P.3.5 Process validation and/or evaluation

- A description, documentation and results of the validation and/or evaluation studies should be provided for critical steps or critical assays used in the manufacturing process (e.g. validation of the sterilization process or aseptic processing or filling). Detailed information should be provided in 3.2.P.3.4 for critical steps and all intermediates that are controlled.
- The critical process control values from relevant batches to support numeric ranges, limits, etc., for critical process controls should be included in 3.2.P.3.4 and
- If a process test takes the place of an end-product test it should be listed in the specification (3.2.P.5.1) and specified as a process test.

The concept of quality by design and associated topics such as design space, quality target profile, edge of failure and control strategy was introduced in the revision of the ICH guideline on pharmaceutical development (Q8R1) (4) which will be discussed later in this chapter. Due account should be taken of the guidance in ICH Q6A: Specifications: Test Procedures and

Acceptance Criteria for New Drug Substances and New Drug Products: Chemical Substances (5) in some parts of the information on manufacture, particularly in 3.2.P.3.4 Control of critical steps and intermediates.

There are relevant guidelines relating to the manufacture of the dosage form available within the regions:

Europe

- Committee for Proprietary Medicinal Products (CPMP) (which was the predecessor to the current Committee for Human Medicinal Products, CHMP) Note for guidance on manufacture of the finished dosage form (6),
- CPMP Note for guidance on limitations to the use of ethylene oxide in the manufacture of medicinal products (7),
- CPMP Note for guidance on quality of water for pharmaceutical use (8) and the related CHMP reflection paper on the use of Water for Injection[s] prepared by reverse osmosis (9),
- CPMP Note for guidance on process validation (10),
- CPMP Annex II to Note for guidance on process validation Nonstandard processes (11),
- CPMP Note for guidance on parametric release (12) and the related CHMP concept paper on the revision of the Guideline on parametric release (13),
- CPMP Note for guidance on the start of shelf life of the finished dosage form, annex to the CPMP Note for guidance on the manufacture of the finished dosage form (14),
- CPMP Note for guidance on the use of Near Infrared Spectroscopy by the pharmaceutical industry and the data requirements for new submissions and variations (15) and the related CHMP Concept paper for the revision of this guideline (16) and
- CPMP guideline: the use of ionising radiation in the manufacture of medicinal products (17) and the related GMP Annex 12 on the Use of ionising radiation in the manufacture of medicinal products (18).

United States

- Center for Drugs and Biologics, Food and Drug Administration (FDA) Guideline for submitting documentation for the manufacture of and controls for drug products (19),
- ANDAs: Pharmaceutical solid polymorphism Chemistry, Manufacturing, and Controls Information (20),
- Powder blends and finished dosage units—stratified in-process dosage unit sampling and assessment (21),
- Immediate release solid oral dosage forms Scale-up and postapproval changes: Chemistry, Manufacturing, and Controls, *in vitro* dissolution testing and *in vivo* bioequivalence documentation (22),
- SUPAR-IR: Questions and answers (23),
- SUPAC-IR/MR: Immediate release and modified release solid oral dosage forms Manufacturing equipment addendum (24),
- SUPAC-MR: Modified release solid oral dosage forms (25),
- Nonsterile semisolid dosage forms Scale-up and postapproval changes: Chemistry, Manufacturing, and Controls, *in vitro* dissolution testing and *in vivo* bioequivalence documentation (26),
- SUPAC-SS: Nonsterile semisolid dosage forms Manufacturing equipment addendum (27),
- Process validation: General principles and practices (28),
- Guidance for industry for the submission documentation for sterilization process validation in applications for human and veterinary drug products (29),
- Guidance for industry Submission of documentation in applications for parametric release of human and veterinary drug products terminally sterilized by moist heat process (30) and
- Guidance for industry PAT—A framework for innovative pharmaceutical development, manufacturing, and quality assurance (31).

Japan

The author was unable to identify Internet links to official (i.e. government) Japanese guidelines available in English translation relating to the manufacture or process validation of pharmaceutical products. The following information source provides general information on data requirements for Japan:

- Information in English on Japan Regulatory Affairs Pharmaceutical administration and Regulations in Japan March 2008 (32) and
- *Drug Approval and Licensing Procedures in Japan 2006* (33).

General

It should also be kept in mind that the relevant requirements of regional pharmacopoeias should be taken into account. This will include, for example, requirements related to residual solvents in finished products [which is also covered by ICH Q3CR3 (34)]. The general notices, general chapters, and general (and relevant specific) monographs of the *European Pharmacopoeia* (Ph Eur), the *Japanese Pharmacopoeia* (JP), and the *United States Pharmacopeia* (USP) apply to pharmaceuticals in the regions concerned and this can influence the manufacturing method and control tests applied to such products.

MANUFACTURE OF PHARMACEUTICAL PRODUCTS

In this section of the chapter a summary will be given of the information requirements for an MAA in each of the three ICH regions.

Europe

The following information is based on the CPMP guideline on manufacture of the finished product (6) unless otherwise indicated.

Based on the European legislation, Directive 2001/83/EC as amended (35), the following information is required to form part of the MAA:

- *Article 8(3)(a)*: Name or corporate name and permanent address of the applicant and, where applicable, of the manufacturer.
- *Article 8(3)(d)*: Description of the manufacturing method.
- *Annex I Part I 1.2 Application form*: The name and address of the applicant shall be given, together with the name and address of the manufacturers and the sites involved in the different stages of the manufacture [including the manufacturer of the finished product and the manufacturer(s) of the active substance(s)], and where relevant the name and address of the importer.
- *Annex I Part I 3.1:* Format and presentation.

The general outline of Module 3 is as follows:

. . .

Manufacture

- Manufacturer(s)
- Description of Manufacturing Process and Process Controls
- Process Validation and/or Evaluation
- Manufacturing Process Development

Manufacturing process of the finished medicinal product

(a) The description of the manufacturing method accompanying the application for Marketing Authorization pursuant to Article 8 (3) (d) shall be drafted in such a way as to give an adequate synopsis of the nature of the operations employed.

For this purpose it shall include at least:

- Mention of the various stages of manufacture including process controls and corresponding acceptance criteria, so that an assessment can be made of whether the processes employed in producing the pharmaceutical form might have produced an adverse change in the constituents.
- In the case of continuous manufacture full details concerning precautions taken to ensure the homogeneity of the finished product.

- Experimental studies validating the manufacturing process where a non-standard method of manufacture is used or where it is critical for the product.
- For sterile medicinal products details of the sterilization processes and/or aseptic procedures used and
- A detailed batch formula.

- The name, address, and responsibility of each manufacturer, including contractors, and each proposed production site or facility involved in manufacturing and testing shall be provided.

It is indicated in the guideline that many aspects of GMP do not need to be described in the MAA and only those aspects of quality assurance relating specifically to the product should be included.

The manufacturing formula should be included with the intended batch size(s) indicated. The names and quantities of all materials used during the manufacturing process should be stated including solvents and other materials removed during the process. Where certain materials will not be used for every batch—for example, pH adjusting agents—this should be indicated (and a range of "0 to x" could be included). Quantities (with upper and lower limits) should be stated for all ingredients (and justified if they are "factorized" or adjusted for potency)—for active ingredients the limits should be 100% ± 5% of the declared content, while for excipients they should be 100% ± 10% of the declared content. Wider ranges should be justified—for example with data from batches at the extremes showing that the product will remain within its shelf life specification during its shelf life. Where overages are present they should be stated (and justified in the pharmaceutical development section of the dossier).

The manufacturing method should be described. Where alternative steps are proposed—for example two sterilization methods for the container—this should be accompanied by data to show that a finished product of the desired quality is produced consistently with all requested variants. The various steps in the manufacturing process should be shown in a flow chart with an indication of control points and (if appropriate) which site is involved (where more than one is used in the manufacturing process). Very detailed descriptions of the manufacturing process should be avoided to reduce the number of variations that will be required in the future. The author's experience is that the amount of information provided is often a matter of debate between the applicant and the assessors. The level of detail required will depend on the specific product and control tests needed to ensure conformity with the finished product specifications. Notwithstanding this approach there will be certain processes that will need to be described in detail—for example sterilization processes and the process controls applied. The need for specific in process controls and tests should be discussed—for example to ensure uniformity of content of the active ingredient in the dosage forms, mixing process controls, etc.

Each site involved in the manufacture of the product should be stated whether they are part of the same company or not. It should be stated which company is responsible for batch release.

Methods of sterilization used for the final product or the primary containers should be justified in the pharmaceutical development section of the dossier. Terminal sterilization of the product in the final container is preferred [and this is stated in the Ph Eur text on Methods of preparation of sterile products (36)]. The Ph Eur text includes recommendations for various methods of sterilization and these should be followed or any deviation identified and justified. Limited validation data will be required if sterilization is by (dry) saturated steam at a minimum of 121°C for 15 minutes or dry heat at a minimum of 160°C for at least two hours. For all other terminal processes in the final container the acceptance limits for in process controls and the maximum permissible bioburden prior to sterilization should be stated. The validation data for Sterility Assurance Level (SAL) should be included in the MAA.

In the case of preparation of sterile products using sterilization by filtration an acceptable bioburden level should be stated in the application—not more than 10 colony-forming units per 100 ml will be acceptable for bulk product prior to filtration. The type of filter and its pore size should be stated—nominal pore sizes of 0.22 μm or smaller are accepted without further justification. Media fill data may be requested by the competent authorities—in practice this is likely to be asked for.

There is a separate guideline on the use of ionizing radiation for sterilization purposes (17) and a related GMP annex to which this guideline cross refers (18). In addition to its use for sterilization of the dosage form irradiation can also be used for bioburden reduction or for other reasons. If irradiation is to be used then the MAA should indicate what it is to be used for and whether it will be applied to starting or packaging materials, intermediate or bulk product, or to the finished dosage form in its final container. The shape, size and composition of the items to be processed should be stated clearly. Production batch size and irradiation batch size should be stated or, in the case of continuous processing, the batch size should be defined in terms of a stated period of time. The minimum dose of radiation intended to be absorbed should be stated. Information on the authorization of the irradiation plant—which might be using gamma or beta radiation, for example—should be provided (although this has been a cause of some difficulties because many of these plants operate under an ISO 9000 quality system rather than under pharmaceutical GMPs). The address of the irradiation plant should be stated. The information needed on the irradiation plant is summarized in Table 1. The guideline also includes information on validation studies that are required—these will be summarized under Validation, below. The GMP Annex discusses a number of issues related more to responsibilities and commissioning of the plant, the premises, documentation and microbiological monitoring. It also includes guidance on dosimetry and process validation. It is required that the dosimeters be traceable to a national or international standard and information is offered on the use of the same instrument to establish the calibration curve of the routine dosimeters and on the

Table 1 Information to be Included in an MAA for Irradiation Plants (15)

Description of the irradiation plant
– Type of plant (radionuclide source or electron generator)
– Builder of the plant
– Continuous or batch mode operation
– Radionuclide authorized and actual activity (GBq) or maximum and minimum electron energy (MeV) of the generator
– Description of the plant, drawings, course of the product in the plant, position and geometry of the irradiation source, and conveyor system including pass source mechanism
Description of the irradiation process
– Description of the material to be irradiated, limits (if any) on bioburden or process intended to limit or control bioburden; action to be taken if bioburden limits are exceeded
– Number, position of irradiation containers relative to the source position during the dwell time and the method of moving them through the chamber
– Material and dimensions of the irradiation container
– Maximum total irradiation time/maximum dwell time of product in the chamber
– Results of dose mapping studies using dummy product; dose mapping using the loading pattern of the product and a sufficient number of dosimeters to show the distribution of the absorbed dose using a representative number of irradiation containers[a]
– Loading pattern of the product for each irradiation container; if mixed loads are allowed the composition of the load; mean load density and acceptable minimum and maximum density[b]
– SOP with at least the following: loading pattern of product(s) within the chamber; type/number/location of routine dosimeters within one irradiation batch/stated period of time; adjustments to be applied to routine dosimeter measurements to convert to the absorbed dose at minimum and maximum positions; stated minimum and maximum absorbed dose[c] including experimentally determined dosimeter errors; whether or not repeated treatment is acceptable and the circumstances in which such retreatment is allowed (and how many times); for electron beam irradiators the electron energy, average beam current, beam width, conveyor speed, with acceptable limits

[a] Separate dose mapping exercises should be applied to each product/category of products and each pathway used for processed products.

[b] Modification of the load pattern may be allowed provided that a new dose mapping exercise is undertaken and that this shows that the minimum and maximum doses are met.

[c] Minimum dose: that required for the intended purpose; maximum dose: that required to prevent unacceptable changes in the product or packaging (or imposed by official restrictions)—a minimum absorbed dose of 25 kGy may be regarded as adequate for sterilizing materials or products with low initial bioburden and no radioresistant spores, or other doses that have been subject to a biological validation.

establishment of the absolute absorbance of each instrument used. There is also discussion of the validation of the process, which will be summarized below.

There is a separate guideline on the use of ethylene oxide sterilization (which is a surface sterilization method that will have limited potential application)—it is made clear that this process is acceptable only if no other method of sterilization is available for the specific product or material (e.g. primary container). In the general guideline on manufacture of the finished product it is indicated that if this method of sterilization is to be used then information will be required on the equipment to be used, the quantitative composition of the gas (mixture) to be used, presterilization bioburden of the material to be sterilized, details of any pretreatment prior to sterilization, details of the cycle used (temperature, gas exposure time and concentration, humidity level during the process) and details of the method used to remove residual ethylene oxide. The conditions of the process should be monitored and appropriate in process controls and limits applied. Process validation data will be required. A suitable sterility assurance level should be demonstrated and the removal of residual ethylene oxide demonstrated. A limit of not more than 1 ppm of residual ethylene oxide (using a suitable method, such as simulated use extraction) and not more than 50 ppm of residual ethylene chlorhydrin (or other halogenated ethylene hydrin) will be accepted provided that a suitable justification has been put forward for the use of ethylene oxide in the first place. Limits for residual ethylene oxide should be included in the finished product release and end of shelf life specifications. The guideline on the use of ethylene oxide in the manufacture of medicinal products (7) points out that the concerns relating to the use of this material include its cytotoxic, mutagenic and carcinogenic properties. This is the reason behind the very restricted limit for the material in the finished product (1 μg/g). Due account should be taken of the potential presence of ethylene oxide residues in raw materials as a result of their manufacture or sterilization. Containers that are sterilized using this method should also be controlled for residues—limits of 1 μg/ml of container volume for ethylene oxide and 50 μg/ml for ethylene chlorhydrin, etc., are suggested.

Where toxic gases or solvents are used in the manufacturing process appropriate limits for residues of these materials in the finished product should be included in the dossier and may be justified in an appropriate place. The justification should consider both toxicological aspects and aspects related to the manufacturing process technology. Appropriate in process controls and acceptance limits should be established. Validation data should be included in the dossier.

The CPMP guideline on manufacture of the finished product indicates that information on the reprocessing of nonbiological medicinal products is considered to be a matter for GMP and does not need to be included in the MAA dossier. However this is inconsistent with the information requirements in the CTD guidelines and may reflect the age of the CPMP document.

Validation data for cleaning of primary packaging materials may be requested by the competent authorities.

Competent authorities may request information on production area, microbiological quality, and levels of particulate matter in the atmosphere.

The grade of water to be used in the manufacturing process is addressed in a guideline (8) and a reflection paper (9). The grade of water that may be used will depend on the type of product being manufactured. For sterile products the following expectations apply:

- Parenteral products, hemofiltration products, peritoneal dialysis solutions, and irrigation products should be made with water for injections Ph Eur prepared by distillation.
- Ophthalmic products, nasal and otic products and cutaneous use products may be manufactured using purified water Ph Eur.

For nonsterile products purified water may be used in most cases.

The grade of water that should be used in pharmaceutical processes is indicated in Table 2.

In the reflection paper, the Quality Working Party has confirmed that it was not (as of March 2008) possible to assume that the quality of water for injections prepared using reverse osmosis is as safe as water prepared by distillation.

If a manufacturer wishes to use near infrared spectroscopy as part of the manufacturing process controls, then appropriate note should be taken of the relevant CPMP guideline (15)

Table 2 Water Grades That can be Used for Manufacturing Processes (5)

	Intended use	
Grade of water	**Nonsterile products**	**Sterile products**
Potable water	Initial rinse, clean in place of equipment, containers, closures	
Purified water (or same quality as used in the manufacture of the product if higher)	Final rinse, clean in place of equipment, containers, closures	Initial rinse, clean in place of equipment, containers, closures[a] (Sterile nonparenteral products) Final rinse, clean in place of equipment, containers, closures[a,b]
Water for injections		(Sterile parenteral products) Final rinse, clean in place of equipment, containers, closures[a,c]

[a] Initial rinse of some containers may not be required—e.g., plastics containers for eye drops; and for blow-fill-seal containers will not be possible.
[b] If equipment is dried after rinsing using 70% ethanol this should have been diluted using water of the same quality as that used for the final rinse.
[c] Where a subsequent depyrogenation step is used it may be possible to use highly purified water Ph Eur subject to justification and validation.

and a reflection paper on the revision of this document (16). The data requirements will depend on whether the method is to be used qualitatively or quantitatively.

It should be taken into account that there is a guideline on the start of the shelf life of the dosage form which is related to the manufacturing process (14). The start of the shelf life should be timed from the date of release of the batch provided that this does not exceed 30 days from the date of production of the batch. If the period is greater than 30 days then the following applies for products other than biologicals/blood products/biotechnology products:

> The date of production of a batch is defined as the date that the first step is performed involving combining the active ingredient with other ingredients. For medicinal products consisting of a single active ingredient filled into a container, the initial date of the filling operation is taken as the date of production.

In addition to the general guidelines referred to above there are other documents that address data requirements for certain specific product types. These include a guideline on the pharmaceutical documentation for medicinal gases (37) and on the pharmaceutical quality of inhalation and nasal products (38).

For medicinal gases the manufacturing method description should include a detailed diagram including the controls applied at various stages. Automated packaging systems or mixing systems and the specification of the equipment should be described and the process validated.

For inhalation and nasal products the declaration of the formulation should include the concentration of the drug substance, the fill amount and the target delivery amount. The manufacturing process should be described for each strength and each container closure system and should include information on all filling and packaging operations. Appropriate validation studies should be undertaken and reported (including homogeneity of the formulation) throughout the filling process, assurance that the correct fill volume or weight is achieved and that the closure system is correctly applied. A process control should be applied for the performance of the actuation release mechanism. Any equilibration period needed before testing should be stated.

United States

The legislative basis for applications for a new drug application and their content is the *Code of Federal Regulations Title 21* § 314.50 (39). The following information is required in an NDA:

- A summary of the chemistry, manufacturing and controls section of the application.
- Chemistry, manufacturing, and controls section—describing the ... manufacture ... of the drug product.
- Drug product—A list of all components used in the manufacture of the drug product (regardless of whether they appear in the drug product); and a statement of the composition of the drug product ... the name and address of each manufacturer of the drug product; a description of the manufacturing and packaging procedures and in process controls for the drug product ... alternative components, manufacturing, and packaging procedures [and] in process controls ... and
- The proposed or actual master production record, including a description of the equipment to be used for the manufacture of a commercial lot of the drug product or a comparably detailed description of the production process for a representative batch of the drug product.

The FDA guidance on submitting documentation for the manufacture of controls for drug products (19) was published in 1987 but does not appear to have been updated, although a number of other documents of relevance were published at least in draft form but subsequently withdrawn.

The list of components used in the manufacture of the dosage form—whether or not present in the finished product—should be provided and the quality designation or grade stated (e.g. USP, National Formulary or American Chemical Society). Established names or chemical names should be used to ensure that the right material is identified. The identity of proprietary mixtures should be stated and the composition included. Any alternatives listed should be shown not to significantly alter the nature of the final product.

A quantitative statement of composition should be supplied and state per unit dose the weight or range for ingredients in the drug product.

A batch formula should be provided for each formulation of the drug product and this should include all ingredients whether or not they are present in the final product.

The manufacturer's name and (physical street) location should be stated. This should include each facility involved in the manufacture of the product including contract packagers and labelers and external quality control laboratories. Suppliers of components used in the manufacture of the drug product should be stated.

The production process for the finished product should be described and a draft or actual master or batch production record and control records (or comparable documents describing the production process of a representative batch) should be submitted. Each production step should be described with the operating conditions, equipment to be used, and in process control points. A flow chart showing material flow, equipment used, and points of sampling would be "helpful."

If it is intended to allow reprocessing of the drug product then proposals covering foreseeable deviations should be included in the application. Unanticipated reasons for reprocessing should be subject to a supplementary application. Only approved reprocessing proposals should be used for product released to the market.

In process controls at various stages of the manufacturing process should be described and should be supported by appropriate data such as batch records.

Where drug substance polymorphic form can affect the manufacturing process—for example by affecting particle shape, flowability or compaction properties—due account should be taken of the guidance on ANDAs: Pharmaceutical solid polymorphism (20).

Many aspects of the manufacturing process, in process controls and reprocessing are covered in the US by the GMP provisions of 21 *Code of Federal Regulations* 210–211 (40), which are beyond the scope of this chapter. There is a specific FDA guidance on Powder blends and finished dosage units—stratified in-process dosage unit sampling and assessment (21), which provides a number of recommendations that could affect the manufacturing method and the conduct of in process controls for affected products. These include:

- Conducting powder blend sampling and analyses,
- Establishing initial criteria for stratified sampling of in-process dosage units and evaluation of test results,
- Analysis of the stratified samples and evaluation data,
- Correlating the stratified sample data with the powder blend data,
- Assessing powder mix uniformity,
- Correlating the stratified sample data with the finished dosage unit data and assess uniformity of content,
- Testing exhibit and validation batches for adequacy of powder mix,
- Testing and evaluating routine manufacturing batches and
- Reporting the use of stratified sampling in the marketing authorization application.

There are also recommendations on the assessment of powder mix uniformity, correlation of this with stratified in-process dosage unit data and the correlation of data from the in process samples and the finished product. The guidance goes on to suggest how manufacturing criteria may be validated and to offer advice on routine manufacturing batch-testing methods.

The FDA has also published a number of documents in the Scale-Up and Postapproval Changes (SUPAC) series (22–26) that cover immediate release and modified release solid dosage forms and nonsterile semisolid dosage forms. The information in these documents should be taken into account when considering making changes to the information submitted in a marketing authorization. In terms of potential impact on manufacture of the dosage form, particular attention should be paid to the Manufacturing Equipment Addenda (24,27). The documents include tables of classified types of manufacturing equipment used for various purposes. To move from one class of equipment to another class for a particular product requires additional information. The recommendations cover particle size reduction, blending and mixing, granulation, separation, mixing, emulsification, deaeration, transfer, drying, holding, coating, unit dosing, printing dosage forms, filling and sealing.

Japan

Limited information is available in English and in the public domain on the specific data requirements for the manufacture of pharmaceutical products for an MAA to be submitted to Japan. Since Japan has adopted the relevant guidelines for the format of the Common Technical Document, it might be assumed that this is the correct format for an application. However, as has been indicated elsewhere in this book, there are some important differences. Much more information is included in the application form for Japan, as the Quality Overall Summary is expected to be much more detailed (and longer) than in the EU and the US.

A summary of the information relating to manufacture of the finished product that is required in an MAA for submission to Japan is summarized in Table 3. This is based on the information in Reference (33). This book also includes some useful examples of the extent of

Table 3 Summary of Information Easily Available on Manufacturing Information to be Included in a Marketing Authorization Application to be Submitted in Japan

Ethical and OTC drugs: Original applications
Manufacturer of dosage form, testing facilities, contract manufacturers
Ingredients, solvents, immediate container components
Most important points (with regards consistency of product) in the manufacturing method in sequence, process, and flow chart in detail with quantitative information on ingredients and yield, process parameters with target or set values for operating conditions, control methods and limits, intermediate specifications
Summary of other points
Items that can be considered as minor changes
Data to show acceptable level of residual solvents (other than for water and ethanol)
Whether or not protection from light is required

information to be included in the application for several types of dosage form and advice on data requirements for the two types of application for change to the submitted information.

PROCESS VALIDATION

Having described the manufacturing method it is then necessary to provide at least some information on the validation of key (critical) processes that could affect the quality of the manufactured product. The amount of information will vary with the type of product but particular attention will be paid to the validation of key manufacturing steps that control the release characteristics of a drug from a modified release product and to methods of preparation of sterile products. The aim of process validation is to demonstrate that the manufacturing process can produce consistently a finished product of the required quality. The provision of three verification batch analyses at full-scale production (usually before the product has been placed on the market or at least a promise for the provision of the data if they are not available at the time the MAA is submitted) is one aspect of process validation, but information based on analytical data alone will often be insufficient to demonstrate to the assessors of the MAA that the manufacturing processes are under adequate control. The amount of additional information expected will depend on the complexity of the product and its manufacturing procedures.

There is no relevant ICH guideline.

Europe

There are two guidance documents available on process validation. The first is a general document (10) and the second is a specific annex to the first concerned with "non-standard" process validation (11) but for the sake of convenience the information in the two documents will be considered in a single narrative. These guidelines could be seen as complementary to the requirements in the guideline on pharmaceutical development (which is discussed elsewhere in this book) as well as the guidelines on the manufacture of the finished product and on the proposed finished product specifications. The guidelines are not intended for application directly to drug substances manufacture nor to the manufacture of biological or biotechnological preparations.

It should be noted that process validation issues often have overlap with GMP: due account should be taken of the relevant GMP guidance documents in the *Rules Governing Medicinal Products in the European Union Volume 4 Guide to Good Manufacturing Practice* (41) and documents of relevance from the Pharmaceutical Inspection Cooperation Scheme, including the recommendations on validation of aseptic processes and guidance on parametric release (42).

Data from studies into laboratory (i.e. one thousandth to one hundredth of the proposed production scale), pilot scale (at least 10% of the proposed production scale) and production scale batch material should form part of the package of information in the MAA. Information on the scale-up process will be useful in demonstrating that increasing the scale of production will not result in a loss of product quality. Where ranges of production lot sizes are proposed in the MAA, then there should be a demonstration that such variation would not affect product quality.

Information on three production scale batches made by the intended manufacturing process should be made available when available. It is expected that during the development of the product and its associated manufacturing processes, appropriate experiments will have been undertaken to challenge the processes to ensure that they are sufficiently robust for routine use. Such data may not be complete at the time that the MAA is submitted but the manufacturer is expected to keep the whole of the information available for verification by inspectors from the supervising authority.

Having once defined the intended manufacturing method for the finished product (which will take into account the properties of the active ingredient, key excipients, the formulation and the potential impact of the manufacturing process on product quality and stability, and how the process has been optimized during development), this method will need to be described in the MAA in sufficient detail. The process validation guideline suggests that the manufacturing process should be described "fully" which would seem to be at variance with the advice in the guideline on manufacture of the finished product (discussed above) which suggests that the manufacturing process not be described in too much detail to avoid unnecessary variation submissions in the future. This is to some extent mitigated by an indication in the change control section of the guideline that "very detailed description of processing instructions and equipment design need not be included in the dossier, otherwise prior approval by variation will be necessary before they can be altered."

Process validation data can be used to justify skip testing and, possibly, the omission of certain procedures from route batch analyses.

Where a "non-standard" manufacturing method is used, the validity of that method should be demonstrated and data provided in the MAA. In these cases data from pilot scale batches may not be considered to be predictive of full-scale production. For some products with such as modified release preparations, more complete information would also be expected in the MAA. "Nonstandard" processes will include:

- Non-pharmacopoeial sterilization methods
- Aseptic processing
- Lyophilization
- Microencapsulation
- Certain mixing and coating procedures and
- Other specialized processes.

This was expanded in the guideline on non-standard process validation to include the following:

- The manufacture of specialized pharmaceutical dosage forms (including metered dose inhalation products; suspensions, emulsions, or other liquid dispersed forms; prolonged release preparations; unit dose preparations with a low content of active ingredient—$\leq 2\%$ of the composition; parenteral depot products based on biodegradable polymers; liposome preparations; and micellar preparations).
- The incorporation of new technology into a conventional process (such as the introduction of a new drying technology for granules that has not been used previously in the pharmaceutical industry).
- Specialized processes involving new technologies or existing processes known or likely to be complex and therefore requiring particular care (lyophilization, microencapsulation, processes where the properties of the active ingredient or key excipients may give rise to issues during scale up or processing—e.g. lubricants, coating agents).
- Nonstandard methods of sterilization (terminal processing by moist heat using non-pharmacopoeial conditions, terminal sterilization by irradiation using less than 25 kGy, aseptic processing, standard methods of sterilization with a request for parametric release).
- Any manufacturing process not previously approved for pharmaceutical products in the EU.

It is, however, possible for deviations from this to be allowed where a manufacturer of specialized products or using nonstandard processes can be exempted from the need to provide additional data at the time of submission provided that sufficient supporting data are provided. This could have to be justified on a case-by-case basis.

Normally, process validation data will be expected to be included in the MAA on three production scale lots or an undertaking for the provision of the data postapproval will be required. In other cases, information will be expected from one or two production scale batches if these are supported by pilot scale product data and by evidence of a history of consistent manufacture of products by essentially similar processes. It is expected that critical phases of the production process would be addressed.

Process validation data are required for each proposed manufacturing site. There should also be clearly defined procedures for change control. In some cases it could be necessary to submit a variation prior to the introduction of changes to the proposed manufacturing procedure. Minor changes to standard operating procedures should not require formal approval but significant changes to some factors (such as mixing times, drying temperatures, new equipment using different operating design and operating conditions) should be subject to a variation and supported by adequate data.

It is necessary to include a process validation scheme in the MAA in those cases where validation data on production scale batches are not included. This should address the formal

Table 4 EU Process Validation Scheme Minimum Content (8)

Information included in the MAA
Short description of the process including summary of critical processing steps or critical parameters to be monitored during validation
Finished product release specification
Analytical methods (reference to MAA)
In process controls and acceptance criteria
Additional testing to be undertaken with appropriate analytical validation and proposed acceptance criteria
Sampling plan
Methods for recording and evaluation of results
Proposed timetable for conduct of studies
Content of report for consideration by supervisory authority
Batch analytical data
Certificates of analysis
Batch production records
Report of unusual findings and modifications or changes found to be necessary, with rationale
Conclusions

process validation studies to be undertaken, usually on three consecutive batches. This information would then be available for examination by the supervising authority. The minimum information to be included in the scheme is indicated in Table 4.

The CPMP guideline on manufacture of the finished product (6) also includes information on validation of manufacturing processes. This addresses in particular the use of nonstandard processes (and mentions novel dosage forms and the manufacture of liposomes). Other issues included are mixing, granulation, emulsification and non-pharmacopoeial sterilization processes. When process validation data have been produced satisfactorily it will still be expected that the relevant characteristic of the product—for example, sterility, content uniformity—should still be included in the release and shelf life specifications.

Validation of irradiation sterilizations processes is discussed in the relevant CPMP guideline (17).

For medicinal gases, process validation is addressed in the CPMP guideline for this type of product (37).

In the guidance documents on process validation, there is a reference to parametric release and there is a specific CPMP guideline on this topic (12) and a concept paper on the revision of that document (13).

Parametric release is defined as:

> a system of release [of the finished product] that gives assurance that the product is of the intended quality based on the information collected during the manufacturing process and on . . . compliance with specific [relevant] GMP requirements.

Parametric release can be used as an alternative to routine release testing for certain specific parameters.

The main guideline concentrates on the use of parametric release as a replacement for the sterility test for sterile products (and the text recognizes the statistical limitations of the sterility test) and refers to the elimination of this test from the required batch release testing dependent upon the successful demonstration of prespecified validated sterilization conditions having been achieved for the batch concerned. This possibility is also recognized in the Ph Eur text on methods of preparation of sterile products.

Parametric release is permissible only for product terminally sterilized in the final container. It could be applied in the case of processes complying with the requirements of the Ph Eur using moist heat dry heat or radiation. In all cases, compliance with GMP requirements for the process and its validation should have been followed.

Heat sterilization processes should be complemented by biological validation. The heat resistance characteristics of the bioburden should also be known.

The application for permission to use parametric release for sterile products manufactured using heat sterilization methods should be accompanied by the following information:

- A description of the sterilization process—type of cycle, load pattern, cycle parameters (time, temperature, pressure, F_0 value).
- A statement as to whether chemical indicators are used.
- In process control methods and specifications to be applied (presterilization bioburden, cycle parameter monitoring, verification of load sterilization).
- A process validation report (heat distribution and penetration studies for at least three runs, microbiological qualification that a sufficient sterility assurance level is achieved at the minimum level of the cycle including information on the type, D value, z value and stability of bioindicators used and bioburden characterization) and
- Packaging integrity data, if applicable.

Factors that will be taken into account in assessing a request for parametric release will include:

- Whether there is sufficient experience of the process by the manufacturer.
- The history of GMP compliance by the manufacturer (including current compliance).
- Whether the manufacturing process has been validated adequately.
- Whether the process is controlled adequately.
- The relationship between end product testing and process monitoring.
- The justification put forward for the acceptance criteria.
- Whether the accept/reject criteria have been based on the acceptance criteria included in validation records.
- Whether clearly defined procedures are specified for reporting and the actions taken on acceptance or rejection.
- Historical batch data using the process concerned.

In the context of a request for parametric release a preapproval GMP inspection will usually be conducted unless the product concerned is similar to another product made at the same production site by using a similar process that is already approved for parametric release.

Parametric release can also be applied to radiation sterilization. The minimum absorbed dose should normally be 25 kGy but lower doses can be justified if the bioburden levels and radiation resistance are checked routinely and if adequate validation data are provided.

If permission has been granted for the use of parametric release then all subsequent decisions for accepting or rejecting batches should be based on the agreed criteria been met—the use of sterility tests cannot overrule those results. Compliance with the requirements of the sterility test (if the product was to be tested) should still form a part of the release and shelf life specification even where parametric release has been agreed.

Other possible applications of parametric release are touched on in the guidance:

- Tablet manufacture—uniformity of mass, crushing strength, disintegration and
- The application of process analytical chemistry methods—near infrared spectroscopy and Raman spectroscopy (often in combination with multivariate analysis) for content of active ingredient, polymorphic form, water content, blending homogeneity, particle or powder properties, film thickness.

Because of the variety of manufacturing processes that could be applied the guideline recommends a case-by-case consideration of the application of parametric release. Its use will be permitted after due consideration of how well founded the manufacturing process is and requires a demonstration of its robustness in the MAA. Where critical parameters are monitored, it must be shown that the validated conditions have been achieved throughout the batch.

In the concept paper for the revision of the main text, it is recognized that parametric release could be rather more widely applied under the terms of the ICH guidelines on pharmaceutical development (Q8R1) (2), risk management (Q9) (43) and pharmaceutical quality systems (Q10) (44) following the introduction of Quality by Design and Process Analytical Technologies. The revision of the main guideline will take into account the new ICH documents and will also consider the requirement for retesting of products imported into Europe from third countries with which no memorandum of understanding is in place. The issue here relates to the application of retesting on importation when the product was released initially using parametric release (real time testing) (provided that parametric release of the same type had been approved in Europe). A discussion of the three ICH document referred to above is included below.

United States

There are two versions of the FDA guidelines for process validation: one published in 1987 (45) and a draft revision published in November 2008 (28). The draft revision of this guideline is intended to bring the advice into line with the ICH Q8, Q9 and Q10 guidelines and the concept of the product life cycle. The scope of the document will be human and veterinary drugs, biological and biotechnological products, finished products, active pharmaceutical ingredients, and the drug component of a drug and medical device combination product. The following is a composite summary of the two documents with precedence given to the 2008 revised draft.

It is indicated that in order for the product to be fit for its intended use, the following principles (as amended by the draft revision of the guideline) should be applied:

- Quality, safety, and efficacy are designed or built into the product.
- Quality cannot be assured adequately just by in process and finished product inspection or testing.
- Each step of the manufacturing process is controlled to assure that the finished product meets all design characteristics and quality attributes including specifications.

Process validation is a key element in assuring that these goals are met.

Process validation is defined in the draft revision as:

> The collection and evaluation of data from the process design stage throughout production, which establishes scientific evidence that a process is capable of consistently delivering quality products.

A written validation protocol should be prepared by the manufacturer specifying the procedures and tests to be conducted and the data to be collected (and for what purpose). The number of replicate process runs to be used should be defined. The processes used should cover the upper and lower processing limits which could pose a threat of process or product failure—that is, worst-case conditions. Key process variables should be monitored, documented and analyzed. In process test data and finished product test data may both be of value in process validation.

The FDA guideline as amended by the 2008 draft revision identifies the following elements of process validation:

- *Process design*—defining the production process that will be reflected in the master production and control records so that the process can deliver consistently product that meets its critical quality attributes, with decisions and justifications being documented sufficiently and reviewed to verify and preserve their use in later stages of the product life cycle; and establishing a strategy for process control such as controls on process control when the control of the process through operational limits and in process monitoring is essential (e.g. when sampling and detectability limitations affect the ability to measure the attribute—e.g. microbial contamination—or where intermediates or products cannot be characterized highly and well-defined quality attributes cannot be identified).
- *Process qualification*—confirming that the process is capable of reproducing commercial manufacture including facility design and qualification of utilities and equipment and performance qualification that requires the preparation of a performance qualification protocol, which is then executed and a report prepared on the outcome (which should be undertaken before commercial release of finished product other than in special situations—e.g. for orphan drugs where there is limited demand for the product or products which have a low production volume).
- *Continued process verification*—which is intended to ensure that the process remains under control during commercial manufacture of the dosage form.

The 2008 draft guideline goes into considerable detail for these various procedures from the standpoint of process validation being a mandatory requirement under the US GMPs.

The original document discusses the acceptability of product testing and gives a number of examples. Those of relevance to the current chapter include

- The limitation of the sterility test for ensuring product sterility.
- The physical and chemical characteristics of compressed tablets and measures to be taken when changing processing conditions—for example for granulation and
- The pH specification of oral, ophthalmic or parenteral solutions.

There is an FDA guidance document on the documentation for sterilization process validation (29). In the case of product terminally sterilized in its final container using moist heat sterilization the following data should be included in the submission but the information requirements can be modified for other sterilization processes such as ethylene oxide and radiation—the draft guideline has considerable text on the requirements in these cases:

- A description of the drug product and container-closure system including fill size and whether secondary packaging is also to be processed.
- A description of the sterilization process applied to the finished product and to components used in its manufacture with details of the autoclave process and performance specifications including cycle type (saturated steam, water immersion, water spray), cycle parameters, and performance specifications (temperature, pressure, time, minimum and maximum F_0 for particular identified autoclaves).
- Loading patterns for the autoclave.
- Methods and controls for monitoring production cycles (e.g. the number and placement of thermocouples, pilot bottles, biological indicators).
- A description of the routine and unscheduled requalification of production autoclaves.
- A description of any proposed reprocessing options.
- Heat distribution and penetration studies and information on the thermal monitors used and the effect of loading pattern on the thermal input.
- Information on the microbiological efficacy of the cycle demonstrating the sterility assurance level (which should be 10^{-6} or better) with information on the bioburden organisms, the specification for bioburden, the identification of biological indicators and their resistance and stability, and the resistance of the biological indicator relative to that of the bioburden and microbiological challenge studies using the most difficult configuration to achieve sterilization.
- Environmental microbiological monitoring information.
- Container closure and package integrity including simulation of the stresses from processing, integrity following maximum exposure (examining separately each barrier layer), information on the sensitivity of the test and demonstrating the integrity of the product over its shelf life).
- Bacterial endotoxins data and information on the test method and its validation.
- Sterility test method and release criteria and
- Details of formal written procedures.

In the case of aseptic fill manufacturing procedures rather more information is required, including:

- Details of the buildings and facilities with a floor plan identifying the location of equipment.
- A description of the overall manufacturing operation including solution filtration, holding periods, and critical operation.
- Sterilization and depyrogenation of containers, closures, equipment, and components.
- Procedures and specifications for media fills (identifying the filling room on the floor plan, identification of the container closure type and size, volume of medium per container, type of medium used, number of units filled, number of units incubated, number of positive units, incubation parameters, date of each media fill, simulations of normal production operations, microbiological monitoring and process parameters).
- Action taken when media fills fail.

- Environmental microbiological monitoring (including methods for airborne microorganisms, microorganisms on inanimate surfaces, microorganisms on personnel, water systems, product component bioburden) with data for yeast, mould, and anaerobic organisms, and the action to be taken when specification limits are exceeded.
- Container closure and package integrity data.
- Sterility testing methods and release criteria.
- Bacterial endotoxin test methods and
- Evidence of formal written procedures.

With regards to stability considerations the guideline refers to the need for data on container–closure integrity and preservative efficacy data as well as pyrogen or endotoxin testing, much of which would be included elsewhere in a European MAA.

There is an FDA guideline on the submission of documentation for applications requesting parametric release for terminally sterilized products using moist heat processes (30). Although the document is written to address parametric release of products sterilized in their final container using moist heat it is indicated that the same principles can be applied to products sterilized using radiation sterilization.

In the absence of permission to use parametric release the FDA requires that sterile products meet certain sterility requirements before being released to the market, which are fulfilled in many cases by applying the sterility test on finished units taken from the production batch. Parametric release is defined as a sterility assurance program where demonstrated control of the sterilization process enables the manufacturer to use defined critical process controls rather than the sterility test. The guideline recognizes that the sterility test has limitations in its ability to detect contamination because of the small numbers of samples taken for testing and the limited ability of the test media to stimulate growth of all potential microorganisms that could be present in the product. In process control data relating to a validated sterilization method can provide more accurate information relating to product sterility. It is indicated in the guidance that a "sterilization load monitor" (a chemical or biological indicator) is to be included in each load to indicate that the sterilization cycle has occurred and that the critical parameters have been met. The load monitor is placed in an appropriate place in the load based on the validation data for the cycle concerned. It is preferred by the FDA that the load monitor uses a direct measurement of lethality delivered to the load.

In order to gain permission to use parametric release a specific application has to be made to the FDA that demonstrates the acceptability of the proposed critical process controls and how they are controlled. The sterilization cycle should be the same as that already approved and described in the application. The application needs to address the control strategy for the terminal sterilization process, the risk assessment, understanding of the process, and prior knowledge and needs to provide appropriate documentation. These aspects are summarized in Table 5.

Japan

Other than references to the need for manufacturing procedures to have been validated the author was unable to trace any specific public domain guidance document relating to process validation in English.

ICH INITIATIVE RELATING TO PHARMACEUTICAL DEVELOPMENT, QUALITY RISK MANAGEMENT, AND PHARMACEUTICAL QUALITY SYSTEM AS IT RELATES TO PRODUCT MANUFACTURE AND VALIDATION

Three documents have been adopted by the ICH (and will progressively be applied by the regions) on the following topics:

- Pharmaceutical development (ICH Q8R1, 13 November 2008) (2)
- Quality risk management (ICH Q9, 9 November 2005) (43)
- Pharmaceutical quality system [ICH Q10, 4 June 2008) (44).

Many of the concepts introduced in these documents overlap in their application and in some cases are outside the scope of this chapter—for example, some aspects of the

Table 5 Content for Dossiers for Requests for Parametric Release of Products Terminally Sterilized in Their Final Containers Using Moist Heat in the United States

Topic	Information requirement
Control strategy for the terminal sterilization cycle	Rationale for the methods used for monitoring and controlling the process used for the product release (*critical process parameters*)
	Rationale for the selection of the critical process parameters
	Description of the acceptance criteria for parametric release
	Description of the drug product and container-closure system including any secondary packaging to be part of the parametric release program
	Description of the proposed production loading patterns and verification that they are within the validated limits for the sterilization cycle or a statement that they have not changed since the last validation and approval
	Description of the microbiological monitoring plan for the product and its components prior to terminal sterilization with emphasis on spore detection and heat resistance of the product bioburden or a statement that the plan has not been changed since it was last validated and approved
Risk assessment, process understanding, and prior knowledge	Consistency of performance of the terminal sterilization cycle within the approved validated limits
	Discussion of any changes made to the original submission concerning the production terminal sterilization cycle, production loading patterns, and container-closure system including any secondary packaging and an assessment of the risk to sterility associated with such changes
	Experience with the proposed or similar products and container-closure system, the overall risks to sterility and the steps taken to assess and control those risks
	Discussion of the overall prior knowledge and production and testing experience of the manufacturer relevant to the drug product this is to be subject to parametric release
Documentation for the parametric release process	Application or supplement numbers and approval dates for the currently approved sterilization cycle
	Identification of the critical process parameters for the products proposed for parametric release including the minimum and maximum acceptance limits (which should be within the limits that have been validated and approved for sterility assurance for the products concerned)
	Acknowledgment that the parametric release system is the primary release test and that tests based on drawing samples from the finished product lot will not be used to overrule the results obtained—in the event of a lot failing to meet the parametric release criteria the sterilizer load will be rejected and will not be released until there is provision for resterilization (in which case issues related to stability and container-closure integrity become relevant)
	Acknowledgment that regardless of the batch release technique used that any specimen testing for sterility will meet the criteria for sterility

(*Continued*)

Table 5 Content for Dossiers for Requests for Parametric Release of Products Terminally Sterilized in Their Final Containers Using Moist Heat in the United States (*Continued*)

Topic	Information requirement
Documentation for the parametric release process (*continued*)	Description of the sterilization load monitor including an indication of the type of monitor to be used, how it will be used and analyzed, what functions are being measured by the monitor and the rationale for the location of the monitor in the load If indirect monitors are used it is recommended that justification be included for the classification of the indirect indicator that is being used—see ISO 1140 Revision of the certificates of analysis for batch release to indicate that parametric release is the method used to provide assurance of the requirements for sterility

pharmaceutical development process and applications of quality systems to GMPs. In this chapter, attention will be concentrated on those aspects relevant to manufacture, in process control and process validation.

These guidelines introduced a number of new terms, some of which are listed below:

- *Design space*—The multidimensional combination and interaction of input variables (e.g. material attributes) and process parameters that have been demonstrated to provide assurance of quality. Working within the design space is not considered as a change. Movement out of the design space is considered to be a change and would normally initiate a regulatory postapproval change process. Design space is proposed by the applicant and is subject to regulatory change and approval.
- *Formal experimental design* (= design of experiments)—A structured, organized method for determining the relationship between factors affecting a process and the output of that process.
- *Process analytical technology* (PAT)—A system for designing, analyzing, and controlling manufacturing through timely measurements (i.e. during processing) of critical quality and performance attributes of raw and in process materials and processes with the goal of ensuring final product quality.
- *Process robustness*—Ability of a process to tolerate variability of materials and changes of the process and equipment without negative impact on quality.
- *Control strategy*—A planned set of controls, derived from current product and process understanding that ensures process performance and product quality. The controls can include parameters and attributes related to drug substance and drug product materials and components, facilities and equipment operating conditions, in process controls, finished product specifications and the associated methods and frequency of monitoring and control.
- *Critical process parameter*—A process parameter whose variability has an impact on a critical quality attribute and therefore should be monitored or controlled to ensure that the process produces the desired quality.
- *Critical quality attribute*—A physical, chemical, biological or microbiological property or characteristic that should be within an appropriate limit, range or distribution to ensure the desired product quality.
- *Proven acceptable range*—A characterized range of a process parameter for which operation within this range, while keeping other parameters constant, will result in producing a material meeting relevant quality criteria.
- *Quality by design* (QbD)—A systematic approach to development that begins with predefined objectives and emphasizes product and process understanding and process control, based on sound science and quality risk management.
- *Quality target product profile* (QTPP)—A prospective summary of the quality characteristics of a drug product that ideally will be achieved to ensure that desired quality, taking into account safety and efficacy of the drug product.

- *Real time release testing*—The ability to evaluate and ensure the quality of in process and/or finished product based on process data, which typically include a valid combination of measured material attributes and process controls and
- *Quality risk management*—A systematic process for the assessment, control, communication, and review of risks to the quality of the drug (medicinal) product across the life cycle.

ICH Q8R1 is intended to apply to the development of a product and its manufacturing process. The overall function of these developments is to ensure that the intended performance of the product is delivered consistently. The data gathered may be used to establish the design space, specifications and manufacturing controls. The information can be used as a basis for quality risk management but it is emphasized that quality cannot be tested into products—it should be built in by design. Changes during development to the formulation and manufacturing processes can contribute to the establishment of the design space. In the case of manufacturing processes, critical process parameters are usually identified by determining the extent to which variations can impact the finished product quality. The manufacturer may choose to generate additional information (from formal experimental designs and process analytical technology application) to justify an expanded design space, which could allow a more flexible regulatory approach including allowing improvements to the manufacturing process within the defined and approved design space without further regulatory review and allowing real-time quality control with reduced end product testing.

The fact that the concept of design space has been introduced does not mean that manufacturers are obliged to use it. The conventional approach to product and manufacturing process development could still be applied—see for example Table 6 for a comparison of some aspects of the differences in approach in practice.

A summary of the information is ICH Q8R1 relating to the manufacturing process and in process controls is as follows:

- The selection, control and any improvement of the manufacturing process (intended for commercial production batches) should be explained.
- Critical formulation attributes and available process options should be considered.

Table 6 Alternative Approaches for Product and Manufacturing Process Development (2)

Aspect	Traditional (minimalist) approach	Enhanced quality by design approach
Manufacturing process	• Fixed • Validation based primarily on initial full-scale batches • Focus on optimization and reproducibility	• Adjustable within design space • Life-cycle approach to validation, ideally continuous process verification • Focus on control strategy and robustness • Use of statistical process control methods
Process controls	• In process tests primarily for go/no go decisions • Off-line analysis	• Process analytical technology tools used with appropriate feed forward and feedback controls • Process operations tracked and trend analyses used to support continual improvement efforts post approval
Control strategy	• Finished product quality controlled primarily via intermediates (in process materials) and finished product testing	• Finished product quality ensured by risk-based control strategy for well-understood product and process • Quality controls shifted upstream with the possibility of real-time release testing or reduced finished product testing
Life-cycle management	• Reactive—problem-solving and corrective action	• Preventive action • Facilitation of continual improvement

- The appropriateness of the selected equipment should be discussed.
- Process improvement, process validation, continuous process verification (if appropriate) and process control requirements should be reported including physical, chemical and microbiological attributed as relevant.
- Critical process parameters that need to be monitored and controlled should be identified.
- The choice of sterilization method for primary packaging materials and product should be justified.
- Significant differences between manufacturing processes used for pivotal clinical batches and primary stability lots and commercial material should be discussed.
- Measurement systems that allow monitoring of critical process attributes and process end points should be described.
- The ability to produce an end product of reliable quality and the robustness of the process should be discussed.
- The use of risk assessment tools should be considered to identify and rank parameters such as the process and the equipment used, which have a potential impact on product quality and identification of significant parameters.
- Consideration should be given for establishing mathematical relationships to describe the design space that will result in satisfactory product.
- Consideration should be given to the establishment of unit operation design space(s).
- Consideration should be given to the relevance of the design space defined by the manufacturer to the intended scale of commercial production and
- It would be helpful to establish the edge of failure for process parameters beyond which the relevant quality attributes cannot be met but this is not essential in establishing a design space.

As part of the control strategy the following could be useful:

- Controls for unit operations that have an impact on downstream processing or product quality and
- In process or real-time release testing in lieu of end product testing.

Process performance should be monitored and trend analyses prepared.

The ICH quality risk management guideline (Q9) (43) is intended to offer information on the application of risk management tools to pharmaceutical products and the processes by which they are made. Many of the concepts and procedures are included in documents produced by the International Standards Organization (ISO), International Electrotechnical Commission (IEC), and the World Health Organization (WHO), including:

- ISO/IEC Guide 73,
- ISO/IEC Guide 51,
- IEC 61025 Fault tree analysis,
- IEC 60812 Failure mode and effects analysis,
- WHO Technical Report Series 908 Annex 7 Hazard analysis and critical control points,
- IEC 61882 Hazard operability analysis,
- ISO 14971 Application of risk management to medical devices,
- ISO 7870 Control charts,
- ISO 7871 Cumulative sum charts,
- ISO 7966 Acceptance control charts and
- ISO 8258 Shewell control charts.

The references for these documents are included in ICH Q9. There is also a discussion of the potential application of each of these methods in the text of the guideline, the detail of which is beyond the topic of this chapter.

ICH Q10 (43) describes the application of ISO quality concepts (ISO 9000) with applicable GMPs in a pharmaceutical quality system, which complements ICH Q8 and ICH Q9. This should cover the lifetime of the product including its development, the technology transfer stage, commercial production and product discontinuation. The need for a quality manual or equivalent documentation is emphasized. This should include the company's quality policy, the scope of the pharmaceutical quality system, identification of the pharmaceutical quality system processes with their sequence, linkages and interdependencies, and it is indicated that process maps and flow charts can be useful tools to visualize the pharmaceutical quality systems. Management responsibilities should also be indicated.

Other topics included in the guideline are concerned with the continual improvement of process performance and product quality. This is addressed in considerable detail, much of which is not directly relevant to the current chapter but is more related to GMP compliance.

In 2004 the FDA published a guideline on process analytical technology (PAT) (a framework for innovative pharmaceutical development, manufacturing and quality assurance) (31). PAT was identified as being a system for designing, analyzing and controlling manufacturing through the timely measurement of critical quality and performance attributes during processing with the goal of ensuring final product quality—it can therefore be seen as a precursor to the more broadly based concepts in ICH Q8, Q9, and Q10. With respect to the application of PAT to the manufacturing process, the following potential advantages were identified:

- Reduction in production cycle times by the use of online, in line, and/or at line measurements and controls.
- Prevention of rejects, scrap, and reprocessing.
- Permit real-time release.
- The use of increased automation with greater operator safety and reduction in human error and
- Facilitation of continuous processing.

In order to gain these advantages it is first necessary to gain process understanding (i.e. identification and explanation of all critical sources of variability, control of variability by the process, and accurate and reliable prediction of product quality attributes within the design space, process parameters, and manufacturing and other conditions). The PAT guideline goes on to discuss the principles and tools that can be employed but much of this is outside the scope of the present chapter. Interested readers are invited to read the source document for further information.

REFERENCES

1. Common Technical Document ICH M4 Documents can be found at http://www.ich.org/cache/compo/276-254-1.html. Accessed December 18, 2008. Individual parts of the M4 document can be found at the following addresses: General and Organisational: M4R3 http://www.ich.org/LOB/media/MEDIA554.pdf. General Questions and Answers: http://www.ich.org/LOB/media/MEDIA1189.pdf. Quality Overall Summary and Module 3 (Quality): M4QR1 http://www.ich.org/LOB/media/MEDIA556.pdf. Quality Questions and Answers: http://www.ich.org/LOB/media/MEDIA620.pdf. Regional versions of these documents can be found on the Web sites of the regional agencies.
2. http://www.ich.org/cache/compo/1325-272-1.html#Quality. Accessed December 18, 2008.
3. http://www.ich.org/LOB/media/MEDIA620.pdf. Accessed December 18, 2008.
4. http://www.ich.org/LOB/media/MEDIA4986.pdf. Accessed December 18, 2008.
5. http://www.ich.org/LOB/media/MEDIA430.pdf and http://www.ich.org/LOB/media/MEDIA431.pdf. Accessed December 18, 2008.
6. http://www.emea.europa.eu/pdfs/human/qwp/048695en.pdf. Accessed December 18, 2008.
7. http://www.emea.europa.eu/pdfs/human/qwp/015901en.pdf. Accessed December 18, 2008.
8. http://www.emea.europa.eu/pdfs/human/qwp/015801en.pdf. Accessed December 18, 2008.
9. http://www.emea.europa.eu/pdfs/vet/qwp/2827108en.pdf. Accessed December 18, 2008.
10. http://www.emea.europa.eu/pdfs/human/qwp/084896en.pdf. Accessed December 18, 2008.
11. http://www.emea.europa.eu/pdfs/human/qwp/205403en.pdf. Accessed December 18, 2008.
12. http://www.emea.europa.eu/pdfs/human/qwp/301599en.pdf. Accessed December 18, 2008.
13. http://www.emea.europa.eu/pdfs/human/qwp/56995908en.pdf. Accessed December 18, 2008.
14. http://www.emea.europa.eu/pdfs/human/qwp/007296en.pdf. Accessed December 18, 2008.
15. http://www.emea.europa.eu/pdfs/human/qwp/330901en.pdf. Accessed December 18, 2008.
16. http://www.emea.europa.eu/pdfs/human/qwp/17369806en.pdf. Accessed December 18, 2008.
17. http://www.emea.europa.eu/pdfs/human/qwp/3aq4aen.pdf. Accessed December 18, 2008.

18. http://ec.europa.eu/enterprise/pharmaceuticals/eudralex/vol-4/pdfs-en/anx12en.pdf. Accessed December 18, 2008.
19. http://www.fda.gov/CDER/guidance/old029fn.pdf. Accessed December 18, 2008.
20. http://www.fda.gov/cder/guidance/7590fnl.pdf. Accessed December 18, 2008.
21. http://www.fda.gov/CDER/GUIDANCE/5831dft.pdf. Accessed December 18, 2008.
22. http://www.fda.gov/CDER/guidance/cmc5.pdf. Accessed December 18, 2008.
23. http://www.fda.gov/cder/guidance/qaletter.htm. Accessed December 18, 2008.
24. http://www.fda.gov/Cder/guidance/1721fnl.pdf. Accessed December 18, 2008.
25. http://www.fda.gov/Cder/Guidance/1214fnl.pdf. Accessed December 18, 2008.
26. http://www.fda.gov/cder/Guidance/1447fnl.pdf. Accessed December 18, 2008.
27. http://www.fda.gov/Cder/Guidance/1722dft.pdf. Accessed December 18, 2008.
28. http://www.fda.gov/CDER/GUIDANCE/pv.htm. (1987 version); http://www.fda.gov/Cder/Guidance/8019dft.pdf (2008 draft revision). Accessed December 18, 2008.
29. http://www.fda.gov/CDER/GUIDANCE/cmc2.pdf. Accessed December 18, 2008.
30. http://www.fda.gov/cder/guidance/7046dftrev2.pdf. Accessed December 18, 2008.
31. http://www.fda.gov/CDER/GUIDANCE/6419fnl.pdf. Accessed December 18, 2008.
32. http://www.nihs.go.jp/kanren/yakuji-e0808.pdf. Accessed December 18, 2008.
33. Jiho Inc, Tokyo, Japan. Relevant Information is included on pages 173–174, 183–184, 232–233, 448, 454, 458, and 689–690.
34. Impurities: Guideline for Residual Solvents. http://www.ich.org/LOB/media/MEDIA423.pdf, which applies in the three regions and is additionally applied in Europe by means of general chapter 5.4 in the *European Pharmacopoeia*. There is also an FDA guideline of relevance: Residual solvents in drug products marketed in the United States http://www.fda.gov/CDER/guidance/8179dft.pdf; and the *United States Pharmacopeia* chapter <467>, Residual Solvents is of relevance. Web links accessed December 18, 2008.
35. http://ec.europa.eu/enterprise/pharmaceuticals/eudralex/vol1_en.htm. Accessed December 18, 2008. This page gives access to the Directive and the amending texts. A consolidated version of the legislation is available at http://ec.europa.eu/enterprise/pharmaceuticals/eudralex/vol-1/dir_2001_83_cons/dir2001_83_cons_en.pdf. Accessed December 18, 2008.
36. Ph Eur general text 5.1.1.
37. http://www.emea.europa.eu/pdfs/human/qwp/171900rev1fin.pdf. Accessed December 18, 2008.
38. http://www.emea.europa.eu/pdfs/human/qwp/4931305en.pdf. Accessed December 18, 2008.
39. http://www.accessdata.fda.gov/scripts/cdrh/cfdocs/cfcfr/CFRSearch.cfm?CFRPart=314. Accessed December 18, 2008.
40. http://www.fda.gov/cder/dmpq/cgmpregs.htm. Accessed December 18, 2008.
41. http://ec.europa.eu/enterprise/pharmaceuticals/eudralex/vol4_en.htm. Accessed December 18, 2008.
42. http://www.picscheme.org/publication.php?p=guides and http://www.picscheme.org/publication.php?p = guides. Accessed December 18, 2008.
43. http://www.ich.org/LOB/media/MEDIA1957.pdf. Accessed December 18, 2008.
44. http://www.ich.org/LOB/media/MEDIA3917.pdf. Accessed December 18, 2008.
45. http://www.fda.gov/CDER/GUIDANCE/pv.htm. Accessed December 18, 2008.

16 Drug Product—Excipients

Brian R. Matthews
Alcon Laboratories, Hemel Hempstead, U.K.

INTRODUCTION

What Is an Excipient?

What is an excipient? For the purposes of this chapter, it will be taken to mean any component in a pharmaceutical formulation other than the active ingredient or a prodrug of the active ingredient. It will therefore include any solvent or diluent (liquid or solid) and any additive in the formulation that is intended to perform a particular function. It will include materials added to the formulation that act as stabilizer, thickener, surfactant, antifoaming agent, solubilizer, plastcizer, lubricant, buffer, pH adjustment agent, adsorbent, binding agent, suspending agent, coating agent, moisture-adjusting agent, filler, colouring matter, flavouring agent, fragrance, tonicity-adjusting agent, emulsifier, viscosity-adjusting agent, thickening agent, dispersing agent, propellant, disintegrating agent, antioxidant, antimicrobial preservative, permeation enhancer, constituents of the coating of tablets or the shell of capsules, adhesive components for transdermal delivery systems and compression aids, for example. It will also cover materials added to the formulation during manufacture but which are not present in the final dosage form supplied to the market (such as water in granulation processes or volatile solvents).

The term "excipient" is preferred to "inactive ingredient" since many excipients have an action—for example, antimicrobial preservation or prevention of oxidative spoilage of the formulated product or enhancement of absorption of the active ingredient. Unfortunately, excipients can also be responsible for adverse reactions (e.g. azo dyes and certain antioxidants).

The Global Nature of Pharmaceutical Excipients

The manufacturers of materials used as pharmaceutical excipients are often involved in the global market for these chemicals. It is rare for the pharmaceutical industry to represent the largest part of the market in a particular material. In many cases the pharmaceutical uses of a material represent a very small part of its global sales volume. This is true for cellulose derivatives and the newer propellants used for pharmaceutical aerosol presentations, for example. The total volumes produced of some cellulose derivative materials is measured in many thousands of tonnes per annum, with a small percentage of this being targeted to the pharmaceutical industry. Sometimes—as with replacement propellants for the chlorofluorocarbons—manufacturers make pharmaceutical grades in special facilities. In other cases, the pharmaceutical market is offered materials that have been developed primarily for some other use. When additional demands are made on the excipient producers with respect to data on or specifications for those materials there is often a reluctance to comply and indeed it may result in the withdrawal of the offer to sell the material concerned to the pharmaceutical sector. This is in spite of the common practice among many producers of undertaking additional testing above pharmacopoeial requirements to ensure suitability for specific uses. The increasing interest in functionally-related characteristic or performance standards is another area where demands are being extended to excipient producers. The approach for implementation of these initiatives in the *United States Pharmacopeia/National Formulary* (USP/NF) and the *European Pharmacopoeia* (Ph Eur) is different and it would be helpful if these could be harmonized.

One of the factors to consider here is that it is not uncommon for excipients to be used for a number of different purposes, and the inclusion of a functionally-related characteristic procedure in a monograph may be relevant for some but not for all uses. Additional testing can make the difference between profitability and loss to a producer since the market in excipients is often characterized by tight margins—so any additional tests need to be necessary and appropriate, and introduced by consensus.

This could be an example of 20% of the market being responsible for 80% of the costs to the excipient producer. In this context, the potential extension of good manufacturing requirements rather than relying on their existing quality management systems has been seen by some suppliers as an unwelcome development. Some suppliers have simply stopped supplying the pharmaceutical sector altogether with some materials when the weight of additional controls became too burdensome. This is not a theoretical issue—it has happened when suppliers of corn syrup NF, propylene glycol stearate NF and sodium bicarbonate Ph Eur withdrew their products from sale to the pharmaceutical market, for example.

An additional and unwelcome complication in this has been the identification of counterfeiting extending into the field of pharmaceutical excipients on an international basis. Examples that were fresh in the memory at the time of writing (August 2008) included the substitution of protein by melamine and the substitution of diethylene glycol for glycerin. While an initial solution to this has been seen to be the addition of additional tests to pharmaceutical material monographs this would not seem to be a practical answer in the long term given the wide range of materials that could be affected. Certification of foreign suppliers might be a more productive approach but, of course, this has implication for resources available to both agencies and manufacturers. The application of appropriate good manufacturing practices (GMPs) (by consensus) could be another fruitful approach, especially if supported by a risk assessment and management initiative.

Excipients and the Pharmacopoeias

There is a wide range of materials that have been used as excipients in pharmaceutical formulations. Many of them are subject of pharmacopoeial monographs in the Ph Eur, *Japanese Pharmacopoeia* (JP) or USP/NF or in other publications such as the *Japanese Pharmaceutical Excipients Directory* and *Japanese Pharmaceutical Codex* or the *British Pharmacopoeia*, for example. In recent years, there has been an initiative within the International Conference on Harmonization (ICH)/Pharmacopoeia Discussion Group (PDG) to prepare harmonized pharmacopoeial specification for excipients that are common to these three pharmacopoeias. This is extremely useful since it avoids the need to test for compliance with different pharmacopoeial requirements for a product intended for global use. A summary of the current status of this harmonization exercise is included in Table 1. However, the observant reader will have noted that even where there is an adopted harmonized monograph available it is not always implemented in the three pharmacopoeias in the same year; in some cases, it is not even implemented in all three of the pharmacopoeias.

Nomenclature of Excipients

Another observation arising from Table 1 could be, depending on the reader's familiarity with the titles of the monographs in the three cited pharmacopoeias, that the names of the various excipient monographs are not always consistent between the three pharmacopoeias. This can result in a degree of confusion and certainly inconvenience when it comes to labelling of pharmaceutical products intended for more than one of the ICH regions. This topic is outside the scope of this chapter but can be followed up in *Pharmaceutical Excipients—A Manufacturer's Handbook* (3). In particular, this publication has tables of monograph titles from the three pharmacopoeias and related publications with alternate names and Chemical Abstract Numbers for the materials described—the latter is most likely to ensure that the monographs are about the same substance. It must, however, be observed that the availability of International Nonproprietary Names (INNs) (which are generated by the World Health Organization) for excipients is very limited. There are around 50 INNs for excipients. It is a policy of WHO not to produce more. A few additional approved names are included in *British Approved Names*, a priced publication available from The Stationery Office in the United Kingdom.

Has the Excipient Been Used in Pharmaceutical Products Before?

Pharmaceutical Excipients—A Manufacturer's Handbook (1) includes information derived from the 1996 version of the U.S. Food and Drug Administration (FDA) *Inactive Ingredients Guide* (4), which indicates the number of products in which the excipient concerned was present when the list was prepared. Unfortunately, the current version of the FDA *Inactive Ingredients Database*,

Table 1 Status of Harmonization of Excipient Monographs in the Ph Eur, JP, and USP

	Implementation dates		
Excipients	**Ph Eur**	**JP**	**USP**
Alcohol revision 1	2004	2006	2005
Alcohol, dehydrated revision 1	2004	2006	2005
Benzyl alcohol	2002	2005	2005
Benzyl alcohol revision 1	2006		
Calcium disodium edetate	2007		2007
Calcium phosphate dibasic dihydrate	2007	2007	2007
Anhydrous calcium phosphate dibasic	2007	2007	2007
Carmellose calcium revision 1	2004	2005	2005
Croscarmellose sodium	2002	2006	2005
Cellulose acetate revision 1	2004		2005
Cellulose acetate phthalate	2003	2005	
Cellulose, microcrystalline			2005
Cellulose, microcrystalline revision 1	2007	2006	
Cellulose powder			2005
Cellulose powder revision 1	2007	2006	
Citric acid anhydrous	2004	2003	
Citric acid anhydrous revision 1		2006	2005
Citric acid monohydrate	2004	2003	
Citric acid monohydrate revision 1		2006	2005
Ethylcellulose	2007	2006	2007
Hypromellose	2007	2006	2007
Hypromellose phthalate	2007	2007	2009
Lactose anhydrous	2004	2006	
Lactose anhydrous revision 2		2007	2008
Lactose monohydrate	2004	2006	
Methylcellulose	2007		
Methylcellulose revision 1	2007	2006	2009
Methyl parahydroxy-benzoate	2002	2006	2006
Saccharin	2006	2006	2006
Saccharin calcium revision 1			2006
Saccharin sodium revision 1	2006	2006	2006
Sodium chloride	2004	2003	
Sodium chloride revision 2		2006	
Sodium starch glycolate	2005	2006	
Sodium starch glycolate revision 1	2007	2007	2006
Starch, corn	2003	2005	2005
Starch, corn revision 1	2005	2006	
Starch, potato		2005	2005
Starch, wheat		2005	2005
Starch, rice	2008	2009?	2009
Talc	2005		2005
Ethyl parahydroxybenzoate	2002	2006	2005
Propyl parahydroxybenzoate	2002	2006	2005
Butyl parahydroxybenzoate	2002	2006	2005
Povidone	2008	2009?	2009

Source: Projected timetable for publication and implementation of texts signed off by the Pharmacopoeial Discussion Group, November 2007 (2). State of Work of International Harmonization, November 2007 (3).

Note: Work is ongoing on the following harmonization projects: Alcohol revision 2; Alcohol dehydrated revision 2; Butyl parahydroxybenzoate revision 1; Calcium carbonate; Carmellose; Carmellose sodium; Copovidone; Crospovidone; Ethyl parahydroxybenzoate revision 1; Gelatin; Glucose monohydrate; Glucose anhydrous; Glycerin; Glyceryl monostearate; Hydroxyethylcellulose; Hydroxypropylcellulose; Hydroxypropylcellulose, low substituted; Magnesium stearate; Mannitol; PetrolatumPetrolatum, white; Polyethylene glycol; Polysorbate 80; Propyl parahydroxybenzoate revision 1; Propylene glycol; Silicon dioxide; Silicon dioxide, colloidal; Sodium laurylsulfate; Starch, pregelatinized; Stearic acid; Titanium dioxide; Water for injection[s] in containers which is downloadable from the FDA Web pages (5), no longer includes this information, but it does, of course, have rather more up-to-date information on what excipients are used in which dosage forms, and the database is updated four times a year. Unfortunately, this useful resource addresses only those products that have been approved by the FDA.

In Japan, it is understood that a list of excipients that have been used in approved pharmaceutical products is included in the following publications: *Japanese Pharmaceutical Excipients* 2004 and the *Japanese Pharmaceutical Excipients Directory* 1996, which are published by Yakuji Nippo, Tokyo. These documents are not completely up-to-date.

Within the European Union (EU) there is no central repository of information on what excipients are included in which pharmaceutical products. Under EU pharmaceutical legislation it is required that the full composition of pharmaceutical products is declared in the labelling, the patient leaflet and the summary of product characteristics. It would be possible theoretically to trace this information for any product on the European market using the manufacturer's published information (or national compendia of this information—electronic or in print) but in practice this is very difficult to do. Even where there are national compendia with product information on pharmaceuticals in particular markets these are frequently incomplete or out-of-date. The regulatory agencies have this information, of course. Unfortunately, it does not seem to have been made available in a public forum by those agencies, although it might be possible to submit requests to individual agencies to seek the information. Tabulations of excipients in products types approved through the European Union Centralised Procedure were included in *Pharmaceutical Excipients—A Manufacturer's Handbook* (1), although this is now incomplete because additional products have been approved since the data were collected. The author is unaware of any official source of information on which excipients are used in approved pharmaceutical products in Japan that have not been published in the various pharmacopoeial and related texts as monographs. However, this is a topic that cannot be expanded on further in the current chapter. It is an area that could be subject of improvement if the regulatory agencies chose to make the information available, perhaps on their Web pages.

The Use of Novel Excipients

Why is it of any interest whether a potential excipient has been included in a product that has already been approved by a regulatory agency? Put simply, this is because the use of an excipient that has not been approved before will result in that material being viewed with some suspicion by the regulatory agencies and there could be demanding data requirements before it will be approved for use in the specific product concerned. In fact, in many jurisdictions the data requirements for a new excipient will be similar to those of a new drug substance.

However, it is possible to use an excipient that has been used in another sector than the pharmaceutical sector and not have to provide this level of information. Examples would be materials that are allowed for use in foodstuffs (particularly for products for oral dosing, and particularly if the material has an "E-number" in the EU and a recognized quality specification in the national or regional legislation) or ingredients approved for use in cosmetics (particularly for products for topical application). In the latter case, listings in national or regional legislation are the most relevant but other sources such as the *International Cosmetic Ingredient Dictionary and Handbook* from the Personal Care Product Council may also be useful. Nonetheless, there could well be additional data requirements if an ingredient approved for uses that employ different routes of administration from those of the intended pharmaceutical formulation are likely to result in additional nonclinical as well as pharmaceutical data requirements and clinical data would be examined for evidence of safety of the whole product in use.

In the development of novel therapeutic approaches it is possible that it will be necessary to think "outside the box" and try to find a new way to meet a formulation or performance target. This could result in the adoption of novel materials for use in the product. Examples include ethers of β-cyclodextrins as solubilizers, novel penetration enhancers from different suppliers, tocopherol as a solvent and polymeric micelle forming materials. It is also possible to use the properties of one excipient to achieve more than one formulation target. The success of such attempts to introduce improved pharmaceutical products will no doubt depend to some extent on the willingness of regulatory agencies to be flexible in their demands for additional data to support new approaches.

When Is an Excipient not an Excipient?

It is not unknown for a substance included in one pharmaceutical as an excipient to be claimed as an active ingredient in another product. Examples of such materials include albumin, ascorbic acid, barium sulphate, calcium carbonate, glucose, dimeticone, docusate sodium, iodine, magnesium carbonate, magnesium sulphate, mannitol, liquid paraffin, potassium chloride, povidone, simeticone, sodium chloride and water of various qualities. This possibility of dual use can raise questions on the suitability of a pharmacopoeial monograph in ensuring the suitability for the intended use of the material for all possible uses. It is quite possible for regulatory agencies to request additional specification tests in such cases. One example would be the need for an endotoxin control for a material such as sodium chloride when intended for use in an injectable product (whether it is claimed as an excipient or an active ingredient).

It should also be borne in mind that some excipients have inherent activity secondary to their intended use as an excipient. Examples include docusate sodium, xylitol, lactulose, polyethylene glycol, guar gum, sodium phosphate, magnesium oxide, calcium salts, amino acids, etc. Due account of such properties will need to be taken during the development of the product and when preparing the marketing authorisation application for the dosage form concerned.

Excipients and GMPs

As an introduction to the topic, at the time of writing there are no specific requirements within the ICH region for excipients to be manufactured under a specific set of GMP rules. This is, of course, not so for ingredients that can be both excipients and active ingredients. The latter are required to be manufactured under ICH Q7 Good Manufacturing Practice Guide for Active Pharmaceutical Ingredients (6). Within the EU, there is already a commitment to apply some form of GMP to "certain" excipients. There is no indication at the time of writing (August 2008) of what form those requirements will take, nor which excipients will be affected. This would seem most likely to include those excipients derived from materials of human or animal origin, and excipients provided with a claim for sterility (which will not be subject to a further sterilization process during the manufacture of the pharmaceutical product containing them), and specific examples where there have been problems in the past—glycerol and propylene glycol for example. As of June 2009 the Commission has indicated that it intends to concentrate on the application of risk analysis rather than extend the application of GMPs (*per se*) to excipients.

Useful Reference Works

Two particularly useful reference works of interest to readers of this chapter are the *Handbook of Pharmaceutical Additives* (7) and the *Handbook of Pharmaceutical Excipients* (8).

EXCIPIENTS IN THE COMMON TECHNICAL DOCUMENT (CTD)

Module 1

The European Union

The EU application form (Module 1.2) requires the excipients in the formulated product to be declared in Section 2.6.1. This includes the full qualitative and quantitative declaration of active ingredients and excipients including any overages. The names to be used in the listing should follow the order of preference: INN, Ph Eur monograph title, national pharmacopoeia title, common name or scientific name. Materials of animal or human origin have to be identified in Section 2.6.2. (9). Excipients will also need to be listed in the Summary of Product Characteristics, the labelling and the package leaflet that are included in Module 1.3, but these are somewhat outside the scope of this chapter and will not be discussed further here.

The United States
For the United States, FDA form 356h does not require information on excipients included in the product (10).

Japan
For Japan, the full composition should be included in the application form (11). There are certain exceptions for this—for example, for JP-listed hard capsules, tablet sugar coatings. The content statements for certain materials other than the active ingredient need not be quantitative—"*q.s.*" will often suffice for vehicles and bases, pH regulators, solvents, lubricants and sugar-coating components (12). The function of excipients is to be stated by using a defined list of functions (13). JP and other Japanese specifications are expected to be applied—*Japan Pharmaceutical Codex* (the 1997 edition is available in English) (14), *Japan Pharmaceutical Excipients* (2004 edition is available in English) (15), *Specifications and Standards for Food Additives* and the former Japanese *Standards of Cosmetic Ingredients*, etc.

Module 2.3 : Quality Overall Summary—Sections 2.3.P.4 (Control of Excipients) and 2.3.A.3 (Excipients)

Module 3 : Quality—Body of Data Section 3.2.P.2.1.2 (Excipients), 3.2.P.4 (Control of excipients), 3.2.A.2 (Adventitious Agents Safety Evaluation), and 3.2.A.3 (Excipients)

Europe
These notes are based on the recommended format for a Common Technical Document (CTD) application based on the EU guideline (16). Module 2.3.P.4 and 2.3.A.3 are derived from the information in Module 3.2.P.

Before considering the technical aspects of data requirements for drug products in Europe, it should be pointed out that for European submissions it is necessary to provide all required information on excipients in the marketing authorisation application. There is no mechanism for the submission of a drug master file for an excipient. It follows that the excipient is not approved *per se*; it is the product containing the excipient that is approved. The approval of a product containing a particular excipient is no assurance that the same material from the same source will be suitable for use in other formulated products.

The absence in Europe of a system analogous to the FDA Type IV drug master file procedure or the Japanese Drug Master File procedure is seen by many excipient producers and pharmaceutical companies to be anomalous. It has been indicated that this is another disincentive to the introduction of novel excipients since it is not possible to file the proprietary information using a mechanism independent of the pharmaceutical manufacturer.

As has been indicated in other places within this publication, the CTD is a framework on which to hang the data that will be required in the marketing authorisation application for a pharmaceutical product. It does not specify *what* information has to be included. This the potential applicant needs to determine from local guidance and perhaps meetings to discuss specific issues with regulatory agencies to which the submission will be sent. It is also open to seek scientific advice from the Committee for Human Medicinal Products (CHMP) and other agencies. There is an option to use a joint scientific advice procedure with the CHMP and the FDA at the same time. Meetings are also possible with Japanese regulators.

Where a company is planning to use known excipients for their well-established purposes within the usual quantitative levels used within the dosage form concerned, the published guidance should suffice. If it is intended to use a novel excipient or an excipient outside its established purposes or in a significantly different quantity then advice will be useful. In many cases, novel excipients and novel use or higher levels of known materials will result in considerable additional data requirements.

The first necessity is to identify sources of guidance in publicly available documents. In the EU, there are a number of relevant documents, and these are identified in Table 2. It will be seen that there is a wide range of topics covered in the guidance documents, from development issues to labelling issues. Some are of general applicability and others are very specific.

Table 2 Guidance Relevant to Excipients in Marketing Authorisation Applications in the European Union

Guidances	Sections of particular relevance	Observations
Excipients in the dossier for application for Marketing Authorisation of a medicinal product, February 1994	All	Replaced by July 2007 document
Note for guidance on inclusion of antioxidants and antimicrobial preservatives in medicinal products, July 1997	All	Replaced by July 2007 document
Draft guideline on excipients in the dossier for application for Marketing Authorisation of a medicinal product. October 2006	All	Replaced by July 2007 document
Guideline on excipients in the dossier for application for Marketing Authorisation of a medicinal product, July 2007	All	
Note for guidance on quality of water for pharmaceutical use, May 2002	All	Relates to the uses of water of different grades for manufacture of active ingredients and formulated products
Reflection paper on Water for Injection[s] prepared by reverse osmosis, March 2008	All	Relates to the use of this grade of water in the manufacture of injectable pharmaceuticals
Note for guidance on development pharmaceutics, January 1998	Sections 2.2 Excipients and other nonactive constituents, 3.1 Overages, 3.2 Physicochemical parameters, 3.3.1 Components of the formulation [liquid and semisolid formulations], 3.5.1 Transdermal patches	ICH Q8 guideline, November 2005, applies
ICH Q8 Note for guidance on pharmaceutical development, November 2005	Sections 2.1.2 Excipients, 2.2.1 Formulation development	
Draft ICH Q8 Annex to note for guidance on pharmaceutical development [principles of Quality by Design], November 2007	Sections 2.2 Critical quality attributes, 2.4 Design space, 2.5 Control strategy, 3.1 Quality risk management and product and process development	
Notice to applicants, Volume 3B, Guidelines, Excipients in the label and package leaflet of medicinal products for human use, revision 1, July 2003	All	
Guideline on medicinal gases: Pharmaceutical documentation . . ., July 2008	Sections relating to excipients such as P.1, P.2 Mixtures, P.5.1 Specification, etc.	
Note for guidance on quality of modified release products . . ., July 1999	Sections 3.2 Development pharmaceutics of delayed release dosage forms, 4.1 Development pharmaceutics of transdermal patches	
Guideline on the pharmaceutical quality of inhalation and nasal products	Sections 4.2.1.20 Preservative effectiveness/efficacy, 4.4 (Nasal products) excipients, 4.5.1.12 Preservative content	

(*Continued*)

Table 2 Guidance Relevant to Excipients in Marketing Authorisation Applications in the European Union (*Continued*)

Guidances	Sections of particular relevance	Observations
Note for guidance on minimizing the risk of transmitting animal spongiform encephalopathy agents via human and veterinary medicinal products, October 2003; see also Explanatory note for medicinal products for human use on the scope of this guideline, February 2001	Sections 2 Materials, 6.1 Collagen, 6.2 Gelatin, 6.3 Bovine blood derivatives, 6.4 Tallow derivatives, 6.5 Animal charcoal, 6.6 Milk and milk derivatives, 6.7 Wool derivatives, 6.8 Amino acids	Relates to materials derived from ruminants such as sheep, goats and cattle (i.e., animals that are susceptible to infection with transmissible spongiform encephalopathy agent); the document is in need of updating
Gelatin for use in pharamceuticals: Explanatory note on the manufacture of gelatin in relationship to the Note for guidance . . ., December 2000	All	
Position statement on Polysorbate 80, October 1998	All	
Position statement on production of tallow derivatives for use in pharmaceuticals, December 17, 1997	All	
Note for guidance on in-use stability testing of human medicinal products, March 2001		
EMEA public statement on thiomersal in vaccines for human use, March 2004	All	

Notes: In addition to documents with specific guidance relating to excipients, a number of other general guidelines should be taken into account. Relevant recommendations from these documents should be taken into account. The documents include

ICH Q1A(R2) Note for guidance on stability testing: Stability testing of new drug substances and products, August 2003
ICH Q1E Note for guidance on evaluation of stability data, August 2003
ICH Q1C Note for guidance on stability testing: Requirements for new dosage forms, January 1998
Guideline on stability testing: Stability testing of existing active substances and related finished products, December 2003
Guideline on stability testing for applications for variations to a marketing authorisation, May 2005
ICH Q3C(R3) Note for guidance on impurities: Residual solvents, March 1998
ICH Q3B (R2) Note for guidance on impurities in new drug products, June 2006
Guideline on the limits of genotoxic impurities, June 2006 and Questions and Answers on the CHMP guideline on the limits of genotoxic impurities, June 2008

The note for guidance on nonclinical local tolerance testing of medicinal products, March 2001, might be relevant for certain formulations but the topic is outside the scope of this chapter.
Sources: Documents were downloaded from (17) other than the guideline relating to medicinal gases development (18), the guideline on medicinal gases (19), the document on thiomersal in vaccines (20) and documents relating to TSE risk that were accessed via (21). All documents were accessed on August 30, 2008.

Pharmaceutical development is discussed in chapter 14. For the purposes of this chapter it is sufficient to note that the choice of excipients—and in some cases the specific grade of excipient—will need to be justified. The function of each excipient should be stated and in some cases be justified in the marketing application dossier. The amount of each excipient in the formulated product should be justified and, in the formula included in the application form, the function and range of amounts should be stated. In some cases (particularly for functional excipients such as antimicrobial preservatives and antioxidants) appropriate experimental data will often be required to justify their inclusion at stated concentrations. For preservatives, data from preservative efficacy tests (PETs) should be provided based on the methodology included in the Ph Eur. It is usually expected by regulators that products will meet the Ph Eur A criteria for preservative efficacy—other performance standards will need to be justified.

The inclusion of preservatives may be required for products intended to be used on more than one occasion where the product does not have inherent antimicrobial properties. However, it should be noted that the purpose of the preservative system in such cases is to protect the product from contamination introduced *by the user*. It is not intended to cover the introduction of contamination during manufacture—this is an issue related to failure to comply with GMPs.

It should also be noted that the Ph Eur and some general monographs place limitations on the inclusion of antimicrobial preservatives (and sometimes other excipients)—for example, for injections gaining access to the cerebrospinal fluid, for certain ocular routes of administration or for products used in the middle ear. Any deviation from this will need to be justified but it must be observed that since some regulatory agencies insist on compliance with the requirements of the Ph Eur and since the European Pharmacopoeia Commission has insisted in some cases that a product needs to be registered before it will amend the limitations in the general monograph, it is easy to get into a circular argument—the applicant provides the data to justify a deviation from the general monograph requirement but the regulators refuse to approve the product because it is not in compliance with the pharmacopoeial requirement (regardless of the quantity or quality of the data to support the product).

Data from in-use stability tests will be expected to address the suitability of the formulation during simulated use and will need to include data on in-use preservative efficacy (using a method defined by the manufacturer or taking into account the relevant guideline since there is no specific Ph Eur test procedure) and on the stability of antioxidants and other formulation components during simulated patient use. Data may need to be provided to address the suitability of the excipients for attaining the necessary maintenance of suspension properties, rheological properties, tonicity, emulsion and suspension properties, particle size and shape, crystal characteristics, and ease of use (e.g. to ensure the ability to inject the product through a syringe and needle, or redispersion of suspension products).

Data should be provided to confirm the compatibility of the various excipients and the active ingredient(s) in the formulated product with each other. Where the medicinal product is to be used with another product (e.g. a diluent or a solvent), the mutual compatibility of the two products will need to be addressed. The pH of the formulated product might need to be justified in terms of its effect on excipient function being maintained—e.g. the effect of pH on antimicrobial preservative efficacy with certain preservative systems.

When it is chosen to use a novel excipient (or to use a known excipient for a new purpose, in a dosage form using a different route of administration or using a different amount of the excipient from that previously employed) there are likely to be additional data expectations from the regulatory agencies—these will be similar to the data requirements for a new active substance or a new route of administration for a known active ingredient. An in-house specification for such novel and nonpharmacopoeial ingredients should include specifications for

- physical characteristics,
- identification tests,
- purity tests including impurities,
- assay or limit tests and
- quantitative parameters that influence the performance of the dosage form.

Manufacturers may want to use an overage of an excipient in the manufactured product. There is a general antipathy toward this practice and any overages will need to be justified. The approach will depend on whether the overage is proposed based on losses during manufacture or the need to meet a requirement for the level of the excipient during the shelf life. Large overages—usually taken as those in excess of 10%—are unlikely to be allowed. This is because such large overages could present unacceptable hazards to users. It would be better to reformulate the product if such large overages are needed. Other overages will need to be justified in terms of the technical need for the overage and the potential for adverse events arising from their presence or the presence of the degradation products arising from them.

In establishing specifications for excipients, it should be kept in mind that compliance with monograph requirements in the Ph Eur should be given precedence. In the absence of such a specification then national pharmacopoeia specifications of Member States of the EU can be used. Failing this other specifications from official compendia can be used but might need to be justified. USP/NF and JP specifications might be examples of such alternate sources—but it should be noted that translations of the monographs and of the analytical methods (general and specific) will be required if the original text is not in English. Applicants can also suggest specifications based on monographs in the *Food Chemical Codex* (which is now under the care of USP) or legislation (such as for cosmetics and food additives). In the case of colours, the substances should be authorized for use in pharmaceutical products and they should comply with the appropriate purity requirements for colours used in foodstuffs—relevant information can be found in Directive 94/36/EEC (colours for use in foodstuffs) and Directive 95/45/EEC (specific purity criteria concerning colours for use in foodstuffs) as amended (22).

Appropriate account will also need to be taken of physical characteristics related to the intended function in the formulated product—for example particle size and shape, rugosity, viscosity, ability to manufacture the dosage form—in addition to chemical characteristics of the material. Supplementary requirements may need to be included where pharmacopoeial monographs are used to take account of this.

While the use of sterile excipients is not required for nonsterile dosage forms or products that are to be terminally sterilized, for example, it will be expected that there will be adequate control on the presence of microbial contaminants. There is guidance on this in the Ph Eur. In most cases, a control on total viable count will be acceptable but there could be additional expectations on the control of undesirable organisms for particular excipients depending on their source and method of manufacture. In the case of ingredients for use in sterile products manufactured by using aseptic technique the ingredients will need to have been sterilized and appropriate information included in the dossier to describe the sterilization method and its validation (and of course the material specification will need to reflect the need for sterility).

In the case of sterile products manufactured by using aseptic filtration, it will be expected that a tight bioburden control will be applied—a limit of not more than 10^2 colony forming units (cfu) per gram or millilitre may be acceptable for excipients other than water (which will need a much tighter bioburden control such as 10 cfu per 100 ml). It might be possible to negotiate some relaxation of the need for batch testing of some excipients such as strong acids or bases or concentrated bactericide preparations for bioburden.

Where the material has been milled or air-jet milled to achieve a particular particle-size distribution data will be needed on the size reduction method and the control tests to ensure that the specification has been met.

Due account should also be taken of general requirements for the establishment of specifications for impurities, including residual solvents, genotoxic impurities and residues from sterilization processes (e.g. ethylene oxide residues, although it should be kept in mind that these can also be present in certain excipients—such as nonionic surfactants—arising from the method of manufacture of the excipient). In the case of excipients particular attention might need to be paid to a description of the material, its physical characteristics, identification tests and purity tests. For novel excipients due account should be taken of the data requirements for a new chemical entity—this should be modeled on CTD Module 3.2.S.

Special attention should be paid to ingredients that are derived from human or animal materials. Appropriate controls may be required to control potential viral or transmissible spongiform encephalopathy (TSE) contamination. This is discussed further in chapter 20 but it must be observed here that the use of such materials will require additional information. There is specific guidance on collagen, gelatin, bovine blood derivatives, tallow derivatives, animal charcoal, milk and milk derivatives, wool derivatives, amino acids in a TSE guideline from the CHMP with supplementary guidance for gelatin, tallow derivatives, and polysorbate 80 (Table 2). The use of such materials will require additional information to be included in Module 1.

Where possible, it is advisable to use sources of such materials for which a TSE Certificate of Suitability (CEP) is available from the European Department for the Quality of Medicines (EDQM) (23). The use of a CEP is also useful for demonstration of compliance with the non-TSE-related requirements of the relevant monograph of the Ph Eur—these can be confirmed by using the same EDQM database. There is a large number of these general CEPs and they are accepted in lieu of the provision of data in the marketing authorisation application. Excipient suppliers can apply for a CEP using the procedure described on the EDQM Web site. There is a guide to the format and content of applications for a CEP (24).

In establishing the specification for the dosage form it will not be necessary to include identity tests or assays for excipients with the following exceptions:

- Antimicrobial preservatives (although it is not necessary to include a PET in the specification)
- Antioxidants
- Excipients essential to the function of the dosage form and
- Colours (for which an identify test to confirm that an authorized color has been used should be available).

Manufacturers will need to take into account any limitations on the use of particular materials, for example:

- Mercurial preservatives in ophthalmic formulations and injections [although the reported association of neurological development problems with the presence of thiomersal (thimerosal) in vaccines appears not to have been confirmed].
- Antimicrobial preservatives and antioxidants in products for use by children.
- Benzyl alcohol used in injectable products for use on neonates and young children.
- Benzoic acid esters in injectable products.
- Sulphites and metabisulphites if their use can be avoided—for example, by the use of appropriate headspace replacement with inert gases—due to the potential risk of adverse events.
- Antimicrobial preservatives or antioxidants in parenteral products that will gain access to the cerebrospinal fluid.
- Antimicrobial preservatives or antioxidants in parenteral products for administration by the retro-ocular route.
- Azo dyes (colours) in products intended for use in children or sensitized persons.

Where the use of excipients cannot be avoided, it might be necessary to add specific warning statements to the product information.

Perhaps the most widely used excipient used in the manufacture of pharmaceuticals (whether or not it is included in the finished product) is water. There are three pharmacopoeial grades of water in the Ph Eur but the most widely used qualities are purified water and water for injections—the third type is highly purified water (which is produced by using reverse osmosis). It should be noted that within the EU it is not permitted to use reverse osmosis as the production method for water for injections according to published documents. For injection products and irrigation products it is necessary to use water for injections. This is also true for sterile products that are to be freeze-dried. For ophthalmic products, nasal and ear preparations and cutaneous preparations purified water can be used. For non-sterile formulated products purified water can be used.

The United States

There are two differences that are immediately apparent between the EU and the US procedures applying to excipients:

1. It is possible to submit a (Type IV) drug master file to the FDA for excipients. This does not change the status of the material concerned—drug master files are not approved or disapproved, it is the product that contains the excipient that is approved and this should not be taken to mean that the same material from the same source will be suitable for use in another product. A large number (in excess of 690) of Type IV drug master files have been submitted to the FDA (25).
2. There is a relatively new procedure for USP-verified pharmaceutical ingredients. At the time of writing (August 2008) there were relatively few verified ingredients used as excipients—povidone, povidone-iodine, crospovidone, and copovidone. It is not clear what the status of this verification is with regard FDA aceptance.

There is no specific guideline available on the FDA Web pages that relates specifically to excipients, other than advice on the requirements for Type IV drug master files, which

Table 3 Guidance Relevant to Excipients in Marketing Authorisation Applications in the United States

Guidances	Sections of particular relevance	Sources
Submitting supporting documentation in drug applications for the manufacture of drug substances	Information relating to data requirements for new drugs (of relevance to novel excipients and those with particular requirements)	(26)
Liposome drug products	II Chemistry, manufacturing, and controls D Control of excipients: Lipid components	(27)
Metered-dose inhaler (MDI) and dry powder inhaler (DPI) products	III Drug product B Composition C Specification for the formulation components 2 Excipients	(28)
Nasal spray and inhalation solution, suspension and spray drug products—Chemistry, manufacturing, and controls documentation	III Drug product A Formulation components B Formulation composition C Specifications for the formulation components 2 Excipients F Specifications for the drug product 1 Nasal sprays (e) Preservatives and stabilizing excipients assay 2 Inhalation solutions, suspensions, and sprays (e) Preservatives and stabilizing excipient assay IV Drug product characterization studies L Preservative effectiveness and sterility maintenance	(29)
Sourcing and processing of gelatin to reduce the potential risk posed by bovine spongiform encephalopathy (BSE) in FDA-regulated products for human use	All	(30)

is not a primary reference of relevance for this chapter. A number of other guidelines with potentially relevant information are available and are included in Table 3. The ICH general guidance documents relating to stability testing, analytical validation, impurities including residual solvents, specifications and test methods, etc., apply in the US as in the other ICH regions and so the general information of relevance to excipients applies.

It is recommended that applications and notifications to the FDA follow the ICH CTD format.

A submission for marketing approval or a notification to the FDA should include information on the full qualitative and quantitative composition of the product including any materials that are used during production but absent from the finished product. In addition, the function of the materials should be stated and the quality standards with which they comply should be indicated. Where there is no "established name" (i.e. USP/NF name), a complete chemical name should be used—that is, the current Chemical Abstracts Service index name.

Appropriate information should be included in the pharmaceutical development Section (P.2) of the file. This should address the choice of excipients, their concentration and the characteristics that could affect drug product performance or the ability to manufacture it. The proposed range of quantities of excipients to be used should be justified. The ability of functional excipients such as preservatives and penetration enhancers to perform as intended throughout the shelf life proposed should be addressed.

Additional information will be required for novel excipients or those used for a particular route of administration, similar to the type of information needed for a new drug substance. Known noncompendial excipients may be subject to requirements for additional information depending on the function of the material and the route of administration to be employed. It is possible for some excipients to possess inherent pharmacological properties—for example, docusate sodium, caffeine, methionine—and it should be shown that at the concentration used there will be no (unwanted?) pharmacological activity from such materials.

The microbiological characteristics of the excipients should be discussed. The type of testing for microbial quality should be described and justified. In particular the omission of tests for microbial limits should be justified. The selection of antimicrobial preservatives and their effectiveness should be discussed—the latter as part of the pharmaceutical development. Tests and acceptance limits for microbiological attributes of excipients should be included in P.4.1 and P.4.6.

Compatibility should be demonstrated of the product with diluents and reconstitution media or other products with which the product is intended to be used. These studies should include commonly used diluents, etc., even if they are not specifically named in the instructions for use.

Specifications for all the excipients should be provided, discussed and justified. For pharmacopoeial materials the inclusion of additional tests such as particle size, flow properties, specific impurities or function-related criteria should be considered. Other criteria should be compliant with the current version of USP/NF. For nonpharmacopoeial excipients the specification should be based on the principles applied to new drug substances. Additional tests may be required where there are potential safety issues—for example contamination of glycerol or propylene glycol.

Lipid components in liposome drug products are considered to be important in the control of the quality of the liposome drug product. The level of information required is similar to that for a new drug substance. Specific additional information is included in the relevant guidance (Table 3).

For use in inhalation products (including nasal route dosage forms and aerosol and dry powder inhalations) additional attention will need to be paid to the excipient specifications for materials critical to the effectiveness and safety of the dosage form. This will include those materials in suspension formulations that ensure that the suspension and particle size of the active ingredient in the product are suitable. Sources should be identified for critical excipients. Where the material is of USP/NF quality additional specifications may be required—for example particle-size distribution, crystal or amorphous form, foreign particulates, moisture controls, compositional profiling, additional controls for related substances in propellants, surface roughness—and should be related to the materials used in critical clinical trial batches where possible. Toxicological aspects of the suitability of the excipient for the intended use should be considered and the relevant data cross-referred to. Preservatives and stabilizing excipients should be subject to assays in the dosage form specification.

The finished product specification may need to include tests for antimicrobial preservatives and antioxidant content.

Particular information should be supplied for excipients of human or animal origin. The specific guidance on gelatin should be taken into account if relevant. For other materials information will be required on the animal species from which the material is taken or from which it is derived. Specific consideration will need to be given to materials derived from ruminants and whether they are sourced from countries affected by bovine spongiform encephalopathy.

Japan

In Japan there is a strong emphasis on the use of excipients included in the JP or in one of the other Japanese compendia on excipients—*Japanese Pharmaceutical Excipients* 2004 and *Japanese Pharmaceutical Excipients Directory* 1996. It is also important to address appropriately the use of an excipient that has not been used in pharmaceuticals before in Japan or where the amount or route of administration has been changed from those used in previously approved products. The requirements for novel excipients will also extend to materials that have been used in food (32) or cosmetics, or if they have been used in "quasi drugs." This will also extend to the use of new polymers in containers for pharmaceutical products. In the quantities used in the formulated product, the excipients should show no pharmacological activity. The Japanese Drug Master File procedure can be used for excipients. Data requirements for a novel excipient include data on the development of the material and its use in other countries; the manufacturing method, specification, test methods, and stability data; and toxicity data particularly relevant to the route of administration to be used [using good laboratory practice (GLP-) compliant studies] (31).

A number of excipients will have limitations on their use. This will include edetic acid salts for use in parenteral products, polyvinylpyrrolidone, pentachlorophenol, thiourea, mercury-containing compounds and chloroform. Specific rules apply to the use of tar colours in pharmaceuticals—these are included in Ministerial Ordinance 30 of August 1966 (a translation of which could not be traced).

Premixed excipients can be used provided that the specification and composition is clear and there are tests in the specification for identification tests and assays. The Japanese Drug Master File procedure is available for such materials.

REFERENCES

1. Matthews BR. Pharmaceutical Excipients—A Manufacturer's Handbook. River Grove, Illinois: PDA, Bethesda, Maryland and DHI Publishing LLC, 2005.
2. http://www.edqm.eu/medias/fichiers/PDG_Projected_Timetable.pdf. Accessed August 30, 2008.
3. http://www.edqm.eu/medias/fichiers/PDG_State_of_Work.pdf. Accessed August 30, 2008.
4. Inactive Ingredients Guide January 1996. http://www.fda.gov/cder/drug/iig/default.htm. Accessed August 30, 2008.
5. About the Inactive Ingredients Database, http://www.fda.gov/cder/iig/iigfaqweb.htm. Accessed August 30, 2008.
6. http://www.ich.org/LOB/media/MEDIA433.pdf. Accessed August 30, 2008.
7. Ash Michael (Compiler), Ash Irene (Compiler). Handbook of Pharmaceutical Additives, 3rd ed. Aldershot, Hampshire, UK, Synapse Information Resources Inc, 2007.
8. Rowe RC, Sheskey P, Owen SC, (eds.). Handbook of Pharmaceutical Excipients, 5th ed. Washington, DC: Pharmaceutical Press, London and American Pharmacists Association, 2006.
9. Module 1: Administrative information Application form, revision 9, May 2008, http://ec.europa.eu/enterprise/pharmaceuticals/eudralex/vol−2/b/update_200805/applicformrevised_rev9.doc. Accessed August 30, 2008.
10. http://www.fda.gov/opacom/morechoices/fdaforms/356Hes.pdf. Accessed August 30, 2008.
11. An English translation is included in Drug Approval and Licensing Procedures in Japan 2006. Tokyo: Jiho: 595; guidance on completion of the form starts at page 157.
12. Ibid, pages 168–169.
13. Ibid, page 171.
14. Japan Pharmaceutical Codex. Tokyo: Jiho, 1977.
15. Japan Pharmaceutical Excipients. Tokyo: Yakuji Nippo, 2004.
16. Module 2 Common Technical Document Summaries and Module 3 Quality, Notice to Applicants Volume 2B, Rules Governing Medicinal Products in the European Union, Commission of the European Communities, May 2008 (but the sections cited are dated July 2003), http://ec.europa.eu/enterprise/pharmaceuticals/eudralex/vol−2/b/update_200805/ctd_05−2008.pdf. Accessed August 30, 2008.
17. http://www.emea.europa.eu/htms/human/humanguidelines/quality.htm. Accessed August 30, 2008.
18. www.emea.europa.eu/pdfs/human/qwp/171900rev1fin.pdf - 2008−07−17. Accessed August 30, 2008.
19. http://www.emea.europa.eu/pdfs/human/qwp/171900rev1fin.pdf. Accessed August 30, 2008.
20. http://www.emea.europa.eu/pdfs/human/press/pus/119404en.pdf. Accessed August 30, 2008.
21. http://www.emea.europa.eu/htms/human/tse/tse.htm. Accessed August 30, 2008.
22. The simplest access point for information on requirements for food additives in the European Union is http://ec.europa.eu/food/fs/sfp/flav_index_en.html. Accessed August 30, 2008.
23. The database of certificated materials can be accessed at http://extranet.pheur.org/publications/recherches_CEP.shtml; it is possible to limit the search to TSE certificates by checking a box. To obtain a list of materials for which a certificate is available search on "Certificate number" and using the search term "CEP." On August 29, 2008, there were in excess of 360 valid TSE CEPs, but this need to be kept under review since such certificates can be withdrawn by the holder or they may lapse if not renewed.
24. http://www.edqm.eu/medias/fichiers/Content_of_the_Dossier_for_Chemical_Purity_Microbiological_Quality.pdf. Accessed August 30, 2008.
25. http://www.betterchem.com/dmfs_select/type_4_by_subject/default.htm. Accessed August 30, 2008.
26. http://www.fda.gov/cder/Guidance/drugsub.pdf. Accessed August 30, 2008.
27. http://www.fda.gov/CDER/guidance/2191dft.pdf. Accessed August 30, 2008.
28. http://www.fda.gov/cder/guidance/2180dft.pdf. Accessed August 30, 2008.
29. http://www.fda.gov/CDER/GUIDANCE/4234fnl.pdf. Accessed August 30, 2008.
30. http://www.fda.gov/opacom/morechoices/industry/guidance/gelguide.htm. Accessed August 30, 2008.
31. Drug Approval and Licensing Procedures in Japan. Tokyo: Jiho Inc, 2006, 466–467.
32. Specifications and Standards for Food Additives, Ministry of Health Labour and Welfare, Tokyo, 2000, http://www.ffcr.or.jp/zaidan/FFCRHOME.nsf/pages/spec.stand.fa. Accessed August 30, 2008. Monographs are at http://www.ffcr.or.jp/zaidan/FFCRHOME.nsf/pages/spec.stand.fa−c−monographs.

17 Control Tests on the Finished Product

Henk J. de Jong
Senior Advisor Pharmaceutical Sciences to Servier International Research Institute, Courbevoie, France

INTRODUCTION

Since the start of the efforts in the framework of the International Conference on Harmonization (ICH) of Requirements for Registration of Pharmaceuticals for Human use in 1989, a lot has been accomplished. As one of the important results we have the Common Technical Document (CTD) laying out the (identical) structure of the marketing application dossier for the three regions: European Communities, the United States, and Japan. Canada and Australia, who have been observers to ICH, also accept the CTD format, as do many other countries.

LEGAL AND TECHNICAL REQUIREMENTS IN EUROPE

The requirements for the type of studies and details of data that need to be presented still differ from country to country, but ICH guidelines on many topics have created quite a level of harmonization. The European Community Code relating to medicinal products for human use: Directive 2001/83/EC as amended by Directive 2004/27/EC (1) harmonizes, in a mandatory way, the requirements in all European Union (EU) member states and associated countries. The outcomes of the ICH process have been taken into account in this Directive. Because of further progress in science, technology, and medical sciences updates are necessary. The European Medicines Agency (EMEA) and especially its Committee for Human Medicinal Products (CHMP) deliver guidance in the form of documents that reflect the harmonized thinking of the member states.

Postapproval Changes

A review of the Commission Regulations 1084/2003 and 1085/2003 on variations to marketing authorizations was carried out in 2008 and a new Commission Regulation 1234/2008 (2) was adopted to reduce the number of variation procedures (for postapproval changes to the dossier), thus reducing the burden on the agencies, and to focus on those which have a genuine impact on quality, safety, or efficacy. A system of annual reporting will be introduced for certain minor variations. The Commission Regulation 1234/2008 will come into force from 1 January 2010.

European Pharmacopoeia

Another important harmonization activity in a territory even wider than the European Union is the work of the *European Pharmacopoeia*. This is based on a Council of Europe Convention adopted in 1964 (3), with 37 members and 22 observer states plus the World Health Organization taking part in its activities. The monographs and general chapters are mandatory in the countries that signed the convention (including in the EU), and in several cases they are adopted by observer states, for example, Canada.

LEGAL AND TECHNICAL REQUIREMENTS IN THE UNITED STATES

In the United States, the Food and Drug Administration (FDA) is in charge of the registration of medicinal products. The requirements are published in the *Code of Federal Regulations*, and guidance is available from the agency web site (4).

Postapproval Changes

The categories of postapproval changes in the United States are PostApproval Supplements (PAS), Changes being Effected (CBE-0 and CBE-30), and Annual Report changes. The Flexible Regulatory Approaches being piloted by the Center for Drug Evaluation and Research (CDER) will offer the potential to "downregulate" some categories of product changes using a "risk-based" approach to reduce the burden on the agency and industry.

United States Pharmacopeia

The *United States Pharmacopeia* is a private organization, but its monographs and general chapters are considered as official in the United States and in a number of other countries (5). They are accepted in Canada and in Australia, for example.

LEGAL AND TECHNICAL REQUIREMENTS IN JAPAN

In Japan, the Ministry of Health, Labor, and Welfare (MHLW) is in charge of the registration of medicinal products for human use, and the requirements are published in the Notifications and Ministerial Ordinances applied under the Pharmaceutical Affairs Law (PAL; Law No. 145, 1960) (6). Guidance is available through the Pharmaceuticals and Medical Devices Agency (PMDA), an organization also in charge of the evaluation of the dossiers (7). The *Japanese Pharmacopoeia* is an integral part of the Ministry and its texts are mandatory (8).

THE COMMON TECHNICAL DOCUMENT AND DRUG PRODUCT QUALITY

In the CTD, the topics on quality are found in Module 2: the Quality Overall Summary (QOS) and in Module 3: Quality (9).

In Module 2, QOS Drug Product Control is under 2.3.P.5.

This module should present a comprehensive summary of all relevant quality data, giving the evaluator a clear outline about the characteristics of the drug substance, drug product, and packaging. It should allow easily targeting on critical points. These points should be addressed in detail to allow for an appropriate quality assurance and quality control scheme, that is, a proposed set of specifications, test methods, and acceptance criteria. Details are given in chapter 5 of this book.

In Module 3, Drug Product Control is under 3.2.P.5.

This module should provide the detailed data including the validation packages as appropriate. The format of this module is specified as follows:

- 3.2. P.5 Control of Drug Product (name, dosage form)
- 3.2. P.5.1 Specification(s) (name, dosage form)
- 3.2. P.5.2 Analytical Procedures (name, dosage form)
- 3.2 P.5.3 Validation of Analytical Procedures (name, dosage form)
- 3.2. P.5.4 Batch Analysis (name, dosage form)
- 3.2. P.5.5 Characterization of Impurities (name, dosage form)
- 3.2. P.5.6 Justification of Specification(s) (name, dosage form).

It is nowadays well understood that quality cannot be controlled into a product. It is the detailed knowledge about and control of starting materials, processes, and technology (extraction, modification, synthesis, etc.) and equipment of the components that make up the medicinal product as well as the manufacturing process of the final product and its distribution to the user (the patient or health care provider) that will guarantee the quality of a product. Laboratory control of the product is one (important) element in this whole scheme. Laboratory control has two roles: first, inside the process, that is, routinely throughout the manufacture; second, through a "receiving body," for example, a business partner or an official medicines control laboratory (often affiliated with the competent authority in a particular territory).

Currently, various codes of Good Manufacturing Practices (GMP) together with Good Trade and Distribution Practices are the guidance or even legal requirements in place in many countries and versions of these texts are recommended by the World Health Organization (10). In the framework of ICH, a broader concept has been developed as ICH Q10 Pharmaceutical Quality Systems (11). This describes a comprehensive model for an effective pharmaceutical quality system based on International Standards Organization (ISO) quality concepts. It complements the existing national and regional GMP regulations and guidelines. It is intended to foster innovation and continual product improvement.

In this Chapter, the topic Control Tests on the Product will be discussed, focusing on the points necessary for registration, which will allow agency reviewers of dossiers as well as official medicines control laboratories to judge the intrinsic quality aspects and the proposed means of controlling them.

First, the European situation will be detailed, and then some specifics for the Unites States and Japan are outlined.

CONTROL OF DRUG PRODUCT: THE REQUIREMENTS IN EUROPE

General Considerations

The detailed European legislation and guidances regarding medicinal products for human use can be found in Eudralex, especially Volume 1: Pharmaceutical Legislation (12) and Volume 2: Notice to Applicants (13, 14). These texts should be taken together with the ICH Quality Guidelines (15) and the EMEA guidelines on quality topics (16). All these texts are freely available and downloadable from the Internet.

Starting with the data on pharmaceutical development [ICH Q8 (17)] that take into account the detailed knowledge on the ingredients (active ingredient, excipients, processing aids, other components like primary packaging materials and the manufacturing process, with particulars of the equipment used), a good overview of the product and its critical points are given. This is the basis for the setting of specifications and control tests that allow measurements to confirm the quality of the product.

ICH Guidelines on Specifications

In the ICH, Q6 guidance can be found on setting of specifications [Q6A for chemical actives and products (18), Q6B for biological products (19)]. In these guidance documents, a number of frequently used tests are mentioned, for example, dissolution of tablets, content uniformity, tablet hardness, microbial contamination, etc.

Pharmacopoeial Harmonization Via the Pharmacopoeial Discussion Group

Parallel to the ICH process, the pharmacopoeias from Europe (Ph Eur), the United States (USP/NF), and Japan (JP) have created the tripartite Pharmacopoeial Discussion Group (PDG). This group has worked on harmonization of general test methods and excipient monographs (Tables 1–4). Among the general test methods priority was given to the test methods required for Q6 (Table 1). The stages of harmonization within the PDG are as follows (20):

- Stage 1: Identification
- Stage 2: Investigation
- Stage 3: Proposal for Expert Committee Review
- Stage 4: Official Inquiry
- Stage 5A: Provisional Consensus
- Stage 5B: Draft Sign-of Consensus
- Stage 6: Regional Adoption and Implementation
- Stage 7: Interregional Acceptance

Table 1 ICH Q6A General Methods to be Harmonized: Status in November 2008 (45)

Item	Coordinating pharmacopoeia	PDG stage of harmonization
Dissolution	USP	6
Disintegration	USP	5B
Uniformity of content/mass	USP	5B
Tests for specified organisms	EP	5B
Microbial enumeration	EP	5B
Microbial contamination limits for nonsterile products	EP	6
Bacterial endotoxins	JP	5B
Color (instrumental method)	EP	3
Extractable volume of parenterals	EP	6
Test for particulate contamination: subvisible particles	EP	6
Residue on ignition	JP	6
Sterility test	EP	5B

Table 2 Other General Methods Under PDG Harmonization: Status in November 2008 (45)

Item	Coordinating pharmacopoeia	PDG Stage of Harmonization
Analytical sieving	USP	5B
Bulk density and tapped density	EP	5B
Conductivity	EP	2
Gas pycnometric density of solids	EP	6
Flowability (powder flow)	USP	6
Tablet friability	USP	6
Heavy metals	USP	3
Inhalation	EP	4
Optical microscopy	USP	6
Powder fineness	USP	5B
Specific surface area	EP	6
Porosimetry by mercury intrusion	EP	5B
Laser diffraction measurement of particle size	EP	5B
X-ray powder diffraction	EP	5B
Water–solid interaction	EP	5A
Thermal analysis	EP	3
Uniformity of delivered dose of inhalation	EP	2
Calorimetry	EP	4

Because the outcome of the harmonization efforts of PDG are not always easy to interpret for reasons of style and presentational differences in the pharmacopoeial texts, the ICH created the Q4B Expert Working Group. This group is in charge of judging regulatory acceptability (of the PDG harmonization outcomes); the purpose being that interchangeability is obtained. The formal stages of Q4B Expert Working Group (EWG) evaluation of documents submitted by the PDG are as follows:

- Step 1: Each Q4B party independently evaluates the document for regulatory impact. Issues are resolved as necessary by communication with the PDG.
- Step 2: The ICH Steering Committee agrees, based on the report of the Q4B EWG, that there is sufficient regulatory consensus on the technical issues for the draft annex to proceed to Step 3 regulatory consultation and discussion.
- Step 3: Draft Q4B annex is made available for regulatory consultation in the three regions (generally for three months). The Q4B EWG can revise the annex based on comments received and then submits a final text to the Steering Committee.
- Step 4: The ICH Steering Committee adopts the annex and issues it as a stand-alone companion document to the ICH Q4B guideline.
- Step 5: The annex moves to the regional regulatory implementation step.

Although the purpose of the texts of the annex is to establish that pharmacopoeial texts can be considered interchangeable, a particular annex may set conditions for the basis of interchangeability in the EU, United States, or Japan.

Table 3 Methods for Biotechnology Products Being Harmonized by PDG: Status in November 2008 (45)

Item	Coordinating pharmacopoeia	PDG stage of harmonization
Amino acid determination	USP	6
Capillary electrophoresis	EP	5B
Isoelectric focusing	EP	6
Protein determination	USP	6
Peptide mapping	USP	6
Polyacrylamide gel electrophoresis	EP	6

Table 4 Excipient Monographs Being Harmonized by PDG: Status in November 2008 (45)

Item	Coordinating pharmacopoeia	PDG stage of harmonization
Alcohol	EP	6
Dehydrated alcohol	EP	6
Benzyl alcohol	EP	6
Butyl paraben	EP	4
Calcium carbonate	USP	3–4
Calcium disodium edetate	JP	6
Calcium phosphate dibasic (and anhydrous)	JP	6
Carmellose	JP	5B
Carmellose calcium	USP	6
Carmellose sodium	USP	4
Cellulose acetate	USP	
Cellulose acetate phthalate	USP	
Cellulose powdered	USP	
Citric acid, anhydrous	EP	
Citric Acid, Monohydrate	EP	
Copovidone	JP	
Croscarmellose sodium	USP	
Crospovidone	EP	
Ethylcellulose	EP	
Ethyl paraben	EP	
Gelatin	EP	
Glucose monohydrate/anhydrous	EP	
Glycerin	USP	3
Glyceryl monostearate	USP	3
Hydroxyethylcellulose	EP	4–3
Hydroxypropylcellulose	USP	4 rev
Hydroxypropylcellulose, low substituted	USP	4 rev
Hydroxypropylmethylcellulose	JP	6
Hypromellose phthalate	USP	6
Lactose, anhydrous	USP	3
Lactose for inhalation	USP	1
Lactose, monohydrate	USP	6
Magnesium stearate	USP	6
Mannitol	EP	4
Methylcellulose	JP	6
Methyl paraben	EP	4
Microcrystalline cellulose	USP	6
Petrolatum	USP	4 rev
Petrolatum, white	USP	4 rev
Polyethylene glycol	USP	4–3
Polysorbate 80	EP	5B
Povidone	JP	5B
Propylene glycol	EP	4 rev
Propyl paraben	EP	4
Saccharin	USP	6
Saccharin, calcium	USP	6
Saccharin, sodium	USP	6
Silicon dioxide	JP	4 rev
Silicon dioxide, colloidal	JP	4 rev
Sodium chloride	EP	6
Sodium laurylsulfate	USP	3
Sodium starch glycolate	USP	6
Starch, corn	USP	6
Starch, potato	EP	6

(*Continued*)

Table 4 Excipient Monographs Being Harmonized by PDG: Status in November 2008 (45) (*Continued*)

Item	Coordinating pharmacopoeia	PDG stage of harmonization
Starch, pregelatinized	JP	2
Starch, rice	EP	6
Starch, wheat	EP	6
Stearic acid	EP	6
Sucrose	EP	6
Talc	EP	6
Titanium dioxide	JP	5A3
Water for injection in containers	USP	3

Currently 10 out of 11 general methods from Q6A have been harmonized by PDG and 4 have reached Q4B Step 5 "approval." For current updates on these harmonization efforts, the ICH website (15) and/or the websites of the Pharmacopoeias (3,5,8) should be consulted.

From the pharmaceutical development section of the dossier (ICH CTD 3.2 P.2, following the ICH Q8 guidance), the physical, chemical, biological, and pharmacotechnical foundations for the product should be clear and this together with the data on toxicological, pharmacological, and clinical properties of the active ingredient (and the product) will define the quality of a new industrially manufactured (commercial) pharmaceutical product, that is, a medicine. It is a basic necessity to identify clearly the critical characteristics required to attain the therapeutic objective—these are the basis for setting an appropriate quality level. The early development phase leads to the choice of prototype specifications that will be upgraded and refined during the later steps of product development. The quality characteristics to be attained are usually not finalized until the end of clinical development Phase III. Full scale-up to routine manufacturing conditions often necessitates additional adjustments in a formulation/presentation with consequences for the specification test methods and acceptance criteria.

Quality Characteristics Covered by Specifications

The ICH Q6A Specifications guideline (18) categorizes tests and acceptance criteria for new drug products into:

- Universal tests or criteria: These include description, identification, assay for the active substance, and impurities.
- Additional tests and criteria that should be included for particular dosage forms of new drug products. These include tests such as dissolution, disintegration, hardness and/or friability, uniformity of dosage units, water content, and microbial limits for solid oral drug products. For oral liquids, the tests may include uniformity of dosage units, pH, microbial limits, antimicrobial preservative content, antioxidant content, extractables from the container/closure, alcohol content, particle size distribution (for oral suspensions), redispersibility (for oral suspensions), and rheological properties (for viscous solutions or suspensions). For parenteral drug products, the tests may include uniformity of dosage units, pH, sterility, endotoxins/pyrogens, particulate matter, water content, antimicrobial preservative content, antioxidant content, extractables from the container closure, functionality of a delivery system (such as a prefilled syringe or autoinjector cartridge), osmolarity, particle size distribution (for an injectable suspension), and redispersibility.

Description

A qualitative description of the dosage form should be provided, such as its size, shape, and color.

Identification of the Active Ingredient(s)

It should be stated to what extent a specific method is required to ensure the identity of the active. The method should be able to distinguish between compounds of closely related structure that are likely to be present. When the assay method is proposed as one of the methods of identification of the active ingredient (e.g., retention time by liquid chromatography), this has to be justified [see ICH, Q2 Validation of Analytical Procedures (21)]. A combination of two chromatographic procedures or a combination of tests is acceptable, for example, LC/MS or LC/UV diode array.

Assay of Active Ingredient(s)

The test procedure should allow the provision of an accurate statement on the content or potency of the active ingredient(s) in the finished product, and should use a specific, stability-indicating method. For this the knowledge of the purity of the active ingredient, the excipients, their stability, and the stability of the dosage form are important items in setting up the validation of specificity, accuracy, and precision. In Europe, there will need to be separate release and shelf life specifications for the active ingredient assay unless the product is completely stable. Section 3.2.2.5 of Part 1 of Annex 1 of Directive 2001/83/EC as amended requires that "unless there is appropriate justification, the maximum acceptable deviation in the active substance content of the finished product shall not exceed ±5% at the time of manufacture." This is different to the requirements for a US "regulatory" specification that usually allows ±10% limits for the active substance content.

When physicochemical methods cannot provide adequate information on the potency of the product, an in vivo or in vitro biological assay is mandatory (including the use of an appropriate reference material and statistical analysis allowing calculation of confidence limits). When these tests cannot be carried out on the finished product, they may be performed at an intermediate stage but as late as possible in the manufacturing process.

Identification and Assay of the Constituents of the Excipients

Test should be provided to allow identification of the excipients. This is especially important for

- Coloring matter: Colorants should be on the list of authorized materials—Annex 1 of Directive 94/36/EC (22) lists colors permitted for use in foodstuffs. The EMEA Guideline on Excipients in the Dossier for Application for Marketing Authorization of a Medicinal Product (23) states that these colorants in the Annex are permitted for use in medicinal products.
- Antimicrobial agents and/or antioxidant preservatives: The 1997 Committee for Proprietary Medicinal Products (CPMP) Note for Guidance on Inclusion of Antioxidants and Antimicrobial Preservatives in Medicinal Products (24) states that the finished product release specification should include an identification test and limits for any antioxidants and antimicrobial preservatives. The finished product shelf life specification should also include limits for the antimicrobial preservatives present.
- Any excipient liable to affect physiological functions: According to the ICH Q6A guideline limits should be given.
- Any excipient liable to affect bioavailability: According to the ICH Q6A guideline limits should be given unless bioavailability is guaranteed by other appropriate tests.

Purity Tests

Residual Solvents

If necessary the investigation of residual solvents or other process related impurities (such as residues of ethylene oxide sterilant), as well as microbial contamination, should be undertaken and appropriate control tests added to the specification as needed.

Degradation Products of the Active Ingredient

If found during stability investigations that the degradation of the active ingredient may take place during manufacturing and/or storage, the finished product specifications should include test(s) that limits the content of breakdown products. Guidance on suitable limits is defined in the ICH Q3B guidance on Impurities in the Drug Product (25).

General Characteristics of the Dosage Form

These characteristics are in general determined by pharmacotechnical and physical tests with limits of acceptance, relating to the product performance or handling.

The provisions of the general monographs of the Ph Eur, or if there is no relevant monograph then those of a EU Member State pharmacopoeia, are applicable to all products defined therein. If test procedures and limits other than those mentioned in the Ph Eur or of a pharmacopoeia of an EU Member State are used, proof should be supplied that the finished product would, if tested in accordance with those monographs, meet the quality requirements of that pharmacopoeia for the pharmaceutical form concerned.

Typical pharmacotechnical tests for a solid oral dosage form such as an immediate release uncoated tablet are specified in the Ph Eur general monograph 0478 and are

- Uniformity of mass (Ph Eur method 2.9.5)
- Dissolution (Ph Eur method 2.9.3 for tablets and capsules)
- Disintegration (Ph Eur method 2.9.1 for tablets and capsules)

Safety Tests

Safety tests such as sterility, bacterial endotoxin, viral safety, pyrogenicity, or abnormal or specific toxicity (in particular for biological products) should be included in the analytical particulars wherever such tests must be undertaken as a matter of routine in order to verify the quality of the medicinal product.

Microbial Limits

The ICH Q6A Specifications guideline recommends that in general it is advisable to test the drug product unless its components are tested before manufacture and the manufacturing process is known not to carry a significant risk of adding microbial contamination. Q6A includes a decision tree on testing of excipients.

Ph Eur general text 5.1.4 on Microbiological Quality of Pharmaceutical Preparations defines the recommended acceptance criteria for microbiological quality. Drug products are categorized as follows:

- Category 1: Products required to be sterile.
- Category 2: Products for topical use and for use in the respiratory tract, except where required to be sterile, and transdermal patches.
- Category 3: Products for oral and topical administration.
- Category 4: Herbal medicinal products consisting of one or more herbal drugs.

The above Ph Eur general text includes recommendations for limits for total viable aerobe count for the different categories and limits for or absence of specified organisms (such as *Escherichia coli*, *Salmonella*, *Pseudomonas aeruginosa*, and *Staphylococcus aureus*).

Tests on Specific Dosage Forms

The ICH Q6A Specifications guideline (18) includes recommendations for tests to be included in the specifications for specific dosage forms:

- Solid oral drug products (uncoated and coated tablets, soft shell and hard capsule, granules): Dissolution, hardness/friability, disintegration, uniformity of dosage units (including both uniformity of mass and uniformity of content), water content, and microbial limits.
- Oral liquids: Uniformity of dosage units, pH, microbial limits, antimicrobial preservative content, antioxidant preservative content, extractables from the container/closure system, alcohol content, dissolution (where applicable), particle size distribution, redispersibility, rheological properties, reconstitution time, and water content.
- Parenteral drug products: Uniformity of dosage units, pH, sterility, endotoxins/pyrogens, particulate matter, water content, antimicrobial preservative content, antioxidant preservative content, extractables from the container/closure, functionality testing of drug delivery

system (prefilled syringe or autoinjector cartridge), osmolarity, particle size distribution, redispersibility, and reconstitution time (where applicable).

In addition to the tests specified in the Ph Eur general monographs, there are specific EU guidelines that should be consulted regarding the evaluation and testing of specific dosage forms.

Categories of product on which there are CHMP quality guidelines are:

- Inhalation and Nasal Products (26)
- Radiopharmaceuticals (27)
- Modified Release Products—Oral Solid Dosage Forms and Transdermal Dosage Forms (28)

Inhalation and Nasal Products

For pressurized metered dose inhalers, the specification should include assay, moisture content, mean delivered dose, delivered dose uniformity, fine particle mass, leak rate, microbial limits, leachables, and numbers of actuations per container. For device metered dry powder inhalers (DPIs), the specification should include assay, moisture content, mean delivered dose, delivered dose uniformity, fine particle mass, microbial contamination, and number of actuations per container. For multiple dose products for inhalation, the specification should include assay, content uniformity, fine particle mass (for suspensions), microbial limits, leachables, and preservative content (if present).

Radiopharmaceutical Products

For radiopharmaceuticals, the specification should include radiochemical identity and purity, chemical purity, and where relevant specific radioactivity, radionuclidic identity, and purity. For radionuclide generators details on the testing for mother and daughter radionuclides are required. For generator-eluates, tests for specific activity, mother radionuclides, daughter radionuclides and for other radionuclidic and chemical impurities from the generator should be provided. For radiopharmaceutical kits, the specifications of the finished product should include tests on the quality of the product after radiolabeling. Appropriate controls on the identification, radiochemical and radionuclide purity, content of radioactivity, and where relevant the specific activity of the radiolabeled compound should be included.

Modified Release Oral Solid Dose form Products

The finished product specification must contain a validated method and acceptance limits for in vitro dissolution. In general, a minimum of three points should be included in the specification. Establishment of an in vitro–in vivo correlation is encouraged to help set a meaningful specification to confirm consistent therapeutic performance from routine production batches.

Transdermal Patches

Release characteristics of the product should be tested using a suitable dissolution method, such as Ph Eur method 2.9.4 Dissolution test for transdermal patches. The specification should be derived from batches used in clinical trials and showing acceptable in vivo performance.

Analytical Procedures

For each test procedure, the analytical, physical, or biological technique should be described in detail. When a technique is described in the Ph Eur, reference can be made to the relevant text. The applicability of any method has to be justified by following the ICH Q2 guidance on Validation of Analytical Procedures (21).

The following list of information is typically included in any analytical procedure:

- Principle (e.g., separation using isocratic reverse phase LC with a UV detector)
- Sampling details (e.g., number of tablets, capsules to be tested, replicate analyses)
- Equipment and equipment parameters (e.g., instrument type, detector, column and flow rate, temperature, run time, etc.)
- Reagents and grades
- System suitability test

- Preparation of standards
- Preparation of samples
- Procedure
- Calculation of results
- Format of the report of results (e.g., percent label claim, parts per million, etc.)

Validation of Analytical Procedures

The appropriate validation report for each specification test method used to test the drug product should be provided, following the guidance of ICH Q2 Validation of Analytical Procedures (21). Stability data, including data from stress stability studies should be used to support the validation package.

Batch Analysis

This section must include the results for all specifications at release, whether or not they are intended for routine batch-by-batch verification. Where possible results should be provided from consecutive batches corresponding to the intended production process and scale derived from all of the proposed manufacturers and at all manufacturing sites stated in the application for marketing authorization according to the process described in the dossier. Analytical results should be provided for the batches of new drug product used for clinical, safety, and stability testing. In addition, batch analyses should be provided that are representative of the proposed commercial manufacturing process—the minimum requirement is data on two pilot scale and one production scale batch per manufacturer per site. A commitment by the applicant to provide additional data as soon as more industrial scale batches are available may also be useful.

The tabulated batch documentation needs to include:

- Batch identity, strength, and size
- Date of manufacture
- Site of manufacture
- Manufacturing process
- Immediate container closure
- Use of batch
- Reference to the analytical procedures used
- Batch number of the drug substance used in the manufacture of the batch of drug product

Characterization of Impurities

Together with the knowledge gathered on the active ingredient(s) and the excipients in depth analytical investigation work on the finished dosage form can reveal the presence of additional impurities. The main sources are the manufacturing process, equipment, interaction with packaging material, and instability during storage. Impurities should be characterized and limited following the guidance given in the ICH Q3A guideline on Impurities in the Finished Product as defined in Table 5. However, if a degradation product is known to be unusually potent or toxic, the quantitation level should be appropriate and not merely reflect the ICH guideline limits given in Table 5.

Individual named degradation products limited in the specification are referred to as "specified" degradation products. Any specified degradation product can be identified or unidentified. However, if unidentified, their structure will be unknown and their potential for toxicity is difficult to assess.

A new drug product specification should include the following list of degradation products:

- Each specified identified degradation product.
- Each specified unidentified degradation product.
- Any unspecified degradation product with an acceptance criterion of not more than the identification threshold.
- Total degradation products.

Table 5 ICH Recommended Threshold Limits for Degradation Products in New Drug Products (25)

Maximum daily dose (amount administered per day)	Threshold expressed either as % of drug substance or as total daily intake of the degradation product
Reporting threshold	
≤1 g	0.1%
>1 g	0.05%
Identification threshold	
<1 mg	1.0% or 5 μg TDI, whichever is lower
1–10 mg	0.5% or 20 μg TDI, whichever is lower
>10 mg–2 g	0.2% or 2 mg TDI, whichever is lower
>2 g	0.10%
Qualification threshold	
<10 mg	1.0% or 50 μg TDI, whichever is lower
10–100 mg	0.5% or 200 μg TDI, whichever is lower
>100 mg–2 g	0.2% or 3 mg TDI, whichever is lower
>2 g	0.15%

The results of formal stability studies following ICH Q1A Stability guidance (29) will lead to setting of the shelf life and label storage requirements by assessing the performance of the product against its proposed specification during the accelerated and real-time studies.

There may need to be a commitment from the applicant to complete a further postapproval program of stability studies in the proposed container/closure using production scale batches (if the marketing application did not include these data). In this case, the performance of batches against the specification would need to be monitored for Out of Specification (OOS) or Out of Trend (OOT) results, which might need the shelf life and/or specification to be reviewed.

Justification of Specifications

Taking into account all the data available from pharmaceutical development, results on drug substance analyses, excipients analyses, drug product batch analyses, and packaging materials analyses and including the stability data and having demonstrated the critical parameters for the quality of the final product, a proposal can be made for a set of justified specifications with test methods and acceptance criteria.

Justification should also be provided for tests omitted from the proposed specification that are normally performed, are listed in a guideline, or were reported in the drug product batch analyses.

It is important to distinguish acceptance criteria at release (i.e., after manufacturing) and acceptance criteria at end of shelf life. The latter are the values that apply during the "commercial life" of the product. They are also the values that are legally binding in respect to independent quality control by an official medicines control laboratory or in supplier/client business relations. The justification for the acceptance limits should discuss the available batch data, analytical and manufacturing capability, and variability.

Routine Tests and Periodic Tests

Routine tests are tests that are carried out as part of the specification on each industrial batch as explained and justified above.

Periodic tests are tests that can be applied at a reduced frequency (such as every tenth batch) or at a stated time interval (e.g., every month). Their periodicity/frequency should be explained and justified. The level of knowledge about the product, its ingredients and components, as well as the manufacturing process, together with the (demonstrated) control over the product are key ingredients in deciding on the suitability of this "skip test" approach. Skip test or routine test—the rule is always that when tested the product should comply with the approved specifications, test methods, and acceptance criteria.

THE REQUIREMENTS IN JAPAN AND THE UNITED STATES

The requirements for drug product testing in Japan and the United States are not fundamentally different from those in the EU. In all three regions, the most recent scientific and technological knowledge must be referred to and be used by the manufacturer including validation of processes and methods. The differences relate less to the general scientific approach than to the presentation of dossiers and the licensing/approval procedures. This is perhaps why the topic "Control tests on the finished product in the marketing authorization dossier" was not identified as a high priority item by the pharmaceutical industry at the start of the ICH process. However, as already explained above, analytical validation (Q2) and harmonization of specifications and general test methods (Q6) were identified as very useful tools to avoid extra (redundant) testing to meet the requirements of agencies in the different regions.

Japan

In Japan, the new Pharmaceutical Affairs Law went into effect on 1 April 2005. This law takes several steps towards global harmonization, for example, by introducing new GMP requirements. However, it is important to note that the MHLW added specific requirements that differ from global standards. The new Marketing Authorization Holder system separates the responsibilities of manufacturing and product release. See for some details on Standards for Manufacturing Control, Quality Assurance, and Quality Control the MHLW Ministerial Ordinances No: 179 and 136 (30, 31).

The marketing authorization application dossier in Japan is composed of the Application Form, Modules 2 and 3. The Application Form has to be in Japanese and must contain the Analytical Control Methods in JP format and also the commitments as to data to be provided once commercial manufacturing has started (e.g., long-term stability data). Module 2 is the QOS, but it is required to submit much more data and is a larger document than is required in Europe and the United States: it is closer to the former Japanese *Gaiyo*. It has to be presented in Japanese. Module 3 contains the detailed data and can be provided in English. The main documents for review in Japan are the Application form and the Module 2 QOS.

Pharmacopoeial Requirements

The *Japanese Pharmacopoeia* is an important element in the application process: in principle the methods described in the JP under General Notices, General Rules for preparations, General Tests, processes and Apparatus, Reference Standards, Reagents, and Test Solutions should be employed in the "specifications and testing methods." Here one can easily understand the importance of the harmonization work of the PDG and Q4B, making interchangeability of Ph Eur, USP/NF, and JP methods possible.

United States of America

In the United States, the FDA is in charge of evaluation and approval of dossiers for marketing authorization. The basis is the Federal Food, Drug and Cosmetics Act (FDAC Act) as amended through December 31, 2004 (32), and detailed regulations regarding New Drug Applications can be found in the *Code of Federal Regulations*, especially Title 21 Part 314 (21 CFR 314) (33).

The FDA website provides a wealth of guidance for industry information reflecting current agency thinking. The 1987 FDA Guideline for Submitting Documentation for the Manufacture and Controls for Drug Products (34) is still relevant for this chapter. This guideline defines the "regulatory specification" in section F: Specifications and Analytical Methods for the Drug Product as follows: "Regulatory specifications are the defined limits (e.g., physical, chemical, biological, and microbiological) within which test results for a drug substance or drug product should fall when determined by the regulatory methodology." Unlike the EU there is no requirement for separate release and shelf life specifications. However, some international companies do submit two specifications to the FDA where they feel it is better to have a single global set of specifications, and these have been accepted. Other companies have a release specification as an in-house internal specification, which is not officially submitted as part of the marketing authorization dossier.

Other FDA guidelines that are relevant include:

- Development of New Stereoisomeric Drugs (35)
- Dissolution Testing of Immediate Release Solid Oral Dosage Forms (36)
- Metered Dose Inhaler (MDI) and Dry Powder Inhaler (DPI) Drug Products (37)
- Nasal Spray and Inhalation Solution, Suspension, and Drug Products (38)
- Orally Disintegrating Tablets (39)
- Residual Solvents in Drug Products Marketed in the United States (40).

The August 2000 Draft Guidance on Analytical Procedures and Methods Validation (41) provides very detailed guidance on submitting analytical procedures, analytical validation data and samples. This guidance provides details as to the information needed to support LC, GC, spectrophotometric, and capillary electrophoresis (CE) methods. The ICH Q2(R1) Validation of Analytical Procedures mainly refers to chemical test methods; the FDA Draft Guidance provides useful information on analytical validation for optical rotation, particle size analysis, and dissolution.

Information on Manufacture and Control in a Drug Master Files

A special remark has to be made as to the existence of a Drug Master File (DMF) system, unique to the United States. Europe and Japan have provisions for DMFs for active pharmaceutical ingredients; Japan also allows the use of DMFs for novel excipients. In the United States, information included in a drug product DMF to satisfy the documentation needed to evaluate any particular part of the manufacture, and control for a drug product is "acceptable provided the reference is specific, current, and applicable to the drug product described in the application" (42). Such a possibility is not envisaged by the European and Japanese regulations.

Abbreviated New Drug Applications

For abbreviated new drug applications (ANDA), specifications are often defined in a USP pharmacopeial monograph. The specifications and test methods issued in the USP/NF monographs are considered as validated and as regulatory specifications [section 501(b) of the FDAC Act] (32). Additional specifications or alternate test methods may be required as necessary, especially with older drug products when assay methodology given in the compendial monograph is not sufficiently specific. Broader limits than those in the USP/NF monograph will not be approved ordinarily as "regulatory specifications" unless the labeling indicates that the product differs from the official monograph.

The FDA has a Dissolution Methods Database (43) prepared by the Division of Bioequivalence, Office of Generic Drugs, Center for Drug Evaluation and Research. These are recommended dissolution test methods for many of the drug products submitted via an ANDA. However, the FDA will consider other dissolution methods if supported by appropriate data.

Methods Validation Package

One difference between marketing applications in the United States and elsewhere is the FDA requirement to submit a separate Methods Validation Package (34). This package allows an FDA laboratory to check the suitability of the analytical procedures proposed for testing. The package consists of information copied from pertinent sections of the dossier. It includes

1. Tabular listing of all samples to be submitted (drug substance, reference standards, drug product, and blanks)
2. Analytical procedures
3. Validation data
4. Results
5. Composition of the drug product
6. Specifications for the drug substance and drug product
7. Material Safety Data Sheets for all samples, standards and reagents.

FUTURE DEVELOPMENTS

As already mentioned above within the ICH quality topic area some very interesting developments are underway. The Q8 (Pharmaceutical Development), Q9 (Quality Risk Management), and Q10 (Pharmaceutical Quality System) guidelines have been developed and adopted and work has started on Q11 (Development of the active pharmaceutical ingredient). Q11 will be a new tripartite Technical Guidance which is proposed for Active Pharmaceutical Ingredients (APIs) harmonizing the scientific and technical principles relating to the description and justification of the development and manufacturing process of Drug Substances including both chemical entities and biotechnological/biological entities (CTD sections S 2.2.–S 2.6).

These guidelines when fully adopted by the pharmaceutical industry will allow for very different approach to manufacturing process control and batch release. Based on a full understanding of the manufacturing process, continuous in-process testing could enable "real-time release" instead of conventional end-product batch Quality Control testing. This approach relies on an in-depth process knowledge based on modeling and definition of a "design space" that incorporates data on variables like the properties of the drug substance, excipients, physical parameters such as temperature, humidity, mixing speed, etc. This greater knowledge of the product and processes could create more "regulatory flexibility" because the continuous process verification would allow feedback into the model for its further improvement, or allow for changes in scale and site without seeking preapproval by authorities, provided the process remains within the borders of the design space, which was defined in the original marketing authorization application. This in-process continuous testing [via the so-called Process Analytical Technologies (PAT)] will thus, once approved, replace the current end-product testing for routine control (44).

But as said earlier, independent testing, for example, by an official government laboratory should remain possible, thus presentation of conventional test methods will still remain a requirement for dossiers. Acceptance criteria making use of the in-process testing/process monitoring data and parallel conventional testing results will need to be established.

REFERENCES

1. European Directive 2001/83/EC as amended. Off J Eur Communities 2001; L 311:67.
2. European Commission Regulation 1234/2008 Concerning the Examination of Variations. Off J Eur Communities 2008; L 334/7–L 334/24.
3. Convention on the elaboration of a European Pharmacopoeia, Strasbourg, 1964. http://www.edqm.eu/medias/fichiers/1964_PhEur_Convention_English.pdf.
4. FDA website: http://www.fda.gov/default.htm. Accessed May 29, 2009.
5. United States Pharmacopeia website: http://www.usp.org/. Accessed May 29, 2009.
6. Chapter 2 of Pharmaceutical Regulations in Japan, The Pharmaceutical Affairs Law (English translation), March 2009. http://www.jpma.or.jp/english/parj/pdf/2009_ch02.pdf. Accessed May 29, 2009.
7. Pharmaceuticals and Medical Devices Agency (Japanese regulatory agency working with the Ministry of Health, Labor and Welfare). http://www.pmda.go.jp/english/index.html. Accessed May 29, 2009.
8. Japanese Pharmacopoeia, 15th edition, English version, official from April 2006. http://jpdb.nihs.go.jp/jp15e/JP15.pdf. Accessed May 29, 2009.
9. The Common Technical Document for the Registration of Pharmaceuticals for Human Use: Quality M4Q(R1), Quality Overall Summary of Module 2, Module 3: Quality, September 2002, ICH, Geneva. http://www.ich.org/LOB/media/MEDIA556.pdf. Accessed May 29, 2009.
10. Quality Assurance of Pharmaceuticals. A Compendium of Guidelines and Related Materials, Vol. 2, updated ed. Good Manufacturing Practices and Inspection. Geneva, Switzerland: WHO, 2006.
11. ICH Q10 Pharmaceutical Quality System, June 2008, ICH, Geneva. http://www.ich.org/LOB/media/MEDIA3917.pdf. Accessed May 29, 2009.
12. Volume 1 of the publications "The rules governing medicinal products in the European Union" compiles the body of European Union legislation in the pharmaceutical sector for medicinal products for human use, European Commission. http://ec.europa.eu/enterprise/pharmaceuticals/eudralex/vol1_en.htm. Accessed May 29, 2009.
13. Volume 2A of the publication "The rules governing medicinal products in the European Union", Procedures for marketing authorisation, European Commission. http://ec.europa.eu/enterprise/pharmaceuticals/eudralex/vol2_en.htm#2a. Accessed May 29, 2009.

14. Volume 2B of the publication "The rules governing medicinal products in the European Union", Presentation and content of the application dossier, European Commission. http://ec.europa.eu/enterprise/pharmaceuticals/eudralex/vol2_en.htm#2b. Accessed May 29, 2009.
15. ICH Quality Guidelines, ICH, Geneva. http://www.ich.org/cache/compo/276–254-1.html. Accessed May 29, 2009.
16. EMEA Quality Guidelines, EMEA, London. http://www.emea.europa.eu/htms/human/humanguidelines/quality.htm. Accessed May 29, 2009.
17. Guideline—Pharmaceutical Development, November 2008, ICH, Geneva. http://www.ich.org/LOB/media/MEDIA4986.pdf ICH Q8(R1). Accessed May 29, 2009.
18. ICH Q6A Guideline-Specifications: Test Procedures and Acceptance Criteria for New Drug Substances and New Drug Products: Chemical Substances, October 1999, ICH, Geneva. http://www.ich.org/LOB/media/MEDIA430.pdf. Accessed May 29, 2009.
19. ICH Q6B Guideline: Specifications: Test Procedures and Acceptance Criteria for Biological/Biotechnological Products, March 1999, ICH, Geneva. http://www.ich.org/LOB/media/MEDIA432.pdf. Accessed May 29, 2009.
20. European Directorate for Quality of Medicines website, International Harmonisation (Pharmacopoeial Discussion Group). http://www.edqm.eu/site/International_Harmonisation-614.html. Accessed May 29, 2009.
21. Guideline, Validation of Analytical Procedures: Text and Methodology, November 2005, ICH, Geneva. http://www.ich.org/LOB/media/MEDIA417.pdf ICH Q2(R1). Accessed May 29, 2009.
22. European Parliament and Council Directive 94/36/EC of 30 June 1994 on colours for use in foodstuffs. Off J Eur Communities 1994; L237:13–29.
23. CHMP Guideline on Excipients in the Dossier for Marketing Authorisation for a Medicinal Product, July 2007, EMEA, London. http://www.emea.europa.eu/pdfs/human/qwp/39695106enfin.pdf. Accessed May 29, 2009.
24. CPMP/CVMP Note for Guidance on Inclusion of Antioxidants and Antimicrobial Preservatives in Medicinal Products, July 1997, EMEA, London. http://www.emea.europa.eu/pdfs/human/qwp/011595en.pdf. Accessed May 29, 2009.
25. ICH Guideline Q3B(R2) Impurities in New Drug Products, June 2006, ICH, Geneva. http://www.ich.org/LOB/media/MEDIA421.pdf. Accessed May 29, 2009.
26. CHMP Guideline on the Pharmaceutical Quality of Inhalation and Nasal Products, March 2006, EMEA, London. http://www.emea.europa.eu/pdfs/human/qwp/4931305en.pdf. Accessed May 29, 2009.
27. CHMP Guideline on Radiopharmaceuticals, November 2008, EMEA, London. http://www.emea.europa.eu/pdfs/human/qwp/30697007enfin.pdf. Accessed May 29, 2009.
28. CPMP Note for Guidance on Quality of Modified Release Products A: Oral Dosage Forms, B: Transdermal Dosage Forms, July 1999, EMEA, London. http://www.emea.europa.eu/pdfs/human/qwp/060496en.pdf. Accessed May 29, 2009.
29. Q1A(R2) Stability Testing of New Drug Substances and Products, February 2003, ICH, Geneva. http://www.ich.org/LOB/media/MEDIA419.pdf ICH Guideline. Accessed May 29, 2009.
30. MHLW Ordinance No. 179, 2004 under the Pharmaceutical Affairs Law, PMDA (English text). http://www.pmda.go.jp/english/services/reviews/file/050909betsu2.pdf. Accessed May 29, 2009.
31. MHLW Ordinance No. 136, 2004 under the Pharmaceutical Affairs Law, PMDA (English text). http://www.pmda.go.jp/english/services/reviews/file/050909betsu1.pdf. Accessed May 29, 2009.
32. US Federal Food, Drug and Cosmetics Act as amended through December 31, 2004. http://www.fda.gov/opacom/laws/fdcact/fdctoc.htm. Accessed May 29, 2009.
33. Code of Federal Regulations Title 21—Food and Drugs, Chapter I—Food and Drug Administration, Department of Health and Human Services, Subchapter D—Drugs for Human Use, Part 314 Applications for FDA Approval to Market a New Drug. http://www.accessdata.fda.gov/SCRIPTs/cdrh/cfdocs/cfcfr/CFRSearch.cfm. Accessed May 29, 2009.
34. FDA Guideline for Submitting Documentation for the Manufacture and Controls of Drug Products, February 1987, FDA. http://www.fda.gov/cder/guidance/old029fn.pdf. Accessed May 29, 2009.
35. FDA'S Policy Statement for the Development of New Stereoisomeric Drugs, May 1992. http://www.fda.gov/cder/guidance/stereo.htm. Accessed May 29, 2009.
36. FDA Guidance for Industry: Dissolution Testing of Immediate Release Solid Oral Dosage Forms, August 1997. http://www.fda.gov/cder/guidance/1713bp1.pdf. Accessed May 29, 2009.
37. FDA Draft Guidance for Industry: Metered Dose Inhaler and Dry Powder Inhaler Drug Products, Chemistry, Manufacturing and Controls Documentation, October 1998. http://www.fda.gov/cder/guidance/2180dft.pdf. Accessed May 29, 2009.
38. FDA Guidance for Industry: Nasal Spray and Inhalation Solution, Suspension, and Spray Drug Products, Chemistry, Manufacturing and Controls Documentation, July 2002. http://www.fda.gov/cder/guidance/4234fnl.pdf. Accessed May 29, 2009.

39. FDA Guidance for Industry: Orally Disintegrating Tablets, December 2008. http://www.fda.gov/cder/guidance/8528fnl.pdf. Accessed May 29, 2009.
40. FDA Draft Guidance for Industry: Residual Solvents in Drug Products Marketed in the United States, August 2008. http://www.fda.gov/cder/guidance/8179dft.pdf. Accessed May 29, 2009.
41. FDA Draft Guidance for Industry: Analytical Procedures and Methods Validation, Chemistry, Manufacturing and Controls Documentation, August 2000. http://www.fda.gov/cder/guidance/2396dft.pdf. Accessed May 29, 2009.
42. FDA Guideline for Drug Master Files, February 1989. http://www.fda.gov/cder/guidance/dmf.htm. Accessed May 29, 2009.
43. FDA Database of Recommended Dissolution Methods for Drug Products. http://www.accessdata.fda.gov/scripts/cder/dissolution/. Accessed May 29, 2009.
44. EMEA Inspections Working Party Reflection Paper: Chemical, pharmaceutical and biological information to be included in dossiers when Process Analytical Technology (PAT) is employed, March 2006. http://www.emea.europa.eu/Inspections/docs/PATGuidance.pdf. Accessed May 29, 2009.
45. State of Work on International Harmonisation (November 2008). Pharmeuropa 2009; 21(1):142–143.

18 Reference Standards or Materials

Anthony C. Cartwright
Pharmaceutical Regulatory Consultant, Harpenden, U.K.

WHAT ARE REFERENCE MATERIALS?

Reference material is defined by the International Organization for Standardization (ISO) as a material or substance one or more of whose property values are sufficiently homogeneous and well established to be used for the calibration of an apparatus, the assessment of a measurement method, or for assigning values to the substance (1).

A certified reference material (CRM) is a reference material one or more of whose property values are certified by a procedure that establishes its traceability to an accurate realization of the unit in which the property values are expressed. CRMs are issued with a certificate in which their critical parameters and assigned values are fully documented.

A computerized databank "Code of Reference Materials" (COMAR) lists 10000 reference materials from about 200 producers in 20 countries (2).

WHAT ARE REFERENCE STANDARDS?

Reference standards are highly characterized specimens of drug substances, excipients, impurities, degradation products, and dissolution performance calibrators used as reference materials in the pharmaceutical field. It is a general term encompassing reference substances, reference preparations, and reference spectra.

Reference standards are supplied by the official compendia such as the United States Pharmacopeia (USP), European Pharmacopoeia (Ph Eur), Japanese Pharmacopoeia (JP), British Pharmacopoeia (BP), etc., for use in the tests specified in monographs for pharmaceutical substances and drug products (3–7). Pharmaceutical companies and their suppliers have their own in-house reference standards for tests on pharmaceutical substances and drug products not described in official monographs.

WHAT ARE REFERENCE STANDARDS USED FOR?

Reference standards are required for performing in-house or compendial tests and assays. Typical uses include:

- identification tests,
- limit tests for impurities or degradation products,
- assays and purity tests on drug substances and drug products,
- system suitability tests, and
- analytical validation studies (e.g., method selectivity/specificity studies).

Reference standards are also used as calibrators for in vitro dissolution tests, particle count, and melting point.

Reference spectra, such as infrared reference spectra, are used for identification by comparing a spectrum prepared with a reference standard prepared by using a potassium bromide or potassium chloride–pressed disc or liquid paraffin mull against the spectrum made similarly by using an experimental or production batch of the drug substance or excipient.

PRIMARY STANDARDS AND WORKING STANDARDS

A primary standard is one that is accepted without reference to other standards. It must have been unequivocally characterized in terms of chemical structure, have undergone a complete analytical characterization, and its purity must be sufficiently high and stated (typically >99%). Its method of manufacture may differ from that used for routine production as it may, for example, have undergone additional recrystallization and drying. It may be characterized by elemental analysis, Infrared (IR), Ultraviolet (UV), proton Nuclear Magnetic Resonance

(^{1}H-NMR), ^{13}C-NMR, Mass Spectroscopy (MS), and Circular dichroism (CD). Its purity determination may include determination of impurities, moisture content, content of residual solvents, inorganic impurities (heavy metals, sulfated ash, and atomic spectrometry). For primary chemical reference substances used for assay purposes the assigned percentage content is usually calculated from (100 − total percent impurities), where the total impurities comprise organic, inorganic, water, and solvents. Primary chemical reference substances are accompanied by a certificate of analysis.

A secondary standard has closely similar properties to the primary standard that are relevant for the test(s) for which it is established. The degree of testing is not as great to establish a secondary standard. The secondary standard is established by comparison with the primary standard to which it is traceable.

Compendial primary standard reference substances are usually meant to be used by the pharmaceutical company's analytical laboratory for establishing their own in-house working standards.

Industry Approaches to Primary and Secondary Reference Standards for Noncompendial Drug Substances

Pharmaceutical companies have two different approaches for establishing reference standards for noncompendial drug substances such as New Chemical Entities (NCEs), which will be used for assays and purity tests:

- Manufacture of a small batch of primary standard that is fully characterized and documented with a certificate of analysis, which is then used as the basis for establishing successive batches of in-house working standards; or
- Manufacture of a larger batch of primary standard that is fully characterized and documented in a certificate of analysis and which is then used to exhaustion as the only reference standard. When the batch is used up a further batch is made and then fully characterized and documented.

Both of these approaches are acceptable for a chemical noncompendial drug substance.

The data on characterization and properties of the reference standard for an NCE is usually used as the evidence on elucidation of structure in section 3.2.S.3.1 Elucidation of Structure and other Characteristics in the CTD registration filing. This includes copies of the IR, UV, ^{1}C-NMR, ^{14}C-NMR, MS, etc., spectra and the peak assignments.

COMPENDIAL REFERENCE SUBSTANCES AND PREPARATIONS

A number of the major pharmacopoeias such as the Ph Eur, USP, BP, JP, and the Pharmacopée Française (PF) provide very extensive series of reference substances (3–7) and reference spectra. Their on-line catalogues provide lists of the available materials.

These reference substances will have been tested for suitability for the specific purposes indicated in the compendial monographs. Their suitability for any other purpose will not have been evaluated.

For substances used in identification tests, a normal production batch of the substance is often used. It is tested for compliance with the monograph. For substances used in impurity testing (related substances), the reference standard is characterized for identity and purity.

Reference standards used in assays or purity testing are usually evaluated by a number of collaborating laboratories to assign the active content.

Reference standards are packaged in stoppered glass vials or glass ampoules. The catalogue from the relevant pharmacopoeia provides directions for storage (e.g., 5°C ± 3°C) of the material and any retest or expiry data (e.g., use within three months of receipt). For Ph Eur reference standards, no retest or expiry date is given because the EDQM retest program monitors their continued fitness for use.

COMMERCIAL SOURCES OF REFERENCE MATERIALS

Some commercial laboratories also offer both pharmacopoeial Chemical Reference Substances (CRS) and in-house certified reference materials such as impurities (including solvents and solvent mixtures). Both R.T. Corp in the United States and LGC Standards in the United Kingdom offer a large range of pharmacopoeial reference materials (8, 9). LGC Standards also offers in-house certified impurity reference materials and custom synthesis of organic impurities.

BIOLOGICAL REFERENCE PREPARATIONS

These are substances or mixtures of substances intended for use in compendial monograph tests or in-house testing by manufacturers of biological products. They allow the definition of internationally agreed units of defined biological activity to allow comparison of strength of products worldwide. WHO has a catalogue of primary standards that allows the pharmacopoeias or manufacturers to create their own secondary reference preparations. The WHO catalogue (10) includes allergens, antibiotics, blood products, coagulation factors, cytokines, immunoglobulins, vaccines, toxoids, and toxins. European Pharmacopoeia Biological Reference Preparations include secondary standards calibrated in International Units and primary standards calibrated in European Pharmacopoeia Units. The National Institute for Biological Standards and Control (NIBSC) distributes WHO biological reference preparations and also British Standards and other reference materials, including those previously distributed by the Staten Serum Institute, Copenhagen, Denmark.

DISSOLUTION CALIBRATORS

Small differences in the performance of dissolution test equipment (such as variation in stirring speed, environmental vibration, and temperature) can affect the observed dissolution rate. The USP defines dissolution apparatus suitability by using dissolution calibrators (11). The USP has a range of three calibrators—USP Prednisone Tablets Reference Standard (Dissolution Calibrator, Disintegrating), USP Salicylic Acid Tablets Reference Standards (Dissolution Calibrator, Nondisintegrating), and USP Chlorpheniramine Maleate Extended-Release Tablets (Drug Release Calibrator, Single Unit). Although only the FDA requires the use of such calibrators, it would obviously be wise to consider using the calibrators to check the suitability of dissolution equipment being used to generate data for a global registration filing and to maintain a laboratory record of the testing in case of an FDA GMP inspection.

ESTABLISHMENT OF IN-HOUSE CHEMICAL REFERENCE STANDARDS

Assay/Purity Reference Standards

These are usually manufactured using the same method of manufacture as for production batches. However, they are often subject to additional purification steps—either additional recrystallization or drying, sometimes by preparative chromatography. Care needs to be taken not to introduce additional impurities by any of these further processing steps. The batch of reference standard is then fully characterized in terms of its chemical structure and information derived on its purity by both assay and chromatographic or other impurity determination (for starting materials, by-products, intermediates, reagents, catalysts, solvents, etc.). Characterization tests will usually include elemental analysis, IR, UV, ^{1}C-NMR, ^{14}C-NMR, MS, and melting point. Tests for purity will include tests for content of organic impurities by a suitable chromatographic (HPLC) or other technique, moisture content by Karl Fischer titration, loss on drying, residual solvents by gas chromatography, and inorganic impurities (by heavy metals test, sulfated ash, atomic spectroscopy, etc.). Additional tests may include X-ray diffraction and thermogravimetry such as thermogravimetric analysis (TGA) or differential scanning calorimetry (DSC). As previously mentioned, the assigned content is usually calculated using the principle of mass balance (subtracting the total percentage of organic, inorganic, water, and solvent impurities).

Impurity/Degradation Product Standards

Impurity reference standards may be available from one of the pharmacopoeias or a commercial reference material supplier. If not commercially available, a technical grade of the impurity may be available from a chemical supplier, which could then be further purified.

If not commercially available, it will be necessary to manufacture the impurity. A literature search may identify a possible route of synthesis from starting materials sourced from in-house or obtained commercially. The likely steps in the manufacture and testing will include:

1. Sourcing starting materials
2. Analysis to check their specification and suitability
3. Small laboratory scale synthesis
4. QC on the product
5. Larger scale laboratory synthesis
6. Purification (e.g., preparative chromatography as a final stage clean-up, recrystallization, and drying)
7. Characterization and analysis for suitability
8. Packaging into glass ampoules or vials
9. Stability evaluation
10. Documentation with certificate of analysis

INFORMATION ON REFERENCE MATERIALS/STANDARDS IN MODULES 3.2.S.5 AND 3.2.P.6 IN A CTD REGISTRATION FILE

3.2.S.5 Reference Standards or Materials

Information on the reference standards used for testing the drug substance should be provided. This should include both official (compendial) reference standards and in-house standards.

Full information should be included on the characterization of any nonofficial reference standard. For purity (assay) standards this should include:

- A brief description of the method of manufacture of the reference product if it is different from the production process used for the drug substance. Any additional purification procedures (e.g., additional recrystallization or drying) should be described.
- Legible copies of relevant spectra (e.g., IR, UV, ^{1}H-NMR, ^{13}C-NMR, MS) and chromatograms.
- Data to establish the chemical structure such as formula, molecular weight, etc. Information to establish that the structure would include elemental analysis and relevant spectroscopy with detailed interpretation of the data.
- Data establishing purity such as mass balance (calculation of purity from total levels of all impurities), phase solubility analysis, and appropriate thermometric analytical procedures.
- A physical description of the drug substance (physical form and color).
- Appropriate physical data such as melting point, refractive index, optical rotation.
- A detailed description of the analytical procedures used.

A certificate of analysis would normally be provided for each batch of reference standard that had been used in the development program.

For impurity reference standards, appropriate analytical information (but less than the above) would be needed. Impurity reference standards are often made in small quantity for the analytical validation studies on the drug substance and dosage form. For the routine QC testing of impurity levels in batches of the drug substance, the "response factor" of the impurity is often used (based on the relative peak areas of the impurity and the drug substance under the same conditions of test), provided the peak can be identified unambiguously and that it is more than 0.2 and less than 5. This is the expected response of the impurity in relation to a response of unity for the drug substance itself.

3.2.P.6 Reference Standards or Materials

A list of reference standards should be provided in this section for any standards that are unique for testing the drug product. These could include impurities and degradation products from drug substance–excipient interactions, degradation products formed during manufacture, degradation products formed during product storage, any excipient impurity, or any impurity leached from the container–closure system that is included in the drug product specification, for example, vinyl chloride monomer or di(2-ethylhexyl) phthalate (DEHP) plasticizer extractable from PVC containers into liquid pharmaceutical products.

Information should be provided on the manufacture and characterization of any drug product reference standards.

REFERENCES

1. ISO Guide 30: 1992. Terms and Definitions Used in Connection with Reference Materials.
2. ISO/REMCO List of Producers of Certified Reference Materials and Computerized Databank "Code of Reference Materials" (COMAR). http://www.bam.de/comar. Accessed March 2008.
3. Catalogue Chemical Reference Substances, Infrared Spectra, Biological Reference Preparations, European Department for Quality of Medicines, Strasbourg. http://www.edqm.eu/medias/fichiers/CRS.pdf. Accessed March 2008.
4. USP Reference Standards and Technical Data Sheets. http://www.usp.org/products/referenceStandards/catalog.html. Accessed March 2008.
5. British Pharmacopoeia Commission Laboratory, Catalogue of Reference Substances. http://www.bpclab.co.uk/cat.html. Accessed March 2008.
6. Japanese Pharmacopoeia Reference Standards. http://www.sjp.jp/eng/04/010103.htm. Accessed March 2008.
7. Substances de Reference de la Pharmacopée Française. http://afssaps.sante.fr/htm/pharma/lise-substances-reference-france.pdf. Accessed March 2008.
8. R.T. Corp Catalogue of Pharmaceutical Certified Reference Substances. http://www.rt-corp.com/products/C28.aspx. Accessed March 2008.
9. LGC Standards Reference Materials listings. http://www.lgcstandards.com/BrowseProducts.aspx. Accessed March 2008.
10. WHO Catalogue of International Biological Reference Preparations. http://www.who.int/bloodproducts/catalogue/en/index.html. Accessed March 2008.
11. Foster T, Brown W. USP Dissolution Calibrators: Re-examination and Appraisal. Dissolution Technol 2005; 15(1):6–8.

19 Drug Substance and Drug Product Stability

Anthony C. Cartwright
Pharmaceutical Regulatory Consultant, Harpenden, U.K.

3.2.S.7 DRUG SUBSTANCE STABILITY AND 3.2.P.8 DRUG PRODUCT STABILITY

These sections in a Common Technical Document (CTD) registration file include the stability protocols, the results of the stability studies, the protocols for any future stability testing, commitment to carry out future stability testing, and analysis of the stability data to indicate the conclusions as to retest date (for the drug substance) or shelf life (for the drug product). These sections also include the proposed text of the storage conditions for the data sheet/Summary of Product Characteristics for healthcare professionals and the Patient Information Leaflet (PIL) for patients.

The CTD includes these three sections in the drug substance 3.2.S.7 Stability module:

3.2.S.7.1 Stability Summary and Conclusions—A summary of the results from the forced degradation studies, stress studies, accelerated testing, and long-term real-time testing, together with conclusions as to retest date and storage conditions for the drug substance.
3.2.S.7.2 Postapproval Stability Protocol and Stability Commitment—Postapproval stability testing protocol and the commitment as to any future drug substance batch testing.
3.2.S.7.3 Stability Data—Detailed results of the stability studies on the drug substance, with analytical methods and their validation, results of studies to determine the degradation pathways of the drug substance; the accelerated and real-time storage testing, together with the analysis of the stability data that led to the proposed shelf life and storage conditions.

The CTD includes three similar sections in the drug product 3.2.P.8 Stability module:

3.2.P.8.1 Stability Summary and Conclusions
3.2.P.8.2 Postapproval Stability Protocol and Stability Commitment
3.2.P.8.3 Stability Data

ICH STABILITY GUIDELINES

Stability testing was the first "Quality" topic on the ICH work program in 1990 as there were considerable differences between the EU, United States, and Japanese requirements at the time. Drug products are often marketed in several strengths and package types. The need for testing combinations of strength, packaging, batch, storage conditions, and test intervals leads to an expensive stability chamber investment and analytical program. When this was combined with the complications of different testing requirements for different countries, the amount of testing needed was enormous. Thus, as stability testing is one of the major costs in developing a new product, and, as many subsequent changes to a drug product also have to be supported with new stability data, it was important to minimize these costs and avoid delays by developing an internationally harmonized approach to stability testing.

The first ICH guideline on stability testing is the Q1A Stability Testing of New Drug Substances and Products guideline (1) adopted by the ICH Steering Committee in October 1993. This is the "parent" guideline to which all of the later ICH and national/regional guidelines refer, and contains all of the basic requirements-–the core stability package. This guideline has been subject to two major revisions and the current version at the time of writing is now denoted Q1A(R2). The subsequent ICH guidelines are

- Q1B: Stability Testing—Photostability Testing of New Drug Substances and Products (2)
- Q1C: Stability Testing for New Dosage Forms (3)

- Q1D: Bracketing and Matrixing Designs for Stability Testing of New Drug Substances and Products (4)
- Q1E: Evaluation of Stability Data (5).

STORAGE TEST CONDITIONS FOR DIFFERENT CLIMATIC ZONES

Haynes in his seminal 1971 paper (6) devised the "virtual temperature" concept to reflect annual temperature variations in various locations in the world, and which could then be used as the basis for standardization of stability testing. Haynes produced the following formula:

$$T_k = \frac{\Delta E/R}{-\ln\left(\dfrac{e^{-\Delta E/RT_1} + e^{-\Delta E/RT_2} + \cdots + e^{-\Delta E/RT_{12}}}{12}\right)}$$

where T_k is the mean kinetic temperature (MKT), which is the equivalent of Haynes' virtual temperature in degrees Kelvin; ΔE is the activation energy in kJ/mol; R is the gas constant in kJ/(mol · K); and $T_{1\text{–}12}$ is the mean temperature for a single month.

Thus, from the mean temperature data for each month the mean kinetic temperature for the whole year can be calculated. This enables an isothermal testing condition to be devised, which is equivalent in its chemical reaction kinetic effect to storage at a variety of temperatures throughout the year. The kinetic mean temperature takes account of the effect of "excursions" of temperature on the chemical degradation kinetics of the drug substance and the drug product. These excursions of temperature are those experienced during the hotter times of the year, when the temperatures will be considerably greater than the annual kinetic mean temperature.

Using the concept developed by Haynes, Grimm (7–9) utilized the climatic zones classification originally devised by Futscher and Schumacher (10) for the purposes of practical pharmaceutical product testing, and calculated the temperature and humidity conditions for each zone. These zones are

Zone I: Temperate zone
Zone II: Mediterranean/subtropical zone
Zone III: Hot dry zone
Zone IV: Hot humid/tropical zone

The conditions originally proposed by Grimm are shown in Table 1, and these were utilized by the ICH Topic Group to devise the first version of the Q1A stability guideline.

In the EU, the northern countries (i.e., Belgium, Denmark, Sweden, Finland, Germany, Hungary, the Netherlands, the Baltic countries, and the United Kingdom) all fall into Zone I. The Mediterranean EU countries (e.g., Spain, Portugal, and Greece) all fall into Zone II. In continental North America, the northern states of the United States (e.g., Illinois) fall into Zone I, and the southern states (e.g., Texas and California) fall into Zone II. Most of Japan falls into Zone II. Thus, stability data generated under Zone II long-term storage conditions provides data suitable to support the manufacture, distribution, and sales of pharmaceutical products in North America, the EU, and Japan. As approximately 85% of the sales of New Active Substance (NAS) products comprise sales in the three ICH regions of continental North America, Europe, and Japan, it is clear that Zone II storage provides data suitable for the majority of the developed world markets.

The first version of the ICH Q1A stability guideline thus adopted the storage conditions shown in Table 2 for testing pharmaceutical products.

Table 1 World Climatic Zones According to the Original Grimm Analysis (8, 9)

Climatic conditions	Zone I	Zone II	Zone III	Zone IV
Mean annual temperature (°C)	20.5	20.5–24	>24	>24
Kinetic mean temperature (°C)	21	26	31	31
Mean annual relative humidity (%)	45	60	40	70

Table 2 Original ICH Q1A Stability Test Storage Conditions

Test condition	Storage conditions (temperature and relative humidity)
Long-term real time	25°C ± 2°C/60% RH ± 5% RH
Accelerated	40°C ± 2°C/75% RH ± 5% RH
Intermediate	30°C ± 2°C/60% RH ± 5% RH

STORAGE TEST CONDITIONS FOR CLIMATIC ZONES III AND IV

Work on stability testing of pharmaceutical products was started by the WHO in 1988 and the guideline entitled *WHO Guidelines on Stability Testing for Well Established Drug Substances in Conventional Dosage Forms* was adopted in 1996 (11).

In 2000, discussions began between the ICH Topic Working Group and the WHO on modifications to the WHO guideline, particularly for the long-term storage condition for climatic Zones III and IV. The condition 30°C/60% RH (the intermediate ICH test condition) was proposed rather than 30°C/70% RH. The WHO consulted widely on this proposal.

In 2001, as a result of further consultation, it was proposed to change the real-time storage conditions for Zone IV from 30°C/70% RH to 30°C/65% RH. The ICH Working Group agreed to amend its intermediate testing condition for Zones I and II countries to 30°C/65% RH as shown in Table 3. The WHO expert committee then agreed to change the storage conditions for Zone IV countries in the WHO guideline to 30°C/65% RH.

The Association of South East Asian Nations (ASEAN) countries are all in climatic Zone IV and they had expressed concern as to whether the conditions in the ICH and WHO guidelines met their needs, as they felt that the climatic conditions in some of their member countries such as Indonesia were more extreme in terms of humidity. The ASEAN countries met in Jakarta in January 2004 and concluded that they would require long-term testing at 30°C/75% RH. As a result of the ASEAN decision, WHO circulated a further document in early 2004 with a view to modify the WHO guideline, and this was discussed in December 2004 at a meeting involving ASEAN, ICH, Mercosur, and other countries. It was agreed to add a new climatic Zone IVb (30°C/75% RH) to accommodate hot and very humid areas of the world. The 40th WHO Expert Committee on Specifications for Pharmaceutical Preparations meeting in Geneva in October 2005 decided to split climatic Zone IV into two zones—IVa as the standard long-term testing condition and IVb for the hot and humid countries. The WHO has circulated the current adopted conditions (Zone IVa at 30°C/65% RH and Zone IVb at 30°C/75% RH) to all of its member states and asked them to state the long-term storage conditions required for marketing in their country. Zahn and colleagues (12, 13) have devised a formula to calculate safety margins for temperature and partial vapor pressure, and have proposed long-term stability testing conditions for each selected country, taking into account the worst case for each region as well as its classification into Zone IVa or IVb.

ASEAN Guideline

The ASEAN guideline (14) was adopted in February 2005 and makes recommendations on the core stability package for drug products. The guideline covers NCEs, generics, and variations, but it excludes vitamin and mineral preparations. It summarizes the information to be submitted in a marketing authorization for drug products in the ASEAN countries and includes an example of a stability testing protocol, a report format, a reduced design, and on extrapolation of data. The general storage conditions are summarized in Table 4.

Table 3 Current ICH Q1A General Stability Test Storage Conditions (1)

Test condition	Storage conditions (temperature and relative humidity)
Long-term real time	25°C ± 2°C/60% RH ± 5% RH
Accelerated	40°C ± 2°C/75% RH ± 5% RH
Intermediate	30°C ± 2°C/65% RH ± 5% RH

Table 4 ASEAN Country Stability Test Storage Conditions (14)

Test condition	Storage conditions (temperature and relative humidity)
Long-term real time	30°C ± 2°C/75% RH ± 5% RH
Accelerated	40°C ± 2°C/75% RH ± 5% RH

Eastern Mediterranean Region Guideline

The Eastern Mediterranean Regional WHO Committee in its 53rd session meeting in February 2006 devised draft guidelines on stability testing of active substances and pharmaceutical products (15). The consultation involved experts from Bahrain, Egypt, Jordan, the Islamic Republic of Iran, Kuwait, Lebanon, Morocco, Oman, Pakistan, Saudi Arabia, Sudan, and the United Arab Emirates. The recommended general storage conditions are shown in Table 5.

Brazil

The Director of the Brazilian Health Surveillance Agency published the Guide to Stability Studies in November 2004 (16). The storage conditions for testing of solid, semisolid, liquid, or gaseous pharmaceutical forms in semipermeable packaging are as defined in Table 6.

China

The latest Chinese Pharmacopoeia 2005 (CP 2005) gives guidance on stability testing for marketing in China (17). China belongs in general to climatic Zone II. The storage test conditions are set out in Table 7.

NEW DRUG APPLICATIONS

Purpose of Stability Testing

The purpose of stability testing is to provide data on how the quality of a drug substance or a drug product varies with time under the influence of environmental factors such as temperature, humidity, and light. These data enable the establishment of a retest period for the drug substance (the period of time during which batches of the drug substance are expected to remain within specification) or the shelf life for the drug product (the time period during which batches of the drug product are expected to remain within specification) and the recommended storage conditions.

Drug Substance Testing

Drug substance testing consists of:

- Stress testing to identify likely degradation products, establish the chemical degradation pathways, the intrinsic stability of the molecule, and to validate the stability-indicating nature of the analytical procedures.
- Formal stability studies on batches of the drug substance whose quality is representative of material made to the final commercial manufacturing process.

Stress Testing

Stress testing will normally be carried out on a single batch of the drug substance. Bakshi and Singh (18) have reviewed and classified the steps involved in stress testing and development of

Table 5 European Mediterranean Region of WHO Recommended Stability Test Storage Conditions (15)

Test condition	Storage conditions (temperature and relative humidity)
Long-term real time	25°C ± 2°C/60% RH ± 5% RH or 30°C ± 2°C/65% RH ± 5% RH
Accelerated	40°C ± 2°C/75% RH ± 5% RH
Intermediate	30°C ± 2°C/65% RH ± 5% RH

Table 6 Brazilian Recommended Stability Storage Conditions (16)

Pharmaceutical form	Label storage condition (°C)	Packaging	Accelerated testing (temperature and RH)	Long-term testing (temperature and RH)
Solid	15–30	Semipermeable	40°C ± 2°C/75% RH ± 5% RH	30°C ± 2°C/75% RH ± 5% RH
Solid	15–30	Impermeable	40°C ± 2°C	30°C ± 2°C
Semisolid	15–30	Semipermeable	40°C ± 2°C/75% RH ± 5% RH	30°C ± 2°C/75% RH ± 5% RH
Semisolid	15–30	Impermeable	40°C ± 2°C	30°C ± 2°C
Liquids	15–30	Semipermeable	40°C ± 2°C/75% RH ± 5% RH	30°C ± 2°C/75% RH ± 5% RH
Liquids	15–30	Impermeable	40°C ± 2°C	30°C ± 2°C
Gases	15–30	Impermeable	40°C ± 2°C	30°C ± 2°C
All pharmaceutical forms	2–8	Impermeable	25°C ± 2°C	5°C ± 3°C
All pharmaceutical forms	2–8	Semipermeable	20°C ± 2°C/60% RH ± 5% RH	5°C ± 3°C
All pharmaceutical forms	−20	All	−20°C	−20°C

suitable analytical procedures. For our purposes, the steps in stress testing can be conveniently classified as follows.

Step 1: Evaluation of the Drug Substance Structure to Assess Likely Degradation Products
The functional groups and key components in the drug structure will give some useful clues as to likely degradation mechanisms. Oxidation reactions are probably the most common—for example, thiols and thioethers are susceptible to oxidation. Olefins, aryl acetic acids, and N-oxides can be subject to photodegradation. Hydrolytic reactions are also common for drug degradation--amides, esters, lactams, and lactones are susceptible to hydrolysis. Other degradation reactions include radical reactions (chemistry of a free radical species), condensation, reduction, isomerization, addition, and substitution. If a drug substance is chemically related to an existing drug substance, this may provide valuable hints to the degradation behavior of the new drug and also suggest possible analytical methods. Consulting reference sources such as the various volumes of *Analytical Profiles of Drug Substances and Excipients* (19), Baertschi (20), Connors et al.'s book *Chemical Stability of Pharmaceuticals* (21), or Xu and Trissel's book *Stability-Indicating HPLC Methods for Drug Analysis* (22) can be a very useful first step. Pole et al. (23) have given details of the construction of an expert system, which is used in-house to predict possible degradation products from new drug substances in development.

Table 7 Recommended Storage Conditions for China (17)

Test condition	Storage conditions (temperature and relative humidity)
Long-term real time	25°C ± 2°C/60% RH ± 10% RH (note difference in allowed fluctuation in humidity compared to ICH limits of ±5%)
Accelerated	40°C ± 2°C/75% RH ± 5% RH
Intermediate	30°C ± 2°C/65% RH ± 5% RH
For products sensitive to temperature	4–8°C (note difference to ICH conditions of 2–8°C)

Notes: Three batches of product should be tested at 40°C ± 2°C/75% RH ± 5% RH for the duration of 6 mo, testing at 1st, 2nd, 3rd, and 6th mo (more testing points than in the ICH guideline). For drug products in semipermeable containers, the accelerated condition is 40°C ± 2°C/20% RH ± 2% RH (ICH: not more than 25% RH).

Step 2: Collecting Information on Physicochemical Properties of the Drug Substance
Information on parameters such as pK_a, log P, solubility in a range of solvents, and UV spectrum will help in method development of a Stability Indicating Analytical Method (SIAM).

Step 3: Stress Studies
Hydrolytic behavior can be studied by refluxing the drug in solution or suspension in water (neutral conditions), 0.1 N hydrochloric acid (HCl), and 0.1 N sodium hydroxide (NaOH) for eight hours, or longer if no degradation is seen. Oxidation is studied by using hydrogen peroxide in the range of 0.01% to 1.0%. Photodegradation is normally studied by exposing the drug substance and drug solutions or suspensions in a light stability chamber to give a combination of ultraviolet (UV) and visible light. For each stress condition, four samples should be generated—blank solution, the stressed blank solution, the initial sample of the drug solution/suspension, and the stressed sample of the drug solution/suspension. The analytical comparison of these samples enables the assessment of possible degradation.

Solid-state thermal stress degradation studies are carried out by heating powder samples in a tube at temperatures 10°C above the accelerated testing conditions (e.g., 50°C, 60°C, and 70°C).

Step 4: Analytical Development and Optimization of a SIAM
A chromatographic method (e.g., reversed phase HPLC) is usually employed, which will separate the active substance from impurities and degradation products, and produce good resolution and well-separated peaks. The method is optimized by suitable choice of column, solvents, and buffers in the mobile phase, etc.

Step 5: Identification of Degradation Products
Liquid chromatography-Mass Spectroscopy (LC-MS) or tandem LC-MS-MS can be used to identify the possible degradation products by determining their molecular weight and fragmentation information. Liquid chromatography-Nuclear Magnetic Resonance (LC-NMR) can be used to provide further identification.

Step 6: Determination of Mass Balance
A key issue in the stress stability studies is mass balance--whether the change in assay in the stress stability samples is explained by the increase in degradation products. This should be discussed in the stability report.

A lack of mass balance may be due to co-elution of degradation product with the main active substance peak, different response factors for the drug substance and degradation products to the detector system, due to the formation of nonchromophore degradation products, or due to insoluble degradation products. Where there is a lack of material balance, further analytical tests should be carried out to try to identify the reasons and to find any "hidden" degradation products. Diode array detection is useful to check whether the main drug substance peak is hiding any co-eluting degradation product and similarly for other peaks.

Formal ICH Stability Studies

Selection of Batches
Formal stability studies on a new drug should be carried out on three batches. These batches should be on a minimum of pilot scale and use the same synthetic route as proposed for commercial manufacture. The quality of these batches must be representative of future commercial production batches. The data on these batches comprises the primary stability data.

Data from supporting stability studies can also be provided and this may comprise data on earlier synthetic route material, laboratory scale batches, and batches stored in containers other than those proposed for routine production batches. It is useful to include such data particularly where the primary data are of limited duration, and it is possible to claim a longer retest date by showing that the rate and extent of degradation is similar between the primary and supporting batches of drug substance.

Container Closure System

The drug substance should be packaged in a container closure system that simulates the packaging proposed for commercial storage. Where the proposed commercial bulk drug substance container closure is a polyethylene (PE) bag in a drum, it is possible to simulate this packaging but care should be taken not to assume that small PE bags sold for laboratory use will be the same as the larger commercial bags, because they may contain added components such as mould release agents, which can interfere with the assay. It is often better to use a section cut from the corner of a larger commercial bag for the stability studies.

Specification

The samples of drug substance on stability test should be analyzed before being placed on the study using the proposed specification and test methods. The tests to be applied during the formal stability studies should include those in the specification that are susceptible to change and will include physical tests (appearance, optical rotation), chemical tests (assay, degradation products), biological tests, and microbiological attributes as appropriate to the drug substance under test. The analytical procedures for the assay and degradation products should have been shown to be stability indicating. All of the analytical procedures should have been validated. Where different analytical procedures have been used during the development of a drug substance to those proposed in the final specification, it may be necessary to include a further analytical validation report in the 3.2.S.7.3 Stability section of the registration file.

Storage Conditions and Testing Frequency

The storage conditions and testing frequencies recommended in the ICH Q1A guideline are summarized in Tables 8 to 10. The purpose of accelerated testing is to evaluate the effects of short-term excursions outside the label storage, for example, during international shipping and distribution.

The tolerances around the storage temperatures and humidities were chosen to allow for the temperature accuracy and gradients within commercial fan/blower air circulation humidity test chambers and rooms. The actual temperature and humidity should be monitored throughout the stability storage period, usually by connecting the controllers to a remote computer system for data logging. Commercial chambers are fitted with safety cut off thermostats and an audiovisual alarm to indicate any malfunction. Short-term spikes due to opening doors to add or remove samples are inevitable but any larger excursions (such as those caused by a power failure) should be described in the stability report.

Where companies carry out 25°C/60% RH testing as the long-term storage condition and 40°C/75% RH as the accelerated condition, they usually put samples on the intermediate 30°C/65% storage test condition as well, but only start testing samples from the intermediate

Table 8 ICH Formal Stability Studies—the General Case (1)

Test condition	Storage conditions (temperature and relative humidity)	Minimum time period needed for registration file (mo)	Testing frequency
Long-term real time	25°C ± 2°C/60% RH ± 5% RH or 30°C ± 2°C/65% RH ± 5% RH	12	Every 3 mo during first year, then every 6 mo during second year, then yearly
Accelerated	40°C ± 2°C/75% RH ± 5% RH	6	0, 3, and 6 mo
Intermediate	30°C ± 2°C/65% RH ± 5% RH	6	0, 6, 9, and 12 mo from a 12-mo study

Notes: Long-term stability studies can be carried out at either 25°C ± 2°C/60% RH ± 5% RH or 30°C ± 2°C/65% RH ± 5% RH. If 30°C ± 2°C/65% RH ± 5% RH is chosen as the long-term storage condition, there is no intermediate condition testing. If long-term testing is carried out at 25°C ± 2°C/60% RH ± 5% RH, and a significant change in the drug substance occurs during the accelerated testing, then additional testing at the intermediate testing condition is needed).

Table 9 ICH Formal Stability Studies—Drug Substances Intended for Refrigerated Storage (1)

Test condition	Storage conditions	Minimum time period needed for registration file (mo)	Testing frequency
Long term	5°C ± 3°C	12	Every 3 mo during first year, then every 6 mo during second year, then yearly
Accelerated	25°C ± 2°C/60% RH ± 5% RH	6	0, 3, and 6 mo

condition if there is a significant change under the accelerated test condition. A significant change is where a batch of drug substance falls outside its proposed specification.

Stability Commitment

Where the available long-term stability data do not cover the entire proposed retest period, a commitment should be included to continue the stability studies until they cover the retest period.

Where the stability data includes only pilot scale batches of drug substance, a commitment should be given to carry out stability studies on three production batches on long-term stability studies through the proposed retest period. If some production batch stability data is available, a commitment should be given to continue testing until data on a total of three batches are available for the proposed retest period.

Evaluation

The purpose of the formal stability study is to determine a retest period and label storage instructions based on the data from a minimum of three batches. The retest period so generated will be applicable to all future batches of the drug substance manufactured. The batch variability affects the confidence that any individual batch will remain in specification over its retest period.

Where the drug substance is very stable and batches show little degradation and minimal variability, the data may not need any formal statistical analysis. However, a justification for omitting statistical analysis should be included in the report.

Where the drug substance batches some changes during storage then it is useful to carry out statistical analysis of the data and the options for this are described later in this chapter.

Labeling

Each country/region has its own specific storage labeling requirements for the bulk drug substance, for example, in the EU there is a guideline (24), and the USP has recommended wording for substances sold in the United States.

Drug Product Testing

Drug Substance–Excipient Compatibility Studies

Part of the development of the drug product is a series of drug substance–excipient compatibility studies using binary mixtures of the drug substance with candidate excipients. These will show

Table 10 ICH Formal Stability Studies: Substances Intended for Freezer Storage (1)

Test condition	Storage conditions	Minimum time period needed for registration file (mo)	Testing frequency
Long term	−20°C ± 5°C	12 mo	Every 3 months during first year, then every 6 mo during second year, then yearly

major incompatibilities. If there are findings that relate to excipients, which are actually used in the drug product, these should be reported in the 3.2.P.8 Drug Product Stability module.

Formal Storage Studies

Selection of Batches

Data from stability studies should be reported from at least three batches of the drug substance, packaged in the same container(s) and closure(s) as proposed for marketing. Two of the three batches can be pilot scale and one can be smaller (laboratory scale), if justified. For a capsule or tablet drug product, pilot scale is defined as one-tenth of commercial scale or 100,000 (whichever is the larger). Where possible drug product batches should be manufactured using different batches of the drug substance. Stability studies should be reported for each individual strength and container size unless a bracketing or matrixing design is used. These data comprise the primary stability data package.

Supporting stability data can also be supplied such as drug product to earlier formulations or using different container closure systems. Where an oral dose form is to be marketed in both the United States and Europe, the marketing pack is likely to be different (e.g., blister packs in Europe, HDPE bottles for the United States) and thus stability in the "other" container closure can be used as supportive data.

Container Closure System

Stability testing should be carried out using the dosage form packaged in the primary container proposed for marketing. In some cases, the product should also be packed in its secondary container.

During the ICH discussions prior to the adoption of Q1A, the Japanese industry, and MHLW advocated the use of "naked product testing"--testing the product outside its primary packaging--reasoning that if the product withstood this stress testing it would usually be more stable in the protective primary packaging.

Storage Conditions and Testing Frequency

The storage conditions and recommended testing frequencies for drug product testing are essentially the same as for the drug substance:

General case—Table 8,
Drug products for refrigerated storage—Table 9, and
Drug products for freezer storage--Table 10.

If long-term storage is carried out at 25°C/85% RH and "significant change" in the drug product occurs after storage at 40°C/75% RH, then additional testing at the intermediate (30°C/65% RH) condition is needed. A minimum of six months data is needed. As for drug substance testing it is usual for companies to place samples of the drug product on test in 30°C/65% RH storage, but only to start testing samples if there is a significant change in drug product samples stored at 40°C/75% RH. A significant change is a 5% change in assay from the initial value or any other failure to meet the acceptance criteria in the proposed specification (e.g., change in appearance, change in physical attributes, failure to meet criteria for dissolution testing for 12 dosage units).

For drug products in impermeable container--where sensitivity to moisture or solvent loss is not an issue—the drug product can be packaged in impermeable containers and, for stability testing, these can be stored under controlled temperature without any control over humidity (e.g., 25°C instead of 25°C/60% RH for long-term storage, and 40°C instead of 40°C/75% RH for accelerated storage).

For drug products packaged in semipermeable containers (drug products in plastic bags or pouches, LDPE ampoules, plastics bottles and vials), water loss is a potential issue in addition to the usual physical, chemical, and microbiological stability parameters. The recommended storage conditions are shown in Table 11. If significant change in moisture content occurs (5% or more) after storage at 40°C/75% RH, then testing at the intermediate condition is not needed if the only change is water loss. However, data are needed to show the water loss through the

Table 11 ICH Formal Stability Studies—Drug Products Packed in Semipermeable Containers (1)

Test condition	Storage conditions (temperature and relative humidity)	Minimum time period needed for registration file (mo)	Testing frequency
Long-term real time	25°C ± 2°C/40% RH ± 5% RH or 30°C ± 2°C/35% RH ± 5% RH	12	Every 3 mo during first year, then every 6 mo during second year, then yearly
Accelerated	40°C ± 2°C/not more than 25% RH ± 5% RH	6	0, 3, and 6 mo
Intermediate	30°C ± 2°C/65% RH ± 5% RH	6	0, 6, 9, and 12 mo from a 12-mo study

Notes: Long-term stability studies can be carried out at either 25°C ± 2°C/40% RH ± 5% RH or 30°C ± 2°C/35% RH ± 5% RH. If 30°C ± 2°C/35% RH ± 5% RH is chosen as the long-term storage condition, there is no intermediate condition testing).

shelf life when stored at 25°C/40% RH. Changes of 5% or more for small containers of 1 ml or less (e.g., single dose eye-drop products) over three months storage at 40°C/not more than 25% RH can be accepted, if justified.

The ICH Q1A guideline offers an alternative approach to studying at the reference RH values given in Table 11. This is to perform the studies at another RH value and then calculate the water loss at the reference RH (e.g., by determining the permeation coefficient for the container closure system, or by using the calculated water loss rates at two relative humidity conditions at the same temperature).

Stability Commitment

As with the drug substance there needs to be a commitment to make available data on three production scale batches of the drug product to the end of shelf life if these data are not included in the registration file.

Where the stability data include only pilot scale batches of drug product, a commitment should be given to carry out stability studies on three production batches on long-term stability studies through the proposed shelf life. If some production batch stability data is available, a commitment should be given to continue testing until data on a total of three batches are available for the proposed shelf life and to submit information on any data showing deviation from the agreed specification.

Evaluation

The purpose of the formal stability study is to determine a shelf life and label storage instructions based on the data from a minimum of three batches. The shelf life so derived will be applicable to all batches of the drug product manufactured in the future. The batch variability affects the confidence that any individual batch will remain in specification.

Where the drug product is very stable and batches show little degradation and minimal variability, the data may not need any formal statistical analysis. However, a justification for omitting statistical analysis should be included in the report.

Where the drug product batches do show some changes during storage, then it is useful to carry out statistical analysis of quantitative attributes of the stability data (such as the assay, degradation products), and the options for this are described later in this chapter.

Statements/Labeling

As with the drug substance, storage statements for labeling are defined nationally/regionally and are based on the evaluation of the stability data—both from the long-term storage results and from the effect of excursions and spikes of temperature as predicted from the accelerated storage test results.

EXISTING DRUG APPLICATIONS

It was not possible to agree comprehensive ICH guidelines on existing drug substances and products as the Food and Drug Administration (FDA) had its own requirements for generic drug products that it wished to retain. Each ICH region has its own national guidelines on these products, as do the ASEAN countries for drug products, the WHO Regional Office for the Eastern Mediterranean (EMRO) and the WHO itself for most other countries (11). All of the regional and national guidelines are based on the ICH Q1A guideline. The storage conditions for each country were reviewed earlier in this chapter.

EU Guideline Requirements for Existing Drug Substances and Finished Products (25)

Drug Substances

For pharmacopoeial drug substances, an applicant can either specify that the batch of drug substance complies with a European Pharmacopoeia (Ph Eur) or European national pharmacopoeia monograph, which has been shown to be suitable (i.e., which controls relevant degradation products) just prior to manufacture of the drug product, or can fix a retest date based on the results of long-term and accelerated testing. Stress testing is not needed on the drug substance if the degradation products are specified in the official monograph or where there is data in the scientific literature to show the proposed mechanism of degradation. If neither is available stress testing should be carried out on the drug substance.

For existing drug substances, which are not described in the Ph Eur or any European national pharmacopoeia, stress testing may be needed if there is no literature data, and also formal long-term and accelerated testing as specified in the ICH Q1A guideline.

The EU guideline offers two options for formal stability testing of existing drug substances:

- testing two production batches of the drug substance for a minimum of six months or
- testing three pilot scale batches of the drug substance for a minimum of six months.

The requirements for storage conditions, container/closure, testing frequency, stability specification, significant change, the stability commitment, and evaluation are the same as in Q1A.

Drug Product

The EU guideline offers two options for testing drug products containing existing drug substances.

For conventional dosage forms (immediate release dosage forms, solutions) containing drug substances known to be stable, a minimum of six months stability data on at least two pilot scale batches is needed.

For critical dosage forms or when the drug substance is known to be unstable, a minimum of 12 months stability data on three primary batches is needed; two of the batches should be at least pilot scale, one can be at laboratory scale.

The requirements for storage conditions, container/closure, testing frequency, stability specification, significant change, the stability commitment, and evaluation are the same as in Q1A.

ASEAN Guideline

The ASEAN guideline (14) makes no recommendations on testing the drug substance, and the recommendations for numbers of batches, batch sizes, and length of testing generic products and variations are identical to those stated in the EU existing drug products guideline. The long-term storage condition is, as previously mentioned, 30°C/75% RH. The guideline includes some detailed lists of stability test parameters for different dosage forms such as tablets, capsules, emulsions, oral solutions and suspensions, oral powers for reconstitution, metered-dose inhalers, nasal aerosols, nasal sprays, ophthalmic and otic products, suppositories, small-volume parenterals, large-volume parenterals, and transdermal patches.

Moisture protection of solid dosage form pharmaceutical products against high humidity is obviously crucial in these Zone IVb countries, and the guideline includes a very useful table of examples of different types of packaging material and the moisture permeability of different thickness materials.

The ASEAN guideline also includes an example of a suggested protocol for stability studies and a stability report format.

WHO Stability Guideline

The 1996 WHO guideline is currently under revision. A series of revisions were circulated in 2007 and 2008 (26), and the new text will be adopted in early 2009. The original WHO 1996 guideline provided a text for stability testing of well-established drug substances in conventional products; the new draft text covers both new drug products and existing drugs. It will include a list of long-term stability testing conditions that are required by each WHO Member State.

Drug Substance

The 2008 draft guideline recommends testing three primary batches of active pharmaceutical ingredient (API) manufactured at a minimum of pilot scale. The long-term storage conditions in the general case can be 25°C ± 2°C/60% RH ± 5% RH or 30°C ± 2°C/65% RH ± 5% RH or 30°C ± 2°C/75% RH ± 5% RH depending on the climatic conditions at which the API is intended to be stored. A minimum of 12 months long-term data is recommended. The intermediate test condition (if needed) is 30°C ± 2°C/65% RH ± 5% RH, with a minimum of six months data, and the accelerated test condition is 40°C ± 2°C/75% RH ± 5% RH, with a minimum of six months data. For existing active substances that are known to be stable, data from at least two primary batches should be provided. The guideline also includes recommendations for APIs intended for storage in a refrigerator or a freezer. The guideline recommends labeling statements that should be used depending on the results of the stability testing. If a product has been shown to be stable at 25°C/65% RH (long term) and 40°C/75% RH (accelerated), the draft guideline suggests labeling "Do not store above 25°C."

Drug Product

The draft guideline recommends studies on a minimum of three primary batches of the drug product in the proposed container closure system(s). Two of the three batches should be at least pilot scale and one can be smaller, if justified. The batches should be made if possible using different batches of API. The storage conditions will depend on the climatic conditions (i.e., Zone II, III, IVa, or IVb) at which the drug product is intended to be stored, and are the same as listed above for the drug substance.

In the case of conventional dosage forms with APIs that are known to be stable, data from at least two primary batches should be provided. Stability studies should be performed on each individual strength, dosage form, and container type and size of the finished product unless bracketing or matrixing is applied.

Long-term testing should cover a minimum of six months (for a finished product containing an API known to be stable) or 12 months duration at the time of submission, and should continue for a period of time sufficient to cover the proposed shelf life.

As with the ICH guideline, the WHO draft guidance also covers the requirements for products in impermeable containers, finished products in semipermeable containers, those intended for storage in a refrigerator, and those intended for storage in a freezer. The WHO draft also includes recommendations similar to those in the EU for in-use stability testing.

Annex 1 of the draft guideline lists the long-term testing conditions for the WHO Member States (countries) arranged by WHO Region (Africa, Americas, Eastern Mediterranean, Europe, South East Asia, and Western Pacific). Annex 2 of the draft guideline includes a useful list of examples of testing parameters for different dosage forms (tablets, capsules, emulsions, oral solutions, powders and granules for oral solution or suspension, metered dose inhalers, nasal sprays, topical, ophthalmic and otic preparations, suppositories, small-volume parenterals, large-volume parenterals, and transdermal patches). Annex 3 of the draft guideline is a model stability protocol and report of an active pharmaceutical ingredient. Annex 4 is a model stability protocol and report of capsules/tablets.

PHOTOSTABILITY TESTING

Work on the ICH Q1B Photostability guideline began in 1993 after the Q1A guideline had been accepted by the ICH Steering Committee. Prior to this a variety of light sources and exposures were used for photostability testing. The development of the guideline included a considerable amount of laboratory work in Japan and the United States to recommend suitable test conditions. The ICH Q1B guideline was adopted in 1996.

The guideline recommends a systematic approach to photostability testing on a new drug substance and product. The stages are as follows:

- Stress tests on the drug substance.
- Tests on the drug product outside of its immediate pack.
- If necessary, test on the drug product in its immediate pack.
- If necessary, test on the drug product in the marketing pack.

Commercial photostability test chambers follow one of the two ICH options for light sources to be used for stability testing.

Option 1: Any light source that will produce an output similar to the outdoor daylight Commission Internationale de l'Eclairage (CIE) D65 standard as defined in ISO 10977 (1993), or the indoor standard ID65. Lamps could be fluorescent with both UV and visible outputs, xenon, or metal halide.

Option 2: A combination of two light sources—A cool white fluorescent lamp designed to produce a similar output to ISO 10977, and a near UV fluorescent lamp having a spectral distribution from 320 to 400 nm with a maximum energy emission between 350 and 370 nm.

The exposure for confirmatory studies is not less than 1.2 million lux hours and an integrated near UV energy of not less than 200 watt hours per square meter. The exposure can be monitored with by using a chemical actinometer system, or by using suitable calibrated radiometers/flux meters. The ICH guideline gives details of a quinine actinometry standard and its use was established by Yoshioka et al. (27).

Drug Substance Testing

Forced Degradation Testing

This is part of stress testing of the drug substance in powder and in solution/suspensions. The design of these studies is not specified in the ICH guideline, as these are designed to deliberately degrade the sample.

Confirmatory Testing

These studies are under milder conditions than the stress testing above. The overall light exposure should be not less than 1.2 million lux hours, with a UV exposure of not less than 200 watt hours per square metre.

A sample of a batch of drug substance can be spread in a 3 mm or less layer in a Petri dish covered with plastic film or similar, covered with a plastic film and then placed in the photostability chamber. It is often useful to have a "dark control"—a sample covered in foil and placed in the chamber alongside the other sample. Both samples can then be inspected for appearance, and tested for assay and degradation products.

The results of the confirmatory testing will provide information on steps that need to be taken during manufacture and storage of the bulk drug substance and manufacture of the drug product.

Drug Product

A sequential test is carried out on the drug product—testing firstly the unpackaged product, then if necessary the product in its immediate container, then if necessary the product in its final marketing container.

A single batch of drug product should be tested using the confirmatory testing conditions. Tablets or capsules should be spread in a single layer (e.g., in a Petri dish covered with plastic film). Again, it may be useful to have a dark control sample of product (e.g., a sample covered in foil).

The exposed sample should be examined for any change in appearance, and tested for assay and degradation products.

The results of the photostability testing will show if any special labeling is needed.

IN-USE STABILITY TESTING

In-use stability data should be provided for multidose liquid products (e.g., eye drops, syrups), which may be affected by repeated opening and closing and removal of part of the contents. The repeated opening and closing may result in microbiological contamination and may affect the physicochemical stability of the product in the container.

The 2001 Committee for Human Medicinal Products (CHMP) Note for Guidance on *In-Use Stability Testing of Human Pharmaceutical Products* (28) contains detailed requirements for such products. It recommends testing two batches, at least one of which should be tested towards the end of its shelf life. Where the product has a number of container sizes, the size chosen for the in-use stability study should be the one that poses the greatest susceptibility to possible change.

The in-use test is carried out by withdrawing a sample equivalent to the dose at regular intervals (simulating the way that a patient might withdraw doses) and then testing the appropriate physical, chemical, or microbiological parameters over the proposed in-use shelf life, particularly appearance, assay, degradation products, level of antimicrobial preservative(s), level of any antioxidant, and pH.

An antimicrobial preservative efficacy study can be carried out at the end of the proposed in-use shelf life by bulking the remaining samples from one batch and then carrying out the challenge study. A suitable test is the Ph Eur general test 5.1.3 Efficacy of Antimicrobial Preservation. This test is more stringent than the equivalent USP test and therefore a product that complies with its requirements will be suitable for global marketing.

The findings of the in-use stability study should be reflected in the in-use storage warnings and an in-use shelf life should be proposed (although pharmacopoeial requirements should also be taken into account). The in-use warnings and recommended period of use should be stated on the container, the patient leaflet, and the data sheet/Summary of Product Characteristics. It may be useful to include space for the user to write a "use by" date on the container label--this can be a requirement in some jurisdictions.

BRACKETING AND MATRIXING STABILITY DESIGNS

A full stability study design is one where all design factors are tested at all time points. Thus, for a new drug product with three strengths, two different container closures, two sizes of each container closure, and three batches needed for each strength, the total number of combinations would be $3 \times 2 \times 2 \times 3 = 36$. A reduced testing design is one where all of the design factor combinations are not tested at all time points.

Bracketing designs are those where only the extremes of certain design factors (e.g., strength, container size) are tested at all time points. Thus, for a range of strengths, only the lowest and highest might be tested. Matrixing designs are ones where a selected subset of samples are tested at any one time point (sample pull point) and at later time point another subset of samples are tested with the intention that all variables are monitored during the study.

A reduced design is applicable mainly to reasonably stable products with little expected variability in the analytical data. Where a reduced design is being applied to a range of strengths of a product, the composition of the range of strengths should be broadly similar (i.e., none of the strengths should contain different major excipients or significantly different ratios of excipients), although minor differences in composition in terms of color or flavor are generally allowable. A justification for using a reduced testing design should be included in the stability report.

Bracketing Designs

An example of a bracketing design is shown in Table 12. In this example, only 12 out of 27 possible combinations are being tested, and thus the cost of the bracketed design stability study is only 44% of the potential cost of a full stability study.

Table 12 Example of a Bracketing Design

	Strength of tablet (mg)								
	5			10			25		
Container size	**Batch 1**	**Batch 2**	**Batch 3**	**Batch 1**	**Batch 2**	**Batch 3**	**Batch 1**	**Batch 2**	**Batch 3**
HDPE bottle 50 tablets	T	T	T				T	T	T
HDPE bottle 100 tablets									
HDPE bottle 250 tablets	T	T	T				T	T	T

Each batch of product on test should be tested at each time point as in a full design. *Abbreviation*: T, sample tested.

For bracketing studies, only the extremes in terms of strengths (lowest and highest) or pack sizes will be tested; thus, the assumption is that the stability of the extremes can represent the intermediates.

If the stability of one of the extremes in a bracketing design proves to be different, then the shelf life of the intermediate strength or pack size should be based on the stability of the least stable extreme.

Matrixing Designs

A matrixing design should be applied so that as far as possible each combination of factors is tested to the same extent over the duration of the study. An example of a matrixing design based on time points for two different product strengths is given in Table 13. It should be noted that the matrixing design is chosen so that at 12 months and 3 years the complete set of samples is analyzed, thus if the registration file is submitted with only 12 months data, there are sufficient data points to make an analysis of any data trends.

An example of a matrixing design based on three container sizes is shown in Table 14. Other examples of design are given in the ICH Q1D guideline (4).

There are risks in using reduced testing designs such as matrixing designs based on factors other than time points, and this may mean that there would be less precision in shelf-life estimates and a shorter predicted shelf life. In addition, the matrixing design may have insufficient statistical power to detect main factors such as batch to batch or different container size variability in stability. It may also not be possible to pool data to produce a combined estimate of shelf life.

STATISTICAL EVALUATION OF STABILITY DATA

The text of the Q1A guideline refers to statistical assessment of stability data and sets out some general principles as to how such evaluation might be carried out using regression analysis to establish a retest date (drug substance) or shelf life (dosage form). The 2003 ICH Q1E guideline on Evaluation for Stability Data (5) now defines detailed methods for carrying out such assessment.

Table 13 Example of One-Half Reduction Matrixing Design on Time Points

		Time point (mo)							
Strength (mg)	**Batch**	**0**	**3**	**6**	**9**	**12**	**18**	**24**	**36**
	1	T	T		T	T		T	T
50	2	T	T		T	T	T		T
	3	T		T		T	T		T
	1	T		T		T		T	T
100	2	T	T		T	T	T		T
	3	T		T		T		T	T

Abbreviation: T, sample tested.

Table 14 Example of a Matrixing Design for a Tablet Product with Three Strengths and Three Container Sizes

Strength	**25 mg**			**50 mg**			**100 mg**		
Container size	**50 tablets**	**250 tablets**	**1000 tablets**	**50 tablets**	**250 tablets**	**1000 tablets**	**50 tablets**	**250 tablets**	**1000 tablets**
Batch 1	T1	T2	T3	T2	T3	T1	T3	T1	T2
Batch 2	T2	T3	T1	T3	T1	T2	T1	T2	T3
Batch 3	T3	T1	T2	T1	T2	T3	T2	T3	T1
Key									

	Time point (months)							
	0	**3**	**6**	**9**	**12**	**18**	**24**	**36**
T1	T		T	T	T	T	T	T
T2	T	T		T	T		T	T
T3	T	T	T		T	T		T

Abbreviation: T, sample tested.

A systematic approach should be made to the evaluation of stability test data, assessing the totality of the physical, chemical, and microbiological data to derive the retest date or shelf life. The ICH Q1E guideline includes a decision tree in its Appendix A, which outlines a stepwise approach to stability data and when and how much extrapolation can be considered for the proposed retest data or shelf life.

In general, chemical parameters such as the assay, levels of degradation products, and content of antimicrobial preservative can be assumed to follow zero-order kinetics during long-term storage (29). Thus, these data can be evaluated using linear regression and poolability testing. Qualitative attributes (e.g., appearance) and microbiological parameters are not suitable for statistical analysis. Although the guideline mentions that attributes such as pH and dissolution "can be applied if appropriate," the author's experience is that this is seldom the case.

Statistical analysis is particularly useful to allow extrapolation of the long-term stability data to allow a retest date or shelf life to be claimed, which is beyond the period covered by the available data. This assumes that the pattern of degradation is the same throughout the shelf life and would need to be confirmed from data generated as part of the 3.2.S.7.2 and 3.2.P.8.2 Stability commitments. The guideline permits a greater degree of extrapolation to be made when statistical analysis is carried out than when no such analysis is done. Thus, there is a clear incentive to carry out such analysis and the ready availability of suitable software programs makes it reasonably straightforward.

Where statistical analysis is carried out, a tabulated summary of the analysis should be included, with a note of the procedures used, and the assumptions made should be stated. Where graphical output from the statistical analysis of the long-term storage data is available, consideration should be given to include this in report as graphs make it easier for the agency reviewer.

Cases Considered in the Q1E Guideline

The guideline considers two main cases:

No Significant Change at the Accelerated Condition

Here, the retest period or shelf life depends on the nature of the long-term data. If there is little or no change over time in the long-term data and the product remains well within the acceptance criteria of the specification, a statistical analysis is normally considered unnecessary in the EU. However, a justification should be provided and FDA reviewers often seem to need a statistical analysis to show that there is no significant changes in assay, potential degradation products, etc.

Where there is change over time in the long-term data, statistical analysis is useful. Where there are differences between batches or other factors (strength, container size, fill), the proposed retest period or shelf life should not exceed the shortest period represented by the least stable batch or any other factor. Where such differences arise, the report should try to provide a rationale for them. Where the data are amenable to statistical analysis, the extrapolation of the shelf life can be up to twice the duration of the study, but not more than 12 months beyond the data period covered in the report.

Significant Change at the Accelerated Condition

Where a significant change occurs at the accelerated condition, the retest period or shelf life depends on the results of the testing at the intermediate condition. If there is no significant change at the intermediate condition, the retest date or shelf life can be extrapolated beyond the period covered by the long-term data. If there is significant change at the intermediate condition, the shelf life should not be extrapolated beyond the period covered by the long-term data.

Where there is no significant change at the intermediate condition and if statistical analysis is carried out on an amenable dataset, then the long-term data can be extrapolated up to one and half times the period covered by the long-term data, but not more than six months beyond.

Statistical Data Analysis Procedures

Regression analysis is used to evaluate the stability data for a quantitative attribute such as the assay or levels of degradation products. In most cases, for the limited change in assay likely in a drug substance or drug product, zero-order kinetics can be assumed and linear regression analysis used. However, the guideline indicates that transforms of the data (e.g., a logarithm) can be carried out to enable a linear regression.

The usual approach to determine the chemical shelf life is to analyze a quantitative attribute such as the assay or level of degradation product(s) with respect to time and determine the time at which the 95% confidence limit for the mean intercepts the proposed specification acceptance criterion. For an attribute that decreases with respect to time (e.g., the assay), the lower one-sided 95% confidence limit should be used. For an attribute that increases with respect to time (e.g., a degradation product), the upper one-sided 95% confidence limit should be used. For an attribute that could increase or decrease (e.g., the assay of a product in a semipermeable container), two-sided 95% confidence limits should be calculated.

Where stability data are available on a number of batches, each batch can be analyzed separately to determine whether the data on each batch supports the proposed shelf life overall. However, if one of the batches shows a predicted shelf life shorter than that proposed it may be advantageous to see if all of the individual batch data can be combined (pooled) to give a better overall estimate. Before pooling the data on several batches, a test is needed to find out whether the regression lines for individual batches have the same slope and intercept at the initial time point. Analysis of covariance (ANCOVA) is used to test the differences in slopes and intercepts, testing each using a significance level of 0.25. If the ANCOVA rejects the hypothesis that the slopes are equal, the data cannot be combined and then the shelf life needs to be based on the stability data of the least stable batch. If the slopes are equal but the intercepts are not equal then the data can be combined to estimate the common slope. If the ANCOVA shows no significant differences between slopes and intercepts for the batches, the data can be pooled to give an overall estimate of shelf life.

Tests for poolability of data can be applied to data on laboratory batches of drug product or studies in different packaging materials together with the formal studies to establish shelf life. As the laboratory batch stability data will usually be for a longer period than the formal ICH studies if the data can be pooled, they support the argument for extrapolation of the shelf life. Similarly, a test of poolability can be applied to the data from the ICH formal studies and the studies on the three production batches. If the data can be pooled, they support the shelf life allocated on the basis of the formal ICH studies.

The ICH guideline gives guidance on analysis of more complex situations with a number of different factors (different strengths, different containers, different container sizes, etc.) and also for bracketing and matrixing designs. Chow has published a textbook on statistical design and analysis of stability data (30).

Mass Balance

The Q1E guideline indicates that "the adequacy of the mass balance should be assessed" without giving any details as to how this might be done. An adequate mass balance means that the quantitative decrease in assay of the drug substance in a drug substance or product during manufacture and over the proposed retest date or shelf life is explained by the total quantitative increase in levels of the degradation products. However, it is not simply a matter of adding all of the levels of individual degradation products as the degradation products will usually have a different molecular mass (MM) to that of the parent drug. The total percentage of degraded drug substance D determined from the levels of individual degradation products A, B, and C can be defined by the following equation:

$$\text{Total \% Degraded Drug} = \frac{\%A \times (MM_D)}{(MM_A)} + \frac{\%B \times (MM_D)}{(MM_B)} + \frac{\%C \times (MM_D)}{(MM_C)}$$

where MM_D is the molecular mass of the drug substance, MM_A is the molecular mass of the degradation product A, MM_B is the molecular mass of the degradation product B, and MM_C is the molecular mass of the degradation product C.

Where there is an apparent lack of mass balance factors affecting, this should be considered and explained. This may be due to co-elution of an unknown degradation product with the main active substance peak, different response factors for the drug substance and degradation products to the detector system, or to the formation of degradation products not containing a chromophore (if relevant to the type of signal detection being used in the analytical method). A significant level of unidentified degradation product(s) raises issues because their chemical structure will be unknown as will their potential toxicity. The levels of degradation products that need to be reported, identified, and biologically qualified are set out in Attachment 1 to ICH guideline Q3B(R2) *Impurities in New Drug Products* (31) for different maximum daily doses of a drug substance.

Manufacturing Overage

If the statistical analysis of stability data for different batches is carried out, there is a consistent trend for the intercept of the assay of the drug product being below 100%; this may indicate instability during the manufacturing stages and during the initial storage before the release assay. For products with marginal stability (i.e., with a calculated shelf life of 18 months or less), it may be worthwhile considering adding a manufacturing overage. This would need to be justified in the stability report. Where the data on individual batches, packaging, strengths, etc., can be pooled to give a common estimate of the intercept; this can be used to calculate the overage.

The manufacturing overage can be calculated as follows:

$$\text{Manufacturing Overage} = \left(\frac{100 \times 100}{\text{Intercept \% Assay}}\right) - 100$$

Statistical Analysis Software

There are a number of commercially available software programs that facilitate statistical analysis and graphical presentation of stability data. The FDA provides the STAB program for free download (32), which is written to carry out expiration dating based on linear regression analysis using SAS/PC software. The UK Pharmaceutical Analytical Sciences Group (PASG) makes

Table 15 Commercial Stability Test Statistical Analysis Software

Software	Supplier
Stability System	Scientek Software Inc (34)
SQL*Stability (part of SQL*LIMS software package)	Applied Biosystems (35)
JMP 7	JMP business division of SAS (36)

Table 16 Types of Out of Trend (OOT) Data (38, 39)

Type of OOT data	When observed	Significance and actions
Analytical alert	Single aberrant result, but within specification (outside normal analytical or sampling variation)	Investigate
Process control alert	Succession of data shows an atypical pattern possibly caused by laboratory or manufacturing changes, but without potential Out of Specification (OOS) situation	Investigate reasons for the alert and report internally
Compliance alert	OOT result indicates potential for OOS results to occur before end of shelf life	Investigate to confirm and then notify regulatory agencies

available a free Microsoft Excel macro routine for estimating retest date or shelf life (33). Some other commercial software programs are summarized in Table 15 (34–36).

MANAGING OUT OF TREND STABILITY DATA

The US FDA in its Establishment Inspection Reports has documented the importance of identification of Out of Trend (OOT) data on marketed products. It is an ICH requirement to provide stability data on commercial scale manufacturing batches of drug product stored until the end of the proposed shelf life if these are not available in the registration file. It is a GMP requirement in Europe (37) and in the United States to examine the stability of one commercial batch of drug product in the proposed packaging configuration each year. If an OOT result occurs during the stability storage of such batches, this may have a significant impact because the shelf life allocated to the entire year's production could be questioned, depending on the OOT alert level. The issues involving OOTs have been reviewed at two major US meetings in 2003 and 2005 involving PhRMA member companies, and the results of their deliberations have been reported (38, 39). The authors of the reports classify OOT data as summarized in Table 16. They recommend that companies develop Standard Operating Procedures (SOPs) for review and investigation of OOT stability data. The SOPs could include topics such as setting alert limits, how to use limits, investigation and internal notification procedures, limits for truncated data (such as data which may be higher or lower than ICH degradation data reporting thresholds), and periodic review of limits.

REFERENCES

1. ICH guideline Q1A(R2) Stability Testing of New Drug Substances and Products, February 2003, ICH. http://www.ich.org/LOB/media/MEDIA419.pdf. Accessed September 9, 2008.
2. ICH guideline Q1B Stability Testing: Photostability Testing of New Drug Substances and Products, November 1996, ICH. http://www.ich.org/LOB/media/MEDIA412.pdf. Accessed September 9, 2008.
3. ICH guideline Q1C Stability Testing of New Dosage Forms, November 1996, ICH. http://www.ich.org/LOB/media/MEDIA413.pdf. Accessed September 9, 2008.
4. ICH guideline Q1D Bracketing and Matrixing Designs for Stability Testing of New Drug Substances and Products, February 2002, ICH. http://www.ich.org/LOB/media/MEDIA414.pdf. Accessed September 9, 2008.
5. ICH guideline Q1E Evaluation of Stability Data, February 2003, ICH. http://www.ich.org/LOB/media/MEDIA415.pdf. Accessed September 9, 2008.
6. Haynes JD. Worldwide Virtual Temperatures for Product Stability Testing. J Pharm Sci 1971; 60(6):927–929.
7. Grimm W, Schepky G. Stabilitätsprufung in der Pharmazie, Theorie und Praxis. Aulendrof, Germany: Editio Cantor Verlag, 1980.
8. Grimm W. Storage conditions for stability testing–long term testing and stress testing. Drugs Made in Germany, 1986; 29:39–47.

9. Grimm W, ed. In: Stability Testing of Drug Products. Stuttgart: Wissenschaftliche Verlagsgesellschaft mbH, 1987.
10. Futscher N, Schumacher P. Climatic zones of the earth. Pharm Ind 1972; 34:479–483.
11. WHO Guidelines for Stability Testing of Pharmaceutical Products Containing Well Established Drug Substances in Conventional Dosage Forms. In: WHO Expert Committee on Specifications for Pharmaceutical Preparations, Thirty-fourth report, Annex 5, Geneva, World Health Organisation 1996, Geneva, and as revised at the thirty-seventh and fortieth meetings of the WHO Expert Committee on Specifications for Pharmaceutical Preparations: WHO Technical Report Series No. 908, p. 13 (2003) and WHO Technical Report Series No. 937, p 12 (2006).
12. Zahn M, Källberg PW, Slappendel GM, et al. A risk-based approach to establish stability testing conditions for tropical countries. J Pharm Sci 2006; 95(5):946–965.
13. Zahn M, Källberg, PW, Slappendel GM, et al. A risk-based approach to establish stability testing conditions for tropical countries [Erratum]. J Pharm Sci 2007; 96(8):2177.
14. ASEAN Guideline on Stability Study of Drug Product, 22 February 2005, 9th ACCSQ-PPWG Meeting, Philippines, February 21–24, 2005.
15. WHO Regional Committee for the Eastern Mediterranean, Draft Regional Guidelines on Stability Testing of Active Substances and Pharmaceutical Products, EM/RC53/12, Jeddah, August 2006.
16. Resolution—RE No. 1 of July 29, 2005. Publication of the Guide for Stability Studies. Brazilian Federal Register 08/01/2005.
17. Zahn M. Stability testing: New requirements in China. Regul Aff J: Pharma 2005; 16(11):5.
18. Bakshi M, Singh S. Development of validated stability-indicating assay methods—Critical review. J Pharm Biomed Anal 2002; 28:1011–1040.
19. Analytical Profiles of Drug Substances and Excipients, edited by Harry Brittain, various volumes to 2008. St. Louis, MO: Elsevier BV.
20. Baertschi SW. Pharmaceutical Stress Testing. New York: Marcel Dekker Inc, 2005.
21. Connors GI, Amidon GL, Stella VJ, eds. Chemical Stability of Pharmaceuticals. New York: Wiley, 1986.
22. Xu QA, Trissel LA. Stability-Indicating HPLC Methods for Drug Analysis, 3rd ed. Washington, D.C.: American Pharmaceutical Association, 2008.
23. Pole DL, Ando HY, Murphy ST. Prediction of drug degradants using DELPHI: An expert system for focusing knowledge. Mol Pharm 2007; 4(4):539–549.
24. CHMP Guideline on Declaration of Storage Conditions: A: In the Product Information of Medicinal Products, B: For Active Substances. Annex to Note for Guidance on Stability Testing of New Drug Substances and Products. Annex to Note for Guidance on Stability Testing of Existing Active Substances and Related Finished Products. CPMP/QWP/609/96 Rev 2. Revised September 2007. EMEA. http://www.emea.europa.eu/pdfs/human/qwp/060996.en.pdf. Accessed September 10, 2008.
25. CPMP Guideline on Stability Testing: Stability Testing of Existing Active Substances and Related Finished Products, CPMP/QWP/122/02, rev 1 corr. December 2003. EMEA. http://www.emea.europa.eu/pdfs/human/qwp/012202.en.pdf. Accessed September 10, 2008.
26. Draft WHO Guideline on Stability Testing of Active Pharmaceutical Ingredients and Finished Pharmaceutical Products, Working Document QAS/06.179/Rev. 3, 2008, WHO, Geneva.
27. Yoshioka S, Ishihara Y, Terazono T,et al. Quinine actinometry as a method for calibrating ultraviolet radiation intensity in light stability testing of pharmaceuticals. Drug Dev Ind Pharm 1994; 20(13):2049–2062.
28. CPMP Note for Guidance on In-Use Stability Testing of Human Medicinal Products, CPMP/QWP/2934/99, February 2001, EMEA. http://www.emea.europa.eu/pdfs/human/qwp/293499en.pdf. Accessed September 10, 2008.
29. Carstensen JT. Stability and dating of solid dosage forms. In: Carstensen JT, ed. Pharmaceutics of Solids and Solid Dosage Forms. New York, NY: Wiley-Interscience, 1977:182–185.
30. Chow S-C. Statistical Design and Analysis of Stability Studies. New York, NY: Chapman and Hall/CRC Biostatistics Series, 2007.
31. ICH Guideline Q3B(R2) Impurities in New Drug Products, June 2006, ICH. http://www.ich.org/LOB/media/MEDIA421.pdf. Accessed September 10, 2008.
32. FDA SAS Drug Formulation Stability Program, Division of Biometrics, CDER, FDA, March 1992. http://www.fda.gov/cder/sas. Accessed September 10, 2008.
33. Downloadable Excel Routine for Estimating Shelf Life, Pharmaceutical Analytical Sciences Group. http://www.pasg.org.uk/excel.htm. Accessed September 10, 2008.
34. Stability System Statistical Program, Scientek Software Inc. http://www.stabilitysystem.com/statistical_data_analysis.htm. Accessed September 10, 2008.
35. SQL*LIMS software includes SQL*Stability, Applied Biosystems. http://www.sqllims.com. Accessed September 10, 2008.
36. JMP7 Software, SAS Institute. http://www.jmp.com/applications. Accessed September 10, 2008

37. Eudralex, The Rules Governing Medicinal Products in the European Union, Volume 4, EU Guidelines to Good Manufacturing Practice Medicinal Products for Human and Veterinary Use, Part 1, Chapter 6: Quality Control, October 2005, European Commission. http://ec.europa.eu/enterprise/pharmaceuticals/eudralex/vol-4/pdfs-en/2005_10_chapter_6.pdf. Accessed September 10, 2008.
38. PhRMA CMC Statistics and Stability Expert Teams, Identification of Out-of-Trend Stability Results—A Review of the Potential Regulatory Issue and Various Approaches, Pharmaceutical Technology, April 2003.
39. PhRMA CMC Statistics and Stability Expert Teams, Identification of Out-of-Trend Stability Results, Part II. Pharmaceutical Technology, October 2005.

20 Transmissible Spongiform Encephalopathy Agent and Adventitious Agent Requirements for Non-Biological Pharmaceutical Products

Brian R. Matthews
Alcon Laboratories, Hemel Hempstead, U.K.

INTRODUCTION

In the context of this chapter "biological" products are those which are based on or frequently contain materials of human or animal origin. These products not only include vaccines, blood products, and immunological products but also extend to products of biotechnology. These product types might be expected to be made in such a way that the possibility of contamination with viral or other contaminants [including transmissible spongiform encephalopathy (TSE) agents] derived from the animal- or human-sourced materials would need to be taken into account during their development, production and release testing. However, these will not be the aspects that will be discussed in this chapter, instead this chapter will address the issues that need to be taken into account during the development processes and included in the marketing authorization application (MAA) for "conventional" pharmaceuticals containing in the main small molecule active pharmaceutical ingredients (although the concepts discussed could also be applied to materials such as fermentation-produced antibiotics and similar semisynthetic materials).

In terms of an International Conference on Harmonization (ICH) Common Technical Document formatted application (1) for an MAA information related to "adventitious agents"—which includes viruses and TSE agents--may be required in the application form [for example, in 2.6.2 of the European Application Form which is to identify animal-sourced materials from species susceptible to TSE and from other animals or human-sourced materials with instructions to include copies of any *European Pharmacopoeia* (Ph Eur) Certificates of Suitability (CEPs) at Annex 5.12 (2)], and in modules 2.3 and 3.2 the following information is required:

> 2.3.A.2 Adventitious Agents Safety Evaluation
> A discussion on measures implemented to control endogenous and adventitious agents in production should be included. A tabulated summary of the reduction factors for viral clearance from 3.2.A.2 should be provided.
> 3.2.A.2 Adventitious Agents Safety Evaluation
> Information assessing the risk with respect to potential contamination with adventitious agents should be provided in this section (3).

This information was amended as follows by the Questions and Answers document on the formatting of M4Q (4):

> The detailed information regarding the routine manufacturing control of adventitious agents, such as bacteria, mycoplasma, and fungi, typically using well-established (e.g., pharmacopoeial) analytical procedures, should be provided in the appropriate sections within Module 3.2.S and 3.2.P. If well-established (e.g., pharmacopoeial) analytical procedures are not used, more detailed information regarding the analytical procedure(s) used should also be included in 3.2.S and 3.2.P. With respect to other non-viral adventitious agents such TSE agents (prions) the detailed information should be placed in 3.2.A.2.

For materials derived from species that are liable to TSEs then TSE-related issues should be considered. For any material derived from animals then possible sources of contaminating organisms should be considered--bacteria, mycoplasma, parasites, viruses, etc.

For more information on the background to TSE risks in Europe see *BSE/TSE risks in Europe* (5).

TRANSMISSIBLE SPONGIFORM ENCEPHALOPATHY

TSEs affect certain mammals, animals and humans. Examples of the conditions involved in animals include

- scrapie in sheep,
- chronic wasting disease in certain deer and elk,
- transmissible mink encephalopathy,
- feline spongiform encephalopathy and
- bovine spongiform encephalopathy (BSE).

In humans there are a number of TSEs that have been described including

- Creutzfeldt–Jakob disease (CJD) (sporadic, familial and variant forms),
- fatal insomnia (sporadic and familial forms),
- Gerstmann–Straussler–Scheinker syndrome,
- kuru and
- iatrogenic transmission of TSEs has been reported—in the pharmaceutical sector there have been cases of CJD transmission *via* contaminated pituitary–derived human growth hormone and from the use of human *dura mater* in surgery.

TSEs are caused by unconventional infectious agents currently thought to be infectious proteins (apparently without nucleic acid) known as "prions" which do not share the normal properties of viruses or bacteria.

All of these conditions affect the brain of the host, forming vacuoles (and giving a spongiform appearance to the brain on *postmortem* examination) and present a variety of physical symptoms prior to diagnosis—each version of human TSE presents with its own typical set of symptoms. Once a TSE has been diagnosed the progression is to an inevitably fatal outcome and no therapeutic regime is available at the time of writing (December 2008).

TSE agents are resistant to inactivation by processes that are effective against bacteria, mould, and yeast. The Advisory Committee on Dangerous Pathogens (ACDP), a body offering advice to the English Department of Health, has indicated

> TSE agents are particularly resistant to standard physical and chemical methods of inactivation and decontamination. . . . The standard laboratory autoclave regimen of 121°C for 15 minutes is ineffective, and autoclaving at 134°C to 137°C held for 3 minutes cannot be relied upon to remove the infectious agents (6).

A wide range of agents has been tested for inactivation of TSE agents. The studies are difficult to undertake (and involve a bioassay in a susceptible animal model) and may need a model TSE agent. The TSE agent also needs to be handled in a Level 3 biohazard facility. Agents that have some efficacy include 20,000 ppm of chlorine from hypochlorite or 2 M sodium hydroxide for one hour. Some authorities advise the use of autoclaving at 121°C in sodium hydroxide for decontamination of surgical instruments. Processes that have been found to be ineffective include 10,000 ppm chlorine from hypochlorite, alcohols, ammonia, β-propiolactone, chlorine dioxide, formalin, glutaraldehyde, hydrochloric acid, hydrogen peroxide, iodophors, peracetic acid, phenolic disinfectants, sodium dichloroisocyanurate, ethylene oxide, formaldehyde, dry heat and incineration below 1000°C, ionizing radiation at the dose level normally employed for sterilization, UV or microwave exposure and moist heat processes at 121°C or 134°C. From this it will be seen that it is difficult to inactivate TSE agents and so a greater emphasis has to be placed on avoiding their presence in the first place. If claims are made for effectiveness in the manufacturing process for a particular product in the inactivation of TSE agents then appropriate data will need to be included in the MAA to demonstrate the log reduction factor that it achieved in prion levels (usually in spiked material validation studies).

TSE Issues in the Manufacture of Pharmaceutical Products

Materials of animal and human origin are used in the manufacture of a wide range of pharmaceutical products and they may be used in the production of ingredients for use in such products. An obvious material is human albumin that might be included in vaccines and blood products. Fetal calf serum might also be used in the production of some products. Gelatin is of animal origin—for example, derived from skin or bone of pigs or bovine animals (or in some cases from fish) using a variety of processes. Collagen is derived from animal species. Tallow derivatives come from tallow (a product which comes from animal carcasses and which is made in enormous quantities) and these materials include some stearates, oleates, etc., which could be used as ingredients in products in their own right or they may be used in the processes used to make ingredients. Amino acids can be derived from materials of animal origin. Animal materials may be used in the production of microbiological media, which might be used in the production of some "biological" medicinal products but may also be used in such processes as aseptic filling validation broth fill trials. If such materials are used in or during the manufacture or even within the manufacturing plant for pharmaceutical products it will be necessary to provide suitable information to confirm that they are suitable for their intended use.

For use in Europe, the European Department for the Quality of Medicines and Healthcare (EDQM, a part of the Council of Europe) introduced the Ph Eur Certificate of Suitability (CEP) scheme for materials with a monograph included in the Ph Eur but this was later extended to include certification of materials with regard TSEs (including bovine spongiform encephalopathy or BSE certification) where the substances concerned were not necessarily included in the Ph Eur. There is a specific procedure to be followed and there is a guideline on data requirements and presentation of the information (7). The wording on the EDQM introductory web page to describe this process indicates that an applicant can apply for a CEP for all material(s) derived from animals that are susceptible to the transmission of TSE agents as described in general chapter 5.2.8 of the *European Pharmacopoeia*, which is *verbatim* the EMEA Note for Guidance on *Minimizing the risk of transmitting animal spongiform encephalopathy agents via medicinal products* (8). An equivalent service does not appear to be available as part of the *United States Pharmacopeia* (USP) Verified Pharmaceutical Ingredient scheme and there does not appear to be an equivalent service from the *Japanese Pharmacopoeia* (JP).

Regulatory agencies have published a number of relevant guidelines that should be taken into account. These include:

Europe

- Minimizing the risk of transmitting animal spongiform encephalopathy agents *via* human and veterinary medicinal products (8) and a proposed revision of this document issued in 2004, but on which no action seems to have been taken (9), and an Explanatory Note for medicinal products for human use on the scope of the guideline (10),
- Position paper on reestablishment of working seeds and working cell banks using TSE compliant materials (11),
- Public report on risk and regulatory assessment of lactose and other products prepared using calf rennet (12),
- Questions and answers on BSE and vaccines (13),
- Public statement on the evaluation of BSE risk via the use of materials of bovine origin in or during the manufacture of vaccines (14),
- Explanatory note: Gelatin for use in pharmaceuticals (15),
- Position statement on polysorbate 80 (16),
- Position paper on production of tallow derivatives for use in pharmaceuticals (17),
- Investigation of manufacturing processes for plasma-derived medicinal products with regard to vCJD risk (18) and
- Position statement on Creutzfeldt–Jacob Disease and plasma-derived and urine-derived medicinal products (19).

In addition, manufacturers will need to take into account the BSE status of countries from which animal materials will be sourced. The scheme for the listing of these countries has changed relatively recently and the primary source of information is in the form of a Commission Decision, which is updated from time to time in line with recommendations from the World Organization of Animal Health (OIE). The most recent document published by the Commission is

- Commission Decision of 30 October 2008 mending the Annex to Decision 2007/453/EC, establishing the BSE status of Member States or third countries or regions thereof according to their BSE risk (20).

This is based on information from the OIE:

- BSE status of members (21).

United States

There are very few documents adopted and published on the Internet by the FDA relating to the use by manufacturers of materials of animal origin in pharmaceutical products. Such issues are usually discussed in meetings with the FDA—for example, IND meetings, end of Phase II meetings, etc. The only current document that the author could identify that is of direct relevance is:

- The sourcing and processing of gelatin to reduce the potential risk posed by BSE in FDA-regulated products for human use (22).

There was some additional advice in a guideline on Chemistry Manufacturing and Controls information for the drug substance but the document concerned has been withdrawn. For the sake of completeness a summary of the requirements of the earlier document--which is no longer available on the FDA website--is included in Annex 1 of this chapter.

It should be noted that some aspects of the control on animal-sourced materials within the United States also fall to the Department of Agriculture and specifically the Animal and Plant Health Inspection Service (APHIS) which publishes a list of counties and their BSE status.

- Countries/regions affected with BSE (23).

Japan

The author could identify no documents of relevance in English that were available *via* the Internet other than General Information Chapter 20 of the Japanese Pharmacopoeia XV, "Qualification of animals as origin if animal-derived medicinal products provided in the General Notices of Japanese Pharmacopoeia and other standards" (24). In addition, the information on "Standards for bioderived materials" in *Drug Approval and Licensing Procedures in Japan 2006* (25, pp. 131–144) will be of relevance (as will information included on pages 174, 184, 428, etc.). This includes information on the countries from which ruminant animal–derived materials must be sourced (25, p. 137).

Other Sources of Information

The World Health Organization has worked for some time on the distribution of infectivity in various tissues and materials of animal origin. The most recent information that has been published is:

- WHO Guidelines on Tissue infectivity distribution in transmissible spongiform encephalopathies--updated information on major categories of infection (26).

There is available a series of International Standards [published by the International Standards Organization (ISO) in 2007] relating to the control of viral and BSE risks associated with medical devices but which might also be useful for pharmaceutical manufacturers in suggesting ways to approach the issues concerned:

- ISO 22442 Medical devices utilizing animal tissues and their derivatives--Part 1: Application of risk assessment,
- ISO 22442 Medical devices utilizing animal tissues and their derivatives--Part 2: Controls on sourcing, collection, and handling and
- ISO 22442 Medical devices utilizing animal tissues and their derivatives--Part 3: Validation of the elimination and/or inactivation of viruses and transmissible spongiform encephalopathy (TSE) agents (27).

An Approach to Dealing with Animal-Derived Materials with a Potential Associated TSE Risk

Manufacturers should first determine whether they have any materials that are used within their manufacturing processes--whether they will be included in the finished pharmaceutical product--that are derived from animal materials. The most effective means of doing this is to determine all materials that are used in their processes and check with the suppliers (and preferably the primary manufacturers) of the materials concerned to determine whether they contain or are exposed to materials of animal (or human) origin during their preparation. This will require specific questions to be put to suppliers. In some cases the source of such TSE risks can be quite remote--for example, ingredients such as tallow derivatives in fermentation procedures used in the production of ingredients; the use of animal-derived enzymes in stereo-selective syntheses. If any materials are found to be derived from animal (or human) origin it will then be necessary to determine whether any of them could represent a TSE risk.

If TSE risks are identified then it will be necessary first to determine whether the use of the materials concerned is necessary and whether they could be replaced by a non-TSE risk inherent product (preferably one of nonanimal origin). If they can be replaced serious consideration should be given to their replacement. If they cannot be replaced then a justification for their continued use should be documented. This is likely to need to be submitted in the MAA.

Having identified that a material with a TSE risk is necessary and cannot be replaced then further thought will need to be given to mitigating the risks concerned. It would be worthwhile asking the supplier whether any affected materials are subject to a Ph Eur CEP--in Europe at least this will ease the burden on the pharmaceutical manufacturer regarding the generation of additional information. Of course, it will be necessary to confirm (on a regular basis) that the CEP remains valid.

If there is no CEP then TSE-related issues can be addressed in three phases:

- identifying the risks and mitigating them,
- controlling the source and processing applied to the animal materials and
- validating processes for the elimination or inactivation of any TSE materials present.

The risks associated with the material may be linked to the animal species from which it is derived. TSE-susceptible species are obviously the highest risk. If a TSE susceptible species is to be used then the geographical source of the animals could be limited to countries where there is no relevant TSE history or where TSE risk factors are under control. Information on preferred countries can be found on the OIE web pages and in other places (20,21,23). Countries from which the materials are sourced should have controls on the possible use of high-risk tissues and materials in the feed chain for susceptible species and should have controls on the use of such "specified risk materials." Feeding history of the animals might become important—it should be possible to confirm that the animals concerned have not been fed with ruminant-derived meat and bone meal. Then there is the issue of the potential infectivity of particular tissues. Information on this can be obtained from the WHO (26). It would be advantageous to source materials from categories of tissue that indicate that the level of TSE agent present is very low. This might require the use of materials from different organs or tissues to minimize the risk. Another option might be to use animals from a closed herd to source the materials used in the pharmaceutical product or its manufacturing process. The potential for the material coming into contact with humans should be assessed. The lower that possibility, the lower the perceived risk. If the substance does come into contact then the quantity involved and the number of exposures could become important. The route of administration will also be an important factor--injection routes will overcome any potential species barrier. In some cases it may be that there is specific guidance on the conditions under which certain derivatives may be used—for example, gelatin (8,15,22,27--see Annex C to Part 1); collagen (8,27--see Annex C to Part 1); bovine blood derivatives (8,27--see Annex C to Part 1), tallow derivatives (8,16,17,27--see Annex C to Part 1); animal charcoal (8,27--see Annex C to Part 1); milk and milk derivatives (8,27--see Annex C to Part 1); wool and its derivatives (8,27—see Annex C to Part 1); and amino acids (8,27--see Annex C to Part 1). Where there is such advice it should be determined whether any necessary conditions have been met and this information should be included in the MAA.

The source--including species (and in some cases strain) and age of animals (since TSE titres are likely to be lower in younger animals) and country of origin—will by now have been determined. It is important to take into account this complete history of the animals concerned including their conception, birth and rearing. It will now be necessary to consider other issues such as the method of slaughter used (preferably avoiding penetrative stunning methods) and the manner in which the animal tissues or organs are harvested. Appropriate measures should be in place to avoid cross-contamination of the animal materials as they are processed. The animals used should be "fit for human consumption" under the relevant food safety legislation in the country of origin. This will usually require some form of veterinary inspection pre- and postmortem. Certification should be sought for the status of the animals as fit for human consumption.

Having recovered the materials it should then be determined how they will be processed to produce the material that will be used in the manufacturing process for the pharmaceutical or included in the pharmaceutical finished product. If methods are used which are claimed to inactivate or remove TSE agents then information will be needed to validate the claims made. In many cases it will be found that the ingredients or materials will not withstand such processes in which case reliance will need to be placed primarily on the control of the source of the materials concerned. Where the materials can be subject to processing then information will need to be gathered to justify the choice of processing and its efficacy. The first phase is to undertake a literature review to identify relevant information which can then be assessed for relevance. If no relevant literature can be found then a relevant study will need to be designed and undertaken to provide information on the efficacy of the processes concerned. Factors to consider will include

- the risk concerned,
- identify relevant model TSE agent(s),
- develop a rationale for the particular combination of model agents to be used,
- identification of the manufacturing stages claimed to remove or eliminate TSE agents and
- calculation of the reduction factors achieved.

This information together with the details of the method and the results will need to be included in a report to be included in the MAA.

ADVENTITIOUS AGENTS

As indicated above "adventitious agents" is often taken to include a limited range of potential contaminating agents that might be found in starting materials or possibly introduced in manufacturing processes for pharmaceutical products. For this part of the chapter the term will be taken to include microbial contamination sources involving bacterial, moulds, yeasts, viruses, mycoplasma, rickettsia and other organisms. In some cases, the potential for contamination with other pathogenic species such as parasites should also be taken into account. As with the earlier part of the chapter, the scope will be limited to "conventional" pharmaceuticals.

Bacteria, Moulds and Yeasts

The principles of good pharmaceutical manufacturing practice should be applied for all pharmaceutical products. This should include appropriate measures to ensure that microbial contamination is not present to an unacceptable extent in raw materials and that such contamination should not occur to an unacceptable extent during all stages of the preparation and manufacture of pharmaceutical products (including the handling of raw materials to be included in those products).

In the case of bacteria, moulds and yeasts, there are pharmacopoeial requirements for nonsterile and sterile products. An example would be the requirements of the *European Pharmacopoeia* for *Microbial quality of pharmaceutical preparations* (general text 5.1.4 of the Ph Eur) (28). For products that are not required to be sterile there are recommended acceptance criteria for microbiological quality. There were, at the time of writing (December 2008), two sets of criteria—one from the Ph Eur and one from a harmonization process with the USP and JP. Since the latter are to replace the former in the foreseeable future they are worth repeating here. There are separate limits for total aerobic microbial count (TAMC) and for total combined yeasts/moulds count (TYMC). For substances that are used in pharmaceutical products, the recommended limits are 10^3 colony forming units (CFU) per gram or per milliliter for TAMC and 10^2 CFU per gram or per milliliter (TYMC), although it is indicated that the limits actually applied in a particular case should be subject to a risk-based assessment and should take into account the processes that will be applied in making the finished product. There are separate recommended limits for formulated products with TAMC and TYMC values, and for some products there are additional controls for specific microorganisms and it is indicated that due account should be taken of the significance of other microorganisms present in the product (dependent on the intended route of administration and how it is used and the target patient population and any other medicines likely to be used in that population (e.g., immunosuppressive agents), the nature of the product and its ability to support microbial growth and whether there is an antimicrobial preservation system present in the product and potential comorbidity in the patient population--for example, disease, wounds, and organ damage. Table 1 is a summary of the requirements.

In the case of products that are required to be sterile (such as parenteral products and eye preparations) there should be adequate controls on the bioburden of raw materials and intermediate stages of manufacture of the product. The bioburden controls will need to be tailored to the intended method of manufacture. For products sterilized in the final container using a high-lethality process (such as dry saturated steam at 121°C with a holding time of 15 minutes) the bioburden limits can be higher than for a process such as sterile filtration through a sterilizing grade filter (less than 0.22 μm nominal pore size) followed by aseptic processing. In the EU bioburden limits of 100 CFU per gram or milliliter could be acceptable for raw materials. The expected limits for bioburden on presterilization bulk liquid products will depend on the manufacturing process--a limit of 10 CFU per 100 ml could be appropriate for solutions to be sterile filtered. In addition to bioburden there could be a need for determination of the resistance of the organisms present to the lethal process--in the case of low-lethality dry saturated steam processes heat resistance may need to be monitored (for example using a boil test) and radiation resistance might need to be monitored for low-lethality irradiation processes.

There is advice in the pharmacopoeias relating to methods of preparation of sterile products and this should be taken into account--for example, Ph Eur general text on *Methods of preparation of sterile products* (5.1.1) and *Application of the* F_0 *concept to steam sterilization of aqueous preparations* (5.1.5) (29). There is also a series of International Standards on methods of sterilization and their validation which could be taken into account (30).

Other Adventitious Agents

Where materials of animal or human origin are used as components of or are utilized in the manufacturing process for human pharmaceuticals, there is a possibility that the raw material could be contaminated by viruses or other organisms. The potential contaminants might be known or there may be no methods by which their presence can be determined at the time of use. This is most likely to be an issue for products such as blood products, immunological products, and vaccines (which are outside the scope of this chapter)--examples of problems associated with such contaminants have included SV40 in polio vaccines, West Nile virus contamination and hepatitis virus contamination of blood products. It can also be relevant for other types of product and where such starting materials are used it will be necessary to undertake a risk evaluation and to identify any means that can be used to mitigate any risks associated with their use.

As far as the author is aware there are no specific guidelines for manufacturers to take into account the assessing potential risk from adventitious contamination other than for "biological" and "biotechnological" products—for example, ICH Q5AR1 *Viral safety evaluation of biotechnology products derived from cell lines of human or animal origin* (31) and the CPMP *Note for guidance on virus validation studies: the design, contribution, and interpretation of studies validating the inactivation and removal of viruses* (32). Readers might also find ISO 22442 Parts 1 and 3 (27) to be helpful in suggesting an approach to the control of risks associated with adventitious agents. There are no guidelines of which the author is aware relating to the control of risks related to parasites.

In order to assess risks it is necessary to consider the information available in the literature and then to undertake a gap analysis for any additional data needed for a specific risk and product. Where it is necessary to undertake studies into the elimination or inactivation of viruses it is common to undertake studies with a variety of RNA and DNA enveloped and nonenveloped model organisms and information on this is included in ICH Q5AR1 (31), the relevant CPMP guideline (32), and ISO 22442–3 (27).

Table 1 Acceptance Criteria for Microbiological Quality of Nonsterile Pharmaceutical Products (28)

Route of administration	Total aerobic microbial count (CFU/g or CFU/ml)	Total yeasts/ mould count (CFU/g or CFU/ml)	Specified microorganisms
Oral, nonaqueous product	10^3	10^2	Absence of *Escherichia coli* (1 g or 1 ml)
Oral, aqueous product	10^2	10^1	Absence of *Escherichia coli* (1 g or 1 ml)
Rectal	10^3	10^2	
Oromucosal; gingival; cutaneous; nasal; auricular	10^2	10^1	Absence of *Staphylococcus aureus* and of *Pseudomonas aeruginosa* (1 g or 1 ml)
Vaginal	10^2	10^1	Absence of *Staphylococcus aureus* and of *Pseudomonas aeruginosa* and of Candida albicans (1 g or 1 ml)
Transdermal patch	10^2 per patch including adhesive layer and backing	10^1 per patch including adhesive layer and backing	Absence of *Staphylococcus aureus* and of *Pseudomonas aeruginosa* (one patch)
Inhalation (liquid preparation for nebulization)	10^2	10^1	Absence of *Staphylococcus aureus* and of *Pseudomonas aeruginosa* and of bile-tolerant gram-negative bacteria (1 g or 1 ml)
Ph Eur: Oral, containing raw material of natural origin[a] for which antimicrobial treatment is not feasible and for which the competent authority accepts raw material TAMC exceeds 10^3 per g or per ml	10^4	10^2	Not more than 10^2 CFU bile-tolerant gram-negative bacterial per g or per ml Absence of *Salmonella* (10 g or ml) Absence of *Staphylococcus aureus* and of *Pseudomonas aeruginosa* (1 g or 1 ml)
Ph Eur: Herbal products containing solely one or more herbal drugs (whole, reduced, or powdered)			
To which boiling water is added before use	10^7	10^5	Not more than 10^2 CFU *Escherichia coli* per g or per ml Not more than 10^3 CFU bile-tolerant gram-negative bacterial per g or per ml
To which boiling water is not added before use	10^5	10^4	Absence of *Escherichia coli* (1 g or ml) Absence of *Salmonella* (10 g or ml)

[a] Animal-, vegetal- or mineral-sourced materials.

REFERENCES

1. http://www.ich.org/cache/compo/276-254-1.html. Accessed December 27, 2008.
2. http://ec.europa.eu/enterprise/pharmaceuticals/eudralex/vol-2/b/update_200805/applicformrevised_rev9.doc. Accessed December 27, 2008.
3. http://www.ich.org/LOB/media/MEDIA556.pdf. Accessed December 27, 2008.
4. http://www.ich.org/LOB/media/MEDIA620.pdf. Accessed December 27, 2008.

5. Matthews, BR. BSE/TSE risks in Europe. RAJ 2001; 12(5):374–387, 436–442 and 12(6):534–531.
6. See http://www.advisorybodies.doh.gov.uk/acdp/tseguidance/. Accessed December 27, 2008. The quotation is from Appendix C.
7. http://www.edqm.eu/medias/fichiers/Content_of_the_Dossier_for_a_Substance_for_TSE_Risk_Assessment.pdf. Accessed December 27, 2008.
8. http://www.emea.europa.eu/pdfs/human/bwp/TSE%20NFG%20410-rev2.pdf. Accessed December 27, 2008.
9. http://www.emea.europa.eu/pdfs/vet/regaffair/041001en.pdf. Accessed December 27, 2008.
10. http://www.emea.europa.eu/pdfs/human/bwp/049801en.pdf. Accessed December 27, 2008.
11. http://www.emea.europa.eu/pdfs/human/press/pp/2231402en.pdf. Accessed December 27, 2008.
12. http://www.emea.europa.eu/pdfs/human/bwp/033702en.pdf. Accessed December 27, 2008.
13. http://www.emea.europa.eu/pdfs/human/bwp/081901en.pdf. Accessed December 27, 2008.
14. http://www.emea.europa.eu/pdfs/human/press/pus/047601en.pdf. Accessed December 27, 2008.
15. http://www.emea.europa.eu/pdfs/human/press/pus/430600en.pdf. Accessed December 27, 2008.
16. http://www.emea.europa.eu/pdfs/human/press/pos/195298en.pdf. Accessed December 27, 2008.
17. http://www.emea.europa.eu/pdfs/human/press/pp/116397en.pdf. Accessed December 27, 2008.
18. http://www.emea.europa.eu/pdfs/human/bwp/513603en.pdf. Accessed December 27, 2008.
19. http://www.emea.europa.eu/pdfs/human/press/pos/287902enfin.pdf. Accessed December 27, 2008.
20. http://eur-lex.europa.eu/LexUriServ/LexUriServ.do?uri = OJ:L:2008:294:0014:0015:EN:PDF. Accessed December 27, 2008.
21. http://www.aphis.usda.gov/import_export/animals/oie/. Accessed December 27, 2008.
22. http://www.fda.gov/opacom/morechoices/industry/guidance/gelguide.htm. Accessed December 27, 2008.
23. http://www.aphis.usda.gov/import_export/animals/animal_import/animal_imports_bse.shtml. Accessed December 27, 2008.
24. Japanese Pharmacopoeia XV, pages 1734–6, http://jpdb.nihs.go.jp/jp15e/JP15.pdf. Accessed December 27, 2008.
25. Drug Approval and Licensing Procedures in Japan 2006, Jiho, Tokyo, Japan, ISBN 978-4-8407-3646-0.
26. http://www.who.int/biologicals/BS%202078%20TSE.pdf. Accessed December 27, 2008.
27. Available from national standards bodies or *via* the ISO web site: http://www.iso.org/iso/search.htm?qt=22442&sort=rel&type=simple&published=on. Accessed December 27, 2008.
28. The text included in the British Pharmacopoeia 2009 at Volume IV, Appendix XVI D (pages A426–A428) was used. London: The Stationery Office.
29. The text included in the British Pharmacopoeia 2009 at Volume IV, Appendix XVIII (pages A468–A471) was used. London: The Stationery Office.
30. The following documents are potentially of interest and relevance in the context of sterilization methods and their validation:
    ISO 11135–1:2007 Sterilization of health care products—Ethylene oxide—Part 1: Requirements for development, validation, and routine control of a sterilization process for medical devices
    ISO/TS 11135–2:2008—Part 2: Guidance on the application of ISO 11135–1
    ISO 11137–1:2006 Sterilization of health care products—Radiation—Part 1: Requirements for development, validation, and routine control of a sterilization process for medical devices
    ISO 11137–2: (in draft): Part 2: Establishing the sterilization dose
    ISO 11137–3:2006—Part 3: Guidance on dosimetric aspects
    ISO 17664:2004 Sterilization of health care products—Moist heat—Part 1: Requirements for development, validation, and routine control of a sterilization process for medical devices.
    ISO/TS 17665–2: Part 2: Guidance on the application of ISO 17665–1
    ISO/DIS 20857: Sterilization of health care products—Dry heat—Requirements for development, validation, and routine control of a sterilization process for medical devices.
    These documents can be purchased from the International Standards Organization or from national standards bodies—e.g., via http://www.iso.org/iso/search.htm?qt=sterilization&published=on&active_tab=standards. Accessed December 27, 2008.
31. http://www.emea.europa.eu/pdfs/human/ich/029595en.pdf. Accessed December 27, 2008.
32. http://www.emea.europa.eu/pdfs/human/bwp/026895en.pdf. Accessed December 27, 2008.

Annex 1: Summary of information requirements on viral adventitious agents and TSE agents from the draft Guidance for Industry on Chemistry, Manufacturing, and Controls for the Drug Substance issued by the FDA in January 2004.

Note: The draft guideline from which this information is extracted was withdrawn by the FDA. It has not been replaced (as of December 2008) and the original document is no longer available on the FDA's website.

Additional information may be appropriate in the section of the application on facilities and equipment if contamination with viral adventitious agents or TSE agents is a possibility.

If drug substances are manufactured in the same areas as a drug substance used in a drug product described in an application that could result in contamination or cross-contamination with viral adventitious agents and TSE agents then information will be required on design features of the manufacturing facility that are intended to prevent such cross-contamination.

Information will also be required on product contact equipment and whether this is for dedicated or general use together with information on its preparation, cleaning, sterilization and storage.

An assessment of the risk of cross-contamination with viral adventitious agents and TSE agents should be provided. Information may be required on viral safety evaluation studies. In the case of semisynthetic or synthetic drug substances a reduced testing program could be proposed and justified.

Routine manufacturing controls on bacteria, mycoplasma and fungi should be described. These should generally involve the use of well-established (pharmacopoeial) analytical procedures or full details of alternative procedures should be supplied.

TSE and other nonviral adventitious agents should be addressed in detail. Certification relating to the use of bovine-derived materials as coming from acceptable countries should be discussed.

Viral safety validation studies should be discussed. The proposed viral safety tests should be discussed and justified. Appropriate test results should be submitted. Particular attention may need to be paid to the results from scaled-down models of the intended manufacturing procedure, the viral removal or inactivation processes and equipment and the adequacy of those processes.

21 Nonclinical Testing Strategy

Klaus Olejniczak
Federal Institute for Drugs and Medical Devices, Bonn, Germany

Rolf Bass
Charité-Universitätsmedizin Berlin, Berlin, Germany

SUMMARY

Nonclinical studies with and without the use of animal testing methods have the purpose of limiting risks whenever a new substance is to be tested or used therapeutically in a medicinal product in humans. Such studies should be designed and applied strategically in such a way as to achieve early, (almost) risk-free, and ethically unproblematic transition from nonclinical testing to clinical use under economically feasible conditions. For this purpose, the legitimate interests of trial subjects, patients, pharmaceutical manufacturers (usually the sponsors), the scientific community, as well as competent regulatory authorities have to be understood and harmonized to achieve a course of action that is medically and ethically correct (with respect to animal testing and use in humans), and to adhere to legal requirements. The development of a new medicinal product is thus guided and restricted: its progress depends on the outcome of various nonclinical and clinical investigations. The sequential order, the test details and their logic, and the duration of testing need to be agreed among all parties concerned with product development. Chronological overlaps rather than shortcuts need to be planned carefully. As pharmaceutical developments are moving toward "personalized medicines," it becomes clear that each new active substance is a separate entity, whose investigations have to be strategically designed. The "flowcharts" described in this chapter support the development and application of product-specific testing strategies. They provide a platform for transparent use of decision-tree methods. The "flowchart" system allows for a structured and science-based discussion among nonclinical and clinical investigators as well as between sponsor and competent authority. Each test system available can be probed in principle to determine whether or not it is required.

INTRODUCTION

Certain in vivo and in vitro studies are referred as "nonclinical." They address pharmacological (mechanism of action), pharmacokinetic (behavior of substance and its metabolites), and toxicological (untoward and unexpected action) issues. They are carried out and followed up within the framework of medicinal products development, with and without the use of animal testing methods. They are chosen and applied within a "testing strategy," which will need to be adapted (changed, widened, or shortened), throughout product development. Application of proper testing strategies should allow for carrying out risk assessment for a new substance before it is used in man or before it will be authorized as a medicinal product.

Following the thalidomide disaster, nonclinical studies have been developed more widely, introduced, and legally required to limit the risk of clinical trial subjects and patients whenever they are exposed to a developmental or therapeutic compound. This desired principle of maximum risk exclusion is valid from the first orientating use in humans (healthy subjects/patients) onward throughout its establishment in routine therapy. Thus, nonclinical studies constitute a rolling program to be applied in a stepwise fashion. Results are expected to offer as much safety as possible during every phase of the use in humans. Nonclinical studies begin before the substance is used in healthy volunteers and are completed before an application for marketing authorisation can be submitted successfully. Whereas the average overall development time for new medicinal products ranges from 5 to 15 years, nonclinical studies are actually performed only part-time during these years. Thus, they are not the rate-limiting step of drug development, but it remains crucial to have available those nonclinical results that are needed at a certain point in time of clinical development. If this is to be attempted successfully, a "testing strategy" must be available. The development of such nonclinical testing strategy is a process (1), (2) during which the legitimate interests of the following groups must be duly considered:

Subjects and patients: They have a right to the highest possible level of safety, and in the case of the patients, they have an interest in that new and promising medicinal products become available as quickly as possible.

Manufacturers of medicinal products: They have an interest in developing their product in a way that carries as few risks as possible and is fast and economically advantageous in the global market.

Scientific community: This community consists of competent scientists at universities, industry, independent and state-controlled (contract) research facilities and authorities that work in the areas of medicinal products research and development, and in associated areas. This community, by publishing and discussing scientific work, reveals the state of scientific knowledge relevant at the time when the medicinal product is developed.

Regulatory competent authorities: They are obliged, in the interest of maintaining and fostering individual and public health, to guarantee sufficient control during and after fast, ethical, and economical development of a product. Balancing the impact of other relevant regulations and recommendations (e.g., laws on animal protection) is also under the control of the authorities. The apparently varying requirements of the groups mentioned above are not basically contradictory, but rather complementary and overlapping. However, identical and similar requirements may be weighted differently depending on their human and animal ethical origin. Competent authorities are available throughout the European Union (EU). They reside in the Member States (and associated countries), are networked within the EMEA in London, and reside at the EU level at the EMEA (requiring the European Commission for legalizing their decisions).

HISTORICAL DEVELOPMENT OF NONCLINICAL TESTING

After the thalidomide disaster, nonclinical studies developed from "nice to have" to "must have." At the same time, the nonclinical toolbox became bigger and bigger. All its (growing) contents had to be applied for each case and under every circumstance. As the substances to be developed into medicinal products used to be of chemical nature, it was practical and understandable to apply the same tools for all: "one size fits all." As a consequence, tests that were scientifically unnecessary (or have become outdated) have been performed in order to obtain marketing authorisations until today. Competition between pharmaceutical companies and introduction of new classes of products (e.g., biotechnologically derived products that started in the 1980s) created an understanding that nonclinical testing should be done according to need and value and not according to toolbox opportunities. This could only be handled by applying the best strategy for using which tests and when. Individual strategies have already been applied by companies/sponsors for their own internal decision making in the past—with or without impact on the clinical trial authorization and the marketing authorisation conditions. In this chapter, a generally applicable "testing strategy" is set out and described, which is neither company nor product specific. It is thus suitable and recommended for use by all involved parties, and, namely, under those conditions when arguments and reasons are to be exchanged between these parties.

NETWORKING OF ESSENTIAL AREAS OF BIOLOGICAL RESEARCH IN THE PROCESS OF MEDICINES DEVELOPMENT

It has already been pointed out that the design of a testing strategy is a continuous process, which, regarding its contents and chronological order, is generated on the basis of ever-growing scientific knowledge in the areas relevant for the medicinal product under development, and on the increase in data for the developmental product. Fundamental agreement and understanding among the scientists working in product development must be achieved. They can be chemists/pharmacists, biologists, biochemists, biometricians, veterinarians, and physicians working in different fields. Those fields cover everything from chemical synthesis to the various phases of clinical investigation as well as the patient-oriented work of the health-care practitioner. Intensive communication and exchange of information during the whole development process is an essential prerequisite for success.

INTERTWINING OF NONCLINICAL AND CLINICAL STUDIES

Chronologically, certain nonclinical trials usually precede certain clinical trials. Chronologically, certain nonclinical trials usually precede marketing authorisation. The specific contents and experimental endpoints of nonclinical studies can be deduced from the scientific questions arising from the clinical trials and the marketing authorisation planned. Generally, within the nonclinical chain of experiments the characterization of the pharmacodynamic effect and pharmacokinetic behavior is followed by an investigation into toxicodynamic effects before the substance can be used in humans at all. Results obtained from the use in humans usually have an effect on the kind and content of further nonclinical studies and their logical and chronological position in the development process. Thus, the first cycle of accumulating knowledge and its impact on the design of the remaining nonclinical program is completed. Such cycle can and will be repeated as a spiral as development progresses. This is depicted in Figure 1, which gives an overview of the most important areas of research concerned and their integration. Each cycle will allow for new, improved, and more valuable interpretation bridges.

In addition to the well-established fields of nonclinical testing, we are now facing a number of additional areas that will need to be included in the toolbox for their strategically planned use. Among these are the following:

- How to improve predictability of nonclinical safety evaluation?
- Use of biomarkers and surrogate markers for establishing (nonclinical) safety (for use instead of or in combination with established test systems)

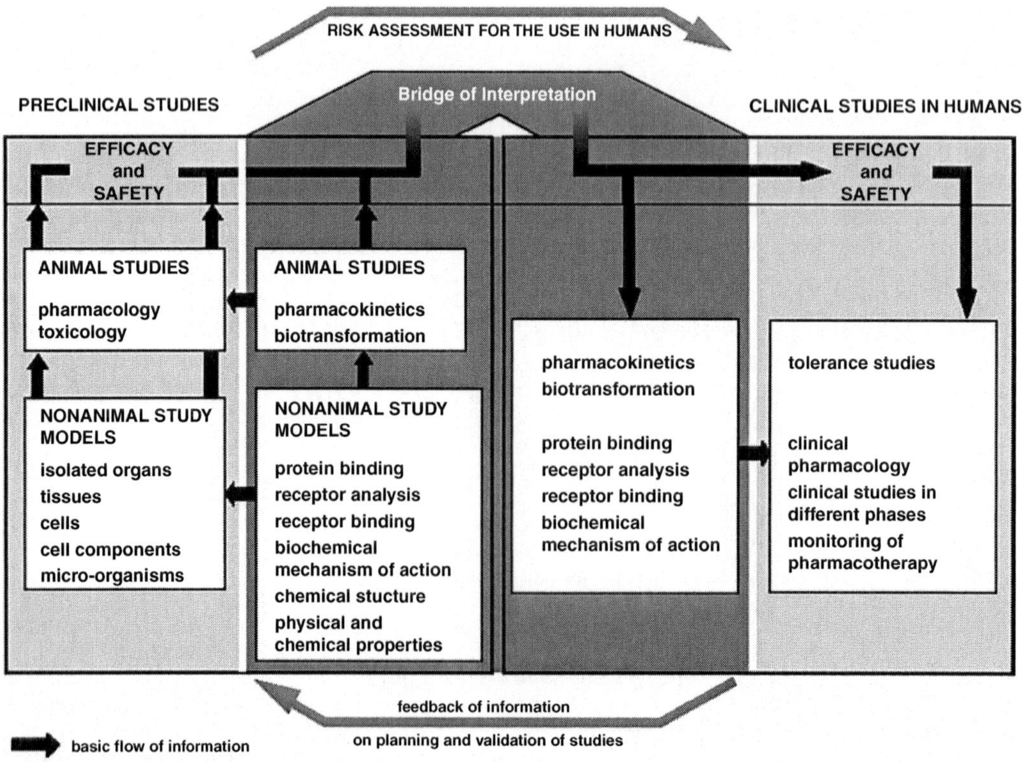

Figure 1 Intertwining of the most important biological fields of research of nonclinical and clinical studies to assess safety and efficacy.

- Strengthening the monitoring of risk/benefit for improving (clinical) safety (look-back and feedback procedures and strategies for resolving clinical questions and issues).

These additional areas are inherent in the intertwining of nonclinical and clinical studies depicted in Figure 1, and the normal routes taken for the flow of information are indicated.

EXPLANATION

Before a potential medicinal product is used in humans **for the first time,** there are, with regard to risk assessment, exclusively results available from nonclinical investigations and possibly hints on potential efficacy in humans stemming from experience with more or less related compounds. At this point in the development program, no reliable statement can be made about the similarity of the testing models to the situation in humans. As a result, the investigator must use the potential medicinal product in a variety of testing models (various models of animal species, application forms and experimentation, and different duration of studies) in order to increase the chances of having included relevant models for the situation in humans. Their actual relevance can only be stated **after** the first studies in humans have been undertaken. With an increase in scientific knowledge resulting from various steps of the clinical trials, certain results obtained in the nonclinical stage will invariably lose their meaning for risk assessment.

On the other hand, through the flow of information from clinical studies, those nonclinical models can also be identified which are particularly appropriate for risk assessment on the basis of their similarity to the use in humans. Results from these tests together with those from the first stage of clinical investigation form the basis of risk assessment, which must always be carried out before the next phase of clinical trials can be entered. The spiral is now exchanged for a staircase with step-by-step process, where the steps and landings are formed by newly added nonclinical and clinical data.

Before entering the staircase, it is necessary to construct a risk assessment from the most sensitive nonclinical testing models. This reflects the zero knowledge base and follows medicinal and ethical reasons. This rule remains valid until it can be shown convincingly that the models have no or only limited biological impact on risk assessment.

The process thus described and depicted in Figure 1 (i.e., drawing conclusions from nonclinical test results with a view to possible results and risks for humans, and the feedback from relevant information obtained from clinical trials) turns the development and application of nonclinical testing strategies into a complex and dynamic process beyond fixed plans or checklists.

The actual design process necessitates the ability to criticize and reanalyze the planning, undertaking, evaluation, and assessment of results from nonclinical or clinical trials. It is acknowledged that any individual kind of investigation may be of limited relevance. Prudence is needed to understand how results obtained can influence the type and extent of subsequent nonclinical and clinical studies. The design of testing strategies must, therefore, be accompanied by a high sense of responsibility reconciling the patients' and doctors' requests for new safe medicines as quickly as possible, the opportunities to improve the scientific state of the art, and the requirement of protecting laboratory animals. Adhering to this principle will reveal, after thorough evaluation, which practical steps **must** be taken regardless of all the steps that **could** be taken (toolbox)—for each step of development.

CURRENT BASIS FOR THE DESIGN OF NONCLINICAL TESTING STRATEGIES

European Legislation and Guidelines

The current legal basis in the EU concerning nonclinical requirements in medicinal products testing is laid down in Directive 2001/83, as amended, on the approximation of the laws of Member States relating to analytical, pharmacotoxicological and clinical standards (also described as pharmaceutical/biological **quality**, nonclinical **safety,** and clinical safety and **efficacy**), and protocols in respect of the testing of medicinal products. The nonclinical requirements are further specified in the guidelines (see Appendix 1), which have been developed by the Committee for Human Medicinal Products (CHMP) Safety Working Party—also in conjunction with the International Conferences of Harmonisation of Technical Requirements

Table 1 Possibly Available Clinical Data to Be Considered

- Drugs for diseases predominantly or exclusively affecting pediatric patients
- Duration of pediatric treatment
- Age of pediatric population
- Primary pharmacodynamics in target organs/tissues with significant postnatal development
- Safety data in adults and the pediatric population in relevant age groups
- Results from pediatric treatment with a drug of similar chemical structure and of the same pharmacological class
- Medicinal product intended to treat serious or life-threatening diseases
- Relevant pharmacokinetic (ADME) data

of Pharmaceuticals for Human Use (ICH) process. These guidelines, while not directly legally binding on the involved parties, set forth the scientific state of the art usually to be applied in drug development. The EU guidelines available in mid-2008 are given in Appendix 2 (http://www.emea.europa.eu/htms/human/humanguidelines/nonclinical.htm). Tables 1 and 2 should be used in conjunction with Fig. 15 (p. 317).

International Developments and Implications

ICH reflects the trilateral negotiations between Japan, the United States, and the EU which, in October 1989, led to the initiation of the ICH process hosted by the International Federation of Pharmaceutical Manufacturers Associations (IPFMA). This ongoing work has the following objectives:

- To identify and eliminate the differing technical requirements in the three states/regions
- To avoid repetition of all kinds of tests
- To accelerate development of medicinal products, thus giving patients quicker access to new medicinal products without negatively affecting quality, safety, and efficacy

The expert working group for the field of (nonclinical) safety consists of a representative of each member authority [Ministry of Health Labour and Welfare (MHLW), U.S. Food and Drug Administration (FDA), and the EU and each regional trade association [Japan Pharmaceutical Manufacturers Association (JPMA), Pharmaceutical Research and Manufacturers of America (PhRMA) and European Federation of Pharmaceutical Industries Association (EFPIA)] (3–6). Whereas ICH started out with creating and harmonizing technical test areas (harmonization of toolbox contents), it has since moved into semistrategic areas providing guidance on which tests should normally be applied and when. Any individual test strategy would thus be derived by choosing appropriate tests from the toolbox according to the guidance on their applicability. Although there is in place an updating process for adjusting ICH Guidelines to the moving target state of the art, the guidelines may not reflect the actual requirements for proper strategies on a case-by-case basis. This implies that the development of individual testing strategies may be guided by what is available from ICH but often will need to be changed according to more specific needs and development stages. ICH Guidelines available in mid-2008 are given in Appendix 1 (http://www.emea.europa.eu/htms/human/ich/ichsafety.htm and http://www.emea.europa.eu/htms/human/ich/ichmulti.htm).

Table 2 Possibly Available Nonclinical Data to Be Considered

- Extent of relevant data from existing animal studies
- Adverse and irreversible reaction observed
- Target organs/tissues identified
- Mechanism of action
- Ratio between exposure resulting in the nonclinical effect and the human adult exposure, low or high
- Pharmacokinetic data show exposure of organs with significant postnatal development, relevant for the intended age group
- Data from pre- and postnatal toxicity studies, in light of the extent of pup exposure relative to the expected therapeutic concentrations
- Juvenile animal data from a drug of similar chemical structure and of the same pharmacological class

From Legislation and Guidelines to Testing Strategies

The top layer of strategic guidance is set out below. Such strategic guidance is supported further in the EU and in the United States, where "Scientific Advice Procedures" are in place. Strategies can be discussed between regulatory authorities and companies/sponsors on a case-by-case basis. In addition, attention must be drawn to a guideline on single dose (acute) toxicity. This guideline has not been approved as an official ICH Guideline but has been published as background paper in the proceedings of ICH 1 in Brussels (7).

Over time, the areas addressed and the content of approved guidelines, drafts, and concept papers have grown immensely. It is highly recommended to study them in detail. Whatever the workload required to fully comprehend what is currently available (and understood by the competent authorities to reflect the current state of the scientific art), this will enable and assure their proper reflection in the development of individual testing strategies. Guidelines not only set out when they would usually have to be applied for test performance but also provide hints about their limitations in applicability. Scientific reasoning believed to be sound may be submitted to the competent authorities for scrutiny, discussion, and advice (using national or EMEA advice procedures). As the listing of guidelines in the EU is somewhat difficult to comprehend, such listing (with source information and a cut-off date of end 2008) is given in Appendix 2. The same sources should be searched in order to find out about any new or changed guidance after the cut-off date.

DEVELOPMENT OF INDIVIDUAL TESTING STRATEGIES

General Concept and Prerequisites

Introduction

An adequate nonclinical testing strategy must be designed to allow for a reliable risk assessment regarding the use of a new medicinal product in humans in every phase of the clinical trials as well as during therapeutic use, as far as this is possible on the basis of nonclinical trials. This implies that initially the *assessment of potential risks* will be broadly based and will become more and more focused during clinical development. In order to fulfill this requirement the three aspects described in Figure 2 must be considered.

Explanation

Consideration of Main Features of Risk as Arising from Planned Clinical Studies

The identification process of a *potential risk* must be based on the situation envisaged for the clinical study scheduled to be performed. Therefore, planned studies in humans must be characterized with regard to the type of study aimed for, the group and type of persons, and the

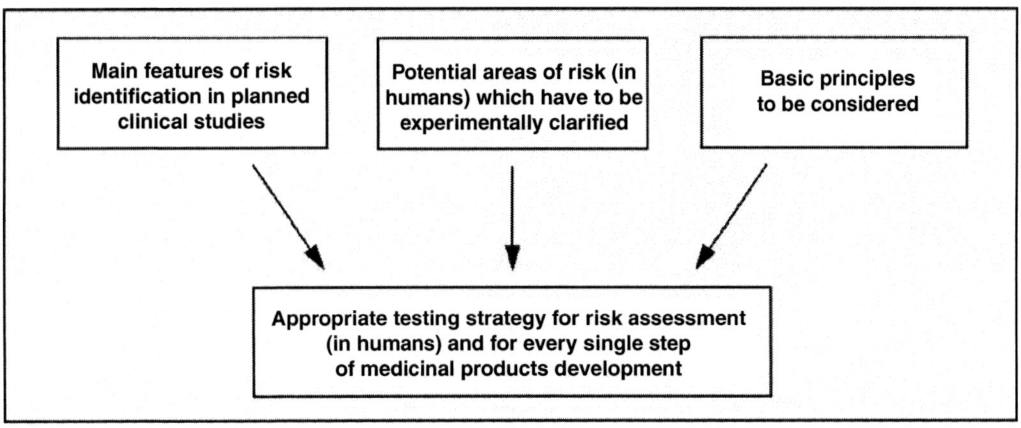

Figure 2 Concept for the development of individual testing strategies.

stage of illness that is sought to be treated, as well as the planned dose (single and cumulative) and the intended duration of treatment.

With regard to the types of studies aimed for it is necessary to differentiate at least between studies in humans on pharmacokinetics, dose titration, human pharmacodynamics, and safety. With regard to group/type of persons to be enrolled in a study, they can either be healthy subjects or patients whose sex, age, etc., must be addressed/known to make feasible assessment of potential risks through nonclinical studies. With regard to stage and severity of illness, these are important criteria to determine whether it is necessary to undertake special in-depth nonclinical trials referring to an individual study situation.

The following criteria of clinical studies have been agreed and established to determine choice and specification of toxicity studies: duration of treatment and doses to be administered. These usually need to be covered (and often in multiples of such doses) in the nonclinical study to allow for proper assessment of potential risk and the evaluation of safe clinical testing conditions.

It is expected that a thorough analysis of answers to these questions will lead to an all-inclusive identification of obvious risks for the groups of persons that are to be treated under certain conditions (all-inclusive meaning a strict conservative identification pinpointing even the smallest of potential risks arising from the nonclinical situation).

Potential Areas of Risk for Humans Which Have to Be Clarified Experimentally

In an attempt to cover all relevant areas of potential risk, it is necessary to consider an exhaustive list of adverse reactions that can be expected to occur in principle and to compare them with the adverse reactions observed during/after clinical trials. Consequently, identified and potentially meaningful risks have to be investigated experimentally.

Typical areas of possible damage in humans (according to the nomenclature of animal experimentation) are

- acute toxicity,
- repeated-dose toxicity,
- adverse effects on male or female fertility,
- embryotoxicity/postnatal adverse events,
- genotoxicity,
- tumorigenicity,
- safety pharmacology,
- sensitization/immune suppression and stimulation, and
- local and other particular adverse events.

Basic Principles to Be Considered

Following identification of those potential risks which would need to be investigated experimentally, nonclinical trials are planned, undertaken, and evaluated. The binding principles and norms to be considered refer to the present state of scientific knowledge in methodology, knowledge on mechanisms, knowledge on and observance of pharmacokinetic/toxicokinetic aspects, and biometric requirements. At the same time, ethical norms of human and animal testing have to be adhered to.

Technical Details of Planned Experiments

The identification of a potential risk for humans, which has to be simulated in an experimental trial situation for full clarification, as well as the consideration of technical and ethical norms, leads to decisions about the undertaking of general and special experiments. Such assumption of possible risk will determine a number of details of tests to be performed in the following:

- Basic suitability of the test system and practicability of individual studies
- Type of study (in vitro, in vivo, area of investigation, etc.)
- Animals (number and species)
- Routes of administration
- Duration of treatment and follow-up observation periods

Use of Flowcharts

Introduction

The nonclinical testing program has to be linked closely to the whole clinical testing program. The resulting coordination plan will reflect general requirements for the program and the individual requirements of the medicinal product under development. Therefore, the overall process, from the initial selection of hopefully suitable substances to the final submission of an application for marketing authorisation, needs to be divided into a number of steps (both the jumping of steps and their treadmill-like reiteration have to be envisaged) (Fig. 3): choice

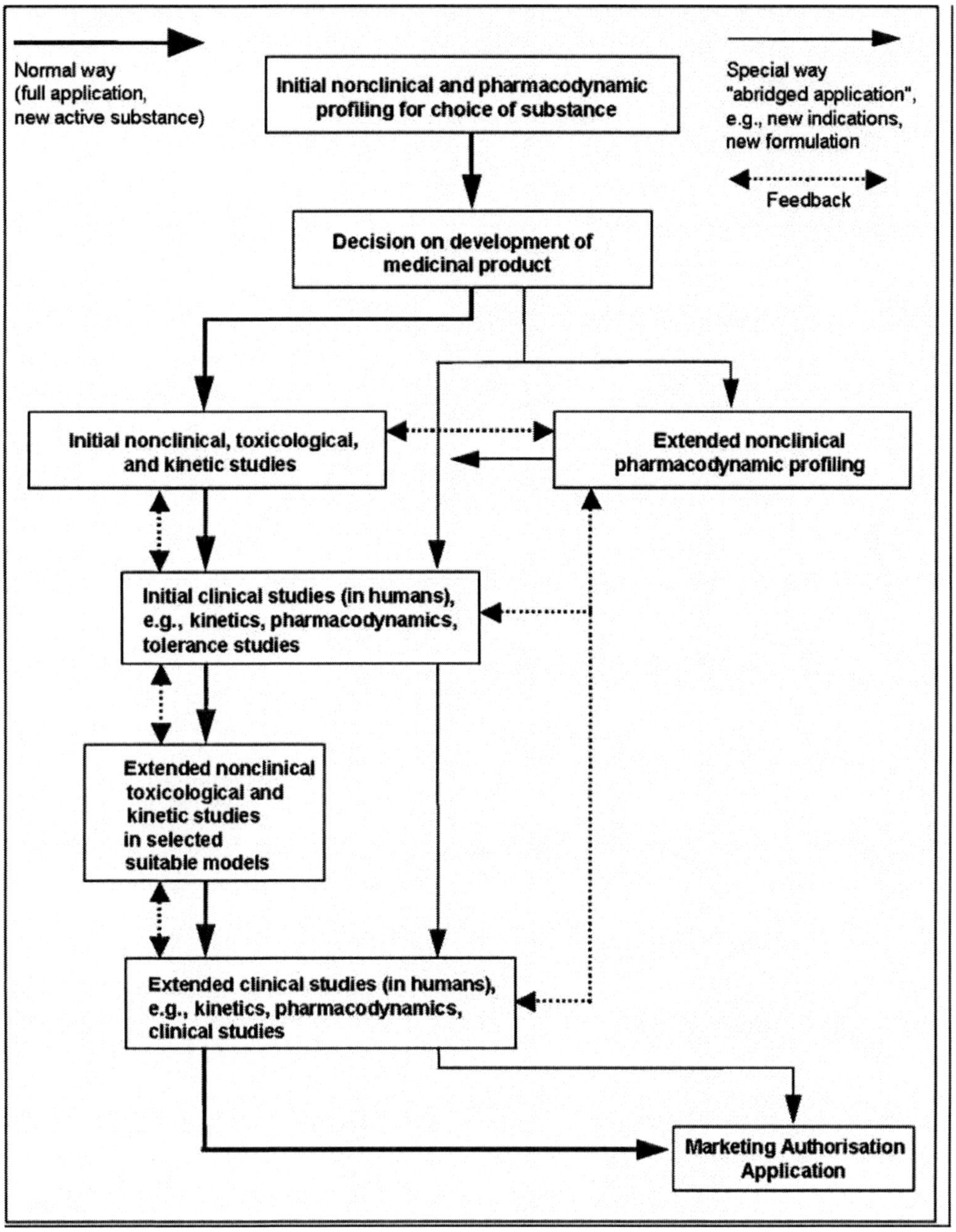

Figure 3 Coordination of nonclinical and clinical studies.

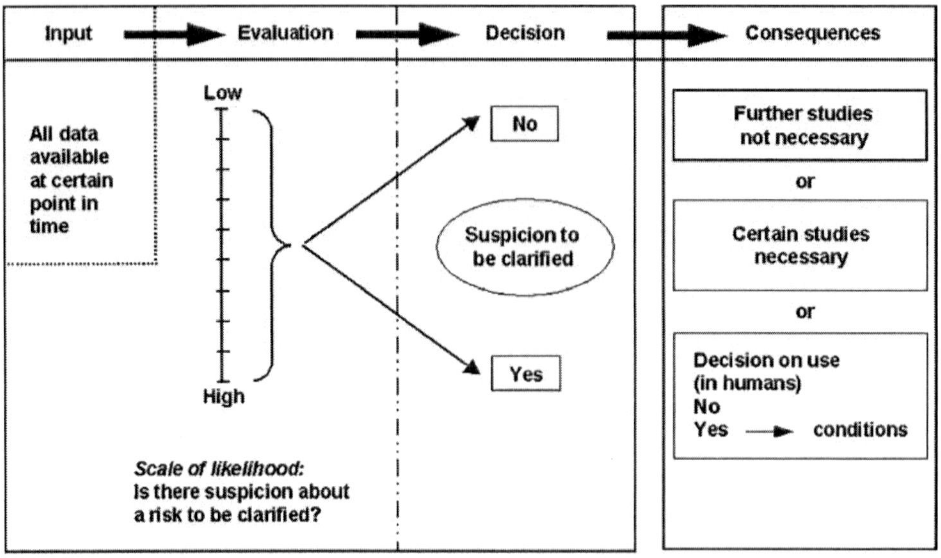

Figure 4 Flowchart on the development of nonclinical testing strategies.

of substance, initial toxicological and kinetic studies, extensive characterization of the pharmacodynamic profile, followed by introductory clinical studies, and subsequently followed by extensive studies in all nonclinical and clinical areas.

Although such coordination programs differ in detail, they are developed and applied by the whole of the pharmaceutical industry. It must be underlined that continuous feedback is necessary, vertically (between toxicology and clinic and from clinic to toxicology) as well as horizontally (between toxico- and pharmacodynamics, experimental and clinical kinetics) (Fig. 3). It is presupposed that the main path of development of new active substances (Fig. 3) will be applied unless serious reasons suggest differently. Abridging is also indicated, for example, in the case of new forms of administration, and possibly also for new clinical indications.

Principle of the Flowcharts

The principle of the flowcharts suggested for determining and agreeing nonclinical testing strategies lies in the stepwise use of a procedure allowing for logically deducted "yes"/"no" decisions (Fig. 4). In order to facilitate the understanding of this principle, the terms "risk" and "probability" must be defined.

- "Risk" means that something undesired can occur (because there is a certain likelihood of occurrence that has not (yet) been sufficiently excluded).
- "Probability" means that the evaluation of occurrence of a potential risk will remain relative. It cannot be evaluated toward an absolute degree of certainty, that is, its likelihood can be called/classified anything from "low" to "high," but it can never be fixed at "zero" or "100%." Therefore, nothing can ever be excluded or foretold with certainty.

Experience and scientific knowledge are necessary for risk identification. If the assessment of a graded risk shows that the probability of the risk occurring is too high to be considered irrelevant for the clinical situation (i.e., the suspicion having arisen requires experimental clarification), then a suitable nonclinical experiment must precede testing in humans. It is understandable that during medicinal product development there will be a continuous increase in the amount and type of those available data (Fig. 4), which suggest that a certain degree of risk is sufficient to raise a suspicion.

This means that due to a mere lack of information a "Yes" decision referring to an unclear situation is more likely to be taken in the early phase of development than in a later phase.

The flowcharts (Fig. 4 and following) primarily serve the purpose of discussion preceding the decision making. Furthermore, they facilitate the communication of such a decision, once it has been taken, to others. The aim is to deliver instructions for decision making based on scientific knowledge by considering all available data and the scientific knowledge at that point in time when a decision on certain nonclinical trials as a precondition of clinical trials is to be made, for example, before healthy subjects are to be exposed to an active substance for the first time. The same aim applies at later points in time, when decisions on additional nonclinical trials are to be made as a precondition of later clinical trials, for example, before patients are to be exposed.

The "decision" whether there is a risk, which must be evaluated on the "scale of likelihood" to determine whether it could lead to such a degree of suspicion that would require scientific clarification, must be answered on the basis of all available information. The "evaluation of the degree of risk," which ranges from "low" to "high," can only lead to one of the two following "decisions":

- *No*, there is no suspicion requiring experimental clarification
- *Yes*, there is suspicion requiring experimental clarification

The impact of these structured decisions on the requirement for particular nonclinical studies as well as on the decision about the *possible, conditional, or prohibited use* of the developmental compound in humans will become evident and can be categorized with the help of special flowcharts covering all typical areas of possible damage to humans as reflected in nonclinical studies. The different testing areas are usually of interest during various phases of medicinal products development, that is, they will have to be covered and decided sooner or later during development.

Areas of Investigation Represented in Flowcharts

There are flowcharts for the following areas of investigation:

- Single-dose toxicity
- Repeated-dose toxicity
- Reproductive toxicity
- Genotoxic potential
- Tumorigenic potential
- Safety pharmacology
- Delayed ventricular repolarization (QT interval prolongation)
- Juvenile toxicity

Single-Dose Toxicity

The information available in the early stages of product development (when it is usually decided to carry out acute toxicity studies) is normally scarce. Presuming possible acute toxicity in humans with the product under investigation will very probably lead to the conclusion that there is sufficient suspicion to require experimental clarification. In these cases, an adequate testing method must be chosen and applied (choice of adequate acute toxicity testing methods be it choice of species, escalating or single doses, nonlethal or lethal doses, etc., are not addressed specifically in this chapter). However, the discussion leading to the decision that single-dose toxicity testing must be performed will provide guidance on the types of questions to be answered experimentally (Fig. 5). This will provide almost automatically guidance toward choosing proper approaches to single-dose toxicity testing.

Repeated-Dose Toxicity

Under application of the above-described procedures, a decision will be taken on whether there is justified suspicion of toxicity in humans from repeated administration in clinical trials, which will require experimental clarification by means of (sub)chronic toxicity testing in the experimental environment (Fig. 6).

As a decision regarding the repeated clinical use of a new active substance must usually be taken at a point in time when there is only insufficient information to argue in favor of the claim "no suspicion in humans," this gap must be bridged by experimental data. The conditions

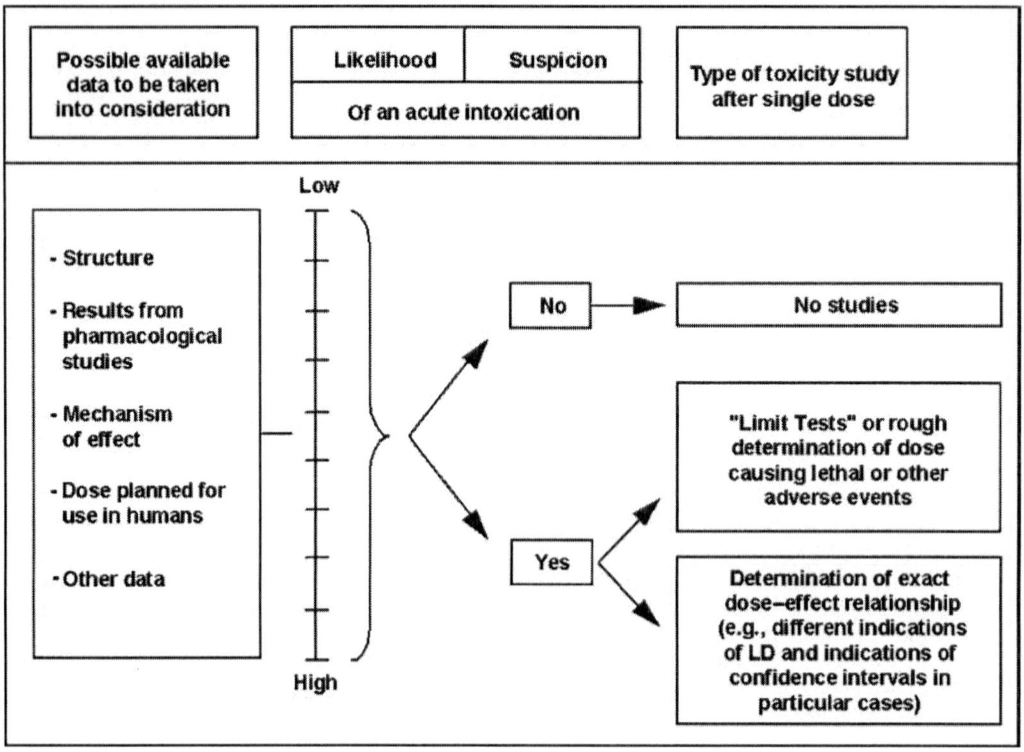

Figure 5 Single-dose toxicity.

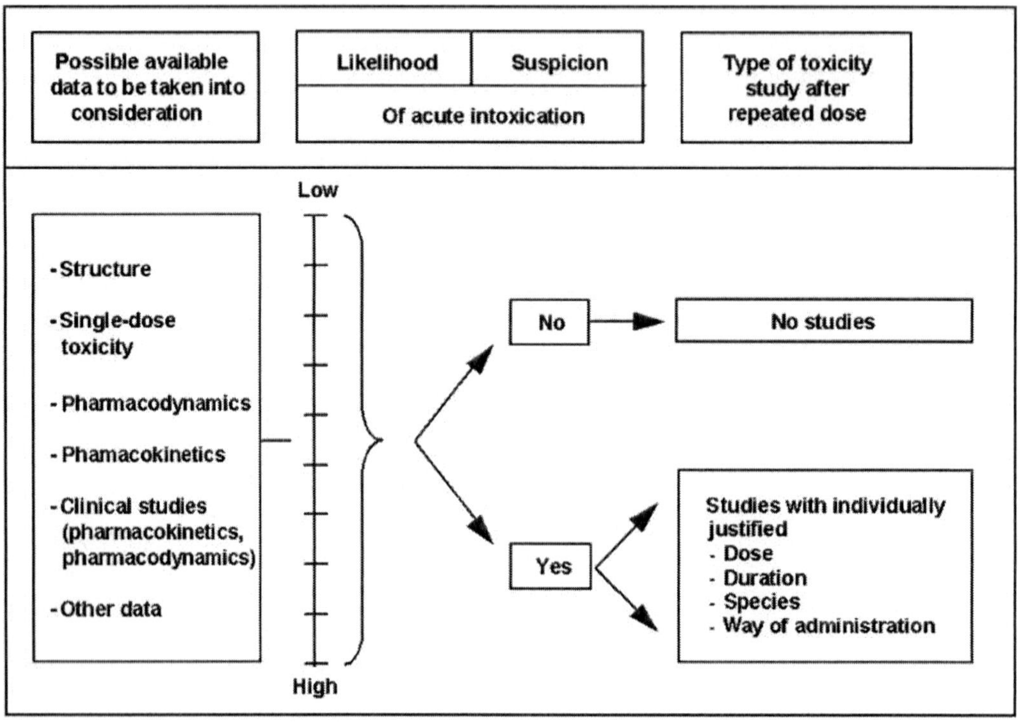

Figure 6 Repeated-dose toxicity.

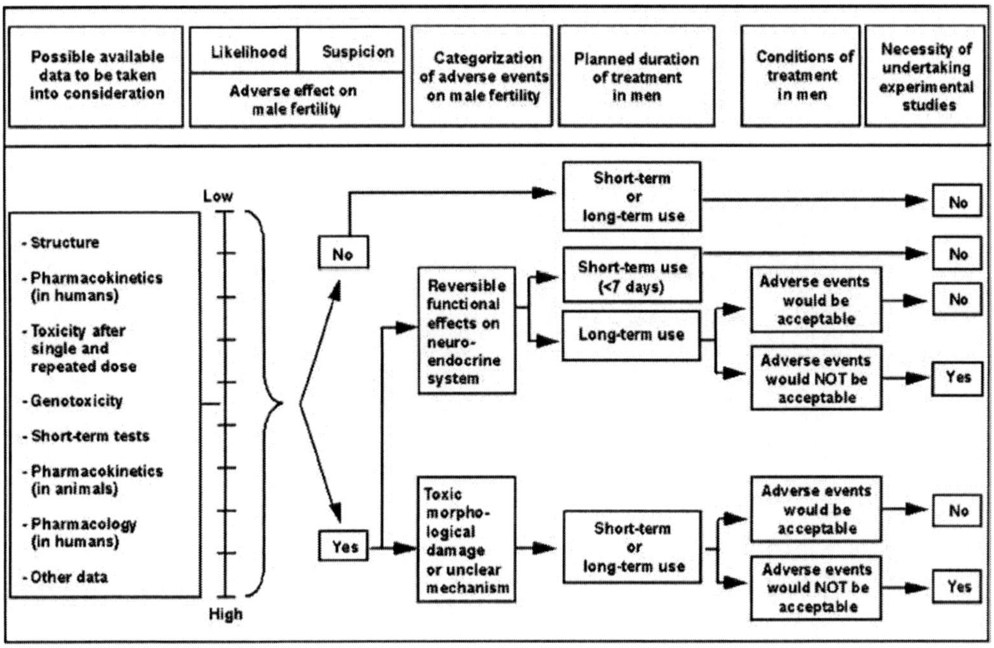

Figure 7 Reproductive toxicity, effects on male fertility, and studies usually in rats.

of the study must be laid down in harmonized consideration of the features of the individual substances to be investigated and the applicable guidelines.

Reproductive Toxicity

Owing to the variety of forms of potential toxicity that can occur in humans, namely, male fertility disorders (Fig. 7), female fertility disorders (Fig. 8), embryonic or fetal toxicity (Fig. 9),

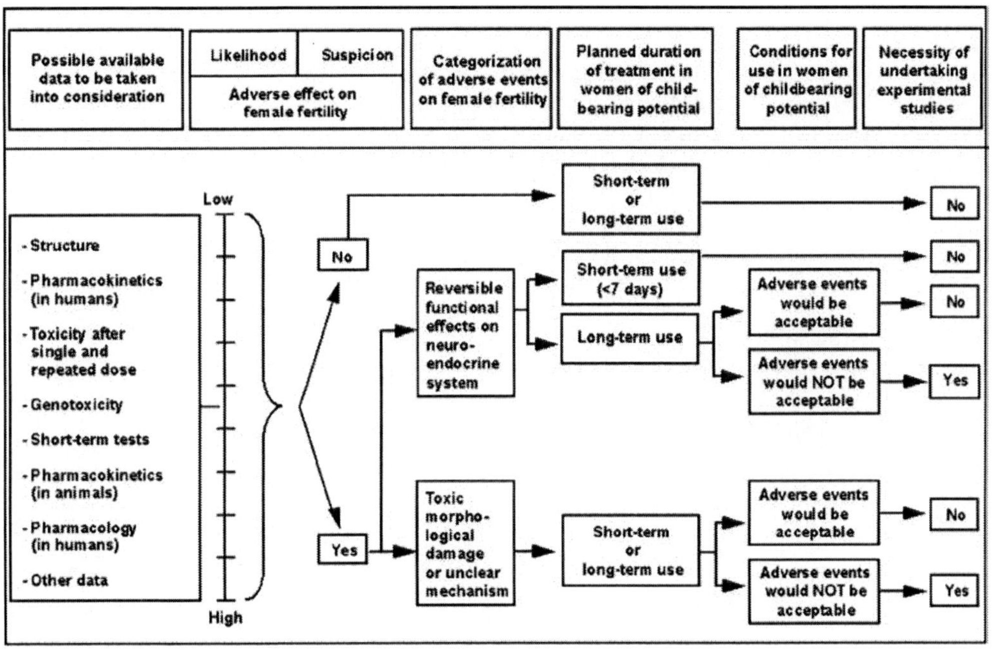

Figure 8 Reproductive toxicity, effects on female fertility, and studies usually in rats.

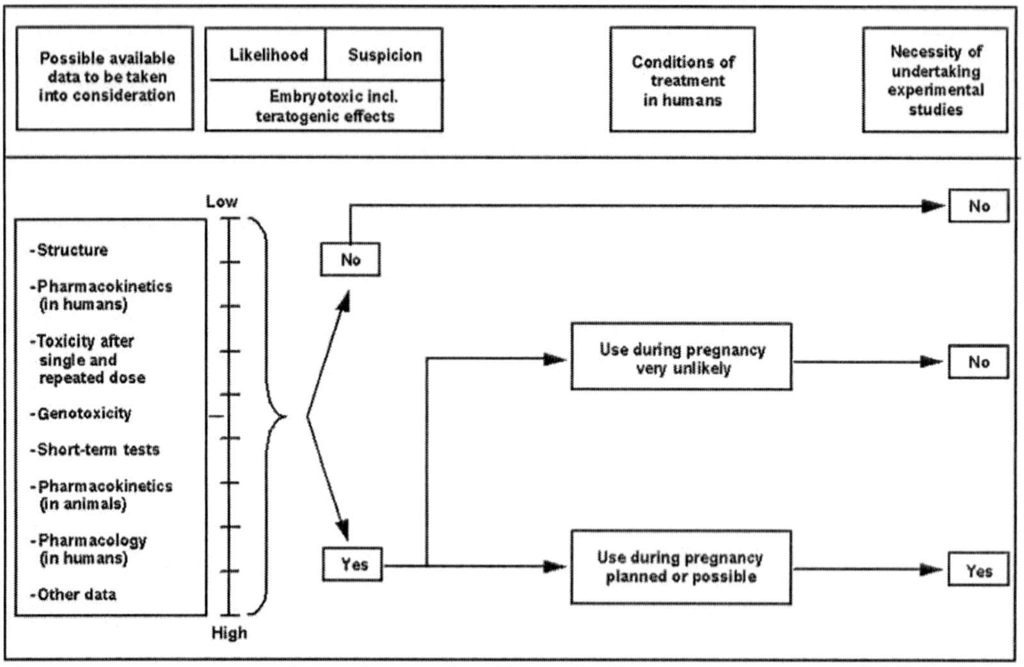

Figure 9 Reproductive toxicity, studies on embryonal toxicity and pharmacokinetic investigations on placental crossing, and studies usually in rats/rabbits.

and postnatal development disorders (Fig. 10), individual flowcharts have been developed for these areas to facilitate decisions on the topics concerned.

If a "yes"/"no" decision on a justified suspicion has been taken, there follows a range of further instructions:

- Qualification of suspicious adverse effects on fertility, such as "reversibly functional" or "toxic" or "unclear" (Figs. 7 and 8)
- Duration of proper use as either "short-term use" or "long-term use" (Figs. 7 and 8)
- Conditions for the use in humans with "undesired effects are acceptable" or "undesired effects are not acceptable" (Figs. 7 and 8), or "use during pregnancy is very unlikely" or "use during pregnancy is foreseen or possible"

These points lead to more than one possibility for a "yes"/"no" decision about the necessity of experimental studies. The availability of certain data that might change the decision from "yes" to "no" or vice versa based on suspicion must be analyzed on a case-to-case basis. The utilization of data derived from (sub)chronic toxicity studies for the understanding of male fertility disorders is one example.

In addition to the scientifically justified and "neutral" decision-tree method, which is represented in the flowcharts, also emotional reasons have come into play for undertaking experimental studies. The consistent use of the flowcharts should help in avoiding decisions and actions that are medically and ethically unjustifiable.

Genotoxic Potential

This is the only flowchart where sequential generation of data has been incorporated into the decision-making process in order to specify the extent to which results can provide indications of mutagenic effects occurring when administering therapeutic doses: "basic data" and "additional data". The following two options for the standard battery are considered

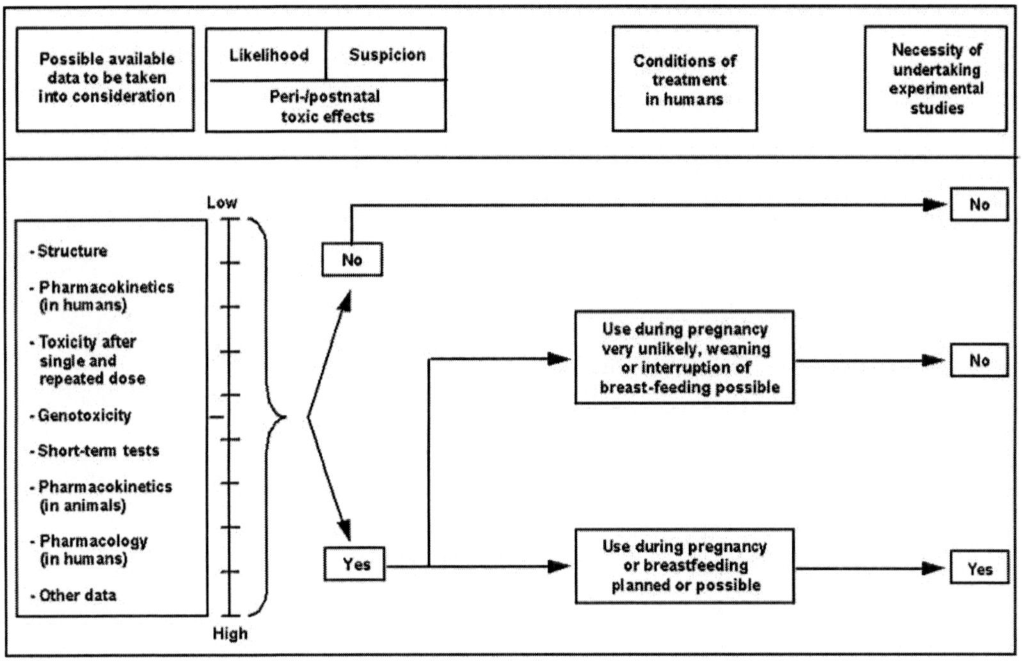

Figure 10 Reproductive toxicity, studies on peri-/postnatal toxicity and pharmacokinetic investigations on transition into mother's milk and studies usually in rats.

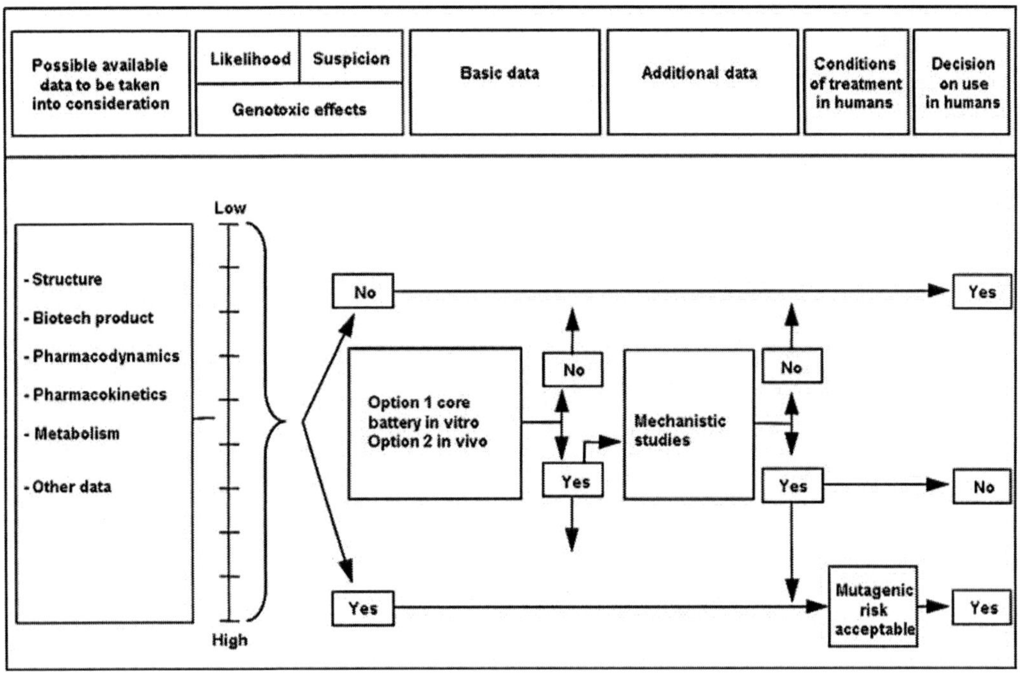

Figure 11 Genotoxicity.

equally suitable:

Option 1

- A test for gene mutation in bacteria
- A cytogenetic test for chromosomal damage (the in vitro metaphase chromosome aberration test or in vitro micronucleus test), or an in vitro mouse lymphoma *tk* gene mutation assay
- An in vivo test for genotoxicity, generally a test for chromosomal damage using rodent hematopoietic cells, either for micronuclei or for chromosomal aberrations in metaphase cells

Option 2

- A test for gene mutation in bacteria
- An in vivo assessment of genotoxicity with two tissues, usually an assay for micronuclei using rodent hematopoietic cells and a second in vivo assay

Under both standard battery options, the in vivo genotoxicity assays can often be integrated into repeated-dose toxicity studies. The value of in vivo results is directly related to the demonstration of adequate exposure of the target tissue to the test substance (Fig. 11).

As is the case with the procedure described for reproductive toxicity, acceptance of effects is a criterion requiring clarification. The final "yes"/"no" decision, however, will affect the use in humans (or not). In all other flowcharts it is decided whether experimental studies are necessary.

Tumorigenic Potential

At the point a decision on suspicion requiring experimental clarification must be taken in order to assess a product's tumorigenic potential in humans, a lot of experimental and clinical data is normally available. Nevertheless, evaluation of these data toward "no suspicion" requiring clarification remains very difficult. Thus, economic criteria continue their role in the decision-making process.

It must be underlined that in addition to the neutral criteria suggesting "suspicion" used so far (e.g., chronic use, suspicion arising from genotoxicity studies, chronic toxicity studies, structure-and-effect relationships) more "modern" criteria remain desirable to scientifically and rationally justify the suspicion of tumorigenic potential of an active substance in humans. Such "modern" criteria have not been generally accepted to date. The currently valid situation suggesting "suspicion" is hard to understand. It seems hardly justified to infer sufficient suspicion solely from long-term use, that is, more than six months. Still, ICH partners have agreed on this procedure (lowest common denominator).

It seems apt to also apply the arguments already listed for reproductive toxicity in this field, that is, to reach a decision about the necessity of tumorigenicity studies depending on the acceptance of an alleged tumorigenic potential of the medicinal product and the indication proposed (Fig. 12).

To aggravate matters further the meaning of the various toxicological studies in use is seen as contradictory. Although it has been confirmed that many long-established studies are insufficient for the purpose of risk assessment, their replacement by other studies is very problematic. The difficulties with justifying scientifically a suspicion as well as with evaluating results from so-called, short-term studies for the investigation of tumorigenic potential very often do not lead to logical arguments in favor of or against typical long-term studies.

Safety Pharmacology

Safety pharmacology investigations are meant to answer questions regarding adverse events that cannot be fully addressed in the framework of toxicopharmacodynamic studies. In the foreground, there are effects on vital functions that are meaningful when it comes to assessing the safety of a medicinal product. Safety pharmacology investigations that are carried out at an early stage of medicinal products development can influence the further development (Fig. 13).

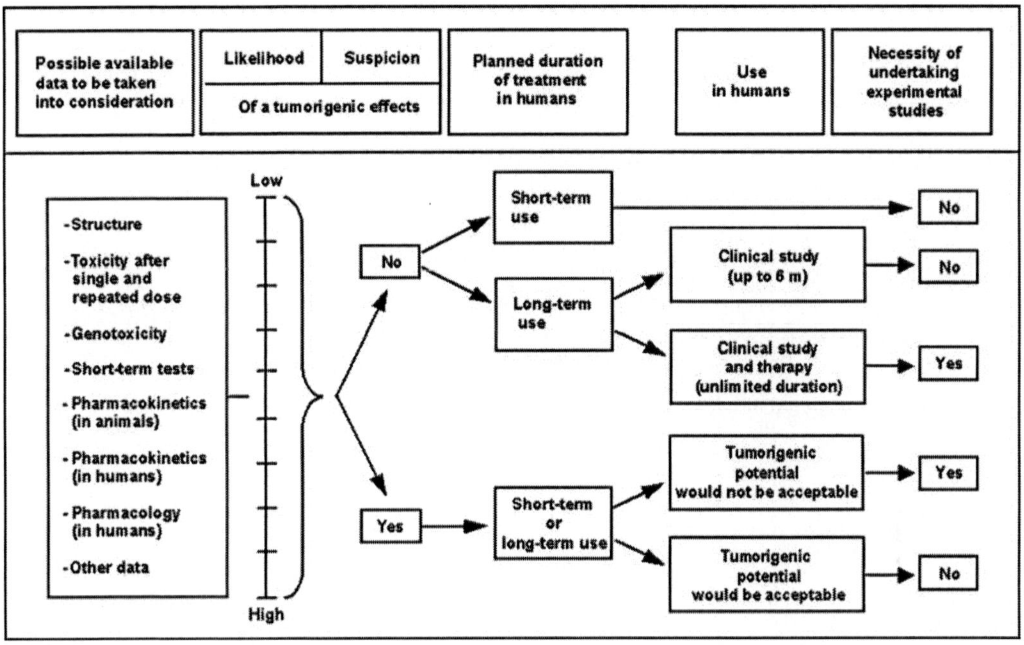

Figure 12 Studies on tumorigenicity.

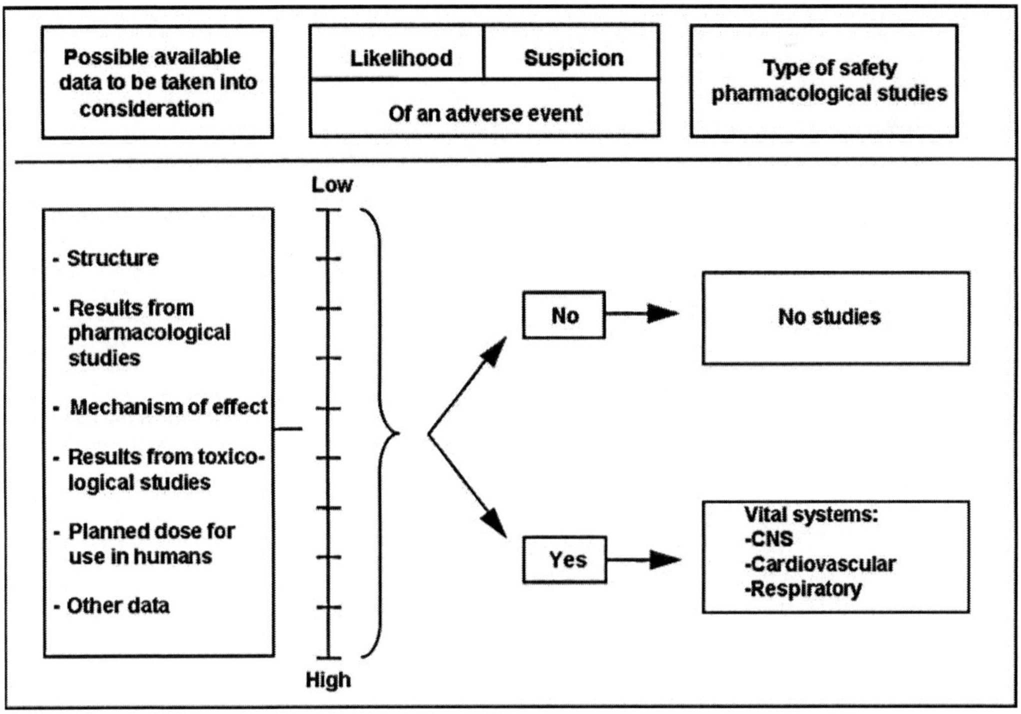

Figure 13 Safety pharmacology.

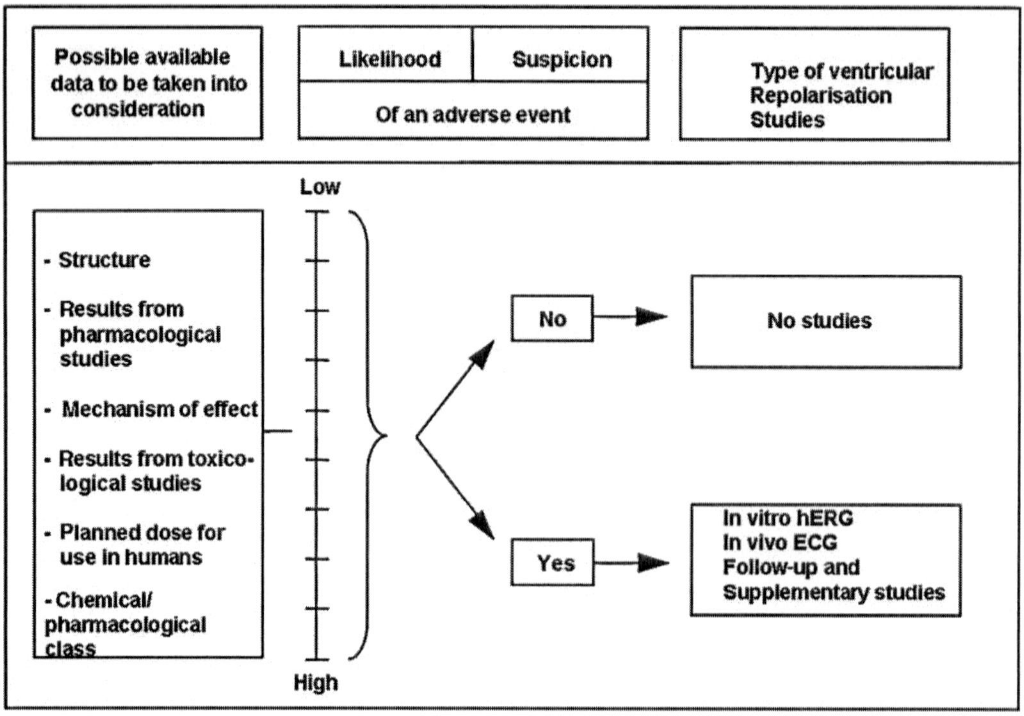

Figure 14 Delayed Ventricular Repolarization (QT Interval Prolongation).

Delayed Ventricular Repolarization (QT Interval Prolongation)

The flowchart for delayed ventricular repolarization (QT interval prolongation) (Fig. 14) extends and complements the flowchart in Figure 13 safety pharmacology. The testing strategy for assessing the potential of a test substance to delay ventricular repolarization is a major part of safety pharmacology development. In vitro and in vivo assays are complementary approaches. Therefore, according to current understanding, both assay types should be conducted. In vitro electrophysiology studies can explore potential cellular mechanisms that might not be evident from in vivo data. Although delay of repolarization can occur through modulation of several types of ion channels, inhibition of IKr is the most common mechanism responsible for pharmaceutical-induced prolongation of QT interval in humans.

In vivo models that possess the full complement of molecular, biochemical, and physiological systems can also be informative with regard to the response in humans to the test substance. Carefully designed and conducted in vivo studies allow evaluation of parent substance and metabolites and can enable estimation of safety margins. In vivo ECG evaluations provide information on conduction properties and noncardiac influences (e.g., autonomic nervous system tone).

The study results can be used to elucidate the mechanism of action and, when considered with follow-up and supplemental studies, estimate risk for delayed ventricular repolarization and QT interval prolongation in humans.

Nonclinical Studies in Juvenile Animals

Following international developments and introduction of the EU legislation concerning need and opportunities to therapeutically use new pharmaceuticals not only in tested adult populations but also in clinically tested pediatric populations (to be differentiated into different age groups), discussion has arisen about the need to perform nonclinical special/additional studies in juvenile animals (Fig. 15). Whereas for many scenarios results obtained in adult humans will serve the purpose of designing pediatric clinical trials under special circumstances, such clinical investigations will need to be preceded by animal studies. The conduct of studies in juvenile animals should be considered when human safety data and previous animal studies

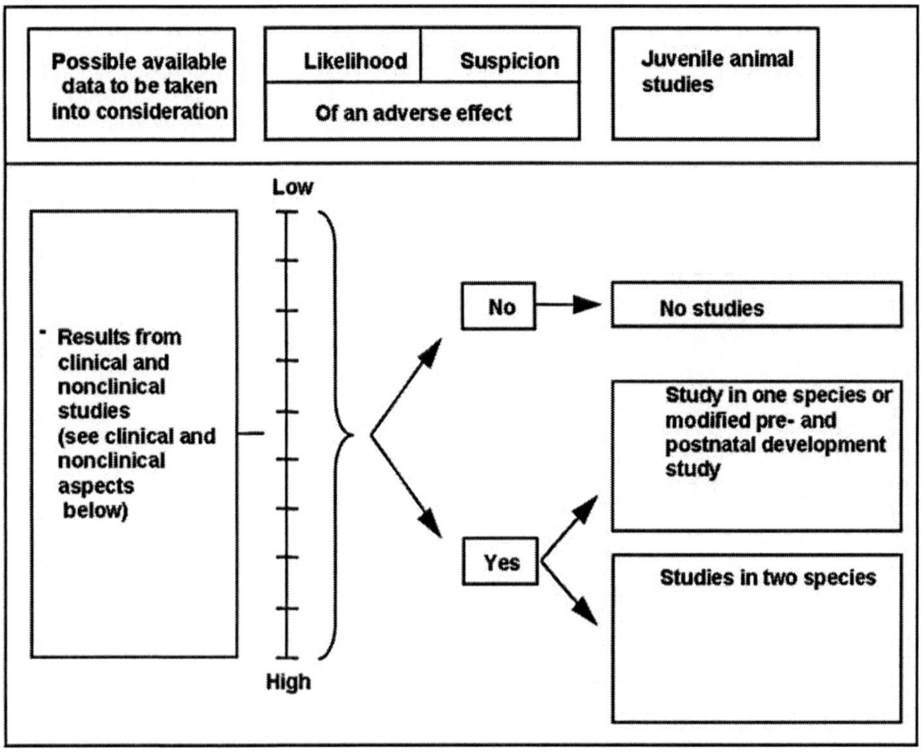

Figure 15 Nonclinical studies in juvenile animals.

are considered insufficient for a safety evaluation in the intended pediatric age group. Before performing a juvenile animal toxicity study, it should be considered whether a developmental toxicity issue could be addressed in a modified pre- and postnatal development study in rats. Testing of juvenile toxicity in one appropriate species using both sexes will normally be sufficient.

Thus, a special flowchart has been designed and is now included. As the possibly available data to be taken into consideration for the design and need of juvenile animals is extremely variable, their clinical and nonclinical aspects have been compiled separately.

Outlook

The flowcharts presented and described present the main areas of concern to be addressed and resolved during development of a medicinal product.A development team may find that the flowcharts do not fully take into account their specific product and they may thus wish to supplement the flowchart series by adding additional areas of risk to be addressed or by becoming more specific in those areas already presented. The area of risk described as "phototoxicity" is currently not included (the reason being that the guideline is under revision). Certain areas of risk such as drug-induced liver injury (DILI) and other organ toxicities, which remain difficult to assess, have come into scientific focus recently. Quite naturally, the desire is to understand and detect such toxicities in relation to drug intake in humans but depending on the stage of development and knowledge, nonclinical investigations may be the proper choice. Here test systems such as biomarkers are creating new opportunities for widening the toolbox of available test systems.

REFERENCES

1. Olejniczak K, Günzel P, Bass R. Preclinical testing strategies. Drug Inf J 2001; 35(2):321–333.
2. Günzel P. Grundlagen und allgemeine Methoden des Tierversuchs in der Toxikologie. In: Hess R, ed. Arzneimitteltoxikologie, Anforderungen—Verfahren—Bedeutung. New York: G Thieme Verlag Stuttgart, 1991:32–49.
3. D'Arcy PF, Harron DWG. Proceedings of the First International Conference on Harmonisation. The Queen's University of Belfast, Northern Ireland, UK, 1992.

4. D'Arcy PF, Harron DWG. Proceedings of the Second International Conference on Harmonisation. The Queen's University of Belfast, Northern Ireland, UK, 1994.
5. D'Arcy PF, Harron DWG. Proceedings of the Third International Conference on Harmonisation. The Queen's University of Belfast, Northern Ireland, UK, 1996.
6. D'Arcy PF, Harron, DWG. Proceedings of the Fourth International Conference on Harmonisation. The Queen's University of Belfast, Northern Ireland, UK, 1998.
7. Cordier A. Single dose toxicity: industries perspectives. In: D'Arcy PF, Harron DWG, eds. Proceedings of the First International Conference on Harmonisation. Northern Ireland, UK: The Queen's University of Belfast, 1992:189–191.

Appendix 1 ICH Guidelines Important for the Planning of Nonclinical Testing Strategies

Field	ICH topic	Title	CPMP doc. ref.	Status
Safety	S 1 A	The Need for Carcinogenicity Studies of Pharmaceuticals	CPMP/ICH/140/95	Step 5
	S 1 B	Testing for Carcinogenicity of Pharmaceuticals	CPMP/ICH/299/95	Step 5
	S 1 C (R2)	Dose Selection for Carcinogenicity Studies of Pharmaceuticals	CPMP/ICH/383/95	Step 5
	S 2 (R1)	Guidance on Genotoxicity Testing and Data Interpretation for Pharmaceuticals Intended for Human Use	CHMP/ICH/126642/08	Step 3
	S 3 A	Toxicokinetics: A Guidance for Assessing Systemic Exposure in Toxicology Studies	CPMP/ICH/384/95	Step 5
	S 3 B	Pharmacokinetics: Guidance for Repeated-Dose Tissue Distribution Studies	CPMP/ICH/385/95	Step 5
	S 4	Duration of Chronic Toxicity Testing in Animals (Rodent and Nonrodent Toxicity Testing)	CPMP/ICH/300/95	Step 5
	S 5 (R2)	Detection of Toxicity to Reproduction for Medicinal Products & Toxicity to Male Fertility	CPMP/ICH/386/95	Step 5
	S 6	Preclinical Safety Evaluation of Biotechnology-Derived Products	CPMP/ICH/302/95	Step 5
	S 7 A	Safety Pharmacology Studies for Human Pharmaceuticals	CPMP/ICH/539/00	Step 5
	S 7 B	The Nonclinical Evaluation of the Potential for Delayed Ventricular Repolarization (QT Interval Prolongation) by Human Pharmaceuticals	CPMP/ICH/423/08	Step 5
	S 8	Immunotoxicity Studies for Human Pharmaceuticals	CHMP/ICH/167235/04	Step 5
	S 9	Nonclinical Development of Oncological Therapeutics		Step 2
Quality	Q 3 A (R2)	Impurities in New Drug Substances	CPMP/ICH/2737/99 Rev. 2	Step 5
	Q 3 B (R2)	Impurities in New Drug Products	CPMP/ICH/2738/99	Step 5
	Q 3 C (R3)	Impurities: Residual Solvents	CPMP/ICH/283/95	Step 5
Multidisciplinary	M 3 (R2)	Nonclinical Safety Studies for the Conduct of Human Clinical Trials for Pharmaceuticals	CPMP/ICH/286/95	Step 5

Appendix 2 EU Guidelines Important for the Planning of Nonclinical Testing Strategies

Field	Title	CPMP doc. ref.	Status
Toxicology	Single-Dose/Acute Toxicity	CHMP/SWP/302413/08	Concept Paper
	Single-Dose Toxicity	3BS1A	Adopted Guideline
	Repeated-Dose Toxicity	CHMP/SWP/488313/07	Draft Guideline
	Repeated-Dose Toxicity	CPMP/SWP/1042/99	Adopted Guideline
	Limits of Genotoxic Impurities	CPMP/SWP/5199/02 EMEA/CHMP/SWP/431994/2007Revision 1	Adopted Guideline Question & Answer
	Reflection Paper on the Assessment of the Genotoxic Potential of Antisense Oligodeoxynucleotides	CHMP/SWP/199726/04	Adopted Guideline
	Position Paper on the Genotoxic and Carcinogenic Potential of Phenolphthalein	CPMP/818/97	Adopted Guideline
	Carcinogenicity Evaluation of Medicinal Products for the Treatment of HIV Infection	EMEA/194898/2006	Adopted Guideline
	CHMP SWP Conclusions and Recommendations on the Use of Genetically Modified Animal Models for Carcinogenicity Assessment	CPMP/SWP/2592/02 Rev 1	Adopted Guideline
	Carcinogenic Potential	CPMP/SWP/2877/00	Adopted Guideline
	Points to Consider on the Nonclinical Assessment of the Carcinogenic Potential of Human Insulin Analogues	CPMP/SWP/372/01	Adopted Guideline
	Risk Assessment of Medicinal Products on Human Reproduction and Lactation: From Data to Labeling	EMEA/CHMP/203927/05	Adopted Guideline
	Need for Nonclinical Testing in Juvenile Animals on Human Pharmaceuticals for Pediatric Indications	CHMP/SWP/169215/05	Adopted Guideline
	Points to Consider on the Need for Assessment of Reproduction Toxicity of Human Insulin Analogues	CPMP/SWP/2600/01	Adopted Guideline
	Nonclinical Local Tolerance Testing of Medicinal Products	CPMP/SWP/2145/00	Adopted Guideline
Other Toxicity	Nonclinical Guideline on Drug-Induced Hepatotoxicity	CHMP/SWP/150115/06	Draft Guideline
	Nonclinical Investigation of the Dependence Potential of Medicinal Products	CHMP/SWP/94227/04	Adopted Guideline
	Need for Revision of the Note for Guidance on Photosafety Testing	CHMP/SWP/534549/07	Concept Paper
	Photosafety testing	CPMP/SWP/398/01	Adopted Guideline
	Background to the CPMP Position Paper on Possible Preclinical Studies to Investigate Addiction and Dependence/Withdrawal Related to the Use of Selective Serotonin Uptake Inhibitors (SSRIs)	EMEA/CPMP/2278/00	Adopted Guideline

(*Continued*)

Appendix 2 (*Continued*)

Field	Title	CPMP doc. ref.	Status
	Background to the CPMP Position Paper on Selective Serotonin Uptake Inhibitors (SSRIs) and Dependency/Withdrawal Reactions	EMEA/CPMP/2775/99	Adopted Guideline
	Replacement of Animal Studies by In Vitro Models	CPMP/SWP/728/95	Adopted Guideline
	In Vitro Investigation of Mitochondrial Toxicity of Anti-HIV Nucleoside Reverse Transcriptase Inhibitors	CHMP/SWP/8212/07	Draft Guideline
Pharmacokinetics	Pharmacokinetics and Metabolic Studies in the Safety Evaluation of New Medicinal Products in Animals	3BS11A	Adopted Guideline
General Guidelines	Medicinal Gases: Pharmaceutical Documentation (including recommendation on nonclinical safety requirements for well-established medicinal gases)	CPMP/QWP/1719/00	Adopted Guideline
	Strategies to Identify and Mitigate Risks For First-In-Human Clinical Trials with Investigational Medicinal Products	CHMP/SWP/28367/07 EMEA/295190/2007 Corr.	Adopted Guideline Overview of Comments
	Nonclinical Studies Required Before First Clinical Use of Gene Therapy Medicinal Products	CHMP/GTWP/125459/2006	Draft Guideline
	Scientific Requirements for the Environmental Risk Assessment of Gene Therapy Medicinal Products	CHMP/GTWP/125491/06	Draft Guideline
	Specification Limits for Residues of Metal Catalysts	CPMP/SWP/QWP/4446/00	Adopted Guideline
	Environmental Risk Assessment of Medicinal Products for Human Use	CPMP/SWP/4447/00	Adopted Guideline
	Development of a CHMP Guideline on the Nonclinical Requirements to Support Early Phase I Clinical Trials with Pharmaceutical Compounds	EMEA/CHMP/SWP/91850/06	Concept Paper
	Annex Guideline on Similar Biological Medicinal Products Containing Biotechnology-Derived Proteins as Active Substance: Nonclinical and Clinical Issues—Guidance on Similar Medicinal Products containing Recombinant Erythropoietins	EMEA/CHMP/94526/05	Adopted Guideline

Appendix 2 (*Continued*)

Field	Title	CPMP doc. ref.	Status
	Quality, Preclinical, and Clinical Aspects of Gene Transfer Medicinal Products—Annex on Nonclinical Testing for Inadvertent Germline transmission of Gene Transfer Vectors	EMEA/273974/05	Draft Guideline
	Nonclinical Documentation for Mixed Marketing Authorisation Applications	CPMP/SWP/799/95	Adopted Guideline
	Nonclinical Development of Fixed Combinations of Medicinal Products	CHMP/SWP/258498/05	Adopted Guideline
	Evaluation of Control Samples for Nonclinical Safety Studies: Checking for Contamination with the Test Substance	CPMP/SWP/1094/04	Adopted Guideline
	Environmental Risk Assessment for Human Medicinal Products Containing or Consisting of GMOs	3BR1A	Adopted Guideline
	Development of a CHMP Guideline on Nonclinical Testing for Inadvertent Germline Transmission of Gene Transfer Vector	CPMP/SWP/110180/04	Concept Paper
	Position Paper on the Nonclinical Safety Studies to Support Clinical Trials with a Single Micro Dose	CPMP/SWP/2599/02	Adopted Guideline
	Dossier Structure and Content for Pandemic Influenza Vaccine Marketing Authorisation Application	CPMP/VEG/4717/03	Adopted Guideline
	EMEA Public Statement on Thiomersal in Vaccines for Human Use—Recent Evidence Supports Safety of Thiomersal-Containing Vaccines	CPMP/VEG/1194/04 v02*	Adopted Guideline
	Comparability of Medicinal Products Containing Biotechnology-Derived Proteins as Active Substance—Annex on Nonclinical and Clinical Issues	CPMP /3097/02*	Adopted Guideline
	Limitations to the Use of Ethylene Oxide in the Manufacture of Medicinal Products	CPMP/QWP/159/01	Adopted Guideline
	Preclinical Pharmacological and Toxicological Testing of Vaccines	CPMP/SWP/465/95	Adopted Guideline
	Preclinical Evaluation of Anticancer Medicinal Products	CPMP/SWP/997/96	Adopted Guideline
	Investigation of Chiral Active Substances	3CC29A	Adopted Guideline

(*Continued*)

Appendix 2 (*Continued*)

Field	Title	CPMP doc. ref.	Status
Herbal Medicinal Products	Selection of test materials for genotoxicity testing for Traditional Herbal Medicinal Products/Herbal Medicinal Products	EMEA/HMPC/315413/08	Concept Paper
	Assessment of Genotoxicity of Herbal Substances/ Preparations	EMEA/HMPC/107079/07 EMEA/HMPC/232888/2008	Adopted Guideline Overview of Comments
	Nonclinical Documentation for Herbal Medicinal Products in Applications for Marketing Authorization (Bibliographical and Mixed Applications) and in Applications for Simplified Registration	EMEA/HMPC/32116/05 EMEA/HMPC/132154/2006	Adopted Guideline Overview of Comments

22 Pharmacology

James W. McBlane
Licensing Division, Medicines and Healthcare Products Regulatory Agency, London, U.K.

INTRODUCTION

Pharmacology arguably provides the most fundamental descriptor of a drug—classification of the mechanism of action of the chemical substance—and in the context of medicinal products, this relates to how the therapeutic action arises. Not only are drug formularies ordered by similarity of pharmacological action, but the names given to drugs are also designed to reflect similarity of pharmacological action. For example, any drug ending in -olol will be a member of the class of beta-adrenoceptor antagonists (e.g., atenolol, propranolol) and any drug ending in -mab will be a monoclonal antibody (e.g., rituximab). The system for International Nonproprietary Names for pharmaceutical compounds is overseen by the World Health Organization and further details can be found at http://www.who.int/medicines/services/inn/en.

The use of medicinal products to treat disease is underpinned by the concept that drugs produce their effects by binding to specific targets and that these targets are constituents of cells that typically respond to natural chemicals to bring about an alteration in cellular function. Prior to the development of organic chemistry from the middle of the 19th century, chemical synthesis of drugs was not considered to be likely to result in material with therapeutic activity. Although remedies were widely known (e.g., opium, ergot, belladonna), the concept that activity resided in the chemical nature of the active substance did not exist. Development of the ability to synthesize simple chemicals such as nitrous oxide, ether, and chloroform led to the recognition of the anesthetic effects of these agents, first in human subjects, with demonstration of similar effects in animals following. In this period, many alkaloids were extracted from plant sources and found to have pharmacodynamic activity (e.g., cocaine, nicotine, morphine) and efforts to chemically modify these resulted in agents with a similar, but slightly different, pharmacological profile. For instance, pethidine and methadone were derived from morphine: pethidine retains analgesic activity yet has reduced respiratory depressant activity and methadone lacks antidiarrheal activity.

Considering that pharmacopoeial texts were in existence for hundreds of years prior to this period [the first London pharmacopoeia was published in 1618 (1)], it is surprising to think that the tenet that drugs act by interacting with receptors was introduced barely 100 years ago. The German scientist, Ehrlich, suggested that the actions of drugs can be understood in terms of conventional chemical interactions between the drug substance and a biological target (2), and the British physiologist, Langley, described how a chemical combination between the drug and a constituent of the cell causes a detectable effect. Langley used the term "receptive substance" to describe this cellular constituent, particularly with regard to the action of nicotine on muscle contraction, writing that "both nicotine and curari combine with a 'receptive' substance in muscle, i.e., with some substance which is not the actual contractile molecule, though capable of acting upon it" (3). The receptor concept was firmly anchored in pharmacological thinking by the work of Clark, who in describing experiments examining the effects of acetylcholine on frog hearts, wrote "a demonstrable action may be produced on the heart when only 20,000 molecules per cell are fixed" (4). The excitatory and inhibitory effects of acetylcholine were recognized as arising from different receptors and Ahlquist proved that different actions of several sympathomimetic amines arise through stimulating subclasses of adrenoceptors: he termed these alpha and beta (5). From these early explorations into classification of receptors based on observable changes in physiological function, modern pharmacology recognizes a large number of targets. These are set out in the *Guide to Receptors and Channels* (6). This classification includes seven sections: the seven-transmembrane receptors, transmitter gated channels, ion channels, catalytic receptors, nuclear receptors, transporters, and enzymes (Table 1). Pharmacology can be considered as the study of receptors (7) and today is as reliant on molecular biology as the

Table 1 Summary of Receptor Classification

7 Transembrane receptors (7 TM)	5-HT; adenosine; adiponectin; adrenoceptors; angiotensin; bombesin; bradykinin; cannabinoid; chemokine; cholecystokinin; dopamine; endothelin; $GABA_B$; ghrelin; glucagon; glutamate; histamine; leukotriene; melanocortin; melatonin; motilin; muscarinic acetylcholine; neuropeptide Y; nicotinic acid; opioid and opioid-like; orexin; parathyroid hormone and parathyroid hormone-related peptide; platelet-activating factor; prostanoid; somastatin; vasopressin
Transmitter gated channels	$5\text{-}HT_3$; $GABA_A$; glutamate; glycine, nicotinic acetylcholine; PX2, zinc-activated
Ion channels	Acid sensing; aquaporins; voltage-gated calcium; CatSper; chloride; connexins and pannexins; cyclic nucleotide gated; epithelial sodium; hyperpolarization-activated, cyclic nucleotide-gated; IP3; potassium, ryanodine, voltage-gated sodium; transient receptor potential (TRP) cation
Catalytic receptors	ErbB family; GDNF family; natriuretic peptide; neurotrophin; toll-like family; tumor necrosis factor (TNF) family
Nuclear receptors	Liver X and farnesoid X; peroxisome proliferator-activated; retinoic acid, retinoid X and retinoic acid-related orphan; steroid hormone; thyroid hormone; vitamin d and pregnene X and constitutive androstane
Cell surface transmitter transporters	Choline; GABA; glutamate; glycine; monoamine; nucleoside
Enzymes	Adenosine metabolizing enzymes; adenylyl cyclizes; amino acid hydroxylases; cycoloxygenases; decarboxylases; endocannabinoid metabolizing enzymes; inositol monophosphatase; lipoxygenases; nitric oxide synthetase; phosphatidylcholine-specific phospholipase d; phosphodiesterases; phospholipase A_2; phosphoinositide-specific phospholipase C; protein serine/threonine kinases; soluble guanylyl cyclase

This is not a comprehensive list. Further details can be found in Ref. (6).

early pioneers were on characterizing physiological effects (8). Additionally, drug discovery is also heavily dependent on computerized data analysis to generate new leads (9,10).

There are very few exceptions to the general rule that drugs produce their effect by specific chemical interactions with cellular constituents. The classes of general anesthetics, where physicochemical characteristics of lipid solubility of the agent seem more important than specific chemical structure, osmotic laxatives, where the desired effect is derived from their capacity to retain fluid in the gastrointestinal tract, volume replacement agents used where there is blood loss potentially leading to hemorrhagic shock and chelating agents, acting to sequester target ions, could all be considered to be examples where the therapeutic benefit is brought about by nonreceptor-mediated means. This description might also apply to gene therapy products. Table 2 lists further examples.

It is not always the case that the target of a drug's effect exists in the mammalian body: for instance, where bacteria, fungi, or viruses are causing the disease for which the patient will take the drug, then bacterial, fungal, or viral molecular targets can readily be perceived to be the intended target of a drug. In theory, selective action against such nonmammalian targets, as compared to mammalian targets, should enhance the apparent safety of the medicinal product: in practice, such selectivity remains difficult to achieve.

The majority of medicinal products contain conventional chemical or biological substances that produce their pharmacological effects by acting on defined mammalian targets. This chapter discusses the type of information that is typically expected to be presented in an application for a new medicinal product characterizing its pharmacological actions. In most instances, much of this information will be available prior to conducting clinical trials.

Let us begin with, as an example, a novel substance that is a new chemical agent, detailing the type of pharmacodynamic studies that might be expected for such an agent and then discuss differences that might be anticipated with other types of medicinal product, such as novel biological substances, biosimilar products, and generic medicinal products (where the active ingredient is already commercially available in existing products).

Table 2 Agents That are Thought Not to Act by Targeting a Specific Receptor

Type of agent	Mechanism of action	Example
Antacids	Neutralize stomach acid	Aluminium hydroxide
Defoaming agents	Reduce bubble surface tension	Simeticone
Mucosal protection	Physically coats ulcer	Sucralfate
Bulk forming laxative	Increase fecal mass	Ispaghula
Osmotic laxative	Increase water content of large bowel	Lactulose
Bile acid sequestrant	Anion exchange resin	Cholestyramine
Chelating agents	Iron chelation	Desferrioxamine
Electrolytes	Correction of imbalance	Sodium chloride
Plasma substitutes	Volume replacement	Dextran
Alteration of urine pH	Acidify or alkalinize urine	Potassium citrate
Keratolytics	Skin removal	Salicylic acid
Modification of wax	Soften wax to facilitate removal from ear	Almond oil
Emollients	Skin hydration	Soft paraffin

PRIMARY PHARMACODYNAMICS

Primary pharmacodynamics is the term given to the collection of studies that determine the pharmacological effects of the drug as they relate to the therapeutic action of the medicinal product. In vitro pharmacology studies delineate the interaction of the drug substance with its target receptor and characterize the basic pharmacological action (e.g., agonist, antagonist), usually in terms of quantifiable biochemical and/or electrophysiological changes. In vivo studies are conducted to explore whether the drug has the desired effect, either in normal animals or in animals that have had their physiology altered to mimic human disease states to better determine the therapeutic potential of the molecule.

In Vitro Pharmacology

The objectives of in vitro primary pharmacology studies can be stated as

- to describe the binding of the drug to the target of interest and to other possible targets, and
- to demonstrate whether an agonistic, antagonistic, or other such effect arises from such binding.

Primary pharmacodynamics typically starts with an assessment of binding of the drug to its target receptor. The target receptor is typically expressed in a suitable cell line (11,12) and binding of the molecule is determined at the receptor in question. The principal measure quantified in such experiments is the dissociation constant (K_d), which is based on the assumption that there is a chemical reaction taking place between the drug and the receptor, which can be described by the equation:

$$\text{Drug} + \text{Receptor} \xrightarrow{k_1} \text{Drug/Receptor complex} \xrightarrow{k_2} \text{Drug} + \text{Receptor}$$

K_d is the ratio of k_2 to k_1 and is a quantitative measure of the propensity of a drug to cease binding to the defined target and corresponds to the concentration of drug that results in half of the receptors being occupied. There is an inverse correlation between K_d and affinity of the drug for the target, such that the lower the K_d value, the greater is the affinity of the drug. Plasma concentration is usually taken as a surrogate of the drug concentration at the site of the receptor. The validity of this assumption is rarely tested, but where the plasma concentration approximates to the K_d, 50% receptor occupancy is estimated. It is important to note that receptor-binding screens give no indication of whether the drug is an agonist or an antagonist. Indeed, they give no indication that a drug will have pharmacological activity; thus, the term ligand is usually applied, as it does not imply some action resulting from the binding in the way that the term "drug" does. Where binding is covalent, there is no dissociation and binding is irreversible. This is rare for medicinal products, however. An irreversible effect can equate to irreversible toxicity, such as occurs with organophosphate nerve agents. In medicine, it is generally a desirable trait of a drug's actions that its effects are reversible.

As an example of the above, consider the development of a novel opiate. Availability of an effective analgesic without the undesirable properties of morphine (such as respiratory depressant activity and tolerance) would be a significant therapeutic advance (13). There are multiple opiate receptors, principally δ, κ, and μ and although the known pharmacological effects of opiates cannot be simply attributed to action at each subtype of receptor, there is evidence of differential functional interactions with other brain systems (14,15). Human recombinant δ, κ, and μ receptors can each be expressed in separate cells and displacement of binding of a specific ligand for each receptor by the novel opiate can be measured and K_d values at the three subtypes of opiate receptor so determined. A comparison of the relative affinity for binding at each receptor subtype may indicate which receptor is likely to be targeted first when dose escalating in initial clinical trials.

It is usual to determine binding at receptors from human and animal species, particularly from those species that are used in the in vivo pharmacological and toxicological studies. Where there are major differences in receptor binding between species, this calls into question the extent to which the pharmacological and toxicological tests are relevant for predicting effects in humans. In practice, it is typically the case that the difference in binding between human and animal receptors is modest and can be estimated in quantitative terms (e.g., as less than a 10-fold difference) and this difference can be quantitatively taken into account when considering cross-species effects. For instance, in the example of a novel opiate, the K_d value at human μ receptors may be fourfold less than that at both rat and mouse μ receptors, such that this increase in affinity might be expected, all other things being equal, to result in greater potency of analgesic activity in humans than in animals.

The consequences of receptor binding need to be further described to determine whether the ligand is an agonist, an antagonist, or can be described in other terms such as a partial agonist or inverse agonist, or if the drug acts to influence binding of another substance that brings about the pharmacological action: this last class refers to drugs acting as allosteric agonists (16). An example of allosteric agonism is the action of benzodiazepines at the $GABA_A$ receptor (17). Activation of the γ-aminobutyric acid $(GABA)_A$ receptor by GABA results in a measurable increase in chloride ion flux through the ion channel into the cell. Thus, GABA is an agonist at $GABA_A$ receptors. Benzodiazepines bind to a different site on the $GABA_A$ receptor than does GABA and, in their own right, benzodiazepines do not activate the chloride channel; however, they induce a change in the $GABA_A$ receptor, increasing its affinity for GABA. The resulting increase in binding of GABA to the $GABA_A$ receptor results in greater chloride ion influx. There are many benzodiazepine-containing medicinal products and these have a range of medical uses, including the treatment of anxiety, insomnia, and epilepsy.

Whereas the majority of drugs are perceived to be relatively pure agonists or antagonists, it is possible that a spectrum of activity can exist—a drug may appear to be an agonist in some circumstances and an antagonist in other circumstances (18,19). An agonist is defined as an agent that, once bound to its target, elicits a response. The effects of an agonist can be measured and quantified and in test systems a maximum effect can be described. A partial agonist has similar features, except that the maximum effect that it can induce is reduced in comparison to that of a full agonist. The range of partial agonist action can be from almost full agonism, where near maximal responses can be measured, to minimal agonism, where a small but measurable response can be measured. An antagonist is defined as an agent that once bound to its target does not produce a measurable effect, but can be shown to inhibit responses induced by an agonist. It is clear that in some circumstance, partial agonists can produce effects similar to an antagonist as they prevent a full agonist inducing the maximum effect (20). This is not just a theoretical point: in asthma, β2 agonists can be used as rescue medication in potentially life-threatening situations. The effect of a long-acting partial agonist at β_2 receptors could be to inhibit the agonistic action of such rescue medication (21).

An inverse agonist is an agent that can induce a clear and measurable response, but the effect induced is opposite to that of the natural agonist. Partial inverse agonism can also be understood as a possible effect of a drug acting at a receptor to which it has bound.

Concerning receptor occupancy and drug action, it might be reasonable to assume that the effect of a drug is proportional to the number of receptors occupied: this is the simple occupation theory and can be applied reasonably well to antagonists (22). However, where agonistic actions are under consideration, experimental observations indicate that maximal effects can be produced where there is only occupation of a small proportion of available receptors and further that different agonists can have a different degree of receptor occupancy despite producing effects of similar magnitude. The simple depiction above that a drug binds to a receptor to elicit a response is, in fact, a considerable simplification (23). For agonists, measurement of binding affinity may be confounded by a conformational change arising from the binding (24).

The action of antagonists can be further described in terms of being either competitive or noncompetitive (25,26). The former compete with the agonist such that in the presence of increasing amounts of a competitive antagonist, increasing amounts of agonist are required to produce the same magnitude of pharmacological response: in effect, the dose–response curve is shifted to the right. A plot of the concentration of agonist needed to produce the same biological response against the concentration of antagonist is known as a Schild plot. Schild plots allow determination of the pA_2 value, the estimated equilibrium dissociation constant. More simply, pA_2 is the dose of antagonist that requires a doubling of agonist concentration to produce the same magnitude of effect. By contrast, noncompetitive antagonism results when the antagonist inhibits the consequence of agonist binding by binding equally well to the free receptor or to the receptor–agonist complex. Increasing the concentration of agonist does not affect the block induced by a noncompetitive antagonist and this therefore represents a fundamentally different type of pharmacological antagonism.

In some circumstances, receptor kinetic studies can be considered to underlie the claim that a new product will offer additional therapeutic benefit to existing treatments (27). For instance, in chronic obstructive pulmonary disease, muscarinic antagonists can produce alleviation of symptoms by reducing mucus secretions. Tiotropium has a long dissociation half-life at the muscarinic M_3 receptor of over 30 hours, giving it a long pharmacodynamic effect (28) that is of benefit to patients, as this drug can be taken once daily. It should be noted that this is different to a long pharmacokinetic action. It may be advantageous in the marketplace for new drugs to differentiate themselves by more detailed research into receptor kinetics, although debatable whether it is necessary for all such data to be reviewed in a marketing authorization application. This could depend on the extent to which the applicant seeks to make claims based on such profiling.

In an application for a new product, it is generally not necessary to explore molecular pharmacological actions beyond characterization at the fairly basic level presented here. Selectivity of binding for the intended target should be shown and a quantitative estimate of binding should be presented. The basic mode of action as an agonist or antagonist or other variant should be demonstrated by appropriate experimental means relevant to the particular molecule: these are likely to include a range of different biochemical and electrophysiological effects. These effects should be placed in the context of basic scientific research that demonstrates the relevance of the intervention to the disease for which the medicinal product is to be used. In very rare instances, it may not be possible to determine how, in molecular terms, a drug produces its effects, yet demonstration of clinical efficacy and safety can be shown. In such cases, regulators could be reluctant to approve the product, as a lack of this fundamental knowledge undermines the scientific basis of the systems for registering medicinal products. However, provided reasonable science-based experimentation is shown to have been applied, the lack of firm proof of a molecular mechanism of action need not be an absolute bar to product approval.

In Vivo Pharmacology

The objectives of in vivo primary pharmacology studies can be stated as

- to show that administration of the drug to animals produces an effect of relevance to the therapeutic action of the drug in patients, and
- to quantify the dose and exposure at which such activity is observed.

These data can be considered together with toxicity data to demonstrate that a beneficial effect can be expected in the absence of an undesirable toxic effect.

In vivo pharmacology studies are also conducted with the purpose of proving that there is sufficient evidence to support belief in the potential activity of the drug to justify testing in human subjects. This will include defining doses and concentrations at which effects are evident: where the intended clinical route of administration has not been used in animals, a kinetic argument to justify human doses can be applied. It is necessary to consider the extent to which the species used in in vivo studies is relevant for predicting effects in humans: this will vary from case to case, but where there is a substantial difference between humans and animals in receptor binding, target distribution, pharmacokinetic, or metabolic fate of the drug, or where there are innate biological differences in the systems under study, these place limitations on the ability to extrapolate the findings from in vivo pharmacological studies to humans (29). Many diseases have been mimicked experimentally in animals, and where there are established methods, these can be useful to indicate therapeutic potential of a novel agent, including that relative to established treatment. In some circumstances, there are major limitations on the ability to generate pharmacological data from in vivo experiments in animals: for instance, hepatitis C infection is difficult to study in laboratory animals as chimpanzees are the only known species other than humans to develop disease following infection (30). It may be acceptable to proceed to human trials in the absence of in vivo pharmacology studies and in these cases, there will be supporting data from in vivo pharmacokinetic and toxicological studies.

It should be noted that in vivo pharmacodynamic studies are distinct from those in which safety pharmacology endpoints are assessed (31,32). Also of note in testing novel agents for pharmacological activity is that studies in whole animals are only permitted where no other means of assessing activity is available and that the potential benefit of conducting the experiment must be balanced against the discomfort that is inevitably experienced by animals subjected to such experiments (33).

Although such data are reviewed by regulatory authorities to determine whether a product may be licensed or a clinical trial conducted, these studies are also applied to selection of which among a range of candidate drugs should transfer from laboratory-based research to the development of a potential medicinal product. Pharmacological activity is one factor influencing decisions that have to be made within pharmaceutical companies to determine whether to develop a drug into a medicinal product, but there are several other factors and the relative importance of each will vary depending on the specifics of each case. These other factors include kinetic and safety profiling, development of a suitable dosage form, establishing manufacturing capacity at a suitable scale to meet market needs and estimates of the chance of development leading to a commercially viable medicinal product. An applicant for a marketing authorization is not required to demonstrate why a particular molecule was selected for development over others, but the pharmacological profile of the agent selected should be sufficiently detailed to give a scientific basis for the reasons why the drug is therapeutically useful. One exception to this, however, is where an enantiomer is selected for development: the reasons for the choice of enantiomer are of interest and justification should be provided.

There is no generic set of studies to characterize sufficiently the pharmacological effects of a novel drug, as these must depend on the intended medical use of the product, on the features of the drug and on the applicability, or otherwise, of existing methods for detecting effects that might be suggestive of therapeutic action in patients. However, illustrative examples of a novel analgesic and a novel anticancer agent are given below to suggest the type of data likely to be adequate in these circumstances.

Development of an Analgesic Agent

In vitro testing of a novel analgesic might well include demonstration of

- inhibition of binding of the ligand at the target receptor in human and animal species and
- demonstration of agonist, antagonist, or similar activity consequent upon binding by measurement of diverse biochemical changes.

In vivo testing of a novel analgesic often includes the assessment of a motor response to a nociceptive stimulus. It might include the following:

- Use of thermal, electrical, mechanical, or chemical stimuli to induce pain and demonstration of inhibition of the pain response by the active agent by time taken to respond (examples include tail flick, paw withdrawal, hot plate, intradermal injection of formalin).
- Demonstration that the animal is capable of responding in the presence of treatment (for instance, if the drug is a sedative, delays in responding in a hot plate test cannot be attributed solely to an analgesic effect).
- Estimates of drug concentration in vivo at effective analgesic doses and comparison with in vitro receptor binding data.

Reviews of animal tests used in testing novel analgesics can be found in the scientific literature (34–36) as can be found in studies of individual drugs (37–39). Further, with regard to predicting an active human dose, it can be understood that the degree of pain suffered influences the magnitude of detectable analgesic action. Consequently, an assumption that an effective dose in such animal testing is similar to that of a human therapeutic analgesic dose equates to an assumption that the degree of pain involved is the same: this clearly is open to question. Across several tests, it is likely that different estimates of an effective dose will be derived. These allow estimates of potential target concentrations for human dosing. In a marketing authorization application, it is appropriate to provide an integrated discussion of human analgesic doses and concentration and in vitro effects at human target receptors and acknowledge any differences identified in animal testing. Whereas the in vivo animal data provide evidence of analgesic activity suitable to select among a range of drug candidates and to support clinical trials, such data are largely superseded by clinical data at the time of a marketing authorization application. Nevertheless, such data can usefully indicate potential analgesic activity of novel compounds of the same class, where in vitro and in vivo data from animals can be analyzed with human efficacy data (40).

Development of an Anticancer Agent

In vitro testing of a novel anticancer agent might well include demonstration of

- inhibition of binding of the known ligand in cells expressing the target receptor,
- inhibition of a biochemical change induced by the natural ligand (e.g., tyrosine phosphorylation),
- inhibition of tumor cell growth of various types of cancer, and
- (relative) lack of effect on normal cells and on tumor cells known not to express the target.

In vivo testing might well include demonstration of

- a higher concentration of drug in tumor than in other tissues in tumor-bearing mice;
- inhibition of tumor growth, or eradication of tumor, in xenotransplant experiments in nude mice;
- influence of dosing schedule on pharmacokinetics and pharmacodynamics in xenotransplant experiments in nude mice;
- effects of combination treatment with novel agent and established treatments for target disease; and
- changes in relevant biochemical markers of drug activity following dosing to nude mice bearing tumors.

Human tumor tissues are used in these tests, and tissues from multiple types of tumor are likely to have been tested. Demonstration of a tumor-specific or tumor-selective effect is sought. This should demonstrate the suitability of the drug in treating the type of tumor for which the product is to be commercialized. Additionally, quantitative estimates of active doses and drug concentrations should be estimated based on both in vitro and in vivo data, and the possibility of increased efficacy or toxicity, in the event of combination of the novel treatment with existing treatments for the disease in question, should be explored.

SECONDARY PHARMACODYNAMICS

Secondary pharmacodynamics describes inherent pharmacological actions of a drug that do not contribute to its primary potential therapeutic benefit. Such effects may detract from the potential benefit of the drug by causing unintended, undesirable but predictable effects, or they may pose no concerns at the doses that are intended to be given to humans. They can also suggest alternative uses. For instance, erythropoietin may have potential benefit postinfarct that is distinct from its erythropoietic effects (41,42). Secondary pharmacodynamic effects can also be identified once a product is in clinical use and can significantly alter the patient population in which the product is used. An example of this is the use of the antimalarial drug chloroquine in patients with active rheumatoid arthritis (43,44).

Typically, secondary pharmacodynamic studies include a screen to identify binding at a range of other receptors, transporter proteins, ion channels, and enzymes, with follow-up investigations to determine the potential significance for understanding the action of the drug and for predicting possible adverse events at any target, where binding is found at concentrations potentially relevant to human use. The binding screen, with appropriate functional testing, is a simple means of establishing an apparent claim for a new molecule of the expectation of superiority over existing treatments. For instance, a novel agent with potential in treating schizophrenia may be identified as acting as an antagonist at a specific type of dopamine receptor; similar drugs may also have activity at a range of 5-HT receptors and it could be postulated that such actions do not contribute to the therapeutic action of these drugs. In a binding screen for the novel drug, lack of interaction at 5-HT receptors could be perceived as a potential advantage, or, could be associated more with the control of the negative symptoms of schizophrenia rather than the positive symptoms. The selection of which targets to include in a broad screen of receptor binding must be driven by the nature of the molecule and what the primary pharmacodynamics suggest about its action.

SUMMARY OF PRODUCT CHARACTERISTICS (LABELING)

The importance of the label, or, in Europe, the Summary of Product Characteristics (SmPC) is emphasized elsewhere in this volume. Section 5 of the SmPC contains a summary of the pharmacodynamic, pharmacokinetic, and preclinical safety data that are considered useful to prescribers. As the clinical data on the product are the most relevant to the prescriber, it predominates in this section. However, a description of the mechanism of action as derived from the primary pharmacodynamic studies is included, usually without quantitative details. Where particular features of the pharmacological profile are potentially significant for the prescriber, this information should be included. For instance, the product called Champix™ in Europe, Chantix™ in the United States, contains varenicline, a partial agonist at $\alpha_4\beta_2$ neuronal nicotinic acetylcholine receptors, and is indicated for smoking cessation. The partial agonist nature of its action means that it can stimulate these receptors with a consequential alleviation of craving and feelings of withdrawal, but it has less propensity to activate reward systems that are stimulated by full nicotinic agonists (45). This is of relevance in the use of the product, as it is initiated while patients continue to smoke. The presence of the drug blocks the full agonist action of nicotine at this receptor, so allowing patients to stop smoking with a greater chance of success than when only a placebo is used. This information can be contrasted with products that contain nicotine, SmPCs for which typically describe only that it is an agonist at central and peripheral nicotinic receptors. In some instances, the exact molecular mechanism causing the therapeutic effect is not known, although effects that are likely to be relevant can be summarized. This is so for the antiviral agent ribavirin, used in treating hepatitis C and respiratory syncytial virus bronchiolitis, and the SmPCs for ribavirin-containing products reflect this.

BIOTECHNOLOGY PRODUCTS

Biotechnology products include active agents of diverse natures, including proteins, monoclonal or, more rarely, polyclonal antibodies, antibody fragments, peptides, antigens for use in vaccines, and gene therapy and cellular products. Oligonucleotide-based products may be considered as having features typical of biotechnology products, for instance in their selectivity of action, but may be chemically synthesized. This indicates that the distinction between chemical and biotechnological products is not absolute.

Pharmacological testing is an area where the differences between biotechnology-derived and chemical drugs are modest, when compared to, for instance, production processes. For biotechnology products, it is necessary to describe binding at the pharmacological target and

Table 3 Monoclonal Antibody Products

Product name	Target	Immunoglobulin	Nature of proof of in vivo pharmacological activity	Indication
Avastin	VEGF	Humanized IgG1	Experiments in nude mice with human tumor xenografts	Colorectal, breast, non small cell lung and renal cancers
Erbitiux	EGFr	Chimaeric IgG1	Experiments in nude mice with human tumor xenografts	Colorectal and head and neck cancers
Herceptin	HER-2	Humanized IgG1	Experiments in nude mice with human tumor xenografts	Breast cancer
Humira	TNFα	Human IgG1	Experiments in transgenic mice expressing human TNFα	Rheumatoid arthritis; psoriatic arthritis
MabCampath	CD52	Humanized IgG1	Experiments in normal cynomolgus monkeys	B cell chronic lymphocytic leukaemia
Mabthera	CD20	Chimaeric IgG1	Experiments in normal cynomolgus monkeys	Non Hodgkin's lymphoma; rheumatoid arthritis
Raptiva	CD11a	Humanized IgG1	Experiments in mice with a rodent surrogate antibody & with efalizumab in psoriatic skin grafted to nude mice	Psoriasis
Simulect	IL-2r	Chimaeric IgG1	No *in vivo* pharmacology	Prophylaxis of acute organ rejection
Soliris	C5 protein	Humanized IgG2/4	Experiments in mice using a rodent surrogate antibody	Paroxysmal nocturnal haemoglobinuria
Synagis	RSV protein	Humanized IgG1	Experiments in virus-infected cotton rats	Respiratory syncytial viral infection
Tysabri	α4 integrin	Humanized IgG4	Experiments in guinea pig autoimmune encephalitis	Multiple sclerosis
Vectibix	EGFr	Fully human IgG2	Experiments in nude mice with human tumor xenografts	Colorectal cancer
Xolair	IgE	Humanized IgG1	Skin reactivity testing in ragweed-sensitized monkeys	Asthma
Zenapax	IL-2r	Humanized IgG1	Allograft transplant experiments in monkeys	Prophylaxis of acute organ rejection

Abbreviations: EGFr, epidermal growth factor receptor; HER2, human epidermal growth factor receptor 2 protein; IL-2r, interleukin 2 receptor; VEGF, vascular endothelial growth factor.

to characterize whether the binding results in agonistic or antagonistic effects. For monoclonal antibodies, both properties can be expressed in a single product: the F_{ab} region of an IgG_1 immunoglobulin may result in pure blocking activity at its target, with the F_c region activating cell-mediated cytotoxicity. This is a well-established action of several licensed medicinal products that contain monoclonal antibodies.

In vivo testing of biotechnology products has the same aim as that for chemical agents, namely, to provide sufficient evidence for the potential activity of the drug and to indicate what dose and concentration is associated with such effects. Quantitative estimates of potentially therapeutically active doses and concentrations should be provided. In some instances, there are existing experimental systems for detecting pharmacodynamic activity in rodents, whereas a novel biotechnological product is not active in rodents. Clearly, if this is the case, pharmacological testing in rodents is not relevant. Table 3 lists a series of monoclonal antibody-containing products and summarizes the type of in vivo pharmacological studies that were used.

Pharmacological testing of biotechnology products should support the conclusion that the species used in toxicity studies are pharmacologically responsive to the product. In some instances, biotechnology products are not pharmacologically active in laboratory animal species, or are only active in great apes. This can pose significant challenges to being able to define pharmacological effects and doses and concentrations at which effects might be observed in humans. In vitro testing in human cells should be considered and there may well be instances where this will suffice in terms of establishing the relevance of the pharmacodynamic action of the drug. Where there is no responsive species with the molecule to be tested, evidence of pharmacodynamic activity can be from a combination of approaches that might include the use of homologous products in rodents or data from transgenic animals, again usually rodents, which have the target either removed or overexpressed. A similar situation can arise for vaccines if a therapeutic action in animals cannot be demonstrated, as the species do not develop the disease for which the vaccine is being developed. In these instances, demonstration of a specific immune response to the vaccine in the species used in toxicity testing is expected. Finally, for gene therapy products, it is expected that evidence of a similar action as anticipated in humans is shown in animals or in vitro and where this is relevant, that tissue-specific expression of the transgene is demonstrated (46,47).

As biotechnology products tend to be created with much higher specificity of action than chemical agents, it is consequently appropriate that the secondary pharmacology binding screen, as was described for new chemical agents, is not conducted. Instead, a cross-reactivity study to assess comparative binding to human and animal tissues is an appropriate way to identify possible unintended effects. Cross-reactivity studies are used to select the species chosen for toxicity testing based on relevance of predicting effects in humans.

For a product composed of the patients own cells, which may or may not have been genetically, or otherwise, manipulated *ex vivo*, testing of the medicinal product in animals may be of little relevance. The product to be given cannot be tested as it is derived from the patient to be treated. Administration of human cells to animals may result in immune rejection, making testing scientifically invalid. In such circumstances, demonstration of relevant pharmacodynamic activity using syngeneic cells in animal testing can demonstrate that the methods used do result in a product with the desired effect. This can establish sufficient evidence of pharmacodynamic activity.

BIOSIMILAR PRODUCTS/FOLLOW-ON BIOLOGICS

A biosimilar product is one that contains as an active agent, a substance that is similar to one in an already approved medicinal product (48). The term "biosimilar" is used in Europe, whereas in the United States, these are termed "follow-on-biologics." Unlike generic chemical drug substances, biosimilar products will contain subtle structural differences from an existing approved product, arising out of differences in manufacture. It is therefore necessary to demonstrate similarities between the existing product and the product that is claimed to be biosimilar. The necessary pharmacological testing relates to binding to the target receptor, demonstration of lack of binding at similar receptors (to prove equal specificity between the two products), and additional measures relating to biochemical pharmacological actions that show both the similarity between the two products and capable of detecting differences, should these exist. These data should be generated prior to commitment to full-scale development of the product, as if there is a failure to demonstrate similarity of pharmacodynamic activity, it is not rational to expect that the remainder of the testing program will result in an adequate demonstration of biosimilarity. For several classes of product [e.g., insulins (49), colony stimulating factors (50)], there are guidance notes published by the European Medicines Agency detailing what studies are required and these should be followed and any deviations justified.

GENERIC PRODUCTS AND LINE EXTENSIONS

Where the product applied for is a generic version of a product already approved, that is to say, the new product is of the same dosage form and strength as an existing, licensed product, and if the active ingredient is a chemical as compared to a biological, there is no need for experimental pharmacological data to be submitted. It is necessary, however, to review up-to-date thinking with respect to how the drug's effects arise to ensure that the rational basis for efficacy of the medicinal product is maintained. This expectation should be met in the Nonclinical Overview, Section 2.4 of the Common Technical Document.

For line extension applications, where, for instance, a different formulation is introduced, again there is no need to include any new pharmacological data, as cross-reference to an existing approval suffices. It is sometimes the case that a medicinal product is developed that is not identical to the originator product but will be the same active agent used in the same patient population. Examples of this include changes to the formulation to support a different route of administration, such as in the development of an inhaled antibiotic for use in controlling bacterial infections in patients with cystic fibrosis, or the introduction of an injectable form of antipsychotic agent that has been available as a tablet for oral administration. In relation to the content of a marketing authorization application, proof of activity is accepted based on the existing product: results of studies should be submitted where they provide additional information that is of potential use in characterizing the effects of the product. This could be demonstrating the reasons for selecting doses used in toxicity studies in animals and clinical studies with the new product.

GOOD LABORATORY PRACTICE

There is no need for exploratory pharmacology studies to be conducted in compliance with Good Laboratory Practice (GLP). GLP is applied to preclinical safety studies.

GUIDELINES

There are relatively few guidelines from regulatory authorities of direct relevance to pharmacological testing: guidelines tend to concentrate on detailing expectations for safety testing (51,52). However, guidelines are in place detailing the expectation of the type of pharmacodynamic data that should be presented for anticancer agents (53) and also in specialist cases where the medicinal product is to be used in the event of an outbreak of pandemic influenza (54) or in cases of bioterrorist attack with, for instance, smallpox (55). In these situations, direct evidence of efficacy cannot be obtained from clinical trials, and the doses determined for licensing, while these must be supported by trials in humans that show immunoactivity and acceptable safety, are also dependent on data from therapeutic testing in animals exposed to infective agents.

Where guidelines have been published by the European Medicines Agency, these are public statements of the expectation of what types of studies should be conducted. Assessors in national agencies expect that such studies will be presented. As drug development is an international endeavor, guidance from other national regulatory agencies or from other organizations, such as the World Health Organization, may well have influenced the studies conducted to characterize the pharmacology of the agent and where this is so, it can be helpful to describe this in European marketing authorization applications.

CONCLUDING REMARKS

In an application for a new medicinal product, the objective of presenting the pharmacology data is to describe the data proving the mechanism of action of the drug and to relate this to its therapeutic action. For products that contain a new drug, in vitro studies should demonstrate specificity of binding to the target(s) of relevance, selectivity of binding with respect to other potential targets, and the nature of pharmacological effect, that is, whether agonist, antagonist, or other. In vivo pharmacology should be conducted with the purpose of establishing an action relevant to the intended therapeutic use, either in normal animals or in animals with experimentally induced disturbances in their physiology. Estimates of the doses necessary in human patients should be presented based on these considerations. For new products that contain a known drug, it should generally not be necessary to present new experimental pharmacological data, but a discussion on the mechanism of action should be presented. However, biosimilar drugs represent an exception to this, as comparative pharmacological data are likely to be required.

REFERENCES

1. British Pharmacopoeia website: http://www.mhra.gov.uk/Howweregulate/Medicines/Inspectionandstandards/BritishPharmacopoeia/index.htm. Accessed August 20, 2008.

2. Ehrlich P. Croonian lecture: On immunity with special reference to cell life. Proc Royal Soc Lond 1899–1900; 66:424–448.
3. Langley JN. On the contraction of muscle, chiefly in relation to the presence of 'receptive' substances. Part 1. J Physiology 1907; 36:347–384.
4. Clark AJ. The reaction between acetyl choline and muscle cells. J Physiology 1926; 61:530–546.
5. Ahlquist RP. A study of adrenotropic receptors. Am J Physiol 1948; 153:586–600.
6. Alexander SPH, Mathie A, Peters JA. Guide to receptors and channels. Br J Pharmacol 2008; 153(Suppl 2):S1–S209.
7. Rang HP. The receptor concept: Pharmacology's big idea. Br J Pharmacol 2006; 147:S9–S16.
8. Rubin RP. A brief history of great discoveries in pharmacology: In celebration of the centennial anniversary of the founding of the American Society of Pharmacology and Experimental Therapeutics. Pharmacol Rev 2007; 59:289–359.
9. Ekins S, Mestres J, Testa B. In silico pharmacology for drug discovery: Methods for virtual ligand screening and profiling. Br J Pharmacol 2007; 152:9–20.
10. Ekins S, Mestres J, Testa B. In silico pharmacology for drug discovery: Applications to targets and beyond. Br J Pharmacol 2007; 152:21–37.
11. Gamper N, Stockand JD, Shapiro MS. The use of Chinese hamster ovary (CHO) cells in the study of ion channels. J Pharmacol Toxicol Meth 2005; 51:177–185.
12. Thomas P, Smart TG. HEK293 cell line: A vehicle for the expression of recombinant proteins. J Pharmacol Toxicol Meth 2005; 51:187–200.
13. Groer CE, Tidgewell K, Moyer RA, et al. An opioid agonist that does not induce μ-opioid receptor—arrestin interactions or receptor internalization. Mol Pharmacol 2007; 71:549–557.
14. Bailey CP, Smith FL, Kelly E, et al. How important is protein kinase C in μ-opioid receptor desensitization and morphine tolerance? Trends in Pharmacol Sci 2006; 27:558–565.
15. Zhao Z-Q, Gao Y-J, Sun Y-G, et al. Central serotonergic neurons are differentially required for opioid analgesia but not for morphine tolerance or morphine reward. Proc Natl Acad Sci 2007; 104:14519–14524.
16. Langmead CJ, Christopoulos A. Allosteric agonists of 7TM receptors: Expanding the pharmacological toolbox. Trends Pharmacol Sci 2006; 27:475–481.
17. Barnard EA, Skolnick P, Olsen RW, et al. International Union of Pharmacology. XV. Subtypes of γ-aminobutyric $acid_A$ receptors: Classification on the basis of subunit structure and receptor function. Pharmacol Rev 1998; 50:292–313.
18. Negus SS. Some implications of receptor theory for *in vivo* assessment of agonists, antagonists and inverse agonists. Biochem Pharmacol 2006; 71:1663–1670.
19. Parra S, Bond RA. Inverse agonism: From curiosity to accepted dogma, but is it clinically relevant? Curr Opin Pharmacol 2007; 7:146–150.
20. Urban JD, Clarke WP, von Zastrow M, et al. Functional selectivity and classical concepts of quantitative pharmacology. J Pharmacol Exp Ther 2007; 320:1–13.
21. Battram C, Charlton SJ, Cuenoud B, et al. *In vitro* and *in vivo* pharmacological characterization of 5-[(*R*)-2-(5,6-diethyl-indan-2-ylamino)-1-hydroxy-ethyl]-8-hydroxy-1H-quinolin-2-one (Indacaterol), a novel inhaled beta2 adrenoceptor agonist with a 24-h duration of action. J Pharmacol Exp Ther 2006; 317:762–770.
22. Stephenson RP. A modification of receptor theory. Br J Pharmacol 1956; 11:379–393.
23. Strange PG. Agonist binding, agonist affinity and agonist efficacy at G protein-coupled receptors. Br J Pharmacol 2008; 153:1353–1363.
24. Colquhoun D. Binding, gating, affinity and efficacy: The interpretation of structure–activity relationships for agonists and of the effects of mutating receptors. Br J Pharmacol 1998; 125:924–947.
25. Verzijl D, Storelli S, Scholten DJ, et al. Noncompetitive antagonism and inverse agonism as mechanism of action of nonpeptidergic antagonists at primate and rodent CXCR3 chemokine receptors. J Pharmacol Exp Ther 2008; 325:544–555.
26. Wyllie DJA, Chen PE. Taking the time to study competitive antagonism. Br J Pharmacol 2007; 150:541–551.
27. Swinney DC. Applications of binding kinetics to drug discovery: Translation of binding mechanisms to clinically differentiated therapeutic responses. Pharm Med 2008; 22:23–34.
28. Disse B, Reichl R, Speck G, et al. Ba 679 BR, a novel long-acting anticholinergic bronchodilator. Life Sci 1993; 52:537–544.
29. Mestas J, Hughes CCW. Of mice and men: Differences between mouse and human immunology. J Immunol 2004; 172:2731–2738.
30. Bukh J. A critical role for the chimpanzee model in the study of hepatitis C. Hepatology 2004, 39; 1469–1475.
31. ICH Topic S7A Safety Pharmacology Studies for Human Pharmaceuticals.

32. ICH Topic S7B The Nonclinical Evaluation of the Potential for Delayed Ventricular Repolarisation (QT Interval Prolongation) by Human Pharmaceuticals.
33. UK Home Office http://scienceandresearch.homeoffice.gov.uk/animal-research/animal-testing-faqs/. Frequently Asked Questions About the Use of Animals in Scientific Procedures. Accessed August 20, 2008.
34. Gil DW, Cheevers CV, Donello JE. Transient allodynia pain models in mice for early assessment of analgesic activity. Br J Pharmacol 2008; 153:769–774.
35. Le Bars D, Gozariu M, Cadden SW. Animal models of nociception. Pharmacol Rev 2001; 53:597–652.
36. Negus SS, Vanderah TW, Brandt MR, et al. Preclinical assessment of candidate analgesic drugs: Recent advances and future challenges. J Pharmacol Exp Ther 2006; 319:507–514.
37. Clark P, Rowland SE, Denis D, et al. MF498 [*N*-{[4-(5,9-diethoxy-6-oxo-6,8-dihydro-7H-pyrrolo[3,4-g]quinolin-7-yl)-3-methylbenzyl]sulfonyl}-2-(2-methoxyphenyl)acetamide], a selective E prostanoid receptor 4 antagonist, relieves joint inflammation and pain in rodent models of rheumatoid and osteoarthritis. J Pharmacol Exp Ther 2008; 325:425–434.
38. Rowland SE, Clark P, Gordon R, et al. Pharmacological characterization of a selective COX-2 inhibitor MF-tricyclic, [3-(3,4-difluorophenyl)-4-(4-(methylsulfonyl)phenyl)-2-(5H)-furanone], in multiple preclinical species. Eur J Pharmacol 2007; 560:216–224.
39. Sullivan NR, Leventhal L, Harrison J, et al. Pharmacological characterization of the muscarinic agonist (3R,4R)-3-(3-hexylsulfanyl-pyrazin-2-yloxy)-1-aza-bicyclo[2.2.1]heptane (WAY-132983) in *in vitro* and *in vivo* models of chronic pain. J Pharmacol Exp Ther 2007; 322:1294–1304.
40. Huntjens DRH, Spalding DJM, Danhof M, et al. Correlation between *in vitro* and *in vivo* concentration–effect relationships of naproxen in rats and healthy volunteers. Br J Pharmacol 2006; 148:396–404.
41. Moon C, Krawczyk M, Paik D, et al. Erythropoietin, modified to not stimulate red blood cell production, retains its cardioprotective properties. J Pharmacol Exp Ther 2006; 316:999–1005.
42. Riksen NP, Hausenloy DJ, Yellon DM. Erythropoietin: Ready for prime-time cardioprotection. Trends Pharmacol Sci 2008; 29:258–267.
43. Freedman A. Chloroquine and rheumatoid arthritis: A short-term controlled trial. Ann Rheum Dis 1956; 15:251–257.
44. Popert AJ, Meijers KAE, Sharp J, et al. Chloroquine diphosphate in rheumatoid arthritis. Ann Rheum Dis 1961; 20:18–35.
45. Hogg RC, Bertrand D. Partial agonists as therapeutic agents at neuronal nicotinic acetylcholine receptors. Biochem Pharmacol 2007; 73:459–468.
46. Committee for the Medicinal Products for Human Use (CHMP). EMEA/CHMP/GTWP/12459/2006. Guideline on the Non-clinical Studies Required Before First Clinical Use of Gene Therapy Medicinal Products.
47. Committee for the Medicinal Products for Human Use (CHMP). CHMP/BWP/3088/99 Note for Guidance on the Quality, Preclinical and Clinical Aspects of Gene Transfer Medicinal Products.
48. Committee for Medicinal Products for Human Use (CHMP). CHMP/437/04 Guideline on Similar Biological Medicinal Products.
49. Committee for medicinal products for human use (CHMP). CHMP/BMWP/32775/2005 Annex to Guideline on Similar Biological Medicinal Products Containing Biotechnology-Derived Proteins as active substance: Non-clinical and clinical issues. Guidance on Similar Medicinal Products Containing Recombinant Human Soluble Insulin.
50. Committee for Medicinal Products for Human Use (CHMP). CHMP/BMWP/31329/2005. Annex to Guideline on Similar Biological Medicinal Products Containing Biotechnology-Derived Proteins as Active Substance: Non-clinical and clinical issues. Guidance on similar medicinal products containing recombinant granulocyte-colony stimulating factor.
51. Committee for the medicinal products for human use (CHMP). EMEA/CHMP/SWP/294648/2007. Guideline on Strategies to Identify and mitigate risks for First-in-Human Clinical Trials with Investigational Medicinal Products.
52. ICH Topic M3(R2). CPMP/ICH/286/95. Non-clinical Safety Studies for the Conduct of Human Clinical Trials for Pharmaceuticals.
53. Committee for the Medicinal Products for Human Use (CHMP). CPMP/SWP/997/96. Note for Guidance on the Pre-clinical Evaluation of Anticancer Medicinal Products.
54. Committee for Medicinal Products for Human Use (CHMP). CHMP/VEG/4717/03. Guideline on Dossier Structure and Content for Pandemic Influenza Vaccine Marketing Authorisation Applications.
55. Committee for Medicinal Products for Human Use (CHMP). CHMP/1100/02. Note for Guidance on the Development of Vaccinia Virus Based Vaccines Against Smallpox.

23 Nonclinical Pharmacokinetics and Toxicokinetics

Richard J. Weaver
Servier R&D, Slough, Buckinghamshire, U.K.

Roeline Jochemsen
Servier International Research Institute, Courbevoie, France

INTRODUCTION

Pharmacokinetics is arguably one of the most important disciplines in pharmaceutical development as it permits "windows of opportunity" to measure drug and/or metabolite concentrations and link these measurements with pharmacological effect. In today's pharmaceutical industry, the science of pharmacokinetics has evolved such that, when used in support of drug safety testing, it forms an indispensable discipline in the pursuit of new candidate drugs. The earliest concepts of pharmacokinetics probably began with Torsten Teorell, a Swedish physiologist, who in 1937 described the movement of drugs through various compartments of the body (1,2), but it was not until 1953 that the term "Pharmacokinetics" was first used by Friedrich Dost in his text "Kinetik de Konzentrationsabläufe in der Kreislaufflüssigkeit" (3). Despite these earliest descriptions, the use of pharmacokinetics as a routine discipline to study the movement of drugs in and out the body had to wait a further two decades until the development of specific, sensitive, and accurate chromatographic methods for the quantitative analysis of drugs became widely available for use in drug research.

As the study of pharmacokinetics began to attract growing interest throughout the mid-20th century onward, the study of drug metabolism as a subdiscipline of pharmacokinetics in its own right, attracted considerable attention as drug-metabolizing enzymes were identified increasingly in the metabolic elimination of drugs. However, unlike pharmacokinetics, studies in drug metabolism appear to have originated in Germany as early as the 18th century where pioneering chemists administered chemical substances to animals to observe physiological effects and to collect excreta to establish if the original chemical was excreted unchanged or changed as a result of metabolism. Studies in metabolism continued throughout Europe during the 18th and early 20th century, but it was Tecwyn Williams' pioneering studies in drug metabolism, developed from his earlier work on terpene glucuronides, that gave most impetus to our present day understanding of drug metabolism (4). In the ensuing years, the identification of drug-metabolizing enzymes, transport proteins, and the kinetics governing drug–drug interactions (DDIs) have each fuelled in turn important developments in our understanding of the many factors governing the pharmacokinetic characteristics of drugs.

As the importance of drug metabolism and pharmacokinetics began to grow during the 1960s, regulatory authorities began asking for species comparisons of drug levels to help explain differences in observed toxicity, particularly in the aftermath of the thalidomide tragedy, to establish certain basic safety procedures. Since this time, regulatory expectations have kept pace with many of the scientific and technical advances, often as a consequence of the pharmaceutical industry's own endeavours towards achieving greater understanding of drug properties and the biological processes that act upon them.

This chapter brings together the regulatory requirements issued by the International Conference on Harmonization (ICH) initiative on the studies and protocols perceived to be necessary for nonclinical pharmacokinetic investigations of small molecular entities by the pharmaceutical industry. Examples in support of the practical implementation of these studies and discussion items are cited from guidelines, scientific journals, and concept papers.

REGULATORY REQUIREMENTS

Prior to the ICH, regional authorities in the United States, Japan, and across Europe had established their own guidelines on nonclinical safety data required prior to the start of clinical investigations. Although these guidelines were often quite detailed in their requirements on study design and conduct of nonclinical studies, they were rarely congruent between different regional authorities, and consequently, gaining acceptance of a single safety package by different regional authorities was largely very difficult. Since the creation of the ICH, tripartite discussions between the industry and authorities in the United States, Japan, and Europe have enabled guidelines to be issued that are applicable in all three regions and beyond. Current ICH guidelines now require toxicokinetic and ADME (absorption, distribution, metabolism, and excretion) information that encompasses, as part of a submission to regulatory authorities, a good understanding of the pharmacokinetics of the drug in animal species at several carefully selected dose levels that are relevant for the evaluation of human safety. In general, there are two types of guidelines in the conduct of nonclinical work. The first relates to the types of studies required and the second relates to protocol requirements for the studies themselves. Reference to the use of toxicokinetic or pharmacokinetic assessment of drug and/or metabolites is mentioned in many of the 13 Safety and Multidisciplinary Guidelines now issued under the umbrella of the ICH. Indeed, such is the importance of pharmacokinetic information on drug and/or metabolites the ICH have issued two guidelines specifically on this topic. Within the ICH, topics on safety are carcinogenicity (S1), genotoxicity (S2), toxicokinetics and pharmacokinetics (S3), reproductive toxicology (S5), safety pharmacology (S7), and timing of nonclinical safety studies required for clinical studies (M3). Each of these topics refers, to varying degrees, the use of pharmacokinetic and metabolism information required for supporting the interpretation of findings within each of the respective disciplines on drug safety.

ICH guidelines on the application of pharmacokinetics in safety evaluation testing have been available for a several years. In contrast, there are notably fewer details within the ICH initiative on the conduct of ADME studies. This is perhaps a consequence of the diversity and breadth of ADME techniques, studies, and protocols available to the investigator that, at least for now, places a greater challenge to achieving tripartite agreement within an acceptable timeframe for adoption by the three main regional authorities. For guidance on ADME studies, it is possible to turn to other sources of information such as regional guidelines, scientific discussion papers, which provide insights to the reader on the current views of pharmaceutical and regulatory opinion leaders within their respective field(s) of expertise.

Regulatory agencies delineate guiding principles on such issues as pharmacokinetics because the work required to support safety evaluation is highly dependent on each drug and the results obtained. ICH guidance documents are designed to be sufficiently flexible to allow the adoption of alternative pharmacokinetic approaches providing they are supported by adequate scientific justification. Another reason for the lack of prescriptive detail is if a guideline is carefully followed then the Regulatory Authorities must accept the information generated and cannot necessarily ask for more work to be done. ICH guidelines can be found on the ICH Home Page (5) and ICH guidelines together with regional guidelines can be located for the United States on the Food and Drug Administration (FDA) Home Page (6) and for Europe on the European Medicines Agency (EMEA) Home Page (7). For Japan, it is difficult to find English versions of regional guidelines on the web, but the translation of Japanese guidelines can be obtained through commercial sources such as the IDRAC regulatory database (8). Tables 1 and Table 2 summarize guidelines pertaining to nonclinical ADME and toxicokinetic investigations as they related to the expectations of the three main regulatory agencies under the umbrella of the ICH initiative. An appreciation of the timing of these studies is summarized in Figure 1, although this will vary between companies and the drug under investigation. These have been divided into toxicokinetic and ADME categories of drug development, and within the text these requirements are discussed together with suggestions on how studies could be designed and undertaken to obtain this information. A glossary of terms used throughout this chapter can be found in Table 3.

Table 1 Summary of Guidelines for Nonclinical Pharmacokinetics and Metabolism

Methods	MHW 1998	Of sufficient validity and reproducibility to accomplish ADME objectives.
Pharmacokinetics	ICH3B MHW 1998	Blood, plasma, bile, urine, feces: AUC, Cl, F, Cl_R, $t_{1/2}$, absorption, C_{max}, T_{max}, V.
Pharmacodynamics	ICHS5(R2) ICHM3(R2)	Permits the interpretation of pharmacokinetic and toxicokinetic data and establishment of dose–response relationships. Can be used to identify role of pharmacologically active metabolites. PK/PD can help support decisions for selecting first dose to human.
	EMEA 1997	When similar mechanisms and/or effects are found in animals and in humans, animal studies and/or *in vitro* studies can be used to describe the pharmacokinetic and pharmacodynamic interaction profile of a drug.
Species	ICH3A	Species should be selected whenever possible with regard to pharmacodynamic and pharmacokinetic principles (see also section toxicokinetics).
Animal numbers		Minimum number to achieve meaningful interpretation of ADME data (see also section toxicokinetics).
Radiolabel		Position of radiolabel should be metabolically stable and, where possible, close to the pharmacologically active moiety of the molecule. Radiochemical purity, stability, and position of the label from the supplier should be available.
Routes and dose		Route of administration should relate to clinical use (see also section toxicokinetics). Appropriate dose levels applicable to pharmacological, toxicological, and clinical endpoints.
Sampling		Sufficient to describe pharmacokinetic parameters.
Absorption	ICH S1C(R2)	Rate and extent of absorption, and characterization of intrinsic and vehicle effect of absorption. Absolute absorption from incorporation of IV dose, including dissolution, absorption site, pH, and food effect.
		Saturation of absorption is an approach that could be useful for high-dose selection for carcinogenicity studies. The mid and low doses selected for the carcinogenicity study should take into account saturation of metabolic and elimination pathways.
	EMEA 1997	Drug–drug interactions might affect fraction absorbed, first-pass metabolism. For example, P-glycoprotein may limit absorption of a drug. Conversely, concomitant drug therapy may inhibit/induce P-glycoprotein activity, and thereby modulates extent of exposure to drug(s). Interactions and clinical relevance should be discussed with respect to dose/concentration effect. Interactions observed in *in vitro* should be confirmed by *in vivo* studies.
	ICH S3A	Where data indicate that absorption of a compound limits exposure to parent compound and/or metabolite(s), the lowest dose level of the substance producing the maximum exposure should be accepted as the top dose level to be used (when no other dose-limiting constraint applies).
Distribution	EMEA 1997 MHW 1998 ICH S3B	Single-dose tissue distribution studies form part of the nonclinical program. Time course, potential accumulation in organs and tissues can be used for dosimetric calculations to support radioactive studies in human. Circumstances under which repeated-dose tissue distribution studies should be considered include (*i*) apparent half-life of the test compound (and/or metabolites) in organs or tissue exceeds the apparent half-life of the elimination phase in plasma, (*ii*) when steady-state levels of a compound/metabolite in the circulation are higher than those predicted from single dose, (*iii*) histopathological changes not observed from single dose tissue distribution studies, and (*iv*) developed for site-specific delivery.

Table 1 (*Continued*)

		Repeat dose studies may be achieved using radiolabeled compounds or alternative methods of sufficient sensitivity and specificity. Duration of dosing should be selected based on the blood/plasma concentration of the compound and/or its metabolites reaching steady state. Consideration should be given to measuring unchanged compound and/or metabolites in organs and tissues in which extensive accumulation occurs or if it is believed that such data may clarify mechanisms of organ toxicity.
Protein binding	ICH S1C(R2)	Interspecies protein binding is taken into consideration when estimating relative exposure. Although *in vivo* determinations of unbound drug might be the best approach, *in vitro* determination of protein binding using parent and/or metabolites as appropriate (over the range of concentrations achieved *in vivo* in rodents and humans) might be used in the estimation of AUC unbound. When protein binding is low in both human and rodents, or when protein binding is high and the unbound fraction of drug is greater in animal species than in human, the comparison of total plasma concentration of drug is appropriate. When protein binding is high and the unbound fraction is greater in human than in animal species, the ratio of the unbound concentrations should be used.
Metabolism	ICHM3(R2)	*In vitro* metabolic data for animals and human should be evaluated prior to initiating human clinical trials.
	ICH S1C(R2)	It is recommended that *in vivo* metabolism be characterized in animal species and human. In the absence of appropriate *in vivo* data, *in vitro* metabolic data can provide appropriate support for the similarity of metabolism across species.
	ICH S2(R1)	The liver is typically the preferred tissue, but choice of *in vivo* tissue and assay should be based on factors such as any knowledge of the metabolism *in vivo*, and of the exposed tissue thought to be relevant. If the *in vivo* and *in vitro* results do not agree, then the difference should be considered/explained on a case-by-case basis, for example, difference in metabolism, rapid excretion may occur *in vivo*, etc. Analysis of the metabolite profiles in preclinical species in microsomes or hepatocytes, or *in vivo*, can help determine the relevance of the results and follow-up studies will usually focus on *in vivo* data.
		When *in vitro* positive results are seen only in the presence of the S-9 activation system, it should first be verified that metabolic activation is responsible and not some other difference in conditions. The follow-up strategy is then aimed at determining the relevance of any reactive metabolites produced *in vitro* to conditions *in vivo*, and will generally focus on *in vivo* studies in liver.
	ICH S3A	There may be circumstances when measurement of metabolite concentration in plasma or other body fluids is especially important in the conduct of toxicokinetics. For example, (*i*) when the administered compound acts as a "pro-drug" and the delivered metabolite is acknowledge to be the primary active entity, (*ii*) when the compound is metabolized to one or more pharmacologically or toxicologically active metabolites, which could make a significant contribution to tissue/organ response, (*iii*) when the administered compound is very extensively metabolized and the measurement of plasma or tissue concentrations of a major metabolite is the only practical means of estimating exposure following administration of the compound in toxicity studies.
	ICH S3B	Consideration should be given to measuring unchanged compound and/or metabolites in organs and tissues in which extensive accumulation occurs or if it is believed that such data may clarify mechanisms of organ toxicity.

(*Continued*)

Table 1 (*Continued*)

	ICH S7A	During early development, sufficient information (e.g., comparative metabolism) may not always be available to rationally select or design the study in accordance with study endpoints. Generally, any parent compound and its major metabolite(s) that achieve, or are expected to achieve systemic exposure in human should be evaluated in safety pharmacology studies. If the major human metabolite(s) is (are) found to be absent or present only at relative low concentrations in animals, assessment of the effects of such metabolite(s) on safety pharmacology endpoints should be considered. Additionally, if metabolites from humans are known to substantially contribute to the pharmacological actions of the therapeutic agent, it may be important to test such active metabolites.
	ICH S8	If the compound and/or its metabolites are retained at high concentrations in cells of the immune system, additional immunotoxicity testing should be considered.
	EMEA 1997 FDA 1997	Use of *in vitro* studies to characterize drug interactions for elimination pathways representing >30% of total clearance. Nature of interaction is important, that is, competitive, noncompetitive inhibition, or induction. These studies are recommended before initiation of Phase II clinical studies.
	FDA 2008	At least one animal species forms drug metabolites equal or greater than human exposure during toxicity testing of parent drug. Further investigation should be a single metabolite form >10% of parent drug systemic exposure at C_{SS}. Direct dosing of metabolite if not formed in animal species (21 CFR part 58). >10% of metabolite relative to bioavailable dose should be addressed on a case-by-case basis.
Excretion	ICH S2(R1) EMEA 1997 MHW 1998	If the *in vivo* and *in vitro* results do not agree, then the difference should be considered/explained on a case-by-case basis, for example, difference in metabolism or rapid excretion of a compound may occur *in vivo*, etc. Primary routes of excretion in urine and feces should be characterized. Major route of biliary elimination, enterohepatic circulation can be investigated. Interactions affecting renal excretion can be studied *in vitro*. pH effect may be significant if p*K*a of a drug is 7.5–10.5 for bases and 3.0–7.5 for acids.
Balance studies		Material balance should be performed normally over 7 days incorporating feces, urine, expired air using radiolabel drug. Recovery of drug and drug-related material should be quantified. Samples from these studies permit profiling of metabolites and direct a comparison of rate, routes, and amount of metabolites formed in toxicology species. These are compared with human to established safety cover.
Fetal uptake and milk transfer	ICH S3A	In pregnant and lactating animals, the treatment regimen during the exposure period should be selected on the basis of the toxicological findings and on pharmacokinetic and toxicokinetic principles. Consideration should be given to the possibility that the pharmacokinetic parameters will differ in pregnant and nonpregnant animals. Toxicokinetics may involve exposure assessment of dams, embryo, fetuses, or newborn at specific days (see section toxicokinetics). Secretion in milk may be assessed to define its role in exposure of newborns. In some situations, additional studies may be necessary or appropriate in order to study embryo/fetal transfer and secretion in milk. Consideration should be given to the interpretation of reproductive toxicity tests in species in which placental transfer of the substance cannot be demonstrated.
Repeat dose	ICH S1C(R2)	Ideally, animal species/strains with metabolic profiles as similar as possible to human should be studied. Changes in metabolite profile or alterations in metabolizing enzyme activities (induction or inhibition) over time should be understood to allow for appropriate interpretation of studies. Criteria for comparisons of AUC in animals and human require that similarity/differences of between animal species and human are documented.

Table 1 (*Continued*)

	EMEA 1997	Confirm whether drug induces or inhibits it own elimination, or potentially of other drugs to be administered concomitantly in drug therapy. Use of radiolabel material or specific assays can be used to qualify differences in metabolite profile between single and repeat dose.
		Where high levels of material are identified, it may be necessary to identify presence of drug and/or identity of metabolite(s).
	ICH 7A	Safety pharmacology studies are generally performed by single-dose administration. When pharmacodynamic effects occur only after certain duration of treatment, or when results from repeat-dose nonclinical studies or results from use in humans give rise to concerns about safety pharmacological effects, the duration of the safety pharmacology studies to address these effects should be rationally based.
Physicochemical		Specifications and stability data.

ICH S1C(R2): Dose selection for Carcinogenicity Studies of Pharmaceuticals (43)
ICH S2(R1): Guidance on Genotoxicity Testing and Data Interpretation for Pharmaceuticals Intended for Human Use (101)
ICH S3A: Toxicokinetics: A Guidance for Assessing Systemic Exposure in Toxicity Studies (18)
ICH S3B: Pharmacokinetics: Repeated Dose Tissue Distribution Studies (102)
ICH S4: Duration of Chronic Toxicity Testing in Animals (Rodent and Nonrodent Toxicity testing) (103)
ICH S5(R2): Detection of Toxicity to Reproduction for Medical Product and Toxicity to Male Fertility (92)
ICH S7A: Safety Pharmacology Studies for Human Pharmaceuticals (104)
ICH S8: Immunotoxicity Studies for Human Pharmaceuticals (78)
EMEA 1997: Note for guidance on the investigation of drug interactions (44)
FDA 2008: Guidance for Industry: Safety Testing of Drug Metabolites (39)
MHLW 1998: Japanese guidelines for nonclinical kinetics (47)

DRUG DISCOVERY

Progress in molecular pharmacology and bioanalytical chemistry along with more recent developments in the technology platform of genomics, proteomics, and metabolomics ("omic"-technologies) together with increasing global competition are markedly reshaping the paradigm of the drug discovery process. Throughout much reshaping of this process, pharmacokinetics has remained an integral component of drug discovery with its notable support in the lead optimization and selection of new candidate drugs. Early pharmacokinetic studies are not themselves governed by regulations but it is felt useful to describe how they influence the choice of a candidate drug and assist in the design of future pharmacokinetic studies to be undertaken later in drug development. It is also important to mention that the first choice of candidate drug selected from a drug discovery program may not necessary be the most pharmacologically potent compound, because the comparative safety and pharmacokinetic profile of the drug must be taken into consideration. It is this integrated approach toward pharmacokinetics, safety, and pharmacology in drug discovery that highlights the importance of a partnership between Drug Metabolism and Pharmacokinetic, Safety Assessment, and Medicinal Chemistry disciplines.

Pharmacokinetics enables the investigator to assess the temporal behaviour of a drug in the body with respect to the absorption, distribution, metabolism, and excretion (ADME) of a drug. When a particular aspect of pharmacokinetics is studied, whether by *in silico*, *in vitro*, or *in vivo* techniques, they are normally referred to as ADME studies. During the early stages of drug discovery, the ADME of a compound can be assessed by using a variety of *in vitro* models with a limited number of *in vivo* animal studies to support final candidate selection. In the last 15 years, many cell- or tissue-based *in vitro* models have been adopted as part of the research and development of novel drug candidates. These models offer opportunities to investigate the likely ADME characteristics of a drug in terms of the different biological barriers and metabolic processes that a drug has to overcome to finally reach its pharmacological site of action. The utility of these models permits compounds that are poorly or erratically absorbed with or without high first-pass metabolism to be identified rapidly and characterized. Information from these studies can also help underpin *in silico* structure–activity relationships

Table 2 Summary of ICH Toxicokinetic Guidelines

Studies	ICH3A	Toxicokinetics may be used in acute and repeated dose toxicity studies, reproductive, genotoxicity and carcinogenicity studies, and toxicity studies with toxicodynamic endpoints.
		In designing protocols, account should be taken of ADME data and species selection.
Methods	ICH3A	Methods should be specific, accurate, and precise, and of adequate sensitivity for the range of concentrations expected. Matrix effects should be considered to account for endogenous components that may interfere with bioanalysis and interpretation of results. Attention should be paid to chiral drugs, whether a racemate or single enantiomers to account for differences in pharmacokinetics and/or enantiomeric conversion.
Toxicokinetics	ICHS3A	AUC, C_{max}, T_{max}, C_{ss}.
Physicochemical		Specification and stability data is available.
Species	ICH 3A ICH S1C(R2)	Carcinogenicity: Ideally, rodents/strains with metabolic profiles as similar as possible to human. Rodent pharmacokinetic data are derived from the strains used for the carcinogenicity study.
	ICH S2(R1)	Genotoxicity: Both rats and mice are appropriate for detection of clastogens—bone marrow.
	ICH S5(R2)	Reproduction: Studies should be conducted in mammalian species. If it can be shown by means of pharmacokinetic, pharmacological and toxicological data that the species selected is a relevant model to human—a single species can be sufficient. Study of fertility and early embryonic development to implantation—at least one species, preferably rat. Study on pre- and postdevelopment, including maternal function in at least one species, preferably rat.
		Study for effects on embryo–fetal development—usually two species: one rodent, preferably rat, one nonrodent, preferably rabbit (lagomorph).
	ICH S8	Immunotoxicity: Rodent and nonrodent species and strains should be consistent, where possible, with the standard toxicity studies in which an adverse immune effect was observed. Usually, both sexes should be used in these studies, excluding nonhuman primates.
Animal numbers	ICH S1C(R2)	AUC values in rodents are usually obtainable using a small number of animals, depending on the availability of pharmacokinetic data.
	ICH S2(R1)	Determined by current recommendations for the micronucleus assay (OECD) or genotoxicity assays.
	ICH 3A	Systemic exposure in toxicity studies should be measured in an appropriate number of animals and dose groups, the design of which can be selected using pharmacokinetic data from same animal species, and can be performed either in all or a representative proportion of the animals used in the main study, representative subgroups, satellites groups, or separate studies. Number of animals should be the minimum consistent with generating meaningful toxicokinetic data.
	ICH S5(R2) ICH S7A	Numbers per sex per group should be sufficient to allow a meaningful interpretation of reproductive and safety pharmacology studies.
Route	ICH S2(R1) ICH S7A	In general, the expected clinical route should be use.
	ICH S5(R2)	Similar to the intended clinical usage.
	ICH S8	Consistent with the standard toxicity studies for assessment of immunotoxicity.
Dose	ICH S1C(R2)	Dose-ranging studies should be conducted for both male and females for all strains and species to be tested in the carcinogenicity bioassay. Dose selection is generally determined from a 90-day study. Traditionally, the maximally tolerated dose (MTD) is the standard high dose selection for carcinogenicity studies but other endpoints based on pharmacological properties and toxicological profile of the test compound are also accepted.

Table 2 (*Continued*)

	ICH S2(R1)	A preliminary range-finding assay over a broad range of concentrations for genotoxicity testing. Intended versus actual concentration(s) should be confirmed. For *in vivo* studies, dose/exposure relationship should be determined. The top dose recommended for genotoxicity assays is a limit dose of 2000mg/kg if tolerated, or maximum tolerated dose defined, for example, for the micronucleus assay (OECD 474). For multiple administration studies, the top dose is limited to 1000mg/kg for studies of 14 days or longer, if tolerated.
	ICH S3A	The setting of dose levels in toxicity studies is largely governed by toxicology findings and pharmacodynamic responses of the test species (see toxicokinetics table).
	ICH S3B	Dose level(s) should be chosen to address the problem that led to the consideration of the repeated dose tissue distribution study (see also distribution—below).
	ICH S5(R2)	The choice of the high dose in reproductive studies should be based on data from available studies such as, pharmacology, acute and chronic toxicity, and pharmacokinetic studies. It is recommended to use at least three dose levels. Fourth dose may be applicable to avoid excessive dosage intervals. Such a strategy should define exposure versus "'no-observable-adverse-effect level" for reproductive aspects.
	ICH 7A	Safety pharmacology studies should be design to define the dose–response relationship of the adverse effect observed. The highest dose should be a dose that produces moderate adverse effects. *In vitro* studies should be design to establish a concentration–effect relationship. The upper limit of this range may be influenced by physicochemical properties.
	ICH S8	Multiple dose levels are recommended in order to determine dose–response relationships and the dose at which no immunotoxicity—high dose should be above the no-observed-adverse-effect level (NOAEL) but below a level inducing changes secondary to stress.
Sampling	ICH S3A	In each study, the number of time points should be justified on the basis that they are adequate to estimate exposure. Sampling should not be so frequent as to interfere with the normal conduct of the study, or cause physiological stress to the animal. If samples are taken from main study animals, it should be considered whether samples should be taken from all the dosed animals and the controls. Justification of time points should be based on kinetic data gathered from earlier toxicity studies, pilot, or range-finding studies.
	ICH S7A	The time points for safety pharmacology measurements should be based on pharmacodynamic and pharmacokinetic considerations.
Metabolites	ICHS3A	It may be necessary to measure metabolites where: (*i*) parent drug can not be measured, (*ii*) a prodrug is administered, (*iii*) metabolite(s) are pharmacologically or toxicologically active, (*iv*) there is extensive metabolism, or human specific metabolite(s) are administered to assess exposure-safety in toxicology species.
Exposure	ICHS3A	In toxicity studies, systemic exposure of drug and/or metabolites should be measured in an appropriate number of animals and dose groups. Concomitant toxicokinetics may be performed either in all or a representative proportion of the animals used in the main study or in special satellite groups.
Control samples	ICHS3A	Control samples should be assayed to support the interpretation of data.

(*Continued*)

Table 2 (*Continued*)

Acute dose	ICHM3(R2) ICHS3A	Plasma samples may be taken from single-dose toxicity studies in very early phase studies and stored for later analysis, if necessary. Appropriate stability data for the analyte in the matrix samples would then be required. Alternatively, additional toxicokinetic studies may be carried out after completion of an acute toxicity study.
Repeat dose	ICH S2 (R1)	In principle, genotoxicity studies of any length may be appropriate provided the top dose/exposure is adequate.
	ICH S3A	The design of the toxicokinetic studies may consist of exposure profiling or monitoring at appropriate dose levels at the start and towards the end of the treatment period of the first repeat-dose study. The treatment regimen and species should be selected whenever possible with regard to pharmacodynamic and pharmacokinetic principles. Toxicokinetics should be incorporated appropriately into the design of the studies. It may consist of exposure profiling or monitoring at appropriate dose levels at the start and toward the end of the treatment period of the first repeat-dose study. The procedure adopted for later studies will depend on the results from the first study and any changes in the proposed treatment regimen.
	ICH S4	Duration of chronic toxicity testing in animals: Rodents—6 mo duration; nonrodents—9 mo duration.
Fertility	ICH S5(R2)	It is preferable to have some pharmacokinetic information before initiating reproduction studies. At the time of study evaluation further information on pharmacokinetics in pregnant or lactating animals may be required according to the results obtained. The usual frequency of administration is once daily but consideration should be given to use either more frequent or less frequent dosing to ensure adequate exposure.
		Usually, two species: one rodent, preferably rat; one nonrodent, preferably rabbit (lagomorph). Pharmacokinetic investigations in pregnant and lactating animals may pose problems due to the rapid changes in physiology. It is best to consider this as a two- or three-phase approach. In planning studies, pharmacokinetic data (often from nonpregnant animals) to provide information on the general suitability of the species, and to assist in study design and choice of dose(s). During a study kinetic investigations can provide assurance of accurate dosing or indicate marked deviations from expected patterns.
		Consideration should be given to the possibility that the pharmacokinetics will differ in pregnant and nonpregnant animals. Toxicokinetics may involve exposure assessment of dams, embryos, fetuses, or newborn at specified days. In some situations, additional studies may be necessary or appropriate in order to study embryo/fetal transfer and milk secretion.
Genotoxicity	ICH S2(R1)	The value of *in vivo* results is directly related to the demonstration of adequate exposure of the target tissue to the test compound—particularly true for negative *in vivo* test results when *in vitro* test(s) have shown convincing evidence of genotoxicity, or in the absence of data from an *in vitro* mammalian cell assay. Evidence of adequate exposure could include toxicity in the tissue in question, or toxicokinetic data. If dose/exposure is not appropriate, an acute *in vivo* study should be performed to optimize dose selection based on exposure/toxicity.
		Recommendations for determining whether the high dose in a toxicity study (typically rat) is appropriate for micronucleus analysis and other genotoxicity evaluations may apply if: (*i*) maximum feasible dose (MFD) is achieved based on physicochemical properties of the drug, (*ii*) dose of 1000mg/kg for study of 14 days or longer is achieved, (*iii*) plateau/saturation in exposure, accumulation, or substantial reduction in exposure to parent drug with time, (*iv*) top dose is $>50\%$ of the top dose that would be used for acute administration, that is, close to the maximum dose.

Table 2 (*Continued*)

Carcinogenicity	ICH S1C(R2)	One of several approaches can be used for dose selection in carcinogenicity studies, these include (*i*) toxicity-based endpoints, (*ii*) pharmacokinetic endpoints, (*iii*) saturation of absorption, (*iv*) pharmacodynamic endpoints, (*v*) maximum feasible dose, (*vi*) limit dose, and (*vii*) additional endpoints. The selection of a high dose which represents a 25:1 exposure ratio of rodent to human plasma AUC of parent compound and/or metabolites is considered pragmatic. The maximum feasible dose to dietary administration is considered to be 5% of diet. It is appropriate to limit the high dose to 1500mg/kg/day where the maximum recommended human dose does not exceed 500mg/day. If the human dose exceeds 500mg/day, the high dose can be increased up to the maximum feasible dose.

ICH S1C(R2): Dose selection for Carcinogenicity Studies of Pharmaceuticals (43)
ICH S2(R1): Guidance on Genotoxicity Testing and Data Interpretation for Pharmaceuticals Intended for Human Use (101)
ICH S3A: Toxicokinetics: A Guidance for Assessing Systemic Exposure in Toxicity Studies (18)
ICH S3B: Pharmacokinetics: Repeated Dose Tissue Distribution Studies (102)
ICH S4: Duration of Chronic Toxicity Testing in Animals (Rodent and Nonrodent Toxicity Testing) (103)
ICH S5(R2): Detection of Toxicity to Reproduction for Medical Product & Toxicity to Male Fertility (92)
ICH S7A: Safety Pharmacology Studies for Human Pharmaceuticals (104)
ICH S8: Immunotoxicity Studies for Human Pharmaceuticals (78)

(SAR) from screening chemical entities with simple measures of partition coefficient, solubility, pK_a, molecular weight, and protein binding, particularly when comparisons are made within a chemical series. During late discovery and early development, these same *in vitro* models are also gainfully employed in studying drug delivery systems for improved delivery and enhanced absorption, and to support more detailed mechanistic studies to understand the likelihood of drug–drug interactions that can occur during absorption and distribution.

The metabolism of a drug plays a pivotal role in the elimination of most drugs. The majority of drugs undergo metabolism by oxidation, reduction, or hydrolysis (Phase I) followed by conjugation (Phase II). Characterization of the human drug-metabolizing enzymes involved in metabolic elimination of a drug is an important component of drug discovery as it permits an understanding of first-pass effects in relation to bioavailability, variability associated with dose–exposure relationships, and the role of drug-metabolizing enzymes to support the risk-benefit analysis of a drug in different populations and individuals. Closely linked with understanding the metabolic fate of a drug are studies to investigate the formation of reactive intermediaries of drug metabolism that may be implicated in toxicity. Where the formation of reactive intermediaries is suspected, studies designed to investigate their formation can be established by using various *in vitro* techniques coupled with metabolite identification. The direct measurement of chemically reactive intermediates arising from Phase I metabolism is rarely possible per se because of their short half-lives in biological systems. However, the presence of stable metabolites arising from the nucleophilic trapping of intermediates by nucleophiles is widely used as an indirect measure of their formation. The detection of these metabolites is not in itself a measure of potential toxicity, as abundance, rate of formation, and dose are also claimed to play an important role in determining the likelihood of reactive metabolite toxicity. Nevertheless, the detection and measurement of these metabolites enables risk-benefit management strategies to be built into drug discovery and future development programs.

Compared to the complexities of the *in vivo* situation, *in vitro* models offer a relatively fast, convenient approach with cost advantages most of the time! The use of *in vitro* and *in silico* ADME assays enables the investigator to select lead candidate drugs with often more detailed *in vitro* and *in vivo* animal studies adopted during late discovery and lead optimization. At some point, however, a decision is taken to move the drug candidate into the development phase with the initiation of nonclinical toxicity studies to establish drug safety before clinical trials can commence.

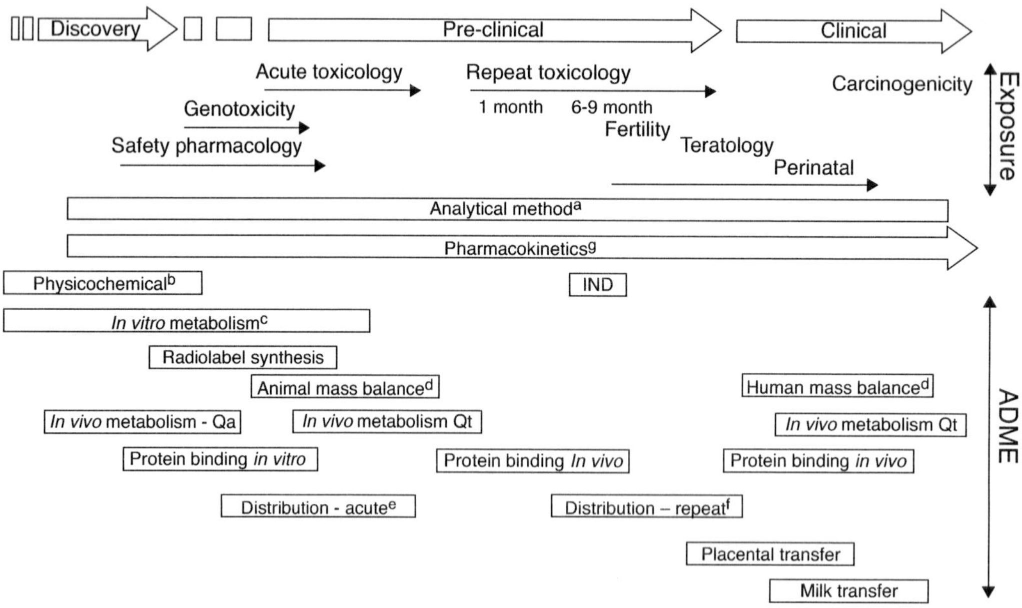

Figure 1 Phases of drug development: Nonclinical pharmacokinetic and ADME studies. The majority of nonclinical studies are assumed to be performed by Phase 1: Qa, qualitative; Qt, quantitative. (a) Assay is refined with improvement of sensitivity, specificity, etc.; (b) QSAR to support ADME and candidate selection; (c) cell or tissue-based assays, inhibition, and induction assays—ADME characterization; (d) validation of toxicology species and safety cover in human; (e) specify tissues exposed to drug/drug-related material; (f) Examine retention of drug/drug-related material and related changes in metabolite profile; (g) measurement of parent and/or metabolites to relate exposure with PD effect. If *in vitro* or animal studies raise questions, then Phase 0 exploratory studies in human can commence after 1-wk or 2–4 wk toxicity studies.

ADME STUDIES

Regulatory Requirements

The Clinical Trials Directive (CTD) (5) for the commencement Phase I studies has been standardized across the EU, together with the package of nonclinical studies required to commence clinical trials themselves. The critical path to initiating Phase I studies is the package of work required on the safety of the drug candidate. Although the ICH M3 guidelines make reference to the use of ADME studies in support of drug safety, guidance on ADME studies are largely absent probably because the type and extent of ADME studies required to support the development of a drug will depend on the properties of the drug itself. Yet despite the absence of ICH guidelines on the use of ADME studies in safety assessment, these studies undoubtedly play a central role in the understanding and knowledge of how animals used in pharmacology and toxicity testing handle a drug in terms of its ADME characteristics, which in turn influences the *in vivo* pharmacokinetic profile of a drug.A recent examination of 54 Investigator Brochures submitted between 1997 and 2007 for first-in-human (FIH) studies showed regional differences in the amount of ADME work reported by pharmaceutical companies from Europe, the United States, and Japan. Although the number and types of ADME studies were broadly similar between Europe and United States, there were notably more studies undertaken by Japanese companies. In the past 10 years, the average number of ADME study-types contained within an Investigators Brochure was also found to increase year-on-year from an average of 2 in 1997 to 10 in 2007, the increase of which was attributed to a greater number of *in vitro* studies (6).

ADME studies for regulatory purposes are carried out for four main reasons: (*i*) to validate the choice of species used in safety testing, (*ii*) to compare the pharmacokinetics and metabolism in animals with that in human to aid in the interpretation of results, (*iii*) to help in the design of

Table 3 Glossary of Terms and Abbreviations

ICH	International Conference for Harmonization
FDA	Food & Drug Administration
GLP	Good Laboratory Practice
CTD	Common Technical Document
ADME	Absorption, Distribution, Metabolism, and Excretion
AUC	Area under blood/plasma concentration time curve
T_{max}	Time to maximum blood concentration
C_{max}	Maximum blood or plasma concentration
MTD	Maximum tolerated dose
fu	Unbound drug level in plasma
CL	Total body clearance
CL_R	Urinary clearance
$t_{1/2}$	Half-life of elimination
C_t	Blood or plasma concentration at a specific time point
C_{ss}	Blood or plasma concentration steady state after repeat dosing
$C_{ss\ max}$	Maximum blood or plasma concentration achieved at steady state
$C_{ss\ min}$	Minimum blood or plasma concentration at steady state
F	Bioavailability
V	Volume of distribution
R_{ac}	Accumulation ratio
DDI	Drug-drug interaction
IVIVE	*In vitro-in vivo* extrapolation
PBPK	Physiologically based pharmacokinetics
TK	Toxicokinetics
TKTD	Toxicokinetic–toxicodynamic
(Q)WBA	(Quantitative) Whole Body Autoradiography
GRAS	Generally recognized as safe
HPLC	High-Pressure Liquid Chromatography
pK_a	Acid dissociation constant

toxicity studies, and (*iv*) to identify any unusual or potentially problematic aspects of the ADME characteristics of a drug, for example, accumulation in sensitive tissues or prolonged half-life. It is on this last point that recent ICH S3B guidelines ask for information on the distribution, accumulation, and retention of the drug and its metabolites after a single dose, and to define the situations in which repeat dose studies are requested. There are five main reasons for undertaking repeat dose distribution studies in animals prior to regulatory submission, namely, (*i*) to determine the extent of localization in tissues, (*ii*) to identify where there is retention of drug and the possibility of covalent binding; (*iii*) to help explain specific organ toxicities; (*iv*) to validate certain toxicity studies, for example, segment II and fetal uptake, and (*v*) to determine quantitative radioactive tissue exposure prior to administration of radiolabeled drug to humans. In addition, knowledge of where and in what concentration drug and drug-related material are located may provide useful information when designing or interpreting results from pharmacology studies.

The timings of ADME studies will vary from one company to another depending on their philosophy of how quickly the company wishes to get into human subjects. Figure 1 assumes that much of the preclinical work is undertaken prior to Phase I. However, some companies will undertake very limited preclinical investigations with a minimum number of *in vitro* metabolic studies together with toxicokinetic evaluation of animal plasma data. Certainly, delaying ADME studies until Phase I can have its merits by first permitting a confirmation of the dose–exposure range in humans and thus knowledge of the likely dose to be used in patients. The risk, on the other hand, is to not identify a problem with the drug at an early stage prior to human testing, and consequently, there may be insufficient information on how to manage the development program if problems should arise.

In the last 10 years, there have been significant advances in our understanding of the role of transport proteins in absorption, distribution, and excretion (ADE) of many drugs, and these are reviewed in detail elsewhere (7). Localized in cell membranes, these transport proteins influence the rate and extent of uptake and efflux of drugs across enterocytes, apical and basolateral membranes of hepatocytes, and endothelial cells of the blood–brain barrier. Many of the transport proteins are subject to induction and inhibition by drugs, or are genetically polymorphic and are therefore synonymous with drug-metabolizing enzymes in their potential to alter the pharmacokinetic profiles of drugs over time. Characterizing the role of transport proteins and potential drug–drug interactions (DDIs) are increasingly included as part of the ADME characterization of drugs. In the short period since their discovery, FDA guidelines on the evaluation of transport proteins have been issued and specifically mention P-glycoprotein (Pgp), but also allude to further investigation into the role of other transport proteins, as required (8,9). Whether the investigation of transport proteins is conducted during drug discovery or development is a matter for individual companies, but what is increasingly clear are regulatory expectations to assess the impact of transport proteins on drug disposition as part of the overall safety evaluation of new drugs. European guidelines on interaction studies contain some limited guidance on the evaluation of drug transport proteins. These guidelines mention the importance of considering interactions with Pgp at the level of drug absorption as well as interactions for drugs that are actively secreted in urine or bile. Therefore, while regulatory guidelines are limited, characterizing the role of transport proteins and potential for drug interactions are expected increasingly in the submission for marketing authorization of a new drug. At present, the European guidelines are under revision with future guidelines likely to incorporate more on the evaluation of transport proteins and quite possibly a mention of altered tissue exposure as a consequence of transport proteins involved in the tissue distribution of drugs (10).

Analytical Requirements

Bioanalytical methods available for routine ADME analysis of small molecules broadly comprises the use of high-performance liquid chromatography (HPLC) coupled with UV, fluorescence, mass spectrometry, or radiochemical detection. Other less frequently used methods for small-molecular-weight entities include GLC and receptor-based immunological methods where analysis by other means is not possible. For most comparative ADME studies, radiolabeled drugs are commonly used.

The use of a radiolabeled molecule enables the drug and its metabolites to be followed throughout the body and excreta, and allows for a material mass balance to be calculated. The two more commonly used radionucleotides are ^{3}H and ^{14}C. ^{14}C is commonly used because it is more stable, has a more efficient detection, and can produce collectable $^{14}CO_2$ if the molecule is degraded. On the other hand, ^{3}H-containing compounds are easier to synthesis and have a higher specific activity, but a shorter path length of activity, which makes it more ideally suited for use in early exploratory-based studies or when drugs with high pharmacological potency are dosed at $\mu g/kg$ body weight levels and the lower specific activity of ^{14}C will not permit adequate detection by conventional liquid scintillation counting. Furthermore, ^{3}H half-life is considerably shorter than ^{14}C, more unstable, and ^{3}H label can undergo exchange with water when specific synthesis is not undertaken. The position of the radiolabel is vital and where possible should be close to the pharmacologically active moiety of the drug which in a position within the molecule which is stable from metabolic or chemical degradation. For larger molecules that are prone to metabolism and where the active portion is not known or spread over a large part of the molecule, two radiolabels are required either as a dual labeled entity or as separate syntheses, which can be combined to give a stoichiometric mix of the two radiolabeled forms. It is important to confirm the chemical and radiochemical purity prior to use, and syntheses, which are less than 98% radiochemically pure (depending on the individual levels and types of impurities), would normally require further purification before use. In general, synthesis should commence as soon as possible in the development process particularly where ADME data are required to support decision-making prior to further development. In practice, however, radiochemical synthesis often occurs once the decision is made on taking the drug into development and in some instances can be delayed until the first clinical results are obtained, for example, until there is proof of exposure in human.

General Considerations

An ADME package is investigated typically in the rat and a nonrodent species, dog or nonhuman primates, the choice of which may be dependent on the findings of the metabolic *in vitro* species comparison or on company policy because many do not wish to use nonhuman primates and because of large interanimal variation. Some work may also be undertaken in mice and rabbits later in development to support carcinogenicity studies and ICH Stage B (Segment II) teratology studies. *In vitro* metabolic comparisons may also provide an indication of the need for specific studies in these species.

The strains, weights, and ages of animals should be as similar as possible to those used in the safety studies and within reason, should be housed under similar conditions. Often in these early studies, both sexes are examined but normally there will be too few animals to see any statistical differences unless the numbers or differences are very large ($\geq$5). Sex differences that occur frequently in rodents, particularly rats, are rarely seen in other species and the value of examining this aspect in single dose ADME studies is questionable because it will become apparent in early toxicokinetic studies. The number of animals used for each study will also vary from one company to another because none of the guidelines for pharmaceuticals provide an indication of animal numbers required for ADME work. Table 4 suggests the animal numbers required for each type of study, but for blood pharmacokinetics more may be required depending on the pharmacokinetic characteristics of a compound and the type of sampling technique adopted. For the larger species, 10 to 12 repeat serial samples of 1 to 2 ml can be achieved without any deleterious effect on the animal, while for the rodents serial sampling can be problematic if anything more than 0.5 ml is taken. The U.K. Home Office requirements state that no more than 1.1 ml per 100 g body weight or 15% of the blood volume can be taken over any one month, making pharmacokinetic and metabolite profiling from a single animal problematic unless the analytical method is particularly sensitive. This difficulty can be overcome by destructive sampling while composite sampling, two to three samples from each animal but at different times spread out over the total collection period, overcomes this need for an excessive number of animals. A simple formula to calculate the number of animals to be used for composite sampling is

$$\text{No. of animals} = \frac{T}{S} \times R$$

where T is the number of time points, S the number of samples to be taken from each animal, and R the number of repetitions at each time point.

Guidelines require that at least one of the doses used in these studies should be similar to that used in humans to allow a comparison with Phase I studies and the human excretion balance study. If the ADME animal studies start too early in the development, it can be argued that the exposure in humans will not be known unless some physiologically based pharmacokinetic PBPK modeling and use of allometric scaling can provide an indication of the expected therapeutic dose in humans, as described above. Therefore, while a single dose is all that is required normally, a mid and high oral dose, and IV dose in animals can provide information on the ADME of a drug over the intended therapeutic dose and important additional information on route dependent metabolism and nature of metabolites formed. Any metabolites of particular importance can then be further investigated during toxicokinetic monitoring of safety studies. Normally, doses for ADME studies will be expressed, as with toxicity studies, in terms of mg/kg and for anticancer drugs as mg/m^2.

In Vitro–In Vivo Extrapolation

The ultimate objective during preclinical development is to predict from simple *in vitro* ADME and *in vivo* animal experiments the ADME profile of a drug in humans. Considerable efforts on the part of academia and industry to develop *in vitro* and *in silico* models have enabled a greater understanding of the pharmacokinetic parameters, scaling factors, and physicochemical properties of drugs that permit acceptable *in vitro–in vivo* extrapolation (*IVIVE*) (11,12). As a result, there are many good examples of successful *IVIVE*, but equally there are examples where quantitative *IVIVE* are poorly predictive of the *in vivo* situation. Despite the difficulties

Table 4 Summary of ADME Studies in Support of Safety Assessment

Study	Animal	Animal number	Sex	Route	Dose	Analysis	Blood/plasma	Urine	Feces	Other
Candidate selection	R	2–3/dose	M	po, IV	P	Cold	4–5 Samples	-	-	-
Plasma binding	R, Mi, Rb, D, Mo	*In vitro*, *In vivo*, ($n = 2$–4)	M(F); M(F)		P-T	Cold or radiolabel	Plasma B:P ratio	-	-	-
Mass balance	R, Mi, Rb, D, Mo	3–6/dose; 3–6/dose; 3–6/dose	M[a]	po, IV	low[b] mid[b] high[b]	Radiolabel	6–14 Samples metabolite profile	0–8, 8–24, . . . , 168 hr or >95% total	0–8, 24, . . . , 168 hr or >95% total	Carcass cage wash CO_2
Biliary	R	3–4/dose	M[a]	IV	P-T	Radiolabel	Metabolite profile	00–8, 8–24 hr	0–24	Bile hourly, 8–24 hr
Pharmacokinetics	R, Mi	1–3/time pt	M/F	po, IV	D-R	Cold	3/time pt/animal	-	-	
Distribution (single dose)	R, Pg, A	1/time pt	M	po	P	QWBA radiolabel		-	-	Tissue exposure assessment
Distribution (repeat dose)	R	1–2/time pt	M	po	P (T)	QWBA cold or radiolabel	Sample at Css	-	-	Tissue exposure assessment
Fetal uptake	R,Rb	2–3/time pt	F	po	P (T)	QWBA radiolabel	Rat – D15; rabbit-D18; 3–4 time pt metabolite profile	-	-	Placenta, fetus
Milk transfer	R	2–3/time pt	F	po	P	Radiolabel or cold	C_{max} and 3–4 time pt metabolite profile	-	-	Milk

Abbreviations: Pg, Pigmented; A, Albino; P-T, pharmacological–toxicological range of dose; b:p, blood:plasma ratio.
[a] Male and female animals if gender differences are observed.
[b] Most bioavailable form.

associated with *IVIVE*, investigators believe that *in vitro* data can be gainfully employed to help in not only decision-making prior to nonclinical development, but also in clinical development when detailed studies using radiolabel drug are possible.

IVIVE for many drug-metabolizing enzymes, particularly the cytochrome P450 enzymes, are well established, but for a number of non-cytochrome P450 drug-metabolizing enzymes and transport proteins, the results are currently somewhat difficult to compare to human *in vivo* data. This is largely due to only an intrinsic clearance (Clint) value being derived from *in vitro* studies and an absence of or poorly defined *IVIVE* scaling factors. Drug elimination measurements *in vitro* are reported as intrinsic clearance from which pharmacokinetic parameters can be estimated such as bioavailability (F%) when values of fraction absorbed become available. The value of these predictions, however, rely heavily on the accuracy of *IVIVE* parameters and although predictions of *in vivo* clearance and metabolic F% can be reasonably good for cytochrome P450-dependent clearance, for other non-P450 mechanisms of elimination, use of *IVIVE* to predict the *in vivo* situation can be rather poor (13,14). Where possible, intrinsic data generated from *in vitro* studies are used increasingly in conjunction with PBPK methodology to facilitate pharmacokinetic predictions in human. It is by use of PBPK coupled with Clint and tissue partition experiments (*kp*) that additional primary parameters such as volume (*v*) can be obtained, thereby refining the prediction of *in vivo* animal and human pharmacokinetic parameters (15–17).

Suspensions of hepatocytes are widely used for the assessment of hepatic uptake, metabolism, and inhibition studies within several hours of isolation, but internalization of canalicular transport proteins and limited cell viability do not allow assessment of biliary clearance and induction processes. For these studies, the use of cultured hepatocytes and/or cell-lines is required. Maintaining expression of proteins in isolated hepatocytes presents some methodological challenges, where the isolation of hepatocytes levels of proteins responsible for the elimination of drugs can rapidly decline unless specialized cocultures and collagen-sandwich techniques are employed to maintain cell polarity and basolateral-cannicular configuration of cells. The utility of these particular cultures permits the study of basolateral uptake and canalicular efflux transport processes. Rat hepatocytes are widely used simply because of the availability of fresh tissue, but where human tissue can be sourced, human hepatocytes cultures enables the scientist to study the interplay between drug transport and metabolic processes that occurs in *in vivo* human.

Other *in vitro* models include the use of cell-lines, for example, Madin-Darby Canine Kidney (MDCK II) and human colonic adenocarcinoma (Caco-2) cells with or without transfection of drug-metabolizing enzymes or transport proteins. These particular types of cell-based models are grown on semipermeable membranes to cell confluence and enable measurements of transcellular passage of drugs from the apical-basolateral and basolateral-apical membranes. The unfortunate drawback with these models is that they do not express the full complement of *in vivo* metabolic enzymes and transport proteins, which, coupled with a lack of suitable scaling factors, can limit the scientist to the use of indirect or relative measures of activity against references drugs by which to extrapolate results to *in vivo* human.

Absorption

Regulatory expectations necessitate rate and extent of drug absorption by the intended therapeutic route in human to be characterized to ensure that adequate exposure in toxicity studies has been achieved (18). For the majority of drugs in clinical use this requires rates and extent of absorption following oral administration to be characterized, but where other routes of administration are selected on the basis of pharmacological and/or pharmacokinetics characteristics of the drug, intravenous dosing is recommended as the second route of administration.

The utility of *in vitro* models allows the scientist to not only characterize the permeability of drugs across a variety of cell membranes, but also to characterize the mechanisms of drug absorption to improve delivery and enhanced absorption (19). Compared to the complexities of the *in vivo* situation, *in vitro* models offer important insights to the mechanisms affecting good absorption, but equally have given arise to several misconceptions that assume, for example, drug absorption, permeability, and bioavailability are equivalent. Despite many factors that can affect rate and extent of drug absorption (such as intestinal pH for ionizable drugs, presence of transport processes and the presence or absence of food), it appears that interspecies differences in drug absorption itself are relatively uncommon (20,21). Cao et al. (22) examined 14 drugs and 3 "drug-like" compounds with different absorption mechanisms, and 48 drugs for a comparison of oral bioavailability. Results showed similar transport protein expression and absorption profiles between rat and human.By contrast, marked species differences in the expression and pattern of intestinal drug-metabolizing enzymes in rat and human markedly affect the ability to predict oral bioavailability.

Provided that a reasonable proportion of the dose is eliminated in the urine then the use of radiolabeled drug enables an estimate of the extent of absorption, and the measurement of unchanged drug in urine gives a minimum estimate of the amount of parent drug absorbed. Providing there is no route-dependent metabolism, for example, due to saturation of a pathway after oral absorption, a comparison of radioactivity in urine after oral and IV dosing gives a better measure of absorption. Estimates of drug absorption can be improved yet further where measurement of radioactivity in bile together with urine will give the best estimate of the amount of drug absorbed. Normally, assays of unchanged drug are necessary to define plasma profiles, rate of absorption, area-under-curve (AUC), and bioavailability, providing that IV administration is also undertaken. For large animals, this can be analyzed at the same time as the radioactivity, providing a direct comparison between cold and radioactive levels. For rodent species this is not normally possible and separate kinetic studies would need to be undertaken unless composite or destructive sampling is used (Table 4). If the drug is extensively metabolized or is a pro-drug, then measurement of metabolite levels may be an acceptable measure of absorption.

In single dose pharmacokinetic studies, it is possible to investigate the range of doses and sampling times to be used in toxicity studies to explore the possibility of nonlinearity of absorption and other aspects affecting drug absorption. Specific animal absorption studies can also be undertaken to investigate formulations proposed for the clinic. The rat is commonly used in these types of studies as available techniques permit automated sampling of small sample volumes at appropriate time points from a single animal. The net result offers a well-defined plasma profile using fewer animal numbers (23).

Distribution

Studies are performed in animals that allow tissue distribution data to be related as closely as possible with the low dose used in toxicity studies and the therapeutic dose. This ensures that tissue distribution data can confidently be used to support the evaluation of toxicological findings and generate human radiolabeled dose estimates. For practical and ethical reasons, rodents are used in tissue distribution studies, but larger species may be used for specific purposes. The albino Wistar or Sprague Dawley rodent strains used in toxicity testing can be selected for tissue distribution studies, thereby providing a link between tissue distribution of a drug and drug-related material, and toxicokinetic findings. However, where regulatory authorities also request the evaluation of melanin binding, alternative strains such as the pigmented Long Evans rat are required to investigate binding to tissues such as the eye and skin. Studies designed to assess tissue distribution and melanin binding of drugs in pigmented animals are required to satisfy radiolabeled dosimetry calculations for radiolabeled studies in human. Where extensive melanin binding over time is demonstrated, then the amount of radioactivity that can be safely administered to humans may be considerably reduced, as outlined below.

There are two methods available for the assessment of tissue distribution—Quantitative Whole Body Autoradiography (QWBA) and Quantitative Tissue Dissection (QTD). QWBA is widely accepted for the assessment of tissue distribution in rodents and occasionally small nonrodent species. Typically, animals are administered a radioactive dose of between 4 and 15 MBq/kg body weight and at predetermined times following administration are sacrificed and sagittal sections of between 20 and 40 μm prepared, exposed to phosphoimaging plates along with radioactive standards to permit quantitative measurement of radioactivity in different tissues. The length of time sections are exposed will depend on the distribution and elimination characteristics of the drug and the amount of radioactivity given. QWBA images are normally readable between 4 and 10 days depending on the amount of radioactivity dosed. Quantitative tissue dissection (QTD) is much less commonly used than QWBA as the approach is extremely labor intensive, but still offers perfectly acceptable method for tissue measurements of radioactivity, particularly where larger animal species are used to address specific issues not otherwise answered by the use of rodent species alone.

QWBA provides a rapid assessment of absorption and distribution patterns into various tissues following administration of the dose. On the other hand, excretion balance studies enable routes of elimination to be explained but where percentage recovery of dose is poor then QWBA can assist with identifying less expected routes of elimination, such as skin or saliva. High levels of dose associated with a particular tissue may warrant further investigation to determine if accumulation is associated with tissue specific uptake and/or metabolism of the drug. QWBA can also help differentiate distribution associated with metabolite formed *in vivo* following administration of parent drug and possible differences in the pattern of distribution following administration of metabolite itself. When assessing toxicity or therapeutic benefit, QWBA provides a method by which to assess retention of radiolabel in specific organs linked with toxicity findings, or demonstrate desired tissue exposure/penetration of drug thereby linking exposure and therapeutic activity, for example, tumor penetration of anticancer drugs. QWBA is also used to provide information to support reproductive toxicology studies by measurement of radioactivity in different tissues such as the placenta and fetus.

Guidelines suggest repeat dose tissue distribution studies may be achieved by use of radiolabeled drug or "cold" analysis. If radioactivity is used during the assessment of tissue distribution after repeat administration, then it is likely to provide a better method compared to the isolation and identification of a large number of metabolites with use of "cold" assays of adequate sensitivity and selectivity towards the analytes of interest. Whether using radiochemical detection or cold analysis, guidelines recommend one-week administration providing metabolites levels are assessed at steady state. The ICH Guidance also states that "studies may be requested on a case by case basis," and therefore within the CTD overview of the written summary a positive comment on why such studies are not required may be useful or expected.

The amount of nonclinical data required before radiolabeled material can be administered to human is quite variable and depends on the country in which the study is performed and regional Ethics Committee requirements. In the United Kingdom, before labeled studies are undertaken in humans, approval must be obtained from the Administration of Radioactive Substances Advisory Committee (ARSAC), which uses the WHO guidelines (24) on the use of ionizing radiation in humans (25). Simply, administration of radioactive drugs to human volunteers is acceptable providing that the Total Body Exposure comes within WHO Category I, equivalent to <0.5 mSv, or with the ICRP Category IIa of between 0.1 and 1mSv and complies with the principles of "as low as reasonably achievable." Where the dose of radioactivity equates to less than 0.1 μSv (200 nCi, or less), then ARSAC approval is not required for these studies. Measurement of the exposure to ionizing radiation is complex and many assumptions are made. Nevertheless, a tissue distribution study in a pigmented rodent species, normally the rat, and a mass balance study in the same species permits a calculation (equations published by ICRP) of the acceptable dose of radioactivity to human. For most drugs, this calculation allows administration of radioactivity between 0.93 and 1.85 MBq (~25–50 μCi) as a single dose to human male volunteers aged between 20 to 60 years and postmenopausal or infertile women depending on the drug's indication. It is the comparison of material mass balance and quantitative metabolite profiling in animal species and human that enables an investigator to establish that safety cover of drug and drug-related material has been achieved in the animal species selected for toxicity testing.

While QWBA offers an important tool in the safety evaluation of drugs, it is unfortunately limited to the visualization and quantification of total radiolabeled material and cannot therefore distinguish between levels of drug and metabolites in the various tissues of interest. Fortunately, newer techniques are beginning to emerge that may eventually offer the ability to detect distribution of drug and drug-related material in different tissues by coupling laser-desorption with matrix-assisted laser desorption/ionization-time of flight mass spectrometry (MADLI-TOF) (26). Autoradiography permits a two-dimensional view of tissue exposure, which limits the precision of tissue structure–exposure relationship. However, three-dimensional imaging methods such as Single Photon Emission Computed Tomography (SPECT) and Positron

Emission Tomography (PET) offer potentially exciting prospects for their use in nonclinical safety evaluation. And while three-dimensional imaging is at present apparently limited to human studies, due to cost and availability, it is conceivable that these techniques can eventually be applied routinely in nonclinical investigations.

Plasma Binding

Large species differences in the plasma binding of drugs can be observed with anywhere up to a 20-fold difference in free drug levels in animals compared to humans (27). These differences can potentially alter the calculated AUC if not corrected for free drug and measures of total body and renal clearances, and uptake into tissues will be incorrect leading to apparent nonlinearity in kinetics. A combination of *in vitro* and *in vivo* studies over the pharmacological and toxicity ranges permits the evaluation of free drug concentration (*fu*) relative to total drug and metabolite concentrations in plasma. Where high plasma binding is established, more meaningful data can be obtained from *ex vivo* measurements, particularly if competition for plasma binding between parent drug and metabolites are suspected. The correction of free drug and/or metabolite concentration becomes particularly important where concentration dependent and/or species differences in plasma binding are known. As a general rule, where a drug shows high plasma binding or large differences in the extent of *fu* between animals and human, it is more judicious to determine safety margins based on the unbound plasma concentrations of drug and relate these to pharmacodynamic and toxicodynamic effects.

Binding is typically expressed as a percentage of the *fu* and it is generally accepted that only the free drug is pharmacologically or toxicologically active. Assessment of *fu* and the blood:plasma ratio (*B:P*) are required for registration and generally use radiolabeled drug. However, radiochemical purity rarely exceeds >98%, and for highly plasma bound drugs (>99%), it is problematical to distinguish if the unbound fraction remaining is drug or an impurity associated with the synthesis. In these instances, determining *fu* by use of a "cold" assay and highly pure drug is required. Species differences in *fu* and *B:P* are normally undertaken in *in vitro* by spiking drug into plasma over the concentration ranges expected in *in vivo* animal and human studies. The *fu* is then normally determined by use of techniques such as ultracentrifugation and equilibrium dialysis (28,29), of which the latter is generally preferred.

The three most important plasma proteins responsible for binding drugs are albumin, α-1-acid glycoprotein, and lipoproteins. The amount of drug binding to these proteins is dependent on the number of binding sites, plasma protein concentration, and association constant of drug to protein. α-1-Acid glycoprotein accounts for the binding of most basic drugs, while albumin accounts for most of the binding of neutral and acidic drugs (30). Lipoproteins are a heterogeneous group of proteins of which very low density lipoproteins, low-density lipoproteins, and high-density lipoproteins are the most important, and bind not only certain drugs such as imipramine, cyclosporine, and tetracyclines as well as a range of other drugs that exhibit some degree of lipophilicity (31). The enantiomeric selectivity of plasma proteins to bind drugs is discussed below. When all available binding sites for a drug to a protein are occupied, then binding is saturated and *fu* will increase with increasing dose. Where nonlinear binding occurs above a certain concentration, and saturation in animal species and human are quite different, then it becomes pivotal to understand the relationship between *fu* and total drug concentration when determining safety cover. Nonlinear binding tends to be encountered in high-dose toxicity studies and is rarely characterized within the pharmacological and mid-range toxicological doses for the vast majority of drugs.

Reproduction and ADME Studies

The ICH S5 guidance consolidates a strategy for reproductive toxicity testing and defines stages of reproduction to reflect possible human exposure and identify stages of risk. Dose should reflect other pharmacological and toxicological investigations to determine if reproductive risks are higher, lower, or equal with other toxicological findings. The ICH guidance identifies six stages of the reproductive life cycle, which historically fall into one of three phases of testing, namely, Segment I, Segment II, and Segment III. In general, three phases of the reproductive process are evaluated. The first phase, ICH Stage A (Segment I) evaluates the effect of the drug on fertility, early implantation, and embryogenesis. The second phase, ICH Stage B (Segment II), is the teratology study or developmental toxicity study and performed in two species. The third phase, ICH Stage C (Segment III) evaluates late gestation, parturition, and lactation. In terms of the ADME of a drug, there are two important questions to address in developmental studies. First, whether drug and/or metabolites cross the placenta to reach the developing embryo/fetus and second, whether pups are exposed to drug and/or metabolites during lactation.

Placental Transfer

The placental transfer of drug and/or metabolites can be examined either in the rat and/or in rabbit at a time after organogenesis and before term, approximately day 15 and day 18 after conception, respectively. Radiolabeled drug may be given throughout pregnancy, but administration for two to three days prior to necropsy is normally sufficient for accumulation to occur in the fetus. At necropsy the animals are either prepared for QWBA or fetuses taken out for radioactivity measurement. In the absence of pharmacokinetic data, dosing of nonpregnant animals can assist in selecting appropriate dose in developmental studies, but normally information from one-month toxicokinetic data is available before these studies are initiated. Pharmacokinetic data in both nonpregnant and pregnant animals can be used to compare pharmacokinetic information between pregnant and nonpregnant animals in terms of C_{max} and AUC.

As a satellite study to the ICH Stage B (Segment II) teratology study, fetal uptake investigations are used only as a method to show that drug and/or its metabolites can cross the placenta and that the developing fetuses have been adequately exposed. Indeed, some investigators may feel that a single sample at C_{max} may be sufficient. However, since exposure in pups does not necessarily parallel that of the mother, additional time points may be warranted to assess AUC (32). For example, in a recent study, the infusion of morphine-3-β-glucuronide to steady-state showed fetal–maternal ratio of less than unity and a bidirectional clearance across the placenta, but negligible clearance or sequestration by the fetal intestine (33). If there is no discernable level of drug or metabolite, for example, less than 1% of the levels in the blood of the dam, then the teratology study may need to be evaluated carefully with respect to the extrapolation of the findings to humans where the drug may cross the placenta and further work or comment in the CTD overview could be required.

Milk transfer

Guidelines ask for information on the exposure of lactating animals to drug and/or metabolites. It is not clear, however, whether this is to validate ICH Stage C (Segment III) postnatal development studies by showing that the pups are exposed to drug secreted in milk, or if it is to provide information on the likely exposure to breast-fed human infants. Using mated rats, 14 to 17 day *post partum*, the time of maximum milk production, approximately 1.0 to 0.5g/day, can be collected as 100 μl samples spread over four to five collections in 24 hours without upsetting the physiological system. This number of samples may not be necessary if only an estimate of milk uptake is required but the problem with, for instance, one single sampling time at C_{max} in maternal plasma is to know whether this is also C_{max} for milk (34). When plasma samples are collected at the same time as milk, the relative uptake can be calculated using various methods including PBPK modeling for predicting milk:plasma ratio to support prediction of infant exposure (35). It has been shown that in general for compounds, which are weakly bound in plasma and are primarily ionized, for example, antipyrine, sulfapyridine, or ethanol, there also appears to be a good relation between what is found in rat and in human milk. However, for a number of drugs correlations between the species becomes difficult and cannot always be corrected by using the pK_a of the drug, pH differences in milk and plasma, and relative milk and plasma free drug levels. This is particularly true for the antiviral drugs ganciclovir and zidovudine that accumulate in breast milk by passive processes but for aciclovir active transport processes reportedly contribute to the accumulation of aciclovir in milk (36).

Metabolism

Regulatory Requirements

A drug metabolism program encompasses a broad range of studies to support full drug development. Use of *in vitro* studies to assess the drug–drug interaction potential of drug and/or metabolites and to identify the enzymes involved, as defined by the FDA's Guidance document (8) on drug–drug interaction studies form an integral part of the ADME work. Guidelines first suggest using *in vitro* studies to assess the effect of drugs on metabolic pathways and where DDIs are characterized to follow-up with appropriately designed *in vivo* studies. Identifying the major metabolic pathways of a drug and exploring the effect of the drug on the metabolism of other drugs and vice versa are two primary goals of the *in vitro* studies. It is these results that help guide the design of future *in vivo* human DDI studies and permits the investigator to take decisions on which metabolites are required to be followed when designing future toxicity and clinical studies.

The question of what, when, and whether to measure metabolites in the safety assessment of drugs has been a subject of much debate for many years. The publication of a paper "metabolites in safety testing" by the Pharmaceutical Research and Manufacturers of America (PhRMA) group in 2002 (37), with input from industry, academia, and regulatory agencies, prompted further attention on the debate on metabolites in safety testing (38), and in 2005, the FDA issued draft guidance for industry, as an overlay on an historical background of the ICH S3A guidance on toxicokinetics issued in 1995. In 2008, the FDA's Guidance for Industry: Safety Testing of Drug Metabolites was issued (39), thereby formalizing the FDA's position on circumstances where the safety of human metabolites are to be further investigated. Although there are no specific guidelines issued by the ICH, there is nevertheless a common agreement between different regulatory agencies that knowledge of the metabolic profile and the identification of metabolites for each animal species and human are required. However, despite recent FDA guidelines and several expert opinions on metabolite safety testing, views and options on what to follow still remain to be universally agreed and it therefore remains a future challenge for the ICH process. In the meantime, it is clear that pragmatic and scientific approaches are required for the safety assessment of drugs during each phase of drug development.

Design of Metabolism Studies

Availability of *in vitro* metabolism data generated during drug discovery provides important information on the likely *in vivo* metabolic fate of a drug. More importantly, a good indication of the species differences and drug-metabolizing enzymes coupled with the availability of bioanalytical methods from drug discovery can greatly assist with the metabolic part of the ADME package. Availability of authentic metabolites synthesized for testing pharmacological activity can greatly facilitate the characterization and quantification of metabolites formed in animal and human ADME studies.

In vivo metabolic studies are carried out by use of radiolabeled drug as part of the ADME package in toxicological species with at least one dose approximating to the human therapeutic dose and by the therapeutic route of administration.A plasma profile at C_{max} but also at other time points to reflect the elimination process with a urinary profile over the time period of the major excretion phase is normally required. Unless the majority of the dose is excreted in urine then some degree of metabolite profiling of feces is also required. HPLC linked with radiochemical detection is used to separate and quantify metabolites. Some companies will investigate the metabolism of their drug at the higher doses used in toxicity testing to show whether there is the possibility of saturation and nonlinearity. These additional investigations may help to explain changes in rates and routes of elimination in nonclinical ADME work. However, these additional investigations may prove little value when trying to extrapolate findings to human because the doses will be much higher than those used in the clinic and in any case toxicokinetic studies will provide levels of the drug and important metabolites under actual conditions of long-term repeat-dose toxicity studies.

Species differences in metabolism may provide a rationale for the differences in toxicity (40,41) even if sometimes the link between the two is unsubstantiated and rather tenuous. Nevertheless, the identification and characterization of metabolites permits the investigator to take appropriate decisions on which metabolites further work is required.

Enzyme (Transport) Induction

The ICH Guidance requires that "changes in metabolite profile or alternations in metabolizing enzyme activities over time should be understood to allow for appropriate interpretation of studies." It is therefore necessary to show if the experimental drug can cause hepatic microsomal enzyme induction or more rarely suppression of drug-metabolizing enzyme activities. Some strong inducers of cytochrome P450 such as phenobarbitone can induce levels of enzyme activity over a period of up to three days, while 3-methylcholanthrene, on the other hand, can have a maximal effect within only 24 hours. Thus, to cover all types of induction profile, drug is administered for periods no shorter than seven days with subsequent measurement of drug levels in plasma and comparing the AUC after the first and last dose. A quantitative change of the parent drug AUC after single and repeat-dose toxicity studies may suggest autoinduction or inhibition of enzyme activity as a consequence of changes in the constitutive expression of drug-metabolizing enzymes involved in elimination of drug. To characterize changes in the constitutive expression of enzyme activity, livers from untreated animals can be compared with treated animals. *Ex vivo* analysis of the hepatic drug-metabolizing enzymes, cytochromes P450 but also other Phase I and Phase II enzymes are made by use of *in vitro* probe substrate assays. Another approach is to incorporate an assessment of enzyme activities with four-week toxicity studies because toxicokinetics will probably be undertaken along with histopathology of various tissues. If changes are seen then further studies may be initiated. The range of suggested probes substrates for the measurement of enzyme activities are many, but characterization of the selectivity and specificity of probes substrates is paramount if there is any chance of relating changes in enzyme activity in animal species with that of human. Indeed, regional guidelines are issued, which recommend probe substrates for use in the characterization of enzyme activities, in particular the cytochromes P450 (42–44).

To characterize changes in the elimination of parent drug, livers can be examined from control and treated animals to study changes in the intrinsic clearance of a drug and to profile for changes in the pattern of metabolites formed. The consequence of induction of drug-metabolizing enzyme activity does not necessarily mean that observed toxicity will be reduced. Indeed, the opposite may occur if the toxicological effect resides in a metabolite itself. Inhibition of enzyme activity is perhaps more rarely observed but can occur rapidly, even after a single dose. Under these circumstances, changes in AUC and liver enzyme activity can identify this possibility during the same experiment described for induction.

Where time-related changes in plasma AUC and/or changes in amount of drug and/or metabolites excreted in urine occur, which are not attributable to metabolic elimination, then other mechanisms of elimination such as transport proteins need to be investigated. Changes in the expression of transport proteins affecting drug distribution and excretion are reported for a few drugs. Characterization of the role of transport proteins becomes particularly important where drug combinations give rise to the potential for DDIs. It is only with the characterization of the functional and regulatory mechanisms associated with transport proteins involved in the elimination of drugs that can help minimize unfavorable pharmacokinetics, toxicity, and DDIs during clinical use of a drug.

An important point to recognize is that induction or inhibition of enzyme or transport activity in toxicological species does not always translate equivocally to the human situation. Clear species differences in the regulatory mechanisms governing constitutive expression and induction of similarly orthologous enzymes and/or transport proteins across species can be markedly different. And the measurement of enzyme induction or inhibition at mid- and high-dose ranges used in toxicity testing is not necessarily reflected at the intended therapeutic dose. Nevertheless, if induction or inhibition of enzyme and/or transport proteins is observed in toxicological species, it will be necessary to confirm that these same effects are observed in human, first by use of cultured human hepatocytes and/or other *in vitro* methods of assessment with possible confirmatory studies *in vivo* human (45).

Excretion

Mass Balance

The use of radiolabeled material in ADME studies serves many purposes in the safety assessment of drug entities. Arguably, the most important of these radiolabeled ADME studies is the excretion balance study, also referred to as the "mass balance" study. Excretion of the drug and metabolites from the body can occur by many routes, including urine, feces, bile, and expired air, and the use of radioactivity in "mass-balance" excretion studies allows a quantification of these various processes with time. The excreta and blood samples collected in such studies are used to determine the rate and routes of elimination, and for the identification and quantitative profiling of metabolites and parent drug. Studies are commonly performed earlier in the drug development program, although some companies prefer to delay studies until later in development when chances of marketing the drug become clearer. Unfortunately, the later these studies are conducted, the less likely information on the metabolism of the drug can be used to aid development and facilitate the design of future studies around drug–drug interaction strategies, bioanalytical target, and special population studies.

The main reasons for identifying and characterizing the metabolites are due to their potential for activity, toxicity, inhibition, and displacement. When an active metabolite is formed, the relative potency and pharmacokinetic profile needs to be determined. Similarly, if a toxic metabolite is suspected, safety studies have to be evaluated for the presence of the metabolite in toxicology species. The characterization and quantification of the metabolite needs to be established and if necessary administer the metabolite itself, albeit this approach may not emulate the true tissue distribution of the metabolite formed *in vivo* (46). The consequence of notable differences between tissue distribution from the metabolite formed *in vivo* and metabolite administered by po or IV can therefore be substantially misleading, particularly when assessing pharmacologically or toxicologically active metabolites.

The regulatory guidelines covering acceptable percentage recovery of dose are rather vague as to what is actually required, but there is a necessity to show that at least 90% to 95% of the administered dose can be accounted for by the end of a minimum of seven days collection but longer if necessary (47). In practice, recovery tends to be rather variable, and in a recent review on a retrospective analysis of 27 proprietary compounds, it was conclude that absolute dose recovery should not be used as a primary objective for the aforementioned criteria of identifying and quantifying important routes of excretion and metabolites formed (48). This can result in protracted collection times for certain drugs with long half-lives, where in the authors experience daily collection of urine and feces can take up to three months to achieve a good balance recovery. Experimentally, an accurately measured radiolabeled dose is administered to rodents and placed in metabolic cages where urine and feces can be collected separately and all expired air is collected by using trapping agents for $^{14}CO_2$. This type of experiment can also be undertaken for the larger animals, but if the radiolabel has been placed in a stable position and no volatile component has been found in rodents, it is unlikely that collected expired air will be necessary. Excreta, together with cage washes, enable the total excreted drug-related material to be measured and compared to the dose administered. Any discrepancies would normally be accounted for by retention in the tissues, which can be also measured by counting the residual radioactivity in the carcass, but volatile metabolites that are not trapped with the CO_2 can exist.

Excreta collected from mass balance studies allow the comparative species differences in the speed of total elimination of parent drug from the body and the relative ratio between urinary and fecal excretion. High fecal elimination, particularly after intravenous dosing, would suggest biliary elimination, but not always because some high-molecular-weight lipid soluble compounds are extensively secreted directly into the gut (49). For comparison, human mass balance is also undertaken following ARSAC approval. It is only when metabolite profiles in human are compared with toxicology species and adequate exposure to metabolites in animal species can be demonstrated, will animal species selected for toxicity testing be largely validated for safety cover in human. Reporting of results is currently a topic of debate where recent FDA guidelines recommend that metabolites are expressed as a percentage of parent rather than total exposure to drug-related material (37,38). It is the authors' view that absolute values expressed in terms of concentration and amount prove more informative than relative abundance when assessing safety cover across species. In any case, a pragmatic and scientific approach in the assessment of which metabolites require further investigation is on a project-by-project basis for establishing a strategy for monitoring metabolites.

In certain circumstances, it may be necessary to evaluate metabolite profiles following repeated dosing when (*i*) single dose tissue distribution suggests the apparent half-life of the drug and/or metabolites in organs or tissues exceeds the apparent half-life of the elimination phase in plasma; (*ii*) steady state levels of drug/metabolite in circulation are higher than those predicted from single dose kinetic studies; and (*iii*) histopathological changes are not predicted from short-term toxicity testing. Normally one week is sufficient providing metabolites are measured at steady state. Radiolabeled drug is dosed for the duration of the study and plasma samples collected at the beginning and end of the study are profiled. Where "cold analysis" of parent drug and metabolites describe the majority of the drug and drug-related material, then an assessment of the AUC and C_{max} may prove sufficient to establish that safety cover is achieved following repeat dosing.

Biliary Excretion

Biliary excretion can have an important impact on the systemic exposure and toxicity of certain drugs particularly when drugs excreted into the bile often have some degree of reabsorption from the gastrointestinal tract. Enterohepatic recycling of parent drug sometimes with hepatic conjugation and intestinal deconjugation is often associated with multiple peaks and a longer apparent half-life in the plasma concentration time-profile. High-molecular-weight lipophilic drugs are typical candidates for enterohepatic circulation, where physicochemical characteristics, transport across sinusoidal and canniculae membranes, metabolism by hepatic and gut microflora, and reabsorption by gut-wall transport all contribute to the extent of enterohepatic recycling of the drug.

The guidelines suggest that when biliary excretion is thought to be high then enterohepatic recycling should be investigated. Practically, this is probably necessary only when there is more than 30% of radioactivity found in the feces after an intravenous administration and may provide limited information on enterohepatic recycling alone. Biliary collection can provide a cleaner source of metabolites compared with extracting feces. Notable species differences exist in the biliary excretion of drugs with a molecular threshold, which is typically 250 to 350 in rodents and 500 to 600 in human. Also, the rate of flow relative to body weight is approximately 20 times faster in rodents (50) compared to human and some notable gender and species differences have also been reported (51). Thus, any results obtained in small animals may bear little relationship to those found in human. From a practical perspective, biliary studies are routinely performed in rodent species and in dog or monkey where the ADME profile and other physiological considerations necessitate further investigation in nonrodent species. With any biliary study, special precautions to ensure correct bile flow and bile salt replacement are necessary. Enterohepatic circulation can be measured semi-quantitatively by taking the bile from a donor animal with a biliary fistula, which has been administered the drug and introducing it to a recipient animal via the intestine, and measuring blood or urine in this second animal. If it is thought that this information is really necessary then the dog or monkey, depending on the metabolic profile and other physiological considerations, can be used. No less challenging are biliary studies in humans but several *in vivo* approaches are available to the investigator (52). Despite the fact that biliary studies are very time-consuming, approximations of the fraction of the drug, which is excreted in the bile and subsequently reabsorbed (F_a), can be derived from the formula:

$$F_a = \frac{1 - \mathrm{AUC}/\mathrm{AUC}_{(can)}}{F_b}$$

where AUC and $\mathrm{AUC}_{(can)}$ are the areas under the blood level time curve after intravenous administration of the same dose to control animals and to biliary cannulated animals without recycling, and F_b is the fraction of the drug in the body, which is excreted in the bile.

TOXICOKINETICS

Definition

The ICH has issued a tripartite definition of toxicokinetics describing it as "the generation of pharmacokinetic data, either as an integral component in the conduct of nonclinical toxicity studies, or in specifically designed supportive studies, to assess systemic exposure." For the purpose of this chapter, toxicokinetics is defined as a measure of drug exposure in animals used in toxicity testing, while pharmacokinetics, on the other hand, helps to describe fully the *in vivo* ADME properties of a drug. Toxicokinetics, therefore, acts as a bridge between animal and human studies and should only be used for interpretation of safety data and not to provide detailed mechanistic explanations that can be derived from appropriately designed ADME studies. This does not mean that information on the ADME of a drug in animals and humans should not be used in the design or interpretation of toxicity studies, but rather that there is a clear distinction between the information expected by the regulatory authorities to support toxicological findings.

General Requirements

The ICH harmonized tripartite guidance for the assessment of systemic drug exposure in toxicology studies was issued in 1995 (18). It requires two key objectives of toxicokinetic studies to be fulfilled. The primary objective is "to describe the systematic exposure achieved in animals and its relationship to dose level and the time course of the toxicity study." The secondary objectives include the choice of species and regimens, design of toxicity studies, and to relate the exposure achieved in these studies to the toxicological and clinical results. As with many ICH Guidelines, no prescriptive details are issued in order that flexibility is maintained to allow for a case-by-case approach based on good science. Assessment of exposure by way of measuring plasma C_{max} and/or C_t and AUC on total or unbound drug concentrations is the most commonly used measures of exposure.

Toxicokinetic monitoring during toxicity testing should establish what level of exposure has been achieved during the course of the study and may serve to alert the toxicologist to nonlinear dose-related changes in exposure that may have occurred. It is this toxicokinetic information that is supposed to give a better indication of the interspecies comparison on drug safety. The route of administration should be based on the intended clinical route of administration. Where the intended clinical route of administration changes for a drug, it will be necessary to establish if, by changing the clinical route, the safety margin of the drug also changes.A comparison of the systemic exposure of parent drug and/or metabolites by the existing and proposed routes is expected, together with some conformation that there is no change in metabolic route. Where a change in exposure and/or metabolic fate of the compound is observed, the safety assessment must be reviewed in light of new data. In setting dose levels, the effect of gender must be considered, particularly in rodent species where higher exposure in female rats is not uncommon (53).

The timely integration of toxicokinetic assessment into a nonclinical toxicology programs promotes optimum design of future safety studies, and if these are appropriately monitored could avoid the duplication of studies, which in turn will reduce animal numbers required to support the safety program—an overall objective of the ICH initiative. The following more detailed description of the practical aspects of undertaking toxicokinetic studies should be read in conjunction with Table 2, which summarizes the present status of ICH Guidance documents for the toxicokinetic assessment of new drug entities.

Analytical Requirements

It is generally accepted that the bioanalytical methods used to measure the analyte(s) of interested should be validated in accordance with a universally accepted approach to assay validation, such as the FDA Guidance for Industry on Bioanalytical Method Validation (54).A validated method should include analyte stability, specificity, precision, accuracy, and sensitivity. For acute toxicity or dose-ranging studies, the method may not necessarily be optimal in terms of sensitivity and not fully validated, but nevertheless sufficient to obtain basic information on exposure. In later stages of development, it is expected that the method will be refined to obtain the maximal amount of information on the concentration–time profile of drugs. The FDA Guidance can be implemented for both GLP and non-GLP work. The ICH guidelines require that toxicokinetic data submitted for regulatory approval are validated and that studies comply with GLP for all matrices in all species, and ensure that analytically methods are robust for the bioanalysis of samples. Similarly, assay specificity of the assay will enable drug and matrix to be assayed without interference by endogenous components present in matrices. Few novel drugs are now developed as a racemate, but whether developing a racemate or single enantiomer, it is necessary to assess exposure to enantiomers of the parent drug and possible enantiomeric conversion of metabolites. Where matrix effects can affect the interpretation of concentration data, Incurred Sample Reanalysis (ISR) of samples should be incorporated into a development program for the ISR of each species used in toxicity and safety pharmacology testing (55).

Sampling

The ICH3A guideline states that for each study "the number of time points should be justified on the basis that they adequately estimate exposure." The justification of sampling times selected for toxicity studies can be gathered from early pharmacokinetic studies or dose-ranging studies in the same species.

In the authors' laboratories, single dose pharmacokinetic studies at two doses (low and high) are performed. A matrix approach is employed using three animals/sex/dose/time points for rodent studies.A total of nine different sampling times covering 24 hours following administration are typically selected to define the pharmacokinetic profile. The individual time points can then be put together to define the average plasma concentration profile. These single dose pharmacokinetic studies provide valuable information about the concentration–time profile of a drug and help support the selection of sampling times in early toxicity studies. As toxicokinetics data becomes available from subsequent studies, it can be combined with data generated from other studies using a population approach. The population model can then be used to support the design of long-term studies and facilitate the interpretation of data from these rodent studies (56).

Indeed, population approaches can facilitate the evaluation of concentration–time data not only from composite sampling in rodents, but also in large animal species to support the interpretation of exposure and toxicological outcomes (57–59). Unfortunately, expertise in modeling is required for population approaches and at present there is a lack of trained people making it difficult to use this approach systematically during nonclinical phase of drug development. In preparing the dossier for submission, all preclinical data for each species can be combined in a population model to facilitate the identification of important covariates and obtaining a better understanding of the pharmacokinetic characteristics of the drug in animal species. Nonclinical population analysis has successfully been applied within the authors' laboratory in two recent submissions for marketing approval. Despite the utility of this approach for analyzing data, a noncompartmental calculation of exposure parameters in individual studies will be necessary in the absence of validated population software.

Animal Numbers

Given the above criteria, it is perhaps not surprising that the actual number of animals required for a study is not given in the ICH guidance because this will depend on the species, type of study, frequency of sampling, and assay sensitivity. Nevertheless, some general suggestions can be made as shown in Table 5. Animal husbandry and sampling requirements normally require setting-up a satellite group for rodent species in toxicokinetic assessment in parallel with animals for toxicological assessment. In contrast, samples from nonrodent species are taken from animals for toxicological assessment as the required sample volumes are easily obtainable and animal numbers would not, therefore, normally exceed three to five per dose group. The volume and frequency of samples taken for bioanalysis, particularly in small animal species, impact on study design and animal numbers used. The UK Home Office stipulates that up to 15% of the blood volume can be taken from any one animal in any four-week period. For rats and dogs, this represents 1.0 to 1.3 ml/100 g body weight. The recently reported application of dried blood sample analysis for the toxicokinetic analysis of samples to support safety testing using small microliter volumes of blood offers promising alternatives to minimize sample volumes and in turn animal numbers required for the assessment of drug exposure (60).

Setting Dose Levels

Toxicokinetics and knowledge of dose–exposure relationships with time is important in supporting the selection of dose levels. For this reason, three dose levels are expected: low, intermediate, and high. The low dose is preferably nontoxic and represents a dose that is intended to equal or

Table 5 Summary of Toxicokinetic Studies Needed to Satisfy ICH Guidelines

Study	Species	Animals	Dose	Time	Blood/plasma	Parameter
Acute[a] (exploratory)	Rat or dog/monkey	Composite 3–5 ♂, ♀ / dose (n = 9–18) Intended clinical route +/- IV 1–3 ♂, ♀ / dose (n = 6–10)	Escalating (3–5) (controls)	D_1	8–10 samples (depending on $t_{1/2}$) up to 24/72 hr	C_{max}, T_{max}, C_{24}, AUC, $t_{1/2}$
Acute[b,c] (MTD)	Rat/mouse Dog/monkey	Composite 2–5 ♂, ♀ / time (n = 24–40). 1–3 ♂, ♀ / time / dose (n = 6–10)	Escalating (3–5)	D_1–D_t	0, C_{max}, t_2, t_3, t_4 8–11 samples	AUC, C_{max}, C_{24} Linearity[d]
Subacute[d] 1–2 wk	Rat Dog/monkey	Repeat 3–5 ♂, ♀ / dose (n = 36–60) 1–2 ♂/time/dose	3 Doses	D_1, D_{14}	C_{max}, t_2, (t_3), t_4	AUC, C_{max} Linearity[d] R_{ac}
1 mo	Rat Dog/monkey	Composite 3–5 ♂, ♀ / dose (n = 9–18) 3–5 ♂, ♀/ dose (n = 9–12)	3 Doses + (controls)	[D_1], D_{28}	C_{max}, t_2, t_3 (t_4–t_5), t_6	AUC, C_{max} Linearity[d] R_{ac} Induction/inhibition—*ex vivo*[e]
3 mo	Rat Dog/monkey	Composite 3–5 ♂, ♀ / dose (n = 9–18) 3–5 ♂, ♀/ dose (n = 9–12)	3 Doses + (controls)	[D_1], [D_{28}], D_{84}	C_{max}, t_2, t_3 (t_4–t_5), t_6	AUC, C_{max}, Linearity[d] R_{ac}
6 mo 9 mo (12 mo)	Rat Dog/monkey	Composite 3–5 ♂, ♀ / dose (n = 9–18) 3–4 ♂, ♀/ dose (n = 9–12)	3 doses + (controls)	[D_1], [D_{84}], D_{168}, [D_{252}]	C_{max}, t_2, t_3 (t_4–t_5), t_6	AUC, C_{max}, Linearity R_{ac}

Dietary study[f]	Rat/mouse	Repeat 5 ♂ / dose ($n = 15$) Destructive 5/time (mouse) ($n = 90$) Composite 5/time ($n = 30$)	3 Doses	D_1, D_7, [D_{14}]	t_1, t_2, (t_3–t_5), t_6	AUC, R_{ac}
Carcinogenicity	Rat/mouse	Repeat 5 ♂, 5 ♀/ dose (satellite)[g]	3 Doses + (controls)	M_3, M_6, M_{12}	C_{max}, t_2, (t_3–t_5), t_6	AUC, C_{max} Linearity[d]
Segment I[h] (fertility)	Rat or mouse	♂ - Repeat 3–5 / dose ♀ - Repeat 3–5 / dose (Satellite)[g]	Toxicology[i]	- D_{-1}, D_{-14} - GD_{-1} – $-_{14}$, GD_6 - necropsy[h]	C_{max}, t_2, (t_3–t_5), t_6	AUC, C_{max}, R_{ac}
Segment II[g] (teratology)	Rat or mouse Rabbit	♀ - Repeat 3–5 / dose (Satellite)[g] ♀ - Destructive 3–5 / time	Toxicology[i]	- GD_6 GD_{17} - GD_7, GD_{18}	C_{max}, t_2, (t_3–t_5), t_6	AUC, C_{max}, R_{ac}

Route of administration assumes the intended clinical route.
[a]Single dose or acute toxicity studies used to support single dose human studies (e.g., microdose). See also regional guidelines.
[b]Revision of CPMP/ICH/286/95 (July 2008; Step 2) proposes availability of acute toxicity data to support "consequences of human overdose"—data should be available prior to Phase 3.
[c]Data not necessary for Phase 1 studies if followed by repeat dose study. See also regional guidelines.
[d]Can be used to support single or repeat dose exploratory studies up to 14 days.
[e]Other repeat-dose studies can also be used.
[f]The absorption of food from the diet may also be incorporated into MTD or palability studies.
[g]Samples can be taken from concomitant study, but satellite groups may be preferred.
[h]Data may be available from other toxicokinetics.
[i]Choice of dose will depend on toxicity—if none, use top dose only and frequency in toxicity guidelines.
Abbreviation: D_1–D_t, first and last day. GD, gestation day.

just exceed the maximum exposure expected in patients where a pharmacological response is expected but no-observed-adverse effect is found. The intermediate dose normally represents a multiple of the exposure at the low- or high-dose levels depending upon the objectives of the toxicity study, and the high dose achieves a toxicological endpoint producing the maximum exposure in animals. Where the physicochemical properties of a drug limits the exposure, it is important to confirm that less than proportional increase in exposure is not attributed to increase in drug clearance through autoinduction, see above. More than proportional increase in exposure with dose should not necessarily result in limiting dose because these finding are important and often helpful with understanding the relationship between dose and systemic exposure. Low- and mid-dose levels provide information to assist with assessing the relevance of exposure in animals with exposure in human. The low and mid doses are selected by use of pharmacokinetic, pharmacodynamic, and ADME data on the linearity of dose–response, possible saturation of elimination pathways, therapeutic dose, and mechanistic information.

Determination of Metabolites

Strategies adopted to measure the exposure and pharmacokinetics of metabolites will depend on the metabolism of the drug. In practice, quantifiably important metabolites are identified during preclinical ADME phase of drug development, but the relative importance of these metabolites in terms of human exposure has yet to be characterized. Unequivocally, the human mass balance study will provide answers on which of the metabolites formed, if any, are worth measuring. Only when the human excretion balance data becomes available can the assessment of which metabolites to follow in toxicology species be made. When the decision is taken to measure metabolite exposure, then the question of method validation, stability, and matrix effects need to be addressed using the same approach required to assess exposure to parent drug. Where possible, it is desirable to measure both parent drug and metabolite in the same assay. An assessment of which metabolites to follow depends on whether metabolites are pharmacologically active and the abundance of metabolites measured not only in plasma but also other biological matrices.

Complicating Factors in Exposure Assessment

Whether plasma concentration or AUC of parent and/or metabolites are used for the assessment of exposure, the intended therapeutic dose in human should be considered, such that relevant exposure and multiple exposure levels are achieved in animal toxicity studies. Monitoring of toxicity studies should establish levels of exposure during the course of the study to compare dose and time-related changes in exposure. Toxicity or toxicodynamic effects of the drug or lack thereof are supported by evidence of exposure. If toxicokinetic data were explicit, then the link with toxicity findings and relationship to human exposure would be simple to interpret. Unfortunately, there are several caveats worth noting, such as species differences in protein binding, uptake into tissues, pharmacological activity, and metabolic fate of drugs.

The ICH guidelines suggest that "it may be more appropriate to relate exposure to free (unbound) concentrations for highly plasma bound compounds." But clearly, if there are significant species differences in plasma binding, this factor becomes an important consideration in the calculation of comparative exposure, particularly because for many drugs and their metabolites, the degree of plasma binding will change with dose, concentration, and even the age of the animal. Intuitively, blood concentrations should be expressed as free levels because it is this fraction that distributes into tissues and receptors. Therefore plasma binding studies should either be done in all species over the full range of concentrations expected in toxicity studies so that each concentration can be individually "corrected," or free drug levels should be measured *ex vivo* from all or a representative number of samples. The measurement of total and free drug concentrations presents an additional challenge to the measurement of total drug concentration alone. For those drugs that are highly bound, that is, >98%, and where there may be sources of errors such as high affinity for the apparatus, it may be difficult to measure accurately free drug levels and small apparent differences of 1% to 2% can have a disproportionate influence on the calculation of exposure and safety margins. Similarly, repeated sampling to obtain ex vivo data may alter the protein binding results due to physiological changes (61). Converting measures of exposures to free levels do not take into account that there may be species differences in tissue uptake of the free drug or tissue/receptor protein binding. For example, it has been shown that there are large species differences in the uptake of fluoxetine into the brain (62–64). Similarly, large differences in the uptake into the brain have been observed for the highly bound drug zosuquidar, despite having similar unbound levels in blood (65). In these cases, blood levels, whether they are free or bound, would be of limited use in the assessment of tissue exposure, particularly where transport proteins are involved in the uptake and efflux of drug and drug-related material in tissues. For other classes of drugs, such as the benzodiazepines and NSAIDs, the use of unbound concentrations has generally provided a better measure of comparative tissue exposure, albeit there is no a priori reason not to suspect the involvement of transport proteins for some or all of these drugs as well. To overcome the problem of relative species differences in tissue uptake, the ICH guidelines suggest that plasma concentration and the volume of distribution at steady state should be calculated. This would provide a measure of total body burden and take into account toxicity, which is more likely to be associated with levels found in organs rather than blood. For specific organ toxicity, physiological modeling of species differences may give a better indication of relative exposure (17,66), but may not predict as well where active transport processes are likely to be implicated in tissue distribution (67).

The measurement of drug and/or metabolite levels for the assessment of drug toxicity and its extrapolation from animals to human is arguably tenuous where, for some drugs such as the NSAIDs, the exposure safety margin is less than unity (68). It is also argued that toxicity, in particular carcinogenicity, often bears no relationship between drug or metabolite levels and observed toxicity where the latter is dependent on the formation of short-lived reactive "free radical type" or electrophilic intermediate metabolites (69,70), which are normally impossible to measure in body fluids (71). The use of comparative toxicokinetics also relies on the assumption that the animal species share similar mechanisms of pharmacological or toxicological activity and yet there are examples where interspecies differences in sensitivity are reported (72,27). However, when predictions are examined closely, for example, potential teratogenic, embryotoxic, and chemical compounds assessed for developmental toxicity, the prediction for compounds in general are good with the majority correctly classified, even if the toxicity was not identical in each of the species examined (73,74). Of course, extrapolation of animal data to human can be envisaged to be greatly improved with the use of mechanistic PKPD (75,76).

Single-Dose Toxicity Studies

The primary objective of single-dose ranging or dose-ranging studies is to evaluate the *in vivo* safety assessment of potential drug candidates for future drug development. Data from *in vitro* and pharmacokinetic studies gives important ADME information on likely dose regimens in toxicity studies. These preliminary studies are used practically for the purpose of drug discovery, lead-optimization, and candidate selection.A validated analytical method is not a requirement where single-dose toxicity data are used to support the refinement of dose, sampling times, and formulation of future repeat-dose studies intended for regulatory submission.

Pharmacokinetic data provide an early perspective on the likely safety profile of a drug but should not be considered exclusive of other data, which reflect the specific activity, potency, or metabolism of the drug. In these early studies, it is normally the profile of the parent drug that is assessed unless developing a pro-drug or there is a priori information arising during the discovery phase to suggest the formation of quantifiably important active metabolites.

Repeat-Dose Toxicity

The majority of toxicokinetic investigations are undertaken as part of a four-week, three-month, and six to nine month repeat-dose toxicity study in the rodent and nonrodent species. Wherever possible, single-dose toxicokinetics should be used to support the design of repeat-dose toxicity studies, thereby ensuring dose and dose formulation, and sampling times are optimized against available information to ensure adequate drug exposure.

ICH S3 guidelines suggest that samples should be taken at the beginning of the study, which in practice should be the first day of administration to permit a "baseline" measure of exposure before autoinduction or inhibition of elimination takes place. The end of treatment would normally be taken at the time of sacrifice but for practical reasons could be within the last week of dosing, depending on the length of the study. Clearly, as suggested by the present guidelines (18,43,77), two sampling times for a 12-month study would in itself be insufficient. However, it could be expected that the sampling times may be built up from previous studies. Thus, during the four-week investigations, day 1 and day 28 samples are taken. For six-month studies, only day 28 and day 84 samples are taken and the previous data used to complete the information (Table 5). The simple use of data from other studies can thus be used for the longer term studies, thereby reducing the number of samples and animals. Toxicity studies with endpoints, such as the proposed ICH S8 guidance (78), require similar considerations to the dose–response relationship assessments for general toxicity studies covered by ICH S3 guidance document.

Drug levels in the body will increase over and above the levels found after the first dose to a degree, which is dependent on the relative frequency of dosing in comparison to the half-life of the drug. Thus, a drug with a half-life of only one hour will accumulate if administered every 15 minutes, while a drug with a half-life of six hours will not accumulate if administered daily. Two measures of accumulation are used: the Accumulation Ratio, which measures how much a drug builds up in the body after repeated dosing to steady state; and the Accumulation Factor, which measures how much accumulation occurs above or below that expected from single-dose considerations. The Accumulation Ratio or Index (R_{ac}) can be measured by comparing the AUC over a dosing interval on Day 1, with the AUC measured over the same time interval at steady state, or the ratio of C_{min} compared with C_{min} on the first day of dosing, or by the equation

$$R_{ac} = \frac{1}{1 - e^{-kt}}$$

where k is the elimination constant and t the dosing frequency. However, this method relies on an accurate measure of k, which may change with repeat dosing. If this figure is greater than 1, then the drug accumulates, and when multiplied by dose provides an estimate of the total amount of drug in the body at steady state. The Accumulation Factor is a measure of nonlinearity after repeated dosing and is calculated as the ratio of infinite AUC after the first dose, compared with the AUC during a dosing interval at steady state. If greater than unity, it would suggest enzymatic inhibition, while if less than unity, it could infer induction.

Additional Studies

Where the intended clinical route of administration or the physicochemical property of the drug changes with, for example, the change in salt then toxicokinetic bridging studies are required. Similarly where drug combinations are known to affect the AUC of one or both drugs, then an assessment of exposure is required. Development of drugs for pediatric use require additional testing prior to approval (79,80). Where juvenile animals are used, it is common to compare findings with toxicokinetic data in adult animals. Regulatory guidance (81) indicates that assessment is case-by-case but that data from animal species requires careful interpretation with respect to its relevance in humans (82,83,84).

Safety Pharmacology Studies

The ICH S7 safety pharmacology guidelines require *in vivo* cardiovascular, CNS, and respiratory assessments and the measurement of plasma levels at predefined sampling times in each of these studies to relate response with systemic exposure in the animal species. Safety pharmacology studies intended for submission to regulatory authorities are usually performed early in drug development. Protocols adopted for regulatory submission are often based on preliminary safety pharmacology findings with the studies conducted during the discovery phase of a compound. Selection of relevant animal models can be based on pharmacodynamic response, pharmacokinetic profile, species, and strain. The selection of sampling times should be based on available *in vitro* and *in vivo* pharmacodynamic and pharmacokinetic data. Where feasible, the intended route of administration should be used and the number of animals should be sufficient to permit adequate sampling (see the discussion under toxicokinetics earlier in this chapter) to determine dose–response relationship and if necessary to design follow-up studies. *In vitro* tests intended for regulatory submission should, in the absence of a pharmacological effect, determine the actual concentration of compound in the test system to prove that adequate concentrations have been reached during the course of the study. Metabolites that are suspected to achieve systemic exposure in humans and contribute to pharmacological activity require suitably designed *in vitro* studies to assess pharmacological activity.

Genotoxicity

Two *in vitro* studies and one *in vivo* study are usually required to support drug development with respect to potential genotoxicity. The *in vivo* test is usually a rodent micronucleus test or chromosomal aberration test. For these *in vivo* studies, proof of absorption is required to demonstrate that adequate exposure has been achieved. For *in vitro* investigations, it is particularly important to confirm that the intended drug concentrations are actually achieved in the studies, particularly for drugs with known stability, solubility, and nonspecific binding issues. ICH guidelines (85,86) expect the investigator to demonstrate exposure to drug with toxicokinetic or genotoxicity data. Consideration may need to be given to use of activated *in vitro* tests using S9 rat fractions where this species is shown not to have a similar metabolic pattern to that seen in humans. In such circumstances, it would appear appropriate to substitute rat S9 for human S9 material to ensure that *in vitro* tests are evaluated in the presence of metabolites, which are unique to human.

Carcinogenicity Studies and Dose-Setting

The ICH S3A guidance on toxicokinetics in carcinogenicity studies places particular emphasis on species and strains, dose selection, and routes of administration when designing dose-ranging studies. The data generated from dose-ranging studies are then used to assist in the design of the main carcinogenicity study. In addition to the ICH toxicokinetic guidelines, ICH S1C(R2) guidance outlines six criteria on high-dose selection for carcinogenicity studies: the Maximum Tolerated Dose (MTD), 25-fold AUC ratio, pharmacodynamic effects, saturation of absorption, maximum feasible dose, and upper dose limit of 1500 mg/kg body weight. While ICH S1C retains the view that a MTD of perhaps 100 times the human dose is a useful parameter, it also recognizes that a several kinetics endpoints are also worthy of consideration based on the following: (a) the steady state kinetic data obtained in subchronic or dose-ranging studies should be derived from the same strain and the same route of administration as those used in carcinogenicity studies; (b) these previous kinetic studies should be of sufficient duration to show nonlinearity over dose and time; (c) the metabolic profile should be qualitatively similar in humans and the test species; (d) measurement of plasma concentrations in animals and humans should be based on equivalent analytical methods; and (e) species differences in protein binding should be taken into account when calculating exposure.By consideration of these kinetic endpoints, it has been suggested that a 25-fold ratio of $AUC_{unbound}$ for animals/human would be acceptable. This figure of 25-fold systemic exposure is based on a review of more than 120 submissions to the FDA, where 30% had ratios between 0.1 and 1, 30% had ratios 1 to 10, and 30% had ratios 10 to 200, although few had ratios greater than 100. Overall, approximately 20% of these compounds had unbound AUC ratios greater than 30. Other studies confirm AUC ratios of between 5- and 10-fold (87). As previously discussed this does rely significantly on plasma binding, and species differences may be difficult to measure when the free level is low.

Many physiological changes occur during a two-year lifetime study in rats or mice, which in turn may affect the exposure of the animals to the drug. For example, animals may get larger with subsequent change in exposure based on mg/kg dosing, and during the course of a lifetime study, the clearance of certain drugs can be affected as animals become geriatric with a reduction in renal and hepatic clearance. Other factors, such as stress caused by the pharmacological or toxicological effect of the drug and animal husbandry, can also have profound effects on drug clearance (88,89). ICH S3A guidance does not consider exposure assessment essential beyond six months. However, only by monitoring of exposure can the possibility of modifying the dose be validated during the study. Adoptive strategies by pharmaceutical companies monitor exposure at 1, 13, 26, and 52 weeks. The ICH S1B guideline recommends only one long-term carcinogenicity study, normally in the rat, provided that the study is complemented by a second short- or medium-term study using one of the alternative *in vivo* mouse models for carcinogenicity studies. These alternative studies are expected to provide information on the mechanism of tumor formation and progression, yet the validity of these alternative models has been questioned (90).

For practical reasons, it is desirable to administer drug in the diet for life-time carcinogenicity studies. However, designing studies to incorporate dose administration in the diet raises a number of practical issues itself. First, there is the timing of sampling where rodents tend to eat at night during the dark period. There is also the problem of some drugs being affected by circadian rhythm that may affect the measurement of AUC depending on when blood samples are taken. Within the authors' laboratory, we have investigated this question using short (paracetamol), intermediate (antipyrine), and long (phenylbutazone) half-life drugs, and taking nine samples throughout a 24-hour period (91). Although plasma levels were found to be higher during the dark period, a good measure within 15% of the expected concentration could be obtained by only measuring samples at 07.00, 10.00, and 16.00 hours, together with the assumption that the plasma concentration at steady state (C_{ss}) at 7-hour time point will equal the concentration at 24 hours. This provides four points for calculating AUC_{24}. A good fit was obtained between measured concentrations and simulations using the pharmacokinetic parameters of each compound, and the food consumption profile reflected the importance of the diurnal feeding pattern. Further confidence can be put on these calculations by relating the drug levels to measured food intake over the 12-hour period, and with this additional information a good estimate of AUC (AUC_{24}) at steady state can be obtained with only one determination at 7 hours at the time when the lights are switched on.

$$AUC_{24} = \frac{C_{LO}}{FF} \times 12$$

where C_{LO} is the median concentration from the animals at lights on at the end of feeding in the morning and FF is the fraction of the 24-hour food eaten during the previous 12-hour period. Thus, dose selection and sampling times from dose-ranging studies are pivotal to ensure appropriate exposure is achieved and reflects the intended clinical use and exposure patterns in human.

Reproductive Studies

ICH S5 guidance requires pharmacokinetic information to support exposure assessment in pregnant and lactating animals (92). Kinetic assessments are taken from rat fertility Stage A (segment I), rat and rabbit embryo–fetal development Stage B (segment II), and rat peri- or postdevelopment Stage C (segment III). In setting dose levels it is desirable to have information on the pharmacokinetic characteristics of the drug, either from ADME and or dose-setting studies, because it may be necessary to adjust either the study design and/or the dose schedules. Exposure data required to set the high dose level is determined by maternal toxicity. Where distribution can be shown to be similar, one route of administration that provides the highest exposure is adequate. Rapid changes in animal physiology during pregnancy can affect both distribution and metabolic fate of a drug and accountability for possible changes are reflected in the three phases of reproductive studies.

Concomitant sampling in animals during the course of the study is not advised because this could interfere with the outcome but single samples can be taken at necropsy, and with some planning, the composite sampling procedure could be used allowing C_{max} and an AUC to be measured. Alternatively, one satellite study could be planned for all the necessary time points but using the same group of pregnant animals throughout segment I to III, with the male rat data for the 80 day repeat dosing coming from the three-month toxicity study. Toxicokinetic evaluation for segment III studies normally takes place in embryo–fetal studies at the beginning and at the end of gestation with samples taken from either the main toxicity study or from satellite animals. Normally, the rat is used for all these studies but the mouse may be used instead for Segment I or II, and the rabbit as a nonrodent species for Segment II. The extent of investigation will greatly depend of the class of drug, relative exposure, and known ADME characteristics.

Exploratory Studies and Special Animal Populations

In most development programs, additional studies are warranted to assist in data interpretation from specific ADME and/or toxicity findings. For example, additional studies may be requested to assess class effect of certain drugs or modification of a protocol to assist with understanding metabolic fate of a drug. Within ICH M4S(R2), ADME necessary for extrapolation to humans covers: (a) the methods of analysis, (b) absorption, (c) distribution, (d) metabolism and interspecies comparison, (e) excretion, (f) pharmacokinetic drug interactions, and (h) specialized pharmacokinetic studies. Under the umbrella of this guidance document, specialized studies to support the safety evaluation of drugs are discussed.

Exploratory clinical studies intended to facilitate go/no-go decisions are permitted using a much reduced toxicological program when the clinical exposure is a "microdose," usually defined as 100 μg dose and only using excipients that are on the generally recognized as safe (GRAS) list (81,93). Within the ICH M3 guidelines, there are two approaches that are described under the conduct of "exploratory clinical studies." The extent of toxicological work depends on whether single or multiple microdose(s) are administered to humans, which in turn determines use of an extended single-dose toxicity study in one species or seven-day toxicity study, respectively. Conformation of exposure coupled with relevant pharmacokinetic data permit exploratory clinical studies to be performed, which are commonly referred to as Phase 0 studies.

STEREOISOMERISM AND DRUG DEVELOPMENT

Besides the properties of lipophilicity, solubility, and p*K*a, there exist in some molecules a further physiochemical characteristic, namely, chirality. Most often these chiral drugs, referred to as isomers or enantiomers, can exhibit markedly different characteristics of disposition and metabolism. Pairs of enantiomers are not distinguishable by lipophilicity or p*K*a, but are readily distinguishable by the body's own enzymes and receptors. It is the body's ability to distinguish enantiomeric pairs that can give rise to marked differences in pharmacokinetic and pharmacodynamic response.

While guidelines do not prescribe the selection of either a racemate or a single enantiomer for development, applicants are expected to provide a scientifically based justification for the development of a racemate (94,95,96). EEC Directive (95) on "preclinical pharmacokinetics" suggests that if a racemate is to be developed, both enantiomers should be "studied separately as early as possible to assess relevance of stereoisomerism for its effect *in vivo* to assess the balance of benefit versus risk." Where a single enantiomer is intended for development, the possibility of the *in vivo* formation of the other enantiomer should be investigated. ICHQ6A guidelines on the specifications and test procedures for new drug substances recommends that when developing a single enantiomer the other enantiomer should be considered as an impurity. There is a general agreement across the three geographical regions to suggest consistency in approach towards chiral drug development (94,96,97). Therefore, whether developing a racemate or single enantiomer, chiral analytical methods will need to be developed to measure the enantiomers so that the possibility of inversion, racemization, or epimerization can be investigated.

In practice, one of two situations occur *in vivo*: (a) little or no interconversion of the enantiomer occurs, or (b) interconversion is faster than the distribution or elimination rates of either enantiomer. Where interconversion is rapid, it may be possible to study the racemate as the active substance provided absorption from the gut occurs by passive diffusion because lipophilicity and ionization are indistinguishable by the body. However, where drug transport proteins may be involved, enantiomeric selectivity by transport proteins can be expected. Similarly, distribution of a drug involves crossing membrane barriers, and if transport proteins are suspected to be involved, then differences can arise in the pharmacokinetics of the enantiomers. The plasma proteins to which drugs bind are enantioselective and therefore the free fraction of the enantiomers can be quite different. Although differences in absorption and distribution between enantiomers have been reported, the number of drugs where enantiomeric differences in absorption and distribution are notably fewer (98,99). In contrast to absorption and distribution, drug-metabolizing enzymes exhibit stereoselectivity toward the elimination and inversion of many drugs with examples from NSAID, β-blockers, and anticoagulant classes of drugs. Thus, it would be necessary to look for interconversion in all species used in development at various doses. Like metabolism, renal and biliary clearance shows stereoselectivity toward a number of drugs but just like absorption drugs renally cleared by filtration with no active secretion will show no difference between enantiomers. As metabolic interconversion can be measured with some success using *in vitro* metabolic screens, the results from these *in vitro* studies can make an important starting point for helping to build a strategy for nonclinical and clinical development programs.

Racemate

There are circumstances when developing a racemate is sensible or even preferred, for example, rapid *in vivo* isomerization, technical difficulties in synthesizing individual isomers, commercial patents and costs, and synergistic benefit of both enantiomers used in combination. Assuming that no unexpected toxicities are found, which cannot be explained by an exacerbation of the pharmacology of the isomers, the pharmacokinetic and metabolic evaluation is relatively straightforward. Because the majority of the safety and clinical studies would be undertaken on the racemate, the analysis of the blood levels using chiral assays to quantify the individual isomers when administered together in pivotal studies would be sufficient. This would provide a measure of the relative exposure of both isomers in the animals used in toxicity studies in comparison to that found in human, and would overcome any interspecies stereoselective differences in metabolism and kinetics.

Racemate–Enantiomer Switch

Differences between regulatory requirements for the development of a new single enantiomer from an approved racemate currently exist, with the FDA seeing this as an "abbreviated" submission, while the CHMP view it as a "new active substance" requiring a "complete new application." Fortunately, the development of a single enantiomer can utilize as much of the dossier on the corresponding racemate for developing a single enantiomer. Measurement of the exposure to the single isomer in pivotal safety studies undertaken with the racemate or those set up to investigate repeated dosing in comparison with the steady state in humans would confirm whether there were interspecies differences in the relative pharmacokinetics of the isomers. These studies would provide the necessary "bridging" studies to link the complete racemate data with data required on the single enantiomer to confirm safety margins in human.A suitable pharmacokinetic and toxicokinetic bridging package may comprise an acute toxicity study with the selected enantiomer and racemate as the control, repeat dose study in a single species, and an assessment of pre- and postnatal development. The precise amount of work required will depend greatly on the amount of available data and the extent to which the synthesis and impurity profiles differ between products, but most certainly there will be a requirement to investigate the possibility of inversion or racemization of the enantiomer.

GOOD LABORATORY PRACTICE

The regulatory expectation is that analytical methods used to determine blood and plasma concentrations of drugs achieve adequate sensitivity and precision, and are validated for the purposes of sample analysis. For the analysis of samples from toxicity studies, it is expected that the methods employed conform to Good Laboratory Practice (GLP). GLP requires a clear auditable record of activities, sample storage, and sample stability associated with the study. Although toxicokinetic assessments are undertaken to GLP standards, there remains no such directive for ADME studies (see above). Despite the lack of regulatory requirements for ADME studies, many companies nevertheless undertake pivotal ADME studies in conformity with the "principles of GLP."

REPORTING RESULTS

Statistical Evaluation

ICH guidelines do not require complex statistical analyses of data because of small animal numbers used in generating toxicokinetic data. Nevertheless, it is expected that in the submission for a marketing authorization, any discussion will come to some conclusions on whether, for example, there is a gender difference or whether the data are nonlinear with respect to time. In certain instances, associated with the uncertainty of data, limited statistical data analysis may be used to support optimal study design where, for example, nonlinearity at the high dose is not clear because of high variability.

Reports

A full account of the bioanalysis and interpretation of toxicokinetic data is required to support the interpretation of the toxicology findings. As a minimum an outline of a validated analytical method is required to be reported or referenced, along with the toxicokinetic phase and toxicology findings. For regulatory submission, toxicokinetic data is tabulated to reflect the relationship between exposure in toxicology species and humans. The AUC and C_{time} (C_{max}) and methods of calculating exposure ratios should be incorporated in the nonclinical pharmacokinetic section and also in the toxicology sections of the submission. Throughout the submission, adequate cross-referencing between the nonclinical pharmacokinetic and toxicology sections is required.

Where possible, the toxicokinetic–toxicodynamic (TK/TD) relationships can be used to derive the optimal amount of information reported on the drug. Toxicodynamic parameters include organ, body weight, and histopathology findings. While limited sampling may hamper TK/TD interpretations, the value of TK/TD information can be enormous in identifying, for example, time-dependent toxicity, tissue exposure using PET technology, and probable active metabolites.

Overviews and Tabulated Summaries

Regulatory Authorities in the three main geographical regions require several overviews. One of these reports examines the nonclinical toxicology of the drug and is typically 30 pages long, summarizing the toxicology findings in relation to the intended clinical use. As an integral part of regulatory submission, the overview is not a factual summary but a critical evaluation of the strengths and weaknesses of the submission.

The ICH "Common Technical Document," or global dossier, provides a harmonized format for new product applications for drug approval in the three regions. The electronic Common Technical Document (eCTD) has been promoted as a means for the submission of the dossier for marketing authorization by, for example, the FDA, the EMEA, and the NIHS. The submissions are as per the ICH guideline ICH M4S(R2). In total, there are five modules of which Module 4 relates to nonclinical study reports (4.0); Module 2 comprises seven parts of which 2.6 relates to the nonclinical summary and 2.4 relates to the nonclinical overview. Further detailed description of the CTD and eCTD processes is discussed elsewhere in this book.

Nonclinical section of the CTD requires that data must be tabulated from pharmacokinetic and toxicokinetic studies as part of submission. The Nonclinical Written Summaries (100–150 pages) are required to provide more extensive summaries and discussion of the nonclinical information on pharmacology, pharmacokinetics, and toxicology. Thirty-eight templates are provided for the preparation of the Nonclinical Tabulated Summaries, and 31 example tables are provided. The Nonclinical Overview (Expert Report) should present an integrated and critical assessment of the pharmacokinetic, pharmacological, and toxicological evaluation of the product, and generally should not exceed 30 pages.

THE FUTURE

The ICH will continue to play an important role in the way drugs are developed, and the collaboration between academic, industrial, and regulatory institutions will continue to be required to streamline areas such as pharmacokinetic, ADME, and toxicokinetic studies to ensure that scientific best practice prevails.

A review of 2006–2008 literature shows an increasing use of pharmacokinetic and toxicokinetic data in conjunction with suitable biomarkers to achieve greater insights into mechanisms of pharmacological and toxicological action and establishment of exposure–response relationships. Besides classical biomarkers (such as measures of heart rate, hematological, and biochemical parameters), novel biomarkers are appearing that rely on "omic" platforms technologies. The application of these "omic" technologies is offering exciting prospects for the identification of endogenous biomarkers of drug response, disease, and disease progression. With these newer technologies, novel biomarkers are available increasingly for identification of disease risk, diagnosis, and intervention (100). Furthermore, linking individual biomarkers through use of mechanistic PKPD modeling approaches will inevitably support better prediction of biological response and its extrapolation from animals to human. Clearly, the next steps must be a consensus on integrating these newer technologies to support the drug development process.

CONCLUSIONS

The process of harmonization has engendered considerable alignment of accepted procedures in drug development among the regulatory authorities of Europe, Japan, and the United States. ICH Guidance documents, whether issued in final form or still in draft, have enabled scientists in pharmacokinetic, bioanalysis, and metabolism departments to become more aware of exactly what is required for a submission rather than of perceived differences between various countries and regulatory agencies.

This chapter has attempted to clarify not only the present situation with respect to the existing expectations from ICH but also highlight a number of important considerations associated with generating ADME and toxicokinetic data and the importance of these disciplines in supporting the safety evaluation of new drug entities.

ACKNOWLEDGMENTS

We are indebted to the many academic and industrial scientists with whom we have had the pleasure of working with, and throughout this chapter, we have tried to reflect many enlightened discussions on pharmacokinetics and toxicokinetics, and their application in drug discovery and development.

REFERENCES

1. Teorell T. Kinetics of distribution of substances administered to the body. I: The extravascular modes of administration. Arch Intern Pharmacodyn 1937; 57:205–225.
2. Teorell T. Kinetics of distribution of substances administered to the body. II: The intravascular modes of administration. Arch Intern Pharmacodyn 1937; 57:226–240.
3. Dost FH, ed. Der Blutspiegel: Kinetik der Konzentrationsablaufe. Leipzig, Germany: Georg Thieme, 1953.
4. Williams RT. Detoxification Mechanisms. London: Chapman & Hall, 1959.
5. Common technical document for the registration of pharmaceuticals for human use: Organization of common technical document. ICHM4. 2004. Accessed at www.emea.eu.int. Accessed September 2008.
6. Turcan RG, Lancaster DJ. ADME strategies to support first-in-man studies. In: ISSX North America, 2008. San Diego, CA: Drug Metabolism Reviews, 2008.
7. Ayrton A, Morgan P. Role of transport proteins in drug discovery and development: A pharmaceutical perspective. Xenobiotica 2008; 38(7):676–708.
8. FDA. Guidance for Industry: Drug Metabolism/Drug Interaction Studies in the Drug Development Process: Studies In Vitro. 1997.
9. Zhang L, Zhang YD, Strong JM, et al. A regulatory viewpoint on transporter-based drug interactions. Xenobiotica 2008; 38(7–8):709–724.
10. Edhom M. Pharmacokinetic considerations regarding drug transporter proteins—A regulatory perspective. In: Uppsala: Drug Transport and Delivery: Impact on Drug Discovery and Development. EUFEPS, 2008, 8–10 October.
11. Shiran MR, Proctor NJ, Howgate EM, et al. Prediction of metabolic drug clearance in humans: In vitro-in vivo extrapolation vs allometric scaling. Xenobiotica 2006; 36(7):567–580.
12. Rostami-Hodjegan A, Tucker GT. Simulation and prediction of in vivo metabolic drug clearance from in vitro data. Nat Rev 2007; 6(2):140–149.
13. Fagerholm U. Prediction of human pharmacokinetics: Biliary and intestinal clearance and enterohepatic circulation.J Pharm Pharmacol 2008; 60(5):535–542.
14. Weaver RJ. Assessment of drug–drug interaction: Concepts and approaches. Xenobiotica 2001; 31(8/9):499–538.
15. Bogaards JJP, Hissinka EM, Briggs MM, et al. Prediction of interindividual variation in drug plasma levels in vivo from individual enzyme kinetic data and physiologically based pharmacokinetic modeling. Eur J Pharm Sci 2000; 12:117–124.
16. Nestorov I. Whole-body physiologically based pharmacokinetic models. Expert Opin Drug Metab Toxicol 2007; 3(2):235–249.

17. Dixit R, Riviere J, Krishnan K, et al. Toxicokinetics and physiologically based toxicokinetics in toxicology and risk assessment. J Toxicol Environ Health 2003; 6(1):1–40.
18. Note for guidance on toxicokinetics: A Guidance for Assessing Systemic Exposure in Toxicology Studies. ICHS3A. CPMP/ICH384/95, 1995. Accessed Sep. 5, at www.emea.eu.int. Accessed September 2008.
19. Pang KS. Modelling of intestinal drug absorption: Roles of transporters and metabolic enzymes (for the Gillette Review Series). Drug Metab Dispos 2003; 31:1507–1519.
20. Clarke B, Smith DA. Pharmacokinetics and toxicity testing. CRC Crit Rev Toxicol 1984; 12:343–385.
21. Xianhua C, Gibbs S, Lanyan F, et al. Why is it challenging to predict intestinal drug absorption and oral bioavailability in human using rat model. Pharm Res 2006; 23(8):1675–1686.
22. Cao X, Gibbs ST, Fang, L, et al. Why is it challenging to predict intestinal drug absorption and oral bioavailability in human using rat model. Pharm Res 2005; 23(8):1675–1696.
23. Bohs C, Cregor M, Gunaratna G, et al. Culex automated blood sampler part II: Managing freely-moving animals and monitoring their activity. Curr Sep 2000; 18(4):147–151.
24. WHO. Use of Ionising Radiation and Radionuclides on Human Beings for Medical Research, Training, and Non-medical Purposes. Report of a WHO Expert Committee. Technical Report Series 611 Geneva, 1977.
25. HPA. Notes for Guidance on the Clinical Administration of Radiopharmaceuticals and Use of Sealed Radioactive Sources. www.arsac.org.uk. Health Protection Agency 2006. Accessed September 2008.
26. Cornett DS, Reyzer ML, Chaurand P, et al. MALDI imaging mass spectrometry: Molecular snapshots of biochemical systems. Nat Methods 2007; 4:828–833.
27. Smith DA. Integration of animal pharmacokinetics and pharmacodynamic data in drug safety assessment. Eur J Drug Metab Pharmacokinet 1993; 18:31–39.
28. Oravcova J. Drug–protein binding studies: New trend in analytical and experimental methodology.J Chromatogr B Biomed Appl 1996; 677:1–28.
29. Wright JD. Measurement and analysis of unbound drug concentrations. Clin Pharm 1996; 30(6):445–462.
30. Meyer MC, Guttman DE. The binding of drugs by plasma proteins. J Pharm Sci 2008; 57(6):895–918.
31. MacKichan JJ. Influence of protein binding and use of unbound (free) drug concentration. In: Burton ME, Shaw LM, Schentag JJ, Evans WE, eds. Applied Pharmacokinetics and Pharmacodynamics: Principles of Therapeutic Drug Monitoring, 4th ed. Baltimore, MD: Lippincott Williams & Wilkins, 2005.
32. Yoon M, Barton HA. Predicting maternal rat and pup exposures: How different are they? Toxicol Sci 2008; 102(1):15–32.
33. Garland M, Abildskov KM, Kiu T-W, et al. Placental transfer and fetal elimination of morphine-3-β-glucuronide in the pregnant baboon. Drug Metab Dispos 2008; 36:1859–1868.
34. Newport DJ, Pennell PB, Calamaras MR, et al. Lamotrigine in breast milk and nursing infants: Determination of exposure. Pediatrics 2008; 122(1):223–231.
35. Clewell RA, Gearhart JM. Pharmacokinetics of toxic chemicals in breast milk: Use of PBPK models to predict infant exposure. Environ Health Perspect 2002; 110(6):333–337.
36. Alcorn J, McNamara PJ. Acyclovir, ganciclovir and zidovudine transfer into rat milk. Am J Microbiol 2002; 46(6):1831–1836.
37. Baillie TA, Cayen MN, Founda H, et al. Drug metabolites in safety testing. Toxicol Appl Pharmacol 2002; 182:188–196.
38. Smith DA, Obach RS. Seeing through the MIST: Abundance versus percentage. Commentary on metabolites in safety testing. Drug Metab Dispos 2005; 33:1409–1417.
39. FDA. Guidance for Industry: Safety Testing of Drug Metabolites. 2008.
40. Calabrese EJ. Animal extrapolation and the challenge of human heterogenicity. J Pharm Sci 1986; 75:1041–1046.
41. Wilkinson GR. The role of active metabolites in drug responsiveness and toxicity: Encainide and profafenone. In: Yacoby AS, Skelly JP, Batra VK, eds. Toxicokinetics and New Drug Development. New York, NY: Pergamon Press, 1989:97–108.
42. Worboys PD, Carlile DJ. Implications and consequences of enzyme induction on preclinical and clinical drug development. Xenobiotica 2001; 31(8/9):539–556.
43. Dose selection for carcinogenicity studies of pharmaceuticals and limit dose. ICHS1C(R2), 2008. Accessed at www.emea.eu.int. Accessed September 2008.
44. Note for guidance on the investigation of drug interactions. CPMP/EWP/560/95. 1998. Accessed at www.emea.eu.int. Accessed September 2008.
45. Pichard-Garcia L, Weaver RJ, Eckett, N, et al. The olivacine derivative S 16020 (9-hydroxy-5,6-dimethyl-*N*-[2-(dimethylamino)ethyl)-6H-pyrido(4,3-B)-carbazole-1-carboxamide) induces CYP1A

and its own metabolism in human hepatocytes in primary culture. Drug Metab Dispos 2004; 32(1):80–88.
46. Prueksaritanont T, Lin JH, Baillie TA. Complicating factors in safety testing of drug metabolites: Kinetic differences between generated and preformed metabolites. Toxicol Appl Pharmacol 2006; 217(2):143–152.
47. Guidelines on non-clinical pharmacokinetic studies. Notification No. 496. 1998. Accessed at www.nihs.go.jp/mhlw. Accessed September 2008.
48. Roffey SJ, Obach RS, Gedge JI, et al. What is the objective of the mass balance study? A retrospective analysis of data in animal and human excretion studies employing radiolabelled drugs. Drug Metab Rev 2007; 39:17–43.
49. Dayton PG, Israili ZH, Henderson JD. Elimination of drugs by passive diffusion from blood to intestinal lumens: Factors influencing non-biliary excretion by the intestinal tract. Drug Metab Rev 1983; 14:1193–1206.
50. Altman PL, Dittmer DS, eds. Biological Handbook Blood and Other Body Fluids. Newark, NJ: American Society for Experimental Biology, 1971:363–409.
51. Levine WG. Biliary excretion of drugs and other Xenobiotics. Ann Rev Pharmacol Toxicol 1978; 18:81–96.
52. Ghibellini G, Leslie EM, Brouwer LR. Methods to evaluate biliary excretion of drugs in humans: An updated review. Mol Pharm 2006; 3(3):198–211.
53. Igarashi T, Sekido T. Case studies for statistical analysis of toxicokinetic data. Regul Toxicol Pharmacol 1996; 23:193–208.
54. FDA. Guidance for Industry: Bioanalytical Method Validation. 2001.
55. Viswanathan CT, Bansal S, Booth B, et al. Workshop/conference report—Quantitative bioanalytical methods validation and implementation: Best practices for chromatographic and ligand binding assays. AAPS J 2007; 9(1):E30–E42.
56. Bouzom F, Laveille C, Merdjan H, et al. Use of non-linear mixed effect modeling for the meta-analysis of preclinical pharmacokinetic data: Application to S 20342 in the rat. J Pharm Sci 2000; 89(5):603–613.
57. Burtin P, VanBree J, Mentre F, et al. A prospective evaluation of sparse sampling for toxicokinetics in the course of drug development. 613. In: Annual Meeting of the Population Approach Group in Europe; Frankfurt, Germany: 1995.
58. Tse FLN JR. Serial versus sparse sampling in toxicokinetic studies. Pharm Res 1996; 13(7):1105–1108.
59. Kendall L, Miller M, Kimberly, L, et al. Minimalist sampling strategies for toxicokinetics: A retrospective analysis. In: AAPS Workshop on BE, BCS and Beyond, North Bethesda, MD, 2007.
60. Barfield M, Spooner N, Lad R, et al. Application of dried blood spots combined with HPLC-MS/MS for the quantification of acetaminophen in toxicokinetic studies. J Chromatogr 2008; 870(1):32–37.
61. Hulse M, Feldman S, Brucker, JV. Effect of blood sampling schedules on protein binding in the rat. J Pharm Exp Ther 1981; 218:416–420.
62. Caccia S, Fracasso C, Garattini S, et al. Effects of short and long-term administration of fluoxetine on the monoamine content of rat brain. Neuropharmacology 1992; 31:343–347.
63. Renshaw PF, Guimaraes A, Fava M, et al. Accumulation of fluoxetine and norfluoxetine in human brain during therapeutic administration. Am J Psychiatry 1992; 149(11):1592–1594.
64. Hirano K, Kimura R, Sugimoto Y, et al. Relationship between brain serotonin transporter binding, plasma concentration and behavioral effect. Br J Pharmacol 2005; 144:695–702.
65. Anderson BD, May MJ, Jordan S, et al. Dependence of Nelfinavir brain uptake on dose and tissue concentrations of the selective P-glycoprotein inhibitor zosuquidar in rats. Drug Metab Dispos 2006; 34:653–659.
66. Anderson ME. Physiological modeling of organic compounds. Ann Occup Hyg 1991; 35(3):309–321.
67. Dobson PD, Kell DB. Carrier-mediated cellular uptake of pharmaceutical drugs: An exception or the rule? Nat Rev 2008; 7:205–220.
68. Dollery C. The clinical Pharmacologist's view: Drug discovery and early development. In: Davies K, ed. Experimental Therapeutics. New York: Taylor & Francis, 2003:3–24.
69. Lecoeur S, Bonierbale E, Challine D, et al. Specificity of in vitro covalent binding of tienilic acid metabolites to human liver microsomes in relationship to the type of hepatotoxicity: Comparison with two directly hepatotoxic drugs. Chem Res Toxicol 1994; 7:434–442.
70. Li AP. A review of the common properties of drugs with idiosyncratic hepatotoxicity and the "multiple determinant hypothesis" for the manifestation of idiosyncratic drug toxicity. Chem Biol Interact 2002; 142(1–2):7–23.
71. Monro AM. What is an appropriate measure of exposure when testing drugs for carcinogenicity in rodents? Toxicol Appl Pharmacol 1992; 112:171–181.

72. Monro AM. Interspecies comparisons in toxicology: The utility and futility of plasma concentrations of the test substance. Regul Toxicol Pharmacol 1990; 12(2):137–160.
73. Jelovsek FR, Mattison DR, Chen JJ. Prediction of the risk for human developmental toxicity. Obstet Gynecol 1989; 74:624–636.
74. Olsen H, Betton G, Robinson D, et al. Concordance of the toxicity of pharmaceuticals in human and in animals. Regul Toxicol Pharmacol 2000; 32:56–67.
75. Breimer DD, Danhof M. Relevance of the application of pharmacokinetic-pharmacodynamic modeling concepts in drug development. The "wooden shoe" paradigm. Clin Pharmacokinetics 1997; 32:259–267.
76. Danhof M, Alvan G, Dahl SG, et al. Mechanistic-based pharmacokinetic/pharmacodynamic modeling—A new classification of biomarkers. Pharm Res 2005; 22:1432–1437.
77. Guideline on Repeated Dose Toxicity. EMEA/CHMP/SWP/488313/2007. 2008. Accessed at www.emea.eu.int. Accessed September 2008.
78. Immunotoxicity Studies for Human Pharmaceuticals. ICHS8. 2005. Accessed at www.emea.eu.int. Accessed September 2008.
79. Pediatric Rule CFR 314. 1999. Accessed at www.fda.gov/cder/pediatric. Accessed September 2008.
80. Clinical Investigation Of Medicinal Products in the Pediatric Population. CPMP/ICH/2711/99. 2001. Accessed at www.emea.eu.int. Accessed September 2008.
81. Guidance on Non-Clinical Safety Studies for the Conduct of Human Clinical Trials and Marketing Authorization for Pharmaceuticals. ICHM3(R2). CPMP/ICH/286/95. 2008. Accessed at www.emea.eu.int. Accessed September 2008.
82. Schwartz S. Providing toxicological support for reproductive toxicology studies in pharmaceutical development. Arch Toxicol 2007; 75(7):381–387.
83. Robinson K. Evaluations of organ system development in juvenille toxicology testing. Reprod Toxicol 2008; 26(1):51–53.
84. Seely JC. Toxicology and pathology considerations for the design of juvenille animal studies. Lab Anim (NY) 2008; 37(5):206–209.
85. Guidance on Specific Aspects of Regulatory Genotoxicity Tests for Pharmaceuticals. S2A. 1995. Accessed at www.emea.eu.int. Accessed September 2008.
86. Genotoxicity: A Standard Battery for Genotoxicity Testing for Pharmaceuticals. S2B. 1997. Accessed at www.emea.eu.int. Accessed September 2008.
87. Morgan D. The applications of toxicokinetic data to dosage selection in toxicology studies. Toxicol Pathol 1994; 22:112–123.
88. Pollack GM, Browne JL, Marton J, et al. Chronic stress impairs oxidative metabolism and hepatic excretion of model xenobiotic substrates in the rat. Drug Metab Dispos 1991; 19(1):130–134.
89. Saranteas T, Mourouzis C, Dannis C, et al. Effect of various stress models on lidocaine pharmacokinetic properties in the mandile after masseter injection. J Oral Maxillofac Surg 2003; 62(7):858–862.
90. Jacobs A. Prediction of 2-year carcinogenicity study results for pharmaceutical products: How are we doing? Toxicol Sci 2005; 88(1):18–23.
91. Jochemsen R, Bazot D, Brillanceau MH, et al. Assessment of drug exposure in rat dietary studies. Xenobiotica 1993; 23(10):1145–1154. Accessed September 2008.
92. Detection of Toxicity to Reproduction for Medicinal Products & Toxicity to Male Fertility. ICHS5(R2). 2005. Accessed at www.emea.eu.int. Accessed September 2008.
93. Guidance for Industry, Investigators, and Reviewers: Exploratory IND studies. 2006. Accessed at http://www.fda.gov/cder/guidance/index.htm. Accessed September 2008.
94. FDA Policy Statement for the Development of New Stereoisomeric Drugs. 1992. Accessed at www.fda.gov/cder/guidance/stereo.htm. Accessed September 2008.
95. Investigation of Chiral Active Substances. 1994. Accessed at http://www.emea.europa.eu/pdfs/human/qwp/3cc29aen.pdf. Accessed September 2008.
96. Shimazawa R, Nagai N, Toyoshima S, et al. Present state of new chiral drug development and review in Japan. J Health Sci 2008; 54(1):23–29.
97. Investigation of Chiral Active Substances. 1997. Accessed at http://www.emea.europa.eu/pdfs/human/qwp/3cc29aen.pdf. Accessed September 2008.
98. Lee EJD, Williams KM. Chirality: Clinical pharmacokinetics and pharmacodynamic considerations. Clin Pharmacokinet 1990; 18:339–345.
99. Walle T, Webb JG, Bagwell EE, et al. Stereoselective delivery and actions of beta-receptor antagonists. Biochem Pharmacol 1988; 37:115–124.
100. Altar CA. The biomarkers Consortium: On the critical path of drug discovery. Clin Pharmacol Ther 2008; 83(2):361–364.

101. Guidance on Genotoxicity Testing and Data Interpretation for Pharmaceuticals Intended for Human Use. 2008. Accessed at www.emea.eu.int. Accessed September 2008.
102. Guidance for Repeated Dose Tissue Distribution Studies. ICHS3B. 1994. Accessed at www.emea.eu.int. Accessed September 2008.
103. Duration of Chronic Toxicity Testing in Animals (Rodent and Nonrodent Toxicity Testing). S4. CPMP/ICH/300/95. 1998. Accessed at www.emea.eu.int. Accessed September 2008.
104. Safety Pharmacology Studies for Human Pharmaceuticals. ICHS7A. 2000. Accessed at www.emea.eu.int. Accessed September 2008.

24 Single and Repeat Dose Toxicity

Barry S. Levine
Levine & Associates, LLC, Chicago, Illinois, U.S.A.

INTRODUCTION

The development of new chemical entities (NCEs) as potential therapeutic agents is dependent on successful clinical trials, which ultimately demonstrate efficacy and safety in patient populations under controlled conditions. To initiate Phase I clinical trials for NCEs intended for multiple dosing, appropriate repeat dose toxicity studies in two species, one of which is a nonrodent (in addition to genotoxicity and safety pharmacology studies), are necessary to submit petitions to worldwide regulatory agencies. While the determination of the acute toxicity and associated median lethal dose (LD_{50}) of environmental and industrial chemicals is useful for assessing consumer and worker protection, risk–benefit analysis, etc., this information is less important for understanding the potential safety of most drugs under development. Exceptions would be for products being developed for single dose or infrequent administration such as medical imaging agents (including both contrast agents and diagnostic radiopharmaceuticals) (1), some therapeutic radiopharmaceuticals (2), and gene therapy medicinal products (3). For those products, comprehensive single dose toxicity studies in two species (one of which is a nonrodent) would be necessary prior to a Phase I clinical trial.

For NCEs intended for multiple dosing, repeat dose toxicity studies should mimic the dosing regimen of the planned clinical trials. Repeat dose toxicity studies should also identify target organs, define dose–response relationships (including a no observed adverse effect level, NOAEL), determine reversibility of toxicity, and establish useful biomarkers for monitoring potential toxicity in subsequent clinical trials. As the clinical development of an NCE progresses, additional longer term repeat dose toxicity studies of at least similar duration to those planned for clinical study will be necessary. The design of these longer term toxicology studies should incorporate relevant endpoints based on the results of early clinical trials. This "ping-pong" approach results ultimately in a more thorough understanding of the efficacy, safety, and potential risks of the NCE under development. Well-designed and executed repeat dose toxicology studies throughout an NCE's development are critical to the success of the clinical program as well as the successful regulatory approval of the new therapeutic agent.

Published guidelines of the requirements to support a clinical trial or marketing authorization application for a candidate drug are available from the International Conference on Harmonization of Technical Requirements for Registration of Pharmaceuticals for Human Use (ICH), the European Medicines Agency (EMEA), and the U.S. Food and Drug Administration (FDA) Web sites. Guidelines from other countries such as Canada and Australia are also available, and are typically based on, or are very similar to, the ICH Guidelines. In the United States, the Pharmaceutical Manufacturers Association (PMA) (now known as the Pharmaceutical Research and Manufacturers of America—PhRMA) published guidelines in 1977 (4), which included a description of studies the FDA expected to be conducted to support clinical studies. In Japan, a detailed guideline was published by the Ministry of Health, Labour, and Welfare (MHLW) in 1990/91 (5,6). Summaries of the history and published guidelines for single dose and repeat dose toxicity requirements are subsequently described in the "Background" section.

SINGLE DOSE TOXICITY

Background

In 1927, the concept of a median lethal dose from a single exposure to a chemical was introduced to standardize doses of certain natural products such as insulin and digitalis extracts (7). Since that time, a number of papers have been published refining the procedures used to estimate this parameter (8–10). Because of numerous variables encountered in toxicology research, for example, species, strain, sex, age, the determination of a precise LD_{50} with its associated 95% confidence limit is not considered critical or necessary. Those variables among others such as husbandry, body weight, dose volume, vehicle, etc., are discussed later in this chapter.

In 1997, the U.S. National Institute of Environmental Health Sciences, one of the National Institutes of Health, established the Interagency Coordinating Committee on the Validation of Alternative Methods (ICCVAM). The responsibilities of ICCVAM were to (1) coordinate interagency technical reviews of new and revised toxicological test methods with regulatory applicability, including alternative test methods that reduce, refine, or replace the use of animals, and (2) coordinate cross-agency issues relating to validation, acceptance, and national and international harmonization of new, modified, and alternative toxicological test methods.

In 2006, ICCVAM published a report based on collaborations between the U.S. National Toxicology Program Interagency Center for the Evaluation of Alternative Toxicological Methods (NICEATM) and the European Centre for the Validation of Alternative Methods (ECVAM) (11). The report concluded that in vitro cytotoxicity test methods were not sufficiently accurate to replace animals for acute toxicity testing in order to determine regulatory hazard classifications, based on IC_{50}–LD_{50} regressions. Each of the two in vitro assays (BALB/c 3T3 mouse fibroblasts and normal human epidermal keratinocytes) predicted the Globally Harmonized System of Classification and Labelling of Chemicals (12) for only about 30% of the chemicals ($n = 72$). However, the report indicated that the in vitro assays could be used to estimate starting doses for acute oral toxicity studies in rats. Replacement of rat LD_{50} studies by in vitro assays was recently reviewed, which included a discussion of regulatory status and limitations such as potential ethical issues precluding rigorous validation procedures (13).

In 1981, the Organization for Economic Co-Operation and Development (OECD) published guidelines on toxicology testing of chemicals (14). Test Guideline 401 set forth procedures for acute toxicity testing and the determination of the LD_{50} of chemicals. The guideline indicated that

1. the rat is the preferred species,
2. at least five animals/sex/dose level should be used, and
3. at least three dose groups should be included in the study.

Although both sexes were not mandated, males and females were required if a limit test of 2000 mg/kg was first conducted. Emphasis of the guideline was on dose–mortality relationships as the data should be sufficient to calculate an LD_{50} by standard methods (8–10).

Several years later (1987), a guideline based on European Economic Community [later the "EC" and then European Union (EU) as of 1993] Directive 75/318/EEC as amended was issued by the European Union entitled "Single Dose Toxicity" (15). It is currently listed as EMEA Guideline 3BS1A. Two mammalian species were required, both of which could be rodents. Both sexes were necessary for the first study. However, only one sex was subsequently needed if no apparent sex difference occurred in the first investigation. Two routes of administration were required, one being the clinically intended route. In 1996, the FDA Center for Drug Evaluation and Research (CDER) issued a Guidance for Industry entitled "Single Dose Acute Toxicity Testing for Pharmaceuticals" that indicated that acute toxicity studies are usually necessary for any drug intended for human use (16). The FDA Guidance indicated that rodents and a nonrodent species (where reasonable) should be used, and information obtained from those studies would be useful in

1. choosing doses for preliminary repeat dose studies,
2. providing preliminary information on target organs of toxicity,
3. possibly demonstrating delayed toxicity,
4. generating information related to accidental poisoning, and
5. assisting in selection of an initial dose for Phase I clinical trials.

In 2001, attendees at the OECD Joint Meeting of the Chemicals Committee and Working Party on Chemicals, Pesticides and Biotechnology agreed that the LD_{50} for acute oral toxicity (OECD Test Guideline 401—previously discussed) should be abolished. It was subsequently deleted from the OECD manual in 2002. Three alternative OECD methods (Test Guidelines 420, 423, and 425) adopted in the 1990s were available to provide sufficient information to replace Test Guideline 401 (14). In 1998, all three methods were revised to consider further reduction in the numbers of animals while improving performance characteristics. The updated methods were adopted by the OECD Council in 2001. Although the revised alternative methods still require the use of animals, their numbers are significantly reduced compared to the conventional LD_{50} test. Moreover, one of the alternative tests (Test Guideline 420) does not require death as an endpoint; whereas for the other alternatives (Test Guidelines 423 and 425), the expected number of deaths is typically not more than three. All three methods provide information on hazardous properties and allow the chemical to be classified for acute toxicity according to the Globally Harmonized System of classification and labeling of chemicals (12).

The principle of Test Guideline 420: Acute Oral Toxicity—Fixed Dose Procedure was first suggested by the British Toxicology Society in 1984 based on the administration of a series of fixed dose levels (17). The approach avoided using death of animals as an endpoint and relied instead on signs of toxicity. Following U.K. (18) and international (19) in vivo validation studies, the procedure was adopted as a Test Guideline in 1992 by the OECD Council. Test Guideline 423: Acute Oral Toxicity—Acute Toxic Class Method, first described in 1986 (20), was based on U.K. (21) and international (22) validation studies published in 1992 and 1995, respectively. The results of those studies showed that this method allows for allocation to toxicity classes as for the Fixed Dose Procedure. OECD Test Guideline 425: Acute Oral Toxicity: Up-and-Down Procedure permits estimation of an LD_{50} with a confidence interval. In 1985, this procedure was proposed for use in acute toxicity testing (23) and confirmed in a follow-up study in 1987 (24).

In 1997, the ICH adopted the S6 guideline entitled "Preclinical Safety Evaluation of Biotechnology-Derived Pharmaceuticals" (25). The S6 guideline addressed the nonclinical studies necessary to develop biopharmaceuticals such as proteins and monoclonal antibodies. Acute toxicity studies are indicated (which can be included as a component of safety pharmacology or pharmacology studies). The guidance mentioned that commonly used species for toxicity testing may not be appropriate for biopharmaceuticals since species or tissue specificity may not be relevant. In June 2008, a revised version of the S6 guideline (S6R1) was published as a Final Concept Paper, although acute toxicity testing was not addressed (26).

In the same year that the ICH S6 guideline was initially adopted (1997), ICH also published the M3 guideline entitled "Maintenance of the ICH Guideline on Non-Clinical Safety Studies for the Conduct of Human Clinical Trials for Pharmaceuticals." The original M3 guideline is no longer available on the ICH Web site. This guideline indicated that both single and repeat dose toxicity studies were necessary for marketing approval of a pharmaceutical. More specifically, the acute toxicity of a pharmaceutical should be evaluated in two mammalian species prior to human exposure, although a dose escalation study would be acceptable as an alternative to a classic acute toxicity study design. Although no changes to acute toxicity assessments were included in the guideline's first revision, that is, ICH M3(R1) published in 2000 (27), significant changes to the draft second revision, ICH M3(R2), were published in 2008 (28). These changes included deletion of "acute toxicity" in the "Scope of the Guideline" section and an expanded section on "Acute Toxicity Studies" requirements in a subsequent section. In essence, separate acute toxicity studies are not required if appropriate data are obtained from other studies such as a non-GLP pilot study to select doses for a repeat dose range-finding study in rodents. In addition, for animal studies in which (1) exposure is at least 50-fold the intended clinical dose (C_{max} or AUC), (2) saturation of exposure occurs, or (3) a maximum feasible dose is given, separate acute toxicity studies are not needed. Formal GLP single dose toxicity studies, similar in design to repeat dose studies, would be needed to support products intended for single dose exposure in humans. The draft-revised guideline (M3[R2]) still referred to the usefulness of acute toxicity information in terms of understanding potential overdose situations prior to Phase III studies.

In 2008, the results of a European initiative to conduct an evidence-based review of acute toxicity studies and assess the value of the data generated were published (29). Eighteen companies studied 74 compounds in the project. The retrospective analysis indicated that acute toxicity data were not used to

1. terminate drugs from development,
2. directly support dose selection for GLP repeat dose toxicity studies in animals, or
3. set doses in the first clinical trials in humans.

Their use in treating a drug overdose was also questionable. Acute toxicity data are not generally used to identify target organs unless the drug is for single administration to humans (and in this instance, extensive single dose toxicity studies would be conducted in accordance with GLPs). No additional information was gleaned from parenteral route rodent studies when the clinically intended route was oral. The project concluded that acute toxicity data were extremely limited in use. A strict GLP single dose toxicity study, therefore, may not be necessary from a regulatory perspective. A non-GLP acute toxicity study in rats as subsequently described in the "Single Dose Toxicity Studies in Rodents" section may be sufficient to provide limited dose-mortality data as well as provide information to select dose levels for the 14-day dose range-finding (DRF) study in rats.

In 2008, the EMEA published a "Concept Paper on Single Dose/Acute Toxicity" (30). Referencing the European Initiative study mentioned above (29), the Concept Paper indicated that data from acute toxicity studies are not used to support the planning and conduct of early clinical trials. The EMEA Committee for Medicinal Products for Human Use (CHMP) Safety Working Party, which is responsible for preparing the Agency's opinions on all questions concerning medicinal products for human use, recommended revising the existing guideline on single dose toxicity based on the limited usefulness of acute toxicity studies for the safety assessment of medicinal products. The Concept Paper indicated that the final guideline would most likely be published in the second half of 2009.

Thirty-four Investigator Brochures used to support First-In-Man studies during 1997 to 2006 were reviewed recently (31). From around 2003, companies generally performed an acute toxicity study in the rat by the proposed clinical route of administration and only occasionally by a second route such as intravenous. In a few instances, no single dose toxicity studies were conducted and the IND was accepted by the regulatory agencies. Single dose nonrodent studies were typically not conducted but were sometimes included in the one- or two-week DRF phase as an ascending dose design. The acute rat study was typically of limited design with two to three animals/sex/dose, potentially increased to $n = 5$. The studies were generally used to select doses for the 14-day DRF study and not to assess organ system toxicity, etc.

In 2004, the EMEA published a "Position Paper on the Conduct of Nonclinical Studies to support Clinical Trials with a Single Microdose" (32). A microdose was defined as less than 1/100th of the dose calculated to yield a pharmacologic effect of the drug based on primary pharmacodynamic data obtained in vitro and in vivo. The extent of required studies should be proportional to the nature and scope of the clinical trial. CHMP/EMEA proposed that certain deviations from ICH and EMEA Guidelines to support pre-Phase I clinical trials may be justified scientifically. Two routes of administration in animals (the intended route in man and IV) should be used, and study designs for single dose studies should include early (day 3) and day 15 necropsies. The corresponding stable isotope test substance should be used for extended single dose toxicity studies on radiopharmaceuticals. A few years later (2006), the FDA published a similar document entitled "Guidance for Industry, Investigators, and Reviewers: Exploratory IND Studies" (33). It was more detailed than the EMEA document with regard to preclinical study requirements to support microdose studies in humans. An exploratory IND study was defined as a clinical trial that is conducted early in Phase I, involves limited human exposure, and has no therapeutic or diagnostic intent (e.g., screening or microdose studies). The guidance described toxicology studies necessary for

1. clinical studies of PK or imaging,
2. clinical trials to study pharmacologically relevant doses, and
3. clinical studies of mechanisms of action related to efficacy.

The document stressed that preclinical testing programs for exploratory IND studies can be less extensive than for traditional IND studies.

Acute/single dose toxicity studies are considered to be of limited usefulness in the overall development of NCEs intended for multiple dose administration in humans. For NCEs being developed for single exposure such as diagnostic agents, GLP single dose toxicity studies would be necessary in two species. Experimental designs would be similar to those used for repeat dose toxicity programs subsequently discussed (except for single dose exposure). Microdose clinical programs would also require single dose toxicity studies of sufficient design to support an early Phase I program. Although abbreviated single dose toxicity studies in rodents can assist in the selection of doses for a 14-day non-GLP DRF study, available rodent pharmacology data may be used to design a 3- to 4-day pilot study for the same purpose of dose level section for the 14-day DRF study.

REPEAT DOSE TOXICITY

Background

As mentioned in the "Single Dose Toxicity" section, the Pharmaceutical Manufacturers Association (PMA), now known as the Pharmaceutical Research and Manufacturers of America (PhRMA), published guidelines in 1977, which included a description of studies the FDA expected to support clinical studies (4). A few years later (1981), the previously mentioned OECD toxicology test guidelines were published, which included detailed study outlines for repeated dose 28-day oral toxicity study in rodents (Test Guideline 407), repeated dose 90-day oral toxicity study in rodents (Test Guideline 408), repeated dose 90-day oral toxicity study in nonrodents (Test Guideline 409), chronic toxicity studies (Test Guideline 452), and combined chronic toxicity/carcinogenicity studies (Test Guideline 453) (13). The 28- and 90-day test guidelines were revised in 1995 and 1998, respectively. For the rodent studies, the revisions included more emphasis on neurologic effects. They also clarified the use of a second, specific nonrodent species. The OECD Guidelines were fairly comprehensive and included specific recommendations for clinical pathology tests, organ weights, and tissues for necropsy/histopathology.

In 1989, the European Community (EC) Committee for Proprietary Medicinal Products (CPMP) published a draft document on nonclinical testing strategies that attempted to consolidate EC requirements (34). Previously, in the EC, each member state had its own requirements. In 1991, regulatory authorities and trade organizations from the United States, Europe, and Japan initiated a cooperative effort through ICH to develop harmonized guidance documents regarding the safety evaluation of pharmaceutical agents (35).

In 1994, the ICH published S3A entitled "Note for Guidance on Toxicokinetics: The Assessment of Systemic Exposure in Toxicity Studies" (36). Its primary objective was to describe systemic exposure in animal toxicity studies and its relationship to dose and time. Secondary objectives were to relate exposure to toxicity findings, support choice of species and treatment regimen in nonclinical toxicity studies, and provide information in the design of subsequent nonclinical toxicity studies. The guidance also noted that determination of metabolites may be necessary when the drug

1. is a prodrug,
2. is metabolized to pharmacologically and toxicologically active metabolite(s), and
3. undergoes extensive metabolism, which requires metabolite measurements to accurately assess exposure.

In 1997, as previously described, ICH published M3 entitled "Maintenance of the ICH Guideline on Non-Clinical Safety Studies for the Conduct of Human Clinical Trials for Pharmaceuticals." The guideline harmonized international standards for nonclinical safety studies to support human clinical trials of a given scope and duration. This included the timing of appropriate nonclinical studies [e.g., acute, repeat dose, safety pharmacology, absorption-distribution-metabolism-excretion (ADME), reproductive/developmental toxicity, local tolerance, genotoxicity, carcinogenicity] in relation to various clinical trials. The guideline noted that the duration of the animal repeat dose toxicology studies conducted in two mammalian species (one a nonrodent) should be equal to or exceed the duration of the clinical trials up to the maximum recommended duration for the repeat dose toxicity studies to support clinical trials and marketing authorization.

In 2000, the M3 guideline was revised (M3[R1]) with minimal changes (28). In 2008, however, significant changes were proposed in a published draft, M3(R2) (29). The proposed

Table 1 Duration of Repeated Dose Toxicity Studies to Support the Conduct of Clinical Trials and Marketing in Europe, United States, and Japan[a]

Maximum duration of clinical trial	Minimum duration of repeated dose toxicity studies to support clinical trials	
	Rodents	**Nonrodents**
Up to 2 wk	2 wk	2 wk
Between 2 wk and 6 mo	Same as clinical trial	Same as clinical trial
>6 mo	6 mo	9 mo
Duration of indicated treatment	Duration of repeated dose toxicity studies to support marketing in all regions	
Up to 2 wk	1 mo	1 mo
>2 wk to 1 mo	3 mo	3 mo
>1 to 3 mo	6 mo	6 mo
>3 mo	6 mo	9 mo

[a]ICH M3(2) (29).

changes harmonized the scope and duration of the nonclinical safety studies to support Clinical Trials versus Marketing Authorization among the various regions of Europe, Japan, and the United States, as shown in Table 1. Specific recommendations included limit doses for repeated dose toxicity studies of 2000 and 1000 mg/kg/day for rodents and nonrodents, respectively, if there are significant margins to clinical exposure and the clinical dose does not exceed 1 g/day. Dose levels providing a 50-fold margin of exposure to the clinical regimen generally would be considered acceptable as the maximum dose for repeat dose toxicology studies in any species. Clinical exposure at a therapeutic dose, however, is usually not known at the earlier phases of development. Also noted in the proposed draft update was that six month rodent and nine month nonrodent studies would generally support dosing for longer than six months in clinical trials. This had been previously mentioned in the ICH S4 Guideline (see subsequent discussion). The proposed revised guideline also clarified toxicity requirements for juvenile animal toxicity studies to support product use in pediatric populations. When the drug was initially developed for adults with previously conducted repeat dose toxicology studies in two appropriate species, one rodent species would generally be considered adequate. When the drug is being developed for primary use in a pediatric population and existing studies do not address adequately developmental concerns for target organs, a nonrodent (e.g., dog) chronic toxicity study may be necessary to cover the full development period in this species.

In 1998, ICH S4 was published entitled "Duration of Chronic Toxicity Testing in Animals Rodent and Non-Rodent Toxicity Testing" (37). This document indicated that for rodents chronic toxicity studies of six months' duration, already accepted in the three major regions (the European Union, Japan, and the United States), were sufficient. However, for nonrodent chronic toxicity studies, 6 versus 12 months' study duration was described briefly with a consensus opinion that 9 months' duration would be sufficient. The ICH expert working group review of 6- and 12-month nonrodent chronic toxicity studies from the European Union, Japan, and the United States, which formed the basis of the 9 month duration proposal by the ICH regulatory review group, was published subsequently (38). However, it is advisable to interact with regulatory agencies on this topic for a given NCE. For example, the FDA may still require a one year nonrodent chronic toxicity study for a first-in-class agent.

To address the worldwide increase in biologics development, the ICH published the S6 guideline in 1997 entitled "Preclinical Safety Evaluation of Biotechnology-Derived Pharmaceuticals" (39). The guidance applies to products derived from characterized cells through the use of a variety of expression systems including bacteria, yeast, insect, plant and mammalian cells. Intended indications for the products include in vivo diagnostic, therapeutic and prophylactic uses. Active substances include proteins and peptides, their derivatives and products of which they are components. They could be derived from cell cultures or produced using recombinant

DNA technology including production by transgenic plants and animals. The guidance pointed out that conventional approaches may be inappropriate for biopharmaceuticals due to unique and diverse structural and biological properties that may include species specificity, immunogenicity and unpredicted pleiotropic activities. Further, the relevant animal species for toxicology testing of monoclonal antibodies are those that express the desired epitope and demonstrate a similar tissue cross-reactivity profile as for human tissues. Safety evaluation programs should normally include two relevant species. However, in some cases, one species may suffice, for example, when only one relevant species is identified or where the biologic activity of the biopharmaceutical is well understood. In addition, even where two species may be necessary to characterize the toxicity in short-term studies, the use of only one species may be justified for subsequent long-term toxicity studies if the toxicity profile in the two species is similar in the short term. In certain cases, studies performed in animal models of disease, gene knockout(s) or transgenic animals may be an acceptable alternative to studies in normal animals.

The continued development of biotechnology-derived products prompted the EMEA to publish a "Concept Paper on the Development of a Committee for Proprietary Medicinal Products (CPMP) Guideline on Comparability of Biotechnology-Derived Products" in 1998 (40). It addressed issues related to changes made in the production process of biotechnology-derived products either during its development or after marketing authorization has been granted. A similar guideline entitled "FDA Guidance Concerning Demonstration of Comparability of Human Biological Products, Including Therapeutic Biotechnology-Derived Products" was published in 1996 (41). In addition to manufacturing changes made on a given product by its developer, the EMEA and FDA began to consider the regulatory requirements for the development of generic biologics. Biosimilars, or follow-on proteins, are new versions of existing biotechnology-derived pharmaceuticals whose patents have expired. They are large, complex molecules produced by living organisms. No two cell lines, developed independently, can be considered identical. Biopharmaceuticals may be quite sensitive to manufacturing process changes. The follow-on manufacturer does not usually have access to the originator's molecular clone and original cell bank, nor to the exact fermentation and purification process. There may therefore be differences in impurities and breakdown products, which could have clinical safety implications. This has created a concern that copies of biologics might perform differently from the original branded version of the drug.

Similar concerns also apply to any production changes by the maker of the original branded version. In the European Union, a specially adapted approval procedure has been developed for certain protein drugs termed "similar biological medicinal products," that is, biosimilars. This procedure is based on a thorough demonstration of comparability of the similar product to an existing approved product. Several EMEA Guidelines on this subject exist. The EMEA "Guideline on Similar Biological Medicinal Products containing Biotechnology-Derived Proteins as Active Substance: Non-Clinical and Clinical Issues" (published in 2006) includes current nonclinical requirements (pharmacotoxicological assessment) for biosimilar product approval (42). The guideline indicates that the studies should be comparative in nature, designed to detect differences in response between the similar biological product and a reference medicinal product and not just the response per se. Nonclinical toxicity should be determined in at least one repeat dose toxicity study. Toxicokinetic measurements in the study should include determination of antibody titers, cross-reactivity, and neutralizing capacity. The duration of the study(ies) should be sufficiently long to allow detection of relevant differences in toxicity and immune responses between the similar biological medicinal product and the reference medicinal product. Normally, other routine toxicological studies such as safety pharmacology, reproduction toxicology, and mutagenicity and carcinogenicity studies are not required for similar biological medicinal products. In addition to the general biosimilar guideline on nonclinical (and clinical) issues, the EMEA has also published nonclinical/clinical guidelines for specific products. These include erythropoietins, low-molecular-weight heparins, interferon-α, granulocyte-colony stimulating factor, human growth hormone, and insulin. The EMEA Web site for Specific Guidelines for Human Medicinal Products, Multidisciplinary Guidelines should be consulted for specific Biosimilar guidances (43).

In the United States, the FDA refers to biosimilars as follow-on protein products (FOPPs). As of 2009, the FDA considered FOPPs as new biologics and therefore subject to Biologic

Licensing Applications regulatory filings. Thus, biosimilars are subject to nonclinical requirements in accordance with the ICH S6 guideline. The FDA is awaiting new legislation to address FOPP registrations in a manner similar to biosimilar marketing authorizations in the Europepan Union.

In 2008, the EMEA published a draft guidance entitled "Guideline on Repeated Dose Toxicity" (44). The guidance addressed the choice of species, group size, husbandry, dose regimen, duration of dosing, route of administration, potential target organs, dose–response relationships, and reversal of toxicity. Also mentioned was that the substance used in the repeat dose toxicity studies should present a pattern of impurities similar to the product intended for humans (clinical trials and marketing). Marketed products with different impurities would need to be qualified in accordance with current ICH Q3 Guidelines for drug substance and drug product impurities (45,46). The repeat dose toxicity guideline also addressed new excipients (see below) that would need to be studied for toxicity. In addition, the species chosen (two, one of which is a nonrodent) should be based on pharmacodynamics, toxicokinetics, and biotransformation. Exposure to major human metabolites should be ensured which has recently been addressed by the FDA document "Guidance for Industry: Safety Testing of Drug Metabolites" (47). This guidance addresses safety testing for human metabolites formed at greater than 10% of parent drug exposure at steady state. Toxicology requirements could include a repeat dose toxicity study with supporting toxicokinetics in one species, a developmental toxicology study (potentially in one species), and in vitro genotoxicity assays (with a subsequent carcinogenicity study if the genotoxicity results are positive).

In 2005, the FDA published "Guidance for Industry: Nonclinical Studies for the Safety Evaluation of Pharmaceutical Excipients" (48). For new excipients that are intended for use in drug products that are intended for extended use, repeat dose toxicology studies are required in two species (one of which is a nonrodent). In addition, safety pharmacology, genotoxicity, and developmental and reproductive toxicology studies would be needed (and possibly carcinogenicity studies). The guideline paraphrases the draft M3(R2) document regarding acute toxicity studies (29). They would not be necessary if alternative studies are performed, such as a rodent pilot dose-escalation study used to select dose levels for the rodent repeat dose toxicology study. The guideline also indicates the duration of repeat dose toxicology studies necessary to support the NCE based on the intended duration of use.

In 2006, the FDA published "Guidance for Industry: Nonclinical Safety Evaluation of Drug or Biologic Combinations" (49). This guidance provided recommendations on nonclinical approaches to support the clinical study and approval of fixed-dose combination products (FDCs, i.e., one dosage form), copackaged products, and some adjunctive therapies (collectively referred to as "combinations"). The guidance indicated that combination toxicology studies should include several dose levels of the combination and a high dose of each drug alone. Dose combinations should allow for additive or synergistic effects without unacceptable toxicity in the high dose groups. One species may be acceptable if toxicity in the species has high concordance with human toxicity, the toxicities are similar among species, or one species is a more relevant model for human risk based on pharmacokinetics, ADME, or pharmacodynamics. For combinations where one drug is an NCE and one has been marketed previously, a bridging study of up to 90 days with the combination in the most appropriate species should be performed. The guidance also indicated that mechanistic studies should be conducted if the mechanism(s) of the drug interaction(s) identified in a bridging study (e.g., PK, PD, or overlapping toxicity) is not apparent. Since the drug ratio may change during drug development, the toxicity studies should be designed to provide adequate margins of safety for future clinical studies.

Recently (2008), the EMEA published a similar guidance entitled "Guideline on the Non-Clinical Development of Fixed Combinations of Medicinal Products" (50). It indicated that for any nonclinical combination study, doses should be based on interspecies differences in PK and PD in order to encompass the clinical situation as much as possible, in terms of systemic exposure of animals to the individual drugs and in relation to the PD effects (while avoiding high dose effects that may be irrelevant to human safety assessment). When the fixed combination of drugs under development includes compounds for which there is sufficiently documented human experience of their individual and combined use, safety studies in animals are not generally required. For the safety evaluation of a fixed combination of drugs not approved as combination therapy, toxicology (and safety pharmacology) studies may be necessary based on anticipated reactions between the two drugs and the range of concentrations and exposures covered in the previously conducted studies for each agent. For NCEs to be used in a fixed combination, or for combined therapy, one approach would be to conduct a nonclinical program for each of the new substances, together with combination bridging studies. Alternatively, an extensive nonclinical program could be conducted on the combination together with a limited number of studies on each NCE. For repeat dose toxicology combination studies, one species may be sufficient. For a fixed combination containing approved compounds for long-term use, a three month combination toxicity study in one species, supported by toxicokinetic measurements, would be considered sufficient.

In 2006, the FDA published "Guidance for Industry: Nonclinical Safety Evaluation of Pediatric Drug Products" (51). The guidance discusses the role and timing of animal studies in the nonclinical safety evaluation of therapeutics intended for treatment of pediatric patients. Juvenile animal studies may assist in identifying postnatal developmental toxicities that are not assessed adequately in reproductive toxicity assessments and that may not be adequately and safely tested in pediatric clinical trials. Organ systems at highest risk for drug toxicity are those that undergo significant postnatal development including brain, kidneys, lungs, immune system, reproductive system, skeletal system, and GI system. Studies conducted in juvenile animals to support the safety of pediatric therapeutics include modified repeat dose toxicology studies in an appropriate species, studies designed to address a specific safety concern, or modified peri- and postnatal developmental study protocols. Toxicologic endpoints should include measurement of overall growth, clinical observations, organ weights, gross and microscopic evaluations, assessment of sexual maturation, and neurobehavioral testing.

Several other guidelines regarding repeat dose toxicology studies exist within the FDA and the EMEA Web sites. The home pages for the FDA Center for Drug Evaluation and Research (CDER) Guidance Documents (52), and the EMEA Scientific Guidelines for Human Medicinal Products, Non-Clinical Guidelines (53) and Multidisciplinary Guidelines (43) should be consulted. For example, the EMEA document "Evaluation of Control Samples in Nonclinical Safety Studies: Checking for Contamination with the Test Substance" indicates that for pivotal studies, which include toxicokinetic evaluations, control samples should be collected and analyzed (54). For nonrodents, all control samples should be analyzed. For rodents, at least the control samples collected in the proximity of T_{max} should be studied. The FDA 2008 draft guidance entitled "Nonclinical Safety Evaluation of Reformulated Drug Products and Products Intended for Administration by an Alternate Route" provides repeat dose toxicology recommendations for a product to be given by an alternate route of administration for which the product was not previously approved (even if there is no change in the composition of the formulation) and for a proposed new formulation (55). Other EMEA Guidelines address the nonclinical requirements for the development of vaccines (56) and gene therapy products (3), both of which can require repeat dose toxicity studies.

EXPERIMENTAL DESIGNS

Single Dose Toxicity Studies in Rodents

It has been suggested anecdotally and occasionally in published guidelines (29,31) that an understanding of the acute toxicity of a drug is useful in establishing treatment following an overdose. However, this information is rarely used in the emergency room. In most cases, supportive measures are used to treat patients after accidental or intentional poisoning. The primary use of acute rodent (typically the rat) toxicity data is to aid in selection of dose levels for a short-term, for example, 14 day, repeat dose range-finding (DRF) study. The results of the 14 day rodent DRF study would typically be used to select doses for a GLP-compliant 4-week toxicity study in that species and for a 14 day nonrodent DRF study (typically using canines or nonhuman primates). For most drugs, single dose toxicity study data have little relevance in selecting an initial dose for Phase I studies in humans. An exception is cytotoxic anticancer drugs for which there is a well-defined paradigm relating acute toxicity to an appropriate initial clinical dose. This is described in the 2008 ICH S9 Draft Consensus Guideline entitled "Nonclinical Evaluation of Anticancer Pharmaceuticals" (57). As stated, a common approach is

Table 2 Acute Oral Toxicity Study in Rats

GLP status	Possibly
Species	CD rats (or other appropriate strain)
Age	Approx. 7- to 8-wk old at dosing
Sex	Males and females
Quarantine period	1 wk (approx.)
Route of administration	Gavage for oral compounds
Dose levels	10 mg/kg 100 mg/kg 500 mg/kg Vehicle Control (not routinely necessary unless vehicle is unusual)
No. of animals/sex/group	3
Dosing regimen	Once by gavage Based on initial results: 1. additional dose levels may be added 2. additional animals may be dosed to result in 5 rats/sex/group
Measurements	
Body weight	Pretest, day 1, twice weekly thereafter, and day 15
Mortality check	Twice daily, at least 6 hr apart
Clinical signs	At least three times daily on day 1 after dosing (~1–2, 3–4 and 4–6 hr after dosing) and once daily in the AM thereafter
Necropsy	Day 15 (gross necropsies, but tissues not typically collected)
Statistical analyses	Body weights (Mean ± SDs)

to set a start clinical dose at 1/10th the Severely Toxic Dose in 10% of the animals (STD_{10}) in a rodent single dose toxicity study. If the nonrodent is the more sensitive species, then 1/6th the Highest Non-Severely Toxic Dose (HNSTD) is considered an appropriate start dose. The HNSTD is defined as the highest dose level that does not produce evidence of lethality, life-threatening toxicities, or irreversible findings. Another therapeutic area in which single dose studies have a direct relevance to clinical development is general anesthetics, which are typically used in single-exposure mode and have relatively low therapeutic ratios (58).

Based on the previous discussion, two scenarios emerge for the conduct of acute-/single dose toxicity studies in rodents (preferably rats) during the drug development process. When the drug is to be given clinically on a daily basis over an extended period of time, an acute toxicity study in a rodent species may be useful in selecting doses for the subsequently conducted 14-day rodent DRF study. Under this scenario, the study design for a non-GLP study is provided in Table 2. It is similar to the previously discussed OECD Test Guideline 420: Acute Oral Toxicity—Fixed Dose Procedure, although the dose range has been compressed and lowered based on anticipated biologic activity of NCEs under development to treat disease. Alternatively, a 3- to 4-day repeat dose pilot study in rats (doses based on rodent pharmacology study data) can be conducted as a prelude to the 14-day DRF study, obviating the need for an acute toxicity study.

In a second scenario, typically, the drug would be given once clinically such as for medical imaging agents, therapeutic radiopharmaceuticals, and gene therapy medicinal products. Accordingly, the GLP-compliant rodent and nonrodent toxicology studies for IND submission would be single dose administration. Prior to a Phase II study, short-term repeat dose studies in both species would be necessary. The single dose toxicity study designs should be comprehensive, similar to those which are subsequently described for repeat dose toxicity studies.

Repeat Dose Toxicity Studies in Rodents and Nonrodents

Repeat dose toxicology studies typically include 1- to 2-week pilot DRF studies, 4- and 13-week subchronic studies, and 6- to 12-month chronic studies. For this latter category, generally 6 month rat studies and 9- to 12-month dog studies are conducted. While a 9 month study is consistent with the current ICH S4 Guideline (37), studies on a first-in-class agent may require a 12-month study for FDA submissions. Examples of 13-week repeat dose toxicology study designs in rats and dogs are shown in Table 3 and Table 4, respectively. A discussion on study design factors follows.

Table 3 Thirteen-Week Oral (Gavage) Toxicity and Toxicokinetics Study in Rats with a Four-Week Recovery Period

GLP status	Yes
Species	CD rats (or other appropriate strain)
Age	Approx. 7- to 8-wk old at dosing
Sex	Males and females
Dose levels	Vehicle control Low dose Mid dose High dose
No. of animals/sex/group	10 for core toxicity animals (all groups) 9 for TK satellite animals[a] (drug-treated groups) 5 for recovery animals (high dose and vehicle-control groups)
Dosing regimen	Once daily in the AM by gavage for 13 weeks
Quarantine period	1 to 2 weeks
Recovery period	Four weeks from the end of the 13-week treatment period
Measurements	
Test article analyses	ID and stability (purity before and after the treatment period)
Dosage formulation analyses	One set of formulations in each of weeks 1, 5, 9, and 13
Body weight	Week 1, day 1, weekly thereafter and at termination
Food consumption	Weekly starting week 1
Mortality check	Twice daily, at least 6 hours apart.
Clinical signs	Once daily, ~1 to 2 hr after dosing (unless T_{max} differs)
Clinical observations (physicals)	Weekly, starting week 1
Clinical pathology	Weeks 5, 9, and 13 and at the end of the recovery period
Toxicokinetics	Conducted in the 9 satellite rats/sex/drug-treated groups on day 1, and in weeks 4 and 13. Typical time-points are 0, 0.5, 1, 2, 4, 8, 12 and 24 hr postdosing. Blood is collected at 3 time-points per rat; therefore, blood is collected at alternate time-points from 2 to 3 groups of 3 rats/sex/dose level (*n* of 3).
Ophthalmology	All animals pretest and in week 13 and at the end of the recovery period. The cornea, iris, lens, fundus, and anterior and posterior chambers are examined.
Necropsy	Week 14 and at the end of the recovery period for all survivors, and for all found dead and moribund kill animals.
Organ weights	Major organs
Histopathology	All tissues in all control and high dose core study animals at the end of the treatment period. Target tissues/organs examined in mid- and low-dose core study animals and in recovery animals.

[a]Satellite animals for plasma drug levels, and discarded from study after their last blood samples, are collected in week 13.

Dose Selection

Dose selection is a critical aspect of repeat dose toxicity study design. The high dose level should be selected to produce frank target organ or systemic toxicity. If this cannot be achieved, the maximum dose, as limited by nonspecific effects, pharmacologic effects (e.g., hypoglycemia) or the volume to be administered should be used. The low dose should be the highest dose that is not expected to cause toxic effects, that is, the NOAEL. At this dose, pharmacodynamic effects related to the desired therapeutic effect may be seen, based on the animal model chosen. Exposure (C_{max}, AUC) and dose (mg/kg/day) at the low dose level should ideally be a small multiple of that anticipated when the drug is given to humans. The intermediate (mid) dose is often set geometrically or arithmetically between the low and high dose levels. This should assist in the further characterization of the dose–response relationship of any effects that may be seen in the study.

Species and Strain/Breed

The majority of guidelines for repeat dose toxicology testing require the use of two species, one of which is a nonrodent. GLP toxicology studies conducted for regulatory submissions are generally performed in rats and a large animal species such as dogs or nonhuman primates.

Table 4 Thirteen-Week Oral Toxicity and Toxicokinetics Study in Dogs with a Four-Week Recovery Period

GLP status	Yes
Dose levels	Vehicle control Low dose Mid dose High dose
No. of animals/sex/group	Four for core toxicity animals (all groups) Three for recovery animals (high dose and vehicle control groups)
Dosing regimen	Once daily by gavage or capsule/tablet administration for 13 weeks
Quarantine period	Approximately two weeks
Recovery period	Four weeks following the end of the 13 week treatment period
Measurements	
Test Article analyses	ID and stability (purity before and after the treatment period)
Dosage formulation analyses	One set of formulations in each of weeks 1, 5, 9, and 13
Body weight	Week 1, day 1, weekly thereafter and at termination
Food consumption	Weekly (24-hr assessment) starting week 1
Mortality check	Twice daily, at least 6 hr apart
Clinical signs	Once daily, ~1 to 2 hr after dosing (unless T_{max} differs)
Clinical observations (physicals)	Weekly, starting week 1.
Clinical pathology	Pretest and weeks 5, 9, and 13, and end of the recovery period
Toxicokinetics	Conducted on day 1, and in weeks 4 and 13. Typical time-points are 0, 0.5, 1, 2, 4, 8, 12, and 24 hr following dosing.
Ophthalmology	All animals pretest and in week 13 and at the end of the recovery period. The cornea, iris, lens, fundus, and anterior and posterior chambers are examined.
Necropsy	Week 14 and at the end of the recovery period for all survivors, and for all found dead and moribund kill animals.
Organ weights	Major organs
Histopathology	All tissues in all core study animals at the end of the treatment period. Target tissues examined in all recovery animals.

In vitro hepatic metabolism systems during the drug discovery/early development phases are often used to determine which large animal species metabolizes the NCE most like humans. Alternatively, dogs can be initiated in IND-directed studies with the possibility of shifting to nonhuman primates if early clinical trials indicate that metabolites formed in vivo favor this species rather than canines.

There are numerous strains of rats and mice typically used for pharmacologic and toxicologic research. For rats, Sprague-Dawley, CD, F344 and Wistar strains are often used for GLP toxicology programs. For mice, CD-1, BALB/c, B6C3F1, and Swiss Webster strains are often used. Beagles are almost exclusively used when canines are the species selected. For nonhuman primates, either Rhesus or Cynomolgus species are generally used, although some investigators favor the Marmoset based on its size (i.e., because less drug is used).

Research animals should be housed and cared for in accordance with local guidelines such as the U.S. "Guide for the Care and Use of Laboratory Animals" (59), the U.K. Home Office "Code of Practice for the Housing and Care of Animals" (60), and the "Canadian Council on Animal Care Guideline: Procurement of Animals used in Science" (61). A review of the European Union animal care standards and legislation was published recently (62). Accreditation by the Association for Assessment and Accreditation of Laboratory Animal Care International (AAALAC) also ensures adequate animal care and use programs. Certified diets, which indicate upper limits for various contaminants, are required for GLP-compliant programs. Animals can be identified by a variety of procedures including tattoos, ear tags, and subcutaneously implanted microchip transponders. These latter devices allow for the use of a hand-held scanner, which can be linked electronically directly to computer programs. These microchips can also relay certain physiologic information such as body temperature.

Age

For GLP single dose and repeat dose toxicology programs, rats should be 7 to 8 weeks of age and dogs should be 7 to 9 months of age at the start of dosing. In a recent Society of Toxicologic Pathology Position Paper, it was strongly recommended that rodents be at least 9 weeks old and dogs at least 9 months of age at the end of study (sexually mature) as it may not be possible to assess testicular toxicity in immature animals (63). This was recently reiterated for dogs that should be at least 10 months old at necropsy for accurate assessment of drug-related effects on spermatogenesis (64). This avoids low testes weights and microscopic testicular findings such as hypospermatogenesis and atrophy/hypoplasia, which are more common in younger versus older dogs. In the case of nonhuman primates where males do not mature until after 4 years of age (65), supply problems dictate that most studies are conducted with immature animals. Thus, adverse effects on spermatogenesis may or may not be detected in this age of animal.

Sex

Both males and females are, with few exceptions, included in repeat dose toxicology programs. An exception would include the testing of NCEs being developed for one sex only, such as hormonal agents (e.g., estrogens and compounds with estrogen-like activity).

Drug Administration

For rats, a gavage needle is generally used to deliver drugs orally. For dogs, oral drug administration can include gavage via a gastric tube as well as administration of the drug formulated into tablets or as a simple mixture with an excipient (e.g., lactose) in gelatin capsules. The choice of method is influenced by the characteristics of the drug and dose formulation, and experience with the methods in particular laboratories. For primates, nasogastric administration is usually employed, although tablets can be given orally via the back end of a gavage tube, the tip of which can be cut diagonally to allow placement of the tablet prior to instilling the gavage tube flared-end first.

Number of Animals and Groups

For rodent (typically rat) GLP toxicology studies, each core study group should have at least 10 animals/sex. This is typical for 4- to 13-week studies whereas 15 rats/sex/group are often employed for 26 week, that is, chronic, toxicity studies. For nonrodents, three to four animals/sex/group are used for GLP toxicology programs of 4 to 13 weeks' duration. For a chronic dog toxicity study (9 to 12 months in duration), five animals/sex/group should be used. Additional groups/animals would be required for ancillary studies, for example, interim necropsy(ies), toxicokinetic sampling for rats, recovery phase, etc. For recovery phases, it may only be necessary to include animals from the high dose group (and corresponding controls) to demonstrate the reversibility of induced changes. Additional animals for toxicokinetics would not generally be necessary for canine or nonhuman primate studies.

The majority of GLP repeat dose toxicology studies should include at least three dose groups and a vehicle-control group as previously discussed. If an unfamiliar vehicle is used, an additional untreated control group may be necessary. For some studies, the inclusion of a reference/comparator (positive or negative control) group may be useful (or required, e.g., for biosimilar toxicity studies) to aid in the evaluation of any changes induced by the NCE.

Control group data should be compared with corresponding data from previous studies to ensure that the results are consistent with the experience of the laboratory for animals of that age, sex, strain, and species. If a control data set is outside the expected range or shows unexpected variability, the reasons should be identified. For certain data sets, such as clinical pathology tests, it is important to establish that the assay procedures gave the expected results with known controls and standards. Quality control procedures should also be in place for various phases of all data collection processes.

Drug Substance

The drug substance, that is, "test article" in GLP vernacular, should be well characterized in accordance with GLP regulations as discussed in chapter 10. Briefly, this includes identification, purity, and stability. The same batch should ideally be used throughout a study. This may not be feasible for longer term studies and for studies with large animal species. Although GMP material is not absolutely necessary for GLP studies, the impurity profile of the drug used in toxicology studies should be similar to that used in clinical programs. This will ensure that the toxicologic potential of any impurity subsequently administered to man has been adequately assessed. If the impurity profile changes, for example, if a new synthetic route is employed, then new impurities may need to be qualified. ICH Q3A (45), Q3B (46), and Q3C (66) address drug substance, drug product, and residual solvent impurities, respectively.

Drug Formulation

Dose formulations for oral toxicology studies use various vehicles such as water, carboxymethylcellulose (CMC), methylcellulose, Tween 80, and mixtures thereof (e.g., 0.5% CMC/0.4% Tween 80). Dose formulations should be well characterized prior to their use in accordance with GLP regulations. Formulations should be evaluated for homogeneity and stability (both chemical and physical) for the full range of concentrations, storage conditions, and duration of use to be employed in the animal studies. Any change in formulation between studies should be accompanied by a kinetic and biological comparison to assure that expected drug exposures are achieved.

For intravenous administration studies, blood compatibility of a sterile formulation is a prerequisite. For repeated use formulations, injection site tolerance is also necessary. These properties are desirable for formulations for other parenteral (e.g., intramuscular or subcutaneous) routes to avoid potential confounding effects associated with infection and local inflammation.

Recovery Phase

The potential reversibility of any drug-induced changes is of considerable importance in assessing the significance of toxicologic effects. Accordingly, recovery phase observation periods should be included in some repeat dose toxicity studies. The duration of the recovery phase seldom exceeds the duration of drug administration but should reflect the cellular kinetics and control mechanisms that may be involved in the reversal of induced changes. For example, a hormonal agent that leads to a disruption of the oestrus cycle in dogs may require a prolonged period of observation to demonstrate a return to normal cycling. In this example, hormonal measurements to demonstrate return to normal could be employed as a surrogate endpoint, which may significantly reduce the duration of the recovery period. For many studies, one month is an acceptable recovery period. This may need to be lengthened to allow for clearance of drugs with a long half-life.

Statistical Considerations

The statistical power of a study is generally limited by study design. The primary variable is the number of animals/sex/group. For smaller species, for example, rodents, used in groups of 10/sex, the use of conventional parametric or nonparametric analyses or tests for trends as appropriate provides a reasonable assurance that differences observed between groups of animals are not due to chance. For larger species, that is, canines and nonhuman primates, typically three to four animals/sex/group are used. This limits the statistical power of the various analyses. However, the judicious use of animals precludes increasing their numbers. Statistical analyses of data from large animal studies can be used to compare results

1. across groups during the treatment period and
2. between treatment periods and baseline measurements.

Statistical procedures can also be used to ensure random allocation of animals to treatment groups during the quarantine/pretest period. This procedure can be totally random, or more frequently, a restricted randomization procedure can be based on some parameter such as body weight. For large animals, it is also advisable to avoid litter mates (at least of the same sex) within a treatment group. In some instances, cages or pens may be arranged to minimize any positional effects that could influence the study. A replicate system is commonly employed to achieve this. A replicate contains representative animals from each group/sex and may be considered to be a cross-section of the study. To minimize experimental or environmental factors during various procedures (e.g., blood sampling, ECG recording, necropsy), randomization procedures can be used. Drug administration, however, should be given on a group basis to minimize potential errors. Also histopathologic evaluation of tissues by the pathologist is typically performed by a read of the control tissues first to identify any normal variations, that is, background for that shipment of animals. If potential target organs are subsequently identified in the treated groups, the slides should be "blinded" and reread to verify treatment-related effects.

The occurrence of statistically significant effects should be only part of the process of identifying drug-induced changes. Drug-related effects are not always statistically significant. Conversely, statistically significant differences from the control group occurring in the drug-treated groups are not always indicative of a drug-related effect. Examples include statistically significant effects occurring at low- and mid-dose groups but not at the high dose group, and significant effects seen for parameters with very small variability such as serum electrolytes.

Measurements

Clinical Observations

Body weight (BW) and food consumption (FC) measurements provide an assessment on the general health of the animals. BWs should be assessed shortly after the animals arrive in the vivarium to determine the suitability of the shipment. If the shipment is large, a reasonably random sample may be sufficient. BWs should be collected a few days prior to treatment initiation for use in randomization to treatment groups, on day 1 (first day of treatment), and weekly thereafter. For a short-term study such as two weeks in duration, biweekly BW measurements may be appropriate.

FC measurements are typically collected as a range for rodents, that is, days 1 to 8. For canines, a shorter range (three to four days) or a 24-hour assessment is often performed. While rodents are fed ad libitum, a fixed daily amount of food such as 350 or 400 g is typically provided to dogs. In this latter instance, nonparametric statistical procedures should be used as an upper limit precludes normally distributed data. Rodents may be fed either pellets or a powdered diet, the latter of which may allow for more accurate measurements. Dogs are typically provided pellets. For nonhuman primates, quantitative assessment of FC is difficult. A specific number of biscuits is often provided. At the end of the feeding period, semiquantitative assessments can be recorded as none, about half, or all of the feed having been consumed. Group-housed animals obviously make assessments of individual FC measurements somewhat problematic.

Daily clinical signs are assessed in GLP toxicology studies, typically one to two hours after treatment (unless the C_{max} is known to occur at a different time postdosing). Mortality checks should be performed twice daily, at least 6 hours apart. Physical examinations are often conducted at least once weekly, typically when BWs are collected. The examination should include an assessment of general behavior, skin, fur, mucous membranes, eyes, respiration, CNS, and somatomotor function.

Clinical Pathology

Clinical pathology measurements provide information on organ system function, electrolytes, and carbohydrate and protein metabolism. Excellent reviews on the principles of clinical pathology pertaining to animal toxicology studies can be found in toxicology (67,68), animal clinical chemistry (69), and animal hematology (70) texts. Measurements should generally include routine hematology parameters (RBC indices, WBC total and differential counts, platelet counts, and coagulation parameters), serum indices of liver and kidney function, serum electrolytes, lipids, and urinalyses assessments (specific gravity, pH, "dipstick" semiquantitative measurements, etc.).

In 1996, a Position Paper from The Joint Scientific Committee for International Harmonization of Clinical Pathology Testing, American Association for Clinical Chemistry Division of Animal Clinical Chemistry was published (71). It indicated that clinical pathology parameters should be assessed for rodents and nonrodents at the end of the treatment period and at the end of a recovery period if included. For nonrodents, an interim sampling period should also be included. Clinical pathology testing in rodent studies includes one or more interim sampling period(s) based on study length, drug studied, toxicity issues, etc. Specific tests were recommended in the publication including two scientifically appropriate blood tests each for hepatocellular evaluation and for hepatobiliary evaluation.

Table 5 Clinical Pathology Parameters Assessed in Repeat Dose Toxicity Studies

Clinical chemistry	Hematology and coagulation	Urinalysis
Alanine aminotransferase (ALT)	Erythrocyte count	Appearance
Albumin	Hematocrit	Color
Albumin/globulin ratio (calc.)	Hemoglobin	Semiqualitative
Alkaline phosphatase	Mean corpuscular hemoglobin (MCH)	Bilirubin
Aspartate aminotransferase (AST)[a]	Mean corpuscular hemoglobin concentration (MCHC)	Glucose
Bilirubin, total	Mean corpuscular volume (MCV)	Ketone
Calcium	RBC morphology	Leukocytes
Chloride	Reticulocyte count	Nitrite
Cholesterol	Leukocyte count, total and differential	Occult blood
Creatinine	Platelet count	pH
Globulin (calculated)	Prothrombin time	Protein
Glucose	Activated partial thromboplastin time	Urobilinogen
γ-Glutamyltranspeptidase	Fibrinogen (possibly)	Specific gravity
Inorganic phosphate		Microscopic examination of spun sediment
Potassium		Volume (rodents)
Protein, total		
Sodium		
Sorbitol dehydrogenase (rats)		
Triglycerides		
Urea nitrogen (BUN)		

[a]AST is not specific for hepatotoxicity in rats.

In 2005, the Regulatory Affairs Committee of the American Society for Veterinary Clinical Pathology published a Position Paper on the use of preclinical clinical pathology endpoints in the assessment of drug-induced hepatotoxicity in animals (72). It supported the Division of Animal Clinical Chemistry publication cited above and further clarified the interpretation of appropriate clinical pathology indicators of hepatic injury, etc. In 2008, the EMEA published a draft document entitled "Non-clinical Guideline on Drug-Induced Hepatotoxicity" (73). It discussed drug-related effects in hepatocellular and hepatobiliary clinical pathology parameters in rodent and nonrodent toxicity studies, individual animal results from nonrodent toxicology studies, and mechanistic in vitro and in vivo models. In a comprehensive retrospective examination of published data, the guideline indicated that nonclinical concordance to a clinical hepatotoxic event was lowest with a rodent finding only, increased when the finding was in nonrodent only, and was greatest when a finding was found in both species.

Serum troponins (cardiac Tn; cTn) have been recently studied as potential biomarkers of drug-induced cardiac toxicity in animal toxicology studies based on their usefulness in clinical cardiology (74,75). Troponin complex is a heteromeric protein consisting of three subunits, which is important in the regulation of skeletal and cardiac muscle contraction. Recent validation studies in rats, dogs, and monkeys should allow for its usefulness in safety assessment programs (75).

Bone marrow cytology slides should be prepared from each animal at termination in addition to blood and urine sampling. Based on the above-mentioned publications and the author's experience, clinical pathology parameters that should be routinely assessed in repeat dose toxicity studies are shown in Table 5.

Ophthalmologic Examinations

Ophthalmologic examinations should be conducted in all main study and recovery rodents and nonrodents in repeat dose toxicity studies. An indirect ophthalmoscope and a slit lamp are routinely used. The eyes should be examined before the start of the study, at study termination, and at the end of the recovery period if included. The cornea, iris, lens, fundus, and anterior and posterior chambers are routinely examined.

Electrocardiography

An assessment of cardiovascular function from an ECG can be obtained routinely from dogs and primates. Blood pressure from large animals can also be collected but the results are highly variable. Technical difficulties preclude the routine collection of ECG data from rodents. Data should be collected at prestudy, study termination, and the end of the recovery period if included. Interim periods may be appropriate for longer term studies. The timing of ECG collection during the repeat dose treatment period should consider whether the observed effects are acute, that is, collected at $\sim T_{max}$, or cumulative, that is, collected prior to dosing.

Toxicokinetics/Plasma Drug Levels

The toxicokinetics and metabolic fate of an NCE can aid in understanding differences in toxicologic responses observed under different conditions of exposure to the drug and its metabolites. Exposure achieved in repeat dose toxicology studies can be established by monitoring plasma levels of the drug and metabolite(s) to estimate C_{max} and AUC. These data should be collected at the start, possibly mid-study and the end of the treatment period. Changes in exposure over time can occur as a result of enzyme induction, changes in clearance, or bioaccumulation. Exposure at the no observed toxic effect level can allow comparison with that expected or achieved in humans at steady state at therapeutic dose levels. The occurrence of nonlinear kinetics may signal enzyme induction, enzyme inhibition, or saturable absorption or clearance. Species differences in protein-binding and receptor properties may potentially confound interspecies comparisons. Differences in affinity for specific organs or tissues may result in higher levels than those measured in plasma, which may or may not be correlated with observed effects. For a drug with slow clearance and long half-life, exposure increases (C_{max} and AUC) during the initial phases of a study may lead to the occurrence of effects only after several doses have been administered.

Occasional differences in sensitivity between male and female rats seen in drug toxicology studies are often related to exposure. Female rats may show higher AUC and C_{max} levels than males as a result of lower hepatic CYP450 levels. Female rats have 10% to 30% less total CYP450 activity compared with male rats (76). There is therefore a slower rate of detoxification of the NCE. This phenomenon is typically not observed in dogs or nonhuman primates.

In repeat dose toxicity studies, usually the drug is administered once daily unless the clinical regimen dictates more frequent dosing. For NCEs with a long half-life, slow clearance, or prolonged absorption or distribution phases, less frequent administration may be appropriate.

Organ Weights

Organ weights are measured routinely in repeat dose toxicity studies to aid in the interpretation of gross and microscopic evaluations. The Society of Toxicologic Pathology recently published a Position Paper entitled "Organ Weight Recommendations for Toxicology Studies" (77). It recommended that liver, heart, kidneys, brain, testes, and adrenal glands be weighed for all repeat dose toxicology studies. Thyroid gland and pituitary glands should also be weighed (preferably postfixation) for all species except mice. Reproductive organs are most valuable when weighed in sexually mature animals. The paper also suggests that epididymides and prostate weights should be assessed in rat studies and possibly in nonrodents (on a case-by-case basis).

Organ/body weight ratios are typically evaluated statistically. Organ/brain weight ratio analyses, however, may be more useful when terminal body weights are affected by drug treatment. Drugs that alter body weight do not generally alter brain weight (78). Organ/brain weight ratio analyses may also be appropriate for nonrodents where body weights may vary more within treatment groups than in rodents, that is, brain weights help to normalize animal-to-animal variability (79).

Apparent drug-related effects on female reproductive organs (weights and histopathology) may represent secondary effects of treatment on the reproductive cycle rather than a direct drug-related effect. Splenic weights in nonrodents can be affected by the method of anesthesia and exsanguination as blood pooling can occur following splenic contraction

Table 6 Organs for Weighing in Repeat Dose Toxicity Studies

Liver	Spleen
Kidneys	Thymus
Heart	Thyroid/parathyroid glands[a]
Adrenals	Pituitary gland[a]
Brain	Lungs
Testes	Ovaries

[a]Postfixation weight.

secondary to barbiturate anesthesia. The Society of Toxicologic Pathology recommends that organ weights not be collected from moribund or found dead animals as there are too many extraneous variables including the lack of concurrent controls (80). Based on the above discussion, Table 6 contains a list of organs for weighing in rat and nonrodent repeat dose toxicity studies.

Pathology

In repeat dose (and appropriate single dose) toxicity studies, all nonrodents and all core study and recovery rodents are necropsied at protocol-indicated time-points. Extensive necropsies are performed and numerous tissues/organs are collected and fixed in 10% neutral buffered formalin (NBF). Exceptions include special fixatives for certain tissues such as testes (Bouin's; picric acid) and eyes (Davidson's—4% Paraformaldehyde in 0.1 M Phosphate Buffer, or 4% glutaraldehyde/10% NBF). Paraformaldehyde is also used for in situ (perfusion) fixation of the central nervous system.

The previously discussed 1981 OECD toxicology test guidelines specified tissues to be collected for histopathologic evaluation. Many pharmaceutical companies follow these guidances, which are modified in-house as necessary. The various subsequent guidelines from ICH, EMEA, and FDA did not generally indicate specifics with regard to protocol design. In 2003, the Society of Toxicologic Pathology published a Position Paper regarding recommended tissues for histopathologic examination in repeat dose toxicity and carcinogenicity studies (81). For nonrodents, all collected tissues from all groups sacrificed at treatment termination should be examined. For rodents, all tissues from all animals in the high dose and control groups sacrificed at treatment termination and any unscheduled deaths should be examined. Target organs/tissues identified in the high dose group should be examined from the mid- and low-dose group animals. For all rodent and nonrodent studies, target organs/tissues should be examined in all recovery group animals. Based on the above discussion, Table 7 contains a comprehensive list of tissues for histopathologic evaluation in rat and nonrodent repeat dose toxicity studies.

"Omics"

Traditional toxicology study designs may not address adequately mechanistic issues, which can provide insight into the relevance of findings from animal studies to humans. Recent technological advances in genomics (toxicogenomics, proteomics, and metabolomics) have resulted in new tools for studying such issues (82–84). The resulting information can be used to understand the mechanisms of and biomarkers for selected aspects of a toxicity profile. The screening of genes for changes in expression patterns in response to drug treatment (toxicogenomics) has already been used in various investigations to clarify toxicologic observations in animals and their potential relevance to clinical studies. It is anticipated that future studies in proteomics and metabolomics will also result in useful information to aid in the prediction of clinical toxicity from nonclinical toxicology studies.

Variables in Toxicology Testing

There are numerous variables in biologic research that can influence the results. In pharmacology and toxicology research, many of these variables have been well studied. Variables inherent to the animals under study include species, strain, breeding facility, animal husbandry, age, sex, and body weight. Age and body weight can be significant variables as older/heavier rats have

Table 7 Tissues for Histopathologic Evaluation in Repeat Dose Toxicity Studies

Adrenal gland	Pancreas
Aorta	Parathyroid gland
Bone w/bone marrow[a]	Peripheral nerve, e.g., sciatic
Brain (fore, mid, and hind)	Pituitary
Cecum	Prostate
Cervix	Rectum
Colon	Salivary gland
Duodenum	Seminal vesicle[b]
Epididymis	Skeletal muscle
Esophagus	Skin
Eye	Spinal cord (cervical, thoracic, and lumbar)
Gallbladder[c]	Spleen
Harderian gland[d]	Stomach
Heart	Testis
Ileum	Thymus
Jejunum	Thyroid gland
Kidney	Tongue
Liver	Trachea
Lung/Bronchi	Urinary bladder
Lymph node(s)	Uterus
Mammary gland	Vagina
Ovary	Other organs or tissues with gross lesions
	Tissue masses

[a]Femur for rats; rib or sternum for nonrodents.
[b]N/A for dogs.
[c]N/A for rats.
[d]N/A for nonrodents.

a slightly greater adipose tissue concentration, thereby resulting in a shift in the disposition of drugs between aqueous and lipid-soluble compartments in the body (85,86). Breeding facilities and animal husbandry practices can also influence the effects of drugs in animals. The same strain of rat from two separate breeders, and even from separate facilities from the same breeder, can show quantitative but typically not qualitative differences in response to drug treatment. In principle, species should be selected that exhibit the pharmacologic response characteristics of the drug (not always possible to establish) and have a metabolic profile similar to humans. In practice, the choice of rodent species is inevitably confined to rats and mice, although other laboratory rodents could be considered.

Drug-related reductions in food intake can result in secondary responses usually considered as a direct drug effect. Moderate caloric restriction of about 20% to 30% enhances survival in long-term rodent studies, reduces the onset and severity of certain spontaneous nonneoplastic lesions, and delays the onset of certain tumors such as those of the mammary and pituitary gland (87,88). The Society of Toxicologic Pathology accordingly published a Position Paper on diet as a variable in rodent toxicology and carcinogenicity studies, which recommended dietary caloric restriction or variably formulated diets whenever scientifically appropriate (89). The effects of food intake on the testis and accessory sex glands were recently examined in maturing rats (90). The study was designed to assess the effects of food restriction in rats during different periods of maturation on testicular histology, reproductive organ weights, and plasma hormones. Food restriction (21 g/day; 20–36% reduction compared to ad libitum) for two or six weeks caused a reduction in body and reproductive organ weights, lower plasma testosterone levels, and degeneration of pachytene spermatocytes in stage VII. The duration of lower food intake (6 weeks vs. 2 weeks) was the greatest determinant of testicular degeneration whereas a younger initial age of lower food intake (6 or 8 weeks vs. 10 or 12 weeks) was most influential on testosterone levels.

Stress is a confounding variable in toxicology research and can be related to housing practices (single vs. multiple housing), "high dose" effects, dose volume, vehicle, and blood volume removal for small animals. For example, group-housed mice have been shown to be significantly more sensitive to the acute toxicity of d-amphetamine than animals housed individually (91,92). It was subsequently hypothesized that chronic stress may increase the responsiveness of the brain to the pharmacologic and neurotoxic effects of amphetamines (93). It has also been suggested that individual housing, that is, isolation, may actually be more stressful than group housing as individual caging of male rats produces a stress response associated with increased serum corticosterone (CORT) levels (94). Individually housed rats have demonstrated an increase in pituitary tumors (*pars distalis* adenoma) and a decrease in Leydig interstitial cell tumors compared to group-housed animals (94). This decrease in Leydig cell tumors may be related to a reduction in testosterone secretion as increases in serum CORT inhibit testosterone biosynthetic enzyme activity (95).

Dose formulation volume can also influence the results of toxicology studies. In 1995, the Toxicology Subcommittee of the Association of the British Pharmaceutical Industry presented its view regarding dose formulation volume limits, expressed as common practice rather than absolute maxima (96). In 2001, the results of an initiative between the European Federation of Pharmaceutical Industries Association (EFPIA) and the European Centre for the Validation of Alternative Methods (ECVAM) were published, which described "good practices" related to dose volume administration, vehicles, and removal of blood volumes (97). In both papers, dose volumes of 10 to 20 ml/kg for the oral administration of drug formulations to rats were recommended. It is the experience of this author that daily administration of 20 ml/kg of aqueous vehicles to rats over extended periods of time poses the potential for aspiration (see below). Oral toxicity studies in rats probably represent the largest subset of general toxicology studies as well as being the most sensitive to experimental variables such as dose volume.

The relationship between gavage administration of various vehicles and induction of the stress response in rats, as defined by increased plasma CORT levels, was recently examined (98). Male rats received a single gavage administration of various vehicles containing Higgins India Ink, blood was collected one hour after dosing for measurement of plasma CORT, and necropsies were performed four hours postdosing. Corn oil at 20 ml/kg, but not the same dose volume of 1% methylcellulose/0.2% Tween 80 or water, elevated plasma CORT levels (with volume-dependent increases at higher volumes). This was not due to aspiration, which occurred after the administration of 20 ml/kg of water or 1% methylcellulose/0.2% Tween 80 but not corn oil (possibly related to vehicle viscosity). The stress response produced by corn oil also occurred after gavage administration of sesame, soybean and peanut oils. The results suggested that dose volumes for gavage administration in the rat generally should not exceed 10 ml/kg per dose administration.

In rats, stress has been shown to be associated with thymic involution, leucopenia, adrenocortical hypertrophy, and suppression of immune function (99–102). Those observations in rodent toxicology studies might incorrectly be ascribed to drug effect(s) as opposed to responses secondary to elevated serum CORT levels.

Stress in rodent toxicology studies can also be secondary to blood collection (and other routine handling) procedures including the use of anesthetics (103–105). Blood volume withdrawal is a critical factor in rodent repeat dose toxicology studies. Blood volumes necessary for periodic clinical pathology tests and for repetitive plasma drug level measurements for toxicokinetics (often up to eight time-points over 24 hours for IV administered drugs) are generally insignificant for canines and nonhuman primates. For rats, however, separate satellite groups for toxicokinetics, including subsets for separate groups of time-points, is typically used as repetitive bleeding over a 24-hour period can result in anemia followed by bone marrow regenerative activity. This can be problematic for drugs that are cytotoxic to the bone marrow such as certain classes of anticancer agents, resulting in confounding hematology data.

CONCLUSIONS

A satisfactory assessment of the toxicologic potential of an NCE requires evaluation of the entire nonclinical program. This typically includes at least two repeat dose toxicity studies of differing length in rodents and nonrodents by the intended clinical route. These data are supplemented by relevant information from pharmacology, pharmacokinetics, ADME, safety pharmacology, developmental and reproductive toxicity, genotoxicity and carcinogenicity studies, etc. Additional investigative toxicology studies are frequently conducted in animals with the objective of elucidating the pathogenesis of lesions or effects that have been identified as drug-related. Data arising from such investigative programs can be paramount in satisfying regulatory authorities' concerns that the toxicologic potential is acceptable in the context of the expected benefits from the therapeutic use of the drug.

Consideration of the results obtained during repeat dose toxicity studies is greatly facilitated by an understanding of the pharmacokinetics and metabolism of the drug in the animal species and man. In many instances, species differences in sensitivity to the pharmacologic action or toxicologic effects of a drug are due to differences in tissue or receptor exposure secondary to differences in the absorption, distribution, metabolism, and elimination of the drug. The closer the disposition characteristics of the drug in one or more species used in the toxicology studies are to man, the more confident will be the extrapolation of the animal data to humans.

In 2002, the relative predictiveness of animal toxicology studies with regard to human toxicity (HT) identified during clinical development was reported (106). Data were obtained from 12 pharmaceutical companies and encompassed 150 NCEs. The results showed that the HT concordance rate was 71% for toxicities seen in both rodent and nonrodent studies whereas toxicities in rodents or nonrodents alone were predictive for only 43% or 63% of HTs, respectively. Accordingly, repeat dose toxicity studies in two species, one of which is a nonrodent, are warranted. In concert with worldwide guidelines from ICH, FDA, EMEA, etc., comprehensive nonclinical development programs, which include appropriately designed repeat dose toxicity studies in multiple species, coupled with good scientific judgment, should allow for a thorough understanding of the potential toxicity profile of an NCE in Man.

REFERENCES

1. FDA Guidance for Industry; Developing Medical Imaging Drug and Biological Products; Part 1: Conducting Safety Assessments, June 2004. http://www.fda.gov/cder/guidance/5742prt1.pdf. Accessed February 2009.
2. FDA Draft Guidance for Industry; Nonclinical Evaluation of Late Radiation Toxicity of Therapeutic Radiopharmaceuticals,June 2005. http://www.fda.gov/cder/guidance/6815dft.pdf. Accessed February 2009.
3. EMEA Guideline on the non-clinical studies required before first clinical use of gene therapy medicinal products,March 2007. http://www.emea.europa.eu/pdfs/human/genetherapy/12545906en.pdf. Accessed February 2009.
4. Pharmaceutical Manufacturers Association. Guidelines for the Assessment of Drug and Medical Device Safety in Animals. Washington, DC: Pharmaceutical Manufacturers Association, 1977.
5. Drug Approval and Licensing Procedures in Japan. Tokyo: Yakugo Jiho, 1991.
6. Guidelines for toxicity studies of drugs manual 1990. Tokyo: Yakuji Nippo, 1991.
7. Trevan JW. The error of determination of toxicity. Proc R Soc Lond B Biol Sci 1927; 101B:483–514.
8. Litchfield JT, Wilcoxon FJ. A simplified method of evaluating dose-effect experiments. J Pharmacol Exp Ther 1949; 96:99–113.
9. Miller LC, Tainter ML. Estimation of the ED50 and its error by means of logarithmic-probit graph paper. Proc Soc Exp Biol Med 1944; 57:261–264.
10. Weil CS. Tables for convenient calculation of median effective doe (LD50 or ED50) and instruction in their use. Biometrics 1952; 8:249–263.
11. ICCVAM Test Method Evaluation Report (TMER): In Vitro Cytotoxicity Test Methods for Estimating Starting Doses For Acute Oral Systemic Toxicity Testing. NIH Publication No. 07–4519. http://iccvam.niehs.nih.gov/docs/acutetox_docs/BRD_TMER/TMERmain_Nov2006.pdf. Accessed February 2009.
12. OECD Series on Testing and Assessment Number 33. Harmonized Integrated Classification System for Human Health and Environmental Hazards of Chemical Substances and Mixtures, August 14, 2001. http://www.oecd.org/dataoecd/48/51/37182285.pdf. Accessed February 2009.
13. Ukelis U, Kramer PJ, Olejniczak K, et al. Replacement of *in vivo* acute oral toxicity studies by *in vitro* cytotoxicity methods: Opportunities, limits and regulatory status. Reg Toxicol Pharmacol 2008; 51:108–118.
14. OECD Guidelines for the Testing of Chemicals Section 4: Health Effects, 1981. http://oberon.sourceoecd.org/vl=1542855/cl=24/nw=1/rpsv/cw/vhosts/oecdjournals/1607310x/v1n4/contp1–1.htm. Accessed February 2009.

15. CPMP Guideline on Single dose toxicity, August 1987. http://www.emea.europa.eu/pdfs/human/swp/3bs1aen.pdf. Accessed February 2009.
16. FDA CDER Guidance for Industry: Single Dose Acute Toxicity Testing for Pharmaceuticals. August 1996. http://www.fda.gov/downloads/Drugs/GuidanceComplianceRegulatoryInformation/Guidances/ucm079270.pdf. Accessed February, 2009.
17. British toxicology society working party on toxicity. Special report: A new approach to the classification of substances and preparations on the basis of their acute toxicity. Human Toxicol 1984; 3:85–92.
18. Van den Heuval MJ, Dayan AD, Shillaker RO. Evaluation of the BTS approach to the testing of substances and preparations for their acute toxicity. Human Toxicol 1987; 6:279–291.
19. Van den Heuval MJ, Clark DG, Fielder RJ. The international validation of a fixed-dose procedure as an alternative to the classical LD50 test. Fd Chem Toxicol 1990; 28:469–482.
20. Roll R, Hoffer-Bosse T, Kayser D. New perspectives in acute toxicity testing of chemicals. Toxicol Lett 1986; 31(suppl):86.
21. Schlede E, Mischke U, Roll R, et al. A national validation study of the acute-toxic-class method—An alternative to the LD50 test. Arch Toxicol 1992; 66(7):455–470.
22. Schlede E, Mischke U, Diener W, et al. The international validation study of the acute toxic class method (oral). Arch Toxicol 1995; 69:659–670.
23. Bruce RD. An up-and-down procedure for acute toxicity testing. Fundam Appl Toxicol 1985; 5:151–157.
24. Bruce RD. A confirmatory study of the up-and-down method for acute oral toxicity testing. Fundam Appl Toxicol 1987; 8:97–100.
25. ICH Guideline S6: Preclinical Safety Evaluation of Biotechnology-Derived Pharmaceuticals, July 1997. http://www.ich.org/LOB/media/MEDIA503.pdf. Accessed February 2009.
26. ICH Final Concept Paper—S6(R1): Preclinical Safety Evaluation of Biotechnology-Derived Pharmaceuticals (Revision of the ICH S6 Guideline), June 2008. http://www.ich.org/LOB/media/MEDIA4733.pdf. Accessed February 2009.
27. M3(R1): Maintenance of the ICH Guideline on Non-Clinical Safety Studies for the Conduct of Human Clinical Trials for pharmaceuticals, November 2000. http://www.ich.org/LOB/media/MEDIA506.pdf. Accessed February 2009.
28. ICH M3(R2): Guidance on Nonclinical Safety Studies for the Conduct of Human Clinical Trials and Marketing Authorization for Pharmaceuticals, June 2008. http://www.ich.org/LOB/media/MEDIA4744.pdf. Accessed February 2009.
29. Robinson S, Delongeas JL, Donald E, et al. A European pharmaceutical company initiative challenging the regulatory requirement for acute toxicity studies in pharmaceutical drug development. Regul Toxicol Pharmacol 2008; 50(3):345–352.
30. EMEA CHMP Concept Paper on Single Dose/Acute Toxicity, June 2008. http://www.emea.europa.eu/pdfs/human/swp/30241308en.pdf. Accessed February 2009.
31. Baldrick P. Safety evaluation to support first-in-man investigations II: Toxicology studies. Reg Toxicol Pharmacol 2008; 51:230–236.
32. EMEA CHMP Position Paper on Non-Clinical Safety Studies to Support Clinical Trials with a Single Microdose, June 2004. http://www.emea.europa.eu/pdfs/human/swp/259902en.pdf. Accessed February 2009.
33. FDA CDER Guidance for Industry, Investigators, and Reviewers: Exploratory IND Studies, January 2006. http://www.fda.gov/cder/guidance/7086fnl.pdf. Accessed February 2009.
34. Rules Governing Medicinal Products in the European Community (1989). Vol. III Guidelines on the quality, safety and efficacy of medicinal products for human use. Office for Official Publications of the European Communities, Brussels. ISBN: 9282596192.
35. D'Arcy PF, Harron DW, eds. Proceedings of the First International Conference on Harmonization. Belfast, UK: Queen's University, 1991.
36. ICH Guideline S3A: Note for Guidance on Toxicokinetics: The Assessment of Systemic Exposure in Toxicity Studies, October 1994. http://www.ich.org/LOB/media/MEDIA495.pdf. Accessed February 2009.
37. ICH Guideline S4: Duration of Chronic Toxicity Testing in Animals (Rodent and Non-Rodent Toxicity Testing), September 1998. http://www.ich.org/LOB/media/MEDIA497.pdf. Accessed February 2009.
38. DeGeorge JJ, Meyers LL, Takahaski M, et al. The duration of non-rodent toxicity studies for pharmaceuticals. Toxicol Sci 1999; 49:143–155.
39. ICH Guideline S6: Preclinical Safety Evaluation of Biotechnology-Derived Pharmaceuticals, July 1997. http://www.ich.org/LOB/media/MEDIA503.pdf. Accessed February 2009.
40. EMEA CPMP Concept Paper on the Development of a Committee for Proprietary Medicinal Products Guideline on Comparability of Biotechnology-Derived Products, June 1998. http://www.emea.europa.eu/pdfs/human/bwp/111398en.pdf. Accessed February 2009.

41. FDA Guidance Concerning Demonstration of Comparability of Human Biological Products, Including Therapeutic Biotechnology-Derived Products, April 1996. http://www.fda.gov/cder/guidance/compare.htm. Accessed February 2009.
42. EMEA CHMP Guideline on Similar Biological Medicinal Products containing Biotechnology-Derived Proteins as Active Substance: Non-Clinical and Clinical Issues, February 2006. http://www.emea.europa.eu/pdfs/human/biosimilar/4283205en.pdf. Accessed February 2009.
43. EMEA Guidelines for Human Medicinal Products, Multidisciplinary Guidelines. http://www.emea.europa.eu/htms/human/humanguidelines/multidiscipline.htm. Accessed January 2009.
44. EMEA CHMP Guideline on Repeated Dose Toxicity, February 2008. http://www.emea.europa.eu/pdfs/human/swp/48831307en.pdf. Accessed February, 2009.
45. ICH Guideline Q3A(R2): Impurities in New Drug Substances, October 2006. http://www.ich.org/LOB/media/MEDIA422.pdf. Accessed February 2009.
46. ICH Guideline Q3B(R2): Impurities in New Drug Products, June 2006. http://www.ich.org/LOB/media/MEDIA421.pdf. Accessed February 2009.
47. FDA CDER Guidance for Industry: Safety Testing of Drug Metabolites, February 2008. http://www.fda.gov/cder/guidance/6897fnl.pdf. Accessed February, 2009.
48. FDA CDER/CBER Guidance for Industry: Nonclinical Studies for the Safety Evaluation of Pharmaceutical Excipients, May 2005. http://www.fda.gov/cder/guidance/5544fnl.pdf. Accessed February 2009.
49. FDA CDER Guidance for Industry: Nonclinical Safety Evaluation of Drug or Biologic Combinations, March 2006. http://www.fda.gov/cder/guidance/6714fnl.pdf. Accessed February 2009.
50. EMEA CHMP Guideline on the Non-Clinical Development of Fixed Combinations of Medicinal Products, January 2008. http://www.emea.europa.eu/pdfs/human/swp/25849805enfin.pdf. Accessed February 2009.
51. FDA CDER Guidance for Industry: Nonclinical Safety Evaluation of Pediatric Drug Products, February 2006. http://www.fda.gov/cder/guidance/5671fnl.pdf. Accessed February 2009.
52. Center for Drug Evaluation and Research, Food and Drug Administration Guidelines. http://www.fda.gov/cder/guidance. Accessed February 2009.
53. EMEA Scientific Guidelines for Human Medicinal Products, Nonclinical Guidelines. http://www.emea.europa.eu/htms/human/humanguidelines/nonclinical.htm. Accessed February 2009.
54. EMEA CHMP Guideline on Evaluation of Control Samples in Nonclinical Safety Studies: Checking for Contamination with the Test Substance, March 2005. http://www.emea.europa.eu/pdfs/human/swp/109404en.pdf. Accessed February 2009.
55. FDA CDER Guidance on Nonclinical Safety Evaluation of Reformulated Drug Products and Products Intended for Administration by an Alternate Route (Draft Guidance), March 2008. http://www.fda.gov/cder/guidance/7686dft.pdf. Accessed February 2009.
56. EMEA CPMP Note for guidance on Preclinical Pharmacological and Toxicological Testing of Vaccines, December 1997. http://www.emea.europa.eu/pdfs/human/swp/046595en.pdf. Accessed February 2009.
57. ICH Guideline S9 Nonclinical Evaluation for Anticancer Pharmaceuticals,November 2008. http://www.ich.org/LOB/media/MEDIA4917.pdf. Accessed February 2009.
58. Brunton LL, Lazo JS, Parker KL, eds. General anesthetics. In: Goodman & Gilman's The Pharmacological Basis of Therapeutics, 11th ed. New York: McGraw Hill, 2006.
59. Institute of Laboratory Animal Care, Commission for Life Sciences Guide for the Care and Use of Laboratory Animals, National Research Council, 1996. http://oacu.od.nih.gov/regs/guide/guide.pdf. Accessed February 2009.
60. UK Home Office for the Code of Practice for the Housing and Care of Animals, March 2005. http://scienceandresearch.homeoffice.gov.uk/animal-research/publications-and-reference/publications/code-of-practice/code-of-practice-housing-care/?view=Standard&pubID=428573. Accessed February 2009.
61. Canadian Council on Animal Care Guideline: Procurement of Animals used in Science, 2008. http://www.ccac.ca/en/CCAC_Programs/Guidelines_Policies/PDFs/Procurement_Guidelines_EN.pdf. Accessed February 2009.
62. Anderson DB. Standards of Accommodation and Care for Animals used in Scientific Procedures in Europe. AATEX 2007; 14(special issue):51–55.
63. Lanning LL, Creasy, DM, Chapin RE, et al. Recommended approaches for the evaluation of testicular and epididymal toxicity. Toxicol Pathol 2002; 30(4):507–520.
64. Goedken MJ, Kerlin RL, Morton D. Spontaneous and age-related testicular findings in Beagle dogs. Toxicol Pathol 2005; 36:465–471.
65. Mitchell DJ, Creasy DM. Sexual maturity and age of animals in preclinical studies to support first in man studies. Abstracts of Papers, Annual Meeting of the Society of Toxicology, Charlotte, NC, March 25–29, 2007. Toxicologist 2007; 96(1):58.

66. ICH Guideline Q3C(R3) Impurities: Guideline for Residual Solvents, revised November 2005. http://www.ich.org/LOB/media/MEDIA423.pdf. Accessed February 2009.
67. Hall RL, Everds NE. Principles of clinical pathology for toxicology studies. In: Hayes WA, ed. Principles and Methods in Toxicology, 5th ed. New York: Informa Healthcare, 2008:1317–1358.
68. Levine BS. Animal clinical pathology. In: Derelanko MJ, Hollinger MA, eds. Handbook of Toxicology, 2nd ed. Boca Raton, FL: CRC Press, 2002:741–768.
69. Loeb WF, Quimby FW. The Clinical Chemistry of Laboratory Animals, 2nd ed. Philadelphia, PA: Taylor & Francis, 1999.
70. Feldman BF. Zinkl JG, Jain NC. Schalm's Veterinary Hematology, 5th ed. Baltimore, MD: Lippincott Williams & Wilkins, 2000.
71. Weingand K, Brown G, Hall R, et al. Harmonization of animal clinical pathology testing in toxicity and safety studies. Fundam Appl Toxicol 1996; 29(2):198–201.
72. Boone L, Meyer D, Cusick P, et al. Selection and interpretation of clinical pathology indicators of hepatic injury in preclinical studies. Vet Clin Pathol 2005; 34:182–188.
73. Draft EMEA CHMP Non-Clinical Guideline on Drug-Induced Hepatotoxicity, January 2008. http://www.emea.europa.eu/pdfs/human/swp/15011506en.pdf. Accessed February 2009.
74. Wallace KB, Hausner E, Herman E, et al. Serum troponins as biomarkers of drug-induced cardiac toxicity. Toxicol Pathol 2004; 32:106–121.
75. Schultze AE, Konrad RJ, Credille KM, et al. Ultrasensitive cross-species measurement of cardiac troponin-I using the erenna immunoassay system. Toxicol Pathol 2008; 36:777–782.
76. Kedderis GL, Mugford CA. Sex-dependent metabolism of xenobiotics. Drug Metab Rev 1998; 30:441–498.
77. Sellers RS, Morton D, Michael B, et al. Society of toxicologic pathology position paper: Organ weight recommendations for toxicology studies. Toxicol Pathol 2007; 35:751–755.
78. Wilson NH, Hardisty JF, Hayes JR. Short-Term, subchronic, and chronic toxicology studies. In: Hayes WA, ed. Principles and Methods in Toxicology, 5th ed. New York: Informa Healthcare, 2008:1223–1264.
79. Baily SA, Zidell RH, Perry RW. Relationship between organ weight and body/brain weight in the rat: What is the best analytical endpoint. Toxicol Pathol 2004; 32:448–466.
80. Michael B, Yano B, Sellers RS, et al. Evaluation of organ weights for rodent and non-rodent toxicity studies: A review of regulatory guidelines and a survey of current practices. Toxicol Pathol 2007; 35:742–750.
81. Bregman CL, Adler RR, Morton DG, et al. Recommended tissue list for histopathologic examination in repeat-dose toxicity and carcinogenicity studies: A proposal of the society of toxicologic pathology (STP). Toxicol Pathol 2003; 31:252–253.
82. Foster WR, Chen SJ, He A, et al. A retrospective analysis of toxicogenomics in the safety assessment of drug candidates. Toxicol Pathol 2007; 35:621–635.
83. Wetmore BA, Merrick BA. Toxicoproteomics: Proteomics applied to toxicology and pathology. Toxicol Pathol 2004; 32:619–634.
84. Xu EY, Schaefer WH, Xu Q. Metabolomics in pharmaceutical research and development: Metabolites, mechanisms and pathways. Curr Opin Drug Discov Devel 2009; 12:40–52.
85. Cavallini I, Marino MA, Tonello C, et al. The hydrolipidic ratio in age-related maturation of adipose tissues. Biomed Pharmacother 2006; 60:139–143.
86. Thomas MA, Rice HB, Weinstock D, et al. Effects of aging on food intake and body composition in rats. Physiol Behav 2002; 76:487–500.
87. Keenan KP, Ballam GC, Soper KA, et al. Diet, caloric restriction, and the rodent bioassay. Toxicol Sci 1999; 52(suppl):24–34.
88. Keenan KP, Laroque P, Ballam GC, et al. The effects of diet, *ad libitum* overfeeding, and moderate dietary restriction on rodent bioassay: The uncontrolled variable in safety assessment. Toxicol Pathol 1996; 24:757–768.
89. Nold JB, Kennan KP, Nyska A, et al. Society of toxicologic pathology position paper: Diet as a variable in rodent toxicology and carcinogenicity studies. Toxicol Pathol 2001; 29:585–586.
90. Rehm S, White TE, Zahalka EA, et al. Effects of food restriction on testis and accessory sex glands in maturing rats. Toxicol Pathol 2008; 36:687–694.
91. Kataoka Y, Gomita Y, Fukuda T, et al. Effects of aggregation on methamphetamine toxicity in mice. Acta Med Okayama 1986; 40:121–126.
92. Vargas-Rivera J, Ortega-Corona BG, García-Pineda J, et al. Influence of previous housing history on the toxicity of amphetamine in aggregate mice. Arch Invest Med (Mex) 1990; 21:65–69.
93. Matuszewich L, Yamamoto BK. Chronic stress augments the long-term and acute affects of methamphetamine. Neuroscience 2004; 124:637–646.
94. Nyska A, Leininger JR, Maronpot RR, et al. Effect of individual versus group caging on the incidence of pituitary and Leydig cell tumors in F344 rats: Proposed mechanism. Med Hypotheses 1998; 50:525–529.

95. Hardy MP, Gao HB, Dong Q, et al. Stress hormone and male reproduction. Cell Tissue Res 2005; 322:147–153.
96. Hull RM. Guideline limit volumes for dosing animals in the preclinical stages of safety evaluation. Human Exp Toxicol 1995; 14:305–307.
97. Diehl KH, Hull R, Morton D, et al. A good practice guide to the administration of substances and removal of blood, including routes and volumes. J Appl Toxicol 2001; 21:15–23.
98. Brown AP, Dinger N, Levine BS. Stress produced by gavage administration in the rat. Contemp Top Lab Animal Sci 2000; 39:17–21.
99. Gruver AL, Sempowski GD. Cytokines, leptin, and stress-induced thymic atrophy. J Leukoc Biol. 2008; 84:915–923.
100. Anisman H. Cascading effects of stressors and inflammatory immune system activation: Implications for major depressive disorder. J Psychiatry Neurosci 2009; 34:4–20.
101. Harvey PW, Everett DJ, Springall CJ. Adrenal toxicology: A strategy for assessment of functional toxicity to the adrenal cortex and steroidogenesis. J Appl Toxicol 2007; 27:103–115.
102. Nyska A, Hester SD, Cooper RL, et al. Single or group housing altered hormonal physiology and affected pituitary and interstitial cell kinetics. J Toxicol Sci 2002; 27:449–457.
103. Balcombe JP, Barnard ND, Sandusky C. Laboratory routines cause animal stress. Contemp Top Lab Anim Sci 2004; 43:42–51.
104. Altholtz LY, Fowler KA, Badura LL, et al. Comparison of the stress response in rats to repeated isoflurane or CO_2: O_2 anesthesia used for restraint during serial blood collection via the jugular vein. J Am Assoc Lab Anim Sci 2006; 45:17–22.
105. Abelson KS, Adem B, Royo F, et al. High plasma corticosterone levels persist during frequent automatic blood sampling in rats. In Vivo 2005; 19:815–819.
106. Olsen H, Betton G, Rovinson D, et al. Concordance of the toxicity of pharmaceuticals in humans and in animals. Regul Toxicol Pharmacol 2000; 32:56–76.

25 Genotoxicity

Peter Kasper
Federal Institute for Drugs and Medical Devices, Bonn, Germany

INTRODUCTION

Genotoxicity testing of chemicals was developed during the 1950s and 1960s and much attention was given during this period to the development of appropriate test models for the detection of germ line mutagens that can induce heritable mutations. During the 1970s, the development of the so-called "Ames test," a bacterial gene mutation test, initiated a new understanding of the relevance of mutagens for humans. Early testing of hundreds of chemicals demonstrated a strong correlation of mutagenic activity in *Salmonella* with carcinogenicity in rodents (1,2), and hence, the primary purpose in mutagenicity testing changed to screening for carcinogens. However, it was already recognized at that time that a wider range of genetic alterations than those detectable in a bacterial gene mutation test may be involved in carcinogenesis, in particular effects leading to chromosomal damage. Thus, a battery approach comprising of tests that detect both gene mutation and chromosomal damage was recommended. This initial concept still underlies the testing batteries currently in use.

In the early 1980s, new legislation required systematic testing of chemical substances for genotoxic effects, and the Organization for Economy Cooperation and Development (OECD) developed a number of test guidelines in order to ensure minimum standards for test performances. At the same time, regulatory guidelines including those addressing genotoxicity testing of pharmaceuticals were implemented in different regions worldwide. However, there were many differences in the regulatory testing requirements in the regions. In this regard, the International Conference on Harmonisation of Technical Requirements for Registration of Pharmaceuticals for Human Use (ICH) was a major breakthrough as in the late 1990s this project achieved international consensus for the requirements of genotoxicity testing of pharmaceuticals (3). Two ICH genotoxicity guidelines were released: S2A, "Guidance on Specific Aspects of Regulatory Genotoxicity Tests" (4) and S2B, "Genotoxicity: A Standard Battery for Genotoxicity Testing of Pharmaceuticals" (5). Since then the recommended standard testing battery for pharmaceutical registration in the European Union (EU), Japan, and the United States is

(i) a test for gene mutation in bacteria,
(ii) an in vitro test with cytogenetic evaluation of chromosomal damage with mammalian cells *or* an in vitro mouse lymphoma assay, and
(iii) an in vivo test for chromosomal damage using rodent hematopoietic cells.

A process of revision of the ICH genotoxicity guidelines was started in 2006 after an additional decade of accumulating experience and knowledge specifically with testing pharmaceuticals. The use of newer test methods (such as the in vitro micronucleus assay, blood as a target tissue for micronucleus analysis in the rat, and the Comet assay for DNA strand breaks) and a more efficient use of resources (for example, by integrating genotoxicity endpoints into existing toxicology studies) are some of the key topics in the updating process of the guideline. The two existing ICH Guidelines S2A and S2B have been merged into one draft document titled "ICH S2R1: Guidance on Genotoxicity Testing and Data Interpretation for Pharmaceuticals intended for Human Use" (6). The revised guideline is expected to be implemented in the regulatory framework in the three ICH regions in the course of 2009. The principles of genotoxicity testing as outlined in this chapter are mainly based on the new version of the revised ICH S2 Guideline.

GENERAL PRINCIPLES

The terms "genotoxicity" and "mutagenicity" are often used interchangeably. However, genotoxicity as is used in this chapter is a broader term that refers not only to the induction of mutations but also includes processes that provide evidence of interaction with DNA or damage to DNA, for example, DNA adduct formation, sister chromatid exchange (SCE), DNA strand breaks, or DNA repair processes such as unscheduled DNA synthesis (UDS). It is important to note that the detection of such effects does not in itself provide direct evidence of inherited mutations.

It is generally agreed that, for a full evaluation of the genotoxic potential of a substance, data are required that address the three levels of mutations, namely

1. gene mutations, which are changes in the genetic code,
2. structural chromosome aberrations, which entail loss or translocation of chromosomal segments, and
3. genome mutations, which involve changes in chromosome numbers.

The different types of mutations are induced by different modes of action, although in practice most mutagens that induce gene mutations are also inducing structural chromosome aberrations. Since the different mutational endpoints currently cannot be covered by a single test model, it is necessary to combine several different assays for adequate genotoxicity assessment. There is a large variety of tests available that detect these mutagenic endpoints directly or indirectly by measuring effects that are precursors to these endpoints or are indirectly related to them (i.e., indicator tests). For regulatory purposes, a systematic "testing strategy" is usually advised, that is, an ordered approach is used employing a limited number of well-defined tests that complement each other in terms of endpoints.

A factor of critical importance in genotoxicity testing is the incorporation of adequate metabolizing systems. Since target cells for in vitro genotoxicity tests possess only a limited capacity for metabolism of xenobiotics, induced rat liver S9 mix is generally applied as an exogenous activation system for detecting potential promutagens. This approach has several limitations as standard S9 preparations contain a very limited selection of liver enzymes in active form and lacks, in particular, the detoxifying components of in vivo metabolism.

CURRENTLY USED TEST MODELS

Many test methods currently in use for regulatory purposes have an officially adopted OECD Guideline for the testing procedure, but for some, in particular, newer test methods such as the Comet assay or transgenic rodent gene mutation tests, this is not the case. A very valuable source of internationally harmonized test protocols including those for new test models is the International Workshop on Genotoxicity Testing (IWGT). Their published recommendations serve as important supplement to established regulatory guidelines (7). While both OECD and IWGT documents provide testing recommendations for chemical compounds irrespective of their intended use, certain attributes of pharmaceutical testing justify specific modifications for application to drugs. The ICH guidance therefore shows some differences from OECD and IWGT recommendations.

In Vitro Tests

Most short-term tests in bacteria and mammalian cell cultures have been designed primarily for hazard identification. Highly sensitive in vitro systems are preferred in order to ensure that there is a high probability of detecting a genotoxic potential. The advantage of this approach is that negative results will usually give sufficient evidence that the test compound does not induce genotoxic effects under in vivo conditions. However, weak effects or effects occurring only under nonphysiological conditions may not represent biologically relevant findings. The interpretation of findings from in vitro tests can therefore create a problem if one does not bear their high sensitivity in mind.

In Vitro Test for Gene Mutation in Bacteria

In vitro tests for gene mutation in bacteria are by far the most widely employed genotoxicity assays. All of the current guidelines require a bacterial mutagenicity test as part of the initial test battery. The test uses several strains of *Salmonella typhimurium* and *Escherichia coli,* which, as a consequence of mutations in genes coding for enzymes for the biosynthesis of an amino acid (histidine or tryptophan, respectively), have lost the ability to grow on agar lacking this amino acid. The principle of the test is that it detects the potential of the test article to induce reverse mutations in the histidine or tryptophan loci back to the normal function, that is, the ability to synthesize the amino acids. Revertants are able to grow in the absence of histidine or tryptophan and form visible colonies. The mutagenicity of a test substance is proportional to the number of colonies observed.

Each of the bacterial tester strains contains one of a number of specific mutations in the amino acid coding genes and can be reverted by either a base-pair substitution or by frameshift mutations dependent upon the strain concerned. Therefore, the type of mutation induced by a mutagenic test compound can be elucidated by examining the pattern of reversion response in several tester strains. In addition, the routinely used test strains have several features that make them more sensitive for the detection of mutations including increased cell permeability to large molecules and elimination of DNA repair systems or enhancement of error-prone DNA repair processes.

The ICH S2R1 Guideline recommends the use of a set of tester strains comprising the base substitution strains TA1535, TA100, and TA102 and the frameshift strains TA1537 (TA97 or TA97a) and TA98. Strain TA102 contains AT base pairs as the site of mutation as compared with the other *Salmonella* strains that detect GC base-pair damage. Instead of TA102, the *E. coli* strains WP2 uvrA or WP2 uvrA (pKM101), which also detect AT site mutations, may be employed. TA102 is the only excision-repair proficient strain and thus able to detect cross-linking agents such as mitomycin C, which require an intact excision-repair pathway to generate mutations. Because the different strains detect different types of mutations, it is likely that a particular mutagen will be positive in some strains and negative in other strains.

There are two common procedures used to conduct the bacterial gene mutation test: the plate incorporation method and the preincubation method. In the plate incorporation assay, the bacterial test strain, test material, and metabolic activation system (S9 mix) are added to a small volume of molten soft agar and plated immediately onto minimal medium. In the preincubation assay, the test compound is exposed to the bacteria in liquid medium for 20 to 60 minutes before addition of the top agar and plating. This method is reported to be more sensitive than the plate incorporation variant toward a number of different classes of genotoxic compounds (8).

Further information and advice on the protocols is given in the OECD Guideline (9) and the IWGT report (8). One difference of the new S2R1 ICH Guideline from the OECD and IWGT recommendations is that, based on experience with testing pharmaceuticals, it is no longer required to do two independent bacterial mutation tests. A single test is sufficient when it is clearly negative or positive and carried out with a fully adequate protocol including all strains with and without metabolic activation, a suitable dose range that fulfills criteria for top dose selection, and appropriate positive and negative controls. Also, for testing pharmaceuticals, either the plate incorporation or the preincubation method is appropriate for this single experiment. Equivocal or weak positive results may indicate the need to repeat the test, possibly with a modified protocol such as appropriate spacing of dose levels.

In Vitro Mammalian Cell Gene Mutation Assays

The common gene mutation assays employing mammalian cells detect forward mutations, that is, mutations that are associated with a loss of function (10). The target genes for measuring mutations in these assays are coding for enzymes involved in nucleic acid biosynthesis, the two most popular ones being thymidine kinase (TK) and hypoxanthine-guanine phosphoribosyl transferase (HPRT). Mutations in these genes abolish the enzyme functions and the ability to incorporate the respective nucleotides into the DNA. Because of this defect, mutants are also unable to incorporate toxic nucleotide analogues. Hence, cells deficient in TK function due to mutations in this gene are resistant to the cytotoxic effects of the pyrimidine analogue trifluorothymidine (TFT). Thymidine kinase proficient cells are sensitive to TFT, which causes the inhibition of cellular metabolism and halts further cell division. Thus, mutant cells are able to proliferate in the presence of TFT whereas normal cells, which contain thymidine kinase, are not. Similarly, cells deficient in HPRT are selected by resistance to 6-thioguanine. The mammalian cell lines most extensively employed for in vitro gene mutation assays are V79 and CHO Chinese hamster cell lines using HPRT gene as the genetic marker and the mouse lymphoma cell line

L5178Y using the heterozygous $TK^{+}/^{-}$ gene. The latter model is commonly referred to as "mouse lymphoma assay" (MLA).

The basic principle of the test method is common to the different cell test models. The cells are typically exposed to the test compound for three to six hours in the absence and the presence of a metabolic activation system followed by a splitting of the treated cultures to determine cytotoxicity and to allow phenotypic expression prior to mutant selection. A phenotypic expression period is necessary to eliminate the preexisting gene product. Cytotoxicity is usually determined by measuring the relative cloning efficiency (survival) or relative total growth of the cultures after the treatment period. Mutant frequency is determined by seeding known numbers of cells in medium containing the selective agent to detect mutant cells and in medium without selective agent to determine the cloning efficiency (viability). After a suitable incubation time, colonies are counted. The mutant frequency is derived from the number of mutant colonies in selective medium and the number of colonies in nonselective medium.

The mouse lymphoma TK assay can detect a wide range of genetic alterations including gene mutations, large-scale chromosomal aberrations, as well as recombination and mitotic nondisjunction (11,12). Therefore, it is much more sensitive to "genotoxic" compounds than other mammalian cell mutation assays such as the HPRT models, which are sensitive to gene mutations in a strict sense (including small deletions). Furthermore, the L5178Y cells used for the assay do not have a wild-type p53 gene; both alleles of the p53 gene are mutated (13). This means that the cells do not undergo apoptosis readily following primary DNA damage and, therefore, the damage cells survive to be detected as mutations. Two types of TFT-resistant mutant colonies are selected and these are designated large colonies and small (slow-growing) colonies. Molecular analysis has indicated that the large colonies tend to represent events within the gene such as base substitution, small deletions, and small insertions, whereas small colony mutants often involve large genetic changes frequently visible as chromosome aberrations (11). For the evaluation and interpretation of positive results in the MLA, it is of importance to provide a differentiation of large and small colonies. More guidance documents have been compiled by an IWGT MLA expert workgroup as to how assays should be evaluated for acceptability and for determining whether the response is positive or negative (14–18).

The ICH S2 Guideline does not request specifically a mammalian cell gene mutation assay as part of the standard testing battery. However, due to the broad spectrum of genetic events detected by the MLA, including chromosomal aberrations, the guideline recommends to include either an MLA or a chromosomal aberration test or a micronucleus test as part of the standard battery.

In Vitro Tests for Chromosomal Damage

The term "chromosomal damage" or "chromosome aberration" includes both alterations of chromosome structure and of chromosome number. Chemicals that induce structural chromosome aberrations are termed clastogens whereas those that induce numerical chromosome aberrations (gain or loss of single chromosomes) are termed aneugens. Testing for chromosomal damage can be done with two different approaches: either directly by cytogenetic analysis of metaphase chromosomes or indirectly by measurement of micronuclei.

Cytogenetic Metaphase Analysis

Cytogenetic analysis of chromosomes is usually done with cultured human peripheral lymphocytes or established cell lines, usually Chinese hamster fibroblasts such as CHO, V79, or CHL cells (19). The test generally involves exposure to the test material in the absence and the presence of an exogenous metabolic activation system for a few hours. After a period equivalent to about 1.5 normal cell-cycle lengths, the cells are fixed and prepared for chromosome analysis. Before fixation, the cells are exposed to a metaphase-arresting spindle poison such as colcemide in order to increase the number of metaphase cells available for evaluation. The focus of analysis is on easily detectable damages to the chromosomes such as terminal deletions, breaks, and exchanges. Such gross chromosome changes are likely to be lethal to the cell during following cell division and therefore the toxicological relevance of this endpoint is sometimes questioned. However, it must be pointed out that the observed and recorded chromosome aberrations are used as biomarkers of the presence of nonlethal, more subtle changes such as reciprocal translocations or small deletions, which cannot be observed readily with standard staining procedures but may play an important role in human genetic diseases and carcinogenesis.

Formation of DNA double-strand breaks is an essential step in the formation of chromosome aberrations. While this can result from exposure to reactive DNA-damaging compounds, there is a variety of indirect mechanisms that can also lead to changes of chromosome structure some of which are unlikely to be relevant under in vivo conditions (see "High False-Positive Rate with In Vitro Mammalian Cell Tests and How to Avoid It" section).

Although cytogenetic metaphase analysis allows for chromosome counting, this test is not designed to measure numerical aberrations and is not used routinely for that purpose. An increase in polyploidy, which is easily detectable and usually recorded, may indicate that a chemical has the potential to induce numerical aberrations. More appropriate tests such as micronucleus assay would be needed to clarify the biological meaning of such a finding.

In Vitro Micronucleus Test

The in vitro micronucleus test is a genotoxicity test for the detection of micronuclei (MN) in the cytoplasm of interphase cells. MN are small extranuclear chromatin-containing bodies, which are considerably smaller than the main nucleus but still can be easily identified by light microscopy after staining. MN are formed during cell division by acentric chromosome fragments (i.e., those lacking a centromere) or whole chromosomes that are not included in the main nuclei following DNA replication and nuclear division.

The in vitro micronucleus assay has been evaluated widely in international collaborative studies (20), is considered validated by ECVAM (21), and an OECD Guideline is in preparation (22). One important consideration in the performance of the in vitro micronucleus test is ensuring that the cells being scored have completed mitosis during the exposure period, otherwise false-negative results may be produced. The most efficient protocol is to use the actin polymerization inhibitor Cytochalasin B (cytoB). The addition of the cytoB prior to the targeted mitosis allows for the identification and selective analysis of micronucleus frequency in cells that have completed one mitosis because such cells are binucleate (20,23). The use of cytoB as a cytokinesis blocker is mandatory when human lymphocytes are used because cell cycle times can be quite variable within cultures and among donors and not all lymphocytes will respond to phytohemagglutinin used as mitogenic stimulus. When cell lines are used, other methods can be employed to determine if the cells being scored have divided such as the formation of clones when cells from permanent cell lines are treated and scored in situ on a microscope slide (24,25) or the measurement of population doubling (20). When there is an increase in micronucleus formation, different procedures can be used, such as immunochemical labeling of kinetochores or hybridization with centromeric/telomeric probes (fluorescence in situ hybridization; FISH), to determine if the increase was the result of clastogenic or aneugenic events.

The MN test is a more practical technique than metaphase analysis for chromosomal damage and, moreover, has the advantage to detect the activity of both clastogenic and aneugenic chemicals. Based on the advantages and the extensive experience with the in vitro micronucleus assay, it is now considered as acceptable mammalian cell test option in the revised ICH S2 Guideline.

High False-Positive Rate with In Vitro Mammalian Cell Tests and How to Avoid It

A severe problem, which emerged during recent years in regulatory testing, is the high rate of positive findings with the in vitro mammalian cell tests recommended in the ICH testing battery, that is, the tests for chromosome breakage and the MLA. The interpretation of the relevance of many of these in vitro findings has been debated frequently and extensive in vivo and mechanistic follow-up studies are usually required. Experience in the pharmaceutical industry is that in most cases follow-up tests demonstrate that the agents are not directly DNA damaging or mutagenic in other systems and no genotoxicity is demonstrable in vivo at maximal testable doses. Several recent reviews confirm oversensitivity and lack of specificity of these in vitro test models (26,27). A recent analysis of data of several hundred chemicals has highlighted the very low specificity of mammalian cell tests to predict the rodent bioassay results for carcinogenicity (26). When a standard two or three in vitro genotoxicity tests were performed, at least 80% of compounds gave a false-positive result. A similar outcome was obtained in an analysis by the U.S. Food and Drug Administration of an even larger database of chemicals (27).

There is now a large body of evidence demonstrating that the extreme experimental conditions currently recommended for the in vitro genotoxicity models are seen as a potential source of nonrelevant findings, for example, a high level of cytotoxicity, precipitating concentrations, and concentrations up to 5000 μg/ml or 10 mM that are very unlikely to be reached in vivo. Most often, positive results under such conditions do not reflect a DNA-damaging capability of the test compound but are a secondary response to perturbation of cell physiology. Since the genotoxicity seen under these high-dose conditions occurs through mechanisms that are not operating at lower doses, there is a threshold and the results are "nonrelevant" to in vivo conditions. The need to reconsider the conditions under which in vitro mammalian cell tests are conducted, as well as the evaluation of positive findings in these assays and their impact on risk assessment, has been explored recently by several organizations, such as the IWGT (28), the European Centre for the Validation of Alternative Methods (ECVAM) (29), and International Life Sciences Institute/Health and Environmental Sciences Institute Subcommittee on the Relevance and Follow up of Positive Results in In vitro Genetic Toxicity Testing (30).

The revised ICH S2 Guideline provides several recommendations directed at avoiding extreme culture conditions in mammalian cell tests that could lead indirectly to DNA damage, thus reducing the occurrence of nonrelevant positive results. The guideline includes

1. a reduction of the recommended top concentration from 10 mM to 1 mM when not limited by cytotoxicity or solubility;
2. omission of the statement that the concentrations used should extend into the range with visible precipitate, if needed, to obtain toxicity and recommendation instead that the top concentration analyzed should be at the apparent limits of solubility of the test compound in culture; and
3. reinforcement of the upper toxicity limits for all in vitro mammalian cell assays and emphasizes not to exceed these limits, that is, it should be avoided to exceed a reduction of about 50% in cell growth for in vitro cytogenetic assays for metaphase chromosome aberrations or for micronuclei, or a reduction of about 80% in Relative Total Growth (RTG) for the mouse lymphoma assay.

The use of appropriate measures of cytotoxicity is of importance since cytotoxicity estimates are known to differ depending on the endpoint used (31). Simple cell counts underestimate real toxicity and can result in selecting inappropriately high concentrations for testing with the potential for generation of irrelevant positive results. For in vitro cytogenetic assays, the ICH S2R1 Guideline recommends measurement of cell growth relative to control to assess toxicity such as relative population doubling (32) while RTG is the agreed standard of cytotoxicity measurement to be used for concentration selection and data evaluation in the MLA. For lymphocyte cultures, analysis of mitotic index (MI) for metaphase aberration assays or replication index (RI) based on cytokinesis block for in vitro micronucleus assays is considered appropriate measures to estimate cytotoxicity. Usually, a preliminary range-finding assay over a broad range of concentrations is performed to obtain an appropriate toxicity range. In the following main genotoxicity assay, the use of multiple concentrations that are closely spaced (less than twofold dilutions) would be advisable. Extra concentrations may be tested but usually not all need to be evaluated for genotoxicity. It is not intended that multiple experiments be carried out to reach exactly 50% reduction in growth or exactly 80% reduction in RTG.

In Vivo Tests

In vivo genotoxicity tests are included in most regulatory test batteries for two purposes. The first is to follow up positive results obtained in vitro in order to determine whether in vitro genotoxic activity can be expressed in the whole animal. There are several reasons why activity shown in vitro may not be observed in vivo such as lack of absorption, inability of the active metabolite to reach DNA, or rapid detoxication and elimination. The second purpose is to ensure that mutagens that are difficult to detect in vitro but may cause genotoxic effects in animals are recognized. In fact, the ICH S2 Guideline mentions that there is a small but significant number of genotoxic carcinogens (such as procarbazine, hydroquinone, urethane, and benzene) that are detected reliably by the bone marrow tests for chromosomal damage but have yielded negative or conflicting results in the in vitro tests (6). In a recent survey of industry data, some further examples were identified (33). The reasons why these compounds appear to be more readily detected in vivo than in vitro include metabolic differences, the influence of gut flora, higher exposure in vivo compared to in vitro, and effects on pharmacology, in particular folate depletion or receptor kinase inhibition (33). It is for these reasons that the ICH S2R1 Guideline has included in vivo testing as part of the core testing battery.

Tests for the Detection of Chromosomal Damage In Vivo

Testing of chromosomal damage induced in animals is usually performed in bone marrow of both rats and mice (34,35). Measurement of MN rather than direct cytogenetic analysis of chromosomal damage is the endpoint of choice because of better detection of aneugenic compounds. Another reason is the simplicity of scoring MN in the cytoplasm of the polychromatic erythrocytes from which the main nucleus has been extruded during maturation. The bone marrow micronucleus assay has been used routinely for decades and is currently the most widely used in vivo genotoxicity test in regulatory testing.

Peripheral erythrocytes in blood instead of bone marrow have been proposed as another appropriate target for MN analysis. A potential advantage of this approach in mice is that micronucleated erythrocytes reach a steady-state frequency in blood, thus allowing sampling at a single time (36–38). However, using peripheral blood erythrocytes from rat can be problematic since the rat spleen selectively removes micronucleated red blood cells from the circulation (39). The outcome of a comprehensive collaborative study (40) suggests that rat blood cells could be suitable for use in the micronucleus assay if scoring is restricted to reticulocytes, that is, the youngest fraction of polychromatic erythrocytes. Under these conditions, immature erythrocytes are scored prior to being captured by the spleen. The subpopulation of young reticulocytes can be identified by their staining characteristics, for example, by use of acridine orange supravital staining method (41) or, more elegantly, by targeting the transferrin receptor CD71, a specific cell surface marker that is active for only a short time after a reticulocyte enters the peripheral blood from the bone marrow. More recently, the use of flow cytometry scoring methods has increased further the efficiency and reproducibility of the rat peripheral blood assay (42–45). This method is now considered acceptable for routine regulatory use (46) and ready to replace the traditional bone marrow assays. There are several advantages of using peripheral blood sampling in the rat over the bone marrow–based MN studies:

- Less labor-intensive
- Require only microliter quantities of blood samples, which can be obtained without sacrificing the animals
- Would allow integration of this assay into routine toxicology and pharmacokinetic studies
- This combination would also allow kinetic aspects and general toxicological observations to be used for the interpretation of genotoxicity data
- Flow cytometry methods can simultaneously provide information about quantity of DNA present in the observed MN, thereby providing an indication of the mechanism by which MN are induced

Based on these recent advances in the development and evaluation in using rat peripheral blood reticulocytes as target cells for the micronucleus assay, the ICH S2R1 Guideline now accepts this method for assessing chromosomal damage in vivo as an alternative to the previously recommended MN test in rat bone marrow and in mouse bone marrow or blood. Moreover, the ICH Guideline encourages to integrate in vivo genotoxicity assays into repeat-dose toxicity studies so that it is not necessary to conduct an independent study for this purpose.

One concern when integrating genotoxicity measurement into repeat-dose toxicity studies is a possible reduced sensitivity as typically the top dose would be lower under these conditions than in acute studies. It is for this reason that the ICH S2R1 Guideline defines criteria to determine whether the top dose in multiple administration studies is appropriate for genotoxicity evaluation. Any one of the criteria listed below is sufficient to demonstrate that the top dose in a toxicology study (typically in rats) is appropriate for MN analysis and for other genotoxicity evaluation:

- Maximum feasible dose (MFD) based on physicochemical properties of the drug in the vehicle, provided the MFD in that vehicle is similar to that achievable with acute administration
- Limit dose of 1000 mg/kg for studies of 14 days or longer, if this is tolerated
- Maximal possible exposure demonstrated either by reaching a plateau/saturation in exposure or by compound accumulation
- Top dose is $\geq$ 50% of the top dose that would be used for acute administration, if such acute data are available

Selection of a top dose based only on an exposure margin, that is, multiple over clinical exposure, is not considered sufficient justification.

Test for Detection of DNA Strand Breaks In Vivo (Comet Assay)

The in vivo Comet assay, also referred to as the single cell gel electrophoresis (SCGE) assay, is a rapid technique for measuring DNA damage in eukaryote cells (47,48). This assay has gained wide popularity over recent years. The basic principle of the Comet assay is the migration of DNA in an agarose matrix under electrophoretic conditions. When damage to DNA occurs, a cell gets the appearance of a Comet with a head (the nuclear region) and a tail containing DNA fragments migrating in the direction of the anode. Increased DNA migration can indicate the induction of different types of DNA damage, such as single- and double-strand breaks, alkali-labile sites, and incomplete excision repair sites. Under certain conditions, the assay can also detect DNA–DNA and DNA–protein cross-linking, which appears as a relative decrease in DNA migration compared with concurrent controls. Several commercial and freeware image analysis systems are available for assessing DNA migration in single cells. The most common parameters analyzed are percentage DNA in the tail, tail length, and tail moment, which in the most commonly used version is defined as the product of the amount of DNA in the tail and the mean distance of migration in the tail (47).

Measurement of cytotoxicity is an important issue in the Comet assay since it is vital to distinguish DNA strand breakage that may be a direct consequence of genotoxic insult from DNA degradation resulting from cell death. The IWGT Comet expert workgroup considers a histopathological examination to assess levels of necrosis and apoptosis as the "gold standard" when positive results are seen in an in vivo Comet assay (49). Other methods for measuring cytotoxicity include tests for membrane integrity or metabolic competence and the frequency of cells with low-molecular-weight DNA using the neutral diffusion assay (49).

The major advantages of the Comet assay include its applicability to any tissue of interest provided that a single cell suspension can be obtained, the detection of multiple classes of DNA damage, the general ease of test performance, and the generation of data at the level of the single cell. As a consequence of its methodological advantages and the ongoing efforts in developing a standard protocol and acceptance criteria for the assay through IWGT working parties and international Comet assay workshops, the Comet assay is gaining increasing acceptance also for regulatory purposes (50). The primary use is that of a supplemental assay for mechanistic investigations such as tissue-specific genotoxic activity including "site-of-contact" genotoxicity, that is, cases of high local versus low systemic exposure (51). It is also proposed to complement the in vivo micronucleus assay to follow up positive findings from in vitro studies (52,53).

In a recent published evaluation of in vivo results from Comet assay, rat liver UDS test and transgenic gene mutation tests were examined for rodent carcinogens that were negative in the MN test but positive in the Ames test indicating genotoxicity is likely to contribute to tumorigenicity (53). The Comet assay was the most sensitive test in detecting such potential genotoxic carcinogens suggesting its suitability as an appropriate complement to the in vivo MN test. In the revised ICH S2R1 Guideline, the Comet assay is included as preferred to the previously recommended (in ICH S2B) rat liver UDS assay when a second in vivo test is required (see below).

Other In Vivo Test Models

Besides the test models for chromosomal damage and the Comet assay, the ICH S2R1 Guideline mentions other in vivo test models, which might be used under specific circumstances, including the rat liver UDS test, transgenic animal models for measuring gene mutations, or DNA-binding tests.

The rodent liver UDS assay is an established approach for investigating genotoxic activity in the liver and for which there is an OECD Guideline (54). The endpoint measured is indicative of DNA damage and subsequent repair in liver cells. Since the liver is usually the major site of metabolism of absorbed compounds, this assay is particularly appropriate for investigating compounds that require metabolic activation to express genotoxic activity.

Transgenic rodent gene mutation models have the advantage that they can be applied to all tissues from which sufficient DNA can be extracted and that they measure gene mutation rather than interaction with DNA (55,56). Mutations can be analyzed upon the efficient retrieval of reporter genes from the rodent genomic DNA into a suitable bacterial host. Test models using *lacI*, *lacZ*, *cII*, or *Spi-* as reporter genes are considered sufficiently developed to be regarded as suitable test assays (55).

Test models for measurement of DNA adducts use either ^{32}P-postlabeling or covalent binding to DNA (57). These approaches measure exposure, uptake, and reactivity to DNA rather than mutagenicity, but they are useful in considering in vivo mechanisms in combination with other data (58). The ^{32}P-postlabeling assay is a sensitive method for measuring DNA adducts that does not require the test compound to be radiolabeled. The method involves numerous steps including digestion of DNA followed by ^{32}P-labeling of adduct nucleoside 3-monophosphates and detection of labeled adducts, for example, by chromatography and autoradiography. The sensitivity of the assay can be further increased by adduct enrichment techniques, which remove normal nucleotides from the digest before ^{32}P-labeling. Another method for measuring DNA adducts is to use radiolabeled compounds and measure covalent binding to DNA. This is a well-established technique but it does need radiolabeled compounds. The biological consequences of low levels of DNA adducts observed in these assays are often difficult to interpret (57,58).

Demonstration of Target Tissue Exposure in In Vivo Studies

The value of in vivo results is related directly to the demonstration of adequate exposure of the target tissue to the test compound. This is especially true for negative in vivo test results when in vitro tests have shown convincing evidence of genotoxicity. In this case, the ICH S2R1 Guideline asks for exposure data from measurements of either toxicity in the tissue in question or toxicokinetic data. The assessments should be made at the top dose or other relevant doses using the same species, strain, and dosing route used in the genotoxicity assay. Data on tissue toxicity in cytogenetic assays in bone marrow or blood can be obtained by measuring changes in the proportion of immature erythrocytes among total erythrocytes in the tissue used in the MN test or by measuring a significant reduction in mitotic index for the chromosomal aberration assay. For other in vivo genotoxicity assays, toxicity in the liver or tissue being assessed should be done, for example, by histopathological evaluation or blood biochemistry toxicity indicators. Measurement of exposure of bone marrow or liver to drug-related material can be done either in blood or in plasma. The bone marrow is a well-perfused tissue and levels of drug-related materials in blood or plasma are generally similar to those observed in bone marrow. Liver is expected to be exposed for drugs with systemic exposure regardless of the route of administration. Drug-related material can also be measured directly in target tissue or by autoradiographic assessment of tissue exposure. When genotoxicity is measured in repeated-dose toxicology assays, exposure information is generally available as part of the toxicology assessment.

If systemic exposure is similar to or lower than expected clinical exposure, alternative strategies may be needed such as

(i) use of a different route of administration
(ii) use of a different species with higher exposure, and
(iii) use of a different tissue or assay.

If adequate exposure cannot be achieved, for example, with compounds showing very poor target tissue availability, conventional in vivo genotoxicity tests may have little value and it may be appropriate to base the evaluation only on in vitro testing. Examples of such compounds are some radioimaging agents, aluminium-based antacids, some compounds given by inhalation, and some dermal or other pharmaceuticals applied topically. In some cases, evaluation of genotoxic effects at the site of contact may be warranted, although such assays have not yet been used widely.

If in vitro tests do not show genotoxic potential, then in vivo (systemic) exposure can be assessed by any of the methods above or can be assumed from the results of standard absorption, distribution, metabolism, and excretion (ADME) studies in rodents done for other purposes.

THE ICH STANDARD TEST BATTERY FOR GENOTOXICITY

The ICH S2R1 Guideline defines as minimum criteria for a standard genotoxicity testing battery for pharmaceuticals the following features. It should include

(i) an assessment of mutagenicity in a bacterial reverse mutation test as this test has been shown to detect relevant genetic changes and the majority of genotoxic rodent and human carcinogens and
(ii) an evaluation of genotoxicity in mammalian cells in vitro and/or in vivo.

Acceptable in vitro mammalian cell systems are the in vitro metaphase chromosome aberration assay, the in vitro MN assay, and the mouse lymphoma L5178Y cell *tk* gene mutation assay. These three assays are currently considered equally appropriate and therefore interchangeable for measurement of chromosomal damage. The preferred in vivo study in the standard test battery is either an analysis of MN in rodent erythrocytes in blood or bone marrow or of chromosome aberrations at metaphase in bone marrow cells, which may be supplemented by a liver Comet assay. Based on these general considerations, the following two options for the standard battery are considered equally suitable:

Option 1

- A test for gene mutation in bacteria
- A cytogenetic test for chromosomal damage (the in vitro metaphase chromosome aberration test or in vitro micronucleus test), or an in vitro mouse lymphoma *tk* gene mutation assay
- An in vivo test for genotoxicity, generally a test for chromosomal damage using rodent hematopoietic cells either for MN or for chromosomal aberrations in metaphase cells.

Option 2

- A test for gene mutation in bacteria
- An in vivo assessment of genotoxicity with two different tissues, usually an assay for micronuclei using rodent hematopoietic cells and a second in vivo assay.

Typically for the second part of the in vivo assessment under Option 2, a DNA strand break (Comet) assay in the liver would be the preferred test model. Under both standard battery options, the ICH Guideline expresses it preferable that the in vivo genotoxicity assays be integrated into repeat-dose toxicity studies. Under Option 2, it is important to demonstrate that the top dose in the repeat-dose toxicity study is appropriate. Only if the repeat-dose toxicology study does not meet one of the dosing criteria described above (see "Tests for the Detection of Chromosomal Damage In Vivo" section), an acute in vivo study can be performed to optimize dose selection based on exposure/toxicity. Alternatively, an in vitro mammalian cell assay can be used (Option 1). Under Option 1 when the in vitro mammalian cell assay is negative and the in vivo genotoxicity test is integrated into a multiple administration toxicology study, the top dose does not need to meet the criteria as under Option 2 but generally is considered appropriate when the toxicology study meets the criteria for an adequate study to support human clinical trials. This may differ from dose selection criteria in the OECD Guideline for the in vivo MN assay.

The justification for considering Option 2, which does not include an in vitro mammalian cell test equivalent to Option 1, is that the two in vivo assays of Option 2 fulfill the recommendations to follow up a positive result in a mammalian cell test (see section on followup strategies). Thus, it is assumed that the omitted in vitro mammalian cell test, if done, could have been positive. For compounds that give negative results, the completion of either option of the standard test battery usually will provide sufficient assurance of the absence of genotoxic activity and no additional tests will be needed. Compounds that give positive results in the standard test battery may, depending on their therapeutic use, need to be tested more extensively (see section on follow-up strategies).

The standard battery does not require an independent test designed specifically to test for aneugenic potential. However, information on numerical changes can be derived from the mammalian cell assays in vitro and from the MN assays in vitro or in vivo, for example, elevations in the mitotic index, polyploidy, and MN induction. There is also experimental evidence that spindle poisons can be detected in the MLA. The preferred in vivo cytogenetic test under Option 2 where there is no in vitro mammalian cell test is the MN assay (not a chromosome aberration assay) to include more direct capability for detection of potential aneugenic compounds.

There may be circumstances where modifications to the standard test battery are advisable. For instance, in cases where compounds are highly toxic to bacteria (e.g., some antibiotics), any one of the in vitro mammalian cell assays should be done, that is, Option 1 is followed. The bacterial reverse mutation (Ames) test should still be carried out because mutagenicity can occur at lower, less toxic concentrations and thus needs to be investigated. Also for compounds bearing certain specific structural alerts modification to standard protocols can be appropriate. The choice of additional test(s) or protocol modification(s) depends on the chemical nature, the known reactivity, and any metabolism data on the structurally alerting compound in question.

Although one of the aims of genotoxicity testing is to assess the ability of chemicals to induce heritable mutations, the test battery for pharmaceuticals does not request tests, specifically, for detection of germ cell mutations. Results of comparative studies have shown that in a qualitative sense most germ cell mutagens are likely to be detected as genotoxic in somatic cell tests so that negative results of in vivo somatic cell genotoxicity tests generally indicate the absence of germ cell effects.

Additional Testing of Compounds Giving Positive Results in the Standard Test Battery (Follow-up Strategies)

The discussion on follow-up studies for positive in vitro results in the ICH S2R1 Guideline does not address in detail the situation of a positive finding in the Ames bacterial mutation assay. This is a specific feature of pharmaceutical testing since, in practice, a positive Ames test will almost always result in termination of a drug candidate's development unless there is a clear unmet medical need. For positive results from in vitro mammalian cell assays for which there is insufficient weight of evidence to indicate lack of relevance (see "Evaluation of Individual Tests" section), the guideline recommends two different approaches: either to do two appropriate in vivo assays or to do experiments to provide mechanistic information underlying the observed in vitro effect.

Negative results in two appropriate in vivo assays with adequate justification for the endpoints measured and usually demonstration of exposure would be sufficient to demonstrate absence of a clinically relevant genotoxic activity for mammalian cell in vitro positives. For instance, a combination of clastogenicity test in blood or bone marrow and a DNA strand break (Comet) test in liver would be sufficient in many cases to follow up in vitro positives, but the choice of in vivo tissue and assay should be based on factors such as any knowledge of the potential mechanism of the metabolism in vivo and of the exposed tissues thought to be relevant.

Usually, investigations for gaining an understanding of mechanisms underlying the observed in vitro effect can be carried out using further in vitro tests. Many examples of accepted mechanisms by which a chemical may demonstrate a genotoxic effect (in vitro) that is either not relevant for humans or has a No Observed Effect Concentration below which genotoxicity would not be induced have been described (30,59–61) including the following:

- Inhibition of DNA polymerases
- Inhibition of gyrases/topoisomerases

- Imbalance of DNA precursors
- Energy depletion
- Lipid peroxidation
- Disruption of normal defense mechanisms, such as glutathione depletion
- Production of active oxygen species

The ICH S2R1 Guideline does not prescribe any methods by which to obtain convincing evidence of such indirect modes of genotoxic action. Appropriate approaches should be selected according to available published data on a case-by-case basis (30,61). There might also be some suggestions from the intended use or pharmacological class of a pharmaceutical what kind of molecular or cellular targets may be affected. Additional experimental evidence that a test compound or its metabolites shows no reactivity with DNA and is not a direct DNA-damaging agent would further support existing of an indirect mechanism, for example, negative findings from mutation/DNA damage tests others than Ames test or structural considerations. If the mechanistic information and weight of evidence supports the lack of relevant genotoxicity, only a single in vivo test is needed with appropriate evidence of exposure to confirm lack of in vivo genotoxic activity.

In case of a positive result in the in vitro MN assay, mechanistic investigations are aimed at determining whether micronucleus induction is primarily because of chromosome loss or of chromosome breakage. This could include staining of micronuclei to determine whether centromeres are present (see section "In Vitro Micronucleus Test" section). An alternative approach is to carry out an in vitro or in vivo assay for metaphase structural aberrations. If negative, this would infer that MN induction is related to chromosome loss.

When positive results are seen only in the presence of the S9 activation system, it should first be verified that metabolic activation is responsible and not some other difference in conditions, for example, low or no serum in the S9 mix resulting in higher exposure to the free drug compared with usually $\geq 10\%$ serum in the nonactivated incubations. The follow-up strategy is then aimed at determining the relevance of the results in vitro to metabolic conditions in animal studies and in humans. For instance, analysis of the metabolite profile in the genotoxicity test incubations for comparison with known metabolite profiles in nonclinical species, or in preparations from humans, can help determine the relevance of test results (62). Liver-specific in vivo genotoxicity testing and in vitro studies with human S9 or other human-relevant activation systems are further appropriate options to follow up in vitro positives in the presence of S9.

Follow-up genotoxicity testing in appropriate models may also be needed for compounds that were negative in the standard test battery but which have shown increases in tumors in carcinogenicity bioassay(s) with insufficient evidence to establish a nongenotoxic mechanism. To help understand the mode of action, additional testing can include modified conditions for metabolic activation in in vitro tests or can include in vivo tests measuring genetic damage in target organs of tumor induction (63).

GUIDANCE ON EVALUATION OF TEST RESULTS

The process of evaluating genotoxicity data for the purpose of hazard identification and human safety assessment usually involves three sequential steps:

1. Evaluation of individual tests in order to assess the test compound's ability to produce a genotoxic response in any test.
2. Evaluation of the complete data set in order to assess whether or not the test compound is a relevant mutagen in vivo.
3. Evaluation of potential risk in order to assess whether the test compound, depending on the perceived exposure, poses any adverse health risk.

Evaluation of Individual Tests

The data from individual study reports should first be evaluated against current acceptance and quality criteria. Since the primary source of data is usually data submission to regulatory bodies, the test protocols largely follow internationally accepted guidelines (OECD, ICH) and thus comply with basic requirements for scientific reliability. Moreover, testing in compliance with the principles of Good Laboratory Practice ensures the generation of high quality and reliable test data with respect to study documentation, consistency with Standard Operation Procedures, identity of testing material, etc. Evaluation of genotoxicity test results can be more difficult when data from the literature are used. In general, the basic principles set forth by international guidelines apply also for such data not generated primarily for regulatory purposes. Studies that do not meet minimum standards of scientific reliability both with regard to protocol performance and to documentation should be excluded from the evaluation process.

Current testing guidelines include criteria on how to determine a test result as positive or negative (or equivocal) for a particular test method. Thus, the evaluation whether an individual test provides a negative or a positive outcome will usually not pose a problem. The more crucial point is the assessment of biological relevance of a positive result, in particular, in case of isolated in vitro findings. Basic criteria that should be taken into consideration in the data assessment include the type of studies with positive findings (indicator or mutagenicity studies), the magnitude (increase over background) and type of response (dose–response relationship), dose range (positive response in μg or ng concentration range rather than mg), knowledge about direct or indirect modes of genotoxic action, level of cytotoxicity at which genotoxic effects occur, and the adequacy of in vitro metabolic pathways employed. The ICH S2R1 Guideline provides some examples of in vitro positive responses that may be considered nonrelevant for the in vivo situation:

- Small increases that are statistically significant compared with the negative or solvent control values but are within the confidence intervals of the appropriate historical control values for the testing facility.
- Weak/equivocal responses that are not reproducible.
- Positive responses in mammalian cell genotoxicity assays under conditions of strong cytotoxicity only (in the MLA increases at $\geq$80% reduction in RTG, in in vitro cytogenetics assays when growth is suppressed by $\geq$50%).
- Positive response due to an artifact of a high-dose protocol, for example, increased osmolality, extremes of pH.
- Positive response due to bacterial-specific metabolism, for example, activation by bacterial nitroreductases.
- Artifactual increases in colonies in the bacterial mutation assay due to contamination with amino acids when testing a peptide.

If any of the above conditions apply, the weight of evidence indicates a lack of genotoxic potential and a single in vivo test as described in Option 1 of the test battery is sufficient.

Evaluation of the Complete Data Set

The battery testing approach using assays that measure different endpoints at different levels of organization and therefore with different levels of sensitivity and specificity is primarily designed to reduce the risk of false-negative results for compounds with genotoxic potential. A positive result in any assay for genotoxicity does not necessarily mean that the test compound poses a genotoxic/carcinogenic hazard to humans. Evaluation of positive genotoxicity results should be based on a rigorous weight-of-evidence approach that includes consideration of the results of all genotoxicity data and other relevant information. In many instances, genotoxicity data profiles are typically characterized by the conflict of positive in vitro results and negative in vivo findings. In general, in vivo studies are of higher significance than in vitro studies and are thus given a central role in the evaluation process of positive results and the question of how likely an agent is to be genotoxic in humans. However, experiences show that exceptions from this general rule exist, that is, a simple overruling of in vitro positives by in vivo negatives is not always justified, particularly when there is strong evidence from in vitro assays that the chemical is mutagenic (64). Evaluation of contradictory results should therefore be evaluated with great care. As mentioned before, the ICH S2R1 Guideline requests negative results in two appropriate in vivo assays with adequate justification for the endpoints measured and demonstration of exposure to demonstrate absence of in vivo activity. In some cases, the discordant results between in vitro and in vivo may simply be due to the lower exposure in in vivo tests. Further support of lack of relevance of such high-dose in vitro response for risk assessment would be provided by demonstrating that the underlying mechanism does not involve reactivity of a chemical or its metabolites with DNA but some other indirect mechanisms of genotoxicity that operate only under certain concentrations. Examples of mechanisms only operating at high doses because of saturation of detoxification pathways or disruption of homoeostatic mechanisms are described in the previous section.

Other toxicological data may also provide valuable supportive evidence in this evaluation process. In particular, negative results from carcinogenicity studies would support the view that a compound has no relevant mutagenic potential in vivo. Transgenic mouse models, such as the p53 heterozygote strain, may provide even more specific information as the p53 model has been found useful in the identification of carcinogens operating by a genotoxic mode of action (65).

It is important to also evaluate critically positive (or equivocal) findings from in vivo studies with regard to toxicological relevance as some in vivo genotoxicity tests can give misleading positive results that do not indicate true genotoxicity. For example:

- Increases in micronuclei can occur due to disturbance in erythropoiesis, stress, or hypo- and hyperthermia (66).
- Small increases in DNA adduct level might be without adverse consequences and data should be interpreted in the light of the known background level of endogenous adducts (57).
- Indirect toxicity-related effects might induce positive findings in the DNA strand break assays, for example, Comet assays (49).

Evaluation of in vivo genotoxicity data should therefore always take into account all the toxicological and hematological animal findings.

Risk Evaluation

An evaluation of the potential mutagenic risk to humans is generally required only for compounds found to be mammalian in vivo mutagens rather than compounds that have given positive results only in vitro. Chemicals that are recognized as in vivo somatic cell mutagens will be assumed to be both potential genotoxic carcinogens and potential germ cell mutagens. In the process of low-dose-extrapolation and determination as to whether a mutagen presents a risk of causing genetic damage (either in somatic or in germ cells) in humans exposed to this compound, there should be consideration particularly of the mechanisms by which the mutations were induced. For some in vivo mutagens, there may be sufficient evidence for the existence of a threshold, for example, from toxicokinetic studies and from mechanistic studies indicating that the observed genotoxicity results from interaction with non-DNA targets (see above for examples). Safe exposure limits in these cases may be derived from applying no observed adverse effect level (NOAEL) from appropriate animal studies and the use of uncertainty factors. If insufficient evidence for a threshold is available, a nonthreshold approach is usually applied, that is, it is assumed that such mutagens have the potential to cause genotoxic effects at any level of exposure. It is clear that any full (quantitative) assessment of a possible cancer risk on the basis of in vivo mutagenicity data alone is not feasible but would need data from long-term rodent carcinogenicity studies. In this case, the knowledge of potential mutagenicity would determine the model used to calculate the risk, and it is generally assumed that the carcinogenicity dose response is linear and without a threshold for chemicals tested positive for in vivo mutagenicity (67).

REFERENCES

1. Ames BN, Durston WE, Yamasaki E, et al. Carcinogens are mutagens: A simple test system for combining liver homogenates for activation and bacteria for detection. Proc Natl Acad Sci USA 1973; 70:2281–2285.
2. McCann J, Choi E, Yamasaki E, et al. Detection of carcinogens as mutagens in the Salmonella/microsome test: Assay of 300 chemicals. Proc Natl Acad Sci USA 1973; 70:2281–2285.
3. Müller L, Kikuchi K, Probst G, et al. ICH-Harmonized guidances on genotoxicity testing of pharmaceuticals; Evolution, reasoning and impact. Mutat Res 1999; 436:195–225.

4. International Conference on Harmonisation of Technical Requirements for Registration of Pharmaceuticals for Human Use (ICH). S2A: Guidance on specific aspects of regulatory genotoxicity tests for pharmaceuticals. 1995.
5. International Conference on Harmonisation of Technical Requirements for Registration of Pharmaceuticals for Human Use (ICH). S2B: Genotoxicity: A standard battery for genotoxicity testing of pharmaceuticals. 1997.
6. International Conference on Harmonisation of Technical Requirements for Registration of Pharmaceuticals for Human Use (ICH). S2(R1): Genotoxicity: A standard battery for genotoxicity testing of pharmaceuticals. 2008.
7. Kirkland D, Hayashi M, Jacobson-Kram D, et al. Editorial: The international workshop on genotoxicity testing (IWGT): History and achievements. Mutat Res 2007; 627:1–4.
8. Gatehouse D, Haworth S, Cebula T, et al. Report from the working group on bacterial mutation assays: International workshop on standardisation of genotoxicity test procedures. Mutat Res 1994; 312:217–233.
9. OECD Guideline 471. Bacterial Reverse Mutation Test. Paris: Organisation for Economic Cooperation and Development (OECD), 1997.
10. OECD Guideline 476. In vitro Mammalian Cell Gene Mutation Test. Paris: Organisation for Economic Cooperation and Development (OECD), 1997.
11. Moore MM, Clive D, Hozier JC, et al.Analysis of trifluorothymidine-resistant (TFT) mutants of L5178Y/TK± mouse lymphoma cells. Mutat Res 1985; 151:161–174.
12. Applegate ML, Moore MM, Broder CB, et al. Molecular dissection of mutations at the heterozygous thymidine kinase locus in mouse lymphoma cells. Proc Natl Acad Sci USA 1990; 87:51–55.
13. Clark LS, Hart DW, Vojta PJ, et al. Identification and chromosomal assignment of two heterozygous mutations in the Trp53 gene in the L5178Y Tk± 3.7.2C mouse lymphoma cells. Mutagen 1998; 13:427–434.
14. Moore MM, Honma M, Clements J, et al. Mouse lymphoma thymidine kinase locus gene mutation assay: International workshop on genotoxicity test procedures workgroup report. Environ Mol Mutagen 2000; 35:185–190.
15. Moore MM, Honma M, Clements J, et al. Mouse lymphoma thymidine kinase gene mutation assay: Follow-up international workshop on genotoxicity test procedures, New Orleans, Louisiana, April 2000. Environ Mol Mutagen 2002; 40:292–299.
16. Moore MM, Honma M, Clements J, et al. Mouse lymphoma thymidine kinase gene mutation assay: International workshop on genotoxicity tests workgroup report—Plymouth, UK 2002. Mutat Res 2003; 540:127–140.
17. Moore MM, Honma M, Clements J, et al. Mouse lymphoma thymidine kinase gene mutation assay: Follow-up meeting of the international workshop on genotoxicity testing—Aberdeen, Scotland, 2003—Assay acceptance criteria, positive controls, and data evaluation. Environ Mol Mutagen 2006; 47:1–5.
18. Moore MM, Honma M, Clements J, et al. Mouse lymphoma thymidine kinase gene mutation assay: Meeting of the International Workshop on Genotoxicity Testing, San Francisco, 2005, recommendations for 24-h treatment. Mutat Res 2007; 627:36–40.
19. OECD Guideline 473. In Vitro Mammalian Chromosome Aberration Test. Paris: Organisation for Economic Cooperation and Development (OECD), 1997.
20. Kirsch-Volders M, Sofuni T, Aardema M, et al. Report from the in vitro micronucleus assay working group. Mutat Res 2003; 540:153–163.
21. Corvi R, Albertini S, Hartung T, et al. ECVAM Retrospective validation of the in vitro micronucleus test (MNT). Mutagen 2008; 23:271–283.
22. OECD Draft Guideline 487. In Vitro Mammalian Cell Micronucleus Test (Mnvit). Paris: Organisation for Economic Cooperation and Development (OECD), 2007.
23. Fenech M. The advantages and disadvantages of cytokinesis-blood micronucleus method. Mutat Res 1997; 392:11–18.
24. Kersten B, Zhang J, Brendler-Schwaab SY, et al. The application of the micronucleus test in Chinese hamster V79 cells to detect drug-induced photogenotoxicity. Mutat Res 1999; 445:55–71.
25. von der Hude W, Kalweit S, Engelhardt G, et al. In vitro micronucleus assay with Chinese hamster V79 cells—results of a collaborative study with in situ exposure to 26 chemical substances. Mutat Res 2000; 468:137–163.
26. Kirkland D, Aardema M, Henderson L, et al. Evaluation of the ability of a battery of three in vitro genotoxicity tests to discriminate rodent carcinogens and non-carcinogens. I. Sensitivity, specificity and relative predictivity. Mutat Res 2005; 584:1–256.
27. Matthews EJ, Kruhlak NL, Cimino MC, et al. An analysis of genetic toxicity, reproductive and developmental toxicity and caricnogenicity data, II. Identification of genotoxicants, reprotoxicants and carcinogens using in silico methods. Regul Toxicol Pharmacol 2006; 44:97–110.

28. Thybaud V, Aardema M, Clements J, et al. Strategy for genotoxicity testing: Hazard identification and risk assessment in relation to in vitro testing. Mutat Res 2007; 627:41–58.
29. Kirkland D, Pfuhler S, Tweats D, et al. How to reduce false positive results when undertaking in vitro genotoxicity testing and thus avoid unnecessary follow-up animal tests: Report of an ECVAM Workshop. Mutat Res 2007; 628:31–55.
30. Thybaud V, Aardema M, Casciano D, et al. Relevance and follow-up of positive results in in vitro genetic toxicity assays: An ILSI-HESI initiative. Mutat Res 2007; 633:67–79.
31. Fellows MD, O'Donovan MR. Cytotoxicity in cultured mammalian cells is a function of the method used to estimate it. Mutagenesis 2007; 22:275–280.
32. Greenwood SK, Hill RB, Sun JT, et al. Population Doubling: A simple and more accurate estimation of cell growth suppression in the in vitro assay for chromosomal aberrations that reduces irrelevant positive results. Environ Mol Mutagen 2004; 43:36–44.
33. Tweats DJ, Blakey D, Heflich RH, et al. Report of the IWGT working group on strategy/interpretation of regulatory in vivo tests. II. Identification of in vivo-only positive compounds in the bone marrow micronucleus test. Mutat Res 2007; 627:92–105.
34. OECD Guideline 475. Mammalian Bone Marrow Chromosome Aberration Test. Paris: Organisation for Economic Cooperation and Development (OECD), 1997.
35. OECD Guideline 474. Mammalian Erythrocyte Micronucleus Test. Paris: Organisation for Economic Cooperation and Development (OECD), 1997.
36. MacGregor JT, Wehr CM, Henika PR, et al. The in vivo erythrocyte micronucleus test: Measurement at steady state increases assay efficiency and permits integration with toxicity studies. Fund Appl Toxicol 1990; 14:513–522.
37. MacGregor JT. Series: Current Issues in Mutagenesis and Carcinogenesis, No. 23 Letter to the Editor. Micronucleus assay protocols. Mutat Res 1991; 259:123–125.
38. Hayashi M, MacGregor JT, Gatehouse DG, et al. In vivo rodent erythrocyte micronucleus assay. II. Some aspects of protocol design including repeated treatments, integration with toxicity testing, and automated scoring. Environ Mol Mutagen 2000; 35:234–252.
39. Schlegel R, MacGregor JT. The persistence of micronucleated erythrocytes in the peripheral—circulation of normal and splenectomized Fischer 344 rats—implications for cytogenetic screening. Mutat Res 1984; 127:169–174.
40. Hamada S, Sutou S, Morita T, et al. Evaluation of the rodent micronucleus assay by a 28-day treatment protocol: Summary of the 13th collaborative study by the collaborative study group for the micronucleus test (CSGMT)/Environmental Mutagen Society of Japan (JEMS)—Mammalian Mutagenicity Study Group (MMS). Environ Mol Mutagen 2001; 37:93–110.
41. Hayashi M, Morita T, Kodama Y, et al. The micronucleus assay with mouse peripheral blood reticulocytes using acridine orange-coated slides. Mutat Res 1990; 245:245–249.
42. Dertinger SD, Bishop ME, McNamee JP, et al. Flow cytometric analysis of micronuclei in peripheral blood reticulocytes: I. Intra- and interlaboratory comparison with microscopic scoring. Toxicol Sci 2006; 94:83–91.
43. MacGregor JT, Bishop ME, McNamee JP, et al. Flow cytometric analysis of micronuclei in peripheral blood reticulocytes: II. An efficient method of monitoring chromosomal damage in the rat. Toxicol Sci 2006; 94:92–107.
44. Cammerer Z, Elhajouji A, Suter W. In vivo micronucleus test with flow cytometry after acute and chronic exposures of rats to chemicals. Mutat Res 2007; 626:26–33.
45. Witt KL, Livanos E, Kissling GE, et al. Comparison of flow cytometry and microscopy-based methods for measuring micronucleated reticulocyte frequencies in rodents treated with nongenotoxic and genotoxic chemicals. Mutat Res 2008; 649:101–113.
46. Hayashi M, MacGregor JT, Gatehouse DG, et al. In vivo erythrocyte micronucleus assay. III. Validation and regulatory acceptance of automated scoring and the use of rat peripheral blood reticulocytes, with discussion of non-hematopoietic target cells and a single dose-level limit test. Mutat Res 2007; 627:10–30.
47. Tice RR, Agurell E, Anderson D, et al. Single cell gel/Comet assay: Guidelines for in vitro and in vivo genetic toxicology testing. Environ Mol Mutagen 2000; 35:206–221.
48. Hartmann A, Agurell E, Beevers C, et al. Recommendations for conducting the in vivo alkaline Comet assay. 4th International Comet Assay Workshop. Mutagenesis 2003; 18:45–51.
49. Burlinson B, Tice RR, Speit G, et al. Fourth international workgroup on genotoxicity testing: Results of the in vivo Comet assay workgroup. Mutat Res 2007; 627:31–35.
50. Cimino MC. Comparative overview of current international strategies and guidelines for genetic toxicology testing for regulatory purposes. Environ Mol Mutagen 2006; 47:362–390.
51. Brendler-Schwaab S, Hartmann A, Pfuhler S, et al. The in vivo Comet assays: Use and status in genotoxicity testing. Mutagenesis 2005; 20:245–254.

52. Pfuhler S, Albertini S, Fautz R, et al. Genetic toxicity assessment: Employing the best science for human safety evaluation part IV: Recommendation of a working group of the Gesellschaft fuer Umwelt-Mutationsforschung (GUM) for a simple and straightforward approach to genotoxicity testing. Toxicol Sci 2007; 97:237–240.
53. Kirkland D, Speit G. Evaluation of the ability of a battery of three in vitro genotoxicity tests to discriminate rodent carcinogens and non-carcinogens III. Appropriate follow-up testing in vivo. Mutat Res 2008; 654:114–132.
54. OECD Guideline 486. Unscheduled DNA synthesis (UDS) test with mammalian liver cells in vivo. Paris: Organisation for Economic Cooperation and Development (OECD), 1997.
55. Thybaud V, Dean S, Nohmi T, et al. In vivo transgenic mutation assays. Mutat Res 2003; 540:141–151.
56. Lambert IB, Singer TM, Boucher SE, et al. Detailed review of transgenic rodent mutation assays. Mutat Res 2005; 590:1–280.
57. Phillips D, Farmer PB, Beland FA, et al. Methods of DNA adduct determination and their application to testing compounds for genotoxicity. Environ Mol Mutagen 2000; 35:222–233.
58. Sander M, Cadet J, Casciano DA, et al. Proceedings of a workshop on DNA adducts: Biological significance and applications to risk assessment. Toxicol Appl Pharmacol 2005; 208:1–20.
59. Scott D, Galloway SM, Marshall RR, et al. Genotoxicity under extreme culture conditions. A Report from ICPEMC Task Group 9. Mutat Res 1991; 257:147–204.
60. Galloway SM, Miller JE, Armstrong MJ, et al. DNA synthesis inhibition as an indirect mechanism of chromosome aberrations: Comparison of DNA-reactive and non-DNA-reactive clastogens. Mutat Res 1998; 400:169–186.
61. Kirkland D, Aardema M, Banduhn N, et al. In vitro approaches to develop weight of evidence (WoE) and mode of action (MoA) discussions with positive in vitro genotoxicity results. Mutagenesis 2007; 22:161–175.
62. Ku WW, Bigger A, Brambilla G, et al. Strategy for genotoxicity testing—Metabolic considerations. Mutat Res 2007; 627:59–77.
63. Kasper P, Uno T, Mauthe R, et al. Follow-up testing of rodent carcinogens not positive in the standard genotoxicity testing battery: IWGT workgroup report. Mutat Res 2007; 627:106–116.
64. Brambilla G, Martelli A. Failure of the standard battery of short-term tests in detecting 20 some rodent and human genotoxic carcinogens. Toxicology 2003; 196:1–19.
65. MacDonald J, French JE, Gerson R, et al. The utility of transgenic mouse assays for identification human carcinogens—A basic understanding and path forward. Toxicol Sci 2004; 77:188–194.
66. Tweats DJ, Blakey D, Heflich RH, et al. Report of the IWGT working group on strategies and interpretation of regulatory in vivo tests. I. Increases in micronucleated bone marrow cells in rodents that do not indicate genotoxic hazards. Mutat Res 2007a; 627:78–91.
67. Dearfield KL, Moore MM. Use of genetic toxicology information for risk assessment. Environ Mol Mutagen 2005; 46:236–245.

26 Carcinogenicity

Gerd Bode
Lecturer and Consultant, Goettingen, Germany

INTRODUCTION

The greatest concern for pharmaceutical or chemical compounds is the drug-dependent development of tumors. This induction of cancers was from the beginning of the International Conferences on Harmonization (ICH) in the focus of the discussions of the participating experts. The objectives in the ICH Safety Expert Working Party were to improve the sensitivity of preclinical testing and provide better tools for a valid and early assessment of a tumorigenic potential without increasing the number of animals necessary for the evaluation. Accordingly, the ICH Expert Working Groups created guidelines for carcinogenicity testing (Table 1).

These guidelines recommend when there is a need or not for conducting carcinogenicity studies, which strategy should be chosen, which criteria should be used for selecting the high dose, and how to limit the amount of compounds being administered. This chapter deals with the complex issues associated with the potential for tumorigenesis and how to cope with it.

The testing for the potential to induce the growth of tumors is usually not possible in humans. Such experiments are forbidden ethically because the risk for patients is unacceptable. Human epidemiological studies provide signals for such undesirable tumorigenic reactions but the valid evaluation is only possible by applying nonclinical in vitro and in vivo methods.

The development of tumors is always a long-term effect. Tumors do not appear suddenly but in most cases need many months in rodents and years in humans before they become clinically apparent. Tumors most often develop in older age when the repair mechanisms against mutagenic or epigenetically altered cells are less effective. The appearance of tumors with increasing age is the justification for long-term animal studies where compounds are being administered in a lifelong study, usually 24 months in rats and mice. Strong carcinogens are easy to detect in genotoxicity assays and even in repeat-dose studies over a period of many months by revealing tumors or at least precancerous lesions. Weak carcinogens use the long-term applications of the compounds in question and reveal their inherent risk better at the time when spontaneous tumors, mostly correlated with age, appear. This experience is the basis for the request for lifelong bioassays.

Many long-term animal studies have proven to be predictive for human conditions. Within the ICH discussions some deficiencies of these bioassays have been identified—there are too many positive results, often contradictory outcomes, and a poor concordance of organ sites. As a consequence ICH raised the question: if the long-term testing in two species could be reduced without compromising human safety? Accordingly, ICH Expert Working Group members recommended alternatives suitable for an assessment of tumorigenesis.

GENERAL REGULATORY BACKGROUND

Treatment with compounds associated with carcinogenic potential for banal indications is unacceptable. For severe indications such as life-threatening cancer diseases, the treatment with carcinogenic compounds often does not increase the overall risk of the underlying disease.

- The term *carcinogen* denotes a chemical substance or a mixture of chemical substances, which induce cancer or increase its incidence (1). An alternate definition is that carcinogenic substances are ones that
  - induce tumors (benign or malignant),
  - increase their incidence or malignancy, or
  - shorten the time of tumor occurrence

when they are inhaled, injected, dermally applied, or ingested (2).

Table 1 ICH Carcinogenicity Guidelines

ICH guideline code/number	Title and EMEA reference number	Implementation status in ICH regions
S1A	Guideline on the need for carcinogenicity studies of pharmaceuticals (CPMP/ICH/140/95)	Step 5
S1B	Guideline on the need for carcinogenicity studies of pharmaceuticals (CPMP/ICH/140/95)	Step 5
S1C	Dose selection for carcinogenicity studies of pharmaceuticals (CPMP/ICH/383/95)	Step 5
S1C(R)	Addendum to "dose selection for carcinogenicity studies of pharmaceuticals": addition of a limit dose and related notes of pharmaceuticals (CPMP/ICH/366/95)	Step 5

Carcinogens are classified according to their mode of action as genotoxic or nongenotoxic carcinogens. Genotoxic carcinogens initiate carcinogenesis by direct interaction with DNA resulting in DNA damage or chromosomal aberrations that can be detected by genotoxicity tests as indicated in a document from the Organization for Economic Cooperation and Development (OECD) (3). Nongenotoxic carcinogens are agents that, at least initially, interact indirectly with DNA. These indirect modifications to DNA structure, amount or function may result in altered gene expression or signal transduction (3). In animal studies, most potent mutagens are also found to be carcinogenic (2). Substances that induce tumors in animals are considered as suspected human carcinogens until convincing evidence is presented to the contrary (1).

Regarding animal testing, a number of regulatory recommendations and advisory documents are available.

The conventional test for carcinogenicity is the long-term rodent carcinogenicity bioassay described in Ref. (4). Table 2 describes dose-range-finding studies for carcinogenicity.

Table 3 provides the general design of long-term assays.

The objective of these tests is to observe test animals for a major portion of their life span for the appearance of neoplastic lesions during or after exposure to various doses of a test substance by an appropriate route of administration. The study is usually conducted using two species. Rats and mice of both sexes are typically the selected models. The animals are dosed by oral, dermal, or inhalation exposures based upon the expected type of human application. Dosing typically lasts around two years. General animal health features are monitored throughout the study, *intra vitam*, there is a focus on palpable masses but the key assessment resides in the full histopathological analysis of the animal tissues and organs when the study is terminated.

Two endpoints in animal bioassays, carcinogenicity and chronic toxicity, can be combined to reduce animal use as described in Ref. (5), but this practice is less common for pharmaceutical compounds.

Table 2 Dose-Range-Finding Studies for Carcinogenicity Testing

Parameter	Recommendations
Usual length of studies	• 3 mo for long-term studies • 1 mo for neonatal or transgenic mice
Range of different dose levels	Often 5
Focus of studies	Toxicity endpoints, determination of MTD (maximum tolerated dose)
Profiling of area under curve (AUC)	• e.g., 1 and 3 mo for rats, • e.g., 1 and 4 wk for alternatives Transgenic = wild type useful for kinetic data

Table 3 General Design Parameters for Long-Term Carcinogenicity Studies

Parameter	Recommendation
Species	Rats, mice, rarely hamster
Duration	24 mo for rats, 24 mo for mice
Route	As in clinical use, in feed, water, by gavage, inhalation
Dose levels	Mostly three, mostly with a factor (e.g., 1:3:9)
Animal numbers	50/group/sex
Numbers of histological slides	Ca 25,000
When results to be available	Before NDA/MAA submission
Cost	€1.0–1.5 million

Tumorigenicity studies are the longest (lifelong or two year duration) and most expensive (approximately $/€ 1.5 to 4 million) nonclinical studies. Therefore they should be well designed and conducted in such a way that they indicate clearly any risk involved. Such studies should not be repeated and should support the market authorization by the international regulatory bodies. These objectives explain why a great number of international and regional guidelines exist today and provide best advice for the experimental researchers.

This chapter focuses predominantly on the guidelines of the International Conferences of Harmonization (ICH). The author was Topic Leader for ICH for 16 years. For the complex information needed for optimal design and evaluation of carcinogenicity studies other guidance documents like those from the European Medicines Evaluation Agency (EMEA) are also being considered.

On the highest ICH level, there are altogether four guidelines [S1A, S1B, S1C and S1C(R) (6,7,8,9)] which were all implemented during 1995–97 in the ICH regions (USA, Europe, and Japan) and distributed worldwide via the World Health Organization (WHO) (Table 1; ICH Carcinogenicity Guidelines).

ICH GUIDELINES FOR CARCINOGENICITY

S1A: Guideline on the need for carcinogenicity studies of pharmaceuticals (6)

This guideline was adopted in 1997 and has been implemented in all regions. The objectives of carcinogenicity studies are

- to identify a tumorigenic potential in animals
- to assess the relevant risk for humans.

Any cause for concern derived from laboratory investigations, animal toxicology studies, and data in humans may lead to a need for carcinogenicity studies.

Carcinogenicity studies should be performed for any pharmaceutical whose expected clinical use is continuous for at least six months (Table 4).

Certain classes of compounds may not be used continuously over a minimum of six months but may be expected to be administered repeatedly in an intermittent manner. For pharmaceuticals used frequently in an intermittent manner during the treatment of chronic or recurrent conditions, carcinogenicity studies are generally needed. Examples of such conditions include allergic rhinitis, depression, anxiety, and delivery systems with prolonged exposure.

Pharmaceuticals administered infrequently or for short duration of exposure (e.g., anesthetics and radio-labeled imaging agents) do not need carcinogenicity studies unless there is cause for concern.

Table 4 When Are Carcinogenicity Studies Needed?

Factor	Issue
Length of use of medicinal product in man	Continuous use for 6 months or longer
Toxicology findings causing concern	• Positive genotoxicity • Results of short-term carcinogenicity tests • Results of exposure levels in specific organs

Carcinogenicity studies may be recommended for some pharmaceuticals if there is concern about their carcinogenic potential. Concern could derive from the following information:

1. Previous demonstration of carcinogenic potential that is considered relevant to humans in the product class;
2. Structure–activity relationship suggesting carcinogenic risk;
3. Evidence of preneoplastic lesions in repeated dose toxicity studies; and
4. Long-term tissue retention of parent compound or metabolite(s) resulting in local tissue reactions or other pathophysiological or pathomorphological responses.

How Should Genotoxic Compounds Be Tested?
Unequivocally genotoxic compounds are presumed to be trans-species carcinogens, implying a hazard to humans. Such compounds need not be subjected to long-term carcinogenicity studies. However if such a drug is intended to be administered chronically to humans, a chronic toxicity study (up to one year) is recommended by ICH Guideline S1A (6) to detect early tumorigenic effects. In practice, this option has not been used since 1997. The main concern of the pharmaceutical industry against this option is the lack of historical comparative data from competitor drugs. The tumor evaluation was always based on two-year data. It is considered that the one-year studies will not detect the tumorigenic potential and power appropriately.

When Should Carcinogenicity Studies Be Available?
Carcinogenicity studies usually need to be completed before application for marketing approval. However completed rodent carcinogenicity studies are not needed in advance of the conduct of large-scale clinical trials unless there is concern for the patient population. For products to be used in the pediatric population, for example, carcinogenicity testing is required before long-term exposure to children.

For pharmaceuticals developed to treat serious life-threatening morbidities without available adequate treatment conditions carcinogenicity testing need not be conducted before market approval. Under such conditions these studies should be conducted postapproval. This speeds up the availability of pharmaceuticals for such severely debilitating diseases especially where no satisfactory alternative therapy exists.

In instances where the life expectancy in the indicated population is short (i.e., less than two to three years) no long-term carcinogenicity studies may be required. For example, oncolytic agents intended for treatment of advanced systemic disease do not generally need carcinogenicity studies. In cases where the therapeutic agent for cancer is generally successful and life is significantly prolonged, there may be requirements to provide knowledge about the tumorigenic risk. When such pharmaceuticals are intended for adjuvant therapy in tumor-free patients or for prolonged use in noncancer indications, carcinogenicity studies are usually needed.

Pharmaceuticals showing poor systemic exposure from topical routes in humans may not need studies by the oral route to assess the carcinogenic potential to internal organs.

Carcinogenicity studies are not generally needed for endogenous substances when given as essentially replacement therapy. Here physiological compounds such as erythropoietin are being administered to compensate for a deficiency and to lift their availability to approximately physiological levels. This holds true particularly when there is previous clinical experience with similar products (e.g., animal insulins, pituitary-derived growth hormone, and calcitonin).

ICH S1B: Testing for carcinogenicity of pharmaceuticals (7)
ICH S1B was implemented in 1997. This is a recommendation to facilitate the decision on which strategy should be adopted if carcinogenicity studies need to be performed. The options are either to follow the conventional strategy by conducting two long-term assays over a period of two years in two rodent species or, on the other hand, a combination of one long-term study in one rodent species and a midterm study often by using transgenic animals.

Historically the regulatory requirements for the assessment of the carcinogenic potential were the conduct of long-term carcinogenicity studies in two rodent species, usually the rat and the mouse. It was the mission of ICH to examine whether this practice could be reduced without compromising human safety. The discussion in the Expert Working Group soon revealed that

Table 5 ICH S1B: Basic Principle for Testing the Carcinogenic Potential

Study	Rationale for study
Long-term rodent carcinogenicity study, plus	——
One short- or medium-term study	• Supplements the long-term carcinogenicity study and • Provides additional information not readily available from the long-term assay

the rule of testing in two species should be the standard also in the future. The American consumer societies demanded to continue with this same standard of safety assessment.

As the new experimental approach to testing for carcinogenic potential, a basic scheme (Table 5) was set up to comprise one long-term rodent carcinogenicity study and one other study of the type that supplements the long-term carcinogenicity study and provides additional information that is not readily available from the long-term assay.

At the request of ICH, the International Life Sciences Institute (ILSI) and Health and Environmental Science Institute (HESI) coordinated experts for a large-scale research program (Alternatives to Carcinogenicity Testing Technical Committee) to characterize and evaluate selected transgenic rodent models (Table 6) proposed for use in human cancer risk assessment (10).

None of these models were considered sufficient as stand-alone assays. Most assays could detect genotoxic compounds, evaluated already within a genotoxicity test battery, but a better detection especially of nongenotoxic carcinogens is still needed (11). At a 2003 ILSI-HESI workshop on the use of transgenic animals for carcinogenicity testing, regulators concluded that these assays should be integrated with traditional test methods (12,13). The regulators considered the $p53^{+/-}$ and Tg.RasH2 models useful in providing data for regulatory purposes and the Tg.AC model useful in evaluating dermal products. The US Food and Drug Administration (FDA) considered also the neonatal mouse model. This model seems less favored in other regions today although it offers advantages with its administration of the test compound only twice on day 8 and 15 and a subsequent observation period of 12 months.

The species selected should be appropriate based on considerations of the primary pharmacodynamics of the compound, on findings from repeat-dose toxicity studies, on data on metabolism [see also guidelines S1C and S3A (8,14)], toxicokinetics [see also guidelines S1C, S3A, and S3B (8,14,15)] and route of administration (e.g., less common routes such as dermal and inhalation). In the absence of clear evidence favoring one species, it is recommended that the rat be selected for the long-term study.

Additional tests may be either short- or medium-term in vivo rodent test systems, which are usually in the mouse. These models of carcinogenesis may use transgenic or neonatal rodents. The guideline included at the time of its finalization in 1995 additional models of initiation or promotion in rodents. These models are considered today as useful for hepatic carcinogenesis or adequate mechanistic studies but not as assays appropriate as general screen for drug-induced carcinogenesis.

A long-term carcinogenicity study in a second rodent species (e.g., mouse) is still considered acceptable.

Table 6 ICH/S1B: New Approaches for Testing the Carcinogenic Potential

Studies	Mechanism studied
Transgenic mouse assays	
• p53 ±	• Inactivated tumor suppressor gene
• Tg.AC model	• Activated oncogenes
• Tg.RasH2 model	• Activated oncogenes
• XPA ± deficient model	• Inactivated DNA repair gene
Neonatal rodent tumorigenicity model	——

The guideline describes important information about the new models in the following "Notes."

Note 1 informs about another short-term study, the SHE assay (Syrian Hamster Embryo Cell Transformation Assay), The SHE assay is a short-term in vitro assay that assesses a substance's carcinogenic potential based on its ability to cause morphologic transformation of SHE cells.

Note 2 informs about conditions to limit testing with a single species. If the findings of a short- or long-term carcinogenicity study and of genotoxicity tests and other data indicate that a pharmaceutical clearly poses a carcinogenic hazard to humans then a second carcinogenicity study would not usually be useful.

Note 3 provides details about the short- or mid-term models as summarized in Table 6, which provides a representative list of approaches that may meet these criteria.

These new models have been accepted by regulatory agencies. The opinions of FDA and EMEA about these assays are identical and their acceptability for regulatory purposes has been often expressed. The Expert Working Group of ICH S1B discussed very intensively the optimization of the carcinogenicity assessment and recommended consideration of nonanimal alternative methods in addition to new animal models in the Appendix of ICH S1B (7).

Such nonanimal methods include cell-based assays and computational prediction models. It is recognized today that genotoxicity assays can be used to classify possible carcinogenic substances and further in vitro methods (e.g., cell transformation and gap junction intercellular communication) can identify possible carcinogens including nongenotoxic carcinogens. The advantage of such in vitro methods is that they are significantly faster and less expensive but current methods are not yet considered sufficient to serve as full animal replacements and are not recognized by agencies as validated alternatives.

The most widely used of the cell transformation assays (CTA) is the SHE assay which is believed to detect early steps of carcinogenesis by observing the cytotoxicity of test substances by measuring effects on morphology, colony-forming ability, and/or growth rate (16). A European Committee on the Validation of Alternative Methods [ECVAM (17), prevalidation project on SHE] is under way.

Another assay may be the gap junction intercellular communication (GJIC) method based on the disruption of the intercellular exchange of low-molecular-weight molecules through the gap junctions of adjacent cells. This disruption can result in abnormal cell growth and behavior (2). The assay seems good for screening for nongenotoxic carcinogens and tumor promoters but still needs to be validated.

Further (quantitative and qualitative) structure–activity relationship models [(Q)SARs and SARs] and expert systems have been developed to predict carcinogenicity. Several recent publications have reviewed models such as TOPKAT, CASE, and DEREK used by regulatory authorities (18,19). In general the computation of carcinogenicity is complex and predictive capability has been limited. The US FDA funded the development of MultiCASE based on data from regulatory submissions and it was reported to have improved predictivity (18).

As a general conclusion, because of the multiple stages and longer time periods needed for the expression of the carcinogenesis due to the multiple mechanisms and for example the necessary metabolic conversion of some substances, existing cell-based assays can be used only in a tiered testing scheme or test battery as a partial replacement for the animal bioassays (2).

ICH S1C: Dose selection for carcinogenicity studies of pharmaceuticals (8)

This guideline was adopted and implemented in the ICH regions in 1997. It focuses basically on the selection and justification of the high dose in animal studies.

Traditionally, carcinogenicity studies for chemical agents have relied upon the maximally tolerated dose (MTD) as the standard method for high dose selection.

The MTD is generally chosen based on data derived from repeat-dose toxicity studies of three months' duration.

Ideally the doses selected for rodent bioassays for nongenotoxic pharmaceuticals should provide an exposure to the agent and adequate margin of safety, no significant chronic

Table 7 ICH/S1C: Guideline on High-Dose Selection for Carcinogenicity Studies, Options for the Dose Selection

Maximum tolerated dose (MTD)
25-fold AUC ratio (rodent:human)
Dose-limiting pharmacodynamic effects
Saturation of absorption
Maximum feasible dose

physiological dysfunctions and be compatible with good survival rates of animals in the designed studies.

The guideline calls for a flexible approach to dose selection. The guideline proposes five different approaches (Table 7).

These criteria for the selection of high-dose levels include the most frequent aspect of toxicity-based endpoints (MTD: maximal tolerated dose or minimal toxic dose); for well-tolerated compounds: pharmacokinetic endpoints by applying a dose which represents 25-times the human exposure; further limitation of resorption called saturation of absorption; then limitation of tolerance by the primary pharmacodynamics and finally by limiting the dose with the principles of the maximum feasible dose.

1. The ICH Expert Working Group on Safety has agreed to continue to use the MTD as the most frequent option for high-dose selection for carcinogenicity studies. The MTD is defined as the top dose or maximum tolerated dose that produces a minimum toxic effect over the course of the carcinogenicity study. Factors to consider are alterations in physiological functions which would alter the animal's normal life span or interfere with interpretation of the study. Such factors include no more than 10% decrease in body weight gain relative to controls, or target organ toxicity or significant alterations in clinical pathological parameters.
2. The second option is the systemic exposure in rodents of a 25-times multiple of the human AUC (area under the curve) (at the maximum recommended daily dose). This may be an appropriate endpoint for dose selection for carcinogenicity studies for nongenotoxic pharmaceuticals with good general tolerance and lack of MTD values.
3. The third option to select the high dose is based on saturation of absorption measured by systemic exposures. This is the case when the absorption is satisfied and any higher administration of dose levels would not lead to a higher AUC. The mid and low doses selected for the carcinogenicity study should take into account saturation of metabolic and elimination pathways.
4. Option number 4 considers the pharmacodynamics of the compound in question for high-dose selection. The high dose selected should produce a pharmacodynamic response in dosed animals of such a magnitude as would preclude further dose escalation. However the dose should not produce disturbances of physiology or homeostasis. Examples include hypotension, inhibition of blood clotting, or insulin-like effects. Here, as in the conditions above, the high dose selected should always allow a survival of two years and therefore include age-related higher sensitivity periods to neoplastic processes.
5. The maximum feasible dose by dietary administration was considered to be 5% of diet. Many scientists considered this amount of drug to be too high. Therefore a new and more reasonable solution was formulated in the Guideline ICH/S1C(R), (9) which is discussed below.

ICH S1C (R): Addendum to dose selection for carcinogenicity studies of pharmaceuticals (9)—Addition of a limit dose and related notes of pharmaceuticals

Addendum S1C (R) was adopted in 1998. It focuses on the limit dose criteria. The R in this guideline is actually misleading because S1C does not address this aspect. There is no real revision—it is a new concept to facilitate the maximum dose application and prevent that more compound is being administered than needed.

The recommended conclusion is

In determining the high dose for carcinogenicity studies, it may not be necessary to exceed a dose of 1500 mg /kg/day. This limit dose applies only in cases, where there is no evidence of genotoxicity, and where the maximum recommended human dose does not exceed 500 mg/day.

In general this guideline recommends careful consideration of the maximum amount to be administered. In daily practice during the last ten years this guideline describes rare cases.

Evaluation and Classification of Results from Carcinogenicity Assays

The ICH and drug regulatory authorities provide guidance on testing for the carcinogenic potential of new drugs. The focus of ICH guidance is on how to design the tumorigenic studies. Useful detailed recommendations can be found also in OECD guidelines or in the EMEA Note for Guidance on Carcinogenic Potential (20). The OECD TGs 451, 452, and 453 (4,5,21) provide detailed information for conducting carcinogenicity and chronic toxicity studies mostly for the purpose of chemicals but also very useful for pharmaceutical compounds as well. The OECD *Guidance Notes for Analysis and Evaluation of Chronic Toxicity and Carcinogenicity Studies* [ENV/JM/MONO (2002)] (22) provides broad guidance on approaches to hazard assessment and on some of the problems and pitfalls that may arise during an assessment.

The EMEA Note for Guidance on Carcinogenic Potential was issued by the Committee for Proprietary Medicinal Products (20) in 2003; very relevant information is expressed there on how to evaluate the results from carcinogenicity rodent tests.

As general principles in this EMEA note preneoplastic and neoplastic lesions should be described in conventional histopathological vocabulary according to internationally accepted classifications like ILSI, STP, RENI (20), or others. The general reading of the slides should ideally be performed by one pathologist, and a peer review of 10% of all animal slides by pathologists not involved in the specific study is recommended. The data should be presented for each treatment or control group. Table 8 provides information on the number of animals examined, the number of animals with tumors (separating benign from malignant tumors), time of unscheduled death, and time of appearance of any mass and its progress.

Answers need to be provided for any dose relationship, treatment relationship, tumor incidences higher than in historical controls, rare tumors, and location of tumors (including any at the site of injection). Finally, a "weight of evidence" should be provided with a focus on the pharmacological effects, the genotoxicity findings, any preneoplastic lesions discovered in repeat-dose studies, changes in husbandry, diet, background health status of the animals, etc. Finally, this Note offers a list of tissues to be examined histopathologically for all species.

The counting of the different types of tumors and their statistical analysis is a relevant part of the assessment but it is very important also that the experimenter tries to investigate the mechanism leading to tumorigenesis. There are genotoxic or nongenotoxic mechanisms. Species-specific metabolism or sensitivity may cause tumors in animals but not in humans. Intensive search for biomarkers of cancer may help understand cellular changes in carcinogenesis. Embryonic and tumorigenic signaling pathways (23) are in discussion. These pathways represent potential new therapeutic targets as well as potential new biomarkers for detecting the early stages of cell transformation in vitro.

Table 8 Evaluation of Carcinogenic Potential

Item	Issue
Information on tumor	Benign vs. malignant, incidence, local induction, and latency; frequent vs. rare tumors
Kinetics	AUC comparison of rodent and humans; dose-dependency; control group data to evaluate contamination
Ancillary data	Genotoxicity or mechanistic studies; species-specificity; historical data, databanks; literature
Overall consideration	Use results from any tests as part of the overall "weight of evidence" for carcinogenic potential

While the genotoxic mechanism is relatively well understood today for tumorigenesis, epigenetic mechanisms are still intensively investigated. *Epigenetic* refers to mechanisms that alter gene expression without actual changes to the gene/DNA sequence. DNA methylation is an example of epigenetic mechanism involved in carcinogenesis that causes heritable changes in gene expression independent of altering the DNA sequence. Scientists have shown that DNA methylation is an important component in a variety of chemical-induced toxicities including carcinogenicity and should be assessed in the overall hazard assessment (24,25). DNA methylation evaluated along with other in vitro test data has been found useful in prioritizing chemicals for further assessment. DNA methylation coupled with cytotoxicity and genotoxicity assays strengthened the prioritization and was especially useful in identifying genotoxins that were toxic at nonlethal concentrations (26).

The ECVAM Task Force on Carcinogenicity (17) tries to develop a testing strategy that can detect both genotoxic and nongenotoxic carcinogens. The task force has founded a collaborative network of scientists to conduct work aimed at developing an integrated strategy based on the use of cell transformation assays by mouse fibroblast (Balb/3T3) in combination with in vitro testing (Micronuclei and Comet assays). ECVAM is also conducting cell-based studies to determine the molecular signatures of carcinogenic chemicals.

Numerous further efforts are under way to address problems like false-positive results in current in vitro genotoxicity tests, species and gender differences, refinement of extrapolation of animal data to human settings, more and earlier knowledge on the different steps of tumorigenesis, etc. Initiatives include work being done by the following organizations: European Centre for Ecotoxicology and Toxicology of Chemicals (ECETOC), ECVAM, European Cosmetic Toiletry and Perfumery Association (COLIPA), ICH, and ILSI-HESI. The experimenter is advised to observe carefully these new developments. Many of these new models are being used for internal decision-making processes (e.g., in order to prioritize lead compounds among many candidates). Efforts will continue to validate or at least evaluate these new approaches, but without recognition and approvability of these new models by regulators one cannot yet deviate from the ICH recommendations.

ICH will follow intensively any new and better approaches to test the potential for developing tumors. In 2009a revised version of the guidelines for genotoxicity will be available, reanalyses of data banks will take place, the usefulness of alternative models such as the transgenic mouse will be reconsidered and there will be a strong effort to optimize the predictability further.

REFERENCES

1. United Nations Economic Commission for Europe (UNECE). Globally Harmonized System of classification and labelling of chemicals (GHS). Part 3, Health and environmental hazards. Chapter 3.6, Carcinogenicity. 2004:167. http://www.unece.org/trans/danger/publi/ghs/ghs_rev00/English/GHS-PART-3e.pdf. Accessed January 26, 2009.
2. Maurici D, Aardema M, Corvi R, et al. Carcinogenicity. Altern Lab Anim 2005; 33(suppl. 1):177–182.
3. Organisation for Economic Co-operation and Development (OECD). Detailed Review Paper on Cell Transformation Assays for Detection of Chemical Carcinogens. DRP No 31, revised 4th version, August 2007. http://www.olis.oecd.org/olis/2007doc.nsf/LinkTo/NT00002FOA/$FILE/JT03230941.pdf. Accessed July 2009.
4. OECD Guidelines for Testing Chemicals, Test No 451, Carcinogenicity. http://www.oecdbookshop.org/oecd/display.asp?K = 5LMQCR2K7MQ4&m = 26&dc = 52&mw = 1&lang = en&st_01 = ser-00611e1&sf_01 = seriesidentifier&plang = en; http://www.oecdbookshop.org/oecd/display.asp?K = 5LMQCR2K7MJG&m = 28&dc = 52&mw = 1&lang = en&st_01 = ser-00611e1&sf_01 = seriesidentifier&plang = en. Accessed January 26, 2009.
5. OECD Guidelines for Testing Chemicals, Test No 453, Carcinogenicity. http://www.oecdbookshop.org/oecd/display.asp?K = 5LMQCR2K7MJG&m = 28&dc = 52&mw = 1&lang = en&st_01 = ser-00611e1&sf_01 = seriesidentifier&plang = en. Accessed January 26, 2009.
6. ICH Guideline S1A: Note for Guidance on the Need for Carcinogenicity Studies of Pharmaceuticals. http://www.ich.org/LOB/media/MEDIA489.pdf. Accessed January 26, 2009.
7. ICH Guideline S1B: Note for Guidance on Carcinogenicity: Testing for Carcinogenicity of Pharmaceuticals. http://www.ich.org/LOB/media/MEDIA490.pdf. Accessed January 26, 2009.

8. ICH Guideline S1C: Note for Guidance on Dose Selection for Carcinogenicity Studies of Pharmaceuticals. http://www.ich.org/LOB/media/MEDIA491.pdf. Accessed January 26, 2009.
9. Addendum S1C (R): Note for Guidance on Dose Selection for Carcinogenicity Studies of Pharmaceuticals: Addition of a Limited Dose and related Notes (CPMP/ICH/366/96). http://www.ich.org/LOB/media/MEDIA3560.pdf. Accessed January 26, 2009.
10. Robinson DE, MacDonald JS. Background and framework for ILSI's collaborative evaluation program on alternative models for carcinogenicity assessment. International Life Sciences Institute. Toxicol Pathol 2001; 29:13–19.
11. Goodman JI. A perspective on current and future uses of alternative models for carcinogenicity testing. Toxicol Pathol 2001; 29(suppl.):173–176.
12. ILSI-HESI. Workshop on the Utility of Transgenic Assays for Risk Assessment. Washington, DC: Oral Personal Communication, 2003.
13. MacDonald J, French JE, Gerson R, et al. The utility of transgenic assays for risk assessment. Toxicol Sci 2004; 77(2):188–194.
14. ICH guideline S2A: Specific aspects of regulatory genotoxicity tests for pharmaceuticals (ICHS2A). http://www.ich.org/LOB/media/MEDIA493.pdf. Accessed January 26, 2009.
15. ICH guideline S2B: Genotoxicity: a standard battery for genotoxicity testing of pharmaceuticals (ICHS2B). http://www.ich.org/LOB/media/MEDIA494.pdf. Accessed January 26, 2009.
16. Combes R, Balls M, Curren R, et al. Cell transformation assays as predictors of human carcinogenicity. ECVAM Workshop Report 39. Altern Lab Anim 1999; 27:745–767.
17. ECVAM. Genotoxicity and carcinogenicity. Altern Lab Anim 2002; 30(suppl. 1):83–93.
18. Cronin MTD, Jaworska JS, Walker JD, et al. Use of QSARs in international decisions-making frameworks to predict health effects of chemical substances. Environ Health Perspect 2003; 111:1391–1401.
19. OECD Report on the Regulatory Uses and Application in OECD Member Countries of (Quantitative) Structure–Activity Relationship (QSAR) Models in the Assessment of New and Existing Chemicals. ENV/JM/MONO (2006) 25, Series on testing and Assessment, No 58. February 2008. http://www.oecd.org/dataoecd/55/22/38131728.pdf. Accessed July 2009.
20. EMEA Note for Guidance on Carcinogenic Potential (CPMP/SWP/ 2877/00 from 25 July 2002). http://www.emea.europa.eu/pdfs/human/swp/287700en.pdf. Accessed January 26, 2009.
21. OECD Guidelines for Testing Chemicals, Test No 452, Chronic toxicity studies. http://www.oecdbookshop.org/oecd/display.asp?K = 5LMQCR2K7MLW&lang = EN&sf1 = SeriesIdentifier&st1 = SER-00611E1&m = 27&dc = 52&plang = en. Accessed January 26, 2009.
22. OECD Guidance Notes for Analysis and Evaluation of Chronic Toxicity and Carcinogenicity Studies (ENV/JM/MONO(2002)19. http://www.olis.oecd.org/olis/2000doc.nsf/LinkTo/env-jm.mono(2002)19. Accessed January 26, 2009.
23. Hendrix MJ; Seftor EA, Seftor RE, et al. Reprogramming metastatic tumor cells with embryonic microenvironments. Nat Rev Cancer 2007; 7:246–255.
24. Watson RE, Goodman JI. Epigenetics and DNA methylation come of age in toxicology. Toxicol Sci 2002; 67:11–16.
25. Moggs JG, Goodman JI, Trosko JE, et al. Epigenetics and cancer: implications for drug discovery and safety assessment. Toxicol Appl Pharmacol 2004; 196:422–430.
26. Watson RE, McKim JM, Cockerell GL, et al, The value of DNA methylation analysis in basic, initial toxicity assessments. Toxicol Sci 2004; 79:178–188.

27 Reproductive and Developmental Toxicity

John Baldwin
Kelvin Associates, Campton, U.K.

The purpose of testing new potential medicinal products for reproductive toxicity is to identify possible hazards to human reproduction. First-pass studies are conducted in animals of reproductive age and their offspring and, if effects are seen, more refined follow-up studies are designed on a case-by-case basis to characterize further those effects and to provide an indication of the degree of potential risk to humans.

There is a plethora of regulatory guidelines and guidance notes that describe in some way the regulatory attitudes to reproductive toxicology testing. The best place to start any consideration of regulatory requirements in this area is the International Conference on Harmonisation of Technical Requirements for Registration of Pharmaceuticals for Human Use (ICH) guideline S5(R2) (1). There are separate guidelines that discuss requirements for biotechnology-derived products, vaccines, anticancer medicinal products, herbal medicines, etc., and yet others that discuss the timing of studies in relation to the clinical programme as well as risk–benefit considerations. These other guidelines will be reviewed further on in this chapter.

The S5 Guideline was first introduced as an approved Step 4 document (S5A) in June 1993; it was devoted primarily to a consideration of reproductive toxicity studies in females. In November 1995, the corresponding guideline for studies in males (S5B) was finally approved. There had been some interagency disagreement about the conduct of male fertility studies, particularly concerning the duration of treatment, hence the delay in approval; even with the supposedly harmonised guideline, some disagreement still remained until recently. The next step in the development of the ICH S5 Guideline was the publication of S5(R2) in November 2005, which saw the two separate guidelines amalgamated and its contents harmonised to a far greater extent.

The S5 Guideline, and good common sense, requires that any programme of reproductive toxicity studies should allow exposure of the novel chemical to all stages of development throughout one complete life cycle, that is, from conception in one generation through to conception in the following generation, or from any stage in one life cycle through to an identical stage in the next cycle. This life, or reproductive, cycle is subdivided into the following stages:

A. Premating to conception (adult male and female reproductive functions, development and maturation of gametes, mating behaviour, and fertilisation)
B. Conception to implantation (adult female reproductive functions, preimplantation development, and implantation)
C. Implantation to closure of the hard palate (adult female reproductive functions, embryonic development, and major organ formation)
D. Closure of the hard palate to the end of pregnancy (adult female reproductive functions, foetal development and growth, and organ development and growth)
E. Birth to weaning (adult female reproductive functions, neonate adaptation to extrauterine life, and preweaning development and growth)
F. Weaning to sexual maturity (postweaning development and growth, adaptation to independent life, and attainment of full sexual function).

The guideline requires that all stages of this cycle are covered in one or more suitable investigations. Where more than one investigation is used, there must be an overlap between studies so that no gaps are left between key stages. In theory, it would be possible to conduct just one study where it is known that there are no significant pharmacokinetic changes over prolonged administration and it could also be argued that reproductive hazards due to long-term exposure to low doses may be represented better by a one or two generation study approach. However, this does present problems where a chemical has an effect on some stage of reproduction that may produce "knock on" difficulties in interpreting effects on subsequent stages of the reproductive cycle. If the chemical affected fertility, for example, then there may not be adequate numbers of pregnant animals or fetuses to assess properly developmental toxicity in utero; effects in utero leading to malformation or effects on embryonic/fetal growth or survival may reduce the ability of the study to detect postnatal effects, and so on. In practice, it is usually more appropriate to conduct a number of overlapping studies. Guideline S5(R2) describes what is termed "the most probable option" and is the option that is favoured by most investigators other than in rare circumstances. This "most probable option" consists of three studies that individually examine effects on

- fertility and early embryonic development,
- pre- and postnatal development, including maternal function, and
- embryo–foetal development.

It should be emphasized that this strategy may represent only a starting point in the investigation of reproductive toxicity. If no biologically significant effects are seen in this battery of studies and the studies have been conducted to a suitably robust design in the appropriate species, then no further investigations would be required. If effects were to be seen, then the findings might lead to a second level of examination to provide a more in-depth assessment. This would be on a case-by-case basis but some examples of possible follow-up strategies will be presented in this chapter.

STUDY OF FERTILITY AND EARLY EMBRYONIC DEVELOPMENT TO IMPLANTATION

This study is designed to investigate effects on stages A and B of the reproductive cycle described above. It should detect toxic effects/disturbances resulting from treatment from before mating (males/females) through mating and implantation. For females, this includes effects on the estrous cycle, tubal transport, implantation, and development of the preimplantation embryo. For males, it will permit detection of functional effects (e.g., on libido or epididymal sperm maturation) that may not be detected by histological examination of the male reproductive organs. This study can be run as a male fertility study (treated males paired with untreated females), a female fertility study (treated females paired with untreated males), or a combined-sex study where both sexes are treated. This will depend on the nature of the compound and, to some extent, the constraints of the manufacturer's development plan for the product. Clearly, a male fertility study will suffice where the drug is intended to be given therapeutically only to males and vice versa for females.

The development plan for the novel pharmaceutical can be a decisive factor. ICH Guideline M3 provides recommendations for which nonclinical studies need to be conducted to support the various phases of clinical trials and to support a marketing application. The current version of this guideline, M3(R1) (2), allows men to be included in Phase I and II trials in the European Union (EU) and the United States prior to the conduct of the male fertility study so long as an adequate evaluation of the male reproductive organs has been conducted in the repeat-dose toxicity studies and that no effects were found. Females of childbearing potential could also be included without fertility studies so long as developmental toxicity studies had been conducted (the European Union) or so long as women trialists used adequate methods of contraception (the United States). A male fertility study, or a fertility study in both sexes, should be completed prior to the initiation of Phase III trials. For Japan, however, the existing regulations require that an assessment of male fertility should be conducted before including men in clinical trials and an assessment of female fertility and developmental toxicity should be conducted prior to women of childbearing potential entering trials. This has led, in some cases, to companies understandably performing separate sex fertility studies to meet the Japanese requirements and to expedite their clinical trials. The M3 Guidelines are under revision at the time of writing and an updated version was released for consultation in July 2008 as M3(R2) (3). The new revision harmonises the requirements such that Phase I and II trials in all marketing areas (including Japan) can be conducted prior to performing a male fertility study. The requirement that male reproductive organs should be evaluated in the repeat-dose toxicity studies, and that there should be no effects on these organs, remains. Women of childbearing potential can now be included in Phase I and II trials in all regions without a fertility study being conducted. Fertility studies for males and females are now required for all regions only prior to Phase III trials and, if the M3 revisions are implemented, it would make good sense to conduct a combined study in treated males and females.

The fertility study needs to be conducted only in one mammalian species, and invariably, the species chosen is the rat. The main consideration, though, as for all reproductive studies, is whether the chosen species is relevant in terms of kinetic, pharmacological, and toxicological comparisons such that results can be extrapolated to humans. By the time that fertility studies are conducted, repeat-dose toxicity studies of at least one month's duration with toxicokinetics should have been conducted and pharmacokinetic data in humans should also be available. This information should help determine the suitability of the chosen species and provide invaluable information for selecting dose levels rather than having to perform preliminary dose range-finding studies. The mouse would make a useful alternative species to the rat and the rabbit should not be overlooked: the ability to conduct longitudinal semen analysis in rabbits without using invasive techniques is an obvious advantage. Where all the above circumstances permit, the ideal situation is to use a species and strain that has also been used in the toxicological assessment of the test substance.

Whatever species is used, the basic principle of the study design is that males and females are dosed for a suitable period prior to pairing with treated or untreated partners. During this period, females, if rodents are used, are assessed for the regularity of the estrous cycle. Dosing continues throughout pairing until mating has occurred (this may not happen for a number of days and is detected, in rodents, by the presence of a copulation plug or sperm in a vaginal smear), with treatment of the females continuing until just after implantation and treatment of the males continuing until termination of the study in case further mating trials are required. If rats are used, a typical study design would entail treatment of the females for two weeks before pairing, through the pairing period until shortly after implantation: Day 7 of pregnancy is the most likely choice. Animals are then left untreated until mid pregnancy, days 13 to 15 usually, when they are killed. Note that, at this stage of pregnancy, embryos/foetuses are not really amenable for examination for abnormalities and weight and the only observation required on these individuals is to count them and to deduce the number that are alive and dead by reference to the number of corpora lutea and implantation sites in the uterus. Males are treated for four weeks prior to the start of the pairing period and then continuously until postmortem examination of the females has been completed and the litter data evaluated in case further matings are needed. The investigator should also concentrate on evaluating the parent animals for signs of general toxicity, which might indirectly lead to an effect on reproduction. Thus, animals are observed regularly for signs of reaction to treatment, they are weighed, and food and water intakes are recorded. At termination, the parents will be subjected to a full macroscopic examination, and testes, epididymides, ovaries, and uteri from all animals will be taken and preserved for possible histological examination and evaluation on a case-by-case basis. Sperm count in epididymides or testes, as well as sperm viability, may also be assessed.

Guideline S5 suggests that the number of animals per sex per group should be sufficient to allow meaningful interpretation of the data. This is rather hard to interpret in numerical terms. In practice, most laboratories will use around 20 to 24 males and females per group and this is invariably adequate to pick up all but the rarest of events. When both sexes are being dosed or are of equal consideration in separate male and female studies, the preferred mating ratio is 1:1 since this is the safest option in respect of obtaining good pregnancy rates and avoiding incorrect analysis and interpretation of results. Three active dose groups plus a vehicle control group are invariably used. The choice of the high dose should be based on data from all available studies (pharmacology, acute and chronic toxicity, and kinetic studies). The results of a two- to four-week repeat-dose toxicity study should be available and this will provide a close approximation to the duration of treatment in fertility and other reproductive studies. Strategies to be employed in the selection of dose levels are discussed later in this chapter.

This is the first tier assessment of fertility and the final tier if there are no untoward findings. Where there are findings of concern, then further investigations may be required. The easiest first step is where there are signs of impaired fertility in a study. Where males and females have been treated, it is usually difficult to determine whether that impaired fertility was due to the male, the female, or both. By not killing males, and continuing to treat them until all results are available, it is possible to conduct further mating trials using those males and additional untreated females brought onto study. Subsequent follow-up studies will obviously involve separate sex fertility studies but may also home in on a specific feature of the reproductive cycle such as uterotubal transport, sperm transport, etc.

STUDY OF EFFECTS ON EMBRYO–FOETAL DEVELOPMENT

This study is designed to detect adverse effects on the pregnant female and development of the embryo and foetus following exposure of the female from implantation to closure of the hard palate (i.e., stages C to D of the reproductive cycle listed above). It should detect altered embryonic/foetus growth, structural changes (i.e., malformation), and embryo–foetus death. The study should also be able to delineate whether pregnant animals respond to the test drug in a different way to their nonpregnant counterparts.

Two mammalian species should usually be tested: one should be a rodent and the other a nonrodent. The most frequently used species are the rat and rabbit but other species can and are used. In the same way as the fertility study, the most important consideration is to select relevant species where data (principally pharmacokinetic) can be extrapolated to humans. By the time these developmental toxicity studies are conducted, a considerable amount of toxicokinetic data will have been generated from the repeat-dose toxicity programme and pharmacokinetic data in humans will also be available to make cross-species comparisons that should validate the test species as suitable for extrapolation to humans. Rats are used for pragmatic reasons whereas rabbits have been used largely for historical reasons in that embryos exposed to thalidomide show limb malformations (although not the phocomelia seen in humans). On the plus side, since rabbits have been used so much, a large amount of background data has been generated in this species. Since rabbits are not commonly used for repeat-dose toxicity testing, they will have to be validated as a relevant species in a different way, probably by use of toxicokinetic assessments as part of range-finding studies in this species. If the rat is used, it makes good scientific sense to use the same strain as used for the repeat-dose toxicity study—in fact, it makes bad scientific sense to do anything else. If pharmacology and pharmacokinetics in one or other species shows an excellent match with humans, then this can be an adequate justification for using that species alone for developmental toxicity testing. Where either the rat or the rabbit are deemed to be unsuitable, Guideline S5 discusses the use of alternative species such as mice, guinea pigs, hamsters, ferrets, dogs, domestic and mini pigs, and nonhuman primates (and it is best to refer to the guideline for more detail).

Whichever species is used, the basic principle of the study design is that pregnant females are dosed for a period that extends from implantation to the closure of the hard palate. If using the rat as the experimental species, this would be from days 6 to 17 of pregnancy and in the rabbit it would be from days 6 to 19. During the study, the parent females should be observed regularly for signs and mortalities and body weights and food intake should be closely monitored. Other observations may be included on a case-by-case basis, especially if they have proven useful as markers in the repeat-dose or other toxicity studies.

Parent females should be killed and examined about one day prior to parturition. Convenient times would be gestation days 21 for rats and 29 for New Zealand white rabbits (other rabbit strains may differ slightly). They should be given a thorough macroscopic postmortem examination and organs showing macroscopic change should be preserved and processed for possible microscopic examination (and control organs should be taken for comparison). Other examinations required include counting the number of corpora lutea and the numbers of live and dead implantations, weighing foetuses (placental weights may also be useful), and examination of all foetuses for viability and the presence of abnormalities. Detection of abnormalities should include external and internal examination and the latter will include a combination of techniques to determine the presence of visceral (including brain) and skeletal defects. Where foetuses are subjected to a variety of examination techniques, they should be identified individually so that findings can be correlated. It is permissible to just examine high-dose and control groups for abnormalities in the first instance, and if no relevant intergroup differences are seen, then examination of the low- and intermediate-dose foetuses can be omitted but it would be advisable to store the fixed specimens for possible later examination.

The general principles relating to the number of animals to be used and the selection of dose levels (and number of dose groups) are the same as already described for the fertility study. Commonly, for rats and rabbits, three treatment groups and a control group are used, each containing around 20 to 25 pregnant females.

Clearly, if the study has been conducted to high standards at adequate dose levels under optimal conditions and there are no signs of embryo–foetal toxicity or teratogenicity, then these definitive studies will have served their purpose. These studies represent the first (and maybe the only) tier approach. Where untoward effects are seen, then one may wish to explore them further. An obvious (but often overlooked) situation is where the only observation of note is a high incidence of embryonic or foetal death. The possibility that death may be due to malformation and the dead foetuses could be masking a teratogenic effect must not be ignored. Exploration of the effects of lower doses will be required. Sometimes, it may be necessary to simply explore additional intermediate-dose levels to help characterise dose–response relationships to assist in risk assessment or to obtain a better understanding of the relationship between maternal effects (toxicity or pharmacological effects) and responses in the embryo–foetus.

Nonanimal alternatives (e.g., whole embryo culture) are available but are generally used only in preliminary investigations, for example, to screen a range of analogue substances. They also have a role to play in follow-up investigations where they might be used to reveal specific mechanisms of toxicity, to establish concentration–response relationships, to select "sensitive periods," or to detect effects of defined metabolites. The definitive, regulatory studies must, however, be conducted in whole animals since the alternatives lack the complexity of the developmental processes and the dynamic interchange between the maternal and the developing organisms cannot provide assurance of the absence of effect nor provide perspective in respect of risk/exposure.

The regulatory attitude to bringing women of childbearing potential into clinical trials has shown considerable regional variation within the existing relevant guideline (ICH Guideline M3(R1) (2). In the European Union, it has been required that embryo–foetal development studies in two species should be completed prior to Phase I trials or prior to including women of childbearing potential at whatever stage of the development programme, with female fertility studies needing to be conducted prior to Phase III. In Japan, the female fertility and the embryo–fetal development studies have been required prior to the inclusion of women of childbearing potential in clinical studies. In the United States, it has been the case that women of childbearing potential may be included in early, carefully monitored studies without reproduction toxicity studies provided appropriate precautions are taken to minimise risk and as long as patients are monitored regularly on trial to ensure that they have not become pregnant. There must be informed consent from the patient. Assessment of female fertility and embryo–foetal development is required prior to Phase III trials.

The new M3 Guideline revisions (3) now provide a similar approach across all regions. So long as the revisions are ratified, and that seems most likely, women of childbearing potential may be included in clinical trials where

- effective precautions are taken to prevent pregnancy,
- study entry is only after a confirmed menstrual period,
- there is pregnancy testing during the trial,
- there is informed consent,
- there is intensive control of pregnancy risk over short duration clinical trials (such as two weeks) or where trials (or exposure for compounds with long half-lives) are of longer duration and where there is a predominance of the disease in women,
- the mechanism of action of the agent is fully understood, and
- there is difficulty in conducting developmental toxicity studies in an appropriate animal model.

For clinical studies involving relatively few women (up to 150) and that are of relatively short duration (up to three months), those studies can be conducted prior to full assessment of developmental toxicity but preliminary animal studies whose findings give no cause for concern are generally considered desirable. These preliminary studies are described below. In the United States, definitive embryo–foetal development studies need not be conducted until prior to Phase III. In the European Union and Japan, other than the situations described above, definitive nonclinical developmental toxicity studies should be completed prior to women of childbearing potential being exposed to the drug.

The preliminary embryo–foetal studies mentioned above should be conducted in two species using six dams per group treated over the period of organogenesis, with adequate dose levels and including external, visceral, and skeletal examinations; these studies should be conducted under good laboratory practice conditions.

Prior to the inclusion of pregnant women in clinical trials, all the reproduction toxicity studies and the standard battery of genotoxicity tests should have been conducted. In addition, safety data from previous human exposure are generally needed. This requirement is unchanged by the new revisions to the M3 Guideline. If the use of the drug in pregnant women is contraindicated, then developmental toxicity (and fertility) studies should still be conducted.

The M3(R2) revision now contains a section for bringing women who are not of childbearing potential into trial (i.e., permanently sterilised, postmenopausal). Postmenopausal is defined as 12 months with no menses without an alternative medical cause. They can be included without any reproduction toxicity studies if appropriate repeat-dose toxicity studies, including an evaluation of the female reproductive tract, have been conducted.

STUDY FOR EFFECTS ON PRE- AND POSTNATAL DEVELOPMENT INCLUDING MATERNAL FUNCTION

This study is designed to detect adverse effects on the pregnant/lactating female and on development of the embryo–foetal and the offspring following exposure of the female from implantation through to weaning. All treated females are allowed to give birth and rear their young to weaning and the offspring are then selected to continue on study untreated until sexual maturity or, more particularly, until they have demonstrated that they are capable of producing a subsequent generation of offspring. The experiment therefore covers stages C to F listed in the ICH S5(R2) Guideline. There is emphasis first of all on the dam to ensure that she maintains pregnancy and delivers her young, and that the duration of gestation is unaffected and that parturition proceeds normally. Maternal function (e.g., milk production, suckling, dam/pup interaction) is also assessed. Even though effects on embryonic and fetal development will have been evaluated in the embryo–foetal development study, treatment during the major period of organogenesis is repeated in this study with foetuses being delivered naturally rather than derived by Caesarean section. Although fetuses will have been examined thoroughly for the presence of malformation in the embryo–foetal study, it is naive to assume that every abnormality will have been detected. Some defects are very subtle and there are situations where, for example, small interventricular septal defects have gone unnoticed in the embryo–foetal study but have led to postnatal offspring death in the pre- and postnatal study because the defects were sufficiently major to be incompatible with life. Other defects, such as biochemical or functional deficits, cannot be detected by anatomical examination and their presence can only be determined by allowing the foetus to be born and examining its development in detail throughout its life span. Some functional changes, such as behavioural abnormalities, may only be detected later in life using special methodology. It should also be realised that not all development takes place in utero—the CNS continues to develop after birth and a considerable portion of functional development occurs postnatally whilst the offspring can continue to be exposed to the test article by transfer via breast milk.

Only one species is normally tested and the most usual, and preferred, species is the rat. However, the rat is not always a relevant species and can, in some cases, be classed as a highly irrelevant species, as discussed later. Mice can be used as an alternative and it is possible to conduct such studies in rabbits. Because of the need to retain offspring to maturity to assess their reproductive function, those species with a long maturation period such as dogs and most nonhuman primates cannot readily be used for reasons of pragmatism.

If the rat is used, then treatment will commence around day 6 of gestation and continue uninterrupted throughout pregnancy and through the postnatal period until weaning of the offspring at around day 21 postpartum. Around 20 to 24 pregnant females should be used in each treatment group. Observations on the dam will include signs and mortalities, body weight, food intake, duration of pregnancy, the normality of parturition, maternal instinct, and maternal/pup interaction. The aim is to detect enhanced toxicity relative to that in nonpregnant females and to determine whether there are any effects that might impact on the well-being of the offspring. Examination of the offspring will include the detection of pre- and postnatal death, altered growth and development, and functional deficits in offspring including behaviour, sexual maturation and reproductive performance. Where the drug is known to produce target organ toxicity such as kidney or liver damage in the adult, it may be sensible to extend the battery of observations in the offspring to include specific organ function tests.

Some laboratories cull the offspring at birth to provide, as far as possible, standardised litters of, say, four males and four females per litter. This practice is frowned upon by others and, if culling is to form part of the experiment, it should be justified adequately. It seems somewhat counterproductive to purposely exclude offspring that might otherwise provide valuable information—it is just like throwing data away randomly.

Just as with the other reproductive studies, the standard experiment that has been described may be the only examination of pre- and postnatal toxicity that is required. That is, of course, if no untoward effects are detected. Adverse effects may lead the investigator into a second tier of studies, especially if those adverse effects cannot be readily interpreted from the available data. The presence of postnatal offspring death could be due to exposure of the developing embryo–fetus in utero or the offspring postnatally via milk. It may also be due to complications arising from a difficult delivery or to failure of the dam to produce milk, suckle, or bond with the offspring. The cause or causes may not be readily obvious. Follow-up studies may involve cross-fostering where offspring of treated dams are fostered to control mothers and vice versa and this relatively simple procedure can often be successful in separating prenatal from postnatal influences. Where difficult delivery is thought to be the etiological factor, then treatment can be stopped several days before expected parturition and then resumed after birth. Dams can be observed to see if there are any obvious effects on maternal behaviour and suckling can be determined by the presence of milk in the relatively transparent abdomen of the young animal. Pharmacokinetic studies can be carried out to quantitate the degree of drug transfer in the milk and this, coupled with neonatal acute toxicity studies, can give a good idea whether findings are due to a direct effect of the drug on the offspring. Young individuals can be far more susceptible to a drug than their adult counterparts. These are just some of the options available.

The pre- and postnatal development study is required in all regions for marketing approval but it makes good sense to conduct it earlier if there is reason to believe from the compound's pharmacological activity or by analogy to related compounds that the novel pharmaceutical may produce effects at this stage of the reproductive cycle.

There will be situations where there will be good reasons to modify this general strategy. For example, when developing new drugs intended to be administered only to females for gynecological problems, for example, to regularize menstruation, adverse effects on female reproduction would be highly undesirable and it would be prudent to determine whether this might be the case at any early stage in the compound's development. Similarly, for drugs intended only for males, there would be little sense in conducting reproductive toxicity studies in females. There will also be situations where omission of all reproductive toxicity studies could appear justified, for example when treating life-threatening/terminal illnesses. This should be done only with appropriate regulatory approval, following an appropriate Scientific Advice procedure, for example.

FURTHER CONSIDERATION OF SOME EXPERIMENTAL VARIABLES INVOLVED IN REPRODUCTIVE TOXICITY TESTING

Choice of Species

Studies should be conducted in mammalian species. It is generally desirable to use the same species and strain as in other toxicological studies. The number of species to be used and the most commonly used species have already been discussed.

Although the rat is the most frequently used species for all reproductive toxicity studies, there will be occasions where this species is inappropriate. Where the pattern of metabolism in rats and humans is dissimilar, an alternative species should be sought. Female rats are highly susceptible to nonsteroidal anti-inflammatory agents during late pregnancy and they are also unsuitable for testing dopamine agonists or compounds reducing circulating prolactin levels; one effect of such drugs in rats (but not humans) is to inhibit lactation and offspring may fail to gain weight or even die from starvation. In both cases, the rabbit may be a more appropriate choice. The rabbit, on the other hand, is unsuitable for testing many antibiotics since they may destroy the gastrointestinal microflora essential for well-being in this species. This species also has a low tolerance to angiotensin-converting enzyme inhibitors. Mice are useful in many instances, but they do show an exaggerated stress response leading to high circulating corticosteroid levels and stressful procedures or marked maternal toxicity can lead to high instances of cleft palate that are not directly related to the drug itself.

Selection of Dose Levels

When designing reproductive toxicity studies, and particularly when selecting dose levels, it is important to take into account all existing data on toxicity, pharmacodynamics, kinetics, and similarity to other compounds with respect to structure–activity relationships where available. Where preliminary dose-finding studies have been conducted, these will be pivotal in selecting appropriate dose levels for the main studies; all preliminary studies must be reported as part of the submission package. Dose levels should be selected to avoid suffering in the experimental animals. Three treatment groups plus a control group are usually employed. The highest dosage should normally produce some degree of (but not excessive) maternal toxicity, embryo–foetal toxicity, or exaggerated pharmacological response. Reduced body weight or food intake may be useful markers but some maternal effects may be less obvious such as changes in hematology or clinical chemistry. The high dose may also be restricted by the physicochemical characteristics of the test article or by nonlinear dose-limiting kinetics. Regardless, in most circumstances, a limit dose of 1000 mg/kg/day is usually considered adequate for reproductive toxicity studies. The lowest dose should usually be a small multiple of the anticipated human dose or achieve an exposure that is a small multiple of the expected human exposure, with an intermediate dose somewhere in between. In some cases, it may be necessary to have closer dose intervals to evaluate dose–response relationships. The calculation of a no-effect level is desirable, but the demonstration of a toxic effect, where one exists, is essential, and it may be necessary to use more than three dose levels or to perform separate studies with differing objectives.

Numbers of Animals on Study

Studies should use the minimum number of animals necessary to achieve the overall objectives but not so few animals as to render the study uninterpretable leading to a need to repeat the experiment. Typical numbers of rodents used in the regulatory standard studies have been discussed elsewhere in this chapter. Group sizes of 20 to 25 rats, mice, or rabbits are normal. There is no scientific justification why these group sizes should be reduced for other species, although for financial and ethical considerations, far fewer primates tend to be used. In fact, for species with small litter sizes and where there is little background information, there are compelling reasons for using even larger group sizes. Furthermore, where a study is designed specifically to detect an event that occurs at a low incidence, more animals may be required and input from a statistician in the study design phase is clearly desirable.

Route of Administration and Dose Frequency

The route or routes of administration used should be similar to those intended to be used in humans. One route of administration may be acceptable if there is a similar kinetic profile for the different routes. There are examples of full reproductive toxicity packages on intravenous products being submitted where the studies were conducted by the oral route despite limited bioavailability. Such data would be of extremely limited value and perhaps even worthless. Animals should usually be treated once daily but this may be modified to more or less frequent administration depending on kinetic variables.

Toxicokinetic Assessment Within Reproductive Studies

It is highly desirable to have some pharmacokinetic information before reproduction studies commence. This will aid in the choice of appropriate species, study design, and dosing schedules. At this stage, the information does not have to be derived from pregnant or lactating animals. Where appropriate, however, toxicokinetic information should be obtained from animals that are actually on the reproduction studies or in satellite groups run concurrently with those studies. One reason for conducting toxicokinetic assessments is to demonstrate adequate systemic exposure, especially where this might be questioned because of absence of pharmacological response or toxic effects. A more meaningful use of these data is to quantitate exposures achieved at different stages of the reproductive process. Some information on absorption, disposition, and clearance should be available prior to clinical studies in order to predict margins of safety based upon exposure and dose. Exposure, in terms of AUC, will be required to calculate safety margins by comparison with human exposure, and peak plasma level determination may help to determine what blood level is required to produce a particular toxic effect if that effect is related to blood level rather than exposure.

For males and nonpregnant animals on fertility studies, adequate information from the repeat-dose toxicity studies should already be available and no further investigation would normally be required. Toxicokinetics should form part of studies involving pregnant and lactating animals and it should be appreciated that kinetics may differ in pregnant, nonpregnant, and lactating animals. Toxicokinetics may involve exposure assessment of dams, embryos, foetuses, or offspring. Secretion in milk may be assessed to define its role in the exposure of newborns. In some situations, additional studies may be necessary or appropriate in order to study embryo–foetal transfer and secretion in milk. Toxicokinetic data obtained from reproductive toxicity studies can be very helpful but it may also have limitations. In developmental toxicity studies, for example, it is quite common to obtain a full kinetic profile on the last day of dosing (usually day 17 of pregnancy in rats) but these data may not be too helpful in explaining effects on the embryo occurring earlier in pregnancy.

ICH Guideline S3A (4) provides guidance for toxicokinetic assessment in all types of toxicity studies.

Interpretation and Assessment of Data

Appropriate statistical analysis of data generated from reproductive toxicity studies should be conducted. The outcome of all reproductive studies should be related to all other pharmacological and toxicological data available to determine whether there are potential reproductive risks to humans that are of more concern than other toxicities that may have been detected, that is, is the no-effect level for reproduction lower than for other significant findings or, more importantly, how does the systemic exposure at the no-effect level compare with systemic exposure in humans at the clinical dose. The role of other toxicity studies in these interpretations should not be underestimated. Repeat-dose toxicity studies, for example, can provide important information regarding potential effects on reproduction, particularly male fertility.

The EMEA has published guidelines on risk assessment for reproduction (5). These contain useful information on the statements that need to be included in the product labeling including the Summary of Product Characteristics. There is also a schematic that defines the thought processes leading to contraindication of a drug in pregnancy. (It is best to refer to these guidelines rather than summarising them here.)

SPECIAL REQUIREMENTS FOR VARIOUS MEDICINAL PRODUCTS

Special Requirements for Paediatric Products

Guideline S5(R2) (1) recommends that, where medicinal products may be used in infants and juveniles, special nonclinical studies (case-by-case designs) involving direct treatment of offspring at ages to be specified should be considered. This is outside the remit of this chapter. There are, however, certain expectations regarding the need for reproductive toxicity studies in this situation. Where a paediatric formulation is developed as a line extension, then the full battery of reproductive toxicity studies will already be available. Where a drug is developed specifically for paediatric use, then these reproductive studies will still be required. This may seem a little unnecessary but effects on reproductive organs are just as much a concern for young as well as adult individuals and a fertility study will cover this. The rationale behind requiring the other studies is that here developing, immature individuals are exposed to the drug at all stages up to at least weaning and the findings can be of help in determining whether there are specific risks for the young, developing child.

Special Requirements for Biotechnology-Derived Pharmaceuticals

These are covered in ICH Guideline S6 (6). The guideline applies to products derived from characterised cells through the use of expression systems such as bacteria, yeast, insect, plant, and mammalian cells. The intended indications may include in vivo diagnostic, therapeutic, or prophylactic uses. The active substances include proteins and peptides, their derivatives, and products of which they are components. Examples include but are not limited to: cytokines, plasminogen activators, recombinant plasma factors, growth factors, fusion proteins, enzymes, receptors, hormones, and monoclonal antibodies. The principles outlined in ICH S6 may also apply to recombinant DNA protein vaccines, chemically synthesised peptides, plasma-derived products, endogenous proteins extracted from human tissue, and oligonucleotide drugs. These guidelines do not apply to antibiotics, allergenic extracts, heparin, vitamins, cellular blood components, conventional bacterial or viral vaccines, DNA vaccines, or cellular and gene therapies.

The need for reproductive/developmental toxicity studies is dependent upon the product, clinical indication, and intended patient population. Where extensive information regarding the potential reproductive toxicity of a particular class of compounds (e.g., interferons) already exists and where the only relevant species is the nonhuman primate, mechanistic studies indicating that similar effects are likely to be caused by a new but related molecule may obviate the need for formal reproductive/developmental toxicity studies. A full justification will need to be provided. The dosing schedule and other aspects of study design may be modified where there is a long elimination half-life or where there are issues of immunogenicity. Where there are prolonged immunological effects, as with certain monoclonal antibodies, the study design may need to incorporate monitoring of immune function in the neonate.

One particular problem with biotechnology-derived materials is the choice of species to be used for reproductive toxicity studies. Only relevant species should be used. Irrelevant species may give grossly misleading results. Relevant species are ones in which the test material is pharmacologically active due to activity at the receptor or an epitope (in the case of monoclonal antibodies). Also, monoclonal antibodies need to demonstrate a similar tissue cross-reactivity profile as for human tissues. The use of surrogate antibodies in more conventional species and the use of transgenic animal models may also be appropriate.

Special Requirements for Vaccines

Fertility studies are not usually required as long as adequate examination of reproductive organs, including detailed histopathology, has been conducted and there are no untoward findings. Since most vaccines are administered during childhood, embryo–foetal development studies would not normally be required. They would only become necessary if a vaccine was to be given to women of childbearing age. Some existing vaccines, although safe for nonpregnant women, may cause foetal infection or malformation or abortion in women who are pregnant. Such vaccines include those for Rubella, Venezuelan Equine Encephalitis, and Varicella, among others. The reader should refer to the relevant guideline (7).

Special Requirements for Combination Products

When the single components have been tested adequately, and the reproductive/developmental toxicity profiles of these compounds are sufficiently characterised, additional studies with the combination may not be warranted. However, the decision will depend on the nature and properties of the individual components and of their potential interactions. The reader should refer to the relevant guidelines (8).

Special Requirements for Anticancer Medicinal Products

Studies of toxicity to reproduction are not required since cytotoxic/cytostatic drugs are assumed to cause reproductive disturbances. Pregnant women may nevertheless be treated with these agents and, therefore, studies elucidating the potential for reproductive toxicity are encouraged. For further information, the reader is referred to the appropriate guideline (9).

Special Requirements for Herbal Medicines

Fertility studies are required only where literature reports give rise to a cause for concern (e.g., the observation of hormone-like actions) or where the product is explicitly indicated in pregnancy, for example, where the medicine is traditionally used to regulate fertility. The reproductive toxicological potential with regard to embryo–foetal and peri- and postnatal development should be assessed taking exposure data into consideration. Any marketing authorization application for a herbal medicine should take into account all such background information and intended use. Tests in animals are not necessary if one of the following criteria is fulfilled:

- Results from postmarketing studies or epidemiological data of adequate power or postmarketing safety studies are available.
- The assessment of the results of a systematic and comprehensive scientific literature search and postmarketing experience does not identify a positive signal of reproductive toxicity, and the herbal medicinal product is not intended to be used during pregnancy and lactation.
- Results from investigations in pregnant women and neonates are present.
- The medicinal product is not intended to be used in women of childbearing potential.

For herbal medicinals, some, or even much, of the data may be old and unreliable and should be verified as adequate. Where this cannot be achieved, further literature surveys or even new reproductive studies in animals may be required unless the applicant can provide suitable justification to the contrary.

For full details, see the guideline covering herbal medicine applications (10).

CONCLUSION

The regulatory approach to reproductive and developmental toxicity testing is based around new medicinal products being administered to test animals in a way that ensures that all stages of the reproductive cycle are covered. This is achieved by one or more experiments that collectively cover mating, pregnancy, and lactation in the adult and prenatal, postnatal, and postweaning development of the offspring to include an assessment of reproductive performance in the first filial generation. The "most likely option" of three overlapping reproductive studies has been presented for first-tier experimentation but the relevant guidelines and experience show that a flexible approach is allowed and encouraged where appropriate. To demonstrate this, separate guidelines have been developed to cover the special cases such as biotechnology-derived products. No guideline can provide sufficient information to cover all contingencies and flexibility and a sound knowledge of the principles of reproductive and developmental toxicity coupled with a willingness to comply with the general spirit and intent of the guidelines is far better than slavish compliance to what some may perceive erroneously as regulatory dictates. Although it has been mentioned elsewhere in this chapter, it does no harm to reinforce the need not to design, conduct, and interpret reproductive and developmental toxicity studies in isolation. Consideration should be given at every stage to all other aspects of the medicinal product that might have some bearing on reproductive function, aspects such as its pharmacology, pharmacokinetics and metabolism, toxicity to other systems, and intended therapeutic use, amongst others.

REFERENCES

1. ICH Harmonised Tripartite Guideline S5(R2): Detection of Toxicity to Reproduction for Medicinal Products and Toxicity to Male Fertility. Parent Guideline dated June 24, 1993 (Addendum incorporated in November 2005). http://www.ich.org/LOB/media/MEDIA498.pdf. Accessed November 19, 2008.
2. ICH Harmonised Tripartite Guideline (M3(R1): Maintenance of the ICH Guideline on Non-Clinical Safety Studies for the Conduct of Human Clinical Trials for Pharmaceuticals: Parent Guideline dated November 7, 1996 (Addendum incorporated in November 2000). http://www.ich.org/LOB/media/MEDIA506.pdf. Accessed November 19, 2008.
3. Guidance on Nonclinical Safety Studies for the Conduct of Human Clinical Trials and Marketing Authorization for Pharmaceuticals: Revision M3(R2) to M3(R1) Guideline, issued for consultation July 17, 2008. http://www.ich.org/LOB/media/MEDIA4744.pdf. Accessed November 19, 2008.
4. ICH Harmonised Tripartite Guideline S3A: Note for Guidance on Toxicokinetics: The Assessment of Systemic Exposure in Toxicity Studies: Issued October 27, 1994. http://www.ich.org/LOB/media/MEDIA4744.pdf. Accessed November 19, 2008.

5. Guideline on Risk Assessment of Medicinal Products on Human Reproduction and Lactation: From Data to Labelling. EMEA/CHMP/203927/2005. http://www.emea.europa.eu/pdfs/human/swp/20392705en.pdf. Accessed November 19, 2008.
6. ICH Harmonised Tripartite Guideline S6: Preclinical Safety Evaluation of Biotechnology-Derived Pharmaceuticals: Issued July 16, 1997. http://www.ich.org/LOB/media/MEDIA503.pdf. Accessed November 19, 2008.
7. Note for Guidance on Pre-Clinical Pharmacological and Toxicological Testing of Vaccines. CPMP/SWP/465/95. http://www.emea.europa.eu/pdfs/human/swp/046595en.pdf. Accessed November 19, 2008.
8. Guideline on the Non-Clinical Development of Fixed Combinations of Medicinal Products. EMEA/CHMP/SWP/258498/2005. http://www.emea.europa.eu/pdfs/human/swp/25849805en.pdf. Accessed November 19, 2008.
9. Note for Guidance on the Pre-Clinical Evaluation of Anticancer Medicinal Products. CPMP/SWP/997/96. http://www.emea.europa.eu/pdfs/human/swp/099796en.pdf. Accessed November 19, 2008.
10. Guideline on Nonclinical Documentation for Herbal Medicinal Products in Applications for Marketing Authorisation (Bibliographical and Mixed Applications) and in Applications for Simplified Registration. EMEA/HMPC/32116/2005. http://www.emea.europa.eu/pdfs/human/hmpc/3211605en.pdf. Accessed November 19, 2008.

28 Safety Pharmacology Testing: Past, Present, and Future

Paul Baldrick
Scientific and Regulatory Consulting, North Yorkshire, U.K.

INTRODUCTION AND DEFINITIONS

In vitro and in vivo animal pharmacology studies (to show that a drug has efficacy and examining high-dose pharmacological effects) are a usual component of drug safety assessment, but until fairly recently the relationship between what was termed primary pharmacology, secondary pharmacology, and safety pharmacology or even general, ancillary, high-dose, or "regulatory" pharmacology was unclear. However, in recent years, regulatory guidance (ICH S7A) has appeared, which makes the situation clear, with the following definitions:

- Primary pharmacology (pharmacodynamic) studies are those on the mode of action of a drug in relation to its desired therapeutic effect.
- Secondary pharmacology (pharmacodynamic) studies are those on the mode of action and/or effects of a drug not related to its desired therapeutic target.
- Safety pharmacology studies are those that investigate the potential undesirable pharmacodynamic functions of a drug on physiological functions in relation to exposure in the therapeutic range and above. There is a need to:
  - relate the relevance of these undesirable properties to human safety.
  - evaluate adverse findings in toxicology/clinical studies.
  - investigate the mechanisms of adverse pharmacodynamic effects.

It should be noted that occasionally, primary and secondary pharmacology studies singly or together are still sometimes referred to as "General pharmacology" studies.

HISTORY

Until fairly recently, studies examining pharmacodynamic actions of a drug outside its main primary pharmacological activity were performed as part of the general pharmacology evaluation, normally in exploratory pharmacology laboratories. Unlike toxicology studies, which were "regulated" from the early 1980s, there was no requirement for compliance with the requirements of Good Laboratory Practice (GLP), for example, protocol-driven studies, with standard operating procedures in place, evidence of adequate staff training, calibration of equipment, etc. In the absence of specific regulatory guidance, each company tended to perform its own in-house battery of tests. However, possibly as a means of standardizing testing for these additional pharmacology studies, the then Japanese Ministry of Health and Welfare (JMHW) produced a document entitled "Guideline for General Pharmacology" in 1991 (1,2). This document divided testing into LIST A studies (needed to get a general picture of pharmacological actions and essential for all new therapies) and LIST B studies (for further evaluation of effects indicated by LIST A studies). The LIST A and B studies were broken down into body functions (and focused on single physiological systems or organs); details are given in Table 1.

In the early 1990s, many companies [especially in Europe and the United States (US)] started to put more emphasis on studies not directly associated with the primary pharmacology of a drug to give extra assurance for detection of high-dose pharmacological effects and the term "safety pharmacology" started to appear. Furthermore, despite its title, some companies started relating the Japanese guideline to a safety pharmacology testing package despite the fact that it was more suited for pharmacological, and not safety, profiling. Indeed, a number of the Japanese guideline tests did not provide any pharmacological information that was particularly pertinent to human safety. Additional confusion was that some companies (especially in Europe) assumed

Table 1 Japanese List A and B General Pharmacology Testing

Pharmacological action	Example of study (and identification)
General signs and behavior	Irwin behavioral screen (A1)
Central nervous system	Effects on spontaneous motor activity, barbiturate sleep time, convulsion response, pain response or body temperature/EEG, spinal reflex, or conditioned avoidance responses (A2 and B1)
Somatic nervous system	Neuromuscular junction effects (B2)
Autonomic nervous system and smooth muscle	Effects on isolated ileum/pupil size or nictating membrane response (A3 and B3)
Respiratory and cardiovascular systems	Effects in anesthetized animal preparation/vagal stimulation, carotid artery occlusion, or isolated cardiac effects (A4 and B4)
Gastrointestinal system	Gastrointestinal transport or gastric emptying effects/response on gastric juice secretion or excised gut movements (A5 and B5)
Water and electrolyte excretion	Effects on urinary volume and electrolyte concentration (A6)
Others	Blood coagulation, platelet aggregation, or hemolytic action (A7 and B6)

that, although not mentioned in the Japanese guideline, the LIST A studies should be conducted to GLP. This was because safety pharmacology studies ("pharmacodynamic studies designed to test the potential for adverse effects") were considered as safety tests which in Europe needed to conform to GLP (3).

Because of uncertainty, global discussions among companies occurred on whether these "special" pharmacology studies should be subjected to GLP along with discussions on the need for a separate identity for safety pharmacology assessment outside the confines of general pharmacology. A milestone was the inaugural conference on General Pharmacology/Safety Pharmacology in the US in late 1993. A further step was an industry survey published in 1995 (from 16 US and 10 United Kingdom pharmaceutical organizations), which showed that 62% of companies were performing safety pharmacology studies to GLP (4). Company clinicians were most often cited as the primary customer. The high-dose level in these studies was a Maximum Tolerated Dose (MTD) for 38% of companies with others using multiples of pharmacological or estimated therapeutic doses. The studies were performed in non-GLP pharmacology units by discovery research staff and in GLP-development groups, generally in toxicology. The survey showed that many companies were using the Japanese guidelines only as an outline or guide for study design. However, despite this finding, it was also apparent that some companies at this time were performing all Japanese LIST A studies as part of their early drug safety evaluation process, as a "tick box" activity (as well as a genuine evaluation for potential safety signals in the clinic) (Baldrick P, personal observation).

More recently, "safety pharmacology" testing was proposed as an International Conference on Harmonization (ICH) topic in 1998. As a result, harmonization has occurred to formally identify safety pharmacology testing with the introduction of the regulatory guideline "Note for Guidance on Safety Pharmacology Studies for Human Pharmaceuticals" (ICH S7A) (5), which became operational in June 2001. In order to standardize and define a further aspect of safety pharmacology testing not fully covered by this guideline, namely, specific in vitro and in vivo cardiovascular testing/risk assessment, a further document "The Nonclinical Evaluation of the Potential for Delayed Ventricular Repolarization (QT Interval Prolongation) by Human Pharmaceuticals" (ICH S7B) (6) was developed and became operational in November 2005. This document updated earlier European "Points to Consider" (7) and Canadian (8) guidance on the topic; both these latter guidance documents have been subsequently withdrawn. Interestingly, the ICH S7B document took a few years before adoption, as there was (and still is) a lack of global scientific agreement on the best in vitro and in vivo tests, to determine for repolarization-associated ventricular arrhythmia (see later).

Table 2 Safety Pharmacology Testing

Test needed	Body system	Example of study
Core battery	Central nervous system	General and behavior study in the rodent (generally a modified Irwin behavioral screen)
	Cardiovascular system	In vitro assessment for QT prolongation (usually a hERG assay) Conscious/anesthetized study in the non-rodent
	Respiratory system	"Head out" or whole body plethysmograph study in the rodent
Follow-up and supplemental battery	Central nervous system	Motor activity in monitoring chambers, rotarod test or body temperature in the rodent
	Cardiovascular system	Action potential duration changes in cardiac tissues (e.g., dog Purkinje fibre)
	Respiratory system	In depth plethysmograph study to measure pulmonary resistance, elastance and lung compliance
	Renal/urinary system	Effect on urine volume and electrolytes in the rodent
	Gastrointestinal system	Intestinal transit or gastric emptying study in the rodent

SAFETY PHARMACOLOGY TESTING

General/Guidelines

ICH S7A

The ICH S7A guideline defines a "core battery" of necessary studies to be performed before First In Man (FIM) clinical trials are conducted. These studies relate to functions that are acutely critical for life and comprise the central nervous system (CNS), the cardiovascular system, and the respiratory system (Table 2). The guideline also defines follow-up and supplemental studies, dependent on the pharmacological properties or chemical class of the drug, including further studies on the CNS, cardiovascular, or respiratory function as well as renal/urinary function and the autonomic nervous system (Table 2). Because of their nature, safety pharmacology studies may not be needed for biotechnology-derived (see later) or cytotoxic anticancer drugs (except perhaps for those with a novel mechanism of action).

The ICH S7A guideline gives information on experimental design for safety pharmacology studies including the use of appropriately sized dose groups to allow meaningful interpretation of results and use of the expected clinical route of administration. It is recognized that positive controls are not necessary for well-characterized test systems (which also reduces animal use). High-dose selection for in vivo testing resulted in much discussion during the development of the guideline (Baldrick P, personal observation). However, it is stated that doses should include and exceed the primary pharmacodynamic or therapeutic range with provision of moderate adverse effects at the highest tested dose (although it is recognized that effects in the toxic range may confound study interpretation). For in vitro testing, it is noted that the upper limit for testing may be influenced by physicochemical properties of the material under test.

Other aspects of the guideline indicate that the studies are generally by single-dose administration and although performed on the parent compound, consideration for testing of major metabolites and isomers (for racemic compounds) needs to occur. A facet of safety pharmacology testing is that the data are superseded by human data fairly rapidly and so studies should be done early in the development process if they are to be useful. Such thinking is endorsed in ICH S7A, which clarifies that "core battery" studies should be performed prior to first administration in humans as should appropriate follow-up or supplemental studies. However, it may be that the some of the latter studies are performed later in development when results do not directly impact on human safety but are needed to better understand mechanisms underlying particular pharmacodynamic effects. Finally, all standard safety pharmacology studies should be performed to GLP, although the guideline acknowledges that this may not be possible for nonstandard studies of unique design. However, noncompliance with GLP needs to be adequately justified.

ICH S7B

The ICH S7B guideline describes a nonclinical strategy for assessing the potential of a drug to delay ventricular repolarization in humans. A key part of this assessment is the measure of prolongation of the electrocardiogram (ECG) QT interval as a surrogate measure of the ability of nonantiarrhythmic drugs to delay cardiac reporlarization with potential development of ventricular tachyarrhythmia including torsade de pointes (TdP). The QT interval (time from beginning of the QRS complex to end of T wave) of an ECG is a measure of the duration of ventricular depolarization and repolarization. Ventricular repolarization is determined by the duration of the cardiac action potential, which involves the activities of various membrane ion channels (notably those for potassium, sodium, and calcium) and transporters. Although delay of repolarization can occur through modulation of several types of ion channels, it has been shown that the rapidly and slowly activating components of the delayed rectifier potassium current have the most influential role in determining the duration of the action potential and thus the QT interval (ICH S7B). QT interval prolongation arises from drug inhibition of the delayed rectifier potassium channel.

In describing both in vitro and in vivo study types to assess for a potential risk for delayed ventricular repolarization and QT interval prolongation, ICH S7B has some overlap with ICH S7A as in vivo cardiovascular function studies are described in both documents, although more emphasis is made for ECG evaluation in ICH S7B. A slightly confusing statement in the latter guideline is that these studies "should be considered" prior to first drug administration in humans, while ICH S7A clearly states that core battery safety pharmacology studies including cardiovascular function evaluation need to occur beforehand. In any case, companies tend to perform at least some in vitro work (see later) before clinical entry.

The ICH S7B guideline gives details on study design indicating that assay methods and endpoints should be valid and robust, standardized, reproducible, and relevant for assessing human risk. For in vitro studies, a positive control substance needs to be included to demonstrate responsiveness, and concentrations used should span a broad range, covering and exceeding the anticipated maximum therapeutic plasma concentration. The guideline indicates that, once available, all data need to be included in an integrated risk assessment (along with information on the chemical/pharmacological class of drug) to evaluate whether there is any evidence for potential delayed ventricular repolarization and QT interval prolongation. Finally, it should be noted that the ICH S7B guideline should be used in conjunction with a clinical one (9), which addresses evaluation of QT interval prolongation and proarrthymic potential.

For assessment for potential delayed ventricular repolarization, ICH S7B mentions the use of four functional levels of testing (which also includes in vivo testing) to assess for effects as follows:

- Ionic currents measured in isolated animal or human cardiac myocytes, cultured cardiac cell lines, or heterologous expression systems for cloned human ion channels.
- Assay of action potentials in isolated cardiac preparations (e.g., Purkinje fibers, papillary muscle, intact heart) or specific electrophysiology parameters in anesthetized animals.
- ECG parameters in conscious or anesthetized animals (it is mentioned that mouse or rat are not useful for assessment as ionic mechanisms of repolarization differ in these species).
- Proarrhythmic effects measured in isolated cardiac preparation or animals.

A recommended package of studies is not given in ICH S7B, although it is suggested that initial evaluation should come from an in vitro assay examining ionic current as well as in vivo QT evaluation. For the former, examination for inhibition of the delayed rectifier potassium channel current is mentioned using a human ether-a-go-go-related gene (hERG) assay.

The hERG encodes for a protein that is thought to be involved in modulating this potassium channel.

In Vitro Methods

Initial Model

Currently, the most popular in vitro study performed by companies prior to initial entry into the clinic is the hERG assay. The hERG assay is an electrophysiological (whole-cell patch clamp) method normally using heterologous human embryonic kidney cells or Chinese hamster ovary cells into which the human hERG channel has been encoded. In a typical study design, the cells are placed in a chamber (perfused with vehicle, ascending concentrations of drug and/or positive control for, for example, 10 minutes per treatment) from which microelectrode recordings are made following voltage pulsing. Inhibition is measured as an effect on the amplitude of the potassium current evoked by a voltage step (e.g., from +20 to −50 mV), the so-called tail current, which is analogous to the plateau-repolarization phase of the cardiac action potential. Modifications on voltage pulsing (e.g., ramping) can give further information analogous to action potential repolarization.

The assay is usually carried out in three to five separate cell preparations; a modification to the design is to evaluate each treatment (vehicle/positive control and each drug concentration) in separate preparations. Positive control drugs can include dl-sotalol or terfenadine, which result in marked hERG channel inhibition. Other methods to examine hERG function include ligand-binding or rubidium efflux studies. The literature supports the utility of the hERG assay in that clinical torsadogens have all been shown to be hERG positive with no false negatives (10). The model also remains the only potassium channel target linked to delayed repolarization and a propensity towards proarrhythmia (11).

Follow-up assessment studies to provide a greater depth of understanding or additional information on potential delayed ventricular repolarization and QT prolongation include measurement for effects on the cardiac action potential duration (APD) using dissected tissue Purkinje fibers or papillary muscles (usually from dogs). However, the literature also reports use of Purkinje fibers from, for example, guinea pigs, rabbits, and pigs; papillary muscles from, for example, guinea pigs, rabbits, and cynomolgus monkeys; and dissociated cardiac (ventricular) myocytes from, for example, guinea pigs and rabbits. As in the hERG assay, the tissue is placed in a chamber with perfusion of vehicle, drug, and/or positive control, and microelectrode recordings (e.g., for up to 30 minutes) are made following stimulation of the preparation. A drug's effect on APD (which includes sodium, calcium, and potassium channel effects) is measured at various percentage of repolarization (e.g., APD_{40}, APD_{70}, or APD_{90}). The maximum velocity of the action potential upstroke (V_{max}), action potential amplitude, and resting membrane potential can also be made. For dog Purkinje fiber assessment, a total of six separate fibers are usually used. Positive control drugs can include dl-sotalol or astemizole, which result in prolongation of APD. Recent work with rabbit Purkinje fibers has defined an "effect" as $\geq$10% prolongation of APD at APD_{90} (12).

The concentration of drug used in these in vitro studies is obviously dependent on solubility and potency but values in the range sub-μM up to 100 μM are commonly used. It is stated that concentrations studied in hERG and APD models should cover a two to three log unit range with the highest concentrations used being at least 10- to 30-fold higher than the anticipated plasma or tissue level necessary for therapeutic activity (13). Others have endorsed use of a wide concentration range (at least 1–100-fold the primary pharmacological potency or expected free plasma peak concentration in man) to allow assignment of an in vitro therapeutic window for a compound (14). In the risk assessment process, examination of various marketed drugs indicated that if the therapeutic ratio for in vitro (and in vivo) experiments was >30 for a given compound, then it was unlikely to have the potential for significant QT prolongation or TdP in the clinic (14). Further work, with a much larger drug database, also suggested that if the margin between either the hERG IC_{50} value or concentration producing a 10% increase in APD_{90} and anticipated C_{max} (unbound drug) is <30-fold, then this is a signal for possible arrhythmogenic activity, while a 10% to 20% inhibition of the delayed rectifier potassium channel current in a hERG assay may not be of negligible physiological significance (and may translate into detectable increases in APD_{50} and in vivo QT interval) (15).

A further publication has suggested that 20% inhibition of binding to the hERG channel is a sensitive predictor for compounds to prolong the QT interval in vivo (16). Such calculations, however, can only be theoretical until FIM dosing occurs. In the latter case, calculation of safety margins needs to take total and free drug plasma concentration into account (11,14,17). Indeed, it has been suggested that although in vitro data are generally expressed as IC_{50} values, use of IC_{10-20} values (to correlate with clinical free concentration of drug) may more accurately reflect cases in which compounds are likely to cause clinically significant QT prolongation and TdP (14). Indeed, clinical experience has indicated that drugs that bind to hERG channels in concentrations close to or lower than unbound therapeutic plasma concentrations have a proarrhythmic risk (18).

Utility of Model

Although commonly used in safety pharmacology packages, there are concerns with the predictability of the hERG assay. In a recent assessment of 14 candidate drugs, which were examined using this assay, 10 showed weak inhibition, dose-dependent inhibition, or high concentration inhibition (although this finding did not prevent the drugs being taken into the clinic as there was no effect on in vivo ECG in any of these cases) (19). Various definitions of concentrations giving concern are given in the literature:

- Compounds with little hERG blockage at 1 μM will have relatively low risk of having a QT issue, whereas compounds showing substantial blockade at 1 μM may be viewed as being potentially troublesome (20).
- When the hERG binding potency is <5 μM, there is a risk of inducing QT prolongation, which becomes more significant when the potency is <1 μM (21).
- A compound giving IC_{20} and IC_{50} values of 0.5 and 5 μM is considered as a moderately potent hERG inhibitor (22).
- A ≥20% inhibited hERG current at a concentration of ≤3 μM has been defined as "hERG inhibition" while "no effect" was associated with <20% inhibition at ≤3 μM (or effects only occurred at a high concentration of >3 μM) (12).
- In the candidate selection process by one company, compounds with hERG IC_{50} values of >10 μM are usually progressed, while those with values of <1 μM are usually withheld; compounds in between these values need to undergo in vivo testing before it is determined whether they are progressed or not (23).

It has been pointed out that as a large percentage of compounds can interact with the hERG channel when the concentration is high enough, many compounds will demonstrate some blocking activity and that very few compounds will exhibit no interaction at high concentrations (20). For this reason, it is suggested that the IC_{50} value should be referenced to a realistic drug concentration (17). It is also reported that the predictive value of the hERG assay is limited by a high false-positive rate, a lack of consensus on the definition of a negative study, and interlaboratory variations (10). Indeed, the literature reports considerable variation in IC_{50} values for the hERG assay, generally of about threefold when two laboratories were directly compared (24). Although such a difference was considered reasonable by the latter authors for an in vitro assay, it does have an important impact on using such data to predict clinical safety ratios. Discrepancies among labs in IC_{50} values can arise from differences in protocols and method criteria, for example, block may be strongly temperature and frequency dependent (17). Unlike other in vitro models such as the Purkinje fiber study, the hERG assay does not assess for prolongation of the action potential from effects on sodium and calcium channels.

Use of follow-up in vitro studies such as the Purkinje fiber assay is not without issues. It has been reported that results from this model can be confounded by unexpected APD_{90} prolongation observed with vehicle-treated fibers, which reduce sensitivity (24). Also, testing to the limit of solubility can lead to electrically unresponsive fibers, making the role of data obtained at exceedingly high concentrations highly questionable in risk assessment. Although canine Purkinje fiber tissue is often used for APD measurement, a species comparison (guinea pig, rabbit, dog, swine, goat, and sheep) showed that rabbit Purkinje fibers constitute the most sensitive model for detecting drug-induced, potential long APD, and proarrhythmogenic effects in vitro (25). It may be that rabbit Purkinje preparations are more sensitive due to better drug access (17). Further work with rabbit Purkinje fibers, papillary muscles, and ventricular trabeculae showed different responses among these three tissues to the same ion channel blockers/activators, highlighting that the choice of cardiac tissue used for in vitro testing could affect results and thus the risk assessment process (26). Other work has shown that the guinea pig ventricular myocyte model may be more sensitive than the dog Purkinje assay at detecting action potential prolongation with compounds that inhibit multiple ion channels (27). A recent inclusion by some investigators in these models is calculation of triangulation (defined as the difference between APD_{40} value in dog or APD_{30} value in guinea pig and APD_{90} value) as a further prediction for repolarization change and development of early after-depolarizations (EADs). However, it has been pointed out that such use is still relatively new (24), as is an understanding of how it might be related to an increased risk of arrhythmias (27). However, work with rabbit Purkinje fibers and guinea pig papillary muscle has shown that actions potential assays are useful for assessing risk of QT interval prolongation when APD_{90} and triangulation are considered (28,29).

To correlate with the hERG assay, inhibition of the delayed rectifier potassium current should result in increased APD_{90}. However, it has been shown that a poor correlation between the results of hERG and repolarization (dog Purkinje fiber) assays can occur, suggesting that the former can oversimplify drug effects on the complex repolarization process for compounds demonstrating multichannel block (e.g., inhibitory actions on sodium and calcium channels affect the expected prolonging effects on APD_{90} caused by hERG/delayed rectifier potassium current inhibition) (30). Indeed, in a recent comparison of hERG versus APD assessed in rabbit Purkinje fibers, it was found that of 92 hERG channel inhibitors, 55.4% prolonged APD, 28.3% had no effect and 16.3% actually shortened APD (12). Of 70 compounds with no effect on hERG channels, 54.3% did not affect APD, 25.7% prolonged APD, while 20% significantly shortened APD.

Overall, however, despite the limitations of hERG channel testing described above, this assay is commonly the only in vitro one performed before FIM studies begin (in line with ICH S7B) and providing a "clean" in vivo ECG result occurs does not prevent regulatory approval to allow clinical dosing to commence even when evidence of inhibition is seen. Its continued use as one important part of a multifactorial integrated risk assessment including in vitro and in vivo components to evaluate for proarrhythmic risk continues to be mentioned in the literature [e.g., (11,24)]. However, maybe it will become standard for future FIM packages to be accompanied by an APD assay. There seems to be a growing appreciation that the hERG assay alone, even in combination with in vivo nonrodent data, is not sufficient (in terms of precision and/or accuracy) to predict proarrhythmic risk with recommendation for supplementary data from other investigations such as a Purkinje fiber study (28,30,31).

Further Models

It has been pointed out in the literature that QT interval prolongation itself is not a perfect surrogate for prediction of TdP (17,32,33). Thus, although current in vitro models have utility in predicting the risk of QT interval prolongation in man, they are less good at predicting proarrhythmia risk. As a result, an array of other in vitro and ex vivo models have been developed. Electrophysiological markers in these studies include triangulation of the action potential, transmural dispersion of ventricular repolarization, EADs, and ectopic beats. Models (many of which have shown evidence of predictivity for potential delay in ventricular repolarization in humans) are described in the literature [e.g., (32,34–37)] and include:

- Isolated Langendorff-perfused heart (Screenit) model: Typically, rabbit (but sometimes guinea pig or rat) hearts are isolated and perfused with electrode recording of repolarization changes/abnormalities from measurement of ECG and monophasic action potentials (MAPs). Changes in APD [reduction in triangulation, reverse-use dependence and instability (TRI)] can be calculated. The rabbit model can be adapted so that bradycardia [achieved by chronic atrioventricular (AV) block] and hypocalcemia are used to induce arrhythmic episodes.

- Arterially-perfused left ventricular (Wedge) preparation: A full thickness slab of canine (but sometimes rabbit) ventricular wall is immersed in a physiological bath and perfused through its natural vessels with recording of transmembrane action potentials from electrodes as well as measurement of a pseudo-ECG. Recent species comparison work with this model has shown that in the presence of dofetilide the rabbit (vs. the dog and monkey) was the most sensitive species for predicting arrhythmogenic potential (38).
- Furthermore, it has recently been shown that the use of the isolated Langendorff heart under either sinus rhythm or pacing and beat-to-beat electrical alternans in the same guinea pig model may be a useful tool in early cardiac risk assessment (39,40).

CNS Methods

Initial Model

ICH S7A mentions the need to assess for motor activity, behavioral changes, coordination, sensory/motor reflex responses, and body temperature using, for example, a modified Irwin test or a functional observational battery (FOB). Follow-up assessment parameters include behavioral pharmacology, learning and memory, ligand-specific binding, neurochemistry, visual, auditory, and/or electrophysiology examinations. Currently, CNS assessment takes the form of a general activity and behavior study based closely on that originally developed by Irwin (41) for psychoactive compounds. For this reason, the test is sometimes called a modified Irwin behavioral screen. Examples of parameters measured (of which there can be up to 50) are given in Table 3. There is some overlap in the general activity and behavior study and what is known as the FOB. Such testing (with its evaluation of autonomic, neuromuscular, sensiomotor, and behavioral aspects) was first introduced by the Environmental Protection Agency in the US in the mid-1980s for examination of neurotoxicity by chemicals [e.g., (42,43)]. The FOB assesses many of the CNS activities given in Table 3 and nowadays, for some companies, the two tests represent the same thing. However, other companies keep the two evaluations separate by performing not only a general activity and behavior study as part of their safety pharmacology core battery but also include a FOB testing regimen into the repeat-dose toxicology studies in the rodent. Recent work with a range of positive control compounds has also validated FOB testing in the dog (44,45) and the monkey (46,47).

In a typical general activity and behavior study, groups of animals receive a single dose of either vehicle or three dose levels of drug with assessment for onset, duration, and reversibility of findings at, for example, 30, 60, 120, 240, and 360 minutes (oral drug) or 0–5, 15, 30, 60, and 120 minutes (intravenous drug) postdose. Changes can be measured using integers (+ or − representing present or absent may also be used) so that parameters in normal animals are scored using an integer that allows for increases and decreases to be recorded in

Table 3 Example of Parameters Measured in a CNS Safety Pharmacology Study

Activity	Parameter
Home cage	Dispersion (0), respiratory rate (0), locomotor activity (3), restlessness (0), fighting (0), alertness (3), apathy (0), exophthalmus (−)
Open field (arena)	Transfer arousal (3), spatial locomotion (2), ptosis (0), startle (2), tail elevation (1), piloerection (1), touch response (2), positional passivity (N/A), catalpsy (−), righting reflex (0), visual placing (3), grip strength (3), pinna (+), corneal (+), pain response (2), wire maneuver (0)
Manipulative (handling)	Cyanosis (0), cutaneous blood flow (3), body temperature (N), body tone (3), pupil size (N), light-pupil response (4), lachrymation (0), grooming (3), red staining (0), salivation (0), provoked biting (0)
General	Fearfulness (2), irritability (0), abnormal gait (0), abnormal body carriage (0), tremors (0), twitches (0), convulsions (−), bizarre behavior (0), writhing (0), vocalization (N/A), diarrhea (−), number of defecations (N/A), number of urinations (N/A), moribund (−), lethality (−), abnormalities detected (−)

Example of scoring system with value in parenthesis = scoring integer for normal animal.
Abbreviations: N/A, not applicable; +/−, sign present/absent; N, normal.

drug-treated animals (Table 3). Other scoring systems can be used such as 0/1 or 4 for a normal animal depending on the parameter assessed (48). Typically, observation sessions last for three minutes per animal or longer. It is common to use six animals/group (and usually males), although some workers have indicated eight rodents/group as a minimal number to enable an accurate assessment (taking into account interanimal variability) of neurobehavioral changes (48). If used, typical positive control compounds are chlorpromazine hydrochloride (giving depressor effects), amphetamine sulfate (giving stimulant effects), diazepam (giving anxiolytic and sedative effects), or clonidine hydrochloride (giving sedative effects). Although originally developed in the mouse, the general activity and behavior study is now largely performed in the rat. Indeed, in a recent assessment of 26 candidate drugs for which this study was performed, the rat was used on 21 occasions (19).

Utility of Model

Positive results are often seen in the general activity and behavior study. However, although high-dose CNS effects were seen on 12 occasions in an evaluation of 26 candidate compounds, none were of concern in moving the drugs concerned into the clinic (19). In an examination of new drugs approved during 1987 to 1992 in Japan, it was found that 61% of 89 drugs gave a finding when examined in a general activity and behavior study (49). Also, although findings were not specified, an assessment of core battery studies showed that 6/20 candidate drugs gave CNS effects (50). In an analysis of 50 compounds, it was found that all of the materials had an effect of some sort (51). Commonest effects were decreases in bodyweight, decreases or increases in rectal temperature, open field effects (changes in line crossings, time to exit the centre circle or supported rears), and changes in tail flick latency or landing foot splay. Examination of related FOB evaluation has concluded that it is a valid screen for neurotoxicity screening (42). Indeed, it has been shown to detect neurological dysfunction produced by chemicals known to be neurotoxic. However, it has been cautioned that for some tests in the Irwin battery or FOB (e.g., grasp irritability, hindlimb foot splay), there is no obvious clinical neurological counterpart (52), which makes interpretation of findings difficult. Furthermore, clinical symptoms of, for example, nausea, sweating, headache, vertigo, or dizziness cannot be predicted from these models.

Further Models

A range of other models exist to further study for potential CNS effects (Table 2). Examples include:

- Motor/locomotor activity: For example, photobeam breaks/video-image analysis by rodents in monitoring chambers to assess horizontal, ambulatory, and vertical activity.
- Motor coordination: Rotarod performance of a rodent on either a fixed speed or accelerating rod regimen or beam walking tasks.
- Body temperature: Assessment of rectal temperature in rodents (although usually included in the general activity and behavior study nowadays and not as a stand-alone investigation).

Various other measures for potential CNS effects were included in the Japanese LIST A package described earlier (Table 1) but are rarely performed for modern drugs (unless there is a specific concern). Examination can include assessment of antiepileptic, analgesic, hypnotic, anxiolytic, neuroleptic, antidepressant, or anti-inflammatory activity with well-known models described in the literature [e.g., (37)] and including:

- Anti- or proconvulsant effects (e.g., following pentylenetetrazol or electrical induction) in the rodent as measured by diminution or escalation of convulsive threshold; in vivo electroencephalogram (EEG) measurement in rats/rabbits or in vitro brain slice (hippocampal) electrophysiology can also occur. The utility of measuring EEG to evaluate for proconvulsant effects in conscious dogs by means of telemetry has recently been demonstrated (53).
- Analgesic effect on writhing (following acetic acid induction) in the rodent as measured from the number of writhes.
- Pain response/nociception (e.g., latency in tail flick response to a focused infrared light source, paw lick/removal from a hot plate or tail pinch response) in the rodent.

- Barbiturate sleep time (following administration of pentobarbital or hexobarbital) in the rodent as measured by the onset of sleep and duration of sleeping time (as indicated by loss or reappearance of the righting reflex); a positive result can indicate a potential for the compound under investigation to produce drug interactions.
- Acoustic function: Startle response measured by a motor reflex evoked by a suprathreshold auditory stimulus in the rodent.

Cognitive Function (Learning, Memory, and Attention)

Animal tests for potential effects on cognitive function, as evaluated by learning and memory, are usually only applicable to drugs developed for a CNS indication and often overlap with secondary pharmacology evaluation. Learning, memory, and attention studies include spatial navigation of mazes, associations with shock, conditioned responses, and appetite-motivated operant responses. A wide range of models exist [e.g., (37)] and a few well known ones include:

- Passive avoidance: Rodents receive an adverse stimulation in a recognizable environment and on a later occasion show that they remember by not going there.
- Morris water maze (for spatial memory assessment): Rats are timed in finding a submerged platform. A radial maze model is also available.
- Operant responses: Food reward in rats following lever pressing under a discrimination regimen.

It is important to remember that false-positive results can occur in these studies at higher drug dose levels due to clinical signs (e.g., sedation or impairment of motor function).

Cardiovascular System Methods

Initial Model

ICH S7A mentions the need to assess for blood pressure, heart rate, and ECG and suggests a preference for use of unanesthetized animals. Follow-up assessment parameters include cardiac output, ventricular contractility, vascular resistance, and the effects of endogenous and/or exogenous substances on cardiovascular responses. Currently, in vivo safety pharmacology cardiovascular assessment is performed in either an anesthetized or a conscious (telemetry) nonrodent (normally dog or cynomolgus monkey) study. Routine cardiovascular studies in the rat (e.g., anesthetized arterially cannulated or tail cuff sphygmomanometry preparations to measure blood pressure, heart rate, and ECG) are no longer performed, although some companies perform investigative studies in conscious rats. It has been shown that the rat has limited value for predicting electrophysiological effects on the human myocardium due to the relatively brief period of ventricular repolarization in this species (54). If an anesthetized nonrodent preparation is used, it is possible to include assessment for effects on respiratory function as well (see later). If a conscious animal study is utilized, assessment of respiratory function is generally examined in a separate study, such as whole body plethysmograph in the rat (see later). Methodology on the actual design of cardiovascular studies varies among companies. However, basic designs can take the following form:

- Cardiovascular effects in the anesthetized cannulated dog or monkey: Animals are anesthetized and maintained in this state throughout the study. Two groups, each comprising two male and two female dogs, receive either three ascending intravenous (or rarely, intraduodenal) doses of drug or three doses of vehicle (at e.g., 2 ml/kg) by infusion through the left jugular vein with an interval of at least 25 minutes between each dose administration. Heart rate, blood pressure (femoral artery), systemic resistance (contralateral femoral artery), left ventricular pressure, and its derivative dP/dt (left ventricle via the left carotid artery) and lead II ECG are measured. Other measurements such as cardiac output and pulmonary artery pressure can be included in this model. As this study is usually performed by the intravenous route (and not the clinical oral route), toxicokinetic sampling is needed to allow comparison with any study finding to blood levels of drug.
- Cardiovascular effects in the conscious dog or monkey: Animals are surgically implanted with transducers (e.g., in subcutaneous area or abdominal cavity) with catheterization to

allow measurement of blood pressure (femoral), ECG (thoracic inlet and cranial to last rib), and body temperature. Following recovery from surgery, animals are administered vehicle or drug (usually orally) in a Latin square design such that each dose level is given on each of four dosing days, no animal receives the same dose more than once and no two animals have the same dosing sequence during the study. Measurement (from four to eight dogs) occurs from a telemetry pen with radio-transmitted signals detected by receivers in the pen and sent to a commercial acquisition and analysis system. Although sometimes performed, toxicokinetic sampling is not necessarily required as extrapolation of exposure can be made to Day 1 values from the repeat-dose nonrodent toxicology study.

It should also be remembered that further information on ventricular events are also available from the ECG evaluation that is routinely made in nonrodent repeat-dose toxicology studies. Provided that assessment is made using specifically trained staff and in properly trained animals (e.g., two weeks recording pretraining) with standardized recording times (e.g., C_{max}), information can be used to support findings seen in the safety pharmacology work. Although the dog and monkey are the most popular species, it has been shown that the conscious miniature pig can also be used for assessment of drug-induced QT interval prolongation (55).

Utility of Model

In a recent assessment of 34 candidate drugs, in vivo safety pharmacology cardiovascular assessment occurred in both anesthetized and conscious nonrodent studies (19). Although high-dose effects on heart rate (usually increased) and blood pressure (usually decreased) were occasionally seen (for 15 drugs), they did not prevent taking the affected drugs into the clinic. No effect on ECG was seen for any drug. Although findings were not specified, an assessment of core battery studies showed that 7/20 candidate drugs gave cardiovascular effects (50). Predictiveness for positive finding has been suggested in the literature; in that, if the margin between the concentration producing a 10% increase in QT for an in vivo (dog) study and anticipated C_{max} (unbound drug) is <30-fold, then this is a signal for possible arrhythmogenic activity (15).

A great deal has been published on which model is best but it appears that anesthetized or conscious methods are used by different companies and each has merits over the other. In a recent evaluation using various reference drugs, and with the exception of one drug that showed lower sensitivity in anesthetized dogs, both models detected QT interval prolongation with similar sensitivity (56). However, the authors recommended the conscious telemetered model due to its ability to allow extended continuous ECG recording, need for a limited number of dogs (which can be reused on many occasions), and exclusion of potential effect of premedication and anesthetics on study parameters. Despite this conclusion, it was recognized that the anesthetized model has utility in the presence of dose-limiting factors such as convulsions or vomiting interfering with ECG recording.

In another recent evaluation using various reference drugs in both models, QT interval prolongation was detected, but it was concluded that the sensitivity and specificity of the anesthetized preparation was superior to that of the conscious preparation "for detecting whether a drug will cause QT interval prolongation in humans" (57). The authors also point out benefits of the anesthetized model as no implantation surgery is needed and there is a shorter experiment period. More support for the anesthetized animal has occurred due to improved sensitivity as a result of reduced variability of heart rate and QT interval (20). Additionally, other workers have commented that the anesthetized model appears to be more sensitive for slight chronotropic and hemodynamic effects and that a more in-depth evaluation of cardiovascular function can occur (58,59). Finally, it should be pointed out that the conscious model allows rechallenge for reproducibility in the same animal at a later time point if an unexpected finding is seen.

Increasing heart rate can decrease QT interval, while decreasing heart rate can increase QT interval, and proper correction of QT for heart rate in calculation for potential QT interval prolongation (QTc) is vital. A large literature occurs on the subject. It is now accepted that correction using Bazett's formula (QTcB) is not suitable for dogs, as it overcorrects at rapid heart rates, even though 41% of laboratories were using it as recently as 1999 (54). Other suggested methods for dogs are the methods of Fridericia (QTcF) or Van de Water (QTcVdW), with the latter method also recently recommended for toxicology studies (60). QTcB or QTcF can both be used for primates. A recent evaluation of data from conscious (telemetry) dogs and monkeys recommended the optimal correction formula as being Van de Water's and Bazett's, respectively (61). If miniature pigs are used, correction using Fridericia's formula is recommended (55). ICH S7A mentions the use of Bazett or Fridericia formulae to correct QT interval for heart rate, although it also states that the choice of correction formula should be justified in order to avoid ineffective evaluation of risk of QT interval prolongation. Some companies are using their own in-house formula generated with data from QT-heart rate relationships in individual animals under drug-free "resting" conditions [e.g., (24,54,62)]. As there is no universal correction factor, it has been suggested that each laboratory should establish its own formula based on dog strain, the methodology, and the experimental conditions under which data are generated (60,63). The controversy over best method is likely to continue, especially with the recent comment that conscious telemetrized dog studies utilizing Fridericia's or Bazett's correction factors are able to differentiate between drugs that pose a proarrhythmic risk and those that do not (64).

It is common to utilize up to eight nonrodents for cardiovascular studies with mixed findings on an ideal number reported in the literature. In assessing numbers used in telemetry dog work, a double Latin square design with $n = 8$ gave at least an 80% power of detecting at least a 7% increase in QT interval and a 5% increase in QTcF interval versus a single Latin square design with $n = 4$, which reduced sensitivity only slightly to 12% and 8%, respectively (64). Recent assessments using compounds positive and negative in causing QT interval prolongation and ventricular arrhythmias in clinical use have shown a >10% increase in QTcF interval for positive compounds using a Latin square design with $n = 4$ conscious dogs; however, findings of up to 10% were also seen among negative compounds (65). Similar results were seen from calculation of QTcB interval using a dose-escalation or Latin square design with $n = 4$ conscious monkeys, although a few of the positive controls were not detected (66). Using the same materials, clear prolongation of >10% was seen for QTcF interval with positive compounds using $n = 4$ anesthetized dogs with only minor changes with negative compounds (67). Further recent work using data from groups of four to eight conscious dogs, has shown that changes of 5% to 10% can be detected using appropriately corrected QT (QTcVdW) (60). However, by using a standard formula for power calculations, it is reported that group sizes of $n = 4$ may be inadequate to detect changes in QT interval of 10% in conscious dogs, but may detect changes of around 20% (54).

Further Models

As described earlier, as a result of current thinking on whether QT interval prolongation itself is a perfect surrogate for ventricular arrhythmia, various other in vivo cardiovascular models have been re-visited or developed. Models (which have shown evidence of predictivity for potential delay in ventricular repolarization in humans) are described in the literature [e.g., (32,34,35,37)] and include:

- Anesthetized animal (e.g., dog): MAP measurement.
- Methoxamine-pretreated anesthetized rabbit model: Methoxamine (an α_1-adrenoceptor agonist) results in greatly increased sensitivity to repolarization-altering effects for some drugs (34,68).
- Chronically AV node-blocked dog model: Bradycardia is induced after, for example, radiofrequency ablation; the structural, mechanical, and electrical remodeling due to the block increases the susceptibility of the heart to arrhythmias. It has been pointed out that experience for this model is not large (34,35).
- Cardiac electrical alternans model: Measurement of beat-to-beat alternations/variations of cardiac APD at rapid frequencies measured from MAPs from the left ventricle of anesthetized guinea pig heart. The method is claimed to discriminate between proarrhythmic and nonarrhythmic hERG blockers (32).

Orthostatic Hypotension

Measurement for the potential of a drug to cause orthostatic hypotension effects (precipitous fall in blood pressure seen clinically upon assumption of a standing position) can be assessed in conscious rodents on a tilt table, by examination of supine and erect blood pressure, and by heart rate changes in conscious monkeys or specific alternation of rhythm in temperature-stressed rats (69).

Respiratory Function Methods

Initial Model

ICH S7A mentions the need to assess for respiratory rate and other measures of respiratory function such as tidal volume or hemoglobin oxygen saturation. Follow-up assessment parameters include airway resistance, compliance, pulmonary arterial pressure, blood gases, and blood pH. Currently, in vivo safety pharmacology respiratory assessment is performed either in conjunction with cardiovascular function in an anesthetized nonrodent (normally dog or cynomolgus monkey) study or as a stand-alone plethysmography study (usually in the rat). A typical design for the former study has been described earlier with measurement of the following study parameters: respiratory flow, tidal volume, respiration rate, and minute volume (calculated by multiplying tidal volume by breathing rate). In a typical plethysmography study, groups of animals receive a single dose of either vehicle or three dose levels of drug and are placed within (whole body) or "head out" using a plethysmography chamber (for which pressure changes due to thoracic movements of the animal are a direct measure of lung airflow) with measurement for respiration rate, tidal volume, and minute volume made. Other parameters such as inspiratory and expiratory flow, inspiratory and expiratory time, and total breath time can be (but are not usually) reported. Interestingly, recent work in rats with "head out" plethysmography has shown that addition of inspiratory and expiratory time as well as mid-tidal expiratory flow to routine measurements can increase the sensitivity of this respiratory model (70). Typically, measurements occur to set predose baseline levels and then continuously for a four-hour postdose period (although the timing of measurements can depend on the kinetics of the drug). Commonly, six animals/group (and usually males) are used, although a group size of eight has also been suggested for safety pharmacology studies (71). If used, typical positive control drugs can be morphine (which depresses minute volume by reducing respiratory rate without affecting tidal volume) or theophylline (which increases minute volume by increasing tidal volume without affecting respiratory rate) (72). Bethanecol (which causes respiratory depression as decreased respiratory rate and minute volume and increased tidal volume) has also been used as a positive control in rat plethysmography studies (73). In a recent assessment of 34 candidate drugs, respiratory findings (usually depressed respiratory rate) were seen for five drugs (using anethetized non-rodent or plethymography studies), but they did not prevent taking the affected drugs into the clinic (19). Although findings were not specified, an assessment of core battery studies showed that 2/18 candidate drugs gave respiratory effects (50). The importance of assessing for respiratory function effects is shown by findings with an immunosuppressant drug "Compound A" in development for organ transplantation (74). This compound binds to sphingosine 1-phosphate receptors, which occur in smooth muscle including bronchioles. A study using whole body plethysmography in rats showed altered respiratory function indicating that the drug may cause bronchoconstriction in the clinic, which may well limit development of this class of compounds.

As well as in rodents, respiratory measurements can be made for nonrodents either as "head out" or "head in" using a plethysmography chamber. A facemask with a pneumotachometer or a head only flow helmet can also be used to measure ventilatory patterns in nonrodents. Measurement of arterial gases (partial pressure of arterial CO_2 or O_2) and arterial pH can be made in nonrodents as part of the assessment for ventilatory effects.

Utility of Model

The merits of different methods are discussed in the literature. It is pointed out that whole body rodent plethysmograph measurement offers the advantage of no restraint but is limited versus "head out" measurement in that the volume of each breath is determined indirectly from relatively minor increases in chamber pressure (caused by humidified air exhaled from the lungs of the enclosed animal) and that tidal volume calculation can suffer from variables such as chamber temperature and humidity (72,73). An obvious advantage of these models versus the anesthetized nonrodent preparation is that anesthetics, which can affect ventilatory response in their own right, are not involved. It has been shown that although both models responded well to positive control compounds, the conscious dog (using a facemask method) showed greater sensitivity for respiratory effects (respiratory rate, tidal volume, and minute volume) when compared to the anesthetized dog (using intubated pneumotachometry), with speculation that the reduced response in the latter case may be due to an interfering effect by the anesthetic drug (75). Interestingly, other work has shown that when tidal midexpiratory flow is used as a respiratory endpoint, lower sensitivity and greater variability were seen in the conscious versus anesthetized rodent model (71). However, it was concluded that noninvasive methodology is appropriate for quick and repeatable screening of respiratory function in conscious rodents. Finally, the utility of different species has been compared in rat, dog, and monkey using "head out" plethysmography, masks, and flow helmets, respectively, at equivalent doses of various positive control drugs (76). All three species were shown to be adequate for respiratory safety pharmacology evaluation as assessed from respiratory rate, tidal volume, and minute volume with the rat less sensitive (likely due to its less developed respiratory system) and the monkey giving responses closest to humans.

Further Models

Other safety pharmacology studies are available if further examination of potential respiratory findings or specific effects on the mechanical function of the respiratory system are considered necessary [e.g., (37,71,72)]. Models include:

- Pulmonary resistance, elastance, and lung compliance measurements in anesthetized or conscious animals recorded by plethysmography and from pressure-sensitive catheters with or without a telemetry transmitter. Anesthetized models use designs of orotracheally intubated or tracheostomized intubated spontaneously breathing animals or ventilated animals. Typical parameters measured can be lung airflow, transpulmonary pressure, and lung volume changes.
- Determination of whether effects are due to the respiratory muscle itself or neural function change by examination of effects on, for example, isolated guinea pig trachea or phrenic nerve preparations.

Renal/Urinary Function Methods

Initial Model

ICH S7A mentions assessment for urinary volume, specific gravity, osmolality, pH, fluid/electrolyte balance, proteins, cytology, and blood chemistry determinations such as blood urea nitrogen, creatinine, and plasma proteins. For most drugs, however, these parameters can be readily assessed in repeat-dose toxicology studies. When there is a need to evaluate for specific effects in a safety pharmacology study, the most commonly used renal function study involves dosing groups of rats (e.g., six males) with control and three levels of drug immediately followed by supplementation with water or physiological saline (e.g., 10–30 ml/kg) with assessment of urine volume and selected urinary electrolytes (e.g., sodium, potassium, and chloride), for example, for up to six hours postdose (dependent on the drug's kinetics) in metabolism cages. Osmolarity, uric acid, protein, and calcium measurement can also be made. If used, a typical positive compound is furosemide, although hydrochlorothiazide is also suggested (77).

Further Model

A range of other models exist to further study for potential renal effects [e.g., (37,77)]. Examples include:

- Clearance in rodents by evaluation of a drug's performance in the presence of inulin (glomerular filtration rate) or para-aminohippurate (renal plasma flow).
- Clearance studies can also be performed in anesthetized dogs.
- Renal function can further be assessed in spontaneously hypertensive rats.

Autonomic Nervous System Methods

ICH S7A mentions assessment for binding to receptors relevant for the autonomic nervous system (ANS), functional responses to agonists or antagonists in vivo or in vitro, direct stimulation of autonomic nerves and measurement of cardiovascular responses, baroreflex testing, and heart rate variability. Receptor binding studies against a range of potential drug target sites are commonly performed for compounds but are generally considered as part of the primary pharmacology screen (or secondary pharmacology evaluation). In vitro studies can assess for interaction with autonomic transmitters and, as well as investigating a potential effect on the gastrointestinal system (see later), the application of a drug to isolated guinea pig ileum can also be used to assess for ANS (and smooth muscle) response.

Unless there is a specific need, stand alone in vivo safety pharmacology studies to measure potential effects on the ANS are not usually performed with modern drugs. Furthermore, assessment is likely to be included under other areas of functional testing (e.g., cardiovascular or behavioral testing). Indeed, it has been pointed out that cardiovascular endpoints are useful parameters to assess effects of compounds on ANS function (78). Not only can a direct effect on cardiovascular indexes be made but antiadrenergic, anticholinergic, and ganglionic-blocking properties may also be examined from blood pressure and heart rate responses to ANS stimulants. Measurement of baroreceptor reflex has also been described in the literature by infusion of a pressor agent (e.g., phenylephrine) to conscious animals and recording of reflex changes in blood pressure/heart rate [e.g., (79,80)]. Although the presence of anesthetic may affect results, baroreflex testing has also been described in anesthetized rats and dogs [e.g., (37,81)]. The modified Irwin screen/FOB test and general toxicology studies can pick up ANS effects on, for example, pupil response, palpebral closure, lachrymation, and salivation.

Gastrointestinal System Methods

Initial Model

ICH S7A mentions assessment for gastric secretion, gastrointestinal injury potential, bile secretion, transit time in vivo, ileal contraction in vitro, and gastric pH measurement and pooling. Standard models for assessment for potential effects on the digestive system are intestinal motility/transit or gastric emptying in mice or rats. A range of methods are available for these models, but motility is typically examined as distance traveled along the gut of a charcoal or dye meal (e.g., 0.2–0.3 ml for mouse and 1–2 ml for rat) given after administration of the drug. Comparison of distance traveled from the stomach is expressed as a percentage of the length of the whole intestinal tract at sacrifice versus vehicle and/or positive control groups usually occurs. Gastric emptying can be measured as the percentage change in stomach weight following drug and then drinking water (e.g., 20 ml/kg) administration with stomach removal (after clamping to prevent leakage of the contents into the esophagus or duodenum) after a further defined period. The timing between charcoal/dye meal or drinking water administration after drug dosing and then gastrointestinal/stomach evaluation is dependent on the kinetics of the drug but periods of 30 minutes are common. Typically, for both models, groups of animals (e.g., six per group and usually males) receive a single dose of either vehicle or three dose levels of drug. It has been suggested that due to the number of possible variables, test groups of eight or more are required to show significant changes in gastrointestinal motility in the mouse (82). If used, morphine or atropine is frequently used as positive control compounds. Modifications in the design of these studies are covered in the literature [e.g., (83,84)] but include measurement of the amount of dye (phenol red or carmine dye) remaining in the stomach after a liquid meal, the amount of charcoal remaining in the stomach after a charcoal meal, the amount of beads or pellets remaining in the stomach or small intestine after a meal with such materials, or following the fate of radioisotope administration. Furthermore, charcoal or dye transport and stomach weighing can be included in the one study. Fed or fasted animals can be used; a recent evaluation of gastrointestinal transit in the rat following a charcoal meal (30–120-minute posttest compound with animals assessed 30 minutes later) showed that although both models were effective, transit was greater in animals that had been fasted versus those that had been fed (85).

Utility of Model

Although widely used, caution needs to be applied to the results of these simple models of gastrointestinal transport as a means of detecting potential effects of a new drug on the digestive system. This concern was highlighted in a study on intestinal motility (assessed by the transit of carmine dye in the mouse) and gastric motility (assessed by stomach weight in the rat after water loading) using a range of potent pharmacological agents known to affect gastrointestinal function as well as assessment of both forms of motility in the guinea pig (83). Although activity was seen with codeine, diazepam, atropine, and cholecystokinin-8 (which inhibited function), neither mouse nor rat model gave consistent results with metoclopramide, bethanechol, cisapride, deoxycholate, carbachol, and domperidone, which questions the robustness of these models. Furthermore, the guinea pig "combination" model was less sensitive for gastrointestinal effects when compared to the individual models in mice and rats. Similar inconsistent results were seen in other work in which gastric emptying (assessed by stomach weight in the mouse after administration of hydrated chow as a test meal) was measured; although inhibitory compounds (atropine and morphine) were effective, various stimulants (carbacol and metoclopramide) did not significantly affect the rate of emptying (82). The study authors commented that the method used could only be recommended for detecting the actions of inhibitory agents. However in contrast, a similar gastric emptying model (again assessing stomach weight in the mouse but after use of semisolid, paste-like test meal) showed both stimulatory and inhibitory activity with various positive control compounds (86). Finally, a good knowledge of a drug's pharmacology is often needed to explain findings in safety pharmacology studies. Thus, in addition to causing expected vasodilator effects (decreased blood pressure and increased heart rate in rodents and nonrodents) through phosphodiesterase type 5 (PDE5) inhibition, sildenafil also reduced intestinal transit in rodents (Sildenafil EPAR) (87). The latter finding is a reflection of PDE5 inhibitory action on smooth muscle.

Further Models

A range of other models exist to further study for potential gastrointestinal effects [e.g., (37,84)]. Examples include:

- Gastric acid secretion: Rats are anesthetized and a ligature placed around the duodenum just below the pylorus sphincter following midline incision. Following recovery, animals receive drug and at a suitable time (e.g., five hours) are euthanized with the abdomen opened, the esophagus clamped, and the stomach removed with measurement of contents volume, electrolyte concentration, and acidity. Another model is the rat gastric fistula in which gastric lumen intubation occurs under anesthesia; following recovery, gastric juice is collected for up to three hours. This model can also be performed in the dog.
- Fecal output/colonic propulsive motility: Measured as fecal pellet output (along with composition) in the rodent. Another model is to measure glass bead expulsion following insertion into the distal colon of mice or rats (this is also a good model to assess for constipation).
- Ulcerogenic effect: Assessed in the fasted rat from examination of the stomach and small intestine (for hyperemia, hemorrhage, and ulcers), perhaps in comparison with a drug (e.g., indomethacin) known to have this effect.
- Emetic effect using ferrets (can also be used to assess for CNS effects).

Other more sophisticated methods for examining the gastrointestinal tract are described in the literature and include endoscopic biopsy and imaging techniques (88). Another study related to the gastrointestinal tract is examination of submaximal contractions (tone and basal) with a range of drug concentrations using guinea pig ileum in the presence of acetylcholine, histamine, 5-hydroxytrypamine, and barium chloride. This study was commonly performed as part of the Japanese LIST A package described earlier but is rarely performed for modern drugs (unless there is a specific concern).

Other System Methods

ICH S7A mentions assessment for effects on, for example, dependency potential or skeletal muscle, immune, and endocrine functions. Following are some examples of available assessment models.

Abuse/Dependence Potential

For some CNS-active drugs there is a need to examine the potential for drug abuse liability/dependence. A recent preclinical guideline in Europe (89) describes relevant drugs as opiates, CNS stimulants, sedatives, and anxiolytics and various other classes with examination using a tiered approach. The first tier involves determination of any signal from receptor-binding studies from known targets involved in drug dependence or in vivo pharmacology studies followed up by a second tier of specific behavioral pharmacology studies if a signal is seen. For the latter, various models are available (usually in the rat and primate), which involve multiple endpoints (e.g., motor function, learning, memory, motivated behavior, and drug withdrawal behavior). The work can be undertaken alongside clinical development. Different studies cover reinforcing effects (study of self-administration), drug discrimination (overlap of interoceptive stimulus effects), and physical dependence potential (assessing for presence of a withdrawal syndrome) (90). Study examples are

- Self-administration model: Depression of levers (rodent and nonrodent) and nose poke (rodents) that result in delivery of positive control (e.g., cocaine or heroin) through chronically indwelling venous catheters (measured as change in level pressing/nose poke upon substitution with test drug).
- Drug discrimination model: Two lever choice model in which animal learns that operation of one lever will produce a reward (e.g., food pellet) if the animal is pretreated with a positive control (e.g., cocaine), but operation of the other lever will not (and that operation of this latter lever produces food if the drug is not present) with measurement of response for the test drug when training is complete.
- Physical dependence model: One study design with, for example, an opiate drug, is to give vehicle, various levels of drug, and positive control (morphine) repeatedly to groups of rats followed by a challenge with the opioid antagonist (naloxane) to precipitate possible withdrawal syndrome. The morphine-naloxane treatment results in signs including writhing, squealing, diarrhea, teeth chattering, ptosis, and "wet dog" shakes (91). Further assessment for withdrawal (discontinuation) syndrome can occur by following animals for an additional dose free period.

Antigenicity Testing

Although also a consideration for immunotoxicity assessment, specific antigenicity tests [passive cutaneous anaphylaxis (PCA) and active systemic anaphylaxis (ASA)] are sometimes included as part of a safety pharmacology package. The two most well-known tests are

- Antigenicity in the mouse using the PCA test in the rat: Typically, groups of mice (e.g., five males) are dosed with vehicle, various levels of drug, and positive control (human serum albumin; HSA), weekly on up to five occasions with removal of blood approximately 10 days after the last sensitization dose. Two routes of administration are used—the intended clinical route of administration and a route that is most likely to initiate an antigenic response. Concurrent dosing of an adjuvant (e.g., 2% aluminium hydroxide) to some groups is also included to further increase the antigenic potential of the drug. Extracted sera is then diluted and intradermally dosed to the shaved flank of rats (6 sites per animal) followed 48 hours later [immunoglobulin E (IgE)] or 4 hours later (IgG) by intravenous challenge to the rat with drug (with or without HSA) mixed with Evan's blue dye. The intradermal injection sites are then examined 30 minutes later for "blueing." A positive response is an indication of the ability of the drug to produce IgE and/or IgG antibodies in the mouse.
- Antigenicity in the guinea pig using the ASA and PCA tests: Typically, groups of guinea pigs (e.g., 10 control and 5 test males) are dosed with vehicle, various levels of drug, and positive control (HSA), up to three times weekly for four weeks with removal of blood approximately seven days after the last sensitization dose. PCA testing follows the method described above for rats (only using fresh guinea pigs). For ASA testing, the originally treated guinea pigs receive an intravenous challenge of drug or HSA approximately 14 days after the last sensitization dose. Assessment for the presence of anaphylaxis (e.g., scratching, red ears, unrest, tremor, sneezing) is made for up to 210 minutes postchallenge (although an effect is usually seen within a few minutes of challenge).

Altered Hormone or Other Mediator Release

Measurement of potential effects on hormones or other parameters can be included in repeat-dose toxicology studies as part of the clinical pathology battery. However, some companies also perform specific safety pharmacology evaluation with some examples as follows:

- Altered hormone levels: Measurement of hormone (e.g., blood prolactin, thyroid hormone, or testosterone level) in rat; estrogenic studies in ovarectomized rats are also used.
- Gastric acid suppression inducing secretion of gastrin: Blood gastrin measurement in rat.
- Histamine release examination: Measurement in rat or an in vitro comparison using a rat peritoneal mast cell versus human mast cell model.

Blood Coagulation

There may be a need to assess whether, for example, an intravenously administered drug has a local effect on blood. Typical studies used are described in the literature [e.g., (37)] but commonly include:

- In vitro platelet aggregation (e.g., human blood).
- In vitro hemolytic potential (e.g., rabbit or human blood).

Local Tolerance

For drugs administered by injection, there is a need to assess for local dose site reaction (e.g., inflammatory response in the muscle or subcutis). These "local tolerance" type investigations may be included into toxicology studies but, more typically, are performed as a stand-alone study (e.g., single dosing in the rabbit) with histopathological examination of the dose site. Most companies, however, tend to classify local tolerance studies as part of the toxicology, not safety pharmacology package.

BIOLOGICAL DRUGS

In addition to small molecules, ICH S7A also applies to biotechnology-derived products (biologicals). Specific reference is also given in a further guidance "Preclinical Safety evaluation of Biotechnology-derived Pharmaceuticals" (92). These documents highlight that, as such materials often achieve highly specific receptor targeting, it is often sufficient to evaluate safety pharmacology endpoints as part of toxicology and/or pharmacodynamic and clinical testing. In reality, the need and extent of safety pharmacology testing for biological drugs is very compound-specific given the range and diversity of these therapeutic proteins, which include monoclonal antibodies (mAbs) as well as recombinant hormones, interferons, interleukins, growth factors, enzymes, blood factors and thrombolytics and vaccines, and cell/gene therapy and antisense molecules.

Examples of the types of studies performed with some recently approved mAbs and recombinant proteins are given in Table 4. For many of these compounds, it can be seen that the need for stand-alone safety pharmacology work does not exist or is greatly reduced compared to small, chemically synthetized molecules with the actual extent of the testing program, the subject of much debate within project teams, and/or with regulatory agencies. Indeed, it has recently been clarified that hERG channel evaluation is not appropriate for mAbs based upon the high selectivity of these compounds for their therapeutic targets and the extremely low probability for off-target interactions with binding sites of the channel (93). In recent years (and when scientifically justified), the inclusion of safety pharmacology endpoints into toxicology studies has become more frequent. Parameters that can be routinely incorporated into such studies in the primate include analysis of body temperature, heart and respiratory rates, blood pressure, and ECGs. Potential effects on the CNS (clinical signs assessment) and renal system (urinalysis) can also be assessed. This approach has further gained acceptance as the high-dose levels used in these toxicology studies are usually only at a 10- to 30-fold margin over the therapeutic level of drug (in terms of bodyweight) and so are much more relevant than the much higher safety margins seen with small molecules (see later). Obviously if an effect is seen, a specific safety pharmacology study may then be needed. Also, additional testing may be needed if there is a mechanistic basis for a potential cardiovascular effect (e.g., a mAb directed against a target expressed on cardiac myocytes or within the vascular system) (93).

Table 4 Examples of Safety Pharmacology Testing with Recently Approved Biological Drugs (mAbs and Various Recombinant Proteins)

Drug	Indication	Safety pharmacology package
Bevacizumab (Avastin) [Humanized MAb]	Metastatic carcinoma of colon or rectum	None (safety pharmacology endpoints evaluated in toxicology work with cynomolgus monkey)
Cetuximab (Erbitux) [Chimeric Mab]	Mentastics colorectal cancer	Cardiovascular and respiratory function in anesthetized cynomolgus monkey Safety pharmacology endpoints also incorporated into toxicology work with cynomolgus monkey
Eculizumab (Soliris) [Humanized MAb]	Paroxysmal nocturnal hemoglobinuria	None (drug has minimal cross-reactivity with nonprimate and primate with toxicology work using a surrogate murine antibody)
Efalizumab (Raptiva) [Humanized MAb]	Chronic plaque psoriasis	None (safety pharmacology endpoints evaluated in toxicology work in primate, which is only species found to cross-react and/or surrogate rodent antibody)
Natalizumab (Tysabri) [Humanized MAb]	Relapsing multiple sclerosis	Cardiovascular and respiratory function in anesthetized dog Safety pharmacology endpoints also incorporated into toxicology work with cynomolgus monkey
Panitumumab (Vectibix) [Fully humanized MAb]	Colorectal cancer	Combined CNS, cardiovascular, and respiratory function in (telemetry) conscious cynomolgus monkey Safety pharmacology endpoints also likely to have been included into toxicology work with cynomolgus monkey
Ranibizumab (Lucentis) [Humanized MAb]	Age-related macular degeneration	None (safety pharmacology endpoints evaluated in toxicology work with cynomolgus monkey)
Alglycosidase (Myozyme)	Glycogen storage disorder – Pompe disease	Cardiovascular and respiratory function (+ assessment of general condition and body temperature) in dog Cardiovascular function in cynomolgus monkey NOTE: No specific evaluation of CNS was performed as it was stated that the protein was unlikely to cross the blood–brain barrier
Antithrombin alfa (Atryn)	Hereditary antithrombin deficiency	None (safety pharmacology endpoints evaluated in toxicology work in primate—cardiovascular and respiratory + rat—CNS and respiration)
Galsulfase (Naglazyme)	Mucopolysaccharidosis	None (safety pharmacology endpoints evaluated in toxicology work in dog and cynomolgus monkey)
Idursulfase (Elaprase)	Mucopolysaccharidosis (Hunter syndrome)	Cardiovascular and respiratory function (+ assessment of clinical signs) in cynomolgus monkey

(*Continued*)

Table 4 Examples of Safety Pharmacology Testing with Recently Approved Biological Drugs (mAbs and Various Recombinant Proteins) (*Continued*)

Drug	Indication	Safety pharmacology package
Palifermin (Kepivance)	Oral mucositis	Modified Irwin screen in mouse Locomotor activity, barbiturate sleep time + writhing and convulsive activity in mice and body temperature in rat Contraction of isolated guinea pig ileum Cardiovascular and respiratory function in anesthetized rhesus monkey Gastrointestinal transit in mouse Renal effects in saline-loaded rat
Parathyroid hormone (Preotach)	Osteoporosis	hERG and dog Purkinje assays Cardiovascular and function in anesthetized and conscious rat + anesthetized dog No effect on cardiovascular function in toxicology work in monkey

Source: From Ref. 104.

Sometimes a specific safety pharmacology study is needed when a healthy animal does not express the target receptor. An example is that for examination for potential hypertensive effects (which can be found in patients with chronic renal failure following use of recombinant human erythropoietin). No such effect was seen with the recombinant protein in routine safety pharmacology or general toxicology studies, but an effect on cardiovascular regulatory control can be observed in spontaneously hypertensive rats (which more closely mimic the clinical situation) (94).

CURRENT PRACTICE AND THE FUTURE

Current Practice

The ICH safety pharmacology guidance has now been fully implemented by pharmaceutical companies and Contract Research Organizations worldwide. To comply with the need for studies to be performed to GLP, most companies have specialized, fully compliant safety pharmacology units often separate from the toxicology function. With the introduction of a core battery approach for safety pharmacology evaluation, it has been shown that the number of studies performed has been greatly reduced, notably in Japan (50). Interestingly, although performed prior to first human exposure, it has been shown that there is still variation in the timing of studies. Thus, some companies perform safety pharmacology studies at an early stage (precandidate selection and even at lead selection or optimization) or later (post-candidate selection) (50). In the former situation, the work may not be to GLP (e.g., due to lack of a relevant clinical formulation), which results in some repetition of studies.

Current safety pharmacology practices (as for 2007) have recently been benchmarked with results, largely for small molecules, available from 125 companies (95). It was reported that almost all companies included cardiovascular, CNS, and respiratory studies in the safety pharmacology package submitted to regulatory agencies, approximately 40% of the companies also submitted studies on gastrointestinal and renal function. The main in vitro cardiovascular study was the hERG assay (56/57 companies) with less common evaluation (36/57 companies) in APD assays using rabbit, guinea pig, or dog tissue. In fact, APD assays were performed as follow-up work to clarify ambiguous results in the hERG assay or to define mechanisms of action. The preferred in vivo cardiovascular model was the telemetered dog. The process of "frontloading" (safety pharmacology studies performed under non-GLP conditions during lead optimization before selection of a candidate drug), largely for cardiovascular and CNS studies, was often seen (43/55 companies).

A common question is what is a suitable high-dose level for in vivo safety pharmacology studies? ICH S7A indicates that doses should exceed the primary pharmacodynamic or therapeutic range and in the absence of an adverse effect on the safety pharmacology parameters evaluated in the study, the highest tested dose should be a dose that produces moderate adverse effects. In a recent assessment of 34 candidate drugs, it was found that a range of different approaches was used including basing values for rodent CNS and conscious cardiovascular studies on selected multiples of the pharmacodynamic effect, using data from pharmacokinetic studies, using MTD values based on single dose (rodent) or range-finding (rodent and nonrodent) toxicity studies or limit doses (19). A level of up to 30-fold the "therapeutic" dose has been suggested for conscious dog telemetry work, if the tolerability is thought to be good (20). Whether considered high enough nowadays, others have stated that for dog telemetry assessment the highest dose should be at least 10-fold higher than the plasma concentration expected to afford the desired therapeutic effect in man (13). As the clinical dose route is commonly not used, some pilot work in anesthetized animals may occur to select the high dose for cardiovascular assessment in this model. Other workers report that for rodent safety pharmacology CNS evaluation, the high dose is routinely set at either the MTD or the maximum feasible dose with a maximum of 2000 mg/kg (51). This gave values of 3.3- to 1000-fold the therapeutic dose. It could be argued that the latter safety margins give more of a toxicological evaluation than that of a high-dose pharmacological one. Indeed, other work (using levels of 10- to 100-fold the clinical dose for in vivo work) showed that the higher the test dose, the higher the incidence of false-positive results (i.e., findings in animals but not in the clinic) (49). For respiratory safety pharmacology studies, it has been suggested that the high dose should be based on acute toxicology findings and be a dose that produces evidence of toxicity, but not findings that may compromise respiratory function measurements (72). However, in the absence of toxicology data, it is stated that multiples of the pharmacological dose should be used and if toxicity is not limiting, the highest multiple should be at least 100 times the pharmacologically active dose. A recent evaluation of industry practice has shown that the highest dose tested for in vivo safety pharmacology studies gave a multiple (most often 10, 30, 100 times) of the expected clinical concentration or one that matches the plasma concentration with moderate adverse effects (95). In the latter case, the dose producing such effects was determined in separate studies. For in vitro studies, the target maximal concentration was reported to be the limit of solubility or, a multiple of the expected clinical concentration (30–100 or 100–300 times) with two-thirds of companies using unbound (free) and one-third using total, drug concentration.

A continuing debate is whether stand-alone safety pharmacology testing could be reduced or modified by inclusion of endpoints in toxicology studies. It is true that neurobehavioral/FOB measurements can be included in, for example, one-month toxicology studies in the rodent, as could plethysmographic assessment. Nonrodent toxicology studies routinely include assessment of ECG. However, a range of issues occur including the use of dose levels used in safety pharmacology versus toxicity testing, the timing of studies, actual interpretation of "pharmacological" effects, possible development of tolerance with repeat dosing, and the need for appropriately trained staff. A further major consideration is overcomplicating repeat-dose toxicity studies, which are already labor intensive. Thus, measurement of both safety pharmacology and toxicology endpoints could be compromised if included in one study.

Finally, on the topic of current practice from its inaugural meeting in 1993, the General Pharmacology/Safety Pharmacology discussion group developed into the Safety Pharmacology Society in 2000/2001 and has had annual update meeting on the topic ever since then (96).

Future Developments

Despite the development of the "core battery," the field of safety pharmacology is still dynamic with companies looking at ways to improve existing study models and developing others to help understand high-dose pharmacological effects caused by new drugs. To finish this chapter, a few areas of potential future developments are summarized as follows:

- In light of earlier, fast screening to avoid drug attrition and to reduce animal use, the development of new in vitro models (e.g., evaluation in human stem cell–derived cardiomyocytes) would be very useful. Indeed, it has been pointed out that innovative in vitro techniques would contribute significantly to the early safety pharmacology evaluation of new drugs (97).

- A potential change is to integrate neurobehavioral and cardiovascular assessments into one study in the nonrodent. Recently, it has been shown that examination of dogs for blood pressure and ECG changes by telemetry along with a detailed examination of 44 behavioral observations/measurements following administration of positive control drugs resulted in expected changes for both systems (98). Thus, a separate behavioral study in the rodent may not be needed in the future?
- In the last few years, advances in electronic monitoring allows for both cardiovascular and respiratory function assessment in conscious (telemetry) nonrodents with the latter measured from intrathoracic pressure and respiratory rate. Thus, a separate study, such as whole body plethysmograph in the rat, may not be needed in the future?
- Although a range of models exists, currently no consensus on a preferred "best" approach on addressing risks for repolarization-associated ventricular tachyarrhythmia is available. Maybe in the next few years a standard industry position on the most appropriate studies will be developed?
- The use of existing or new models to assess for the potential of drugs to cause QT shortening is also a consideration. Recent evidence suggests that just as prolongation of QT interval can be proarrhythmic, shortening of QT interval can also lead to atrial fibrillation and sudden cardiac death (12,33,99). As described earlier, testing in rabbit Purkinje fibers has revealed cases of APD shortening (12). The situation is also confused as it has been pointed out that ICH S7B does not mention QT interval shortening or its potential consequences.
- New nonmammalian models to assess for human cardiac proarrhythmia are being examined. Recent work has shown that ECGs can be reliably obtained from anesthetized zebra fish (100). QT prolongation was seen in the presence of various drugs causing such an effect in humans; the model also showed bradycardia for some of the compounds.
- Although with possible use as an early screening tool, other recent literature has supported the use of the zebra fish for assessment of the cardiovascular system as well as the CNS (measurement of locomotor activity, convulsant and proconvulsant liability, cognitive impairment, drug dependence potential plus impaired visual, and auditory functions), renal system, and gastrointestinal function (101). Indeed, a larval zebra fish locomotor assay for assessing the seizure liability of drugs (through recording of high-speed movements) has been validated using a range of positive control compounds (102).
- Finally, a future development may be safety pharmacology studies in juvenile as well as adult animals to aid prediction of potential unforeseen adverse findings in the pediatric human population. A recent comparison of CNS function in neonatal (two weeks of age), juvenile (four weeks of age), and adult rats (eight to nine weeks of age) using a range of psychoactive drugs has shown that although responses were qualitatively similar, quantitative differences were seen at different ages (103).

REFERENCES

1. Anon. Guidelines for General Pharmacology Required for Application for Approval of Manufacturing (Import) of New Drugs (1991): Notification YAKUSHIN-YAKU No. 4, Director of New Drug Division, Pharmaceutical Affairs Bureau, Japanese Ministry of Health and Welfare, Japan, 1991.
2. Anon. Guidelines for General Pharmacology. In: Drug Appraisal and Licensing Procedures in Japan 1992. Tokyo, Japan: Yakugyo Jiho Company Limited, 1993:137–140.
3. Anon. Applicability of Good Laboratory Practices. Committee for Proprietary Medicinal Products, Directorate-General for Internal Market and Industrial Affairs, 1992.
4. Sullivan AT, Kinter LB. Status of safety pharmacology studies in the pharmaceutical industry. Drug Dev Res 1995; 35:166–172.
5. ICH S7A. Safety Pharmacology Studies for Human Pharmaceuticals, International Conference on Harmonisation (ICH). www.emea.europa.eu/pdfs/human/ich/053900en.pdf. Accessed July 2008.
6. ICH S7B. The Nonclinical Evaluation of the Potential for Delayed Ventricular Repolization (QT Interval Prolongation) by Human Pharmaceuticals, International Conference on Harmonisation (ICH). www.europa.eu/pdfs/human/ich/042302en.pdf. Accessed July 2008.

7. Committee for Medicinal Products for Human Use (CHMP) Points to Consider on the Assessment of the Potential for QT Interval Prolongation by Non-cardiovascular Medicinal Products (CPMP/SWP/986/96), 1997.
8. Therapeutic Products Directorate Guidance Document (Health Canada). Assessment of the QT Prolongation Potential for Non-antiarrhythmic Drugs, 2001.
9. ICH E14. The Clinical Evaluation of QT/QTc Interval Prolongation and Proarrhythmic Potential for Non-Antiarrhythmic Drugs, International Conference on Harmonisation (ICH). www.emea.europa.eu/pdfs/human/ich/000204en.pdf. Accessed July 2008.
10. Shah RR. Drug-induced QT interval prolongation-regulatory guidance and perspectives on hERG channel studies. Novartis Found Symp 2005; 266:251–280.
11. Gintant GA, Su Z, Martin RL, et al. Utility of hERG assays as surrogate markers of delayed cardiac repolarisation and QT safety. Toxicol Pathol 2006; 34:81–90.
12. Lu HR, Vlaminckx E, Hermans AN, et al. Predicting drug-induced changes in QT interval and arrhythmias: QT-shortening drugs point to gaps in the ICHS7B Guidelines. Br J Pharmacol 2008; 154:1427–1438.
13. Crumb W, Cavero I. QT interval prolongation by non-cardiovascular drugs: Issues and solutions for novel drug development. Pharm Sci Technol Today 1999; 2(7):270–280.
14. Webster R, Leishman D, Walker D. Towards a drug concentration effect relationship for QT prolongation and torsades de pointes. Curr Opin Drug Discov Dev 2002; 5(1):116–126.
15. Redfern WS, Carlsson L, Davis AS, et al. Relationships between preclinical cardiac electrophysiology, clinical QT prolongation and torsade de pointes for a broad range of drugs: Evidence for a provisional safety margin in drug development. Cardiovasc Res 2003; 58:32–45.
16. Doherty C, Claiborne C, Glyptis T, et al. Is 20% inhibition of hERG sufficient for prediction for in vivo QT interval prolongation? Toxicol Sci 2004; 78(1-S):(Abstract no: 998).
17. Brown AM. hERG Assay, QT liability and sudden cardiac death. In: Morganroth J, Gussak I, eds. Cardiac Safety of Noncardiac Drugs, Practical Guidelines for Clinical Research and Drug Development. New York, NY: Humana Press, 2005:67–81 [Chap. 4].
18. De Bruin ML, Pettersson M, Meyboom RHB, et al. Anti-HERG activity and the risk of drug-induced arrhythmias and sudden death. Eur Heart J 2005; 26:590–597.
19. Baldrick P. Safety evaluation to support First-In-Man investigations I: Kinetic and safety pharmacology studies. Regul Toxicol Pharmacol 2008; 51:230–236.
20. Guth BD, Germeyer S, Kolb W, et al. Developing a strategy for nonclinical assessment of proarrhythmic risk of pharmaceuticals due to prolonged ventricular repolarization. J Pharmacol Toxicol Methods 2004; 49:159–169.
21. Murphy SM, Palmer M, Poole MF, et al. Evaluation of functional and binding assays in cells expressing either recombinant or endogenous hERG channel. J Pharmacol Toxicol Methods 2006; 54:42–55.
22. Misner DL, Ly J, Guo L, et al. The importance of early characterization of metabolites to access potential cardiovascular effects. Toxicol Sci 2008; 102(1):339 (Abstract).
23. Yao X, Anderson DL, Ross SA, et al. Predicting QT prolongation in humans during early drug development using hERG inhibition and an anaesthetized guinea-pig model. Br J Pharmacol 2008; 154(7):1446–1456.
24. Hanson LA, Bass AS, Gintant G, et al. ILSI-HESI cardiovascular safety subcommittee initiative: Evaluation of three non-clinical models of QT prolongation. J Pharmacol Toxicol Methods 2006; 54:116–129.
25. Lu HR, Marien R, Saels A, et al. Species plays an important role in drug-induced prolongation of action potential duration and early afterdepolarizations in isolated Purkinje fibres. J Cardiovasc Electrophysiol 2001; 12:93–102.
26. Lu HR, Vlaminckx E, Gallacher DJ. Choice of cardiac tissue in vitro plays an important role in assessing the risk of drug-induced cardiac arrhythmias in human: Beyond QT prolongation. J Pharmacol Toxicol Methods 2008; 59:1–8.
27. Terrar DA, Wilson CM, Graham SG, et al. Comparison of guinea-pig ventricular myocytes and dog Purkinje fibres for in vitro assessment of drug-induced delayed repolarization. J Pharmacol Toxicol Methods 2007; 56:171–185.
28. Ducroq J, Printemps R, Guilbot S, et al. Action potential experiments complete hERG assay and QT-interval measurements in cardiac preclinical studies. J Pharmacol Toxicol Methods 2007; 56:159–170.
29. Hayashi S, Kii Y, Tabo M, et al. QT PRODACT: A multi-site study of in vitro potential assays on 21 compounds in isolated guinea-pig papillary muscles. J Pharmacol Sci 2005; 99:423–437.
30. Martin RL, McDermott JS, Salmen HJ, et al. The utility of hERG and repolarization assays in evaluating delayed cardiac repolarization: Influence of multi-channel block. J Cardiovasc Pharmacol 2004; 43(3):369–379.

31. Pugsley MK, Curtis MJ. Safety pharmacology methods: Anticipating the transition from long QT (LQTS) to short QT syndromes (SQTS). J Pharmacol Toxicol Methods 2007; 56:87–90.
32. Lawrence CL, Pollard CE, Hammond TG, et al. Nonclinical proarrhythmia models: Predicting Torsades de Pointes. J Pharmacol Toxicol Methods 2005; 52:46–59.
33. Shah RR. Interpretation of clinical ECG data. In: Morganroth J, Gussak I, eds. Cardiac Safety of Nonclinical Drugs, Practical Guidelines for Clinical Research and Drug Development. New York, NY: Humana Press, 2005:259–298.
34. Carlsson L. In vitro and in vivo models for testing arrhythmogenesis in drugs. J Intern Med 2006; 259:70–80.
35. Fenichel RR, Malik M, Antzelevitch C, et al. Drug-induced Torsade de Pointes and implications for drug development. J Cardiovasc Electrophysiol 2004; 15(4):475–495.
36. Steidl-Nichols JV, Hanton G, Leaney J, et al. Impact of study design on proarrhythmia prediction in the SCREENIT rabbit isolated heart model. J Pharmacol Toxicol Methods 2008; 57:9–22.
37. Vogel HG, ed. Drug Discovery and Evaluation: Pharmacological Assays, 2nd ed. New York, NY: Springer, 2002.
38. Chen X, Cordes JS, Sun Z, et al. Species comparison-sensitivity of the ventricular wedge preparation to arrhythmogenic drugs. J Pharmacol Toxicol Methods 2007; 56(2):e27 (Abstract).
39. Guo L, Dong Z, Guthrie H. Validation of guinea pig Langendorff heart assay to evaluate cardiovascular toxicity of drug candidates. Toxicol Sci 2007; 96(1):78 (Abstract).
40. Guo L, Qian J. Beat-to-beat electrical alternans in guinea pig Langendorff hearts as a predictor of cardiotoxicants. Toxicol Sci 2008; 102(1):475 (Abstract).
41. Irwin S. Comprehensive observational assessment: 1a. A systemic quantitative procedure for assessing the behavioural and physiological state of the mouse. Psychopharmacologia 1968; 13: 222–257.
42. Moser VC. Approaches for assessing the validity of a functional observational battery. Neurotoxicol Teratol 1990; 12:483–488.
43. Moser VC. Applications of a neurobehavioral screening battery. J Am Coll Toxicol 1991; 10(6):661–669.
44. Gad SC, Gad SE. A functional observational battery for use in canine toxicity studies: Development and validation. Int J Toxicol 2003; 22:415–422.
45. Wasielewski JA, White JC, Newton PE, et al. Development and validation of a canine functional observational battery (FOB) for use in toxicity studies. Toxicol Sci 2002; 66(1-S):Abstract.
46. Deguchi Y, Anraku R, Nagata R, et al. New assessment of neurobehavioural changes using functional observational battery (FOB) in monkeys. J Pharmacol Toxicol Methods 2005; 52:215 (Abstract).
47. Gauvin DV, Baird TJ. A functional observational battery in non-human primates for regulatory-required neurobehavioural assessments. J Pharmacol Toxicol Methods 2008; 58:88–93.
48. Moscardo E, Maurin A, Dorigatti R, et al. An optimised methodology for the neurobehavioural assessment in rodents. J Pharmacol Toxicol Methods 2007; 56:239–255.
49. Igarashi T, Nakane S, Kitagawa T. Predictability of clinical adverse reactions of drugs by general pharmacology studies. J Toxicol Sci 1995; 20:77–92.
50. Valentin J-P, Bass AS, Atrakchi A, et al. Challenges and lesions learned since implementation of the safety pharmacology guidance ICH S7A. J Pharmacol Toxicol Methods 2005; 52:22–29.
51. Redfern WS, Strang I, Storey S, et al. Spectrum of effects detected in the rat functional observational battery following oral administration of non-CNS targeted compounds. J Pharmacol Toxicol Methods 2005; 52:77–82.
52. Ross JF, Mattsson JL, Fix AS. Expanded clinical observations in toxicity studies: Historical perspectives and contemporary issues. Regul Toxicol Pharmacol 1998; 28:17–26.
53. Durmuller N, Guillaume P, Lacroix P, et al. The use of the dog electroencephalogram (EEG) in safety pharmacology to evaluate proconvulsant risk. J Pharmacol Toxicol Methods 2007; 56(2):234–238.
54. Hammond TG, Carlsson L, Davis AS, et al. Methods of collecting and evaluating non-clinical cardiac electrophysiology data in the pharmaceutical industry: Results of an international survey. Cardiovasc Res 2001; 49:741–750.
55. Kano M, Toyoshi T, Iwasaki S, et al. QT PRODACT: Usability of miniature pigs in safety pharmacology studies: Assessment for drug-induced QT interval prolongation. J Pharmacol Sci 2005; 99:501–511.
56. Ollerstam A, Visser SAG, Duker G, et al. Comparison of the QT interval response during sinus and paced rhythm in conscious and anesthetized beagle dogs. J Pharmacol Toxicol Methods 2007; 56:131–144.
57. Omata T, Kasai C, Hashimoto M, et al. QT PRODACT: Comparison of non-clinical studies for drug-induced delay in ventricular repolarization and their role in safety evaluation in humans. J Pharmacol Sci 2005; 99:531–541.
58. Oueller M, Authier S, Nelson A, et al. Cardiovascular safety pharmacology evaluation in non-human primates: comparison of conscious and unconscious models. Toxicol Sci 2006; 90(1):446 (Abstract).

59. Authier S, Tanguay JF, Fournier S, et al. Conscious and anesthetized non-human primate safety pharmacology models: Hemodynamic sensitivity comparison. J Pharmacol Toxicol Methods 2008; 58:94–98.
60. Tattersall ML, Dymond M, Hammond T, et al. Correction of QT values to allow for increases in heart rate in conscious Beagle dogs in toxicology assessment. J Pharmacol Toxicol Methods 2006; 53:11–19.
61. Soloviev MV, Barrett RM, Birchmeier PA, et al. Comparison of two major in vivo cardiovascular safety pharmacology models. J Pharmacol Toxicol Methods 2007; 56(2):e27 (Abstract).
62. Spence S, Soper K, Hoe C-M, et al. The heart rate-corrected QT interval of conscious Beagle dogs: A formula based on analysis of covariance. Toxicol Sci 1998; 45:247–258.
63. Hanton G, Nahas K, Priou C, et al. The QT interval in the dog ECG: Importance in preclinical toxicology and relationship to heart rate. Toxicol Methods 2001; 11:21–40.
64. Chiang AY, Bass AS, Cooper MM, et al. ILSI-HESI cardiovascular safety subcommittee dataset: An analysis of the statistical properties of QT interval and rate-corrected QT interval (QTc). J Pharmacol Toxicol Methods 2007; 56:95–102.
65. Toyoshima S, Kanno A, Kitayama T, et al. QT PRODACT: In vivo QT assay in the conscious dog for assessing the potential for QT interval prolongation by human pharmaceuticals. J Pharmacol Sci 2005; 99:459–471.
66. Ando K, Hombo T, Kanno A, et al. QT PRODACT: In vivo QT assay with a conscious monkey for assessment of the potential for drug-induced QT interval prolongation. J Pharmacol Sci 2005; 99:487–500.
67. Tashibu H, Miyazaki H, Aoki K, et al. QT PRODACT: In vivo QT assay in anesthetized dog for detecting the potential for QT interval prolongation by human pharmaceuticals. J Pharmacol Sci 2005; 99:473–486.
68. Roden DM, Viswanathan PC. Genetics of acquired long QT syndrome. J Clin Invest 2005; 115(8):2025–2032.
69. Pugsley MK, Curtis MJ. Safety pharmacology on focus: New methods developed in the light of the ICH S7B guidance document. J Pharmacol Toxicol Methods 2006; 54:94–98.
70. Petit-Turcotte C, Authier S, Legaspi M, et al. Respiratory safety pharmacology: Relevant parameters. Toxicol Sci 2008; 102(1):42 (Abstract).
71. Hoymann HG. Invasive and non-invasive lung function measurements in rodents. J Pharmacol Toxicol Methods 2007; 55:16–26.
72. Murphy DA. Comprehensive non-clinical respiratory evaluation of promising new drugs. Toxicol Appl Pharmacol 2005; 207:S414-S424.
73. Harris D, Graham M, Price J, et al. Respiratory function in rats restrained for extended periods: Assessment of the effects of bethanecol. J Pharmacol Toxicol Methods 2005; 52:83–89.
74. Travis JJ, Zingaro GJ, Briscoe RJ. Spingosine 1-phosphate receptor activation elevates PENH in conscious rats using whole body plethysmography. J Pharmacol Toxicol Methods 2005; 52(1):218 (Abstract).
75. Authier S, Legaspi M, Gauvin D, et al. Validation of respiratory safety pharmacology models: Conscious and anesthetized beagle dogs. J. Pharmacol Toxicol Methods 2008; 57:52–60.
76. Authier S, Legaspi M, Fournier S, et al. Respiratory safety pharmacology: Comparison of conscious Beagle dogs, cynomolgus monkey and Sprague-Dawley rats. Toxicol Sci 2008; 102(1):43 (Abstract).
77. Chiu PJS. Models used to assess renal function. Drug Dev Res 1994; 32:247–255.
78. Bunting PB, Siegl PKS. Models used to assess cardiovascular function in general pharmacology. Drug Dev Res 1994; 32:256–259.
79. Holmberg MJ, Zucker IH. Increased left ventricular pressure attenuates the baroreflex in unanesthetized dogs. Am J Physiol 1986; 251(1 Pt 2):R23–R31.
80. Little RA, Randall PE, Redfern WS, et al. Components of injury (haemorrhage and tissue ischaemia) affecting cardiovascular reflexes in man and rat. Q J Exp Physiol 1984; 69:753–762.
81. McMahon NC, Drinkhill MJ, Hainsworth R. Reflex vascular responses from aortic arch, carotid sinus and coronary baroceptors in the anaesthetized dog. Exp Physiol 1996; 81:397–408.
82. Yeung C-K, McCurrie JR, Wood D. A simple method to investigate the inhibitory effects of drugs on gastric emptying in the mouse in vivo. J Pharmacol Toxicol Methods 2001; 45:235–240.
83. Baldrick P, Bamford DG, Tattersall ML. An assessment of two gastric transport models currently used in safety pharmacology testing. Hum Exp Toxicol 1998; 17:1–7.
84. Jacoby HI. Safety pharmacology and the GI tract. In: Gad SC, ed. Toxicology of the Gastrointestinal Tract. Boca Raton, FL: CRC Press, 2007:53–80 [Chap. 3].
85. Mittelstadt SW, Hemenwey CL, Spruell RD. Effects of fasting on evaluation of gastrointestinal transit with charcoal meal. J Pharmacol Toxicol Methods 2005; 52:154–158.
86. Osinski MA, Seifert TR, Cox BF, et al. An improved method of evaluation of drug-evoked changes in gastric emptying in mice. J Pharmacol Toxicol Methods 2002; 47:115–120.

87. Sildenafil European Public Assessment Report (EPAR). www.emea.europa.eu/humandocs/PDFs/EPAR/viagra/113698en6.pdf. Accessed July 2008.
88. Harrison AP, Erlwanger KH, Elbrond VS, et al. Gastrointestinal-tract models and techniques for use in safety pharmacology. J Pharmacol Toxicol Methods 2004; 49:187–199.
89. Guideline on the Non-clinical Investigation of Dependence Potential of Medicinal Products (CHMP/SWP/94227/2004), 2006. www.emea.europa.eu/pdfs/human/swp/9422704en/pdf. Accessed July 2008.
90. Ator NA, Griffiths RR. Principles of drug abuse liability assessment in laboratory animals. Drug Alcohol Depend 2003; 70:S55–S72.
91. Algate DR, Beard DJ. The physical dependence liabilities of oxybutynin and morphine in rats. Arzneimittelforschung/Drug Res 1985; 35(10):1577–1579.
92. ICH S6. Preclinical Safety evaluation of Biotechnology-Derived Pharmaceuticals, International Conference on Harmonisation (ICH). www.emea.europa.eu/pdfs/human/ich/o30295.pdf. Accessed July 2008.
93. Vargas HM, Bass AS, Breidenbach A, et al. Scientific review and recommendations on preclinical cardiovascular safety evaluation of biologics. J Pharmacol Toxicol Methods 2008; 58:72–76.
94. Green MD. Problems associated with the absence of activity in standard models of safety pharmacology used to assess biologic products. Int J Toxicol 1997; 16:33–40.
95. Lindgren S, Bass AS, Briscoe R, et al. Benchmarking safety pharmacology regulatory packages and best practice. J Pharmacol Toxicol Methods 2008; 58:99–109.
96. Safety Pharmacology Society Homepage. www.safetypharmacology.org. Accessed July 2008.
97. Wakefield ID, Pollard C, Redfern WS, et al. The application of in vitro methods to safety pharmacology. Fundam Clin Pharmacol 2002; 16:209–218.
98. Tontodonati M, Fasdelli N, Moscardo E, et al. A canine model used to simultaneously assess potential neurobehavioural and cardiovascular effects of candidate drugs. J Pharmacol Toxicol Methods 2007; 56:265–275.
99. Bjerregaard P, Gussek I. Short QT syndrome. Ann Noninvasive Electrocardiol 2005; 10:436–440.
100. Park D, Seng WL, Li C, et al. ECG assessment of drug-induced proarrhythmia in adult zebrafish. Toxicol Sci 2008; 102(1):338 (Abstract).
101. Barros TP, Alderton WK, Reynolds HM, et al. Zebrafish: An emerging technology for in vivo pharmacological assessment to identify potential safety liabilities in early drug discovery. Br J Pharmacol 2008; 154:1400–1413.
102. Winter MJ, Redfern WS, Hayfield AJ, et al. Validation of a larval zebrafish locomotor assay for assessing the seizure liability of early-stage development drugs. J Pharmacol Toxicol Methods 2008; 57:176–187.
103. Himmel HM. Safety pharmacology assessment of central nervous system function in juvenile and adult rats: Effects of pharmacological reference compounds. J Pharmacol Toxicol Methods 2008; 58:129–146.
104. European Public Assessment Report (EPAR). www.emea.europa.eu/htms/human/epar/a.htm. Accessed July 2008.

29 Immunotoxicology

Danuta Herzyk
Merck Research Laboratories, West Point, Pennsylvania, U.S.A.

INTRODUCTION

Immunotoxicology is a scientific discipline focused on the evaluation of potential adverse effects of xenobiotics (chemicals, drugs, and biotechnology-derived products) on the host immune mechanisms that can lead to harmful changes in the immune system. Unintended immunosuppression can lead to an increased susceptibility to infectious diseases and tumorigenesis and unintended immunostimulation can lead to the induction of hypersensitivity reactions, or an increased incidence of autoimmune disease.

The immune system is a tightly regulated and very complex network of various lymphoid and other cell types interacting by cell-to-cell contact and communicating via soluble mediators such as cytokines. This multicellular organ system consists of granulocytes, macrophages, lymphocytes, and dendritic cells with various functions and unique phenotypic characteristics. The cells that constitute the immune system are of hemopoietic origin residing in the peripheral blood, lymphatic fluid, and organized lymphoid tissues, including bone marrow, spleen, thymus, lymph nodes, tonsils, and mucosa-associated lymphoid tissue (1). Since the immune system is very dynamic and involves cell proliferation, differentiation, activation, and maturation, it is vulnerable to agents that disrupt any of these cellular processes. At the same time, immune cells are in a constant state of self-renewal that contributes to the functional immunological reserve. An intact and functional immune system that includes local and systemic, innate and acquired, as well as humoral and cell-mediated immunity is required to protect from and eliminate infectious and neoplastic disease. Apart from defending the host against invasion by infectious and opportunistic microorganisms and spontaneously arising neoplasm, this network of cells and soluble mediators must discriminate self from non-self and be able to react to non-self with a variety of defensive responses (2). In addition, the immune system can develop a response to a chemical or drug or their reactive metabolite that might bind to or alter a host protein, resulting in an allergic or autoimmune response.

Over the past two decades, there has been a growing awareness that pharmaceuticals can cause a variety of immunologically mediated adverse effects in patients. It is also well established that the immune system of experimental animals, although exhibiting some obvious differences from that of humans, is sufficiently similar that data obtained from lower species are instructive of a potential human response (3) and these data can be used for risk assessment. Furthermore, recognition by regulatory agencies that the immune system is an important as well as sensitive target organ for drug-induced toxicity has contributed to the growth of immunotoxicology as a discipline of toxicology. Immunotoxicology testing has emerged as a significant adjunct to routine safety evaluation of pharmaceuticals in development and has been addressed by guidelines issued by several regulatory authorities including the European Medicines Agency (EMEA) (4), the U.S. Food and Drug Administration (FDA) (5), and the International Conference on Harmonisation (ICH) (6). There is an agreement that reliable immunotoxicity methods are needed to address potential adverse effects on a target organ that is as dynamic and complex as the immune system. The main goal of the immunotoxicological evaluation of pharmaceuticals is to differentiate between true immunotoxic effects and a small decrease in one or several immunological functions that may not result in increased disease susceptibility due to immunological reserve (7). Regulatory approaches to immunotoxicity have evolved with the science, and all immunotoxicity testing strategies to date have recognized the complexity of the immune system as a target organ and that no single immune parameter can be used with sufficient confidence to test for the hazard of immunotoxicity.

GUIDANCE ON IMMUNOTOXICOLOGY TESTING FOR SAFETY ASSESSMENT OF PHARMACEUTICALS

The latest ICH S8 Guideline "Immunotoxicity Studies for Human Pharmaceuticals" (6) is focused on providing recommendations on nonclinical tests and approaches to the evaluation of immunotoxicity of new drug candidates (excluding biopharmaceuticals), which is defined as unintended immunosuppression and immunoenhancement. According to this guideline, evaluation of potential adverse effects on the immune system should be incorporated into routine safety assessment conducted during drug development. Evaluation of immunotoxicity includes endpoints on standard toxicology studies (STS) regarding hematology and lymphoid tissues (organ weights and histopathology) and additional immunotoxicity studies conducted based on a weight of evidence (WoE) review of multiple factors.

WoE review is a process to establish a need for specialized immunotoxicology studies. Factors to consider that might prompt additional immunotoxicity studies are the following:

- Findings from STS.
- The pharmacological properties of the drug.
- The intended patient population.
- Structural similarities to known immunomodulators.
- The disposition of the drug.
- Clinical information.

The document provides a decision tree as illustrated in Figure 1.

The ICH S8 Guideline also recommends a selection of immunotoxicology assays in the following:

- T-cell-Dependent Antibody Response (TDAR)
- Immunophenotyping
- Natural Killer (NK) Cell Activity Assays
- Host Resistance Studies
- Macrophage/Neutrophil Function
- Assays to Measure Cell-Mediated Immunity

The top three tests (TDAR, NK cytotoxicity, and immunophenotyping) listed in the guidance were selected based on demonstrated concordance of these tests (TDAR or NK combined with immunophenotyping) with models of host resistance to infections (8). They may be

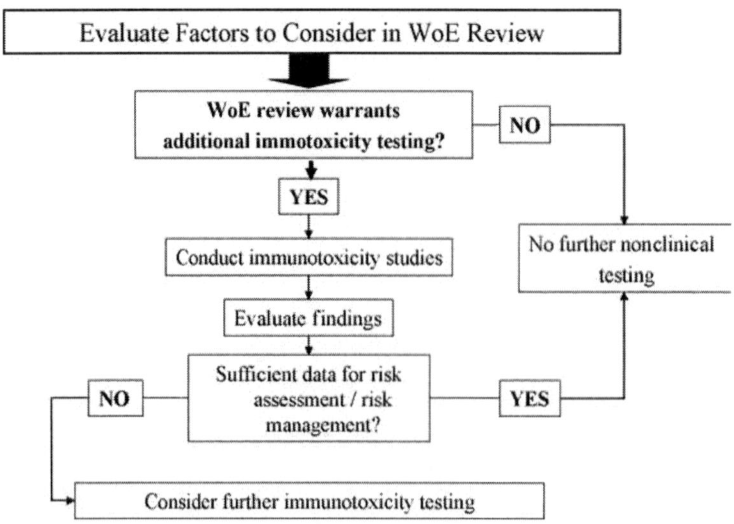

Figure 1 Schematic illustration of decision tree for immunotoxicity testing of pharmaceuticals based on ICH S8 Guidance.

viewed as core functional immunotoxicity studies and are most commonly used by immunotoxicologists. The other tests listed in the guidance (macrophage/neutrophil function assays, cell-mediated immunity assays, and host resistance models) are performed less frequently in immunotoxicology laboratories and are viewed as a second tier of testing to aid in risk assessment of immunotoxicity of pharmaceuticals.

EVALUATION OF POTENTIAL IMMUNOTOXICITY BASED ON STS

The evaluation of clinical pathology parameters, particularly hematology data, is crucial in WoE review of data obtained from standard toxicity testing for potential immunotoxicity signals. Clinical pathology data (hematology and serum biochemistry) often provide the earliest indication of effects of drugs or chemicals on immune status as reviewed in detail by Evans (9). The clinical pathology data collected in a standard toxicity study should be interpreted in context of other data, that is, clinical signs, exposure to the pharmaceutical and its metabolites, and histopathology findings to determine a potential concern for immunotoxicity. If a cause for concern is raised, subsequent immunotoxicology studies should be considered based on specific findings. For example, observed pancytopenia, abnormal hematopoietic cell morphology, or abnormal maturation of one or more hematopoietic lineages may suggest the need for bone marrow culture, which could be used to assess relative liability in human versus animal hematopoietic lineages. Lymphopenia may suggest a need to conduct immunophenotyping via immunohistochemistry in tissues or flow cytometry of peripheral blood or tissue cells. Functional immune tests, for example, TDAR, NK activity, and B- and T-cell proliferation, are often warranted in situations of lymphoid depletion in the absence of clear evidence of systemic toxicity/stress. Abnormal globulin values may prompt the need to conduct TDAR to determine whether the decrease in globulins represents a diminished ability to respond to challenge with infectious agents. Clinical pathology data are important in the WoE assessment of immunotoxic potential of pharmaceuticals and could be useful in guiding the extent of monitoring clinical patients.

Standard histopathology practices applied throughout the entire drug development process have been generally successful at identifying potential immunotoxicants. However, with the emergence of variety of immunomodulatory agents, the differentiation between intended and unintended excessive immunomodulation becomes more difficult and characterization of subtle effects is often warranted for risk assessment of immunotoxicity as reviewed in detail by Healy (10). In recent years, many scientific debates have taken place to address some concerns around inconsistent description of pathology examination and terminology used to characterize alterations of the immune system in toxicology studies. These discussions resulted in the generation and publishing of a few comprehensive reviews dedicated to "Best Practices" applied in histopathology evaluation of lymphoid organs (11,12). The ICH S8 Guidance contains a list of tissues that should be evaluated specifically by histological examination for signs of immunotoxicity: thymus, spleen, lymph nodes draining the anatomical site of route administration and drug exposure, at least one lymph node distal to the site, bone marrow, Peyer's patch for drugs administered by the oral route, and bronchus- and nasal-associated lymphoid tissues for drugs administered by the inhalation or nasal route. For intravenously administered drugs, the spleen is considered to be the draining lymph node equivalent. It is also recommended to use a semiquantitative description of observed morphological changes ("enhanced histopathology") reflective of the fact that lymphoid tissues demonstrate potential cellular functional changes. For example, drug-related increases in lymphoid germinal centers typically result from some "immune activation" and can indicate increased susceptibility to infections and hypersensitivity/autoimmunity. Thus an additional immunohistochemistry evaluation can be very helpful in determining if increased expression of markers associated with inflammation could indicate potential adverse drug reactions not generally thought of as immunotoxicity. Overall, histopathology may be one of the more sensitive tests for early effects on the immune system. However, based on the nature of the pathology and the biology of the particular cell population and organ affected, it should be determined if the change is of sufficient magnitude to constitute immunotoxicity and to pose a risk to the patient. Based on such determination, additional specialized immunotoxicology studies may be warranted.

TYPE AND PURPOSE OF SPECIALIZED IMMUNOTOXICOLOGY STUDIES

While strong immunotoxicity signals can be detected by hematology and lymphoid tissue histopathology evaluation on standard toxicity studies, potential moderate immunotoxicity, as a consequence of immune dysregulation by drugs, will manifest only at the functional level during an immune response to a challenge with an antigen (e.g., foreign protein, pathogen, toxin, etc.). Thus evaluation of the functional immune system requires studies involving "activated" immune cells, organs, or entire hosts in response to encountered "enemy." Beyond the conventional immune parameters incorporated into general toxicology studies, that is, lymphoid organ histopathology and hematology, there is a wide array of immune function assays, immune parameters, and analytical methods to select from to address specific questions/concerns of a drug's immunotoxic potential. Establishing the strategy, study design, and timing of immunotoxicity evaluations during preclinical drug development require careful consideration of the pharmacologic properties and toxicity profile of the drug under investigation.

Evaluation of Drug Effects on Immune Cell Phenotypes

Immunophenotyping refers to identification of specific cell types using tagged antibodies to specific cell surface antigens defined as clusters of differentiation (CD). The tags used are often fluorescent dyes applied in flow cytometry. Evaluation of subsets of leukocytes in peripheral blood and lymphoid tissues is broadly used in immunotoxicology and provides information on drug effects on the immune system on in vivo, ex vivo, and in vitro studies. Lymphocyte subset analysis is one of many extremely versatile applications in flow cytometry technology being used in both preclinical and clinical research as reviewed in detail by Narayanan et al. (13). Lymphocyte immunophenotyping provides information about the types of cells that could be altered following exposure to a drug. Reagents are now available to assess lymphocyte subpopulations in many of the standard toxicology species including mice, rats, dogs, and nonhuman primates (marmosets, cynomolgus, and rhesus monkeys). The standard set of markers used for lymphocyte immunophenotyping generally includes markers for total T and B cells, $CD4^+$ T-helper (T_h) cells, and $CD8^+$ T-cytotoxic/suppressor ($T_{cyt/sup}$) cells. In addition to these subsets, reagents are available to monitor NK cells in mice, rats, and monkeys. The same lymphocyte populations can be evaluated in the peripheral blood of patients in clinical trials to help determine the relevance of animal findings for humans and to establish a biomarker for immunomodulatory effects. The use of lymphocyte immunophenotyping in risk assessment, considerations for study and assay design, and relevance of animal data to effects in humans are reviewed in detail by Iciek (14).

Evaluation of Immunotoxicity Potential by Immune Function Tests

Innate Immunity

Innate immunity is considered to be the body's first line of defense against pathogens and malignant transformation and initiates within minutes to hours following exposure to pathogens. This phase of the immune response consists of both cellular and soluble mediators including phagocytic cells (macrophages and neutrophils), NK cells, and molecules such as complement proteins and cytokines, respectively. During the safety evaluation of therapeutics, toxicity signals related to innate immunity may be observed and there are specific assays available to determine which aspect of innate immunity is affected. These assays include NK cell, macrophage, and neutrophil function assays.

NK Cell Activity Tests

NK cell cytolytic activity is a key mechanism of host defense against tumor and virally infected cells. Drug-induced alterations in NK cell function can lead to increased incidence of neoplasia and infections. Regulatory guidelines on immunotoxicity testing recommend the NK cytotoxicity assay to determine potential drug effect on innate immunity as a follow-up to immunophenotyping studies that demonstrate a change in the number of NK cells or if the routine toxicity study demonstrates increased viral infection rates or in response to other factors (6). Incorporation of NK cytotoxicity measurements in preclinical testing is performed on a case-by-case basis primarily driven by mechanism of drug action (i.e., modulation of inflammatory cytokines).

The guidance specifically mentions the chromium-51 (^{51}Cr) release assay as a means to assess NK cell function but suggests that new methods that involve nonradioactive labels can be used if adequately validated. NK cytotoxicity test may be used in rodents, nonhuman primates, and dogs. The main differences in these species-specific assays are the tumor target cells used. The YAC-1 cell line is established as target cells for killing by NK cells in rodent assays (15), the K562 cell line is used for NK cells in primate assays (16), and the adenocarcinoma cell line (CTAC) is applied in canine NK cell assays (17). A novel flow cytometry–based method for evaluation of NK kill function described by Marcusson-Ståhl et al. (18) has comparable sensitivity to the ^{51}Cr-release method. The assay involves staining target cells with the nonradioactive dye carboxyfluoresceine succinimidyl ester (CFSE) followed by staining with propidium iodide (PI) and analysis by flow cytometry. NK cell assays performed in rats are based on spleen-derived NK cells while NK cell assays in nonhuman primates and humans use peripheral blood as a source of NK effector cells. These tests may be incorporated into general toxicity studies, allowing for simultaneous measurement of specific immunotoxicological endpoints within the same study as toxicological endpoints.

Phagocytosis and Microbicidal Tests

The role of phagocytosis is primarily the removal of microorganisms and foreign bodies but also the elimination of dead or injured cells. Phagocyte defects are associated with varied pathological conditions. Mononuclear leukocytes can bind to and subsequently phagocytose foreign particles through a number of membrane receptors. Because phagocytosis is a very complex process, a variety of ex vivo and in vitro assays have been designed to evaluate this cell function. If in the initial evaluation of a test article it is determined that follow-up immunotoxicology testing is required and the concern involves phagocytes, changes in many different aspects of phagocyte function, including phagocytosis and intracellular killing, can be evaluated. New methods, for example, phagocytosis of fixed and fluorescein-labeled *Escherichia coli* or flow cytometry–based ingestion of labeled beads, are increasingly used to measure phagocytosis.

The intracellular killing of microbial pathogens (microbicidal activity) is part of the innate immunity performed by granulocytes and mononuclear phagocytes as the result of the oxidative burst in these cells. The metabolic activation of phagocytes induced by the ingestion of bacteria, fungi, or latex beads, or by stimulation with zymosan or phorbol myristate acetate (PMA), results in the release of varied substances such as acidic hydrolases, lysozyme, superoxide anion, and reactive oxygen intermediates. Defects in oxidative metabolism of neutrophils and macrophages lead to impaired ability of these cells to kill microorganisms. Such impaired function is manifested in an increased frequency of skin or tissue infections. The assessment of bacterial killing involves measuring the viability of bacteria by various methods, for example, absorbance measurements using tetrazolium MTT [3-(4, 5-dimethylthiazolyl-2)-2, 5-diphenyltetrazolium bromide] dye, growth on agar plates, thymidine uptake, or staining with fluorescent dyes following incubation with phagocytic cells (19). Tumor cell killing may be assessed both in vivo (20) and in vitro where tumor cells are incubated with phagocytic cells and their destruction is assessed by measurement of released radioactivity or chemical dye used to label tumor cells (21,22).

Evaluation of NK cell function and macrophage tumoricidal activity using in vitro tests may be, particularly, important in assessment of immunomodulatory biopharmaceutical agents as discussed later in the chapter.

Adaptive Immunity

T-Cell-Dependent Antibody Response Test

TDAR to an antigen is a comprehensive immune function assay evaluating various aspects of immune responses including antigen processing and presentation, B and T lymphocyte interactions, antibody production, and cytokine-dependent isotype class switch (i.e., IgM- to IgG-specific antibody response). The use of a TDAR assay is endorsed by regulatory agencies as "default" immune function test for evaluating the immunotoxicity potential of new drug candidates because any compound-induced alteration in antigen processing, presentation, cell proliferation, differentiation and secretion of interleukins, antibody production, and cytokine-dependent isotype class switch is thought likely to modify the response. Originally, this test was configured as an IgM-/complement-dependent ex vivo plaque forming cell (PFC) assay in mice, with sheep red blood cells (SRBC) as both the immunogen and the target for complement-mediated lysis. Subsequently, an enzyme-linked immunosorbent assay (ELISA-) based format was developed by using SRBC as both immunogen and antigen in the antibody detection system (23). In recent years, however, TDAR assay development and application has refocused on the use of keyhole limpet hemocyanin (KLH) as a T-cell-dependent antigen (24–26). KLH is a soluble and highly immunogenic protein and provides a better standardized and characterized reagent compared to SRBC. Antibody responses to KLH, including anti-KLH IgM and anti-KLH IgG, are measured by ELISA.

The widely used KLH TDAR test is typically conducted as a 28-day repeated-dose study in rats administered a test article at multiple dose levels. The dose selection of the drug is usually guided by results from initial standard toxicity studies. To avoid immunotoxicity secondary to overly toxic doses, a high dose in the TDAR study should not exceed a maximal-tolerated dose of the drug. Typically, rats are immunized with KLH by intravenous administration on day 14 of the study and the primary anti-KLH IgM and anti-KLH IgG-specific antibody response is measured in serum collected on days 19 and 29, respectively (24). The TDAR test as a stand-alone functional evaluation is mainly used for hazard identification, that is, unintended immunosuppression. However, hazard identification alone may be insufficient for risk assessment of novel drugs, especially immunomodulators, in development. Additional in vitro and ex vivo tests can be incorporated into TDAR studies to elucidate mechanisms of immunotoxicity and provide helpful information for better human risk assessment. Such additional parameters that may be incorporated into the KLH–TDAR model include ex vivo lymphocyte proliferation in recall response to KLH, cytokine protein expression, and immunophenotype analysis, as reviewed in detail by Gore (27).

Nonhuman Primate TDAR

Development of drug candidates that are highly specific to primates and have a potential to adversely affect the immune system (e.g., immunomodulatory biopharmaceuticals) may require preclinical evaluation in a nonhuman primate model of immune function. The KLH–TDAR model used in rodents has been adapted to a nonhuman primate model (NHP TDAR). Unlike rodent TDAR, the NHP TDAR uses KLH with adjuvant to elicit a strong response and to reduce high interindividual variability in consideration of the relatively fewer number of monkeys assigned to toxicology studies. Single immunization with adjuvant by the intramuscular route demonstrated high titer anti-KLH IgM and IgG antibody responses with low interindividual variability (26). In the interest of reducing nonhuman primate usage on toxicology studies, NHP TDAR evaluations are typically built into other toxicology studies to characterize immunotoxicity signals identified either in previous studies or based on a priori knowledge of immunomodulatory activity of the drug candidate.

Cell-Mediated Immunity: Delayed Type Hypersensitivity Test

While core functional immunotoxicity studies, mainly the TDAR models, are designed to evaluate various effector cells of the systemic immune response working in concert, the cell-mediated immune response in local tissue is not assessed. A conventional test system for localized cell-mediated immune reaction is delayed type hypersensitivity (DTH) test. DTH is a response to an antigen (microorganisms and protein extracts) processed by antigen-presenting cells and presented to T-helper lymphocytes in local tissue, for example, skin. Only T lymphocytes previously exposed to an antigen (sensitized) can respond by a characteristic inflammatory reaction (mediated by certain cytokines and chemokines) upon subsequent exposure to the same antigen (challenge) within 24 to 72 hours at the exposure site. As described by Price (28), conventional DTH involves local intracutaneous injection of microorganism (e.g., bacteria, fungi, or their extracts) into sensitized host and evaluation of local induration and swelling. DTH, as the evaluation of host ability to mount an effective cellular immunity, is used in clinical practice for testing the immune competence in immunosuppressed patients (e.g., posttransplantation, HIV-infected) and as a diagnostic test for the presence of antigen-specific immunity to microbial agents (e.g., tuberculin test) or for the detection of potential allergies in symptomatic patients. Development of certain immunomodulatory drugs may justify inclusion of DTH test for immunotoxicology risk assessment.

Animal Models of Host Resistance

When immunotoxicity of a drug candidate is identified or anticipated based on the mechanism of action, its effect may result in an impaired clearance of an infectious agent, increased susceptibility to opportunistic infections, exacerbation of latent viral infections, ineffective immunization, or unintended immunostimulation. Host resistance models evaluate the functional integrity of the immune system and provide perspective on the extent of immunotoxicity vis-à-vis the functional immunological reserve as reviewed in detail by Burleson (29). A statistically significant decrease in several immunological functions may not always translate to biologically significant impairment resulting in increased disease susceptibility. Conversely, several small, not statistically significant, decreases in one or more immunological function may result in additive or synergistic effects that together affect susceptibility to opportunistic microorganisms (30). The biological significance of effects detectable in any selected immunological test may vary depending on a number of factors including the virulence of the infecting microorganism and the immunological status of the host. When there are concerns about drug-induced increased susceptibility to specific types of infection in patients, the conduct of host resistance studies in relevant animal models should be considered to aid in risk assessment of novel drug candidate.

Models of Host Resistance to Infections

The influenza infection model is a well-characterized test for assessment of the overall status of the immune system. Therapeutic agents can be evaluated for immunotoxicity by using the influenza host resistance model in either mice or rats (31,32). In this model, clearance of virus requires an intact and functional immune system involving innate immunity (NK cell activity, macrophage activity, and cytokine activity) as well as adaptive immunity (cytotoxic T lymphocyte activity and antibody production).

The *Streptococcus pneumoniae* infection model is designed for evaluation of innate immunity, particularly effects on function of neutrophils and macrophages. *S. pneumoniae* has been used in mice and rats as a pulmonary infection following intranasal infection (33). Bacterial clearance is evaluated by determining the number of colony forming units (CFU) per gram of lung tissue. Strong immunosuppressive drugs, for example, dexamethasone, are known to impair innate immunity and to delay bacterial clearance in the *S. pneumoniae* model. *Pseudomonas aeruginosa* can also be used as a pulmonary bacterial host resistance assay to evaluate therapeutics when immunotoxicity toward neutrophils, macrophages, and TNF-α production is suspected (33,34).

S. pneumoniae infection model represents means of testing clearance of bacteria encapsulated with polysaccharides in a T-cell-independent manner. Capsular polysaccharide antigens belong to a group of thymus-independent Type 2 antigens (TI-2) (35) and are highly dependent on the presence of functional marginal zone B cells (36) consisting of a distinct naive B cell lineage as well as memory B cells that are considered a critical host defense mechanism directed against encapsulated blood-borne pathogenic microorganisms. The *S. pneumoniae* marginal zone B cell model has been characterized in mice and Sprague Dawley rats with a systemic blood-borne infection by intravenous inoculation. Bacteria are quantified by determining the number of CFU in the spleen, liver, lungs, and blood over a two-week period. The measurements of cytokine production, lymphocyte immunophenotes, and antistreptococcal antibody levels are also incorporated in this model (33).

The *Listeria monocytogenes* infection model represents systemic infection (primarily in the liver and spleen) and can be useful in evaluation of adverse effects on neutrophils and Kupfer cells/macrophages residing in the liver and the spleen. NK cells and T lymphocytes also play a role in bacterial clearance of *L. monocytogenes* (29).

Latent viral infection models are warranted to evaluate potential viral reactivation. Reactivation of latent viruses may be a serious problem in immunosuppressed patients and may involve cytomegalovirus (CMV), herpes virus (HSV), or the human BK and JC polyoma viruses. Murine cytomegalovirus (MCMV) latent virus reactivation and rat cytomegalovirus (RCMV) infection are well-characterized models and have been used for immunotoxicity testing by evaluating the effect on clearance of the virus in a primary infection (37). Using the MCMV latent viral model, lymphocyte depletion studies have been conducted and revealed an hierarchy of immune control functions dependent on key cell populations, $CD8^+$, NK, and $CD4^+$ cells accordingly. Reactivation of the infection was rarely observed when only one cell type was depleted but was evident where depletion included multiple cell types (38). The MCMV reactivation model is viewed as suitable test to assess immunosuppressive therapeutics with the potential to cause serious recrudescence of latent viral disease (29).

Host Resistance Against Tumors

Effects of drug candidates on resistance to transplantable tumors could be considered as means to address immune surveillance based on concerns driven by mechanism of action of some immunomodulatory agents. Tumor host resistance models, including the B16F10 and PYB6 experimental metastasis studies using syngeneic C57BL/6 mouse, have been adopted in immunotoxicity evaluation (39). In the B16F10 melanoma model, B16F10 tumor cells are injected intravenously and tumor nodules (metastasis) develop in lungs within three weeks after the inoculation. The nodules are counted visually on the surface of the lungs using dissecting microscope. An alternative method for the B16F10 model uses the uptake of a thymidine analogue ^{125}I-5-iodo-2′-deoxyuridine (^{125}IUdR) in the lungs as a measure of tumor growth. The model can be easily expanded by the phenotypic evaluation of leukocytes infiltrating the tumor site to provide additional characterization of drug-induced immunotoxicity and to better understand the mechanisms of drug effects on host resistance to tumors. Such evaluations should be considered when the proposed models and data to be generated are deemed to be relevant and meaningful in human risk assessment.

In general, host resistance models are used for target-specific immunotoxicity testing of drug candidates.

Assessment of Hypersensitivity Potential of New Chemical Entities

According to the FDA Guideline "Immunotoxicology Evaluation of Investigational New Drugs," testing for hypersensitivity potential is required for drugs administered topically and by inhalation (5).

Contact Hypersensitivity for Dermal Drugs

Drugs intended for topical application should be tested for sensitization potential in an appropriate nonclinical model. Regulatory agencies (EPA and FDA), based on the recommendations of the Interagency Coordinating Committee on the Validation of Alternative Methods (ICCVAM), have considered the murine local lymph node assay (LLNA) an acceptable method to screen for allergic contact sensitization (40). The standard assay uses CBA mice treated with test article by topical administration to the dorsum of the ears bilaterally. Measurement of the degree of cell proliferation in response to the treatment is quantified by incorporation of ^{3}H-thymidine into DNA of replicating lymphocytes in draining lymph nodes. The response is compared to appropriate concurrent positive and negative controls. Interpretation is based on a statistically significant difference in the stimulation index (SI) between the test and negative control groups. Assay performance requires that the positive control's SI be greater than 3.0.

Respiratory Hypersensitivity for Inhaled Drugs

Drugs that are intended for delivery by the inhalation route may require an evaluation for their potential to induce respiratory hypersensitivity. Since a standardized validated nonclinical model for evaluating respiratory hypersensitivity potential of drugs does not exist, stepwise approaches are being employed to identify and characterize this risk (41). The most common strategy for pharmaceuticals is to first identify immune sensitization hazard potential using contact hypersensitivity tests (e.g., LLNA). If the test article is positive in LLNA and is considered to have the necessary chemical attributes to elicit an immune response, further evaluation of risk to produce pulmonary allergies in patients should be considered. Since the route and level of exposure are critical factors in induction of respiratory hypersensitivity, positive skin sensitizers may be subsequently tested in a guinea pig or rat inhalation challenge models (42) to better characterize potential risk.

IMMUNOTOXICITY TESTING OF BIOPHARMACEUTICAL PRODUCTS

Biotherapeutic agents often target the immune system and the desired pharmacologic effect is to modulate the immune response in various diseases. However, due to the complexity of the immune system regulation, there can also be undesirable effects of the immune modulation, which would constitute immunotoxicity. Evaluation of potential immunotoxicity is likely to be conducted as an integral component of overall pharmacology as opposed to separate studies driven by observed "cause for concern." This approach is advocated by the ICH S6 Guideline "Preclinical Safety Evaluation of Biotechnology-Derived Pharmaceuticals" (43). In general, the major concerns for risk associated with use of biological immunomodulators are acute reactions or cytokine release syndrome or chronic immunosuppression leading to opportunistic infection, chronic infection, and malignancy, as reviewed in detail by Bussiere and Mounho (44).

At one end of the immunotoxicity spectrum is a concern about acute systemic effects consistent with adverse immunostimulation. Such a case is illustrated by the recent example of severe toxicity in humans described as "cytokine storm syndrome" and caused by an experimental drug (TG-1412, agonistic anti-CD28 monoclonal antibody) that stimulates T cells (45). This example also heightens the importance of adequate immunotoxicity assessment of novel drug candidates using well-designed and target-driven experimental systems rather than routine testing approaches.

At the other end of the immunotoxicity spectrum is immunosuppression caused by biopharmaceutical agents that may have the potential to lead to tumor promotion or progression. A concern about potential tumorogenicity for such drugs is driven by exaggerated and prolonged immunomodulation that could lead to potential impairment in host defense against neoplasm. As reviewed by Bugelski et al. (46), in contrast to small chemical drugs that can be tumor initiators/promoters through the interaction with cellular DNA or other chromosomal material biopharmaceuticals, due to their large size, are not expected to pass through the cellular and nuclear membranes of an intact cell. Therefore, they and their metabolites (oligopeptides and amino acids) are unlikely to be tumor initiators. An evaluation of potential decreased immune surveillance caused by protein therapeutics should be considered in the context of pharmacological data and additional immunotoxicology testing. A battery of in vitro immune function tests that address major immune effector functions (e.g., NK cell assay, macrophage tumorocidal assay, cytotoxic T lymphocyte assay) can be useful in characterization of the hazard posed by this type of biotherapeutics and enable risk assessment for patients.

Assessment of Drug Immunogenicity Potential

According to the ICH S6 Guideline (43), immunogenicity evaluation is essential to the development of protein therapeutics. The drug-induced immune reaction can manifest as an antidrug antibody response that may alter pharmacokinetic and pharmacodynamic properties of the drug by formation of drug: antidrug antibody complexes. Many biotechnology-derived pharmaceuticals intended for human use are immunogenic in animals. Therefore, measurement of antibodies associated with administration of these types of products should be performed when conducting repeated-dose toxicity studies in order to aid in the interpretation of these studies. Antibody responses should be characterized (e.g., titer, number of responding animals, neutralizing, or nonneutralizing antibodies) and their appearance should be correlated with any pharmacological or toxicological changes. Specifically, the effects of antibody formation on pharmacokinetic/pharmacodynamic parameters, incidence and severity of adverse effects, complement activation, or the emergence of new toxic effects (e.g., pathological changes related to immune complex formation and deposition) should be considered when interpreting the data.

SUMMARY OF IMMUNOTOXICOLOGY EVALUATION

Specialized immunotoxicology studies are based on a "tool box" concept. Some functional immunotoxicology tests are fairly standardized and considered as core assays (i.e., TDAR, NK cytotoxicity, immunophenotyping) that can be applied consistently and conducted to GLP compliance. Many additional tests may be considered and conducted based on specific drug and target characteristics. While testing for potential immunotoxicity of all pharmaceuticals is recommended by regulatory guidelines, the approaches for drugs not intended to affect the immune system are better established than approaches for immunomodulatory agents. The latter include new chemical entities and biopharmaceuticals designed to produce some degree of "immunomodulation" (immunosuppression and anti-inflammatory properties). Therefore, their immunopharmacology and potential immunotoxicity are intertwined posing a greater challenge in both hazard identification and risk assessment during the drug development process.

REFERENCES

1. Dean JH, House RV, Luster MI. Immunotoxicology: Effects of and Response to Drugs and Chemicals. In: Hayes AW, ed. Principles and Methods of Toxicology, 5th ed. Philadelphia, PA: Taylor & Francis, 2007:1761–1793.
2. Paul WE. Fundamental Immunology, 4th ed. Philadelphia, PA: Lippincott Williams & Wilkins, 1999.
3. Haley PJ. Species differences in the structure and function of the immune system. Toxicology 2003; 188:49–71.
4. CPMP—Committee on Proprietary Medicinal Products. Note for guidance on repeated-dose toxicity 2000, CPMP/SWP/1042/99 July 2000. http://www.emea.europa.eu/pdfs/human/swp/104299en.pdf.
5. FDA: Guidance for industry: Immunotoxicology evaluation of investigational new drugs October 2002. http://www.fda.gov/Drugs/GuidanceComplianceRegulatoryInformation/Guidances/ucm065014.htm.
6. International Conference on Harmonisation of Technical Requirements for Registration of Pharmaceuticals for Human Use. ICH Harmonised Tripartite Guideline: Immunotoxicity Studies For Human Pharmaceuticals S8, May 2006. (CHMP/ICH/SWP/167235/2004; ICH S8)
7. Gore ER. Immune function tests for hazard identification: A paradigm shift in drug development. Basic Clin Pharmacol Toxicol 2006; 98:331–335.
8. Luster MI, Portier C, Pait DG, et al. Risk assessment in immunotoxicology. I. Sensitivity and predictability of immune tests. Fundam Appl Toxicol 1992; 18:200–210.
9. Evans E. Clinical pathology as crucial insight into immunotoxicity testing. In: Herzyk DJ, Bussiere JL, eds. Immunotoxicology Strategies for Pharmaceutical Safety Assessment. New Jersey: Wiley & Sons, 2008:13–26.
10. Haley P. Histomorphology of the immune system: A basic step in assessing immunotoxicity. In: Herzyk DJ, Bussiere JL, eds. Immunotoxicology Strategies for Pharmaceutical Safety Assessment. New Jersey: Wiley & Sons, 2008:27–54.
11. Haley P, Perry R, Ennulat D, et al. STP Position Paper: Best practices guideline for the routine pathology evaluation of the immune system. Tox Path 2005; 33:404–407.
12. Maronpot RR. A monograph of histomorphologic evaluation of lymphoid organs. Tox Path 2006; 34(5):409–696.
13. Narayanan P, Capocasale RJ, Li N, et al.. Application of flow cytometry in drug development. In: Herzyk DJ, Bussiere JL, eds. Immunotoxicology Strategies for Pharmaceutical Safety Assessment. New Jersey: Wiley & Sons, 2008:141–159.
14. Iciek L. Evaluation of drug effects on immune cell phenotypes. In: Herzyk DJ, Bussiere JL, eds. Immunotoxicology Strategies for Pharmaceutical Safety Assessment. New Jersey: Wiley & Sons, 2008:103–123.
15. Friberg D, Bryant JL, Whiteside TL. Measurements of natural killer (NK) activity and NK-cell quantification. Methods Enzymol 1996; 9:316–326.
16. Condevaux F, Guichard J, Forichon A, et al. Compared effects of morphine and nickel chloride on NK cell activity *in vitro* in rats and monkeys. J Appl Toxicol 2001; 21:431–434.
17. Gondolf C, Burkhardt E, Failing K, et al. A new colorimetric method for measuring cell-mediated cytotoxicity in dogs. Vet Immunol Immunopathol 1996; 55:11–22.
18. Marcusson-Ståhl M, Cederbrant K. A flow-cytometric NK-cytotoxicity assay adapted for use in rat repeated dose toxicity studies. Toxicology 2003; 193:269–279.
19. Hampton MB, Winterbourn CC. Methods for quantifying phagocytosis and bacterial killing by human neutrophils. J Immunol Methods 1999; 232:15–22.
20. Ou DW, Shem M-L, Luo Y-D. Effects of cocaine on the immune system of Balb/c mice. Clin Immunol Immunopathol 1989; 52:305–312.
21. Betjes MGH, Havenith CEG, van de Loosdrecht AA,et al. Methods for studying immuno-effector functions and antigen presenting activity of human macrophages. J Immunol Methods 1994; 174:215–222.
22. Hubbard AK. Effects of xenobiotics on macrophage function: Evaluation *in vitro*. Methods 1999; 19:8–16.

23. Temple L, Kawabata TT, Munson AE, et al. Comparison of ELISA and plaque-forming cell assays for measuring the humoral immune response to SRBC in rats and mice treated with benzo[a]pyrene or cyclophosphamide. Fundam Appl Toxicol 1993; 21:412–419.
24. Gore ER, Gower J, Kurali E, et al. Primary antibody response to keyhole limpet hemocyanin in rat as a model for immunotoxicity evaluation. Toxicology 2004; 197:23–35.
25. Finco-Kent D, Kawabata TT. Development and validation of a canine T-cell-dependent antibody response model for immunotoxicity evaluation. J Immunotoxicol 2005; 2:197–201.
26. Piccotti JR, Alvey JD, Reindel JF, et al. T-cell-dependent antibody response: Assay development in cynomolgus monkeys. J Immunotoxicol 2005; 2:191–196.
27. Gore ER. Functional cellular responses and cytokine profiles. In: Herzyk DJ, Bussiere JL, eds. Immunotoxicology Strategies for Pharmaceutical Safety Assessment. New Jersey: Wiley & Sons, 2008:127–140.
28. Price K. Cellular immune response in delayed-type hypersensitivity test. In: Herzyk DJ, Bussiere JL, eds. Immunotoxicology Strategies for Pharmaceutical Safety Assessment. New Jersey: Wiley & Sons, 2008:87–101.
29. Burleson GR, Burleson FG. Animal models of host resistance. In: Herzyk DJ, Bussiere JL, eds. Immunotoxicology Strategies for Pharmaceutical Safety Assessment. New Jersey: Wiley & Sons, 2008:163–177.
30. Burleson GR. Pulmonary immunocompetence and pulmonary immunotoxicology. In: Smialowicz R, Holsapple MP, eds. Experimental Immunotoxicology. Boca Raton, FL: CRC Press, 1996:113–135.
31. Burleson GR. Immunotoxicology: Influenza virus host resistance model for assessment of immunotoxicity, immunostimulation, and antiviral compounds. In: Burleson GR, Dean JH, Munson AE, eds. Immunotoxicology: Past, Present, and Future, Vol 2. New York: Wiley-Liss, 1995:181–202.
32. Hammerbeck DM, Burleson GR, Schuller CJ, et al. Administration of a dual toll-like receptor 7 and toll-like receptor 8 agonist protects against influenza in rats. Antiviral Res 2007; 73:1–11.
33. Burleson GR, Burleson FG. Influenza virus host resistance model. Methods 2007; 41:31–37.
34. Gosselin D, DeSanctis J, Boule M, et al. Role of tumor necrosis factor alpha in innate resistance to mouse pulmonary infection with *Pseudomonas aeruginosa*. Infect Immun 1995; 63(9):3272–3278.
35. Pillai S, Cariappa A, Moran ST. Marginal zone B cells. Annu Rev Immunol 2005; 23:161–196.
36. Guinamard R, Okigaki M, Schlessinger J, et al. Absence of marginal zone B cells in Pyk-2-deficient mice defines their role in the humoral response. Nat Immunol 2000; 1:31–36.
37. Ross PS, van Loveren H, de Swart RL, et al. Host resistance to rat cytomegalovirus (RCMV) and immune function in adult PVG rats fed herring from the contaminated Baltic Sea. Arch Toxicol 1996; 70(10):661–671.
38. Polić B, Hengel H, Krmpotić A, et al. Hierarchical and redundant lymphocyte subset control precludes cytomegalovirus replication during latent infection. J Exp Med 1998; 88:1047–1054.
39. McCay JA. Immunotoxicology: Syngeneic tumor cell models: B16F10 and PYB6. In: Burleson GR, Dean JH, Munson AE, eds. Immunotoxicology: Past, Present, and Future, Vol 2. New York: Wiley-Liss, 1995:143–157.
40. Basketter DA, Evans P, Fielder RJ, et al. A Local lymph node assay—validation, conduct and use in practice. Food and Chemical Toxicology 2002; 40:593–598.
41. Maier CC. Nonclinical models to assess respiratory hypersensitivity potential. In: Herzyk DJ, Bussiere JL, eds. Immunotoxicology Strategies for Pharmaceutical Safety Assessment. New Jersey: Wiley & Sons, 2008:257–270.
42. Arts JHE, Kuper CF. Animal models to test respiratory allergy of low molecular weight chemicals: A guidance. Methods 2007; 41:61–71.
43. International Conference on Harmonisation of Technical Requirements for Registration of Pharmaceuticals for Human Use. ICH Harmonised Tripartite Guideline: Preclinical Safety Evaluation of Biotechnology-Derived Pharmaceuticals S6, July 1997. http://www.ich.org/LOB/media/MEDIA503.pdf.
44. Bussiere JL, Mounho B. Differentiating between desired immunomodulation and potential immunotoxicity. In: Herzyk DJ, Bussiere JL, eds. Immunotoxicology Strategies for Pharmaceutical Safety Assessment. New Jersey: Wiley & Sons, 2008:191–198.
45. Suntharalingam G, Perry MR, Ward S, et al. Cytokine storm in a Phase 1 trial of the anti-CD28 monoclonal antibody TGN1412. New Eng J Med 2007; 355:1018–1028.
46. Bugelski PJ, Sachs C, Cornacoff J, et al. Immunomodulatory biopharmaceuticals and risk of neoplasia. In: Cavagnaro JA, ed. Preclinical Safety Evaluation of Biopharmaceuticals: A Science-Based Approach to Facilitating Clinical Trials. New Jersey: Wiley & Sons, 2008:601–632.

30 Local Tolerance and Other Toxicity Studies

Andrew Makin

LAB Research Denmark (Formerly Scantox), Hestehavevej, Lille Skensved, Denmark

INTRODUCTION

Drugs are administered to humans by a variety of routes depending on drug type, disease type, and the kinetic profile of the drug. Although it is clearly important to have characterized the safety and efficacy of the product in all cases, it is important to know that local effects of the drug product at the site of administration are not adverse.

In this chapter, I will discuss the various aspects of local tolerance related to medicinal products based on effects at the site of administration. I will avoid discussing in detail evaluation of subsequent systemic effects following administration because these aspects are covered by other chapters. It is, however, the case that local tolerance is more than just single-dose skin and eye irritation studies in rabbits and so there is discussion of repeat-dosing studies and some reference to the local tolerance aspects of the regulatory toxicity studies.

In many cases, local tolerance issues are addressed in the routine toxicology studies as a normal part of assessing reactions at the site of administration. So, for instance, in a repeated-dose intravenous toxicity study in animals, a standard part of the study protocol would be regular clinical inspection of the dose sites for signs of reaction to the treatment followed by sampling representative sections at the necropsy and subsequent histopathological evaluation. There are, however, cases where this is not done and where specific studies focused just on the local tolerance issues are required. Another area where local tolerance studies are appropriate is where drugs of known or established systemic toxicity are developed for new routes of administration. So, for example, where an antibiotic drug with a well-characterized systemic toxicity profile is adapted for topical use, the focus of any additional testing will be on the local effects at the site of application. This assumes that some basic bioavailability studies have been performed and that they have shown that the topical application of the product gives a lower systemic exposure than using the original route of administration. Additionally, it may be appropriate to evaluate local tolerance in relation to a site of administration that is not the intended site—for instance, for a drug that is intended to be administered intravenously, the possible adverse effects of unintentional peri-venous or intra-arterial administration should be evaluated.

Where studies are designed to look specifically at local tolerance, rather than general toxicity, it is important to remember that what is being tested is the final formulation or product that is intended to be given to humans. Whereas in a standard toxicology study local tolerance is investigated as a routine, those studies often employ higher doses or different formulations than will be used in the clinic. In a specific study of local tolerance, the normal toxicology endpoints associated with high-dose toxicity following administration of the active drug may be of little interest, while direct comparison with the probable human dosing is the principal aim. For this reason, it is important that, as far as practicably possible, the final formulation intended for human use should be employed in the local tolerance studies. Adjusting the formulations (and particularly altering characteristics of the vehicle or excipients) will render the tests meaningless unless, of course, the purpose is to compare the local effects of different preparations prior to proceeding to use them in humans.

A final example of where local tolerance studies could be of interest are where minor changes in the human formulation need to be evaluated for local effect and where full formal toxicology studies are considered unnecessary. This last example could also cover situations where the drug is delivered by a device (e.g., a needle-free injection device) and where investigation of the effects of the changed delivery system is of the greatest interest.

REGULATORY BACKGROUND

The clearest regulatory guidance on local tolerance is found in the European Agency for the Evaluation of Medicinal Products (EMEA) Note for Guidance (1). This guidance outlines the basic principles associated with local tolerance testing, and in this chapter, it is used as the guide to what dosing routes should be considered for local tolerance testing as well as to explore the more practical aspects of meeting it. In addition to the published guideline from the EMEA, the United States Food and Drug Administration (FDA) has issued a draft guidance concerning testing of products following reformulation for new routes of administration (2). It is emphasized that at the time of writing (November 2008) this guidance was drafted and was only circulated for discussion purposes. However, it does reflect the way FDA is thinking on this topic and several of the recommendations in the guidance have relevance to local tolerance testing.

Local tolerance of drugs and formulated drug products is rarely studied in complete isolation. As mentioned earlier, mostly the local tolerance of the drug by the intended clinical route is studied in the standard nonclinical toxicology studies in animals. Local tolerance studies can also be associated with elucidating potential effects after unintentional administration of a product to a site that is not the expected site of administration (e.g., performance of an ocular irritation study for a product that is intended for dermal application) or where a drug of known toxicity is reformulated for delivery by another route (e.g., where an antibiotic of known oral toxic profile is reformulated for dermal application). Reformulation of drugs can also lead to issues in relation to the excipients used in the vehicle. Where novel excipients are used, the regulatory agencies are required to see a full evaluation of the excipients themselves, both systemic, and for local tolerance safety. If the excipients have not been used before or have only been used for other routes or dose forms, this can be quite burdensome.

In this chapter, all of these aspects of local tolerance are discussed, so that we are not only concerned with direct effects at the site of administration but also for the wider implications of administration of drug products at various sites. Reformulation of drugs in itself can give rise to the need for further testing, particularly if the combination of drug and vehicle could have an effect on toxicity (e.g., where a product for dermal application is formulated with vehicle constituents that enhance absorption). As stated I have tried to cover local tolerance in its wider aspects. I have tried to explore some of the key points related to what testing is required and then to look, where appropriate, at the practicalities of such testing, study design, and so on.

STUDY DESIGN PRINCIPLES

Study designs are often not specified in the pharmaceutical testing guidelines [in contrast with the Organization for Economic Cooperation and Development (OECD) test guideline series for chemicals] (3). In most cases, it is up to individual laboratories to employ protocols that are suitable for the purpose based on good scientific principles and on the ability of the study design to provide sufficient data to answer the questions posed. An example of the variability of study design for local tolerance testing is given in the paper by Jochims et al. (4) where in a survey of nine separate laboratories a wide variety of study designs for such studies was found. There were variations in species used for specific tests, species, number and sex of animals, use of controls, endpoints, etc. One can assume that each of the protocol variants is correct in its own way and therefore one can conclude that there cannot be a definitive guidance that can be given to cover all types of local tolerance study. However, there can be assumed to be several design principles that should be taken into consideration, as discussed below.

Species Selection

Species selection depends greatly on the test being performed and the availability of other data on the drug being tested. For certain tests, the species to be used is described in the guidance (e.g., rabbit for in vivo ocular irritation, mouse for the Local Lymph Node Assay). For others, the choice of species could be dependent on being able to extrapolate results from the species used in the standard toxicology program. For others again it might be more appropriate to use a completely different species, for example, for a drug developed using the dog as the principal nonrodent species, it would be more appropriate to switch to the pig or minipig where the drug is reformulated for dermal use. A sensible precaution would be to make sure that the laboratory has some background experience in the study type and the species chosen.

For many local tolerance studies, particularly where the focus is mostly on local tolerance rather than on a wider investigation of general toxicology following administration by a new route, it may be sufficient to test products in a single species (assuming the use of the species most sensitive to the effects of the drug) and in a single sex (where gender differences in response are not anticipated). Numbers of animals to be used should, as always, be minimum consistent with being able to analyze the data produced.

Frequency and Duration of Administration

Several factors affect choice of frequency and duration of dose administration. Most obviously there should be a relation to the proposed human exposure. However, comparative pharmacokinetic or other metabolism data (bioavailability or bioequivalence of the drug after administration by the new route) might also indicate the need for additional studies to characterize further the toxicology of the drug prior to human administration. Altered absorption or elimination of a drug following a change of dose route could result in an altered toxicological profile and this should be taken into consideration when designing local tolerance studies.

Reversibility

Reversibility of effects should be investigated as with all pharmaceutical evaluation. Often with short local tolerance studies reversibility is a standard parameter (i.e., single dose followed by an observation period). For repeat-dose studies, where investigation of local tolerance is a primary endpoint, addition of groups of animals to look at reversibility should be considered. As mentioned later, there can be case where the dose formulation and route could be expected to cause some local tissue damage (e.g., intramuscular injection of a depot formulation) and the reversibility period would need to be sufficiently long to be able to at least clearly demonstrate that reversibility is in progress. Additionally, one could design several recovery points in the study in order to be able to follow the process of repair/recovery.

Preparation to be Used

As far as possible, a study specifically designed to demonstrate local tolerance should use the proposed human formulation. Practical considerations may mean that this cannot be the case but the predictivity of tests that do not use the final formulation is not as great as that of tests that do. As discussed elsewhere in this chapter local tolerance issues can be associated not only with the active drug substance but also with other ingredients in the formulation, some of which might be active in their own right. Therefore, while it can be interesting to know the cases where there are local tolerance problems and how much of this problem is due to the active drug, it is equally important to know if the formulation itself has an effect. The synergy between the various excipients used in some formulation can be important in modifying the behavior of the drug substance. Clearly, it is important that study designs include groups that receive the vehicle alone (in some cases, the control can also be within the same animal), and, where appropriate, untreated controls as well. There may be reasons why this cannot be done; for instance, a particular dose vehicle that is not proposed for the clinical formulation might be appropriate to perform a Local Lymph Node Assay. In such cases, the altered vehicle should be taken into consideration when interpreting the results of the studies and/or when evaluating the formulation prior to entering into clinical trials.

Choice of Dose

As discussed elsewhere the choice of dosage will depend somewhat on the other endpoints of the study. Therefore, if the study is a routine toxicology study, the dosage selection will be aimed at evaluating multiple endpoints associated with administration of high doses of the drug as well as local tolerance issues. Where the study is specifically designed to look at local tolerance as the principal endpoint, it would be more natural to use the proposed clinical formulation in the test as far as is practical.

Route of Administration

As far as possible the proposed human route of administration should be used in local tolerance studies. Otherwise there would be little meaningful data that could be extracted from the studies.

Animal Welfare

As with all experimentation involving the use of animals, consideration should be given to the principles of Reduction, Refinement, and Replacement. Where there is already existing information about a product, this should be used as far as is possible to define whether it is even necessary to perform a local tolerance study. If a study is necessary, the available data should be used to help in defining the ideal design of study. In vitro studies are available in some cases and these can reduce or eliminate the need for testing in animals, and at the very least can screen out those compounds that are likely to have the worst effects in vivo.

SPECIFIC DOSAGE ROUTES

Dermal Application

The skin is a good example of an organ that is exposed to medicinal products both as the therapeutic target and as an administration site for products whose target is remote (e.g., transdermal patches). This is an area where there are significant local tolerance issues and, from the experimental point of view, there are many study design challenges.

Dermal application of test substances in animal studies is broadly used in two contexts. First, it is used for testing pharmaceutical products where the intended route of human exposure is dermal; and second, for testing any product (principally chemicals) where there may be unintentional human dermal exposure. In either case the principles are the same.

The nature of the products to be applied is very variable—for example, creams, gels, ointments, other forms of liquid delivery system, patches, etc. However, as mentioned elsewhere in this chapter, the basic fundamental rule in animal studies where the objective is to test possible adverse effects of the product is that as far as can be achieved the product should be identical to that of intended to be used in humans. Dermal products generally, in addition to the active (drug) ingredient, contain a wide range of other constituents such as absorption enhancers, preservatives, buffer systems, emulsifiers, solubilizers, solvents, cosolvents, etc., which means that the actions and effects of the drug substance could be dependent on the concentrations and proportions of these other constituents. Therefore, as an example, if the proposed human formulation is 1% of the drug substance in a gel there would be little meaningful data that could be obtained from testing a 5% concentration in animals because the balance of ingredients would be incorrect. In addition, if the higher drug concentration or altered balance of constituents has an effect on the skin that would not be seen with other proportions, this might cause differences at the site of application that make any other results from the study irrelevant. When selecting dosages in the early stages of development of a product, it might be appropriate to test a range of concentrations of the formulation in order to choose the final clinical formulation. In addition, it could be the case that there is the intention of producing a range of concentrations for testing in the clinic. However, the most common approach is to take a single formulation with the highest intended human concentration and to administer varying amounts to the different groups of animals. This has to be considered carefully and some degree of judgment will have to be applied in the individual situations, particularly in studies where the systemic toxicity also has to be evaluated. The key point would be that over the duration of the application period, the absorption of the drug substance, or at least contact of the skin with the drug substance, should be maximized. A relatively soluble drug in a gel formulation will readily absorb into the skin, while a more complex ointment will not and, in the latter case, simply putting a greater volume on to the test site may not provide any benefits in relation to the study outcomes.

Experience in performing a great number of dermal studies in animals says that the two most frequently asked questions are: how much drug substance I can administer? and what should be the size of the dosing area? The reason for this is obviously to attempt to maximize the dosing in order to obtain the greatest possible safety margin. The answers to the two questions are linked but when considering them I need to go back to basic toxicological principles and think carefully about what I am doing in my animal studies. The purpose of nonclinical studies for any pharmaceutical product, regardless of dose route, is to provide information on the likely toxic (adverse) effects of the drug and to establish a dosage relationship so that the findings can be used to provide a guide to selection of safe doses in humans. In an oral study, for instance, one can simply give higher dose volumes or volumes with increased concentrations in order to increase dose levels. With dermal application there are certain practical limitations and the application should be appropriate to the species of animal that is being used. One can be faced with problems in study design in that, for instance, in human use a dermal product could be used on large areas of the body including in clinical trials where up to 70% of the body surface might be used. For purely practical reasons it is not possible to cover large areas of animal skin regularly on a daily basis. Furthermore, as mentioned above, the option of increasing the drug concentration is not always a possibility in dermal studies so that the dose can be varied either by varying the amount applied (but see the previous comment about ensuring maximal absorption of the drug at the application site) or by varying the area to which it is applied. One should, however, be pragmatic about this. I would not advise varying both area and volume in the same study, because this makes interpretation of results difficult.

The question of dosing area is interesting and there is in fact not much in the way of helpful guidance in the regulatory documents. The best one can find is that the guidelines in the OECD series (3) refer to 10% of the animal's surface area as being the maximum one should use and I normally use this. There are various mathematical formulas available for calculating an animal's surface area based on its bodyweight and these are used to calculate the area, which is then marked on the animal. The area is recalculated and remeasured regularly during studies to ensure that I am still using the correct area as the animal grows.

The amount of material that can be applied to an animal varies depending on the nature of the material but as a rule of thumb a practical maximum will be *ca* 1 g/kg bodyweight, assuming one is using the maximum dose area (10% of the surface area). Regardless of what the recommended human dose is, I find that for practical reasons in the animal room it is much easier to dose as g/kg bodyweight than to express doses as, for instance, ml/cm^2. The nature of the formulation has a great influence on how much can be applied. Gels are normally rapidly absorbed and are easy to apply and the maximum amounts can be given in this form. Ointments tend to be thick and poorly absorbed and while they can be applied in large amounts as quite thick layers, the poor absorption means that in effect putting on a thick layer makes no sense because most of the applied material never comes into contact with the animal. Watery liquids are the most difficult to apply as they tend to run off, and care and patience are required in their application to ensure that the material is dosed evenly over the application area.

I find that by far and away the most common method used in selection of dose levels in order to achieve the aim of getting a series of doses in regulatory toxicology studies is to keep the dosing area constant and to vary the volume of material applied to that area. Again with reference to introduction of variables, this makes the final interpretation of the results easier.

Species selection in dermal testing is an interesting topic and I also find that, paradoxically, this also leads to discussion of dosing route in toxicology studies with compounds that are intended for human dermal application. Dose route is discussed later and I will deal with species selection first.

It is a general regulatory requirement that nonclinical toxicology studies are performed in two species, one of which is a nonrodent. It is clearly important that the species used should be appropriate for the drug and the route of administration. Rats are unquestionably the most commonly used rodent species in safety studies and therefore they should be chosen for dermal products. From the range of nonrodent species that is available, the minipig or pig should be the first choice. The reason for this is that, of all the experimental species in current use, the structure and nature of porcine skin is the most similar to that of humans. The similarity is shown in both the physical characteristics (thickness, density of hair growth) and in the finer structure (close adherence to the underlying structures, rete ridges, presence of Langerhans cells, etc.). As an example, the range of thickness of the epidermis of a human is 50 to 120 μm, while that of a minipig is 70 to 140 μm, so there are clear similarities.By comparison the range of epidermal thickness in a rat is 10 to 20 μm, which is outside of the range for both the other species. Although not specified in the guidelines, based on experience in the laboratory and communication with others working with dermal products, I believe that most regulatory authorities expect to see that minipig or pig has been used in any package of studies intended for regulatory submission.

As mentioned previously, the rat would normally be selected as the rodent species and this leads in to the interesting question of dose selection. The FDA draft guidance previously referred to (2) specifically refers to the use of nonrodent species for dermal local irritation testing of drugs that have not been tested previously by this route. Inclusion of both untreated and vehicle-treated controls is recommended. Taking into consideration the purpose of the nonclinical safety studies, one of the key outcomes is to be able to get a prediction of likely adverse effects in humans and to define a dose–response relationship. The nature of many products for dermal application is that they are often intended to work at or close to the site of application. Therefore, it is normal that systemic exposure is low. An approach that is often used, and one that is accepted by regulatory authorities, is to use the rat to give information on systemic exposure and to use only the clinical dose route in the minipig. For practical reasons this makes sense because rat skin is a poor model of human skin and therefore the value of using this species to define, for example, local tolerance is questionable. Therefore, the subcutaneous or oral routes can be used for dosing in the rat. The dose formulation in this case should be appropriate to the dose route, for example, a solution for subcutaneous injection, and I would not advocate using the clinical formulation in this case. The subcutaneous route gives the advantage of presenting a kind of "worst-case scenario," which is representative of what would happen if all of the drug substance were to become available systemically after reaching the subcutis (e.g., if the dose formulation had been applied to broken or abraded skin). Some people use the oral route and this is perfectly acceptable. However, the systemic exposure would normally be expected to be lower following oral administration compared to subcutaneous, although other factors could come into play, such as liver metabolism, first-pass effects, and so on that could make the systemic exposure pattern different after oral dosing in comparison to subcutaneous. The main point is to know the relative bioavailability by the different routes of administration.

Among the other questions frequently asked when designing studies is whether the application site should be occluded or not during the exposure period. Clearly, if the human exposure would include occlusion (and by this is meant a dressing that excludes light and air), then this should be mimicked in the animal model. In other cases, I always recommend some form of light occlusion because the alternative is that the animal could interfere with the dose site and thus render ultimate interpretation of results difficult. The best recommendation is to cover the treated site with a couple of layers of light gauze held in place by an net "body stocking," which does not exclude the light and which allows the skin to "breathe" while making it less likely that the animal can interfere with the site. Experience says that if the material is applied and the animal is left free to roam around its environment without any form of protection, then there will be loss of the applied dose on the sides of the pen or cage, that if irritated the animal will be able to scratch or rub itself possibly causing damage to the treated skin, and that the local irritation results will be almost impossible to interpret.

I am also faced from time to time with the question as to whether the skin of the animals should be intact or if I should use some form of abrasion. I believe that in repeat-dose studies abrasion should be avoided for the reasons that (a) abrading the skin will lead to changes (physical damage, seen clinically as erythema, edema, or erosion) that could affect the interpretation of results, rate of absorption of the test article, etc.; and (b) that in the case of animals it is virtually impossible to abrade the skin repeatedly in a uniform manner, which also impacts adversely on the ability to interpret results. In single-dose local tolerance studies, particularly in rabbits, abrasion has been widely used in order to get a better contact with and penetration of the test formulation. In principle, for a single-dose study, abrasion is far easier to control because it only needs to be done once and only on untreated skin so that it can be done relatively uniformly. Tape stripping also provides a method of getting a uniform removal of the outer skin layers, but again this is only recommended for a one-off treatment prior to the first application of the test formulation.

Where products are formulated for dermal administration, including patches, consideration should be given to performance of photoirritation studies if the product absorbs light in the ultraviolet or visible spectrum. Phototoxicity is the subject of another chapter in this book and is not discussed further here.

Acute Dermal Irritation/Corrosion

For products that are expected to come into contact with the skin, either intentionally or unintentionally, it has been traditional to perform an in vivo test of irritation/corrosivity, either according to the OECD Test Guideline 404 (5) or other similar guidelines. In principal, the test is designed for chemicals rather than pharmaceuticals and therefore in the case of pharmaceuticals it could be considered that this test would be more suited to those drugs that are not intended for dermal application, but there could well also be applications for such a test in the discovery/screening phase of pharmaceutical development. Although this test is more appropriate for chemicals, there are some principles involved in it that are applicable in all cases of safety testing. The most important of these is that before engaging in any test of irritation or corrosion, there should be a thorough review of existing data and the decision to proceed with tests on animals should be based on a weight-of-evidence approach that says that no testing is required if the available data already points to a positive result. Complementing this test are alternatives, in vitro and *ex vivo*, that can significantly contribute to the weight of evidence, and that can also have wide application in early discovery/screening of drugs with potential for dermal application.

In the in vivo test, a single animal, usually a rabbit, is exposed to the test substance for four hours and thereafter the test site is observed for up to 14 days, and reactions are recorded. Only if there are not any signs of corrosivity should the substance be tested on a further two rabbits and the test is then complete. For in vitro testing, there are many tests used for early studies and it is not the intention here to go into this subject. However, there are two tests in the OECD series that have been accepted from the regulatory point of view and these are worth a mention. The two tests are the Transcutaneous Electrical Resistance Test (TER) (6) and the Human Skin Model Test (7). The TER test is not a true in vitro test because the skin has to be obtained from rats, although the animals themselves are not used in the experiments. The Human Skin Model Test uses commercially available models such as EpiDerm™ or EPISKIN™. The purpose here is not to go into detail about the test designs but to highlight that there are nonanimal alternative methods available for testing. Unlike some other alternative tests (see, for instance, the *ex vivo* ocular irritation/corrosivity tests), a negative result in either of the two in vitro corrosivity tests does not need to be confirmed by an in vivo test.

Ocular Tolerance

Ocular administration as a standard route of dosing in the in vivo testing strategy would be appropriate for any product that is intended to be administered to the eye or parts of the eye. Therefore, the local tolerance of such products would be evaluated during the routine toxicology studies. The EMEA guidance note (1) indicates that single- and repeat-dose studies up to 28 days in duration in the rabbit would be acceptable. However, products that are intended to be administered to the eye regularly over longer periods of time in clinical use should be tested for appropriate periods. Again the routine toxicology studies would be the best place to do this. The question is as to whether the ocular contact is intentional or unintentional. For intentional eye contact, the ocular route would be used in a full two-species testing strategy.

Ocular tolerance studies are also appropriate for products that are not intended to come into contact with the eye but which, by their use, might do so anyway. These kinds of products could be gels or creams for dermal use, medicinal shampoos, etc. As with testing for skin corrosivity, the guidance argues for a weight-of-evidence approach before placing anything into the eye of an animal. Therefore, available data should be evaluated in a stepwise fashion, including existing human and animal data, structure/activity relationships, physicochemical properties, results of in vivo, in vitro, or *ex vivo* tests and results of dermal application studies. Thereafter, and only when it is clear that severe effects are not expected, the in vivo test in rabbits should be considered. The recommendation for in vivo testing of such products is a single-dose study in animals with observation for up to 72 hours postdosing. A good strategy to use in such testing is to follow the OECD protocol (test guideline 405) (8). Although the OECD guidelines have been developed principally for the testing of chemicals, they are useful background documents for testing of human pharmaceuticals as well. The single-dose test is a minimal procedure in rabbits where one eye of a single animal is treated while the other eye acts as control. Assuming that the reaction in the first animal is not too severe, a further two animals are dosed in the same way to confirm the finding.

Recently, several alternative tests have been evaluated and of there, two in vitro (or rather *ex vivo*) tests have been introduced and approved for regulatory acceptance. At the time of writing (November 2008), they have been formally accepted by the U.S. authorities (9) and OECD protocols were in preparation. The two tests, the Bovine Corneal Opacity and Permeability (BOCP) and Isolated Chicken Eye (ICE) assays, use tissues taken from animals at slaughter. They do not remove completely the requirement to perform the in vivo test in rabbits but they provide an additional screen for those compounds that are likely to cause the most severe pain and discomfort. Compounds that give a positive result in the BOCP or ICE assays are regarded as damaging to the eye and no further testing will be required. Where a negative or equivocal result is obtained in these tests the in vivo test would still be needed.

The BCOP test is based on using bovine eyes (obtained from slaughter animals) and exposing them to chemical substances (including pharmaceuticals) to evaluate effects on opacity and permeability. The eyes used should be freshly collected from the animals and used while the cells are still viable. Isolated corneas from the bovine eyes are placed in chambers where both the epithelial and endothelial surfaces of the cornea can be brought into contact with the test substance. After exposure the corneal opacity is measured as the amount of light that passes through the corneas. Following this the permeability is measured by evaluating the barrier properties of the epithelium to sodium fluorescein. Lastly, histopathology can also be incorporated into the test in order to further elucidate any findings seen.

The ICE test is based on similar principles to the BCOP and in addition it includes the corneal swelling parameter. The test takes corneas from enucleated chicken eyes, again from slaughter animals. In this test, the exposure to the test substances is over a short period, followed by rinsing and thereafter the endpoints are evaluated.

As mentioned above, the important thing to note about both of these in vitro tests is that the results are used only to identify positive substances. A negative result does not imply that the substance tested is not irritant or corrosive and therefore such a result should be followed up by performing an in vivo study. However, these tests do have the positive effect that some animal testing is no longer necessary [the United States National Institute of Environmental Health Sciences (NIEHS) estimates that the number of live animal tests could be reduced by 10%] and that those substances that are tested in live animals are less likely to cause suffering.

For longer studies, both for toxicological and local tolerance purposes using the ocular route, more detailed examinations including PK evaluation, detailed ophthalmoscopy (including slit-lamp), fluorescein staining, histopathology (including detailed evaluation of cell type and number), and other examinations should be considered. It should be remembered that even though a drug is only intended to be placed on the eye, it could result in systemic exposure.

Otic Tolerance

If a drug is to be administered to the ear and/or is expected to penetrate the tympanic membrane or to reach the middle or inner ear, this is an indication for detailed evaluation of the potential effects on the ear. In addition to detailed histopathology, other functional tests, such as the auditory brainstem response, should be performed.

Parenteral Routes

Parenteral is used here to mean any systemic exposure route other than the gastrointestinal tract and is taken to mean also those routes that do not just use some sort of topical administration. In the following sections, various parenteral routes are described individually. There may well be other potential routes of administration that are not discussed, but it is clear that there are several key principles that apply to all routes and that, with common sense, a reasonable study design could be developed to look at any of them.

Subcutaneous and Intramuscular

It is not uncommon to see companies trying to develop long-acting, slow-release, or depot formulations of some types of drug. Often these are injected either intramuscularly or subcutaneously. Studies should be designed appropriately but should in any case be performed in an appropriate species. Often it is better to use larger nonrodent animals such as dogs for evaluation of the effects of intramuscular deposit formulations. Studies can be performed often in a single sex. If the objective is simple local tolerance, then individual animals can act as their own controls, receiving the test formulation in one injection while receiving vehicle alone at a second site. If the study has more endpoints, for example, kinetics following the slow or extended release of the drug, then a separate control group will be needed. Histopathological evaluation is important in these studies. Intramuscular or subcutaneous injection is inevitably associated with some local tissue damage, including inflammation, necrosis, fibrosis, etc., and it is important to characterize the extent, time course, and resolution of this damage in relation to human safety. Furthermore, it should be considered to allow sufficient time to be able to follow this time course, which could be several months. In this case, it could be appropriate to have small groups of animals that are killed at different times following the last dose in order to characterize the progress of the recovery phase properly.

Intra-arterial, Peri-venous

As stated earlier the local effects of such administration are normally evaluated in the routine nonclinical toxicology studies where the clinical route of administration is followed. However, there are cases where it may be appropriate to design smaller studies to investigate specific aspects of local tolerance to the product. One example is where the effects following unintentional administration are examined. Therefore, with drugs that are intended to be given intravenously, it can be of interest to know the likely effects of unintentional peri-venous or intra-arterial administration.

The rabbit ear is commonly used for evaluation of intravenous, peri-venous, or intra-arterial injection. Small numbers of animals can be used and, because the purpose of the study is local tolerance rather than systemic toxicity, each animal can be used as its own control by administering the test formulation to one ear and the control or vehicle formulation to the other ear of the same animal, thus minimizing animal use. Frequency and duration of dosing will depend on the intended human exposure, but a simple experiment to look at the effects of potential effects of accidental misdosing could be assumed to require only a single dose followed by an observation period of two to four days followed by macroscopic and histopathological examination of the injection sites. The observation period could be prolonged in case of marked adverse effects in order to be able to follow and characterize the course of recovery.

Inhalation

If the drug has not been tested previously by the inhalation route, then testing in animals by this route should be performed. Inhalation toxicology is a specialized area that is covered by another chapter in this book and therefore is not discussed in further detail here except to say that studies of up to six months duration in animals might be required, although if the compound is not carcinogenic by the oral route, and if there is no finding that gives cause for concern in the six-month study, then an inhalation carcinogenicity study is not needed. This is also an area where thorough evaluation of the earlier data could be advantageous including whole body autoradiography studies to have a knowledge of the tissue distribution following administration by various routes.

Intranasal

If a drug has not previously been administered intranasally, the amount of local tolerance testing that is required will be dependent on available information from earlier studies. For instance, systemic exposure following administration to the mucous membranes of the nose would be expected to be rapid and extensive plus any nonabsorbed drug could be expected to be swallowed. Therefore, information on oral toxicity would also be important. In any case, the endpoints in a nasal application study would have to include histopathology of the nasal passages. Administration of the formulations themselves is often relatively straightforward and applicators such as those used in humans can be employed or adapted for animal use. Care needs to be taken with animal species that are obligate nose-breathers to ensure that they can continue to breathe during and after the dosing process. Large animals such as dogs and minipigs are often used in such studies because they approximate better to the size and anatomy of the human.

Intra-oral

This refers to drugs that are applied directly to the mouth or associated areas (sublingual, buccal, etc.). If the drug (and the formulation to be used) is already tested for safety using the oral route, then the focus will be on local tolerance. Animal studies can be carried out but it may be enough to conduct appropriate monitoring in the clinical studies. If animal studies are performed, then there are certain practical considerations. Principally, there is a problem of retaining the drug in the mouth. This is particularly the case with dispersible tablets. One approach is to use an animal model where retention of materials in the mouth is not such an issue (i.e., the hamster cheek pouch). In a study in hamsters, study specific considerations could be consideration of inclusion of only one sex (if there is no indication from other studies that there is expected to be a gender difference in the toxic profile) and inclusion of a group receiving the vehicle alone as well as a group receiving the clinical (human) formulation. The duration of dosing would be appropriate to the clinical situation. The study endpoints would include clinical observation of the application sites, and histopathology certainly of the application sites but also consideration should be given to extending examination to the rest of the gastrointestinal tract plus to any other tissues that may be expected to be target organs. Where the material to be applied is in the form of a cream or paste, this can be more easily applied and larger animals, for example, dogs, can be used. Evaluation of the local effects could also include fluorescein staining of the oral cavity to look for local damage as well as histopathology, where appropriate.

Rectal

FDA guidance indicates that no additional testing is required for this route. This would assume that the toxic profile of the drug was already adequately characterized.

Vaginal

If a drug is to be administered by the vaginal route, then local tolerance testing should include a test for hypersensitivity [normally the Local Lymph Node Assay (LLNA)]. Reproductive and development toxicity studies should be considered depending on the evaluation of risk to reproduction based on previous studies. Vaginal administration of drugs may well give rise to a different exposure profile to the unborn animal and should be thoroughly evaluated before human exposure via this route.

Intravesicular, Intra-urethral, Intracavernosal

Clearly there is an advantage in choosing larger animals (rabbits or larger) for this kind of study for practical reasons. Common to all of these routes of administration is to have information on the potential for effects on reproduction including male fertility. Otherwise, all of the other key points relating to formulation, frequency of dosing, etc., should be considered. Endpoints such as histopathology will be important in studies involving routes such as these—remembering that the dosing procedures can be associated with mechanical damage to sensitive tissues. The use of controls is essential in this regard. It may be an idea to include some limited measurement of pharmacokinetics in the studies to monitor systemic exposure. The extent of the investigations required will depend on the availability of existing data on systemic toxicity.

Intrathecal, Epidural

Again, as with all other routes of administration, the extent of testing necessary will depend on available information about the drug product. However, any studies should include evaluation of neurotoxicity (including potential clinical effects as well as extended pathological evaluation) and it will probably be necessary to investigate the pharmacokinetics of the drug in cerebrospinal fluid. Again, depending on the endpoints, it might be that there is an advantage in using larger animals such as dogs for studies like this. Particularly, they can be useful in relation to being able to monitor basic neurological and behavioral responses posttreatment, such as normality of gait, reflexes, and so on. In addition to sampling cerebrospinal fluid for drug levels, it might also be useful to look at systemic exposure.

Sensitization

Products applied to the skin or mucous membranes (e.g., rectum or vagina) have the potential to be sensitizing and they should therefore be tested in an appropriate system. There are two basic systems available, one using guinea pigs and the other using mice. The guinea pig tests have been used for many years and are being replaced by the Local Lymph Node Assay in mice; however, for completeness all the tests are described here, starting with the "older" tests and moving on to discussion of the replacement.

Guinea Pig Maximization Test (GPMT) and the Buehler Test

The original animal model of sensitizing potential is in the guinea pig, of which there are two tests—the Guinea Pig Maximization Test (GPMT) and the Buehler Test. The tests are described in OECD Guideline 406 (10). In both cases, the principle of the test is the same, that is, the test animals (usually 20 treated animals and up to 10 controls) are exposed to the test article (the induction exposure) followed by a period of 10 to 14 days where any immune response to the treatment can develop after which a challenge dose is administered. In the GPMT, the first induction doses are administered intradermally given in combination with Freunds Complete Adjuvant. A few days later, a patch of skin is treated to create a local irritation following which the test material is applied to the irritated area loaded onto a filter paper and held in place for 48 hours by an occlusive dressing. After 10 to 14 days, the animal is exposed to the test article again for 24 hours and after a further period the skin is observed for evidence of erythema and/or swelling. This method is most suitable for chemicals and other similar products that can be easily formulated for the intradermal induction dose. In the Buehler Test, the product is applied dermally and this test has often been regarded as most suitable for pharmaceutical products because it enables use of the formulated human product (cream, gel, ointment, etc.). In this case, the animal is given the induction dose and is dosed again after about one and two weeks. Two weeks later, the challenge dose is given and thereafter the skin is examined. The Buehler Test has been regarded as being less sensitive than the GPMT because the use of adjuvant and intradermal injection in the latter increased the chance of getting a positive result, or rather said, reduce the chances of getting a false-negative result.

The Local Lymph Node Assay

Although well established, the tests in guinea pigs have been criticized, particularly from an animal welfare perspective. In more recent times, the Local Lymph Node Assay (LLNA) in mice has been developed and protocols are described in OECD Test Guideline 429 (April 2002) (11) and by the Interagency Coordinating Committee on the Validation of Alternative Methods (ICCVAM) whose protocol was published in January 2001 (9).

From the regulatory perspective, the LLNA is regarded as an equal test to the guinea pig tests and the results of the LLNA can be accepted without further confirmatory tests. The test is shorter than the guinea pig tests and looks at the induction phase of sensitization. The LLNA does not replace the use of animals in sensitization testing but it is a Reduction alternative using a minimum of five groups of four mice (vs. 30 guinea pigs) and it is a Refinement alternative, focusing on the induction phase therefore eliminating the need to challenge the animals to elicit the hypersensitivity reactions. The use of an adjuvant is not needed in the LLNA.

The principle of the LLNA is that the induction phase of sensitization involves a proliferation of lymphocytes in the lymph nodes draining the site of application and that this proliferation is proportional to the dose. Therefore, the test measures not only the sensitizing potential but also gives a dose response, and hence the potency. Animals (three treated groups, a negative control, and a positive control) are treated for three consecutive days by application of the material to the dorsal aspect of the ears (Days 1 to 3). On Day 6, the animals are given an intravenous injection of tritiated thymidine, and after five hours, the animals are killed and the auricular lymph nodes are harvested. The principle of the test is that, if the test material is a sensitizer, the cells in the lymph nodes will have been stimulated to proliferate and the radiolabeled thymidine will have been taken up by these cells. The radioactivity for all groups is measured and a Stimulation Index (SI) is calculated. The SI of the controls is one, and if the SI of any of the treated groups is at least three, then the substance can be considered to be a potential sensitizer. The positive control group should always give a result over three otherwise there can be doubt over the conduct of the test.

The LLNA is specifically designed to test the sensitizing potential of small molecules. As part of the validation of the test, a series of vehicles/solvents was approved and this is listed in the guidelines. However, it may be appropriate to test the test substance in a clinically relevant vehicle/solvent—for a drug intended for dermal application this could be the intended human formulation.

SUMMARY

Local tolerance studies are in fact a broad range of different types of study in which local tolerance might not be the only endpoint. However, common to all such studies is that the potential for effects at the site of application, either adverse or otherwise, is one of the main endpoints. Such studies can be specifically designed in their own right, and can be of variable duration, and with various endpoints. Alternatively, local tolerance endpoints can be included in standard toxicology studies. Finally, there may be a need to compare between different dose and delivery forms for established drug products where systemic toxicity is not of interest. There is no specific study design for a local tolerance study and a great deal of experience needs to be used in selecting the best design for a specific need. The basic starting point has to be to use as much of the data that might already be available for a product before starting on a study or studies. After employing a weight of evidence approach and utilization of alternative and validated in vitro methods to filter out those compounds that are least likely to be tolerated, one could continue to the animal studies. In general one should consider carefully how to minimize the use of animals and, as far as possible, the studies should be designed to be as close as possible to the proposed human use of the drug or formulation. There is much in local tolerance testing that is relevant to all types of chemical, including pharmaceuticals, and there is much information in the chemical testing guideline series that can be used to help inform protocol design. One should also remember that in studies that are specifically designed to investigate local tolerance one should, however, try to get as much other information as possible at the same time. Therefore, when dosing animals by different routes, it is always the case that one should be looking for data on bioavailability, recognizing that toxicity could be manifested at a site remote from the application site.

REFERENCES

1. European Agency for the Evaluation of Medicinal Products, Committee for Proprietary Medicinal Products, Note for guidance on Non-clinical Local Tolerance Testing of Medicinal Products, CPMP/SWP/2145/00, February 2001.
2. United States Food and Drug Administration, Draft Guidance, Nonclinical Safety Evaluation of Reformulated Drug Products and Products Intended for Administration by an Alternate Route, March 2008.
3. Organisation for Economic Cooperation and Development Test Guideline Series. Available on OECD website http://www.oecd.org/document/. Accessed on 15 June, 2009.
4. Jochims K, Kemkowski J, Nolte T, et al. Local tolerance testing of parenteral drugs: How to put into practice. Regul Toxicol Pharmacol 2003; 38:166–182.
5. OECD Test Guideline 404, Acute Dermal Irritation/Corrosion, April 2002.
6. OECD Test Guideline 430, In vitro Skin Corrosion: Transcutaneous Electrical Resistance Test (TER), April 2004.
7. OECD Test Guideline 431, In vitro Skin Corrosion: Human Skin Model Test, April 2004.
8. OECD Test Guideline 405, Acute Eye Irritation/Corrosion, April 2002.
9. Interagency Coordinating Committee on the Validation of Alternative Methods Working Group Recommended Protocol for the Murine Local Lymph Node Assay, ICCVAM, January 2001. http://iccvam.niehs.nih.gov. Accessed on 15 June, 2009.
10. OECD Test Guideline 406, Skin Sensitisation, July 1992.
11. OECD Test Guideline 429, Skin Sensitisation: Local Lymph Node Assay, April 2002.

31 Bioavailability and Bioequivalence Studies

Iain J. McGilveray
McGilveray Pharmacon Inc, Ottawa, Ontario, Canada

INTRODUCTION

The scope of this chapter will not include biologics for which there are guidance documents on comparability (1), which can include bioavailability (BA) comparisons. The chapter will be concerned with chemical entities for which full marketing authorization applications (MAAs) will be made according to regional or national regulations and ICH guidelines and documentation. (The term "guideline" is used for all the regulatory guidance documents quoted.) The scope will be for such MAAs that will be referred to in this chapter using the U.S. Food and Drug Administration (FDA) nomenclature, New Drug Applications (NDAs), rather than abbreviated or abridged applications for generic drug products, termed ANDAs (Abbreviated New Drug Applications). However, the same bioequivalence (BE) standards are usually applied to the pivotal BE study between clinical trial formulations and the to-be-marketed (or market image) formulations.

As with pharmacokinetics, there is no specific ICH guideline for bioavailability (BA) and bioequivalence (BE), but the main results are to be summarized in the Common Technical Document (2) Module 2.7.1 *Summary of biopharmaceutic studies and associated analytical methods.* This is a comprehensive overview of BA, comparative BA, BE, and in vitro dissolution studies, for the new drug entity and formulations, with the associated validated analytical procedures. Most of the content of this summary is predicated from the ICH guideline, *Pharmaceutical Development*, Q8 (3) and, although this guideline indicates that it does not apply to contents of submissions for drug products during clinical research stages of development, the principles in the guideline are important to consider during those stages. In addition, this guideline states that "a summary of formulations used in clinical safety and efficacy and any relevant bioavailability studies should be provided. Any changes between the proposed commercial formulation and those formulations used in pivotal clinical batches and primary stability batches should be clearly described and the rationale for changes provided." This information would be described in the above-mentioned Summary 2.7.1 and is also a key to the life-cycle management described in ICH guidelines Q9 and Q10, "Quality Risk Management" and "Pharmaceutical Quality System", respectively (4,5).

There are elements of chemistry and manufacturing, such as formulation and dissolution, detailed in other parts of the ICH Module 2 and the Phase 1 clinical studies of bioavailability are detailed in Module 5. The in vitro and in vivo studies are conducted according to several ICH guidelines, such as Q1A (R2), B-F (stability) (6), Q2(R1) (Analytical procedures) (7), Q3A(R2), Q3B(R2) and Q3C(R3) (Impurities) (8), Q6A (Specifications) (9), Q7 (Good Manufacturing Practice) (10), as well as in E3 Structure and content of clinical study reports (11) and E6 (Good Clinical Practice; GCP) (12).

Regional Guidances

Regional and individual country guidances (or guidelines) for BA and BE are manifold, but in many cases countries (such as those in the Pan American Health Organization) follow the FDA general guideline (13) with some variations noted below for Europe and Japan. The Australian TGA and the Association of South East Asian nations (ASEAN) have adapted the European Medicines Evaluation Agency (EMEA) guidelines (14,15) and Canada has some particular variations in standards that will be noted as appropriate.

It should be appreciated that in "Europe," the European Union comprises 27 countries and the European Economic Area (EEA) comprises the EU + Norway, Iceland, and Liechtenstein, bringing the number to 30 countries. Marketing authorization can be applied for in those countries using the centralized procedure for a new drug (and other categories) via the EMEA.

Other countries such as Croatia and Turkey are adapting their pharmaceutical legislation so that it conforms to that in the EU. Thus, the EMEA guidelines are important throughout Europe.

As most new drugs are introduced in Europe, the United States, or Japan, the key regions involved in ICH, most other countries devote more rules for documentation to BE of generic FDA, ANDAs, than to bioavailability in NDAs. Also, there is no specific ICH guideline on pharmacokinetics and little has been provided by the FDA (16) or in Europe (17). However, there is a Japanese guideline, which is quite comprehensive (18). Nonetheless, the investigation of the characteristics of absorption, distribution, metabolism, and elimination of each new drug by different routes of administration is important in all regions and parts on the basis of bioavailability determination.

Japanese Pharmacokinetic Guideline

One of the important recommendations in the Japanese guideline from the Ministry of Health, Labor, and Welfare (MHLW) (18) is that a pharmacokinetic study may be needed for extrapolating foreign data to the Japanese population. The scope of any pharmacokinetic study involving Japanese individuals who live in Japan or abroad should be determined for each drug while considering its physical and chemical properties, previous pharmacokinetic data, and intended clinical use. To extrapolate from foreign data, it is necessary to compare the data on ethnically diverse populations with the data on Japanese populations living in Japan or living overseas. This would usually be one or more comparative studies, similar to bioequivalence studies described later in this chapter.

World Health Organization (WHO) Guidelines

The WHO has also worked in the area of BA and BE with two documents (19,20). Harmonization of BA and BE approaches is desirable and follows a major ICH mandate to reduce unnecessary studies. The ICH and WHO have worked toward standardization of nomenclature, agreement of test procedures and standards, and ongoing revisions of regional guidelines, which should reflect these initiatives.

Biopharmaceutics

Biopharmaceutics, a term coined by Wagner in 1961 (21), was defined as the study "encompassing the relation between the nature and intensity of the biological effects observed in animals and man and (*i*) the nature of the form of the drug (ester, salt, complex, etc.); (*ii*) the physical state, particle size, and surface area; (*iii*) the presence or absence of adjuvants with the drug; (*iv*) the type of dosage form in which the drug is administered; and (*v*) the pharmaceutical processes used to make the dosage form."

Gibaldi (22) describes "biopharmaceutics as a major branch of the pharmaceutical sciences concerned with the relationship between the physicochemical properties of a drug in dosage form and the pharmacologic, toxicologic, or clinical response observed after its administration."

Biopharmaceutics is the bridge discipline between formulation description, usually assessed by chemistry and manufacturing divisions of chemists in regulatory agencies and pharmacology/clinical data, assessed by pharmacologists, medical, and clinical scientists.

Exposure

Biopharmaceutics is concerned with the effects of these physicochemical and pharmacology factors on exposure, which is defined by ICH (23) as

> Exposure is represented by pharmacokinetic parameters demonstrating the local and systemic burden on the test species with the test compound and/or its metabolites. The area under the matrix level concentration–time curve (AUC) and/or the measurement of matrix concentrations at the expected peak-concentration time C_{max}, or at some other selected time $C_{(time)}$, are the most commonly used parameters. Other parameters might be more appropriate in particular cases.

This is one of the ICH Drug Safety guidelines, which link animal toxicology information to Phase 1 pharmacokinetic studies in humans.

DEFINITIONS OF BIOAVAILABILITY

There does not appear to be a common definition of bioavailability in the ICH documents, and so the FDA, the EMEA, and Japanese MHLW definitions are given.

FDA

At the FDA "Bioavailability" is defined (24) in the Code of Federal Regulations (CFR) § 320.1 as "the rate and extent to which the active ingredient or active moiety is absorbed from a drug product and becomes available at the site of action."

EMEA

The EMEA guideline states "Bioavailability means the rate and extent to which the active substance or active moiety is absorbed from a pharmaceutical form and becomes available at the site of action." (14)

This guideline also provides a more practical definition: "Bioavailability is understood to be the extent and rate at which a substance or its active moiety is delivered from a pharmaceutical dosage form and becomes available in the general circulation."

It continues with the comment, "It may be useful to distinguish between 'absolute bioavailability' of a given dosage form as compared with that (100%) following intravenous administration (e.g., oral solution) and the 'relative bioavailability' as compared with another form administered by the same or another non-intravenous route (e.g., tablets vs. oral solution)."

Japan

The Japanese MHLW guideline (25) defines bioavailability as "the rate and extent of absorption of parent drugs or active metabolites from a dosage form into the systemic circulation."

Link to Exposure

The FDA general guideline (13) notes the relationship to exposure:

> As part of Investigational New Drugs (INDs) and New Drug Applications (NDAs) for orally administered drug products, BA studies focus on determining the process by which a drug is released from the oral dosage form and moves to the site of action. BA data provide an estimate of the fraction of the drug absorbed, as well as its subsequent distribution and elimination.BA can be generally documented by a systemic exposure profile obtained by measuring drug and/or metabolite concentration in the systemic circulation over time. The systemic exposure profile determined during clinical trials in the IND period can serve as a benchmark for subsequent bioequivalence (BE) studies.

The guideline (13) also notes that "From a pharmacokinetic perspective, BA data for a given formulation provide an estimate of the relative fraction of the orally administered dose that is absorbed into the systemic circulation when compared to the BA data for a solution, suspension, or intravenous dosage form." [This is similar to the expectation in the EMEA guideline (14).] "In addition, BA studies provide other useful pharmacokinetic information related to distribution, elimination, the effects of nutrients on absorption of the drug, dose proportionality, linearity in pharmacokinetics of the active moieties, and, where appropriate, inactive moieties.BA data can also provide information indirectly about the properties of a drug substance before entry into the systemic circulation, such as permeability and the influence of presystemic enzymes and/or transporters (e.g., P-glycoprotein)."

Bioequivalence Definitions

The United States

The CFR § 320 (24) also defines BE for the United States as "the absence of a significant difference in the rate and extent to which the active ingredient or active moiety in pharmaceutical equivalents or pharmaceutical alternatives becomes available at the site of drug action when administered at the same molar dose under similar conditions in an appropriately designed study."

EMEA

In the EMEA guideline (14,15), BE is defined as "Two medicinal products are bioequivalent if they are pharmaceutically equivalent or pharmaceutical alternatives and if their bioavailabilities after administration in the same molar dose are similar to such degree that their effects, with respect to both efficacy and safety, will be essentially the same."

Japan

The MHLW guideline states, "Bioequivalent products are defined as drug products having the same bioavailabilities" (25).

Pharmaceutical Equivalents and Pharmaceutical Alternatives

"In the United States, drug products are considered pharmaceutical equivalents if they contain the same active ingredient(s), are of the same dosage form, route of administration, and are identical in strength or concentration" (26).

The EMEA guideline notes that "medicinal products are pharmaceutically equivalent if they contain the same amount of the same active substance(s) in the same dosage forms that meet the same or comparable standards" (14,15).

In the United States, "drug products are considered 'pharmaceutical alternatives' if they contain the same therapeutic moiety, but are different salts, esters, or complexes of that moiety or are different dosage forms or strengths (e.g., tetracycline hydrochloride, 250 mg capsules vs. tetracycline phosphate complex, 250 mg capsules; quinidine sulfate, 200 mg tablets vs. quinidine sulfate, 200 mg capsules). Data are generally not available for FDA to make the determination of tablet to capsule bioequivalence. Different dosage forms and strengths within a product line by a single manufacturer are thus pharmaceutical alternatives, as are extended-release products when compared with immediate-release or standard-release formulations of the same active ingredient" (26).

The EMEA definition is similar. "Medicinal products are pharmaceutical alternatives if they contain the same active moiety but differ in chemical form (salt, ester, etc.) of that moiety or in the dosage form or strength" (14,15).

However, the EMEA guideline notes that " 'essentially similar' products can be considered equivalent" and, unlike in the United States, for immediate-release products, different oral forms such as tablets and capsules can be bioequivalent (14,15). This also pertains in Canada (27).

This issue is more important for the development of generic products (ANDAs) than during new drug (NDA) development when development formulations are compared with pivotal bioequivalence studies of clinical trial and to-be-marketed lots (which could be capsules vs. tablets). In the latter cases, bioequivalence evidence would be accepted in all jurisdictions.

FACTORS AFFECTING ABSORPTION AND BIOAVAILABILITY

In the first edition of this book, there were many examples of the attributes of the active pharmaceutical ingredient, formulation, and manufacturing process, which could influence absorption and other pharmacokinetic processes (absorption, distribution, metabolism, and elimination/excretion; ADME) measures and hence bioavailability. These are now covered in chapter 32 on Pharmacokinetics of this edition.

Some Key Factors Influencing Bioavailability

In brief, molecular size and shape, degree of ionization, relative lipid solubility of ionized and unionized forms, and its binding to serum and tissue proteins are the key drug properties that influence its movement and availability at sites of action (28).

For most orally administered drugs, absorption is mainly via passive membrane transport but some drugs are transported actively, such as by glucose and amino acid transporters involved in the uptake of endogenous compounds. The rate of gastric emptying is an important influence on absorption as most drugs cross the membranes of the mucosal epithelium in the high surface area villi of the small intestine. The drug may be metabolized in the small intestine, but in most cases the drug passes to the liver for metabolism or biliary excretion before entering the systemic circulation. This sequence of absorption to appearance in the systemic circulation allows for the process termed "first pass" metabolism. A drug may be absorbed quantitatively and have a low bioavailability, and it is important to differentiate between absorption and bioavailability. The amount of drug reaching the circulation depends on the administered dose but often mainly on the fraction of dose (F) that is absorbed and escapes first-pass elimination. Bioavailability is represented by this fraction, F (28).

Bioavailability comparisons of intravenous and oral formulations are also applied to some nonoral routes of administration, such as sublingual, transdermal, and rectal doses (28).

Nonlinear Pharmacokinetics

Changes in ADME parameter disproportionate with administered dose are termed nonlinear pharmacokinetics. This can occur by saturation of absorption, distribution (protein binding), metabolic enzymes, or excretion processes, but is important to quantify for BA/BE as will be described later.

Newer Knowledge of Factors Influencing Bioavailability

The most important new knowledge affecting BA/BE is the Biopharmaceutics Classification System (BCS) of categorizing orally administered drugs and drug products and the newer Biopharmaceutics Drug Disposition Classification System BDDCS (29), as well as the increasing information on the influence of transporters in absorption (30,31).

Biopharmaceutics Classification System

As the BCS is applied widely in justifying waivers for BE studies, it is summarized here. The FDA has a guideline on BCS (32) and it is being applied in part to the revision of the EMEA guideline (15).

The BCS is a scientific framework for classifying drug substances based on their aqueous solubility and intestinal permeability. When combined with the dissolution of the drug product, the BCS takes into account three major factors that govern the rate and extent of drug absorption from immediate-release (IR) solid oral dosage forms: dissolution, solubility, and intestinal permeability (33).

According to the BCS drug substances are classified as follows:

Class 1: High Solubility—High Permeability
Class 2: Low Solubility—High Permeability
Class 3: High Solubility—Low Permeability
Class 4: Low Solubility—Low Permeability

In addition, immediate-release solid oral dosage forms are categorized as having rapid or slow dissolution. Within this framework, when certain criteria are met, the BCS can be used as a drug development tool to help sponsors justify requests for waivers of products from BE studies (biowaivers).

The FDA BCS guideline (32) provides the class boundaries for solubility, permeability (or absorbability), and product dissolution as well as the test conditions considered acceptable. These are

Solubility of drug substance: A drug substance is considered *highly soluble* when the highest dose strength is soluble in 250 ml or less of aqueous media over the pH range of 1 to 7.5.

Permeability of drug substance: In the absence of evidence suggesting instability in the gastrointestinal tract, a drug substance is considered to be *highly permeable* when the extent of absorption in humans is determined to be 90% or more of an administered dose based on a mass balance determination or in comparison to an intravenous reference dose. Alternatively, nonhuman systems capable of predicting the extent of drug absorption in humans can be used (e.g., in vitro epithelial cell culture methods). Examples are given in the guideline (32).

Dissolution from drug product: An IR drug product is considered *rapidly dissolving* when no less than 85% of the labeled amount of the drug substance dissolves within 30 minutes, using *U.S. Pharmacopeia* (USP) Apparatus I at 100 rpm (or Apparatus II at 50 rpm) in a volume of 900 ml or less in each of the following media: (*i*) 0.1 N hydrochloric acid or Simulated Gastric Fluid USP without enzymes; (*ii*) a pH 4.5 buffer; and (*iii*) a pH 6.8 buffer or Simulated Intestinal Fluid USP without enzymes.

Applications of BCS

In brief, bioawaivers for BE are possible for BCS Class 1 drug products IR formulations provided they are not narrow therapeutic index drugs. There are movements to extend the waivers to Class 3 drugs, which are dissolution limited (34–36). Some pharmaceutical experts also suggest the addition of highly soluble, highly metabolized drugs (>90%) to the BCS Class 1 (29,37).

The EMEA guideline (14) in paragraph 4.2 also refers to bridging of some of the development formulations including waivers for BE studies by application of in vitro information based on the BCS classification referred to in paragraph 5.1.1 of the guideline. The draft EMEA guideline (15) has more detail in the appended decision trees, but there appears to be less emphasis in the EMEA documents on the permeation boundaries than in the FDA guideline.

The Japanese guidelines rely heavily on dissolution information for such bridging as different strengths (38) and formulations (39), but depend more on rate of dissolution and less on the permeability aspects of BCS (40).

DETERMINATION OF BIOAVAILABILITY AND BIOEQUIVALENCE

The U.S. CFR (24) provides the most comprehensive definition of bioequivalence and the approaches toward determining BA and BE. "'*Bioequivalence*' means the absence of a significant difference in the rate and extent to which the active ingredient or active moiety in pharmaceutical equivalents or pharmaceutical alternatives becomes available at the site of drug action when administered at the same molar dose under similar conditions in an appropriately designed study."

Approaches to Determine Bioavailability and Bioequivalence

The CFR (24) states that "the following in vivo and in vitro approaches, in descending order of accuracy, sensitivity, and reproducibility, are acceptable for determining the bioavailability or bioequivalence of a drug product."

In Vivo Test Using Blood Samples

"An in vivo test in humans in which the concentration of the active ingredient or active moiety, and, when appropriate, its active metabolite(s), in whole blood, plasma, serum, or other appropriate biological fluid is measured as a function of time. This approach is particularly applicable to dosage forms intended to deliver the active moiety to the bloodstream for systemic distribution within the body."

In Vitro Test

An in vitro test that has been correlated with and is predictive of human in vivo bioavailability data.

In Vivo Test Urine

"An in vivo test in humans in which the urinary excretion of the active moiety, and, when appropriate, its active metabolite(s), are measured as a function of time. The intervals at which measurements are taken should ordinarily be as short as possible so that the measure of the rate of elimination is as accurate as possible. Depending on the nature of the drug product, this approach may be applicable to the category of dosage forms described above in the paragraph 'In vivo test using blood samples'. This method is not appropriate where urinary excretion is not a significant mechanism of elimination."

Pharmacodynamic In Vivo Test

"An in vivo test in humans in which an appropriate acute pharmacological effect of the active moiety, and, when appropriate, its active metabolite(s), are measured as a function of time, if such effect can be measured with sufficient accuracy, sensitivity, and reproducibility. This approach is applicable to the category of dosage forms described above in paragraph 'In vivo test using blood samples' only when appropriate methods are unavailable for measurement of the concentration of the moiety, and, when appropriate, its active metabolite(s), in biological fluids or excretory products but a method is available for the measurement of an appropriate acute pharmacological effect. It may be particularly applicable to dosage forms that are not intended to deliver the active moiety to the bloodstream for systemic distribution."

Clinical Trials
"Well-controlled clinical trials in humans that establish the safety and effectiveness of the drug product, for purposes of measuring bioavailability, or appropriately designed comparative clinical trials, for purposes of demonstrating bioequivalence. This approach is the least accurate, sensitive, and reproducible of the general approaches for measuring bioavailability or demonstrating bioequivalence. For dosage forms intended to deliver the active moiety to the bloodstream for systemic distribution, this approach may be considered acceptable only when analytical methods cannot be developed to permit use of one of the approaches outlined above in the paragraph 'In vivo test using blood samples' and when the in vitro test or pharmacodynamic approaches described above are not available. It may also be considered sufficiently accurate for measuring bioavailability or demonstrating bioequivalence of dosage forms intended to deliver the active moiety locally, for example, topical preparations for the skin, eye, and mucous membranes; oral dosage forms not intended to be absorbed, for example, an antacid or radiopaque medium, and bronchodilators administered by inhalation if the onset and duration of pharmacological activity are defined."

In Vitro Test
"A currently available in vitro test acceptable to FDA (usually a dissolution rate test) that ensures human in vivo bioavailability."

The use of plasma concentrations after oral administration of systemically absorbed drugs is the most documented approach to determine BA and BE, as evidenced by the FDA (13) and EMEA guidelines (14,15). These also require that the approach be used for nonoral products with systemic action, such as transdermal and rectal applications. For oral and nonoral administration of drugs intended for local action the pharmacological (pharmacodynamic) or clinical trial approaches above may be required. The European guidelines (14,15) note that whenever systemic exposure resulting from locally applied, locally acting medicinal products (e.g., nasal, inhalation, dermal, or vaginal) entail risk of systemic adverse reactions, systemic exposure should also be measured.

APPLICATIONS OF BIOEQUIVALENCE

While BA determination is part of the characterization of new chemical entities for NDAS, BE determination in comparative studies serves a variety of purposes. As is noted in the draft EMEA guideline (15), during development of a new chemical entity, bioequivalence may be used to bridge data among different formulations, for example, between a formulation used in the pivotal studies and the formulation to be marketed. The document also suggests that wider bioequivalence acceptance limits may be justified than with generic comparisons based on clinical and other data submitted. Other uses of BE information, which the EMEA document describes, are applications for extensions, such as additional dosage forms or strengths and variations for change in composition or the manufacturing process. In addition, the draft guideline notes that the methodology of bioequivalence studies can be used to assess differences in the pharmacokinetic parameters in PK studies such as drug–drug or food–drug interactions or to assess differences in subsets of the population. However, only the food–drug interaction application will be discussed later in this chapter.

These applications are similar to those described in the FDA general guideline (13): "BE documentation can be useful during the IND or NDA period to establish links between (*i*) early and late clinical trial formulations; (*ii*) formulations used in clinical trial and stability studies, if different; (*iii*) clinical trial formulations and to-be-marketed drug product; and (*iv*) other comparisons, as appropriate. In each comparison, the new formulation or new method of manufacture is the test product and the prior formulation or method of manufacture is the reference product. It is recommended that the determination to redocument BE during the IND period be generally left to the judgment of the sponsor, who can wish to use the principles of relevant guidelines to determine when changes in components, composition, and/or method of manufacture suggest further in vitro and/or in vivo studies be performed." However, the FDA appears to be less flexible than the Europeans for BE standards: "we recommend that similar approaches to measuring BA in an NDA generally be followed in demonstrating BE for an NDA or an A/NDA."

Dose-Proportionality, Linearity of Pharmacokinetics

A particular application of comparative bioavailability is dose-proportionality or linear pharmacokinetics, described above under section "nonlinear pharmacokinetics." In fact, there is little guidance on how to measure this, although finding nonlinearity has implications for conduct and design of comparative BA or BE studies. The FDA has an old guideline on *Content and format of the human and BA section of an application* (16), which indicates that studies to support and NDA plasma drug levels will be used in ADME studies and "Of particular interest are changes in kinetic parameters with dose (i.e., dose-dependent kinetics) within the recommended clinical dosing range."

The EMEA PK guideline (17) also indicates that the study of PK should examine linearity. With multiple dose studies "accumulation kinetics of the substance should be verified experimentally: different doses should be included in one study to determine dose dependence and to decide whether changes from linearity to nonlinearity occur at dosage levels, which are normally used." However, there are few details on interpretation or presentation of the results in either of these guidelines.

The Japanese PK guideline (18) has some advice:

> "information as to whether the pharmacokinetic parameters of a test drug (C_{max}, AUC, etc.) are proportional to dose over the dose range relevant to clinical use (including the anticipated clinical dose) is considered important. In this case, one should perform the linearity evaluation as follows: begin with plotting the study results against the dose, with careful observation of the pattern, then estimate the regression line and assess the linearity of the line. If linearity exists between the dose and the parameter, various kinds of pharmacokinetic analyses will be performed easily. Although different methods have been used to evaluate proportionality between AUC and dose in the test range, including the (anticipated) clinical dose, there is probably no general consensus on the criterion to evaluate linearity, including extrapolation to the origin, because of differences in pharmacokinetics of test-drug, size, and precision of the study, therapeutic indication, etc. Accordingly, linearity should be evaluated with the proper criteria set for each individual case by the most appropriate method selected, depending on the properties of the test drug. Consequently, if a pharmacokinetic parameter can be practically evaluated as linear for dose, it is acceptable to consider that further analysis is allowed to proceed, assuming linearity. When linearity is not recognized, the degree of nonlinearity is important in the dose range used for clinical practice, and a discussion of clinical relevance will be necessary."

There is a Canadian draft policy *Bioequivalence Requirements: Drugs Exhibiting Non-Linear Pharmacokinetics* (41) that indicates that "a drug will be considered to exhibit nonlinear pharmacokinetics if this is indicated in the peer-reviewed scientific literature or the approved labeling for the drug. However, the drug may be treated in the same way as those exhibiting linear pharmacokinetics, if evidence is provided to show that dose-normalized AUC deviate (increase or decrease) by less than 25% over the labeled dose range for the proposed indication."

DESIGN AND CONDUCT OF STUDIES

Both the FDA and the EMEA guidelines provide advice on the conduct of BA and BE studies in which blood (plasma or serum) drug measurements are applied. This approach is particularly applicable to dosage forms intended to deliver the active moiety to the bloodstream for systemic distribution within the body (14,24) and is the best characterized of the methods described above under section "approaches to determine bioavailability and bioequivalence."

In general, the applicable sections of ICH E2 guideline *Structure and content of clinical study reports* are applied (11). As these are Phase 1 studies, patients are not usually being treated and the main "efficacy measure" is the plasma concentration of parent drug and/or metabolites. The report would begin with the title page and synopsis of study (summary description with objective and results). This is followed by details of the ethical oversight by an independent ethics committee, which should approve the protocol and any amendments before the study is initiated. Subject information and consent, investigators and study administrative structure, study objectives, investigational plan, overall study design, and description of the plan would follow. Although each region has Good Clinical Practice (GCP) requirements, these are reflected in the ICH E6 GCP guideline (12), which is accepted by all the regions.

Design

Both the FDA and the EMEA guidelines remark on designs for BA and BE studies. The emphasis is actually more on BE study design, but the EMEA guideline notes (14) that, since BA studies are comparative in nature, the design sections for BE apply to BA studies as well with any necessary adaptations in accordance with the specific aim of the study. The FDA general guideline (13) states that "to measure product quality BA and establish BE, reliance on PK measurements may be viewed as a bioassay that assesses release of drug from the drug product into the systemic circulation. A typical study is conducted as a crossover study. In this type of study, physiological variables are assumed to have less inter-occasion variability compared to the variability arising from formulation. Nonreplicate crossover designs are recommended for BE studies of immediate-release and modified-release formulations."

The FDA has a web listing of draft guideline of individual product bioequivalence recommendations (42). In general, these indicate that for ANDAs of these drugs crossover studies should be completed with healthy human male and female volunteers both in fasted and fed conditions. It is likely that for formulation changes for NDAs, for which a biowaiver could not be justified (see section "biowaiver" below) the same approach would be expected.

The EMEA guideline (14), in section 3.1, notes that the study should be designed in such a way that the formulation effect can be distinguished from other effects. If the number of formulations to be compared is two then a two-period, two sequence crossover design is often considered the design of choice.

Add-on or Sequential Designs

The FDA guideline does not allow for such designs and indeed notes that "statistical analysis of subgroups is not recommended" (13).

The EMEA draft guideline (15) is somewhat contradictory on this type of design. If the failed trial was ambiguous, for example, the confidence intervals were wide and were consistent with both possible bioequivalence and lack of bioequivalence, then a subsequent positive study can be convincing. However, it is not acceptable to pool two ambiguous studies to reach a positive conclusion.

However, from this guideline, it is acceptable to use a two-stage approach when attempting to demonstrate bioequivalence. An initial group of subjects can be treated and their data analyzed. If bioequivalence has not been demonstrated, an additional group can be recruited and the results from both groups combined in a final analysis. If this approach is taken, appropriate steps must be taken to preserve the overall type I error of the experiment. The analysis of the first-stage data should be treated as an interim analysis and both analyses conducted at adjusted significance levels (with the confidence intervals accordingly using an adjusted coverage probability which will be higher than 90%). The plan to use a two-stage approach must be prespecified in the protocol along with the adjusted significance levels to be used for each of the analyses.

The Japanese guideline (25) also allows for add-on studies stating that if bioequivalence cannot be demonstrated because of an insufficient number of subjects in the original study, an add-on subject study can be performed using not less than half the number of subjects in the initial study.A sample size of 20 ($n = 10$/group) for the initial study and pooled size of 30 for initial plus add-on subject study may suffice if test and reference products are equivalent in dissolution and similar in average AUC and C_{max}.

The Canadian general guideline (43) allows pooling of two or more studies, provided that the same protocol is used for all studies and two consistency tests are done on the studies to ensure that pooling is meaningful. The first test is the test of equality of the residual mean squares and the second test is the formulation by study interaction. Details of the calculations are given in the guideline.

Replicate Designs

The FDA general guideline (13) remarks that replicate study designs may offer several scientific advantages compared to nonreplicate designs and they have noted that the approach is now favored for highly variable drugs (HVD), discussed later in this chapter.

The EMEA draft guideline (15) also indicates that a three- or four-period replicate design is an acceptable approach for HVD. This is also noted in the Japanese guideline (25).

Multiple-Dose Steady-State Designs

For the United States, the CFR § 320.27(a) (3) (24) notes that "A multiple-dose study may be required to determine the bioavailability of a drug product in the following circumstances: (*i*) there is a difference in the rate of absorption but not in the extent of absorption; (*ii*) there is excessive variability in bioavailability from subject to subject; (*iii*) the concentration of the active drug ingredient or therapeutic moiety, or its metabolite(s), in the blood resulting from a single dose is too low for accurate determination by the analytical method; and (*iv*) the drug product is an extended-release dosage form."

However, the BA/BE general guideline (13) indicates that it "recommends single-dose pharmacokinetic studies for both immediate- and modified-release drug products to demonstrate BE because they are *generally* more sensitive in assessing release of the drug substance from the drug product into the systemic circulation. If the multiple-dose study is chosen, attainment of steady-state must be documented."

The EMEA draft guideline (15) states that, in general, single-dose studies will suffice, but there are situations in which steady-state designs may be required. Examples are provided as quoted below:

> In case of dose or time-dependent pharmacokinetics, resulting in markedly higher concentrations at steady state than expected from single-dose data, a potential difference in AUC between formulations may be larger at steady state than after single dose. Hence, a multiple-dose study may be required in addition to the single-dose study to ensure that the products are bioequivalent regarding AUC also at steady state. However, if the single-dose study indicates very similar PK profile for test and reference (the 90% confidence interval for AUC is within 90–111%), the requirement for steady-state data may be waived.

In certain cases when a single-dose study cannot be conducted in healthy volunteers due to tolerability reasons, and a single-dose study is not feasible in patients, conduct of a multiple-dose study in patients may be acceptable.

A multiple-dose study as an alternative to a single-dose study may also be acceptable if problems of sensitivity of the analytical method preclude sufficiently precise plasma concentration measurements after single-dose administration. As C_{max} at steady state may be less sensitive to differences in the absorption rate than C_{max} after single dose, bioequivalence should, if possible, be determined for C_{max} after the single-dose administration (i.e., after the first dose of the multiple-dose study) as a measure of peak exposure while extent of exposure can be based on demonstration of bioequivalence of AUC at steady state. In steady-state studies, the administration scheme should preferably follow the highest usual dosage recommendation."

The Japanese guideline (25) indicates that "In principle, bioequivalence studies should be performed by single-dose studies. Multiple-dose studies may be employed for drugs that are repeatedly administered to patients."

The Canadian BA/BE guideline for modified-release products (44) requires multiple, as well as single-dose studies, for drugs that accumulate.

Parallel Group Designs

These will be discussed under section "long half-life drugs" below.

Population PK Designs

This is a specialized approach that allows investigation of PK, including BA/BE, in patients. It is now much used in the clinical efficacy stages of new drug development, but its application in BE is still in its infancy (45).

The European draft guideline (15) is silent on population pharmacokinetics, while the earlier guideline (14) notes that experience with this approach is limited and "no specific recommendation is given".

Selection of Subjects

Recommendations on subject selection are similar in most jurisdictions, and the FDA, EMEA, and Japanese MHLW guidelines indicate that, for most drugs, normal, healthy volunteers over the age of 18 years of both genders should be recruited. The European guideline (14) suggests 55 years of age as the upper limit and that the risk to women of childbearing potential be considered on an individual basis. The FDA recommends (13) that in vivo BE studies be conducted in individuals representative of the general population, taking into account age, sex, and race. Also, if the drug product is intended for use in both sexes, the sponsor should attempt to include similar proportions of males and females in the study, and if the drug product is to be used predominantly in the elderly, it is recommended that the sponsor attempt to include as many subjects of 60 years of age or older as possible.

As the ICH Structure and Content of Clinical Study Reports (11) and GCP (12) guidelines detail, appropriate medical screening for health, demographic information including body weight, laboratory tests and any special tests appropriate to the drug's safety profile or therapeutic class, inclusion and exclusion criteria should be included in the protocol and adhered to by sponsors. The EMEA guideline (14) states that "subjects should preferably be nonsmokers and without a history of alcohol or drug abuse. If moderate smokers are included (less than 10 cigarettes per day), then they should be identified as such and the consequences for the study results should be discussed."

In Japan, the guideline (25) also notes that healthy adult volunteers should be employed.

However, if the use of the drug is limited to a special population and test and reference products show a significant difference in dissolution results, even with one of the several recommended dissolution tests, the in vivo test should be performed using subjects from the target population. When test and reference products showed a unique significant difference in dissolution at around pH 6.8 by the dissolution test or between pH 3 and 6.8 for products containing basic drugs, then subjects with low gastric acidity (achlorhydric subjects) should be employed, unless the application of the drug is limited to a special population. With regard to the use of foreign sourced information on bioavailability, a Japanese document (46) suggests that "on the whole, it is preferable if a study conducted in Japan is included in the application."

The FDA general guideline (13) does not remark on genetic phenotyping but for safety reasons this may be expected according to GCP. In the case of NDAs, during development, BA studies should explore the possibility of genetic differences in metabolism.

The EMEA guideline (14) notes that phenotyping and/or genotyping of subjects should be considered for exploratory bioavailability studies and all studies for which a parallel group design is applied. If a metabolic polymorphism is found for a drug, studies could be performed in panels of subjects of known geno- or phenotype. The draft guideline (15) echoes this for safety reasons. The Japanese guideline (25) notes that if the clearance of drug differs to a large extent among subjects due to genetic polymorphism, subjects with higher clearance should be employed.

Number of Subjects

The FDA general guideline (13) notes that pilot studies may be applied with a small number of subjects to assess variability and refers to the guideline on *Statistical Approaches to Establishing Bioequivalence*, Appendix C, which, as well as some tables (in Appendix C of the guideline), states that "sample sizes for average BE should be obtained using published formulas" (47).

The EMEA guideline notes (14) that "the number of subjects required is determined by

(a) the error variance associated with the primary characteristic to be studied as estimated from a pilot experiment, from previous studies or from published data;
(b) the significance level desired;
(c) the expected deviation from the reference product compatible with bioequivalence (delta); and
(d) the required power.

The clinical and analytical standards imposed may also influence the statistically determined number of subjects. However, generally the minimum number of subjects should be not smaller than 12 unless justified."

The Canadian general guideline (43) provides a formula for calculating acceptance probabilities to estimate sample sizes of sufficient power. Again, the minimum number is 12 subjects.

The Japanese guideline (25) notes that a sufficient number of subjects for assessing bioequivalence should be included and also provide literature references for power calculations. However, as noted above under "Add-on or sequential designs" the Japanese guideline also allows for add-on groups.

Reference and Test Lots

For the NDA pivotal BE study during development, the Q8 ICH Development Pharmaceutics guideline (3) remarks that "changes between the proposed commercial formulation and those formulations used in pivotal clinical batches and primary stability batches should be clearly described and the rationale for the changes provided." The FDA has IND guidelines that indicate that at Phase 3 considerable information on the clinical trial lots will be available, including stability and dissolution. Manufacture at that stage would be expected to meet Good Manufacturing Practice (GMP) requirements. This would be the reference formulation product.

The EMEA guideline (14) notes that "The test products used in the biostudy must be prepared in accordance with GMP-regulations. In the case of oral solid forms for systemic action, the test product should usually originate from a batch of at least 1/10 of production scale or 10,0000 units, whichever is greater, unless otherwise justified. The production of batches used should provide a high level of assurance that the product and process will be feasible on an industrial scale; in case of production batch smaller than 10,0000 units, a full production batch will be required. If the product is subjected to further scale-up this should be properly validated. Samples of the product from full production batches should be compared with those of the test batch, and should show similar in vitro dissolution profiles when employing suitable dissolution test conditions." This agrees with the FDA Scale Up and postapproval (SUPAC) guidelines (48,49). In Japan, the batch size of 1/10 of production scale or 10,0000 units is also accepted (25).

Standardization of the Study

The FDA general guideline (13) includes an attachment with some advice on conduct of BA/BE studies, recognizing that the elements can be adjusted for certain drug substances and drug products, including

(i) the test or reference products can be administered with about 8 oz (240 ml) of water to an appropriate number of subjects under fasting conditions;
(ii) generally, the highest marketed strength can be administered as a single unit. If warranted for analytical reasons, multiple units of the highest strength can be administered, providing the total single dose remains within the labeled dose range;
(iii) an adequate washout period (e.g., more than five half-lives of the moieties to be measured) would separate each treatment; and
(iv) before and during each study phase, it is recommended that subjects be allowed water as desired except for one hour before and after drug administration, be provided standard meals no less than four hours after drug administration, and abstain from alcohol for 24 hours before each study period and until after the last sample from each period is collected.

The draft EMEA guideline (15) notes that the test conditions should be standardized in order to minimize the variability of all factors involved except that of the products being tested. Therefore, standardization of the diet, fluid intake, and exercise are recommended.

Preferably subjects should be fasting at least during the night prior to administration of the products. The time of day for ingestion should be specified and, as fluid intake may profoundly influence gastric passage for oral administration forms, the volume of fluid (at least 150 ml) should be constant (which is a slight difference from FDA recommendations). All meals and fluids taken after the treatment should also be standardized with regard to composition and time of administration during the sampling period. The subjects should not take other medicines during a suitable period before and during the study and should abstain from food and drinks, which may interact with circulatory, gastrointestinal, liver, or renal function (e.g., alcoholic or xanthine-containing beverages or certain fruit juices). As the bioavailability of an active moiety from a dosage form could be dependent upon gastrointestinal transit times and regional blood flows, posture and physical activity may need to be standardized. It is noted that an adequate washout period should separate the treatments.

The Japanese guideline (25) notes that drugs are usually given to subjects with 100 to 200 ml water (normally 150 ml) after fasting for more than 10 hours. Fasting lasts for at least four hours postdose. Generally, one dose unit or a clinical usual dose should be employed. A higher dose, which does not exceed the maximal dose of the dosage regimen, may be employed when analytical difficulties exist, such as a high detection limit. In crossover studies, washout periods between administration of test and reference products should usually be more than five times the elimination half-life of the parent drug or active metabolites.

Sampling Times

Blood, Plasma Serum Samples

The FDA general guideline (13) recommends that under normal circumstances blood, rather than urine or tissue, be used. In most cases, the parent drug or its metabolites are measured in serum or plasma. However, in certain cases, whole blood may be more appropriate for analysis. The guideline recommends that blood samples be drawn at appropriate times to describe the absorption, distribution, and elimination phases of the drug. For most drugs, it recommends that 12 to 18 samples, including a predose sample, be collected per subject per dose. This sampling can continue for at least three or more terminal half-lives of the drug. The exact timing for sample collection depends on the nature of the drug and the input from the administered dosage form. The sample collection can be spaced in such a way that the maximum concentration of the drug in the blood (C_{max}) and terminal elimination rate constant (λ_z) can be estimated accurately. At least three to four samples can be obtained during the terminal log-linear phase to obtain an accurate estimate of λ_z from linear regression. It is recommended that both the actual clock time when samples are drawn and the elapsed time related to drug administration be recorded.

The European draft guideline (15) also recommends blood as the matrix of choice and notes that a sufficient number of samples adequately describing the complete plasma concentration–time profile should be collected. The sampling schedule should include frequent sampling around C_{max} to provide a reliable estimate of peak exposure. The sampling schedule should be planned to avoid C_{max} being the first point of a concentration time curve. The sampling schedule should also cover the plasma concentration time curve long enough to provide a reliable estimate of the extent of exposure, which is achieved if AUC_t is at least 80% of AUC_∞. At least three to four samples are needed during the terminal log-linear phase in order to reliably estimate the terminal rate constant (which is needed for a reliable estimate of $AUC_{0-\infty}$.).

The Japanese guideline (25) is similar for sampling but would allow fewer sampling points. Sampling points should be at least seven, including zero time, one point before C_{max}, two points around C_{max}, and three points during the elimination phase. Sampling should be continued until AUC_t is over 80% of AUC_∞ (normally more than three times the elimination half-life) after t_{max}..

Urine Samples

In the U.S. CFR Sec. 320.26 (24), there is a paragraph (d) *Collection of urine samples*. "When comparison of the test product and the reference material is to be based on cumulative urinary excretion–time curves, unless some other approach is more appropriate for valid scientific reasons, samples of the urine should be collected with sufficient frequency to permit an estimate of the rate and extent of urinary excretion of the active drug ingredient or therapeutic moiety, or its metabolite(s), measured." An example of the application for BE comparison from urinary excretion is given for the specific guideline for the in vivo bioequivalence study for slow-release potassium chloride tablets (50). Samples are required to be collected at 0–1, 1–2, 2–4, 4–6, 6–8, 8–12, 12–16, and 16–24 hours after dosing. Creatinine clearance is also measured to assure that adequate urine is collected. The parameters evaluated in each subject to calculate the mean for each product are cumulative urinary excretion over 0 to 24 hours, the maximum rate of urinary excretion (R_{max} also known as U_{max}), the time of maximal urinary excretion (T_{max}), and a rate profile.

From the European draft guideline (15), it is noted that the use of urinary excretion data as a surrogate for a plasma concentration may be acceptable in determining the extent of exposure in case it is not possible to measure reliably the plasma concentration–time profile of parent compound. However, the use of urinary data has to be justified carefully when used to estimate peak exposure. If a reliable plasma C_{max} can be determined, this should be combined with urinary data on the extent of exposure for assessing bioequivalence.

The Japanese guideline notes that generally blood samples should be employed (25). Urine samples may be used if there is a rationale. When urine samples are used, they "should be collected in the same manner as blood samples."

Analytes to be Measured

Parent Drug or Metabolite

Although for ANDAs BE is usually based on parent drug, the FDA guideline (13) recommends that, for BA studies in NDAs, determination of moieties to be measured in biological fluids take into account both concentration and activity. *Concentration* refers to the relative quantity of the parent drug or one or more metabolites in a given volume of an accessible biological fluid such as blood or plasma. *Activity* refers to the relative contribution of the parent drug and its metabolite(s) in the biological fluids to the clinical safety and/or efficacy of the drug. It also recommends that, for BA studies, both the parent drug and its major active metabolites be measured, if analytically feasible (13).

For BE studies measurement of only the parent drug released from the dosage form, rather than the metabolite, is generally recommended (13). The rationale for this recommendation is "that concentration–time profile of the parent drug is more sensitive to changes in formulation performance than a metabolite. Exceptions to this are if parent drug occurs in too low a concentration to be measured and if a metabolite may be formed as a result of gut wall or other presystemic metabolism. If the metabolite contributes meaningfully to safety and/or efficacy, it also recommend that the metabolite and the parent drug be measured but the BE would be based on parent, with metabolite data supportive" (13).

The EMEA guideline (14) has rather extensive recommendations for measuring parent compound and metabolite(s), which depend on the contribution of parent compound and metabolite(s), respectively, to activity. These appear to be different from the FDA recommendation and clearly could impact on design of studies for some products in the two jurisdictions.

The EMEA draft guideline (15) notes that "In principle, evaluation of bioequivalence should be based upon measured concentrations of the parent compound. The reason for this is that C_{max} of a parent compound is usually more sensitive to detect differences between formulations in absorption rate than C_{max} of a metabolite."

"Also for inactive pro-drugs, demonstration of bioequivalence for parent compound is the preferred option when the pharmacokinetics of pro-drug and active metabolite(s) is linear. In this situation, the active metabolite does not need to be measured. However, in case the pro-drug or active metabolites display nonlinear pharmacokinetics (or it is difficult to conclude linear pharmacokinetics from available data), it is recommended to demonstrate bioequivalence for the main active metabolite." In this situation, it is acceptable to demonstrate bioequivalence for the main active metabolite without measurement of parent compound.

"The use of a metabolite as a surrogate for an active parent compound can only be considered if the applicant presents convincing arguments demonstrating that it is not possible to reliably measure the parent compound after single-dose administration or at steady state. However, as C_{max} of the metabolite is usually less sensitive to differences in the absorption rate than C_{max} of the parent drug, bioequivalence should, if possible, be determined for C_{max} of the parent compound as a measure of peak exposure while extent of exposure can be based on demonstration of bioequivalence of AUC of metabolite. Furthermore, when using metabolite data as a substitute for parent drug concentrations, the applicant should present any available data supporting the view that the parent drug exposure will be reflected by metabolite exposure and that the metabolite formation is not saturated at therapeutic doses."

"In exceptional cases, bioequivalence of active metabolite(s) may need to be demonstrated in addition to parent drug (15). This is applicable if the metabolite has a major contribution to clinical efficacy of an active substance and metabolite concentrations may reflect differences in formulation, which may not be detected in parent compound, such as drugs with linear pharmacokinetics for parent compound, and where the active metabolite shows nonlinear pharmacokinetics caused by significant saturation of formation and/or elimination."

"When evaluating the significance of the contribution of an active metabolite to the clinical efficacy, available information on differences in AUC and pharmacodynamic activity between parent compound and metabolite should be taken into account. Depending on how pharmacodynamic activity has been determined, differences in protein binding between parent compound and metabolite may also need to be taken into account."

The Japanese guideline is not clear on this topic. Parent drugs should, in principle, be measured (25). Major active metabolites may be measured instead of the parent drug if it is rational to do so. The active metabolite is accepted for pro-drugs.

Enantiomers

The FDA guideline states that "for BA studies, measurement of individual enantiomers may be important" (13). For NDAs early studies should document clearly the importance of each enantiomer in terms of pharmacokinetics and pharmacodynamics. Based on this information the guideline indicates that achiral methods can be applied to BE studies for later comparison studies of development, to-be-marketed formulations, and ANDAs, unless the enantiomers show different pharmacodynamic and/or pharmacokinetic properties or the minor enantiomer provides most of the primary efficacy and safety activity or the absorption of the two enantiomers is nonlinear (enantiomer ratio changes with change in rate of input). The European guidelines do not mention the NDA aspect, but for generics (ANDAs), the same considerations are noted, concerning application of achiral methods, as in the FDA guideline (14,15). The Japanese guideline is similar to both FDA and EMEA guidelines on this aspect (25,51).

BIOANALYTICAL METHODOLOGY

Since 1990, there were a number of international workshops (52–54) on the bioanalytical validation of methods for PK, BA, and BE studies, with all jurisdictions expecting the same requirements. FDA has a separate guideline (55). The European draft guideline (15) summarizes these expectations. In brief, the main characteristics of bioanalytical methods to be documented to ensure acceptance are

- stability of stock solutions and analytes in the matrix under storage and process conditions for the duration of analytical study periods,
- specificity (selectivity),
- accuracy,
- precision,
- limit of quantitation (and range), and
- response function.

Recently, the concern for reproducibility of results with incurred samples (study samples that are drawn from dosed subjects) has been controversial (56) and it appears that the FDA will require studies to document an appropriate number in each BE study, with stated tolerances (possibly two-thirds to be within 20% of average) (57).

PHARMACOKINETIC PARAMETERS

The pharmacokinetic parameters BA–BE to be listed for single- and multiple-dose studies are similar in all jurisdictions.

Single Dose

In the FDA general guideline (13), the following pharmacokinetic information from single-dose studies is recommended for submission:

- plasma concentrations and time points;

- subject, period, sequence, treatment;
- peak exposure is to be assessed by measuring the peak drug concentration (C_{max}) obtained directly from the data without interpolation; and
- AUC_{0-t}, $AUC_{0-\infty}$, C_{max}, T_{max}, λ_z, and $t_{1/2}$.

The measures of total exposure are area under the plasma/serum/blood concentration–time curve from time zero to time t (AUC_{0-t}) is the required measure of total exposure, where t is the last time point with measurable concentration for individual formulation; ($AUC_{0-\infty}$) is the area under the plasma/serum/blood concentration–time curve from time zero to time infinity, where

$$AUC_{0-\infty} = AUC_{0-t} + \frac{C_t}{\lambda_z}$$

C_t is the last measurable drug concentration and λ_z is the terminal or elimination rate constant (also known as k_{el}) calculated according to an appropriate method. It is recommended that the terminal half-life ($t_{1/2}$) of the drug also be reported and *inter* subject, *intra* subject, and/or total variability, if available.

The parameters are identical in the EMEA draft guideline (15). "In studies to determine bioequivalence after a single dose, AUC_t, $AUC_{0-\infty}$, C_{max}, and t_{max} should be determined. Additional parameters that may be reported include the terminal rate constant, λ_z, and $t_{1/2}$."

In the Japanese guideline (25), there is also guideline on urine parameters. "When blood samples are used, AUC_t and C_{max} should be subjected to the bioequivalence assessment in single-dose studies." "C_{max} is an observed value and AUC is calculated by using the trapezoidal integration method. If F can be estimated by deconvolution, then F can be used for AUC. Parameters such as $AUC_{0-\infty}$, t_{max}, MRT (Mean Residence Time), and k_{el} (equivalent to λ_z) should be submitted as reference data." This guideline also notes "When urine samples are used, A_{et}, $A_{e\tau}$, $A_{e\infty}$, and U_τ are employed instead of AUC_t, AUC_τ, AUC_∞, C_{max}, and C_τ."

Multiple Dose

For multiple-dose studies, the FDA recommends (13) "that the measurement of total exposure be the area under the plasma, serum, or blood concentration–time curve from time zero to time tau (τ) over a dosing interval at steady state ($AUC_{0-\tau}$ or AUC_τ), where tau is the length of the dosing interval. C_{min} (concentration at the end of a dosing interval), C_{av} (average concentration during a dosing interval), degree of fluctuation [$(C_{max}-C_{min})/C_{av}$], and swing [$(C_{max}-C_{min})/C_{min}$], if steady-state studies are employed."

In the draft EMEA guideline (15), "AUC_τ, $C_{max,ss}$, $C_{min,ss}$, $t_{max,ss}$, and fluctuation should be determined for studies to determine bioequivalence at steady state." The C_{max} and C_{min}, as listed above by FDA for steady state, are identical to $C_{max,ss}$ and $C_{min,ss}$.

The Japanese guideline (25) recommends measurement of AUC_τ, C_{max}, and C_τ that is also used as a reference parameter. It would seem that all jurisdictions would require the same level of detail in reporting data to the standard of ICH E6 and GCP guidelines (11,12). The Japanese guideline (25) lists the following results:

i. Individual subject data
   - Tables should be provided showing drug levels in biological fluids at each sampling time for each subject with each treatment.
   - Tabulation of primary and secondary pharmacokinetic parameters from each individual with each treatment, including C_{max}, C_τ, AUC_t, AUC_τ, AUC_∞, λ_z, or k_{el}, t_{max}, and MRT.
   - The correlation coefficient for determining k_{el} should be reported together with time points used for its calculation for each subject with each treatment.
   - Tables with the ratios of C_{max} and AUC_t of test product to those of reference product in each individual.
   - Figures comparing individual drug level–time profiles of the two products drawn on a linear/linear scale. (The Canadian guideline (43) also requires semi-log plots and a note of the time point from which the terminal slope is calculated.)

ii Averages and standard deviations
Tables should be provided showing
- averages and standard deviations of raw data of drug levels in biological fluids at each time point;
- averages of the primary and secondary pharmacokinetic parameters C_{max}, C_{τ}, AUC_t, AUC_{τ}, AUC_{∞}, k_{el}, t_{max}, and MRT.
- the ratios of average of C_{max} and AUC_t of test product to those of reference product should be reported.

Also there should be figures provided comparing average drug level-time profiles of the two products drawn on a linear/linear scale.

iii. Statistical analysis and equivalence assessment (discussed in the next section below)

iv. Analysis of pharmacokinetic parameters
If deconvolution is used, the program, algorithm, pharmacokinetic models, and fitting information should be listed.

v. Others
Information on dropouts (data, reasons), monitoring, and records of health of subjects.

STATISTICAL ANALYSIS

Analysis

The FDA general guideline (13) notes that the following be obtained for AUC_t, $AUC_{0-\infty}$, and C_{max}: Geometric and arithmetic means, ratio of means and confidence intervals. Logarithmic transformation is to be provided for these measures for BE demonstration. To compare measures in single-dose crossover studies, the guideline states that "data have been analyzed using an average BE criterion. This guideline recommends continued use of an average BE criterion to compare BA measures for replicate and nonreplicate BE studies of both immediate- and modified-release products." Details are given in the statistical guideline (47) and application to replicate designs in recent literature (58,59). The statistical guideline also documents handling of population bioequivalence data.

The EMEA guidelines (14,15) have the same requirements as FDA for nonreplicate crossover designs and states that "the pharmacokinetic parameters under consideration (AUC_t, $AUC_{0-\infty}$, and C_{max}) should be analyzed using ANOVA (or equivalent parametric method). A confidence interval for the difference between formulations on the log-transformed scale is obtained from the ANOVA model. This confidence interval is then back-transformed to obtain the desired confidence interval for the ratio on the original scale. The data should be transformed prior to analysis using a logarithmic transformation.A nonparametric analysis is not acceptable. The precise model to be used for the analysis should be prespecified in the protocol."

The Japanese general guideline (25) has similar provisions but adds that the acceptable ranges for other parameters such as t_{max} should be determined for each drug.

Acceptance Limits

The three regions have the same general acceptance limits for BE studies. In studies to determine bioequivalence after a single dose, the parameters to be analyzed are AUC_t and C_{max}. For these parameters, the 90% confidence interval for the ratio of the test and reference products should be contained within the acceptance interval of 80% to 125%. Confidence intervals should be presented to two decimal places. To be inside the acceptance interval the lower bound should be $\geq$80.00 and the upper bound should be $\leq$125.00. The FDA recommends no rounding of calculated values (13).

An exception to the C_{max} BE standard is provided in Canada (43) with the ratio log transformed of means (point estimate, not confidence interval) between 80.00% and 125.00%. The Canadian guideline also requires results to meet the acceptance criteria for both observed and content (or potency) corrected values.

For studies to determine BE at steady state, AUC_{τ}, $C_{max,ss}$, and $C_{min,ss}$ should be analyzed using the same acceptance interval as stated above (13–15).

SPECIAL TOPICS

Long Half-Life Drugs

The FDA has the clearest explanation of BA/BE studies of this group of drugs. The guidance states, "In a BA or pharmacokinetic study involving an oral product with a long half-life drug, adequate characterization of the half-life calls for blood sampling over a long period of time" (13). This would be investigated in the development stages of an NDA. For BE studies in later development or ANDAs, "a nonreplicate, single-dose, crossover study can be conducted, provided an adequate washout period is used. If the crossover study is problematic, a BE study with a parallel design can be used. For either a crossover or parallel study, the sample collection time should be adequate to ensure completion of gastrointestinal transit (~2–3 days) of the drug product and absorption of the drug substance." The parameters C_{max}, and a suitably truncated AUC, usually AUC truncated at 72 hours ($AUC_{0\text{–}72}$ hours in place of $AUC_{0\text{–}t}$ or $AUC_{0\text{–}\infty}$.), can be used to characterize peak and total drug exposure, respectively. The guideline notes "for drugs demonstrating high intrasubject variability in distribution and clearance, AUC truncation warrants caution. In such cases, it is recommended that sponsors and/or applicants consult the appropriate (FDA) review staff." The EMEA (14,15) and Japanese guidelines (25) have similar truncation recommendations.

Narrow Therapeutic Range Drugs

The FDA general guideline provides examples of narrow therapeutic range drugs (NTR). These are digoxin, lithium, phenytoin, theophylline, and warfarin (13). Clearly, the dose response for NDAs will characterize drugs to be added to the list. There is no narrowing of acceptance range recommended by the FDA for those drugs. However, in general, no waiver of BE studies would be considered and fed as well as fasted studies will be expected.

The EMEA guideline (14) remarks that "in specific cases of products with a narrow therapeutic index, the acceptance interval may need to be tightened." It is not considered possible to define criteria to define NTR drugs and this would be a case-by-case consideration from a 'full application' for changing limits for both AUC and C_{max} or for AUC only. When tightening of limits is considered "the acceptance interval for concluding bioequivalence should generally be narrowed to 90% to 111%. In individual cases alternative or additional requirements might be set" (15).

The Canadian guideline "Bioequivalence guidelines, Critical Dose drugs" (60) (similar to the FDA NTR list but including immunosuppressants such as cyclosporine) requires BE studies to be completed in the fasted and fed states, and the following acceptance standards are used:

- the 90% confidence interval of the relative mean AUC of the test to reference formulation should be within 90.0% to 112.0% and
- the 90% confidence interval of the relative mean measured C_{max} of the test to reference formulation should be between 80.0% and 125.0%.

"Steady-state studies are not required for 'critical dose drugs' unless warranted by exceptional circumstances. If a steady-state study is required, the 90% confidence interval of the relative mean measured C_{min} of the test to reference formulation should also be between 80.0% and 125.0%."

The guideline requires observed and potency corrected values to meet these acceptance standards.

Japanese guidelines (38,39) also provide a list of NTR drugs for which in vivo methods are required and waivers of changes by in vitro tests is not permitted. The BE standard is not altered.

Rapid Onset Drugs

The importance of early exposure would be determined during clinical efficacy studies for an NDA. The FDA exemplifies drugs requiring rapid onset of analgesic effect or to avoid an excessive hypotensive action of antihypertensive drug (13). Partial area estimation is recommended to the population median of the reference formulation peak concentration time (t_{max}) with at least two quantifiable samples collected before peak time.

There is a similar section in the EMEA draft guideline, but this notes that the time point for truncating the AUC should be specified and justified in the study protocol. Presumably this would be established for a new drug (MA).

The Canadian agency has a separate guideline for this (61). To date the standard has only been applied to two drugs, gel formulations of ibuprofen and tablet formulations of sumatriptan. In addition to the BE standards usually applied, the relative mean (point estimate) $AUC_{Reftmax}$ (reference t_{max}) of the test to reference formulation should be within 80% to 125%, where $AUC_{Reftmax}$ for a test product is defined as the area under the curve to the time of the maximum concentration of the reference product, calculated for each study subject. Other drugs for which this additional standard would be required will be considered from development and clinical efficacy information.

Highly Variable Drugs

FDA (13) allows replicate designs (described above under DESIGNS) to be applied for highly variable drugs (HVD) (commonly defined as comparisons when the within subject variability of C_{max} or AUC is 30% or greater). They have also published recently on applications of this approach (58,59). This uses a scaling approach based on the within subject variance of the reference. The statistical analysis may yield wider confidence limits than the general standard (of 80–125%) permits, but an additional safeguard is applied limiting the point estimate (test/reference geometric mean ratio) of AUC and C_{max} to fall within [0.80, 1.25]. Several generic applications are under review and it would also apply to major preapproval changes in NDAs.

The European draft guideline (15) notes that "in certain cases, C_{max} is of less importance for clinical efficacy and safety compared with AUC. When this is applicable, the acceptance criteria for C_{max} can be widened to 75% to 133% provided that all of the following are fulfilled: the widening has been prospectively defined in the study protocol; it has been prospectively justified that widening of the acceptance criteria for C_{max} does not affect clinical efficacy or safety; the bioequivalence study is of a replicate design where it has been demonstrated that the within-subject variability for C_{max} of the reference compound in the study is >30%. However, this approach does not apply to AUC. It is acceptable to apply either a three-period or a four-period crossover scheme in the replicate design study."

This is a major difference between the FDA and Europe because the FDA accepts scaling to the reference for widening the 90% confidence interval for AUC and does not suggest a limit on such scaling for C_{max}.

The Japanese guideline (25) has the same general BE standard as described by the FDA and Europe. However, add-on studies are permitted and wider limits for C_{max} are allowed for "less potent, mild effect" drugs. Additionally, if the study confidence intervals are outside of 80% to 125% limits, BE may still be claimed provided that the study meets all three conditions listed below:

1. the total number of subjects in the initial BE study is no less than 20 ($n = 10$/group), or pooled sample size of the initial and add-on studies is no less than 30;
2. the differences in average values of logarithmic AUC and C_{max} between two products are between log (0.9) and log (1.11); and
3. dissolution rates of test and reference products are determined to be equivalent under all dissolution testing conditions specified in the guideline.

Solutions

All jurisdictions waive bioequivalence studies, for simple, solutions, elixirs, syrups tinctures, or other formulations in which the drug is solubilized. However, as the FDA general guideline notes (13), certain excipients such as sorbitol or mannitol can reduce permeability and interfere with absorption of some drugs, and this would have to be investigated during development. Intravenous solutions are also exempt but other injectables such as intramuscular or subcutaneous may require an in vivo study if there are changes in composition. BE standards would be identical to those for solid oral dose forms.

Combination Products

Clearly considerable clinical trial work would be essential for developing an NDA for a new combination product, which may consist of existing marketed single component products or completely new agents or a new agent with an existing marketed product. The EMEA draft guideline on fixed-dose combination products provides PK information that is expected for combination products (62). Drug–drug interaction studies should be completed at appropriate doses, although enhancement of the PK profile of one drug on the other may be a rationale. The dose response of various fixed doses must be mapped. For pharmacokinetics, this is usually completed with the drugs administered separately and then the "free" combination of the two drugs is administered in separate dosage forms.A separate study of the to-be-marketed fixed dose combination (at appropriate strengths) compared to the same doses of free combination would complete the BE data package. If the clinical trials were completed with the free combination, then the final formulation should be shown to be bioequivalent. According to dosage instructions fed and fasted studies may be required.

Oral Suspensions and Suppositories

In general, at the NDA stage BA and BE studies should be provided for these dosage forms and linked with any dissolution studies for waiver of different strengths or formulations. BE standards would be identical to those for solid oral dose forms.

Food Effect Studies

The FDA has the most comprehensive guideline (63). It would be expected that food effect BA studies would usually be conducted for new drugs and drug products during the IND period to assess the effects of food on the rate and extent of absorption of a drug when the drug product is administered shortly after a meal (fed conditions) as compared to administration under fasting conditions. This would be part of the development plan to study formulation contributions to food effects. The guideline notes that food effects would be expected to be less with BCS Class 1 drugs but for BCS Class 2, 3, and 4 and for all modified-release drug products, food effects are most likely to result from a more complex combination of factors that influence the in vivo dissolution of the drug product and/or the absorption of the drug substance and will be difficult to predict without conducting a BE study. Unless there are safety concerns, in general, the highest strength of a drug product intended to be marketed should be tested in (pivotal) food-effect BA and fed BE studies. Although different meals can be investigated, one of the meals for the food-effect BA studies should be the high-fat and high-calorie test meal described in the guideline. An equivalence approach is recommended for food-effect BA (to make a claim of no food effects) and fed BE studies, analyzing data using an average criterion. Thus, to claim that there is no food effect, the 90% confidence interval around the geometric mean ratio of fed/fasted observations for AUC_t and C_{max} should be within 80% to 125%. This result, or if a food effect is found, the magnitude should be described on the clinical pharmacology and dosage and administration sections of the labeling.

The EMEA pharmacokinetic guideline (17) notes, "For medicinal products intended to be orally administered, it is important to study the effects of food on absorption." The Modified Release (MR) guideline (64) also indicates that the effect of food on the BA of oral modified-release dosage forms must be investigated. In this case, a high fat meal is usually investigated but other types of meals will be accepted if justified.

In the Japanese guideline (25), the food challenge depends on the prescribing information. If this specifies clearly that the drug should be administration after a meal, or if administration after fasting results in significantly lower bioavailability or in a higher incidence of serious adverse events, then the study should be performed with dosing after a meal.A low fat meal is preferred (700 kcal or less and the rate of fat energy to the total energy is 20% or less).

MODIFIED-RELEASE DOSAGE FORMS

The nomenclature for different types of MR dosage forms includes "delayed-release" (DR) and extended, controlled, or prolonged-release. DR is described among USP (65), FDA (13), and EMEA quality modified-release (66) guidelines as dosage forms that release the drugs at a time later than immediately after administration, yielding a lag time in appearance of drug in plasma. Usually, this is because of enteric-coating to protect the drug from stomach acidity or the stomach from the drug. In the USP, all other types of MR appear to be covered by the term extended-release (ER), although FDA also refers to "controlled-release".

In the EMEA MR quality guideline (66) "prolonged-release" is noted as equivalent to "extended-release" as dosage forms showing a slower release than that of conventional (immediate-release) dosage forms administered by the same route-–achieved by special formulation design and/or manufacturing methods. These are listed with transdermal systems in the EMEA MR efficacy guideline (64) and it is also noted that "most items" are also applicable to implants and intramuscular or subcutaneous depot formulations.

Delayed-Release (Enteric-Coated) Dose Forms

There is not much advice given for DR products in the European or FDA guidelines but the FDA indicates that "In vivo tests for delayed-release drug products are similar to those for ER products" (13). In contrast, the EMEA guideline (66) notes that for DR formulations "BE is assessed using the same main characteristics and statistical procedure as for immediate-release formulations with emphasis on the DR characteristics," thus indicating that single-dose fasted studies would be expected and that food effect studies would also be necessary. Appendix I of the guideline recommends the studies expected for either an NDA introduced as an MR or as an extension with an MR product, after an IR formulation was approved.

Extended-Release (Controlled-Release) Dose Forms

Although the FDA guideline no longer requires multiple-dose studies for BE studies of generic MR products (13), there are two types of MR that would be submitted as NDAs and in common with EMEA would require extensive clinical efficacy and pharmacokinetic studies.

ER as a New Drug First Application

A new chemical entity may be introduced as an MR dosage form requiring a full safety, efficacy, and quality data package. This a Type 1 ER.

ER When the Drug Has Been Approved as an IR Form and the ER is an Extension

In the case when the new chemical entity was first introduced as a conventional (immediate-release) product, the ER product application may be supported with fewer toxicology studies, clinical trials, and with extensive pharmacokinetic study (Type 2 ER). The FDA recommends that such products be submitted as NDAs (13). Both the FDA and the EMEA guidelines have similar requirements for Type 1 and 2 ER products.

The purpose of an *in vivo* BA study for which a controlled-release claim is made is to determine if all of the following conditions are met:

- the drug product meets the controlled-release claims made for it,
- the BA profile established for the drug product rules out the occurrence of any dose dumping, and
- the drug product's steady-state performance is equivalent to a currently marketed conventional-release product and the drug product's formulation provides consistent pharmacokinetic performance between individual dosage units.

While fewer animal toxicology studies may be expected for the immediate- to extended-release situation in the case of new nonactive ingredients (such as controlling release), animal safety studies may have to be carried out.

The claim may be either to reduce the dosage interval or to suppress the peak exposure (less fluctuation) than for an immediate-release product (or both) and this would have to be supported by clinical trial data, which would also be required for any additional indication for the drug in an application.

The reference in bioavailability studies of Type 2 ER products allows proving of the controlled-release claims made for the drug product, such as a solution or suspension of the active drug ingredient or therapeutic moiety or a currently marketed immediate-release drug product containing the same active drug ingredient or therapeutic moiety and administered according to the dosage recommendations in the labeling.

The FDA guideline (13) suggests the following BA studies be completed for the NDA of an ER product:

- a single-dose, fasting study on all strengths of tablets and capsules and highest strength of beaded capsules, and
- a single-dose food-effect study on the highest strength and a steady-state study on the highest strength.

If there are lower strengths, these may be approved on the basis of dissolution and other in vitro measures covered later under SUPAC. This is especially relevant to beaded granules that are of the same composition from dose to dose.

The EMEA ER guideline (efficacy) (66) is concerned about pharmacokinetic linearity in bridging to other strengths. If the drug substance is shown to have linear pharmacokinetics, it is necessary to establish similar total exposure between the MR and IR formulation at one dose level following multiple-dose administration. In the case of nonlinear pharmacokinetics, the steady-state ER to MR comparison should be completed at the highest and lowest dose level. Also, dose proportionality for different strengths should be appropriately addressed.

Food Challenge Studies of ER Products

For food challenge studies, the FDA food guideline (63) requires that a single-dose fasted fed study be carried out for the NDA or all MR products. In addition "for NDAs, the labeling of certain drug products (e.g., controlled-release capsules containing beads) can recommend that the product be sprinkled on soft foods, such as applesauce, and swallowed without chewing. For the labeling to indicate that the drug product can be sprinkled on soft foods, additional in vivo relative BA studies should be performed by sprinkling the product on the soft foods to be listed in the labeling (test treatment) and comparing it to the product administered in the intact form (reference treatment), and then administering both on an empty stomach." Usually, these are for product used in pediatric patients.

The EMEA MR guideline (efficacy) (66) has a decision tree appended with the recommended studies to investigate the effect of food on absorption from MR dosage forms. These are usually single-dose healthy subject studies comparing fasted and high-fat meal with appropriate immediate-release formulation as reference. The final crossover study for an NDA type application is comparison of the clinical trial form fasting with the to-be-marketed in fasted and fed conditions.

In general, the standards for BE in the NDAs of such products would be the same as described above under "Acceptance limits," except for drug types discussed above under "special topics."

PHARMACODYNAMIC AND CLINICAL TRIAL ESTABLISHMENT OF BE

Manufacturers and regulators have struggled for 30 or more years with drug products whose BE cannot be assessed reliably by pharmacokinetic or surrogate in vitro tests, and for NDAs, this occurs with late-stage changes in formulation or with aerosols when fluorocarbon propellant agents had to be replace. These are case-by-case situations. For pharmacodynamics, there are four examples, three from the FDA website.

Topical Steroids

The vasoconstrictor test (skin blanching) or Stoughton–McKenzie method is described in an FDA guideline (67) with great detail on the methodology. Although introduced in 1995, it has not been updated and the standard is for the ratio of Area Under Effect Curve (AUEC) comparison of new and old (reference) formulations of skin blanching on the forearm of volunteers. The 90% confidence interval between 80% and 120% is mentioned as a possible standard, but it is noted that it may have to be wider.

Albuterol (Salbutamol)

Bronchodilation and bronchoprovocation (methacholine or histamine challenge) have been used to measure the pharmacodynamic effects from albuterol both for generic comparisons and formulation changes (68). In general, the 90% confidence interval of the AUEC has to be widened to 67% to 150%. Together with extensive in vitro testing of aerosol quality, these pharmacodynamic tests have been used in approvals by FDA.

Canada also has a guideline "to establish equivalence or relative potency of safety and efficacy of a second entry short-acting beta$_2$-agonist metered dose inhaler" and includes the bronchodilation and bronchoprovocation studies (69). This guidance requires an 80% to 125% standard to be met for ratio of test to reference AUEC.

Inhaled Corticosteroids

Health Canada also issued a draft guideline for Inhaled Corticosteroid Products (70), which proposes anti-inflammatory markers as primary endpoints. The main measure is the effect on serum cortisol levels (SCO) and the standard proposed is that the 90% confidence interval of the relative mean serum cortisol area under the 24-hour curve (SCO-24 AUC) of the test to reference product should lie between 80% and 125%.

Nasal Aerosols and Nasal Sprays for Local Action

The FDA has a specific draft guidance on bioavailability and bioequivalence studies for nasal aerosols and nasal sprays for local action (71). The endpoints for the equivalence and efficacy analyses are expressed as mean change from baseline (pretreatment) of the total nasal symptom score (TNSS) expressed in absolute units rather than percent change from baseline over seven days.

Clinical Trials

One of the approaches to determine bioequivalence noted earlier is by well-controlled clinical trials. The EMEA guideline (14) also notes that human studies with clinical or pharmacodynamic endpoints can be envisaged "as long as they are appropriately justified and/or validated." Such studies have been discussed from the ANDA viewpoint by Hendy (72), who remarks that "there is limited regulatory guidance." Many of these products are topical products for a variety of diseases as well as optical or otic products. In general, formulation performance (BE) is not a problem for an NDA where the final formulation would likely have been studied in clinical trials. However, if there is a late, significant change in formulation, the same issues would have to be addressed as for an ANDA. The Nasal Aerosol guideline is similar to the specific guideline that is required for such studies, and the FDA draft guidelines on allergic rhinitis and *acne vulgaris* are examples of potential clinical endpoint requirements that would have to be discussed case by case with regulatory agencies. Patient numbers would be large, and greater numbers would be required to show superiority of two active treatments than active versus placebo.

In Vitro Test

An in vitro test that has been correlated with and is predictive of human in vivo bioavailability data, and a currently available in vitro test acceptable to the FDA (usually a dissolution rate test) that ensures human in vivo bioavailability are BE approaches, also noted above. Initially, this was a measure for drugs approved before the U.S. regulations changed in 1962, such as prednisone, but recently use of the BCS has extended this approach for biowaivers.

BIOWAIVERS

This section will briefly discuss scale-up, preapproval formulation or process changes, and strength changes that may be required in NDAs and situations when biowaivers may be justified. This would be presented in CTD summary section 2.7.1 and provide information on dissolution experiments during pharmaceutical development. The type of information required is described by FDA in the SUPAC IR (48) and MR guideline (49) documents as well as those for dissolution testing of IR (73) and ER (74) dosage forms. The dissolution approach is also summarized in Appendix III of the draft EMEA document (15). It is also extensively applied by the Japanese in different strengths (38) and formulation change (39). There are considerable differences among regions in the granting of biowaivers and this has been discussed in detail by Gupta et al. (75).

Immediate Release

Preapproval Changes

The FDA approach depends largely on the BCS approach described earlier. Notably, the EMEA draft guideline (15), the FDA BCS document (32), and the Japanese documents (38,39) note that NTR or critical dose drugs are not eligible for BCS waivers.

The Japanese approach does not apply the BCS, but formulations are categorized based on dissolution speed (46). The rationale suggests that differences in bioavailability between formulations are due to the differences in formulation characteristics, such as particle size of the pharmaceutical substance, compounding, and manufacturing processes. It is considered that formulations, which dissolve in the same way in the digestive tract, are unlikely to have differences in bioavailability. However, to test equivalence of dissolution with a diverse range of conditions encountered in different digestive tracts, equivalence of dissolution profiles in dissolution tests is required under many conditions. The Japanese guidelines are also concerned about achlorhydric patients and thus higher pH values are investigated.

The European draft guideline (15) notes that although BCS-based waivers described therein are intended only to address the question of bioequivalence test and a reference (ANDAs), the approach can be applied to show bioequivalence between early clinical trial and to-be-marketed products and in the case of variations (changes) that may require bioequivalence testing. The latter would include strength changes and formulation or process changes during development of NDAs. In general, the drug substance should exhibit high solubility and complete absorption [BCS class 1; with very rapid (>85% within 15 minutes) in vitro dissolution characteristics and no absorption interference by excipients].

BCS-based biowaivers may also be applicable for an immediate-release drug product of limited absorption (BCS Class 3) with the same constraints noted above for the BCS Class 1 situation.

Generally, the risks of an inappropriate biowaiver decision should be more critically reviewed (e.g., site-specific absorption, risk for transport protein interactions at the absorption site, excipient composition, and therapeutic risks) for products containing BCS Class 3 than for BCS Class 1 drug substances. As yet the U.S. FDA guideline for biowaivers is directed at BCS Class 1 drugs, and several ANDAs have been approved based on the guideline (34) without in vivo studies but there has been discussion of extension to BCS Class 3 (35).

Extensive dissolution comparisons are directed at justifying biowaivers, as well as the usual chemistry and manufacturing control (CMC) specifications, controls, and stability for drug substance and product. Again, while the FDA SUPAC and dissolution guidelines include great detail on the levels of requirements, these are well summarized in the European draft guideline (15).

Different Strengths

The FDA (13), European (14), and Japanese guidelines (25,38) are similar concerning BE testing of different strengths in both NDA and ANDA situations. Usually, the highest strength is tested in vivo (unless there are subject safety concerns), and in vivo BE demonstration of one or more lower strengths can be waived based on the demonstration of BE with the highest strength and dissolution tests (as mentioned above). For biowaivers in these circumstances, the following conditions should be met:

- clinical safety and/or efficacy studies including data on the dose and the desirability of the higher strength should be documented,
- pharmacokinetics should be linear (dose proportional) over the therapeutic dose range,
- the higher strength should be similar proportionally to the lower strength, and
- the same dissolution procedures should be used for both strengths and similar dissolution profile results obtained.

The SUPAC IR guideline (48) indicates the tolerances for acceptable variations in excipients for decisions on proportionality of formulations. The EMEA draft document (15) provides a decision tree to decide when waivers may be appropriate or when additional BE studies by strength may be required. The Japanese change document (38) has considerable differences from FDA in the tabulation of acceptable variations. In particular, the degree of change in the core and the coating layer must be calculated separately for coating products, including that for sugar-coated tablets. The rationale is that some coating layers have a greater effect on dissolution, and with coating layers, thickness matters more than weight.

Changes or Different Strengths of Extended-Release Products

For slowly dissolving drugs, and particularly ER products, it may be possible to develop an in vivo/in vitro correlation (IVIVC) based on in vivo plasma level profiles compared to in vitro dissolution profiles, which can be applied to biowaivers of lower strengths and certain scale-up and preapproval changes (or variations) in NDAs.

The most useful IVIVC is termed a Level A correlation by both FDA (49,75) and EMEA (15) documents. The specification is based on 1:1 correlation between the in vivo and in vitro dissolution profiles. The in vivo profile is usually obtained from the plasma concentration profile by deconvolution, and the in vitro profile may be based directly on release at different time points or can be described by mathematical models, such as the Weibull function. A correlation of this type is generally linear and represents a point-to-point relationship between in vitro dissolution and the in vivo input rate (e.g., the in vivo dissolution of the drug from the dosage form). In a linear correlation, the in vitro dissolution and in vivo input curves may be directly superimposable or may be made to be superimposable by the use of a scaling factor. Nonlinear correlations, while uncommon, may also be appropriate.

The most commonly seen process for developing a Level A IVIVC is to

(a) develop formulations with different release rates, such as slow, medium, fast;
(b) obtain in vitro dissolution profiles and in vivo plasma concentration profiles for these formulations; and
(c) estimate the in vivo absorption or dissolution time course using an appropriate deconvolution technique for each formulation and subject.

These three steps establish the IVIVC model. Details of establishing the IVIVC and validating tolerances are given in the U.S. guideline document (75).

The FDA SUPAC MR guideline (49) notes that for modified-release beaded capsules, where the strength differs only in the number of beads containing the active moiety, a single-dose fasting BE study be carried out only on the highest strength with waiver of in vivo studies for lower strengths based on dissolution profiles. For MR tablets, when the drug product is in the same dosage form but in a different strength, when it is proportionally similar in its active and inactive ingredients, and when it has the same drug release mechanism, an in vivo BE determination of one or more lower strengths can be waived based on dissolution profile comparisons, with an in vivo study only on the highest strength. Clearly, this is facilitated if an IVIVC has been validated.

OTHER ISSUES

The different BE approaches among regions to highly variable drugs and the use of dissolution and BCS in changes or strengths during development have been noted above.

Many of the issues of BE testing are concerned with ANDA problems and some of those are apparent in the EMEA inspectors draft document (76), which noted that the problems were "manifold, diverse, and observed in many aspects of the clinical trials." These included poor documentation of study subjects, ethics committees approvals, bioanalytical data, quality of test and reference products, protocol violations, safety reporting (adverse events), study reports, and archiving (lost or incomplete). Inspection reports from the FDA also provide evidence of shortcomings.

The reference or comparator product choice remains a problem for ANDAs around the jurisdictions and, although less of a problem for NDAs, sometimes "formulation creep" and stability problems during development can bring problems to BA/BE comparison of clinical trial products to the market image. This can be especially troublesome if the original product is no longer available for reference.

Outliers, which skew BE comparisons, remain problematic when studies fail narrowly when they are included, but there is no obvious reason to exclude their values.

The FDA guideline (13) notes the problem of complex mixtures (i.e., active ingredients that are mixtures of multiple synthetic and/or natural source components). In such situations it is recommended that BA and BE studies be based on a small number of markers of rate and extent of absorption.

This is somewhat related to the problems of BA/BE assessment of endogenous drug substances (such as minerals and hormones) administered exogenously. Usually, in an ANDA problem, the increased development of biotechnology-derived drugs has renewed attention to characterizing the PK of such drugs for NDAs. Colocci et al. (77) have described approaches and much depends on whether the baseline is stable or the endogenous active substance can be suppressed safely to differentiate from exogenous dosing.

As noted the approach for orally administered systemically available drugs is well accepted in all regions, but use of pharmacodynamic studies or clinical trials to support changes during development requires consultation with each agency.

There is a useful website (78) for following updates of biopharmaceutics guidelines in many jurisdictions internationally, although not all are available in English translations.

REFERENCES

1. ICH Harmonised Tripartite Guideline. Comparability of Biotechnological/Biological Products Subject to Changes in Their Manufacturing Process Q5E, Step 4 Version, 18 November 2004. http://www.ich.org/cache/compo/276–254-1.html. Accessed November 14, 2008.
2. ICH Harmonised Tripartite Guideline. The Common Technical Document for the Registration of Pharmaceuticals for Human Use Efficacy—M4E(r1) Clinical Overview and Clinical Summary of Module 2 module 5: Clinical Study Reports, Step 4 Version Dated 12 September 2002, page 6, 2.7.1 Summary of Biopharmaceutic Studies and Associated Analytical Methods. http://www.ich.org/LOB/media/MEDIA561.pdf. Accessed November 14, 2008.
3. ICH Harmonised Tripartite Guideline. Pharmaceutical Development Q8, Step 4 Version 10 November 2005. http://www.ich.org/cache/compo/276–254-1.html. Accessed November 14, 2008.
4. ICH Harmonised Tripartite Guideline. Quality Risk Management, Q9, Step 4 Version 9 November 2005. http://www.ich.org/cache/compo/276–254-1.html. Accessed November 14, 2008.
5. ICH Harmonised Tripartite Guideline. Pharmaceutical Quality System, Q10, Step 4 Version, 4 June 2008. http://www.ich.org/cache/compo/276–254-1.html. Accessed November 14, 2008.
6. ICH Harmonised Tripartite Guideline. Stability Testing of New Drug Substances and Products, Q1A(R2), Step 4 Version, 6 February 2003. http://www.ich.org/cache/compo/276–254-1.html. Accessed November 14, 2008.
7. ICH Harmonised Tripartite Guideline. Validation of Analytical Procedures: Text and Methodology Q2(R1) Step 4, Parent Guideline, 27 October 1994 (Complementary Guideline on Methodology, 6 November 1996 Incorporated in November 2005) http://www.ich.org/LOB/media/MEDIA417.pdf. Accessed November 14, 2008.
8. ICH Harmonised Tripartite Guideline. Impurities in New Drug Products, Q3B(R2) Step 4, 2 June 2006. http://www.ich.org/LOB/media/MEDIA421.pdf. Accessed November 14, 2008.
9. ICH Harmonised Tripartite Guideline. Specifications: Test Procedures and Acceptance Criteria for New Drug Substances and New Drug Products: Chemical Substances, Q6A, Step 4, 6 October 1999. http://www.ich.org/LOB/media/MEDIA430.pdf. Accessed November 14, 2008.
10. ICH Harmonised Tripartite Guideline. Good Manufacturing Practice Guide for Active Pharmaceutical Ingredients, Q7 Step 4, 10 November 2000. http://www.ich.org/LOB/media/MEDIA433.pdf. Accessed November 14, 2008.
11. ICH Harmonised Tripartite Guideline. Structure and Content of Clinical Study Reports, E3, Step 4, 30 November 1995. http://www.ich.org/LOB/media/MEDIA421.pdf. Accessed November 14, 2008.
12. ICH Harmonised Tripartite Guideline. Guideline for Good Clinical Practice, E6(R1), Step 4, 10 June 1996. http://www.ich.org/LOB/media/MEDIA482.pdf. Accessed November 14, 2008.
13. Guidance for Industry Bioavailability and Bioequivalence Studies for Orally Administered Drug Products—General Considerations, U.S. Department of Health and Human Services, Food and Drug Administration Center for Drug Evaluation and Research (CDER), March 2003. http://www.fda.gov/cder/guidance/5356fnl.htm. Accessed November 14, 2008.
14. Committee for Proprietary Medicinal Products (CPMP) Note for Guidance on the Investigation of Bioavailability and Bioequivalence. January 2002. http://www.emea.europa.eu/pdfs/human/qwp/140198enfin.pdf. Accessed November 14, 2008.

15. Committee for proprietary medicinal products (CPMP) *DRAFT* Guideline on the Investigation of Bioequivalence. Released for Consultation, July 24, 2008. http://www.emea.europa.eu/pdfs/human/qwp/140198enrev1.pdf. Accessed November 14, 2008.
16. Guideline for the Format and Content of the Human Pharmacokinetics and Bioavailability Section of an Application. U.S. Department of Health and Human Services, Food and Drug Administration Center for Drug Evaluation and Research (CDER), February 1987. http://www.fda.gov/cder/guidance/old071fn.pdf. Accessed November 14, 2008.
17. EMEA, Pharmacokinetic Studies in Man, Directive 75/318/EEC as Amended, Date of First Adoption February 1987, Date of Entry into Force October 1988. http://www.emea.europa.eu/pdfs/human/ewp/3cc3aen.pdf. Accessed November 14, 2008.
18. Clinical Pharmacokinetic Studies of Pharmaceuticals, Ministry of Health, Labour, and Welfare, Japan, 1 June 2001, http://www.nihs.go.jp/phar/material/material2/ClPkEng011122.pdf. Accessed November 15, 2008.
19. WHO 1996. Multisource (Generic) Pharmaceutical Products: Guidelines on Registration Requirements to Establish Interchangeability. Technical Report Series, No 863, Annex 9, pp. 115–139. http://www.who.int/medicinedocs/en/d/Js5516e/#Js5516e. Accessed November 15, 2008.
20. WHO Technical Report Series, 937, WHO Expert Committee on Specifications for Pharmaceutical Preparations, 40th Report, Geneva 2006, Annex 7, pp. 347–390. whqlibdoc.who.int/trs/WHO_TRS_937_eng.pdf. Accessed November 15, 2008.
21. Wagner JG. Biopharmaceutics: Absorption aspects.J Pharm Sci 1961; 50(5):359–387.
22. Gibaldi M. Biopharmaceutics and Clinical Pharmacokinetics, 4th ed. Malvern, PA: Lea & Febiger, 1991:preface, vii.
23. ICH Harmonised Tripartite Guideline. Note for Guidance on Toxicokinetics: The Assessment of Systemic Exposure in Toxicity Studies S3A, Step 4 Version, 27 October 1994. http://www.ich.org/LOB/media/MEDIA495.pdf. Accessed November 28, 2008.
24. Code of Federal Regulations, title 21, Part 320 http://www.access.gpo.gov/nara/cfr/waisidx_02/21cfr320_02.htm. Accessed November 28, 2008.
25. Japan NIHS, Ministry of Health Labor and Welfare, Division of Drugs, Guideline for Bioequivalence Studies of Generic Products,December 1997. http://www.nihs.go.jp/drug/be-guide(e)/Generic/be97E.html. Accessed November 28, 2008.
26. Approved Drug Products with Therapeutic Equivalence Evaluations, "Orange Book" 28th edition, February 21, 2008, pp vi–vii, US Department of Health and Human Services, Food and Drug Administration, Center for Drug Evaluation and Research, Office of Science, Office of Generic Drugs. Electronic edition. http://www.fda.gov/cder/ob. Accessed November 29, 2008.
27. Canadian Food and Drug Regulations, Part C, Drugs, Division 8, Paragraph C.08.001.1 "Pharmaceutical Equivalent". http://laws.justice.gc.ca/en/showdoc/cr/C.R.C.-c.870/bo-ga:l_C-gb:s_C_08_001//en#anchorbo-ga:l_C-gb:s_C_08_001. Accessed November 29, 2008.
28. Buxton ILO. Pharmacokinetics and Pharmacodynamics [Chapter 1]. In: Brunton LL, Lazo JS, Parker KL, eds. Goodman & Gilman's The Pharmacological Basis of Therapeutics, 11th ed. New York: McGraw Hill Inc, Medical Publishing Division, 2006:1–39.
29. Benet LZ, Amidon GL, Barends DM, et al. The use of BDDCS in classifying the permeability of marketed drugs. Pharm Res 2008; 25(3):483–488.
30. Schinkel AH, Jonker JW. Mammalian drug efflux transporters of the ATP binding cassette (ABC) family: An overview. Adv Drug Deliv Rev 2003; 55(1):3–29.
31. Nakamura T, Yamamori M, Sakaeda T. Pharmacogenetics of intestinal absorption./ Curr Drug Deliv 2008; 5(3):153–169.
32. Guidance for Industry Waiver of In Vivo Bioavailability and Bioequivalence Studies for Immediate-Release Solid Oral Dosage Forms Based on a Biopharmaceutics Classification System U.S. Department of Health and Human Services Food and Drug Administration Center for Drug Evaluation and Research (CDER) August 2000. http://www.fda.gov/cder/guidance/3618fnl.htm. Accessed, November 29, 2008.
33. Amidon GL, Lennernäs H, Shah VP, et al.A Theoretical Basis for a Biopharmaceutics Drug Classification: The Correlation of In Vitro Drug Product Dissolution and In Vivo Bioavailability. Pharm Res 1995; 12:413–420.
34. Polli JE, Abrahamsson BS, Yu LX, et al. Summary workshop report: Bioequivalence, biopharmaceutics classification system, and beyond. AAPS J 2008; 10(2):373–379.
35. Stavchansky S. Scientific perspectives on extending the provision for waivers of in vivo bioavailability and bioequivalence studies for drug products containing high solubility-low permeability drugs (BCS-Class 3). AAPS J 2008; 10(2):300–305.
36. Karalis V, Macheras P, Van Peer A, et al. Bioavailability and bioequivalence: Focus on physiological factors and variability. Pharm Res 2008; 25(8):1956–1962.

37. Fagerholm U. Evaluation and suggested improvements of the Biopharmaceutics Classification System (BCS).J Pharm Pharmacol 2007; 59(6):751–757.
38. Japan NIHS, Ministry of Health Labor and Welfare, Division of Drugs, Guideline for Bioequivalence Studies of Oral Solid Preparations with Different Strengths. July 24, 2006.
39. Japan NIHS, Ministry of Health Labor and Welfare, Division of Drugs. Guideline for Bioequivalence Studies of Oral Solid Formulations with Formulation Changes. July 24, 2006.
40. Japan NIHS, Ministry of Health Labor and Welfare, Division of Drugs, Q&A on Guideline for Bioequivalence Test on Oral Solid Preparation with Different Drug Contents, Guideline for Bioequivalence Test on Oral Solid Preparation which the Formulation has been changed. July 24, 2006.
41. Health Canada, Therapeutics Product Directorate, Draft policy Bioequivalence Requirements: Drugs Exhibiting Non-linear Pharmacokinetics. http://www.hc-sc.gc.ca/dhp-mps/prodpharma/applic-demande/pol/nonlin_pol-eng.php. Accessed December 1, 2008.
42. Guidance for Industry: Individual Product Bioequivalence Recommendations. U.S. Department of Health and Human Services Food and Drug Administration Center for Drug Evaluation and Research (CDER) May 2007. http://www.fda.gov/Cder/Guidance/bioequivalence/default.htm. Accessed December 1, 2008.
43. Health Canada, Therapeutics Product Directorate. Conduct and Analysis of Bioavailability and Bioequivalence Studies—Part A: Oral Dosage Formulations Used for Systemic Effects. http://www.hc-sc.gc.ca/dhp-mps/prodpharma/applic-demande/guide-ld/bio/bio-a-eng.php. Accessed December 1, 2008.
44. Health Canada, Therapeutics Product Directorate. Conduct and Analysis of Bioavailability and Bioequivalence Studies—Part B: Oral Modified Release Formulations. http://www.hc-sc.gc.ca/dhp-mps/prodpharma/applic-demande/guide-ld/bio/bio-b-eng.php. Accessed December 1, 2008.
45. Colucci P, Marier J-F, Ducharme MP. Population pharmacokinetic approaches for assessing bioequivalence. In: Kanfer I, Shargel L, eds. Generic Drug Product Development: Bioequivalence Issues. (Drugs and the Pharmaceutical Sciences, v.180). New York: Informa Healthcare USA, Inc, 2008; 143–183.
46. Japan NIHS, Ministry of Health Labor and Welfare, Division of Drugs, Guideline for Bioequivalence Studies of Generic Products, Q & A Document,November 2006.
47. Guidance for Industry Statistical Approaches to Establishing Bioequivalence. U.S. Department of Health and Human Services Food and Drug Administration Center for Drug Evaluation and Research, January 2001. http://www.fda.gov/CDER/GUIDANCE/3616fnl.htm. Accessed December 1, 2008.
48. Guidance for Industry Immediate Release Solid Oral Dosage Forms Scale-Up and Postapproval Changes: Chemistry Manufacturing, and Controls, *In Vitro* Dissolution Testing, and *In Vivo* Bioequivalence Documentation, (SUPAC-IR) U.S. Department of Health and Human Services Food and Drug Administration Center for Drug Evaluation and Research, November 1995. http://www.fda.gov/CDER/GUIDANCE/#chemistry. Accessed December 2, 2008.
49. Guidance for Industry SUPAC-MR: Modified Release Solid Oral Dosage Forms Scale-Up and Post approval Changes: Chemistry, Manufacturing, and Controls; In Vitro Dissolution Testing and In Vivo Bioequivalence Documentation. U.S. Department of Health and Human Services Food and Drug Administration Center for Drug Evaluation and Research, September 1997. http://www.fda.gov/CDER/GUIDANCE/#chemistry. Accessed December 2, 2008.
50. Potassium Chloride (Slow-Release Tablets and Capsules) In Vivo Bioequivalence and In Vitro Dissolution Testing. U.S. Department of Health and Human Services Food and Drug Administration Center for Drug Evaluation and Research, June 1994 http://www.fda.gov/CDER/GUIDANCE/#Biopharmaceutics. Accessed December 2, 2008.
51. Shimazawa R, Nagai N, Toyoshima S, et al. Present state of new chiral drug development and review in Japan.J Health Sci 2008; 54(1):23–29.
52. Shah VP, Midha KK, Dighe S, et al. Analytical methods validation: Bioavailability, bioequivalence and pharmacokinetic studies. Conference report. Eur J Drug Metab Pharmacokinet 1991; 16(4):249–255 [also in Pharm Res 1992; 9:588–592].
53. Shah VP, Midha KK, Findlay JW, et al. Workshop report bioanalytical method validation—A revisit with a decade of progress. Pharm Res 2000; 17(12):1551–1557.
54. Viswanathan CT, Bansal S, Booth B, et al. Quantitative bioanalytical methods validation and implementation: Best practices for chromatographic and ligand binding assays. Pharm Res 2007; 24(10):1962–1973.
55. Guidance for the Industry. Bioanalytical Method Validation. U.S. Department of Health and Human Services Food and Drug Administration Center for Drug Evaluation and Research (CDER) and Center for Veterinary Medicine (CVM), May 2001. http://www.fda.gov/CDER/GUIDANCE/4252fnl.htm. Accessed December 2, 2008.
56. Rocci ML Jr, Devanarayan V, Haughey DB, et al. Confirmatory reanalysis of incurred bioanalytical samples. AAPS J 2007; 9(3):E336–E343.

57. Subramaniam S. Progress in incurred sample reproducibility: Agency perspective. Round Table Presentation at AAPS Annual Meeting, Atlanta, Georgia, November 2008 (FDA Office of Compliance).
58. Davit BM, Conner DP, Fabian-Fritsch B, et al. Highly variable drugs: Observations from bioequivalence data submitted to the FDA for new generic drug applications. AAPS J 2008; 10(1):148–156.
59. Haidar SH, Davit B, Chen ML, et al. Bioequivalence approaches for highly variable drugs and drug products. Pharm Res 2008; 25(1):237–241.
60. Health Canada, Therapeutics Product Directorate. Guidance for Industry Bioequivalence Requirements: Critical Dose Drugs. http://www.hc-sc.gc.ca/dhp-mps/prodpharma/applic-demande/guide-ld/bio/critical_dose_critique-eng.php. Accessed December 2, 2008.
61. Health Canada, Therapeutics Product Directorate Notice to industry: Bioequivalence Requirements for Drugs for Which an Early Time of Onset or Rapid Rate of Absorption Is Important (rapid onset drugs) http://www.hc-sc.gc.ca/dhp-mps/prodpharma/applic-demande/guide-ld/bio/notice_rapidonset_avis_apparitionrapide-eng.php. Accessed December 2, 2008.
62. European Medicines Agency Committee for Medicinal Products for Human Use (CHMP) Draft Guideline on Fixed combination medicinal products, January 2008. http://www.emea.europa.eu/pdfs/human/ewp/024095en.pdf. Accessed December 2, 2008.
63. Guidance for Industry Food-Effect Bioavailability and Fed Bioequivalence Studies. U.S. Department of Health and Human Services Food and Drug Administration Center for Drug Evaluation and Research (CDER), December 2002. http://www.fda.gov/CDER/GUIDANCE/5194fnl.htm. Accessed December 2, 2008.
64. European Medicines Agency Committee for Medicinal Products for Human Use (CHMP Note for Guidance on modified release oral and Transdermal Dosage Forms: Section II (Pharmacokinetic and Clinical Evaluation) January 2000. http://www.emea.europa.eu/htms/human/humanguidelines/efficacy.htm. Accessed, December 4, 2008.
65. United States Pharmacopeia, 32, Definition of Modified, Extended and Delayed-Release Products. Rockville, MD: United States Pharmacopeia.
66. European Medicines Agency Committee for Medicinal Products for Human Use (CHMP) Note for Guidance on Quality of Modified Release A: Oral and B: Transdermal Dosage Forms: Section I (Quality) January 2000. http://www.emea.europa.eu/htms/human/humanguidelines/quality.htm. Accessed December 4, 2008.
67. Corticosteroids, Dermatologic (topical) In Vivo Topical Dermatologic Corticosteroids, *In Vivo* Bioequivalence. U.S. Department of Health and Human Services Food and Drug Administration Center for Drug Evaluation and Research (CDER), June 1995. http://www.fda.gov/cder/guidance/index.htm#Biopharmaceutics. Accessed December 2, 2008.
68. Stewart BA, Ahrens RC, Carrier S, et al. Demonstration of in vivo bioequivalence of a generic albuterol metered-dose inhaler to Ventolin. Chest 2000; 117(3):714–721.
69. Health Canada, Therapeutics Product Directorate. Guidance for Industry Guidance to Establish Equivalence or Relative Potency of Safety and Efficacy of a Second Entry Short-Acting $Beta_2$-Agonist Metered Dose Inhaler (MDI). http://www.hc-sc.gc.ca/dhp-mps/prodpharma/applic-demande/guide-ld/inhal-aerosol/mdi_bad-eng.php. Accessed December 2, 2008.
70. Health Canada, Therapeutics Product Directorate. Guidance for Industry Draft Guidance. Submission Requirements for Subsequent Market Entry Inhaled Corticosteroid Products for Use in the Treatment of Asthma. http://www.hc-sc.gc.ca/dhp-mps/prodpharma/applic-demande/guide-ld/inhal_corticost-eng.php. Accessed December 2, 2008.
71. Draft Guidance, Bioavailability and Bioequivalence Studies for Nasal Aerosols and Nasal Sprays for Local Action. U.S. Department of Health and Human Services Food and Drug Administration Center for Drug Evaluation and Research (CDER), April 2003. http://www.fda.gov/CDER/GUIDANCE/#Biopharmaceutics. Accessed December 3, 2003.
72. Hendy C. Bioequivalence using clinical endpoint studies. In: Kanfer I, Shargel L, eds. Generic drug product development: Bioequivalence issues. (Drugs and the Pharmaceutical Sciences, v.180), New York: Informa Healthcare USA, Inc, 2008; 71–96.
73. Guidance for Industry. Dissolution Testing of Immediate Release Solid Oral Dosage Forms U.S. Department of Health and Human Services Food and Drug Administration Center for Drug Evaluation and Research (CDER), August, 1997. http://www.fda.gov/CDER/GUIDANCE/#Biopharmaceutics. Accessed December 3, 2008.
74. Guidance for Industry. Extended Release Oral Dosage Forms: Development, Evaluation, and Application of In Vitro/In Vivo Correlations. Department of Health and Human Services Food and Drug Administration Center for Drug Evaluation and Research (CDER), September 1997. http://www.fda.gov/CDER/GUIDANCE/#Biopharmaceutics. Accessed December 3, 2008.
75. Gupta E, Barends DM, Yamashita E, et al. Review of global regulations concerning biowaivers for immediate release solid oral dosage forms. Eur J Pharm Sci 2006; 29(3–4):315–324.

76. EMEA Inspectors Working Group Draft Reflection Paper on Advice to Applicants/Sponsors/CROs of Bioequivalence Studies. http://www.pdfdownload.org/pdf2html/pdf2html.php?url = http%3A%2F%2Fwww.emea.europa.eu%2FInspections%2Fdocs%2F46897507en. Accessed December 3, 2008.
77. Colocci P, Paternak-Di Marco M, Potvin D, et al. Bioequivalence assessment of endogenous drug substances. In: Kanfer I, Shargel L, eds. Generic Drug Product Development: Bioequivalence Issues. (Drugs and the Pharmaceutical Sciences, v.180). New York: Informa Healthcare USA, Inc, 2008; 233–256.
78. http://bebac.at/Guidelines.htm. Accessed December 16, 2008.

32 Pharmacokinetics in Man

Don J. Nichols and Don K. Walker
Pfizer Global Research and Development, Sandwich, Kent, U.K.

INTRODUCTION

In this chapter, we aim to provide an outline of the pharmacokinetics (PK) that would normally be part of new drug development for an oral formulation. We discuss the in vitro evaluation of the metabolism of drugs as a means of identifying potential routes of metabolism in vivo and predicting potential drug–drug interactions (DDIs). The integration of PK and pharmacodynamics (PD) as a means of understanding the dose-concentration–response relationship and the application of population PK and PD will be addressed.

PK data provide a scientific basis for interpreting clinical data on safety and efficacy by relating them to the dose, dosage regimen, and dose form. For example, a given dose of a drug does not produce the same effect in different individuals: part of this variability can be explained by differences in the absorption, distribution, metabolism, and excretion between individuals. Understanding these processes is the goal. Another significant source of variability between individuals is the different response observed for a given plasma concentration. This response may be a biomarker or surrogate PD endpoint or a direct outcome measure of safety or efficacy. For PK data to be useful in understanding the variability in dose response, there should be a predictable relationship between the plasma drug concentration and the effect. False conclusions can be obtained if the link between PK and PD is not established and decisions are based solely on the PK information.

The theme of this approach is therefore one of integration: it is inadequate to conduct a comprehensive series of PK studies defining the PK parameters if this information is not translated into the dosage regimen design in Phase 2 and 3 clinical studies and into understanding the variability in response. It is thus important to provide interpretation to the PK findings that are relevant in clinical practice.

PRECLINICAL SAFETY DATA

In order to provide assurance of human safety it is a requirement, in all but the most exceptional circumstances, for a pharmaceutical agent to be tested in animal species before and during the clinical phases of drug development. For general toxicity testing, the safety assessment will be performed in a rodent and nonrodent species. The selection of the nonrodent species has been the subject of considerable scientific debate over recent decades. While the most generally used species is the dog, alternatives include cynomolgus monkeys, marmosets, minipigs, and ferrets. The preference for using dogs in these studies is largely driven by experience, ethical considerations, and regulatory acceptance; however, reasons for selecting an alternate species include similarity to human metabolism, relevant physiology, or relevant pharmacological response.

TOXICOKINETICS

Toxicokinetic (TK) data are a vital component in the interpretation of nonclinical safety studies. International guidelines on the conduct of TK studies were issued in 1995 after several years of scientific discussion and debate (1). TK is defined as PK data generated either as an integral component of nonclinical toxicity studies or in specially designed supportive studies. The data provided may be used in the interpretation of toxicology findings and their relevance to human safety issues.

The primary objective of TK is to describe the systemic exposure achieved in animals and its relationship to dose level and the time course of the toxicity study. In addition, a number of secondary objectives have also been defined (i) to relate the exposure achieved in the toxicology studies to toxicological findings and to assess the relevance of these to human safety, (ii) to

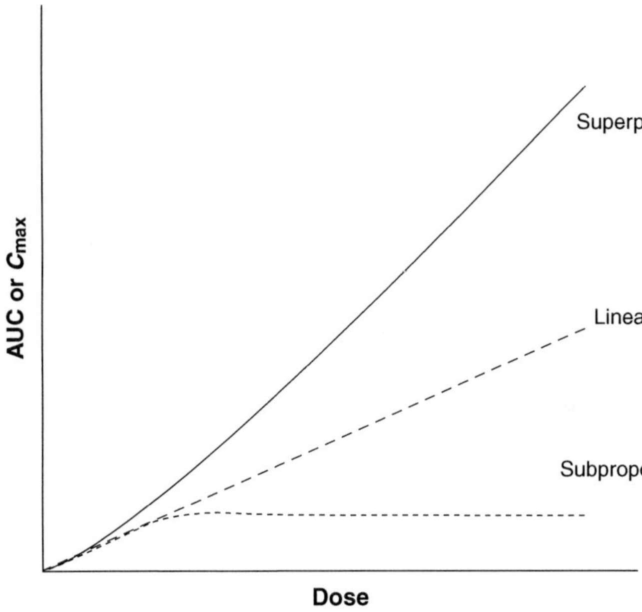

Figure 1 Possible relationships between administered dose level and systemic exposure that may be observed through toxicokinetics.

support the choice of species and dosage regimen used in the toxicology program, and (iii) to contribute to the design of the subsequent toxicology program.

It is vital that PK data are available in early toxicology studies in order to relate systemic exposure to the administered dose (2). TK measurements usually consist of plasma (or whole blood or serum) concentrations of parent compound and/or pharmacologically or toxicologically relevant metabolites with actual protocol details decided on a case-by-case basis. As the focus is primarily on the interpretation of toxicity results (and not on the characterization of basic PK parameters) the number of sampling points is generally small (typically four to five time points). Area under the plasma concentration–time curve (AUC) and maximum observed plasma concentration (C_{max}) are the most commonly used parameters in assessing TK studies. These parameters allow an assessment of systemic exposure at different doses over different study durations and in the different species that may be used. For some compounds, it may be more appropriate to calculate exposure based on the (plasma protein) unbound concentration. This is particularly relevant for highly protein-bound compounds that show variation in free fraction between species.

Examination of AUC and C_{max} allows two key questions to be answered:

1. What are the plasma concentrations of drug present in each species?
2. How do the concentrations relate to dose size?

Simple graphical examination of AUC or C_{max} values against dose allows consideration of the relationship between systemic exposure and dose size. The relationship between dose and systemic exposure can be linear, subproportional or superproportional with regard to AUC or C_{max} (Fig. 1).

The observation of linear increases in systemic exposure with increasing dose indicates that absorption and clearance processes remain constant over the administered dose range. Subproportional increases are most commonly an indication of saturation of the absorption process, generally as a result of limited compound solubility. This may often be foreseen in the drug development process through knowledge of the physicochemical characteristics of the compound. This situation requires caution in interpretation of data from higher dose levels particularly when setting safe dose limits for humans. In such situations, unless formulation can be used to achieve greater absorption, the high dose in further toxicology studies will be limited to the lowest dose producing the maximum exposure (when no other dose-limiting constraint applies). Other potential reasons for subproportional increases in exposure include autoinduction (on multiple administration) or saturation of plasma protein binding. Superproportional increases in exposure are generally due to saturation of clearance due to the limited capacity of drug metabolizing enzymes. In such circumstances rapid onset of PD or toxic effects may occur and the dose–response curve becomes exaggerated with narrow separation between desirable and undesirable effects. While TK provides value in understanding the results of the toxicology study limitations in any cross-species, interpretation need to be recognized. Factors contributing to interspecies differences in response have been reviewed (3) and highlight the confounding factors of species differences in metabolism and disposition, which complicate the relationship between parent drug exposure and effect. However, these differences may be relatively trivial compared to interspecies differences in PD response. Thus for a really meaningful cross-species comparison of desired and undesired effects and their relative exposures, it is necessary to develop some PK-PD understanding of the compound in the animal species and in man.

PLASMA PROTEIN BINDING AND FREE FRACTION

As previously stated, there are occasions when it is important to consider the free drug exposure when comparing systemic exposure of a compound across species. It is generally considered that it is the free fraction of drug in plasma which is available for interaction with receptors (4) and thus it is this that needs to be considered for comparisons of toxicity and efficacy across species. Plasma protein binding may be determined by a number of methods; however, the most popular and reliable remains equilibrium dialysis whereby plasma is dialyzed against isotonic physiological buffer and the free fraction equilibrates across the semipermeable membrane (5). For highly protein bound drugs, differences in free fraction can become very important in understanding variation in pharmacological response between species. This is illustrated by the experimental antimuscarinic agent zamifenacin. This compound was very highly protein bound in all species with values for plasma protein binding of 99.8% in rat and 99.98% in man, reflecting a 10-fold difference in free fraction from 0.002 to 0.0002. Only through consideration of this difference in free drug exposure is it possible to rationalize species differences in PK-PD (6).

PRECLINICAL METABOLISM

Prior to the first clinical studies with a new drug candidate, in vitro and in vivo metabolism studies are performed as part of the overall understanding of compound disposition and to identify potential issues that may arise during drug development. A number of objectives for these studies can be identified:

1. To identify potential active metabolites that may contribute to the pharmacological activity of the compound
2. To compare the metabolic fate of the compound in animals and humans to ascertain the suitability of toxicology species or identify potential unique human metabolites that may require further investigation
3. To identify the mechanisms and enzymes involved in the clearance of the compound in order to assess possible DDI liabilities through coadministered agents
4. To investigate the potential of the new compound as an inhibitor of drug metabolizing enzymes and thus liability to cause DDIs when used in combination with other drug therapies.

ACTIVE METABOLITES

Active metabolites may be identified through the conduct of preclinical in vitro and in vivo investigations, which demonstrate the presence of a metabolite with structural similarity to the parent which is then found to possess pharmacological activity. Alternatively these may be suspected as a result of well-designed PK-PD studies, which give rise to the following observations:

1. The drug shows conventional rapid equilibrium with the receptor in in vitro testing but demonstrates a duration of action that exceeds its PK profile and may show anticlockwise hysteresis in the PK-PD relationship
2. The drug shows greater potency than expected based on in vitro studies. The difference may be highlighted as particularly unusual by comparison with structural analogues of similar pharmacological activity
3. The drug demonstrates greater potency for a given plasma concentration after oral administration than would be expected based on simple comparison with the PD and plasma concentrations after intravenous administration. Such phenomena will be characterized by significant first-pass effects.

Whether active metabolites are identified through structural information or via PD observations, once their presence is realized then there will be a requirement to assess their contribution to the overall drug activity. This will necessitate synthesis of the authentic compound which can then be assessed for pharmacological activity in vitro and possibly also in vivo in preclinical species. The use of in vivo studies allows direct administration by the intravenous route thus facilitating pharmacological assessment in the absence of parent compound, something that would be difficult to achieve clinically.

PRECLINICAL METABOLITE IDENTIFICATION

Knowledge of the general metabolic fate of a new compound in animal species and in man can provide confidence that the animal species to be used in toxicology studies will underwrite clinical safety in terms of providing exposure to relevant metabolites. Should the situation arise where one or more animal species do not exhibit a particular metabolic pathway that appears to be important in human, a decision may be taken to use alternative species in toxicology testing. This can avoid the need for separate and specific safety assessment of metabolite(s) further down the drug development path (7). It should be recognized that metabolite profiles are qualitatively (if not quantitatively) similar in animals and man in the majority of cases. This can be attributed to the similarity of cytochrome P450 enzymes (CYPs) across the species that are responsible for many of the important primary clearance pathways of xenobiotics. Homologues of most of the isoforms of human CYPs are present in animals (8) and catalyze similar metabolic transformations. Examples of CYP reactions leading to the formation of active metabolites in man, which also occur in animals include CYP2D6-mediated alprenolol 4-hydroxylation, encainide O-demethylation and propafenone 5-hydroxlation and CYP3A4-mediated quinidine N-oxidation, diltiazem N-demethylation, and cyclosporin N-demethylation. All of these reactions are observed in man and dog.

Over the past 25 years, the focus of metabolite consideration has switched from excreted metabolites to those present in the systemic circulation. This is largely attributable to the advances in bioanalytical technology which enable the detection and quantitation of ever lower concentrations of drug and metabolites in biological matrices. There is also clearly a logical relationship between metabolites present in the circulation and the probability of their contribution to on- or off-target pharmacology. Early knowledge of metabolic pathways is most readily accessible using in vitro systems. In vitro systems, such as liver microsomes that contain the important CYP enzymes, are extensively used to provide rapid access to metabolic routes in animals and man prior to human exposure to new chemical entities. Such knowledge can inform the planning and interpretation of clinical studies. The application of these techniques is included in FDA guidance to the industry (9).

IDENTIFICATION OF CLEARANCE MECHANISM AND DDI LIABILITY

Understanding the mechanisms responsible for drug clearance permits an assessment of DDI liability that may be encountered in clinical use. Such interactions are generally metabolism based and involve the inhibition (or induction) of drug metabolizing enzymes by coadministered drugs. Such knowledge prior to a decision to progress to clinical development can be extremely useful in the overall assessment of the potential utility and acceptability of a new drug candidate. Obviously an estimate of the role of metabolic clearance in the overall clearance of the drug must be made to allow this assessment. This estimate can initially be based on physicochemical and/or animal PK data (2). The use of in vitro systems, including human liver microsomes and recombinant CYP enzymes, permits characterization of the enzymes responsible for compound metabolism and therefore an indication of potential impact by coadministered agents that are known to be inhibitors or inducers of these enzymes. The understanding of metabolism-based DDIs has progressed enormously over the past 20 years, following on from examples such as the potentially fatal interaction between ketoconazole and terfenadine. In vitro studies have proved their ability to predict such liabilities and thus have recognized value with regulatory agencies (10). The finding from in vitro studies that particular CYP enzymes are not involved in the clearance of a compound can provide good rationale for not requiring specific DDI studies with inhibitors of these enzymes during the early phase of clinical development (11). Sophisticated modeling and simulation software is available that allows the prediction of in vivo drug clearance based on in vitro data as well as the impact of coadministration with CYP inhibitors (12) and can thus inform early clinical development strategy.

In addition to concerns over DDI risks with a new molecule as a potential victim of enzyme inhibition are the concerns over the potential of a new molecule to inhibit the metabolism of other drugs. Relatively straightforward assays are available using competent in vitro metabolizing systems and probe substrates of the various CYP enzymes that will allow such an assessment to be made (13). Assessment of a new compound to induce drug metabolizing enzymes is also possible using in vitro systems and may inform the need for subsequent in vivo investigation (14). Marked species differences occur in the induction of drug metabolizing enzymes, particularly in the CYP3A family, and the presence or absence of induction in preclinical species during toxicology studies should not necessarily be seen as indicative of enzyme induction in man (15).

ROLE OF TRANSPORT PROTEINS

Over the last decade, there has been a significant increase in the characterization and understanding of the role of transport proteins in drug disposition. Transport proteins are now recognized to play a role in the absorption, distribution, and elimination of many drug molecules by facilitating their transport across tissue membranes (16). The best characterization of the transport proteins is P-glycoprotein (P-gp), which is present in various tissues and may prevent absorption from the gastrointestinal tract, facilitate biliary and renal excretion, and prevent distribution into the CNS. In vitro systems, such as Caco-2 monolayers, have been extensively used to examine the absorption potential of drug candidates. These have the ability to identify compounds that may be expected to show P-gp limited absorption in man. An example where nonproportional PK in man have been ascribed to P-gp limited absorption is the CCR5 antagonist, maraviroc. Preclinical studies including Caco-2 monolayers demonstrated the potential role of P-gp in limiting compound absorption, which was subsequently confirmed during dose escalation studies in man (17). While of less profound impact than CYP-mediated DDIs, transporter-mediated DDIs have been described and again the potential for these can be assessed using in vitro techniques (18).

BIOANALYTICAL METHODS

Development of suitable bioanalytical methods to determine accurately the concentrations of drug or metabolites in biological media provides the basis of all pharmacokinetic (and PK-PD) analysis. Guidelines have been issued for the validation of these methods (19) which were based upon the deliberations of workshops in the 1990s (20). It is necessary to quantitate the active components, which may include active metabolites, with precision, accuracy, and reliability at concentrations that are therapeutically relevant and over a sufficient concentration range to adequately describe the pharmacokinetic profile of the drug or metabolite (typically about five half-lives).

The requirement for development of metabolite assays depends on both the pharmacological activity of the metabolite and the systemic concentration that is achieved. While there is an expectation that all significant circulating drug metabolites will be characterized to some degree this does not require the quantitation to be performed to a high level of accuracy and reproducibility for all components. Developing fully validated bioanalytical methods for multiple analytes should not be undertaken lightly due to the considerable resource implications and technical challenge, especially in situations where the analytes may differ in physicochemical properties and require multiple internal standards (21). At one extreme, it is readily seen to be appropriate to develop a bioanalytical method to quantify the active principle generated from a prodrug, such that pharmacokinetic characteristics can be determined and applied to the rationalization of pharmacological activity. At the other end of the spectrum, it is not necessary to develop an assay for a metabolite that shows only 10% of the activity and occurs at only 10% of the concentration relative to the parent compound. The challenge is in determining whether a metabolite is contributing to the pharmacological activity in man without embarking on in vivo pharmacokinetic studies with the metabolite, although clues are given as outlined above (see "Active Metabolites" section).

An ever present issue for bioanalytical methods is the stability of the analyte from the point of collection until quantitation. Many drugs degrade in plasma even under normal frozen storage conditions (typically −20°C) or colder. Analyte stability is therefore a recognized component of bioanalytical method validation and may necessitate addition of stabilizers or extreme sample handling procedures to avoid degradation. Equally challenging is the potential for metabolites to degrade to parent compound and thus provide erroneous data for quantities of parent substance in samples. This is a recognized problem with some glucuronic acid conjugated metabolites and has also been observed with lactone metabolites (22). Provided such issues have been identified, the problem can often be overcome either by acidification of samples at the time of collection. Obviously, if the issue is not appreciated, there is the potential to generate erroneous and variable data. However, following discussion between industry and regulatory agencies guidelines have recently been issued on the reanalysis of incurred study samples that will identify and mitigate such a problem (23).

PHARMACOKINETIC PARAMETERS

For most Phase 1 PK studies, noncompartmental analysis is the typical way to conduct pharmacokinetic analysis of the drug concentrations. The primary parameters are the maximum concentration (C_{max}), the time of the maximum concentration (T_{max}), the area under the concentration–time curve from zero time to the time of the last plasma concentration (AUC_t) or zero time to infinity (AUC) or over the dosing interval at steady state (AUC_τ), and the terminal phase rate constant (k_z) [and hence terminal half-life, ($t_{1/2z}$)]. Details of the calculation of these parameters can be found in pharmacokinetics textbooks (24).

Two other parameters are extremely useful in understanding the physiological processes involved in drug distribution and elimination, particularly if calculated following intravenous administration. The clearance (CL) of a drug describes its overall elimination from the body: it is defined as the volume of plasma (or blood) irreversibly cleared of drug per unit of time. This parameter can be related to hepatic blood flow rate for a drug cleared by the liver or renal blood flow rate for a drug cleared by the kidney. For renally cleared drugs, other flow terms such as glomerular filtration rate are also available for determining extent of tubular secretion. The volume of distribution at steady state (V_{ss}) and other volume terms describe the distribution of drug in the body: a high volume of distribution indicates an uptake into a tissue or tissues, and a low volume of distribution suggests limited tissue distribution. Similarly, this term can be set against physiological volumes such as blood volume, total body water, etc., in defining the terms high and low. Renal clearance (CL_r) can be calculated from urine and plasma drug concentration and provides a means of estimating metabolic or non-renal clearance, CL_{nr}: that is, $CL = CL_r + CL_{nr}$. The time constant used most widely is terminal half-life. Because it is a hybrid parameter of CL and V_{ss}, it reflects both distribution and elimination. For dosage regimen design terminal half-life can be misleading. A long terminal half-life may reflect accumulation in tissues which may or may not be of consequence for efficacy. Peak-to-trough ratio at steady state can aid dosage regimen design in these cases.

The extent of bioavailability (F) of an oral formulation is determined by evaluating the AUC, AUC_t, and/or the AUC_t in comparison with that from an oral solution, a reference formulation, or an intravenous administration. The rate of absorption is normally evaluated by C_{max} and T_{max} pairs. Thus, if the drug is absorbed faster, C_{max} will be higher and T_{max} shorter. Using the Wagner-Nelson, Loo-Reigelman, or other deconvolution methods can be helpful in elucidating the input rate, particularly for controlled release formulations.

Population model based analyses are becoming commonly used throughout the development process to integrate all the PK and PD data from Phase 1 through Phase 2 and 3. These typically use structural or compartmental models, for example, two compartmental model for PK and E_{max} model for PD. There are broadly two approaches to modeling PK-PD relationships; one utilizes empirical models such as linear or E_{max} models. These models are flexible and can be applied to many PK-PD data but typically provide no mechanistic understanding and hence extrapolation outside of the data domain is risky. An alternative approach is to utilize more mechanistic and pharmacologically realistic models, such as the indirect effect models (25). Indirect response models are more mechanistic, can be reused, can sometimes provide insights on likely model behavior beyond what is possible with empirical models and are increasingly applied in drug development programs.

PHASE 1 STUDIES

Phase 1 studies are conducted throughout drug development and are designed to investigate the PK and PD of the drug. The PK objectives of the Phase 1 studies have been identified (26):

1. To characterize the PK of the drug after a single and multiple doses and as a function of dose size including time to steady state
2. To characterize the pharmacological effects and their relationship with dose and plasma concentration
3. To assess the suitability of the animal models used in toxicology studies with respect to the exposure to the drug and its metabolites
4. To evaluate the bioavailability of dosage forms and examine the effects of food and clinical variables on the rate and extent of absorption
5. To identify special populations or clinical conditions that result in altered PK and PD requiring dose adjustment during clinical use.

The objectives of some early clinical studies in man are to allow progression of the compound into further development particularly in patients. Such studies may or may not be pivotal for final regulatory submission. Usually, if there are good quality PK and PD data from these studies, they will contribute to the overall database for the compound.

The subjects in most of the Phase 1 PK studies are normally healthy males between the age 18 and 55 years who are within normal body weight ranges and have normal blood hematology and biochemistry. Females may be included if the reproductive toxicology information is available. In the case of genetic polymorphism in clearance, it is wise to take this into consideration in selecting subjects because the PK may be different in the two (or more) groups. In some situations, for example where the drug is toxic, patients may have to be used.

SINGLE DOSE TOLERATION, SAFETY, AND PK STUDY

This is the first dose study in man and, while safety and toleration are the primary concerns, understanding the exposure is key during escalation to relate exposures in preclinical studies to human exposure. This study is sometimes the only study that explores the full range of doses up to the maximum tolerated dose. The starting dose should be based on the minimal anticipated biological effect level (MABEL) rather than a fraction of the no adverse effect level (NOAEL). Typical practice is to base the starting dose, the dose expected to provide efficacy and the maximum dose on the unbound exposure in the animal studies and the prediction (and emerging data) of the PK in man, and particularly AUC and C_{max}. Thus, the MABEL dose would be the dose that is predicted to give unbound human exposures that are associated with minimal pharmacological activity in animals. Similarly, the maximum dose in man would be limited by the unbound exposure (AUC and C_{max}) observed in the most sensitive species used for animal safety studies at the NOAEL (or a fraction or multiple of this depending on the nature of the safety findings).

The usual course is to escalate the dose at reasonable increments (two- to threefold) to some maximal dose based on toleration or other considerations. Ideally, the dose should be increased until side effects or toleration are deemed prohibitive. Dose increases should be dependent on the safety, toleration, and PK from previous doses. It is not uncommon for the PK to show nonproportionality with dose: subproportional increases due to solubility or dissolution limited absorption or superproportional increases due to saturation of first pass (metabolism and/or transporters).

The study could be conducted with a solution, suspension, or capsule if the drug is administered orally. Solution administration is preferred, however, in that it overcomes potential problems with digestion and dissolution of capsules and the dissolution from suspensions. The PK information available may be limited at the lower, nontherapeutic doses because of limitation in the assay. It is possible to modify the blood sampling schedule as the study progresses so that information gathering will be optimized. The importance of this PK study is to gather data that will impact on the design of the subsequent multiple dose toleration and PK studies. It will also provide the PK over a range of doses that may never again be explored formally.

It is not uncommon to conduct a leg in this first study where the dose is given in the fed state. Where it is predicted that food may affect the absorption of drug (27), it is desirable to understand the magnitude of this effect as early as possible. With the increasing proportion of small molecule drugs having poor solubility/dissolution, the effect of a high fat meal in increasing absorption may be critical to achieving adequate exposure. The alternative approach in this situation is to use a formulation optimized for dissolution and a leg of this study can evaluate the effect of these formulations.

SINGLE AND MULTIPLE DOSE TOLERATION, SAFETY, AND PK

The design of this study will be dependent on the results from the first single-dose study. The length of the study will be typically 10 to 14 days. Ideally, the study should be conducted with the preparation that will be used in the Phase 2 studies and this is typically a tablet or capsule. Ideally, there should be a full PK profile after the first dose, which will include 10 to 15 samples over a period of at least three half-lives. In reality, this may require a washout of several days and for practicality reasons a minimal sampling over the dosing interval will allow estimation of the accumulation ratio (but not the linearity ratio). Dosing will resume using the desired regimen as dictated by the PK from the first study for 10 to 14 days after the last plasma sample from the first dose profile. Further profiles over the dosing interval could be taken during the multiple dosing regimen to establish when steady state is achieved. Alternatively, predose samples should be taken throughout the dosing to establish when steady state is achieved. After the last dose, samples for a full PK profile should be collected over five half-lives. The doses used in this study should escalate to the maximum tolerated dose defined in the first study and should include at least two other lower doses including the perceived therapeutic range. This type of study may provide a means of assessing enzyme-inducing potential. This can be accomplished by measuring cortisol/6-OH-cortisol ratio in urine predrug and at steady state in both treated and placebo subjects. If metabolites have been identified as active all efforts should be made to quantitate them in the plasma. Urine could be collected after the first and last dose and renal clearance (of drug and metabolites, if available) could be determined if appropriate.

DOSE PROPORTIONALITY: LINEARITY AND NONLINEARITY WITH DOSE

It is necessary to demonstrate that the PK are linear with dose and time. Linearity implies that as the dose is increased, AUC and C_{max} increase in direct proportion. If the increase in AUC or C_{max} is disproportionate, then the pharmacokinetics are nonlinear; with regard to dose, this nonlinearity may be a disproportionate increase or decrease. The first study in man should be designed to provide at least preliminary information on linearity with dose. If the dose escalation and single/multiple dose studies have not provided definitive evidence of dose proportionality it will be necessary to conduct a formal study. If the earlier studies have provided indicators of linearity, the dose proportionality study may be deferred until Phase 3 and may not be needed if there is sufficient confidence in prior studies. In a dose proportionality study, the dose range used should encompass the anticipated therapeutic range.

BIOAVAILABILITY AND BIOEQUIVALENCE

Details of bioavailability and bioequivalence are outlined in the chapter on Bioavailability and Bioequivalence Studies. For the development of a new chemical entity the formulation used in the early clinical studies is not usually the final formulation that would be marketed. Ideally, the formulations used in the early studies can be linked to those used in pivotal Phase 3 trials by comparative bioavailability studies and the final formulation developed for marketing is used in Phase 3. Sometimes this is not possible and a formal bioequivalence study of the final marketed formulation and the formulation used in Phase 3 is required. The advantage of being able to link the bioavailability of early formulations to the later ones is that the PK data generated early in development can be easily linked to the doses used later.

The design of bioavailability and bioequivalence studies is relatively standardized and is aimed at separating the treatment (i.e., formulation effect) from other effects.

The development of an intravenous (IV) formulation is also important for the determination of IV pharmacokinetics and absolute bioavailability. Characterizing absolute bioavailability is generally considered necessary to obtain registration in all territories.

FOOD EFFECTS ON BIOAVAILABILITY

It is now possible to predict when a food effect study may not be necessary (27) such that this study can be delayed until after proof of concept. This study should be done using the solid form to be used for Phase 2/3. If a food effect is predicted, its outcome could impact on Phase 2/3 study design and it should be completed prior to start of clinical studies. Using a two-way crossover design the dosage form should be administered after an overnight fast and after high fat breakfast. The dose should be the maximum tolerated from the first study in man unless a large increase in exposure is anticipated. If a significant food effect is observed, it is necessary to investigate it further and initially the study should be repeated with a solution to determine whether the effect is intrinsic to the drug or a formulation effect.

ALTERNATIVE ORAL FORMULATIONS

Alternative oral formulations to an immediate release oral tablet or capsule are normally developed to alter the shape of the plasma concentration–time profile. The most common is a prolonged (sustained or controlled) release formulation designed to flatten the plasma concentration–time profile. Thus for a short half-life drug the peak to trough ratio can be reduced, which can be beneficial if the therapeutic window is narrow, and the dosage interval extended with a controlled release formulation. Typically a controlled release formulation for a new chemical entity would be developed in parallel with or after the immediate release formulation. The goal would be to produce a formulation with equivalent AUC values for the two forms but with a reduced degree of fluctuation in plasma concentrations. The controlled release formulation would not necessarily have to have the same amount of drug as several doses of an immediate release formulation administered over the same dosing interval. This could occur if the controlled release formulation had a higher first-pass metabolism than the immediate release formulation. For a new chemical entity, the controlled release formulation should be studied to determine single- and multiple-dose pharmacokinetics (and particularly the rate, pattern, and extent of absorption) and the fluctuation of peak and trough concentrations in comparison with an immediate release formulation or a solution. The effect of food needs to be characterized and, if the immediate release formulation shows nonlinearity with dose or time or is poorly characterized, the dose dependency needs to be determined (28). Determining regional absorption, particularly colonic absorption, can be useful in understanding whether appropriate bioavailability of a controlled release formulation is due to the formulation or due to the inherent nature of the drug. If bioavailability is equivalent to an immediate release formulation the implication is that there is good colonic absorption and a study would only confirm this observation. An in vitro dissolution/in vivo bioavailability relationship needs to be explored as this will facilitate the understanding of critical formulation characteristics that are likely to have an effect in vivo without the need for testing for all these variable in vivo. This correlation could be of the in vitro dissolution curve with the in vivo absorption curve (which is generated by deconvolution of the plasma concentration data). The relationship could be validated by preparing batches of product with different release rates and determining the absorption characteristics in vivo.

ALTERNATIVE ROUTES OF ADMINISTRATION

There are several reasons for developing drugs that are not administered orally. Some drugs are administered nonorally for biopharmaceutical reasons. Others are dosed closer to the site of action and do not require systemic absorption and distribution for pharmacological activity. The most common alternative routes of delivery are intravenous, intramuscular, subcutaneous, topical (transdermal), ocular, rectal, sublingual, intranasal, and inhalation. Others include vaginal and intra-articular. Drugs are given intramuscularly (e.g., spectinomycin), subcutaneously (e.g., insulin, sumatripan), and rectally because either there is poor and/or erratic absorption from the gastrointestinal tract or extensive first-pass metabolism. Some drugs given by these routes also provide a sustained delivery (e.g., medroxyprogesterone acetate). Transdermal delivery of drug using patches offers bypass of first-pass metabolism or poor gastrointestinal absorption, the potential for sustained delivery thus reducing dosing frequency (e.g., hormone replacement therapy) and flattening the plasma concentration–time profile which can reduce the incidence of side effects (e.g., scopolamine). Ocular (e.g., pilocarpine), intra-articular (e.g., methylprednisolone acetate), topical (e.g., minoxidil, betamethasone), and inhalation products (e.g., budesonide) are given in an attempt to target the site of activity thereby avoiding systemic exposure and consequent unwanted side effects.

The PK and bioavailability of drug given by alternative routes of delivery should be evaluated where possible. Where systemic delivery is not intended, the extent of systemic absorption should be evaluated. There are presently no agreed guidelines for the evaluation of the bioavailability of compounds at the site of action where the drug is targeted to these sites, and clinical or PD endpoints would likely be appropriate in these situations.

METABOLISM

The primary purpose for elucidating the metabolic fate of a drug substance in man is to provide assurance that human safety is supported by toxicological testing. In general, safety considerations throughout a drug development program include comparison of systemic exposures of the drug in the toxicology species with systemic exposure in humans to assess the potential risk and to guide clinical safety monitoring. This paradigm is appropriate when the metabolic profile is similar in the animal species and in man; hence comparison of parent drug exposure is also representative of various metabolites. An issue arises when there is a lack of concordance between human metabolites and those in the toxicology species such that humans may be exposed to a clinically relevant metabolite, which has not been present in nonclinical safety studies. Such situations, as previously discussed, are relatively rare given the broad similarity in biotransformation reactions across the species, coupled with the generally higher doses encountered in the toxicology testing which provide cover for quantitative differences in proportions of individual metabolites. Clearly of prime importance is that any metabolites that cause a significant proportion of the dynamic effects of the drug in man are present in the safety species. An example illustrating this is the β-adrenoceptor antagonist acebutolol, which forms an active metabolite, diacetolol, by acetylation (29). This metabolite is present at higher concentrations than the parent in man and possesses a longer elimination half-life. The metabolite is not formed in the dog to any significant degree and hence the dog is not an ideal species for safety assessment of this compound.

Metabolism studies in toxicology species and man continue to be deemed necessary to underwrite the safety assessment program. The actual requirements of these studies and what constitutes a significant metabolite or a significant difference in metabolites between species has been the subject of debate over several decades. This debate was brought to a head through a cross-industry group discussion and subsequent publication (30) that looked to set guidelines for what constituted a major metabolite and what safety assessment was then required. The ensuing scientific debate on this topic encompassed the relative value of considering metabolite risk in terms of absolute abundance rather than traditional percentage and the specific safety implications of metabolite location, be that within the circulation or excreta (31). The FDA issued a guidance document in 2008 (32) which suggests that all human metabolites present at more than 10% systemic exposure relative to parent compound at steady state are subjected to adequate safety assessment. This safety assessment may be achieved by exposure in at least one of the toxicology species that is similar to or greater than human exposure (in absolute amount terms) or failing this it may require separate administration of the metabolite within the safety program.

Aside from the interspecies comparisons of drug metabolism, perhaps of more significant benefit is the application of the detailed understanding of drug disposition and metabolism gained in animals and man to assess likely consequences in various clinical situations. For example, the preclinical studies with tolterodine indicated the formation of the active metabolite, 5-hydroxymethyl tolterodine, which contributed to the pharmacological profile. Clinical studies demonstrated that this pathway was mediated by the polymorphic enzyme CYP2D6 in man resulting in marked variability in exposure between poor metabolizers (PMs) and extensive metabolizers (EMs). Despite this difference the PD effects of tolterodine are similar in the two groups due to the combined contribution of parent compound and metabolite, with parent compound contributing most of the pharmacological effect in PMs and metabolite most of the effect in EMs (33). Detailed understanding of more complex drug disposition may only be achieved through access to preclinical studies. For example, the identification of enterohepatic circulation, whereby a compound undergoes glucuronide conjugation, biliary excretion into the gastrointestinal tract, hydrolysis to parent compound and reabsorption, can only be readily determined in animal studies. Knowing that this process can take place permits recognition of the potential consequence of reduced renal function in man, whereby decreased elimination of glucuronide in urine has the potential to increase systemic parent drug levels due to an increased proportion of the glucuronide undergoing enterohepatic circulation. Hence, while parent compound may not be eliminated by renal clearance per se, reduced renal function will impact exposure and potentially cause drug accumulation. An example where this has been implicated is provided by oxazepam (34).

RADIOLABEL BIOTRANSFORMATION STUDY

This may also be referred to as a human mass balance study, reflecting that one of the purposes of such studies is to account for all of the administered radioactive dose within the excreta. The other principal purpose is to delineate the metabolic pathways and clearance mechanisms which, for the reasons outlined above, make this study an important component of nearly all small molecule drug development programs. While the human biotransformation study is regarded as the definitive study in terms of allowing categorical identification of the metabolic products, modern analytical technology allows some preliminary information to be obtained without radiolabeled material both from in vitro and in vivo studies. For this reason, conduct of the study may be delayed within overall drug development but the information should be available prior to the start of Phase 3 trials.

Prerequisites for the human biotransformation studies are a quantitative whole-body autoradiography study in rodents which permits the calculation of the acceptable human radioactive dose and preclinical mass balance studies to confirm appropriate positioning of the radioisotope within the molecule. The human studies themselves are normally conducted in a small number of volunteers (six or less) and for most compounds these will be healthy male subjects. The study may be conducted in females of non–child-bearing potential when the target population is predominantly female. If a compound is known to be metabolized by a polymorphic enzyme then parallel groups may be utilized to investigate genotype differences. Generally, studies are open-label single-radiolabeled dose study design. Multiple radiolabeled dosing may be considered if auto-induction is known to occur, although this may compromise the assessment of mass balance or quantitative metabolite profiling. Studies are normally performed at a specialist contract research organization facility equipped for dealing with radioactivity and holding an appropriate license. The compound is usually administered by the intended clinical route and a solution formulation will often be used for ease of dealing with radiolabeled dose preparation. Carbon-14 (^{14}C) is the most commonly used radioisotope for radiolabeled studies owing to its chemical stability, long decay half-life (5730 years), and the ubiquity of carbon in drug molecules. Tritium (^{3}H) may also be used but often suffers from tritium exchange, whereby radiolabel is transferred to the water pool, compromising the mass balance element of the study (35). Other radioisotopes that may be used if the element is present within a molecule include ^{32}P, ^{35}S, ^{36}Cl, and ^{131}I but these suffer from short or excessively long decay half-lives, which complicate their application. In situations where administration of radioactivity is deemed inappropriate, the use of stable isotopes may be considered (36).

The design of the study should allow for calculation of the total radiolabeled dose recovered in urine and feces and the rate at which the radiolabel appears in the excreta. In addition, the concentration of radioactivity and parent drug in the circulation should be analyzed to compare the relative abundance of metabolites to parent compound. A comparison of the rate of disappearance of radioactivity from the circulation and appearance in excreta provides information on the importance of circulating materials in terms of their proportion of the metabolic products from the administered dose (2). Metabolites often have lower volumes of distribution than the parent drug, hence may represent large plasma concentrations but a relatively small proportion of the administered dose. The plasma and excreta from the study are then available for detailed characterization of the metabolic products using appropriately pooled samples from each matrix. In exceptional cases, a radiolabel study may not be required in man for compounds which are well absorbed (or intravenously administered) and completely cleared (>90%) as unchanged drug in the urine. This obviates the need for a radiolabeled mass balance study as adequate information is available from the unlabeled studies.

PHARMACOKINETICS IN SPECIAL POPULATIONS AND SITUATIONS

In some cases, it will be necessary to assess the PK in special populations and situations prior to clinical studies in these populations: for example when recruitment could be compromised by the lack of such information. Examples of special populations situations may include diabetics, patients with heart failure or other forms of cardiovascular disease, patients with renal insufficiency, the elderly and children, and when the test product is to be coadministration with other drugs. For almost all drugs at some point, it will be necessary to compare the PK in the elderly and the young and to investigate the effects of renal and hepatic insufficiency on drug PK. However, these studies can often be delayed until the end of Phase 2 when information on therapeutic efficacy and dose range are available.

DRUG INTERACTION STUDIES

In the last 20 years, there has been a shift in the driver for conducting drug interaction studies much of it outlined in regulatory guidelines (9,11). Today, most of the rationale for understanding PK drug interaction studies comes from an in vitro knowledge of metabolism and drug transporters and routes of clearance. A new drug candidate can be an inhibitor or inducer of drug metabolizing enzymes, particularly P450 enzymes or transporters such as P-glycoprotein (PGP), and the clinical concern would be their effects as "perpetrators" of PK effects on other drugs, particularly if these drugs have a narrow therapeutic window. Alternatively, new drug candidates can be "victims" of existing drugs which are inhibitors or inducers of enzymes or transporters.

The timing of these studies will be highly drug-dependent and will likely be driven by the in vitro metabolism data. If there are no strong clinical safety concerns, they may be deferred until the start of Phase 3. If the lack of interaction data could possibly interfere with efficient recruitment into the Phase 2 program, relevant studies should be considered. The specific type of study will be compound-dependent but will probably include, as a minimum, interaction studies with the enzyme responsible for the metabolism. For example, if the drug is metabolized predominantly by CYP3A4 then an interaction study with the potent CYP3A4 inhibitor, ketoconazole, will provide information on the magnitude of the increase in exposure. Similarly, if low concentrations of drug represent a risk, as with anti-infectives, a study with an inducer such as rifampicin would be indicated.

If the in vitro data indicate the possibility of the drug being a perpetrator of interactions, then studies with model substrates would be indicated. For example, if the drug was possibly a CYP3A4 inhibitor at therapeutic doses then a study to evaluate the effects on midazolam would be indicated. The timing of these studies will also be dependent on the knowledge of the clinical dose range of interest. If the drug candidate is a victim and has linear PK, the effect of an inhibitor can be evaluated at a dose below the clinical dose range to ensure exposures do not increase above a level that would be of clinical concern. If the drug candidate has a nonlinear PK due to first pass metabolism or transporter saturation, then the magnitude of the interaction with an inhibitor may also be dose dependent: at low doses the bioavailability is lower and an inhibitor will lead to a greater increase than at higher doses where the bioavailability is higher. Where the drug is an expected perpetrator the interaction will likely be dose-dependent so that the drug candidate needs to be tested at the highest clinical dose.

While the ability to predict PK interactions has improved over the last 20 years, there is sometimes a clinical concern with drugs that have narrow therapeutic index and where there is a high level of coadministration in the target population. Sometimes the need for data can be met by population analysis of patient data. In other circumstances Phase 1 studies are still conducted.

The ideal design for an interaction study is a crossover design. Alternative designs include a parallel group study with a placebo arm or a fixed sequence study. When examining the effect of a new drug on the pharmacokinetics of another drug, steady state should be attained with the new drug and maintained throughout the sampling period of the test compound. In most cases, the test compound can be administered as a single dose. If safety and toleration is also an endpoint, it may be desirable to conduct the study with both compounds at steady state.

Alternatively, when examining the effect of another agent on the new drug kinetics, the design should be reversed. The most likely example of this type of study will be a ketoconazole interaction study, in which case the ketoconazole should be administered throughout the sampling period for assessment of both absorption and disposition effects. A single dose for the new drug may be adequate if it has linear PK and if there is no interest in the safety and toleration of the combination.

PHARMACOKINETICS IN THE ELDERLY

All PK parameters may be altered in the elderly. The aim of studies is to characterize the PK (and the PK-PD relationship if useful) in the elderly subjects. A major problem with typical formal studies in the elderly (i.e., intense study of a limited number of healthy elderly) is that the individuals are chosen according to relatively stringent inclusion and exclusion criteria and do not necessarily represent the sort of elderly patients who would be likely to receive the drug (37). There are both practical and ethical difficulties in conducting studies in frail elderly volunteers. An alternative way to obtain information on the pharmacokinetics in the elderly is to use the population PK approach, see later). This involves taking a limited number of blood samples from both young and elderly patients during Phase 2 and 3 clinical trials. Large age-associated differences in PK would be detected; small differences are unlikely to be of medical importance. A formal study could normally be done during Phase 2, but its priority will be dependent on the expected patient population (e.g., dementia) and the therapeutic index. Young and elderly subjects (normal volunteers or patients) could be studied as two parallel groups in the same study with sufficient numbers to permit meaningful conclusions.

PHARMACOKINETICS IN RENAL INSUFFICIENCY

A study comparing the PK in renally impaired patients with a group of subjects with renal function similar to the proposed patient group should be conducted if the drug, or an active metabolite, has significant renal clearance. Because severe renal impairment may affect hepatic metabolism, a study is also needed for drugs with significant metabolism if they have a narrow therapeutic window (38). A study may not be necessary if the drug is metabolized but has a wide therapeutic window, is gaseous or volatile, or if intended only for single-dose usage. A study should be done after there is knowledge of circulating metabolites. If relevant (i.e., active) the metabolites should be assayed in plasma and urine. Subjects should be selected to allow examination of the full range of renal function. Also, the effect of dialysis on drug kinetics should be examined in patients routinely undergoing dialysis. Where possible serum protein binding measurements should be made in all subjects. A suitable measure of renal insufficiency, typically creatinine clearance (CLcr) evaluated by the Cockcroft and Gault method from serum creatinine measurements, should be evaluated as well as the drug PK to allow correlation of CLcr with drug clearance. For those drugs intended for indications in which renal insufficiency is prevalent a steady state study should be considered. If there is a clinically significant change in the PK due to renal impairment, this study should provide the guidelines for dosage adjustments. The effect of renal insufficiency can also be picked up in population PK analyses in Phase 2or 3. In general for drugs with a high renal clearance, there is a strong correlation with CLcr and, after accounting for this covariate, they show low between-subject variation in exposure.

PHARMACOKINETICS IN HEPATIC INSUFFICIENCY

Regulatory guidance (39) recommends a PK study in patients with impaired hepatic function if hepatic metabolism and/or excretion accounts for a substantial portion (>20% of the absorbed drug) of the elimination of a parent drug or active metabolite. The guidance also recommends a hepatic impairment study even if the drug and/or active metabolite is eliminated to a lesser extent (<20%), if it is a drug with a narrow therapeutic range. If the metabolism of the drug is unknown the FDA recommends that the drug be considered extensively metabolized.

For some drugs, hepatic functional impairment is not likely to alter PK sufficiently to require dosage adjustment. In such cases, a study to confirm the prediction is generally not important. These cases include the following: renally cleared drugs, drugs with <20% metabolism and a wide therapeutic window, gaseous or volatile drugs, or drugs intended for single-dose usage (40). Patients with hepatic failure or insufficiency and age and weight matched volunteers should be administered a single dose of the drug. Although there is a wide range of hepatic disease, this study is typically conducted in cirrhotic patients. There should be a full evaluation of hepatic function. Classification is using the Child Pugh scale. All three categories could be studied but normally only Child Pugh A and B category patients are studied. A reduced design can be used with only moderate hepatic impairment with an acceptance that the dosing instructions in mildly impaired will follow that for the moderate and that the severely impaired patients will be contraindicated in the product labeling. Plasma protein binding measurements should be made in all subjects if appropriate. This study is usually not required prior to Phase 3.

CHIRAL DRUGS

Many of the biological processes in the body are able to distinguish between two enantiomers so that the absorption, distribution, metabolism, elimination, and pharmacology are different (41). Giving a racemic compound (i.e., an equal mixture of the two enantiomers) in this situation can be equivalent to administering a mixture of two different compounds. Where racemic drugs are being developed the PK of the enantiomers need to be evaluated. Thus, a stereoselective assay will need to be developed if the PK of the two isomers is the same. Later development work could be conducted with either a chiral assay or an assay that monitors one of the enantiomers.

PEDIATRIC PHARMACOKINETICS

Most development program need to consider the use of drugs in children. The PK in children changes with age and is typically different from adults so examination in different age groups is necessary. Pediatric PK studies are typically conducted after completion of Phase 3 in adults. The population PK sampling is often used, particularly in the very young children, although specific studies in older pediatric patients with more extensive sampling is also possible. In general, population modeling approaches are used to compare the PK to adults.

POPULATION PHARMACOKINETICS AND PHARMACODYNAMICS

Most PK studies are conducted under highly controlled conditions in a population of healthy male and female subjects of restricted age and weight. The drug, when marketed, will be used in a much more heterogeneous population of patients who will possibly be taking concomitant therapy, may have diseases other than the one being treated and will vary widely in body weight, age, and physiological function. Collecting a full set of PK samples in a large number of patients will often be impractical. Population PK, which is now an established component of drug development programs, involves taking a limited number of blood samples for drug assay from patients in clinical studies. The samples are taken at different but not necessarily fixed time points throughout the dosing interval to provide information on the plasma concentrations and allows estimation of PK parameters for the population of interest. For this population PK approach to be successful, it is necessary to have a basic understanding of the PK of the drug. The PK parameters (usually apparent clearance and volume of distribution) for the population studied and the between-subject variability can be estimated and the influence of covariates examined by exploratory data analysis. These covariates could include demographic data (age and race), pathophysiological conditions (e.g., renal function as evaluated by CLcr), concomitant medication or disease condition.

The sparse data collected during these types of population PK studies typically require nonlinear mixed effects modeling methods of data analysis in order to maximize the information yield. The most widely used method for formal population analysis is nonlinear mixed effects modeling as implemented by the computer program NONMEM (42). With this approach a compartmental (exponential) model is fitted to the data with fixed effect parameters. However random effects are also modeled to allow estimation of the intersubject variability and residual variability (43). Covariates can be incorporated into the model to determine whether they have a significant effect on the overall model fit. Data quality needs to be controlled and this requires considerable planning and coordination in major global Phase 2/3 clinical trials. The impact of poor compliance is harder to control but also represents a significant problem. Because population analyses may involve extensive modeling and exploratory data analysis, it is important to document all the analyses conducted. To assist independent review, it is particularly important to document objectives of the analyses, the hypotheses tested, the models used, the assumptions of the analyses, the adequacy of models, and the goodness of fit. Ideally the final model should be validated with a new data set obtained from existing but independent data or from a prospective study (44). This modeling approach to PK-PD has been extended to outcome efficacy and safety in what has been termed model-based drug development (45). This approach attempts to integrate all available relevant data into informative models which can then be used to optimize dose selection and clinical study design and quantify decision making. The investigation of the population PK-PD should involve data generated throughout the development program. Thus data generated in volunteers can be used to define the basic PK and PK-PD models, Phase 2a will define the PK in the target patients group and potentially the exposure–response relationship for patient-specific biomarkers, outcome efficacy, and safety. Typically dose-response studies are conducted in Phase 2b and the models for PK, efficacy and safety can be refined (40). This allows dose (concentration) efficacy and safety simulations of potential Phase 3 dosing regimens, simulations of the effects against comparators and clinical trial simulations of outcomes for different Phase 3 study designs. In Phase 3, three or four PK samples are taken over the duration of therapy and population PK analysis can explore the effects of covariates including DDIs in a wider patient group and with larger numbers of patients.

INTEGRATION: PHARMACOKINETICS AND PHARMACODYNAMICS

A modeling- and simulation-based approach to clinical drug development allows the integration of all pertinent data and the conversion to knowledge (45). When integrated these models should provide important scientific insights to support the safety and efficacy profile of new medicines. During all stages of drug development taking a more quantitative model-based approach for making drug development decisions efficiently based on all the pertinent data available is advantageous. Although modeling and simulation have been used for PK-PD analyses for years, and is used in many other industries, the broader use in drug development for safety and efficacy evaluation is more recent and limited (e.g., it may not be integrated in the strategy and decision-making). Broader implementation of this more quantitative approach to drug development represents one of the opportunities to address the "critical path" problem in drug development.

CONCLUSIONS

We have presented an analysis of the PK information required for development of pharmaceutical agents which is in line with our understanding of current regulatory attitudes. Obviously each drug should be treated on an individual basis and guidelines must not be allowed to become tramlines. The regulatory package should provide a complete model for the disposition and PD action of a drug. To achieve such a goal, rigid adherence to a checklist of studies is a hindrance.

REFERENCES

1. EMEA. CPMP/ICH/384/95 ICH Topic S3A. Toxicokinetics: a Guidance for Assessing Systemic Exposure in Toxicity Studies. Issued by The European Medicines Agency, 1995. http://www.emea.europa.eu/pdfs/human/ich/038495en.pdf. Accessed June 2, 2009.
2. Smith DA, Humphrey MJ, Charuel C. Design of toxicokinetic studies. Xenobiotica 1990; 20:1187–1199.
3. Smith DA. Integration of animal pharmacokinetic and pharmacodynamic data in drug safety assessment. Eur J Drug Metab Pharmacokinet 1993; 18:31–39.
4. Trainor GL. The importance of plasma protein binding in drug discovery. Exp Opin Drug Discov 2006; 2:51–64.
5. Banker MJ, Clark TH, Williams JA. Development and validation of a 96-well equilibrium dialysis apparatus for measuring plasma protein binding. J Pharm Sci 2003; 92:967–974.
6. Beaumont KC, Causey AG, Coates PE, et al. Pharmacokinetics and metabolism of zamifenacin in mouse, rat, dog and man. Xenobiotica 1996; 26:459–471.
7. Davis-Bruno KL, Atrakchi A. A regulatory perspective on issues and approaches in characterizing human metabolites. Chem Res Toxicol 2006; 19:1561–1563.
8. Smith DA. Species differences in metabolism and pharmacokinetics: Are we close to an understanding? Drug Metab Rev 1991; 23:355–373.
9. FDA. FDA Guidance for Industry. Drug Metabolism/Drug Interaction Studies in the Drug Development Process: Studies www.fda.gov/downloads/Drugs/GuidanceComplianceRegulatoryInformation/Guidances/ucm072104.pdf. Accessed June 2, 2009.
10. Peck CC, Temple R, Collins JM. Understanding consequences of concurrent therapies. JAMA 1993; 269:1550–1552.
11. FDA. FDA Guidance for Industry. Drug Interaction Studies—Study Design, Data Analysis and Implications for Dosing and Labelling www.fda.gov/downloads/Drugs/ GuidanceComplianceRegulatoryInformation/Guidances/ucm072119.pdf. Accessed June 2, 2009.
12. Ozdemir M, Crewe K, Tucker G, et al. The impact of ketoconazole dosage regimen on midazolam clearance and its prediction using PBPK modelling. J Clin Pharmacol 2006; 46, 1065.
13. Hutzler JM, Messing DM, Wienkers LC. Predicting drug-drug interactions in drug discovery: where are we now and where are we going? Curr Opin Drug Discov Devel 2005; 8:51–58.
14. Sinz M, Wallace G, Sahi J. Current industrial practices in assessing CYP450 enzyme induction: preclinical and clinical. AAPS J 2008; 10:e-pub.
15. Martignoni M, Groothuis GMM, de Kanter R. Species differences between mouse, rat, dog, monkey and human CYP-mediated drug metabolism, inhibition and induction. Exp Opin Drug Metabol Toxicol 2006; 2:875–894.
16. Ayrton A, Morgan P. Role of transport proteins in drug absorption, distribution and elimination. Xenobiotica 2001; 31:469–497.
17. Walker DK, Abel S, Comby P, et al. Species differences in the disposition of the CCR5 antagonist, UK-427857, a new potential treatment for HIV. Drug Metab Dispos 2005; 33:587–595.
18. Keogh JP, Kunta JR. Development, validation and utility of an in vitro technique for assessment of potential clinical drug-drug interactions involving P-glycoprotein. Eur J Pharm Sci 2006; 27:543–554.
19. FDA. FDA Guidance for Industry. Bioanalytical Method Validation 2001. www.fda.gov/downloads/Drugs/GuidanceComplianceRegulatoryInformation/Guidances/ucm070107.pdf. Accessed June 2, 2009.
20. Shah VP, Midha KK, Findlay JWA, et al. Bioanalytical methods validation—a revisit with a decade of progress. Pharm Res 2000; 17:1551–1557.
21. Patorini E, Locatelli M, Simoni P, et al. Development and validation of a HPLC-ESI-MS/MS method for the determination of 5-aminosalicylic acid and its major metabolite *N*-acetyl-5-aminosalicylic acid in human plasma. J Chromatogr B 2008; 872:99–106.
22. Jemal M, Xia Y-Q. LC-MS development strategies for quantitative bioanalysis. Curr Drug Metab 2006. 7:491–502.
23. Rocci ML, Devanarayan V, Haughley DB, et al. Confirmatory reanalysis of incurred bioanalytical samples. AAPS J 2007; 9, E336–E343.
24. Rowland M, Tozer TN. Clinical Pharmacokinetics. Concepts and applications, 2nd ed. Philadelphia, PA: Lea and Febinger, 1989.
25. Gabrielsson J, Weiner D. Pharmacokinetic and Pharmacodynamic Data Analysis: Concepts and Applications. Stockholm: Swedish Pharmaceutical Press, 1997.
26. Peck CC, Barr WH, Benet LZ, et al. Opportunities for integration of pharmacokinetics, pharmacodynamics and toxicokinetics in rational drug development. Pharm Res 1992; 9. 826–833.
27. Custodio JM, Wu C-Y, Benet LZ. Predicting drug disposition, absorption/elimination/transporter interplay and the role of food on drug absorption. Adv Drug Deliv Rev 2007; 60:717–733.

28. FDA. FDA Guidance for Industry. Extended Release Oral Dosage Forms: Development, Evaluation, and Application of In Vitro/In Vivo Correlations 1997. www.fda.gov/downloads/Drugs/GuidanceComplianceRegulatoryInformation/Guidances/UCM070239.pdf. Accessed June 2, 2009.
29. Ruelius HW. Extrapolation from animals to man: predictions, pitfalls and perspectives. Xenobiotica 1987; 17:255–265.
30. Baillie TA, Cayen MN, Fouda H, et al. Contemporary issues in toxicology: drug metabolites in safety testing. Toxicol Appl Pharmacol 2002; 182:188–196.
31. Smith DA, Obach RS. Seeing through the mist: abundance versus percentage. Commentary on metabolites in safety testing. Drug Metab Dispos 2005; 33:1409–1417.
32. FDA. FDA Guidance for Industry. Safety Testing of Drug Metabolites 2008, www.fda.gov/downloads/Drugs/GuidanceComplianceRegulatoryInformation/Guidances/ucm079266.pdf. Accessed June 2, 2009.
33. Brynne N, Dalen P, Alvan G, et al. Influence of CYP2D6 polymorphism on the pharmacokinetics and pharmacodynamics of tolterodine. Clin Pharmacol Ther 1998; 68:529–539.
34. Odar-Cederlof I, Vessman J, Alvan G, et al. Oxazepam disposition in uremic patients. Acta Pharmacol Toxicol 1997; 40:52–62.
35. Shaffer CL, Gunduz M, Thornburgh BA, et al. Using a tritiated compound to elucidate its preclinical metabolic and excretory pathways in vivo: exploring tritium exchange risk. Drug Metab Dispos 2006; 34:1615–1623.
36. Browne TR, Szabo GK, Ajami A, et al. Performance of human mass balance studies with stable isotope-labelled drug and continuous flow-isotope ratio mass spectrometry: a progress report. J Clin Pharmacol 1998; 38:309–314.
37. Thompson AH, Tucker GT. Gerontokinetics – a reappraisal. Br J Clin Pharmacol 1992; 33: 1–2.
38. FDA. FDA Guidance for Industry. Pharmacokinetics in Patients with Impaired Renal Function 1998. www.fda.gov/downloads/Drugs/GuidanceComplianceRegulatoryInformation/Guidances/ucm072127.pdf. Accessed June 2, 2009.
39. FDA. FDA Guidance for Industry. Pharmacokinetics in Patients with Impaired Hepatic Function: Study Design, Data Analysis, and Impact on Dosing and Labeling 2003. www.fda.gov/downloads/Drugs/GuidanceComplianceRegulatoryInformation/Guidances/ucm072123.pdf. Accessed June 2, 2009.
40. FDA. FDA Guidance for Industry. Exposure-Response Relationships—Study Design, Data Analysis, and Regulatory Applications 2003. www.fda.gov/downloads/Drugs/GuidanceCompliance RegulatoryInformation/Guidances/ucm072109.pdf. Accessed June 2, 2009.
41. Ariens EJ. Stereochemistry, a basis for sophisticated nonsense in pharmacokinetics and clinical pharmacology. Eur J Clin Pharmacol 1984; 26:663–668.
42. Beal S, Sheiner LB. The NONMEM system. Am Stat 1980; 34:118.
43. Whiting B, Kelman AW, Grevel J. Population pharmacokinetics. Theory and clinical application. Clin Pharmacokin 1986; 11:387–401.
44. FDA. FDA Guidance for Industry. Population Pharmacokinetics 1999. www.fda.gov/downloads/Drugs/GuidanceComplianceRegulatoryInformation/Guidances/ucm072137.pdf. Accessed June 2, 2009.
45. Lalonde RL, Kowalski KG, Hutmacher MM, et al. Model-based drug development. Clin Pharmacol Ther 2007; 82:21–32.

33 Pharmacodynamics

Corinne Seng Yue
Faculté de Pharmacie, Université de Montréal, Montréal, Québec, Canada

Pina D'Angelo
Cetero Research, Cary, North Carolina, U.S.A.

Murray P. Ducharme
Cetero Research, Cary, North Carolina, U.S.A. and Faculté de Pharmacie, Université de Montréal, Montréal Québec, Canada

INTRODUCTION

There has been a growing trend in the last 20 years to consider more and more the pharmacological activity of a drug in addition to its pharmacokinetic behavior. The pharmacological activity of a drug, its efficacy and toxicity, is commonly referred to as its pharmacodynamics (PD), while what is happening with its concentration time profile or its systemic exposure is referred to as its pharmacokinetics (PK). Another simple way of describing PK and PD is that the PK is what the body does to the drug (how it absorbs, distribute, and eliminates it), while PD is what the drug does to the body (how it decreases blood pressure, how it kills bacteria causing infections, etc.). For decades now, ever since the work of Shannon on antimicrobials (1), we know that PK and PD are linked. It is not enough to just discuss PK or PD separately but one needs to consider them together. Indeed, a fundamental principle of clinical pharmacology is that there is always a relationship between the concentrations of a drug at its sites of efficacy and/or toxicity and its efficacy and/or toxicity.

Nowhere is it more evident than in the drug development process that we have been considering PK differently from PD for too long. Despite technological advances and increased medical knowledge, the attrition rate for drugs entering Phase II of the drug development process is as high as 62%, while it is approximately 45% for compounds entering Phase III (2). In the early 1990s, failure was mostly related to pharmacokinetic causes (poor characterization of pharmacokinetic properties or low bioavailability). However, improvements in our understanding of pharmacokinetics and in the quality of the tools used for pharmacokinetic analyses led to less pharmacokinetic-related failure. Now, at the beginning of the 21st century, attrition rates remain high but, rather than being attributable to pharmacokinetics, failure is often related to safety and efficacy outcomes and it occurs at the end of the drug development process (Phase III). Some causes of drug failure include poor efficacy, unacceptable toxicity, or the selection of an inappropriate dose or dosing regime. These could be avoided in a significant part by better studying the PK/PD of a drug, not just its PK and PD in isolation. This is one of the recommendations arising from the white paper from the FDA entitled "Challenge and Opportunity on the Critical Path to New Medical Products" published in 2004 (3). By studying the PK/PD of a compound more carefully, proof of concept studies can be performed early in Phase I, allowing one to either kill a compound sooner or to study it in Phase II/III with a more appropriate regimen that will ensure efficacy and minimize toxicity (4).

Among the challenges faced in drug development and in the quest toward personalized medicine is the variability in drug response. As clinicians well know two patients who are given the equal doses of the same drug under identical conditions may have completely different responses to the drug. Drug response is influenced by many factors related to the drug itself and to the conditions surrounding drug administration such as food intake but, more importantly, it is also influenced by patient's characteristics. Intersubject genetic variability can translate into differences in drug transport (influx and efflux transporter proteins), volumes of distribution and elimination (metabolizing enzymes). Conditions influencing drug response can even vary within an individual from one occasion to another, which further adds to the disparity in drug response (5,6).

Although clinicians and those involved in the drug development process are concerned with the pharmacokinetic behavior of molecules, it is ultimately their pharmacodynamic properties that matter to patients. In light of the staggering failure rates in Phases II and III of drug development and the high costs associated with the development of a new compound and, considering the complexity of drug response, it is therefore crucial that we refine the tools and approaches that are used throughout the drug development process (7).

This chapter starts with a review of the relationship between PK and PD. We will then describe how one studies the pharmacological activity of a drug (its PD), links it to its pharmacokinetic properties, and what critical activities/descriptions need to be done within the drug development process to optimize the development of drugs. Topics such as PK/PD modeling will also be explored, as well as the importance of evaluating PK/PD throughout the drug development process.

PHARMACOKINETIC AND PHARMACODYNAMIC RELATIONSHIPS

The assumption that there is a relationship between drug dose and effect underlies all principles of clinical pharmacology. Our improved understanding of PK has allowed us to further refine this relationship to relate dose to concentration, which is in turn related to response. The science of PK explains how the same dose of a medication given to different individuals can elicit different responses because of differences in drug exposure. It is this relationship between drug concentration and response that forms the foundation for most clinical trials. For instance, by relating drug effect to its plasma concentrations, we can presume that similar concentrations will produce a similar response in different individuals. This forms the foundation of bioequivalence studies, which seek to demonstrate equivalent efficacy through equivalency in pharmacokinetic parameters. Therefore, a clear understanding of this relationship is vital to the drug development process.

The PD effect of a drug can be classified as direct or indirect and reversible or irreversible. This means that four different cases can occur. An example of a direct reversible effect is with theophylline and bronchodilation. Theophylline acts within the lungs and, the lungs being extremely well-perfused organs, the bronchodilation is directly related to the concentrations of the drug in the plasma (i.e., direct effect) and its effect is short lived, so the bronchodilation is maximal at the time of maximum drug concentration (C_{max}) and minimal at the time of minimum drug concentration (C_{min}). The effect is therefore reversible. The extreme opposite example would be of an anticancer agent such as cyclophosphamide and the number of cancer cells. Cyclophosphamide needs to be transformed to an active metabolite before acting (i.e., indirect) and the killing of cancer cells is obviously an irreversible effect. Of course this relationship can be altered by simply studying the effect of the active metabolite instead of the parent drug for the PK part and on the killing activity of cells instead of the number of cancer cells for the PD part.

As previously mentioned, a fundamental principle of clinical pharmacology is the link between the PK of the drug and its PD. Figure 1(A) describes this relationship between PK and PD schematically. The link may not always be easy to describe but it is always there. A common relationship is a direct relationship between the daily exposure to the drug characterized by its area under the plasma concentration–time curve (AUC) and the inhibitory or stimulatory effect on a pharmacological phenomenon. An example is presented in Figure 1(B) for an antiallergic compound where the exposure to the drug inhibits the appearance of symptoms such as nasal congestion.

Direct Response

For a drug to exert its pharmacological effect, it must bind to the necessary receptors to activate the electrophysiological events or biochemical cascades that will produce the pharmacological response. However, before reaching its site of action, the drug must undergo absorption, disposition, and possible metabolism. In other words, the PK properties of the drug are intrinsically linked to the pharmacological response. A pharmacological response may be poor not because the drug is ineffective but simply because the drug is unable to reach its site of activity.

At steady-state the average concentration of free (unbound) drug in systemic circulation (i.e., the compartment from which drug concentrations are being measured) will be in

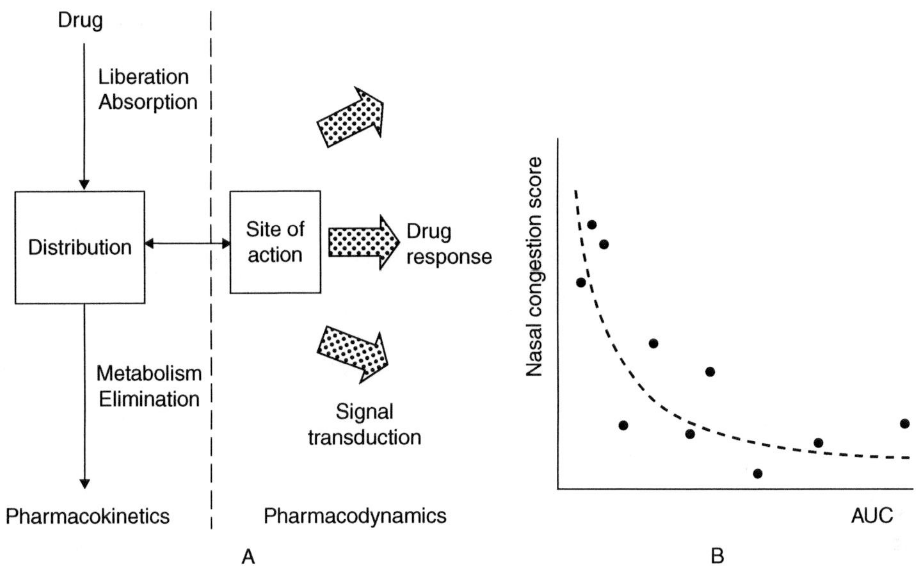

Figure 1 Relationship between pharmacokinetics and pharmacodynamics. AUC, Area under the plasma concentration-time curve; solid points, individual subject data; dashed line, predicted exposure-response relationship.

equilibrium with those at the site of action (e.g., in some particular tissue or organ). This assumption allows us to relate drug plasma or blood concentrations to the pharmacodynamic response.

If we assume that at a basic level, a drug exerts its effect by binding to a receptor, we can apply receptor theory to better describe drug effects (8,9). Equation 1 describes the formation of the receptor–drug complex:

$$[\mathrm{R}] + [\mathrm{D}] \overset{K_\mathrm{d}}{\longleftrightarrow} [\mathrm{R}] \cdot [\mathrm{D}] \tag{1}$$

The free drug ([D]) binds to the available receptor sites ([R]) to form a receptor–drug complex ([R]·[D]). The balance between the free drug and the complexed drug is determined by the dissociation constant K_d, but the formation of the complex is also limited by the quantity of free drug that is available as well as the number of available receptor binding sites. The formation of receptor–drug complexes is dependent therefore upon the affinity of the drug for the receptor (a drug-specific characteristic) as well as the number or density of receptors in the given tissue.

E_{max} and Sigmoidal E_{max} Models

Receptor theory suggests that drug response that is related to receptor binding is limited by several factors. It is conceivable that there would therefore be a maximal response, the point at which the maximal number of receptor–drug complexes is present. This is known as the E_{max} response model, and this phenomenon is characterized by Equation 2.

$$\text{Response} = \frac{E_{\max} \cdot C_\mathrm{p}}{\mathrm{EC}_{50} + C_\mathrm{p}} \tag{2}$$

The parameter E_{max} provides an estimate of the maximum response associated with the drug, while the EC_{50} parameter represents the plasma concentration associated with 50% of this maximal response. A variation of this concentration–response curve is known as the sigmoidal E_{max} response model. Both models are depicted in Figure 2.

The sigmoidal E_{max} model produces a profile that is more "S-shaped" than the simple E_{max} model. In mathematical terms, the sigmoidal E_{max} response is described by Equation 3,

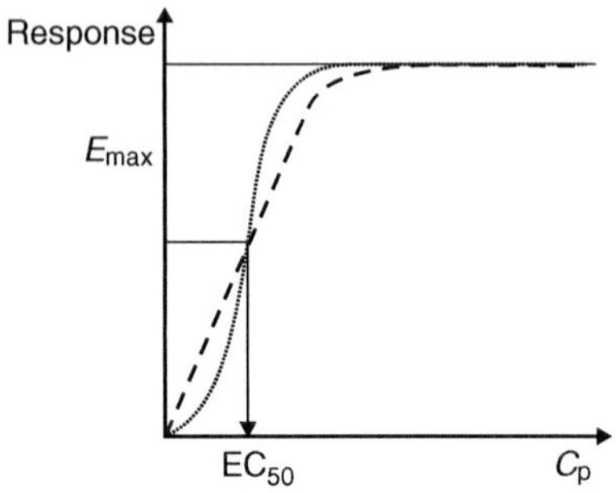

Figure 2 Ordinary and sigmoidal E_{max} concentration response profile. C_p, plasma drug concentration; E_{max}, maximal response; EC_{50}, plasma drug concentration associated with 50% of the maximal response; dotted line, sigmoidal E_{max} model; dashed line, ordinary E_{max} model.

where the exponent γ (also known as the Hill factor) governs the steepness or the shallowness of the central portion of the curve.

$$\text{Response} = \frac{E_{\text{max}} \cdot C_{\text{p}}^{\gamma}}{\text{EC}_{50}^{\gamma} + C_{\text{p}}^{\gamma}} \tag{3}$$

The ordinary E_{max} model and sigmoidal E_{max} model are intuitively appealing because they indicate that there is no effect when drug levels are equal to zero and that, above a certain concentration threshold, increases in concentration cause no further increment in effect (10).

Although the ordinary or sigmoidal E_{max} models are hyperbolic in nature, it must be pointed out that, within certain concentration ranges, the relationship between the concentration and response is linear. For the sigmoidal E_{max} model, there is a log-linear relationship between concentration and response between 20 to 80% of the maximal effect while, below 20% of the maximal response, the relationship is linear.

While Figure 2 and Equations 2 and 3 depict a response that is agonistic, these models can also be used to describe antagonistic drug responses. These equations can also be modified to take into consideration a baseline response that may be present. For instance, Equation 3 can be simply rewritten to include a baseline response, R_0, which increases as a result of drug exposure.

$$\text{Response} = R_0 + \frac{E_{\text{max}} \cdot C_{\text{p}}^{\gamma}}{\text{EC}_{50}^{\gamma} + C_{\text{p}}^{\gamma}} \tag{4}$$

These equations, describing response as a function of plasma concentration, assume that the response is directly related to the drug concentration in systemic circulation. In these cases, a change in plasma concentration would be associated with an immediate response. Maximal drug response would concord with the occurrence of peak plasma drug concentrations. In addition, each drug concentration is associated with only one response level.

Time-Dependent Models

In many cases, however, a delay exists between the change in plasma drug concentrations and the associated drug response. Also the same drug concentration can be associated with different responses. These situations are easily observed in hysteresis and proteresis figures where concentration and response are plotted with respect to time, which is indicated by arrows. Hysteresis is derived from Greek and means "arrives later." Such a curve is, therefore, associated with a greater effect later in time for similar systemic concentrations. A proteresis curve is seen when greater effects of the drug are seen early on (the word "proteresis" comes from the Greek term for "arrives early"). Such a curve may be seen if one follows the efficacy of a drug where a tolerance effect is seen for example.

Figure 3(A) illustrates a hysteresis plot that suggests that there is some delay between the observed response and the drug concentration level. The response occurring at time A is smaller than the response occurring later at time B. This is the situation seen with the majority of drugs. The systemic plasma concentrations are not in "equilibrium" with the concentrations of the drug at the site of activity. The drug needs to "travel" from the systemic compartment to the site of

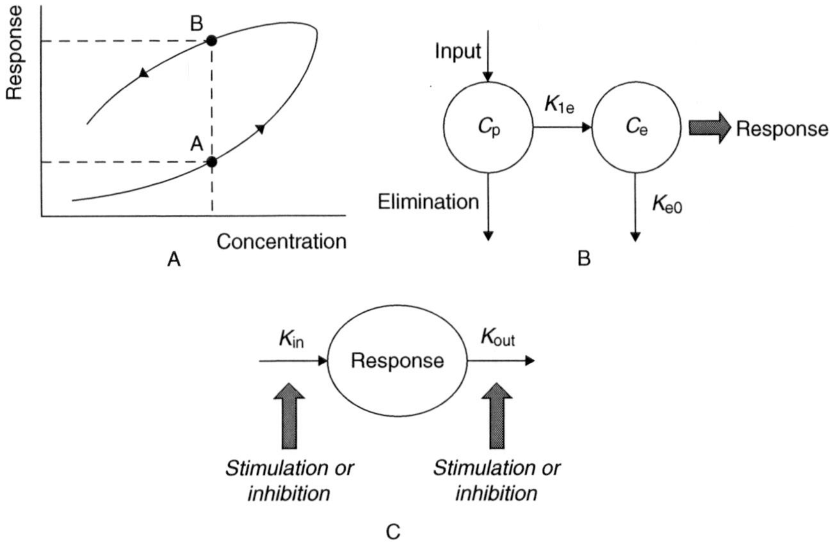

Figure 3 Time-dependent models. A, response associated with a given concentration at an early time; B, response associated with the same concentration at a later time; C_p, plasma drug concentration; C_e, drug concentration in the effect compartment; k_{1e}, first order rate constant governing transfer of drug into the effect compartment from the plasma compartment; k_{e0}, first-order rate constant describing the irreversible disappearance of the drug from the effect compartment; k_{in}, zero-order constant for the production of the response; k_{out}, first order rate constant for loss of the response.

activity. As the systemic concentrations of the drug rise early on, very little drug has arrived yet at the site of activity. So the concentrations at the site of activity are small relative to the systemic ones. Similarly, as the systemic concentrations decline rapidly the concentrations at the sites of activity do not decline at the same rate. The "active" concentrations of phenytoin, for example, are not directly those seen systemically but those in the central nervous system and that we could not measure directly. There are a myriad of other reasons for seeing a hysteresis loop besides the fact that concentrations of a drug need to travel to the efficacy site. These could include the fact that we are measuring only the parent drug concentration when metabolites are active, that we are measuring the concentrations of a racemate instead of solely the active enantiomer, or that the drug under study has nonlinear protein binding characteristics. Conversely a proteresis loop can suggest also a myriad of hypotheses, a common one being that tolerance to the drug effect or receptor desensitization occurs such that the magnitude of response associated with a given systemic concentration decreases with time (11).

To account for these time-dependent responses, many response models have been developed from the now classic effect compartment (or link) model to the most recent indirect modeling approach.

Link or Effect Compartment Model

One of the first models to be proposed to explain the delay for the systemic concentration to arrive at the sites of activity was the link model. It included an artificial delay compartment where its concentrations could be in equilibrium with those at the biophase and resulting in a "collapsing" of the hysteresis curve. The link model attributes the delay in drug response to a process whereby the drug must attain the biophase from systemic circulation (12–14). This process is influenced by many factors including, but not limited to, drug transport across various membranes, organ/tissue perfusion, and drug solubility in tissue (9). Therefore the observed delay between the pharmacological response and drug concentration levels are related to the time necessary for equilibration of concentrations in the plasma and effect compartments (15). This model is depicted in Figure 3(B).

The link model assumes that the amount of drug transferred from the pharmacokinetic compartment to the effect compartment is negligible so that the PD does not affect the PK. The drug concentration in the plasma compartment is represented by C_p while C_e reflects the drug level in the effect compartment. The transfer of drug into the effect compartment is governed by the first-order rate constant k_{1e}, which, for lack of identifiability and to minimize the impact of PD on PK, is fixed to an extremely low value. In other words the parameter k_{1e} cannot really be "identified" because no "sample" is actually drawn from the effect compartment, therefore it is fixed to a value that is physiologically sound. The other parameter, k_{e0}, is the first-order rate constant describing the irreversible disappearance of the drug from the effect compartment. Thus the temporal displacement of the pharmacological effect is dependent solely on modeling the parameter k_{e0}. Finally, the drug response can be related to its concentration in the effect compartment (C_e) through the ordinary and sigmoidal E_{max} response models, where C_p in Equation 2 and Equation 3 is replaced by C_e. This model was associated with many drawbacks including a nonphysiological elimination, albeit extremely small, from the effect compartment instead of from the central compartment as well as inconsistency in its ability to correctly predict steady-state efficacy from the fitting of single dose data. Subsequent improvements were proposed and made to this model but none as dramatic as the indirect response modeling technique.

Indirect or Turnover Model

A major advance was made in our ability at relating the PK with the PD of drugs with the indirect modeling technique proposed by Jusko (16,17). In this approach, concentrations of drugs inhibit or induce a formation or an elimination of an effect. As the name suggests indirect models relate drug concentrations to their pharmacological effect by an indirect mechanism. Let us consider the example of warfarin where the drug, rather than act directly on the PD response marker, is acting on precursors eventually decreasing the rate of formation of vitamin K–dependent clotting factors thereby achieving its anticoagulant effect.

The basic indirect or turnover model is depicted in Figure 3(C), where k_{in} is the zero-order constant for production of the response while k_{out} represents the first-order rate constant for loss of the response. An increase in response can be achieved by either stimulation of the response production (k_{in}) or by inhibition of the loss of response (k_{out}). Conversely, a decreased response can be attributed to an inhibition of k_{in} or a stimulation of k_{out}. These four situations correspond to the four basic indirect models (16,17). Of course response can be described by a combination of these four basic models. As with the link model and indirect model, drug effect on k_{in} or k_{out} can be described by an E_{max} or sigmoidal E_{max} function.

These indirect models are understandably mechanistic in nature. That is some appreciation for the mechanism of action or true physiological effect of the drug is necessary in order to implement these models.

Irreversible Response Models and Disease Modifying Modeling

The direct response model and link model assume that drug effect is reversible and dependent on a relatively fast interaction between the drug and receptor (12). However some drugs exert a pharmacological effect that is irreversible such as antibacterials and chemotherapeutic agents, which result in cell death. Models can be elaborated for cell-cycle non-specific chemotherapeutic drugs, accounting for natural cell turnover as well as drug effect (13).

Sometimes the disease can be modified by the drug activity hence the term disease-modifying modeling. This will be touched upon later in this chapter.

ASSESSING PHARMACOLOGICAL RESPONSE

Markers of Pharmacodynamic Response

Pharmacodynamic response encompasses not only the desired pharmacological effect of the drug but also includes unwanted effects or toxicity. These responses can be categorized broadly as quantal or continuous variables. Quantal responses are discrete categorical responses that do not belong to a continuum of responses. Such responses include dichotomous responses (awake/asleep) or polychotomous responses (grades of hematological toxicity) (14). Continuous responses can include laboratory values such as triglyceride levels or other clinical results such as heart rate.

The models described in the previous section provide a framework that allows us to characterize and understand drug response in relation to drug concentrations. However the selection of an appropriate marker for pharmacodynamic response is crucial in the establishment of an accurate PK/PD model.

Pharmacodynamic response markers can be categorized as biochemical markers (angiotensin I, plasma renin), clinical results (blood pressure, heart rate), or clinical outcomes/endpoints (life or death, cure or failure) (18). The term "biomarker" is often used to refer to physical signs or laboratory values that have a putative diagnostic or prognostic value (19,20). Surrogate endpoints are a subgroup of biomarkers that can be used to replace clinical endpoints, which are clinically meaningful measures of "how a patient feels, functions, or survives" (19). Clinical endpoints can be subdivided further into two categories: intermediate endpoints (such as increased exercise tolerance) or ultimate outcomes (for instance, decreased risk of stroke).

The ideal biomarker possesses certain characteristics. Firstly, it must be clinically relevant in the sense that it must be physiologically or mechanism based. There must be a link between this biomarker and the pharmacological mechanism of the drug. The biomarker must also be objective and reproducible so that the information provided by the biomarker is universally accessible. It must also be practical and relatively easily obtainable, minimizing invasive procedures for the patient. A good biomarker must also be sensitive and selective to the pharmacological effects of the drug as far as possible. This can be a challenge when biomarkers are also influenced by disease state or other physiological processes (18,20,21). A biomarker that is gradual is also desirable, as it will allow the characterization of incremental changes in response (22).

In most cases it is the ultimate outcome (increased survival or decrease of the risk of irreversible morbidity) that is of interest. However, because of time constraints, surrogate endpoints that are a substitute for clinical endpoints provide an alluring alternative. For instance the desired clinical outcome of a drug may be a decrease in mortality and it may be difficult to wait for information on patient mortality. Therefore, it would be more practical to associate drug levels with a surrogate biomarker that is easily and quickly detectable (18,21). Cases have been reported where the use of a surrogate endpoint proved to be successful as a predictor of clinical outcome (e.g., HMG-CoA reductase inhibitors) while, in other cases (e.g., antiarrhythmic agents) the surrogate endpoint was inadequate. Caution must be employed with the use of surrogate endpoints because there is the danger that they might not be valid and that they fail to predict rare, negative side effects that can lead to overall negative clinical outcome (23). Indeed, because various mechanisms can contribute to disease progress and to drug response, it is rare that a single biomarker can account for all drug effects. In many cases, the measurement of multiple markers of pharmacological response can provide more insight into the PK/PD relationship than the assessment of a single biomarker (24).

The validation or evaluation of a biomarker as a surrogate endpoint can be a long and extensive process but it is an essential step if one wants to use these biomarkers in pivotal Phase III efficacy studies. The process has been described elsewhere but essentially biomarker validation must include data from informative studies that relate drug exposure to response, provide insight on the mechanism of action of the drug and quantify the magnitude of drug response (18,20,25). This is often a retrospective process (19). To date very few biomarkers are recognized as validated surrogate endpoints for pivotal Phase III studies.

Utility of Biomarkers

Despite the many factors that must be considered with the use of biomarkers their utility extends well beyond their roles as measures of efficacy in proof-of-concept studies or as substitutes for clinical outcome measures.

By their very definition, biomarkers can provide vital information on a patient's health status as well as the effects of disease progression on their clinical state (14). For example, clinical laboratory tests such as glycemia or glycosylated hemoglobin (hemoglobin A_{1c}) are measured to follow the status of diabetic patients. Biomarkers are used daily by clinicians in order to guide their therapeutic decisions. One such example is serum creatinine that can be measured to perform dosage adjustments in renally impaired patients. In a clinical trial setting baseline biomarker values can indicate a patient's preintervention status and these values can then be compared to values obtained posttreatment.

Biomarkers can provide useful knowledge about the clinical pharmacology of a drug and increase our understanding of disease mechanisms. They can provide information that can be used to guide dose selection and escalation such that interindividual variability in response is minimized (19,26). Their use can also be expanded to explore the impact of covariates, investigate drug–drug interactions, and assess the effects of renal or hepatic impairment on various pharmacological effects.

The insights on drug efficacy, safety, and disease afforded by biomarkers are beneficial at early stages of drug development (24). The information that they supply can form the basis for selection of lead compounds in later stages of drug development (19) or for the discontinuation of the development of other compounds.

Clearly biomarkers not only inform us about a drug's efficacy but also reveal data pertaining to subject safety. Frequently this application of biomarkers is undervalued. Because of their very nature biomarkers can detect subclinical adverse events or laboratory values that could be predictive of significant clinical adverse events (26). Therefore biomarkers could act as predictors of clinical risk or benefit.

Important Elements of Study Design

In order to assess the pharmacodynamic properties of a drug robustly studies must be designed properly. One must employ an appropriate biomarker and take into consideration several other factors. Firstly, because the pharmacokinetics and pharmacodynamics of a molecule are intrinsically linked, all of the design elements necessary to properly characterize the pharmacokinetics of a compound must be respected.

Essentially two important questions must be answered by clinical pharmacology studies. The first objective is to determine if the drug reaches its site of action and the second is to evaluate if the drug exerts its intended effect (27).

Two types of studies can be performed to assess PK/PD relationships (25). The first type of study is the most commonly used. The investigational product is given to subjects at varying doses, plasma concentration levels are determined and these are related to response. In the second type of study subjects are randomly assigned to various plasma concentration levels and are titrated to achieve these levels. These plasma concentration levels are then correlated with response. Potential problems with the first type of study design include the possibility of confounding factors that affect both the PK and PD and therefore misconstrue the PK/PD relationship. For instance, if certain subjects had increased absorption of the drug, and associated increased exposure and response, there may be a factor causing both the increased absorption and increased response. This problem can be avoided when the PK/PD relationship is observed in the same individual over time or when the second design strategy is employed.

Dose–response studies, in which multiple doses of a drug are compared to a placebo or control, can provide important information regarding clinically relevant doses. However, in cases where there is a poor correlation between dose and exposure it may be wise to assess response with respect to systemic exposure. Ideally, concentrations at the presumed effect site should be measured but, due to sampling and assay limitations, this is not always possible so plasma concentrations are often used instead (10). Applying the PK/PD modeling techniques presented earlier in this chapter enables us, however, to circumvent these limitations.

Information obtained from single- and multiple-dose studies should be used in order to better appreciate if response is subject to time-dependency. Time-related changes in response, such as tolerance or induction, might be observed. To determine whether the drug response is direct or indirect, drug response should be measured at the same time as drug concentration levels. This would allow the determination of the shape of the concentration–response curve (single curve, hysteresis, or proteresis).

In order to capture the potency (EC_{50}) and maximal effect (E_{max}) of a drug it is important to obtain response data associated with a range of drug concentrations (28). As illustrated in Figure 2, depending on the portion of the response curve that is being examined, different concentration–response relationships are noted. Within certain ranges the relationship between concentration and response is linear or log-linear while at the higher doses the response attains a plateau. Although certain situations (narrow therapeutic index, severe toxicity) may prevent the administration of a large dose range to test a wide range of concentrations, it is necessary to bear in mind this concept. Classically, people were studying the dose–response relationship resulting in a limited amount of information capturing the true relationship thereby resulting in an uncertain relationship. This is why assessing the PK with the PD of the drug and evaluating the concentration–response relationship (instead of dose–response) using all available discrete concentrations and efficacy time-points makes much more sense. By analyzing all available data in this way, enormous information can be gained from the comodeling of PK and PD because rising and decreasing concentrations give information throughout the concentration–response relationship instead of the limited information available per patient with the dose–response modeling.

At the preclinical stage, to maximize the information obtained in animals, one must ensure that plasma concentrations are measured and that the pharmacological model occurs in the animal species (27). If possible measures of the potential biomarker must also be obtained.

ROLE OF PK/PD MODELING AND SIMULATIONS

Using data obtained from well-designed exposure-response studies PK/PD relationships can be established. Although exposure and response can be related through the use of a single measure of drug exposure, such as the area under the plasma concentration–time curve, it is much more interesting to link the two through the relationships described previously, which account for temporal changes in drug exposure (29).

A powerful and instructive tool in constructing and making use of the exposure–response curve is what we call PK/PD modeling and simulations. Modeling and simulations allow us to make maximal use of the available information even in cases where little information can be obtained (e.g., due to sparse sampling or due to restrictions because of assay limitations). Although these models do not necessarily prove causality or provide insight into the mechanisms behind the concentration response, they can still help us understand the nature of the pharmacological response.

Population PK/PD modeling is essential to the determination of robust concentration–response relationships. The mathematical principles that underlie PK/PD modeling, as well as the procedures involved in modeling, have been reviewed extensively in other publications and are beyond the scope of this chapter (29–31). However we will review some of the main aspects of PK/PD modeling and some important considerations.

As in any sort of modeling, the first step in PK/PD modeling is the definition of the problem. There must be a good understanding of the study design and of the available data and the modeling objective must be clearly defined. Any assumptions that are being made must then be stated, such as hypotheses regarding mechanism of action, placebo response, appearance of tolerance, and pharmacokinetic parameters, to name a few. The modeling process can then be undertaken, selecting the best model based on the putative mechanism of action and drawing on available information on the drug's pharmacology and system's biology (14,22,25,28). Finally the model must be validated through simulations or by using internal or external data.

The population PK/PD approach also allows us to identify and quantify factors that contribute to intra- and intersubject variability in response, through the construction of a concentration–response model. The various applications of PK/PD models throughout the drug development process are examined in the following section. As mentioned previously every concentration and observational data should be fitted so that the maximum amount of information can be obtained to construct the exposure–response relationships.

PK/PD IN THE DRUG DEVELOPMENT PROCESS

The utility of PK/PD modeling and simulations throughout the drug development process has been the focus of many reports and discussions (21,25,27,29,32–40). In this section we will attempt to highlight some of the key applications of PK/PD modeling and simulations at different stages of the drug development process.

The Learn–Confirm Paradigm

An important shift in the drug development paradigm was fuelled by Sheiner's "learn–confirm" concept (41). This novel notion views the phases of drug development as successive learn–confirm cycles. In the learning phases the goal of study design is to glean as much information as possible about exposure–response relationships and mechanisms of action. The confirmatory portions of the cycle serve to provide further evidence to support specific claims, for instance regarding risk–benefit ratios or optimal doses. Thus, Phase I studies are studies designed to learn about tolerated doses and explore dose–response relationships, while Phase IIa studies are meant to confirm that selected doses are associated with the desired response. The learn–confirm cycle begins anew with Phase IIb studies where researchers learn about the drug's behavior in the intended patient population. Finally Phase III and IV studies aim to confirm the dose and dosing regimen selected based on the Phase IIb studies as well as provide further supportive evidence on the drug's postulated efficacy. This understanding of the learning and confirming paradigm will allow the reader to better appreciate how PK/PD modeling and simulations can be used in both the learning and confirming stages of drug development, making it a very versatile tool.

Nonclinical Stage

At the nonclinical stage compounds are screened for efficacy and safety through in vitro and animal studies (38). Performing PK/PD modeling using animal data can provide some insight into the nature of the exposure–response shape as well as estimates of potency (EC_{50}) or maximal effect ($E_{\max}$) (36). Information on pharmacokinetic properties (absorption, metabolism, elimination) that could impact pharmacological response can also be obtained at this stage. Modeling can also make use of existing clinical data on similar compounds. Despite its utility in screening potential drug candidates one of the main uses of PK/PD modeling at the nonclinical stage remains the selection of doses for first-time human administration (27). Different modeling approaches can be used for dose range projection including population PK/PD analyses of sparse nonclinical data, allometric scaling to predict human PK, and efficacy scaling (33). Indeed the selection of an appropriate dosing range for humans can, at times, be more of an art than a science. Therefore a tool that provides a scientific rationale for dose selection is much needed.

Phase I

The PK/PD model developed at nonclinical stages can be further refined, using data obtained from Phase I studies conducted in small numbers of healthy volunteers. Phase I studies offer the first opportunity to characterize the PK/PD of a compound in humans. At this stage of drug development PK/PD modeling can serve several purposes. Using a PK/PD model, a link can be established between animal and human data.

In this first phase of clinical development, a PK/PD model can describe complex exposure–response relationships, such as those which involve nonlinearity, as well as make use of scarce data (related to sparse sampling or assay limitations) (42). PK/PD modeling can make use of data available from all Phase I studies in order to get robust estimates. Modeling can also reveal if there are any deviations from dose-proportionality, and to describe possible time-related phenomena such as tolerance or sensitization.

Modeling and simulations can also be used to select doses and dosing regimens for subsequent studies as well as to optimize sampling schemes and study designs. For instance, based on a developed PK/PD model, several dosing schemes can be simulated in order to select the range of doses or types of regimens that would provide optimal response (43).

Another important application of PK/PD modeling is that it can support proof-of-concept (POC) claims by demonstrating that the drug acts on its targeted mechanism, thereby leading to the desired short-term outcome. PK/PD models can, in fact, be used to support claims of efficacy throughout drug development (25).

Phase II

Phase II studies are conducted in a larger number of subjects and information can therefore be obtained to better understand intersubject variability. As in Phase I studies, PK/PD modeling

Table 1 Key Features of the Classical PK/PD Approach Compared to the Learn–Confirm Method

Classical approach	Learn–confirm approach
- Only a small number of regimens can be assessed	- PK/PD model can be used to simulate and predict results for many regimens
- A limited set of outcomes are evaluated	- Multiple outcomes can be evaluated
- Dosage individualization cannot be undertaken	- The assessment of inter- and intraindividual variability, as well as the impact of covariates, through PK/PD modeling promotes dosage individualization.
- Manner of selecting minimal effective dose is illogical	- Able to provide objective basis for decision-making and dosage selection
- Unable to extrapolate beyond bounds of the study design	- Can provide credible extrapolations
- Associated with a high late-phase failure rate	- Reduction of failure rates in late phases of drug development

can be used for further evidence of POC as well as in the selection of dosing regimens, sampling schedules, and study design.

Information gleaned from Phase II studies can therefore be used to optimize the design of future trials, as well as further enrich the understanding of a compound's attributes that will distinguish it from competitor drugs (33).

Phases III and IV

In later stages of clinical development, PK/PD models can be used to understand the impact of covariates on drug response. Because Phase III studies are conducted in larger numbers of patients from the target population, they provide key information regarding the effect of patient characteristics and different pathological states on drug response. In addition, disease–drug models can be developed.

Models can further confirm the dose–exposure–response relationship in the target population. In some cases, the use of exposure–response information coupled with a single pivotal clinical trial can be sufficient evidence of effectiveness (25). This demonstrates how a thorough understanding of the exposure–response relationship can even obviate the need for additional studies.

Overall Benefits Associated with PK/PD Modeling

Beyond their use at specific stages of the drug development process, the construction of PK/PD models allow researchers to integrate data from all phases of drug development for future use, such as the determination of a different dose, dosing regimen, or route of administration (33, 34,44).

It is obvious that the use of PK/PD modeling and simulations throughout the drug development process can decrease the number of failed studies and aid us to better understand the results that are obtained (37). Even in "confirming" phases of drug development a researcher may be confronted with unexpected results that can only be explained through PK/PD modeling.

The use of disease state information, combined with PK/PD models and knowledge of important patient characteristics, will lead to a better prediction of the range of possible outcomes and to a more efficient drug development process.

Table 1 compares the classical approach where PK and PD are studied separately and linked by dose and response to more holistic "learn-and-confirm" methods described in this chapter. It summarizes the important advantages associated with the latter approach.

Case Studies

Now that we have seen how PK/PD modeling can be applied throughout the drug development process and how it is superior to the classical sequential approach, we will examine specific examples of how this can be done.

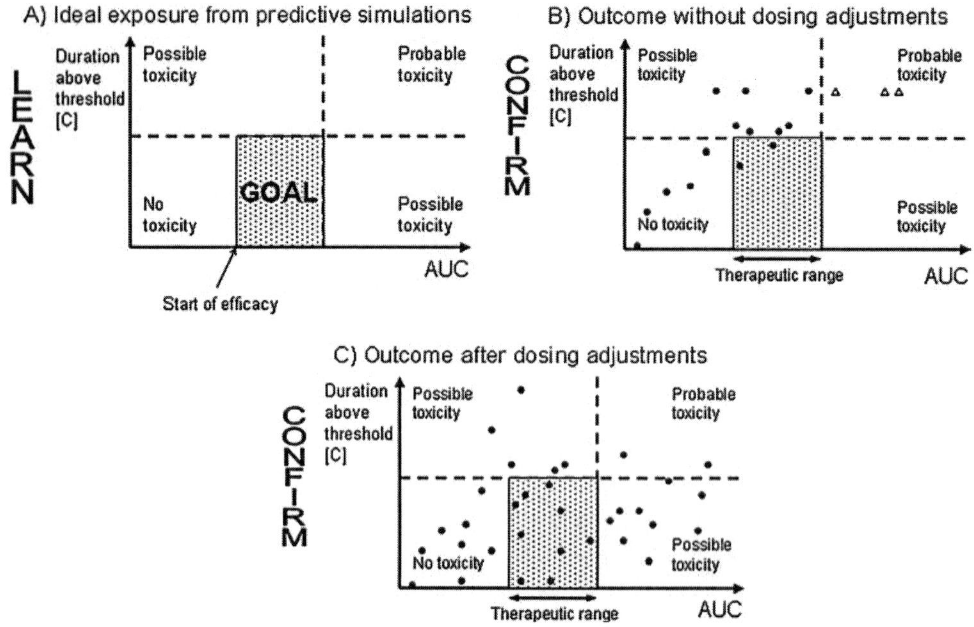

Figure 4 Relationship between exposure and response for an oncology agent. AUC, area under the plasma concentration-time curve; [C], drug plasma concentration; dashed lines, delimiters for probabilities of experiencing toxicities; dotted area, target exposure based on efficacy and toxicity; solid circles, individual patient responses; white triangles, patients experiencing dose-limiting toxicities.

Oncology

An example of the utility of PK/PD modeling in the development of an anticancer drug is illustrated in Figure 4.

Panel A of Figure 4 depicts for an immediate-release oncology drug the relationship between (1) the total exposure (AUC), (2) the predicted toxic and efficacious AUC, and (3) the duration of time above which its levels are greater than a threshold "active" concentration. The threshold concentration and AUC were determined from PK/PD modeling by linking the exposure of the drug with a toxicity biomarker, the absolute neutrophil counts (ANC). The link was made based on all data obtained from three different Phase I studies, using an indirect response modeling technique as presented earlier in this chapter where the systemic concentrations of the anticancer agent inhibit directly the formation of the neutrophil precursors in the bone marrow. Estimates resulting from this PK/PD modeling exercise are presented in this figure and illustrate the first part of the "learn and confirm" paradigm where information about the activity of the drug as it relates to its exposure is acquired.

Therefore based on what was learned from the analysis of pooled data from different Phase I studies, ranges of exposure that ensured efficacy while minimizing toxicity [Fig. 4(A)] were proposed with a potential optimal exposure range represented by the dotted area. Once an ideal exposure range was proposed from the PK/PD modeling of this anticancer agent (through the learning phase) exposure to the same active compound but from a different slow-release formulation (the previous modeling was done with an immediate release dosage form) was plotted on the proposed graphical relationship [Fig. 4(B)]. One can see that the actual exposures of patients receiving this new sustained release formulation were not resting in the targeted area proposed by the initial PK/PD modeling. In fact only two patients had their exposure in the "ideal" goal range—the majority received inefficacious regimens and three subjects appeared in the area thought to be associated with definite toxicity. Upon further assessment it was found that these three patients were actually the only ones having experienced dose-limiting toxicity with the agent, and their exposures are therefore represented by white triangles on the figure. This confirmed the utility of the proposed exposure–toxicity relationship.

Following this observation new dosing regimens were proposed for this sustained release formulation. This was to maximize the likelihood that patients would be more likely to receive dosing regimens that would be efficacious while minimizing definite toxicity. This was also part of the "confirm" phase of the exercise, illustrated in Figure 4(C), where actual exposures in patients are plotted again on the proposed exposure–toxicity relationship figure and show that a greater proportion of subjects received active and less toxic regimens.

As illustrated by this example, the first dosing adjustment based on only one cycle of learn-and-confirm PK/PD modeling, dramatically reduced the number of patients who experienced toxicity while increasing the number of patients who responded to the drug. This is but one example of how proper characterization of the PK/PD relationship of a drug can aid researchers in selecting adequate doses and dosing regimens for investigational products and ensure that an efficacious and safe dosing regimen will be tested in Phases II and III.

Cardiology

This example demonstrates how the early development of a PK/PD model combined with PK/PD knowledge of drugs acting similarly allowed a company to halt the development of a particular compound thereby saving precious time and money.

The drug in question was a cholesterol absorption inhibitor designed to be administered alone or with a HMG-CoA reductase inhibitor or a statin. A PK/PD model was developed for this drug in order to link its plasma concentrations with its effects on low-density lipoprotein cholesterol. In addition, models were developed for five statins that were on the market as well as for the drug combination of statin and cholesterol absorption inhibitor (35).

The PK/PD models thus constructed enriched researchers' understanding of the investigational drug as well as a competitor compound and therefore contributed to the "learning" phase of the drug development process. The models were then used to simulate the exposure–response relationships associated with these drugs and various drug combinations. Subsequently, Phase II data became available to "confirm" and modify the previously developed models.

Finally, the updated model was used to acquire even more information about the compound and its projected benefits. This learning phase answered questions pertaining to comparability with a competitor compound, expected response when co-administered with statins and the reliability of these predications. Ultimately this led to the demise of the compound but resulted in substantial savings for the drug company compared to the case where they would have learned this only after performing Phase III studies.

Infectious Diseases

Interferons are used to treat various viral infections and malignancies. Despite their therapeutic benefits their short half-life requires frequent administration (three times per week) and they can be highly antigenic. PEGylation of interferons is thought to increase the circulating half-life as well as decrease immunogenicity. In this example, a PK/PD model was constructed to relate the exposure to PEG-modified interferon α-2a to its effect on the induction of the production of MX protein (45). Because of their many effects, MX proteins were considered to be a useful PD probe. The goal of model development was to provide information to improve dosing strategies as well as guide the drug development of future modified molecules.

The PK/PD model was based on data from a randomized single ascending dose study that included 45 healthy adult male subjects receiving one of four subcutaneous doses of PEG-modified interferon α-2a or interferon α-2a. The PK of the interferon products, described by a one-compartment model with first-order absorption and elimination, was related to the PD through an indirect model. The drug stimulated the production of MX protein (stimulation of k_{in}) via an E_{max} function.

The simulations obtained from the PK/PD modeling exercise indicated that, although the addition of a PEG moiety to interferon α-2a did indeed prolong the half-life of the drug, the PD properties associated with the PEG-modified interferon α-2a would still necessitate a twice-weekly dosing regimen in order to attain a comparable response to the unmodified product. This was a far cry from the anticipated once-weekly dosing for the PEG-modified product and these predictions were confirmed by two Phase II trials.

In conclusion, PK/PD modeling demonstrated that the PEG-modified interferon α-2a provided little therapeutic benefit over its unmodified counterpart which proved to be consistent with Phase II findings. These findings contributed to the decision to discontinue the development of this product for this indication.

ADVANCED TECHNIQUES AND CONCEPTS

Although the goal of this chapter was to familiarize the reader with fundamental principles and applications associated with pharmacological response, as well as PK/PD modeling, this section will briefly touch upon relevant concepts and approaches that are still in their infancy.

Placebo Response and Disease Modeling

Pharmacological response is not only subject to the factors governing the occurrence of the drug–receptor interaction (e.g., pharmacokinetic variability, inhibitory competition with other substrates, etc.) but it is affected by an array of postreceptor phenomena. Response can be a consequence of signal transduction and a cascade of downstream events occurring as a result of the drug–receptor interaction. In many cases, as a natural protective reflex, the body can compensate for the effects of certain drugs by counteracting their effects. A typical example of this is the baroreceptor reflex which maintains blood pressure.

Similarly pathophysiological changes that result from disease or comorbid conditions can also affect drug response. Knowing this, a control group or placebo group is included in clinical trials in an attempt to control for the effects of disease or changes which are unrelated to the treatment under investigation.

Clearly it is important to dissociate drug effects from confounding disease or placebo effects. More and more PK/PD models incorporate this concept into their framework in order to better capture the pharmacological response. As shown in Equation 4, drug response can be separated into a baseline component (disease or placebo portion) and the actual effects related to the drug. However, this baseline component is rarely static and the modeling of the disease effects as a function of time can prove to be a daunting task (46).

An early model for Alzheimer's disease was formulated in the early 1990s which showed that tacrine exerted a symptomatic effect on the disease since discontinuation of the drug resulted in continuing disease progression (47). Many other placebo and disease models have since been created including models for diabetes, schizophrenia, and Parkinson's disease (48).

The conception of additional disease models and the creation of a placebo response database are needed (26) but will require many collaborative efforts among the pharmaceutical industry, academia, and regulatory bodies.

Clinical Trial Simulation

Clinical trial simulations involve the replication of a particular trial design in order to assess the probability distribution of the possible outcomes. These simulations necessarily involve PK/PD models that may or may not include a disease progression component or a compliance model. Trial simulations must also account for variability associated with population kinetics, dynamics, and compliance (49).

Clinical trial simulations aim to reduce costs associated with the testing of various hypotheses, improve decision making, and the overall efficiency of drug development (26). They allow researchers to investigate different scenarios under certain assumptions made with the information on hand, such as "what happens to response if noncompliance increases by 10%?" Researchers can therefore conduct a multitude of "virtual" trials thereby identifying strengths and weaknesses associated with different study designs (50). By examining the outcome associated with different situations, researchers are then able to select optimal doses, dosing regimens, and study designs (51).

The results of clinical trial simulations can therefore provide a rational and scientific basis for making decisions (49). In essence, they help researchers design trials that are more likely to attain desired outcomes or prove particular hypotheses (50).

Mechanistic PK/PD Modeling

Early attempts at PK/PD modeling largely involved empirical models which were further refined by the advent of the link models and indirect models. Despite the improvements associated with the use of such models many experts believe that the predictive power of PK/PD models can benefit from a more mechanistic approach. This approach applies current PK/PD modeling methods from a more physiological perspective.

Mechanistic PK/PD modeling, also known as physiologically based PK/PD modeling, is still at its early stages and has been extensively reviewed in other publications (22,52). Essentially it distinguishes drug attributes (octanol-water partition coefficient, dissolution constants, ionization) from patient characteristics (organ volume and perfusion, homeostatic feedback mechanisms, transduction). This approach integrates drug-specific information with our knowledge of receptor theory, biology, pathophysiology, and system dynamics. Proponents of mechanistic PK/PD models suggest that physiologically based PK/PD models can provide better models for animal–human scaling or to determine effects in pediatric populations (52). Mechanistic PK/PD modeling can also help researchers better understand factors that can affect efficacy and toxicity, such as the expression of efflux transport proteins.

Despite its promising features, the use of mechanistic PK/PD modeling is not very widespread due in part to the complexity associated with this approach. Nonetheless continuing advances in bioinformatics will likely make this type of analysis more accessible and popular.

CONCLUSIONS

It is evident that an improved understanding of the pharmacodynamics of a drug through the use of tools such as modeling and simulations can benefit commercial drug developers throughout the drug development process. However, it is crucial that these tools and principles be applied at the beginning of the drug discovery process rather than as an afterthought during clinical trial stages. Modeling and simulation analyses are often complex and time-consuming, and therefore these processes must be undertaken as early as possible in the drug development process.

In the end a better appreciation for pharmacological responses, both desired and unwanted ones, will not only make the drug development process more efficient but will benefit patients who will be better served by a larger array of therapeutic options and optimal usage of these treatments.

REFERENCES

1. Shannon JA. The study of antimalarials and antimalarial activity in the human malarials. Harvey Lect 1946; 41:43–89.
2. Kola I, Landis J. Can the pharmaceutical industry reduce attrition rates? Nat Rev Drug Discov 2004; 3:711–715.
3. FDA. Innovation or Stagnation? Challenge and opportunity on the critical path to new medical products. March 2004. http://www.fda.gov/oc/initiatives/criticalpath/whitepaper.pdf. Accessed September 15, 2008.
4. DiMasi JA. The value of improving the productivity of the drug development process: faster times and better decisions. Pharmacoeconomics 2002; 20(suppl 3):1–10.
5. Lin JH. Pharmacokinetic and pharmacodynamic variability: a daunting challenge in drug therapy. Curr Drug Metab 2007; 8:109–136.
6. Undevia SD, Gomez-Abuin G, Ratain MJ. Pharmacokinetic variability of anticancer agents. Nat Rev Cancer 2005; 5:447–458.
7. Rawlins MD. Cutting the cost of drug development? Nat Rev Drug Discov 2004; 3:360–364.
8. Danhof M, de Lange EC, Della Pasqua OE, et al. Mechanism-based pharmacokinetic-pharmacodynamic (PK-PD) modeling in translational drug research. Trends Pharmacol Sci 2008; 29:186–191.
9. Kern SE, Stanski DR. Pharmacokinetics and pharmacodynamics of intravenously administered anesthetic drugs: concepts and lessons for drug development. Clin Pharmacol Ther 2008; 84: 153–157.
10. Williams RL. Dosage regimen design: pharmacodynamic considerations. J Clin Pharmacol 1992; 32:597–602.

11. Cawello W, Antonucci T. The correlation between pharmacodynamics and pharmacokinetics: basics of pharmacokinetics–pharmacodynamics modeling. J Clin Pharmacol 1997; 37:65S–69S.
12. Csajka C, Verotta D. Pharmacokinetic-pharmacodynamic modelling: history and perspectives. J Pharmacokinet Pharmacodyn 2006; 33:227–279.
13. Jusko WJ. Pharmacodynamics of chemotherapeutic effects: dose-time-response relationships for phase-nonspecific agents. J Pharm Sci 1971; 60:892–895.
14. Mick R, Ratain MJ. Statistical approaches to pharmacodynamic modeling: motivations, methods, and misperceptions. Cancer Chemother Pharmacol 1993; 33:1–9.
15. Abdel-Rahman SM, Kauffman RE. The integration of pharmacokinetics and pharmacodynamics: understanding dose-response. Annu Rev Pharmacol Toxicol 2004; 44:111–136.
16. Dayneka NL, Garg V, Jusko WJ. Comparison of four basic models of indirect pharmacodynamic responses. J Pharmacokinet Biopharm 1993; 21:457–478.
17. Sharma A, Jusko WJ. Characteristics of indirect pharmacodynamic models and applications to clinical drug responses. Br J Clin Pharmacol 1998; 45:229–239.
18. Colburn WA, Lee JW. Biomarkers, validation and pharmacokinetic-pharmacodynamic modelling. Clin Pharmacokinet 2003; 42:997–1022.
19. Biomarkers and surrogate endpoints: preferred definitions and conceptual framework. Clin Pharmacol Ther 2001;69:89–95.
20. Lesko LJ, Atkinson AJ Jr. Use of biomarkers and surrogate endpoints in drug development and regulatory decision making: criteria, validation, strategies. Annu Rev Pharmacol Toxicol 2001; 41:347–366.
21. Dingemanse J, Appel-Dingemanse S. Integrated pharmacokinetics and pharmacodynamics in drug development. Clin Pharmacokinet 2007; 46:713–737.
22. Mager DE, Jusko WJ. Development of translational pharmacokinetic-pharmacodynamic models. Clin Pharmacol Ther 2008; 83:909–912.
23. Temple R. Are surrogate markers adequate to assess cardiovascular disease drugs? JAMA 1999; 282:790–795.
24. Colburn WA. Biomarkers in drug discovery and development: from target identification through drug marketing. J Clin Pharmacol 2003; 43:329–341.
25. FDA Guidance for Industry: Exposure-Response Relationships: Study Design, Data Analysis, and Regulatory Applications, CDER/CBER, FDA. April 2003. http://www.fda.gov/cder/guidance/5341fnl.pdf. Accessed September 15, 2008.
26. Rolan P, Atkinson AJ Jr, Lesko LJ. Use of biomarkers from drug discovery through clinical practice: report of the Ninth European Federation of Pharmaceutical Sciences Conference on Optimizing Drug Development. Clin Pharmacol Ther 2003; 73:284–291.
27. Cohen A. Pharmacokinetic and pharmacodynamic data to be derived from early-phase drug development: designing informative human pharmacology studies. Clin Pharmacokinet 2008; 47: 373–381.
28. Mager DE, Wyska E, Jusko WJ. Diversity of mechanism-based pharmacodynamic models. Drug Metab Dispos 2003; 31:510–518.
29. Minto C, Schnider T. Expanding clinical applications of population pharmacodynamic modelling. Br J Clin Pharmacol 1998; 46:321–333.
30. Ette EI, Williams PJ. Population pharmacokinetics II: estimation methods. Ann Pharmacother 2004; 38:1907–1915.
31. Steimer JL, Mallet A, Golmard JL, et al. Alternative approaches to estimation of population pharmacokinetic parameters: comparison with the nonlinear mixed-effect model. Drug Metab Rev 1984; 15:265–292.
32. Chaikin P, Rhodes GR, Bruno R, et al. Pharmacokinetics/pharmacodynamics in drug development: an industrial perspective. J Clin Pharmacol 2000; 40:1428–1438.
33. Chien JY, Friedrich S, Heathman MA, et al. Pharmacokinetics/Pharmacodynamics and the stages of drug development: role of modeling and simulation. AAPS J 2005; 7:E544–E559.
34. Gobburu JV, Marroum PJ. Utilisation of pharmacokinetic-pharmacodynamic modelling and simulation in regulatory decision-making. Clin Pharmacokinet 2001; 40:883–892.
35. Lalonde RL, Kowalski KG, Hutmacher MM, et al. Model-based drug development. Clin Pharmacol Ther 2007; 82:21–32.
36. Meibohm B, Derendorf H. Pharmacokinetic/pharmacodynamic studies in drug product development. J Pharm Sci 2002; 91:18–31.
37. Perez-Urizar J, Granados-Soto V, Flores-Murrieta FJ, et al. Pharmacokinetic-pharmacodynamic modeling: why? Arch Med Res 2000; 31:539–545.
38. Rajman I. PK/PD modelling and simulations: utility in drug development. Drug Discov Today 2008; 13:341–346.

39. Sheiner LB, Steimer JL. Pharmacokinetic/pharmacodynamic modeling in drug development. Annu Rev Pharmacol Toxicol 2000; 40:67–95.
40. Zhang L, Sinha V, Forgue ST, et al. Model-based drug development: the road to quantitative pharmacology. J Pharmacokinet Pharmacodyn 2006; 33:369–393.
41. Sheiner LB. Learning versus confirming in clinical drug development. Clin Pharmacol Ther 1997; 61:275–291.
42. Aarons L, Karlsson MO, Mentre F, et al. Role of modelling and simulation in Phase I drug development. Eur J Pharm Sci 2001; 13:115–122.
43. Park K, Verotta D, Gupta SK, et al. Use of a pharmacokinetic/pharmacodynamic model to design an optimal dose input profile. J Pharmacokinet Biopharm 1998; 26:471–492.
44. FDA Guidance for Industry: Providing Clinical Evidence of Effectiveness for Human Drug and Biological Products, CDER/CBER, FDA. May 1998. http://www.fda.gov/cder/guidance/1397fnl.pdf. Accessed September 15, 2008.
45. Nieforth KA, Nadeau R, Patel IH, et al. Use of an indirect pharmacodynamic stimulation model of MX protein induction to compare in vivo activity of interferon alfa-2a and a polyethylene glycol-modified derivative in healthy subjects. Clin Pharmacol Ther 1996; 59:636–646.
46. Mould DR, Denman NG, Duffull S. Using disease progression models as a tool to detect drug effect. Clin Pharmacol Ther 2007; 82:81–86.
47. Holford NH, Peace KE. Results and validation of a population pharmacodynamic model for cognitive effects in Alzheimer patients treated with tacrine. Proc Natl Acad Sci U S A 1992; 89:11471–11475.
48. Chan PL, Holford NH. Drug treatment effects on disease progression. Annu Rev Pharmacol Toxicol 2001; 41:625–659.
49. Lockwood PA, Cook JA, Ewy WE, et al. The use of clinical trial simulation to support dose selection: application to development of a new treatment for chronic neuropathic pain. Pharm Res 2003; 20:1752–1759.
50. Girard P. Clinical trial simulation: a tool for understanding study failures and preventing them. Basic Clin Pharmacol Toxicol 2005; 96:228–234.
51. Bonate PL. Clinical trial simulation in drug development. Pharm Res 2000; 17:252–256.
52. Danhof M, de Jongh J, De Lange EC, et al. Mechanism-based pharmacokinetic-pharmacodynamic modeling: biophase distribution, receptor theory, and dynamical systems analysis. Annu Rev Pharmacol Toxicol 2007; 47:357–400.

34 Statistical Concepts in the Design and Analysis of Clinical Trials

Richard Kay

Independent Statistical Consultant, rkstatistics, Great Longstone, Bakewell, U.K.

INTRODUCTION

Statistical concepts clearly play a key role in both the design of clinical trials and the subsequent analysis of the clinical trial data. The purpose of this chapter is to provide a broad overview of the statistical methodology used in trials. In recent years, we have also seen the publication of numerous regulatory guidelines that directly relate to statistical issues. These build on the foundations of the general ICH E9 Guideline "Statistical Principles in Clinical Trials" (1). The material that follows in this chapter is very much motivated by what these guidelines say, and the way that methodology has developed and been applied in the pharmaceutical industry.

The chapter starts with some discussion of randomization and then presents the ideas surrounding hypothesis testing and confidence intervals. The important design issue of the sample size calculation and its association with power is then discussed followed by intention to treat and missing data and adjusting the analysis for baseline factors through, for example, analysis of covariance (ANCOVA). The analysis of survival data and a discussion of Kaplan-Meier curves and the hazard ratio follow. The important issue of multiplicity is always a major concern of regulators and a discussion of this follows. Interim analyses, data monitoring committees, and adaptive design are addressed next followed by equivalence and noninferiority. The chapter concludes with a discussion on the role of the statistician in clinical trials.

Many topics are covered but I am also conscious that, because of constraints on space, there are some omissions. There is no discussion on nonparametric methods, in part because these methods increasingly seem to gain little favor within the regulatory setting. Meta-analysis is a further omission neither have I included any discussion on Bayesian methods. The application of Bayesian methods in the pharmaceutical industry has increased in recent times, in particular in relation to dose finding and adaptive designs, and maybe when the time comes to write an update of this section this topic will have sufficiently settled down to be able to say something useful. The interested reader is referred to the book by Spiegelhalter et al. (2) for a fairly recent and comprehensive coverage.

RANDOMIZATION

Randomization is the cornerstone of the clinical trial where patients are randomized to one of the two or more treatments being compared. This process ensures that there is no bias in the allocation of patients to the treatment groups and underpins the validity of the statistical comparisons of the treatments.

Unrestricted and Block Randomization

Unrestricted, or simple, randomization is based on a computer-generated random list of treatment allocations with the next patient recruited being assigned to the group corresponding to the next allocation. Whilst this works acceptably well in large and small trials, it may result in treatment groups that are unbalanced in terms of numbers of patients. From a statistical perspective it is usually best to have equal numbers, or approximately equal numbers, of patients in each of the treatment groups and block randomization is used to ensure this. Block randomization places a restriction on the construction of the randomization list. For example, with two treatment groups A and B each set of four allocations would be constrained to contain precisely two allocations to A and two allocations to B. If the trial then recruits a multiple of four patients then there would be exactly equal numbers of patients allocated to the two treatment groups.

If the total number of patients recruited is not a multiple of four then there will be approximate balance in terms of the sizes of the groups.

Stratified Randomization

Randomization will produce groups, at least in the long run, that are balanced in terms of the kinds of patients recruited—the "mix" of patients. By chance, however, imbalances in the patient mix could occur which could skew the subsequent treatment comparisons. Although there are methods available that could account for such imbalances at the data analysis stage (see Adjusting the Analysis, below), these methods add complexity and it is worthwhile at the design stage to avoid any major imbalances of this kind. Stratified randomization is used to force the balance with regard to the mix of patients recruited into the treatment groups. Following identification of important baseline factors predictive of patient outcome, randomization is stratified according to the levels of these baseline factors, that is, separate randomization lists are produced for the various combinations of baseline factors. For example, if age (<65, ≥65) and baseline risk (high, low) are thought to be important predictors of outcome, based on historical data, then the stratified randomization would involve four strata: (<65, High), (<65, Low), (≥65, High), (≥65, Low), with a distinct randomization list produced for each of these strata. The stratum for each patient entering the trial would then be identified based on baseline data and the next allocation in the corresponding list would indicate the treatment group for that patient. Following such a scheme would provide balance in relation to age and baseline risk. Combining stratified randomization and block randomization, where each randomization list has the block structure, ensures balance in terms of both mix and number of patients.

It is usually the case in a multicenter trial, where the centers are fairly large, to stratify the randomization by center. It is not that center per se is important in relation to predicting outcome but it is more that "center" may hide several factors, potentially unquantifiable, that could impact on outcome: for example the skill of the surgeon, the quality of ancillary care, the subpopulation of patients recruited in the different centers, and so on.

> It is advisable to have a separate random scheme for each centre, i.e., to stratify by centre or to allocate several whole blocks to each centre. Stratification by important prognostic factors measured at baseline (e.g., severity of disease, age, sex, etc.) may sometimes be valuable in order to promote balanced allocation within strata; .... [ICH E9: Statistical Principles for Clinical Trials (1)]

For practical and statistical reasons only a limited number of baseline factors can be accounted for effectively through stratified randomization. With many factors, and therefore a large number of strata, many of the strata could end up with only a handful of patients in them. With a block of size four, there is then no guarantee that balance, both with regard to numbers and mix, will be achieved. In practice this is not too restrictive. Providing that balance can be achieved on, say, center and the two most important predictors of outcome then this is usually enough to ensure a valid analysis and comparison of treatments.

Dynamic Allocation

There may, however, be a desire to go beyond this and achieve balance on the basis of several baseline factors. Dynamic allocation is a methodology that, at least in theory, can be used to achieve such balance. In this methodology a randomization list is not prespecified. Allocation for the next patient entering the trial is based on the characteristics of that patient and the current mix of patients across the treatment groups. So, for example, suppose that at a particular stage in the trial, treatment group A contains proportionately more males than treatment group B and that the next patient to enter is male. Biasing the randomization (say 80–20) to allocate this patient to treatment group B would then help to redress the existing gender imbalance. Dynamic allocation can take account of many factors simultaneously providing these are predefined. On the basis of each of these factors, the current balance in the trial is assessed and the new patient, depending on their values for the identified factors, is allocated to that treatment group that reduces the overall imbalance. If a random element in this allocation is not used so that the patient is simply placed in the treatment group that is optimal for reducing the imbalance then the procedure is known as minimization. In the ICH E9 document regulators insisted however that a random element be used.

> Dynamic allocation is an alternative procedure in which the allocation of treatment to a subject is influenced by the current balance of allocated treatments and, in a stratified trial, by the stratum to which the subject belongs and the balance within that stratum. Deterministic dynamic allocation procedures should be avoided and an appropriate element of randomisation should be incorporated for each treatment allocation. [ICH E9: Statistical Principles for Clinical Trials (1)]

There has, however, been considerable debate about the validity of this procedure subsequent to this recommendation and in relation to the properties of standard statistical procedure in the analysis of data from trials designed in this way. As a consequence regulators now advise against the use of dynamic allocation.

> techniques of dynamic allocation such as minimisation are sometimes used to achieve balance across several factors simultaneously. Even if deterministic schemes are avoided, such methods remain highly controversial. Thus applicants are strongly advised to avoid such methods. [CPMP: Adjustment for Baseline Covariates (3)]
>
> See Ref. (4) for a comprehensive review and discussion of dynamic allocation.

HYPOTHESIS TESTING AND CONFIDENCE INTERVALS

Constructing Hypotheses

Hypothesis testing, with the associated p value for assessing statistical significance, provides the primary method for evaluating the relative effectiveness of our treatments. In comparisons based on detecting the superiority of a test treatment to a control treatment the null hypothesis (H_0) will be an appropriate specification of "equality" of the treatments while the alternative hypothesis (H_1) will capture "difference." If the endpoint is continuous, for example the fall in cholesterol level between baseline and end of treatment, then H_0: $\mu_1 = \mu_2$, states that the mean fall in each of the two treatment groups is the same while the alternative H_1: $\mu_1 \# \mu_2$, states that the two means are different. If the endpoint is binary, for example cure at the 21-day test of cure visit, then the null and alternative hypotheses will specify equality and inequality respectively for the cure rates θ_1 and θ_2. The μ and θ parameters that are being compared here relate to what would happen in the populations as a whole from which the patients are being drawn where they are to be treated by either the test or the control treatment.

The *p* Value

The data from the trial will provide evidence, in the form of sample means $\overline{x}_1$ and $\overline{x}_2$ for continuous endpoints, and sample cure rates r_1 and r_2 for binary endpoints and in each case an associated p value comparison. The p value is the probability associated with the observed difference in the sample means (or in the cure rates for the second example) where the means μ_1 and μ_2 are truly identical. More specifically the p value measures the probability associated with a difference at least as big as $|\overline{x}_1 - \overline{x}_2|$ in either direction when the null hypothesis H_0: $\mu_1 = \mu_2$ is true. If the p value is sufficiently small, and conventionally we use ≤ 0.05 as our operational definition of sufficiently small, then we have enough evidence to suggest that the null hypothesis is false—the data is not consistent with "equal" treatments. On this basis we reject the null hypothesis and accept the alternative hypothesis as true and talk in terms of statistically significant differences, at the 5% level, between the treatment means. Alternatively, if $p > 0.05$, we do not have enough evidence to reject the null hypothesis and cannot conclude that the treatment means are different. We then talk in terms of "nonsignificant differences."

One-Sided and Two-Sided Tests

The alternative hypotheses specified above are two-sided in the sense that they do not prespecify a direction for the difference. The subsequent p value is calculated by considering differences that could occur in either direction, $\mu_1 > \mu_2$ or $\mu_1 < \mu_2$. These so-called two-sided or two-tailed p values are required by regulators who are interested in detecting differences in either direction—that is that the test treatment is either better or worse than the control. A sponsor may argue that only differences in favor of the test treatment are of interest and therefore prefer the use of p values that are one-sided and calculated by considering only differences in the direction of interest. As a result of the symmetry that exists in most settings the one-sided p value will invariably be exactly one-half of the two-sided p value. However, the cutoff for statistical significance with one-sided p values is set at 0.025 rather than 0.05, and so in terms of drawing a statistically significance inference there will be no difference between the use of two-sided and one-sided p values.

Simple Test Procedures

When statisticians talk in terms of undertaking a statistical test they are simply referring to the method that will be used to calculate the p value. The particular statistical test that we use will depend primarily on two things—the type of endpoint that we are dealing with and the nature of the comparison. For example, for a continuous endpoint in a parallel group study the two-sample t-test (also called the unpaired t-test) provides the appropriate methodology for the p value calculation. For a binary endpoint in a parallel group setting, it will be the Pearson chi-square test that labels the method of calculation for the p value. For a continuous endpoint in a crossover trial it is the paired t-test that gives the appropriate p value. Later we will talk in detail about the analysis of time to event data, and it is the log-rank test that will give us the p value for the comparison of treatments.

Confidence Intervals

The analysis of data invariably involves constructing confidence intervals in addition to calculating p values. A confidence interval provides a range of plausible values for a particular parameter that are supported by the data. Consider the example from the previous section where we were looking to compare the mean fall in cholesterol level between a test treatment group (mean μ_1) and a control group (mean μ_2). The parameter of major interest is $\mu_1-\mu_2$, the additional benefit afforded to the test treatment over and above the benefit seen in the control group. A 95% confidence interval for μ_1-μ_2 is constructed to give a range of values for this treatment difference, based on the data, and we can be 95% confident that the true difference lies within this range.

It is standard practice to use 95% confidence intervals and this is in line with the use of the 5% level of significance in significance testing. There is in fact a mathematical connection between confidence intervals and p values in this setting. If the 95% confidence interval for μ_1-μ_2 does not include zero, then the p value for the treatment comparison will be <0.05. Conversely if the confidence interval includes zero then the p value will be >0.05. In other settings, for example for the comparison of two cure rates, this is an approximate relationship rather than a mathematical one. If the 95% confidence interval for the difference in the cure rates includes/excludes zero the p value will be $>0.05/<0.05$. In this and many other settings, the closeness of the approximation will depend on the method of calculation for the confidence interval and the method for the calculation of the p value.

POWER AND SAMPLE SIZE

Type I and Type II Errors

Unfortunately the hypothesis-testing environment with the p value determining the conclusion does not always lead us to draw the correct conclusion. Sometimes we are misled by the data. There are two potential mistakes: the false-positive and the false-negative. It can happen that when the null hypothesis is true the data by chance give $p \leq 0.05$. Under these circumstances we would reject the null hypothesis and falsely conclude that μ_1#μ_2. This is the false-positive and is termed the type I error. In contrast it can also happen that even though μ_1#μ_2 we fail to get $p \leq 0.05$ and are therefore unable to reject the null hypothesis and conclude that the treatments are different. This is the false-negative and is termed the type II error. Type I and type II errors are sometimes labeled the α and β errors, respectively.

There is a very important theorem in statistics, the Neyman-Pearson lemma, which tells us that with a fixed sample size it is not possible to eliminate the potential for making these two

Table 1 Power for the Two-Sample *t*-Test, $\alpha = 0.05$ (two-sided), sd = 1.0 mmol/l

$\mu_1 - \mu_2$	Power, $n = 40$/group	Power, $n = 80$/group
0.25	19.7%	34.9%
0.50	59.8%	88.2%
0.75	91.2%	99.7%
1.00	99.3%	100.00%

mistakes. If we try to bring either one of them down then the other one is increased so that the best we can do is to control them. The type I error is fixed at 0.05. If the null hypothesis is true then we will make the false-positive mistake on 5% (1 in 20) occasions. That this mistake occurs on 5% of such occasions is entirely determined by the use of 0.05 as the cut-off for statistical significance. We will see $p \leq 0.05$ purely by chance when $\mu_1 = \mu_2$, 5% of the time.

Control of type II error is generally thought through in terms of the linked concept of power. This will be covered in the next section.

Type II Error and Power

The connection between type II error (expressed here as a percentage) and power is Power = 100% – type II error. So, for example, if the type II error is set at 10% then power = 90%. Loosely speaking type II error is missing a true difference while power is capturing that true difference. To control type II error at, say, 10% we design our trial in such a way that it gives power equal to 90%.

It is possible to calculate power in advance of running a study by speculating about what might happen. Suppose that we are planning a parallel group clinical trial in cholesterol lowering where the primary endpoint is the fall in LDL cholesterol from baseline to the end of a 12-week treatment period with 40 patients per treatment group. The treatment means are to be compared using the two-sample *t*-test. Assume finally that historical data has provided an estimated value for the inherent patient-to-patient variation through the standard deviation of the primary endpoint and this is equal to 1.0 mmol/L. Loosely speaking the standard deviation is the "average" distance of each of the patient observation from the corresponding mean value. Table 1 provides calculated values for power for various true differences in the treatment means. So for example if the true difference in the treatment means were 0.50 mmol/L then this trial, with 40 patients per group, would only have power equal to 59.8%—that is a 59.8% probability of coming out with a significant *p* value, or the equivalent that there would be a 40.1% probability of coming out with a nonsignificant *p* value, the type II error. In contrast, if the true treatment difference were 0.75 mmol/L then the power would be 91.2% with then only an 8.8% type II error. If in the context of the trial setting a difference of 0.5 mmol/L was a clinically important difference then this trial is underpowered with an unacceptable value for type II error. The only real solution is to increase the level of power by an increase in the sample size. The final column in Table 1 gives the power values for a doubling of the sample size to 80 patients per group. The power now to detect a difference of 0.50 mmol/L has increased to the much more acceptable level of 88.2%, a type II error of 11.8%. All of the other values for power have also increased.

Calculating Sample Size

The considerations of the previous section give us the basis for calculating sample size. We first specify what level of effect we are interested in detecting and then choose that sample size which provides sufficient power at the specified difference. This prespecified level of effect of interest is often based on clinical considerations and is usually termed the clinically relevant difference. It can however also be chosen on either commercial or practical grounds or indeed a combination of all three of these.

Regulators are primarily concerned with the control of type I error. They do not want to registering drugs that do not work. They are less concerned with the control of type II error. That is more the concern of the sponsor who does not want to be failing to detect important differences should they exist. The regulators only therefore make recommendations regarding type II error and power.

> The probability of type II error is conventionally set at 10% to 20%; it is in the sponsor's interest to keep this figure as low as feasible especially in the case of trials that are difficult or impossible to repeat. [ICH E9: Statistical Principles for Clinical Trials (1)]

INTENTION TO TREAT AND MISSING DATA

The Principle of Intention to Treat

Randomization produces groups that are balanced in terms of baseline characteristics and avoids any bias in the allocation of particular kinds of patients to the different treatment groups. At the analysis stage, it is equally important to analyze the data according to the randomization process and in line with this the principle of intention to treat says that the primary treatment comparison should include all patients based on the treatment groups to which they were randomized. By excluding patients who withdraw from treatment, deviate from the trial procedures, take a banned concomitant medication, give missing data, and so on will potentially bias the treatment comparison since there is no guarantee that the resultant "groups" will be alike in terms of the mix of patients. In practice complete compliance with this principle can be very difficult and we will discuss in the next section how compromises can be made to enable an analysis close to the ideal of including all randomized patients to be undertaken.

The Practice of Intention to Treat

Regulators accept that full compliance with the principle of intention to treat is simply not going to be possible and they ask us to provide, for the primary analysis, an analysis based on the so-called full analysis set (FAS). This is the set of patients that gets us as close as we possibly can get to the ideal of including all randomized patients.

> The intention-to-treat principle implies that the primary analysis should include all randomised subjects. Compliance with this principle would necessitate complete follow-up of all randomised subjects for study outcomes. In practice this ideal may be difficult to achieve, ... [ICH E9: Statistical Principles for Clinical Trials (1)]
>
> In this document the term "full analysis set" is used to describe the analysis set which is as complete as possible and as close as possible to the intention-to-treat ideal of including all randomised subjects.

The alternative term modified intention to treat is sometimes used for FAS. In practice the kinds of patients we exclude from the FAS are as follows:

- Patients who do not satisfy the inclusion/exclusion criteria
- Patients who do not take any medication
- Patients who provide no on-treatment efficacy data

Excluding patients who do not satisfy the inclusion/exclusion criteria should not cause bias providing patients in each treatment group receive equal scrutiny for such violations. Further the final two conditions often go together. For each exclusion, it is important to evaluate the potential for bias and possibly undertake sensitivity analyses to ensure that including or excluding certain patients from the FAS does not affect the conclusions materially.

The Per-Protocol Set

A per-protocol analysis which excludes patients who do not comply with the protocol to a defined extent is an attempt to provide an analysis which is closer to the basic scientific question being asked. This is in contrast to the analysis based on the FAS which is more aligned with a broader pragmatic question accepting that, in practice, patients will not comply fully with the rules set down in the protocol regarding the taking of the medication, visit discipline, and so on. It must be recognized however that analyses based on the per-protocol set, where patients are excluded as a result of deviations from the protocol during the course of treatment, are subject to bias and that this bias can very often be substantial. Examples of per-protocol analyses are those involving only completers, patients taking at least X% of the prescribed medication, and responder analyses where patients who respond over an initial period are then included in a subsequent comparison.

Missing Data

The previous sections have discussed the construction of both the FAS and the Per-Protocol Set in terms of which patients are included and which patients are excluded. There remains however the issue of how to include such patients who have missing data. For example if the primary endpoint is the fall in diastolic blood pressure over a 12-week period, how do you include the patient who withdrew from treatment at week 8 because of a serious side effect and who provides no further data? The most common approach with continuous data is to use the principle of last observation carried forward (LOCF). In the above example, the diastolic blood pressure at week 8 would be used as a substitute for the missing value at week 12 to enable that patient to be included in the analysis. One way to think about the validity of this approach is to consider the week 8 value as an appropriate measure of what has been achieved with this patient on the assigned treatment rather than see the week 8 value as some kind of surrogate for week 12. In this way the use of LOCF can be viewed as a pragmatic evaluation of the relative efficacy of the treatments accepting that some patients will not be able to tolerate treatment for 12 weeks. In some therapeutic settings however the use of LOCF is controversial, for example in multiple sclerosis and Alzheimer's.

The CPMP Points to Consider paper on "Missing Data" (5) provides further discussion on the methodologies for accounting for missing data. One particular issue raised concerns the extent to which we "allow" missing data to occur. It is certainly true that the presence of missing data can cause enormous problems for the analysis and the potential for its occurrence should be minimized by

- designing trials in ways that do not place too much burden on patients by allowing drug holidays or dose reductions when patients suffer side effects and
- following up patients who withdraw from medication.

On this second point, it is perfectly possible to have informed consent at two levels, one relating to the data collection and one relating to the medication. Patients can withdraw their consent in relation to the medication without withdrawing their consent to data collection.

> it is extremely important to avoid the presence of unobserved measurements as much as possible, by favouring designs that minimise this problem, as well as strengthening data collection regardless of the patient's adhesion to the protocol and encouraging the retrieval of data after the patient's drop-out. [CPMP: Missing Data(5)]

Considerations above primarily relate to the evaluation of efficacy rather than in relation to the routine collection of safety information. Safety evaluation will be based on the Safety Set—the collection of patients who take at least one dose of study medication, irrespective of whether they have satisfied the inclusion/exclusion criteria or have provided any on-treatment efficacy data.

ANALYSIS ADJUSTED FOR BASELINE FACTORS

Extending the Simple Tests

The simple tests of significance outlined above can be extended in a straightforward way to take account of factors measured at baseline, for example center, baseline risk, age, and so on. There are several reasons why we may want to do this. Firstly, if the treatment groups are imbalanced with respect to these factors then an adjusted analysis can account for this. Secondly, an adjusted analysis can increase the power of the treatment comparison and make it easier to detect differences should they exist. Finally, these adjusted analyses provide a framework within which we can evaluate the homogeneity of the treatment effect across subgroups of patients defined by baseline factors.

For continuous data, the extension of the two sample t-test comparing means that incorporates adjustment for baseline factors is the two-way analysis of variance (ANOVA). For binary data, the corresponding extension of the chi-squared test is the Cochran-Mantel-Haenszel (CMH) chi-squared test. Each of these procedures provides a p value comparison of the treatments that is adjusted for the baseline factors incorporated into the analysis.

For example, suppose that patients are classified into one of three age categories: <50, ≥50 but <65, and ≥65, and we are comparing two treatments A and B. Two-way ANOVA will base a treatment comparison of an "averaging" over the age categories of the three ($\overline{x}_1 - \overline{x}_2$) differences (where $\overline{x}_1$ is the mean on A and $\overline{x}_2$ is the mean on B) in the treatment means. This averaging process will take account of the numbers of patients within each of the age categories so that an age category containing 100 patients will be given more weight than a category with only 10 patients. If there were imbalances in terms of the numbers of A and B patients within each age category coupled with an age effect, for example younger patients responding better than older patients, then a simple two sample t-test could give a biased result and the adjustment will correct for this. In addition, irrespective of imbalances in the patient age mix between groups A and B, ANOVA gives a more efficient analysis, increasing the power to detect treatment differences. Finally, ANOVA enables an evaluation of the consistency of the treatment effect by providing a p value for the treatment by age (treatment × age) interaction comparing the treatment differences across each of the three age categories.

Analysis of Covariance

The methods for adjusting the analysis for baseline factors discussed in the previous section have the limitation that they can only account for a small number of such factors at the same time. For example, with four centers and three age categories there will be 12 levels of the two factors combined and the ANOVA and CMH methodologies then start to break down mathematically, especially with small sample sizes.

ANCOVA for continuous data and logistic regression for binary data are methods that allow the inclusion, at least in theory, of several factors simultaneously. Both of these techniques are based on constructing statistical models that specify the way in which the endpoint being evaluated depends upon treatment and the baseline factors (called covariates). The trial data is then used to estimate the contribution of treatment and each factor simultaneously on predicting the value of the endpoint, and through this, a p value for the contribution of treatment can be obtained. If this p value is ≤ 0.05, then the data are telling us that treatment significantly (at the 5% level) affects outcome—we have detected statistically significant treatment differences having adjusted for the contribution of the covariates.

ANCOVA and logistic regression are both forms of adjusted analysis that have the same advantages over simple tests as do ANOVA and the CMH test. Additionally, as mentioned above, they are able to account for several baseline factors simultaneously.

CLINICAL AND STATISTICAL SIGNIFICANCE

Statistical significance and clinical importance are different things. Statistical significance is achieving $p \leq 0.05$ and with a large sample size this could be obtained when in fact the true difference between the treatments is small—less than what would be agreed as being of importance clinically. In contrast, in a trial with a small sample size achieving $p \leq 0.05$ may be extremely unlikely even when the true treatment differences are large simply as a result of low power. These arguments are well captured by the following quote from Gardner and Altman (6):

> Presenting p values alone can lead to them being given more merit that they deserve. In particular, there is a tendency to equate statistical significance with medical importance or biological relevance. But small differences of no real interest can be statistically significant with large sample sizes, whereas clinically important effects may be statistically nonsignificant only because the number of subjects studied was small.

This particular issue is not well understood. The p value by itself says nothing about the magnitude of the treatment effect. It is simply asking whether there is enough evidence to be able to reject the hypothesis that the treatments are the same. Clinical importance is all to do with the magnitude of the effect. Large differences are important, small differences are unimportant. The best way to gain information about the magnitude of the treatment effect is to look at the observed treatment difference, for example, $\overline{x}_1 - \overline{x}_2$, and the associated confidence interval. Presenting both p values and confidence intervals gives the full picture of both statistical and clinical significance, and this is the recommended approach within the regulatory setting.

SURVIVAL ANALYSIS

Time to Event Data and Censoring

In clinical trials, it is quite common to have an endpoint that is the time to some well-defined event. Examples include time to death, disease-free survival in oncology, time to rash healing in *Herpes zoster*, time to first seizure in epilepsy, time to resolution of symptoms in flu, and time to first elevation of AST or ALT above $3 \times$ ULN. In each case time would be measured from the point of randomization. The feature of data of this kind that separates them from other data types is the occurrence of censoring. Suppose, for example, that we are looking at time to death (survival time) in a two-year oncology study. At the end of the two-year follow-up period some patients will still be alive, and we will not have an outcome measure in those cases. We have partial information in that for those patients still alive at two years we know that survival time is at least two years. These are termed censored observations, and it is inevitable that when looking at time to event there will be patients in whom the event is not observed over the duration of the trial. In a similar way, censored observations arise from patients who are lost to follow-up or who are withdrawn from the study.

When we have a data that gives a mixture of exact and censored event times, it is not possible even to do the simple things, for example to calculate the mean time to event. This whole area tends to go under the heading of Survival Analysis in part because the methodology that was developed many years ago was applied primarily to mortality studies where the endpoint of interest was indeed time to death. Nowadays, however, we apply these same techniques to a whole range of time to event type endpoints.

Kaplan-Meier Curves

Time to event data is most commonly displayed in Kaplan-Meier curves. Figure 1 displays the Kaplan-Meier survival curves from a placebo controlled trial evaluating carvedilol in patients suffering severe chronic heart failure (7) at the first planned interim analysis. The trial was stopped following this analysis.

The Kaplan-Meier method in general estimates the probability of remaining event-free over time based on the patterns of events and censored observations in the data. Where the event is death, as in this example, this is the probability of surviving as a function of time. It is clear from Figure 1 that carvedilol is effective in prolonging survival. The "number of patients at risk" along the foot of figure gives the numbers of patients alive and in the trial (not yet censored) at various time points from the point of randomization in each of the two treatment groups. This allows us to make at least an informal assessment of the amount of information

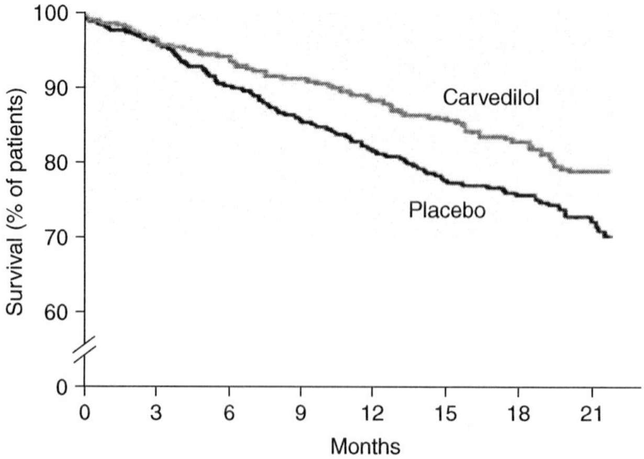

No. of patients at risk

Placebo	1133	937	703	580	446	286	183	114
Carvedilol	1156	947	733	620	479	321	208	142

Figure 1 Kaplan-Meier plots for time to death in placebo and carvedilol groups.

in the curves as time increases. In particular when these numbers become small, it should be recognized that the curves at those time points are based on limited amounts of data and are therefore less precise.

While the Kaplan-Meier method gives a visual impression of the differential rates at which the event of interest is occurring, it does not provide a p value comparison.

The Log-rank and Other Tests

The most common approach for a formal p value comparison for time to event data across two treatment groups is the log-rank test (8). For the data displayed in Figure 1, the log-rank p value was 0.00013, a highly significant result. Survival is significantly prolonged in the carvedilol group. It is also possible to take account of a small number of baseline factors in a stratified analysis. This stratified log-rank test is akin to the ANOVA procedure for continuous (uncensored) data.

The log-rank test is most powerful and therefore most relevant for settings where the Kaplan-Meier curves gradually separate out over time as in Figure 1. However not all patterns seen display this relationship, and it is not infrequent to see two curves that separate out early and converge later in time. Such a pattern would be seen, for example, in looking at time to resolution of symptoms in influenza. In a placebo controlled trial with an effective test treatment, there would be fairly rapid resolution in many patients in the test treatment group over a short period of time, say three to five days, while symptoms would only gradually resolve in the placebo group. Out at 14 days however most patients will be free of symptoms, except perhaps the subset of smokers, and the proportions with symptoms will be approximately the same irrespective of treatment group. The log-rank test is not especially sensitive in terms of detecting departures from the null hypothesis of equal time to event curves of this kind which display large early differences but with the curves coming together again later in time. A better procedure in this case is the Gehan-Wilcoxon test (9). It is important therefore to be able to anticipate the kinds of departures from equality that are anticipated based on experience with the particular endpoint and the treatments being compared and to prespecify the test procedure that is best suited to detecting those treatment differences.

Hazard Ratios

A useful measure of treatment difference in time to event analysis is the hazard ratio. In order to be able to understand what a hazard ratio is, you need to know first of all what a hazard rate is. The hazard rate is the conditional event rate through time. For example, considering death as the event, suppose that in a cohort of 1000 patients 45 die in month 1. The hazard rate for month 1 is 45/1000. Now suppose that 40 patients die in month 2, the hazard rate for month 2 is 40/955, the death rate in month 2 among those patients still alive at the start of month 2. In month 3, suppose that 30 patients die; the hazard rate for month 3 is 30/925 and so on. The hazard rate will of course not necessarily be constant. However, if we have two treatment groups then the ratio of the two hazard rates may well be approximately constant. If this is the case then it is appropriate to calculate a single value for that ratio. This is the hazard ratio, the ratio of the hazard rates, the ratio of the (conditional) death rates across the two groups.

By convention, the hazard rate for the test treatment group is placed in the numerator while the hazard rate for the test treatment group is placed in the denominator. A hazard rate of 1 corresponds to equal treatment groups. A hazard rate < 1 tells us that the hazard rate in the test treatment group is lower than the hazard rate in the control group and vice versa for a hazard rate > 1. In the carvedilol trial discussed earlier, the hazard ratio was 0.65 so that overall the death rate in the carvedilol group was 65% of the death rate in the placebo group, a 35% reduction over the period of follow-up of approximately 21 months.

The Proportional Hazards Model

There may well be interest in adjusting the analysis for several baseline covariates. As with ANOVA, the stratified log-rank test is somewhat restricted in terms of the number of factors that can be incorporated in such an analysis. Also the log-rank procedures, while providing p value comparisons, do not directly provide a measure of the treatment difference in terms of a hazard ratio. The proportional hazards model (10) is a modelling framework, based on the hazard ratio, that allows us to incorporate both treatment and covariates in the analysis. Including covariates in this way gives an adjusted hazard ratio and in particular removes any bias in the estimated treatment differences caused by baseline imbalances in the covariates. The proportional hazards model also allows the investigation of treatment by covariate interactions in order to judge the homogeneity of the treatment effect.

Independent Censoring

The methods of statistical inference set down so far in this subsection, that is the log-rank and Gehan-Wilcoxon tests and the proportional hazards model, assume that the mechanism producing the censored observations is independent of the event process. Usually in clinical trials the majority of the censored observations are due to patients completing planned follow-up without the event occurring and in this case there are no problems arising, the censoring mechanism is independent. If however patients were removed systematically from the study (and therefore censored) because they were at higher (or lower) risk of suffering the event, then the censoring mechanism would not be independent and biased estimates of the hazard ratio and incorrect p values would be produced. An example of this would be patients removed from the study because of serious adverse events where the event of interest is death. It could be argued that those patients are at higher risk of dying than other similar patients remaining in the study. It is important therefore to consider in each case the mechanism producing the censored observation to ascertain whether there could be a link to the risk of the event of interest. The usual way to gauge the possible extent of the bias produced by dependent censoring is to undertake sensitivity analyses. In the earlier example undertaking an analysis where the serious adverse events are treated as if they were deaths, a kind of worst-case assumption, and reanalyzing the data would produce hopefully conclusions that were unchanged from those with such events treated as censorings. If the conclusions are changed then it would be difficult to justify a robust conclusion.

MULTIPLICITY

Inflation of the Type I Error Rate

Under the usual conditions of statistical testing, where tests are conducted at the 5% level, every test has the potential to give a false-positive result. If the treatments truly are identical in terms for example of the mean response then there is a 1 in 20 probability of seeing $p \leq 0.05$ purely by chance. If under these same circumstances several tests are undertaken then each such test has this 5% false-positive possibility and the likelihood of seeing at least one spurious positive result increases as the number of tests increases. There are numerous ways in which this multiple testing (or multiplicity) can arise and the following are some examples:

- Multiple primary endpoints
- Multiple treatment comparisons in a trial with more than two treatments
- Comparing treatments within different subgroups
- Comparing treatments at various time points
- Interim analyses

The problem arises when several analyses are undertaken, each with the potential for the false-positive, and the positive results are then selected ("cherry-picked") as being confirmatory for the superiority of the test treatment. To take an extreme case suppose we had 20 endpoints and only one of those turned out to be statistically significant and in favor of the test treatment then it would clearly be inappropriate to claim this as confirmatory evidence that the test treatment is effective. Remember 1 in 20 p values will be statistically significant purely by chance with "equal" treatments.

Bonferroni Correction

One fairly simple way to account for multiplicity and allow the cherry-picking to take place is to adjust the level at which statistical significance is declared. For example, with five tests, only declaring significance at the 1% level rather than the 5% level would maintain the overall (collective) type I error across all of the five tests at 5%. This is the so-called Bonferroni correction.

For m tests significance is declared only if the p value is less than or equal to the adjusted significance level $= 0.05/m$. The more tests that are undertaken, the greater the price paid for the luxury of being able to cherry-pick the results. The division of 0.05 by m, the number of tests, maintains the overall type I error rate at 5%.

The Bonferroni correction is a fairly crude and conservative approach, and for many situations there are much better strategies for adjusting the significance level to account for multiplicity.

Regulatory View

Regulators are rightly concerned about multiplicity. As has been said earlier, regulators do not want to be registering drugs that do not work and it is important from this perspective that the type I error is strictly controlled.

> When multiplicity is present, the usual ... approach to the analysis of clinical trial data may necessitate an adjustment of the type I error. Multiplicity may arise, for example, from multiple primary variables, multiple treatment comparisons, repeated evaluation over time, and/or interim analyses. Methods to avoid or reduce multiplicity are sometimes preferable when available, such as the identification of the key primary variable (multiple variables), the choice of a critical treatment contrast (multiple comparisons), the use of a summary measure ... (repeated measures). In confirmatory analyses, any aspects of multiplicity which remain after steps of this kind have been taken should be identified in the protocol; adjustment should always be considered and the details of any adjustment procedure or an explanation of why adjustment is not thought to be necessary should be set out in the analysis plan. [ICH E9: Statistical Principles for Clinical Trials (1)]

The regulators point to several ways in which multiplicity can arise and offer solutions. With multiple endpoints, choose a single primary endpoint. This is then the basis for a confirmatory claim, there is no opportunity for cherry-picking and the usual 5% level of significance can be used. When the trial involves multiple treatment groups, restrict the primary comparisons to only those that form part of a confirmatory claim. For example, in a late phase II trial with three test treatment groups (low dose, medium dose, high dose) and a placebo group, the primary comparisons could be restricted to just the three dose levels against placebo. A Bonferroni correction would then allow each of these comparisons to be undertaken with an adjusted significance level equal to $0.05/3 = 0.017$ and claims could be made on those differences that are statistically significant at this adjusted 1.7% level. When the value of the outcome variable is measured repeatedly over time, it is tempting to compare the treatments at each such time point. Regulatory guidance tells us not to do this but to form a summary measure, one value for each patient, by combining the measurements in an appropriate way that has clinical meaning, maybe for example the average of the final three monthly values. Matthews et al. (11) give some good advice with regard to the handling of repeated measures.

The concerns of regulators and the above comments apply only to confirmatory analyses and claims. In exploratory settings, we tend not to worry too much about multiplicity. So for example we may capture the primary analysis strategy within one or a small number of primary endpoints and classify other endpoints of interest as secondary endpoints. In the multiplicity sense, these secondary endpoints are exploratory. If the primary endpoints give negative results then a confirmatory claim cannot be made by promoting a secondary endpoint to become primary. Only if the primary endpoints give a positive result will a confirmatory claim be allowed. An alternative approach to the treatment of secondary endpoints will be covered in the next section.

These considerations of multiple testing also do not apply to the analysis of safety data where p values tend not to be produced unless a specific safety endpoint is the basis of the primary objective of the study and a confirmatory claim in relation to that endpoint is being considered. In such a case, the safety endpoint would be treated in the same way as a primary efficacy endpoint.

Avoiding Adjustment

In the final sentence of the ICH quotation in the previous subsection, there is mention of the possibility that in some circumstances in which adjustment of the significance level may not be necessary. These particular settings have been elaborated in a subsequent CPMP Points to Consider paper on "Multiplicity Issues in Clinical Trials" (12). We will deal with these in the context of multiple primary endpoints, but the ideas can be applied more widely to other areas of multiplicity. There are essentially three situations where adjustment of the significance level would not be needed: composite endpoints, requiring statistical significance on each endpoint, and hierarchical testing.

Composite Endpoints

For a composite endpoint the endpoints of interest, for example, death, nonfatal stroke, myocardial infarction, and heart failure are combined into a single endpoint, and a single p value is produced for this combination. This avoids multiplicity and also increases power since the incidence of each of the components may be quite low. From a regulatory perspective, this composite may be adequate for registration purposes providing there is assurance that an overall difference is not hiding a negative effect in one or more of the components.

> One concern with composite outcome measures from a regulatory point of view is, however, the possibility that some of the treatments in the study may have an adverse effect on one or more of the components, and that this adverse effect is masked by the composite outcome .... [CPMP: Multiplicity Issues in Clinical Trials (12)]

This reassurance can be given by looking at the incidence rates (without p values) for the individual components of the composite. If any of these go in a negative direction for the test treatment then the validity of a claim based on the composite will be undermined.

Significance Required on all Endpoints

If $p \leq 0.05$ is required for each of the specified (multiple) primary endpoints then no adjustment is needed. In this setting the issue of cherry-picking the endpoints that are significant does not arise and the type I error rate is not inflated. In some therapeutic settings, multiple endpoints are required by regulators.

> A significant benefit for both primary endpoints, lung function and the symptom based clinical endpoint, should be demonstrated so that no multiplicity adjustment to significance levels is indicated. [CPMP: Clinical Investigation of Medicinal Products in the Treatment of Asthma (13)]

Hierarchical Testing

In this approach, a prespecified order for the primary endpoints is set down and the significance testing proceeds in that order. Statistical significance can only be claimed down to the first nonsignificant p value. Each test is conducted at the standard 5% significance level, and it can be shown that this procedure maintains the overall type I error at 5%. Suppose for example there are three primary endpoints that are ordered according to their clinical importance. If the p value for the first of those endpoints gives $p \leq 0.05$ then the significance for the second endpoint in the hierarchy can be assessed. If $p > 0.05$ for this second endpoint then a confirmatory claim can only be based on the first endpoint. If $p \leq 0.05$ for the second endpoint however, then the significance for the third endpoint can be assessed and the first two endpoints can be incorporated into a confirmatory claim. Finally, if $p \leq 0.05$ for the third endpoint, this can also be incorporated into the confirmatory claim.

Clearly the hierarchy needs to be prespecified. This approach is also a way of incorporating secondary endpoints into a claim; secondary endpoints can be thought of as simply primary endpoints lower down in the hierarchy.

Subgroup Testing

If specific confirmatory statements are to be made for subgroups then some account of multiplicity will be needed. This could be in the form of a Bonferroni correction, which will allow the cherry-picking to take place. Alternatively organizing the testing in a hierarchy may be useful, for example placing a particular subgroup where the treatment effect is likely to be the greatest at the top of the hierarchy with the comparison for the treatment groups overall as the second evaluation. Testing then proceeds by looking at the subgroup to begin with. If this gives $p \leq 0.05$

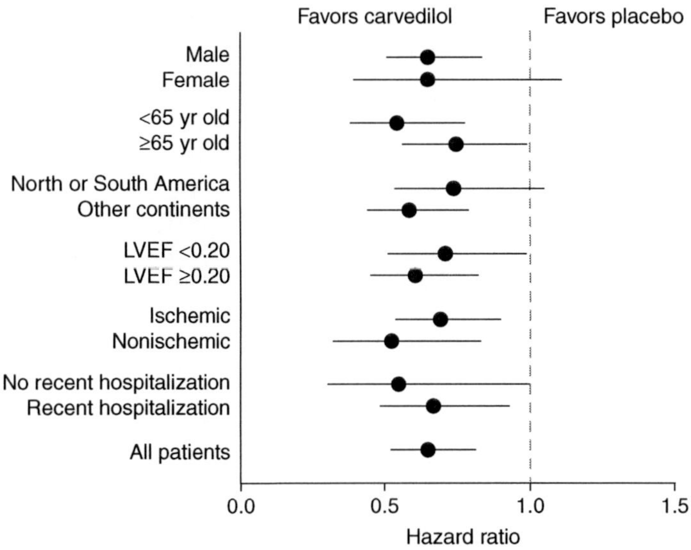

Figure 2 Hazard ratios for carvedilol in subgroups.

then the treatment groups overall can be compared and confirmatory claims made accordingly. If a treatment difference in the subgroup is more likely than a treatment effect overall, taking into account the sample sizes in both the subgroup and overall, then this strategy may stand a better chance of success than undertaking the tests in the reverse order—treatment groups overall followed by the subgroup.

Recovering claims based on subgroups, where these subgroup evaluations are not preplanned and multiplicity is not taken into account, are rarely successful in a regulatory setting. Subgroup evaluation however is often undertaken but simply to judge the homogeneity of the treatment effect. This is an alternative to the investigation of the treatment by covariate interactions within the context of ANCOVA covered previously.

> Marked heterogeneity may be identified by graphical display of the results of individual centres or by analytical methods, such as a significance test of the treatment-by-centre interaction. [ICH E9: Statistical Principles for Clinical Trials (1)]

Figure 2 shows the hazard ratios, together with 95% confidence intervals, for subgroups defined by baseline factors in the carvedilol trial (5) discussed above.

The treatment effect is seen to be fairly consistent across the subgroups investigated. Not all of the 95% confidence intervals exclude the value 1 and so in some subgroups the corresponding p value for the treatment comparison will be nonsignificant at the 5% level. This is of no concern. We are not looking to confirm a significant effect across these subgroups. Rather we are simply confirming a consistent treatment effect with the hazard ratios in each case varying tightly around the overall value of 0.65.

Preplanning

The importance of preplanning the methods of statistical analysis is widely recognized. The definition of the primary endpoint(s), analysis sets, methods for handling missing data, and the precise statistical methods that will be used for the analysis of the primary endpoint will all be set down in either the protocol or in the Statistical Analysis Plan, which will be finalized prior to database lock and the unblinding of the treatment codes. Among other things this provides protection against multiplicity. Having defined the primary endpoint(s), the strategy for handing any multiplicity issues prior to data analysis and the test to be used for analysis there is little room for maneuver with regard to cherry-picking across a series of different analyses. Any analyses that are presented at the regulatory stage and were not preplanned will be viewed if not with suspicion then certainly with concern. These analyses may well be justifiable from a clinical perspective but unless they comply with strict statistical control of the type I error through prespecification, they are unlikely to be successful as the basis for a confirmatory claim.

INTERIM AND ADAPTIVE DESIGN

Interim Analysis for Efficacy

Interim analyses are undertaken when we want to give ourselves the opportunity to stop the trial early, for example for evidence of overwhelming efficacy. Because we are potentially analyzing the data on several occasions, there is again an element of multiplicity in this process and we need to adapt our testing procedures to preserve an overall false-positive rate of 5%. For the purposes of this development, let us assume that we have a planned total sample size of 600 and that we wish to have two interim looks at the data after 200 and 400 patients, respectively, have provided data on the primary endpoint, and the final analysis after 600 patients.

Pocock (14) developed a scheme that divided the type I error equally across the three analyses. Naively using the Bonferroni adjustment with an adjusted significance level of 0.017 ($= 0.05/3$) at each analysis might seem appropriate. It turns out however that paying such a big price at each analysis is not needed. The data sets on which the three analyses are based are overlapping since the data for the first 400 patients contains the data for the first 200 patients and the data for all 600 patients contains the data on the first 400. Pocock showed that the correct adjusted significance level for each analysis is 0.022 and this preserves the overall type I error at 5%.

O'Brien and Fleming (15) proposed an alternative scheme in which the 0.05 is not divided equally across the three analyses but has a very strictly adjusted significance level early on and leaves most of the 0.05 left over for the final analysis. The adjusted significance levels for the O'Brien-Fleming scheme with three analyses are 0.0006 for the first analysis, 0.014 for the second analysis, and 0.045 for the final analysis. This makes sense for two reasons. Firstly, it would be inappropriate to stop a trial very early unless the evidence was absolutely overwhelming and, secondly, the final analysis is the most important analysis and most of the 0.05 should be preserved for that.

If the p value at any particular stage falls below the corresponding adjusted significance level then the evidence is sufficiently strong to recommend stopping the trial. There are, in addition, several other schemes available but the two outlined above tend to be the most common. It is also possible, and sometimes desirable, to have schemes where the analyses are not equally spaced. For instance, in the above example with three analyses at 200, 400, and 600 patients, it may be that stopping very early is inappropriate, irrespective of the strength of evidence, as this would restrict the information needed on safety and secondary endpoints. In this case, the first analysis could be removed and only a single interim after 400 is considered. The adjusted significance levels would change and we would need to use so-called α-spending functions to calculate those levels.

Interim Analysis for Futility

It may also be useful to consider stopping the trial for futility. If at an interim stage an analysis of the data suggests that there is essentially no possibility of demonstrating superiority were the trial to continue then it may be appropriate to stop the trial early and move resources to another project where the chances of success are higher. Decisions of this kind are usually based on considerations of conditional power. Suppose that a trial with 500 patients has been powered at 90% to detect an improvement in cure rates of 25%, from 60% in the placebo group to 85% in the test treatment group. At the interim look based on 250 patients suppose the placebo cure rate is seen to be 65% with the test treatment cure rate at 70%. There is little possibility, therefore, even if the trial data from this point on were favorable for the test treatment, to turn this around and gain a significant difference at the end of the study. The conditional power is the probability of detecting the 25% difference conditional on the observed data on the first 250 patients and can be calculated based on certain assumptions regarding what the true cure rates will be as the trial progresses. Usually the conditional power is calculated "under the current trend," in other words assuming that the true rates from this point on in the trial will be 65% and 70%, respectively, although other assumptions are also possible. If this conditional power were say <20%—based on realistic assumptions about the true rates—then there would seem little point in continuing as there is only a very small chance of success. The choice of cutoff for conditional power, that is the level below which the trial would be stopped for futility, is entirely a commercial decision and is the responsibility of the sponsor.

Data Monitoring Committees

Many long-term morbidity/mortality trials have associated with them an Independent Data Monitoring Committee (IDMC), also referred to as a Data and Safety Monitoring Board (DSMB). These committees have responsibility for looking at the accumulating safety information to protect the safety of the trial participants and can recommend modifying the study if safety is a concern or in extreme cases recommending that the study be stopped. Having an IDMC looking an unblinded data in this way, with the sponsor remaining blind, can avoid bias in the ongoing trial conduct and therefore protect the scientific integrity of the trial. A comprehensive review of the role of IDMCs is given by Ellenberg et al. (16). Both the EMEA [CHMP (17)] and the FDA [FDA (18)] provide guidelines on the setup, responsibilities, and conduct of IDMCs.

If a trial has an interim analysis for efficacy then it is usually most appropriate that these analyses are reviewed within the IDMC, which then make recommendations to the sponsor, possibly through a Steering Committee, regarding stopping or continuing the trial. This again allows the sponsor to remain blind to the unfolding unblinded data to avoid bias.

Adaptive Design

There is a substantial amount of current interest in so-called adaptive designs where the trial design is modified during the course of the trial based on accumulating data. Indeed there is some discussion of these issues in the next chapter (Efficacy and Safety Clinical Studies). In one sense the interim analysis schemes outlined above are adaptive in that the trial can be stopped early based on interim data. Such schemes, however, relate to a simple go/no-go decision and providing that blinding is preserved outside of the IDMC and there is no dissemination of unblinded information then the scientific integrity of the trial is assured, while in addition the scheme controls the type I error.

More broadly adaptive design incorporates the following potential design changes:

- Increase (or decrease) sample size
- Choosing optimal dose—Adaptive Dose Response
- Dropping (or adding) treatment arms in phase II/phase III
- Changing primary endpoint
- Adjusting the population
- Changing randomization ratio.

It is beyond the scope of this chapter to discuss each of these issues in details, so we will only make some general comments. The reader is referred to Kay (19) and Gallo et al. (20) for further details and references therein.

The CHMP has made numerous recommendations in relation to adaptive designs in a recent guideline, CHMP (21) and there are two issues that are of greatest concern to regulators in relation to such designs. Firstly, the design must be able to control in a formal statistical way the type I error at an overall prespecified level, usually 5%. Secondly, there must be control of the dissemination of unblinded information at the stage where the design change is made. Otherwise there could be bias in the future conduct of the trial that undermines its scientific integrity. In relation to control of the type I error, increasingly we are seeing methods in the statistical literature that take care of these things. The control of dissemination is in one sense more difficult, and behavioral issues are often of much more concern to regulators than the statistical complexities. See Gallo (22) for further discussion on this point.

EQUIVALENCE AND NONINFERIORITY

Demonstrating Similarity

In the development so far, we have focused on detecting superiority of a test treatment over a control treatment. The objective with equivalence and noninferiority evaluation is exactly the opposite and we are now looking to demonstrate similarity between a test treatment and an active comparator. Obtaining a nonsignificant p value in a superiority study does not necessarily mean that the treatments are similar. This may be a result simply of low power to detect differences.

> Concluding equivalence or noninferiority based on observing a nonsignificant test result of the null hypothesis that there is no difference between the investigational product and the active comparator is inappropriate. [ICH E9: Statistical Principles for Clinical Trials (1)]

There are two reasons why we may want to conduct either an equivalence trial or a noninferiority trial. On the one hand it may be unethical to conduct a placebo controlled trial, and we are therefore looking to demonstrate the efficacy of the test treatment indirectly by showing it is similar to an active control that we know works. Alternatively we may have a new treatment that is potentially better than an active control because it is safer while we do not expect any change in terms of efficacy. In this case we would of course need to demonstrate the improvement in safety but also we would need to show similarity in terms of efficacy.

In equivalence the objective is to show similarity—that is, that the test treatment is no worse but also no better, in a clinical sense, than the control treatment; while in noninferiority the objective is only to show that the test treatment is no worse than the control treatment. One critical issue is how we define "no worse" and "no better" and this will be discussed later. In a clinical setting, we usually find ourselves dealing with noninferiority and this will be the focus in the remainder of this section. The most common example of equivalence testing is in bioequivalence, although there are some applications outside of that.

Confidence Intervals

The first step in establishing noninferiority is to define what we mean by "no worse." As an example, suppose we are comparing two treatments in terms of their cure rates and for improvements in safety we are prepared to pay a small price in terms of efficacy, say a 5% reduction in the cure rate. In other words from an efficacy point of view, we feel that the test treatment is noninferior to the active control if the cure on the test treatment is no more than 5% below the cure rate on the active control. The value 5% is the noninferiority margin.

Having conducted the trial we compute the 95% confidence interval for the difference (test treatment minus active control) in the cure rates. If the lower limit of this confidence interval is greater than −5% then noninferiority has been established. Note that conventional p values have no role in establishing noninferiority.

Assay Sensitivity

One concern in noninferiority trials is that a conclusion of similarity does not necessarily imply that the test treatment is effective. Similarity can simply mean that the two treatments are equally ineffective rather than equally effective. Assay sensitivity is the ability of the trial to detect differences should those differences truly exist and it is important in any noninferiority trial that the trial has assay sensitivity. This is most easily confirmed by including a placebo arm, which can then be compared directly with both the test treatment and the active treatment. If the active control does not separate from placebo by an anticipated amount then the trial will lack sensitivity. This is somewhat of a circular argument in that one reason why we may be conducting an active control trial is because it is unethical to include a placebo.

If it is not possible to include a placebo then ensuring assay sensitivity is best achieved through good design and a trial conducted to a high standard. In terms of design, for example, we should not use the active control at too low a dose; we should use established sensitive endpoints and include patients at the moderate to severe end of the disease scale. At the extreme if patients are included who are not really ill then the treatments will look similar! With regard to trial conduct, it is important to have good compliance to the medication, few dropouts, and so on, as these aspects could be seen to be weakening any treatment differences that might exist biasing the study toward a conclusion of similarity. In addition the data coming out of the trial can be used to support the presence of assay sensitivity. Going back to the cure rates example we would expect to see a cure rate on the active control in line with historical data. If this is lower, or higher, then the assay sensitivity of the study could be undermined.

The Choice of Margin

The choice of the noninferiority margin is the most difficult aspect in the design of a noninferiority study. The CHMP have produced a paper [CHMP (23)] that gives some recommendations on the thinking behind such a choice. The choice should be based on both statistical and clinical reasoning. The statistical arguments require an evaluation of the effect size of the active comparator treatment obtained by looking historically at placebo controlled trials of that treatment. Suppose either in a single large trial or through a meta-analysis combining a series of those trials the 95% confidence interval for the effect size is (D_1, D_2). The value chosen for the noninferiority margin would need to be somewhat less than D_1 otherwise we open up the possibility of concluding that the new treatment is noninferior to the active control (within D_1 units) yet no different from placebo. Conventionally we take the margin to be equal to one-half or maybe one-third the value of D_1. We can then be assured that, from a statistical perspective, the test treatment will still have benefit over placebo following a conclusion of noninferiority to the active control. Once a preliminary value for the margin has been obtained through such statistical arguments, we then need to appeal to clinical judgment to argue that the chosen margin represents an irrelevant difference from a clinical perspective.

One very important aspect of all of this is that having chosen a value for the margin there needs to be buy-in from the regulators. Regulators need to see the arguments on which the choice has been made, and give their approval.

Switching Between Superiority and Noninferiority

If the 95% confidence interval for the difference between the test treatment and the active control not only excludes the noninferiority margin, −5% in the cure rates example but also excludes 0, there is evidence for the superiority of the test treatment. The question is can we draw the conclusion that the test treatment is not only noninferior but also superior to the active control? The answer is in fact yes and this is made clear in a CHMP guideline that specifically addresses this issue.

> If the 95% confidence interval for the treatment effect not only lies entirely above $-\Delta$ but also above zero, then there is evidence of superiority in terms of statistical significance at the 5% level ($p < 0.05$).
>
> In this case, it is acceptable to calculate the exact probability associated with a test of superiority and to evaluate whether this is sufficiently small to reject convincingly the hypothesis of no difference. . . .
>
> Usually this demonstration of a benefit is sufficient for licensing on its own, provided the safety profiles of the new agent and the comparator are similar. [CPMP: Switching Between Superiority and Noninferiority (24)]

Are we also allowed to switch in the opposite direction? That is in a superiority trial if we have failed to establish superiority is the lesser claim of noninferiority a possibility. Unfortunately the answer here is no for two reasons. Firstly, it is very unlikely in a superiority trial that we will have prespecified a noninferiority margin and a post hoc choice could clearly be data driven. Secondly, switching from noninferiority to superiority can be considered in the context of hierarchical testing where noninferiority has been established in position number 1 in the hierarchy and then superiority is evaluated in position number 2 in the hierarchy. For switching in the opposite direction, we have fallen at the first hurdle in failing to establish superiority and the type I error is not protected.

Given that switching from noninferiority to superiority is possible, but not the reverse, there is one school of thought that says always design trials as noninferiority trials. If the data are strong enough, a conclusion of superiority is then also allowed.

THE ROLE OF THE STATISTICIAN

It is of course important to involve the statistician in the design, analysis, interpretation, and reporting of a clinical trial in the broadest sense. Design is not just about calculating the sample size nor is the analysis just about calculating p values. Statistical thinking is important in relation to all aspects of design, including

- Structuring the randomization (stratification, blocking)
- Choice of endpoints in line with the trial objectives
- Methods to control variability and increase precision (precise definition of endpoints and their measurement)
- Clear identification of the confirmatory elements of the trial, structuring the statistical testing strategy to control multiplicity
- Avoidance of missing data through flexibility in the design and design methodologies to deal with missing data
- Interim analysis and its structure
- Statistical methods to be used for data analysis
- Considerations of power and the resulting requirement for sample size.

These considerations relate to each individual trial. There will be similar considerations at the development plan level and the consistency of endpoint and covariate measurement, the recovery of missing data and the definition of analysis sets will be needed across trials to facilitate integrated summaries.

The statistician will then be involved in writing the statistical methods section of the protocol and casting a critical eye over the complete protocol. The statistical methods section of the protocol will then be expanded to produce the Statistical Analysis Plan. The Blind Review takes place following database lock but prior to breaking the blind and gives an opportunity to define the analysis sets at the patient level, possibly revisit the algorithms for combining centers and for dealing with missing data. The data analysis itself then takes place together with the production of the tables and graphs that will form part of the Clinical Study Report which will then be put together in conjunction with the medical writing team.

Finally, and again at the broader level, the statistician will be a key member of the team interacting with regulators at the stage where the regulatory submission is made. Many of the questions that come back from regulators will have a statistical element.

REFERENCES

1. ICH E9. Note for Guidance on Statistical Principles for Clinical Trials. CPMP/ICH/363/96, 1998.
2. Spiegelhalter DJ, Abrams KR, Myles JP. Bayesian Approaches to Clinical Trials and Health-Care Evaluation. Chichester, UK: Wiley, 2004.
3. CPMP. Points to Consider on Adjustment for Baseline Covariates. CPMP/EWP/2863/99, 2003.
4. Roes KCB. Dynamic allocation as a balancing act. Pharmaceut Stat 2004; 3:187–191.
5. CPMP. Points to Consider on Missing Data. CPMP/EWP/1776/99, 2001.
6. Gardner MJ, Altman DG. Estimation rather than hypothesis testing: confidence intervals rather than p-values. In: Gardner MJ, Altman DG, eds. Statistics with Confidence. London: BMJ, 1989:6–19.
7. Packer M, Coats AJS, Fowler MB,et al. Effect of carvedilol on survival in severe chronic heart failure. N Engl J Med 2001; 344:1651–1658.
8. Peto R, Peto J. Asymptotically efficient rank invariant procedures. J Roy Statist Soc A 1972; 135:185–207.
9. Gehan EA. A generalised Wilcoxon test for comparing arbitrarily single censored samples. Biometrika 1965; 52:203–223.
10. Cox DR. Regression models and life tables. J Roy Stat Soc B 1972; 74:187–220.
11. Matthews JNS, Altman DG, Campbell MJ, et al. Analysis of serial measurements in medical research. Br Med J 1990; 300:230–235.
12. CPMP. Points to Consider Multiplicity Issues in Clinical Trials. CPMP/EWP/908/99, 2002.
13. CPMP. Note for Guidance on Clinical Investigation of Medicinal Products in the Treatment of Asthma. CPMP/EWP/2922/01, 2003.
14. Pocock SJ. Group sequential methods in the design and analysis of clinical trials. Biometrika 1977; 64:191–199.
15. O'Brien PC, Fleming TR. A multiple testing procedure for clinical trials. Biometrics 1979; 35:549–556.
16. Ellenberg SS, Fleming TR, DeMets DL. Data Monitoring Committees in Clinical Trials: A Practical Perspective. London: Wiley, 2003.
17. CHMP. Guidance on Data Monitoring Committees. CHMP/EWP/5872/03, 2005.

18. FDA. Establishment and Operation of Clinical Trial Data Monitoring Committees. 2006, fda.gov/cder/gdlns/clintrialdmc.pdf.
19. Kay R. Statistical Thinking for Non-Statisticians in Drug Regulation. Chichester, UK: Wiley, 2007, Section 14.5.
20. Gallo P, Chuang-Stein C, Dragalin V, et al. Adaptive designs in clinical drug development – an executive summary of the PhRMA working group. J Biopharm Stats 2006; 16:275–283.
21. CHMP. Reflection Paper on Methodological Issues in Confirmatory Clinical Trials Planned with an Adaptive Design. CHMP/EWP/2459/02, 2007.
22. Gallo P. Confidentiality and trial integrity issues for adaptive designs. Drug Inf J 2006; 40:445–450.
23. CHMP. Guidance on Choice of a Non-Inferiority Margin. CPMP/EWP/2158/99, 2005.
24. CPMP. Points to Consider on Switching between Superiority and Non-Inferiority. CPMP/EWP/482/99, 2000.

35 Efficacy and Safety Clinical Studies

David Jefferys
Eisai Europe Ltd, Hertfordshire, U.K.

INTRODUCTION

The development of a pharmaceutical products containing new active substance or a product that is a major line extension is a complex task involving a wide range of disciplines. It may take between 10 and 15 years to develop a new active substance and cost in excess of US $1 billion (1). Many functions have an important role in the successful development of a product but arguably the demonstration of efficacy and safety lies at the very center of drug development.

The other 10 chapters in part 5 of this book cover in depth some of the very important aspects of clinical development. It is right to cover these in separate chapters, but in reality they are interlocking parts in the demonstration of efficacy and safety.

Over the years the word "efficacy" has generated much debate. Is there a difference between efficacy, clinical efficacy, and effectiveness? Some have characterized efficacy as the demonstration of a response compared to placebo, whereas clinical effectiveness could only be demonstrated in comparative studies. The importance of this argument is past since today, if a drug is to be successful and is to be reimbursed, then clinical effectiveness has to be demonstrated.

This chapter will look at various aspects around the design of clinical trials and the need to incorporate the clinical development plan at the earliest possible stage into the overall drug development plan (DDP). Although health technology outcome research (HTOR) will not be specifically discussed, the importance of comparator studies will be considered in the context of obtaining the satisfactory position, reimbursement, and technology assessment for the new medicine.

The development of product specific (so-called vertical) and general (horizontal) guidelines through the International Conference of Harmonisation (ICH) and from the Food and Drug Administration (FDA) in the United States, the European Medicines Agency (EMEA) in Europe, and the Pharmaceutical and Medical Devices Agency (PMDA) in Japan have greatly assisted clinical development. Scientific advice from the major regulatory agencies (FDA, EMEA, and PMDA) is now crucial for the successful development of a medicine. The advice from the regulators has helped to reduce the uncertainty in product development although, on occasion, differences in the advice from the different regions have introduced a new set of issues. Increased transparency through summary bases of approval (SBAs) documents and the European Public Assessment Reports (EPARs) from the EU along with the introduction of clinical trial registration databases has meant that companies are much better informed about the development of competitor products. There is thus a convergence in the clinical development programs.

There is considerable interest in novel and adaptive/innovative designs for clinical programs. All companies are seeking to reduce the time and cost of clinical development. There are also considerable pressures to identify a go/no go decision as early as possible. These initiatives from the industry have been supported by the regulatory authorities. New procedures have been put in place for expedited, accelerated, and conditional approval. The traditional phase I, IIa, IIb, and III clinical trial development programs are being replaced by various innovative and "adaptive" designs, which seek to reduce the number of patients in trials and reduce the overall time in clinical development. These novel approaches are supported through the "Critical Path Initiative" launched by the FDA (2) and by the "Innovative Medicines Initiatives" launched in May 2008 as a public/private partnership between the European Federation of Pharmaceutical Industry Association (EFPIA) and the European Commission (3).

This chapter will explore the new initiatives and consider the present and future environment for clinical trials. Biomarkers are likely to play an increasing part in shaping the clinical development program both to aid the proof of concept and, with appropriate validation, to provide surrogate endpoints.

The introduction of structured scientific advice, points to consider, feedback documents, and comprehensive detailed guidelines have made it much easier to develop a clinical trial program. The current focus is to generate an innovative program to gain time and cost advantages. There are extensive contacts between the major international regulatory agencies and gradually we may see the development of true "global clinical advice" for certain product areas.

The life cycle management (LCM) of products is of the utmost importance and today the clinical development program and the trial program have to be defined at an early stage to fully capture the long-term opportunity for the product. The timing of follow-up clinical trials is crucial. Increasingly, there is an interest in the development of drug/device and device/drug combinations to prolong the life cycle. These products present interesting issues for the clinical trial program.

Some clinical trials are primarily established to demonstrate efficacy, whereas others are principally aimed at generating safety data. This said most trials will generate safety data to a greater or lesser extent. However, the final decision on a drug concerns its benefit–risk profile. This chapter will consider some of the methodologies that can be used to integrate the safety and efficacy data to provide reassurance to regulators that the product has an acceptable profile for the claimed indication.

CLINICAL DEVELOPMENT PLAN

In the development of a new drug, it is essential to generate a DDP, and the section on clinical development is a key component in this document. The DDP should be a high-level document focusing on the overall strategy. It should be aligned across all the functional areas. The clinical development plan commences with the clinical introduction, and then outlines how full development will be taken forward with the appropriate milestones. The plan needs to cover all the phases of the clinical development program and include the proposed postmarketing and postauthorization studies along with the LCM of the compound. It is important that the LCM is considered and set out at an early stage of the overall development. This is necessary to maximize the potential of the drug and to generate additional exclusivity and protection from a second applicant. New legislative provisions in the United States and in the EU mean that the pediatric development program should be considered at the earliest stage of development and be an integral part of the development plan as is described later in section Pediatric Studies. Similarly, health technology assessment and health economic outcome research need to be brought into the overall plan as discussed in section Health Economic Outcome Research.

THE CLINICAL DEVELOPMENT PROGRAM

The current traditional clinical development program is divided into four main phases (although increasingly phase II and phase III are recognized to be subdivided into "a" and "b" sections). This is illustrated in Figure 1.

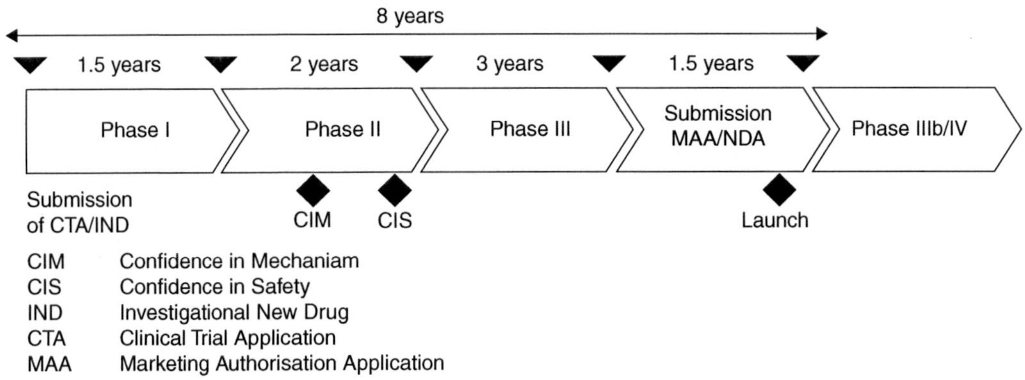

Figure 1 The current development process.

At the end of phase II, there should be a reasonable understanding of the efficacy and identification of major safety issues, with a proof of concept. However, even then there is much uncertainty with a recent study (4) showing in a review of 73 clinical candidate molecules, that 31% failed for safety concerns in phase III development and 50% failed to demonstrate efficacy.

Phase I

The clinical introduction of the new active substance begins with the phase I studies. These are often undertaken in healthy human volunteers. For oncology drugs, the phase I studies are usually undertaken in patients because of the toxicity profile of the drug. The phase I program is undertaken to establish the human pharmacology and the pharmacokinetics of the drug along with the pharmacodynamic profile. In addition, the phase I studies will give an early evaluation of the safety profile and patient tolerance of the drug.

The phase I program has become more extensive and increasingly will overlap the initial phase II (or phase IIa) program. QT interval testing and, where appropriate, an evaluation for drug dependency will be undertaken in the phase I program. Such studies are often performed rather later in the development program but are still considered phase I studies. The requirements for QT interval testing are described in the relevant ICH guidelines S7B and E14 (5,6). The guidelines suggest a sequential evaluation with a hERG gene (human Ether-a-go-go-Related Gene) bioassay followed by preclinical animal studies. If the latter is positive, then a thorough QT test with an active control (moxifloxacin) should be undertaken, supported appropriately with ECG monitoring during the clinical trials. This approach is followed by the European Regulatory Authorities but increasingly the FDA is requiring all new drugs to have a thorough QT test performed notwithstanding the results of the preclinical screening tests. Studies for potential drug abuse dependency will only be required in the light of the human pharmacology and from safety reports of euphoria and mood change in the latter stage development. In oncology phase I studies the maximum tolerated dose and the dose-limiting toxicity will be determined.

Before a phase I study is undertaken an investigator's brochure has to be generated, and in the EU an investigation on medicinal product dossier (IMPD) has to be prepared.

Microdosing

In recent years, there has been increasing interest in the use of microdosing studies. These employ a very low dose of the drug, which is devoid of pharmacological activity. The drug is radiolabeled and sensitive assays are used to detect plasma levels of the administrated drug. Guidelines on the preclinical requirements that have to be undertaken before microdosing studies in humans have been issued by the FDA and the EMEA (7,8). Microdosing studies require approval from the relevant agencies through a clinical trial application or IND.

The advantages of microdosing are that early pharmacokinetic data can be obtained quickly at a minimal risk allowing for a better selection of compounds, and uses only a small quantity of the investigational drug. Development times are reduced and fewer patients are exposed. Microdosing provides an earlier introduction to man and gives much earlier indication of clinical exposure. Microdosing has the opportunity to improve early clinical development.

Safety Issues for First in Man Studies

Following serious safety events in a trial at Northwick Park Hospital in the United Kingdom in 2006, the supervision and the guidelines on the conduct of initial phase I studies have been significantly tightened. New recommendations demand that companies consider not only the safety margin and the no effect level (NOEL), but also the minimally active biological effect level (MABEL). An investigation of the Northwick Park Hospital incident was undertaken by Sir Gordon Duff and his report made a range of recommendations (9). These have now been taken up in guidelines from the FDA and the EMEA (10). There are some differences in the regional recommendations, but essentially they recommend staggered dosing regimes with defined minimum intervals between the dosing of subjects. These are designed to limit the exposure and risk to human volunteers.

In the era of multinational/global clinical trials, it is recommended that companies consider undertaking the phase I program in ethnically diverse populations. Therefore if a company is planning to develop and introduce a drug for the Japanese and South East Asian markets, it would be sensible to consider conducting early phase I studies in subjects from Japan and China or other parts of South East Asia.

Phase II

The clinical studies undertaken in phase II development are intended to demonstrate the activity of the compound and to provide further information on the safety. Phase II studies are undertaken in patients. Frequently, phase II studies are differentiated as phase IIa and phase IIb. The earlier phase II studies are more concerned with demonstration of pharmacodynamic response, whereas later studies focus on the proof of concept and are of a controlled design either against placebo or increasingly interactive comparator. The phase II studies are usually powered to show a trend in efficacy but not to demonstrate a statistically significant effect. The phase II studies are also aimed at determining the appropriate dose range and dosage regimen and to identify a dose-response relationship. Phase II studies often provide a critical decision point for the development program. It is important to ensure that there is adequate preclinical safety cover to allow a relevant range of doses to be explored in the phase II program.

Phase III Development

The phase III program provides the essential approval trials. Phase III protocols are designed to demonstrate both the short- and the long-term safety and efficacy of the product proposed for marketing. Studies have to be adequately powered to demonstrate the primary efficacy point and the appropriate secondary endpoints. The phase III studies are usually randomized, blinded, controlled trials. Many will include a crossover design and frequently will be a multiarmed study to allow more than one dose to be explored. Detailed points on the duration and the number of patients required in a study to demonstrate efficacy and safety will be considered in depth later in the chapter. The ICH has given guidance on when a single pivotal trial is acceptable (11). This guidance is open to different interpretations among the regions and will also be considered further. As the effect size of the response decreases so ever larger number of patients are required in the phase III program to give a robust statistical result. Increasingly, very large-scale trials are being undertaken which include interim analyses and futility analyses.

Phase IV

In phase IV, studies are undertaken after the authorization and the marketing of the pharmaceutical product. These studies are normally used to further define the drug profile and to assess therapeutic value or treatment strategies. Postmarketing surveillance studies are increasingly being required as part of the postauthorization commitment programs imposed at the time of approval by the regulatory authorities. In some diseases, so-called "registry studies" are being requested. These are studies designed to capture and record both adverse reactions and the therapeutic response for all patients. They are particularly useful for drugs indicated for rare diseases, drugs used in selected populations and for biotechnology, and biological and advanced therapy products. For the latter, the interest may be on long-term safety concerns such as the development of resistance through the production of neutralizing antibodies or long-term safety concerns. Registry studies have been employed for many years to assess the safety and the performance of medical devices. Postmarketing studies may also be undertaken using patient record databases such as the GPRD (General Practice Research Database) held by the MHRA or the prescription event monitoring system organized through the drug safety research unit of the University of Southampton in the United Kingdom.

Adaptive Clinical Trials

There is considerable interest in novel or "adaptive" clinical trial design. On occasion, the use and the development of biomarkers is included under this term. While biomarkers may be used as part of adaptive designs, they represent a separate important topic, which needs separate and more detailed consideration (see section Accelerating Proof of Concept—Biomarkers). Adaptive designs are being pursued to reduce the time of drug development and to reduce the number of patients and hence the costs of the clinical programs.

One form of adaptive design is the so-called "group sequential" design, which has been developed to avoid increased type one errors seen with the repeated testing of the treatment effect. The methodologies are reasonably well established for undertaking interim analyses. Interim analyses are used to stop the trial early when efficacy has been adequately substantiated or for futility when a predefined efficacy level is not achieved after a set number of patients have been recruited. In a reflection paper from the Committee on Human Medical Products (CHMP) issued in October 2007 (12), the Committee stresses considerations which need to be put in place to minimize the risks of obtaining a false result with an interim analysis. It has to be remembered that regulatory acceptance of the results from a trial will often require assessment of the secondary endpoints as well as the primary endpoint. Sometimes the secondary endpoints give important clinical information. These results can be lost if the interim analysis is considered positive on only the primary endpoint. An alternative is to use probability density analyses or Bayesian statistics.

Several different types of adaptive design studies are recognized. These include changes or modification of primary endpoints, reassessment of the samples size and switching between superiority and noninferiority assessments. In the latter instance, the trial may have been planned to show noninferiority. However, if an interim analysis is undertaken and this is positive, then one option is to recruit more patients and to switch to a superiority endpoint.

In the majority of trials the randomization ratio is 1:1 but in some instances a 2:1 or 3:1 ratio may be used to reduce the numbers of patients who have to be recruited. With an adaptive design the randomization ratio may be changed after an interim analysis has been performed.

The greatest interest in adaptive design centers on the seamless phase II/phase III trial. If successful, such trials can significantly reduce the number of patients who have to be recruited and avoid the inevitable delay between the completion of the phase II study and the regulatory approval for the commencement of the phase III investigation. The rationale for such studies is that the same patient population will be used in the phase II and the phase III studies. If the same endpoints are being used, then the wish is to perform an interim analysis on the phase II data, and if this is positive then continue the patients into the phase III study with additional patients being recruited. Such an approach can only be used when the authorities are prepared to accept a single pivotal trial for approval.

An adaptive design is one where the statistical methodology allows the design of the study to be modified after an interim analysis. Modifications may include changing sample size, randomization ratio, the number of treatment arms, or the seamless switch of phase II patients into the phase III program.

Figure 2 is taken from a recent Price Waterhouse Coopers report (13) and suggests what the clinical development process might look like in 2020, taking account of recent changes and the impact of adaptive designs.

Accelerating Proof of Concept—Biomarkers

The previous section described the conventional approach for the clinical development of a new drug. The use of adaptive designs was also discussed. Adaptive designs, however, are only part of the current interest in accelerating the proof of concept or, as some refer to it, the new approach to "learn and confirm." There is a strong move to progress beyond the phase

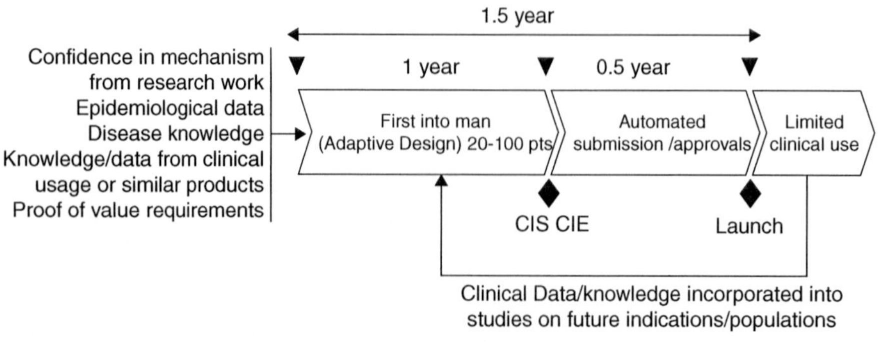

Figure 2 What the development process might be like in 2020.

trial model, with a drive to better understand the mechanisms and action of a drug at the early stage. Approaches include using receptor occupancy studies to better define the dose response, metabolomic profile, and microdosing techniques referred to earlier. Pharmacogenomic profiling is being employed to stratify patient populations with a view to enhancing the probability of demonstrating efficacy, although at the potential cost of restricting the initial approved indication for the product. The development of biomarkers forms an important part of these initiatives.

Biomarkers are being sought at the early stage to help identify and elucidate the mechanism and action of a drug, also to accelerate decisions on the proof of concept (14). A variety of markers are being explored such as FDG-PET imaging for use in Alzheimer's disease and tumor biomarkers for oncology products. Many would like to see biomarker data being accepted as a "conditional proof" of efficacy. The regulatory agencies are resisting this, and both they and those engaged in health technology assessment wish to see the validation of biomarkers before accepting them as a surrogate endpoint for clinical development. The validation of a biomarker is not easy and increasingly this is being taken forward in a precompetitive model by groups or consortia of companies. Recognizing the importance of this work, the EMEA and the FDA have set up a joint task force and more recently the EMEA has launched an initiative on biomarker qualification (15). Interestingly, there is cross representation on this group with FDA officials participating alongside the EMEA.

Biomarkers also have an important role to play in understanding adverse reaction mechanisms and are being used predictively, to identify those at risk of particular adverse reactions. In effect, this is stratification for safety postapproval. Already we have seen such approaches for abacavir and for increasing the safe dosing with warfarin (dicoumarol).

THE DESIGN OF INDIVIDUAL STUDIES

Controlled Trials

The majority of clinical trials will be controlled either against placebo or a comparator. Increasingly three arm studies are being used with both a placebo and an active comparator. The use of an active control will not only provide an assessment of the efficacy but will also provide a validation assay. In the event of a negative trial, the comparator arm can be valuable to look at issues of sensitivity within the study.

Comparative trials are increasingly being required to allow the position of the new treatment to be established and to assess the clinical meaningfulness of the drug. In many instances, a placebo controlled study would be unethical, although rescue therapy and other techniques can allow placebo control studies to be undertaken in more difficult circumstances. The switch to more active comparator trials is partly driven by reimbursement and health economic issues, but also because regulatory authorities wish to see longer duration trials. These are often not possible or practical with a placebo control design.

Uncontrolled Trials

Uncontrolled trials may be used to study treatments in rare diseases. They are also used to demonstrate the safety of the drug in clinical usage. Many trials will have an open label or uncontrolled extension, when those patients on placebo will convert to the active drug. This phase allows an analysis of the maintenance of effect and also valuable information can be obtained from the cohort of patients who switch from placebo to the active substance adding confirmation of efficacy. Additionally, the open label extension will expand the evaluable database for safety.

Crossover Design Studies

In a crossover design study all the patients will receive the drug under investigation and the comparator in a randomized sequence with a suitable washout period (drug free) between the two treatments. The crossover designs theoretically allow a number of patients being studied to be halved. They also have the benefit that each patient acts as his or her own control. The downside is the carry over effect at the crossover point. This may be significant for drugs with a long half-life.

CLINICAL TRIAL PROTOCOL AND DESIGN

The Protocol

The protocol is the crucial document which describes in considerable detail all the aspects of the study. How the results will be analyzed is set out in the statistical analysis plan of the protocol. The protocol has to be submitted to the regulatory authorities as part of an IND application in United States and Japan and as part of the clinical trial application in the EU. In addition the protocol has to be reviewed by the appropriate ethics committee or institutional review body. In the EU, all the clinical trial documentation is brought together in the Investigational Medicinal Product Dossier (IMPD). A relatively new key document is the investigator's brochure (IB), which brings together all the knowledge of the drug at the time of the clinical trial.

Design of the Program

The design of the clinical trial is crucial to the successful development of a new medicine. The process has become much easier over the last 10 years for several reasons. A large number of general (so-called horizontal) and product-specific (vertical) guidelines have been issued by the three major regulatory regions. Most of the key issues have been taken up in ICH guidelines, which are common to the three regions. In addition, the agencies have issued points to consider documents and reflection papers. This means that there is now a large body of knowledge available to those developing a new drug. In addition, the drive for transparency means that detailed summaries are publicly available for all new drugs. Thus a competitor can clearly see the development program for a rival drug, or for a drug in the same class, or with a similar mode of action. This can be a considerable help.

SCIENTIFIC ADVICE DURING DEVELOPMENT

Extensive scientific advice is available to the industry from each of the three major agencies. In the United States the key meetings are the pre-IND meeting, the end of phase II discussion, and the prefiling meetings. In addition, for drugs deemed to be of significant therapeutic interest, the FDA will offer so-called "special protocol assistance" (SPA). Under an SPA, a rolling program of meetings and discussions can be held with the relevant center in FDA and extensive support is proffered by the Agency. In the EU, individual Member States still offer scientific advice, but the predominant route is through the Scientific Advice Working Party of the CHMP. Scientific advice in the EU is free for orphan drugs, whereas for all other drugs it is chargeable. However, there is no limit on the number of meetings that can be requested and often companies will seek repeat meetings up to six times during the product development cycle. It has to be appreciated that scientific advice meetings are, as their name suggests, advisory. The agency and its staff will give their best assistance and where appropriate will work closely with the newly established CHMP scientific advisory groups (SAGs). The latter are composed of external experts.

OBJECTIVES OF TRIAL

It is essential that the objectives of the clinical trial are clearly defined at the outset. The hypothesis to be tested should be clearly defined. This will reflect the results obtained from previous studies (usually the phase II studies) and the criteria set out in guidelines from the major regulatory agencies. Once the objectives have been defined, then the study protocol can be generated. Without clear objectives the design of the trial will be difficult and it is likely to fail.

Study Endpoints

The endpoints of the clinical trial need to be clearly defined. Often the primary endpoints are specified in guidelines from the agencies or agreed in a scientific advice meeting. If not, the endpoints must reflect the desired indications for the product. Extrapolation from a limited database is difficult (section Study Size/Number of Patients).

Study Size and Power Calculation

The power calculation is one of the key issues. All studies need to be adequately powered taking account of their endpoints. The agencies have issued a wealth of methodology and statistical guidance to help appropriate calculation of the number of subjects required.

Study Size/Number of Patients

Estimating the number of patients required to deliver the objectives of a clinical trial is difficult. It is the most frequently asked question in scientific advice meetings to both the FDA and the EMEA. All phase III studies need to be adequately powered taking account of their endpoints. The power calculation needs to reflect the population being studied, compliance and likely withdrawal rates, the use of interim analyses and the frequency of repeat observations. If a noninferiority trial is being conducted the acceptable noninferiority margin has to be defined. Both the FDA and the EMEA have generated a large number of guidelines on methodology and statistical approaches, which should be carefully studied and are very useful in designing the trial. Advice should be sought from the regulatory agencies on the proposed numbers to be included in the trial and the assumptions behind any proposal. It is also helpful to look at recent approvals and other compounds under development, to see the range of numbers being investigated in similar studies (11,16).

The ICH E1 guideline considered the number of patients needed to assess the clinical safety for drugs intended for long-term treatment. The current recommendation is that 1500 patients is an acceptable number with between 300 to 600 patients being exposed for six months. Previously, it had been recommended that 100 patients should be studied for one year which would normally require a similar 1500 patients to be studied to allow such a one-year cohort to emerge. On occasions the evaluable database will be much smaller, particularly when clinical trials are being undertaken for orphan drugs and the treatment of rare diseases as is described in the following section. For other drugs where there is a small effect size, much larger patient numbers will be required to demonstrate efficacy, and hence a larger database will be evaluable for safety. In the area of cardiovascular medicine and diabetes, studies of more than 10000 patients are now being undertaken.

Guidance on Clinical Trials in Small Populations

In conditions where there are small, and occasionally very small, populations novel methodological approaches to study design may be acceptable. The CHMP issued a guideline on clinical trials in small populations in 2006 (17). This guidance notes that there are hierarchies of evidence which range from metaanalyses of randomized controlled clinical trials to published case reports, anecdotal case reports, and opinions of experts in the fields. In very rare diseases, a combined evaluation of single case studies may be the only way to provide evidence. In such situations, the data collection and the treatment conditions should adhere to GCP standards. The use of small trials is clearly assisted if there is a detailed knowledge of the pathophysiology of the disease and the pharmacology of the drug. In undertaking such small studies it is essential that there is strict adherence to the protocol to avoid so-called "bionoise". The regulatory decision on such studies is usually aided by the offer of a postauthorization registry study. This makes it easier for agencies to consider granting a "conditional" approval.

Patient Selection and Stratification

It is important to include as representative a population as possible in the trial. This is both to demonstrate the efficacy and the safety across the population and also because regulatory agencies are becoming less inclined to extrapolate across a population. Thus, if strict inclusion and exclusion criteria are used, then these may be reflected in the subsequent labeling of the product. Regulatory agencies expect to see adequate numbers of elderly patients exposed and an appropriate balance between male and female subjects. They look carefully at the exclusion criteria.

There is an obvious tension between the desire to reduce "bionoise" and to have a clear result with the minimum of number withdrawals, against the need to include "real-life" patients and to give early identification of potential safety concerns. This is typically seen in the need for information on the safety of the drug in patients with impaired renal and hepatic function. Extensive guidance is available from the agency Web sites covering patient selection for clinical trials, but too often this is ignored or inadequate numbers are included in studies. It is tempting to meet the requirement of studies in the elderly by recruiting fit and healthy patients in the age groups 65 to 70 years. The regulators and public health would wish to see an older group of patients studied who will have comorbidities and better reflect the patients who will ultimately be receiving the medication.

Patient stratification within trials is important and can improve the efficiency of the trial. Stratification may be helped by early use of a biomarker to identify those patients more likely to demonstrate a response. Planned stratification allows such response groups to be identified and studied as part of the protocol.

The inclusion of pediatric patients is a crucial consideration and this is addressed in section Pediatric Studies.

POINTS TO CONSIDER IN TRIAL DESIGN

Pivotal Studies

Is one political study acceptable?

Most agencies wish to see two pivotal phase III studies, with one study being confirmatory of the lead pivotal study. ICH guidance helps to define when only a single pivotal study is acceptable. In recent years, with the increasing availability and uptake of scientific advice, this is one of the major questions of discussion with the agencies. The decision to accept a single pivotal study will be influenced by the class of drug and the indication being sought. It will also be influenced by the range and extent of studies that have been undertaken in the phase II program. If there is a well-powered phase II study, then this makes it much easier to accept a single pivotal phase III study. The robustness of the result is also of importance and the agencies would be looking for a value of around 0.01. The agencies will also be checking to see if there is center bias within the study, and whether particular centers or regions are overly influencing the result. On occasions, they may ask for a so-called "destructive" analysis whereby the result is tested with the exclusion of certain positive centers. This will reveal whether the effect on the trend is preserved independent of those centers.

Drug Dosage

Determining the dose and defining the posology in the drug label is arguably one of the most difficult tasks in clinical development. The dose should be defined during the phase II development, but on occasion the optimal dose will be determined during later development. This may involve using two or even three treatment arms at different dose levels. Measurement of drug blood levels during the trial will also assist in optimizing dosage recommendations and will be helpful in investigating drug–drug interactions.

Regulatory agencies will pay particular attention to determination of the starting dose, clear establishment of the maintenance dose, and will also wish to see demonstration of the minimally effective dose through the use of appropriate dose escalation studies. On occasion companies may undertake back-titration studies: these involve progressive reduction in the dose to define the minimally effective and the true maintenance dose.

Study Endpoints—Primary Versus Secondary

In designing a clinical trial, it is necessary to specify the study endpoints and to decide which will be the primary and which the secondary endpoints. It is possible to have more than one (so-called coprimary) endpoints. However, there are risks in using more than one primary endpoints. If the study fails to show efficacy in one of the coprimary endpoints, even if the other is positive, then the FDA will consider this as a failed study. The secondary endpoints are also of importance, particularly if these are more clinically meaningful than the primary endpoint. Example here might be if the use of nerve conduction velocity as the primary endpoint in a study in peripheral neuropathy, whereas the secondary endpoint may be looking at a reduction in the signs/symptoms associated with the neuropathy. The European agencies are more likely to pay attention to the second endpoint. On occasion, companies will receive differing scientific advice and as a result will include one measurement as the primary endpoint, with the other as the secondary endpoint for a different region. This is to avoid selecting coprimary endpoints with the inherent regulatory risks.

The selection of the endpoints is usually as a result of guidance and dialogue with the regulatory agencies, and will also reflect the views of key opinion leaders, product-specific guidelines, and "notes to consider" papers from the regulatory agencies. Key issues in defining study endpoints are whether a responder analysis should be undertaken and the treatment of patient withdrawals and those lost to treatment during the trial.

A major area of debate concerns the selection of a superiority versus a noninferiority endpoint in a trial. This debate is often part of the wider debate on the use of including a comparator arm study. Increasingly companies are undertaking comparator studies with the comparator arm being either used for internal sensitivity validation or to enable specific labeling claims and to secure adequate reimbursement. If the comparator study is undertaken, then the debate is whether the analysis should be of noninferiority or superiority. This is a complex matter outside the scope of this chapter. The balance is often between the risk of failure and the number of patients required and cost of the study program.

In the field of oncology, it is interesting to see that the agencies are requiring increasingly the endpoint of overall survival rather than a significant improvement in the progression free interval.

Duration of the Study

The duration of the clinical trial is another contentious issue. The duration of the trial to demonstrate the efficacy of the drug will clearly depend on the disease being treated. It will also be influenced whether this is a placebo or a comparator controlled study. The design is also influenced by feedback from scientific advice and by product-specific guidelines. For drugs to treat long-term conditions, at least one six-month study is required and usually this will include an additional six-month open label study to gain adequate numbers for safety evaluation. Currently, the European agencies are requiring two six-month studies. We are also seeing the emergence of one- and two-year studies for the specific conditions. Partly this reflects the difference in diseases that are now being investigated as well as the requirement for more long-term data to demonstrate long-term effectiveness and maintenance of the effect.

HEALTH ECONOMIC OUTCOME RESEARCH

Health technology assessment and health economic outcome research is becoming increasingly important for the pharmaceutical industry. Many regions of the world now require such studies to be undertaken and the evidence provided before a drug be given access to the market or be considered for reimbursement. This has implications for the clinical development program. As already discussed, it is one of the drivers for including more comparator studies in the clinical program. As HEOR develops it is recommended that all programs are assessed carefully by health economics teams and a decision is made in late phase I development as to what additional elements or studies should be put into the development program. These may often involve including specific secondary outcome measures to address important economic outcomes. Regulatory agencies are prepared to accept such studies as part of the development program and will allow reference to these studies to be made in the product labeling. Such labeling statements are increasingly important for the positioning and the success of a product. The regulatory agencies are expected to issue further advice on how they will handle such economic labeling claims and their approach to economic outcome research.

MULTINATIONAL CLINICAL TRIALS

Most pharmaceutical companies wish to introduce their new drugs into the global market as rapidly as possible. On occasions, some line extensions are developed for individual markets or regions and some products may be developed with particular markets in mind. The DDP should identify at an early stage the commercial strategy. Wherever possible, companies will seek to have a global development program, and to use the same registration file with multicenter multinational trials. This is relatively easy where there is a common pathophysiology of the disease such as in oncology. In other areas where there may be differences in the pathophysiology between the regions ethnic factors may complicate global development programs in diabetes and cardiovascular disease, making the conduct of such trials more difficult. The E5 ethnic guideline (18) generated by ICH has been extremely influential. For the US and EU markets, the intention should always be to undertake a common development program although sometimes this may be complicated by differing scientific advice from the two regions. On occasions, this will mean that an additional study will be required for one of the two regions. In conducting multicenter multinational trials, it is important to avoid center bias and wherever possible to ensure an equal representation of centers. Should there be a predominance of centers in one region then appropriate statistical analysis of each region should be undertaken.

Studies in Japan and South East Asia are more complex. Clearly, South East Asia is becoming a significantly more important market for the pharmaceutical industry. In Japan, there are currently concerns over differences in the pharmacokinetics and the pharmacodynamics of drugs and also a difference in sensitivity to adverse reactions. In light of this, it is necessary to conduct early phase I studies in the Japanese population. From the results of such studies a decision can be made either to proceed through a conventional development program for Japan, or to use the phase I data and the dose determination to include Japan in the global phase III multinational trials. Traditionally, bridging studies have been used to access the Japanese market, but increasingly companies are switching to the use of multicenter multinational trials. Discussions have recently started between Japan, China, and Korea about the possible acceptance of clinical trials results between these three countries. If the discussions reveal that there is no significant pharmacogenetic difference in the handling of drugs between the populations of the three countries, then the possibility of a mutual recognition agreement and acceptance of data from one of the three countries will be taken forward. This could be a very significant development.

Currently, countries in South East Asia, excluding Japan which has its own requirements and is a founder member of ICH, can be divided into two groups. The first group of countries are China, Korea, India, and Taiwan which require a clinical study to be undertaken before a drug can be approved along with an approval in another major territory. This requirement is usually for a bridging study of around 100 patients. India and Korea and Taiwan will often waive this requirement if adequate numbers from each of their countries have been included in the multinational trial. Other countries in South East Asia rely on the prior approval using the certificate of the pharmaceutical product (CPP) process for the approval. They require that the drug has been evaluated and authorized in one or two major countries prior to granting them a national approval.

The inclusion of emerging markets, and in particular South East Asia, in drug development programs will become increasingly important over the next few years and is a subject which requires close attention by anyone interested in international drug development.

LIFE CYCLE MANAGEMENT STRATEGIES

The LCM of a product is of increasing importance as companies attempt maximize the opportunities for a new molecule. Some companies will introduce a new molecule with a restricted indication and then seek to widen the use and also to explore its efficacy in other treatment areas. Alongside this new formulations and presentations may be developed such as extended or sustained release preparations, transdermal delivery systems and on occasion novel delivery systems involving a combination with a medical device.

For oncology drugs it is not uncommon to find the molecule introduced initially for third-line treatment in a specific tumor such as breast cancer, with subsequent clinical studies being undertaken to extend the use into second- and first-line therapy while at the same time exploring the use in other tumor types. In these circumstances, the safety database is accumulating as more patients are studied and this can provide additional reassurance. A similar approach may be seen with antiepileptic drugs where the initial indication may be as adjunctive therapy for partial seizures and then additional indications such as generalized seizures and first-line therapy in partial seizures will be explored. At the same time, given the pharmacology of antiepileptic drugs new indications may be sought for treatment of neuropathic pain.

To support these indications, clinical efficacy and safety studies are required. If the indication is in a similar therapeutic area, then approval may be granted with one robust clinical study although the regulators still may require to see a second confirmatory study. As for a new drug application, companies are well advised to consult the relevant therapeutic guidelines and to discuss the design program with the major agencies to obtain their agreement in advance.

For a novel formulation or new delivery system, reassurance will be sought on the safety of the product in clinical usage alongside the demonstration of its efficacy. A standard development program will include a full pharmacokinetic and pharmacodynamic profile. Some companies may then proceed directly to phase III development while others will undertake a phase II study to confirm the proof of concept before embarking on the more costly clinical development program.

PUBLICATION STRATEGY

Clinical studies may also be undertaken to support the marketing of a product. These studies will supplement an already approved indication or may be used to generate further interest in the product alongside a publication strategy. It has to be appreciated that not all clinical studies will be used for a regulatory filing, although the safety data will need to be appropriately reported to the regulatory agencies.

POSTMARKETING SAFETY STRATEGY

Postmarketing or phase IV studies are undertaken to provide additional reassurance on the safety of a drug. Many are undertaken as part of postauthorization commitments and will have been required or suggested by a regulatory agency during the initial approval. With a move to conditional licensing for certain medicines, more targeted postmarketing studies are being undertaken. These phase IV studies may be supplemented by registry studies or investigations undertaken using electronic public health records.

PEDIATRIC STUDIES

During the late 1990s, it became apparent that many drugs were not being developed or evaluated for use in children. A paper in 2000 (19) suggested that, in the EU, 50% or more of the medicines used in children had never been studied in this population but only in adults and often in different diseases. This was partly for economic reasons but also reflected a widespread view at the time that it was better to develop a drug first for adults and then to seek to widen its use in children when sufficient safety data had been gathered. This said there were still many areas where drugs were successfully developed solely for use in children. Nevertheless there was a perceived delay in the development of medicines for children, a lack of information on dosage regimes for use in all pediatric age ranges and a lack of suitable formulations of medicines to be used in children. This initially led to legislation being introduced in the United States and this was then followed up by similar legislation in Europe (20,21). While the legislation is broadly similar, there are significant differences in implementation that need to be understood. Both systems offer the "carrot and stick" approach. Companies are required to put forward pediatric development plans for any new drug. These have to be agreed before the application for a new medicine can be submitted. The reward is an extra six months exclusivity or six months of patent life which is granted when the pediatric program is completed. The studies do not have to be successful to obtain the reward; rather they have to be completed as agreed with the relevant authorities. For some indications, a company can obtain a waiver if the drug would not be used for children to a significant extent. On occasions, it is possible to obtain a deferral, whereby the studies can be completed after the initial approval for use in adults. For some drugs, a deferral is not acceptable and the pediatric program has to be undertaken alongside the program for adults. The results, if positive, have to be submitted as part of the full NDA or MAA filing. Clinical studies in children follow the same principles as for those in adults, although clearly there are important issues around obtaining informed consent. Pediatric studies usually have to be undertaken in the following age ranges: neonates, 0 to 2 years, 2 to 6 years, 6 to12 years, and 12 to 18 years. On occasions, it is possible to extrapolate between the age bands using pharmacokinetic modeling. On other occasions, the agencies take a tough line and demand studies in each of the age bands. One of the difficulties that has emerged as a result of the legislation is the limited number of centers and qualified investigators available to undertake such pediatric studies. This is becoming a significant rate-limiting step.

INTEGRATION OF SAFETY AND EFFICACY DATA—THE BENEFIT–RISK PROFILE

Clinical Overview Summary

The clinical overview summary is the critical document within Module 5 of the submission dossier. Prior to the introduction of the common technical document (CTD) through the ICH process (22), the three regions had different requirements for integrating the safety and efficacy data into a summary. The FDA favored the separate integrated summaries for safety and efficacy (the ISS and ISE), while Europe required the production of the clinical expert report. The latter was intended to be a critical analysis of the data with a relatively short report supported by overview tables and cross references to the dossier study summary reports. In Japan, a more factual summary (the Gaiyo) was favored. As a result of hard fought negotiations, ICH adopted the clinical overview which combines elements of the critical review with the previous US integrated summaries. However, the reality is that companies are still generating different styles of reports for the three regions, with a shorter more critical overview being undertaken for Europe, and the extended ISE and ISS being generated for the 20 US submissions. The common technical document guideline (22) makes clear that the clinical overview is meant to be an integrated analysis of all the clinical data giving a particular relevance to the pivotal studies. It should provide a clear set of conclusions and recommendations on the benefit–risk profile of a drug for the claimed uses in the target population.

Benefit–Risk Assessment Methodologies

In a recent reflection paper (23), the CHMP) observed that "the assessment of the benefits and risks in the context of a new drug application is a central element of the scientific evaluation on the market authorization application". In 1998, the Council for International Organizations of Medical Science (CIOMS) stated that "it is a frustrating aspect of benefit–risk evaluation that there is no defined and tested algorithm or summary metric that combines benefit and risk data and that might permit straightforward quantitative comparisons of different treatment options which in turn might aid in decision making" (24). The problem is that within a clinical program, there are often multiple risks and multiple benefits. In the past, reference was made to the so-called "risk–benefit ratio". This implied that there was a fixed quotient, which helped regulators decide on the acceptance or rejection of an application. Clearly, no such ratio or figure existed. The reality is that often a subjective or qualitative judgment is made by groups of experts. This approach has worked well and should not be denigrated or unduly criticized. However, there is currently a move to better structure the consideration of benefit–risk assessment and to be more explicit about setting out the decisions behind the approval of a new drug. There is additionally a need to better describe the uncertainties and data variabilities and how they impact benefit to risk assessment.

One approach has been to consider the use of decision analysis tools including a quantitative approach (25). A nonexhaustive list of examples is referenced in the CHMP reflection paper of 21 March 2008 which is available on the EMEA Web site http://www.emea.europa.eu. Such approaches can be used during the drug development process. The regulators are generating templates which may assist in a more uniform approach to determine the acceptable benefit–risk profile of a new drug application and to help better explain regulatory decisions to the wider public and to stakeholders. These initiatives, coupled with linked work on risk communication, are likely to be key considerations over the next five years.

CONCLUSION

The demonstration of efficacy is becoming ever more difficult. In part, this is a reflection of the more difficult diseases that are being tackled. It is also a reflection of the increasing requirements for the demonstration of comparative efficacy and of cost effectiveness.

Two trends can be discerned in the diseases currently being investigated as "druggable targets." The first is that more intervention studies are being undertaken for which there is a smaller effect size but there will be correspondingly a greater benefit because of a larger population to be treated. This inevitably means that much larger studies are required to demonstrate efficacy. The second trend is that "niche" or rare/orphan diseases are being investigated. As described earlier in the chapter, this raises a different set of problems whereby studies have to be undertaken of small populations.

At the same time, reassurance on safety of new medicines is increasing and there is arguably a greater risk awareness and perhaps risk aversion by both regulators and wider society. Thus, greater reassurance is required from more extensive premarketing safety exposure. This increased premarketing safety data also has to be supported by more extensive phase IV or postmarketing surveillance studies.

These increasing demands are taking place against a background of decreasing productivity in research and development and rapidly escalating cost pressures on the pharmaceutical industry. In an attempt to overcome these difficulties both the industry and the regulators are considering new approaches such as adaptive design, new methods for validating biomarkers and the possibility of introducing "conditional approvals." There are some interesting paradoxes and inevitable tensions in these different approaches but equally this remains an exciting time and a highly innovative one in the area of clinical trial development.

REFERENCES

1. DiMasi J, Hansen R, Grabowski H. The price of innovation: new estimates of drug development costs. J Health Econ 2003; 22:151–185.
2. http://www.fda.gov/oc/initiatives/criticalpath. Accessed in 2008.
3. http://imi.europa.eu. Accessed in 2008.
4. Gordian MA, Singh N, et al. Why drugs fall short in late stage trials. McKinsey Q 2006. http://www.mckinseyquarterly.com/article_abstract_visitor.aspx?av = 1879. Accessed in 2008.
5. ICH S7B. The Non Clinical Evaluation of the Potential for Delayed Ventricular Repolarisation (QT Interval Prolongation) by Human Pharmaceuticals CPMP/ICH/423/02.
6. ICH E14. The Clinical Evaluation of QT/QTs Interval Prolongation and Pro-Arrhythmic Potential for Non Anti-Arrhythmic Drugs. CHMP/ICH/2/04.
7. FDA Guidance for Industry, Investigators and Reviewers Exploratory IND Studies 2006.
8. Concept Paper on the Development of a CHMP Guideline on the Non-Clinical Trials with Pharmaceutical compounds. EMEA/CHMP/SWP/91850/2006.
9. Expert Group on Phase One Clinical Trials Final Report 2006. www.dh.gov.uk.publications. Accessed in 2008.
10. Strategies to Identify and Mitigate Risks for First-in-Human Clinical Trials with Investigational Medicinal Products. CHMP/SWP/28367/07.
11. ICH Topic EI. Population Exposure: the Extent of Population Exposure to Assess Clinical Safety. CHMP/ICH/375/95.
12. Reflections Paper on Methodological Issues in Confirmatory Clinical Trials Planned with an Adaptive Design. CHMP/EWP/2459/02.
13. Pricewaterhouse Coopers. Pharma 2020: The vision – Which path will you take? July 2008. http://www.pwc.com/pharma2020. Accessed in 2008.
14. Report of December 2007 EMEA/EFPIA Biomarker Workshop. www.emea.europa.eu/htms.humans/mes/biomarkers. Accessed in 2008.
15. Biomarker Qualifications Guidance to Applicants EMEA/CHMP/SAWP/72894/2008 – Consultation.
16. Points to consider an application with: (a) Meta-analysis and (b) One pivotal study. CPMP/EWP/2330/99.
17. Guideline on Clinical Trials in Small Populations. CHMP/EWP/83561/2005.
18. ICH Topic E5 (RI). Note for Guidance on Ethnic Factors in the Acceptability of Foreign Clinical Data. CPMP/ICH/289/95.
19. Convoy S, et al. Survey of unlicensed and off label drug use in paediatric wards in European countries. Br Med J 2000; 320:79–82.
20. Best Pharmaceuticals for Children Act. Public Law 107–109, 2002. www.fda.gov.opacom/laws/pharmakids. Accessed in 2008.

21. Paediatric Regulations. Regulations EC 1901/2006.
22. Common Technical Document for the Regulation of Pharmaceuticals for Human Use – Organisation of CTD. CPMP/ICH/2887/99.
23. Reflection Paper on Benefit-Risk Assessment Methods in the Context of the Evaluation of Marketing Authorisation Applications of Medicinal Products for Human Use. EMEA/CHMP/15404/2007.
24. CIOMS Working Group IV. Benefit-Risk Balanced for Marketed Drugs: Evaluating Safety Signals. Geneva: CIOMS, 1998.
25. Mussen F. The Evaluation of Methods for Benefit-Risk Assessment of Medicines and the Development of a New Model Using Multi Criteria Decision Analysis [doctoral thesis]. University of Wales, Cardiff, 2005.

36 Postmarketing Evaluation

H. Guenter Hennings
Regulatory Consultant, Bad Kreuznach, Germany

INTRODUCTION

Once a new medicinal product for human use has been approved by a Health Authority (HA), for example, the US Food & Drug Administration (FDA), the European Commission, or a national European Union (EU) Member State HA, the holder of the marketing authorization (MA) has the right to place the product on the market. Then many years of dedicated efforts in research, development, nonclinical and clinical investigation come to fruition. Developmental activities however continue to some extent for several reasons. Preauthorization research, development, nonclinical and clinical investigation describe at the time of approval a "model" of the medicinal product. The MA Holders (MAHs) may obtain the MA from the HA under conditions that require them to undertake additional studies within a defined time frame and with subsequent reporting obligations.

After placing the product on the market and exposing it to a broader patient population, new and unexpected scientific and medical questions also have to be addressed. Spontaneous adverse reaction reports require scientific clarification. Within the EU, a MAH is obliged to collect emerging scientific information and adverse event/reaction reports from all sources including accessing a widely used systematic literature review and reference databases (e.g., Medline: http://medline.cos.com/) at least once a week. The MAH would expect to initiate appropriate nonclinical and clinical research when significant benefit/risk issues arise through signal detection processes.

The MAH also continues development in order to support broader indication claims and will then obtain additional exclusivity rights, and may also get new formulations approved, which frequently can be protected by patents.

Once launched, the medicinal product becomes available to a broader audience in clinical practice and exposed to competition from other products and therapeutic procedures. Academic researchers will have a desire to understand the clinical properties and start their own research with the new product. This of course includes expansion into new areas, which may involve using the product "off-label".

Clinical studies in order to investigate the pharmacoeconomic profile, that is, demonstrating a favorable cost–benefit balance, may be necessary for reimbursement as well as contributing substantially to the commercial success of the product if health care value can be demonstrated.

The postauthorization evaluation in the areas of quality, safety, and efficacy is a continuum within active life cycle management and forms the response to scientific and medical issues brought to the attention of the MAH during marketing. Such postauthorization clinical studies are often referred to as Phase IV studies.

The intention here is to describe issues and challenges frequently faced by the MAH after the approval and the contribution of postauthorization research and development activities for an active life cycle management. The focus is on clinical postauthorization activities. Quality and nonclinical aspects are only discussed as far they impact on clinical safety and efficacy and the risk–benefit balance. For additional information, the reader should consult Parts III and IV of this book. Emerging or advanced therapy medicinal products, that is, products derived from cell and gene therapy and tissue engineering have not been discussed since these are still in the nonclinical and clinical preauthorization phase of development.

In this text, the European regulatory terms are used mainly and equivalent terms in the United States are briefly explained. However, for simplicity, EU terminology is predominantly used.

QUALITY

After approval of the MA application, changes to the manufacturing and specifications of the active substance and the finished medicinal product become necessary in most cases. Many of these changes, such as additional stability data leading to claims to extend the shelf life, additional or modified analytical procedures, and minor changes of excipients in standard formulations of small molecules, are not expected to influence the safety and efficacy of the medicinal product. Other changes however may have consequences, which require additional medical evaluation and even clinical data in order to demonstrate absence of negative influences on safety or efficacy and thus the risk–benefit balance of the medicine.

The prediction of such influences is not always easy, especially for medicinal products derived from biotechnology and related areas such as gene and cell therapy and tissue engineering, where the accepted notion (for small molecules) that slight changes of manufacturing under well-defined conditions do not affect the safety and efficacy of a product is not necessarily valid.

A few years ago, this was experienced painfully in a case involving epoetin alfa, which is prescribed to patients suffering from anemia, due to kidney failure cancer or HIV. In 1998, a cluster of epoetin-associated pure red-cell aplasia (PRCA) was reported which reached a maximum in 2001 and which was caused potentially by raised immunogenicity. The epoetin alfa accounted for 92% of these cases. Almost all cases involved patients with chronic kidney disease who had received subcutaneous injections of epoetin alfa. The increase in adverse events began when prefilled syringes became available and subcutaneous use increased. At the same time, the manufacturer had replaced human serum with other excipients (polysorbate 80 and glycine). Other contributing factors could have been polysorbate 80, leaching by organic compounds from the rubber plungers and the use of silicone oil as a lubricant in the prefilled syringes. Beginning in 2002 the formulation, primary packaging, storage, and handling of the product were improved and the subcutaneous administration for patients with chronic kidney disease was restricted. Then the exposure-adjusted incidence of epoetin alfa-associated PRCA decreased by 83% worldwide, dropping nearly to pre-1998 levels (1).

The epoetin alfa case demonstrates that small changes to the formulation and/or handling of a product might have serious consequences especially if factors influencing the patient's immune system are involved.

The MAH should therefore always consider the implications of the change and design an appropriate testing program. Simple comparisons at the level of both active substance and finished product to a publicly available standard as a reference (e.g., European Pharmacopoeia [Ph Eur], World Health Organization [WHO]) are generally not sufficient to establish quality, safety, and efficacy for complex medicines based, for example, on biotechnology. Consequently, extensive *comparability exercises* will be required for changes introduced in the manufacturing process of a given product (i.e., changes during development and postauthorization). The respective EU, United States, and International Conference on Harmonization (ICH) guidelines should be considered (2–7) and early scientific advice should be obtained from experienced HAs.

Because of the complexity of biological medicines, the demonstration of *comparability* will be a step-by-step process starting with the quality and the consistency of the medicinal product and of the manufacturing process. Differences in terms of safety or efficacy should be studied for each disease or indication. The nature of the difference from the authorized product should be taken into account especially regarding immunogenicity. A detailed risk management plan (RMP) should be developed in this context (see chap. 39). In the EU, specific guidance has also been developed for *biosimilar* medicinal products, which have been developed by different manufacturers and the guideline should be consulted when *comparability* has to be assessed. Such guidance has been developed for different classes of biotechnology-derived medicinal products, for example, the recombinant forms of somatotropin, human granulocyte colony stimulating factor, recombinant soluble human insulin, erythropoietin, recombinant α-interferon, and low-molecular-weight heparins (8–14).

In the United States, *comparability* also refers to the comparison of a biological product before and after a manufacturing change by the manufacturer. Specific guidance compatible with the ICH and EU guidelines is available (15, 16). The MAH should demonstrate through analytical and functional testing that after the manufacturing change, the product is still comparable and that its safety, identity, purity, or potency has not been affected. The HA has to determine if additional nonclinical and/or clinical studies are required taking into account experience with similar cases. The role of HAs is crucial because they have access to confidential information from other files, which the applicant does not have.

NONCLINICAL INFORMATION

Nonclinical research in pharmacology, toxicology, and related areas continues after authorization for different reasons.

Since nonclinical information may be relevant to the risk–benefit balance of a medicinal product, there may be requirements from HAs to conduct such postauthorization studies as an element of a conditional approval.

The mode of action and general and specific pharmacology studies are performed in order to identify potential new indications or routes of administration. A better understanding of the pharmacokinetic behavior in animals may improve the basis for dosage and posology recommendations. Additional toxicological data such as repeat-dose toxicity, reproductive/developmental toxicity, nephrotoxicity, hepatotoxicity, genotoxicity, carcinogenicity are also often required when indication claims and routes of administration are changed, treatment duration is prolonged or special (vulnerable) patient populations are to be included in clinical trial programs (17).

New signals from adverse event reports, from research performed in academia and by competitors, publications, and other sources could motivate nonclinical research prior to specific clinical investigation.

Manufacturing changes both of the active substance and the finished product (e.g., impurity profiles) may also require specific nonclinical studies. This is of particular importance for biotechnology and advanced therapy medicinal products. An active dialogue with all relevant regulatory authorities, such as the EU Rapporteur in the centralized procedure (CP), is essential in order to avoid major problems for the changed product in the market.

The relevance of nonclinical findings to the use in humans needs to be assessed and appropriately considered for the risk–benefit balance of the medicinal product. This is of special importance when nonclinical safety findings cannot be addressed by clinical research.

CLINICAL POSTAPPROVAL ACTIVITIES

Postauthorization Safety Studies

When the MA has been granted and the medicinal product has been placed on the market for the first time, a vulnerable phase commences so the risk–benefit balance must be monitored continually in order that it can be reconfirmed at certain time points in terms of broader (and off-label) use, which may be beyond direct control of the MAH. With the limited experience from the clinical trial program in mind, a MA will include conditions or restrictions (the "legal status") under which the medicinal product may be made available to patients, where the medicine can be administered (e.g., in a hospital) or by whom it can be prescribed (e.g., by a clinical specialist). Limitations in the safety database relate to the size of the study population, the study inclusion/exclusion criteria, and the precision with which these reflect the normal marketed use of the product.

In the preauthorization phase populations which may not been studied or may have only been studied to a limited degree may include children, the elderly, pregnant or lactating women, patients with relevant comorbidity such as hepatic or renal disorders, patients with differing disease severity, populations carrying known and relevant genetic polymorphism, and patients of different racial and/or ethnic origins. Some differences may also arise through trial setting (e.g., hospital or general practice) rather than through inclusion/exclusion criteria.

Safety signals from spontaneous adverse reaction reports require continuous monitoring. This so-called passive surveillance needs to be substantiated by more formal studies in order to improve risk identification and to confirm, dismiss, characterize, or quantify possible safety concerns. Postauthorization clinical safety studies (PASS) may be conducted by the MAH without consulting a regulatory authority. However, to aid transparency and reduce the chance of a surprise regulatory decision, the MAH should consider strongly whether to use the PASS mechanism as described in Volume 9A of *The Rules Governing Medicinal Products in the European Union* (18), which contains pharmacovigilance guidelines for medicinal products for human use.

PASS studies are "pharmacoepidemiological studies or clinical trials carried out in accordance with the terms of MA conducted with the aim of identifying or quantifying a safety hazard relating to an authorized medicinal product" (19). They are performed in order to assess if changes in the original balance of risks and benefits at the time of authorization can be identified by careful monitoring of any safety signals from such studies. Bearing in mind the significance of these findings, HAs regularly expect expedited reporting and submission of interim and final study reports from such studies. To be a PASS, relevant regulatory authorities at a national or regional level need to consulted and involved in assessing the trial protocols prior to initiation of safety studies. This is particularly relevant for the European Economic Area (EEA) qualified person for pharmacovigilance who has specific accountability for conduct of a PASS. This is especially important if the MAH anticipates labeling changes will be required at the end of the study.

The objective of such studies could be the identification (quality, quantity) of previously unrecognized safety concerns (hypothesis generation), the investigation of potential and identified risks (hypothesis testing), or the confirmation of the known safety profile of a medicinal product under normal conditions of use.

Well-designed and well-executed postauthorization safety studies are especially important for products with a novel chemical structure or novel mode of action. This may be particularly relevant if there is uncertainty as to the clinical relevance of a toxic effect in animals, if adverse events and risk factors identified in clinical trials need better quantification or where safety concerns from spontaneous reporting or off-label use need to be investigated.

Depending on the objectives, a variety of designs are available and continuously developed including clinical trials and observational cohort studies, case-control studies, or registries (20–24).

With the knowledge gained from these activities, the terms of the MA may have to be amended.

The EU RMP contains the *Safety Specification* as a regulatory instrument addressing the important identified risks of a medicinal product and *interalia,* the safety studies to be performed in the context of risk minimization activities.

In the United States, when NDA applicants are required to perform postapproval commitment studies, they have specific reporting obligations under Section 506 B (21 USC 356b) of the Federal Food, Drug, & Cosmetic Act (FDCA; 25). Voluntary studies, which are conducted on an applicant's own initiative (e.g., for the evaluation of a new indication), are not subject to these specific reporting requirements although the NDA applicant has to inform the FDA of the status of the voluntary studies in the annual report.

Postauthorization studies may be performed as *noninterventional (observational)* or *interventional* studies.

Noninterventional (Observational) Safety Studies

These studies use the prescribed product in accordance with the authorized indication and the decision of the treating physician to prescribe the product is separated from the decision to include the patient in the study. Diagnostic or monitoring procedures must be consistent with normal clinical practice as agreed with relevant HAs and ethics committee(s). Epidemiological methods such as cross-sectional, case control, and retrospective and prospective cohort studies do not fall under the legislation for investigational medicinal products.

When adequately planned and conducted, such studies are useful for better investigation of the signals from spontaneous reports or case series. They can also contribute important information concerning safety and efficacy of a medicinal product under normal conditions of use. They may be required by Regulatory Authorities and may be a commitment of the MAH.

Adverse event information from these studies is subject to the usual regulatory requirements for reporting of adverse reactions; for instance, serious adverse reactions arising from such studies should be reported on an expedited basis according to the same criteria and timelines as adverse reactions reported spontaneously by health care professionals. The reports on adverse reactions should also be included in the Periodic Safety Update Reports (PSURs).

The use of observational studies with the prime purpose of promoting the use of medicinal products may raise concerns. If company sales and marketing representatives are involved with recruitment of patients and physicians in such studies, the company exposes itself to considerable regulatory risk and liability.

Health care professionals should only be offered payment or inducement to participate in a postauthorization study when additional time and expenses will be incurred.

Interventional Studies

These studies are clinical trials which have to comply with the specific legislation covering investigational medicinal products (such as the EU Clinical Trial Directive; 26) and require clinical trial authorizations by Member State HAs in the EU or an approved IND by the FDA in the United States.

Conditional MAs

In rare and specific cases of unmet medical needs or the interest of public health, HAs reserve the right to grant MAs on the basis of fewer data than is normally required. Such MAs are subject to specific obligations and commitments on the part of the MAH.

In the EU, these *conditional* MAs are used sparingly and restricted to categories of medicinal products which aim at the treatment, prevention, or medical diagnosis of seriously debilitating or life-threatening diseases (27,28), medicinal products designated as orphan medicinal products in accordance with Article 3 of Regulation (EC) No 141/2000 (29), or medicinal products to be used in emergency situations in response to public health threats either recognized either by WHO or EU (30). Normally, only incomplete clinical data are acceptable in such instances, incomplete nonclinical or pharmaceutical data can only be accepted in emergency situations, in response to public health threats.

Before advising the EU Commission to grant a "conditional MA" for a medicinal product in the Centralized Procedure (CP), the Committee for Human Medicinal Products (CHMP) of the European Medicines Agency (EMEA) also has to assess whether the initial risk–benefit balance of the medicinal product is positive despite the limitations of the clinical data and whether it is likely that the applicant will be in a position to provide the relevant information for a full MA in due course.

Conditional MAs are valid for one year on a renewable basis and can be revoked if the MAH is unable to comply with the obligations imposed or if the risk–benefit balance turns out eventually to be negative.

The summary of product characteristics of the medicinal product and the package leaflet provided to patients and health care professionals must clearly state the conditional nature of the MA.

Clinical trials performed in order to meet the conditions of the MA have to be approved by HAs according to their national legislation and guidelines and EU legislation, for example, Directive 2001/20/EC (26).

Conditional MAs should not be confused with MAs granted in *exceptional circumstances* in accordance with Regulation (EC) No 726/2004 (27) and Directive 2001/83/EC as amended (19). They are granted when the applicant can show that he is unable to provide comprehensive data on the efficacy and safety for rare indications, that the provision of comprehensive evidence is impossible, or it appears unethical to collect such information. Then the applicant has to commit himself to an identified program of studies within a time period specified by the HA; the results of which shall form the basis of a reassessment of the risk/benefit profile. Supply of the medicinal product will be subject to certain restrictions concerning the legal status and obligations concerning the medical information to health care professionals and the patient.

In the United States, the legal basis and the regulatory terms regarding postauthorization (postmarketing) commitments and studies are different, but in practice the instruments and the underlying principles are similar in terms of protecting public health. The MAH may be required by the FDA to conduct postauthorization studies concerning clinical safety, clinical efficacy, clinical pharmacology, or nonclinical toxicology. The FDA will monitor the progress of such studies under Section 506B of the FDCA.

In an annual report, the MAH must address the progress of studies or the reasons for failure to conduct them. The FDA annually reports the status of postauthorization study commitments in the *Federal Register*. Any information necessary to identify the applicant for a study and establish the status of a study and reasons for failure to carry out a study is considered to be public information.

Postauthorization studies can be required in different situations with a certain degree of flexibility on the part of the FDA. A medicinal product, for instance, may have been approved in accordance with the accelerated approval provisions, which regularly demand the postauthorization conduct of studies to verify and describe clinical benefit, risk, and a positive risk–benefit balance. Detailed final guidance describing the content, format, and timing of the postauthorization commitment reports required by the agency (25) is available from the FDA.

The FDA may also require paediatric studies. These studies may be deferred because of concerns about the safety or effectiveness of the drug in paediatric populations and subject to postauthorization commitments. At the time of approval, postauthorization studies can be agreed with the FDA with the intention to provide additional information about product risks, benefits, and/or optimal use or after approval in order to address a safety concern that has been identified during the postapproval use of the drug.

Postauthorization studies concerning the consistency and reliability of product quality and voluntary nonclinical and clinical studies are not subject to the 506B reporting requirements; they will be reported in other sections of the annual report.

In 1997, the FDCA was amended in order to improve the timeliness of completing postauthorization studies and the update of medicinal product information. Since then the FDA has effective instruments monitoring the postauthorization activities of the MAH.

In the EU, the change in the legislation amending Directive 2001/83/EC and the corresponding Regulation 726/2004/EC (27) now allows for the introduction of specific sanctions for noncompliance with postauthorization commitments. The future will tell if the availability of sanctions means the EU European HAs will change their approach for the granting of conditional MAs if they are justified for public health reasons.

Off-Label Use

Once a medicinal product has reached the market, clinicians will be interested in the pathophysiology of disease and the known pharmacologic properties of the authorized product in order to assess if potentially different types of patients would benefit especially when there are few therapeutic alternatives. As physicians are free to prescribe an authorized medicinal product for any other purpose they consider safe and effective, they are not limited to the authorized indications. Thus, off-label use of the medicinal product in new indications or conditions by individual physicians may occur if not induced by a company and may lead to new data to expand the indication. (31).

In the case of well-established medicines, off-label use is often supported by extensive medical literature and experience but formal MAAs may not have been submitted due to lack of patent protection. β-blockers, for instance, were originally approved for hypertension, angina, and cardiac arrhythmias and contraindicated in patients with congestive heart failure. However, academic investigation into their off-label use for this purpose led to formal clinical trials, which eventually supported congestive heart failure as a standard therapy in symptomatic chronic heart failure (32).

The limited safety and efficacy data for off-label use may present a risk for patients such as unknown or unanticipated side effects or a lack of efficacy. The evidence base may range from substantial published literature supporting an acceptable risk–benefit balance to little or no published or unpublished evidence of risk–benefit. If physicians use medicines off-label in cases with little or insufficient scientific evidence they expose the patient and themselves to much greater risks and liability. The treatment may then be considered as medication error.

For obvious reasons, pharmaceutical companies are only allowed to promote a medicinal product within the confines of the authorized indications. However, as part of medical information services, they can on request, disseminate published study results about off-label use from academia and prescribing physicians. This is particularly justifiable if it provides guidance to ensure safe use. Companies need to tread carefully to avoid going too far from acceptable scientific information to illegal promotion of unapproved indication claims. Robust processes based on applicable legislation and guidelines are required to help mitigate the risk of acting unethically or unlawfully as this can have serious and litigious consequences. In 1993, gabapentin (Neurontin™ marketed at that time by Warner–Lambert) was approved in the United States for treatment of seizures. Warner–Lambert promoted gabapentin indirectly for the off-label treatment of pain and psychiatric conditions using medical education and research activities. In a short period of time of approximately five years, the product was widely used off-label promotion. In 2004, Pfizer, who had bought Warner–Lambert, had to admit to charges that Warner–Lambert had violated FDA regulations by promoting the drug for pain, psychiatric conditions, migraine, and other unapproved uses and paid $430 million to the federal government to settle the dispute (33,34).

In the EU, the MAH should assess the potential for off-label use in appropriate regulatory documents such as the RMP and PSURs wherever this is likely. It is of particular importance if the disease or disorder is found in the paediatric population or the medicinal product has an indication restricted to a subset of the population within a disease area or there are situations where the medicinal product must not be given for safety reasons.

The physician's liability concerning off-label use treatment is addressed in the national legislations of the EU Member States and reimbursement restrictions may apply also.

A manufacturer can actively collect information about the off-label use of the medicinal product either using *noninterventional* or preferably through *interventional* (i.e., clinical trials) studies. Clinical trial authorization applications (EU: CTA, US: IND) have to be submitted to and approved by the HAs. Results need to be reported according to the regional reporting requirements such as the annual safety report as required for the EU Clinical Trial Directive (26) or PSURs. Safety issues may be subject to expedited reporting.

Off-label uses reported by physicians in the literature as case reports need to be taken into account and appropriately followed up as the MAH must assess all normal use of the product not just for the authorized indication.

Clinical postauthorization trials are frequently performed on the investigator's own initiative with or without informing the MAH. The involvement of the MAH can raise liability concerns, which need to be addressed carefully in the contract and training programs. Outcomes of such studies may affect the risk–benefit balance of the product. If the MAH becomes aware of such a study, they should try to get as much information as possible and to ensure that the investigators adhere strictly to Good Clinical Practice rules and common scientific standards as described in the Declaration of Helsinki.

Clinical Life Cycle Management: New Indications, New Populations

MAHs for innovative medicines are usually very interested in postauthorization clinical research as it offers the opportunity to receive intellectual property rights in addition to patent protection. Such rights offer exclusive marketing opportunities for some time.

New Medicinal Products

A medicinal product for human use, which has not been approved previously for marketing in the EU, must be supported by a "full" dossier, that is, complete documentation, as specified in the legislation (19) has to be provided. The product benefits from an eight-year period of *data protection* and a ten-year period of *marketing protection* starting on the day of approval. An authorization obtained by the MAH for one or more new therapeutic indications as a line extension with a significant clinical benefit in comparison with existing therapies would result in an additional one-year *marketing protection* period. Thus a maximum of 11 years of protection is possible (8+2+1 rule). Generic products, with or without the new therapeutic indication, may not be placed on the market until expiry of the eleventh year.

It is the responsibility of the HA, that is, the EU Commission for the CP, the Reference Member State (RMS) in cases of the Decentralized or Mutual Recognition Procedure (DP/MRP) or the national HA to provide a clear statement in the MA as to whether the new indication represents a significant clinical benefit in comparison with existing therapies.

Some intellectual property protection is available even for well-established substances: one year of data exclusivity can be granted by HAs, provided that significant nonclinical or clinical studies were carried out for the new indication (19); in this case, the data exclusivity however is only for the data concerning the new indication.

Similarly, a change of classification (prescription, pharmacy only, general sales list) of a medicinal product based on significant nonclinical tests or clinical trials (19) will protect the MAH for one year against generic competition regarding this change of classification.

Nonclinical tests and/or clinical trials are considered significant if they concern new strengths, new routes of administration, new pharmaceutical forms, new indications or sub-populations like elderly, or children and patients with specific medical conditions (hepatic or renal impairment).

Paediatric Medicinal Products

Although children represent a considerable part of the population the majority of medicines prescribed by paediatricians have never been investigated in this population. For a long time, the relatively small market size, the lack of intellectual property protection, practical difficulty in conducting trials, and restrictive pricing/reimbursement policies were major obstacles to significant investment by the pharmaceutical industry in the development of paediatric formulations and indications.

Both in the US and the EU, legislation is in place requiring for new medicines adequate paediatric clinical data for new medicines preauthorization unless waivers have been granted by HAs.

Established medicines pose a particular problem. Many have been used for years to treat children without a sufficient clinical database. Lawmakers in both regions have addressed the issue in similar ways in order to provide incentives to postauthorization research and development, with the United States having a much longer experience than the EU.

In 1997, the U.S. legislators created Section 505A of the FDCA (21 USC 355a), offering an extension of six months in the period during which a product can be marketed exclusively by the originator and with the experience of the initial phase, the "Best Pharmaceuticals for Children Act" was signed into law in January 2002 (35). Detailed guidance and information is available from the FDA (36). Now, in the United States, paediatric information is also available for many well-established and off-patent medicinal products. More accurate paediatric dosing information has been established for around 150 drug products and over 20 new formulations have been developed specifically for children. New or enhanced paediatric safety information is available for 48 drugs commonly used in children (37).

The EU has also recently established specific legislation and guidance on better medicines for children (38). This new legislation entered into force in the EU on January 26, 2007.

Regulatory instruments, processes, and incentives for the industry have been determined and are currently established. A pan-EEA Paediatric Committee (PDCO) has been formed at the EMEA based at Canary Wharf, London and is involved in the review of paediatric investigation plans and paediatric clinical data submitted by the industry. The incentive for industry to provide paediatric data will be a six-month extension to the supplementary protection certificate (where one is available), that is, a six-month patent extension on the active substance. For orphan paediatric medicines, an additional two years of market exclusivity is granted in addition to the regular 10-year EU orphan exclusivity.

For established off-patent products which are available on the market for adults only, without information and formulations appropriate for children of all ages, a new type of MA, the *paediatric-use MA* (PUMA) has been created, which provides 10 years of data protection for innovative studies on new paediatric indications. The PUMA will cover the indication for children and the formulation appropriate for all age groups.

Other instruments comprise improvements in the area of pharmacovigilance and risk management systems, postauthorization of paediatric medicines, an EU inventory of the therapeutic needs, research, and regulatory activities, an EU network of investigators and trial centres for children, a provision on EU funding into research, a system of free scientific advice for the industry, and a public database of paediatric studies.

In the area of paediatric medicine, an active dialogue between the United States and the EU has been started from which the EU can certainly benefit from the United States experience with the implementation of paediatric research with new and well-established medicines. Principles of interaction have been published jointly by the FDA, the EU Commission, and the EMEA (39).

Orphan Medicinal Products

Many medicines for the treatment of rare diseases have been developed starting from off-label uses of established medicines. HAs have developed a set of guidance based on specific legislation in order to provide guidance, supervision, and stimulate the generation of adequate scientific data in this field.

European Union (25,40–42)

A MAH can apply for *designation* as an *orphan medicinal product* (OMP) at any stage of the development of the medicinal product, however only *prior to* the application for a MA in the proposed therapeutic indication.

The Committee for Orphan Medicinal Products (COMP) is responsible for the evaluation of such an application for orphan drug status at the EMEA.

An application for this designation needs to justify the medical plausibility of the proposed orphan condition with a prevalence of not greater than 5 in 10,000 population in the EU, or any subset or rationale excluding a larger population. A subset of a disease which, when considered as a whole, has a prevalence greater than 5 in 10,000, could be considered a valid condition if patients in that subset present distinct and unique characteristic(s). Prevalence is defined as the number of persons with a disease or condition affected at the time of designation in the 27 Member States of the EU.

The MAH has also to provide reasons for the life-threatening or chronically debilitating nature of the condition supported by scientific or medical references.

It is also possible, but rarely used, to apply for a designation as Orphan Medicinal Product, where, without incentives, it is unlikely that the marketing of the medicinal product in the Community would generate sufficient return to justify the necessary investment.

Incentives from the designation as an Orphan Medicinal Product are extensive data protection rights of 10 years, eligibility for fee waivers, and free protocol assistance from the EMEA. The use of the CP is mandatory for Orphan Medicinal Products. Thus, it is possible for a product to be authorized initially through the mutual recognition procedure for one indication and then through the CP for a second indication, which has an orphan designation.

United States

For the last few years, an orphan designation process has been available in the United States based on the Orphan Drug Act (43) which on request by MAHs and approved by the authority, offers tax credit, exemptions from user fees, and seven years marketing exclusivity incentives. Rare disease or condition in the United States means any disease or condition which affects less than 200,000 persons or if the cost of developing a medicinal product for such a disease or condition could not be recovered.

Experience in both the United States and the EU demonstrate that the incentives for orphan disease medicines has encouraged MAHs to invest not only in preauthorization development for new medicines but also in postauthorization studies for well-established use products as an instrument of active life cycle management.

Pharmacoeconomic Studies

Many Member States of the EU face serious problems financing their public health system due to an ageing population, low birth rates, a shrinking workforce, and a flat economy. The cost of medicinal products is an easy budgetary target for politicians and the social insurance organizations since it can contribute to cost savings in the entire health care system.

According to the pharmaceutical legislation in the EU however MAs are granted *only* on the basis of a scientific risk–benefit assessment, which is the evaluation of quality, safety, and efficacy to demonstrate the therapeutic place of a product.

Cost–benefit considerations (*Fourth Hurdle*) are excluded from the review of MAAs for a medicinal product at this stage. Such a cost–benefit evaluation in the different Member States plays a crucial role for the commercial success of the product in the national markets following the initial approval of a medicinal product. Decisions on reimbursement and pricing fall under the jurisdiction of the Member States, who have developed many different controlling instruments. The price and reimbursement negotiations can substantially delay the introduction of the product to the market. National authorities such as the English and Welsh National Institute for Health and Clinical Excellence (NICE: 44), the French Haute Autorité de Santé (HAS, 45), and the German Institut für Qualität und Wirtschaftlichkeit im Gesundheitswesen (IQWiG, 46) are reviewing the use of different criteria for benefit and an different overall focus on existing clinical data and all other additional information provided by the MAH supporting the cost benefit of the medicinal product. Many stakeholders contribute to the outcome of their reviews although controversy may arise when reimbursement is refused.

It would be advisable to perform cost–benefit-related studies in parallel to the major confirmatory trials in Phase III of the clinical development program. Such trials may still be needed for postauthorization evaluation when pricing and reimbursement authorities require special information on the topic. Design and implementation of such studies depends largely on the individual product.

In the United States, reimbursement and price are not controlled as in the EU. The success of a medicinal product however also depends on the view major payers and health maintenance organizations (HMOs) take and if the medicinal product is accepted by the formularies.

Periodic Safety Update Reports

The EU Guidance in Volume 9A "Pharmacovigilance for Medicinal Products for Human Use" defines a PSUR as an update of the worldwide safety experience of a medicinal product to which is to be submitted to HAs at defined time points postauthorization. MAHs are expected to provide succinct summary information together with a critical evaluation of the risk-benefit balance of the product in the light of new or changing information. This evaluation should ascertain whether further investigations need to be carried out and whether changes should be made to the MA and product information.

HAs expect to be informed on postauthorization information from all countries where the product is marketed or investigated in interventional and noninterventional trials.

The EU format and content of PSURs are based on the ICH guideline E2C(R). The reports are required by law even if the medicinal product is not marketed. With the EU pharmaceutical legislation of 2004 (19), the PSUR submissions are required every six months following initial placing of the product on the market in the EU and until two full years of marketing experience in the EU have been gained. Then PSURs should be submitted once a year for two years and thereafter at three-yearly intervals. It should be noted that the time of validity of a MA in the EU has also been changed. Initially, it is granted for five years, it is renewable and then valid indefinitely. At the request of a HA, a PSUR has to be submitted at any time after the granting of the MA (19,25).

In the United States, spontaneous, nonserious, domestic, adverse drug experiences are required to be submitted in a specific format to the FDA in periodic reports (47) under 21 CFR 314.80. These reports must be submitted quarterly for the first three years after application approval and annually thereafter. With the prior agreement of the concerned FDA division, the MAH may use instead PSURs in the ICH format (48), which would facilitate the management of adverse drug experience reporting for globally marketed medicines. The ICH guidance does not, however, establish legally enforceable responsibilities in the United States.

SUMMARY AND CONCLUSIONS

Despite the rise of Asian markets and industry, the pharmaceutical markets of the ICH area (the EU, Japan, and the USA) remain very important and influential in the world and their HAs monitor the quality, safety, and efficacy of medicinal products during clinical development and postauthorization. MAHs therefore have clear obligations to maintain state of the art processes to monitor the safety of their products, which in essence mean a continuous scientific and medical development from beginning to end of the product life cycle.

The early postauthorization phase after the launch of a medicinal product is a vulnerable phase during which the only limited experience gained from the preauthorization nonclinical and clinical research program meets the broader (and possibly off-label) use beyond direct control of the MAH. The product may also be used in populations which have not been studied very intensively such as children, the elderly, pregnant or lactating women, patients with relevant comorbidity such as hepatic or renal disorders, patients with disease severity different from those studied in clinical trials; populations carrying known and relevant genetic polymorphism, and patients of different racial and/or ethnic origins.

During this phase, safety signals from sources such as spontaneous adverse reaction reports should be monitored extremely carefully. Passive surveillance requires substantiation by more formal studies in order to improve risk identification and to confirm, dismiss, characterize, or quantify possible safety concerns.

PASS, which refer to pharmacoepidemiological studies or clinical trials, can be implemented to help identify or better define newly arising and previously unrecognized safety concerns (hypothesis generation), investigate potential and identified risks (hypothesis testing), or confirm the known safety profile of a medicinal product under normal conditions of use. A variety of designs are available including clinical trials, observational cohort studies, case-control studies, or registries.

Appropriate RMPs (see chap. 39) address the important identified risks of a medicinal product and *inter alia* the safety studies to be performed in the context of risk minimization activities.

The other important aspect of postauthorization clinical development aims at a better understanding of the broader use of the product, new indications, and additional populations. The MAH may then receive benefits in the form of tax incentives and additional intellectual property protection. Postauthorization development, mainly but not necessarily just clinical, is thus an important strategic instrument for the active life cycle management of a medicinal product.

Postauthorization, noninterventional (observational), and interventional Phase IV clinical studies set up primarily for marketing purposes can cause concern because it can appear to be irresponsible use of scientific information. Companies need robust processes in place to ensure good governance around study design and conduct. In particular, the use of studies, research, or educational activities, which might be interpreted as being aimed at promoting the use of medicinal products for off-label use indications, can lead to serious sanctions from regulatory authorities.

It is good company practice for sales and marketing representatives not be involved in postauthorization research activities involving patients and physicians as this can look like a promotional exercise. When doing this, the company exposes itself to considerable regulatory risk and liability.

A favorable risk–benefit balance is the prerequisite for obtaining a MA. However, the commercial viability of a new medicine also depends on a cost–benefit evaluation. The MAH should therefore plan pre- and postauthorization studies carefully, in order to get a favorable position in price and reimbursement negotiations with payers, statutory insurances, and their institutions.

REFERENCES

1. Bennett CL, Luminari S, Nissenson AR, et al. Pure Red-Cell Aplasia and Epoetin Therapy. N Engl J Med 2004; 351:1403–1408.
2. European Medicines Agency (EMEA). Comparability of Biotechnological/Biological Products, ICH Topic Q 5 E. Note for Guidance on Biotechnological/Biological Products Subject to Changes in their Manufacturing Process, CPMP/ICH/5721/03, London, June 2005.
3. European Agency for the Evaluation of Medicinal Products (EMEA). Evaluation of Medicines Human for Use. Guideline on Comparability of Medicinal Products Containing Biotechnology-derived Products as Active Substances : Quality Issues, CPMP/BWP/3207/00,. London, December 2003.

4. European Medicines Agency (EMEA). Committee For Human Medicinal Products (CHMP). Guideline On Comparability Of Biotechnology-Derived Medicinal Products After A Change In The Manufacturing Process – Non-Clinical And Clinical Issues, EMEA/CHMP/BMWP/101695/2006, London, July 2007.
5. European Medicines Agency (EMEA), Committee for Medicinal Products for Human Use (CHMP). Guideline on Similar Biological Medicinal Products, CHMP/437/2004, London, October 2005.
6. European Medicines Agency (EMEA), Committee for Medicinal Products for Human Use (CHMP). Guideline On Similar Biological Medicinal Products Containing, Biotechnology-Derived Proteins as Active Substance: Quality Issues, EMEA/CHMP/BWP/49348/2005, London, February 2006.
7. European Medicines Agency (EMEA), Committee for Medicinal Products for Human Use (CHMP). Guideline On Similar Biological Medicinal Products Containing, Biotechnology-Derived Proteins As Active Substance: Non-Clinical And Clinical Issues, EMEA/CHMP/BMWP/42832/2005, London, February 2006.
8. European Medicines Agency (EMEA), Committee for Medicinal Products for Human Use (CHMP). Annex To Guideline On Similar Biological Medicinal Products Containing Biotechnology-Derived Proteins As Active Substance: Non-Clinical And Clinical Issues. Guidance On Similar Medicinal Products Containing Somatropin, EMEA/CHMP/BMWP/94528/2005, London, February 2006.
9. European Medicines Agency, Committee for Medicinal Products for Human Use (CHMP). Annex To Guideline On Similar Biological Medicinal Products Containing Biotechnology-Derived Proteins As Active Substance: Non-Clinical And Clinical Issues. Guidance On Similar Medicinal Products Containing Recombinant Granulocyte-Colony Stimulating Factor London, EMEA/CHMP/BMWP/31329/2005, February 2006.
10. European Medicines Agency, Committee for Medicinal Products for Human Use (CHMP). Annex To Guideline On Similar Biological Medicinal Products containing Biotechnology-Derived Proteins As Active Substance: Non-Clinical And Clinical Issues. Guidance On Similar Medicinal Products Containing Recombinant Human Soluble Insulin, EMEA/CHMP/BMWP/32775/2005, London, February 2006.
11. European Medicines Agency, Committee for Medicinal Products for Human Use (CHMP). Annex To Guideline On Similar Biological Medicinal Products Containing Biotechnology-Derived Proteins As Active Substance: Non-Clinical And Clinical Issues. Guidance On Similar Medicinal Products Containing Recombinant Erythropoietins, London, EMEA/CHMP/BMWP/94526/2005 Corr., March 2006.
12. European Medicines Agency, Committee for Medicinal Products For Human Use (CHMP). Draft Guideline On Similar Medicinal Products Containing Recombinant Interferon Alpha, EMEA/CHMP/BMWP/102046/2006, London, October 2007.
13. European Medicines Agency, Committee for Medicinal Products For Human Use (CHMP). Draft Guideline On Similar Biological Medicinal Products Containing Low-Molecular-Weight-Heparins, EMEA/CHMP/BMWP/118264/2007, London, April 2008.
14. European Medicines Agency, Committee for Medicinal Products For Human Use (CHMP). Draft Concept Paper On Similar Biological Medicinal Products Containing Low Molecular Weight Heparins Non-Clinical Issues, EMEA/CHMP/BMWP/496286/2006, London, 24 January 2007.
15. U.S. Department of Health and Human Services, Food and Drug Administration (FDA), Center for Biologics Evaluation and Research (CBER), Center for Drug Evaluation and Research (CDER). FDA Guidance Concerning Demonstration of Comparability of Human Biological Products, Including Therapeutic Biotechnology-derived Products, Washington DC, April 1996.
16. U.S. Department of Health and Human Services Food and Drug Administration, Center for Biologics Evaluation and Research (CBER), Center for Drug Evaluation and Research (CDER). Guidance for Industry ICH Q5E Comparability of Biotechnological/Biological Products Subject to Changes in Their Manufacturing Process. June 2005.
17. International Conference on Harmonisation of Technical Requirements for Registration of Pharmaceuticals for Human Use, European Medicines Agency. Note For Guidance On Non-Clinical Safety Studies For The Conduct Of Human Clinical Trials And MA For Pharmaceuticals, CPMP/ICH/286/95, ICH Topic M 3 (R2), London, July 2008.
18. European Commission, Volume 9A, Pharmacovigilance for Medicinal Products for Human Use (version April2007). http://ec.europa.eu/enterprise/pharmaceuticals/eudralex/vol-9/pdf/vol9_2007-07_upd07.pdf. Accessed August 2008.
19. European Parliament and Council Directive 2001/83/EC, of 6 November 2001 on the Community code relating to medicinal products for human use, Official Journal of the European Communities of 28 November 2001 L 311/67 & Commission Directive 2003/63/EC of 25 June 2003 amending Directive 2001/83/EC of the European Parliament and of the Council on the Community code relating to

medicinal products for human use Official Journal of the European Communities of 27 June 2003 L 159/46 & Directive 2004/27/EC of The European Parliament and of the Council of 31 March 2004 amending Directive 2001/83/EC on the Community code relating to medicinal products for human use. Official Journal of the European Union 2004; L136/34.
20. International Conference on Harmonisation of Technical Requirements for Registration of Pharmaceuticals for Human Use, European Medicines Agency. ICH Harmonised Tripartite Guideline, Pharmacovigilance Planning, E2E, CPMP/ICH/5716/03, London, December 2004.
21. Mann RD, Andrews EB, eds. Pharmacovigilance. West Sussex: John Wiley and Sons, Ltd, 2002.
22. Strom BL, ed. Pharmacoepidemiology, 4th ed. New York, NY: John Wiley and Sons, Ltd, 2005.
23. Hartzema AG, Porta M, Tilson H, eds. Pharmacoepidemiology: An Introduction, 3rd ed. Cincinnati: Harvey Whitney Books, 1998.
24. Rothman KJ, Greenland S, eds. Modern Epidemiology, 2nd ed. Philadelphia, PA: Lippincott, Raven Publishers; 1998.
25. U.S. Department of Health and Human Services, Food and Drug Administration, Center for Drug Evaluation and Research (CDER), Center for Biologics Evaluation and Research (CBER). Guidance for Industry, Reports on the Status of Postmarketing Study Commitments Implementation of Section 130 of the Food and Drug Administration Modernization Act of 1997, Washington DC, February 2006.
26. Directive 2001/20/EC of The European Parliament And of The Council of 4 April 2001 on the approximation of the laws, regulations and administrative provisions of the MSs relating to the implementation of good clinical practice in the conduct of clinical trials on medicinal products for human use. Official Journal of the European Communities 2001; L121/34.
27. Regulation (EC) No 726/2004 of the European Parliament and of the Council of 31 March 2004 laying down Community procedures for the authorisation and supervision of medicinal products for human and veterinary use and establishing a European Medicines Agency. Official Journal of the European Communities2004; L136/1.
28. Commission Regulation (EC) No 507/2006 of 29 March 2006 on the conditional MA for medicinal products for human use falling within the scope of Regulation (EC) No 726/2004 of the European Parliament and of the Council.
29. Regulation (EC) No 141/2000 of the European Parliament and of the Council, of 16 December 1999, on orphan medicinal products. Official Journal of the European Communities 2000; L18/1.
30. Decision No 2119/98/Ec of The European Parliament And of The Council of 24 September 1998 setting up a network for the epidemiological surveillance and control of communicable diseases in the Community. Official Journal of the European Communities, 1998; L 268/1.
31. Bennett WM. Off-label use of approved drugs: Therapeutic opportunity and challenges. J Am Soc Nephrol 2004; 15:830–831.
32. Jessup M, Brozena S. Heart failure. N Eng J Med U S A 2003; 348:2007–2018.
33. Steinman MA, Bero LA, Chren MM, et al. Narrative review: The promotion of gabapentin: An analysis of internal industry documents. Ann Intern Med 2006; 145(4):284–293.
34. US Department of Justice Press Release: Warner-Lambert to Pay $430 Million to Resolve Criminal & Civil Health Care Liability Relating to Off-label Promotion, 13 May 2004. http://www.usdoj.gov/opa/pr/2004/May/04_civ_322.htm. Accessed August 2008.
35. 107th Congress of the USA, Best pharmaceuticals for children act, Washington, DC January 2002. http://www.fda.gov/opacom/laws/pharmkids/pharmkids.html. Accessed August 2008.
36. U.S. Department of Health and Human Services, Food and Drug Administration, Center for Drug Evaluation and Research (CDER), Center for Biologics Evaluation and Research (CBER). Procedural Guidance for Industry, Qualifying for Pediatric Exclusivity Under Section 505A of the Federal Food, Drug, and Cosmetic Act, Revised, September 1999.
37. Lumpkin M. Food and Drug Administration: Children Are Not Little Adults. www.fda.gov/oc/vonEschenbach/andys_take/default.html. Accessed August 2008.
38. Regulation (EC) No 1901/2006 of the European Parliament and of the Council of 12 December 2006 on medicinal products for paediatric use and amending Regulation (EEC) No 1768/92, Directive 2001/20/EC, Directive 2001/83/EC and Regulation (EC) No 726/2004. Official Journal of the European Union L 378/1. see also: http://www.emea.europa.eu/htms/human/paediatrics/introduction.htm. Accessed August 2008.
39. U.S. Food and Drug Administration, European Commission, Enterprise and Industry Directorate General, European Medicines Agency, Principles of Interactions: Between EMEA and FDA Pediatric Therapeutics, June 2007. http://ec.europa.eu/enterprise/pharmaceuticals/paediatrics/docs/peds_principles-of-interactions.pdf. Accessed August 2008.
40. Commission Regulation (EC) No 847/2000 of 27 April 2000 laying down the provisions for implementation of the criteria for designation of a medicinal product as an orphan medicinal product and

definitions of the concepts 'similar medicinal product' and 'clinical superiority'. Official Journal of the European Communities 2000; L103/5.

41. European Commission, Enterprise Directorate-General, Single market, regulatory environment, industries under vertical legislation, Pharmaceuticals, Brussels, 22 July 2004, ENTR/6283/00 Rev 2, Guideline on the format and content of applications for designation as orphan medicinal products and on the transfer of designations from one MAH to another, 22 July 2004. http://www.emea.europa.eu/htms/human/orphans/guidance.htm. Accessed August 2008.
42. European Agency for the Evaluation of Medicinal Products, Pre-authorisation Evaluation of Medicines for Human Use, Committee for Orphan Medicinal Products (COMP), Points to Consider document on 'Calculation and Reporting of the Prevalence of a Condition for Orphan Designation (COMP/436/01). http://www.emea.europa.eu/htms/human/orphans/guidance.htm. Accessed August 2008.
43. U.S. Department of Health and Human Services, Food and Drug Administration. http://www.fda.gov/orphan/oda.htm. Accessed August 2008.
44. NICE: National Institute for Health and Clinical Excellence (England). http://www.nice.org.uk/. Accessed August 2008.
45. HAS: Haute Autorité de Santé (France). http://www.anaes.fr/anaes/anaesparametrage.nsf/HomePage?ReadForm. Accessed August 2008.
46. IQWiG: Institut für Qualität und Wirtschaftlichkeit im Gesundheitswesen. http://www.iqwig.de/. Accessed August 2008.
47. US FDA: Periodic Safety Reporting Requirements. http://www.fda.gov/cder/aers/chapter53.htm. Accessed August 2008.
48. International Conference on Harmonisation of Technical Requirements for Registration of Pharmaceuticals for Human Use. Guidance for Industry, Addendum to E2C Clinical Safety Data Management: Periodic Safety Update Reports for Marketed Drugs, ICH Topic ICH-E2C(R), U.S. Department of Health and Human Services Food and Drug Administration. February 2004. http://www.fda.gov/cber/gdlns/ichclinsafety.htm. Accessed August 2008.

37 Pharmacovigilance and Risk Management

Brian Edwards, Mary Teeling, and Markku Toivonen
NDA Regulatory Science Ltd, Leatherhead, Surrey, U.K.

RELATION OF PHARMACOVIGILANCE TO RISK MANAGEMENT

No effective medicine is without risk and this is not a static concept. It changes over time requiring active surveillance and activities to maintain an optimal balance of benefits and risks for a medicinal product. The challenge for regulators is to find the right balance between timely availability of new medicines and the limited knowledge on the safety profile at the time they grant a marketing authorization (MA). How can we rise to this challenge? Firstly, it is important to remember that writing a risk management plan (RMP) as part of a submission for a MA or New Drug Application (NDA) is only part of, and not necessarily the same as, managing risk. The most effective way to manage risk is to create an effective pharmacovigilance (PV) system, so-called "routine pharmacovigilance." However to understand what routine means, we should reflect on the current European Medicines Agency (EMEA) and Medicines and Health care Products Regulatory Agency (MHRA) PV inspectorate's definition of critical and major findings after an inspection as this refers not only to safety but also rights or well being of patients (1,2). Thus, there is much more to "routine" PV than simply collecting and reporting adverse events. In turn, safety is achieved by a combination of having a compliant ("safe") PV system and a safe product, which is one with an acceptable benefit–risk profile appropriate to the therapeutic indication and normal use within that territory.

The focus of this chapter will be on examining the scientific and legislative basis for regulatory requirements and expectations for a robust and realistic PV system to manage risk management. Much of the discussion in this chapter will refer to United States (US) and European Union (EU) initiatives with a brief description of the situation in Japan.

Regulatory Expectations for Pharmaceuticals Leading to a Conceptual Framework Reflected in Risk Management Plans

Risk management is not in any sense a new idea. It is a well-established concept for medical devices that forms part of the International Standards Organization (ISO) initiatives (3). ISO defines risk management as the systematic application of management policies, procedures, and practices to the tasks of analyzing, evaluating, and controlling risk. Thus, although there may well be lack of harmonization in the global pharmaceutical system in many areas, when it comes to managing pharmaceutical risk, regulators are harmonized in their prime objective to achieve an acceptable balance of benefits and risks for medicinal products in their own territory of influence so as to improve public health. Ideally, risk management thinking should begin at the earliest stage of product development and continue throughout the life cycle of any given product. This includes proactive risk assessment and development of plans for interventions to manage patient risk. The processes and activities within the system required to manage risk can be referred to as PV and this is consistent with the WHO definition used to define PV in Volume 9A of the *Rules Governing Medicinal Products in the European Union* (4).

THE HEADS OF AGENCIES EUROPEAN RISK MANAGEMENT STRATEGY

In recognition of the scientific and resource needs to coordinate efforts to deliver safe medicinal products, the Heads of Medicines Agencies (HMA) Human *Ad Hoc* Working Group for the European Risk Management Strategy (ERMS) was first established in July 2002. The Working Group produced an initial report on the establishment of the strategy, which was published in January 2003. On May 2, 2005, in Luxembourg, the HMA agreed to the publication of a progress report on the implementation of the ERMS and of an Action Plan. In July 2005, the HMA agreed the membership and remit of a permanent Working Group of the HMA and the EMEA–the ERMS Facilitation Group. A rolling two-year Work Program (2005–2007) was agreed with the intention that the ERMS will achieve high standards of public health protection for all medicines available in the EU (5). The ERMS aims to provide for a more proactive conduct of PV by putting in place measures that allow for the early detection, assessment, minimization, and communication of risks of medicines in Europe throughout their lifecycle. The "Action Plan" states that the initiatives to be taken during the second implementation phase of the ERMS relate to three priority areas, as follows:

- the implementation of new community legislation;
- complementary implementing initiatives to deliver an intensive drug monitoring system; and
- further strengthening of the EUPV System as part of the EU Regulatory System.

The working methodology applied during the further implementation of the ERMS is based on three pillars:

(i) introducing a targeted approach by providing a further level of prioritization in order to adequately tackle the wide range of activities to be undertaken stemming from the "Action Plan,"
(ii) making the best use of the available resources, hence avoiding duplication of work, and
(iii) involving all relevant stakeholders of the EUPV System.

The HMA recognize the need to prioritize initiatives according to those most likely to provide the most significant contribution to the further improvement of patient safety whilst adhering to the timeframe for the implementation of the new legislative provisions. There is due emphasis on the cost–efficiency aspects of the system with work sharing taking place between EU Member States to optimize resources (such as evaluation of periodic safety update reports). In addition, they intend to involve all appropriate stakeholder involvement. The basic science of PV has not been forgotten as regulators are examining how to move up the levels of evidence in order to obtain the best evidence on which to reach decisions about benefit–risk (6). These ideas surrounding what is Best Evidence are to be examined in a Concept Paper, which is still awaited at the time of writing (November 2008). A list of medicines requiring intensive drug monitoring will be published. New methods of risk detection are being explored. To enhance trust in the system, there is an EU Transparency and Communication strategy (7). The HMA and the EMEA agreed on a rolling two-year (2008–2009) work program to further progress the ERMS at their November 2007 meeting (8). The ERMS work program during these two years is focusing on two areas: further improving the operation of the EUPV System and strengthening the science and methodology that underpins the safety monitoring of medicines for human use. The activities that are underway are meant to build on both the progress already made as well as take into account a number of environmental changes which will impact on the implementation of the ERMS, such as the European Commission's Strategy to Better Protect Public Health by Strengthening and Rationalizing EU Pharmacovigilance, the implementation of new EU legislation on paediatric medicines and on advanced therapies, and the increased regulatory cooperation with non-EU regulatory authorities. Updated guidance in the field of PV of medicines monitoring has been made available through a complete revision of Volume 9 of the *Rules Governing Medicinal Products* which is now published as Volumes 9A and 9B, the latter being for veterinary medicines (4). Regulators believe that further implementation of EudraVigilance and the introduction of additional functionalities and activities such as the validation of the EudraVigilance Data warehouse and Analysis System (EVDAS) and statistical signal detection methods in EVDAS will be central to managing medicinal risk. To help with implementation of postauthorization safety studies (PASS), the EMEA has embarked on an ambitious project entitled European Network of Centers for Pharmacoepidemiology and Pharmacovigilance (ENCePP). It is hoped that by concentrating available expertise and research experience in the field of pharmacoepidemiology and PV across the EU, ENCePP will provide a more cost-effective opportunity to facilitate the conduct of multicenter "independent" PASS. The project consists of several phases. The first phase aiming at identifying the various centers across the EU has been finalized. This has resulted in the establishment of a general inventory of 55 centers and a paediatric inventory of 8 centers, following a survey at the level of the Member States. In addition, a third inventory following feedback from the pharmaceutical industry has been established whilst a fourth inventory of pharmacoepidemiological databases and useful patient registries is envisaged. The project is now focusing on the development of a structure and model for the future network (9,10).

Further to a knowledge "gap analysis" performed by the Committee for Human Medicinal Products (CHMP)/Pharmacovigilance Working Party (PhVWP), information had been provided to the European Commission at the beginning of 2007 with respect to the need for additional safety data, the relevant issues to study, the proposed approaches, and study designs as well as the research output. Furthermore, the CHMP/PhVWP has been asked to draw up a list of the top five public health issues in drug safety affecting groups/classes of medicines including off-patent products. This is in addition to possible inclusion of safety topics within the Seventh Framework Program, under the umbrella "postauthorization studies on the safety of medicines" (11).

Further improvement in the quality of the work performed by regulators was another area where HMA progress is required. This concerns both the introduction of a formal peer review system at CHMP level for centrally processed applications and activities to strengthen the methodology for benefit/risk analysis in order to improve the consistency of decision making. Furthermore, the scientific expertise at PhVWP level has been strengthened taking into account the outcome of a gap analysis of the available expertise. The identified missing or insufficient expertise was complemented by introducing new members into the PhVWP in the fields of pharmacoepidemiology, risk management, risk communication, biotechnology/vaccines/emerging therapies, statistics and methodology, paediatrics, and immunology. Discussions are ongoing to further optimize the operation of the PhVWP, primarily in relation to its interaction and output to the CHMP. Furthermore, the interaction between the PhVWP and the Coordination Group for Mutual Recognition and Decentralized Procedures (Human) (CMD(h)) has been strengthened and formalized, building on the work already undertaken through the Best Practice Guide on the cooperation between the PhVWP and the former mutual recognition facilitation group (MRFG)(12). Preparatory steps were also taken for the development of an EU Regulatory System Incident Management Plan for medicines for human use (irrespective of the licensing route) (13).

EU APPROACH TO RISK ASSESSMENT

The existing peer review systems for the scientific work undertaken at CHMP level have been reinforced through a more formal peer review by CHMP members concentrating on the timeframe up to the adoption of the list of questions for centrally processed MA applications. Such peer review includes all aspect of the risk management plans submitted by applicants. The methodology for benefit/risk analysis has been overhauled with the aim of improving the consistency of decision making focused on the development at CHMP level of a report on benefit/risk assessment models and methods. A two-step approach is being applied. The first step will concentrate on integrating the most useful features of the models into CHMP guidelines and assessment report templates. In the second step, further research into the methodology of benefit/risk assessment will be undertaken. This CHMP approach is undergoing external consultation and comments will be assessed. Based on the principle of continuous improvement, a Review and Learning Project has been set up whereby risk management plans for both centrally authorized products and noncentrally authorized products are being reviewed. The first phase of the project has been concluded and has revealed that the EU risk management plans have been of varying quality during the early period of submissions but there has been a successive improvement of the compliance with the available guidance over time. Identified weaknesses have been communicated through discussions between regulators and the pharmaceutical industry.

Work in this field by regulators has primarily focused on streamlining the provision of information to health care professionals by making available guidance on Dear Health care Professional Communications (DHPCs) through Volume 9A of the Rules Governing Medicinal Products (4). In Part IV, Section 2, the principles for the content and format of DHPCs and situations where dissemination of DHPCs should be considered have been described. In addition, work has started to develop an EU Regulatory System Communication Strategy on emerging safety-related issues for medicines for human use. It is expected for such work to have been finalized before the end of 2008 (14).

EU DEFINITION OF POTENTIAL SERIOUS RISK TO PUBLIC HEALTH

However, before contemplating regulatory requirements is there agreement between EU regulators about what constitutes risk? The answer about how risk relates to the use of a medicinal product is defined in Article 1 (28, first indent) of Directive 2001/83/EC (15,16) as "any risk relating to the quality, safety, or efficacy of the medicinal product as regards to patients' health or public health" and in Article 1 (28a) "an evaluation of the positive therapeutic effects of the medicinal product in relation to the risks as defined in point 28, first indent." A risk is defined as the product of the size of a hazard and the probability of the hazard occurring. A "potential serious risk to public health" is where there is a high probability that a serious hazard could result in severe negative effects with regard to the use of the human medicinal product that cannot be prevented, reversed, or avoided. Thus, the regulatory assessment process should identify the potential of a medicinal product to cause harm to public health and that this will actually be expressed in practice. "Serious" in this context is based on ICH criteria in that a hazard could result in death, could be life-threatening, could result in significant disability or incapacity, could be a congenital anomaly/birth defect, or could result in hospitalization or permanent or prolonged signs in exposed humans.

A serious risk to public health in relation to a specific medicinal product can be considered to exist mainly under the following circumstances:

- Efficacy: the data submitted to support therapeutic efficacy in the proposed indication(s), target population(s), and proposed dosing regimen (as defined by proposed labeling), do not provide adequate scientific justification for the claims for efficacy;
- Safety: evaluation of the preclinical toxicity/safety pharmacology and clinical safety data do not provide adequate support for the conclusion that all potential safety issues for the target population have been appropriately and adequately addressed in the proposed labeling or the absolute level of risk from the product is considered unacceptable;
- Quality: the proposed production and quality control methods cannot guarantee that a major deficiency in the quality of the product will not occur, which might have a negative impact on the safety or efficacy of the product;
- Overall risk–benefit: the risk–benefit balance for the product is not considered favorable, taking into account the nature of the identified risk(s) and the potential benefit in the proposed indication(s) and target patient population(s); and
- Product information: information for prescribers and patients could result in serious harm to public health.

Thus, these criteria drive EU regulatory decision making about what is or is not an acceptable balance of benefits and risks.

EUROPEAN COMMISSION STRATEGY TO STRENGTHEN AND RATIONALIZE EU PHARMACOVIGILANCE

In the 2001 review of the EU pharmaceutical legislation, there was no mention of a systematic review of the PV provisions. However, it soon became apparent that regulators were approaching PV quite differently both as to scientific interpretation and provision of resource. The expansion of the EU to 27 Member States has placed additional strain on the EU PV system. In 2005, after a competitive tender, the Commission arranged for the Fraunhofer study of the EU regulatory authorities within the PV system (17). There then followed a period of public consultation and workshops in 2006. In summary, this consultation identified that the EUPV system was complex with duplication of work, lack of clear roles, significant administrative burdens on industry and regulators, inconsistent implementation in Member States which impacts on the internal market, and no fast and coherent EU action in response to drug safety alerts. A Commission strategy was announced in February 2007 with draft proposals for legislative change published in December 2007. The consultation deadline ended on February 1, 2008. Proposals for a new pharmacovigilance Directive amending 2001/83/EC and a Regulation amending Regulation (EC) No 726/2004 were expected late in 2008. Article 101(b) of the proposals provides the legal basis for the Commission to adopt Good

Vigilance Practices (GVP). There are six key proposed changes which will be summarized as follows:

1. Applications for a MA shall contain a summary (not a detailed description) of the PV system which must contain proof of services of a qualified person for pharmacovigilance (QPPV) and contact details for the QPPV. The MA will also contain a statement that the applicant has the means to fulfill its PV obligations with reference to the PV system master file (SMF) (article 8(3). The SMF is currently what is referred to as the Detailed Description of the PV system and should be maintained on site for regulators and inspectors. There will still be a detailed description of the mandatory risk management system which shall be proportionate to the identified and potential risks. This will need to be integrated into the MA. Indeed, the robustness of the risk management system is a key element as Member States assess MA applications (article 21(4)). The MA may be granted subject to conditions included in the risk management system such as postauthorization safety studies, additional adverse drug reaction (ADR) reporting obligations, and conditions or restrictions of use. These conditions should be met within certain deadlines. Products subject to conditions shall be included in a list of intensively monitored products. Further clarification of what constitutes a PASS is also expected.
2. Simplification of adverse reaction reporting is expected with all EU domestic reports going to EudraVigilance only, all serious third country reports going to EudraVigilance and the EMEA will scan literature. Most notable of all, 15 day reporting requirements will apply for all EU source reports. Patients can report suspected ADRs to the MA holder (MAH) for those products on the intensive monitoring list and to relevant national authorities for other drugs.
3. Periodic Safety Update Reports (PSURs) will retain their emphasis on scientific evaluation of benefit–risk with summaries of data and not just routine listings of individual cases. Electronic submission will be used. PSURs may not be required for generic products (article 10), active ingredients in well-established medicinal use for 10 years with an acceptable level of safety (article 10a), homoeopathic medicinal products (articles 13–16), and traditional herbal medicinal products (articles 16a-16i). There will be a new PV committee at the EMEA which will have a coordinating role making public lists of reference dates, managing periodicity, appointing a lead Member State for nationally authorized products, and coordinating the assessment. The committee's conclusions will be made public. In addition, this committee will recommend safety referrals to CHMP with public hearings and have oversight over postauthorization studies.
4. The safety assessment for products authorized nationally will be centralized at EU level (see article 101(k). Mandatory community assessment will be undertaken if certain criteria ("triggers") are met which include a Member State considering suspension or revocation of a MA, marketing or distribution of a product, refusal of renewal of a MA or if the MAH consider withdrawing a product on safety grounds. Additional triggers include a Member State considering necessary a new contraindication or restriction to indications and "serious deficiencies" following an inspection. A detailed procedure will be followed involving a public hearing and assessment by new PV committee. The resulting CHMP opinion will be made public and a binding commission decision will result. The EMEA will coordinate important safety announcements between MS. A safety web portal will make available safety data such as agreed risk management plans, list of all QPPVs, reference dates for PSURs, agreed PASS protocols and the initiation of a Community assessment of safety issue, including data from public hearings.
5. A new presentation of "key safety information" and "how to minimize risks" within the Summary of Product Characteristics is proposed. In the patient leaflet the same data shall be presented in a box surrounded by a black border.
6. Enforcement will be stepped up so that Member States shall ensure that laws, procedures, and resources are in place to allow enforcement of measures included in risk management plans. Inspection reports shall be sent to the EMEA.

Although we cannot say how many of these proposals will be adopted, certain key messages have emerged. There is an increased emphasis on better definition of the impact of risks of a medicine relative to its benefits with rationalization of risk management planning. However, optimal use of resources, improving and simplifying clarity of legal provisions has attained equal importance. Certainly decision making must improve and so the role and remit of the new PV committee at the EMEA will need greater definition. Roles and responsibilities within the system such as the QPPV will be further clarified. To improve trust transparency must be strengthened with public release of more safety assessments.

THE EU RISK MANAGEMENT PLAN: NEEDED FOR MARKETING AUTHORIZATION APPLICATIONS OR NEW SAFETY ISSUES WITH EXISTING PRODUCTS

It was realized in the late 1990s by the CHMP and the EMEA that more active approaches were required for risk identification, evaluation, minimization, and communication (6). The first details about the EU RMP were first published in Guideline on Risk Management Systems for Medicinal Products for Human Use (18) and more recently within Volume 9A of the *Rules Governing Medicinal Products* (3). The EMEA RMP template was published in October 2006 (19). The description of a risk management system should be submitted in the form of an EU Risk Management Plan (EURMP). The RMP guidelines reiterate that the aim of a risk management system is to ensure that the benefits of a particular medicine (or a series of medicines) exceed the risks by the greatest achievable margin for the individual patient and for the target population as a whole. This can be done either by increasing the benefits or by reducing the risks but by its definition risk management focuses on the risk reduction approach. Whenever possible increases in benefits should also be considered and the characteristics of patients most likely to benefit from treatment better defined. Thus, this system consists of a set of PV activities and interventions designed to identify, characterize, prevent, or minimize risks relating to medicinal products including the assessment of the effectiveness of those interventions. Interestingly, this last point relates to what is commonly called "outcomes research."

By means of the development plan which should be linked to the RMP, the MA applicant should have organized development of the product in such a way that they can demonstrate a clear understanding of the population of users that will be using the product consistent with the terms of the MA. This is not just a question of defining the types of patients (that have and have not been included) but also how the product will be used, what predictable and emerging issues will be expected, and clearly describe how assessment of on-going risk–benefit will occur in the target population based on various sources of evidence. Throughout, the argument should be how to assure more effective protection of public health through proper use of the medicine. Such plans may be required pre- or postauthorization depending on when new risks have been identified. The basic components of an EU Risk Management Plans can be broadly summarized as

- Part I
  Safety Specification identifying important risks, potential risks, and missing information
  PV Plan incorporating many ICH E2E concepts (20)
- Part II
  Evaluation of need for risk minimization activities
  Risk minimization plan

The template is more or less in line with ICH E2E apart from certain EU requirements within the Safety Specification covering the potential for overdose, transmission of infectious agents, misuse for illegal purposes, and off-label use at all ages. These different requirements can be captured in global RMPs (20). As described in Section 3.8.1 of Volume 9A of the *Rules Governing Medicinal Products* (3), and so expected within the EU RMP, the MAH is encouraged to consider routinely the likelihood of medication errors with reference to presentation (size, shape, and coloring of formulation and packaging); instructions for use; labeling; whether the product has life-threatening potential if given by wrong route; whether there is a need to visually or physically differentiate strengths of same product; use in those who are visually impaired and the consequences of accidental ingestion; or unintentional use by children. If suspected, ADRs occur as a result of medication errors these should be discussed in updates to the RMP and ways of limiting errors proposed. Cases not associated with ADRs and near misses should only be reported in accordance with national requirements. However, such cumulative information should be discussed in the PSUR. Throughout this process to create the RMP, a recurrent question is the level of detail that is required. We believe the QPPV has a critical role in determining the quality and adequacy of PSUR and RMP content and their involvement in RMP creation is critical to record for inspectors.

CHALLENGES IN WRITING A RISK MANAGEMENT PLAN: WHEN TO START

As described in the Council for International Organizations of Medical Sciences Working Group VI report (CIOMS VI) (21), the expectation is that the RMP starts early on in development. The focus is how to extrapolate from clinical development to normal use once the product has been marketed. Regulators want to understand clearly the epidemiology of the identified target population and epidemiology of disease behind the indication. This epidemiology section should be created early in drug development. From an early stage, such as when a drug is moved into Phase IIb (and certainly by Phase III of drug development), risk assessment should be planned systematically taking into account known and theoretical risks. The safety database should be sized appropriately to detect serious adverse events and coding of adverse events should be accurate and planned ahead. There needs to be discussion of theoretical, potential, and known risks as part of clinical trial design. With creation of a RMP Risk Tracking Document, this can act as an internal document as part of the Compound Strategy Plan. It can incorporate time frames and milestones into the submission timelines. Many MA applicants would want to discuss Phase III results with relevant authorities. There should be a commitment to dialogue with the US Food and Drug Administration (FDA) and other major agencies with frequent reevaluation of safety data. When it comes to benefit–risk assessment then absolute rather than relative risks are always preferred where possible.

RMP ISSUES FOR SAFETY SPECIFICATION

Regulators will expect that the extent of clinical safety experience will be described precisely. Not only should the MA applicant define the number of people that have been exposed but this number should also be expressed in terms of how long they have been exposed. In addition, the characteristics and demographics of patients should be described and discussion should focus on the relevance of clinical trial population to the target population once the product is launched. This means understanding in depth the epidemiology of the proposed indication and the potential for misuse, abuse, and off-label use. It may well be relevant to obtain and extrapolate from information from similar products or disease management programs. Regulators have reported that too many applicants tend to focus more on what is known rather than what is not known. The data are not presented adequately and all too often long and complicated tables are submitted ("data dump") without adequate thought given how to present the data to justify key risk management conclusions. There is too little use of survival analysis techniques such as Kaplan-Meyer plots. Difficulties in interpreting and presenting data seem to impact identification and agreement on what are important risks and what remain potential risks. There may already be health services risk management programs in place for certain products or in certain disease areas which could be referenced. An example includes the risk management programme for methotrexate developed and implemented by the United Kingdom National Patient Safety Agency (22).

KEY ASSESSMENT ISSUES FOR THE PV PLAN

A key question for the MA applicant to address is that having identified and explained the risks earlier on in a plan are they confident that a regulatory assessor can identify them readily in the Safety Specification together with suggestions for how to obtain information on missing populations? Are the study protocols provided and what proportion of these studies will be provided for assessment and in what detail? Before signing off the plan regulators will want milestones set for such a plan. They do not like excessive emphasis on "routine PV" because this term is misleading and means various things to all concerned. They will want key risks to "read across" throughout the PV plan. Further points of contention for the regulator include leaving insufficient time to develop study protocols and reach internal agreement within the applicant's own organization about what could be defined milestones. In essence, the regulatory assessor wants a clear picture about how the applicant intends to advance safety knowledge about the product according to mutually agreed timescales.

PROBLEMS AND PITFALLS OF RISK MINIMIZATION

Applicants need to understand that if no extra measures are required, they need to justify their position adequately. Definition of what are "Potential Risks" is open to varying interpretation so how decisions are reached needs careful documentation. Even potential risks may require precautionary measures and so the distinction from an actual risk is not clear cut. Education and additional forms of communication may be proposed by the applicant as additional risk minimization activities. However, regulators need to be reassured that the applicant understands the difference between "normal promotional activities" and "risk communication activities" and that both sets of activities are separated adequately. Thus, educational materials which might normally be produced are not necessarily risk minimization tools and so should not be included in the RMP routinely. A MA applicant cannot assume that just because risk minimization actions based on, for instance, the U.S. package insert have been agreed with the FDA that this will be sufficient without further justification for EU regulators. According to feedback from the MHRA, applicants all too often do not provide plans for monitoring success and outcomes. Regulators realize that some measures cannot be uniform across all Member States and want to know what the applicant intends to do about this. Hence, the results of feasibility studies are crucial to demonstrate how risk minimization activities can be flexibly implemented across all Member States to meet local requirements based on a combination of cultural, social, and scientific interpretation of risks as well as logistical differences in how health services are delivered. Nevertheless, even when a RMP has been agreed at central level, there may still be national "RMPs." If a product has been authorized centrally or if the execution of an EU-wide RMP might be disrupted (especially if it has been approved by the EU PhVWP), then the applicant, working through their QPPV, has a good chance of resisting regulatory requests for changes to the RMP. Of course operationally, as we have stated, there may well be differences as implementation will be tailored to local medical practice although the core "plan" objectives are the same.

A popular conclusion arising from the RMP is that the MAH will perform solely "routine PV activities," which is assumed to mean those activities referred to in the detailed Description of Pharmacovigilance Systems (DPS), which is a document which also has to accompany all MA applications. The DPS is too detailed to be replicated within the EU RMP although important features might be selected and discussed in further detail (such as selection of adverse events and reaction of special interests and any additional techniques for signal detection). "Routine pharmacovigilance" is an ambiguous term that may mean different things to different EU regulators. The authors suspect that when MA holders say this what they mean is that they will "just collect ADRs and report as necessary and produce PSURs." In the EU "routine pharmacovigilance" means compliance with Volume 9A of the *Rules Governing Medicinal Products*, which is about achieving an acceptable balance of benefit–risk for a product. Thus managing risk is intrinsic to "routine pharmacovigilance." Another conclusion is that "routine" risk minimization is to be applied which in the author's experience normally means the MAH hopes risk can be minimized (or more appropriately mitigated) solely through product labeling.

INTERNAL OPERATIONAL DIFFICULTIES IN WRITING A RISK MANAGEMENT PLAN

Although RMPs should preferably be concise and short, it is difficult to judge how long a RMP should be and how much detail should be included. Currently, there is no "Benefit" Section in the EU RMP Template and yet this has to be summarized somehow to reach a balanced decision about benefit–risk. This may well be the basis of risk mitigation activities if uncertainties around the indication need to be managed. This is particularly relevant when discussing epidemiology as the evidence can be voluminous. Normally, the applicant can just refer to the epidemiology activities it has been involved in rather than attempt to describe the entire knowledge base. When defining the processes for producing a RMP, how do they relate to other safety documentation produced during development such as Annual Safety Reports? The Safety Specification needs to be aligned with the Integrated Safety Summary included in the MA application. Indeed, the RMP may well have to cross-refer to other material in the dossier so that it is important that it is not written in isolation from the entire process of dossier assembly. A global approach to assembling risk documentation has led to the idea of a "Core RMP" based on the European template but with added features of risk MAPs and REMS. This core RMP can then be adapted as required to meet local needs. Principles for risk minimization activities should be the same globally with the realization that tactical details may differ nationally.

RMPs are still comparatively new so it is uncertain how updates to an agreed plan are supposed to work, how version control is maintained, or how an applicant should propose a new RMP actively if the evidence suggests the current RMP is insufficient. In addition, regulators are not the only stakeholders involved in managing product risk and yet it seems that the plan often arises only from a company-regulator dialogue. How to optimize communication to the target population is still a subject of debate both as to who should be informed and whether they should receive the complete plan, summaries, or extracts.

With the increasing requirement for transparency (such as the Freedom of Information Act in the United Kingdom), a MA applicant has to assume that a RMP may be made public. For example, although Novartis referred to their Enhanced PV Plan for Prexige® (23) as confidential this did not stop the MHRA publishing it on their website. In other member states, benefit–risk reports have been obtained under Freedom of Information legislation. Thus, a MA applicant can assume a RMP may well end up in the public domain.

NEXT STEPS ONCE THERE IS REGULATORY AGREEMENT ABOUT A RMP

Once the product is authorized the practical points to consider include

- timing of the EURMP or Risk MAP implementation, its relation to product launch and rollout and who should oversee these activities (inspectors expect a documented process for monitoring postauthorization commitments);
- how can current management tools be adapted to capture risks, trends, and monitor outcomes;
- the beneficial impact of improving data integrity in supporting claims for new products;
- what should be the required dedicated resource, processes, and tools;
- how to access often small target patient populations especially when a disease is rare or the product is authorized only for severely diseased patients; and
- how to create a budget including costs for design and implementation to cover technology, drug supply, manpower, laboratory costs, education material, and monitoring of effectiveness.

It is important to remember that although a RMP might be agreed centrally invariably it will have to be implemented at the local level. This is particularly relevant when it comes to education and risk communication. Regulators want to be reassured that these activities are differentiated from the normal promotional activities. Thus, the team responsible for the RMP will need to think about impact of an agreed risk communication program on sales force training and certification; national and local education programs; global educational materials; Continuing Medical Education; company sponsored physician speakers and/or key opinion leaders; medical information centers; publication strategy and company sponsored web sites. Thus, feasibility studies may well be important for certain risk management activities.

WHAT SAFETY ARRANGEMENTS MUST A MAH OR A CLINICAL TRIAL SPONSOR HAVE IN PLACE IN THE EU?

Under European Community law, the MAH must ensure that it has an appropriate system of PV and risk management in place to assure responsibility and liability for its products on the market. The sponsor of a clinical trial has a similar obligation for investigational pharmaceutical products under review by the Competent Authority (24). Both the MAH and the sponsor must ensure that appropriate action can be taken when necessary. Whatever arrangements are put in place to delegate safety-related activities and share responsibilities, the MAH retains full legal accountability and liability, both civil and criminal, for the appropriate performance of these activities. While responsibility for the activity can be delegated, accountability cannot. Therefore, a MAH must have in place mechanisms to ensure active oversight of all PV activities related to drug products for which the MAH has an active authorization (although the product does not have to have been launched). In addition, any delegation of such activities must be clearly documented and processes established to assure that obligations are being fulfilled.

The following is a summary of the obligations of being a MA holder:

1. A MAH must have permanently and continuously at their disposal a QPPV who will have oversight over maintaining and managing the company's PV system, will be responsible for the management of the central safety database for authorized products, will have oversight of the safety profiles of the products and any emerging safety concerns and authority to provide input into risk management plans and will act as a single point of contact for the regulatory authorities. This can be contracted out or delegated to a marketing partner but the MAH remains fully accountable for performance of their QPPV.
2. The MAH is responsible for ongoing PV evaluation following the grant of a MA so that all information relevant to the benefits and risks of an authorized product are reported fully and promptly to the appropriate regulatory authorities. Therefore, a MAH must have in place a governance system and appropriate resources that ensure compliance with the PV requirements and reporting obligations. In many Member States, pharmaceutical operating companies appoint a local safety officer (LSO) who is responsible for the collection of all product-related safety information received within that Member State. This includes
    (a) adverse events relating to products in development from studies the MAH sponsors, as well as other company activities likely to collect patient-related safety information, such as market research and outcomes studies; and
    (b) spontaneous and solicited adverse reactions from health care professionals, patients, and local literature related to products your company markets.

    The LSO is responsible for transmitting this information to the MA holder's global headquarters and for completing any necessary follow-up or case validation. All expeditable case reports must be reported to health authorities no later than 15 calendar days after receipt of information and in some cases, for life-threatening cases in clinical trials, within seven days. Thus, good local processes for timely review and follow-up are critical. Therefore, the MAH must have in place mechanisms to ensure that the LSO, and any other organization involved in PV compliance, as well as the QPPV, are fulfilling their obligations
3. The MAH must have in place a governance system and appropriate resources to ensure that any entity conducting a clinical trial on their product in a Member State where there is an active MA, either directly or through a third party, have processes in place to ensure adequate communication of all necessary PV information in a timely manner (24). This includes processes to ensure timely notification to regulatory authorities, ethics committees, and investigators, as appropriate, of safety information generated in these clinical studies.
4. The MAH must have in place processes to ensure that product labels for products the MAH markets and investigational brochures for clinical studies for which the company is the sponsor, contain current, accurate, and relevant safety information.
5. The MAH must document, review, and archive all agreements with other pharmaceutical entities or third parties involving the delegation of PV activities, and must put procedures in place to ensure that all obligations are fulfilled.

VOLUME 9A: DETAILED INSTRUCTIONS FOR POSTAUTHORIZATION PHARMACOVIGILANCE

Following an open consultation from Spring 2006 onward, Volume 9A of the *Rules Governing Medicinal Products* was published on January 25, 2007. It took effect immediately although a revised version was issued in March 2007 after minor amendments were made to withdraw the need to report nonserious ADRs from postauthorization studies. The latest version of Volume 9A was published in September 2008 (4). Volume 9A should be read alongside the CIOMS V Working Group report which adds further details, in particular about validation and classification of cases and management of PSURs (25). Volume 9A is presented with an introduction and has four parts:

- Introduction
- Part I—Guidance for Marketing Authorization Holders
- Part II—Guidance for Competent Authorities and the Agency

- Part III—Guidelines for Marketing Authorization Holders, Competent Authorities, and the Agency on Electronic Exchange of PV Information in the EU
- Part IV—Guidelines for Marketing Authorization Holders and Competent Authorities on PV Communication

It applies to all EU and EEA countries (including Iceland, Liechtenstein, and Norway) as well as the additional two countries which joined the EU in 2007 (Bulgaria and Romania). The key new features in Volume 9A are that the MAH must ensure that there is an appropriate risk management system in place and that, when submitting a MAA, a detailed description of the PV system and proof that the services of a QPPV must be in place (this latter document is similar in content to Summary of PV System required for a MHRA inspection) (1). Significant updates to the PV system should be made as type II variations. Because this is a large document, only significant changes will be reviewed.

PART I, SECTION 1.2.1—ROLE & RESPONSIBILITIES OF THE QPPV

The QPPV has a pivotal role in ensuring patient compliance and patient safety. If the QPPV is not medically qualified they should report to (or have access to) a medically qualified person. The QPPVs name should be provided to the competent authorities of the Member States (and the EMEA for Centralized products). In some Member States national law may require a local nominated individual for safety (who should not be called a local qualified person to avoid confusion). The QPPV has two types of responsibilities: direct and indirect. Direct responsibilities include those that cannot be delegated such as establishing, maintaining, and managing the MA holder's PV system; having an overview of safety profiles and emerging safety concerns for all products for which the MAH holds MAs and acting as a single contact point for Competent Authorities at all times. The QPPV has responsibility for certain activities that can be delegated but for which they must maintain "oversight responsibility." These include ensuring processes are in place for collection of adverse reactions at one point in the EU; preparation of individual case reports, PSURs, and reports on postauthorization safety studies; continuous overall postauthorization benefit–risk evaluation; responding fully and promptly to requests from regulatory authorities providing competent authorities with any other relevant information about risk–benefit. The QPPV also has oversight responsibility for the functioning of the PV system in all relevant aspects including quality control and assurance procedures, SOPs, database operations, contractual arrangements, compliance data, audit reports, and PV training of personnel.

PART I, SECTION 1.2.2—ROLE & RESPONSIBILITIES OF THE MAH TOWARD THE QPPV

This section better defines the broad responsibilities of the MAH especially when it comes to supporting the QPPV. The MAH must have an appropriate PV system "in order to assume responsibility and liability for products" and ensure that appropriate action may be taken when necessary and ensure that all information relevant to the risk-benefit balance is reported fully and promptly. As regards the QPPV, the MA holder's role and responsibilities have been clarified as follows as they must provide the QPPV with "adequate" support; "appropriate resources"; "appropriate processes"; "appropriate communication mechanisms"; access to "all sources of relevant information" relating to the QPPVs tasks; full documentation covering all procedures and activities of the QPPV and mechanisms to allow the QPPV to receive or seek all relevant information. Pharmaceutical companies with more than one PV system can have more than one QPPV. However each company (that is MA applicant, MAH or group of MA holders using the same PV system) should appoint one QPPV responsible for the overall PV for all medicinal products. A deputy QPPV is not necessary but adequate back up is essential.

PART I, SECTION 2.3—MONITORING OF COMPLIANCE BY COMPETENT AUTHORITIES

This section gives guidance as to how regulatory authorities should monitor compliance of PV requirements by MA holders. The actions recommended for authorities include

- maintaining a list of all QPPVs;
- checking that PV data are collated and accessible in at least one point in the EU;

- monitor expedited reporting of individual cases (failure to report, timing of submission so that all serious domestic reports must be submitted within 15 days, quality, under reporting, review sample reports);
- monitor PSURs for nonsubmission, incorrect format, omission of information, poor quality reports, cover letter not highlighting differences between Company Core Data Sheet (CCDS) and EU Summary of Product Characteristics (SPC) and previous requests from Competent Authorities not addressed;
- ask questions to MAH (adjusting deadline for request on urgency of issue);
- monitor timely submissions of safety variations applications;
- monitor follow-up measures after authorization (in particular conditional authorizations and authorizations granted under exceptional circumstances); and
- monitor postauthorization safety studies, including interim and final study reports.

PART I, SECTION 4—REQUIREMENTS FOR EXPEDITED REPORTING OF INDIVIDUAL CASE SAFETY REPORTS

In line with ICH E2D (26), the clock for expedited reporting starts ("day 0") as soon as the minimum information has been brought to the attention of any personnel of the MAH or an organization having a contractual arrangement with the MAH, including medical representatives. This same rule applies to worldwide scientific literature. The clock start day for receiving follow-up information is also "day 0." There needs to be "explicit procedures and detailed agreements" between the MAH and the person or organization receiving such cases to ensure that the MAH can comply with his reporting obligations. MAH are now also "encouraged" to report to the EMEA all expected serious adverse reactions occurring outside the EU on an expedited basis (although only serious and unexpected foreign cases remain mandatory). An overview of expedited reporting requirements in Member States is annexed to Volume 9A. The MAH must monitor their own websites for adverse drug reactions (ADRs). The MAH must review ADRs that they happen to identify on other websites to assess whether or not they are expeditable (although there is no expectation that there is a routine search of the internet for these). The MAH should consider utilizing their website for ADR collection such as by adding ADR collection forms and reporting contact details. Part III of Volume 9A is dedicated to electronic reporting which is mandatory for adverse reactions, save in exceptional circumstances. These "exceptional circumstances" are defined as mechanical, program, electronic, or communication failures that prevent electronic reporting.

PART I, SECTION 5—REQUIREMENTS FOR REPORTING IN SPECIAL SITUATIONS

The MAH should report on worldwide experience with the medicinal product in the following "special situations":

- in between application and granting of a MA in that the applicant has responsibility to "immediately" submit information to CA that "could impact" the risk/benefit balance;
- following suspension or withdrawal of the MA for safety or commercial reasons;
- reporting of data on use of medicinal products in children;
- reporting of suspected transmission of infectious agents;
- medication errors that are associated with serious adverse reactions; and
- in the Event of a Public Health Emergency, reporting requirements can be amended.

One significant change was to the timelines for PSURs after initial placing of the product on the EU market, which means granting of a MA. The six-monthly PSUR submissions should be continued until two full years of marketing experience in the EU has been gained, followed by yearly PSURs for the following two years and thereafter PSURs should be submitted at three-yearly intervals. In addition, PSURs should be submitted immediately upon request from a Competent Authority or, for centrally authorized products, from the EMEA. A MAH may make a special request to the Competent Authority for 30 additional calendar days to submit a PSUR. Ideally, this request should be made before the data lock point. The Competent Authority should respond as rapidly as possible. The basis for such a request should be justified and could include

- a large number of case reports for the reporting period but no new significant safety concern, and

- safety concerns raised by Competent Authorities in the previous PSUR or the MAH has identified subsequently for which the MAH is preparing additional or further analysis in the next PSUR.

APPROACH TO RISK MANAGEMENT IN THE UNITED STATES

In response to the Institute of Medicine report "To Err is Human," (27) the FDA convened an expert working group that examined the principles of risk management that have been used in various industries to identify and develop strategies to minimize various risks. In 1999, this FDA Risk Management Task Force recommended that a systems framework should be applied to medical product risk management (28). This systems framework should be based on the principles of risk identification and assessment, risk confrontation, identification of options to manage risk, communication, and finally assessment of results. It would involve not only industry and regulators but also health care professionals and patients with the ultimate goal of maximizing benefit of medications while minimizing risk. Subsequently, the FDA incorporated Risk Management into the Prescription Drug User Fee Act III (PDUFA III) legislation, which became effective in 2002. As part of PDUFA III, funding was provided to the FDA for dedicated risk management resources and it is stipulated that risk management plans may, on a voluntary basis, be a part of New Drug Applications (NDAs)/Biologics Licence Applications (BLAs) submitted to the agency. This led to FDA draft concept papers published in March 2003 followed in May 2004 and March 2005 by the following guidance documents: Premarketing Risk Assessment (29), Development and use of RiskMAP (30), and Good PV Practices and PE Assessments (31).

The FDA defines the process of "Risk Management" as an iterative process for assessing benefit–risk balance, using tools to minimize risk and preserve benefits followed by an evaluation of tools, risks and benefits and then reassessment of benefit–risk balance. Risk Management is thus a cyclical combination of risk assessment and efforts at risk minimization. This should occur both pre- and postmarketing and ideally activities are evidence-based. Risk Minimization Action Plans (RiskMAPs) in the guidance is intended to distinguish risk minimization interventions from overall process of risk management.

The guidance on Development and Implementation of RiskMAPs provides guidance on when and how they are to be used, the importance of defining goals, objectives, tools, and evaluation to minimize known risks of a product while preserving its benefits (31). The factors to take into account about whether a RiskMAP is appropriate include the nature and rate of known risks and benefits for a product, the preventability of risk, the likelihood that measures to mitigate risk will work, and probability of benefit. Thus, there is no ready formula and decisions need to be made on a case-by-case basis. However, unlike in the EU, a RiskMAP is not required routinely for each NDA submission and so should be used judiciously. Thus, the FDAs original intention was that RiskMAPs would be required for a small number of products because of the wish to avoid unduly encumbering drug availability and access to benefits. Should a RiskMAP Program be required, the goal(s) should reflect the ideal outcome of a RiskMAP while being targeted to specific health outcomes related to known safety risks. Of course, goals then have to be translated into "pragmatic, specific, and measurable" objectives. To give an example of how a Goal relates to Objectives, one type of goal would be to eliminate a specific and dangerous drug–drug interaction. The relevant objectives might target physician coprescribing, pharmacist codispensing, and patient information. Different tools may apply for each target.

Risk Minimization Tools refer to processes intended to minimize known safety risks and which have been designed to help achieve one or more objectives that serve the RiskMAP goal(s), which may be considered as intermediate steps. This involves more than routine product labeling *via* the package insert (FDA-approved professional product labeling). However, the FDA maintains that the package insert (PI) will always be the cornerstone of routine risk communication and risk minimization/management efforts. The FDA emphasizes the importance of keeping the PI updated to reflect new benefits or risk concerns and encourages ongoing efforts to make it clear, concise, and focused. The PI cannot be a "RiskMAP" in its own right. Details within the RiskMAP guidance (31) describe how evaluation of effectiveness should proceed to achieve targeted health outcomes, how to evaluate individual performance tools and their acceptability, as well as discussing what compliance with RiskMAP procedures through metrics and milestones might look like. If the outcome measures are difficult to measure directly then the applicant needs to think of suitable surrogate and process measures. There are health care information sources providing evidence based outcome measures in different therapeutics areas (32,33). This should include how to compensate for the invariable limitations of such evaluation methods. There are two basic strategies to mitigate risk which can vary in the scope of their intensity and intervention. Firstly, risk can be reduced by reducing drug exposure either by restricting indication, controlling drug distribution, and optimize dosage regimen. Secondly, the risk of an ADR occurring can be reduced by screening patients at baseline, monitoring precursors of adverse reactions, educating prescribers and patients, asking for informed consent so that a patient should be fully aware how to reduce a risk, introducing independent patient monitoring, and providing hotlines for medical advice.

Risk Minimization Tools Fall in One of Three Categories

1. targeted education and outreach tools which refer to specific, targeted education, and outreach efforts about risks to increase appropriate knowledge and behaviors of key people or groups (e.g., health care practitioners and consumers) that have the capacity to prevent or minimize the product risks of concern;
2. reminder systems supplementing the targeted education and outreach category. Examples include systems that prompt, remind, double-check, or otherwise guide healthcare practitioners and/or patients in prescribing, dispensing, receiving, or using a product in ways that minimize risk; and
3. performance-linked access tools include systems that link product access to laboratory testing. Despite best efforts of all concerned RiskMAPs can introduce significant practical obstacles and challenges (34). How does one judge that education has failed and that it is time to move to controlled distribution? How can you guarantee reliable methods for contraception even in the context of a pregnancy register? What happens to those medicines available from earlier prescriptions? What can be done about drug sharing and drug distribution across borders, which includes parallel distribution and import? Whichever tools are chosen there needs to be evidence-based discussion about their projected likely effectiveness, how they are to be evaluated prior to full implementation (such as in a pilot and/or feasibility studies) so that the relevant FDA reviewer can comprehend the evidence, its acceptance to patients and physicians and so overall feasibility.

There is now a wealth of evidence from other sources about how to assess interventions on health outcomes (32,33). Ultimately, this is what the validity of any RiskMAP measure is dependent on. The applicant needs to think through logistical variability such as delays in RiskMAP implementation and evaluation because of temporal and geographical reasons which may impact the robustness of the RiskMAP system. Milestones for progress reports need to be agreed up front to allow periodic FDA assessment of RiskMAP evaluation. The applicant needs to anticipate any reasons for delayed feedback and what contingencies will be put in place to cope. This often leaves the question of how much effort has to be undertaken to ensure "remaining risk(s)" are adequately mitigated.

Food and Drug Administration Amendments Act: Significant Strengthening of Safety Legislation

One of the more comprehensive changes to the US Food Drug and Cosmetic Act in 40 years was signed into law on September 27, 2007: the Food and Drug Administration Amendments Act (FDAAA) (34). There are significant changes that increase the FDAs authority to require safety changes in labeling require postapproval studies and impose restrictions in distribution and use. Some provisions are effective immediately whereas others have effective dates ranging from 2007 to 2010 depending on the provision. The FDA was obliged to develop a five-year plan by March 31, 2008. As directed by the Act, the FDA has commenced activities designed to modernize the USPV system. The FDA has solicited and implemented proposals for research on the best way to maximize the public health benefit associated with collecting and reporting ADRs throughout product lifecycle. Investment will occur into best epidemiological best practices with guidance development to expand database resources. The FDA strategy for risk management and risk communication is to be developed and more tools validated. User fees may be used to review RMPs. As regards to postmarketing study commitments, the FDA are required to finalize SOPs by end of 2008. More importantly, the FDA can now require a NDA holder or applicant to conduct postapproval studies or clinical trials if required for the following purposes: to assess a known serious risk, to assess signals of serious risk and/or to identify an unexpected serious risk.

The applicant must submit a timetable and report periodically on the study status. Failure to comply with the timetable or other requirement is a violation which is a significant legislative development.

Risk Evaluation and Mitigation Strategies

Whether the FDA or the applicant should think a Risk Evaluation and Mitigation Strategies (REMS) is required is based on the following criteria: the size of patient population, the seriousness of condition being treated, the benefit and duration of treatment, the seriousness of potential AEs, and whether a drug is a New Molecular Entity. This is consistent with the FDAs previous view that a REMS is not required for every drug. Only those products with safety concerns based on the preceding criteria may be required to have REMS as condition for approval for new drugs or a new indication. A REMS may be needed postapproval (prompted by new safety information) at any time if the FDA believes it is required to ensure benefits outweigh risks. The applicant must submit the REMS within 120 days. Timetable assessments follow at 18 months, three and seven years. Other potential contents of the REMS include Medication Guide/Patient Pack Insert (PPI), communication plan with health care providers, and elements to assure safe use (such as has been used for Thalomid® and Accutane®) (34). Until further experience is obtained, it is not clear how the concept of REMS, as presented under FDAAA, is related to the RiskMap and related terminology in the three guidance documents issued in 2005. We assume that this is the same concept but now within a more stringent legal framework. The language on REMS seems to put responsibilities for determining need for risk management more with FDA than the NDA holder or applicant ("the Secretary determines that a REMS is necessary..."). Generic drugs approved under Abbreviated New Drug Applications (ANDAs) require only a "subset" of the REMS requirements, and with regard to restrictions on use and distribution, they must use a "single, shared system" with the innovator company. This raises significant questions around how legal and resource issues will be arranged, especially for the innovator, and what contribution generic companies will make participating in active surveillance programs that might be required under a REMS. In particular, we assume that these generic provisions apply to biosimilar products. As a new tool, the FDA is now permitted to consider requiring the use of a unique symbol in labeling and Direct to Consumer advertisements for newly approved drugs and indications. This seems to be analogous to the inverted black triangle system in the United Kingdom.

Procedure for REMS

There is a complex procedure with specified timelines for REMS, dispute resolution, and public disclosures with referrals possible to advisory committees and workshops. The FDA will have to establish a process for dispute resolution (much of which will be public). A REMS may be applied retrospectively to drugs that already have had an enhanced safety focus unless "deemed" by FDA to currently have a REMS because currently subject to "elements to assure safe use" (which is undefined). This had to be completed by March 25, 2008 and the REMS had to be submitted for applicable products by September 25, 2008.

For those companies with generic drugs they will only be subject to medication guide, patient package insert requirements, and safe-use requirements (if any are imposed on the reference drug). However, as in the EU with a RMP, the FDA may still require a generic product to be included in same REMS system as the reference product unless the FDA determines burden outweighs risk or the generic manufacturer has sought a license from the reference product manufacturer for protected elements of the system unsuccessfully. Interestingly, these requirements may not be used to "delay" generics. The Act provides a system for the FDA to take action across a class of drugs by allowing for international coordination of timing for studies and utilizing the Drug Safety Oversight Board. The FDA can now withdraw or suspend approval and is not required to order an assessment of an approved REMS first. If a company fails to comply with a risk evaluation and mitigation strategy, this is legally interpreted as misbranding.

In addition, violation of a requirement relating to postmarketing studies or labeling is also misbranding. Penalties may range up to $250,000 per violation. For continuing violations, FDA can impose $250,000 for the first 30-day period, doubling for every 30-day period thereafter, up to $1,000,000.

In June 2006, the FDA implemented new labeling requirements for the content and format of FDA-approved prescribing information (package inserts). The new design helps manage the risks associated with prescription drug use by providing the prescribers with the most up-to-date information in an easy-to-read format. The format includes a "Highlights" section that summarizes the most critical prescribing information and a table of contents that refers prescribers to detailed information located in the labeling. In addition, the content of the new package insert includes a list of recent changes made to the labeling including new warnings, a toll-free telephone number for adverse event reporting, and a separate section that contains information for counseling patients about a drug's risks.

BRIEF SUMMARY OF REQUIREMENTS FOR PHARMACOVIGILANCE IN JAPAN

Since October 2001 and the start of Early Postmarketing Phase Vigilance (EPPV), a rigorous program of surveillance is performed under Good Postmarketing Study Practice (GPSP) for each new approval in Japan (35,36). EPPV is not a clinical study but refers to an intensive period of vigilance during first six months after marketing. For the first two months, health care professionals are contacted every two weeks (in principle) and then for the following four months this is to be done once a month. It is a system of encouraged (augmented) data collection although, surprisingly, most companies do not classify case reports received under this system as "solicited." At the conclusion of EPPV, a report is submitted on the results of the data accumulated over that period. Under Article 14–4 of the Pharmaceutical Affairs Law (PAL) the regulatory authority requests a reexamination period which can range anywhere from six years up to ten years (for orphan drugs, paediatric products) or four years for a new chemical entity to allow additional indications and posology. The surveillance required and the data submitted for reexamination must have been obtained according to GPSP. This specifies items that are to be complied with strictly in order to achieve appropriate postmarketing surveillance and studies to be conducted by manufacturers/distributors. "Actual Use Studies" are required which are observational studies of safety and efficacy when used under conditions of the approved package insert. Typically, around 3000 subjects are studied but this varies by indication. In addition, there may be Special Use Studies in populations that may have been difficult to enroll during the development program which may be observational trials performed under GPSP with smaller patient numbers (typically 20–100).

Article 14 to 16 of the PAL allows for the efficacy and safety of an already approved drug to be reassessed in the light of the current status of medical and pharmaceutical sciences. This refers to those medicines which were approved when approval standards and medical practice were different and this provision could help eliminate drugs which may not have adequate efficacy or safety by today's standards. Reporting processes that are part of the Japanese postmarketing surveillance system fall into three categories:

1. company reporting of cases of adverse reactions and infectious diseases (case reports, research reports, and reports by regulatory measures in foreign countries). The scope of expedited reports is not the same as in the EU/US. A J-PSUR (ICH-style PSUR with Japanese information) should be submitted with an "unexpected nonserious" line-listing for domestic reports;
2. periodic Infection Reports of biological products (every six months); and
3. safety Information Reporting System concerning drugs and medical devices from Health Professionals.

There has long been a system of predefined PV planning and risk management such as when a company submits plan for actual use studies, special use studies, etc., as part of the basic plan for postmarketing surveillance (*kihonkeikakusho*) Module 1 of the Common Technical Document (CTD). The equivalent of a Risk Minimization Action Plan is called a *Zenrei–Chousa*. This is a program required as a condition of approval where all exposures to a newly approved drug are monitored for a specific amount of time or up to a number of exposures, which is similar to a Registry. It is more likely to be required if the product is an orphan drug, if there is a perceived need for greater safety monitoring because it is a new chemical entity, if the approval is based largely on foreign data, if there is a likelihood of frequent serious adverse reactions, or if there is a high risk of off-label use. Under this program, the institutions, which can use the product are restricted and must contract to participate in the monitoring program. Finally, there is a Japanese system for the "Safety Responsible Person" (*Soseki-Anseki*), which is akin to the EU QPPV. Thus when submitting a dossier in Japan, it is important to consider the existing regulatory background for a rigorous program of surveillance which is performed under GPSP for each new approval Expedited reporting timelines and requirements in Japan are different from those in the EU/US while several aspects of the current PMS system in Japan fulfill the objectives of risk management and risk minimization.

SO WHERE TO NOW FOR RISK MANAGEMENT?

A more innovative approach to managing risk is needed to help reverse the decline in pharmaceutical research and development productivity which is often attributed to safety issues and bottlenecks caused by risk averse behavior. For many products, routine risk minimization measures (e.g., healthcare professional labeling and education) and postauthorization PV are sufficient with the *caveat* that the system has to be appropriate and robust. There are, however, a considerable number of products for which product specific surveillance and risk minimization activities should be considered (30). Decisions are usually based on data from the clinical development program, postmarketing surveillance, and Phase IV studies extrapolated to the product's intended population and use. Differences in regulatory assessments globally are somewhat inevitable not only because risk is perceived differently in different territories but also because of what can be done in practical terms given the huge differences in healthcare systems. In the EU, a RMP is required as part of the MA Application but we still do not know whether the pharmaceutical sector is (more) effective at managing pharmaceutical risk. Best-practices and standards are still ill-defined and are evolving in an uncoordinated fashion with little in the way of formal benchmarking (although the authors are aware of the World Class International Consulting Ltd (WCI) initiative PV Safety Net (36). One thing is certain: the global spectrum of risk is changing and the need to acquire new tools and thinking with which to harness risk as a force for pharmaceutical growth in the future has never been more urgent. The complexity of modern life, the proliferation of new forms of communication, and the deluge of available data makes this more difficult than ever before. We must be more creative in our thinking about the place of pharmaceutical risk in our lives as individuals, as members of pharmaceutical organizations and as a society.

REFERENCES

1. Procedure for coordinating PhV Inspections requested by the EMEA Ref. INS/PhV/1.
2. Good Pharmacovigilance Practice. http://www.mhra.gov.uk/Howweregulate/Medicines/Inspectionandstandards/GoodPharmacovigilancePractice/index.htm. Accessed September 23, 2008.
3. Application of risk management to medical devices; ISO 14971:2007.
4. Eudralex—Volume 9A of The Rules Governing Medicinal Products in the European Union—Guidelines on Pharmacovigilance for Medicinal Products for Human Use. final version September 2008. http://ec.europa.eu/enterprise/pharmaceuticals/eudralex/vol-9/pdf/vol9a_09–2008.pdf.
5. European Risk Management Strategy (ERMS) facilitation Group Doc Ref EMEA/424988/2005 December 16th 2005 http://www.emea.europa.eu/pdfs/human/phv/42498805en.pdf. Accessed September 23, 2008.
6. Waller PC, Evans SJW. A model for the future conduct of pharmacovigilance. Pharmacoepidemiol drug saf 2003; 12:17–29.
7. Engaging citizens in the discussion about Europe. http://ec.europa.eu/commission_barroso/wallstrom/communicating/policy/index_en.htm. Accessed September 23, 2008.
8. Implementation of the Action Plan to further progress the European Risk Management Strategy: Rolling Two Year Work programme (2008–2009) EMEA/280089/2007. http://www.emea.europa.eu/pdfs/human/phv/28008907en.pdf.

9. EMEA meeting with interested parties and research centres on ENCePP: Summary of the Outcome of the Working Groups EMEA/601105/2007 December 17th 2007. http://www.emea.europa.eu/pdfs/human/phv/60110507en.pdf. Accessed September 30, 2008.
10. Executive Summary Report on EMEA meeting with interested parties and research centres on ENCEPP June 28th 2007 Doc Ref EMEA/601107/2007 December 18, 2007. http://www.emea.europa.eu/pdfs/human/phv/60110707en.pdf. Accessed September 30, 2008.
11. The Innovative Medicines Initiative (IMI) Strategic Research Agenda. http://imi.europa.eu/docs/imi-gb-006v2–15022008-research-agenda_en.pdf.
12. Work programme for the European Medicines Agency 2008. http://www.emea.europa.eu/pdfs/general/direct/emeawp/EMEA_Work_Programme_2008_full.pdf. Accessed September 23, 2008.
13. 2008 Public Status Report on the Implementation of the European Risk Management Strategy 2008. http://www.emea.europa.eu/pdfs/human/phv/4355609en.pdf. Accessed May 30, 2009.
14. Section 3: Communication and information. http://www.hma.eu/uploads/media/Section_3_Communication_and_Information_Summary.pdf. Accessed September 23, 2008.
15. Regulation (EC) No 726/2004 of the European Parliament and of the Council of March 31, 2004.
16. Directive 2004/27/EC of the European Parliament and Council of March 31, 2004 amending Directive 2001/83/EC.
17. Assessment of the European Community System of Pharmacovigilance. Final Report–Final version 25 January 2006 European Commission Enterprise and Industry Directorate-General, Unit F2, Pharmaceuticals Reference: Service Contract No: FIF.20040739 Submitted by the Fraunhofer Institute for Systems and Innovation Research. http://www.isi.fraunhofer.de/t/projekte/medpharm-e-rt-eurovigilance.htm. Accessed October 4, 2008.
18. Guideline on Risk Management Systems for Medicinal Products for Human Use (EMEA/CHMP/96268/2005). http://www.emea.europa.eu/pdfs/human/euleg/9626805en.pdf. Accessed October 4, 2008.
19. Annex C. Template for EU Risk Management Plan (EU-RMP) EMEA/192632/2006 September 27th 2006. http://www.emea.europa.eu/pdfs/human/euleg/19263206en.pdf. Accessed October 4, 2008.
20. Hartford CG, Petchel KS, Mickail H, et al. Pharmacovigilance during the pre-approval phases: an evolving pharmaceutical industry model in response to ICHE2E, CIOMS VI, FDA and EMEA/CHMP Risk Management guidelines Drug Saf 2006; 29(8):657–673.
21. Report of CIOMS Working Group VI Management of safety Information from Clinical trials Geneva 2005. http://www.cioms.ch/.
22. http://www.npsa.nhs.uk/nrls/alerts-and-directives/alerts/methotrexate/ Accessed September 23, 2008.
23. Lumiracoxib Post-Launch Enhanced Pharmacovigilance and Risk Management Programme Final draft 05 Dec 2006. www.mhra.gov.uk/home/idcplg?IdcService=GET_FILE&dDocName=con2022708&RevisionSelectionMethod=Latest. Accessed September 23, 2008.
24. Detailed guidance on the collection, verification and presentation of adverse reaction reports arising from clinical trials on medicinal products for human use April 2004 revision 1 ENTR/CT http://ec.europa.eu/enterprise/pharmaceuticals/eudralex/vol-10/21_susar_rev2_2006_04_11.pdf. Accessed October 4, 2008.
25. Current Challenges in Pharmacovigilance–Pragmatic Approaches Report of CIOMS Working Group V, Geneva 2001. http://www.cioms.ch/.
26. ICH E2D. http://www.emea.europa.eu/pdfs/human/ich/394503en.pdf. Accessed September 23, 2008.
27. Institute of Medicine; Committee on Quality of Health Care in America. In: Kohn LT, Corrigan JM, Donaldson MS, eds. To Err is Human: Building a Safer Health System. Institute of Medicine, 2000. http://www.nap.edu/books/0309068371/html/. Accessed September 23, 2008.
28. Managing the Risks from Medical Product Use—Creating a Risk Management Framework Report to the FDA Commissioner from the Task Force on Risk Management, FDA, 1999 http://www.fda.gov/oc/tfrm/executivesummary.html. Accessed September 23, 2008.
29. Premarketing Risk Assessment http://www.fda.gov/downloads/RegulatoryInformation/Guidances/UCM126958.pdf. Accessed May 30, 2009.
30. Guidance for Industry Good Pharmacovigilance Practices and Pharmacoepidemiologic Assessment http://www.fda.gov/downloads/RegulatoryInformation/Guidances/UCM126834.pdf. Accessed May 30, 2009.
31. Guidance for Industry Development and Use of Risk Minimization Action Plans. http://www.fda.gov/downloads/RegulatoryInformation/Guidances/UCM126830.pdf. Accessed May 30, 2009.

32. http://www.qualitymeasures.ahrq.gov/. Accessed September 23, 2008.
33. http://www.patientsafety.gov/. Accessed September 23, 2008.
34. Casteneda CP, Zeldis JB, Freeman J, et al. RevAssist: A comprehensive risk minimisation programme for preventing fetal exposure to lenalidomide. Drug Saf 2008; 31(9):743–752.
35. Kubota K, Koyama H. Pharmacovigilance and risk management in Japan. In: Pharmacovigilance, 2nd ed. (Mann RD, Andrews EB, eds.) John Wiley & Sons, Ltd., 2007, Bognor Regis, U.K.
36. Pharmaceuticals and Medical Devices Agency. http://www.pmda.go.jp/english/service/safety_info.html Accessed September 23, 2008. Pharmaceutical Administration and Regulations in Japan http://www. nihs.go.jp/mhlw/jouhou/yakuji/yakuji-e0703.pdf
37. About PV Net. http://pvnet.wcigroup.com/PVNetHTML/AboutPVNet.html. Accessed September 23, 2008.
38. Public status Report on the Implementation of The European Risk Management strategy (reporting mid 2005–Mid 2007) Doc ref: EMEA/168954/2007. http://www.hma.eu/uploads/media/ERMS_Status_Report_20070726.pdf. Accessed September 23, 2008.
39. The Food and Drug Administration Amendments Act of 2007 (FDAAA). http://www.fda.gov/CDER/regulatory/FDAAA/default.htm. Accessed October 4, 2008.

38 Pregnancy and Children

H. Guenter Hennings
Regulatory Consultant, Bad Kreuznach, Germany

INTRODUCTION

The *thalidomide* tragedy in the late 1950s and the early 1960s demonstrated the extreme vulnerability of the unborn child and triggered the development of legislation, rules, and guidelines in order to protect women and their unborn children against safety hazards of medicinal products. Since those days the protection of subjects in clinical trials and patients is a crucial element of the development of medicinal products. Special care is taken for vulnerable patient groups such as pregnant women and children.

This protection, however, creates a certain ethical dilemma because this limits clinical research in these two special populations and as a consequence leaves them to treatment based rather on empirical experience than on evidence. The same ethics requiring that new drugs demonstrate quality, safety, efficacy, and a positive risk–benefit balance prevents children, pregnant women, and other vulnerable groups from being exposed to investigational medicinal products.

For some time public discussion has addressed the issue. Legislation and guidelines have been established, especially in the area of pediatric medicine, for the development of adequate medicines earlier in the United States but subsequently also in the European Union (EU).

This chapter outlines the regulatory framework for clinical research in pregnancy and in children without being able to address this extremely important subject completely with all its different perspectives, epidemiological contexts, ethical considerations, and philosophical principles.

PREGNANCY

Sponsors, clinical researchers, regulators, and ethics committees regard pregnancy often as a cause for exclusion in clinical trials in order to protect the fetus against the risks of the investigational medicine. The unborn child is however exposed to many risks; the most important ones are summarized in the following text.

Medicines and Other Agents

There are indeed many ways by which medicinal products administered during pregnancy may affect the fetus. They may have direct lethal, toxic, or teratogenic effects on the embryo; the placental function could be changed with effects on the gas and nutrition exchange between the fetus and the mother. A changed myometrial activity could lead to severe uterine hypertonia resulting in fetal anoxic injury. Metabolism of medicinal products may be altered in unpredictable ways in pregnant women, thus affecting efficacy or safety or both.

The classical examples for suspected teratogenic and embryotoxic medicines are antineoplastic agents, isotretinoin, androgenic hormones and synthetic progestins, thyroid drugs, oral hypoglycemics, narcotics, sedatives, benzodiazepines, tetracyclines, long-acting sulfonamides, and anticoagulants (coumarins) but also agents like alcohol, caffeine, and nicotine.

Frequently, patients do not know about medications they are taking for the treatment of mild conditions , i.e., they consider that over-the-counter (OTC) medicines do not establish a risk for the fetus. But aspirin, for instance, has blood-thinning properties. Thus it may cause problems in the fetus and may also complicate the delivery by increased bleeding and lengthened clotting time. Ibuprofen is metabolized in the liver and can cross into the fetal blood supply and liver.

For many OTC medicines, but also for herbal remedies, tea, and dietary supplements (frequently labeled as "natural"), it is unknown if they are safe during pregnancy.

Background Risk

Many women of reproductive age are not aware that there is a background risk in all pregnancies. Adverse pregnancy outcomes can occur even if the mother takes no medicines during her pregnancy. The rates of this background risk (spontaneous abortion, stillbirth, and serious birth defects, i.e., those resulting in physical or mental disability or death) vary with maternal age and medical conditions. The causes of most birth defects are not known with the exception of genetic syndromes and chromosomal abnormalities. Since all pregnancies have a background risk of birth defect, loss, or other adverse outcome, regardless of exposure, the fetal risk attributed to a medicinal product is its potential to increase the risk of developmental abnormalities above this background risk (1).

Inadvertent Exposure

Inadvertent exposure is also frequent, e.g., approximately half of all pregnancies in the United States are unintended. Many women are exposed to potential dangerous agents including medicines before they know they are pregnant, and the period with the greatest risk to the fetus is when major organs are developing. It would thus be important to inform the mother of known or predicted risks to the fetus from inadvertent exposure. But if there are no data to assess the risk from inadvertent exposure, decisions, which often involve deciding whether to terminate pregnancies, become very difficult (2).

Untreated Disease

Diseases such as chronic hypertension, psychiatric illness, diabetes, cancer, infections, and autoimmune disorders establish a serious risk to the fetus. Under-treating or not treating a condition can then become dangerous, or even life-threatening, to the mother and the child. A few examples may illustrate this.

Diabetes

Uncontrolled diabetes increases the risk of miscarriage, stillbirth, and serious birth defects such as heart or neural tube defects. Controlling diabetes and careful monitoring of blood sugar decrease the risks for the fetus. Today women with diabetes are no longer discouraged from having children.

Taking the right treatment decisions is however difficult, especially if they are not only based on risk–benefit but also on cost considerations. According to the approved Summary of Product Characteristics (SPC) in the United Kingdom , oral antidiabetic medicines such as metformin or glibenclamide should be stopped before conceiving and during pregnancy and patients should switch to (costly) insulin. Contrary to the official labeling, the English and Welsh National Institute for Health and Clinical Excellence (NICE) considers that strong evidence for its effectiveness and safety supports the use of both the oral medications in pregnant diabetic patients (3). Such an inconsistency between the officially approved labeling and the treatment recommendation of a government institution could raise concerns and liability issues.

Chronic Hypertension

Different forms of hypertension such as chronic hypertension, preeclampsia–eclampsia, and gestational hypertension cause frequent and serious problems during pregnancy. Uncontrolled high blood pressure increases the risks of placental problems and fetal growth retardation (4).

Asthma

Uncontrolled asthma can decrease oxygen in the fetal blood possibly impairing fetal growth and survival. Women suboptimally treated for asthma have worse pregnancy outcomes than those who control it with medication (5).

Epilepsy

Nonmedicated individuals with epilepsy may suffer from seizures, which might also limit oxygen supply or cause injury in an unsuspected event. For these types of illnesses, medication should be continued under the supervision of a physician (6).

Depressive Disorders

For some women, untreated maternal depression creates more problems than taking the medication. Depression can lead to poor diet and inability to care for oneself. Untreated depression in pregnant women not only harms the woman but also carries risks of prematurity, growth restriction, and postnatal complications for the fetus. The management of women with depression during pregnancy is based on balancing the potential risk of the symptoms against the potential risks of pharmacotherapy (7).

Malaria

Untreated malaria, mainly a disease in developing countries, is responsible for many maternal and infant deaths. Safety and efficacy of antimalarials is based rather on postmarketing surveillance experience than controlled clinical data (8).

Regulatory Aspects

With the exception of few products developed to treat conditions unique to pregnancy, medicinal products are almost never tested in controlled clinical trials in pregnant women prior to their approval. Human data, if available, come regularly from other sources such as pregnancy exposure registries, cohort studies, case–control studies, case series, and case reports. Animal reproductive toxicity data are currently the major source of the assessment of human teratogenic risk, which impedes adequate counseling of patients about the risk of the use of the medicinal product in pregnancy. However, the predictive value of animal studies for humans is often uncertain and animal models can be misleading both in terms of false-positive or false-negative associations in humans (9).

Teratogenic and embryotoxic findings in experimental animals may let women avoid becoming pregnant, to terminate pregnancies, or not to use the needed medicinal product while pregnant, although the animal data may not be relevant for human beings. Absence of such findings in animals, however, does not indicate that the related risk for the patient does not exist.

United States

In the United States, medicinal products are currently classified in five categories of safety standards for use in pregnancy (1). The classification of a product depends on the animal and human data:

In *category A*, data from controlled clinical studies are available that demonstrate the absence of fetal risks. Very few medicinal products are listed as category A products.

Category B lists medicinal products where animal studies do not indicate risk to the fetus and controlled clinical trials are not available. Products for which animal studies show a risk to the fetus but clinical trials do not would also be classified as category B.

Many medicinal products are listed in *category C* indicating that adequate studies, either animal or human, have not been performed or adverse fetal effects have been shown in animals but human data are not available.

If positive evidence of human fetal risk exists but benefits in life-threatening situations or serious diseases outweigh the risks and safer treatment options are not available, the product is classified in *category D*.

Category X lists products with proven fetal risks that outweigh any possible benefit.

These categories have some practical relevance helping risk–benefit decision making when prescribing medicinal products during pregnancy. But they have also several shortcomings and limitations such as an inconsistency within classes of medications, a forced aggregation of diverse information from various sources into a single class, and a poor discrimination between different medications within a category.

The U.S. Food and Drug Administration (FDA) is currently proposing to amend its regulations in order to create a more consistent format for providing information about the effects of a drug on pregnancy and lactation. It is intended to eliminate the current pregnancy categories A, B, C, D, and X, reorganize format and content of the concerned subsections of the labeling, and provide relevant risk–benefit decision making including clinical information (1).

The new format should help health care providers and women to make better use of medicinal products during pregnancy and/or lactation.

European Union

In the EU, guidance on the exposure to medicinal products during pregnancy and their risk assessment on human reproduction and lactation and the need for postauthorization monitoring is available (10,11). As the basis for labeling, an overall assessment of the medicinal product is required regarding its influence on pregnancy, fertility, and the fetus, which comprises all available knowledge such as the pharmacological activity and results from nonclinical studies and, as available, clinical experience and also knowledge about compounds within the same class. Human experience from exposed pregnancies would supersede the animal data. Therapeutic benefits of the medicinal product should be compared with other treatment options and clinical practices.

The European Medicines Agency (EMEA) guidance defines an integrated approach in risk assessment and draws from there recommendations for labeling texts for the health care professional (physician, pharmacist, and midwife). The EMEA does not recommend a classification scheme as currently used in the United States.

Since the focus of the guidance is on clinical risk assessment, it does not mention explicitly prospective clinical studies to establish treatment options in pregnant women. It acknowledges different sources of information for this risk assessment such as case reports, case series, epidemiological studies, which may originate from registries of pregnancy and congenital malformations, teratogenic effects networks, pharmaceutical companies, and also bibliographical data.

Collection of Human Data: An Ongoing Need

Preauthorized development of medicinal products includes regularly a series of nonclinical studies in several animal species in order to assess the developmental and reproductive toxicity of the compound. In clinical studies, pregnancy is normally an exclusion criterion for enrolment of the patient. Pharmaceutical companies also have to consider the liability aspects of the potential risk of their products to the pregnant women and the fetus.

At the time of marketing authorization, the risk assessment is incomplete regarding potential harmful effects of the medicinal product. Women during their childbearing years, however, commonly need medicinal products. Complete avoidance of the use of pharmaceuticals by pregnant women cannot be expected realistically. Untreated disease can also be detrimental to the well being of mother and fetus.

There are concerns that the current limitations to clinical research in pregnancy may lead to treatment of pregnant patients based rather on empirical experience than on evidence.

Certainly the collection of human reproductive and developmental toxicity data need to be intensified after the drug has been approved as an ongoing process but also whenever appropriate during preauthorization development of medicinal products. Such human data may come from clinical trials, pregnancy exposure registries, or large-scale epidemiologic studies. Increased funding for research may also be considered. The current debate in a sense is similar to the discussion that had started some years ago in the pediatric field and has stimulated clinical research in children especially in the United States.

PEDIATRIC CLINICAL STUDIES

For many years pediatricians noted a lack of suitable information for a majority of marketed medicinal products to treat diseases in children. F.H. Dost (1910–1985), a German pediatrician, developed more than 50 years ago important basic principles of pharmacokinetics in order to improve the treatment of his young patients.

Still today the EU estimates that possibly more than 50% of the medicines used for children in Europe have not been tested adequately in this vulnerable population. This leaves health care professionals frequently with the only option to use products "off-label" or unlicensed medicines with unknown risks of inefficacy and/or adverse reactions. Labeling and dosing recommendations also will not be adequate with the consequence of potential medication and dosing errors, which also includes magistral preparations, which are prepared frequently when appropriate pediatric formulations are not available. Very few publications specifically address the issue. A high rate of underreporting of pediatric adverse drug reactions (ADRs) can be assumed (12).

Best Pharmaceuticals for Children in the United States

In the United States, the FDA and Congress developed several initiatives from the early 1990s to stimulate pediatric research for medicines and improve the availability of adequately tested medicines to patients. The amendments to the Federal Food, Drug, and Cosmetic Act (FDCA) comprised incentives to the pharmaceutical industry by providing additional exclusivity rights when therapies being used in pediatrics were investigated and a mandatory regulation (the Pediatric Rule) that requires pediatric studies for those conditions in which significant use or benefit in pediatrics was expected. The FDA has given the authority to require certain research, when necessary, into drugs used in pediatric patients. The Pediatric Exclusivity Provision and the Pediatric Rule were amended in 2003 and in September 2007 taking into account the (positive) experiences with the previous legislation. A fund managed by the National Institutes of Health (NIH) and the FDA was also created for studies on older products devoid of commercial interest (13–17).

As a result of the legislation, the regulatory requirements and the available incentives resulted in a clear increase in pediatric clinical trials with medicines. The FDA noted that the Pediatric Exclusivity Provision which extends patent protection for six months was particularly successful in stimulating pediatric research. From 1998 to 2003 more studies were performed than in the whole of the preceding 30 years.

The information gathered from these studies has added pediatric information to the labeling for a considerable number of medicinal products. Under- and overdosing, ineffectiveness, and safety issues were discovered and addressed in improved labeling. In March 2007, the FDA had requested for 341 studies, sponsors had submitted 150 study reports, and 128 label changes were implemented. As of January 15, 2008, 88 summaries of medical and clinical pharmacology reviews of pediatric studies were provided by the FDA. The current information is available to the public at the pediatric FDA website (18).

A few examples may illustrate the progress:

- A safe and effective dose for over-the-counter use has been established for children aged six months to two years for ibuprofen, which previously carried no dosing information for children younger than two years old.
- For gabapentin, it was demonstrated that higher doses are needed in children younger than five years to control seizures.
- Adequate studies supported the development of new oral midazolam syrup, which replaced the only available intravenous or intramuscular injection. Dosing recommendations for this sedative had also to be lowered in order to prevent respiratory problems in children with congenital heart disease and pulmonary hypertension.

International Regulatory Cooperation

The experience gained in the United States and the growing interest of the EU in this field led to an agreement between the FDA Office of Pediatric Therapeutics, the Commission of the European Communities (CEC), and the EMEA to expand their current cooperative activities also in pediatrics. It is expected that regular exchange of information on scientific and ethical issues and other information on pediatric development programs and also the submission of global pediatric development plans by applicants will be facilitated in Europe and the United States. The EMEA and the FDA discuss regularly product-specific pediatric developments and waivers from the obligation to perform such pediatric development.

The EU Pediatric Legislation

The EU Pediatric Legislation entered into force on January 26, 2007 (19–21). The Regulation changes the regulatory environment for pediatric medicines in Europe considerably and gives the EMEA new roles and responsibilities for pediatric medicines.

The legislation has similar objectives to its United States counterpart. Improving the health of children in Europe should be achieved by facilitating the research and development, the authorization process, the information on the use, and the availability of high quality pediatric medicines. Most of the future studies in children will be conducted with new medicinal products but older medicines are offered specific incentives also.

Children from newborn to 17 years should, however, not be subjected to unnecessary trials and the delay of the authorization of medicines for use in adults should be avoided.

Implementation of the legislation is based on the following key measures.

Pediatric Committee

The establishment of a Pediatric Committee (PDCO) within the EMEA in London is intended to support the objectives of the legislation by providing opinions on the Pediatric Investigation Plans (PIPs) submitted by sponsors. Other activities comprise the review of modifications of PIP, waivers, the verification of compliance, the establishment of the pediatric research network, and the development of an inventory of pediatric therapeutic needs.

New Medicines

Pediatric investigation plans (PIPs), waivers, and modifications

The early submission of a PIP to the PDCO will be required for new medicinal products and authorised products before their patent protection has expired unless waivers or deferrals have been granted. The requirement for a PIP can be waived in cases where diseases do not affect children such as Alzheimer's or Parkinson's disease. If studies are intended to be deferred for safety and ethical reasons the PIP should justify this.

The PIP defines the nonclinical studies and clinical trials in children and the timing of their requirement. The development of adequate formulations to make use of the drug more acceptable in children should also be addressed. The needs of all age groups of children should be covered. The PDCO may agree or refuse the proposed plan.

At the time of marketing authorization application (MAA) of new medicinal products and line-extensions for existing patent-protected medicines, sponsors are required to provide data on the use in children, resulting from an agreed PIP unless a waiver or a deferral of the timing of the requirement has been granted by the agency. This PIP compliance check will be performed by competent authorities during the validation of a MAA. Procedural advice for validation of new MAAs, the compliance check with an agreed PIP, and submission deadlines for compliance checks are published by the EMEA (22,23).

Class waivers

The requirement to submit a PIP can be waived for classes of medicinal products or individual medicinal products if it is considered that they are ineffective or unsafe in children, if they treat conditions that occur only in adults, or if they do not offer significant benefits over existing treatments. The PDCO has published a list of such conditions, which is regularly updated (24).

Incentives

As a reward for compliance with the requirement, a six-month extension to the supplementary protection certificate, i.e., a six-month patent extension on the active moiety will be granted. Orphan medicines receive additional two years market exclusivity added to the existing ten years awarded under the EU orphan regulation.

A Selection of Guidelines

Quality and pediatric formulations

Younger children especially require age-appropriate formulations and presentations which should allow safe and accurate dose administration. The different age groups such as preterm and term newborn infants, toddlers, children, and adolescents require very different dosage forms, strengths or concentrations to allow simple, accurate, and safe dosing. Preterm newborns with extremely low-birth weights present special pharmaceutical challenges relating mainly to the size of the dose.

The extemporaneous preparation for children from formulations prepared for adults in hospitals or in the physician's practice (e.g., by crushing tablets and suspending them in water and similar manipulations) may create serious safety problems for the patient. Excipients that are potentially toxic or unsuitable for children should be avoided.

A reflection paper of the EMEA addresses these issues and should be taken into account (25).

Nonclinical guidelines

Nonclinical studies in juvenile animals are useful in cases where suspected toxicity cannot be adequately, ethically, and/or safely assessed in pediatric clinical trials. During the preauthorization phase of the development of medicinal products, information from such studies can be obtained on the potentially different safety profiles from those seen in adults, especially if the product will be of importance for the treatment of diseases in children. Pharmacodynamic properties and pharmacokinetic profiles target organ or systemic toxicity relevant for developing systems, and effects on growth and development in the intended age-group studies in juvenile animals may in certain cases be warranted.

EU guidance is available when nonclinical testing on human pharmaceuticals for pediatric indications is needed and provides recommendations on the timing and utility of juvenile animal studies in relation to phases of clinical development (26).

Clinical guidelines

The protection of the vulnerable pediatric population creates a similar ethical dilemma to that discussed for pregnant women. Clinical research was and still is limited in children and as a consequence they were and still are left to treatment based on empirical experience rather than on evidence. The situation has been improved without doubt especially in the United States and also in Europe. Criteria have been developed for a pragmatic approach backed by legislation and guidelines.

The Council of European Convention on Human Rights and Biomedicine (1997) (27) already provides a legal frame for studies with minors that calls for the research to have the potential of real and direct individual benefit to the patient undergoing clinical trials. The ICH guideline E 11 on the clinical investigation of medicinal products in children (28) and the EU Directive on Good Clinical Practice (Directive 2001/20/EC) further specify under which circumstances pediatric clinical trials can be performed (29).

Guidelines addressing methodological and clinical efficacy and safety aspects such as pediatric pharmacokinetics (30) and oncology (31), trials with small sample sizes (32), and pharmacovigilance for medicines used by children (33) complete the picture.

A considerable challenge is the availability of medicinal products for term and preterm neonates as the immaturity and maturation of different organ systems make the patient extremely vulnerable. A proposed guideline of the EMEA provides criteria for the development of medicinal products for use in the neonatal population (34).

Concept and discussion papers of the agency on the impact of medicinal products on brain, lung and heart, and liver and renal immaturity should be considered when the investigation of medicinal products in neonates is necessary (35–38).

Transparency

Pediatric legislation requires that information on pediatric clinical trials have to be publicly available from the clinical trials database EudraCT. The database includes all trials performed with children under a PIP including trials performed outside the EU anywhere in the world (20,39).

The Commission has also released for consultation a proposal for revision of the SPC guideline, which mainly aims to reflect new requirements in relation to the Pediatric Regulation (40).

Further information can be found on specific Web sites of the European Commission, Directorate General Enterprise and Industry (41) and the EMEA (42).

Off-Patent Medicines

Incentives

Off-patent products without information and a formulation appropriate for children of all ages are eligible to receive a ten-year data protection (intellectual-property protection) period as a reward for the development and for new studies that form a base for an innovation (20). The development in children should have followed a PDCO-agreed PIP. Such products will be marketed under a specific Pediatric Use Marketing Authorization (PUMA).

EU-wide collection and assessment of pediatric data

EU Member States represented by their Heads of Medicines Agencies (HMA) have agreed to start a collaborative initiative called the EU Work Sharing Procedure for the assessment of pediatric data in the EU. Assessment Reports will be created, discussed, and approved by the National Competent Authorities on the basis of data on the use of medicinal products in children submitted by marketing authorization holders (MAHs). Assessment Reports will be published and the results of the assessment will be incorporated in SPCs and patient information leaflets (PILs) as needed.

The data expected from the MAHs are/will be most of the time not sufficient to approve an indication for use in children but it is expected that this inventory of information will in any case be useful for treatment improvements and it will help to focus on research and development and authorization activities for individual and classes of medicinal products.

A Best Practice Guide on the EU Work Sharing Procedure in the assessment of pediatric data has been published by the HMA (43).

The EU is pursuing also a network of investigators and trial centers for pediatric research, a system of free scientific advice provided by the EMEA for the industry and the provision of EU funding for the development and authorization of off-patent medicines for children. The funding is provided through the EU Framework Programmers (44). The PDCO is involved in establishing and updating a priority list of off-patent products for which studies are required and where funds should be directed (45).

A New Approach for Industry in Europe

The Pediatric Regulation requires pharmaceutical companies to review their research and development strategies and their allocation of resources. An active dialogue between sponsors and regulators is essential for a successful implementation of development plans. The submission of PIPs and the justification of waiver and deferral proposals need to be planned, agreed, and implemented early in the R&D process and thus becomes of crucial strategic importance. Sponsors and applicants are offered scientific advice free of charge by the EMEA on the design and conduct of trials necessary to demonstrate the quality, safety, and efficacy of pediatric medicines.

In the EU, the legislation has been introduced recently but the industry and regulators have to adjust very quickly. Experience from the United States demonstrates that this is possible and that as a result of the legislation pediatric research and development in new medicines for children is stimulated with favorable effects on the safety and efficacy of medicines for children.

REFERENCES

1. U.S. Department of Health and Human Services, Food and Drug Administration. Proposed Rules, Content and Format of Labeling for Human Prescription Drug and Biological Products; Requirements for Pregnancy and Lactation Labeling. Fed Regist 2008; 73(104):30831–30868, Docket No. FDA–2006–N–0515, 21 CFR Part 201.
2. U.S. National Center for Chronic Disease Prevention and Health Promotion, Centers for Disease Control and Prevention. PRAMS and Unintended Pregnancy. http://www.cdc.gov/PRAMS/PDFs/PRAMSUnintendPreg.pdf. Accessed November 2008.
3. National Collaborating Centre for Women's and Children's Health, commissioned by the National Institute for Health and Clinical Excellence. Clinical Guide: Management of Diabetes and Its Complications from Preconception to the Postnatal Period. http://www.nice.org.uk/nicemedia/pdf/DiabetesFullGuideline. Revised July 2008. Accessed December 2008.
4. Gibson P, Carson MP. Hypertension and Pregnancy. http://emedicine.medscape.com/article/261435-overview. Accessed November 2008.

5. National Asthma Education and Prevention Program (2005). Working Group Report on Managing Asthma During Pregnancy: Recommendations for Pharmacologic Treatment. NIH Publication No. 05–5236. http://www.nhlbi.nih.gov/health/prof/lung/asthma/astpreg/astpreg_full.pdf. Updated 2004. Accessed November 2008.
6. Epilepsy Foundation. Pregnancy and Epilepsy. http://www.epilepsyfoundation.org/answerplace/Life/adults/women/Professional/pregnancy.cfm. Accessed December 2008.
7. Ward RK, Zamorski MA. Benefits and risks of psychiatric medications during pregnancy. Am Fam Physician 2002; 66:629–636.
8. White NJ, McGready RM, Nosten FH. New Medicines for Tropical Diseases in Pregnancy: Catch-22. PLoS Medicine 2008; 5(6): e133.doi:10.1371/journal.pmed. http://medicine.plosjournals.org/perlserv/?request=get-document&doi=10.1371%2Fjournal.pmed.0050133&ct=1. Accessed November 2008.
9. Rogers JM, Kavlock RJ. Developmental toxicity. In: Klaassen CD, ed. Toxicology: The Basic Science of Poisons. 5th ed. New York: McGraw-Hill, 1996:301–332.
10. European Medicines Agency (EMEA), Committee for Medicinal Products for Human Use (CHMP). Guideline on Risk Assessment of Medicinal Products on Human Reproduction and Lactation: From Data to Labelling. EMEA/CHMP/203927/2005. July 2008.
11. European Medicines Agency (EMEA), Committee for Medicinal Products for Human Use (CHMP). Guideline on the Exposure to Medicinal Products During Pregnancy: Need for Post-Authorisation Data. EMEA/CHMP/313666/2005. November 2005.
12. European Medicines Agency (EMEA). Evidence of Harm from Off-label or Unlicensed Medicines in Children. EMEA/126327/2004. October 2004. http://www.emea.europa.eu/pdfs/human/paediatrics/12632704en.pdf. Accessed November 2008.
13. U.S. Department of Health and Human Services, Food and Drug Administration. Regulations Requiring Manufacturers to Assess the Safety and Effectiveness of New Drugs and Biological Products in Pediatric Patients; Final Rule. Fed Regist 1998; 63:66631, 21 CFR Parts 201, 312, 314, and 601. http://www.fda.gov/ohrms/dockets/98fr/120298c.pdf. Accessed November 2008.
14. U.S. Department of Health and Human Services, Food and Drug Administration, Center for Drug Evaluation and Research (CDER) and Center for Biologics Evaluation and Research (CBER). Guidance for Industry, Qualifying for Pediatric Exclusivity Under Section 505A of the Federal Food, Drug, and Cosmetic Act. http://www.fda.gov/cder/guidance/2891fnl.htm. Revised September 1999. Accessed December 2008.
15. U.S. Department of Health and Human Services, Food and Drug Administration. Best Pharmaceuticals for Children Act. 2002; 107:1094. http://www.fda.gov/opacom/laws/pharmkids/contents.html. Accessed December 2008.
16. U.S. Department of Health and Human Services, Food and Drug Administration. Pediatric Research Equity Act, December 2003. http://www.fda.gov/cder/pediatric/S-650-PREA.pdf Accessed December 2008.
17. U.S. Department of Health and Human Services, Food and Drug Administration. Food and Drug Administration Amendments Act of 2007. http://www.fda.gov/oc/initiatives/hr3580.pdf. Accessed December 2008.
18. U.S. Food and Drug Administration, FDA Consumer Magazine, January-February 2003. Drug Research and Children. http://www.fda.gov/Cder/pediatric/index.htm. Accessed December 2008.
19. European Medicines Agency (EMEA). The European Paediatric Initiative: History of the Paediatric Regulation. EMEA/17967/04 Rev 1. Accessed July 2007.
20. Regulation (EC) No 1901/2006 of the European Parliament and of the Council of 12 December 2006 on medicinal products for paediatric use and amending Regulation (EEC) No 1768/92, Directive 2001/20/EC, Directive 2001/83/EC and Regulation (EC) No 726/2004. OJEU L378.
21. Regulation (EC) No 1902/2006 of the European Parliament and of the Council of 20 December 2006 amending Regulation 1901/2006 on medicinal products for paediatric use. OJEU L378:20.
22. European Medicines Agency (EMEA). Procedural Advice for: Validation of New Marketing Authorisation Application-Extension/Variation Application and Compliance Check with an Agreed PIP. EMEA/553631/2007. July 2008. http://www.emea.europa.eu/pdfs/human/paediatrics/55363107en.pdf. http://www.emea.europa.eu/htms/human/paediatrics/pips_procedural.htm. Accessed December 2008.
23. European Medicines Agency (EMEA). Submission Deadlines for 2010–2011 for Submitting Request for Compliance Check. EMEA/PDCO/495530/2008. September 2008. http://www.emea.europa.eu/pdfs/human/paediatrics/49553008en.pdf.Accessed December 2008.
24. Commission of the European Communities (CEC). Guideline on the Format and Content of Applications for Agreement or Modification of a Paediatric Investigation Plan and Requests for Waivers or Deferrals and Concerning the Operation of the Compliance Check and on Criteria for Assessing Significant Studies. Commission Communication (2008/C 243/01). September

2008. http://www.emea.europa.eu/htms/human/paediatrics/classwaivers.htm. Accessed December 2008.
25. European Medicines Agency EMEA, Committee for Medicinal Products for Human Use (CHMP). Reflection Paper: Formulations of Choice for the Paediatric Population. EMEA/CHMP/PEG/194810/2005.July 2006.. http://www.emea.europa.eu/pdfs/human/paediatrics/19481005en.pdf. Accessed December 2008.
26. European Medicines Agency EMEA, Committee for Medicinal Products for Human Use (CHMP). Guideline on the Need for Non-Clinical Testing in Juvenile Animals of Pharmaceuticals for Paediatric Indications. EMEA/CHMP/SWP/169215/2005. January 2008.
27. Council of Europe, European Treaty Series – No. 164. Convention for the Protection of Human Rights and Dignity of the Human Being with Regard to the Application of Biology and Medicine: Convention on Human Rights and Biomedicine. Oviedo, 4.IV.1997. http://conventions.coe.int/Treaty/EN/Treaties/Html/164.htm. Accessed December 2008.
28. European Medicines Agency (EMEA). Note for Guidance, ICH Topic E 11, Clinical Investigation of Medicinal Products in the Paediatric Population, Step 5. CPMP/ICH/2711/99. January 2001.
29. Directive 2001/20/EC of the European Parliament and of the Council of 4 April 2001 on the approximation of the laws, regulations and administrative provisions of the Member States relating to the implementation of good clinical practice in the conduct of clinical trials on medicinal products for human use. OJEC L 121:34.
30. European Medicines Agency (EMEA), Committee For Medicinal Products For Human Use (CHMP). Guideline on the Role of Pharmacokinetics in the Development of Medicinal Products in the Paediatric Population. EMEA/CHMP/EWP/147013/2004/Corr. June 2006. http://www.emea.europa.eu/pdfs/human/ewp/14701304en.pdf. Accessed December 2008.
31. European Medicines Agency (EMEA). Addendum on Paediatric Oncology (CPMP/EWP/569/02) to the Guideline on the Evaluation of Anticancer Medicinal Products in Man. CPMP/EWP/205/95 Rev. 3. http://www.emea.europa.eu/pdfs/human/ewp/056902en.pdf. Accessed December 2008.
32. European Medicines Agency (EMEA), Committee For Medicinal Products For Human Use (CHMP). Guideline on Clinical Trials in Small Populations. CHMP/EWP/83561/2005. July 2006. http://www.emea.europa.eu/pdfs/human/ewp/8356105en.pdf. Accessed December 2008.
33. European Medicines Agency (EMEA), Committee For Medicinal Products For Human Use (CHMP). Guideline on Conduct of Pharmacovigilance for Medicines Used by the Paediatric Population. EMEA/CHMP/PhVWP/235910/2005-rev.1. January 2007. http://www.emea.europa.eu/pdfs/human/phvwp/23591005en.pdf. Accessed December 2008.
34. European Medicines Agency (EMEA), Committee for Medicinal Products for Human Use (CHMP). Guideline on the Investigation of Medicinal Products in the Term and Preterm Neonate. EMEA/267484/2007. October 2007. http://www.fimp.org/sperimentazione/allegati/Guideline_on_the_investigation_of_medicinal_product_in_the_term_and_preterm_neonate_rev1.pdf. Accessed December 2008.
35. European Medicines Agency (EMEA), Committee For Medicinal Products For Human Use (CHMP). Concept paper on the Impact of Brain Immaturity When Investigating Medicinal Products Intended for Neonatal Use. EMEA/181377/2006. June 2006. http://www.emea.europa.eu/pdfs/human/paediatrics/18137706en.pdf. Accessed December 2008.
36. European Medicines Agency (EMEA), Committee For Medicinal Products For Human Use (CHMP). Concept paper on the Impact of Lung and Heart Immaturity When Investigating Medicinal Products Intended for Neonatal Use. EMEA/CHMP/114218/2006. November 2005. http://www.emea.europa.eu/pdfs/human/paediatrics/11421806en.pdf. Accessed December 2008.
37. European Medicines Agency (EMEA), Committee For Medicinal Products For Human Use (CHMP). Concept Paper on the Impact of Liver Immaturity When Investigating Medicinal Products Intended for Neonatal Use. EMEA/CHMP/PEG/194605/2005. July2005. http://www.emea.europa.eu/pdfs/human/paediatrics/19460505en.pdf. Accessed December 2008.
38. European Medicines Agency (EMEA), Committee For Medicinal Products For Human Use (CHMP). Discussion Paper on the Impact of Renal Immaturity when Investigating Medicinal Products Intended for Paediatric Use. CPMP/PEG/35132/03. December 2004. http://www.emea.europa.eu/pdfs/human/paediatrics/3513203en.pdf. Accessed December 2008.
39. Commission of the European Communities (CEC). Draft Guidance on the Information Concerning Paediatric Clinical Trials to be entered into the EU Database on Clinical Trials (EudraCT) and on the Information to be made Public by the European Medicines Agency (EMEA), in accordance with Article 41 of Regulation No. (EC). 1901/2006. February 2008. http://ec.europa.eu/enterprise/pharmaceuticals/pharmacos/docs/doc2008/2008_02/guidance-eudra-ct-paediatric-clinical-trials_02-%202008.pdf. Accessed December 2008.
40. European Medicines Agency (EMEA), Committee for Medicinal Products for Human Use (CHMP). Draft proposal for a Revision of the European Commission Guideline on Summary

of Product Characteristics. EMEA/299527/2007. December 2007. http://www.emea.europa.eu/htms/human/genguidance/regaffairdraft.htm. Accessed December 2008.
41. Commission of the European Communities (CEC). Enterprise and Industry, Medicines for Children. http://ec.europa.eu/enterprise/pharmaceuticals/paediatrics/medchild_en.htm. Accessed December 2008.
42. European Medicines Agency (EMEA). Medicines for Children. http://www.emea.europa.eu/htms/human/paediatrics/introduction.htm. Accessed December 2008.
43. Heads of Medicines Agencies (HMA). Best Practice Guide, Article 45 – Paediatric Regulation, EU Work Sharing Procedure. September 2008. http://www.hma.eu/uploads/media/BPG_Art45_2008_09_Rev0_Clean.pdf. Accessed December 2008.
44. Commission of the European Communities (CEC). Seventh Framework Programme (FP7). http://cordis.europa.eu/fp7/home_en.html. Accessed December 2008.
45. European Medicines Agency (EMEA), Committee for Medicinal Products for Human Use (CHMP). Revised priority List for Studies into Off-patent Paediatric Medicinal Products. EMEA/226983/2008. August 2008. http://www.emea.europa.eu/pdfs/human/paediatrics/22698308en.pdf. Accessed December 2008.

39 Ageing Populations and Development of Medicinal Products

Jean-Marc Husson
European Diploma in Pharmaceutical Medicine, Lyon, France

Jean-Marie Vetel*
Former Head of Geriatric Department, Le Mans Hospital, Sarthe, France

INTRODUCTION

The world population is ageing, especially within the industrialized countries and this tendency will continue especially due to better social conditions, control of chronic diseases, and improvement of health care. A great deal of effort will be needed especially on a social level with this population which, due to increasing extremes of age, will be more and more isolated but also due to the development and daily use of medicinal products/drugs.

Today, the marketing authorization of drugs does not include older people within their specific profile in most cases, with the exception of degenerative central nervous system (CNS) diseases such as Parkinson's or Alzheimer's diseases, bone disorders (e.g., osteoporosis) and, in part, oncology. Products are not always adapted to an elderly patient's profile or to their condition. The specificities of these vulnerable populations must be taken into account not only during the development of drug development plans but also during routine clinical care and drug usage. For example, pathophysiological age should be taken into account more than the calendar age, as well as gender, multiple comorbidity, physical, and psychological profile. Specially adapted pharmaceutical formulations with specific dosage and regimens should be available. Consideration should be given to specific pharmacokinetics and metabolism according to age, drug–drug interaction, contraindications, safety profile, and risk management.

Specific clinical trials must be performed in the elderly, as they are the major consumers of common medicinal products such as analgesics, antihypertensive, anti-inflammatory and antidiabetic drugs. Increasing recognition of the continuing effectiveness of multiple disease modifying therapies for common cardiac- and cerebrovascular diseases in this population will also lead an explosion in therapeutic interventions. We must also recognize problems of compliance with drug regimens in these age groups.

The absence or insufficient recruitment of older people in clinical trials must be addressed but the reasons are multifactorial. The inclusion criteria may be not only clinical and scientific but also social and ethical to a large extent. The level of comprehension in patients and the evaluation tools frequently are not validated in the elderly.

The European Commission and certain Member States such as the Netherlands have already addressed the issue of clinical trials in the ageing population. Extrapolation of the results of clinical trials to daily medical practice and the "proper use" of medicinal products in ageing populations is hazardous due to the absence of stratification of ages, the exclusion of many of the common comorbid conditions, and use of multiple therapies.

This last point must be emphasized. The older person consumes daily many medicinal products, particularly the more common drugs, and this trend increases markedly with age. For example, in France, today on average the number is three and a half different drugs per day for patients aged older than 65 years and eight drugs for patients older than 75 years (1).

In the coming years, we should recognize and will need to modify the situation for all stakeholders, authorities, physicians, health care providers, as the population of more and more aged patients increases. This so-called "Silver Tsunami" will have a major impact on societal health care provision.

* Currently President of the Association of French Gerontologists.

DEMOGRAPHY

Europe

European Countries

Data are available in *Eurostat Statistics in Focus*, 72/2008 (2) in a paper by Giannakouris entitled "Ageing characterises the demographic perspectives of the European societies." These population projections for the period 2008 to 2060 come from a report of what–if scenarios which aim to provide information about the likely future size and structure of the population if current trends are confirmed. They should therefore be considered with caution.

This paper states that the population of the 27 European Union countries (EU27) is "projected to increase from 495 million on January 1, 2008 to 521 million in 2035, and thereafter gradually decline to 506 million in 2060. The annual number of births is projected to fall over the period 2008 to 2060, while at the same time the annual number of deaths is projected to continue rising. From 2015 onward, deaths would outnumber births and hence population growth due to natural increase would cease. From this point onward, positive net migration would be the only population growth factor. However, from 2035, this positive net migration would no longer counterbalance the negative natural change and the population is projected to begin to fall."

The EU27 population is also projected to continue to grow older with the share of the population aged 65 years rising from 17.1% in 2008 to 30.0% in 2060 and those aged 80 and above rising from 4.4% to 12.1% over the same period. This information is from Stat/O8/119, August 6, 2008 (3).

The analysis includes projections about considerable differences between individual Member States. Between 2008 and 2060, the population is projected to rise in 13 Member States and fall in 14. The strongest population growth is projected to be found in Cyprus (+66%), Ireland (+53%), Luxembourg (+52%), the United Kingdom (+25%) and Sweden (+18%), and the sharpest declines in Bulgaria (−28%), Latvia (−26%), Lithuania (−24%), Romania (−21%), and Poland (−18%). In 2060, the Member States with the largest populations will be the United Kingdom (77 million), France (72 million), Germany (71 million), Italy (59 million), and Spain (52 million). (2, 3) Thirty percent of the EU27 population will be aged 65 year or more in 2060.

The EU27 population is projected to become older throughout the projection period due in particular to persistently low fertility and an increasing number of survivors to higher ages. This ageing process will occur in all Member States. In 2060, the share of the population aged 65 or more is projected to range from 23.6% in Luxembourg, 24.7% in the United Kingdom and 25.0% in Denmark to 36.2% in Poland, 36.1% in Slovakia, and 35.0% in Romania.

In consequence, the old age dependency ratio in the EU27, that is, the population aged 65 years and older divided by the working age population, is projected to increase from 25% in 2008 to 53% in 2060. In other words, there would be only two persons of working age for every person aged 65 or more in 2060, compared with four persons to one today. The old age dependency ratio is projected to be more than 60% in Bulgaria, the Czech Republic, Latvia, Lithuania, Poland, Romania, Slovenia and Slovakia and less than 45% in Denmark, Ireland, Cyprus, Luxembourg, and the United Kingdom.

These data may be found in publications from the Statistical Office of the European Commission (Eurostat) and within Eurostat news releases on the Internet (4) and is summarized in Annexes 1 and 2 of this chapter.

France

Whatever the country in the world, life expectancy is different for women and men. In France, according to Institut National de la Statistique et des Etudes Economiques (Insee), the expectancy for women is at birth 83.2 years, at 65 years is 25.6 years, and at 86 years is 6 years. For men at the same ages, the figures are 76.1 years, 17.3 years, and 5.6 years respectively.

The age stratification for France (5), given in Table 1, is impressive. About one-quarter of the French population of 65 million people is above 65 years. The pyramid of ages and their modification according to time (Fig. 1) and people older than 100 years (Table 2) (6) derive from the French Insee databases.

Table 1 Stratification of French population (Total 65 Million People) Over 65 years is Very Close to That for Most Industrialized Countries

Age (years)	French total population (%)	Number
65–74	15.9	9.5 millions
75–79	7	4.2 millions
80–84	4	2.4 millions
85–100	1.5	900,000
over 100		20000 (in 2008)

Japan and the United States

In Japan, life expectancy at birth in 2008 is 85 years for women and 78 for men, the highest in the world. (7)

In the United States, life expectancy at birth for women (80.4 years) and men (75.2 years) lags behind 41 other nations with marked ethnic disparities between black Americans (73.3 years) and white Americans (77.9 years) (8).

DEFINITION OF "GERIATRIC POPULATIONS"

Introduction

The first question to be asked is how should we define an ageing population?

Ageing has health, social, and economic components leading to marked heterogeneity depending upon the genetic make-up, history and health, and lifestyle of each individual.

Do we have specific regulatory texts in France, the European Union, or at international level (Japan, United States, or others) to guide us? The only international accepted text is the International Conference on Harmonization (ICH) E7 guideline published in 1993 (9), which gives the arbitrary age of 65 years as when an individual is defined as old without any more specific stratification according to the ageing process and biomedical function. In some countries, such as France, becoming a senior citizen starts at 60 years of age.

Fragile and Frail Persons

Age is only one component of the profile of older patients. The European Medicines Agency (EMEA) document (10) specifically refers to frail persons. Many factors must be taken into this definition such as chronic disease, comorbidities with or without a physical dependency, degree of activity, intellectual status, social conditions, and family life environment. A precise

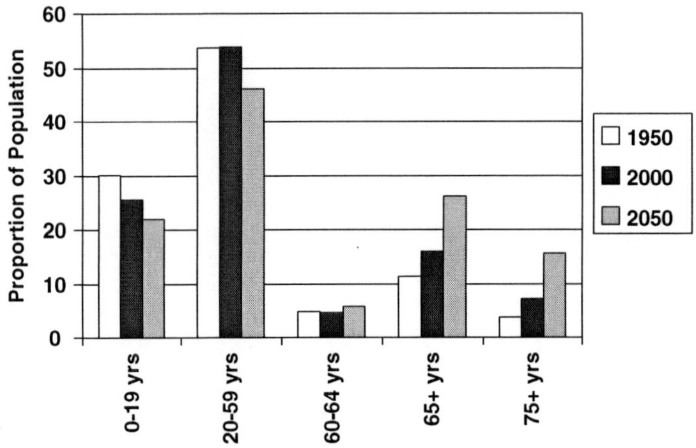

Source: Insee, Demographic projection of population 2005-2050, central scenario

Figure 1 Evolution of population in France which is aging is certainly very close to the population of most industrialized countries (4).

Table 2 The Number of French People More Than 100 Years has Risen (5) from 100 in 1900 to 200 in 1950. In 2008, Their Number is Around 20,000 and will be 46,340 in 2025 and 165,216 in 2050

Years	Number
1900	100
1950	200
1980	1545
2008	20,000
2050	165,216

definition of the frail, weak, and fragile person is to be introduced in a new revision of the ICH E7 guideline, which will be issued at the end of 2009. A person could become frail with ageing even if she/he has a good physiological status. The most vulnerable person is the "very old frail" patient.

Population Profile

Factors which should be considered when faced with an older patient, which may have a very important impact, include items such as gender, number of diseases, matrimonial status, living place, income, education level, and social environment.

AGEING AND DISEASES

Diseases

Ageing is a natural process whatever the species living in our world but a certain number of social, economic, and health factors contribute specifically to its impact in mankind as outlined above.

Descriptive studies have examined frequency and distribution of health problems/ diseases in the population (11–13).

A clinical study on a medicinal product could be global for one indication whatever the type or carried out on a subgroup (cross-sectional) but it would need to include a sufficient number of persons to be representative of the requested population.

The causes of death in patients older than 75 years vary markedly with the gender but often unrecognized morbidity also varies with age and gender. Available data from the French Institut National de la Santé et de la Recherche Médicale (Inserm) (Figs. 2 and 3) show that in France, diseases linked to ageing (13) are very frequent that is, accidents, falls with femoral neck fracture, dementia, depression, suicides, ocular or hearing deficiencies, oral and dental diseases, incontinence, and haemorrhage. These lead to the use of medicinal products including psychotropic agents which could induce iatrogenic diseases or serious adverse events. For example, the use of psychotropic agents is increasingly frequent in both men and women with age.

Health and Social Consequences

The limitation of activity varies greatly in the journey from birth to death and is different for men and women. After 65 years of age, this limitation becomes increasingly important being 27% for men and 30.4% for women overall and with dependency levels of 43.5% for men and 57.4% for women over 85 (Insee)—see Figure 4 (13).

Impact of Comorbid Diseases

The number of active diseases present in older people increases markedly with age.

In France, an investigation in 2003 (13) showed that in retirement homes an average number of seven diseases are found which vary with dependency and age and have an effect on level of individual autonomy. The French Iso Resources Group (IRG) rating scale (14) for the loss of autonomy is distributed over a six point scale ranging from 1/1 with maximum loss to 6/6 with no loss.

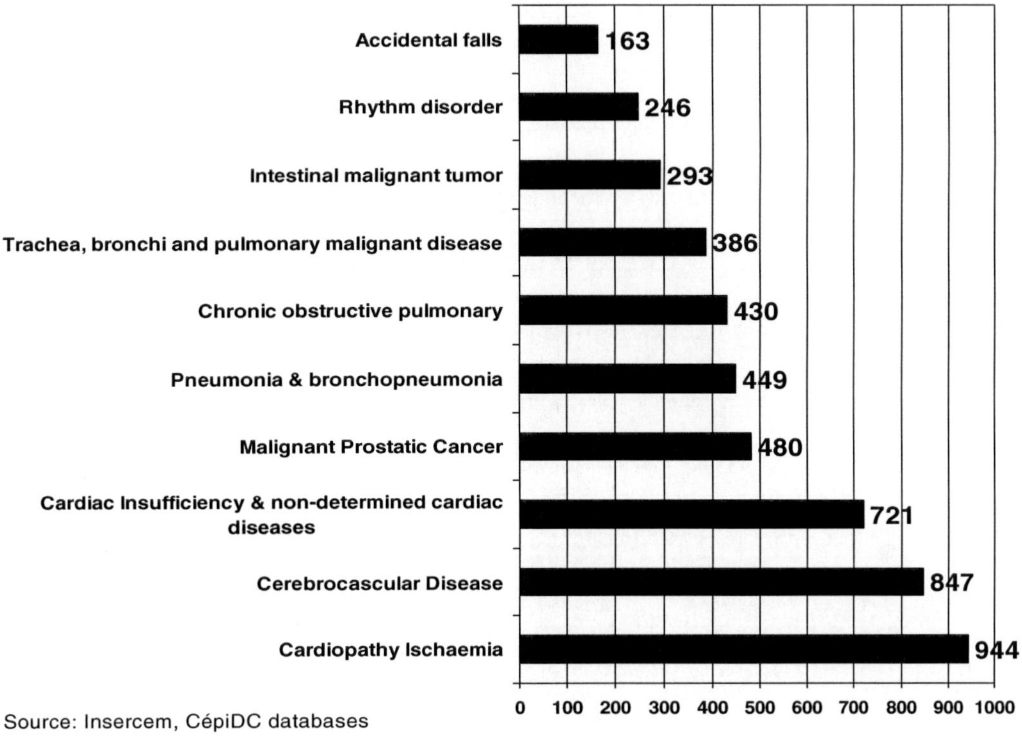

Source: Insercem, CépiDC databases

Figure 2 Main causes of death in France for men aged 75 years or more in 1997.

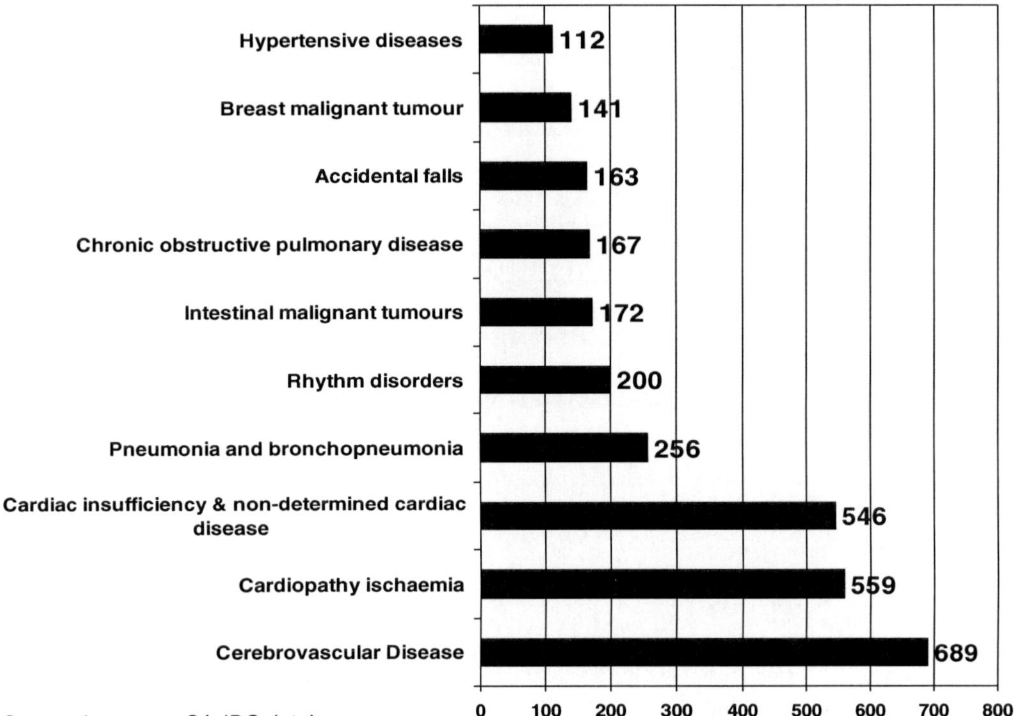

Source: Insercem, CépiDC databases

Figure 3 Main causes of death in France for women aged 75 years or more in 1997.

	Men	Women	Both
Restriction of activity for at least 6 months	10.8%	13.0%	12.0%

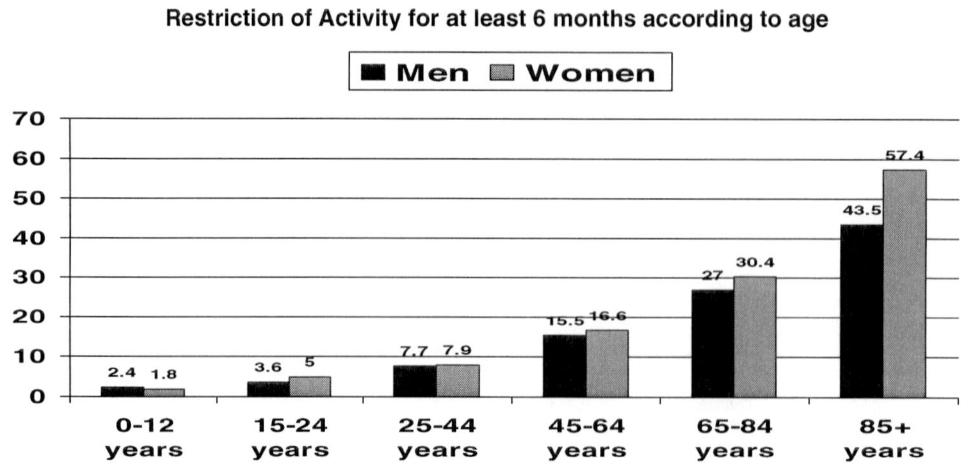

Source: National survey of health, 2002-3, INSEE in Metropolitan France, population aged 18 years or more

Figure 4 Restriction of activity whatever the age in France depending upon gender and ages.

The presence and frequency of comorbid diseases had a marked effect on the degree of individual dependency

- eight diseases are noted for very dependent people (IRG 1 and 2)
- seven diseases are noted for less dependent (IRG 3 and 4)
- six diseases are noted for groups IRG 5 and 6

IRG 1 and 2 have five main pathologies according to age: dementia for 56%, incontinence of urine for 55%, hypertension for 39%, change in behavior in 36%, and cardiac insufficiency in 31%. For IRG 3 and 4, two pathologies are present: hypertension in 50% of people and cardiac insufficiency in 31%, especially above 85 years.

A study published by Dorenlot *et al* in 2005 (15) in the International Journal of Geriatric Psychiatry showed predictive factors of institutionalization after discovery of dementia.

The number of dependent persons will increase dramatically from the years 2000 to 2040 as shown by the Insee data (13).

ICH AND ELDERLY PEOPLE

General Issues

The ICH for the European Union, Japan and the United States covers the different aspects of the development of new medicinal products (chemical or biological products) and particularly involves guidance on issues of quality, safety, and efficacy. It also includes other aspects such as the "pharmacovigilance" issues and the format and content of a dossier—the Common Technical Document (CTD) and the electronic CTD (eCTD). ICH has a fundamental role concerning regulatory issues not only among ICH partners (Europe, Japan, and USA) but increasingly in more and more countries in the world. The guidelines provide a consensus between the three regions: some texts are mandatory while others are merely advisory.

ICH E7 Guideline on the Elderly

The ICH E7 text was published in 1993 (9) and it defined an arbitrary age of 65 years as when an individual becomes old without defining more specific stratification according to extremes of age other than a brief mention of 75 years. Today, it is recognized that age and ageing is more related to pathophysiological status than simply the calendar age. Thus, the EMEA has decided to modify this definition. In December 2006, in a text examining the content of the previous 20 "centralized procedure" dossiers, it had been found that most marketing authorizations (MAs) were poor in dealing with older people. The EMEA text entitled "adequacy of guidance on the elderly" (10) does not give any detailed advice on how to adapt the doses of most drugs with ageing. With the elderly, this is not enough—we need clear data to adapt the doses with increasing age and modifications of metabolic pathways.

E6 Guideline on Good Clinical Practice

The ICH Good Clinical Practice (GCP) guideline E6 (16) is enshrined today in the EU Directive 2005 /28/EC of April 8, 2005. Consequently, it must be taken into consideration for any clinical study in ageing populations as for other type of patients, not only by the pharmaceutical industry but also by all academic and other clinical investigators and their teams, following implementation of a "site management system."

Other ICH guidelines are not mandatory but since they have been adopted by the ICH steering committee and endorsed by the US Food and Drug Administration (FDA), the EMEA, and the Japanese Ministry of Health and Welfare (MHW) at Step 5 they should be taken into consideration. These guidelines are not well accepted by many academic clinical investigators who frequently criticize this approach as overly administrative and too bureaucratic.

Figure 5 shows the ICH Efficacy and Multidisciplinary guidelines and their interrelationship.

According to the GCP texts, clinical investigators have a triple role:

- Protect subjects enrolled in a trial with the support of informed consent
- Insure the quality of data
- Avoid and prevent scientific misconduct, even detecting fraud, especially in case of publications.

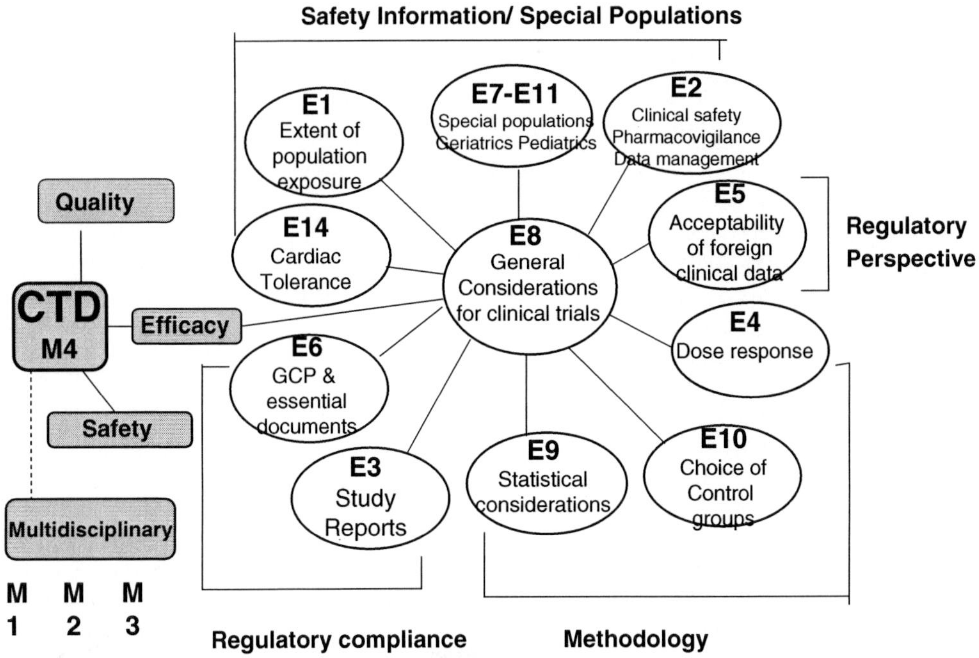

Figure 5 ICH Efficacy and Multidisciplinary guidelines.

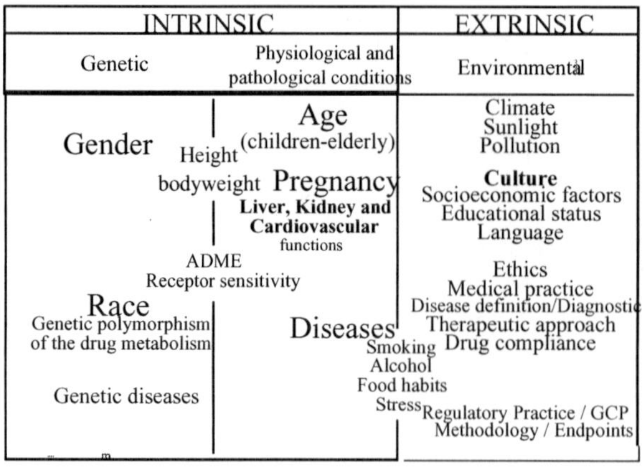

Clinical Studies in Special Populations at Risk

Figure 6 Extrinsic and intrinsic factors which must be taken into consideration for the development (or use) of medicinal products/drugs.

E5 Guideline on Ethnic Factors

Issues around ethnicity will also assume more and more importance in the development of medicinal products for a multitude of reasons:migration of populations, proper use of medicinal products/drugs, cultural differences in lifestyles and attitudes to diseases, ageing and end-of-life care, differing spectra of disease manifestation in ethnic populations, and the application and acceptance of foreign clinical data from different parts of the world in regulatory submissions where research ethical standards have been criticized. This is the origin of the E5 guideline on ethnic factors (17) in order to promote "bridging studies" at the different phases of development.

This is summarized in Figure 6, integrating the many items to be selected during the development process.

In elderly patients, the factors to be considered in this vulnerable population are: age, gender, ethnicity, physical, and psychological profiles, particularly vulnerability of frail and fragile persons. Also to be taken into consideration for the elderly should be the pharmaceutical formulations, routes of administration (e.g., gastrointestinal, inhalation, or transcutaneous absorption), proper dosages and dose regimens, pharmacokinetic/PK specific studies, the search for subtle adverse events, drug/drug interactions, and risk management. Especially, important and often unrecognized are the effects of mild cognitive dysfunction and manual dexterity on drug compliance and therapeutic adherence. These extrinsic factors, and others such as medical practice, education, and nutrition, may be more important than intrinsic factors in the use of medicinal products.

Other Important ICH Efficacy Texts

Many other ICH efficacy texts must be taken into consideration for clinical studies in older populations. The E2 guidelines on pharmacovigilance are mandatory and essential in the use of medicinal product before and after the submission of a marketing authorization application (MAA) or new drug application (NDA) (18).

ICH methodological texts such as the E4 dose response guideline (19) and the E10 guideline on choice of control groups (20) are clear and useful.

WEAKNESS IN EUROPEAN LEGAL AND REGULATORY TEXTS

European Legal and Regulatory Framework Concerning General Clinical Research

There is no law covering clinical research in the European Union or legal text covering specifically physiological or pathophysiological "clinical research" and especially in our case for the elderly. Each European Member State has its own legislation covering in part of or in totality different types of clinical research in humans. The French law 2004 –806 on Public Health Policy with its decree of implementation(2006–477) on "biomedical research" published in August 2006 and its implementation decrees covers any type of clinical research and does not have an equivalent at European level.

European Clinical Research Infrastructures Network

The Regulatory Affairs group of the European Clinical Research infrastructure network (ECRIN) (21) with the support of the European Commission has established an enquiry in ten Member States on academic clinical research programs and this shows that every country is different (21). Clinical research as a whole is not unified in the European Union. More countries from the European Union will be included in ECRIN in the future and this will exacerbate the differences. ECRIN offers a degree of harmonization for many aspects of clinical research programs such as Standard Operating Procedures (SOPs).

This approach is of real importance for European multicenter, multicountry clinical studies in order to facilitate them.

Legal and Regulatory Requirements Covering Clinical Studies on Medicinal Products

Directive 2001/20/EC has been adopted in the 27 European Member States (22) but with obvious differences according to national implementation laws and regulations.

This is particularly true for Ethics Committees for which the support, functioning, and decision making varies markedly in each country. This situation is the same both for adults and older people in contrast to that for children (younger than 18 years of age) who are covered by Regulation EC N° 1901/2006 which applies directly and equally in each member state.

Clinical trial authorizations often have dual requirements (national competent authority and ethics committees) and vary within the European member states. Arrangements for dealing with incapacitated adults and those unable to give "informed consent" are included in Article 5 of Directive 2001/20/EC (22) on clinical trials but interpretation and application vary with different approaches from member state to member state with regard to the specificities.

Legal and Regulatory Requirements Covering Clinical Studies on Medical Devices

For medical devices, there are fewer variations between member states for authorization of clinical trials with Directive 93/42/EC (23). The CE marking can be achieved in specific centers in the European Union and in parallel with clinical trials. Clinical trials are generally less demanding than for "new molecular entities."

ABSENCE OF A REAL REGULATORY FRAMEWORK FOR THE ELDERLY

The ageing population is an orphan population in any country in the world, especially for the development of new medicinal products but also for those studies to be done after the initial marketing authorizations.

Neither Japan nor the United States address the issue of the elderly very well and in 2006 at the request of the U.S. Congress an inspection was performed at the FDA. A Government Accountability Office (GAO) report (24) on the elderly suggested that the U.S. system was insufficiently robust in dealing with clinical trials for the elderly. This led to advice to improve clinical drug development programs for the elderly but so far this is still a wish.

ETHICS AND ELDERLY SUBJECTS/PATIENTS

Few general texts exist on ethics and clinical research in the elderly. Among others only the official texts have a real impact. Few general recommendations exist on ethics and clinical research in the elderly. Official specific geriatric texts are needed to have a real impact. The new text of Declaration of Helsinki published in October 2008 in Seoul (Korea) (25) does not specifically refer to vulnerable populations such as the elderly/older people but says in paragraph 5: "Medical progress is based on research that ultimately must include studies involving human subjects. Populations that are underrepresented in medical research should be provided appropriate access to participation in research."

The Council of Europe (26) does have a specific text for older people called: Parliementary assembly: recommendation 1254 (1994) 1, on the medical and welfare rights of the elderly–ethics and policies.

The Council of International Organization of Medical Sciences (CIOMS) refers to Elderly populations and patients within its "Ethical Guidelines for Biomedical research involving Human Subjects." These guidelines, published in 2002, were produced in conjunction with World Health Organization (WHO) (27).

There are a number of different books on the topic of medical ethics and the elderly in the English language. These include one from the British Medical Association (28) and one edited by Rai (29).

ELDERLY PATIENTS AND THE CONSENT FORM/THE WAY TO ENTER A TRIAL

Capacity for Communication by the Elderly Patient

As already seen, there is a need for more clinical trial research in older people in the European Union. For those people to be included in clinical trials, they must be able to give true informed consent to their participation. This is a challenge especially where they may be fragile or with a psychological handicap. There is thus a need to assess their ability to understand the information given on the nature, risks, and implications of a particular study and to give consent to be signed by the participant before taking part in any clinical trial. This process should be harmonized across the European Union as much of clinical drug development research will be pan-European in the future.

Evaluation of the capacity to make a decision by a person includes three steps (30–36):

Step 1: The patient's participation is entirely voluntary. Evaluation of that capacity by the investigator must show how the participant is able to exhibit that choice with independence.

Evaluation of the consent of an elderly patient to enter a clinical research programs must first search for their capability to communicate with the clinical investigator, especially in case of fragility or senile dementia. Only after this, first approach may the decision-making process and obtaining a valid consent continue.

Normal ageing does not necessarily modify the communication and understanding of information but any pathological process, especially any modification of sense organs associated with ageing, may interfere with the means and innovative techniques may be required.

Step 2 is "comprehension of what is explained by the clinical investigator."

Information given to each individual must be pertinent, simple and clear enough to be understood by the subject in their own native language with verification that everything has been understood and if necessary with a translator.

Step 3 is the "opportunity and freedom to either accept or refuse participation in the proposed clinical research."

This must be explained, the patient given the information in their basic language being the two key elements to understanding before signing a "consent form."

The validity of the decision taken by the subject depends on their capability to communicate and this should also be evaluated.

In practice, it is not difficult to determine if a person has the capability or not to take one or more decisions because the normal ageing process does not modify the capability to communicate or to understand. But as already said a modification of the sense organs, a debilitated state due to difficult social or environmental conditions, a premorbid dementia, a depressive disorder, or aphasia/dysphasia due to cerebrovascular disease may disrupt the ability to communicate and comprehend.

All the mentioned deficits can be evaluated by validated scales. But there is no worldwide agreement on tools to be used in order to define the patient profile and consequently there is no agreement between countries (even in Europe) on the proper way that the decisions should be taken and the consent form signed in order to enter a clinical research project or therapeutic trial. The investigator must insure that the individual subject is able to understand that their agreement could lead to a better or worse condition.

The capability of the elderly to manipulate rationally the given information should be addressed. This part concentrates on the subject's capability to use rational thought to understand the possible risks and benefits they will face during the study. Their capability to remember that decision should be considered. This decision should remain stable and consistent over a certain period of time although variable capacity doesn't necessarily exclude an individual from participation in a clinical trial.

Conclusion in Regard to Patient Consent

The three steps of the consent must be maintained but the content of the "Information text" should be shortened and simplified using language understandable by the elderly patient in their environment.

The procedures used to determine the capability of an old person to take a decision should be reviewed in order to achieve a European consensus and later lead to an international consensus. The "Decision Tree" established by University of California at San Diego (UCSD) Group (37) could be the starting point leading to a European consensus for evaluation tools in order to measure "the capability for decision" by an old person.

The approach concerning Alzheimer's disease is more complicated (36–39) with the participation of a trustworthy person (a "personne de confiance" in French), a substitute person selected by the patient and who could decide instead of the patient if that patient is not able to give consent. This person is a legal representative of the patient and their use is mandatory in France but the concept varies in other European Member States. It has a legal implication but we must say that few persons have selected a "personne de confiance."

A good approach for "decision-making capabilities" is illustrated in Figure 7.

Evaluation Guidelines

There are several guidelines or rating scales to measure "capacities to take a decision" by a subject who have a cognitive diminution (40–42). A subject able to enter a clinical research programs must show to a physician or a member of the team that she/he has a capacity to take a decision.

As already said a good approach to determining capability to take a decision is a "Decision Tree" in this case such as the one established by University of California at San Diego (UCSD) Group (Fig. 7) (37).

Differences in the Approach Relating to Ethics Committees in the European Union

Referring to Ethics Committees, there are a lot of differences between the 27EU Member States as shown by a European Forum for Good Clinical Practice (EFGCP) survey (43). This is one of the main issues for the consent form and ethics as a whole.

In the context of European clinical research, whatever its type, it is not easy to be sure that ethical review process is consistent across the European Union. In fact, evaluation differences are wide especially for multicenter multicountry clinical projects.

The European Directive 2001/20/EC has been differently implemented in different countries, half member states specify that application should be made directly by the sponsor to an ethics committee, whereas the other half specifies that it should be made by the investigator.

Depending upon member states, the way a single opinion is obtained for a multisite application varies in the European Union, some countries designate which committee out of several possible is selected, whereas others have a single committee for the whole country anyway.

Finally, throughout the production of this EFGCP report, it was found that the details of how ethical review was conducted in the various member states varies from one country to another. The most striking differences arise in the areas of training for members of research ethics committees and of quality assurance, assessment, and accreditation of such committees. But this is now under scrutiny in order to achieve a possible harmonization.

Some of these differences are shown in Tables 3 to 6.

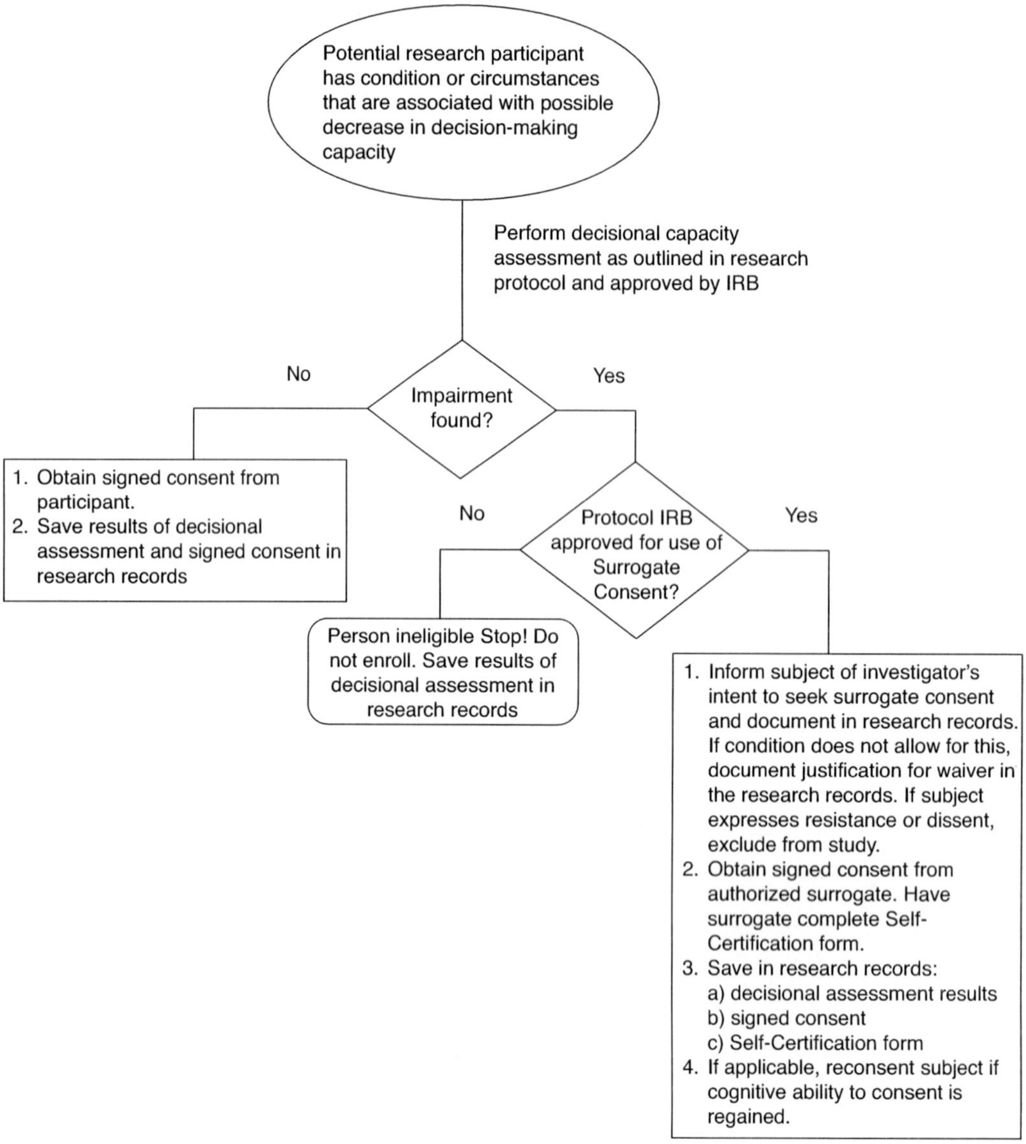

Figure 7 "Capability to take a decision" is a "Decision Tree" established by the University of California San Diego (UCSD) Group.

AGEING POPULATIONS AND CLINICAL RESEARCH

Initiatives leading to the improvement of clinical trials in older patients are one of the wishes of regulatory authorities. They want to rationalize the development plan within this type of population. Frequently, ethics committees do not recognize the necessity of performing clinical trials in older people or they may be completely against them with a paternalistic approach, for example, the elderly should be kept quiet/calm; any disturbance or intrusion should be avoided even if the protocol is necessary and scientifically established.

There is a common therapeutic nihilism in dealing with these age groups which has been refuted by studies such as the recent ones in hypertension in the very elderly. In spite of this experience, often geriatricians want to pursue clinical studies in the elderly in order to understand the role of a particular medicinal product in these populations, only to be obstructed by different stakeholders (44–52).

Table 3 ECRIN Comparison of National Requirements for Clinical Trials on Medicinal Products

		ETHICS COMMITTEE										COMPETENT AUTHORITY										SPONSOR									
Type of trial	**Type of trial therapy**	**AT**	**DK**	**FR**	**DE**	**HU**	**IE**	**IT**	**ES**	**SE**	**UK**	**AT**	**DK**	**FR**	**DE**	**HU**	**IE**	**IT**	**ES**	**SE**	**UK**	**AT**	**DK**	**FR**	**DE**	**HU**	**IE**	**IT**	**ES**	**SE**	**UK**
Medicinal products	Phase 1	✔	✔	✔	✔	✔	✔	✔	✔	✔	✔	✔	✔	✔	✔	✔	✔	✔	✔	✔	✔	✔	✔	✔	✔	✔	✔	✔	✔	✔	✔
	Phase 2	✔	✔	✔	✔	✔	✔	✔	✔	✔	✔	✔	✔	✔	✔	✔	✔	✔	✔	✔	✔	✔	✔	✔	✔	✔	✔	✔	✔	✔	✔
	Phase 3	✔	✔	✔	✔	✔	✔	✔	✔	✔	✔	✔	✔	✔	✔	✔	✔	✔	✔	✔	✔	✔	✔	✔	✔	✔	✔	✔	✔	✔	✔
	Phase 4	✔	✔	✔	✔	✔	✔	✔	✔	✔	✔	✔	✔	✔	✔	✔	✔	✔	✔	✔	✖	✔	✔	✔	✔	✔	✔	✔	✔	✔	✔
Biotherapy	Tissue eng	✔	✔	✔	✔	✔	✔	✔	✔	✔	✔	✔	✔	✔	✔	✔	✔	✔	✔	✔	✔	✔	✔	✔	✔	✔	✔	✔	✔	✔	✔
	Cell therapy	✔	✔	✔	✔	✔	✔	✔	✔	✔	✔	✔	✔	✔	✔	✔	✔	✔	✔	✔	✔	✔	✔	✔	✔	✔	✔	✔	✔	✔	✔
	Gene therapy	✔	✔	✔	✔	✔	✔	✔	✔	✔	✔	✔	✔	✔	✔	✔	✔	✔	✔	✔	✔	✔	✔	✔	✔	✔	✔	✔	✔	✔	✔
Biopharma-ceuticals	Blood-derived	✔	✔	✔	✔	✔	✔	✔	✔	✔	✔	✔	✔	✔	✔	✔	✔	✔	✔	✔	✔	✔	✔	✔	✔	✔	✔	✔	✔	✔	✔
	MAB, prot, peptide	✔	✔	✔	✔	✔	✔	✔	✔	✔	✔	✔	✔	✔	✔	✔	✔	✔	✔	✔	✔	✔	✔	✔	✔	✔	✔	✔	✔	✔	✔
	Oligonucleotide	✔	✔	✔	✔	✔	✔	✔	✔	✔	✔	✔	✔	✔	✔	✔	✔	✔	✔	✔	✔	✔	✔	✔	✔	✔	✔	✔	✔	✔	✔
Vaccines	—	✔	✔	✔	✔	✔	✔	✔	✔	✔	✔	✔	✔	✔	✔	✔	✔	✔	✔	✔	✔	✔	✔	✔	✔	✔	✔	✔	✔	✔	✔
Fixed combination	—	✔	✔	✔	✔	✔	✔	✔	✔	✔	✔	✔	✔	✔	✔	✔	✔	✔	✔	✔	✔	✔	✔	✔	✔	✔	✔	✔	✔	✔	✔
Multimodal	—	✔	✔	✔	✔	✔	✔	✔	✔	✔	✔	✔	✔	✔	✔	✔	✔	✔	✔	✔	✔	✔	✔	✔	✔	✔	✔	✔	✔	✔	✔

Abbreviations: AT, Austria; DK, Denmark; FR, France; DE, Germany; HU, Hungary; IE, Ireland; IT, Italy; ES, Spain; SE, Sweden; UK, United Kingdom.

Table 4 ECRIN Comparison of National Requirements for Clinical Trials on Medical Devices and Other TTT Trials

		ETHICS COMMITTEE										COMPETENT AUTHORITY										SPONSOR									
Type of trial	**Type of trial therapy**	AT	DK	FR	DE	HU	IE	IT	ES	SE	UK	AT	DK	FR	DE	HU	IE	IT	ES	SE	UK	AT	DK	FR	DE	HU	IE	IT	ES	SE	UK
Clinical trial on Medical Device																															
Alone	Authorized	✔	✔	✔	✔	✔	✔	✔	✔	✔	✔	✔	✖	✔	✔	✔	✖	✔	✖	✔	✖	✔	✖	✔	✖	✖	✖	✔	✖	✖	✔
	Not authorized	✔	✔	✔	✔	✔	✔	✔	✔	✔	✔	✔	✔	✔	✔	✔	✔	✔	✔	✔	✔	✔	✔	✔	✔		✔	✔	✔	✖	✔
With Medicinal Product	Authorized	✔	✔	✔	✔	✔	✔	✔	✔	✔	✔	✔	✖	✔	✔	✔	✔	✔	✖	✔	✖	✔	✖	✔	✖	✖	✔	✔	✖	✖	✔
	Not authorized	✔	✔	✔	✔	✔	✔	✔	✔	✔	✔	✔	✔	✔	✔	✔	✔	✔	✔	✔	✔	✔	✔	✔	✔		✔	✔	✔	✖	✔
Other TTT Trials																															
	Radiotherapy	✔	✔	✔	✔	✔	✔	✔	✔	✔	✔	✔	✖	✔	✔	✖	✖	✔	✖	✖	✔	✔	✖	✔	✔	✖	✖	✔	✖	✖	✔
	Surgery	✔	✔	✔	✔	✔	✔	✔	✔	✔	✔	✖	✖	✔	✖	✖	✖	✔	✖	✖	✖	✖	✖	✔	✖	✖	✖	✔	✖	✖	✖
	Transplant	✔	✔	✔	✔	✔	✔	✔	✔	✔	✔	✔	✖	✔	✖	✖	✖	✔	✖	✔	✔	✔	✖	✔	✖	✖	✖	✔	✖	✖	✖
	Transfusion	✔	✔	✔	✔	✔	✔	✔	✔	✔	✔	✖	✖	✔	✔	✖	✖	✔	✖	✔	✔	✖	✖	✔	✔	✖	✖	✔	✖	✖	✔
	Physical therapy	✔	✔	✔	✔	✔	✔	✔	✔	✔	✔	✖	✖	✔	✔	✖	✖	✔	✖	✖	✖	✖	✖	✔	✔	✖	✖	✔	✖	✖	✖
	Psychotherapy		✔	✔	✔	✔	✔	✔	✔	✔	✔	✖	✖	✔	✖	✖	✖	✔	✖	✖	✖	✖	✖	✔	✖	✖	✖	✖	✖	✖	✖

Abbreviations: AT, Austria; DK, Denmark; FR, France; DE, Germany; HU, Hungary; IE, Ireland; IT, Italy; ES, Spain; SE, Sweden; UK, United Kingdom.

Table 5 Comparison of National Requirements for Regulatory Requests for Research Trials—Diagnostic Studies, Nutrition, and Other Clinical Research

		ETHICS COMMITTEE										COMPETENT AUTHORITY										SPONSOR									
Type of trial	**Type of trial therapy**	**AT**	**DK**	**FR**	**DE**	**HU**	**IE**	**IT**	**ES**	**SE**	**UK**	**AT**	**DK**	**FR**	**DE**	**HU**	**IE**	**IT**	**ES**	**SE**	**UK**	**AT**	**DK**	**FR**	**DE**	**HU**	**IE**	**IT**	**ES**	**SE**	**UK**
Diagnostic Studies																															
	In vivo	✔	✔	✔	✔	✔	✔	✔	✔	✔	✔	✔	✔	✔	✔	✖	✖	✔	✖	✖	✖	✔	✖	✔	✔	✖	✔	✖	✖	✖	✖
	In vitro	✔	✔	✔	✔	✖	✔	✔	✔	✔	✔	✔	✖	✔	✔	✖	✖	✔	✖	✖	✖	✔	✖	✔	✖	✖		✖	✖	✖	✖
	Imaging	✔	✔	✔	✔	✔	✔	✔	✔	✔	✔		✔	✔	✔	✖	✖	✔	✖	✖	✔	✔	✖	✔	✖	✖	✔	✖	✖	✖	✖
Nutrition Studies																															
	Nutritional	✔	✔	✔	✔	✔	✔	✔	✔	✔	✔	✔	✔	✔	✖	✖	✔	✔	✖	✖	✖	✖	✖	✔	✖	✖	✖	✔	✖	✖	✖
	Nutrition supplements	✔	✔	✔	✔	✔	✔	✔	✔	✔	✔	✔	✔	✔	✖	✖	✔	✔	✖	✖	✖	✖	✖	✔	✖	✖	✖	✔	✖	✖	✖
Other Clinical Research																															
	CAM	✔	✔	✔	✔	✔	✔	✔	✔	✔	✔	✔	✔	✔	✔	✔	✔	✔	✖	✔	✔	✖	✖	✔	✔	✖	✔	✖	✖	✖	✖
	biobanks	✔	✔	✔	✔	✔	✔	✔	✔	✔	✔	✔	✔	✔	✖	✖	✖	✔	✖	✔	✔	✖	✖	✔	✖	✖	✖	✖	✖	✖	✔
	physiology	✔	✔	✔	✔	✔	✔	✔	✔	✔	✔	✖	✔	✔	✖	✖	✖	✔	✖	✖	✔	✖	✖	✔	✖	✖	✖	✖	✖	✖	✔
	pathophysiology	✔	✔	✔	✔	✔	✔	✔	✔	✔	✔	✖	✔	✔	✖	✖	✖	✔	✖	✖	✔	✖	✖	✔	✖	✖	✖	✖	✖	✖	✔
	psychology	✔	✔	✔	✔	✔	✔	✔	✔	✔	✔	✖	✔	✔	✖	✖	✖	✔	✖	✖	✔	✖	✖	✔	✖	✖	✖	✖	✖	✖	✔

Abbreviations: AT, Austria; DK, Denmark; FR, France; DE, Germany; HU, Hungary; IE, Ireland; IT, Italy; ES, Spain; SE, Sweden; UK, United Kingdom.

Table 6 ECRIN Comparison of National Requirements for Regulatory Requests for Research Trials in Epidemiology

		ETHICS COMMITTEE										COMPETENT AUTHORITY										SPONSOR									
Type of trial	**Type of trial therapy**	**AT**	**DK**	**FR**	**DE**	**HU**	**IE**	**IT**	**ES**	**SE**	**UK**	**AT**	**DK**	**FR**	**DE**	**HU**	**IE**	**IT**	**ES**	**SE**	**UK**	**AT**	**DK**	**FR**	**DE**	**HU**	**IE**	**IT**	**ES**	**SE**	**UK**
Pharmacoepidemiology																															
	Interventional	✔	✔	✔	✔	✔	✔	✔	✔	✔	✔	✔	✔	✔	✔	✔	✔	✔	✔	✔	✔	✔	✔	✔	✔	✔	✔	✔	✔	✔	✔
	Noninterventional	✖	✔	✖	✔	✔	✔	✔	✔	✔	✔	✖	✖	✖	✖	✖	✖	✔	✔	✖	✔	✔	✔	✔	✔	✔	✔	✔	✔	✔	✔
Epidemiology																															
	Interventional	✔	✔	✔	✔	✔	✔	✔	✔	✔	✔	✔	✔	✔		✔		✔	✔		✔	✔	✖	✔	✖	✖	✔	✖	✖	✖	✔
	Noninterventional	✖	✔		✔	✔	✔	✔	✔	✔	✔	✖	✖	✖	✖	✖	✖	✔	✖	✖	✔	✖	✖	✖	✖	✖	✔	✖	✖	✖	✔
Registries of patients		✖	✔	✔	✔	✔	✔	✔	✔	✔	✔	✖	✖	✖	✖	✖	✖	✔	✖	✖	✖	✖	✖	✖	✖	✖	✖	✔	✖	✖	✖

Abbreviations: AT, Austria; DK, Denmark; FR, France; DE, Germany; HU, Hungary; IE, Ireland; IT, Italy; ES, Spain; SE, Sweden; UK, United Kingdom.

Which Medicinal Products Should be Studied in the Elderly?

Balance Between Therapeutic Nihilism and Multiple Therapy

"Just because you are old you should not be treated" is a form of discrimination but the use of too many treatments and their complex interactions produce multiple side effects in patients older than 75 years. The effects of many interventions may be diminished but paradoxically such interventions may also have greater epidemiological impact on populations with more disease prevalence. Thus, there is a balance to be chosen.

Many geriatricians are reluctant to perform clinical studies in elderly patients. We recommend studying only drugs that concern pathologies which are seen specifically in the elderly. Drugs with a narrow therapeutic index due to modification of their characteristics with age, that is, affecting different metabolic systems, kidney, liver, cardiac output, albumin index etc.; or drugs with excessive side effects that could be especially harmful in the elderly (such as dizziness, sleepiness, etc.), should be avoided.

The elderly should be considered as part of an orphan population as was the case with children where medicinal products were commonly used "off-label." Many issues have a common ground with use of medicines in children.

A good approach would be to have a geriatric investigational plan (GIP), not mandatory as with children with the paediatric investigational plan (PIP) but to be able to ask for a "à la carte GIP" referring to the stratification, age, gender, epidemiology, morbidity/ies indications of elderly patients.

Elderly Patients: The Parallel With Children

Older people should be included in their own specific trials with stratified groups according to age and indication. The recruitment of these subjects is not always simple (dependence, fragility, family opposition, etc.). The informed consent is *per se* something specific and Ethics Committees should include in their group at least one geriatrician or member skilled and experienced in dealing with these age groups. Many elements of informed consent vary with ageing, physical and psychological status.

Pharmaceutical forms must be adapted for use in older patients in order to improve compliance with drug therapy.

Specific pharmacokinetic/pharmacodynamic studies should be performed according to age and parameters such as physical/psychological autonomy, renal insufficiency, the absence of mobility, weight, etc.

A long-term follow-up is necessary to look at the absence of efficacy but mainly to investigate any possible drug–drug interactions in order to avoid pharmacovigilance/iatrogenic problems.

Clinical Research on Physiology, PathoPhysiology

In spite of difficulties, it is from time-to-time indispensable to investigate certain aspects of physiology or pathophysiology of very old emerging populations. Here again, it will be difficult to convince the family or Ethics Committees to allow such studies especially in countries, which have no specific regulatory text.

Clinical Studies on Medicinal Products or Medical Devices

Trial Methodology

Elderly people are different from other adults. Therefore, a better knowledge about pharmacotherapy in the elderly is mandatory to establish evidence-based therapeutic guidelines for higher age groups (53).

Recruitment of Patients

As already stated, the recruitment of patients is not simple especially above the age of 75 years. Besides upper age limits, the enrolment of patients in clinical trials is frequently confined to patients with limited comorbidities and comedication, the so-called "elite elderly" patients.

However, higher age is associated with an increased number of chronic conditions, which are usually treated with drugs (54).

The age group between 80 and 85 years especially receives the highest number of daily prescription drugs. The future of clinical trials in elderly patients will be based on two main parameters:

Epidemiology and ageing

As a function of ageing, many parameters should be studied depending on the gender, activity, weight but also renal and hepatic function and volume of distribution depending on drug characteristics. Elderly patients may well have markedly different profile for pharmacokinetics and pharmacodynamics which might make invalid the results of Phase I first in man studies carried out on young adult men.

Later, and depending upon the number and types of drugs taken daily, they could have different type of adverse events, drug–drug interaction, and iatrogenic disease. Providing clinical data for the marketing authorization is one thing but extrapolation to daily care and daily life is another big question.

Pharmacology aspects: pharmacokinetics/pharmacodynamics, drug–drug interaction, and pharmacovigilance

The main issue is the determination of efficient and effective pharmacologic dosing schedules by doing the necessary Pharmacokinetics (PK), Pharmacodynamics (PD), and Drug–Drug Interaction (DDI) studies (55). Very frequently, medical formularies advise reducing the basic dosage schedule and total daily dose without stating the parameters used to make those decisions. This is counterproductive when a specific patient experiences the adverse effects of a particular medicinal product without benefiting from its efficacy.

The study of the relation between PK and PD and any subsequent drug–drug interaction/DDI is mandatory for any new medicinal product and especially so in older patients. This leads to a stratification of ageing populations for any type of clinical development, looking at specific organ impairment (kidney and liver being obvious but cardiac and respiratory also having effects) and metabolic implications such as dehydration, albumin level, gastrointestinal motility, body mass, and fat-free index etc.

This should lead to the construction of a pharmacovigilance plan, a risk management plan for the inclusion in the marketing authorization application, without affecting the efficacy of the drug and the risk to benefit ratio in elderly patients.

For enrolment of elderly patients in clinical trials, sufficient knowledge about pharmacokinetics across all age groups is mandatory as well as information about potential drug–drug interactions as stated in the ICH E7 Guideline (9). However, this guideline resulted in the present situation, if it is known that a certain drug should be used in lower (starting) doses for the elderly or that drug–drug interactions are likely this will probably result in the exclusion of elderly multimorbidity patients, just for convincing safety reasons.

In their analysis of 495 cancer trials Lewis et al. (56) observed that normal renal and hepatic function (relevant for drug excretion) was a strict inclusion criterion for 49% and 60%, respectively, of the studies.

One part of the clinical development program will be the analysis of factors precluding enrolment of elderly patients, also with respect to the exclusion criteria comedication and decreased renal or hepatic elimination capacity.

The aim is to identify solutions and make suggestions for broadening the inclusion criteria with regard to PK and DDI. This may also include methodological aspects of study design (working package "Methodology") such as allowing for lower starting doses in the elderly, more frequent monitoring visits, and dose adaptations for study drugs and comedication throughout the study. After a thorough literature analysis of the most relevant pharmacokinetic aspects, suggestions for future clinical trials will be prepared. This will also include aspects regarding pharmacovigilance with respect to the elderly to be includedfor example, in the pharmacovigilance risk management plans required by the EMEA and national agencies for submission of new drugs.

The choice of the right dosage and dose regimen in older patients is crucial.

Methodology

Besides the rules linked with each therapeutic class, the methodology should aim at using a global geriatric evaluation whatever the main objective. Evaluation tools are specific and validated in many domains such as cognitive function, nutrition, pain, autonomy, and quality of life. The methodologist with the support of a geriatrician should ascertain that methods are compatible with ageing patients, especially if evaluation tools are validated in this patient's population.

Protocol and Case Report Form

The objective is to introduce patients in Phases II and III and to be able to compare them with younger adults using simpler Case Report Form/CRF and electronic Case Report Form/eCRF, but also simplified consent form and information to patients.

Once more the elderly population must be stratified according to their status (institutions or external patients) using evaluation scales,for example, as used in France, especially for people older than 75 years. It is absolutely necessary to have groups of sufficient size in order to be able to derive meaningful conclusions.

Modeling

Modeling must not exclude real clinical trials in the elderly (57). Modeling is a very good tool in order to select the drugs geriatric doses of which can be easily extrapolated and calculated from the adult: loading dose and maintenance doses.

Modeling can also detect drugs that have in the elderly a very different metabolism or pharmacodynamic properties from the adult and that will need specific dose range trials. The ideal objective would be to simulate diseases and their therapeutic approach by using these mathematical models. This does not exclude classical methodologies but could lead to a "personalized, individual approach" concerning therapeutic support. The two approaches are synergistic and complementary.

The optimization of PK/PD could lead to dosage and dose regimen in order to optimize the efficacy and clinical safety. Numeric modeling and theoretical modeling should converge by using mathematical and biological (pathophysiology, phenotype, and genotype) integrations.

Evaluation Tools

If adult evaluation tools are sometimes usable, the tools must have been previously validated in the elderly in many cases. This point is of particular importance to avoid floor effects with rating scales that is, loss of sensitivity for some elderly patients when all the elderly population is at the bottom of the rating scales. Another important point is to define the clinically relevant variation of the rating scale in the elderly.

Oncology and the Elderly as an Example

Preliminary remarks

Many cancer studies have shown many differences in prevalence and responses to treatment according to age. The factors involved are ageing processes, comorbidities, comedications, increase risk of adverse events, and these lead to heterogeneous populations in difficult protocols (58). Generally the need to adapt the PK and PD parameters leads to questions before clinical trials (9,24,55).

Underrepresentation of elderly patients in oncology clinical trials

A retrospective analysis performed by the U.S. National Cancer Institute (NCI) has shown that in 495 Phase II and III oncology clinical trials the elderly were underrepresented, being only 33% of trial participants instead of the 61% of cancer patients older than 65 years in the clinical population.

Twice, in 1989 and 1998, the U.S. FDA insisted that the elderly should be included in clinical trials in order to study DDI and adverse events according to their profile.

Obstacles to the recruitment of elderly patients

The reasons for excluding very old cancer patients from clinical trials were their age, the opposition of investigators, the type of protocols (inclusion/noninclusion criteria), the functional status of patients, organ insufficiencies, and physical dependencies. Other factors were social status and time necessary to include them.

Strategy to improve recruitment

The absence of a biological database concerning this type of elderly population is something which should be taken into account. Biomarkers and safety data are really missing. Survival time and quality of life should be included in databases. The protocols must not exclude systematically comorbidity and elderly patients must be stratified according to ages. By this approach, protocols would be closer to "real life."

Basic and continuous training of physicians (traditional or e-learning) should be done in order to support clinical investigational teams. Beside documentation, case studies would be very helpful.

Medical students and other stakeholders should be educated early on to improve appreciation of the need to include these ageing populations in clinical drug development and encourage inclusion of elderly people in clinical trials.

THE ELDERLY AND DAILY USE OF MEDICINAL PRODUCTS

General Issues

The extrapolation from MAA/New Drug Application (NDA) data to daily life is not easy especially as elderly patients are not included in most clinical trials with the exception of those for some degenerative diseases. The dosage, dose regimen, and duration of treatment are not always well defined.

Efficacy, compliance with drug regimen, and clinical safety are the three main topics to follow. Drug–drug interaction is a difficult issue depending upon the disease, the therapeutic indication, and the type of medicinal product (59).

Efficacy and Safety

Extrapolation to subjects more than 75-year-old is always difficult because, with a few exceptions, they are not included in clinical trials. Some studies have been performed either in specific indications with well-known products (60) or in usual indications such as hypertension in very old patients (61).

Some products, especially in the cardiovascular field, have never proved to be useful (e.g., statins) or can be dangerous (e.g., anti-vitamin K) over a certain age. New indications which may be relevant to these populations are being discovered and will require specific evaluation.

This question is really a problem of therapeutic strategy. Ageing populations are not always compliant especially if they have psychological disturbances.

Compliance with the Drug Regimen

The daily use of medicinal products is difficult especially in the elderly population who were not trained to use them. Ageing populations are using more and more drugs and frequently are not really compliant. The proper use of drugs is very rare (62). This is a critical element to consider especially because the pharmaceutical forms are not always adapted to the needs of this patient group. In French institutions, 30% of medicinal products are crushed or modified in order to give them to hospitalized patients.

Pain is not too badly treated but cardiovascular, osteoporosis, and cancer treatments are regularly forgotten.

Clinical Safety

The elderly patient (as with the child) is fragile and should be followed carefully in order to detect adverse drug reactions. In addition due to age-associated changes in drug metabolism and elimination as well as sensitivity, elderly people have a higher risk for adverse drug reactions then younger patients.

A minor adverse event encountered in a young adult could become a serious adverse reaction due to the age of the patient: for example, dizziness due to short-term orthostatic hypotension could lead to a severe fracture. In Germany, the mean age of elderly patients hospitalized for pharmacovigilance is 70 years (63). Iatrogenic hospitalization is common in all European member states.

The ICH guideline documents E2A to E2E (18) are now implemented in ICH regions and have led to a pharmacovigilance plan and risk management plan for each new product. None of the demands specifically cover the elderly. In the European Union, a "qualified person for pharmacovigilance" is responsible for ensuring that companies follow the pharmacovigilance process sending individual reports for serious Adverse Drug Events (ADEs), annual pharmacovigilance reports, pharmacovigilance plans, or risk management plans for new products as already seen.

Inappropriate Use of Drugs in Elderly Patients

Drugs trials involving geriatric population can also detect drugs with an unfavourable benefit to risk ratio when other safer or more efficient therapeutic alternatives are available. Thus, it would be possible to establish a list of "potentially inappropriate medications in the elderly" and a list of "best choice drugs" (64). This could be a way to reduce adverse effects that are so frequently encountered in geriatric populations. For the moment, it is clear that there is a lack of data and such lists as already exist are only established on the basis of experts' consensus but with no hard evidence-based background. There is thus a risk of selecting or rejecting drugs inappropriately.

CONCLUSIONS

Clinical trials are essential before or after the submission of a MAA/NDA application. They do not include enough patients older than 75 years with a few exceptions.

After approval of the MA, the big issue in assessing the role and value of a new drug is to extrapolate from the registration data and to evaluate whether it is applicable, effective, and safe on an individual basis especially for very old patients with their individual profiles of comorbid disease and multiple therapy which may have marked effects on the characteristics of the medicinal product (65).

But once the drug has been evaluated in the older patient in the short-term, the next problem to solve will be what will be the therapeutic strategy for long-term care. What are the good therapeutic goals at 85 years of age?

Some trials indicate what to do for a few diseases in specific circumstances. For the majority of diseases/symptoms there are no data in elderly patients. For this reason extrapolation from young patients could be dangerous and lead to safety issues.

As an interim conclusion, the health problems of elderly people now mainly concern patients older than 75 years who have a lot of degenerative processes, often neurological. Pathology is multiple and chronic in frail older patients. The problems for the future in every country in the world revolve around health, socioeconomic, and ethics but the silver tsunami is upon us and these issues need to be addressed.

REFERENCES

1. Auvray L, Sermet C. Consommations et prescriptions pharmaceutiques chez les personnes âgées : Un état des lieux. Gérontol Soc 2002; 103:13–27.
2. Giannakouris K. Ageing characterizes the demographic perspectives of the European societies, Eurostat. Statistics in Focus, 72/2008, European Commission. http://epp.eurostat.ec.europa.eu/cache/ITY_OFFPUB/KS-SF-08–072/EN/KS-SF-08–072-EN.PDF. Accessed February 2009.
3. Statistics in "France Metropolis" excluding the four overseas departments : French Guyana, Guadeloupe, Martinique, Réunion, Institut national de la statistique et des études économiques (Insee). http://www.insee.fr. Accessed October 2008.
4. Eurostat Home Page, European Commission. http://ec.europa.eu/eurostat. Accessed February 2009.

5. Beaumel C, Vatan M. La situation démographique en 2006, Mouvement de la population, Institut national de la statistique et des études économiques (Insee). http://www.insee.fr/fr/ppp/bases-de-donnees/irweb/sd2006/dd/pdf/sd2006_brochure.pdf. Accessed February 2009.
6. Age of people of more than 100 years old in France, Institut national de la statistique et des études économiques (Insee). http://www.insee.fr. Accessed October 2008.
7. Japanese Ministry of Internal Affairs and Communications, Statistics Bureau and Statistical Research and Training Institute, Tokyo. http://www.stat.go.jp/english/data. Accessed January 2009.
8. Arias E. United States Life Tables 2004, National Vital Statistics Reports, 56(9), December 28, 2007, US Department of Health and Human Services, National Center for Health Statistics. http://www.cdc.gov/nchs/data/nvsr/nvsr56/nvsr56_09.pdf. Accessed February 2009.
9. International Conference on Harmonization Guideline E7. Studies in support of special populations: Geriatrics, 1993. http://www.ich.org/LOB/media/MEDIA483.pdf. Accessed February 2009.
10. EMEA- CHMP document ref EMEA/498920/2006: Adequacy of guidance on the Elderly regarding medicinal products for human use; December 14, 2006. www.emea.europa.eu/pdfs/human/opiniongen/49892006en.pdf. Accessed February 2009.
11. Gijsen R, Hoeymans N, Schellevis FG, et al. Causes and consequences of comorbidity: A review. J Clin Epidemiol 2001; 54:661–674.
12. Fortin M, Lapointe L, Hudon C, et al. Multimorbidity is common to family practice. Is it commonly researched? Can Fam Physician 2005; 51: 244–245. www.ncbi.nlm.nih.gov/pubmed/16926936. Accessed February 2009.
13. The causes of death above 75 years vary with the gender. Morbidity varies also with age and gender. The number of dependent persons will dramatically increase from 2000 to 2040, Institut national de la statistique et des études économiques (Insee). http://www.insee.fr. Accessed October 2008.
14. AGGIR Scale—Generontological Autonomy, Iso-Resources Group, France. http://ec.europa.eu/employment_social/missoc/2006/02/2006_02_fr_en.pdf. Accessed February 2009.
15. Dorenlot P, Harboun M, Bige V, et al. Major depression as a risk factor for early institutionalisation of dementia patients living in the community. Int J Geriatr Psychiatry 2005; 20(5):417–418.
16. International Conference on Harmonization Guideline E6(R1) Good clinical practice, June 1996. http://www.ich.org/LOB/media/MEDIA482.pdf. Accessed October 2008.
17. International Conference on Harmonization (ICH) Guideline E5(R1) Ethnic factors in the acceptability of foreign clinical data, February 1998. http://www.ich.org/LOB/media/MEDIA481.pdf. Accessed October 2008.
18. International Conference on Harmonization Guideline E2 Pharmacovigilance planning, November 2004. http://www.ich.org/LOB/media/MEDIA1195.pdf. Accessed February 2009.
19. International Conference on Harmonization Guideline E4 Dose response information to support drug registration,March 1994. http://www.ich.org/LOB/media/MEDIA480.pdf. Accessed February 2009.
20. International Conference on Harmonization Guideline E10 Choice of control groups and related issues in clinical trials, Julty 2000. http://www.ich.org/LOB/media/MEDIA486.pdf. Accessed February 2009.
21. European Clinical Research Infrastructures Network (ECRIN); A comparison established in 2007–2008 by the Regulatory Affairs Working Party, on Clinical research in 10 European member states. www.ecrin.org. Accessed December 2008.
22. Directive 2001/20/EC: Clinical trials on medicinal products, April 4, 2001, Article 5: clinical trials in incapacitated adults not able to give informed consent. www.europa.eu/eur-lex/pri/en. Accessed October 2008.
23. Medical devices sector legislation, DG Enterprise and Industry, European Commission. http://ec.europa.eu/enterprise/medical_devices/legislation_en.htmn. Accessed October 2008.
24. United States Government Accountability Office (GAO) report on Prescription Drugs: FDA Guidance and Regulations Relating to Data on Elderly Persons in Clinical Trials, September 28, 2007. http://www.gao.gov/new.items/d0747r.pdf. Accessed October 2008.
25. World Medical Association (WMA), the Declaration of Helsinki, last update Seoul (Korea)- October 2008. http://www.wma.net/e/policy/pdf/17c.pdf. Accessed November 2008.
26. Council of Europe, Parliamentary Assembly Doc. 7170, report of the Social, Health and Family Affairs Committee, Rapporteur: Ms Özver Parliamentary assembly specific text for older people called: Parliamentary assembly: recommendation 1254 (1994) 1, on the medical and welfare rights of the elderly – ethics and policies. http://assembly.coe.int/Documents/AdoptedText/ta94/erec1254.htm. Accessed February 2008.
27. The Council of International Organization of Medical Sciences (CIOMS). Ethical Guidelines for Biomedical Research Involving Human Subjects. Geneva: CIOMS, WHO, 2002. http://www.cioms.ch/frame_guidelines_nov_2002.htm (accessed February 2009).

28. The Ethics of Caring for Older People. British Medical Association, ISBN 978–1–4051–7627–9, January 2009.
29. Rai GS, ed. Medical Ethics and the Elderly, 2nd ed. Radcliffe Publishing, 2004:ISBN 1–85775 851 X.
30. American Psychiatric Association. Diagnostic and Statistical Manual Of Mental Disorders, 4th ed. Washington, DC. http://www.psychiatryonline.com/resourceTOC.aspx?resourceID=1. Accessed February 2009.
31. Hugonot-Diener L, Husson JM. Communicating with the elderly; decision making and informed consent in subjects with frailty or dementia. Res Ethics Rev 2007; 3(3):97–101.
32. Folstein MF, Folstein SE, McHugh PR. Mini Mental State. A practical method for grading the cognitive state of patients for the clinician. J Psy Res 1975; 12:189–198.
33. Roth LH, Meisel A, Lidz CW. Tests of competency to consent to treatment. Am J Psychiatry 1997; 134:28–35.
34. American Psychiatric Association. Defined standards for assessing decision-making capacities. Am. J Psychiatry 1998; 155:11.
35. Palmer BW, Dunn AB, Appelbaum PS, et al. Assesment of capacity to consent to research among older subjects with schizophrenia, Alzheimer diseas e, or diabetes mellitus. Arch Gen Psychiatry 2005; 62:726–733.
36. Williams B, Irvine L, McGinnis A, et al. When "no" might not quite mean "no". The importance of informed and meaningful non consent. BMC Health Serv Res 2007; 7:59.
37. Decision Making Capacity Guidelines. San Diego (UCSD): The University of California, 2008. http://irb.ucsd.edu/decisional.shtml. Accessed December 2008.
38. Schneider LS, Olin JT, Lyness SA, et al. Eligibility of Alzheimer's disease clinic patient for clinical trials. J AmGeriat Soc 1997; 45:923–928.
39. McKhann G, Drachman D, Folstein M, et al. Clinical diagnosis of Alzheimer's disease: Report of the National Institute of Neurological and Communicative Diseases and Stroke/Alzheimer's Disease and Related Disorders Association (NINCDS-ADRDA) work group under the auspices of Department of Health and Human Services task force on Alzheimer's disease. Neurology 1984; 34:939–944.
40. Cummings JL. The Neuropsychiatry of Alzheimer's disease and related dementias. New York, NY: Taylor & Francis, 2005:41.
41. Hugonot-Diener L. La consultation en Gériatrie. Paris: Masson , 2001 mise à jour en 2007.
42. Hugonot-Diener L, Barbeau E, Michel et Robert P. Gremoire. Les tests et échelles de la maladie d[1] Alzheimer et des syndromes apparentés. Collection Greco-Solal. 2008: 308.
43. Wells F. European Forum for Good Clinical Practice / EFGCP/ The procedures for the Ethical Review of Protocols for Clinical Research Projects in the European Union, Internat. J Pharm Med 2007; 21:1.
44. Boissel JP, Gueyffier F, Haugh MC. Response to Inclusion of women and minorities in clinical trials and the NIH revitalization act of 1993. The perspective of NIH clinical trialists. Control Clin Trials 1995; 16:286–288.
45. Gurwitz JH, Col NF, Avorn J. The exclusion of the elderly and women from clinical trials in acute myocardial infarction. JAMA 1992; 268:1417–1422.
46. Bugeja A, Kumar A, Banerjee AK. Exclusion of elderly people from clinical research: A descriptive study of publish reports. BMJ 1997; 315:1059.
47. Hutchins LF, Unger JM, Crowley JJ, et al. Underrepresentation of patients 65 years of age or older in cancer-treatment trials. N Engl J Med 1999; 341:2061–2067.
48. Bayer A, Tadd W. Unjustified exclusion of elderly people from studies submitted to research ethics committee for approval: Descriptive study. BMJ 2000; 321: 992–993.
49. Lee PY, Alexander KP, Hammil BG, et al. Representation of elderly persons and women in published randomized trials of acute coronary syndromes. JAMA 2001; 286:708–713.
50. Masoudi FA, Havraneck EP, Wolfe P, et al. Most hospitalized older persons do not meet the enrollment criteria for clinical trials in heart failure. Am Heart J 2003; 146:250–257.
51. Krumholz HM, Gross CP, Peterson ED, et al. Is there evidence of implicit exclusion criteria for elderly subjects in randomized trials? Evidence from the GUSTO-1 study. Am Heart J 2003; 146:839–847.
52. Mitchell SL, Sullivan EA, Lipsitz LA. Exclusion of elderly subjects from clinical trials for Parkinson disease. Arch Neurol 1997; 54:1393–1398.
53. Schoenmaker N, Van Gool WA. The age gap between patients in clinical studies and in the general population: A pitfall for dementia research. Lancet Neurol 2004; 3:627–630.
54. Adams J, Silverman M, Musa D, et al. Recruiting older adults in clinical trials. Control Clin Trials 1997; 18:14–26.
55. Turnheim K. When drug therapy gets old: Pharmacokinetics and pharmacodynamics in the elderly. Exp Gerontol 2003; 38:843–853.
56. Lewis JH, Kigore ML, Goldmann DP, et al. Participation of patients 65 years of age and older in cancer clinical trials. J Clin Oncol 2003; 21:1383–1389.

57. Boissel JP, Cucherat M, Nony P, et al. Modeling and medical product R&D. In International Pharmaceutical Product Registration, 2nd edition, 2009; pp. 761–772.
58. Witham M, McMurdo M. How to get older people included in clinical trials. Drug Aging 2007: 24(3):187–196.
59. Boyd CM, Darer J, Boult C, et al. Clinical practice guidelines and quality of care for older patients with multiple comorbid diseases. Implications for pay for performance. JAMA 2005; 294:716–724.
60. Masoudi FA. Statins for ischemic systolic heart failure editorial. N Engl J Med 2007; 357:22.
61. Beckett NS, Peters R, Fletcher AE, et al. Treatment of hypertension in patients 80 years of age or older. N Engl J Med 2008; 358:18.
62. Métry J-M, Meyer UA, eds. Drug Regimen Compliance: Issues in Clinical Trials and Patient Management. John Wiley and Sons, Chichester, 1999 [published online April 2002].
63. Thürmann PA, Werner U, Hanke F, et al. Arzneimittelrisiken bei hochbetagten Patienten: Ergebnisse deutscher Studien. In Bundesärztekammer (Hrsg): Fortbildungskompendium Fortschritt und Fortbildung in der Medizin", Band 31, 2007.
64. Fick DM, Cooper JW, Wade WE, et al. Potentially Inappropriate Medications for the Elderly According to the Revised Beers Criteria. Updating the Beers criteria for potentially inappropriate medication use in older adults: results of a US consensus panel of experts. Arch Intern Med 2003; 163:2716–2724.
65. Turnheim K. Drug dosage in the elderly. Is it rational? Drugs Aging 1998; 13:357–379.

Annex 1 Statistical Office of the European Commission

Total population (000s)					
	Population at 1 January			Growth since January 1, 2008 (%)	
	2008	2035	2060	2035	2060
EU27	495,394	520,654	505,719	5.1	2.1
Belgium	10,656	11,906	12,295	11.7	15.4
Bulgaria	7642	6535	5485	−14.5	−28.2
Czech Republic	10,346	10,288	9514	−0.6	−8.0
Denmark	5476	5858	5920	7.0	8.1
Germany	82,179	79,150	70,759	−3.7	−13.9
Estonia	1339	1243	1132	−7.2	−15.4
Ireland	4415	6057	6752	37.2	52.9
Greece	11,217	11,575	11,118	3.2	−0.9
Spain	45,283	53,027	51,913	17.1	14.6
France	61,876	69,021	71,800	11.5	16.0
Italy	59,529	61,995	59,390	4.1	−0.2
Cyprus	795	1121	1320	41.1	66.2
Latvia	2269	1970	1682	−13.2	−25.9
Lithuania	3365	2998	2548	−10.9	−24.3
Luxembourg	482	633	732	31.3	51.7
Hungary	10,045	9501	8717	−5.4	−13.2
Malta	410	429	405	4.5	−1.4
Netherlands	16,404	17,271	16,596	5.3	1.2
Austria	8334	9075	9037	8.9	8.4
Poland	38,116	36,141	31,139	−5.2	−18.3
Portugal	10,617	11,395	11,265	7.3	6.1
Romania	21,423	19,619	16,921	−8.4	−21.0
Slovenia	2023	1992	1779	−1.5	−12.1
Slovakia	5399	5231	4547	−3.1	−15.8
Finland	5300	5557	5402	4.9	1.9
Sweden	9183	10,382	10,875	13.1	18.4
United Kingdom	61,270	70,685	76,677	15.4	25.1
Norway	4737	5634	6037	18.9	27.4
Switzerland	7591	8798	9193	15.9	21.1

Annex 2 Statistical Office of the European Commission

	Older population							
	Percentage aged 65+			Percentage aged 80+			Old age dependency ratio (%)	
	2008	2035	2060	2008	2035	2060	2008	2060
EU27	17.1	25.4	30.0	4.4	7.9	12.1	25.4	53.5
Belgium	17.0	24.2	26.5	4.7	7.4	10.2	25.8	45.8
Bulgaria	17.3	24.7	34.2	3.6	7.1	12.8	25.0	63.5
Czech Republic	14.6	24.1	33.4	3.4	7.9	13.4	20.6	61.4
Denmark	15.6	24.1	25.0	4.1	7.7	10.0	23.6	42.7
Germany	20.1	30.2	32.5	4.7	8.9	13.2	30.3	59.1
Estonia	17.2	22.8	30.7	3.6	6.8	10.7	25.2	55.6
Ireland	11.2	17.6	25.2	2.8	5.0	9.6	16.3	43.6
Greece	18.6	26.3	31.7	4.1	7.9	13.5	27.8	57.1
Spain	16.6	24.8	32.3	4.6	7.2	14.5	24.2	59.1
France	16.5	24.4	25.9	5.0	8.5	10.8	25.3	45.2
Italy	20.1	28.6	32.7	5.5	9.1	14.9	30.5	59.3
Cyprus	12.4	19.0	26.2	2.8	5.3	8.6	17.7	44.5
Latvia	17.3	23.7	34.4	3.6	6.7	11.9	25.0	64.5
Lithuania	15.8	24.3	34.7	3.3	6.4	12.0	23.0	65.7
Luxembourg	14.2	21.3	23.6	3.5	5.8	8.9	20.9	39.1
Hungary	16.2	23.1	31.9	3.7	7.6	12.6	23.5	57.6
Malta	13.8	24.8	32.4	3.2	8.3	11.8	19.8	59.1
Netherlands	14.7	25.9	27.3	3.8	8.0	10.9	21.8	47.2
Austria	17.2	26.1	29.0	4.6	7.2	11.4	25.4	50.6
Poland	13.5	24.2	36.2	3.0	7.7	13.1	19.0	69.0
Portugal	17.4	24.9	30.9	4.2	7.6	12.8	25.9	54.8
Romania	14.9	22.9	35.0	2.8	6.2	13.1	21.3	65.3
Slovenia	16.1	27.4	33.4	3.5	8.4	13.9	23.0	62.2
Slovakia	12.0	23.0	36.1	2.6	6.4	13.2	16.6	68.5
Finland	16.5	26.4	27.8	4.3	9.4	10.8	24.8	49.3
Sweden	17.5	23.6	26.6	5.3	8.1	10.0	26.7	46.7
United Kingdom	16.1	21.9	24.7	4.5	6.7	9.0	24.3	42.1
Norway	14.6	22.6	25.4	4.6	7.1	10.0	22.1	43.9
Switzerland	16.4	25.2	28.0	4.7	7.7	11.1	24.1	48.5

40 Good Clinical Practice

Laura Brown
Laura Brown Training, Chalfont St Peter, Buckinghamshire, U.K.

INTRODUCTION

Good Clinical Practice (GCP) is a standard defining how clinical research should be carried out to ensure that the rights and confidentiality of patients are protected and that the data reported from clinical trials are credible and accurate.

HISTORY OF GCP

The origins of GCP are to be found in concerns about the abuse of human subjects. For example in the United States, in the 1950s, vulnerable subjects such as prisoners, disabled people, and pregnant women took part in studies where they were exposed to radioactive substances without their consent. In 1972, the Tuskegee Syphilis Study was exposed publicly. This study had followed the effects of untreated syphilis in 200 poor black Americans older than 40 years of age, denying them treatment although from 1943 the disease was easily treatable with penicillin (1).

Development of Codes of Ethics for Clinical Research

There had also been ethical concerns since the Nuremberg trials after the Second World War about the experiments carried out by the Nazis on the inmates of the concentration camps at Dachau, Auschwitz, Buchenwald, and Sachsenhausen. This led to the Nuremberg Code of Ethics (2) being formulated in 1947 by the American judges conducting the trial of Nazi physicians. It states

> "The duty and responsibility for ascertaining the quality of the consent rest upon each individual who initiates, directs, or engages in the experiment. It is a personal duty and responsibility which may not be delegated to another with impunity."

Later, this led to the World Medical Association (WMA) producing the Declaration of Helsinki which was adopted at the 18th World Medical Assembly in 1964. The Declaration of Helsinki has had a major impact on how clinical trials are carried out and many of the principles of GCP have their origin in this document.

The Declaration of Helsinki has been amended eight times, most recently in 2008 (3). It is the basis of the guidelines for carrying out biomedical research by the pharmaceutical industry and academic medicine. It is concerned with patient protection as exemplified by obtaining consent and having an independent ethical review of clinical research. The Declaration of Helsinki states in the Introduction that

> "..it is the duty of the physician to promote and safeguard the health of patients, including those involved in medical research."

The principles laid down by the WMA demanded that clinical research must contain the following: good science, sound animal experiments, morally responsible investigators, reasonable risk for subjects, respect for privacy, willingness to stop for hazards, integrity in reporting, and informed consent. These principles are still true of GCP today.

The Thalidomide and Elixir Sulfanilamide Tragedies

In 1962, the thalidomide tragedy resulted in at least 8,000 children being born and surviving with severe birth defects in 46 countries when their mothers had taken the drug during pregnancy (4). The pharmaceutical company, which supplied this drug, did not assess fully the adverse events reported.

Another precursor to the introduction of GCP was the development of a legal regulatory framework and in particular the 1938 Food, Drug and Cosmetic Act in the United States (5). The regulations and guidelines governing clinical trials carried out in the United States have evolved over many decades. The original Food and Drug Act became law in 1906 and required drugs to meet official requirements of strength and purity. There were no requirements concerning safety. It took a major tragedy in 1937 in the United States when "Elixir Sulfanilamide" was linked to over 100 deaths, mainly in children, to change this. The deaths were caused by diethylene glycol (antifreeze), which was used as a solvent in the "Elixir". As a result, the Federal Food Drug and Cosmetic Act was passed in 1938 requiring safety to be assessed by Food and Drugs officials before a drug could be licensed.

In 1962, when the thalidomide tragedy occurred, this was part of the motivation which prompted the US Food and Drug Administration (FDA) to seek to amend the Food, Drugs and Cosmetics Act of 1938 (*via* the Kefauver–Harris Amendments) to require proof of a drug's effectiveness in addition to safety. The United States thought they were exempt from the effects of thalidomide because approval for the dossier was being withheld by the FDA medical reviewer, Dr Frances Kelsey, who had requested further clinical studies. However, more than a 1000 physicians in the United States had used thalidomide. This failure of the U.S. Government to have known about and controlled unlicensed uses, as much as anything else, was what prompted the change in legislation. This act established procedures to gain control over clinical research into new drugs.

There was a similar regulatory aftermath in Europe in the 1960s with the adoption of new national legislation and the establishment of new committees to have oversight of safety and efficacy of medicines. The Committee on Safety of Drugs was set up in the United Kingdom, the College ter Beoordeling van geneesmiddelen (Committee for the Evaluation of Medicines) in the Netherlands, and the Bundesgesundheitsamt (Federal Health Office) in Germany (6). The first of the pharmaceutical Directives (65/65/EC) was adopted in January 1965 for the then six countries of the European Community (France, Federal Republic of Germany, Italy, Belgium, the Netherlands, and Luxembourg). The primary purpose of the Directive was to "safeguard public health".

Investigational New Drug Procedure in the United States

The amended 1938 Act introduced the Investigational New Drug (IND) procedure, which made it mandatory in the United States to notify the FDA before starting a clinical trial. The requirements for the submission of an IND to the FDA were as follows:

- Protocols for all proposed studies
- Identification and qualifications of proposed investigators
- Preclinical data
- Existing human clinical data
- Maintenance of adequate case histories
- Monitoring the progress of studies by the sponsor
- Report by sponsor of findings associated with the test drug.
- Immediate reporting of alarming (i.e., serious and unexpected) adverse events

Background to the United States Introducing GCP

The United States took the lead in introducing GCP because of deficiencies in carrying out clinical research and abuses to vulnerable patients (despite the Nuremberg Code of Ethics in Europe). Uncovering cases of fraud and drug tragedies such as thalidomide were also contributing factors. Although there are various regulations, which contain the principles of GCP in the United States, the term GCP is not used in U.S. legislation.

FDA Inspections and General Accounting Office Findings

The FDA required the sponsor company and the investigator to take on certain responsibilities. The introduction of regulations in 1963 required the investigator to sign Form FDA 1572 accepting legal responsibility for the health and human rights of patients.

The FDA has authority to inspect the sponsor's and investigator's facilities and records to ensure they are carrying out their responsibilities. In the 1970s, the U.S. General Accounting

Office (GAO) published two highly critical reports (in 1973 and 1976) concerning the FDAs inspections (7). The 1976 report concluded

> "As a result of inadequate monitoring of sponsors and clinical investigators, FDA lacks assurance that human test subjects are adequately protected and that the data upon which marketing approval is based is accurate and reliable."

Several cases of fraud were detected at this time and this resulted in an intensified inspection program called the "Bioresearch Monitoring Program" being introduced in 1977. This program was designed to assess ways in which research submitted in support of New Drug Applications (NDAs) could be improved. The following aspects of each study were assessed: design and conduct, including both the scientific value of the study, and the protection for subjects.

The First GCP Proposals

Further efforts were made by the FDA to improve the situation from 1977 and later by publishing proposed regulations in the Federal Register. These are often thought of as the first GCP guidelines.

The FDA proposals were:

- Proposed establishment of regulations on obligations of sponsors and monitors (8)
- Proposed obligations of clinical investigators of regulated articles (9)
- Proposed standards for Institutional Review Boards (IRBs) for clinical investigations (10). This was withdrawn and reissued as Protection of human subjects (11)

Although these proposals were not called GCP guidelines, they contained the elements of GCP and are frequently described as the first GCP guidelines. The proposals met with resistance from investigators and so the FDA decided that guidelines rather than regulations were more appropriate and so neither of the first two proposals became law although parts of the proposals were later included in the IND Rewrite in the *Code of Federal Regulations* (CFR 312 Subpart D) (12) which was proposed in 1982 and became effective in 1985. Compliance is monitored by the FDA in its Bioresearch Monitoring Program.

U.S. GCP REGULATIONS, GUIDELINES, AND INFORMATION SHEETS

In the United States, the term "GCP" is used to refer to a collection of regulations and guidelines that govern how clinical trials are carried out. Adherence to the requirements of GCP is essential if studies are to be used for registration of a new drug in the United States.

The main regulations, guidelines, guidance documents, and information sheets are described below.

Regulations

The CFR is the codification of the rules established by the Executive departments and agencies of the U.S. Federal Government. Title 21 of the CFR is reserved for rules of the FDA.

Those relating to GCP and clinical trials are as follows (9)

Parts 50 (13) and 56 (14)

Protection of Human Subjects; Informed Consent (*January 27, 1981*)

Protection of Human Research Subjects; Standards for Institutional Review Boards for Clinical Investigations (*January 27, 1981*)

Protection of Human Research Subjects; Clinical Investigations Which May Be Reviewed Through Expedited Review Procedure Set Forth in FDA Regulations (*January 27, 1981*)

Protection of Human Subjects; Informed Consent; Standards for Institutional Review Boards for Clinical Investigations (*November 1988*)

Federal Policy for the Protection of Human Subjects (*June 18, 1991*)

FDA Policy for the Protection of Human Subjects (*June 18, 1991*)

Protection of Human Subjects; Informed Consent; Proposed Rule (*September 21, 1995*)

Protection of Human Subjects; Informed Consent, Part II (*October 2, 1996*)

Protection of Human Subjects; Informed Consent and Waiver of Informed Consent Requirements in Certain Emergency Research; Final Rules (*October 2, 1996*)
Protection of Human Subjects; Informed Consent (*December 22, 1995*)
Protection of Human Subjects; Informed Consent Verification; Final Rule (*November 5, 1996*)
Categories of Research That May Be Reviewed by the Institutional Review Board (IRB) Through an Expedited Review Procedure (*November 9, 1998*)
Protection of Human Subjects; Informed Consent, Exception from general requirements (*October 5, 1999*)
Additional protections for children (*66 FR 20589–600, April 24, 2001*)
Exception from General Requirements for Informed Consent (*71 FR 32827, June 7, 2006*)

Part 54 (15)
Financial Disclosures by Clinical Investigators (*63 FR 5233, February 2, 1998*)
Financial Disclosures by a Clinical Investigator (*63 FR 72171–81, December 31, 1998*)

Part 210 (16)
Current Good Manufacturing Practice Regulations and Investigational New Drugs (*63 FR 5233, February 2, 1998*)

Parts 312 (12) and 314 (17)
Proposed New Drug, Antibiotic, and Biologic Drug Product Regulations (*June 9, 1983*)
New Drug and Antibiotic Regulations (*February 22, 1985*)
New Drug, Antibiotic, and Biologic Drug Product Regulations (*March 19, 1987*)
Investigational New Drug Applications and New Drug Applications (*September 8, 1995*)
Investigational New Drug Applications and New Drug Applications (*February 11, 1998*)
Disqualification of a Clinical Investigator (*February 16, 1996*)
Disqualification of a Clinical Investigator (*September 5, 1997*)
Expedited Safety Reporting Requirements for Human Drug and Biological Products (*October 7, 1997*)
Clinical Hold for products intended for life-threatening conditions (*65 FR 34963–71, June 1, 2000*)

Part 320 (18)
Retention of BE and BA testing samples (*April 28, 1993*)

Part 812 (19)
Quality System Regulations (*October 7, 1996*)
Treatment Use of Investigational Devices (*September 18, 1997*)
Withdrawal of Intraocular Lenses Regulation (Part 813) (*January 29, 1997*)
Disqualification of Clinical Investigators (*March 14, 1997*)
Modifications to the List of Recognized Standards (*July 12, 1999*)
Modifications to the Medical Device and/or Study Protocol (*November 23, 1998*)

Part 814 (20)
Medical Devices; Humanitarian Use Devices Part V (*June 26, 1996*)
30-Day Notices and 135-Day PMA Supplement Review (*October 8, 1998*)
Humanitarian Use of Devices (*November 3, 1998*)

Miscellaneous (21)
Part 11 Electronic Records; Electronic Signatures (*March 20, 1997*)

FDA Guidelines and Information Sheets

There are a large number of FDA guidelines, information sheets, and notices relating to GCP in FDA-regulated clinical trials (9). These guidances and information sheets represent the FDAs current view on GCP and the conduct of clinical trials and include the following:

FDA Information Sheet Guidances for IRBs, Clinical Investigators, and Sponsors (22)
Information intended to help IRBs, clinical investigators, and sponsors fulfil their responsibilities to protect human subjects is updated periodically.

FDA/NCI (National Cancer Institute) MOU (memorandum of understanding) Regarding Common Standards-based Data Repository
The purpose of this MOU (23) is to establish a formal collaboration between FDA and the National Cancer Institute (NCI) regarding the creation of a common standards-based data repository to facilitate the electronic exchange and analysis of data from research studies on investigational drugs in a fully secure manner *Federal Register* (April 18, 2007, Volume 72, Number 74).

Financial Relationships and Interests in Research Involving Human Subjects: Guidance for Human Subject Protection
This guidance (24) is for IRBs, investigators, research institutions, and other interested parties as announced in *Federal Register,* Vol. 69, No.92, May 12, 2004, Page 26393.

Guidance for Clinical Investigators, IRBs, and Sponsors Process for Handling Referrals to FDA under 21 CFR 50.54
This guidance (25) is intended to assist clinical investigators, IRBs, sponsors and other interested parties in understanding the FDAs process for handling clinical investigations that include children as subjects and that have been referred to the FDA for review under 21 CFR 50.54. *Federal Register* (December 22, 2006, Volume 71, pp. 77034–77035) (26)

Guidance for Clinical Trial Sponsors: Establishment and Operation of Clinical Trial Data Monitoring Committees
The guidance (27) is intended to assist sponsors of clinical trials in determining when a data monitoring committee (DMC) is needed for study monitoring and how such committees should operate. (*Federal Register* March 28, 2006; Volume 71, Number 59).

Guidance for Industry: Acceptance of Foreign Clinical Studies
This guidance (28) clarified the versions of the Declaration of Helsinki that applies to the FDAs acceptance of a foreign clinical study not conducted under an IND.

Guidance for Industry: Available Therapy
This guidance (29) is intended to provide guidance to industry on the meaning of the term "available therapy" as currently used by the Center for Drug Evaluation and Research (CDER) and the Center for Biologics Evaluation and Research (CBER) of the FDA.

Guidance for Industry: Collection of Race and Ethnicity Data in Clinical Trials
This guidance (30) recommends a standard approach for collecting and reporting race and ethnicity information in clinical trials conducted within and outside the United States for certain FDA regulated products. The recommended standardized approach was developed by the Office of Management and Budget (OMB). The guidance lists the OMB categories for race and ethnicity and describes the FDAs reasons for recommending the use of these categories. In addition, this guidance recommends a format for race and ethnicity data within study data that are submitted in standardized data sets such as the Study Data Tabulation Model or in the electronic Common Technical Document (eCTD).

Guidance for Industry: Computerized Systems Used in Clinical Investigations
This guidance (31) provides recommendations to sponsors, CROs, data management centers, clinical investigators and IRBs regarding the use of computerized systems in clinical investigations

Guidance for Industry: Development and Use of Risk Minimization Action Plans
The guidance (32) provides guidance to industry on risk management activities for drug products, including biological drug products, in the Center for Drug Evaluation and Research (CDER), and the Center for Biologics Evaluation and Research (CBER). The guidance addresses the development, implementation, and evaluation of risk minimization action plans for drug products.

Guidance for Industry Exploratory IND Studies
This guidance (33) describes the preclinical and clinical issues as well as chemistry, manufacturing, and controls information that should be considered when planning exploratory studies including studies of related drugs or biologics under an IND.

Guidance for Industry: Financial Disclosure by Clinical Investigators

This guidance (34) addresses questions received by the FDA concerning the implementation of the final rule on financial disclosure (21 CFR 54). It is intended to ensure that financial interests and arrangements of clinical investigators that could affect the reliability of data submitted to FDA are identified and disclosed to FDA by the applicant (sponsor).

Guidance for Industry: Food–Effect Bioavailability and Fed Bioequivalence Studies

This guidance (35) provides recommendations to sponsors and/or applicants planning to conduct food–effect bioavailability (BA) and fed bioequivalence (BE) studies for orally administered drug products as part of INDs, NDAs, abbreviated new drug applications (ANDAs), and supplemental applications.

Guidance for Industry: Good Pharmacovigilance Practices and Pharmacoepidemiologic Assessment

The guidance (36) addresses good pharmacovigilance practices and pharmacoepidemiologic assessment of observational data.

Guidance for Industry: Guideline for the Monitoring of Clinical Investigations

This guidance (37) considers monitoring of clinical trials.

Guidance for Industry: Guideline for the Study and Evaluation of Gender Differences in the Clinical Evaluation of Drugs

This guideline (38) presents guidance on the FDAs expectations regarding inclusion of both genders in clinical trials during drug development.

Guidance for Industry on Handling and Retention of Bioavailability and Bioequivalence Testing Samples

The guidance (39) is intended to clarify how to distribute test articles and reference standards to testing facilities, how to randomly select reserve samples, and how to retain reserve samples.

Guidance for Industry: IND Exemptions for Studies of Lawfully Marketed Drug or Biological Products for the Treatment of Cancer

This guidance (40) is intended to assist sponsors in deciding whether a study of marketed drugs or biological products for treating cancer falls within the exemption under § 312.2(b)(1) (21 CFR 312.2(b)(1)) from the general requirement to submit an IND application.

Guidance for Industry: Information Program on Clinical Trials for Serious or Life-Threatening Diseases and Conditions (draft guidance)

This draft guidance (41) is intended to assist sponsors who will be submitting information to the Clinical Trials Data Bank. It addresses statutory and procedural issues for submitting information to the data bank.

Guidance for Industry: IRB Review of Stand-Alone HIPAA Authorizations Under FDA Regulations

This guidance (42) provides clarification for IRBs of their responsibilities for reviewing and approving stand-alone authorizations under the Health Insurance Portability and Accountability Act 1996 (HIPAA) Privacy Rule.

Guidance for Industry on Part 11, Electronic Records; Electronic Signatures—Scope and Application

This guidance (21) is intended to describe the FDAs current thinking on the scope and application of Part 11 of Title 21 of the *CFR*; Electronic Records; Electronic Signatures (21 CFR Part 11).

Guidance for Industry and FDA Staff: General Principles of Software Validation (43)

Although the main purpose of this guidance (43) is for medical device software or the validation of software used to design, develop, or manufacture medical devices, it contains useful validation principles that can be applied to software used in the conduct of clinical trials with pharmaceuticals.

Guidance for Industry: Pharmacogenomic Data Submissions

The guidance (44) provides recommendations to sponsors holding INDs, NDAs, and biologics licence applications (BLAs) on what pharmacogenomic data to submit to the agency during the drug development process, the format of submissions, and how the data will be used in regulatory decision making. The guidance is intended to facilitate scientific progress in the area of pharmacogenomics.

Guidance for Industry: Premarketing Risk Assessment

The guidance (45) addresses premarket risk assessment.

Guidance for Industry: Providing Regulatory Submissions in Electronic Format-Human Pharmaceutical Product Applications and Related Submissions Using the eCTD Specifications

This guidance (46) considers issues related to the electronic submission of NDAs, ANDAs, BLAs, INDs, master files, advertising material, and promotional labeling using the electronic common technical document (eCTD) specifications.

Guidance for Industry and Clinical Investigators on the Use of Clinical Holds Following Clinical Investigator Misconduct

This guidance (47) for industry and clinical investigators provides information the FDAs authority to impose a clinical hold on a study or study site if it finds that subjects are or would be exposed to unreasonable and significant risk of illness or injury.

Guidance for Industry on Using a Centralized IRB Process in Multicenter Clinical Trial

This guidance (48) is intended to assist sponsors, institutions, IRBs, and clinical investigators involved in multicenter clinical research in meeting the requirements of 21 CFR Part 56 by facilitating the use of a centralized IRB review process (use of a single central IRB) especially in situations where centralized review could improve efficiency of IRB review.

Guidance for Sponsors, IRBs, Clinical Investigators and FDA Staff on Informed Consent for In Vitro Diagnostic Device Studies Using Leftover Human Specimens that are Not Individually Identifiable

FDA issued this guidance (49) to inform sponsors, IRBs, clinical investigators, and agency staff that the FDA intends to exercise enforcement discretion, under certain circumstances, with respect to its current regulations governing the requirement for informed consent when human specimens are used for FDA regulated in vitro diagnostic device investigations.

ICH Guidances

The following key ICH guidances are listed on the FDAs website:

- ICH E3: Guideline for Industry Structure and Content of Clinical Study Reports (50),
- ICH E5: Ethnic Factors in the Acceptability of Foreign Clinical Data (51),
- ICH E6: Good Clinical Practice: Consolidated Guidance (52), and
- ICH E10: Choice of Control Group and Related Issues in Clinical Trials (53).

Compliance Guidance Manuals

There are a number of compliance-program guidance manuals issued by the FDA (54) which describe procedures to help comply with FDA requirements and inspections. Those covering GCP compliance include

- In-vivo bioequivalence (55);
- Clinical investigators (56);
- Sponsor, Contract Research Organizations (CROs), and monitors (57); and
- IRBs (58).

Institutional Review Boards

CFR 21 Part 56 (14) specifies that all clinical trials must be approved by an IRB before they start. The contents of CFR 21 Part 56 are given in Table 1 An IRB as defined by U.S. law is

> "any board, committee, or other group formally designated by an institution to review, to approve the initiation of, and to conduct periodic review of, biomedical research involving human subjects. The primary purpose of such review is to assure the protection of the rights and welfare of human subjects."

IRBs have been in existence for many years and are familiar with FDA requirements and are liable for audit under the FDAs compliance program. They are accustomed to evaluating protocols, reviewing ongoing progress reports of studies and their responsibilities and working procedures are very similar to the requirements of ICH GCP.

The IRBs decisions must be given in writing. If a study is refused reasons for refusal should be given. The IRB has the right to suspend or stop a study at any time if it is thought

Table 1 Contents of FDA Regulation CFR 21 Part 56—IRBs, Dealing with Institutional Review Boards (14)

Subpart of Part 56	Contents
Subpart A—General Provisions	§ 56.101—Scope.
	§ 56.102—Definitions.
	§ 56.103—Circumstances in which IRB review is required.
	§ 56.104–Exemptions from IRB requirement.
	§ 56.105–Waiver of IRB requirement.
Subpart B—Organization and Personnel	§ 56.107–IRB membership.
Subpart C—IRB Functions and Operations	§ 56.108–IRB functions and operations.
	§ 56.109–IRB review of research.
	§ 56.110–Expedited review procedures for certain kinds of research involving no more than minimal risk and for minor changes in approved research.
	§ 56.111–Criteria for IRB approval of research.
	§ 56.112–Review by institution.
	§ 56.113–Suspension or termination of IRB approval of research.
	§ 56.114–Cooperative research.
Subpart D—Records and Reports	§ 56.115–IRB records.
Subpart E—Administrative Actions for Noncompliance	§ 56.120–Lesser administrative actions.
	§ 56.121–Disqualification of an IRB or an institution.
	§ 56.122–Public disclosure of information regarding revocation.

that a regulation concerning clinical research has been violated or the risk to the subjects has become unacceptable. Approval of a protocol or protocol amendment is by majority vote.

An IRB must have at least five members who are qualified appropriately through experience and expertise and diversity of the members, including consideration of race, gender, cultural backgrounds and, in particular, must be familiar with laws relevant to IRBs. This helps to promote respect for its advice in order to safeguard the welfare and rights of subjects.

IRBs must not consist solely of men or women or people from one profession. At least one member's primary concern must be in a nonscientific area. At least one member must not be affiliated to the institution or be closely related to a member who is. The IRB may seek advice from experts to assist the IRB members reviewing the study documentation but such experts would not vote with the IRB.

A list of IRB members identified by name, earned degrees, representative capacity, indications of experience such as board certifications, licences, etc., sufficient to describe each member's chief anticipated contributions to IRB deliberations and any employment or other relationship between each member and the institution–for example: full-time employee, part-time employee, a member of governing panel or board, stockholder, paid or unpaid consultant–should be available to the sponsor.

IRBs must functions as follows:

- maintain a written constitution and written procedures, including procedures for the regular review of research activity;
- retain and make available to the FDA for inspection, records relevant to the study until three years after completion of the study;
- report findings at properly convened meetings at which a quorum of the members should be present;
- determine how frequently a study should be reviewed if more than an annual review is needed;
- ensure timely reporting to the FDA of any serious or continuing noncompliance by investigators; and
- the IRB and the sponsor share responsibility for ensuring that adverse reactions are reported to the regulatory authority within the required time frame and suitably followed up.

Table 2 Contents of FDA Regulation CFR 21 Part 50—Dealing with Protection of Human Subjects (13)

Subpart of Part 50	Contents
Subpart A—General Provisions	§ 50.1–Scope. § 50.3 –Definitions.
Subpart B—Informed Consent of Human Subjects	§ 50.20–General requirements for informed consent. § 50.23–Exception from general requirements. § 50.24–Exception from informed consent requirements for emergency research. § 50.25–Elements of informed consent. § 50.27–Documentation of informed consent.
Subpart C [Reserved]	–
Subpart D—Additional Safeguards for Children in Clinical Investigations	§ 50.50–IRB duties. § 50.51–Clinical investigations not involving greater than minimal risk. § 50.52–Clinical investigations involving greater than minimal risk but presenting the prospect of direct benefit to individual subjects. § 50.53–Clinical investigations involving greater than minimal risk and no prospect of direct benefit to individual subjects, but likely to yield generalizable knowledge about the subjects' disorder or condition. § 50.54–Clinical investigations not otherwise approvable that present an opportunity to understand, prevent, or alleviate a serious problem affecting the health or welfare of children. § 50.55–Requirements for permission by parents or guardians and for assent by children.

There are some exceptions when an IRB may approve a consent procedure in which the requirement to obtain informed consent may be waived. The U.S. Department of Health and Human Services Regulations (HHS) at 45 CFR Part 46 permit an IRB to waive the requirements to obtain informed consent in the following circumstances (59):

- the research involves no more than minimal risk to the subjects;
- the research will not adversely affect the rights and welfare of the subjects; and
- wherever appropriate the subject will be provided with additional pertinent information after participation.

Informed Consent

The FDA guidelines (CFR 21 Part 50) (13) cover the protection of human subjects including informed consent. The contents of the regulation are given in Table 2.

The FDA guidelines (CFR 21 Part 50) (13), which cover the protection of human subjects, specify that the following items should be included in the informed consent information provided to subjects:

1. that the study involves research and the aim of the research, how long it will take and the procedures that the subject will have carried out;
2. the foreseeable risks and discomforts as well as the potential benefits;
3. alternative treatments;
4. that confidentiality of the subject's records will be maintained and that the subject's records may be inspected by the FDA;
5. that compensation and medical arrangements are available should injury occur;
6. that participation is voluntary and that the subject may withdraw from the study at any time without the penalty of loss of rights to continuing treatment; and
7. an explanation of whom to contact for answers to pertinent questions about the research and the research subjects' rights and whom to contact in the event of a research-related injury to the subject.

Additional elements of consent, which should be provided when appropriate are:

1. unforeseeable risk to the subject (including embryos and foetuses if the subject is or becomes pregnant);
2. circumstances where the subject's participation in the study may be stopped by the investigator;
3. the consequences of the subject withdrawing from the study;
4. any potential additional cost the subject may incur as a result of participating in the study;
5. that new findings will be provided if they could affect the willingness of the subject to continue in the study; and
6. the approximate number of subjects who will be participating in the study.

The informed consent form will be signed by the subject or his/her legally authorized representative and a copy given to the person signing the form. Oral consent is allowed provided that there is a witness to the procedure who will also sign the consent. In these circumstances, a "short form" written consent is used. The IRB must approve a written summary of what is to be said to the subject or his/her representative and the witness will also sign this. A copy of the summary as well as the signed consent form will be given to the subject or his/her representative.

Sufficient time for explanation and questions must be given before the consent is signed. The investigator is responsible for ensuring that the elements of informed consent and what is involved in the study have been understood by the subject or his/her representative.

There are certain circumstances where subjects are allowed to enter studies without giving consent. Such situations include

- life-threatening situations where the study drug is considered appropriate to use;
- subject's inability to communicate or to give legally effective consent;
- insufficient time to obtain consent from the subject's legal representative; and
- no alternative treatment likely to save the life of the subject.

The President May Waive Prior Consent

A rather interesting point is made in the CFR 21 Part 50 (13) concerning where the President may waive consent: Section 50.23: Exception from general requirements:

> "(d)(1) Under 10 USC 1107(f) the President may waive the prior consent requirement for the administration of an investigational new drug to a member of the armed forces in connection with the member's participation in a particular military operation. The statute specifies that only the President may waive informed consent in this connection and the President may grant such a waiver only if the President determines in writing that obtaining consent: Is not feasible; is contrary to the best interests of the military member; or is not in the interests of national security."

Exception from Informed Consent Requirements for Emergency Research

If obtaining informed consent is not feasible and a legally authorized representative is not available, the investigator will summarize efforts made to contact family members. If a legally authorized representative or family member becomes available, they should be informed about the clinical trial and when the subject's condition improves the subject is also to be informed as soon as feasible. If a subject is entered into a clinical investigation with waived consent and the subject dies before a legally authorized representative or family member can be contacted information about the clinical investigation is to be provided to the subject's legally authorized representative or family member if feasible.

Informed Consent Requirements to Enter Children for Research

The following key provisions should be in place when entering children into clinical trials in the United States:

- the IRB must take into account the ages, maturity and psychological state of the children involved;

- taking part in the clinical investigation may provide direct benefit to the child and that this treatment is otherwise not available to the child;
- the clinical investigation should involve no more than minimal risk to the subject; and
- permission is to be obtained from parents–both parents must give their permission unless one parent is deceased, unknown, incompetent or not reasonably available, or when only one parent has legal responsibility for the care and custody of the child if consistent with State law.

Investigator Responsibilities

The responsibilities of the investigator are described in the IND rewrite (12) (Table 3 below). Compliance is assured by FDA inspection.

The responsibilities of the investigator are to:

seek ethical approval for the study;
ensure the study is carried out according to the signed investigator statement (Form FDA 1572, discussed below), protocol and applicable regulations;
protect the rights, safety, and welfare of subjects;
maintain control of the study drug; and
obtain informed consent from subjects.

As part of carrying out studies under an IND, U.S. investigators must sign Form FDA 1572 (60) which is a legal document.By signing this form, the investigator accepts legal responsibility for the health and rights of the subject is obliged to read the investigator's brochure, submit the protocol and any amendments for approval by the IRB, and to be responsible for carrying out the study or supervise staffs who are participating in the study.

Investigators are also responsible for ensuring that the IRB is kept up-to-date about the progress of the study, report serious adverse reactions to the IRB, and sponsor and provide the sponsor with a final report shortly after the investigator has completed the study. Other responsibilities include maintaining drug dispensing records, ensuring the confidentiality of subject's records, and providing a current *curriculum vitae* to the sponsor.

The investigator must also agree to allow FDA inspection. Since such inspections will review all relevant study records as well as the facilities study documents will need to be kept so that they can be available for inspection.

Investigator Record Retention (Source CRF 21 312.16) (12)

An investigator should retain records required to for two years following the date a marketing application is approved for the drug for the indication for which it is being investigated or, if

Table 3 CFR 21 Part 312 – Responsibilities of Sponsors and Investigators (12)

Subpart of Part 312	Contents
Subpart D—Responsibilities of Sponsors and Investigators	§ 312.50–General responsibilities of sponsors. § 312.52–Transfer of obligations to a contract research organization. § 312.53–Selecting investigators and monitors. § 312.54–Emergency research under Section 50.24 of this chapter. § 312.55–Informing investigators. § 312.56–Review of ongoing investigations. § 312.57–Recordkeeping and record retention. § 312.58–Inspection of sponsor's records and reports. § 312.59–Disposition of unused supply of investigational drug. § 312.60–General responsibilities of investigators. § 312.61–Control of the investigational drug. § 312.62–Investigator recordkeeping and record retention. § 312.64–Investigator reports. § 312.66–Assurance of IRB review. § 312.68–Inspection of investigator's records and reports. § 312.69–Handling of controlled substances. § 312.70–Disqualification of a clinical investigator.

no application is to be filed or if the application is not approved for such indication, until two years after the investigation is discontinued and the FDA is notified.

Sponsor Responsibilities

The responsibilities of the sponsor are described in the FDA IND (12) and summarized in Table 3.

The responsibilities of the sponsor are to:

- select qualified investigators;
- provide the selected investigators with the information necessary to carry out the study properly;
- ensure that proper monitoring is carried out including employing appropriately trained monitors;
- ensure that the study is carried out according to the protocol;
- ensure that the investigator(s) and the FDA are informed of new adverse reactions or risks;
- ensure that the study site has sufficient and appropriate staff and facilities to carry out the study;
- maintain an effective IND;
- ensure the investigator has obtained IRB/ethical approval for the study;
- supply the investigational drug to GMP requirements and ensure that appropriate drug accountability records are maintained;
- collect and analyze the study data and produce a final study report;
- obtain financial disclosure information from the investigator. Sufficient accurate financial information is required to allow the sponsor to submit complete and accurate certification or disclosure statements required under Part 54 of the FDA requirements. The sponsor should obtain a commitment from the clinical investigator to update this information promptly if any relevant changes occur during the course of the investigation and for one year following the completion of the study; and
- where a CRO is carrying out any or all of the sponsor's responsibilities, a written contract describing such transfer of responsibilities must be prepared.

Sponsor Forms Required Under an IND: 1571 and 1572

There are two forms that are required under 21 CFR part 312 (12).

Form FDA-1571–"Investigational New Drug Application" (61)

A sponsor who intends to conduct a clinical investigation submits this form to the FDA. It includes

a. a cover sheet containing background information on the sponsor and investigator;
b. a table of contents;
c. an introductory statement and general investigational plan;
d. an Investigator's Brochure describing the drug substance;
e. a protocol for each planned study;
f. chemistry, manufacturing, and control information for each investigation;
g. pharmacology and toxicology information for each investigation; and
h. previous human experience with the investigational drug.

Form FDA-1572—"Investigator Statement" (60)

Before permitting an investigator to begin participation in an investigation the sponsor must obtain and record this form. It includes background information on the investigator and the investigation and a general outline of the planned investigation and the study protocol.

The Investigator's Brochure

An Investigator's Brochure is required if the planned investigation will occur at multiple sites and the sponsor is not a sponsor–investigator. Most academic investigator initiated studies that file for an IND are considered sponsor–investigator studies and will not require an Investigator's Brochure unless the product is supplied by a third party.

The IND regulations refer to the requirements for Investigator Brochure in two complementary sections and both sections apply (12):

- 21CFR312.23 (a) (5)
- 21CFR312.55.

21 CFR312.23 5 (12) states that the Investigator's Brochure should contain the following information:

1. a brief description of the drug substance and the formulation including the structural formula if known;
2. a summary of the pharmacological and toxicological effects of the drug in animals and, to the extent known, in humans;
3. a summary of the pharmacokinetics and biological disposition of the drug in animals and, if known, in humans;
4. a summary of information relating to safety and effectiveness in humans obtained from prior clinical studies. (Reprints of published articles on such studies may be appended when useful); and
5. a description of possible risks and side effects to be anticipated on the basis of prior experience with the drug under investigation or with related drugs and of precautions or special monitoring to be done as part of the investigational use of the drug.

21CFR 312.55 Informing Investigators
To comply with 21CFR 312.55 (12) regarding keeping investigators informed about the study drug the sponsor should keep participating investigators informed of new observations discovered by or reported to the sponsor on the study drug as the trial progresses, particularly regarding to safety findings. The sponsor can distribute this information to investigators by providing them with periodically revised investigator brochures (although reprints or published studies, reports, or letters to clinical investigators or other appropriate means may be used).

The Trial Protocol
In general, protocols for Phase 1 studies are less detailed and more flexible than those for Phase 2 and 3 studies.

Phase 1 protocols should primarily give an outline of the study and provide the following information:

- estimate of the number of patients;
- safety exclusions; and
- a description of the dosing plan including duration, dose or method to be used in determining dose.

A protocol for a Phase 2 or 3 study should enable the sponsor to anticipate that some deviation from the study design may become necessary as the investigation progresses and alternatives or contingencies to provide for such deviation should be built into the protocols at the start of the study.

A protocol should contain the following:

1. A statement of the objectives and purpose of the study.
2. The name and address and a statement of the qualifications (*curriculum vitae* or other statement of qualifications) of each investigator, the name of each subinvestigator (e.g., research fellow) working under the supervision of the investigator, the name, and address of the research facilities to be used and the name and address of each reviewing IRB.
3. The criteria for patient selection and for exclusion of patients and an estimate of the number of patients to be studied.
4. A description of the design of the study including the control group to be used (if applicable) and the methods to be used to minimize bias.
5. The method for determining the dose(s) to be administered, the planned maximum dosage, and the duration of individual patient exposure to the drug.

6. A description of the observations and measurements to be made to fulfil the objectives of the study.
7. A description of clinical procedures, laboratory tests, or other measures to be taken to monitor the effects of the drug in human subjects and to minimize risk.

Sponsor Record Retention

The sponsor should retain the records and reports for the clinical investigation for two years after a marketing application is approved for the drug or if the marketing application is not approved for the drug until two years after the final shipment and delivery of the drug for investigational use.

FDA Acceptance of Foreign Studies

Before 1975, studies conducted outside the United States were only acceptable as supportive data for an NDA. From 1975, the FDA published regulations (CFR 21 Part 312.120) (12), which accepted such studies provided they were as evidence of a major health gain, uncommon disease, or very favorable benefit to risk ratio. The FDA also required the investigator to:

- be well qualified and experienced;
- have adequate facilities to carry out the study;
- maintain detailed records including those of the sponsor (e.g., Case Report Forms (CRFs)); and
- carry out the study in accordance with the Declaration of Helsinki or the local regulations, whichever give the most protection).

It was not until 1985 when the NDA revision of CFR 21 Part 314.106 was published (17) that the FDA accepted non-US studies as the sole basis for drug approval, provided that

- the data was relevant to the U.S. population;
- studies were carried out by investigators of acknowledged competence; and
- the studies were audited by the FDA or by other appropriate means.

The FDA decides which studies are to be considered as pivotal. In practice, pharmaceutical companies normally carry out all studies to be included in an NDA according to US GCP requirements even if the study takes place outside the United States. Today, provided studies are carried out according to the requirements of ICH GCP, it is no longer necessary for studies outside the United States to be conducted under an IND although many sponsor companies still do so.

Amendment of Human Subject Protection; Foreign Clinical Studies Not Conducted Under an IND 312.20

The FDA published an amended rule in April 2008 (62) on acceptance of foreign clinical studies not conducted under an IND (non-IND foreign clinical studies) as support for an IND or application for marketing approval for a drug or biological product. The amended rule replaces the requirement that these studies be conducted in accordance with ethical principles stated in Declaration of Helsinki (Declaration) issued by the WMA with a requirement that the studies be conducted in accordance with GCP including review and approval by an independent ethics committee (IEC). This updates the standards for the acceptance of foreign clinical studies not conducted under an IND and helps ensure the protection of human subjects and the quality and integrity of data obtained from these studies. This rule was effective from October 27, 2008.

Monitor's Responsibilities

The guideline: "Guideline for the monitoring of clinical investigations" (first issued 1988) (37) details how monitoring should be carried out. This is one area in the United States where responsibilities of those involved in clinical research are not governed by the law. The monitor is the link between the sponsor and the investigator and ensures that many of the sponsor's responsibilities are carried out and that the investigator is meeting his/her responsibilities. The monitor visits the investigator site regularly before, during, and at the end of the study. The principles in the guideline are not legal requirements but represent a standard, which is acceptable to the FDA. The guideline stipulates that the sponsor should produce standard-written procedures to be used by monitors.

The monitor must be trained and qualified appropriately, although not necessarily in the therapeutic area under investigation, to be able to monitor the progress of studies. The guideline states that the following would be acceptable monitors depending on the area of study: clinical research associates, physicians, veterinarians, paramedical personnel, nurses, and engineers. The monitor will prepare a report of each visit to the investigator site. The report should include the date of the visit, the name of the monitor who visited the site, the name, and address of the investigator site and findings, conclusions, and any actions needed to deal with any deficiencies.

The monitor will carry out a prestudy visit to ensure that the investigator understands the studies carried out so far on the drug, the trial protocol, and his/her responsibilities including obtaining informed consent. The monitor will also ascertain whether the investigator has adequate facilities, time, and access to a sufficient number of subjects.

The monitor should visit the investigator site frequently during the study to ensure:

- compliance with the protocol;
- that the facilities continue to be suitable for the study;
- that the investigator has not delegated responsibilities to unspecified staff; and
- that reports are accurate, complete, and legible.

During monitoring visits, the monitor should carry out source document verification to verify that the study records are verifiable. The guideline does not refer explicitly to "source document verification". It states:

> "the monitor should compare a representative number of subject records and other supporting documents with the investigator's reports."

The monitor will also review the study records for omissions of data, missing visits, or examinations and reasons given for subjects who do not complete the study.

GCP Compliance: The FDA Bioresearch Monitoring Program

The FDA ensures that investigational studies are carried out according to GCP through inspection of sponsor and investigator sites and IRBs. Bioequivalence, Good Manufacturing Practice (GMP), and Good Laboratory Practice (GLP) inspections are also carried out. The majority of FDA inspections are investigator inspections. If the requirements of the regulations are not met the FDA will apply penalties. Pivotal (i.e., essential) studies carried out under an IND are always inspected. Inspections will also be carried out where inconsistencies or discrepancies are detected in study data or the Agency becomes aware of other issues of concern. This type of inspection is called a "for cause" audit.

The FDA has a Scientific Investigations Division, which is responsible for the compliance program and consists of four subdivisions:

- the Clinical Investigation Branch;
- the Nonclinical Investigations Branch;
- the Institutional Review Branch; and
- the Regulatory Management Branch (this is involved when criminal activity is suspected).

The FDAs compliance program is the FDA Bioresearch Monitoring Program and was set up in 1977, although the FDA has followed a compliance program guidance manual since 1974. The guidance manual (63) consists of more than 32 instructional guides on how to conduct various types of inspections and managing the reporting of inspections. The objectives of the program are:

- to assess adherence to guidelines and assure adherence to the FDA regulations;
- to determine validity of specific studies in support of products pending approval by the FDA; and
- to determine that the rights and safety of subjects participating in clinical trials have been properly protected.

Types of Inspection

The inspections are classified as:

1. study-orientated—covers data verifications for approval of pending submission, pivotal studies, and foreign studies; and
2. investigator-orientated—results from prior knowledge or suspicion of improper study conduct by an investigator.

There are many aspects of the US GCP requirements which have been incorporated into the ICH GCP guideline. This was due to the FDAs representatives exerting great influence on the final content of the guideline. As a result, the final ICH GCP guideline is similar to the U.S. regulations and requirements. The FDA has published the final ICH GCP guideline in the *Federal Register* and so it is an official guideline. However, the ICH document has not replaced the existing FDA requirements nor has it been incorporated into U.S. legislation. It is believed the ICH GCP guideline is consistent with the FDA GCP requirements (which is not surprising considering the influence of the FDA in the process that resulted in the production of the ICH document). Therefore, studies in the United States will continue to be carried out to the FDA requirements.

FDA officials have stated that studies conducted outside the United States and/or studies not conducted under an IND for submission as part of an NDA/product licence application should comply with ICH GCP and so studies carried out to the ICH GCP guideline should be acceptable for an NDA.

NATIONAL AND REGIONAL GCP GUIDELINES AND REGULATIONS

While the concepts of GLP and GMP were quite readily accepted internationally, this was not the case with GCP. In the mid-1980s, GCP guidelines and regulations were developed based on the original FDA proposals but amended to reflect local requirements and cultures. One of the first GCP guidelines to be developed in Europe was the Nordic Guidelines issued in 1989 (102). These guidelines were produced as collaboration by the Council on Medicines in Sweden with the drug regulatory authorities of Denmark, Finland, Iceland, and Norway and have had a significant influence on the development of other GCP guidelines (prior to ICH) including as far away as Australia.

The EU CPMP GCP guidelines (64) were published in July 1990 and took effect from July 1991. These guidelines attempted to harmonize GCP in the European Community. They were developed from existing guidelines including the UKs Association of the British Pharmaceutical Industry (ABPI) guidelines on GCP issued in 1981. The Commission Directive 91/507/EEC (65) was issued on July 19, 1991 and provided a legal basis for GCP with effect from January 1992. In 2001, the Clinical Trial Directive (Directive 2001/20/EC) was published (66) and was implemented in most countries in the EU in 2004. This provided a legal standard for carrying out clinical research in the EU.

With the large number of national and regional GCP guidelines and regulations which existed in the 1980s, the lack of consistency was recognized as an issue for efficient global drug development. To overcome this, GCP became part of the ICH process.

INTERNATIONAL HARMONIZATION OF GCP

The ICH GCP guideline (52) represents the most influential initiative to harmonize GCP between the EU, Japan, and the United States. It was finalized on the May 1, 1996 as the "Good Clinical Practice: Consolidated Guideline" but because of some minor editorial modifications the members of the ICH Expert Working Party responsible for developing the guideline distributed a letter dated June 20, 1996 reissuing the guideline. Apart from changing the name to "Guideline for Good Clinical Practice," improving the presentation of the guideline and changing a few words there are no major changes to the content.

Finalization and Adoption of the ICH GCP Guideline in the European Union, United States, and Japan

It was originally hoped that the ICH GCP guideline would be finalized at the Third International Conference on Harmonization (ICH 3) held at Yokohama, Japan in 1995 but there were several issues to be resolved and the working group deliberately avoided compromises in favor of negotiation and agreement to produce the best possible document. It was particularly difficult working with different regulatory requirements, legal systems, medical practices, administrative procedures, and cultures.

The guideline was approved on the July 17, 1996 by the Committee for Proprietary Medicinal Products (CPMP), the European Community's scientific committee, for implementation from the January 17, 1997. Clinical trials carried out in Europe after this date should have complied with the ICH GCP guideline. Japan implemented the guideline in 1997. Although the FDA intends it to become an official guideline, there are no plans to incorporate it into U.S. legislation. It was published in the Federal Register in May 1997 (67).

The ICH GCP guideline incorporated addenda on the Essential Documents and on the Investigator's Brochure, which had been agreed earlier through the ICH process in 1994 and 1995 respectively. There is provision for amendments to be made to the GCP guideline by each of the three regions to reflect any regional or national differences. However it is hoped, if this facility is used, that any such territory-specific modifications will be kept to a minimum to maintain the consistency of the guideline.

Evolution of the ICH GCP Guideline

The original plan at the start of the ICH process was to produce a series of ten guidelines, each covering a specific GCP topic (including investigator responsibilities, sponsor responsibilities, essential documents, Investigator's Brochures, ethics committees, informed consent, monitoring, final report verification, and auditing). However, after discussion by the members of the ICH Expert Working Party responsible for compiling the guidelines, and because the WHO GCP guideline was finalized at this time as a single GCP guideline, it was decided that the ICH GCP guideline would be more useful to users if each subject was included as a section in a single GCP guideline. As the "Essential Documents" and "Investigator's Brochure" guidelines were already at an advanced stage of preparation and were being prepared separately from the other GCP topics these were included as addenda to the main GCP guideline. The ICH GCP guideline was called a "Consolidated GCP guideline" to reflect this approach. As a result, there are a few inconsistencies between the two documents. For example, the main GCP guideline refers to the investigator documenting the responsibilities of personnel involved with the trial at a site but this document is not mentioned in the "Essential Documents" guideline.

Structure and Content of the ICH GCP Guideline

The ICH GCP guideline (52) is reasonably easy to use, well thought out, and logical. It provides the main users of the guideline with a good understanding of their responsibilities necessary to ensure that the appropriate GCP procedures are carried out and relevant documents are used for clinical trials. After a comprehensive glossary (Section 1), there is a detailed list of 13 principles of GCP (Section 2) which gives general instructions for carrying out clinical trials to GCP (Table 4 below).

One of the aims of the guideline is to harmonize the terminology for those involved with the process and, like the European CPMP GCP guidelines (64), it contains a glossary at the beginning to help the reader with the main body of the document. The glossary (Section 1) contains the term "Independent Ethics Committee" (IEC) (Section 1.27) instead of the normally used "ethics committee." This is to emphasize the independent nature of ethics committee's work. Unlike IRBs, IECs are not normally accountable to any regulatory body. It is also for this reason that the guideline has not used a single term to describe IECs and IRBs.

The next sections (Sections 3–5) are concerned with the activities and responsibilities of the main target groups (IRBs and IECs, the investigator and the sponsor) that the guideline applies to and contains sections concerning the monitor, auditor and CROs. There was more detailed guidance for these target groups than there is in pre-existing GCP guidelines and regulations. There is a suggested structure and layout for protocols and amendments (Section 6) followed by the addenda on the Investigator's Brochure and Essential Documents (Sections 7 and 8).

Although the structure of the ICH GCP guideline was new the concept was consistent with that of previous GCP guidelines and regulations. It was developed taking into account GCP in the EU, Japan, the United States, Australia, Canada, the Nordic countries, and the WHO.

Table 4 The Principles of ICH GCP (52)

Principle Number	Statement
1	Clinical trials should be conducted in accordance with ethical principles that have their origin in the Declaration of Helsinki and that are consistent with GCP and the applicable regulatory requirement(s).
2	Before a trial is initiated, foreseeable risks and inconveniences should be weighed against the anticipated benefit for the individual trial subject and society. A trial should be initiated and continued only if the anticipated benefits justify the risks.
3	The rights, safety, and well being of the trial subjects are the most important considerations and should prevail over interests of science and society.
4	The available nonclinical and clinical information on an investigational product should be adequate to support the proposed clinical trial.
5	Clinical trials should be scientifically sound, and described in a clear, detailed protocol.
6	A trial should be conducted in compliance with the protocol that has received prior IRB/IEC approval/favorable opinion.
7	The medical care given to, and medical decisions made on behalf of, subjects should always be the responsibility of a qualified physician or, when appropriate, of a qualified dentist.
8	Each individual involved in conducting a trial should be qualified by education, training, and experience to perform his or her respective task(s).
9	Freely given informed consent should be obtained from every subject prior to clinical trial participation.
10	All clinical trial information should be recorded, handled, and stored in a way that allows its accurate reporting, interpretation, and verification.
11	The confidentiality of records that could identify subjects should be protected, respecting the privacy and confidentiality rules in accordance with the applicable regulatory requirement(s).
12	Investigational products should be manufactured, handled, and stored in accordance with applicable GMP. They should be used in accordance with the approved protocol.
13	Systems with procedures that assure the quality of every aspect of the trial should be implemented.

This created the greatest potential for future adoption of the guideline in countries outside the three regions of Europe, Japan and the United States.

The FDA, in particular, has had a great influence on the development of the guideline. There are several examples where sections of the guideline originate from FDA GCP requirements. For example, the section concerning sponsor delegation to CROs in Section 5.2 is taken from the FDA regulations given in the CFR 21 Part 312 IND Section 312.52 (12) "Transfer of obligations to a CRO."

There were several new requirements in the chapters referring to IRBs/IECs, investigators and sponsors. The following sections consider some of these.

Informed Consent (Section 4.8.10)

There is a comprehensive list of 20 items to include in the patient consent information, which is more than in the previous European GCP guidelines. This is included in the section on investigators because it is the investigator's responsibility to obtain consent and because investigators in the United States and Japan often compile the informed consent documentation. However it is frequently the sponsor who prepares the informed consent information and will therefore have to ensure that these items are included. The items to include are:

1. to state that the trial involves "research,"
2. purpose of the trial;
3. trial treatments and the probability of the subject being assigned to each treatment;
4. trial procedures including invasive procedures;
5. subject's responsibilities;
6. aspects of the trial that are experimental;
7. the possible foreseeable risks or inconveniences;
8. the expected benefits; if there are none this should be stated;

9. alternative treatments or procedures and their potential benefits and risks;
10. compensation and/or treatment available in the event of injury related to the trial;
11. prorata payment detail, if any;
12. anticipated expenses, if any;
13. that participation is voluntary and the subject may withdraw at any time without penalty or loss of benefits;
14. the signing of informed consent authorizes access to the patient's notes by appropriate sponsor personnel, IRB/IEC and regulatory authority(ies);
15. that the identity of the subject will be kept confidential;
16. that subjects will be informed of any new information that is relevant to the subject's willingness to continue in the trial;
17. details of who to contact for further information about the trial;
18. circumstances and/or reasons the subject will stop participation in the trial;
19. how long the subject will participate in the trial; and
20. number of subjects (approximately) who will participate in the trial.

Independent Data-Monitoring Committee (Section 5.5.2)
Sponsors should consider setting up an

> "Independent Data-Monitoring Committee to assess at intervals the progress of a clinical trial, the safety data, and the critical efficacy endpoints, and to recommend to the sponsor whether to continue, modify, or stop a trial."

The committee is to be used primarily for large multicenter trials, especially where only one study is carried out as pivotal for a marketing licence. The members of this committee must not be investigators participating in the study. Many sponsors currently set up such committees and including this in the guideline reflects current practice.

Electronic Data Handling (Section 5.5.3)
The section of the guideline provides instruction on using electronic trial data handling and/or remote electronic trial data systems.

- validation procedures should be in place and documented;
- SOPs for using these systems should be available;
- an audit trial must be maintained to document any changes which are made;
- a security system to prevent unauthorized access should be maintained;
- a list of personnel authorized to make changes must be maintained;
- adequate back up should be maintained; and
- blinding should be subjected to safeguards (if the study is blinded).

Release of Study Drug (Section 5.14.2)
The trial drug should not be released to the study site until the sponsor has obtained "required documentation" such as IRB/IEC and regulatory approval. Several pharmaceutical companies' existing SOPs include such requirements. This procedure should prevent investigators administering study drugs before the appropriate approvals have been given.

Qualified Monitors and Auditors (Sections 5.18.2 and 5.19.2)
The guideline says that monitors and auditors should be qualified appropriately and asks for their qualifications to be documented. Qualifications of clinical research personnel vary enormously: personnel include physicians, nurses, Ph.Ds, science graduates, and sometimes secretaries with pharmaceutical industry experience who have been promoted to monitors.

Training Courses for Monitors, Auditors etc
Many countries recognize the need for qualifications and a number of initiatives exist [e.g., the postgraduate Diploma in Clinical Science from the University of Wales which has been in operation for over 20 years and the American Association of Clinical Research accreditation system for clinical research associates (CRAs)]. The implementation of GCP requirements has

meant introducing GCP training and although there is a plethora of GCP training initiatives there are no international qualifications or recognized standards for monitors or auditors. Several of the larger pharmaceutical companies and CROs have developed their own internal accreditation and/or competence schemes for CRAs and monitors to meet the requirements of ICH GCP as well as their own internal training needs.

Other personnel involved with clinical research should also be qualified for such a role. For example the guideline states that personnel involved in data handling, verification of data, and statistical analysis should be "appropriately qualified individuals" (Section 5.5.1). For example, Kingston University, United Kingdom, runs a postgraduate course in Clinical Data Management.

Clinical Study Report (Section 5.22)

The mutual recognition of clinical reports in the three regions was one of the most important issues addressed by the harmonization process. The ICH GCP guideline says "clinical reports should meet the standards of the 1995 ICH guideline for Structure and Content of Clinical Study Reports" (68) so that differences between the three regions should have been eliminated.

Protocol and Amendments (Section 6)

This section of the guideline (Section 6) gives topics to include in the protocol. These were similar to those given in the previous CPMP GCP guidelines used in Europe but included one significant addition. Section 6.4.9 makes provision for the "identification of any data to be recorded directly on the CRF...." to be documented in the protocol. This should help ensure that CRFs reflect accurately the requirements of protocols. This section covers items frequently included in protocols. The topics covered are:

- general information (e.g., protocol title, sponsor and monitor details, other relevant names and address such as laboratories);
- background information (e.g., a summary of the nonclinical data, description of the population to be studied);
- trial objectives and purpose;
- trial design;
- selection and withdrawal of subjects;
- treatment of subjects;
- assessment of efficacy;
- assessment of safety;
- statistics;
- direct access to source data;
- quality control and quality assurance;
- ethics;
- data handling and record keeping;
- finance and insurance;
- publication policy; and
- Supplements. (The guideline does not specify what this covers. However, sponsors frequently include a copy of the patient informed consent information, a copy of the Declaration of Helsinki (for European studies), investigator GCP responsibilities, and any study-specific documents.)

There is no specific information concerning protocol amendments. It would have helped to standardize how these are prepared if a suggested template for preparing amendments had been given in the guideline.

Investigator's Brochure (Section 7)

The addendum on the Investigator's Brochure details the minimum information that the Investigator's Brochure should contain (Section 7) and suggests how it should be structured. It contains a summary of the nonclinical studies and clinical data on the investigational product (i.e., the study drug), which is relevant to the study of the drug in human subjects. Its purpose is to provide the investigator(s) and others involved in the study with information to help them understand the background to the development of the drug, including the rationale for the study and the safety profile of the drug as known. The Investigator's Brochure should be reviewed at least annually and in accordance with GCP. If any important new information about the drug is communicated to the investigator(s) or IRB/IEC this should be included in the Investigator's Brochure when it is revised.

The list of suggested contents for inclusion in an Investigator's Brochure is as follows:

Title page
Confidentiality statement
Signature Page
1.0 Table of Contents
2.0 Summary
3.0 Introduction
4.0 Physical, Chemical, and Pharmaceutical Properties and Formulation
5.0 Nonclinical Studies
5.1. Nonclinical Pharmacology
5.2 Pharmacokinetics and Product Metabolism in Animals
5.3 Toxicology
6.0 Effects in Humans
6.1 Pharmacokinetics and Product Metabolism in Humans
6.2 Safety and Efficacy
6.3 Marketing Experience
Summary of Data and Guidance for the Investigator
Appendices (if any).

The sponsor is responsible for ensuring the investigator(s) is provided with an up-to-date Investigator's Brochure. If the study drug is marketed and the pharmacology is well understood by medical practitioners, provided regulatory authorities will allow, an alternative to the Investigator's Brochure such as a product information brochure, package leaflet, or labeling information may be used.

Essential Documents (Section 8)

Section 8, "Essential documents for the conduct of a clinical trial" describes a list of 50 or so documents which are the minimum and which will allow evaluation of the conduct of a trial and the quality of the data produced to comply with GCP. The various documents are grouped into three sections according to the stage of the trial at which they will be needed: before the start of the study, during the study, and at the end of the study. These documents are usually filed in the trial master file, which will be set up for both the investigator and the sponsor to file at the beginning of the study.

Most of the documents will be filed by both the sponsor and investigator. The differences are as follows:

Documents only the sponsor will retain:

- study drug sample label(s) attached to treatment container;
- master randomization list (to document the method of randomization of the trial subjects);
- certificate of analysis of the study drug and for new batches of drug used during the trial;
- monitoring reports including the pre-trial and final monitoring report;
- audit certificate (if available); and
- treatment allocation and decoding documentation.

Documents only the investigator will retain:

- subject enrolment log;
- signed informed consent forms;
- confidential subject identification code list (auditors prefer this to be completed during the trial, rather than at the end); and
- source documents.

It is likely the ICH GCP guideline will be reviewed and updated and this would allow for such inconsistencies to be eradicated.

The ICH process represented a landmark in the progress of achieving greater harmonization of the technical guidelines and requirements for product registration of new drugs. Enormous efforts by representatives from the three regions have been made to reach consensus in order to resolve many technical issues so that harmonization has been increased since it was introduced.

EU CLINICAL TRIALS DIRECTIVE FOR GCP AND CLINICAL TRIALS

The EU Clinical Trials Directive (Directive 2001/20/EC) (66) was adopted in April 2001. This has been the most important piece of legislation to implement GCP and regulate clinical trials in Europe and has given rise to a framework of GCP related documents and includes the following, which are included in Volume 10 Clinical Trial Guidelines of "The rules governing medicinal products in the European Union" (69):

- CT Directive 2001/20/EC (April 4, 2001) (66)
- GCP Directive 2005/28/EC (April 8, 2005) (70)
- CT guidelines on the application for clinical trial
    - Detailed guidance for the request for authorization of a clinical trial on a medicinal product for human use to the competent authorities, notification of substantial amendments, and declaration of the end of the trial (71)
    - Detailed guidance on the application format and documentation to be submitted in an application for an Ethics Committee opinion on the clinical trial on medicinal products for human use (72)
    - Detailed guidance on the European Clinical Trials database (EUDRACT) (first revision, April 2004) (73)
- CT guidelines on Monitoring and Pharmacovigilance
    - Detailed guidance on the collection, verification, and presentation of adverse drug reports arising from clinical trials on medicinal products for human use (April 2006) (74)
    - Detailed guidance on the European database of Suspected Unexpected Serious Adverse Reactions (Eudravigilance–Clinical Trial Module) (April 2004) (75)
- CT guidelines on GCP Inspections
    - Guidance for the preparation for GCP inspections (June 2008) (76)
    - Recommendations on inspection procedures for the verification of GCP compliance (July 2006) (77)
    - Guidance for the conduct of GCP inspections (June 2008) (78). This guidance has Annexes dealing with different types of GCP inspections such as investigator sites, clinical laboratories, computer systems, sponsor and CRO, Phase I units, bioanalytical laboratories
- Other CT guidelines
    - Ethical considerations for clinical trials on medicinal products conducted with the paediatric population (2008) (79) and
- GMP Directive 2003/94/EC (October 2003) (80)

In addition, Annex 13 of Volume 4 of the publications "The rules governing medicinal products in the European Union": Good manufacturing practices of Manufacture of investigational medicinal products was adopted in July 2003 (81).

CONCLUSION

From the dreadful abuses of vulnerable subjects and drug development tragedies such as thalidomide and the discovery of incidents of fraud an ethical and scientific quality standard has evolved for carrying out and reporting clinical research. This is GCP.

GCP originated in the United Stataes and was quickly taken up by other nations resulting in a plethora of guidelines and regulations. With the adoption of the ICH GCP we now have a standard, which has helped facilitate the mutual acceptance of clinical data and to harmonize GCP throughout the world. However, it is still necessary to follow any additional country specific GCP requirements such as those from the transposition of the EU Clinical Trial Directive (2001/20/EC) into European Community Member State legislation and the FDA requirements.

REFERENCES

1. Jones J. Bad blood: The Tuskegee Syphilis Experiment: A Tragedy of Race and Medicine. New York, NY: The Free Press, 1981.
2. Trials of War Criminals Before the Nuremberg Military Tribunals Under Control Council Law No. 10. Vol. 2. Nuremberg, Germany; Washington, DC: US Government Printing Office, 1949:181–182.
3. World Medical Association Declaration of Helsinki—Ethical Principles for Medical Research Involving Human Subjects. http://www.wma.net/policy/pdf/17c.pdf. Accessed December 26, 2008.
4. Sjöstrom H, Nilsson R. Thalidomide and the Power of the Drug Companies. Penguin Special, Penguin Books Ltd., 1972.
5. The 1938 Food, Drug and Cosmetic Act. http://www.fda.gov/oc/historyoffda/section2.html. Accessed December 26, 2008.
6. Cartwright AC. Introduction and history of pharmaceutical regulation. In: Cartwright AC, Matthews BR, eds. In Pharmaceutical Product Licensing: Requirements for Europe. Ellis Horwood, 1991.
7. Farrell F. Drug regulation in the 1990s. In: Lloyd J, ed. ACRPI Handbook of Clinical Research. 1994.
8. Proposed Establishment of Regulations on Obligations of Sponsors and Monitors. Fed Reg 1977; 42:49612.
9. Proposed Obligations of Clinical Investigators of Regulated Articles. Fed Reg 1978; 43:35210.
10. Proposed Standards for Institutional Review Boards for Clinical Investigations. Fed Reg 1978; 43:35186.
11. Protection of Human Subjects. Fed Reg 1979; 44:47688.
12. Code of Federal Regulations,Title 21—Food and Drugs, Chapter 1—Food and Drug Administration, Department of Health and Human Services, Subchapter D—Drugs for Human Use, Part 312C, Investigational New Drug Applications. http://www.accessdata.fda.gov/scripts/cdrh/cfdocs/cfcfr/CFRSearch.cfm?CFRPart=312&showFR=1&subpartNode=21:5.0.1.1.3.4. Accessed December 26, 2008.
13. Code of Federal Regulations, Title 21—Food and Drugs, Chapter I—Food and Drug Administration, Department of Health and Human Services, Subchapter A—General, Part 50 Protection of Human Subjects. https://www.accessdata.fda.gov/scripts/cdrh/cfdocs/cfCFR/CFRSearch.cfm?CFRPart=50. Accessed December 26, 2008.
14. Code of Federal Regulations, Title 21—Food and Drugs, Chapter I—Food and Drug Administration, Department of Health and Human Services, Subchapter A—General, Part 56 Institutional Review Boards. http://www.accessdata.fda.gov/scripts/cdrh/cfdocs/cfcfr/CFRsearch.cfm?CFRPart=56. Accessed December 26, 2008.
15. Code of Federal Regulations, Title 21—Food and Drugs, Chapter I—Food and Drug Administration, Department of Health and Human Services, Subchapter A—General, Part 54 Financial Disclosure by Clinical Investigators. http://www.accessdata.fda.gov/scripts/cdrh/cfdocs/cfCFR/CFRSearch.cfm?CFRPart=54. Accessed December 26, 2008.
16. Code of Federal Regulations, Title 21—Food and Drugs, Chapter I—Food and Drug Administration, Department of Health and Human Services, Part 210 Current Good Manufacturing Practice In Manufacturing, Processing, Packing, or Holding of Drugs; General Part 211 Current Good Manufacturing Practice for Finished Pharmaceuticals. http://www.fda.gov/cder/dmpq/cgmpregs.htm. Accessed December 26, 2008.
17. Code of Federal Regulations, Title 21—Food and Drugs, Chapter I—Food and Drug Administration, Department of Health and Human Services, Subpart D—Drugs for Human Use, Part 314 Applications for Marketing a New Drug. http://www.accessdata.fda.gov/scripts/cdrh/cfdocs/cfcfr/CFRsearch.cfm?CFRPart=314. Accessed December 26, 2008.
18. Code of Federal Regulations, Title 21—Food and Drugs, Chapter I—Food and Drug Administration, Department of Health and Human Services, Subpart D—Drugs for Human Use, Part 320 Bioavailability and Bioequivalence Requirements. http://www1.va.gov/oro/apps/compendium/Files/21CFR320.htm. Accessed December 26, 2008.
19. Code of Federal Regulations, Title 21—Food and Drugs, Chapter I—Food and Drug Administration, Department of Health and Human Services, Subpart H—Medical Devices, Part 812 Investigational Device Exemptions. https://www.accessdata.fda.gov/scripts/cdrh/cfdocs/cfCFR/CFRSearch.cfm?CFRPart=812. Accessed December 26, 2008.
20. Code of Federal Regulations, Title 21—Food and Drugs, Chapter I—Food and Drug Administration, Department of Health and Human Services, Subpart H—Medical Devices, Part 814 Premarket Approval of Medical Devices. http://www.accessdata.fda.gov/scripts/cdrh/cfdocs/cfcfr/CFRSearch.cfm?;CFRPart=814. Accessed December 26, 2008.
21. Code of Federal Regulations 21 CFR Part 11, Electronic Records; Electronic Signatures. http://www.fda.gov/ora/compliance_ref/Part11/ Title 21. Accessed December 26, 2008.

22. US Food and Drug Administration, Guidances, Information Sheets, and Important Notices on Good Clinical Practices in FDA-Regulated Clinical Trials. http://www.fda.gov/oc/gcp/guidance.html. Accessed December 26, 2008.
23. Memorandum of Understanding between the National Cancer Institute and the Food and Drug Administration on Janus Study Data Depository, FDA 225.07.4301. http://www.fda.gov/OHRMS/DOCKETS/98fr/FDA-225–07–4301-mou0001.PDF. Accessed December 26, 2008.
24. Department of Health and Human Services, Final Guidance Document, Financial Relationships and Interests in Research Involving Human Subjects: Guidance for Human Subject Protection. http://www.hhs.gov/ohrp/humansubjects/finreltn/fguid.pdf. Accessed January 26, 2008.
25. FDA Guidance for Clinical Investigators, Institutional Review Boards and Sponsors Process for Handling Referrals to FDA Under 21 CFR 50.54–Additional Safeguards for Children in Clinical Investigations. http://www.fda.gov/OHRMS/DOCKETS/98fr/06d-0172-gdl0002.pdf. Accessed January 26, 2008.
26. Guidance for Clinical Investigators, Institutional Review Boards, and Sponsors; Process for Handling Referrals to Food and Drug Administration Under 21 CFR 50.54: Additional Safeguards for Children in Clinical Investigations, Federal Register: December 22, 2006 (Volume 71, Number 246)], [Notices] [Page 77034–77035]. http://www.fda.gov/OHRMS/DOCKETS/98fr/E6–21950.htm. Accessed January 26, 2008.
27. FDA Guidance for Clinical Trial Sponsors Establishment and Operation of Clinical Trial Data Monitoring Committees. http://www.fda.gov/OHRMS/DOCKETS/98fr/01d-0489-gdl0003.pdf. Accessed January 26, 2008.
28. Guidance for Industry, Acceptance of Foreign Clinical Studies U.S. Department of Health and Human Services, Food and Drug Administration, Center for Drug Evaluation and Research (CDER), Center for Biologics Evaluation and Research (CBER), Center for Devices and Radiological Health (CDRH), Clinical Medical, March 2001. http://www.fda.gov/cder/guidance/fstud.htm. Accessed December 26, 2008.
29. Guidance for Industry, Available Therapy, U.S. Department of Health and Human Services, Food and Drug Administration Center for Drug Evaluation and Research (CDER), Center for Biologics Evaluation and Research (CBER), Clinical Medical, July 2004. http://www.fda.gov/cder/guidance/5244fnl.htm. Accessed December 26, 2008.
30. Federal Register/Vol. 68, No. 20 / Thursday, January 30, 2003 / Notices. http://www.fda.gov/OHRMS/DOCKETS/98fr/03–2162.pdf. Accessed December 26, 2008.
31. Guidance for Industry, Computerized Systems Used in Clinical Investigations, U.S. Department of Health and Human Services, Food and Drug Administration (FDA), Office of the Commissioner (OC), May 2007. http://www.fda.gov/OHRMS/DOCKETS/98fr/04d-0440-gdl0002.pdf. Accessed December 26, 2008.
32. Guidance for Industry, Development and Use of Risk Minimization Action Plans, U.S. Department of Health and Human Services, Food and Drug Administration, Center for Drug Evaluation and Research (CDER), Center for Biologics Evaluation and Research (CBER), Clinical Medical, March 2005. http://www.fda.gov/OHRMS/DOCKETS/98fr/04d-0188-gdl0002.pdf. Accessed December 26, 2008.
33. Guidance for Industry, Investigators, and Reviewers, Exploratory IND Studies, U.S. Department of Health and Human Services, Food and Drug Administration, Center for Drug Evaluation and Research (CDER), Pharmacology/Toxicology, January 2006. http://www.fda.gov/cder/guidance/7086fnl.htm. Accessed December 27, 2008.
34. Guidance, Financial Disclosure by Clinical Investigators, U.S. Department of Health and Human Services, Food and Drug Administration. http://www.fda.gov/oc/guidance/financialdis.html. Accessed December 27, 2008.
35. Federal Register: January 31, 2003 (Volume 68, Number 21), [Notices] Page 5026–5027. http://www.fda.gov/OHRMS/DOCKETS/98fr/03–2214.htm. Accessed December 27, 2008.
36. Guidance for Industry, Good Pharmacovigilance Practices and Pharmacoepidemiologic Assessment, U.S. Department of Health and Human Services, Food and Drug Administration, Center for Drug Evaluation and Research (CDER), Center for Biologics Evaluation and Research (CBER), Clinical Medical, March 2005. http://www.fda.gov/OHRMS/DOCKETS/98fr/04d-0189-gdl0002.pdf. Accessed December 27, 2008.
37. Guidance for Industry, Guideline for the Monitoring of Clinical Investigations, U.S. Department of Health and Human Services, Food and Drug Administration, Office of Regulatory Affairs, January 1988 (Minor editorial and formatting changes November 1998). http://www.fda.gov/ora/compliance_ref/bimo/clinguid.html. Accessed December 26, 2008.
38. Guideline for the Study and Evaluation of Gender Differences in Clinical Evaluation of Drugs, Federal Register 58, 139, 39406. http://www.fda.gov/cder/guidance/old036fn.pdf. Accessed December 26, 2008.

39. Guidance for Industry, Handling and Retention of BA and BE Testing Samples, CDER, Food and Drug Administration, May 2004. htpp://www.fda.gov/OHRMS/DOCKETS/98fr/2002d-0350-gdl0002.pdf. Accessed December 26, 2008.
40. Guidance for Industry, IND Exemptions for Studies of Lawfully Marketed Drug or Biological Products for the Treatment of Cancer, U.S. Department of Health and Human Services, Food and Drug Administration, Center for Drug Evaluation and Research (CDER), Center for Biologics Evaluation and Research (CBER), Clinical Medical, January 2004. http://www.fda.gov/cder/guidance/6036fnl.pdf. Accessed December 27, 2008.
41. Guidance for Industry, Information Program on Clinical Trials for Serious or Life-Threatening Diseases and Conditions, U.S. Department of Health and Human Services, Food and Drug Administration, Center for Drug Evaluation and Research (CDER), Center for Biologics Evaluation and Research (CBER), Procedural, March 2002. http://www.fda.gov/cder/guidance/4856FNL.PDF. Accessed December 27, 2008.
42. FDA Guidance for Industry IRB Review of Stand-Alone HIPAA Authorizations Under FDA Regulations, August 2003. http://www.fda.gov/OHRMS/DOCKETS/98fr/03d-0204-gdl0001.pdf. Accessed December 27, 2008.
43. General Principles of Software Validation; Final Guidance for Industry and FDA Staff, January 11, 2002. http://www.fda.gov/cdrh/comp/guidance/938.html. Accessed December 27, 2008.
44. Guidance for Industry, Pharmacogenomic Data Submissions, U.S. Department of Health and Human Services, Food and Drug Administration, Center for Drug Evaluation and Research (CDER), Center for Biologics Evaluation and Research (CBER), Center for Devices and Radiological Health (CDRH), Procedural, March 2005. http://www.fda.gov/OHRMS/DOCKETS/98fr/2003d-0497-gdl0002.pdf. Accessed December 27, 2008.
45. Guidance for Industry, Premarketing Risk Assessment, U.S. Department of Health and Human Services, Food and Drug Administration, Center for Drug Evaluation and Research (CDER), Center for Biologics Evaluation and Research (CBER), Clinical Medical, March 2005. http://www.fda.gov/OHRMS/DOCKETS/98fr/04d-0187-gdl0002.pdf. Accessed December 27, 2008.
46. Guidance for Industry, Providing Regulatory Submissions in Electronic Format—Human Pharmaceutical Product Applications and Related Submissions Using the eCTD Specifications, U.S. Department of Health and Human Services, Food and Drug Administration, Center for Drug Evaluation and Research (CDER), Center for Biologics Evaluation and Research (CBER), Electronic Submissions, October 2005. http://www.fda.gov/cder/guidance/6766fnl.htm. Accessed December 27, 2008.
47. Guidance for Industry and Clinical Investigators–The Use of Clinical Holds Following Clinical Investigator Misconduct, U.S. Department of Health and Human Services, Food and Drug Administration, Center for Biologics Evaluation and Research (CBER), Center for Drug Evaluation and Research (CDER), September 2004. http://www.fda.gov/OHRMS/DOCKETS/98fr/02d-0320-gdl0002.pdf. Accessed December 27, 2008.
48. Guidance for Industry, Using a Centralized IRB Review Process in Multicenter Clinical Trials, U.S. Department of Health and Human Services, Food and Drug Administration, Good Clinical Practice Program, Office of the Commissioner (OC), Center for Drug Evaluation and Research (CDER), Center for Biologics Evaluation and Research (CBER), Office of Regulatory Affairs (ORA), Procedural, March 2006. http://www.fda.gov/cder/guidance/OC2005201fnl.htm. Accessed December 27, 2008.
49. Guidance on Informed Consent for In Vitro Diagnostic Device Studies Using Leftover Human Specimens that are Not Individually Identifiable–Guidance for Sponsors, Institutional Review Boards, Clinical Investigators and FDA Staff, April 25, 2006. http://www.fda.gov/cdrh/oivd/guidance/1588.html. Accessed December 27, 2008.
50. ICH Tripartite Guideline E3: Structure and Content of Clinical Study Reports, ICH, Geneva, November 1995. http://www.ich.org/LOB/media/MEDIA479.pdf. Accessed December 27, 2008.
51. ICH Tripartite Guideline E5(R1): Ethnic Factors in the Acceptability of Foreign Clinical Data, ICH, Geneva, February 1998. http://www.ich.org/LOB/media/MEDIA481.pdf. Accessed December 27, 2008.
52. ICH Tripartite Guideline E6(R1): Guideline for Good Clinical Practice, ICH, Geneva, June 1996. http://www.ich.org/LOB/media/MEDIA482.pdf. Accessed December 27, 2008.
53. ICH Tripartite Guideline E10: Choice of Control Group and Related Issues in Clinical Trials ICH, Geneva, July 2000. http://www.ich.org/LOB/media/MEDIA486.pdf. Accessed December 27, 2008.
54. Documents, lists, policies, programs, and statements relating to Compliance References used by FDA personnel, Food and Drug Administration. http://www.fda.gov/ora/compliance_ref/. Accessed December 27, 2008.
55. Food and Drug Administration Program, Compliance Program Guidance Manual, Chapter 48 – Bioresearch Monitoring: Human Drugs: In Vivo Bioequivalence, March 2000. http://www.fda.gov/ora/compliance_ref/bimo/7348_001/foi48001.pdf. Accessed December 27, 2008.

56. Food and Drug Administration Program, Compliance Program Guidance Manual, Chapter 48—Bioresearch Monitoring: Human Drugs: Clinical Investigators, September 2000. http://www.fda.gov/ora/ftparea/compliance/48_811.pdf. Accessed December 27, 2008.
57. Food and Drug Administration Program, Compliance Program Guidance Manual, Chapter 48 – Bioresearch Monitoring: Human Drugs: Sponsors, Clinical Research Organizations and Monitors, September 2001. http://www.fda.gov/ora/compliance_ref/bimo/7348_810/48–810.pdf. Accessed December 27, 2008.
58. Food and Drug Administration Program, Compliance Program Guidance Manual, Chapter 48 – Bioresearch Monitoring: Human Drugs: Institutional Review Boards, September 1997. http://www.fda.gov/ora/compliance_ref/bimo/7348_809/irb-cp7348–809.pdf. Accessed December 27, 2008.
59. Office for Human Research Protections (OHRP), Department of Health and Human Services (HHS), Applicability of 45 CFR part 46 to Clinical Investigations, June 2006. http://www.hhs.gov/ohrp/humansubjects/guidance/invitrodev.html. Accessed December 27, 2008.
60. Department of Health and Human Service, Food and Drug Administration, Statement of Investigator, FDA Form 1572. http://www.fda.gov/opacom/morechoices/fdaforms/FDA-1572.pdf. Accessed December 27, 2008.
61. Department of Health and Human Service, Food and Drug Administration, Investigational New Drug Application, FDA Form 1571. http://www.fda.gov/opacom/morechoices/fdaforms/1571es.pdf. Accessed December 27, 2008.
62. Shapiro RS, Rusczek JM. The FDA's New Rule on Foreign Drug Trials, Drinker Biddle Health Law Practice, Client Alert, June 2008. http://www.theworldlawgroup.com/docs%5CFDA%20Drug%20Trials.pdf. Accessed December 27, 2008.
63. FDA Compliance Program Guidance Manual. http://www.fda.gov/ora/cpgm/default.htm#bimo. Accessed December 27, 2008.
64. Legler UF. Experiences with GLP/GCP from the pharmaceutical industry's viewpoint. Methods Find Exp Clin Pharmacol 1993; 15(4):233–236.
65. Commission Directive 91/507/EEC. Official Journal of the European Communities No. 1991; L270/32.
66. Directive 2001/20/EC of the European Parliament and of the Council. Official Journal of the European Communities No. 2001; L121/34.
67. Fed Reg 1997; 62(90):25961–5709E7.
68. ICH Harmonised Tripartite Guideline E3: Structure and Content of Clinical Study Reports, ICH, Geneva, November 1995. http://www.ich.org/LOB/media/MEDIA479.pdf. Accessed December 27, 2008.
69. Volume 10: Clinical trials guidelines of 'The rules governing medicinal products in the European Union'. http://ec.europa.eu/enterprise/pharmaceuticals/eudralex/vol10_en.htm. Accessed December 27, 2008.
70. Commission Directive 2005/28/EC. Official Journal of the European Communities 2005; L91:13–19.
71. Detailed guidance for the request for authorisation of a clinical trial on a medicinal product for human use to the competent authorities, notification of substantial amendments and declaration of the end of the trial, European Commission, October 2005, ENTR/F2/BL D(2003). http://ec.europa.eu/enterprise/pharmaceuticals/eudralex/vol-10/11_ca_14–2005.pdf. Accessed December 28, 2008.
72. Detailed guidance on the application format and documentation to be submitted in an application for an Ethics Committee opinion on the clinical trial on medicinal products for human use, European Commission, February 2006. http://ec.europa.eu/enterprise/pharmaceuticals/eudralex/vol-10/12_ec_guideline_20060216.pdf. Accessed December 28, 2008.
73. Detailed guidance on the European clinical trials database (EUDRACT Database), European Commission, April 2004. http://ec.europa.eu/enterprise/pharmaceuticals/eudralex/vol-10/13_cp_and_guidance_eudract_april_04.pdf. Accessed December 28, 2008.
74. Detailed guidance on the collection, verification and presentation of adverse reaction reports arising from clinical trials on medicinal products for human use, European Commission, April 2006. http://ec.europa.eu/enterprise/pharmaceuticals/eudralex/vol-10/21_susar_rev2_2006_04_11.pdf. Accessed December 28, 2008.
75. Detailed guidance on the European database of Suspected Unexpected Serious Adverse Reactions (Eudravigilance – Clinical Trial Module), European Commission, April 2004. http://ec.europa.eu/enterprise/pharmaceuticals/eudralex/vol-10/22_cp_and_guidance_database_susars16_april_2004.pdf. Accessed December 28, 2008.
76. Guidance Documents containing the Common Provisions of GCP Inspections by Competent Authorities of the Different Member States—Guidance for the Preparation of Good Clinical Practice Inspections, European Commission, Version: 28 May 2008. http://ec.europa.eu/enterprise/

pharmaceuticals/eudralex/vol-10/chap4/guidance_for_the_preparation_of_gcp_inspections.pdf. Accessed December 28, 2008.
77. Recommendation on Inspection Procedures for the Verification of Good Clinical Practice Compliance, European Commission, July 2006. http://ec.europa.eu/enterprise/pharmaceuticals/eudralex/vol-10/v10_inspection-proc.pdf. Accessed December 28, 2008.
78. Guidance Documents containing the Common Provisions on the Conduct of GCP Inspections by Competent Authorities of the Different Member States, European Commission, Version: 28 May 2008. http://ec.europa.eu/enterprise/pharmaceuticals/eudralex/vol-10/chap4/guidance_for_the_conduct_of_gcp_inspections.pdf. Accessed December 28, 2008.
79. Ethical Considerations for Clinical Trials on Medicinal Products conducted with the Paediatric Population, Recommendations of the ad hoc group for the development of implementing guidelines for Directive 2001/20/EC relating to good clinical practice in the conduct of clinical trials on medicinal products for human use, European Commission, Final 2008. http://ec.europa.eu/enterprise/pharmaceuticals/eudralex/vol-10/ethical_considerations.pdf. Accessed December 28, 2008.
80. Commission Directive 2003/94/EC. Official Journal 2003; L262: 22–26.
81. Eudralex VOLUME 4: Good manufacturing practices, Annex 13 Revision 1, Manufacture of investigational medicinal products, European Commission, July 2003. http://ec.europa.eu/enterprise/pharmaceuticals/eudralex/vol-4/pdfs-en/an13final_24–02–05.pdf. Accessed December 28, 2008.

41 Prevention and Detection of Fraud in Clinical Trials

Jane Barrett
MedicoLegal Investigations Ltd, Cambridge, U.K.

INTRODUCTION

Fraud and misconduct in clinical research can be found wherever research takes place whether that research is academic or sponsored by the pharmaceutical industry. Numerically, however, most detected cases have been found in industry-sponsored studies, and it is these cases that have received most publicity. The majority of companies have had Standard Operating Procedures (SOPs) in place for some time to prevent, detect, and investigate research fraud and there is a growing awareness of the need for robust methods to provide similar protection within academic and healthcare institution research.

However, despite apparently careful attention being paid to the prevention and detection of fraud, the majority of cases are still found either through the actions of "whistleblowers" or by complete chance. This chapter will explore the types of fraud, the ways that it can be made more difficult to commit fraud, how detection can be made more likely, and the investigation of fraud and misconduct. In order to both prevent and detect fraud, it is essential that all involved with clinical research accept both the possibility that fraud exists and that the eminence and reputation of the clinical investigator are no safeguards against them being fraudsters.

No matter how suspicions of research fraud or misconduct arise, the handling of all such suspicions must be conducted properly and thoroughly. If the wrong questions are asked or if prearranged procedures are not followed, the person or persons committing the wrongdoing will be given the opportunity to destroy vital information, to correct their errors, and to escape detection and, thus, avoid the confirmation of even well-founded suspicions. If the guilty escape detection and possible punishment, patients' health and well-being are at risk both in the present and the future. Protection of patients must be the first and last aim in preventing and detecting research fraud. There is too much well-publicized evidence of patients being given unnecessary treatment, of being entered into studies despite not fulfilling entry criteria, or in spite of having significant contraindications; patients have been submitted to the risk of taking experimental treatments without the full assessments mandated by the study protocol. We know that at least one drug has been needlessly withdrawn from the market because a study allegedly found it to be teratogenic (see section "Falsification").

Clinical research fraud is treated differently across the world. In the United States, several physicians found guilty of being involved in fraud related to human studies have been given prison sentences (see later for details of one such case). In South Korea too, after one of the most high profile cases of clinical fraud of this century, Hwang Woo-Suk faces a custodial term if finally found guilty of fraud and embezzlement (see later for further details). In Europe, the introduction of the Clinical Trials Directive (1) means that law now dictates that the conduct of clinical trials involving medicinal products must be in accordance with good clinical research practice (GCP). Therefore, anyone making a false declaration in a submission to an ethical review committee or in an application to the National Competent Authority to carry out a clinical trial commits a criminal offence. Strangely though, a person who fabricates data or in any other way commits research fraud in Europe is not automatically regarded as having committed a criminal offence under the legislation supporting the Directive.

We must keep a sense of perspective: Most clinical researchers are honest and lawful people whose focus is on the well-being of their patients. However, to ignore signs of research fraud and misconduct and pretend that they do not exist is to allow (even encourage) these malpractices to continue and thus indirectly to condone bad medicine, bad science, and—above all—abuse of patients. Research misconduct strikes at the very heart of scientific objectivity. It raises doubts about the integrity of science and our trust in the work of others. We must be justified in believing in the reliability of scientific research.

WHAT IS FRAUD?

Behind fraud of any type is the intention to gain advantage for the perpetrator by deception or providing false information. The unique element of fraud in clinical research is that the act of committing it places patients at risk. We often see the words "fraud" and "misconduct" used almost interchangeably but, in most cases, they can be differentiated. In broad terms, research fraud is a behavior that breaches the principles of GCP but, in addition, fraud must have an element of deliberate or reckless action—it can never be an accidental act.

Another way of making a distinction is to look at those affected. Others are always harmed by fraud but with misconduct there may sometimes be no obvious victims. Patients or research subjects are most obviously at risk of harm (physically or emotionally) by fraud.

Others are harmed too. Sponsoring pharmaceutical companies are harmed financially if they have paid for research that they cannot use subsequently because of fraudulent data being generated, and the families of disgraced doctors (who lose their licenses to practice when prosecuted and found guilty of fraud) also suffer financially. Journal editors who unknowingly publish the fraudulent results can be harmed professionally.

All fraud is also misconduct by definition, but misconduct per se is less easy to define. It could be simple lack of attention to detail, for example missing the designated date for patient assessments or completing case report forms carelessly. Errors are common and represent carelessness, pressure of work and time, or badly designed or overcomplicated case report forms. Distinction must be drawn between clinical research that is of poor quality (misconduct) and that which is fraudulent and which, paradoxically, is often of the highest apparent quality.

The transition between carelessness and misconduct and between misconduct and fraud is not clear-cut. An omitted safety assessment might be due to human error by the research team or because the research subject failed to go to the laboratory to have the blood drawn. Alternatively, it might be because a researcher decided not to do it either because the researcher was too busy or decided it was not important. This leads to the possibility, in the next study that the researcher participates in, that he or she would deliberately fabricate larger amounts of data if there were no consequences from his accidental acts in earlier research. Inappropriate delegation of study tasks to an inexperienced member of the study team without input or supervision from the Principal Investigator could also be regarded as serious (but nonfraudulent) misconduct.

The Wellcome Trust, a U.K.-based charity that makes available £600 million every year both in the United Kingdom and internationally to achieve their mission of "supporting and promoting research to improve the health of humans and animals," defines research misconduct as follows (2):

The fabrication, falsification, plagiarism, or deception in proposing, carrying out, or reporting results of research or deliberate, dangerous, or negligent deviations from accepted practices in carrying out research.

It includes failure to follow established protocols if this failure results in unreasonable risk or harm to humans, other vertebrates, or the environment and facilitating misconduct in research by collusion in, or concealment of, such actions by others.

It also includes intentional, unauthorized use, disclosure or removal of, or damage to research-related property of another, including apparatus, materials, writings, data, hardware, or software or any other substances or devices used in or produced by the research.

It does not include honest error or honest differences in the design, execution, interpretation, or judgment in evaluating research methods or results or misconduct unrelated to the research process.

Similarly, it does not include poor research unless this encompasses the intention to deceive.

The Office of Research Integrity, part of the U.S. Department of Health and Human Services, defines research misconduct in the following similar way (3):

Research misconduct means fabrication, falsification, or plagiarism in proposing, performing, or reviewing research, or in reporting research results.

(a) Fabrication is making up data or results and recording or reporting them.
(b) Falsification is manipulating research materials, equipment, or processes or changing or omitting data or results such that the research is not accurately represented in the research record.
(c) Plagiarism is the appropriation of another person's ideas, processes, results, or words without giving appropriate credit.
(d) Research misconduct does not include honest error or differences of opinion.

The underlying principle for both these definitions is that bad ethics make for bad science and clearly research fraud is the prime example of bad ethical behavior. Whatever the explanation, however, research misconduct can always be said to fall short of ethical and/or scientific standards.

TYPES OF FRAUD

There are many different types of fraud (4), some of which will be discussed here.

Fabrication

Fabrication is the deliberate invention of research data or results or the deliberate fabrication of laboratory analyses. The most commonly fabricated clinical trial documents are consent forms and patient diary cards, probably because there are no external checks on them. One of the most famous cases of fabrication occurred in the United Kingdom in the 1990s. An eminent U.K. gynecologist, Malcolm Pearce, published two papers in the *British Journal of Obstetrics and Gynaecology*, of which he was one of the editors (5, 6). The first claimed that he had successfully reimplanted an ectopic fetus, and the second was an extensive series of case studies. The case studies were in a syndrome so uncommon that a tertiary care centre might see only one or two new cases a month, yet Pearce reported on 191 women he had seen over three years--more than five cases per month on average. The editor-in-chief of the journal, Professor Geoffrey Chamberlain, was the head of the department where Pearce worked and was named as coauthor of one of the papers. Chamberlain's role in the work is not known but he was quoted by a newspaper as saying, "The head of department's name is always put on reports out of politeness. I was not part of this work, but I have always trusted Mr Pearce" (7). It is worth noting here that gift authorship is the most common and apparently the most lightly regarded of all the abuses of scientific research. However, when the fraud was discovered through the actions of a whistleblower both Pearce and Chamberlain found their careers effectively ended.

Another once prominent researcher was Robert Fiddes in the United States who conducted over 200 studies for scores of pharmaceutical companies in the early 1990s. In a series of frauds running over several years, he changed patient notes to show falsely that they had particular conditions specified by study protocols and instructed his study staff to buy bacteria from a commercial supplier and send the same bacteria to testing laboratories under the names of patients, claiming to have recruited 25 patients for one study when he had, in fact, recruited just one. He used cervical smears from other sources and entered the results into the patient notes and used blood from employees stating it was from patients. In a sinusitis study, employees' family members were fictitiously enrolled and their records were fabricated. He made his staff wear Holter monitors and took electrocardiograms from them, again claiming that they were taken from patients. Where patients really were entered into studies, he failed to conduct the proper tests and fabricated results instead. Despite inspections from both the concerned pharmaceutical companies and the Food and Drug Administration (FDA), the fraud was only recognized when a former employee blew the whistle. Fiddes was sentenced to 15 months of imprisonment and fined almost a million US$ (8, 9).

More recently, the case of Hwang Woo-Suk in Seoul, South Korea, has drawn international attention to research fraud. In 2004, Hwang announced that his team had created an embryonic stem cell (10). A year later he claimed they had created 11 human embryonic stem cells using 185 eggs (11). This was hailed as a breakthrough as the cells were created with somatic cells from patients of different ages and genders, but it was surprising as it was so sudden and unexpected.

Later that year, one of his coworkers admitted paying women and coercing other researchers to donate their eggs for Hwang's research. Hwang apologized and claimed that he found out about the situation only after it had occurred. He further said that he was not aware of the Declaration of Helsinki, a pivotal element of research ethics. Finally Seoul National University determined that 11 of Hwang's cell lines were fabricated and the journal *Science* retracted both of the published papers (12). Hwang blamed other members of his research project for having deceived him with false data and alleged sabotage of his projects and theft of materials. His trial was ongoing at the time this chapter was written (Autumn 2008) and could result in several years' imprisonment if he is found guilty.

Falsification

Falsification is the deliberate distortion or omission of undesired data or results, including the dishonest misinterpretation of results. Perhaps the best-known example of this is the case of William McBride. McBride, an Australian obstetrician, wrote a letter to the *Lancet* in 1961 in which he suggested that giving the drug thalidomide to pregnant women was causing severe limb deformities in their babies (13). This was before anybody else had published concerns. McBride's hypothesis was based on a limited number of observations but subsequent studies proved him right and thalidomide was removed from the market. Twenty years later he published results of research showing that the active substance in a drug widely used for morning sickness of pregnancy caused birth defects in rabbits (14). The manufacturers removed the drug from the market and lawsuits were prepared. No researchers could reproduce his work but it was only some years later after an exposé by a television investigator and a former research assistant of McBride that it was shown that he had altered research data. The drug, in fact, had no teratogenic effects. McBride was found guilty of scientific fraud by a medical tribunal and removed from the Medical Register (15).

Several different fraudulent practices can be found together in one case such as that of John Anderton, a highly respected physician in Edinburgh with an exemplary career, who falsified data, invented patients, and forged consent. In a pharmaceutical industry-sponsored study, the trial monitor noticed that the signatures of some patients on the consent forms differed from other signatures in their hospital notes. On closer scrutiny, it was also found that electrocardiograms and nuclear medicine investigations seemed to have been reported on forms that had been taken out of use some time before the start of the study. Records had been created documenting treatment effects but they were dated before the patients were actually treated, and there were major discrepancies between the case report forms and letters to patients' general practitioners. Many other dates did not pass the reality test: registers at the hospital did not record patients' attendance for the visits recorded in the study documentation, some patient visits were dated on public holidays when the hospital department was closed, and 13 patient visit dates coincided with the investigator's own vacation. Some of the electrocardiograms and X-rays had clearly been altered to remove dates and patient identification. The most incontrovertible evidence was that patients at Anderton's site apparently had a rise in drug metabolite levels at a time point when all other sites recorded a fall, and they reported a fraction of the number of adverse events reported at other sites. After a disciplinary hearing, Anderton had his name removed from the medical register after being found guilty of serious professional misconduct. He instructed his lawyer to enter a plea of guilty and refused to answer any question related to the charges against him.

Plagiarism

Plagiarism is the deliberate unacknowledged presentation or exploitation of the work and ideas of others as one's own. Worryingly, the culture in schools and universities seems increasingly to accept plagiarism or at least not to penalize it. This may be considered to be as much plagiarism by the teaching professionals as was the original intellectual fraud by the student. The personal experience of the author of this chapter is that at MSc level in Europe and the United States up to one in four students does not perceive that unacknowledged copying of large tracts of text from the Internet is plagiarism. A survey in the United Kingdom published in 2004 seems to confirm this with 16% of respondents admitting that they had plagiarized work more than once and a further 9% stating that they had plagiarized once, most commonly by copying material for essays from the Internet (16). The detection rate was only 3% perhaps showing that while not actively condoned, plagiarism is not always dealt with as firmly as it should be. It is easy to buy from the Internet essays on many school and university subjects. They can even be ordered and customized to be consistent with previous examples of the student's work to avoid raising suspicion by the sudden appearance of an above average piece of work. Submitting another's essay as one's own work is fraud and if it is accepted without comment or penalty by those marking the work, then an opportunity to prevent the formation of the mindset that leads to fraud is lost and at least one barrier to committing further fraud has been removed. There seems to be a very low rate of detection compared with the rate of plagiarism, so it is mandatory that attention is paid to teaching tomorrow's researchers what constitutes good scientific and ethical behavior. It is worrying that 24% of those students who replied to the survey referred to above claimed to have received no guidance as to what constituted plagiarism. Students and scientists entering their research careers must therefore be trained in the concept of research honesty and ethics; it is as important as trial design and methodology.

The Committee on Publication Ethics (COPE) is an independent U.K. Registered Charity concerned with ethical standards in journal publication internationally and provides support and advice to editors, authors, publishers, and others in areas related to publication ethics and fraud. The members of the Committee publish anonymous cases that have been referred to them, and the committee has reviewed 18 cases of alleged plagiarism between 1998 and 2005. One such case was considered in 2004 (17). A review article was sent out to three peer reviewers, one of whom had "serious concerns" about the paper. It transpired that large amounts of the text and most of the references had been copied from a teaching syllabus that the reviewer had written. One of the two review article authors had attended the teaching session and received a copy of the syllabus.

The paper was withdrawn temporarily and the two authors contacted about the alleged plagiarism. They claimed that they had used the teaching syllabus because English was not their first language, but they did not realize that this amounted to plagiarism. They suggested rewriting the paper, but the editors felt that this would be unacceptable and the manuscript was rejected.

Recruiting and Gaining Patients' Consent Without Ethics Approval

Protection of research subjects is the central pillar of ethical clinical research and independent ethical review of research is a critical part of that. The time taken to obtain ethical review can sometimes be time-consuming, however. Using modern scanners and copying equipment it is relatively easy to produce a document that appears to represent Ethics Committee's or Institutional Review Board's approval in order to start a study. Because studies may not start without ethical approval and funding is not forthcoming until it is given, time is probably the driving force behind this type of fraud. The fraudulent preparation of a dishonest piece of paper is further compounded when attempts, often successful, are made to recruit patients and then to induce them to give consent to participation when no genuine ethics approval exists. Subsequent issue of the ethics approval does not make the initial fraudulent conduct justifiable.

Failure To Document Consent Appropriately

In terms of numbers, forged consent forms are among the most common types of research fraud. In one example of consent fraud, a monitor from a Contract Research Organisation (CRO) noticed that patient signatures on consent forms looked similar to each other and also resembled the handwriting of one of the study team. It was eventually proven that none of the patients shown as having been recruited into the study had any knowledge of being in a trial. The study involved women taking hormone replacement therapy over several months with a uterine biopsy being done before and after treatment. All had been supplied with the therapy by the researcher but none had given consent.

Collusion in or Concealment of Research Misconduct by Others

The United Kingdom's governing body for physicians, the General Medical Council (GMC), is just one organization that has made it clear that if a doctor fails to report suspicions or evidence of scientific fraud or misconduct, he or she will also face disciplinary action. This was seen when Professor Timothy Peters knew that a colleague, Anjan Bannerjee, had fabricated patient data in a clinical trial and did not report it. Both men were tried by the disciplinary panel at the

GMC and both were found guilty of Serious Professional Misconduct. Bannerjee's license was suspended and Prof Peters was severely reprimanded, marring an otherwise excellent career.

Despite public encouragement to report suspicions of fraud and misconduct, and notwithstanding legislation in most countries to provide a framework of protection for those wishing to do so, it is not uncommon for those who expose the wrongdoing of others to experience significant prejudice, both personal and professional. The term whistleblower tends to have negative connotations and is increasingly being replaced by such euphemisms as "Open-Practice Policy." Damage to whistleblowers who act in good faith can usually be avoided but it is essential that they and their suspicions are assessed properly and managed by someone experienced in the role. Until recently many SOPs in use in the public arena, such as hospitals and universities, did not provide sufficiently robust protection and, especially, did not allow for a safe route of reporting if the suspected wrongdoer was in a senior position in the organization.

The fear of personal career damage perhaps goes some way to explain why more biomedical fraud is not exposed even though it may have been strongly suspected or even known for certain. This is a particular factor if the scientist with suspicions is in a subordinate position to the person perpetrating the fraud. For example, it was many years after the event that scientist Phil Vardy reluctantly blew the whistle on McBride and then only after he had moved to another city and was working in another institution.

An additional concern is the fear of being shown to be wrong, and this perhaps holds back many from making allegations of research fraud or misconduct. Indeed, the Office of Research Integrity (ORI) in the United States has said that 70% of allegations that they investigate are found to be genuine differences of interpretation (18).

Drug Supplies

Most pharmaceutical industry-sponsored studies require unused medication to be returned and counted as a check that patients have been compliant with their medication. If all the drug containers are clean and unmarked at the end of a study or if all the tubes of cream or ointment have been squeezed in an identical manner, it is almost certain that they have not actually been given to patients. It is possible to use canisters that automatically record the times at which they are opened to counter this, but their use is not yet widespread as they are expensive.

Reuse of Data

Different pharmaceutical companies often conduct almost identical studies on similar drugs in similar patient groups. A recent case in the United Kingdom proved that an investigator had used data from one group of study patients to fulfill fraudulently the needs of a second study. The two sponsor companies knew nothing of each other's study and site staff were instructed to hide documents when monitors from either company visited. One study called for extensive diagnostic imaging and, although this had not been approved by an ethical review committee or mentioned in the patient consent form, the same tests were performed in the other study thus allowing data to be pooled. Patients were subjected, therefore, to tests for which neither they nor the ethics committee had given permission or consent. The same investigator also photocopied case report forms from one pharmaceutical company-sponsored study and used them unchanged in a separate study sponsored by a charitable body.

Another way in which data can be reused is to duplicate data from one patient and attribute it to other patients. For example, before modern technology was available, electrocardiograph recordings were made on a single long strip of heat-sensitive paper, which was cut up and the strips mounted one by one on a sheet of paper. In one study, comparing a calcium channel blocker with a β-blocker in patients with hypertension, an electrocardiogram was taken before and after 12 weeks of treatment. No patient showed the expected variations in heart rate—an improbable, if not impossible, finding. The paper manufacturer's name was repeated along the edge of the paper strip and when the recording for each lead was cut up for mounting, the cut inevitably passed through the name. It became obvious that the investigator must have taken one double-length recording at the first visit and then cut each strip into two before mounting onto two separate sheets and presenting them as the baseline and 12-week recordings. It could not have been argued that this breach of safety assessment was unimportant; it was quite clearly crucial to the safety of patients.

HOW COMMON IS FRAUD?

This is perhaps the most difficult question to answer. Fraud that is detected is easy enough to quantify but it is the undetected fraud that probably forms the vast bulk of the archetypal iceberg. True deliberate fraud is certainly much less common than carelessness, but carelessness in research techniques that is not corrected by sponsor or supervisor can lead to deliberate cutting of corners in the future and perhaps ultimately to outright fraud. More than 70% of the audiences of two separate international clinical research conferences within the last five years indicated they had seen clinical research fraud. Most had implicitly condoned the suspected offences by having taken no action on their suspicions.

The FDA has offered estimates that fraud occurs in around 3% to 5% of clinical trials. Some authorities suggest a rate of 1%, some up to 7% (19). The experience of the author of this chapter suggests a figure around the 3% mark. The number of clinical trials in progress worldwide at any one time must be many thousands, leaving potential for an unacceptably large number of studies putting patients at risk and producing data that are unreliable or even fabricated.

The results of a paper published in 2005 revealed that 33% of early- and mid-career scientists admitted to one or more of ten examples of serious professional misconduct. More than 15% had altered the design, methodology, or results of their research; 12.5% had overlooked research misconduct in others; and almost 8% had in some way "circumvented requirements relating to human subjects" (20, 21).

In recent surveys more than 40% of researchers stated that they were aware of research misconduct but had not reported it (22), and those working in an academic medical setting believed that a junior researcher would probably do nothing if he or she became aware that a principal investigator or research staff member was involved in misconduct (23). Twenty percent of those surveyed had first-hand experience of actual research misconduct within the past year, but only a quarter of them said they would report to the appropriate authorities. So it is clear that both fraud and misconduct are widely seen in research, acknowledged as such, but in the large part ignored. Some of the reasons for this have already been discussed. The fact remains that fraud does matter. Not only does it put the research subjects at risk, it also affects public confidence in clinical research and raises questions about the effectiveness of clinical trial monitoring and supervision.

WHY IS FRAUD COMMITTED?

There are many theories. Given that the creation of fraudulent data probably takes as long as, if not longer than, its legitimate counterpart, motivation can be hard to determine. Peter Jay, one of the founders of MedicoLegal Investigations Ltd (the only Europe-based independent company investigating suspected research fraud and misconduct), suggests that greed, need, and breed are the main underlying reasons. The first is self-explanatory and includes greed for both money and status, and the desire to be invited to join the ranks of key opinion leaders. The second category includes addiction to drugs, alcohol, or gambling and the need for money to fund them. The third group accounts for the craving of the adrenaline buzz achieved by lying, cheating, and deceiving to get something for nothing. Other major motivators include arrogance, with the self-belief that the fraudster is too clever to be caught, and the requirement to achieve an extensive list of publications to further otherwise unachievable and inflated career aims.

PREVENTION

The most important element in the prevention of research fraud is to recognize the possibility of its existence. Unless the possibility of fraud is accepted it will not be seen. Suspicions will be ignored or swept aside as impossible or imagined. In particular, the possibility of eminent and widely respected researchers committing fraud must be embraced as several very high profile doctors have been found guilty of research fraud and misconduct (24). It is not acceptable to hide behind friendship or reputation or to fail to respond to suspicion. It may sound cynical to suggest that in all research it should be up to the researcher to show that he or she has not been fraudulent, but it would certainly be a major step in the prevention of fraud. Prevention is the most effective way of protecting the rights and welfare of the human research subject.

The first step in the prevention of fraud lies with clear statements from both the institution where the research will take place and the body sponsoring the research that suspicions of misconduct will be taken seriously, will be fully investigated, and that energetic prosecution will follow actual evidence or substantial doubt. An atmosphere must be created wherein it is recognized that any expressed suspicions of fraud or misconduct will be taken seriously and followed through in a way that fully protects the person making the allegations. For this to happen requires a robust set of SOPs and all those employed by a pharmaceutical company or in academia must be fully trained in them and their use. A well thought out and well-written SOP makes a statement to would-be fraudulent investigators that such behavior is unacceptable. It gives a framework and protection for those who have concerns that fraud or misconduct has occurred. It allows for a standardized and impartial investigation and, if necessary, an appropriate punishment. Guidelines for the production of these SOPs have been published, for example, by the Department of Health in the United Kingdom (25). The key elements of such SOPs are as follows:

- Definitions of fraud and misconduct; for example, Wellcome Trust, ORI. It is helpful also to say what fraud is not, for example, carelessness.
- Statement that suspected fraud and misconduct will be taken seriously and fully investigated.
- Rules of making allegations: Both parties must be named but confidentiality will be maintained. It must be clear that any person making vexatious or frivolous allegations will be punished and that those making genuine allegations will be protected—even if the allegations prove to be unfounded on complete investigation.
- Policies: Whistleblowers' protection will be invoked for those following the SOP.
- Roles and responsibilities: Specified job titles for those receiving allegations, and investigating and supporting the whistleblower.
- Purpose: To give a framework for both allegations about others and to allow self-reporting.
- Process and investigation: Allegations must be in writing, submitted to specified people, with appropriate people appointed to investigate and follow through.
- Outcome: Decision to take matters further, disciplinary action , sanctions, etc.
- Appeal: Panel of people not involved with the original investigation.
- Documentation: Notes to be made at all stages, in accordance with data protection laws.

The SOP should mandate that allegations of research fraud or misconduct are made to two individuals simultaneously. Within National Health Trusts in the United Kingdom, for instance, the vast majority of SOPs on fraud and misconduct state that one named person, typically the Director of Research and Development, is the only person to be told of allegations. This strategy is risky as collusion should always be regarded as a possibility that also makes more likely the possibility of intimidation of any subordinate person making the allegations. The chair of the relevant ethics committee or independent review board might be an appropriate second person to receive a copy of the allegations. Within industry, the medical director and the head of quality assurance are the most usual persons designated to receive allegations of alleged fraud and misconduct.

Where SOPs exist they must be followed to the letter. The major risk if they are not so followed is that the fraudster will hear of the allegations and take evasive action. They may be able to destroy evidence, to create missing data, or to modify patient records. There are other risks too. There has actually been more than one case of suicide or sudden unexplained death when an investigator, suspected of fraud, believed he was about to be exposed.

When an allegation is made, there must be detailed documentation and the case must be fully investigated by a designated (small) group of people. This group must be independent of the department or departments employing both the accused and the accuser and must have the power to make enquiries and the level of expertise that will give it credibility (26).

The rationale behind the decision as to whether or not to pursue the allegation must be equally well-documented. The default position must always be to pursue the allegation but decisions to do nothing need to be fully justified. The next decision will be when (or, in the case of allegations being found to lack substance or sufficient evidence for further process, if) to inform the person against whom the allegations were made. This will naturally need input from the organization's legal department to avoid possible defamation suits being brought against those making or investigating the allegation.

Other important elements in the prevention of research fraud include research governance. In the United Kingdom, for example, such a framework was published (27) laying down standards and requirements for monitoring in all research involving the National Health Service in England and Wales. The framework's stress on good quality research, the review of research at all stages, and the rights and well-being of research subjects goes a long way towards the prevention of fraud.

Similarly, the guidelines from the Committee on Publication Ethics discussed above would almost certainly have prevented the fraudulent publication of the Pearce paper had it been required to undergo peer review and had the coauthors been obliged to detail their exact involvement.

Outside the pharmaceutical industry many SOPs detail possible sanctions to deal with those found guilty of fraud or misconduct. These can include

- letter of reprimand,
- supervision of future research,
- withdrawal of approval for the research,
- barring the researcher from further research in the organization or from applying for further funding,
- reporting researchers to their professional bodies and ethics committees,
- suspension and/or dismissal, and
- requesting withdrawal or correction of published or pending papers.

The concern with many of these sanctions is that they cannot prevent the researcher moving to another hospital or university or even to another country and starting again. It is even possible that the lessons learned from being caught once will enable avoidance of detection in the future.

IMPACT OF THE EUROPEAN CLINICAL TRIALS DIRECTIVE

As mentioned earlier in this chapter and elsewhere in this book the enactment of the Directive (1) into national law means that throughout the European Union research in human subjects and the protection of those subjects is covered by legislation. Part of this protection allows each Member State's Competent Authority to conduct or commission inspections of research sites and sponsor premises and to effectively demand access to whatever they deem necessary to provide assurance that GCP is being followed and that patients' well-being is foremost. One of the clearly defined objectives of these inspections is to exclude the possibility of research fraud and misconduct. The EMEA is also empowered to trigger inspections, and the FDA has, for decades, carried these out under the Code of Federal Regulations (28) not only in the United States but also anywhere in the world where research that will be used for a license application in the United States is or has been carried out.

Despite the history of inspections, the United States has recorded a considerable number of cases of research fraud and there does not seem to be any compelling evidence that the longstanding history of regulatory inspections has had a discouraging effect on other fraudsters. Perhaps this reflects the arrogance mentioned earlier; nobody believes they will be the one to be caught. Nevertheless it is widely hoped that the new statutory framework in Europe will work to reduce both research fraud and misconduct.

DETECTION

Perhaps the most elementary step in the detection of fraud is for those monitoring it to learn to evaluate study data properly, to learn how to read X-rays and electrocardiograms, and to interpret laboratory data at least well enough to recognize duplication or impossibility and not just to document their presence in the patient files. If a researcher knows that everything that is documented will be reviewed thoroughly, it will act as a deterrent to fraud. Missing data should be pursued energetically and offers by investigators to "find it after you've gone" should be rejected in favor of "finding it while I wait." Sponsors of research must be alerted if investigators demand replacement of study monitors as Fiddes did when he sensed that those allocated to his site were beginning to have suspicions and asking too many questions.

Study monitors are mainly trained to look for errors and missing data but not to detect fraud. It is this that allows fraudsters to avoid detection for so long. Most investigated in the UK have been active for some years before their eventual detection (29, 30).

Many countries have national bodies that are involved in the detection of research fraud and misconduct and offer advice and support. The institutions or companies to whom the allegations of fraud or misconduct are made usually carry the responsibility at least to initiate investigation, however.

The United States was the first country to make public the need for the prevention and investigation of research fraud and misconduct with a national body being established in 1972 to provide guidance, education, and clarification on human research subject protection. In 2000 this became the Office for Human Research Protections (OHRP), and it assesses investigations of cases of alleged noncompliance with regulations and determines any necessary action.

Also in the United States is the ORI, part of the Office of Public Health and Science (OPHS). It promotes worldwide integrity in biomedical and behavioral research projects supported by the OPHS and monitors institutions investigating research, develops policies, and provides training and support to researchers.

The final body in the United States with a role in the prevention and detection of research wrongdoing is the FDA. FDA investigators carry out study-orientated audits on clinical trial data and investigator-orientated inspections as a routine or because asked to do so by a sponsor. If a site fails to comply with regulatory requirements or has submitted false information, the FDA has the power to disqualify the investigator from further research or severely restrict their research activities.

Few European countries have any official sanctions in research fraud, but there have been research misconduct committees in Europe since the early 1990s.

- Finland: National Research Ethics Council, 1991—Produces guidelines for prevention, handling and investigation of alleged scientific dishonesty
- Denmark: Danish Committee on Scientific Dishonesty, 1992—Investigates cases and expresses an opinion
- Norway: National Committee for the Evaluation of Dishonesty in Health Research, 1994—Advises on prevention and investigation of scientific dishonesty
- Sweden: Expert advisory group, 1997, linked to Swedish Medical Research Council—Provides guidance
- Germany: Deutsche Forschungsgemeinschaft, 1999—Advises and assists, produces guidelines
- France: Délégation à l'Intégrité Scientifique, 1999—Focuses on prevention of research fraud and sanctions against individuals or institutions
- United Kingdom: U.K. Panel for Research Integrity in Health and Biomedical Sciences, 2006—Provides Code of Practice, expertise

INDEPENDENT ASSESSMENT OF ALLEGATIONS

For many research institutions and pharmaceutical companies, the investigation of allegations or suspicions of research fraud and misconduct raises significant problems. Their SOPs detail the steps to be taken, but the exact mechanisms of investigation of such allegations are often couched in vague terms such as "suitable procedures" by "appropriate persons." It is important always for investigations to be carried out by those with relevant experience, who are separate from both the department from which the allegations or suspicions arise and the department within which the misconduct is said to have occurred. In many such situations there may be nobody suitable, in which case persons external to the organization can fulfill a very useful role not only to handle whistleblowers and to establish their motivation but also to conduct discreet and impartial enquiries into the allegations. There are significant risks from such enquiries being conducted by the inexperienced as evidence can be lost, whistleblowers can find their careers damaged, and the perpetrators of fraud can escape detection and punishment.

If external bodies are not brought in to initiate investigations into allegations of misconduct, it is vital that the organization carrying out its own enquiries mandate that more than one person is involved at each step of the process. Lack of corroboration can weaken any subsequent disciplinary hearing, and a lone investigator is always open to allegations of malice.

CONCLUSIONS

Over the last decade there has been evidence of an increasing determination on the part of official and governmental bodies to eradicate research fraud and misconduct, but vigilance is still important as there is no evidence of "dramatic decline in the reporting of misconduct cases" (31). Prevention of misconduct will always make more impact on the safety of research subjects than robust procedures for its detection but while fraudulent and careless researchers are still active, the latter continue to be vitally important.

Although nobody would claim that deliberate research fraud is common, it is not possible to make a credible case that it does not exist. To ignore fraud is automatically to condone it. Misconduct is more common and to allow it to continue uncorrected creates an atmosphere in which fraudulent behavior can develop. There have been instances in the past where sponsors of biomedical research have deliberately chosen not to investigate and prosecute and colleagues with strong suspicions have not acted, but fortunately the climate has changed and such decisions are now rare. Ethical review committees too need to accept that fraud and misconduct exist and that they have a role to play in their eradication.

Eradication must be the objective but it will not be an easy process. Increasing levels of research governance in academia and health services will help, certainly, as will determination on the part of the pharmaceutical industry, but any action must be proactive, not reactive. SOPs are only of real benefit if every single person involved in research is trained in their use and uses them every day. They are a useful framework, but they should neither be regarded as safeguards against fraud nor be left as historic documents. They must be reviewed and rewritten on a regular basis. Well-written and appropriately revised, they provide an approved, defined, and thought-out plan of action for those suspecting fraud or misconduct and those called to act on those suspicions.

At the centre of the need to eradicate research fraud and misconduct are the research subjects. The definition of what has happened is irrelevant: the fact is that patients have been harmed. They have been harmed when ethical review has not been obtained, when they have been treated with inappropriate drugs or procedures or investigations, and when their consent has not been sought or has been sought on the basis of false information. They have been harmed when safety data have not been recorded, when drugs have been licensed or withdrawn because of fraudulent data, and when false entries have been made on their doctor's case notes to make them appear eligible for a study.

To even come near to eradicating fraud there must be two elements in conjunction. The first is that everyone involved in biomedical research at any level must share an abhorrence of fraud and a determination to see it wiped out. This must include an absolute acceptance that failure to act is itself misconduct. The second element is that each country must have a truly empowered body to investigate and prosecute research fraud and misconduct.

This is underlined by part of the statement by the Code of Practice Committee of the GMC at the end of a disciplinary hearing against Geoffrey Fairhurst (32), who was found guilty of serious professional misconduct after research fraud. It surely represents the intent of regulatory bodies worldwide.

> "Trust lies at the heart of the practice of medicine. Patients must be able to trust doctors with their lives and wellbeing . . . Medical research must always be conducted with scrupulous honesty and integrity."

There will always be a variety of pressures surrounding biomedical research: pressure to gain publications for researchers, pressure to obtain a license for a new drug for pharmaceutical companies, and pressure to achieve funding for academic units. But there must be another overriding pressure on all sponsors and participants involved in research: the pressure to provide a safe environment for patients. This cannot be assured if fraud and misconduct are permitted to continue.

REFERENCES

1. Directive 2001/20/EC of the European Parliament and of the Council of 4 April 2001 on the approximation of the laws, regulations, and administrative provisions of the Member States relating to the implementation of good clinical practice in the conduct of clinical trials on medicinal products for human use. OJEC 2001; L121;34–44.
2. Definition of research misconduct. http://www.wellcome.ac.uk/About-us/Policy/Policy-and-position-statements/WTD002756.htm. Updated November 2005. Accessed September 15, 2008.
3. Definition of research misconduct. http://ori.hhs.gov/misconduct/definition_misconduct.shtml. Accessed September 15, 2008.
4. University of Abertay Dundee. Code of Good Practice in Research: A Document for All Staff and Students Undertaking Research. http://www.bristol.ac.uk/secretary/studentrulesregs/research-misc.html. Accessed December 30, 2004.
5. Pearce J, Manyonda I, Chamberlain G. Term delivery after intrauterine relocation of an ectopic pregnancy.Br J Obstet Gynaecol 1994; 101(8):716–717.
6. Pearce J, Hamid R. Randomised controlled trial of the use of human chorionic gonadotrophin in recurrent miscarriage associated with polycystic ovaries.Br J Obstet Gynaecol 1994; 101(8):685–688.
7. Sheikh A. Publication ethics and the research assessment exercise: Reflections on the troubled question of authorship.J Med Ethics 2000; 26:422–426.
8. Nordenberg T. Physician sentenced for doctoring drug data. FDA Consumer magazine. www.fda.gov/FDAC/departs/1999/399_irs.html. May–June 1999.
9. Eichenwald K, Kolata G.A doctor's drug studies turn into fraud. New York Times. Section A, New York Edition May 17, 1999.
10. Hwang W, Ryu Y, Park J, et al. Evidence of a pluripotent human embryonic stem cell line derived from a cloned blastocyst. Science 2004; 303:1669–1674.
11. Hwang W, Roh S, Lee B, et al. Patient-specific embryonic stem cells derived from human SCNT blastocysts. Science 2005; 308:1777–1783.
12. Kennedy D. Editorial retraction of Hwang et al. papers. Science 2006; 311:335.
13. McBride W. Thalidomide and congenital abnormalities. Lancet 1961; 2:1358.
14. McBride W, Vardy P, French J. Effects of scopolamine hydrobromide on the development of the chick and rabbit embryo. Aust J Biol Sci 1982; 35(2):173–178.
15. Humphrey G. Scientific fraud: The McBride case—judgment. Med Sci Law 1994; 34(4):299–306.
16. Eaton L.A quarter of UK students are guilty of plagiarism, survey shows. BMJ 2004; 329:70.
17. Plagiarism in a review article 2004 .http://www.publicationethics.org.uk/reports/2005/plagiarism-in-a-review-article-2004/.
18. Martyn C. Fabrication, falsification, and plagiarism. QJM 2003; 96:243–244.
19. Ankier S. Misconduct and fraud in clinical research. In: Medical Law and Research. Middlesex, Ankier Associates, 2004.
20. Martinson B, Anderson M, de Vries R. Scientists behaving badly. Nature 2005; 435:737–738.
21. Wadman M. One in three scientists confesses to having sinned. Nature 2005; 435:718–719.
22. Sheehan J. Fraud, conflict of interest, and other enforcement issues in clinical research. CCJM 2007; 74(2):S63–S67.
23. Pryor E, Habermann B, Broome M. Scientific misconduct from the perspective of research coordinators: A national survey.J Med Ethics 2007; 33:365–369.
24. Barrett J. Fraud and misconduct in clinical research. In: Edwards, Fletcher, Fox, Stonier, eds. Principles and Practice of Pharmaceutical Medicine, 2nd ed. Chichester: John Wiley, 2007:631–641.
25. UK Department of Health. Research Governance Framework for Health and Social Care. http://www.dh.gov.uk/en/Publicationsandstatistics/Publications/PublicationsPolicyandguidance/DH_4108962. 2005.
26. Van Der Weyden M. Preventing and processing research misconduct: A new Australian code for responsible research. MJA 2006; 184(9):430–431.
27. Mayor S. New governance framework for NHS research aims to stop fraud. BMJ 2000; 321:725.
28. Code of Federal Regulations. Title 21; Food and Drugs. Revised April 1, 2000.
29. Wells F. Historical aspects of research misconduct: Europe. In: Wells, Farthing, eds. Fraud and Misconduct in Biomedical Research, 4th ed. London: Royal Society of Medicine Press, 2008:72–85.
30. Barrett J. Conduct of an enquiry into alleged misconduct. In: Wells, Farthing, eds. Fraud and Misconduct in Biomedical Research, 4th ed. London: Royal Society of Medicine Press, 2008:260–273.
31. Farthing M. The role of national advisory bodies. In: Wells, Farthing, eds. Fraud and Misconduct in Biomedical Research, 4th ed. London: Royal Society of Medicine Press, 2008:260–273.
32. Dyer O. GP struck off for fraud in drug trials. BMJ 1996; 312:798.

42 Aspects of Biological and Biotechnological Medicinal Products

Manfred G. Haase
NDA Regulatory Science Ltd, Leatherhead, Surrey, U.K.

INTRODUCTION

This chapter "Aspects of Biological and Biotechnological Medicinal Products" covers primarily quality and quality-related safety issues connected with the following biological/biotechnological manufacturing techniques:

- Growth of strains of microorganisms and eukaryotic cells
- Extraction of substances from biological tissues, including human, animal, and plant tissues (allergens)
- Recombinant deoxyribonucleic acid (rDNA) techniques
- Hybridoma techniques
- Propagation of microorganisms in embryos and animals
- Advanced Therapy medicinal products.

In addition some important nonclinical and clinical aspects will be discussed briefly.

The manufacture of biological/biotechnological medicinal products ("biologics") involves certain specific considerations arising from the nature of the products and the processes used to make them. Furthermore, the ways in which biological/biotechnological medicinal products are produced, controlled, and administered make some particular precautions necessary.

Unlike conventional pharmaceutical medicinal products, which are produced using chemical and physical techniques capable of a high degree of consistency, the production of biological/biotechnological medicinal products involves biological processes and materials such as cultivation of cells or extraction of material from living organisms. These biological processes may display inherent variability so that the range and nature of by-products may be variable. In order to minimize variability (and reduce the opportunity for cross-contamination), steps such as dedicating equipment to specific products should be considered. Moreover, the materials and processing conditions used in these cultivation processes are generally designed to provide conditions for the growth of specific cells and microorganisms. This offers extraneous microbial contaminants the opportunity to grow. Most biological/biotechnological medicinal products cannot be terminally sterilized and processing must therefore be conducted aseptically. In addition, many products are limited in their ability to withstand a wide range of purification techniques particularly those designed to inactivate or remove adventitious viral contaminants. Great care must therefore be taken in the design of the processes, equipment, facilities, utilities, the conditions of preparation and addition of buffers and reagents, and in the training of the operators to minimize contamination (1,2).

Control of biological/biotechnological medicinal products usually involves biological analytical techniques, which have a greater variability than physicochemical determinations. A robust manufacturing process is therefore crucial and in-process controls take on a greater importance in the manufacture of biologics. The complexity and variability of both the biological/biotechnological product and the bioassays used to test them led already at an early stage in the development of the field to the use of reference material (international, national, in-house)—a sample of a product in question of defined potency and purity against which all preparations of that product must be calibrated (1,3).

Growth and innovation will continue in many areas of biological and biotechnological medicinal products, and it is of vital importance to manufacturers, regulators, and the public alike that production and control methods keep pace with current Good Manufacturing Practice (GMP), Good Laboratory Practice (GLP), and Good Clinical Practice (GCP) requirements.

New kinds of biologics are raising new scientific questions and potential public health concerns and effective management of manufacturing, testing, and regulation in this highly dynamic and productive field calls for further harmonization of requirements. Dominant areas of biological/biotechnological medicinal products in the next 10 years will primarily include vaccines, plasma-derived medicinal products, recombinant coagulation factors, allergens, hormones, cytokines and other regulatory factors, monoclonal antibodies, gene therapy medicinal products, and somatic cell therapy medicinal products (1,2).

The nature of biological/biotechnological medicinal products raised, right from their introduction in the 1890s and early 20th century, particular questions in manufacture and quality control and led regulators to the provision of specific rules governing these products. These rules arose from an appreciation of the particular difficulties in assuring the consistency, potency, and safety of these complex, often poorly characterized and not infrequently contaminated products. Two important control components of biologics are the following: the introduction of controls and testing of the total production process, not just at the end product; and the acceptance of calibration against biological standards to improve the accuracy of biological testing. The potency of a biological product usually needs to be assessed in terms of its biological activity rather than in terms of its chemical or physical properties. Biological variability means that this assessment depends on statistical techniques such as bioassays and on the availability of biological standards, which have known properties. Standards can be local (in-house), as might be used within a company for the assessment of a new assay, or regional [e.g., *European Pharmacopeia* (Ph Eur) standards] or international [World Health Organization (WHO) International Standards]. The International Unit (IU) is a unit of measurement for the amount of a substance based on measured biological activity or effect. The precise definition of the IU differs from substance to substance and is established by international agreement for each substance. To define the IU of a substance, the Expert Committee on Biological Standardization of the WHO provides a reference preparation of the substance, arbitrarily sets the number of IUs contained in that preparation and specifies a biological procedure to compare other preparations of that substance to the reference preparation. The goal in setting the standard is that different preparations with the same biological effect will contain a known number of IUs. The WHO Expert Committee on Biological Standardization views and recommendations are published in the WHO Technical Report Series (TRS). They provide updated information on the establishment, discontinuation and replacement of the WHO International Biological Reference Preparations as well on the adoption of Guidelines and Recommendations (3,4).

QUALITY ASPECTS

This paragraph should be read in conjunction with Part 3, Quality Topics in chapters 8 to 21 of this book. The quality requirements for the successful registration of biological and biotechnological medicinal products involve certain specific considerations arising from the nature of the products and the processes and make some particular precautions necessary. These particular precautions relate to, for instance

- starting materials such as cell banks and seed materials;
- control tests carried out at intermediate stages of the manufacturing process;
- safety tests such as sterility, bacterial endotoxin, and pyrogenicity;
- stability indicating tests at intermediate stages of production; and
- treatment of various acquired and inherited pathological dysfunctions with so-called advanced therapy medicinal products (5).

Furthermore, for controls of biological and biotechnological medicinal products that are not specified in the Ph Eur the procedures and the criteria of acceptability published as recommendations in the WHO "Requirements for Biological Substances, Technical Report Series (TRS)" shall serve as guidelines. The following sections inform about quality aspects in a few particular areas.

Production Process

Requirements relating to establishments in which pharmaceuticals are produced (e.g., EU and WHO GMPs) apply generally to the production of all biologics for human use. The way in which biologics are manufactured, controlled, and administered make some particular precautions for their manufacture necessary. Annexes to the principal GMP guidelines take these particular precautions into consideration (1,2). Additional useful information to help applicants prepare marketing authorization applications (MAAs) can be found in numerous product specific recommendations for the production and quality control of biological and biotechnological medicinal products, for example, in the previously quoted TRS of WHO and in Volume 3 "Scientific guidelines for medicinal products for human and veterinary use" of the publication series "The rules governing medicinal products in the European Union" (6).

Many of the general requirements for the quality control of biologics such as potency, abnormal toxicity testing, freedom from contaminants, stability, and freedom from detectable levels of antibiotics apply to most biologics. It is undesirable to use materials and/or excipients in their manufacture which are known to provoke sensitivity reactions in certain individuals, for example, penicillin or other β-lactam antibiotics. Any reagents of biological origin (e.g., sheep erythrocytes, foetal calf serum, bovine serum albumin, human transferrin, insulin, trypsin) used in the generation of the biological products producing cell line and/or during routine production should be free of microbial contamination such as mycoplasma, fungi, and bacteria. Special consideration should be given to possible viral contamination and tests for relevant viruses should be performed, for example, trypsin should be tested for porcine parvovirus and serum of bovine origin for bovine viruses (7,8). In addition, bovine sera and other bovine and ovine (and other species susceptible to TSE agents) derived biologics should comply with the guidance documents for minimizing the risk of transmitting agents causing spongiform encephalopathy via medicinal products and human plasma constituents and in addition with the guidance documents on new variant CJD and plasma-derived medicinal products (9). For plasma-derived medicinal products of human origin for therapeutic use and/or for excipients/stabilizers, the documents stating the requirements and standards of blood products must be taken into consideration (10). The use of transgenic animals to produce biologics requires special production process precautions and the considerations which apply tend to be a blend of those relevant to rDNA products and products from less defined sources (11). Because gene therapy products contain genetic and other biological materials, many of the quality, efficacy, and safety considerations, which apply to rDNA products and other biologics manufactured by modern biotechnological methods, will apply to some stages in their manufacture. Requirements relating to establishments in which biological products are manufactured will apply to the manufacture of gene therapy products as will several of the general recommendations for the quality control of biological products (5,6). Although comprehensive characterization of the final product is essential, considerable emphasis must also be placed on "in-process" control—a concept that has been highly effective in the quality control of vaccines prepared by conventional methods. The basic requirements for manufacture and control of vaccines are essentially the same as for other biological and biotechnological medicinal products. However, the current very promising concept of combined, multicomponent vaccines with numerous new vaccine designs different from monocomponent biological products poses additional issues. Each component must be developed and controlled individually in terms of quality, which includes the pharmaceutical development to establish the correct formulation and the compatibility between the individual components in the combined vaccine, including preservatives, excipients, and adjuvants (12,13). For some combined vaccines, there are already requirements available. For others, they have not yet been published. The methods of preparation vary according to the type of vaccine as described in individual pharmacopoeia monographs, WHO requirements, and European Medicines Agency (EMEA) guidelines. These methods are designed to guarantee the quality of each vaccine component with greatest emphasis on maintaining the appropriate antigenic properties and on ensuring freedom from contamination with extraneous agents.

Cell Substrates and Establishment of the Cell Bank System

A critical part of the quality control of the cell substrates involves the full characterization of cells. Distinguishing genetic, phenotypic and immunological markers of the cell will be useful in establishing the identity of the cells where eukaryotic cells are used. Likewise, where microbial cultures are used, specific phenotypic features which form a basis for identification should be described. It is essential that production of biologics is based on a well-defined Master Cell

Bank (MCB), a culture of fully characterized cells processed together to ensure uniformity and stability, and working cell bank, that is, the bank of cells derived from the MCB, which are used to seed the fermentation manufacturing process. Broad guidance on appropriate standards for the derivation of human and animal cell lines and microbial cells to be used to prepare biologics is provided in ICH guideline "Quality of Biotechnological Products: Derivation and Characterisation of Cell Substrates Used for Production of Biotechnological/Biological Products" (14). Quality concerns for cell-derived biological products have originated from the presence of adventitious contaminants or from innate properties of the cells used to prepare the product which are capable of influencing product quality and safety. There are also quality concerns for rDNA-derived products regarding the expression construct contained in the cell substrate. In the context of this chapter, there is insufficient space to provide details. In summary, ICH guideline "Quality of Biotechnological Products: Derivation and Characterisation of Cell Substrates Used for Production of Biotechnological/Biological Products" (14) deals with source, history, and generation of cell substrates; the cell banking system; cell banking procedures; general principles of characterization and testing of cell banks; cell substrate stability; and tests for karyology and tumorigenicity. This guideline covers cell substrates having a cell banking system and "cell substrate" refers to microbial cells or cell lines derived from human or animal sources that possess the full potential for generation of the desired biological and biotechnological medicinal products for human in vivo or ex vivo use. Cell banks used to prepare gene therapy products or vaccines should follow the recommendations presented in this ICH guideline (14). Some biological products, such as certain viral vaccines, are prepared in primary cell cultures derived directly from animal tissues or organs. Primary cells are not banked and therefore are not discussed in this ICH guideline (14). However, other considerations which may apply to primary cells are discussed further in Appendix 1 of this ICH guideline (14).

Biological Materials used in Production

Careful selection of biological materials used in production is one of the most important criteria for the safety of biological medicinal products. Any material of biological origin (e.g., foetal calf serum, bovine serum albumin, human albumin or transferrin, insulin, trypsin) used in the generation of cell substrates and establishment of the cell bank system and/or during routine production, including for example monoclonal antibodies for purification processes, should be free from microbial contamination such as mycoplasma, fungi, and bacteria. Special consideration should be given to possible viral contamination and tests for relevant viruses should be performed according to available ICH Q5A guidance (15). In addition, materials derived from ruminants should comply with the EU note for guidance "Minimising the Risk of Transmitting Animal Spongiform Encephalopathy Agents via Human and Veterinary Medicinal Products" (9) and materials of human origin should comply with the EU note for guidance "Investigation of Manufacturing Processes for Plasma-Derived Medicinal Products with Regard to vCJD Risk" (16) and/or similar guidance from other regulatory authorities, for example, the "WHO Guidelines on Transmissible Spongiform Encephalopathies in Relation to Biological and Pharmaceutical Products" (17). In order to reduce further the risk of infections in humans via biologics, a clear tendency to use entirely synthetic materials (nonhuman and nonanimal based) can be observed and freedom from biological materials during production is used as a promotional argument by Marketing Authorization Holders (MAHs).

Control of Contamination with Adventitious Agents

The various cell levels, for example, MCB and Manufacturer's Working Cell Bank (MWCB), used in production of biologics should be examined for adventitious agents (viral, bacterial, fungal, or mycoplasmal) in accordance with WHO requirements (18) and/or similar guidance from other regulatory authorities. Special attention should be given to viruses which can contaminate the species from which the cell line has been derived. Appropriate tests to demonstrate the absence of virus contamination should be performed as described in guideline ICH Q5A (15). Cells may have latent or persistent virus infection (e.g., herpes virus) or endogenous retroviruses (e.g., cell lines derived from rodents) which may be transmitted vertically from one cell generation to the next because the viral genome persists within the cell. Such viruses may be constitutively expressed or may unexpectedly become expressed as an infectious virus (15). Furthermore, potential viral contamination may take the form of complete viral genomes or subgenomic fragments resulting in the expression of infectious viral particles. A cell line which produces any virus capable of infecting human cells would be acceptable only in exceptional circumstances. There is special concern about the use of cell lines transformed by the deliberate introduction of Epstein-Barr virus for the production of human monoclonal antibodies (8). The importance of reliable viral validation studies is emphasized. Source cells should undergo suitable screening for extraneous agents and endogenous agents. The choice of viral detection and identification assays is dependent on the species and tissue of origin and recommendations for manufacturers on such assays are provided in guideline ICH Q5A (15). Where materials of bovine or other TSE-relevant animal species have been used in development or manufacture of the product the relevant guidance documents should be consulted (9,17).

The Purification of Biologics and its Validation

Methods used to purify the product and their in-process controls, including their specification limits, should be described in detail, justified, and validated. It is important to ensure that purification procedures do not impair relevant conformational features of the biologics. Consideration should be given to incorporating procedures which inactivate/eliminate potential viral contaminants where such methods will not compromise the biological activity and safety of the product. The capacity of the purification procedure to remove unwanted host cell–derived proteins, nucleic acids, carbohydrates, viruses, and other impurities including product-related proteins should be investigated thoroughly. Any inactivation process used should be shown to be effective and not compromise the biological activity of the product. The reproducibility of the purification process with respect to its ability to remove specific contaminants should also be demonstrated (7). Studies using, for example, a carefully selected group of viruses which exhibit a range of physicochemical features relevant to their behavior on purification [see note for guidance (19) on Virus Validation Studies: "The Design, Contribution and Interpretation of Studies Validating the Inactivation and Removal of Viruses"], host cell proteins, other potential impurities derived from the production process, and DNA intentionally mixed with the crude preparation (spiking) should be undertaken (7). The choice of the nucleotide probe to detect DNA contamination should be relevant to the system used. A reduction factor at each stage of purification for such contaminants and overall should be established by using, if necessary, various concentrations of viruses, host cell proteins, other potential impurities, and DNA in excess of that expected during normal production. Validation of the purification process should also include justification of the working conditions such as column loading capacity, column regeneration and sanitization, and length of use of the columns (7,8). Measures to minimize risks to humans from biological and pharmaceutical products in which ruminant and human materials are used during manufacture are described in "WHO Guidelines on Tissue Infectivity Distribution in Transmissible Spongiform Encephalopathies" (20) and in the EMEA guideline (9) referred to before. Moreover, the EMEA provides detailed recommendations with regards to the use of various specific materials in the manufacture of medicinal products, for example, lactose prepared using calf rennet (21), bovine materials used in the manufacture of vaccines (22), and tallow derivatives (23).

Consistency and Specifications

Evidence should be provided on the consistency of production, for example on at least five consecutive full-scale production batches. This should include information on the bulk harvest and final dosage form as well as on in-process controls. In the case of a production process where multiple harvests are used, batches from different fermentation runs are needed. The studies should include biological, chemical, and immunological methods to characterize and assay the active substance and methods to detect and identify impurities. A selection of the tests used to characterize the purified active substance should be used to confirm the product identity for each batch. The methods employed should include tests for the biological activity as well as physicochemical and immunological methods. The degree of purity desirable and attainable will depend on several factors. These include the nature and intended use of the product, the method of its production and purification, and also the degree of consistency of the production process. The purity of each batch should be established and should be within specified limits.

The analysis should include sensitive and reliable assays for DNA. Strict upper limits should be set for DNA in the product (7,8).

Specifications are chosen to confirm the quality of the biologics drug substance and drug product rather than to establish full characterization and should focus on those molecular and biological characteristics found to be useful in ensuring the safety and efficacy of the product. A specification establishes the set of criteria to which a drug substance, drug product, or materials at other stages of its manufacture should conform to be considered acceptable for its intended use. "Conformance to specification" means that the drug substance and drug product, when tested according to the listed analytical procedures, will meet the stated acceptance criteria. Specifications are critical quality standards that are proposed and justified by the manufacturer and approved by regulatory authorities as conditions of approval.

Specifications are one part of a total control strategy designed to ensure product quality and consistency. Other parts of this strategy include thorough product characterization during development (upon which many of the specifications are based), adherence to GMPs, adherence to a validated manufacturing process, appropriate raw materials testing, defined in-process testing, necessary stability testing, etc. An ICH guidance document provides general principles on the setting and justification, to the extent possible, of a uniform set of international specifications for biotechnological and biological products to support new MAAs (24).

Tests for Potency and Biological Standardization

Potency means the measurement of the units of biological activity of biologicals using a suitably quantitative biological assay based on the attribute of the product that is linked to the relevant biological properties in the clinic. A relevant validated potency assay should be part of the specification for a biologics drug substance and/or drug product. It is an important tool for quality control (24). The concept of the biological standard proved to be the key in establishing a reliable method of assessing potency because biological assays are themselves difficult to standardize, and it is necessary to run parallel assays of the test preparation and a standard preparation of defined potency to get a reasonably accurate result. The WHO has played a major role in achieving international consensus by establishing WHO International Biological Reference Materials, that is, primary standards that ensure the comparability of the activities of biological/biotechnological products worldwide and drawing up internationally agreed requirements and guidelines for potency and also safety measurement of many biologics (4). For Europe, the European Directorate for the Quality of Medicines (EDQM) in Strasbourg is in charge of the organization of the Biological Standardisation Programme. The Biological Standardisation Programme pursues the goal to establish common European reference preparations and to standardize the methods used for the quality control of biologicals (3,4).

Special Considerations for Stability Testing

The guidance stated in the ICH harmonized tripartite guideline entitled "Stability Testing of New Drug Substances and Products" (25) applies in general to biologics. However, biologics have distinguishing characteristics to which consideration should be given in developing a well-defined testing program designed to confirm their stability during the intended storage period. The products are particularly sensitive to environmental factors such as temperature changes, oxidation, light, ionic content, and shear. To ensure maintenance of biological activity and to avoid degradation, stringent conditions for their storage are usually necessary. The evaluation of stability may necessitate the use of complex analytical methodologies. Assays for biological activity, where applicable, should be part of the pivotal stability studies. Appropriate physicochemical, biochemical, and immunochemical methods for the analysis of the molecular entity and the quantitative detection of degradation products should also be part of the stability program whenever purity and molecular characteristics of the product permit use of these methodologies. With these concerns in mind, the applicant should develop the proper supporting stability data for biologics and consider many external conditions that can affect the product's potency, purity, and quality.

Relevant guidance for biologics is given in an annex to the ICH harmonized tripartite guideline "Quality of Biotechnological Products: Stability Testing of Biotechnological/Biological Products" (26). This annex applies to well-characterized proteins and polypeptides, their derivatives and products of which they are components and which are isolated from tissues, body fluids, cell cultures, or produced using rDNA technology. Thus the document covers the generation and submission of stability data for products such as cytokines (interferons, interleukins, colony-stimulating factors, tumor necrosis factors), erythropoietins, plasminogen activators, blood plasma factors, growth hormones and growth factors, insulins, monoclonal antibodies, and vaccines consisting of well-characterized proteins or polypeptides.

Changes to the Manufacturing Process

Manufacturers of biotechnological/biological products make changes to manufacturing processes of products frequently both during development and after approval. Reasons for such changes include improving the manufacturing process, increasing scale, improving product stability, and complying with changes in regulatory requirements. When changes are made to the manufacturing process the manufacturer generally evaluates the relevant quality attributes of the product to demonstrate that modifications did not occur that would adversely affect the safety and efficacy of the drug product.

An ICH "Note for Guidance on Biotechnological/Biological Products Subject to Changes in Their Manufacturing Process" has been proposed (27). The objective of this document is to provide guidance for assessing the comparability of biologics before and after changes are made in the manufacturing process for the drug substance or drug product. Therefore, this guideline is intended to assist in the collection of relevant technical information which serves as evidence that the manufacturing process changes will not have an adverse impact on the quality, safety, and efficacy of the drug product. The guideline does not prescribe any particular analytical, nonclinical or clinical strategy. The main emphasis of the document is on quality aspects. In addition, in Europe Commission Regulations have been adopted concerning the examination of variations to the terms of a marketing authorization. These Regulations contain particular guidance for biologics (28,29).

Plasma Master File and Vaccine Antigen Master File

The process of reviewing the European pharmaceutical legislation resulted in a body of laws in 2004 (5) which contains two new instruments related to marketing authorizations for biological medicines: the Plasma Master File (PMF) and the Vaccine Antigen Master File (VAMF). In the manufacture of plasma derivatives (e.g., human albumin), the same starting material, that is plasma pool, is usually employed for several products. In the case of vaccines, the same active substance, that is vaccine antigen, may be included in several combination vaccine products. The intention behind the introduction of the PMF and the VAMF was to avoid unneccesary and redundant documentation being generated by applicants and to improve and harmonize regulators' assessment by means of procedures for certification of master files on the community level. A VAMF is that part of a vaccine MAA concerning the active substance including information on starting materials and reagents, the production process, specification and routine controls, stability and the viral safety aspects of a vaccine antigen. The PMF contains common information on plasma from collection to plasma pool relevant to the manufacture of all intermediate fractions including cryoprecipitate, all constituents of the excipient, and all active substance(s), which are part of medicinal products or medical devices referred to in Directive 2000/70/EC (6) of the European Parliament and of the Council of 16 November 2000 amending Council Directive 93/42/EC (6) as regards medical devices incorporating stable derivatives of human blood or human plasma. Guidelines for applicants are available that describe the scientific data requirements for the two new procedures (30,31).

Official Control Authority Batch Release

As pointed out previously special considerations apply to the control of certain biologics which do not apply to conventional pharmaceuticals. These arise because biological methods are used in their production which is mostly based on the culture of cells or microorganisms using complex starting materials and because of the intrinsically variable nature of the biological assays used in their testing. Therefore most of the WHO's general and specific guidelines published in the WHO Technical Report Series in relation to biologicals and addressed to national control authorities include the recommendation that the authorities should have access to laboratory facilities for testing of in-process and final product samples. National regulatory systems for certain biologics specify independent testing of all batches before release but there is no globally harmonized approach toward independent testing by national control authorities (3). EU directives now in force include provision of independent (national authority) batch release testing for all vaccines and plasma derivatives as well as selective testing for certain other biotech products. Furthermore, the United States will continue to do lot release testing on vaccines and plasma derivatives (3). Batch control procedures operated by national authorities in many countries have over many years played a vital role in ensuring the quality, potency, and safety of biologics. Independent batch control procedures also do much to generate and maintain public confidence in the effectiveness and safety of biologics. In the European Union a guideline, legally based on Council Directive 2001/83/EC, is applicable on the procedure for official control authority batch release (32). However, there is obviously a need for a rational system, based on internationally agreed criteria, which can be applied globally to help decide where independent batch release testing is still needed (3).

QUALITY ASPECTS IN PARTICULAR AREAS

Vaccines

The biological nature of the starting materials, the complexity of the manufacturing process (e.g., up to two dozen active substances in combined vaccines with adjuvants), and the test methods needed to characterize these active substances and batches of the product raise important quality and quality-related safety questions for vaccines. Subtle changes in the microorganisms/cells used for the production may affect radically the vaccine product. The establishment of seed lot/cell bank systems is therefore essential for vaccine production. Moreover the quality, safety, and potency of vaccines are sensitive to changes in the manufacturing conditions and cannot be assured solely by end product testing but depend on the strict control of the manufacturing process. This includes demonstration of the purity and quality of the starting material, the adequacy of the in-process control testing, testing for process additives, and process intermediates. Consistency of production is of the greatest importance and the demonstration that the product does not differ from vaccine batches that have been shown to be safe and adequately immunogenic in clinical studies is a crucial component of vaccine evaluation, licensing, and batch release (12). The stability evaluation of vaccines is complex as they are very susceptible to inactivation by environmental factors. Therefore the WHO developed a relevant guidance document in 2006 (33). Vaccines have been of outstanding public health benefit and provide excellent prospects for the eradication of many serious infectious diseases. Nevertheless, experience has shown that their use in healthy human beings can also be associated with serious hazards, for example, transmission of infectious agents, anaphylactic reactions, or insufficient protection of recipients against the disease concerned.

To overcome the constraints described above and to cover specific manufacturing aspects of particular vaccines, the WHO develops and publishes recommendations and guidelines on the production and quality control of vaccines in general and on specific issues related to particular vaccines. They are developed and adopted by the WHO Expert Committee on Biological Standardization and published in the WHO Technical Report Series. These documents form the basis for assuring the acceptability of products globally as they specify the need for appropriate starting materials, including seed lot system and cell banks strict adherence to established protocols and tests for purity, potency, and safety at specific steps during production. Regulatory guidelines for vaccines are also developed by other bodies, such as the EMEA—see "The rules governing medicinal products in the EU, volume 3, Scientific guidelines for medicinal products for human use" (6)—and by the Center for Biologics Evaluation and Research (CBER) within the US Food and Drug Administration (FDA) (13,34). In addition, numerous pharmacopoeial requirements (monographs), such as those of the Ph Eur, are also established for vaccines.

Plasma-Derived Medicinal Products

Derivatives of human plasma produced on an industrial scale, such as albumin, transferrin, immunoglobulins, coagulation factors, and other plasma protein solutions, may be used for instance as therapeutic medicines, as starting material for the production of recombinant proteins in the fermentation media or as excipients. The addition of human albumin as stabilizer to a number of vaccines and recombinant medicinal products is still an important issue. The quality and safety of the products rely both upon donor selection and the screening of source materials and on the choice and control of the manufacturing processes, including processes which inactivate or remove microbial contaminants. The transmission of viruses is of particular concern and measures taken to prevent infection by the use of plasma-derived products include selection of donors, screening of individual donations, and starting materials for markers of infection with known viruses and validation of the production process for the inactivation and removal of viruses. GMPs should be adopted as an essential tool of the quality assurance system. In addition, GMPs should be adhered to at all levels of the manufacturing process from the collection of the source materials to the quality control of the final product thus ensuring traceability from the donor to the recipient. Requirements and standards for plasma-derived medicinal products are stated in the monograph "Human Plasma for Fractionation" and in several specific monographs of the Ph Eur (35). Other documents providing important additional guidance and recommendations in the field of plasma-derived products, including Creutzfeldt-Jakob disease, have been published by the Council of Europe (36), the WHO (37), and the EMEA (16).

Recombinant Coagulation Factors

In view of the high rate of transmission of blood-borne viruses by plasma-derived coagulation factor concentrates in the 1970s and early 1980s, there was considerable interest in the possibility of producing coagulation factors by rDNA technology. The structure of the factor VIII gene was elucidated in 1984 followed by the isolation of cDNA clones encoding the complete factor VIII sequence and the in vitro expression of human factor VIII in tissue culture. Since then commercial production of recombinant full-length and modified factor VIII, as well as recombinant factor IX (rFIX) products, have been accomplished for clinical use. Requirements relating to the production and quality control of medicinal products derived by rDNA technology in general will apply to the production of recombinant coagulation factors as will several of the general recommendations for the quality control of biological products. Certain factors may compromise the consistency, safety, and efficacy of rDNA coagulation factors. For example, the occurrence of an antibody against factor VIII, a so-called inhibitor, is the most important complication in hemophilia treatment. Inhibitors occur in up to about 30% of previously untreated patients with severe hemophilia A, usually within the first 100 exposure days. These inhibitors have been observed mainly in previously untreated children and approximately one-third of the inhibitors disappeared on continued treatment with the same product. It now appears that in cases in which inhibitors occur patient-related factors [certain types of mutations in the factor VIII gene, the family history, ethnicity and possibly Human Leukocyte Antigen major histocompatibility complex DR (HLA-DR) constitution] appear to be important determinants of inhibitor development. The immunogenicity of recombinant factor VIII (rFVIII) has to be addressed because in rFVIII products a heterogeneous protein pattern might occur due to differences in the posttranslational modifications of the proteins. Moreover, the choice of the manufacturing procedure will influence the nature, range, and amount of potential impurities in the final product. The purification processes must be shown to be capable of removing these potential impurities. Future applicants for recombinant coagulation factors are reminded of the availability of European Public Assessment Reports (EPARs) at the EMEA Web site (38) which reflect the scientific conclusions reached by the Committee for Medicinal Products for Human Use (CHMP) at the end of the centralized evaluation process. The EPAR provides a summary of the grounds for the CHMP opinion in favor of granting a marketing authorization for a specific medicinal product and an interesting insight into numerous quality, safety and efficacy issues: see for instance the EPARs for Kogenate, Advate, ReFacto and BeneFIX.

Monoclonal Antibodies

Monoclonal antibodies (mAbs) are antibodies with a defined specificity that are identical because they are produced by one type of immune cell that is a clone of a single parent cell. The classical method for obtaining monoclonal antibodies is the hybridoma technology, through fusing B cells from immunized mice with myeloma cells followed by clonal expansion. Monoclonal antibodies are characterized by a specific structure which is based on the immunoglobulin structure and a clearly defined functional activity, which is primarily based on a specific binding characteristic to a ligand (commonly known as the antigen).

Quality requirements for monoclonal antibodies in medicinal products were published in the mid-1990s. In 2004, the CHMP concluded that a complete and general revision and update of the guideline for mAbs was necessary (39). The development of a harmonized ICH guideline together with the United States and Japan has been postponed.

The field of mAb has matured since the mid-1990s. In the last few years, several medicinal products with a monoclonal antibody as the active substance have successfully obtained an MA. Moreover a high percentage of proteins currently in phase II and III clinical studies are mAbs. Monoclonal antibody-related substances are also being developed such as fusion proteins, proteolytic fragments, single-chain variable fragments (scFv) and other fragments, and bispecific monoclonal antibodies (BsAb). It is currently state of the art to humanize monoclonal antibodies by rDNA technology and express the mAb in a recombinant cell culture system. In addition a number of innovative methods to obtain mAbs have been developed, for example those based either on display technology (phage, ribosome, yeast) or genetically engineered mice that produce human antibodies. An updated guideline on "Production and Quality Control of Monoclonal Antibodies and Related Substances" is available as a draft document and planned for coming into operation in 2009 (40). In this draft guideline, specific requirements for monoclonal antibodies for therapeutic and prophylactic use (including ex vivo application) and in vivo diagnostic use are described and, in addition, requirements for mAb used as reagents, especially in the purification of other pharmaceutical products. Reference is made in this guideline to the recently proposed Ph Eur monograph on "Monoclonal antibodies for human use" (41).

Allergen Products

"Allergen product" means (according to EU directives) any medicinal product which is intended to identify or induce a specific acquired alteration in the immunological response to an allergizing agent (5). These products are subject to the requirements of the pharmaceutical legislation as regards the quality, safety, and efficacy testing and marketing authorization. In 1996 the EMEA proposed guidance on allergen products (42) and there is a Ph Eur general monograph (43). Both address the technical quality of allergen products that are based on allergen extracts. In recent years, more and more allergens have been generated by using rDNA technology and improved conventional techniques. Such novel allergens have been evaluated as innovative therapeutic products in clinical trials on specific immunotherapy. An overview on the regulatory system in Europe with special emphasis on allergen products has recently been published (44). A recently published guideline from the EMEA lays down the quality requirements for such novel allergen products of biological origin including allergen extracts from natural origin and allergens produced through rDNA technology used for specific immunotherapy or in vivo diagnosis of immunoglobulin E (IgE)-mediated allergic diseases (45). It applies to all allergen products and their intermediates manufactured by a method involving an industrial process and covers in its scope products obtained from allergen extracts, allergoids (i.e., allergens which are chemically modified to reduce IgE reactivity), conjugates, or allergens manufactured using rDNA technology. This revised guideline does not cover allergenic preparations consisting of synthetic peptides, DNA or RNA constructs and/or cell preparations or low–molecular-weight chemical allergens. This document also provides guidance on the establishment and use of in-house reference preparations, which are used for quality control including the analysis of batch-to-batch consistency. Moreover criteria for the preparation of the serum pools used for potency measurements are defined.

Advanced Therapy Medicinal Products

The treatment of various acquired and inherited pathological dysfunctions in humans calls upon novel concept-based approaches based on the development of biotechnology techniques. The latter involve the use of advanced therapy medicinal products. "Advanced therapy" medicinal products are defined in the EU based on manufacturing processes focused on various gene transfer-produced biomolecules and/or biologically advanced therapeutic modified cells as active substances or part of active substances. Examples are (i) gene therapy medicinal products (human and xenogeneic) and (ii) somatic cell therapy medicinal products (human and xenogeneic). In view of these developments, the European Commission has amended Directive 2001/83/EC (5) which now contains requirements for the presentation of MAAs for such products. Relevant guidelines for production, quality control, and nonclinical and clinical testing have been published or are currently under development (46–50).

Gene Therapy Medicinal Products

Gene therapy medicinal product are defined in the EU as products obtained through a set of manufacturing processes aimed at the transfer of a prophylactic, diagnostic, or therapeutic gene to human/animal cells performed either in vivo or ex vivo, and subsequently expressed in vivo. The gene transfer involves an expression system contained in a delivery system known as a vector which can be viral or nonviral in origin. The vector can also be included in a human or animal cell. Gene therapy medicinal products include naked nucleic acid, complex nucleic acid or nonviral vectors, viral vectors, and genetically modified cells. The EU Directive (5) requires that in Module 3 of the Common Technical Document special attention shall be paid to the following items:

a. Information on the relevant characteristics of the gene therapy medicinal product including its expression in the target cell population. Information concerning the source, construction, characterization, and verification of the encoding gene sequence shall be provided including its integrity and stability. Apart from therapeutic gene the complete sequence of other genes, regulatory elements, and the vector backbone shall be provided.
b. Information concerning the characterization of the vector used to transfer and deliver the gene shall be provided. This must include its physicochemical characterization and/or biological/immunological characterization. For medicinal products that utilize microorganisms such as bacteria or viruses to facilitate gene transfer (biological gene transfer) data on the pathogenesis of the parental strain and on its tropism for specific tissues and cell types as well as the cell cycle-dependence of the interaction shall be provided. For medicinal products that utilize non-biological means to facilitate gene transfer, the physicochemical properties of the constituents individually and in combination shall be provided.
c. The principles for cell banking or seed lot establishment and characterization shall apply to gene transfer medicinal products as appropriate.
d. The source of the cells hosting the recombinant vector shall be provided. For cells of animal origin, detailed information related to the following items shall be provided: sourcing of the animals, animal husbandry and care; transgenic animals (methods of creation, characterization of transgenic cells, nature of the inserted gene), measures to prevent and monitor infections in the source/donor animals, testing for infectious agents, facilities, control of starting and raw materials.
e. The evaluation of viral safety as well as the traceability of the products from the donor to the finished medicinal product are to form an essential part of the documentation to be supplied.

Several guidelines are intended to facilitate the collection and submission of data to support applications for MA within the EU for gene therapy products derived by biotechnology/high technology and intended for medicinal use in man (46–48). These guidelines should be read in conjunction with Directive 2001/83/EC of the European Parliament and of the Council of 6 November 2001 on the Community code relating to medicinal products for human use (5,6) and ICH guidelines (14,15,24) where appropriate.

Somatic cell therapy medicinal products

Somatic cell therapy medicinal products are defined in the EU in the Annex to Directive 2001/83/EC (5) as meaning the use in humans of autologous (i.e., emanating from the patient himself), allogeneic (i.e., coming from another human being), or xenogeneic (i.e., coming from animals) somatic living cells the biological characteristics of which have been altered as a result of their manipulation to obtain a therapeutic, diagnostic, or preventive effect through metabolic, pharmacological, and/or immunological means.

Human somatic cell therapy medicinal products are made of a defined number (pool) of viable cells which are derived from a manufacturing process starting either at the level of organs or tissues retrieved from a human being or at the level of a well-defined cell bank system where the pool of cells relies on continuous cell lines.

The starting materials and each step of the manufacturing process shall be fully documented and characterized including viral safety aspects. This includes the following: organs, tissues, body fluids and cells of human origin, the cell banking systems, ancillary materials, or ancillary medical devices. As regards animal somatic cells, detailed information shall be provided related to the following items: sourcing of the animals, animal husbandry and care, genetically modified animals, measures to prevent and monitor infections in the source/donor animals, testing for infectious agents including vertically transmitted microorganisms, facilities, cell banking systems, control of starting, and raw materials. In addition, the Annex to Directive 2001/83/EC (5,6) describes in detail information needed on

a. the manufacturing process of the active substance(s) and the finished product;
b. the characterization of active substance(s);
c. the pharmaceutical development of finished medicinal product; and
d. traceability of the products from the donor to the finished medicinal product.

Two notes for guidance have been proposed in this field: (i) manufacture and quality control of human somatic cell therapy medicinal products and (ii) points to consider on xenogeneic cell therapy medicinal products, both of which are intended to facilitate the collection and submission of data to support applications for MA within the EU (49,50). They should be read in conjunction with the relevant Directive 2001/83/EC of the European Parliament and of the Council of 6 November 2001 on the Community code relating to medicinal products for human use (5,6) and other specialized guidelines (14,15,18,46) where appropriate. A flexible approach to the control of these products has been adopted so that recommendations can be modified in the light of experience of production and use and of further developments.

Biosimilars

Since 2004 in the EU, a company has been able to choose to develop a new biological medicinal product claimed to be "similar" in terms of quality, safety, and efficacy to an original reference medicinal product which has been granted a marketing authorization in the EU on the basis of a complete dossier. Experience in Europe has shown that it is advisable to stipulate more precisely the cases in which the results of toxicological and pharmacological tests or clinical trials do not have to be provided with a view to obtaining an MA for a biosimilar medicinal product while ensuring that innovative firms are not placed at a disadvantage. Under this scenario, Directive 2001/83/EC lays down the requirements for the MAA based on the demonstration of the similar nature of the two biological medicinal products (5,6). Comparability studies are needed to generate evidence to substantiate the similarity of the nature of the new similar biological medicinal product and the chosen reference medicinal product authorized in the EU in terms of quality, safety, and efficacy.

The general principles to be applied for similar biological medicinal products are addressed in an overarching guideline taking into account the characteristics of the concerned biological medicinal product published by the EMEA (51). The purpose of this guideline is (i) to introduce the concept of similar biological medicinal products; (ii) to outline the basic principles to be applied; (iii) to provide applicants with a "user guide" showing where to find relevant scientific information in the various CHMP guidelines in order for the applicant to be able to substantiate the claim of similarity.

Applicants for similar biological medicinal products who have applied for scientific advice from the CHMP expressed the need for more specific guidance.

The advances as well as the limitations of methods and techniques available today for the full characterization of such biosimilar products have already prompted the CHMP to initiate the preparation of a number of specific guidelines relevant to quality, nonclinical, and clinical issues to be addressed within the development programs of similar biological medicinal products.

In principle the concept of a "similar biological medicinal product" is applicable to any biological medicinal product. The success of such a development approach will depend in practice on the ability of the applicant to characterize the products and therefore to demonstrate the similar nature of the concerned products. The chosen reference medicinal product must have been authorized in the EU on the basis of a complete dossier. The active substance of a similar biological medicinal product must be similar to the active substance of the reference medicinal product in molecular and biological terms. For example, a medicinal product containing interferon α-2a manufactured by Company X claiming to be similar to another biological medicinal product should refer to a reference medicinal product containing as its active substance interferon α-2a. Therefore a medicinal product containing interferon α-2b could not be considered as the reference medicinal product in this example.

The pharmaceutical form, strength, and route of administration of the similar biological medicinal product should be the same as that of the reference medicinal product. When the pharmaceutical form, the strength, or the route of administration is not the same then additional data should be provided in the context of the comparability exercise. Any differences between the similar biological medicinal product and the reference medicinal product will have to be justified by appropriate studies on a case-by-case basis. It should be recognized that, by definition, similar biological medicinal products are not generic medicinal products since it could be expected that there may be subtle differences between similar biological medicinal products from different manufacturers or compared with reference products which may not be fully apparent until greater experience in their use has been established. Therefore, in order to support pharmacovigilance monitoring, the specific medicinal product given to the patient should be identified clearly. While developing a similar biological medicinal product and carrying out the comparability exercise to demonstrate that one product is similar to another already authorized in the EU, some existing ICH and CHMP guidelines may be relevant and should therefore be taken into account (24,26,52). In addition to the guidelines already referred to the CHMP has developed further guidelines on Similar Biological Medicinal Products, for example (i) on quality issues of biosimilars (53); (ii) on nonclinical and clinical issues of biosimilars (54); and (iii) product-specific guidance on nonclinical and clinical issues for biosimilar medicinal products containing recombinant insulins (55), somatotropins (56), granulocyte stimulating factors (57), and epoetins (58).

NONCLINICAL SAFETY ASPECTS

The nature of biologics also raises particular questions of nonclinical safety evaluation. Considerable experience has now been gathered with submission of applications for all kinds of above mentioned biologics by applicants and assessment by regulators. Critical review of this experience has been the basis for development of the ICH Harmonised Tripartite Guideline "Preclinical Safety Evaluation of Biotechnology-Derived Pharmaceuticals" that is intended to provide general principles for designing scientifically acceptable nonclinical safety evaluation programs (52). In addition to the general principles recommended in the ICH Harmonised Tripartite Guideline, the WHO and several regulatory authorities have proposed guidance for special biologics, some of which will further down be briefly discussed in this chapter (13,59–61). Furthermore, an important source of information concerning the state-of-the-art design of nonclinical safety evaluation programs for biologics is published by the EMEA as part of the EPARs following the granting of an MA (38). The legal basis for its development and availability is contained in Article 13(3) of Regulation (EC) No 726/2004 (62). The EPAR results from the scientific committee's review of the documentation submitted by the applicant with respect to quality, nonclinical safety and clinical efficacy and safety and from subsequent discussions held during CHMP meetings. The EPAR is then updated throughout the life of the product as changes are made to the original terms and conditions of the MA.

All regulators throughout the world have, based on the ICH guideline "Preclinical Safety Evaluation of Biotechnology-Derived Pharmaceuticals" (52), adopted a flexible, case-by-case, science-based approach to nonclinical safety evaluation needed to support clinical development and MA, that is, adapted the design of evaluation programs to the products in question such as monoclonal antibodies, interferons, gene therapy products, or vaccines. Existing nonclinical notes for guidance for the safety testing of new chemical medicinal products (see Part 4 of this publication) recommend that studies be conducted on each new substance to investigate pharmacokinetics, pharmacodynamics, acute and chronic toxicity, reproductive effects, and carcinogenicity in suitable experimental systems. These notes for guidance will provide the basic framework for recommendations on testing specific biotechnology-derived products but amendments and alterations are necessary to take account of the particular nature of the biologics considered.

Only some of the studies recommended for chemical substances as presented and discussed in this book in Part 4, the nonclinical chapters are applicable to biologics before the grant of an MA. The primary goals of nonclinical safety evaluation are (i) to identify an initial safe dose and subsequent dose escalation schemes in humans, (ii) to identify potential target organs for toxicity and for the study of whether such toxicity is reversible, and (iii) to identify safety parameters for clinical monitoring. It should be emphasized once again that the ICH Harmonised Tripartite Guideline "Preclinical Safety Evaluation of Biotechnology-Derived Pharmaceuticals" (52) is intended primarily to recommend a basic framework for the nonclinical safety evaluation of biologicals.

Quality-related safety concerns are more important for biologicals than for chemical substances and may arise mainly from the presence of impurities or contaminants. For instance, there are specific potential risks associated with host cell contaminants derived from bacteria, yeast, insect cells, plants, and mammalian cells. The presence of cellular host contaminants can result in allergic reactions and other immunopathological effects. The adverse effects associated with nucleic acid contaminants are theoretical but include potential integration into the host genome and carcinogenicity respectively. For products derived from insect, plant, and mammalian cells or transgenic plants and animals, there may be an additional risk of viral infections. In this respect, it is preferable to rely on purification processes to remove impurities and contaminants rather than to establish a nonclinical testing program for their qualification. In general, the product that is used in the definitive pharmacology and toxicology studies should be comparable to the product proposed for the initial clinical studies. However, it is appreciated that during the course of development programs, changes normally occur in the manufacturing process of biologics in order to improve product quality and yields. This is not the case with conventional products for which the pharmaceutical development is generally terminated when starting the nonclinical testing program. The potential impact of such changes in the manufacturing process of biologics for extrapolation of the findings in animals used in nonclinical studies to humans should be carefully considered. In some cases, additional studies may be needed (e.g., pharmacokinetics, pharmacodynamics, and/or safety).

First-in-Human Clinical Trials

The violent adverse reactions of healthy volunteers who took the new monoclonal antibody TGN 1412 (developed by TeGenero based in Würzburg, Germany) in a phase 1 clinical trial conducted in the United Kingdom which resulted in the hospitalization of six persons on 13 March 2006 occurred due to an unpredicted biological action of the drug. The six healthy volunteers developed a severe cytokine release syndrome. This Mab, originally intended for the treatment of B cell chronic lymphocytic leukemia (B-CLL) and rheumatoid arthritis, was withdrawn from development. This incident led to a profound reconsideration of nonclinical and clinical safety testing requirements of biologics and the proposal of additional guidance in the EU (59). A "Guideline on strategies to identify and mitigate risks for first-in-human clinical trials with investigational medicinal products" has been developed to assist sponsors in the transition from nonclinical to early clinical development (60). It provides criteria to classify new investigational medicinal products according to risk factors including the calculation for the first doses to be given to human subjects, the initial dose-escalation trials, and the management of risk. It also gives guidance on quality aspects, nonclinical testing strategies, and designs for first-in-man clinical trials for investigational medicinal products including the calculation for the first doses to be given to human subjects. It focusses on medicinal products with potential risk factors, that is, drugs where the mode of action, the nature of the target in the body or the limited relevance of animal models for the prediction of toxicological effects in humans raise concerns for potential serious adverse reactions during first-in-man clinical trials. However, it is stressed in the guideline that the ability of nonclinical studies to predict safety issues in humans may be limited because the nature of the target is more specific to humans or because of other factors. Gene and cell therapy medicinal products are excluded and are to be covered by specific guidelines (60).

Nonclinical Safety Aspects in Particular Areas

A wide range of proteins, peptides, and other (even more complex) substances (e.g., gene therapy products) for use as biologicals can now be produced by novel biotechnological processes. Such products require thorough testing by relevant laboratory methods which allow for adequate assessment of safety before clinical investigation.

Bearing in mind the diversity of these products specific aspects of safety testing by product group have been considered by regulators and laid down in particular guidelines. For instance, the WHO and the EMEA have made available guidelines on nonclinical evaluation of traditional live and inactivated vaccines and recombinant vaccines against infectious diseases by outlining the international regulatory expectations in this area (61,63). The WHO recommends that these nonclinical guidelines for vaccine evaluation should be read in conjunction with guidelines on clinical evaluation of vaccines (64) in order to complete the understanding of the whole process of vaccine evaluation. With regards to plasma-derived medicinal products, manufacturers and regulators are faced with different nonclinical safety aspects (e.g., blood can harbour many viruses) and the use of medicinal products derived from human blood or plasma has led to the transmission of severe viral diseases including hepatitis A, B and C, and AIDS caused by the human immunodeficiency virus (HIV). Bacterial contaminants and pyrogens have also been the cause of serious accidents. Thus measures designed to prevent the transmission of such pathogens are essential to ensure the general safety of products derived from human blood and plasma (10). More extensive safety testing is unlikely to be required for plasma-derived products/excipients—for example, no fixed battery of tests is recommended currently. In the past a number of other biologicals administered to humans have also been contaminated with viruses. In several instances, the virus was only identified many years after the product had been introduced into the market since contamination occurred prior to the availability of adequate information concerning the presence of the infectious agents. The primary cause of these viral transmissions has been contamination of the starting or source materials. Therefore several guidance documents discuss the need for, and the contribution of, viral validation studies toward the viral safety of biologicals and provide recommendations on the design of a validation study including selection of viruses to be used and on the interpretation of the data. Similar safety concerns apply to Creutzfeldt-Jacob disease and plasma- and urine-derived medicinal products and excipients (65). A guideline on the nonclinical studies required prior to clinical use of gene therapy medicinal products defines scientific principles and provides guidance to applicants developing gene therapy medicinal products. Its focus is on the nonclinical studies required before the first use of a gene therapy medicinal product in human subjects (46). It has already been noted above that several guideline annexes to the overarching guideline on similar biological medicinal products (51) containing biotechnology-derived proteins as active substance have been proposed with regards to nonclinical and clinical testing issues, for example, for recombinant insulins (55), somatotropins (56), granulocyte stimulating factors (57), and epoetins (58). In all these documents, it is emphasized that they are not intended to provide rigid requirements of universal applicability; but that a degree of flexibility in the application of the recommendations is implicit and that case by case evaluation will be essential.

CLINICAL ASPECTS

Numerous general and therapeutic class guidance documents have been developed in the past years but in contrast to the remarkable differences between biologicals and pharmaceuticals in the area of quality and nonclinical testing the differences are less pronounced in the area of clinical safety and efficacy. Therefore clinical guidelines as discussed in Part 5 of this book, the clinical chapters, and the recommendations for the clinical overview in the Common Technical Document (discussed in chapter 7) apply generally in the same manner to biological and biotechnological medicinal products as to all other pharmaceuticals, and this is in like manner the case for therapeutic class clinical guidelines: Problems specific to most biological and biotechnological medicinal products such as possible transmission of adventitious agents, residual DNA, contaminants from the production process and changes in structure have been discussed in previous sections of this chapter. In most cases, a company wishing to bring biological or biotechnological medicinal products onto the market will have to take into account the general requirements for the clinical dossier and will need to establish the safety and efficacy of the product in the indications requested. However, there are exceptions to this rule and some examples will be discussed below.

Clinical Efficacy and Safety Aspects in Particular Areas

Besides specific requirements for the nonclinical evaluation of vaccines, the WHO and the EMEA have proposed requirements for the clinical evaluation of new vaccines (66,67). The documents provide guidance to manufacturers, clinical researchers, investigators, and regulators for the design of clinical development programs for new vaccines. This guidance is also useful for the generation of clinical data to support changes to the prescribing information in the postmarketing period. Guidance on various aspects for the evaluation of combination vaccines is also available from the FDA "Guidance for industry for the evaluation of combination vaccines for preventable diseases: production, testing and clinical studies" (13).

The public health importance of blood and blood products is recognized world wide and following viral safety concerns relating to plasma-derived medicinal products the EMEA as an example convened a group of experts to address efficacy and safety aspects of both plasma-derived and recombinant blood products. The CHMP Blood Products Working Party (BPWP) has provided globally accepted recommendations on all matters relating directly or indirectly to efficacy and safety aspects of blood products. Specific requirements for the following product classes for which a MAA is to be submitted as well as for authorized products in which a change in the manufacturing process has been made are now available. They concern clinical investigation in the areas of

- human plasma–derived factor VIII and IX products (65)
- recombinant factor VIII and IX products (64)
- human normal immunoglobulin for intravenous administration (68)
- human normal immunoglobulin for subcutaneous and intramuscular use (69)
- plasma-derived antithrombin products (70)
- plasma-derived fibrin sealant/hemostatic products (71)
- human anti-D immunoglobulin for intravenous and/or intramuscular use (72) and
- human plasma-derived von Willebrand factor products (73).

Interestingly, and upon request of allergen product manufacturers, the EMEA decided in 2007 to develop a guideline on the clinical development of products for specific immunotherapy for the treatment of allergic diseases. The EMEA felt that harmonization in this highly controversial therapeutic area was needed. This guideline is intended to address issues of study design, efficacy, and safety for allergen extracts and allergenic proteins being developed for specific immunotherapy of allergic diseases (e.g., the long-term treatment and management of allergic conditions such as rhinoconjunctivitis, mild allergic asthma, and allergic reactions to insect venoms) (74). This guideline covers clinical studies on specific immunotherapy, regardless of the affected organ system [e.g., nose, upper and lower airways, eyes, multiorgan affection (systemic reaction)], the allergen source (e.g., pollen, mites, animal dander, moulds, insect venoms, food), the allergen product (e.g., extracts, purified allergens, modified allergens, adsorbed allergens), or the route of administration (e.g., subcutaneous, sublingual). An excellent overview on the regulatory system in Europe with special emphasis on allergen products has recently been published (44).

Annex I, Part IV to Directive 2001/83/EC as amended (5) of the European Parliament and of the Council provides advice on the content of the clinical documentation for gene therapy and somatic cell therapy products. The efficacy of advanced therapy medicinal products must in general be demonstrated as described in Module 5 of the CTD but the clinical development of advanced therapy medicinal products will have some special features owing to the complex and labile nature of the active substances. The clinical development requires additional considerations because of issues related to viability, proliferation, migration, and differentiation of cells (somatic cell therapy) because of the special clinical circumstances where the products are used and because of the special mode of action through gene expression (somatic gene therapy). Special risks associated with such products arising from potential contamination with infectious agents must be addressed in the MAA for advanced therapy medicinal products.

More and more biological and biotechnological medicinal products have been developed against rare diseases affecting only a few thousand or even fewer than 100 patients. An EMEA guideline considers problems associated with clinical trials when there are limited numbers of patients available to study (75). Under such circumstances a trial enrolling several hundred patients may not be practical or possible. Similarly in pediatric trials, it may be difficult to recruit large numbers of patients. In this guideline, dealing with clinical trials in small populations, several possible strategies are outlined for an approach to trials under such circumstances.

Unwanted Immunogenicity of Protein Therapeutics

Treatment with therapeutic proteins (such as monoclonal antibodies, factor VIII, interferon, and erythropoietin) may induce an unwanted immune reaction, which is sometimes clinically significant. The immune response is influenced by many patient-, disease-, and product-related factors and it is usually impossible to predict the incidence, characteristics, or clinical consequences. Detection and management of the immunogenicity of therapeutic proteins is an essential undertaking during product development and safety surveillance following approval. Questions regarding immunogenicity are often raised during assessment of the MAA procedures, and therefore industry welcomed the initiative of the EMEA to give general guidance for the investigation and assessment of immunogenicity of therapeutic proteins. Some product-specific guidance existed already before, for example, for monoclonal antibodies (8), coagulation factors (65), and biotechnology-derived similar biological medicinal products (54) [i.e., recombinant human insulin (55), somatropin (56), recombinant human granulocyte colony stimulating factor (57), recombinant human erythropoietin (58)]. A guideline on immunogenicity assessment of biotechnology-derived therapeutic proteins has been published (76). The main topics addressed in this guideline are

- factors that may influence the development of an unwanted immune response
- nonclinical assessment of immunogenicity
- development of assays and assay strategies for the assessment of such immune responses
- potential clinical consequences of immunogenicity
- immunogenicity and clinical development, and
- recommendations for risk management plans.

Currently the most significant clinical problem in the treatment of hemophilia patients is the occurrence of neutralizing antibodies (inhibitors) which jeopardize the success of treatment. An EMEA expert meeting on FVIII products and inhibitor development was held in 2006 to discuss the standardization and harmonization of requirements for clinical studies in hemophilia patients. A number of issues were discussed relating to

- current data on FVIII products and inhibitor development
- FVIII inhibitor measurement
- clinical relevance of FVIII inhibitors
- clinical study design
- risk management
- postmarketing studies, and
- pharmacovigilance.

The outcome of these discussions will be the revision of the EMEA/CHMP guidance documents on the clinical investigation of human plasma-derived and recombinant FVIII and FIX products and the corresponding core summaries of product characteristics (77).

In summary, the principal contents of the guideline (76) with regards to the clinical setting for all therapeutic proteins is (i) careful planning of immunogenicity evaluation should include data systematically collected from a sufficient number of patients; (ii) the sampling schedule for immunogenicity evaluation should be standardized, adapted for each product on a case-by-case basis and taking a risk-based approach; (iii) data on the impact on efficacy and safety should be collected in order to fully understand the clinical consequences of the immune response; and (iv) immunogenicity issues should be further addressed in the risk management plan. Although data on possible unwanted immune reactions to therapeutic proteins are required before MA, problems may still be encountered in the postauthorization period. In the MAA, the applicant should therefore include a summary of investigations of immunogenicity in the overview with full cross-reference to the data in the relevant modules (76).

REFERENCES

1. Guidelines for national control authorities on quality assurance for biological products. (WHO Technical Report Series 822, 1992 Annex 2).
2. EudraLex. The Rules Governing Medicinal Products in the European Union, Volume 4 Good Manufacturing Practice. Medicinal Products for Human and Veterinary Use Draft Annex 2: Manufacture of Biological Medicinal Products for Human Use.
3. Review of the Scientific Basis of Standardization and Quality Control/Safety Monitoring of Biological Substances in Medicine. Edited by: National Institute for Biological Standards and Control, October 1996.
4. Guidelines for the preparation, characterization and calibration of national or laboratory working standards and reference reagents for biological substances. (WHO Technical Report Series 800, 1990 Annex 4, Part B).
5. Directive 2001/83/EC of the European Parliament and of the Council of 6 November 2001 on the Community code relating to medicinal products for human use. (Consolidated version: Official Journal No L 81 page 51 20.3.2008).
6. The rules governing medicinal products in the European Union. ec.europa.eu/enterprise/pharmaceuticals/eudralex/index.htm. Accessed June, 2009.
7. Production and Quality Control of Medicinal Products Derived by Recombinant DNA Technology. Eudralex. Notice to Applicants. (3Ab1a).
8. Production and Quality Control of Monoclonal Antibodies. Eudralex. Notice to Applicants. (3AB4a)).
9. Minimising the Risk of Transmitting Animal Spongiform Encephalopathy Agents via Human and Veterinary Medicinal Products. (EMEA/410/01 Rev. 2).
10. Plasma-Derived Medicinal Products. (CPMP/BWP/269/05 Rev. 3).
11. Use of Transgenic Animals in the Manufacture of Biological Medicinal Products for Human use. Eudralex. Notice to Applicants. (3AB7A).
12. Pharmaceutical and Biological Aspects of Combined Vaccines. (CPMP/BWP/477/97).
13. Guidance for industry for the evaluation of combination vaccines for preventable diseases: production, testing and clinical studies. USA, FDA, Center for Biologics Evaluation and Research (CBER), 04/99.
14. Quality of Biotechnological Products: Derivation and Characterisation of Cell Substrates used for Production of Biotechnological/Biological Products. (CPMP/ICH/294/95, ICH Topic Q5D).
15. Quality of Biotechnological Products: Viral safety Evaluation of Biotechnology Products derived from Cell Lines of Human or Animal Origin. (CPMP/ICH/295/95, ICH Topic Q5A).
16. Investigation of Manufacturing Processes for Plasma-Derived Medicinal Products with Regard to vCJD risk. (CPMP/BWP/5136/03).
17. WHO Guidelines on Transmissible Spongiform Encephalopathies in relation to Biological and Pharmaceutical Products. World Health Organization, February 2003. (WHO/BCT/QSD/03.01).
18. General requirements for the sterility of biological substances. In: WHO Expert Committee on Biological Standardization. Twenty-fifth report. Geneva, World Health Organization, 1973, Annex 4 (WHO Technical Report Series, No. 530); and WHO Expert Committee on Biological Standardization. Forty-sixth report. Geneva, World Health Organization, 1998, Annex 3 (WHO Technical Report Series, No. 872).
19. Virus Validation Studies: The Design, Contribution and Interpretation of Studies validating the Inactivation and Removal of Viruses. (CPMP/BWP/268/95 3AB8A).
20. WHO Guidelines on Tissue Infectivity Distribution in Transmissible Spongiform Encephalopathies. World Health Organization, 2006. ISBN 92 4 154701 4.
21. Public Report on Risk and Regulatory Assessment of Lactose and other products prepared using Calf Rennet. (EMEA/CPMP/BWP/337/02).
22. Public Statement on the Evaluation of Bovine Spongiform Encephalopathies (BSE) – Risk via the Use of Materials of Bovine Origin in or during the Manufacture of Vaccines. (EMEA/CPMP/BWP/476/01).
23. Position Paper on Production of Tallow Derivatives for Use in Pharmaceuticals. (CPMP/1163/97).

24. Specifications: test procedures and acceptance criteria for biotechnological/biological products. (CPMP/ICH/365/96, ICH Topic Q6B).
25. Stability Testing of New Drug Substances and Drug Products. (CPMP/ICH/2736/99, ICH Topic Q1A).
26. Quality of Biotechnological Products: Stability Testing of Biotechnological/Biological Products. (CPMP/ICH/138/95, ICH Topic 5QC).
27. Note for Guidance on Biotechnological/Biological Products Subject to changes in their Manufacturing Process. (CPMP/ICH/5721/03- ICH Topic Q5E).
28. Commission Regulation (EC) No 1085/2003 of 3 June 2003 concerning the examination of variations to the terms of a marketing authorisation for medicinal products for human use and veterinary medicinal products falling within the scope of Council Regulation (EEC) No 2309/93. (Official Journal L 159, 27/6/2003 p. 24–45).
29. Commission Regulation (EC) No 1084/2003 of 3 June 2003 concerning the examination of variations to the terms of a marketing authorisation for medicinal products for human use and veterinary medicinal products granted by a competent authority of a Member State. (Official Journal L 159, 27/6/2003 p. 1–23).
30. Scientific Data Requirements for a Vaccine Antigen Master File (VAMF). (CPMP/BWP/3734/03).
31. Scientific Data Requirements for Plasma Master File (PMF). (CHMP/BWP/3794/03 Rev.1).
32. Procedure for Official Control Authority Batch Release (OCABR) of centrally authorised immunological medicinal products for human use and medicinal products derived from human blood and plasma. (PA/PH/OMCL (2000) 83, DEF).
33. Guidelines on stability evaluation of vaccines. (WHO/BS/06.2049).
34. Guidance for industry: Content and format of chemistry, manufacturing and controls information and establishment description for a vaccine or related product USA, FDA, Center for Biologics Evaluation and Research (CBER), 01/97.
35. Monograph: human plasma for fractionation. In: European Directorate for the Quality of Medicines (EDQM), ed. European Pharmacopoeia, 6th ed. Strasbourg: Council of Europe, 2008:2073–2075.
36. Guide to the preparation, use and quality assurance of blood components. http://www.edqm.eu/site/Blood_Transfusion_amp_Organ_Transplantation_Guides-1131.html. Accessed June, 2009
37. Recommendations for the production, control and regulation of human plasma for fractionation. (WHO, Technical Report Series 941, Annex 4).
38. EMEA (European Medicines Evaluation Agency, EU). www.eudra.org/emea.html. Accessed June, 2009.
39. Need to Revise the Guideline on Production and Quality Control of Monoclonal Antibodies. (CHMP/BWP/64/04).
40. Production and Quality Control of Monoclonal Antibodies and Related Substances. (CHMP/BWP/157653/07).
41. Monograph: monoclonal antibodies for human use. In: European Directorate for the Quality of Medicines (EDQM), ed. European Pharmacopoeia, 6th ed. Strasbourg: Council of Europe, 2008:690–693.
42. Allergen Products. (CPMP/BWP/243/96).
43. Monograph: allergen products. In: European Directorate for the Quality of Medicines (EDQM), ed. European Pharmacopoeia, 6th ed. Strasbourg: Council of Europe, 2008:679–680.
44. Lorenz AR, Lüttkopf D, Seitz R, et al. The Regulatory System in Europe with Special Emphasis on Allergen Products. Int Arch Allergy Immunol 2008; 147:263–275.
45. Allergen products: Production and quality issues. (CHMP/BWP/304831/07).
46. Note for guidance on quality, preclinical and clinical aspects of gene transfer medicinal products. (CPMP/BWP/3088/99).
47. Guideline on the quality, preclinical and clinical aspects of medicinal products containing genetically modified cells. (EMEA/GTWP/58311/07).
48. Guideline on clinical monitoring and follow-up of patients exposed to gene therapy/gene transfer medicinal products. (EMEA/GTWP/60436/07).
49. Points to consider on the Manufacture and Quality Control of Human Somatic Cell Therapy Medicinal Products. (CPMP/BWP/41450/98).
50. Points to consider on xenogeneic cell therapy medicinal products. (CPMP/1199/02).
51. Guideline on similar biological medicinal products. (CHMP/437/04).
52. Note for Guidance on Preclinical Safety Evaluation of Biotechnology-Derived Pharmaceuticals. (CPMP/ICH/302/95).
53. Guideline on similar biological medicinal products containing biotechnology derived proteins as active substances: quality issues. (EMEA/CHMP/BWP/49348/05).
54. Guideline on similar biological medicinal products containing biotechnology-derived proteins as active substance: non-clinical and clinical issues. (EMEA/CHMP/42832/05).

55. Committee for Medicinal Products for Human Use. Annex guideline on similar biological medicinal products containing biotechnology-derived proteins as active substance: non-clinical and clinical issues. Guidance on similar medicinal products containing recombinant human insulin. (EMEA/CHMP/32775/05).
56. Committee for Medicinal Products for Human Use. Annex guideline on similar biological medicinal products containing biotechnology-derived proteins as active substance: non-clinical and clinical issues. Guidance on similar medicinal products containing somatropin. (EMEA/CHMP/94528/05).
57. Committee for Medicinal Products for Human Use. Annex guideline on similar biological medicinal products containing biotechnology-derived proteins as active substance: non-clinical and clinical issues. Guidance on biosimilar medicinal products containing recombinant granulocyte-colony stimulating factor. (EMEA/CHMP/31329/05).
58. Committee for Medicinal Products for Human Use. Annex guideline on similar biological medicinal products containing biotechnology-Derived proteins as active substance: non-clinical and clinical issues: guidance on biosimilar medicinal products containing recombinant erythropoietins. (EMEA/CHMP/94526/05).
59. Schneider CK. First-in-human trials with therapeutic proteins: Regulatory rethink? Expert Rev Clin Pharmacol 2008; 1(3):327–331.
60. Guideline on strategies to identify and mitigate risks for first-in-human clinical trials with investigational medicinal products. (CHMP/SWP/28367/07).
61. Guidelines on non-clinical evaluation of vaccines. (WHO, Technical Report Series 927, 2005, Annex 1).
62. Regulation (EC) No 726/2004 of the European Parliament and of the Council of 31 March 2004 laying down Community procedures for the authorisation and supervision of medicinal products for human and veterinary use and establishing a European Medicines Agency (Official Journal L 136, 30/4/2004 p. 1–33).
63. Note for Guidance on preclinical pharmacological and toxicological testing of vaccines. (CPMP/SWP/465/95).
64. Clinical Investigation of Recombinant Factor VIII and IX Products. (CPMP/BPWG/1561/99).
65. Clinical Investigation of Human Plasma Derived Factor VIII and IX Products. (CPMP/BPWG/198/95).
66. Guidelines on clinical evaluation of vaccines: regulatory expectations. (WHO Technical Report Series 924, 2005, Annex 1).
67. Guideline on clinical evaluation of new vaccines. (CHMP/VWP/164653/05).
68. Clinical Investigation of Human Normal Immunoglobulin for Intravenous Administration (IVIg). (CPMP/BPWG/388/95 Rev. 1).
69. Clinical Investigation of Human Normal Immunoglobulin for Subcutaneous and Intramuscular use. (CPMP/BPWG/283/00).
70. Clinical Investigation of Plasma derived Antithrombin Products. (CPMP/BPWG/2220/99)
71. Clinical Investigation of Plasma derived Fibrin Sealant/Haemostatic Products. (CPMP/BPWG/1089/00).
72. Clinical Investigation of Human Anti-D Immunoglobulin for Intravenous and/or Intramuscular Use. (CPMP/BPWG/575/99 Rev. 1).
73. Clinical Investigation of Human Plasma Derived von Willebrand Factor Products. (CPMP/BPWG/220/02).
74. Guideline on the clinical development of products for specific immunotherapy for the treatment of allergic diseases. (CHMP/EWP/18504/06).
75. Guideline on clinical trials in small populations. (CHMP/EWP/83561/05).
76. Immunogenicity assessment of biotechnology-derived therapeutic proteins. (CHMP/BMWP/14327/06).
77. Report of expert meeting on factor VIII products and inhibitor development. (EMEA/CHMP/BPWP/123835/06).

43 Device-Drug Combination Products

Brian R. Matthews
Alcon Laboratories, Hemel Hempstead, U.K.

INTRODUCTION

This chapter will consider the factors associated with combinations of medical devices and pharmaceuticals particularly with respect to nonbiological sourced pharmaceuticals.

The regulation of "device-drug combination products" is not harmonized between the three regions of the International Conference on Harmonization (ICH) nor within the members of the Global Harmonisation Task Force (GHTF). Furthermore, even within the regions and countries, there can be divergent practice in the regulation of these products—the situation within the European Union (EU) is an example of this. Even where the relevant advisory group at the European level (of which more is discussed later) has recommended that a particular product should be regulated under the medical device regime there is no universal adoption of this recommendation between the Member States (or in some cases even within a Member State).

For the purpose of this chapter, it will be assumed that the product being considered consists of at least one component that could in certain circumstances be considered to be covered by pharmaceutical (including biological product) legislation and at least one component that could in certain circumstances be considered to be covered by the medical device legislation. It will also be assumed that the product is intended for use in or on human beings—use on animals can complicate the legislative system that applies in some regions [such as the EU, where the Medical Device Directives (MDDs) do not extend to products used on animals]. Within the EU the "device" component is assumed also to be covered by the MDD (Directive 93/42/EEC as amended) rather than under either the active implantable medical device legislation or the *in vitro* diagnostic MDD for the purposes of this chapter.

To say that the situation is not harmonized could be seen as an understatement. At the time of writing, there was not even a common understanding of what a "device-drug combination product" is.

In the United States (US), the definition (1) of a combination product is as follows:

(1) A product comprised of two or more regulated components, i.e. drug/device, biologic/device, drug/biologic, or drug/device/biologic, that are physically, chemically or otherwise combined or mixed and produced as a single entity;
(2) Two or more separate products packaged together in a single package or as a unit and comprised of drug and device products, device and biological products, or biological and drug products;
(3) A drug, device, or biological product packaged separately that according to its investigational plan or proposed labeling is intended for use only with an approved individually specified drug, device, or biological product where both are required to achieve the intended use, indication, or effect; and where upon approval of the proposed product the labeling of the approved product would need to be changed, e.g., to reflect a change in intended use, dosage form, strength, route of administration, or significant change in dose; or
(4) Any investigational drug, device, or biological product packaged separately that according to its proposed [labeling] is for use only with another individually specified investigational drug, device, or biological product where both are required to achieve the intended use, indication or effect.

These definitions are not intended to imply that a combination product includes products containing two or more different drug components—these are classed as "fixed-combination" pharmaceutical products, and continue to be subject to assessment by the Center for Drug Evaluation and Research. Drug-cosmetic combinations and products containing drugs and dietary supplements are dealt with separately within the Food and Drug Administration (FDA) system.

Within the EU, device-drug combinations are within a broad category of "borderline products." In the draft revision 3 of the relevant guideline (2), it is indicated that "a combination product is considered to be a product which is comprised of two or more differently regulated components." This category is not limited to the borderline between pharmaceuticals (which includes biological medicinal products) and medical devices. Other interfaces can occur with cosmetic products, foodstuffs and personal protective equipment, for example. There is not a simple definition of where the drug-device borderline lies. Instead it is necessary to read the definitions in the pharmaceutical legislation and the medical device legislation within the EU. Thus, if you want to determine whether a product is to be controlled as a pharmaceutical or as a medical device, you have to read these definitions.

Article 1 [of Directive 2001/83/EC as amended, 3].

For the purposes of this Directive, the following terms shall bear the following meanings:

2. Medicinal product:
   a. Any substance or combination of substances presented as having properties for treating or preventing disease in human beings; or
   b. Any substance or combination of substances which may be used in or administered to human beings either with a view to restoring, correcting or modifying physiological functions by exerting a pharmacological, immunological or metabolic action, or to making a medical diagnosis.
3. Substance:

   Any matter irrespective of origin which may be:
   –human, e.g. human blood and human blood products;
   –animal, e.g. micro-organisms, whole animals, parts of organs, animal secretions, toxins, extracts, blood products;
   –vegetable, e.g. micro-organisms, plants, parts of plants, vegetable secretions, extracts;
   –chemical, e.g. elements, naturally occurring chemical materials and chemical products obtained by chemical change or synthesis.

The European definition of a medical device is given as follows.

Definitions, scope [of Directive 93/42/EEC as amended, 4]

Article 1

2. For the purposes of this Directive, the following definitions shall apply:
   a. 'medical device' means any instrument, apparatus, appliance, software, material or other article, whether used alone or in combination, including the software intended by its manufacturer to be used specifically for diagnostic and/or therapeutic purposes and necessary for its proper application, intended by the manufacturer to be used for human beings for the purpose of:
      –diagnosis, prevention, monitoring, treatment or alleviation of disease,
      –diagnosis, monitoring, treatment, alleviation of or compensation for an injury or handicap,
      –investigation, replacement or modification of the anatomy or of a physiological process,
      –control of conception,
      and which does not achieve its principal intended action in or on the human body by pharmacological, immunological or metabolic means, but which may be assisted in its function by such means.

The phrase "principal intended action" is most important in determining whether a particular product is to be controlled under the pharmaceutical regime or the medical device regime.

Given the importance of "pharmacological, immunological or metabolic means," it is perhaps surprising that these terms were not defined in the legislation. However, they were included in MEDDEV 2.1/3 rev 2 (2) in the following terms:

"Pharmacological means" is understood as an interaction between the molecules of the substance in question and a cellular constituent, usually referred to as a receptor, which either results in a direct response, or which blocks the response to another agent. Although not a completely reliable criterion, the presence of a dose-response correlation is indicative of a pharmacological effect.

"Immunological means" is understood as an action in or on the body by stimulation and/or mobilization of cells and/or products involved in a specific immune reaction.

"Metabolic means" is understood as an action which involves an alteration, including stopping, starting, or changing the speed of the normal chemical processes participating in, and available for, normal body function.

"Material" is not defined in the MDD.

There is additional text in Article 1 of the MDD (3) which reads as follows:

3. Where a device is intended to administer a medicinal product within the meaning of Article 1 of Directive 2001/83/EC, that device shall be governed by this Directive, without prejudice to the provisions of Directive 2001/83/EC with regard to the medicinal product.

   If, however, such a device is placed on the market in such a way that the device and the medicinal product form a single integral product which is intended exclusively for use in the given combination and which is not reusable, that single product shall be governed by Directive 2001/83/EC. The relevant essential requirements of Annex I to this Directive shall apply as far as safety and performance-related device features are concerned.

4. Where a device incorporates, as an integral part, a substance which, if used separately, may be considered to be a medicinal product within the meaning of Article 1 of Directive 2001/83/EC and which is liable to act upon the body with action ancillary to that of the device, that device shall be assessed and authorized in accordance with this Directive.

4a. Where a device incorporates, as an integral part, a substance which, if used separately, may be considered to be a medicinal product constituent or a medicinal product derived from human blood or human plasma within the meaning of Article 1 of Directive 2001/83/EC. and which is liable to act upon the human body with action ancillary to that of the device, hereinafter referred to as a 'human blood derivative', that device shall be assessed and authorised in accordance with this Directive.

For the purposes of this chapter the other relevant definitions in EU pharmaceutical legislation (relating to radiopharmaceuticals and immunological products, for example) will not be considered. However, attention has to be drawn to the following legislative text included in a recent revision to the pharmaceutical directive. (Article 2 [of Directive 2001/83/EC as amended) (3)]

2. In cases of doubt, where, taking into account all its characteristics, a product may fall within the definition of a 'medicinal product' and within the definition of a product covered by other Community legislation the provisions of this Directive shall apply.

It is intended that for any particular product, the provisions of *either* the pharmaceutical *or* the medical device regulatory regime should apply. It has been suggested that the meaning of Article 2(2) of Directive 2001/83/EC (as amended) is to be read so that the "principle of non-cumulation only applies if, after a case-by-case assessment, taking into consideration all the characteristics of a product, it is clear that the product in question may fall within the definition of both medical device and medicinal product. In such a case, the provisions of [Directive 2001/83/EC (as amended)] apply" in the draft revision 3 of MEDDEV 2.1 *Guidance on Borderline Products: Medical Products/Medicinal Products and Combination Products*. The draft guidance also goes on to refer to the status of *in vivo* diagnostic agents: "Due to the definition of 'medicinal product', substances used in or administered to human beings to make a medical diagnosis, even if they fulfill their function by physical or chemical means, and not by pharmacological, immunological, or metabolic means in the sense as described above are considered to be medicinal products."

In a presentation, *Combination Products Regulation in Japan* (5) by Tomiko Tawaragi a summary of the situation in Japan was presented. The Japanese Pharmaceutical Affairs Law regulates the following products:

1. **Drugs**
   Articles included in the Japanese Pharmacopoeia
   Articles which are intended for use in the diagnosis, treatment, or prevention of disease and which are not covered by equipment, instrument, etc
   Articles intended to affect the structure or functions of the body and which are not equipment, instruments, etc
2. **Medical devices**
   Equipment or instruments specified by the government ordinance which are intended for use in the diagnosis, treatment, or prevention of disease or intended to affect the structure and functions of the body
3. **Cosmetics**
4. **Quasidrugs.**

It is understood that there is no specific definition of "combination products" in Japanese legislation. In practice, products that consist of a drug and a medical device combined physically or chemically or copackaged are considered to be combination products. This does not extend to products where the instructions for use require certain combinations to be applied—e.g. photodynamic therapy and the associated laser device.

DECISION-MAKING PROCEDURES

Japan

In Japan each potential drug-device combination product is considered as a case-by-case example. The primary mode of action of the combination is the main criterion for deciding whether it should be controlled as a pharmaceutical or a medical device. The specific case will be considered within the Ministry of Health, Labour and Welfare. The Licensing and Evaluation Division, the Medical Devices Evaluation Office, and the Compliance and Narcotics Division will all be consulted. If the product is considered to be a pharmaceutical then the Office of New Drugs of the Pharmaceuticals and Medical Devices Agency (PMDA) will take the lead in its review. If the product is a medical device, the lead will be taken by the Office of Medical Devices at the PMDA. There will be cooperation between these two units during the review process.

A single application (pharmaceutical or medical device) may be used in Japan for a combination product. If the product is considered to be a pharmaceutical then pharmaceutical good manufacturing practices (GMPs) are required to be met. If it is a medical device, then the relevant quality management system and postmarketing surveillance requirements will have to be met.

United States

In the US, the FDA Office of Combination Products has been established to allow a focal point for the discussion of combination product issues between manufacturers and various Agency staff. An FDA Center will be assigned for each combination or other product where the internal FDA jurisdiction is not clear or is in dispute. This office can be contacted to determine which regulatory regime should be applied to a particular product. The contact may be informal or formal. When requesting a formal determination of jurisdiction, the manufacturer concerned should submit at least the following information: a brief description of the product and its major components, the intended uses (clinical and therapeutic claims, types of patient with whom the product will be used) of the product, and explanation of how the product works. The specific requirements are stated thus:

> (c) What to file: an original and two copies of the request for designation must be filed. The request for designation must not exceed 15 pages, including attachments, and must set forth:

(1) The identity of the sponsor, including company name and address, establishment registration number, company contact person and telephone number.
(2) A description of the product, including:
  (i) Classification, name of the product and all component products, if applicable;
  (ii) Common, generic, or usual name of the product and all component products;
  (iii) Proprietary name of the product;
  (iv) Identification of any component of the product that already has received premarket approval, is marketed as not being subject to PMA, or has received an investigational exemption, the identity of the sponsors, and the status of any discussions or agreements between the sponsors regarding the use of this product as a component of a new combination product.
  (v) Chemical, physical, or biological composition;
  (vi) Status and brief reports of the results of developmental work, including animal testing;
  (vii) Description of the manufacturing processes, including the sources of all components;
  (viii) Proposed use or indications;
  (ix) Description of all known modes of action, the sponsor's identification of the single mode of action that provides the most important therapeutic action of the product, and the basis for that determination.
  (x) Schedule and duration of use;
  (xi) Dose and route of administration of drug or biologic;
  (xii) Description of related products, including the regulatory status of those related products; and
  (xiii) Any other relevant information. [6]

There is also a guidance document on how to prepare a request for designation for a particular product (7).

In determining the primary mode of action for a particular product, the following is taken into account: the single mode of action of a combination product that provides the most important therapeutic action of the combination product (8). This is expressed in the *Federal Register* as follows (9):

(m) Primary mode of action is the single mode of action of a combination product that provides the most important therapeutic action of the combination product. The most important therapeutic action is the mode of action expected to make the greatest contribution to the overall intended therapeutic effects of the combination product.

After consideration of the details provided by the manufacturer, there will be a decision on the Center within the FDA that will take the lead on the assessment of the application. This decision will consider not only the information provided by the manufacturer but also wider FDA experience that is of relevance. Where it is not possible to identify a mode of action that is providing the greatest contribution to overall therapeutic effect, then account will also be taken of other combination products with which similar safety and efficacy questions arose and regarding which Center in the FDA has the most relevant expertise.A wide range of guidance documents on various aspects of FDA expectations for combination products may be accessed (10). The decision on jurisdiction must be issued within 60 days. There is the possibility of appealing the outcome, with a response time of 15 days.

European Union

In the EU there is no a simple, well-defined procedure to obtain a definitive decision on the status of a particular product and whether this should be approved for marketing through the pharmaceutical or the medical device regulatory systems. If a product is a medical device then its status is determined by the manufacturer and, in the case of all but the simplest products, there is some involvement with the manufacturer's chosen Notified Body. If there is doubt or disagreement concerning the status of a particular product then the manufacturer or the Notified Body can consult with one or more medical device Competent Authorities within the EU. Where there is a lack of agreement or apparent inconsistency it is also possible to refer issues to the Commission of the European Communities (the Commission) who can then take the issue up with the Member States concerned and/or refer the matter to the Medical

Device Experts Group and its borderline and classification working group, which are advisory bodies consisting of representatives of the Member States and other interested parties (which includes representatives of relevant industry trade associations, Notified Bodies, etc.). Not all of the recommendations made by the Medical Device Experts Group or the borderline and classification group have been published. Some have been included in the revised version of the MEDDEV on borderline issues referred to above.

Within the European system there is an additional requirement for consultation between the Notified Body that is chosen by the manufacturer of the device and the relevant pharmaceutical Competent Authority. This applies to combination products that contain substances that would under other circumstances be considered to be pharmaceutical active ingredients or stable derivatives of human blood products. In the MDD (ibid), the following text is included:

> 7.4. Where a device incorporates, as an integral part, a substance which, if used separately, may be considered to be a medicinal product as defined in Article 1 of Directive 2001/83/EC and which is liable to act upon the body with action ancillary to that of the device, the quality, safety and usefulness of the substance must be verified by analogy with the methods specified in Annex I to Directive 2001/83/ EC. For the substances referred to in the first paragraph, the notified body shall, having verified the usefulness of the substance as part of the medical device and taking account of the intended purpose of the device, seek a scientific opinion from one of the competent authorities designated by the Member States or the European Medicines Agency (EMEA) acting particularly through its committee in accordance with Regulation (EC) No 726/2004 on the quality and safety of the substance including the clinical benefit/risk profile of the incorporation of the substance into the device. When issuing its opinion, the competent authority or the EMEA shall take into account the manufacturing process and the data related to the usefulness of incorporation of the substance into the device as determined by the notified body.
>
> Where a device incorporates, as an integral part, a human blood derivative, the notified body shall, having verified the usefulness of the substance as part of the medical device and taking into account the intended purpose of the device, seek a scientific opinion from the EMEA, acting particularly through its committee, on the quality and safety of the substance including the clinical benefit/risk profile of the incorporation of the human blood derivative into the device. When issuing its opinion, the EMEA shall take into account the manufacturing process and the data related to the usefulness of incorporation of the substance into the device as determined by the notified body.

EXAMPLES OF DECISIONS ON COMBINATION PRODUCTS

The status of a new combination product—whether it is to be processed under the pharmaceutical regulations or the medical device regulations—is often determined taking into account examples that are relevant amongst products that have already been considered by the regulators in the regions concerned. It is therefore useful to be able to see what the outcome has been for other products in different regions. The information that was easily available when the chapter was prepared (August 2008) is summarized in Tables 1 to 4.

DOSSIER REQUIREMENTS FOR DEVICE-DRUG COMBINATION PRODUCTS

European Union Consultation Procedure (Notified Body and Competent Authority)

In the EU, it is necessary for Notified Bodies chosen by the manufacturer of the medical device to consult with a pharmaceutical Competent Authority within a member state, or in some cases, the EMEA on the ancillary substances included in a device-drug combination product. In the case of a medical device incorporating a stable human blood derivative it is mandatory for this consultation to be with the EMEA. There might also be advantages in consulting with the EMEA for a pharmaceutical component that has been approved through the Centralized Procedure. In other cases the Notified Body has freedom of choice of any of the EU pharmaceutical Competent Authorities. However, it is advantageous for the consultation to be with an Agency that has

Table 1 Examples of Determinations of Status of Drug-Device Combinations in Japan

Product	Details	Outcome
Drug-eluting stents	Stent coated with drug-containing polymer	Medical device
Syringe prefilled with pharmaceutical product		Drug product
Heparin-coated catheter		Medical device
Hydroxyapatite artificial bone with biological medicinal product		Drug product
Drug-eluting transdermal patch		Drug product?
Drug-eluting contact lens		Drug product?
Gelatin sponge with biological medicinal product		Drug product?
Iontophoresis system with a drug reservoir	Disposable (single-use) system which is prefilled with a pharmaceutical and which cannot be refilled	Drug product
	Reservoir can be refilled repeatedly with a pharmaceutical; iontophoresis system and the pharmaceutical distributed separately	Iontophoresis system is a device Pharmaceutical component is a drug

Source: Taken from a presentation "Combination products regulation in Japan" made at the 2007 Global Harmonisation Task Force Conference in Washington DC by Tomiko Tawaragi available from www.ghtf.org.

fairly recent experience with the "ancillary substance" (= drug active ingredient) concerned. It is advisable for the Notified Body (ideally with the manufacturer concerned) to hold a meeting in advance with the Agency selected to determine any local requirements for the format and language of the information to be submitted.

The existing guidance on the format and content of consultation applications for the EU is included in MEDDEV 2.1/3 rev 2 (July 2001) (2) but is out of date. It is being updated in the third revision of this document but this has not, at the time of writing, been published officially in hard copy or electronic versions. There is also an MHRA Guidance Note (11). This is also out of date (and refers to the March 1998 version of the MEDDEV referred to earlier). A more up-to-date summary is included in the MHRA's *Special Mail 5* (12), but this does not advise on the detail of the data requirements.A more up-to-date source of guidance is available in the Irish Medicines Board *Guide to Drug-Device Consultations*, published in June 2008 (13). There are also specific EMEA guidance documents relating to procedural aspects and dossier requirements for consultations on ancillary substances (14) and a draft document on drug-eluting stents (15)—this addresses only the nonclinical and clinical aspects of the required dossier. One of the issues is that the MEDDEV initial advice was based on the older format for applications used in the pharmaceutical sector, which has been supplanted by the Common Technical Document (CTD) format. Applications for consultations should now use the CTD format. Table 5 provides a summary of the requirements for the dossier content and structure for EU consultations as of August 2008.

United States

In the US, there three classes of medical devices—I, II and III. These are different from the classes of devices in Europe (classes I, IIa, IIb, and III).

In the US, there are two types of submission for medical devices: Premarket Approval (PMA) submission (applying to Class III devices or high-risk devices that present a significant risk of illness or injury or devices not found to be "substantially equivalent" to Classes I and II devices) and FDA 510(k) (premarket notifications). There is an FDA Web page concerned with the classification (16). A range of devices is exempt from these requirements, and the current listing of such exempted products can be found on the FDA Web site (17).

Table 2 Examples of Jurisdictional Determinations of Lead Centers for Drug-Device Combinations in the United States

A. CBER—Biologics
Interleukin for treatment of HIV and cancer
Hemostasis factor-based hemostatic agent
Protein to maintain normal intestinal microflora
Synthetic protein for treatment of anemia
Plasmid vectored gene silencing to treat hepatitis
Autologous cultured chondrocytes for cartilage repair
Recombinant human protein to treat infection
Enzyme replacement therapy for Fabry disease
Immune globulin treatment for autoimmune skin disease
Recombinant protein to treat bone diseases
Cancer vaccine and antibody for treatment of melanoma
Antigenic modification of autologous blood for allergy treatment
B. CBER—Devices
Device to process autologous cells for wound treatment
Device to facilitate autograft transplantation
Device to concentrate a therapeutic protein from autologous blood
Device to prepare autologous fibrin sealant
In vitro test kit to identify and enumerate CD34+ cells
Device to determine concentration of CD34+ cells in peripheral blood and apheresis products
In vitro urine test to aid in clinical diagnosis of HIV infection
Immunodiagnostic assay for HIV (for individual patient use)
Branched DNA technology–based quantification assay for HIV (for patient management)
Device to inactivate pathogens in blood
Device to modify blood cells
C. CBER—Drugs
Sodium citrate solution to prevent coagulation during blood collection and storage
Citrate-dextrose solution to prevent coagulation during extracorporeal blood processing
Contrast agent-drug combination for tracking cells in cell therapy studies
D. CBER—Combination Products with Biological Product Primary Mode of Action
Biological product with implantable delivery device
Cellular transplant for diabetes treatment
Biological hemostatic agent with reconstitution and delivery devices
Fibrin sealant and delivery catheter
Interferon and injector for treatment of hepatitis C
Injectable protein with delivery device for orthopedic use
Drug/biologic embolization agent
Autologous cellular product and delivery device
Radiolabeled antibody with detector device
Autologous cellular therapy and delivery device to treat cardiovascular disease
Autologous cells and scaffold for orthopedic use
Blood-derived protein and delivery device/dressing for wound treatment
Menstrual tampon with bacteria to restore/maintain natural vaginal flora
Autologous cells and scaffold for organ replacement
Pegylated interferon and ribavirin to treat chronic hepatitis C
Interferon and drug to treat chronic hepatitis C
Blood-derived hemostatic agent and applicator
Device and biologic to separate stem cells for reinfusion after chemotherapy
Tumor-derived antigens for diagnostic use
Recombinant growth factor and delivery vehicle
Monoclonal antibodies and magnetic microspheres to remove tumour cells from bone marrow
Cultured bone marrow cells and bone void filler with handling agent for bone repair
Autologous blood, light source, and photosensitizing agent for immunotherapy
Gene therapy and drug for treatment of malignant glioma
Autologous chondrocytes and scaffold for repair of cartilage defects
Autologous mesenchymal cells and scaffold for diaphragm repair

Table 2 Examples of Jurisdictional Determinations of Lead Centers for Drug-Device Combinations in the United States (*Continued*)

E. CDER—Drugs
Ultrasound contrast agent
Dye mouth-rinse to examine oral tissue
Orally administered drug for ultrasound imaging
Drug used to determine plasma volume
Inhaled chemoirritant to elicit cough reflex
Hormone for the treatment of gastrointestinal disease
Vital dye for ophthalmic use
Aerosolized polysaccharide for pulmonary disorder
Diagnostic drug administered to the bladder
Drug for *in vitro* treatment of sperm
Synthetic protein to treat HIV
Solution used in continuous renal replacement therapy procedures
Vitreous liquefaction agent
Oxygen-carrying liquid drug to treat pulmonary conditions
Antibiotic cream to treat nasal infections
Agent to decontaminate chemical and biological warfare agents
Extended-release synthetic compound for orthopedic use
Isotope-based drug combination to diagnose digestive disease
Topical agent for wound healing
Recombinant hormone for the treatment of obesity
Synthetic protein to treat chemotherapy side effect
Hydroxyapatite-based dentifrice
Solution for use in cardioplegia and preservation of the heart during cardiac surgery
Enzyme to degrade viscoelastic agent during cataract surgery
Synthetic immunomodulator for neurological use
Drug used in therapeutic ultrasound procedures
Polysaccharide for prevention/treatment of bacterial overgrowth
Heat-activated chemotherapy drug
F. CDER—Biological Products
Botulinum toxin for treatment of neuromuscular disorders
Recombinant protein for dental use
Monoclonal antibodies for treatment of blood disorder
G. CDER—Combination Products with Drug Primary Mode of Action
Cytotoxic agent and biological product for targeted cancer treatment
Gynecologic drug and delivery device
Photosensitizing drug and light source for cancer treatment
Drug and transdermal iontophoretic delivery system
Fluoride-containing dental device for anticaries treatment
Fluoride varnish for anticaries treatment
Drug with injector to treat erectile dysfunction
Drug with injector to treat anaphylactic reactions
Drug with implantable delivery device
Antiseptic skin preparation swabs
Peritoneal dialysis solution and administration set
Drug and delivery device for diagnosis of coronary artery disease
Drug and delivery system for cancer treatment
Drug and radiation emitting device to treat cancer
Drug and delivery system for use with vein grafts
Antimicrobial barrier
Spermicidal gel with delivery device
Antibiotic with biodegradable delivery device
Drug reconstitution and delivery system
Contrast agent and saline flush syringes
Chemoablation drug and delivery device
Cytotoxic agent and enzyme for cancer treatment

(*Continued*)

Table 2 Examples of Jurisdictional Determinations of Lead Centers for Drug-Device Combinations in the United States (*Continued*)

G. CDER—Combination Products with Drug Primary Mode of Action (*Continued*)
Drug and drainage device for glaucoma treatment
Drug and delivery system for localized bladder treatment drug and device for *ex vivo* photodynamic cancer treatment
Hormone to induce expression of breast fluids and sample collection device
Enzymatic debriding agent/dressing for wound management
Insulin and inhaler to treat diabetes
Synthetic drug with controlled-delivery device for ophthalmic use
Chemical agent and delivery system for gallstone dissolution
Chlorhexidine gluconate swabs
Radiolabeled contrast agent, production device and administration set
Aerosolized radiolabeled contrast agent and production device
Radiopharmaceutical and carrier for orthopedic use
Drug and delivery system for internal tissue
Photopheresis system and drug to treat skin cancer
Drug, antibody, and enzyme for cancer treatment
PET drug and probe for tumor localization
Topical drug and light source to treat skin cancer
Patient-controlled transdermal system for pain management
Drug and light source for dental use
Drug and collagen combination for pain relief after lumbar surgery
Aerosolized sodium chloride for halotherapy
Prefilled insulin cartridges for use with a dedicated infusion pump
Drug-eluting device to prevent restenosis of vascular grafts
Condom with drug-containing gel to help maintain erection
Cytotoxic agent-monoclonal antibody conjugate for cancer treatment
Cartridges with prefilled syringes for use in specified reusable autoinjector
H. CDRH—Devices
Agent used to prepare skin for adhesion of bandages
Gel used as barrier to prevent surgical adhesions
Hyaluronic acid viscosupplement for treatment of arthritis pain
Absorbable barrier to prevent surgical adhesions
Polysaccharide intended to act as protective barrier
Radiation-containing glass beads for cancer treatment
Device to hyperpolarize gas for use during magnetic resonance imaging
Extracorporeal lipid removal system
Physical barrier for dental use
Oral wound dressing providing physical barrier
Device to provide controlled hypoxia
Needle-free injector for administration of a biological product
Needle-free injector for administration of a drug
Processed animal connective tissue for use in orthopedic surgery
LAL-based *in vitro* test to aid in the diagnosis of fungal infections
Device for disinfection of harvested bone
Collagen/oxidized regenerated cellulose wound dressing
Intraperitoneal drug delivery system
Device that controls delivery of medication from implanted delivery system
Human allograft demineralized bone matrix with additives to make putty or paste
Animal collagen bone void filler
Medical maggots for wound treatment
Solution for rinsing preserved organs prior to reimplantation
Device to produce sterile water for injection
Laser-activated surgical tissue welding tube
Vitreous visualization agent
Ozone production/delivery device for dental use
Device that produces oxygenated water for topical oxygen therapy
Device for pulsatile delivery of therapeutic solutions
Kit for intestinal permeability testing
Peak flow meter accessory for metered dose inhalers

Table 2 Examples of Jurisdictional Determinations of Lead Centers for Drug-Device Combinations in the United States (*Continued*)

H. CDRH—Devices (*Continued*)
Bone and collagen allograft for bone grafting
Device to monitor use of metered dose inhalers
In vitro diagnostic test for hepatitis (for patient management)
Immunodiagnostic assay for hepatitis (for individual patient use)
Branched DNA technology–based quantification assay for hepatitis B (for patient management)
Branched DNA technology–based quantification assay for hepatitis C (for patient management)
Injectable polymer for urological use
Sterile water for inhalation therapy
Sterile water with saline for inhalation therapy
Dressing that binds enzymes from wounds
Adjunct to a PAP machine to treat breathing disorder
Wound cleansing/debriding product
Device that provides humidified hyperbaric oxygen for wound management
Device for mixing and delivery of fibrin sealants
Dual syringe applicator for delivery of fibrin sealants
Reusable autoinjector for use with specified prefilled syringe cartridges
Animal tissue–derived scaffold for ligament replacement
Surfactant-based lubricant to act as adhesion barrier in pelvic and abdominal surgery
Heating/drug delivery device for cancer treatment
Device and inhaled gas mixture to measure cardiopulmonary function
Topical paste for wound covering
Accessories to blood/fluid warming device
Polymeric sealant to seal air leaks in the lung
Computer-based system to help monitor and manage Parkinson's disease
More than minimally manipulated structural human tissue for use in vascular procedures
Device with tracer for evaluation of lung function
I. CDRH—Combination Products with Device Primary Mode of Action
Breath collection/analysis devices and drug for *H. pylori* diagnosis
Laser and photosensitizing agent for antimicrobial use
Dialysate with drug component
Heparin-coated cardiovascular stent
Drug-eluting cardiovascular stent
Radiolabeled drug–eluting cardiovascular stent
Vascular graft with antibiotic
Dental floss with fluoride
Dental floss with antimicrobial agent
Dental filling material with antimicrobial agent
Dental sealant with antimicrobial agent
Dental cement with antimicrobial agent
Periodontal barrier with antibiotic
Adhesive bandage with antibiotic
Physical/chemical embolization product
Cardiovascular stent and injectable drug
Orthopedic prosthesis coated with growth factor
Coronary stent coated with biological product
Dental prophylaxis paste with fluoride and antimicrobial agent
Dental implant coated with autologous cells
Dental implant and synthetic peptide used to promote its integration
Wound dressing impregnated with drug ointment
Occlusive wound dressing with biologic component
Catheter lock/flush solution with antimicrobial and anticoagulant agents
Dental wax with anesthetic for discomfort associated with orthodontic procedures
Dental restorative material with fluoride
Breath collection/analysis devices and drug for diagnosis of lactose maldigestion
Breath collection/analysis devices and drug for diagnosis of gastric emptying disorders
Breath collection/analysis devices and drug to monitor enzyme activity

(*Continued*)

Table 2 Examples of Jurisdictional Determinations of Lead Centers for Drug-Device Combinations in the United States (*Continued*)

I. CDRH—Combination Products with Device Primary Mode of Action (*Continued*)
Breath test kit and drug for use during pharmacological research
Breath collection/analysis devices and drug for diagnosis of intestinal bacterial overgrowth
Breath test and drug for assessment of small intestinal function
Blood test and drug for detection of *H. pylori* in the stomach
Embolization device coated with a radiopharmaceutical drug
Root canal cleanser with antimicrobial agent
Device with enzyme mimic for orthopedic use
Bone void filler with antibiotic
Menstrual tampon with drug to neutralize pH within the tampon
Fluorescent imaging system for use during cardiac surgery
Tubing with drug to reduce discomfort during placement
Device and lithium chloride indicator dilution to monitor cardiac output
Device for delivery and filtration of chemotherapeutic agents to organs
Device and filtration system to administer chemotherapy to limbs
Polymer-astringent granules for wound/laceration management
Laser and photosensitizing agent for dental use
Nonocclusive wound dressing with drug
Wound dressing with analgesic
Wound dressing with antimicrobial agent
Bone void filler with blood component to act as matrix and enhance handling properties
Vascular graft with drug-eluting mesh to maintain graft patency
Breath test kit and drug to measure liver function
Scaffold/matrix for wound treatment
Device and soluble intravenous gases to evaluate pulmonary function
Chemiluminescent light and dye for oral lesion marking
Collagen-thrombin hemostatic agent
Gelatin granules-thrombin hemostatic agent
Breath test kit and drug for diagnosis of insulin resistance
Cryotherapy device and salicylic acid for wart removal
Central venous catheter with antimicrobial coating
Therapeutic ultrasound device and gas microspheres for thrombolytic therapy
Surgical mesh with anesthetic coating
Scented respiratory device

Abbreviations: CBER, Center for Biological Evaluation and Research; CDER, Center for Drug Evaluation and Research; CDRH, Center for Devices and Radiological Health.
Source: From http://www.fda.gov/oc/combination/updates.html and http://www.fda.gov/oc/combination/determinations.html.

This Web page (16) consists of a link to exempted devices in the following categories:

Clinical chemistry and clinical toxicology devices
Hematology and pathology devices
Immunology and microbiology devices
Anesthesiology devices
Cardiology devices
Dental devices
Ear, nose and throat devices
Gastroenterology devices
General and plastic surgery devices
General hospital and personal use devices
Neurological devices
Obstetrical and gynecology devices
Ophthalmology devices
Orthopedic devices
Physical medicine devices and
Radiology devices

Table 3 Specific Examples of Jurisdictional Determinations of Lead Centers for Device-Drug Combinations in the United States

Product	**Determination**
Breath test combination products *Combination product diagnostic breath tests in which the drug component is an isotope-labeled (typically* ^{13}C*) substrate and the device components capture and/or analyze exhaled breath for detection of labeled carbon dioxide or other gases used for breath test combination products for the diagnosis of* H. pylori, *gastric emptying disorders, carbohydrate malabsorption, intestinal bacterial overgrowth, insulin resistance, liver function, for monitoring enzyme activity, for assessment of small intestinal function, and for use in pharmacological research. In some cases, such as diagnosis of* H. pylori, *the substrate is metabolized by bacteria present in the stomach or gut; in others, the substrate is metabolized by the patient when a particular disease or condition is present. Upon metabolism of the substrate, labile metabolites are exhaled that can be uniquely traced to the substrate. Thus, for example, the presence, absence, or rate of release of isotope-labeled carbon dioxide in exhaled breath is intended to be indicative of the presence or absence of the disease or condition in question.*	The primary mode of action (PMOA) of such combination products was attributable to the device components' role in the *in vitro* diagnosis of the disease or condition in question, while the drug component plays a secondary role in acting as the diagnostic substrate. CDRH assigned to be the lead agency center for reviewing these products.
Heparin catheter lock/flush solution	Combination drug-device product. Lead review responsibility CDRH based on determination that the primary mode of action is that of the device part of the combination.
Catheter lock/flush solutions containing an anticoagulant or an antimicrobial in a solution (for example water or saline solution).	The solution component of these products (e.g. water or saline solution) acts by physically occupying space within the catheter and exerting pressure on the patient's circulating blood. In this way, the patient's blood is prevented from backfilling into the catheter and clotting. In acting in this manner, the solution component of the product meets the definition of a device in that it affects the structure or function of the body, and does not achieve its primary intended purposes through chemical or metabolic action. Determined that in these cases, the anticoagulant or antimicrobial component of the catheter lock/flush solutions was intended to affect the structure and function of the body of man by acting chemically on microorganisms and preventing thrombotic occlusions with the catheter. Concluded that such components meet the definition of a drug in that they are intended to affect the structure or function of the body of man. Concluded that the catheter lock/flush solutions consisting of a solution (e.g. water or saline solution) and an anticoagulant or antibiotic are combination products. Primary mode of action of these catheter lock/flush solutions in maintaining catheter patency was attributable to the device component's role in occupying space and applying pressure within the catheter. Concluded that the drug component of the combination product plays a secondary role by, for example, chemically acting on microorganisms and/or preventing thrombotic occlusions within the catheter. Assigned to the CDRH for review and regulation.

Table 3 Specific Examples of Jurisdictional Determinations of Lead Centers for Device-Drug Combinations in the United States (*Continued*)

Product	Determination
Metered dose inhalers (MDIs), spacers and other accessories	MDIs are drug-device combination products. "Stand-alone" spacers: General use spacers are regulated as devices by CDRH. Locking clips: Devices regulated by CDRH. Actuators: An essential part of an MDI. The drug cannot be delivered to the patient without an actuator. Actuators play a crucial role in the delivery characteristics of MDIs. The concept of a general use actuator is not appropriate because of this link to dosing performance: an actuator significantly affects the safety and efficacy of the drug, and must always be studied with a particular MDI drug. Therefore, replacement actuators have been determined to be device components of combination products. Regulated by CDER as new drugs. Spacers incorporating actuators: Designed to replace the actuator of an approved MDI. As with replacement actuators without a spacer, the concept of a general use actuator-spacer is not appropriate because an actuator must be studied with a particular drug. Therefore, actuator-spacers have been determined to be device components of combination products. Primary mode of action of the combination product is attributable to its drug component. Regulated as a new drug by CDER. Dose counters or dose indicators: Have been determined frequently to be device components of combination products. PMOA is attributable to its drug component—dose counters have been regulated by CDER as new drugs. If a "universal" dose counter were developed, i.e. one where data demonstrate that the counter would work regardless of the characteristics of the MDI and would not alter drug delivery from the MDI, and so would not need to be labeled for use with a specific MDI, it would probably be considered a device and regulated by CDRH.
Human demineralized bone matrix (DMB)	Certain components added to DBM products, such as sodium hyaluronate, glycerol, or calcium phosphate, are not sterilizing, preserving, or storage agents. These components meet the definition of a device because they are intended to affect the structure or function of the body by assisting in the filling of bone voids, and they do not achieve their primary intended purposes through chemical or metabolic action. Therefore, in these cases products consisting of DBM and these components are regulated by the CDRH as medical devices.
Dental prophylaxis pastes incorporating a drug component	Such products are combination products. The prophylaxis paste component functions as an abrasive agent intended to clean and polish the teeth, while the drug component has played a secondary role, augmenting the safety and/or effectiveness of the prophylaxis paste. The combination products acted primarily by cleaning and polishing teeth. Concluded that the primary mode of action for the combination product is that of the device component. Regulated by CDRH as medical devices.

Abbreviations: CDER, Center for Drug Evaluation and Research; CDRH, Center for Devices and Radiological Health.
Source: Derived from information accessed *via* http://www.fda.gov/oc/combination/updates.html.

Table 4 Guidance on the Status (Pharmaceutical or Medical Device) of Particular Types of Products in the European Union

Product	Regime applicable	Comments
Bone cement	Medical device	Bone cements containing antibiotic are "medical devices incorporating a medicinal substance with ancillary action"
Dental filling material	Medical device	
Materials for sealing, approximation, or adhesion of tissues (for example cyanocrylates, fibrin-based adhesives not of human origin)	Medical device	Materials of human origin other than stable human blood derivatives are excluded from the scope of the MDD MHRA: Status of biological sealants depends on mode of action
Resorbable materials used in osteosynthesis (for example pins or bone screws manufactured using polylactic acid)	Medical device	MHRA: Resorbable bone plates/polylactic/polyglycolic acid are medical devices
Sutures, absorbable sutures	Medical device	
Soft and hard tissue scaffolds and fillers (for example collagen, calcium phosphate, bioglass)	Medical device	MHRA: Scaffolds containing bioactive materials are likely to be medicinal products Include the following classes of products: a. Hydroxyapatite with/out collagen b. Calcium phosphate with/out collagen c. Bioglas d. Coral e. Cartilage repair systems The status of cartilage repair systems depends on the mode of action MHRA: Human tissue–derived fillers may be regulated as medicinal products, or may come within the Code of Practice for human-derived therapeutic products
Bone void fillers intended for the repair of bone defects where the primary action of the device is a physical means or matrix, which provides a volume and a scaffold for osteoconduction	Medical device	
Intrauterine devices	Medical device	Intrauterine contraceptives containing copper or silver are "medical devices incorporating a medicinal substance with ancillary action." Intrauterine contraceptives whose primary purpose is to release progestogens are medicinal products
Blood bags	Medical device	Blood bags containing anticoagulant or preservation agents are "medical devices incorporating a medicinal substance with ancillary action"

(*Continued*)

Table 4 Guidance on the Status (Pharmaceutical or Medical Device) of Particular Types of Products in the European Union (*Continued*)

Product	Regime applicable	Comments
Systems intended to preserve and treat blood	Medical device	Systems intended for the collection, storage, and preservation of blood or blood components and as an ancillary function, the treatment of blood or blood components where this effect is achieved outside the human body, are classified as devices provided that any residual material is not intended to achieve its intended effect when the blood or cells are reintroduced into the body, e.g. systems incorporating chemicals activated by light to reduce the viral load where the quantity of chemical remaining has no intended effect when transfused. Does not cover substances introduced into an extracorporal circuit
Viscoelastic materials with intended use for mechanical/physical purposes, such as protection of tissues during and after surgery and separation of tissues	Medical device	Such materials are also used as synovial fluid replacements where viscosupplementation provides support and lubrication. Additional pharmacological benefits claimed which are ancillary to the mechanical action do not alter the medical device status. However, certain of these materials such as some hyaluronic acid-based products, where the predominant claims are of a pharmacological nature and not primarily related to any viscoelastic characteristics, are classed as medicinal products MHRA: Viscoelastic products may become medicines if additional claims are made
Gases and liquids for ocular endotamponades	Medical device	MHRA: Ocular endotamponades may become medicines if additional claims are made
Cell separators, including those incorporating antibodies for cell marking	Medical device	
Wound dressings, which may be in the form of liquids, gels and pastes, etc (for example hydrocolloid, hydrogel)	Medical device	Wound dressings, surgical or barrier drapes (including tulle dressings) with antimicrobial agent are "medical devices incorporating a medicinal substance with ancillary action." Wound treatment products comprising a matrix whose primary purpose is the administration of medicinal products, e.g. wound dressings containing an antimicrobial agent where the primary action of the dressing is to administer the agent to the wound for the purpose of controlling infection are medicinal products MHRA: Status depends on manufacturer's claim

Table 4 Guidance on the Status (Pharmaceutical or Medical Device) of Particular Types of Products in the European Union (*Continued*)

Product	Regime applicable	Comments
Hemostatic products, for example patches, plugs, and powders where the hemostatic effect results from the product's physical characteristics, or is due to the surface properties of the material. This includes products containing collagen, or calcium alginate or oxidised cellulose where adhesion of platelets to the surface triggers platelet adhesion and aggregation	Medical device	Hemostatic agents where the primary mode of action is not mechanical, e.g. certain collagens which have a molecular structure capable of a surface-independent demonstrated interaction with platelet receptors, and which achieves platelet adhesion through a pharmacological process are medicinal products
		Hemostatic devices enhanced by the incorporation of collagen, where the primary action of the device is mechanical even though there may be ancillary action because of the presence of collagen having demonstrable action with platelet receptors resulting in platelet adhesion through a pharmacological process are "medical devices incorporating a medicinal substance with ancillary action."
Concentrates for hemodialysis	Medical device	
Pressure-reducing valves and regulators	Medical device	
Irrigation solutions (including those used in the eye) intended for mechanical rinsing	Medical device	If the solution contains a medicinal substance such as chlorhexidine where the principal intended purpose is to provide a local antimicrobial effect, it will be a medicinal product. Solutions incorporating substances for other purposes, e.g. antimicrobial agent for the preservation of the solution remain a medical device Ophthalmic irrigation solutions principally intended for irrigation which contain components which support the metabolism of the endothelial cells of the cornea are "medical devices incorporating a medicinal substance with ancillary action." MHRA: Medical devices for mechanical rinsing purposes but if solution contains a pharmacologically active substance then the product is likely to be a medicinal product. Eye washes for emergency purposes are considered to be medical devices. Eye washes intended for use to alleviate eye irritation resulting from environmental factors such as smoke, dust are not considered to be medical devices
Devices such as catheters, guidewires and stents containing or incorporating radioisotopes where the radioisotope as such is not released into the body, used, for example in cardiology for the prevention of restenosis	Medical device	

(*Continued*)

Table 4 Guidance on the Status (Pharmaceutical or Medical Device) of Particular Types of Products in the European Union (*Continued*)

Product	Regime applicable	Comments
Contact lens care products (disinfecting, cleaning, rinsing, and hydrating solutions, including those which aid the insertion and/or wearing of contact lenses without a therapeutic claim)	Accessory to medical devices	MHRA: Comfort drops: Considered to be accessories to medical devices when specifically intended for use as a result of wearing contact lenses. Considered to be medicinal products if therapeutic claims are made and contain an active ingredient. Drops indicated for relief from purely environmental factors are not considered to be either (medicinal products or medical devices)
Disinfectants specifically intended for use with medical devices (for example endoscopes)	Accessory to medical devices	Multipurpose disinfectants or sterilization agents are not covered by MDD; they are covered by the (EU) Directive on biocides Topical disinfectants (antiseptics) for use on patients are medicinal products. MHRA: a. Topical disinfectants (medicinal products) b. Alcohol preparation pads/swabs (medicinal products) c. Wipes/swabs with medicinal substance (medicinal products) d. Disinfectants specifically intended for disinfecting medical devices (medical devices) Overlap with biocides and medicinal products legislation
Lubricants specifically intended for use together with medical devices (for example for gloves, endoscopes, condoms)	Accessory to medical devices	
Skin barrier powders and pastes or other skin care products specifically intended for use together with ostomy bags	Accessory to medical devices	
Challenge tests specifically intended to assess the tolerance to a given medical device, or its constituents (for example injectable collagen)	Accessory to medical devices	
Gases used to drive cryoprobes and surgical tools	Accessory to medical devices	
Spermicidal preparations	Medicinal product	Condoms coated with spermicides are "medical devices incorporating a medicinal substance with ancillary action." MHRA: Other contraceptive products: Diaphragms—medical devices
Gases intended to be used in anesthesia and inhalation therapy (for example oxygen, medical air supplied in containers) including their primary containers	Medicinal product	These gases are also used in minimal access surgery. However a product intended exclusively for minimal access surgery would be a medical device
Zinc paste for dermatological use	Medicinal product	MHRA: Nonmedicated dermatological creams including those containing zinc oxide (without pharmacological action) are medical devices

Table 4 Guidance on the Status (Pharmaceutical or Medical Device) of Particular Types of Products in the European Union (*Continued*)

Product	Regime applicable	Comments
Water for injections, IV fluids, and plasma volume expanders	"Assimilated to medicinal products"	
Hemofiltration substitution solutions	"Assimilated to medicinal products"	
In vivo diagnostic agents, for example x-ray contrast media, NMR-enhancing agents, fluorescent ophthalmic strips for diagnostic purposes, carrier solutions to stabilize microbubbles for ultrasound imaging	"Assimilated to medicinal products"	
Gases for *in vivo* diagnostic purposes, including lung function, tests, for example carbon dioxide for vascular diagnostic purposes	"Assimilated to medicinal products"	
Solutions for peritoneal dialysis	"Assimilated to medicinal products"	
Antacids	"Assimilated to medicinal products"	
Artificial tears	"Assimilated to medicinal products"	MHRA: Considered to be medical devices if therapeutic claims are made. Products indicated for relief from purely environmental factors are not considered to be devices. Medicated drops are considered to be medicinal products
Fluoride dental preparations and other dental preparations	"Assimilated to medicinal products"	Dental preparations with a typical device mode of action, e.g. cements or varnishes incorporating fluoride, are medical devices, where the fluorine is of ancillary action to that of the device Certain products where the claims are primarily cosmetic in nature and where the fluorine level is less than 0.15% are cosmetic products

(*Continued*)

Table 4 Guidance on the Status (Pharmaceutical or Medical Device) of Particular Types of Products in the European Union (*Continued*)

Product	Regime applicable	Comments
Root canal fillers which incorporate medicinal substances with secondary action	Medical device incorporating a medicinal substance with ancillary action	Temporary root canal fillers incorporating medicinal products, whose primary purpose is to deliver the medicinal product, are medicinal products MHRA: Advice covers a. Pit and fissure sealants (medical devices) b. Root canal sealers: medicated/ nonmedicated (medical devices) c. Root canal dressings (for example polyantibiotic pastes, antiseptics) (medicinal products) d. Pulp capping material e. Dry socket preparations (status will depend on composition of product) f. *In vivo* diagnostics, e.g. disclosing tablets (controlled as medicinal products) g. Hemostatic agents and astringents (status depends on mode of action) h. Aluminium sulphate/salts astringents (medical devices) i. Retraction cords: medicated/ nonmedicated (medical devices) j. Fluoride preparations: tablets, gels, varnishes (medicinal products) k. Hard tissue scaffolds (medical devices) l. Desensitizing agents: physical/ pharmacological (status depends on mode of action) m. Periodontal dressings: medicated/ nonmedicated (medical devices) n. Periodontal antibacterials: gels, ointments, fibers (medicinal products) o. Varnishes: protective/drug delivery (status depends on function and mode of action) p. Toothache preparations (medicinal products) q. Artificial saliva (medical device) r. Mouth ulcer preparations: medicated/ nonmedicated (status depends on mode of action) s. Antibacterial mouthwashes/gels (medicinal product but depends on primary purpose—cosmetic?)
Solutions administered *in vivo* to the local circulation for the cooling of organs during surgery	"Assimilated to medicinal products"	
Agents for transport, nutrition, and storage of organs intended for transplantation		These products are not currently regulated in all Member States as medicinal products. However there was general consensus of public authorities that the medicinal products category is the most appropriate. Some of these products may have a metabolic effect, others however have no such effects

Table 4 Guidance on the Status (Pharmaceutical or Medical Device) of Particular Types of Products in the European Union (*Continued*)

Product	Regime applicable	Comments
Catheters coated with heparin or an antibiotic agent	Medical device incorporating a medicinal substance with ancillary action	
Soft tissue fillers incorporating local anesthetics	Medical device incorporating a medicinal substance with ancillary action	
Bone void filler intended for the repair of bone defects where the primary action of the device is a physical means or matrix, which provides a volume and a scaffold for osteoconduction and where an additional medicinal substance is incorporated to assist and complement the action of the matrix by enhancing the growth of bone cells	Medical device incorporating a medicinal substance with ancillary action	In such cases, the ancillary nature would be determined by the performance of the matrix on its own and the extent of the enhancement of growth because of the presence of the substance. With reference to the overall purpose of the product, where the medicinal substance has such an effect that its ancillary nature cannot be clearly established, then the product should be considered in accordance with the concept of a drug delivery system
Electrodes with steroid-coated tip	Medical device incorporating a medicinal substance with ancillary action	
Drug delivery pump	Medical device	A device which is intended to deliver a medicinal product is itself regulated as a medical device. The medicinal product which the device is intended to administer is subject to the normal procedures for medicinal products. The essential requirements of the MDD apply as far as the device related features of the product are concerned
Implantable infusion pump	Medical device	A device which is intended to deliver a medicinal product is itself regulated as a medical device. The medicinal product which the device is intended to administer is subject to the normal procedures for medicinal products. The essential requirements of the MDD apply as far as the device related features of the product are concerned
Iontophoresis device	Medical device	Single-use disposable iontophoresis devices incorporating a medicinal product are medicinal products A device which is intended to deliver a medicinal product is itself regulated as a medical device. The medicinal product which the device is intended to administer is subject to the normal procedures for medicinal products. The essential requirements of the MDD apply as far as the device related features of the product are concerned

(*Continued*)

Table 4 Guidance on the Status (Pharmaceutical or Medical Device) of Particular Types of Products in the European Union (*Continued*)

Product	Regime applicable	Comments
Nebulizer	Medical device	Nebulizers precharged with a specific medicinal product, and not for universal application, are medicinal products.A device which is intended to deliver a medicinal product is itself regulated as a medical device. The medicinal product which the device is intended to administer is subject to the normal procedures for medicinal products. The essential requirements of the MDD apply as far as the device related features of the product are concerned
Syringe	Medical device	In a kit comprising an insulin pen and insulin cartridges, the pen is subjected to the MDD whereas the insulin cartridge is a medicinal product. A device which is intended to deliver a medicinal product is itself regulated as a medical device. The medicinal product which the device is intended to administer is subject to the normal procedures for medicinal products. The essential requirements of the MDD apply as far as the device related features of the product are concerned
Prefilled syringe	Medicinal product	A device which is intended to deliver a medicinal product is itself regulated as a medical device. The medicinal product which the device is intended to administer is subject to the normal procedures for medicinal products. The essential requirements of the MDD apply as far as the device related features of the product are concerned
Jet injector	Medical device	
Spacer devices for use with metered dose inhalers	Medical device	MHRA: Advice covers a. Prefilled metered dose inhalers b. Chamber spacers for use with metered dose inhalers c. Spinhalers—refillable d. Diskhalers—refillable e. Other empty or refillable inhalers Advice: (b) (c), (d), and (e) may be sold with medication and their performance/drug delivery will be assessed by a drug regulatory authority as part of the medicines marketing authorization application
Aerosols containing a medicinal product	Medicinal product	A device which is intended to deliver a medicinal product is itself regulated as a medical device. The medicinal product which the device is intended to administer is subject to the normal procedures for medicinal products. The essential requirements of the MDD apply as far as the device related features of the product are concerned

Table 4 Guidance on the Status (Pharmaceutical or Medical Device) of Particular Types of Products in the European Union (*Continued*)

Product	Regime applicable	Comments
Port systems	Medical device	
Patches for transdermal drug delivery intended exclusively for use in the given combination and which not reusable	Medicinal product	A device which is intended to deliver a medicinal product is itself regulated as a medical device. The medicinal product which the device is intended to administer is subject to the normal procedures for medicinal products. The essential requirements of the MDD apply as far as the device related features of the product are concerned
Implants containing medicinal products in a polymer matrix whose primary purpose is to release the medicinal product, e.g. plastic beads containing antibiotic for treating bone infections, or a matrix to release osteoinductive proteins into the surrounding bone	Medicinal product	A device which is intended to deliver a medicinal product is itself regulated as a medical device. The medicinal product which the device is intended to administer is subject to the normal procedures for medicinal products. The essential requirements of the MDD apply as far as the device related features of the product are concerned
Light box indicated to treat seasonal affective disorder	Medical device	The manufacturer states that light therapy is "a convenient and effective way of compensating for the lack of light without resorting to medication." The manufacturer also states: "in autumn and winter, the seasons with the least sunlight because the days are shorter, increased symptoms resulting from light depravation may be experienced. Even standard artificial lighting in buildings cannot compensate for a shortage of natural light. The consequences of this may be depression, lack of drive, interrupted sleep and melancholia – the typical autumn winter blues" Statements are effectively claims for treatment of seasonal affective disorder (SAD), which is a generally recognized medical condition.
AB0 and Rhesus (D) blood grouping intended for diet purposes	Not a medical device. Should not use terms defined in the Directives	The manufacturer states the following: HOME-KIT, For *In Vitro* Diagnostic Use, Not for Bed-Side Testing; and currently in the text, which describes interpretation of the result, the manufacturer states *not for clinical use*. The manufacturer has stated that the main reason and purpose of the tests is educational. It is indicated that the product enables the user to ascertain their blood group in order to determine whether a specific (food) diet should be followed. This decision was not related to following a specific diet for medical purposes
Sample receptacles and sampling devices which are intended to be used for the collection by the lay user of samples, which are subsequently examined by third persons	Not a device for self-testing under the *in vitro* diagnostic medical device directive	IVD kits which are being supplied to the public for a variety of medical conditions including tests for food allergies and infections such as *Chlamydia*

(*Continued*)

Table 4 Guidance on the Status (Pharmaceutical or Medical Device) of Particular Types of Products in the European Union (*Continued*)

Product	Regime applicable	Comments
CE marked/labeled microscope slides	General laboratory product unless the manufacturer's intended purpose falls within the definition of an *in vitro* diagnostic medical device	
Product for testing patient reflex cough	The product does not meet the definition of a medical device	Part of a neurological evaluation to assess the patient's laryngeal cough reflex (LCR), neurological airway protection and the vagal (cranial nerve X) component of reflex cough. Uses 1–3 inhalations of a nebulized 20% solution of L-(+)-tartaric acid in a medicinal nonventilatory nebulizer. Tartaric acid acts as a mild irritant to the mucosa of the larynx
Elastoviscous fluids	Data presented do not allow issuing a general statement for the qualification of the elastoviscous fluids	Therapeutic effect through viscosupplementation, a process whereby the physiological and rheological states of the arthritic joint tissues are restored. Viscosupplementation is a treatment to decrease pain and discomfort, allowing more extensive movement of the joint. The product has to be injected in the affected joint to achieve its effect
Hemodialysis water test strips	Such strips are not considered to be "accessories" of a medical device	Products are used together with the hemodialysis machine, one is a residual peroxide reagent strip, which confirms, when used, that the residue of disinfection agents used in the hemodialysis machine have been reduced to safe levels. The other is a reagent strip used to test the water in the hemodialysis machine to ensure that the level of water hardness has been reduced to a level where it is safe to proceed with hemodialysis
Coatings on medical devices		MHRA: a. Antithrombotic coatings gelatin/heparin/protein (medical devices) b. Antibacterial coatings chlorhexidine/benzalkonium chloride/silver/salts/ antibiotics (medical devices unless the primary purpose is to treat infection)
In vivo diagnostic agents		MHRA: a. X-ray and MRI contrast media b. Barium meal c. Other *in vivo* imaging agents d. Labeled urea for *H. pylori* test e. Gases for lung function tests All medicinal products
"Activated" medicinal products		MHRA: a. Medicinal product (medicinal product) b. Activating device, e.g. laser (medical device).

Table 4 Guidance on the Status (Pharmaceutical or Medical Device) of Particular Types of Products in the European Union (*Continued*)

Product	Regime applicable	Comments
Administration devices		MHRA: a. Medicine spoons b. Droppers c. Oral syringes d. Eye baths These products are medical devices even though they may be supplied in the same pack as the medicine, unless they form the closure of the container (e.g. bottle cap/ dropper assembly)
Artificial skin systems		MHRA: Products which do not contain material of human origin will be medical devices. The advanced therapy medicinal product regulations may apply to products of human origin
Head lice products		MHRA: Either medical devices or medicinal products, depending upon their mode of action
Weight loss tablets		MHRA: Status depends on mode of action and claims made: Products that act pharmacologically or metabolically and claim to suppress appetite, burn fat, speed up metabolism, or treat obesity are likely to be regarded as medicinal products. Fat absorption and bulking agents are likely to be regulated as medical devices if making a medical claim (for example treatment rather than just a slimming product). Products not regulated as medical devices or medicinal products are likely to be regulated as food supplements, provided no medicinal claims are made
Active coal or carbon products for treatment of acute poisoning		MHRA: Medicinal products
Products containing lactobacillus for the regulation of vaginal flora		MHRA: Medicinal products
Leeches and maggots		MHRA: Considered to be medicinal products when there is a clear intended medical purpose
Products for the treatment of nicotine addiction		MHRA: Most products intended to treat the addiction to nicotine will be considered to be medicinal products

Source: Based on information included in MEDDEV 2.1/3 rev 2 Demarcation between Directive 90/385/EEC on Active Implantable Medical Devices and Directive 93/42/EEC on medical devices and Directive 65/65/EEC relating to medicinal products and related directives (July 2001), MHRA Medical Device Bulletin No 17 Medical devices and medicinal products (April 2006) and the Manual on Borderline and Classification in the Community Regulatory Framework for Medical Devices (version 1.0, April 2008). Information specifically derived from the MHRA document is identified as such.

For a PMA submission control design documents are required and there is a guideline on the format and data requirements (18). A summary of this information is included in Table 6.

For 510(k) notifications in the US, there is a separate guideline (19). Devices notified under these provisions require the demonstration of "substantial equivalence" to existing devices, should have the same intended use as the existing device, and should have the same

Table 5 Summary of General Data Requirements and Format of Applications for Consultation by Notified Bodies with Pharmaceutical Competent Authorities in the European Union for Medical Devices Containing an Ancillary Medicinal Substance

Section of application	Requirements	Comments
Covering letter		
Comprehensive table of contents		
Section 1	Application form	Specific to the agency concerned
	Product information and labeling	English language text usually Labeling or information to be supplied with the device, particularly as it relates to the ancillary substance *CTD Module 1*
	Explanation that the action of the medicinal substance or human blood derivative incorporated in the medical device is only ancillary to that of the device in line with MEDDEV guidance	General description of the medical device including the manufacturer's claim regarding the purpose of the substance together with a critical appraisal of the results of the risk analysis *CTD Module 1*
	Critical summaries or overview or Expert Reports of the quality, nonclinical, and clinical data for the ancillary substance	
	Tabulated summaries of the clinical studies (efficacy, safety, and individual)	
	Quality Overall Summary sections as relevant for the ancillary substance (Volume 2B – *Presentation and Content of the Dossier*, Common Technical Document, of the *Notice to Applicants*	
Section 2	Quality, nonclinical, and clinical documentation following the headings in MEDDEV 2.1/3 rev 2 for the ancillary substance	*Quality – CTD Module 3* *Nonclinical – CTD Module 4* *Clinical – CTD Module 5*
	Quality documentation *1. For the ancillary substance itself*	Relevant parts of CTD Module 3 (will depend on whether the ancillary substance is an active pharmaceutical ingredient or a formulated product)
		Information on the active substance: an active substance master file (in CTD format) may be submitted
		Reference to monograph for the ancillary substance in the Ph Eur or a monograph in an EU member state pharmacopoeia
	2. For the ancillary substance as incorporated in the medical device	Qualitative and quantitative details of the constituents
		Relevant information on any modification of the ancillary substance during its incorporation into the medical device

Table 5 Summary of General Data Requirements and Format of Applications for Consultation by Notified Bodies with Pharmaceutical Competent Authorities in the European Union for Medical Devices Containing an Ancillary Medicinal Substance (*Continued*)

Section of application	Requirements	Comments
		Description of the method of manufacture
		Control of starting materials: specification for the ancillary substance
		Control tests on intermediate stages of the manufacture
		Control tests on the finished product – qualitative and quantitative tests to control the ancillary substance in the medical device
		Stability – Demonstration that the ancillary substance maintains its function throughout the defined shelf life of the medical device under the recommended storage conditions
		Potential interaction with other materials
		Potential degradation of the ancillary substance
	Nonclinical documentation	Toxicity
		Reference to toxicology profile for known substances
		For new active substances the results of toxicity studies to be provided taking into account CHMP guidelines
		Single-dose toxicity, repeat dose toxicity
		Toxicity and biocompatibility of the medical device from evaluations in accordance with EN ISO 10993 standards may be included
		Reproductive function (as needed)
		Embryo-fetal and perinatal toxicity (as needed)
		Mutagenic potential (as needed)
		Carcinogenic potential (as needed)
		Pharmacodynamics Intended action of the ancillary substance in the context of its inclusion in the medical device
		Pharmacokinetics May not be required in all cases Some or all of the following may need to be addressed: – pattern of local and systemic exposure to the ancillary substance

(*Continued*)

Table 5 Summary of General Data Requirements and Format of Applications for Consultation by Notified Bodies with Pharmaceutical Competent Authorities in the European Union for Medical Devices Containing an Ancillary Medicinal Substance (*Continued*)

Section of application	Requirements	Comments
		– if the level of exposure fluctuates, the maximum level and duration of exposure – if the maximum level of systemic exposure may present a safety concern the maximum peak plasma concentration and information on individual variability – for new active substances information on the release from the medical device and its subsequent distribution and elimination
		Local tolerance Particularly relevant – route of exposure to the ancillary substance may be different from medicinal uses Literature information or results from studies in accordance with EN ISO 10993 to be provided
	Clinical documentation	The usefulness of the ancillary substance in the medical device to be addressed by clinical data (Class III medical device) or by cross-reference to other parts of the dossier Clinical studies: compliance with EN ISO 14155 is appropriate
Section 3	Relevant parts of Module 3 for the ancillary substance following the Common Technical Document, Volume 2B, *Notice to Applicants*	

Note: Special provisions apply to consultations with the EMEA on ancillary human blood derivatives incorporated in medical devices.

technological characteristics as the existing device (or if there are different technological characteristics, these should not raise new questions relating to safety and effectiveness, and the applicant demonstrates that the device would be safe and effective when placed on the market).

Most Class I devices are exempt from the need to submit a 510(k) notification to the FDA unless the product is intended for a use that is of substantial importance in preventing impairment to human health or unless it presents a potential unreasonable risk of illness or injury. The FDA has exempted a total of 574 generic types of Class I devices from the requirement for premarket notification. There are a number of "reserved" Class I devices for which the requirement to submit a notification still applies (20); the provisions apply to many Class II devices (see for exempted Class II devices [21]), while most Class III devices will continue to need a PMA submission until they are reclassified.

Class III devices that require an approved PMA application to be marketed are those

– regulated as new drugs prior to May 28, 1976, also called transitional devices;
– devices found not substantially equivalent to devices marketed prior to May 28, 1976; and
– Class III devices that require a PMA application.

Class III devices that can be marketed with a premarket notification 510(k) are those introduced to the US market after May 28, 1976, which are substantially equivalent to Class III

Table 6 Summary of Information Requirements for a US FDA Premarket Approval Submission

Information requirement	Comments
Cover letter	Identification: Full name and street address of applicant Telephone number with area code FDA facility establishment identifier or registration number Relationship of each manufacturing facility to applicant Date facility sites will be ready for inspection Location and affiliation information to allow determination of facilities to be inspected under the preapproval process
What to send *Initial volume for each facility to include the following information*	Copy of cover letter Description of the device including pictures and where possible the trade name, common name, model number(s), product code, intended use Description of how the device works
Specific information	Design control information Design and development planning including design input, design output, design review, design verification, design validation, design transfer, design changes, design history file
Manufacturing information	
Quality system procedures	
Production flow	
Use of standards	
Purchasing controls	
Production and process controls	
Inspection, measuring, and test equipment	
Process validation	
Receiving acceptance activities	
Final acceptance activities	
Nonconforming products	
Corrective and preventive action (CAPA)	
Compliance files	
Servicing	

Source: Based on quality system information for certain premarket application reviews; guidance for industry and FDA staff, February 2003.

devices introduced to the US market before May 28, 1976, for which a regulation calling for a PMA application has not been published.

For Class II devices that are subject to the requirements for premarket notifications under 510(k), there are two options open to manufacturers. The first is the demonstration of substantial equivalence. There is also the "special 510(k) device modification" option [which relies on the availability of guidance documents, special controls, and recognized standards (22)]. For the second type of product there is an FDA guidance document (23).

FORMAT OF DOSSIERS

United States

An indication of the type of information that might be included in a 510(k) notification is summarized in Table 7. Detailed content will depend on the particular product.

Table 7 Information That Could Be Included in a 510(k) Notification

Cover letter: Name, address, telephone number, contact person, date on which summary prepared
Detailed table of contents
Name of the device (trade name, common name, classification name), establishment registration number
Device classification regulation number and regulatory status (Class I, II, III, unclassified)
Identification of any recognized performance standards, applicable voluntary safety or measurement standards
Identification of the marketed device against to which equivalence is claimed
Description of the device that is subject to the premarket submission (e.g. labeling, promotional information)
Explanation of how the device functions
Scientific concepts forming the basis for the device
Significant physical and performance characteristics of the device (design, materials of construction, physical properties)
Proposed labels and labeling including advertisements
Statement of the intended use of the device (general description of the disease or conditions that will be diagnosed, treated, prevented, cured, or mitigated)
Intended patient population
Explanation as to why any differences between the intended uses of the device and of the comparator product are not critical to the intended uses of the device and why the differences do not affect the safety and effectiveness of the device when used as labeled
Where the device has the same technological characteristics (i.e. design, material, chemical composition, energy source) as the comparator product a summary of the technological characteristics of the new device *vs* those of the comparator device
Where the device has different technological characteristics a summary the characteristics of the new and comparator products
Compliance with performance standards—see http://www.fda.gov/cdrh/stdsprog.html.
Where the determination of substantial equivalence is based on an assessment of performance data: – A brief discussion of the nonclinical tests submitted, referenced or relied on – A brief discussion of the clinical tests submitted, referenced or relied on to include as appropriate a description of the subjects on whom the device was tested, a discussion of the safety and effectiveness data obtained, specific reference to adverse events and complications, and any other information relevant to a determination of substantial equivalence – Conclusions from the nonclinical and clinical tests that demonstrate that the device is as safe, as effective and performs as well as or better than the comparator product
(To include financial certification or disclosure statement relating to clinical studies)
Design control activities summary including – Identification of risk analysis methods used to assess the impact of the modification of the device or its components; results of the analysis – Based on the risk analysis identification of the verification and validation activities needed and the acceptance criteria – Declaration of conformity with the design controls with appropriate statements concerning the conduct of the studies and the results meeting specifications; and concerning the manufacturing facility's conformance with design control procedure requirements
Where reliance is placed on a guidance document or special controls a summary report describing how the guidance or special controls were used to address the risks associated with the particular device type
Where reliance is placed on a recognized standard a declaration of conformity or a statement from the applicant that they intend to conform to a recognized standard and that supporting data will be available before the device is marketed
Where reliance is placed on a nonrecognized standard that has been accepted historically by the FDA a statement that the applicant intends to conform to a recognized standard and that supporting data will be available before the device is marketed
Where reliance is placed on a nonrecognized standard that has not been accepted historically by the FDA a statement that the applicant intends to conform to a recognized standard and that supporting data will be available before the device is marketed

Table 7 Information That Could Be Included in a 510(k) Notification (*Continued*)

Biocompatibility data for all patient-contact materials or certification of identical material/formulation
Safety information
Sterilization and expiration dating information – Sterilization method – Validation method for the sterilization process – Sterility assurance level – Packaging – Freedom from endotoxins/pyrogens – Ethylene oxide residues – Radiation dose – Traditional or nontraditional method

Software documentation and certification (specification/requirements; design; implementation; verification and validation; and maintenance – may need to address level of concern; system and software requirements and design document; hazard analysis document; software development manual/software development plan; software testing, verification and validation document; and software certification)

Source: Based on 21 CFR 807(e) http://www.accessdata.fda.gov/scripts/cdrh/cfdoes/cfcfr/CFRSearch.cfm?fr=807.92 and FDA document *Screening Checklist for all Premarket Notification [510(k)] Submissions*

The FDA introduced a pilot scheme to evaluate the acceptability of the GHTF Summary Technical Document (STED) as an alternative to the existing premarket procedures. The STED document is available at the GHTF Web site (24). The STED could be used for PMA and 510(k) procedures [but not special 510(k) procedures]. It is also necessary to take into account the GHTF document *Essential Principles of Safety and Performance of Medical Devices* (25). Four divisions within the Center for Devices and Radiological Health participated in the pilot programme, although the FDA undertook to consider wider application on a case-by-case basis.

STED format applications should consist of the following components:

Covering letter
Table of contents
Executive summary
Essential principles and evidence of conformity
Device description
Summary documents (engineering, nonclinical, and clinical)
Labeling
Risk analysis
Manufacturing information.

Other provisions of the US system should also be taken into account when preparing a STED application (26).

European Union

In the EU, there is no defined legal requirement for the format used prior to the application of a CE marking to a medical device. As indicated earlier, there are four classes of medical devices in the EU—Classes I, IIa, IIb, and III. Class I devices are placed on the market after the manufacturer has prepared a Declaration of Conformity (i.e. declaration of conformity with the requirements of the national transposition of the MDD and other applicable requirements) and requires no reference to a Notified Body (other than for devices that are sterilized or those that have a measuring function, in which case the Notified Body is involved in those specific aspects). For all other Classes of device there is involvement of a Notified Body. In the case of a Class III device it is necessary to submit for consideration a Design Dossier which includes certain information. In the case of Class II devices, the manufacturer needs to prepare a Design Dossier or a Technical File (depending on the quality system they operate). The items of information that need to be considered for inclusion in the Design Dossier or the Technical File are summarized in Table 8.

Table 8 Typical Content of Technical Documentation for Medical Devices in the European Union

Technical file	Design dossier
Summary document—does not include all of the detailed information but indicates where this can be found	Required specifically for Class III devices; can be applied to other Classes based on the quality system that is operated by the manufacturer To be submitted to the Notified Body Stand alone document – must contain all relevant information Will need to be amended whenever significant new information becomes available

Note: The Notified Bodies have recommended a two-part dossier format, as included in the NB-MED recommendation referenced at the foot of the Table

Application
Made under Annex xx of the MDD (conformity assessment procedure)
Quality system approved by [name of Notified Body]
No application made to another Notified Body
Whether or not the device incorporates a medicinal substance or a stable derivative of a human blood product
Whether the device is presented in a sterile state or is intended to be sterilized prior to use
Whether or not the device has a measuring function
Whether or not the device is intended for use on a single occasion on a particular patient

Name of device

Manufacturer
Name and physical address; where the manufacturer is not based in the EU then the name and address of an Authorized Representative within the EU must be provided
Name and physical address of any facilities involved in the production of the device also to be provided

(Note: "Manufacturer" refers to the person responsible for placing the product on the EU market)

Device description
Include drawings, photographs etc if this make the information easier to follow; include variants intended; include accessories, adaptors and other devices or equipment necessary for the use of the device
Intended use
Short description of the intended purpose or application of the device (patient population, medical conditions in which to be used), whether for professional or lay use; may include instructions for use or device manual
Mode of action
Classification
Using Annex IX of the MDD and taking into account MEDDEV 2.4/1 and additional directives relating to the classification of specific devices
Risk analysis
Using EN ISO 14971or additional methods such as failure mode and effect analysis, fault tree analysis, hazard and operability study . . .
References to standards applied
Harmonized standards (for example EN ISO 13485 Quality System; EN ISO 14971 Application of risk assessment; EN 1041 Information supplied by the manufacturer; EN 980 Symbols used with medical devices; EN ISO 10993 series of standards on biological evaluation of devices; sterilization standards) – note that certain Ph Eur monographs have a status equivalent to harmonized standards

Note: Updates lists of harmonized standards are published regularly by the Commission – e.g. http://eur–lex.europa.eu/LexUriServ/LexUriServ.do?uri=OJ:C:2008:186:0009:0024:EN:PDF, Commission communication in the framework of the implementation of the Council Directive 93/42/EEC concerning medical devices (Official Journal C 186/9 dated 23 July 2008)

Other standards
Note: Compliance with specific standards is voluntary: alternative solutions to the issues concerned can be put forward

Essential Requirements checklist
Indicate which Essential Requirements do and do not apply (with justification) and indication of how relevant requirements are met (tabular format?)
EC Declaration of Conformity

Table 8 Typical Content of Technical Documentation for Medical Devices in the European Union (*Continued*)

Technical file	Design dossier
Composition of device	
Raw material and component specification	
Method of manufacture	
Outline such as flow chart or key steps, list key processes (e.g. blow moulding/extrusion, chemical processing, sterilization)	
Process validation	
To show control and reproducibility of the production of the device	
Method of sterilization and validation data	
In process quality requirements and methods	
Finished device quality requirements and methods	
Shelf life/use by date or lifetime	
How presented to the market—pack sizes and number of devices etc	
Packaging	
Labeling and information supplied with the device—text in one EU language	
Presentation of the device to the market	
Bench testing data	
Nonclinical safety data relevant to degree of body contact and duration of contact—e.g. relevant parts of EN ISO 10993	
Clinical data review from the literature and any clinical studies	
Postmarketing safety experience	
Regulatory approvals	
Include certificates for approval of quality system and approvals from other regulatory authorities	
Markets to which product supplied, date on which first supplied, sales data	

Source: Based on the requirements of the Annexes related to quality systems components in the MDD and Notified Body Group recommendation NB-MED/2.5.1/Rec5 (http://www.team—nb.org/index.php?option=com_docman&task= doc_download&gid=188&Itemid=38).

The Technical File is intended to demonstrate that a device meets all the Essential Requirements (which are included in Annex 1 to the MDD) that are applicable to it. It acts as a signpost to where to find the detailed information. It is potentially subject to review and comment by Notified Bodies (who may want to hold the index and summary) and in some circumstances by Competent Authorities. All of the detailed information is kept by the manufacturer and the Technical File includes references on where to find detailed information.

The Design Dossier is intended to demonstrate that the device was developed in a specified manner and will include a summary of the design process as well as the device itself. All of the information should be included in a single document (which might need to extend to several volumes of information). It will include copies of the *EC Design Examination Certificate* issued by the Notified Body. The Notified Body will retain a copy of the Design Dossier and it will need to be updated whenever necessary should any of the contained information need updating or amending. Due account should be taken of relevant guidance such as the Co-ordination of Notified Bodies Medical Devices (NB-MED) Recommendation NB-MED/2.5.1/Rec5 (27).

Specific requirements relating to the necessary consultation between the Notified Body and the pharmaceutical Competent Authority relating to the ancillary medicinal (or biological medicinal) substance are discussed earlier.

Japan

In preparing the section on the requirements for Japan, it was found that there are few official documents available in English. The official documents from government sources appear not to be available in official English translation.A few documents have been translated by various parties, but the author was not able to confirm the accuracy of those translations. The majority of English translations are available through a limited number of Japanese publishers (and are quite expensive) or through trade associations (which are available to members of the associations concerned). The following is therefore a summary of the understanding of the author but it is of necessity less detailed and without key references in English.

In Japan, the overall regulation of pharmaceuticals and medical devices is the responsibility of the MHLW. The Pharmaceutical and Food Safety Bureau is responsible for policy on pharmaceutical and medical device control. Medical devices are controlled under the Pharmaceutical Affairs Law, the provisions of which were implemented over the period 2003 to 2006. Under the new system, it is no longer assumed that the marketing company in Japan owns the site(s) manufacturing the medical device. For each product the company wishing to market the product in Japan will need to obtain a marketing approval and a marketing licence. Marketing licences are renewable every five years. Manufacturers providing devices to licensed marketing approval holders will themselves need to hold a licence to manufacture the devices.A notification procedure is available for the introduction of minor changes.

There are four classes of medical devices recognized and the system of controls applied differs according to the perceived risks associated with the device. These are related to the four classes of medical device identified in GHTF documents—Class A, B, C, or D, which in Japan are named Classes I, II, III, and IV.

"General medical devices" (*Ippan Iryo Kiki*) (= Class I) are extremely low- (minimal-) risk products. General medical devices will require a manufacturing licence obtained from the prefectural government but do not require marketing approval.A self-declaration/notification procedure is used (*Seihin Tokokede*). An example of a Class I device is an X-ray film.

"Controlled medical devices" (*Kanri Iryo Kiki*) (= Class II) are of low risk where third-party certification (*Ninsho*) is required and predistribution notification is required with certain exceptional cases where there are specially designated maintenance requirements. Controlled medical devices for which there are approved standards (published Japanese Industrial Standards or JIS, such as ISO 13485 and 14971 and IEC 60601, or guidance documents or accepted industry standards) will be subject to third-party certification by organizations subject to MHLW auditing using established ISO and IEC standards. Accreditation to ISO/IEC guides 62 and 65 is required for these third-party organizations who have to be independent of the manufacturers concerned. When an application is submitted by the marketing approval holder to the third-party certification organization the latter determines whether the device is in compliance with the standards listed by MHLW. It was understood at the time of writing that there were in excess of 400 such standards. An English copy of the listing could not be traced.

The certification bodies will also audit manufacturers regularly using quality standards. The application to the third-party notification body requires the following format:

1. An application form (specified by the certification assessment body)
2. Certification application form and necessary attachment documents including documents to demonstrate conformity to the certification criteria (JIS standards) and documents to demonstrate conformity with the Essential Principles
3. GMP conformity inspection/investigation application form and other necessary documents (e.g. quality manual)

The range of products subject to this mechanism includes the following:

Anesthetic/respiratory devices: pulmonary analysis, breathing and anesthetic circuits, humidifiers and components, heat and moisture condensers, single-use anesthesia ventilators, single-use or reusable ventilator connectors, tracheotomy tubes, capnometers

Dental devices: dental equipment, dental materials and accessories, dental retention and support devices

Medical electrical devices: measuring and stimulating equipment (including temperature and blood pressure), audiometric equipment, endoscopes and equivalents and accessories, therapeutic equipment (including heat, UT, microwave, RF), electrosurgical equipment, monitoring, recording and testing equipment

Clinical equipment: sterilizers

Nonactive implantable devices: suprapubic catheters

Ophthalmic devices: measuring equipment, ophthalmic surgical and test equipment

Reusable devices: manual transportable aspirator, pen-type drug injector, blood filter for artificial heart/lung circuit, device for aspiration-type tissue biopsy needle
Single-use devices: needles, tubes, catheters, infusion/transfusion devices, surgical blades, gloves, condoms
Home-use devices: electrical massagers, therapy equipment, inhalers
Hearing aids
Radiological/diagnostic imaging systems: radiography, fluoroscopy, gamma, MRI imaging equipment and accessories
IVDs: reagents.

A comprehensive list of the approved third-party auditors is available (28) but not all of the information on the Web page is in English—e.g. the types of devices the specific organizations have been approved is in Japanese. The listed organizations includes the Japan Association for the Advancement of Medical Equipment, TÜV Sud Japan, TÜV Rheinland Japan, UL Japan, BSI Management Systems Japan, SGS Japan, Cosmos Corporation, Japan Quality Assurance Organization, Spindler Associates Company, Japan Chemical Quality Assurance, Japan Electrical Safety and Environment Technology Laboratories and Fuji Pharmaceutical Company.

Clinical data may be required for certain controlled medical devices. Where there are no approved standards, controlled medical devices will be subject to the procedure for middle-risk "specially controlled medical devices."

The highest category of devices is "specially controlled medical device" (*Kodo-Kanri Iryo Kiki*), and covers two risk groups—middle (= Class III) and high (= Class IV). A government approval (*Shonin*) is required for these products and those controlled medical devices for which standards have not been published. New medical devices and improved medical devices (including new intended uses and products with improved effects or performance) of middle-risk and high-risk devices are required to hold a marketing approval which will be reviewed by the PMDA, which is responsible for the evaluation of safety and effectiveness and reviews manufacturing and import approval applications. Approvals are issued by the MHLW. PMDA is also responsible for the development of standards against which medical devices can be reviewed and the collection and analysis of reports on defective medical devices, and it is responsible for certain good clinical practice and good manufacturing practice (GMP) inspections. Under this arrangement, the prefectural government will undertake GMP inspections of middle-risk device manufacturers, while PMDA will undertake inspections of foreign manufacturers of middle-risk and high-risk devices and domestic manufacturers of high-risk devices. Initial and follow-up GMP inspections will be against ISO 13485. An example of a Class III device is artificial bones. Examples of Class IV devices are pacemakers and artificial heart valves.

Since 2005, the required format for applications in Japan has included the STED.A preliminary programme for the use of the STED was introduced in 2002 and its use was fully implemented in 2005. In addition to the STED, the application form is required to contain information on the design of the device, its structure and dimensions, the materials used, its intended use, effects, performance, specifications and data, operating methods, manufacturing methods, and shelf life (all supported by the appropriate documents).

Clinical data are usually required for high-risk devices. Clinical data are normally required for medium-risk devices identified in the relevant government Notification.

The Japanese Medical Device Nomenclature (JMDN) is used in the regulatory system for the identification of generic nomenclature. Using this nomenclature system there are in excess of 1100 general medical devices, in excess of 1700 controlled medical devices and in excess of 1000 specially controlled medical devices. The JMDN was derived from an earlier version of the Global Medical Device Nomenclature but has not been updated (and therefore a large number of current Global Medical Device Nomenclature codes are not included in the JMDN). This would seem to be a potential area of difficulty since each medical device is required to fall under a JMDN code. There does not appear to be an English version of the JMDN easily available.

With some exceptions the inspection of quality management systems applies to all classes of medical devices.

Table 9 Summary of Dossier Requirements for Class III and IV Devices in Japan[a]

Document	Required information
Application form	Product name, generic name
	Japanese Medical Device Nomenclature code
	Device: design, structure, dimensions, mechanism of action
	Materials of construction
	Intended indication, use, or effects
	Operation/use methods
	Manufacture and quality control methods and information
	Site of manufacture
	Storage conditions, shelf life, life time
	Device specifications
Summary Technical Document (STED or Gaiyo) *(Summaries of information)*	Essential principles General requirements Design and manufacturing requirements
	Applicable standards
	Documents demonstrating conformity to essential principles verification of conformity to performance standards appropriateness of risk management verification of chemical, physical, and biological characteristics verification of removal of infection and contamination appropriateness of labeling, package insert, etc. *An essential principles checklist is a useful document in this context*
	Device description
	Draft labeling
	Risk analysis
	Manufacturing information
	Conformance to quality assurance standard (ISO 13485) appropriateness of design control appropriateness of manufacturing control
Attached documents	Origin
	Development history
	Reason and background for specification
	Use/approvals outside Japan
	Manufacture (and sterilization) and quality control data
	Documentation of conformity with essential principles
	Risk analysis
	Safety data
	Stability data and determination of lifetime
	Performance data
	Risk analysis
	Clinical evidence
	Data on standards and test methods

[a] Also applicable to certain Class II products with no recognized technical standards are available.
Source: For GHTF documents of relevance, see Summary technical documentation for demonstrating conformity to the essential principles of safety and performance of medical devices (STED). http://www.ghtf.org/documents/sg1/sg1final-n11.pdf. Accessed August 29, 2008. Essential principles of safety and performance of medical devices. http://www.ghtf.org/documents/sg1/sg1n41r92005.pdf. Accessed August 29, 2008.

An outline of the format and content of an application for submission in Japan is included in Table 9.

There are Japanese requirements for good clinical practice and good laboratory practice. Clinical trials are subject to a notification procedure. These topics are outside the scope of this chapter and will not be considered further.

There are specific provisions relating to medical devices containing biological materials—including source materials derived from human or animal cells or tissues. These are quite detailed and rather beyond the scope of this chapter.

ACKNOWLEDGEMENTS

I would like to thank my colleagues at Alcon Laboratories for helpful comments during the preparation of this chapter, and particularly, David Krapf, Terry Dagnon, Yuichi Fukazawa, and Yoji Hamano.

REFERENCES

1. 21 CFR 3.2(e). http://www.fda.gov/oc/ombudsman/part3&5.htm. Accessed August 29, 2008.
2. MEDDEV 2.1.3 Rev 3. Guidance on borderline products: Medical products/medicinal products and combination products. http://ec.europa.eu/enterprise/medical_devices/meddev/2_1_3__07–2001.pdf. Accessed August 29, 2008.
3. Consolidated Directive 2001/83/EC of the European Parliament and of the Council of 6 November 2001 on the Community code relating to medicinal products for human use as amended by Directive 2002/98/EC, Directive 2004/24/EC and Directive 2004/27/EC. http://ec.europa.eu/enterprise/pharmaceuticals/eudralex/vol-1/consol_2004/human_code.pdf. Accessed August 29, 2008.
4. Council Directive 93-42/EEC of 14 June 1993 concerning medical devices (consolidated version). http://ec.europa.eu/enterprise/medical_devices/guide-stds-directives/cons_vers_93–42-eec.pdf. Accessed August 29, 2008.
5. Presentation made at the 2007 Global Harmonisation Task Force Conference in Washington, DC. www.ghtf.org. Accessed August 29, 2008.
6. http://www.fda.gov/oc/ombudsman/part3&5.htm. Accessed August 29, 2008.
7. http://www.fda.gov/oc/combination/howtowrite.html. Accessed August 29, 2008.
8. http://www.fda.gov/oc/combination/faqs.html. Accessed August 29, 2008.
9. Federal Register. 2005; 70(164):49848–49862. http://www.fda.gov/OHRMS/DOCKETS/98fr/05-16527.htm. Accessed August 29, 2008.
10. http://www.fda.gov/oc/combination/guidance.html. Accessed August 29, 2008.
11. Guidance for notified bodies: Devices which incorporate a medicinal substance. Consultation with the MHRA. http://www.mhra.gov.uk/home/groups/comms-ic/documents/publication/con007551.pdf. Accessed August 29, 2008.
12. http://www.mhra.gov.uk/Publications/Newsletters/MAIL/CON2024282. Accessed August 29, 2008.
13. http://www.imb.ie/images/uploaded/documents/adv-g0005_guide_to_drug_device_consultations_v0.pdf. Accessed August 29, 2008.
14. EMEA/CHMP/410993/2005, EMEA Recommendations on the procedural aspects and dossier requirements for the consultation to the EMEA by a notified body of an ancillary medicinal substances or an ancillary human blood derivative incorporated in a medical device, April 2008. http://www.emea.europa.eu/htms/human/ancillary/guidance.htm. Accessed August 29, 2008.
15. EMEA/CHMP/EWP/110540/2007, Draft guideline on the development of medicinal substances contained in drug-eluting (medicinal substance-eluting) coronary stents, March 2007. http://www.emea.europa.eu/pdfs/human/ewp/11054007en.pdf. Accessed August 29, 2008.
16. http://www.fda.gov/CDRH/devadvice/3132.html. Accessed August 29, 2008.
17. http://www.accessdata.fda.gov/scripts/cdrh/cfdocs/cfpcd/315.cfm. Accessed August 29, 2008.
18. Quality system information for certain premarket application reviews; Guidance for industry and FDA staff, February 2003. http://www.fda.gov/cdrh/comp/guidance/1140.pdf. Accessed August 29, 2008.
19. The new 510(k) paradigm. Alternate approaches to demonstrating substantial equivalence in premarket notifications, March 1998. http://www.fda.gov/CDRH/ode/parad510.html. Accessed 29 August 2008.
20. http://www.fda.gov/cdrh/modact/fr011400.html. Accessed August 29, 2008.
21. http://www.fda.gov/cdrh/modact/frclass2.html. Accessed August 29, 2008.

22. http://www.fda.gov/cdrh/stdsprog.html. Accessed August 29, 2008.
23. Deciding when to submit a 501(k) for a change to an existing device, 10 January 1997. http://www.fda.gov/cdrh/ode/510kmod.html. Accessed August 29, 2008.
24. Summary technical documentation for demonstrating conformity to the essential principles of safety and performance of medical devices (STED). http://www.ghtf.org/documents/sg1/sg1final-n11.pdf. Accessed August 29, 2008.
25. Essential principles of safety and performance of medical devices. http://www.ghtf.org/documents/sg1/sg1n41r92005.pdf. Accessed August 29, 2008.
26. A pilot program to evaluate a proposed globally harmonised alternative for premarket procedures, 10 November 2005. http://www.fda.gov/cdrh/ode/guidance/1347.html. Accessed August 29, 2008.
27. http://www.team-nb.org/index.php?option=com_docman&task=doc_download&gid=188& Itemid = 38. Accessed August 29, 2008.
28. http://www.jaame.or.jp/jyusho/ninjyu_eng.html. Accessed August 29, 2008.

44 Recommendations for Toxicological Evaluation of Nanoparticle Medicinal Products*

Jean-Roger Claude and Members of Afssaps Working Party†
Toxicology Department, Afssaps, Saint Denis, France

INTRODUCTION

Nanoparticles are defined as particles at dimensions between approximately 1 and 100 nm. They have considerable general industrial applications, and although only a few nanoparticle medicinal products are available at the present time, a spectacular development of these products can be expected over the years to come. In the short and medium term, the main use of nanoparticle medicinal products (NMP) is for vectorization of active principles, corresponding to several products already marketed. Three types of vectors are distinguished at the present time:

- First generation vectors: nanospheres and nanocapsules (the best known and most accessible).
- Second generation vectors: nanoparticles coated with hydrophilic polymers such as polyethylene glycol (PEG), pegylated nanoparticles.
- Third generation vectors, still under development, combining a biodegradable core and a polymer envelope (PEG) with a membrane recognition ligand.

Very generally, nanoparticle medicinal products are colloidal systems, which can be classified into three classes:

(1) biodegradable nanoparticles,
(2) soluble nanoparticles, and
(3) insoluble or slowly soluble nanoparticles.

Biodegradability or solubility is an important property, determining elimination of nanoparticles introduced into the body. Other parameters than the conventional parameters (mass, mass per unit volume) must also be taken into account in order to relate their effects to exposure: for example, the relative size of the particles, their active surface area (total external and possibly internal surface area of exchange), number of particles per unit volume, etc. These parameters largely determine uptake, distribution, and elimination of NMPs. It is also absolutely essential to take into account the formation of nanoparticle agglomerates that subsequently form aggregates; the characteristics and potential risks of which being very different from those of the initial nanoparticles.

In addition to the specific toxicity of the vectorized active principle, the structure in which it is contained could also considerably modify this toxicity. Consequently, it would often be preferable to consider the NMP as a distinct entity that needs to be evaluated as a largely new "total" drug substance. The nanoparticle form can also induce specific risks (formation of agglomerates), transport impurities by adsorption, generate toxic substances by

* This document constitutes the official recommendations of the Agence française de sécurité sanitaire des produits de santé (Afssaps) for the toxicological evaluation of nanoparticle medicinal products, drawn up by the Working group on Nonclinical Innovation (Afssaps, Toxicology Department) chaired by J-R Claude. This document is not a guideline but is designed for giving help to applicants for the preparation of a nonclinical dossier or of a presubmission for a scientific advice.

† Members of the Working Group: J-R. Claude (Chairman), L. Domenjoud, E. Fattal, J. Guillemain, A. Guillouzo, S. Le Crom, A. Le Pape, S. Lerondel, P. Lescuyer, F. Morel, M. Pallardy, G. Peltre, C. Pineau, T. Rabilloud, R. Rahmani, D. Tremblay. Toxicology Department: D. Masset (Head) and S. Personne (Scientific secretary).

degradation or dissolution of the constituents of the NMP, cross physiological barriers (blood–brain, fetoplacental, cell and nuclear membranes, etc.). This illustrates the magnitude of the task of toxicological evaluation of NMPs, especially because part of this field is only very poorly documented, on the one hand, and no reference material is available to evaluate nanoparticles, on the other hand. Under these conditions, to ensure better efficacy (especially for screening) and for ethical reasons (unjustified and extensive use of laboratory animals), the use of validated in vitro methods must be strongly encouraged. These methods must ensure relevant evaluation of genotoxicity, cytotoxicity, free radical formation, biopersistence, phagocytic capacity, etc. It should be stressed that although vectorization of active principles is probably the major potential use of nanoparticles in medicine, other very valuable applications must be considered: tissue engineering and diagnostic tests, for example. Nanoparticle formulations are also widely used in cosmetology (sunscreens) and are a subject of controversy in some countries. Finally, it is also somewhat surprising to find so many scientific publications concerning the occupational and environmental risks of nanotechnologies and nanoparticles, compared to the limited number of documents concerning NMPs.

OBJECTIVE

The recommendations presented in this position paper reflect the opinion of the Afssaps Working Group on New Approaches to Nonclinical Evaluation of the Safety of Health Products. They are therefore opened to reflection and discussion, especially as some proposals are essentially pragmatic, sometimes even empirical. In view of the potentially extensive nature of the field of NMPs, the Working Group has decided to limit its investigations to three sectors already developed or under development.

- Use of NMPs in medical imaging (MRI and ultrasound).
- Vectorization of drug substances (anticancer drugs, antibiotics, antifungal agents, etc.) by introduction of NMPs into the body.
- Use of NMPs by topical routes (skin, lung, eye, etc.) in order to obtain systemic exposure or a local effect.

Recommendations for the toxicological evaluation of NMPs will be formulated in the usual order of other guidelines. The Task Force emphasizes that, in view of the wide range of structures, physicochemical and biological properties, therapeutic uses, etc., case-by-case assessment of the most relevant study program for a given NMP will always be essential.

RECOMMENDATIONS

As indicated above, general scientific and/or regulatory data related to the toxicological evaluation of NMPs are currently lacking. However, we can refer to the European Commission (Health and Consumer Protection) document prepared by the SCENIHR (Scientific Committee on Emerging and Newly Identified Health Risks) entitled "Opinion on the Appropriateness of the Risk Assessment Methodology in Accordance with the Technical Guidance Documents for New and Existing Substances for Assessing the Risks of Nanomaterials" and approved by public consultation in June 2007 (1). Although not specifically dealing with NMPs, this document contains a number of considerations that can apply to NMPs. Also note that the "conventional" toxicological approach proposed by current guidelines for medicinal products in general (ICH, FDA, EMEA) has been accepted up until now for approved NMPs or NMPs currently under evaluation by health authorities. However, some criticisms have been raised concerning the currently available methods of experimental evaluation that are considered to not adequately assess the properties of nanoparticle products. The most frequent criticisms essentially concern pharmacokinetic and toxicokinetic studies, which are considered to not realistically take into account the specificities related to the nanoparticle structure. The relevance of in vitro tests has also been questioned, as the sedimentation rate and diffusion capacity of nanoparticles must modify the conditions of exposure (dose-duration) of the models used (for example, genotoxicity tests). Finally, the lack of data on long-term effects is often emphasized.

Consequently, like certain consumer groups in the United States, we may need to recommend the development of completely new regulations based on "adapted" safety assessment tests for nanomaterials, including NMPs. This maximalist proposal is totally idealistic and scientifically unjustified according to the very great majority of the scientific community. How many years for development and validation would be necessary to achieve such a result? This major revision also does not appear to be justified by the available scientific data. Some manufacturers consulted and most of the task force also consider that toxicological evaluation of NMPs should not be appreciably different from "conventional" evaluation, but with certain specific adaptations (inappropriate nature of repeated-dose studies for NMP used as a single dose in man, such as in medical imaging). The plan adopted for elaboration of these recommendations is based on this approach, that is, adapt the safety assessment strategy, when necessary, without modifying the basic principles.

Pharmacokinetic Studies

Evaluation of absorption, distribution, metabolism, and excretion (ADME) of nanoparticles and of their degradation or solubilization products is essential and must be performed prior to safety studies. The pharmacokinetic properties of nanoparticles are very different from those of conventional molecules but are nevertheless studied in similar ways. Four factors essentially determine the pharmacokinetics of nanoparticles: route of administration, particle size, nature of coating polymers, and animal species. In particular, because of the importance and originality of the biodistribution study, it must be based on relevant methods (radioactive labeling). Scintigraphic imaging techniques and Positron Emission Tomography (PET) are adapted to the biodistribution study of parenterally, pulmonary, or even enterally administered NMPs, and to determination of sites of sequestration and translocation phenomena. Particular attention must be paid to the potential impact of labeling on the properties and outcome of NMPs.

NMPs are frequently administered parenterally (IV, sometimes SC or IM), but also by specific topical routes (lung, skin, eye). NMPs are recognized by the reticuloendothelial system and are phagocytosed by macrophages (liver, spleen, lymph nodes, bone marrow, lungs, etc.). As opsonization phenomena vary enormously according to the animal species, the choice of species possibly predictive for Man is particularly difficult: dogs appear to be less relevant. The use of two animal species (rodent and nonrodent) may therefore be inappropriate. The case of a specific nanoparticle system associated with the active principle raises the question of whether this system should be studied alone and in the presence of the active principle. This question has not yet been resolved, but the list of required studies cannot be unreasonably increased. In the current state of knowledge, apart from NMPs used in medical imaging, little is known about their metabolism and excretion. The formation of degradation products of polymers, for example ("metabolite"), currently remains hypothetical for a large number of compounds. The essential pharmacokinetic studies of NMPs must therefore be based on a case-by-case scientific approach, possibly by referring to the studies conducted on NMP already developed, by resolving any defects of these studies. Observed effects should preferably be expressed per unit surface area rather than unit of mass, as is usually the case, as the smaller the particle, the greater the proportion of atoms exposed to the environment.

Toxicological Studies

In Vitro Toxicity

For the reasons indicated above (rapidity—ethics), and also in view of the absence of *in silico* data, it is highly recommended to develop and validate in vitro methods that are able to provide information on cytotoxicity, phagocytic capacity and macrophage activation, activation of complement pathways, biopersistence, generation of toxic free radicals, topical cutaneous, pulmonary and ocular tolerance (when these routes are used), etc., right from the prerequisite stage. Specific pharmacological tests, especially concerning the action on nerve cells and myocardial fibers, should be considered.

Single Dose Toxicity

Evaluation of single dose toxicity provides a wealth of information on the adverse effects of NMPs also administered as a single dose in man (imaging). These studies should be designed not as acute toxicity studies in which the endpoint is death, but as complete toxicity studies including evaluation of biochemical, hematological, and histological parameters, as in repeated dose studies (extended single dose study). These studies could also be useful to rapidly compare active principles in nanoparticles and conventional forms, intended to be administered repeatedly in Man (anticancer drugs) and for topical uses.

Repeated Dose Toxicity

Repeated dose toxicity cannot be evaluated according to conventional study plans due to the structural and physicochemical differences of the particles, the animal species used, the indications and conditions of administration in therapeutics, etc. It will therefore be recommended to propose case-by-case protocols adapted to the above characteristics, reproducing human exposure conditions as closely as possible; standard protocols inducing massive exposure of the animals and consequently uninterpretable adverse effects should be excluded. Potential target organs or systems, due to the capacity of NMPs to cross physiological barriers, should be investigated as a priority. In particular, potential targets include:

- liver and organs of the reticuloendothelial system (uptake);
- kidney (e.g., possibility of urolithiasis, tubular lesions);
- central nervous system (various mechanisms have been proposed, especially passage across the blood–brain barrier, to evaluate the risk of neuronal degeneration);
- reproductive organs (potential impairment of fertility);
- cardiovascular system (e.g., formation of aggregates); and
- development of inflammatory reactions, which appear to constitute a major risk for the respiratory tract, related to the formation of agglomerates and aggregates, due to its long-term consequences: cancer (DNA damage) and fibrosis (role of cytokines). Pulmonary inflammation also plays a major role in translocation phenomena, leading to exposure of other target organs, especially the brain. There is a high risk of induction of pulmonary intravascular macrophages, because of their phagocytic activity on NMPs or their microaggregates following intravenous administration. This could cause major pulmonary hemodynamic disorders. Scintigraphic and ultrasound imaging techniques are very suitable for in vivo assessment of these phenomena.

Evaluation of systemic exposure during animal studies in order to define safety margins for human exposures obviously remains to be investigated.

Particular Toxicities

Certain forms of toxicity may occur according to the characteristics of the NMP or according to the route of administration and it would be highly recommended to pay particular attention to these aspects.

Immunotoxicity

The immune response to a foreign substance introduced into the body can be globally divided into two compartments: the adaptive response specific to the antigen introduced and the innate immune response not specific to the antigen. The structure and properties of NMPs suggest that these products are able to modify both of these types of immune response. The recognition of NMPs by "scavenger" receptors located on macrophages and polynuclear neutrophils can induce release of cytokines responsible for an inhalation pulmonary inflammatory response. Particulate matter, especially fine particles, is known to possess adjuvant properties that can induce exacerbation or modification of the type of immune response to a given antigen (Th1 response vs. Th2 response). This type of response could possibly induce hypersensitivity or allergic reactions. Finally, uptake of NMPs or their recognition by human dendritic cells could also lead to immunosuppression, and NMPs may also be able to modify self-antigens, inducing autoimmune reactions. Evaluation of the immunotoxic potential of NMPs is therefore recommended, particularly for medicinal products administered by inhalation. This evaluation must be based on appropriate, validated methods. Determination of dermal sensitization by the LLNA (local lymph node assay) test in mice is the test most widely used at the present time for nanoparticle products [OECD guideline 429 (2)], for example, for the various forms of titanium dioxide in cosmetology. The development of cell models, particularly to study the effects on macrophages and polynuclear cells and dendritic cells, should be actively encouraged.

Risks Related to the Formation of Agglomerates

This is a classical risk, identified right from the use of the first NMPs (liposomes). The formation of agglomerates can affect various territories in the body, especially in smaller vessels (peripheral microcirculation, cerebral vessels, etc.) by inducing embolic phenomena. This potential should be evaluated by appropriate techniques, especially histological techniques.

Local Effects

Nanoparticle systems can induce severe irritation and inflammation phenomena via direct mechanisms or mechanisms mediated by signaling pathways. This should be investigated for all routes of administration, particularly topical routes: skin, eye, and especially lung, as indicated above. It is reasonable to propose that OECD guidelines 404 (3) and 405 (4) can be applied to evaluation of the irritating potential on the skin and the eye after single administration in rabbits. The preferred technique to assess local effects on the lung consists of intratracheal administration of the test substance in rats followed by bronchoalveolar lavage (evaluation of biological markers of inflammation in the lavage fluid), evaluation of cellular proliferation, and histological examination. This type of study could be associated with evaluation of reversibility of the observed phenomena (after one to four weeks, for example).

The hemolytic potential by IV administration of NMPs should be evaluated, if this route of administration is used in therapeutics.

Reproduction Toxicity

No published data are currently available concerning the potential effects of nanoparticles on reproduction, fertility, and their teratogenicity, and these aspects need to be evaluated. For example, passage of NMPs across the fetoplacental barrier makes evaluation of embryofetal toxicity and the teratogenic potential essential. The protocols described in current guidelines should therefore be used, but may need to be adapted to NMPs. Maternotoxic and teratogenic effects have been described for medical imaging products (probably due to iron overload).

Genotoxicity

Although the ability of nanoparticles to cross cell membranes has been established, much less is known about their ability to reach the cell nucleus at the appropriate time of the cell cycle to interact directly on DNA, especially during cell division when the nuclear envelope is lost. Various mechanisms can be proposed to explain a genotoxic effect of nanoparticles, which would be fairly unpredictable because of their particular physicochemical characteristics:

- A direct effect by inducing DNA damage, chromosomal alterations, disturbances of cell division (aneugenic potential).
- An indirect effect via alteration of membranes and especially by the formation of free radicals, toxic oxygen metabolites, or lipid peroxidation products, especially related to inflammatory processes.

In practice, the essential question concerns the relevance of the tests recommended by international guidelines to evaluate the genotoxic potential. Conventional batteries of in vitro tests are generally used [Ames, mouse lymphoma assay (MLA), chromosomal aberrations, in vitro micronucleus]. Use of these tests and the conclusions drawn must be particularly cautious, especially as the relevance of these tests is currently controversial. There is a current tendency in favor of in vivo tests, although they are generally criticized for focusing on bone marrow (micronucleus) and liver unscheduled DNA synthesis (UDS test), while the targets of NMPs could also include the gastrointestinal tract, lungs, and skin. From this point of view, the Comet test that can be applied to many tissues appears to be more appropriate.A micronucleus test on rat alveolar epithelial cells has been proposed to test genotoxicity via the respiratory route, and a test on reconstituted human skin has been proposed for products applied to the skin.

Evaluation of the genotoxic potential of NMPs therefore remains absolutely necessary, bearing in mind that although conventional tests must still be performed, they must be modified to improve their relevance in relation to the particular characteristics of the products evaluated.

Carcinogenic Potential

Evaluation of the experimental carcinogenic potential of NMPs is currently a controversial issue.

- On the one hand, it is clear that, in view of their structure, their potential to induce DNA damage and inflammatory reactions and their bioaccumulation, NMPs could induce tumors, especially lung tumors.
- On the other hand, the protocols recommended by guidelines are complex, time-consuming, and poorly adapted to exposure to nanoparticles (metrology, control of exposure, etc.). Furthermore, carcinogenesis studies do not appear to be necessary in view of the current applications of NMPs (single dose in medical imaging, vectorization of anti-cancer drugs).

Consequently, evaluation of the carcinogenic potential of an NMP should not be systematic, and should be justified by detailed consideration of the potential hazard and risk assessment in order to avoid drawing hasty and inappropriate conclusions. An adaptation of current protocols is highly recommended: for example, shorter studies, studies with a limited number of administrations, use of transgenic mice, etc. Particularly relevant evaluation of the genotoxic potential of NMPs would obviously be very useful as part of risk assessment.

CONCLUSION

The recommendations formulated in this document are based on the following concept: evaluation of the safety of NMPs, taking into account scientific and practical considerations such as the need to be immediately operational, must not fundamentally differ from the conventional strategy of safety evaluation of medicinal products. However, the methods of this evaluation must be adapted when necessary and the results must be expressed in relation with the particular characteristics of the nanoparticle structure.

Nevertheless, a much longer term view cannot be excluded and it will also be recommended to apply the conclusions of the document "Nanotechnology: A report of the US FDA Nanotechnology Task Force" published on 25 July 2007 (5), which proposed the following long-term objectives to the FDA:

- Evaluate the adequacy of current testing approaches to assess safety, effectiveness, and quality of products that use nanoscale materials.
- Promote and participate in the development of characterization methods and standards for nanoscale materials.
- Promote and participate in the development of models for the behavior of nanoscale particles in vitro and in vivo.

This ambitious objective could be confided by Afssaps to a Task Force composed of academic scientists, members of regulatory authorities, and obviously manufacturers specialized in this field.

N.B.: This document is not designed to formulate recommendations for the evaluation of environmental toxicity.

REFERENCES

1. Opinion on the Appropriateness of the Risk Assessment Methodology in Accordance with the Technical Guidance Documents for New and Existing Substances for Assessing the Risks of Nanomaterials, Scientific Committee on Emerging and Newly-Identified Health Risks, European Commission Health and Consumer Protection Directorate, 21/22 June 2007. http://ec.europa.eu/health/ph_risk/committees/04_scenihr/docs/scenihr_o_010.pdf. Accessed November 21, 2008.
2. Organisation for Economic Co-operation and Development Test Guideline 429: Skin Sensitisation: Local Lymph Node Assay (Updated Guideline, adopted April 24, 2002), OECD.
3. Organisation for Economic Co-operation and Development Test Guideline 404: Acute Dermal Irritation/Corrosion (Updated Guideline, adopted April 24, 2002), OECD.

4. Organisation for Economic Co-operation and Development Test Guideline 405: Acute Eye Irritation/Corrosion (Updated Guideline, adopted April 24, 2002).
5. Nanotechnology, A Report of the U.S. Food and Drug Administration Nanotechnology Task Force, July 25, 2007, FDA. http://www.fda.gov/nanotechnology/taskforce/report2007.pdf. Accessed November 21, 2008).

45 Modeling and Medical Product R&D

Jean-Pierre Boissel
UMR5558, Institute for Theoretical Medicine, RTH Laennec School of Medicine, Claude Bernard University, Lyon, France

Michel Cucherat
Consultant, Lyon, France

Patrice Nony
Lyon's Public Hospital, UMR5558, Institute for Theoretical Medicine, Lyon, France

François Gueyffier
UMR5558, Institute for Theoretical Medicine, Grange-Blanche School of Medicine, Claude Bernard University, Lyon, France

François-Henri Boissel
NovaPharma, Lyon, France

INTRODUCTION

Drug innovation and development are addressing an increasing number of hurdles that are mostly brought in by several connected challenges. Some of these challenges are

1. the ever growing amount of increasingly diverse evidence in biology and physiopathology,
2. the complexity of still unsolved diseases,
3. the growing constraints from regulations,
4. the increasing awareness of and demand from the citizens, and
5. the ever deepening concerns on public health needs and requirements.

Altogether, these moves have resulted in an entirely new setting.

We postulated that the decreasing efficiency of pharmaceutical R&D is, at least partly, the result of the inability of the current bottom-up approach (see later) to drug development to tackle properly the challenges raised in the above outlined new setting. We believe we need new approaches for pharmaceutical R&D based on new concepts that should be more relevant with the new setting. Such new concepts should emerge from a rethinking of the way therapies work from their biological target to their clinical and population effects.

In this chapter we will propose four complementary concepts that might pave the way for a new approach:

- Systems physiopathology
- The numerical therapeutic model (NTM)
- The effect model (EM)
- The number of spared events (NSE)

All rely on the use of numerical modeling and computing. They illustrate how numerical modeling could be integrated in medical product R&D. EM is the mathematical relation between the probability of occurrence of the outcome without treatment (R_c) and the probability of occurrence of the outcome with the treatment T (R_T). It has been first illustrated and mentioned by Boissel et al. (1) and also developed in more detail by Boissel et al. (2). NTM is the transcription in a series of mathematical and/or logical relations of both the pharmacological properties of the ligand and the disease it is supposed to target. It has been mentioned by Nony et al. (3) and applied by Chabaud et al. (4) and is described more fully later in this chapter.

The Current Bottom-Up Approach for Drug Discovery and Modeling

Discovery of a new drug is a complex trial-and-error learning process whereby research begins with assimilating all available relevant knowledge and continues with uncovering new knowledge generated through the confrontation of a selected new chemical entity with real life. However, this process is difficult to describe in a logical algorithm. Many believe that drug discovery is mostly based on serendipity. The fashionable high-throughput screening of a large number of chemical entities against a number of standardized biological targets is just a technical manifestation of this common belief. However, the decreasing productivity of drug discovery and development and the small number of innovative medicine marketed during the last decades suggest that another approach is worth exploring (5).

In order to tackle the new setting, there is a growing effort to apply numerical modeling and computational power to streamline drug discovery and development combining the chemical and biological domains (6). A number of techniques relying on numerical modeling, neural networks, or multivariate statistical techniques have been devised to solve or improve several steps in drug discovery. On the side of the ligand, commonly used computational approaches include ligand-based drug design with the pharmacophore concept, 3D spatial arrangement of chemical features essential for biological activity, numerical representation of the structure activity and structure–property relationships, structure-based design, or drug-target docking. On the side of the target much hope was put on the deciphering of the human genome, then on all the following "omics." The genomic approach is characterized by the complete assessment of genetic descriptors associated with specific cellular states, diseases, or clinical events. This should have resulted in our ability to derive a precise functional model of signaling pathways involved in normal and abnormal states. It makes more attainable the concept of engineering complex biological processes and leads to the emerging systems biology. It has been suggested that optimized exploitation of the annotated genomic and other omics information through more comprehensive data mining, together with the precise and complete mapping of transcriptional regulatory networks, would allow different and completely unanticipated new perspectives (7).

A step further, the mechanical-based drug discovery has been proposed as an alternative approach. It is based on the elucidation and exploitation of biological, pharmacological, and biochemical mechanisms, combining application of physics-based numerical models, computer simulations, and in vitro or in vivo experiments (5). Included in this new computational approach is physiologically based Absorption Distribution, Metabolism, and Excretion (ADME) modeling (8). As suggested by Ekins et al., all these computational approaches should have more impact if integrated (9).

However, in the mind of these authors, pharmacology modeling relates to ligand to protein binding and also to effect on cellular signaling pathways through models of regulatory networks. Thus they remain within the realm of systems biology, not beyond it. Nevertheless, Ekins et al. opened the door to "systems in silico pharmacology," without defining it further. Following Rognan, we may estimate the score of the number of targets that can lead to new medical products at 3000 (10). Rognan added that only a subset of 300 proteins has been investigated. Again, the focus is put on the targets, and nothing is said on the diseases. We call the molecular level–focused process the "bottom-up" way of drug discovery.

All these thoughts and developments have led to the emergence of a new term—predictive pharmacology (7). Predictive pharmacology is a very attractive concept that fits rather well with all the efforts outlined above. However, in order to emerge as an operational concept, it needs to gain in substance. Its main limitation is the fact that all the new technical computational developments are applied to a discovery process that still relies more or less on serendipity.

We postulated that the core of any drug discovery approach is the choice of the target. Basically the omics-based choice of target still relies more on consideration for correlations derived from annotations than from established causality. Of course, a well-characterized and isolated correlation between a gene transcript and a clinical event certainly restricts the domain of uncertainty and thus the room for serendipity. By chance the choice might be good. This will be checked later in the development. However, there remains a high, although not quantified, degree of chance that it will not be the case. More troublesome is the fact that a good choice is very rare. Most of the time a number of correlations arise form data mining with no information on the causal links with the disease. Further, cutting down a signaling network established on such correlations does not mean that causality will be broken because of hidden redundancy and retroaction elsewhere. Thus we fear with others that predictive pharmacology as it is defined above will provide very few solutions to the problems that discovery and development of new medical products are facing (11).

Beyond the complexity issue, numerical modeling has been introduced in physiopathology and therapeutic research because of the decreasing innovation efficiency in therapeutics and the growing difficulties for physicians to make optimal treatment decision for a given patient through combining the many available modestly efficacious treatments. Altogether these two worrying developments result in a less-than-optimal treatment for patients when compared to the amount and quality of available knowledge.

THEORETICAL BACKGROUND FOR A NEW APPROACH

When working on an innovative medicine, the researcher needs a picture of the mechanism and the epidemiology of the disease it is proposed to treat. Let us assume that the researcher is looking for a treatment that would prevent the occurrence of acute myocardial infarction (AMI). Currently, regarding the mechanism, such a picture is something like "the myocardial cell death is caused by oxygen and nutriments deprivation due to a blockade of arterial blood flow resulting from a thrombosis in a coronary artery." This picture is a discursive model of AMI. It is a model because it reduces what we know to a sentence (or a series of sentences) and also because what we know is far from the true reality of AMI. It is discursive because it is built of words arranged with grammar. It could also be a drawing—a little more convenient to express interaction and causality. Of course, this model of AMI could be much improved by adding more details (e.g., emerging, growing or destabilized atherosclerotic plaque, plaque rupture, artery physics, blood dynamics, etc.). It may need a textbook. However, with the researcher's brain who is designing the research on it, it will be constrained forever by severe limitations. It cannot account for

1. quantitative interactions if their number is over, say, two or three at best,
2. redundancy, which is a signature of living organisms,
3. feedback loops,
4. time, and
5. space.

The discursive model is the traditional approach. Numerical modeling of AMI is the only available approach capable of overcoming these limitations (11–13).

Now let us turn to the side of the outcome of a therapeutic model, whether it is discursive or numerical. The expected effect of a treatment is a decrease of the risk of event(s), i.e., mortality and/or morbidity caused by the disease. Let R_c be the frequency of this event in patients, as they do not receive the treatment T. In the same but treated subjects, the frequency becomes R_T over the same period of time. With others, we postulated that a relation exists between the two frequencies which characterizes the treatment and its efficacy (14–16). It depends on individuals, disease, and the therapeutic objective, i.e., the clinical criterion chosen for efficacy. We called such a relation the effect model of treatment T:

$$R_T = f(R_c).$$

As shown later, the equation may be more complicated. We believe that predicting the effect model is a major concern in designing new therapies. Although it is beyond the focus of this article, we believe further that knowing the effect model is a key for a valid selection of patients to be treated.

FOUR BASIC CONCEPTS

Beyond Systems Biology: Systems Physiopathology

Systems biology has been devised to tackle the barrier of complexity in life science. Actually in terms of complexity, physiopathology is far beyond systems biology. Disease mechanisms are much more complex than anything else. The main reason is that a disease develops throughout a diversity of levels from gene expression up to population (Fig. 1), whereas systems biology limits

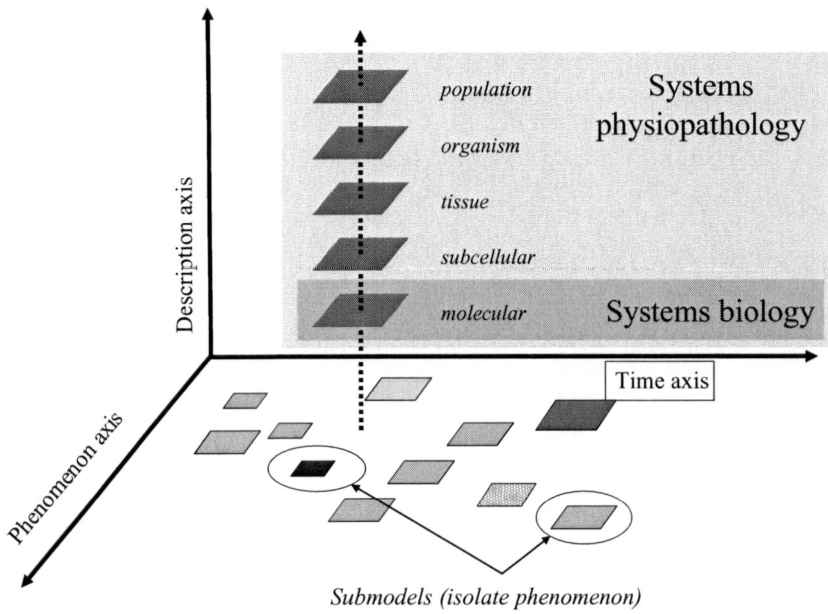

Figure 1 Framework of systems physiopathology compared to systems biology.

itself at a couple of levels (molecules and cell, and sometimes a population of cells). The tissue level is seldom accounted for and neither physiology nor anatomy is paid attention to. Further, the time scales in disease mechanism vary from nanoseconds (chemical interactions) to several decades (evolution to clinical event). In chronic diseases, such as cancer or atherosclerosis, the sequence of events at the molecule, cell, tissue, and target organ levels takes decades to result in death or myocardial infarction.

In addition, new knowledge is emerging at a high rate. This observation has three consequences as follows:

1. the ever extensive amount of knowledge one should incorporate in a numerical model of disease,
2. the need of a careful knowledge management tool, and
3. models, either discursive or numerical should be conceived flexible enough in order to integrate any new relevant knowledge.

The ultimate goal of "in silico" models in physiopathology is to improve health. Through designing numerical models of a disease and running computer simulations we expect a better understanding of the links between its various components, weighing the respective influence of its various factors, deciphering new and more relevant targets for innovative medicines, and paving for a quicker way of testing a new therapeutic intervention at the very beginning of its development before a sizeable amount of resources has been invested (early proof of concept). Altogether these developments should result in improving the way new therapies are discovered and in providing doctors with tools predicting the outcome of combining therapies.

The Numerical Therapeutic Model (NTM)

A Global Account

Activities of a drug or any kind of therapy in the body can be split into four subsystems that can each be modeled separately (Figs. 2 and 3). The unique system entry (input) is the drug administration with amount per dosing, timing of dosing, and cumulative amount. The outputs are the expected clinical effect and the side-effect(s).

Between each subsystem one or more markers of drug activities in the body exist and are accessible to determination: the drug level in relevant fluid (C(t) in Fig. 2) and an intermediary marker (IO(t) in Fig. 2). IO may mediate the drug effect on the disease mechanism and/or in

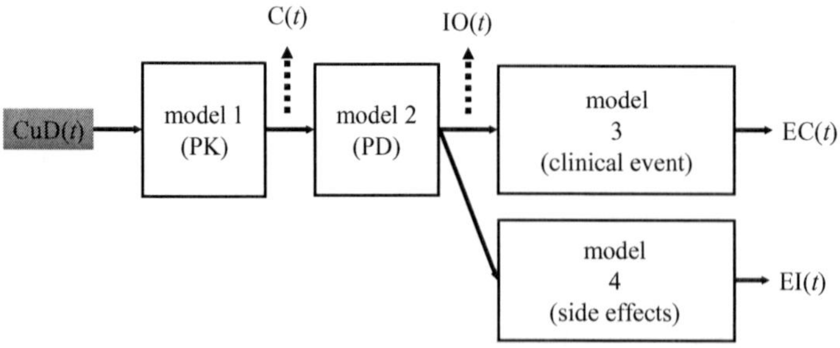

Figure 2 Content of the therapeutic model. Input is a treatment at dose *D* that gives an amount CuD at time *t* delivered to the body. The PK model transforms it in blood level (C(*t*)) through several sequential steps (Fig. 2). In turn, the blood level is transformed during a change of a physiological parameter IO(*t*). If IO(*t*) is the support of the effect of the treatment on the disease, it is quoted as *z*. This variable affects the disease process represented by the clinical model. IO(*t*) or a similar parameter affected by the treatment is the entry into the side-effect model.

the mechanism of side-effects. However, in real medical practice it is more often only correlated with the output. Nevertheless, in the modeling process, it is assumed to bring the drug-related stimulus into the mechanism of the disease. Thus it mediates the drug mode of action. In such a case, and as given below, IO(t) is called z. Each subsystem can be modeled with a phenomenological approach or a mechanistic approach (17). The compartment modeling is an example of the former for the pharmacokinetic (PK) subsystem.

Thus numerical therapeutic modeling, that is, linking "in silico" drug administration at a given dose to the end product (effect on clinical outcome), allows the amount of drug, the drug concentration in biological fluids, mostly the blood, and biomarker(s) of pharmacological activity and clinical effect following drug administration to be integrated in a single global model and computed (1,2).

A Streamlined Formal Therapeutic Model

A Stepwise Process

Each therapeutic model is a cascade of submodels addressing each step of the process that carries the drug potential activity to its final expression (Fig. 2) (4). The sequence of steps and

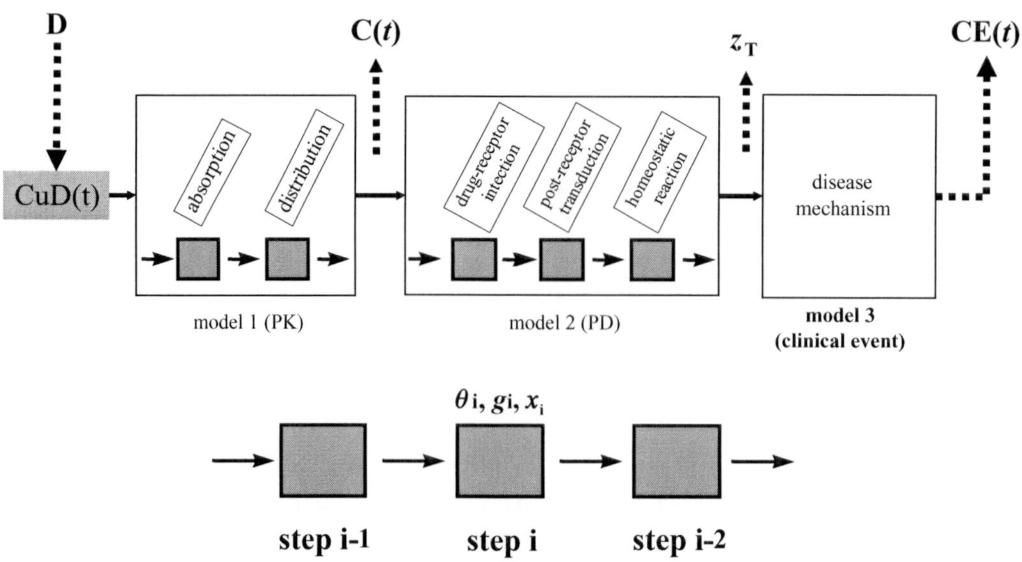

Figure 3 The stepwise therapeutic model. Each step is modeled by one or a few equations based on our understanding of pharmacology and physiology.

their contents can follow the Vengt-Pedersen scheme (18). In turn, each step can be broken-down into substeps. Each step i, or substep ij, is defined by an input, an output, a numerical submodel, which links the output to the input, scale parameter(s) θI, and a single polymorphism, which is expressed by X_i. If a step is composed of substeps, there are as many X as substeps; they are noted as X_{ij}, where j stands for the sub-step. The output of step i-1 is the input of step i. Hence the process is linear in the step domain (i.e., between the steps), although it is not linear within the steps. The treatment is supposed to be a drug. To see in more depth how the steps are structured, let us take the case of drug distribution (step i-2). The input from the previous step (absorption) is the amount of the drug that reaches the systemic circulation, A. Output is the average drug level in blood between two dosings, C_{moy}. The corresponding numerical submodel is given by a classical pharmacokinetic equation (19):

$$C_{moy} = \frac{A}{T \cdot (1 - e^{-\theta_{22}(x_2^{-1})}) \cdot \theta_{21}},$$

where T is the dosing interval, θ_{21} and θ_{22} are the two scale parameters, the latter being the maximum clearance. The patient variable is X_2, the age (20). The last step gives the modified value of z.

More on Clinical Transduction

Model 3 in Figure 2 integrates the effect of the treatment on biomarker z directly into the mechanism of the disease (systems physiopathology). R_c is given by the disease model. R_T is obtained by adding to the disease model the pharmacokinetic and pharmacodynamic submodels. Let us take an example. In this model z is the heart rate. The disease is *angina pectoris*. The therapeutic model has been used in a previous work to predict the clinical efficacy of ivabradine, a drug acting by reducing heart rate (4). A numerical clinical outcome submodel was developed assuming that elevated heart rate during exercise is the cause leading to an angina attack in patients with atherosclerotic stenosis of arterial coronary bed. Exercise induces a rise in heart rate that in turn increases oxygen demand by the heart muscle. When this demand expressed by the coronary reserve exceeds the capacity of oxygen supply by the impaired coronary circulation—the angina threshold—an attack occurs with a chest pain that forces the patient to stop effort. Consecutively the heart rate reduces. The model takes into account the regulatory processes in the cardiovascular system as described by Kappel and Peer (21). Although simplified, it accounts for the current evidence on the mechanism of effort angina. It allowed computing the coronary reserve given the heart rate and physiological parameters specific to patients including the diameter of the lumen of the coronary vessel. Applied to ivabradine with the PK-PD part, it predicts fairly well the dose-effect relationship on markers of heart ischemia during exercise [unpublished]. The PK-PD part of the model was based on data obtained in healthy volunteers with ivabradine (22,23). Details on these data and PK-PD models fitting them and the pathophysiological model for angina are given elsewhere (4).

The Effect Model (EM)

A Few Theoretical Considerations

Now let us turn to the outcome of a therapeutic model, whether it is discursive or numerical. The expected effect of a treatment is a decrease of the risk of event(s) (i.e., mortality and/or morbidity) caused by the disease. In the case of angina pectoris we want to decrease the probability of chest pain. Let R_c be the frequency of this clinical event in patients if they do not receive the treatment T. In the same but treated subjects the frequency becomes R_T over the same period of time. We and others reported empirical and theoretical evidence supporting that a relation between these two frequencies does exist (14–16,24,25). It depends on treatment, disease, and therapeutic objective, i.e., the clinical criterion chosen for efficacy. We called it the effect model of treatment T (14). Actually, four types of effect model have been observed or postulated. Two are linear, the others are curvilinear (2). It can be shown that the EM is fully implicit in the NTM for given drugs and therapeutic objective.

The form of the relation between R_T and R_c is represented by the following equation:

$$R_T = f(R_c, T, x),$$

where T indicates that it is treatment dependent and X is a vector of characteristics of individuals correlated with R_T other than those relevant to R_c (4). X may be phenotype or genotype-derived variables. Some may be altered by the individuals' environment. From this relation, one derives

$$R_c - R_T = g(R_c, T, X). \quad (1)$$

This function gives the absolute benefit, that is, the gain by T expected for a patient $[R_c, X]$.

R_c varies depending on several variables. Accordingly, the gain varies. Variations of R_T with R_c encompassed in the effect model can be called arithmetic interaction as it is represented by a deterministic mathematical relation. The variations of R_T for a given R_c, if not purely random, can be due to polymorphisms and their interactions with the environment. This is the biological interaction (26).

Supportive Empirical Evidence for a Relation Between R_T and R_c

In this section, we present in more details a short sample of cases where the relationship between the two risks has been explored. We reviewed seven different cases based on published or unpublished data. It is not a random sample since we are only seeking to demonstrate that such a relationship does exist. To observe a relationship between R_T and R_c we need to follow up over t untreated and similar treated subjects with relevant recorded baseline data. Data from randomized clinical trials are appropriate because they provide an unbiased estimate of the efficacy, that is, the difference or the ratio of R_T (treated group) and R_c (control group). One can use either summarized, or group, data or pooled individual data (27). As we will see later, with summarized data the number of trials should be large enough although it is not possible to suggest a minimum number since more is always better in this setting. Trials should have been performed in populations with reasonably different risks in the control group. Also, all available information from the studied population should be taken into account (28). In the following, we will use both summarized (first cases) and individual (last case) data.

Angiotensin Converting Enzyme Inhibitors and Heart Failure

Angiotensin-converting enzyme (ACE) inhibitors have been associated with both hemodynamic and symptomatic improvements in the long-term treatment of congestive heart failure (CHF). Garg and Ysuf published in 1995 a meta-analysis of all the available trials evaluating ACE inhibitors *versus* placebo in CHF (29). Their results are based on 7105 patients from 32 randomized trials. Overall, there were 611 deaths among 3870 (15.8%) patients allocated to an ACE inhibitors group and 709 deaths among 3235 controls (21.9%), indicating a statistically significant reduction in mortality. The reduction in mortality was primarily attributable to reduction in deaths due to progressive heart failure with a trend toward reduction in sudden death, fatal myocardial infarction, and strokes. The relation between R_T and R_c for ACE inhibitors in CHF is shown in Figure 4, panel A. A crude linear regression analysis with no attempt to account for regression to mean and measurement errors (30) suggests that the line crosses both axes at the origin. As a consequence, one may assume that the ratio R_T/R_c (the relative risk, given below) is constant for the observed range of R_c. Thus, the risk difference, i.e., $R_c - R_T$, is proportional to R_c.

β-Blocking Agents and Secondary Prevention After Myocardial Infarction

The long-term effect of late intervention with β-blocker treatment on long-term mortality in patients having survived an acute myocardial infarction has been assessed in several randomized trials during the 1980s. Meta-analyses of all trials concluded that the intervention reduced total mortality (31,32). From the data on mortality, standardized rates over one

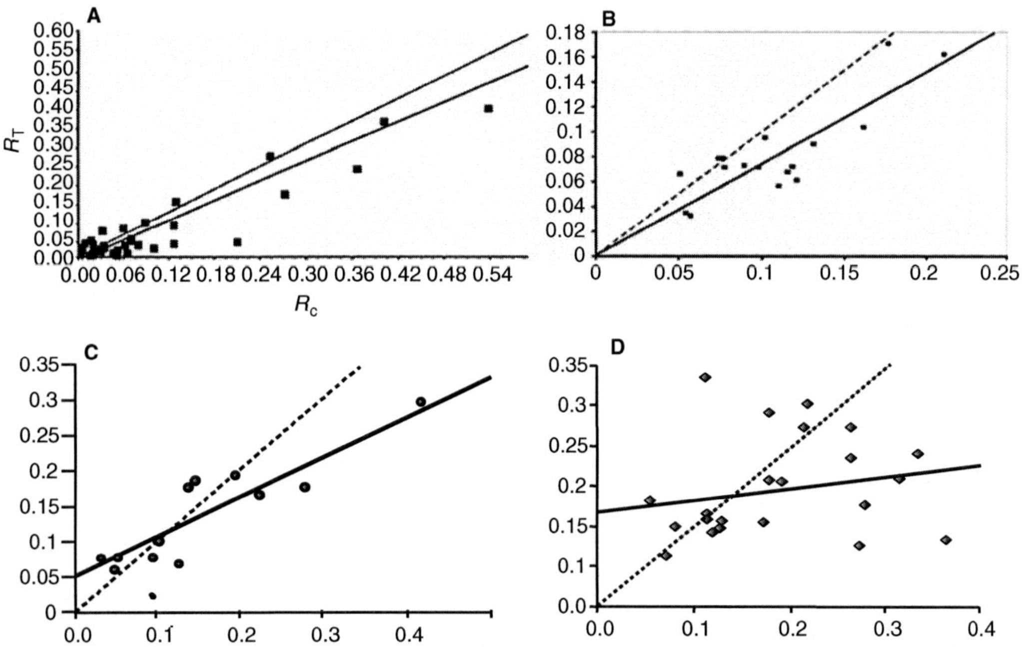

Figure 4 Each dot (solid circle) represents a trial. Both frequencies (R_C, observed frequency in the control groups; R_T, observed frequency in the treated group) were standardized to one year of follow-up. The dash line is the line for $R_T = R_C$. The full line is a linear regression best fit to the data, with no attempt to account for regression to mean and measurement errors because in this work we are not seeking for a precise estimate of the effect model. (Panels: see text).

year were computed. The results are shown on Figure 4, panel B. With the individual data from the nine trials with at least 200 randomized patients, a subgroup meta-analysis has been performed (32). Results are summarized in Figure 5. The graph shows no correlation between the odds ratio for one-year mortality and the one-year risk in the control subgroups. This is another argument for the independence between R_T/R_c and R_c. As for ACE inhibitors in heart failure patients, it seems to be linear with an intercept not significantly different from 0 (i.e., compatible with a multiplicative effect model).

Class I Antiarrhythmic and Secondary Prevention After Yocardial Infarction

The case for class I antiarrhythmic drugs has been published previously (14). Here the relation between R_T and R_c seems to be a straight line again. However, it crosses the y-axis well above the origin (Fig. 4, panel C). According to this model, patients with an untreated risk below 0.12 or so will fare worse on treatment. Thus, there is a threshold between risk values for which the treatment is supposed to be beneficial and those for which it is detrimental. This threshold is natural because it is caused by the interaction between the patients' characteristics and the treatment.

Streptokinase in Acute Myocardial Infarction

Streptokinase at the acute phase of myocardial infarction behaves on total mortality as class I antiarrhythmics (Fig. 4, panel D). Again there is a natural threshold. Threshold can result from an external constraint (33).

Antihypertensive Treatment

The INDANA database established from 1995 to 2000 comprises the individual records of 53,228 patients from 11 major randomized controlled trials designed to assess the preventive value of antihypertensive treatment on cardiovascular risk (34). From patients in the control groups,

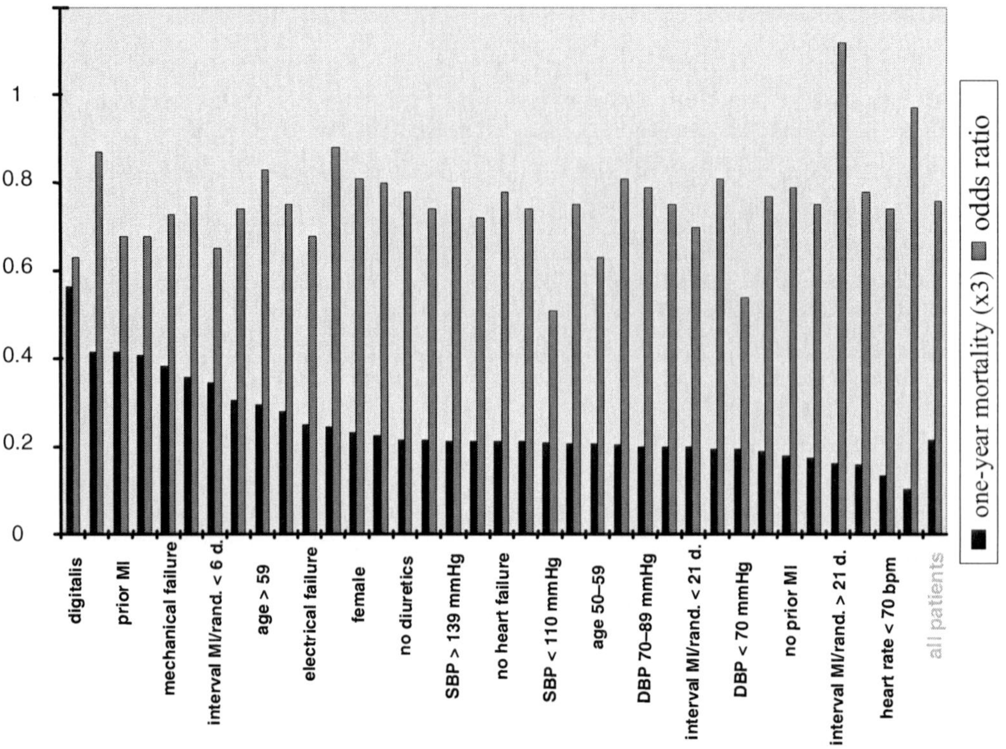

Figure 5 The metaanalyzed one-year mortality in control subgroups of the nine β-blocker treatment randomized trials with at least 200 randomized patients compared to the odds ratio for corresponding subgroups. The post hoc subgrouping according to risk factors of late death after myocardial infarction results in the diversity of R_C. There is no clear relation between odds ratio and R_C as the later decreases from left hand to right hand, the former remains roughly constant.

a Cox model was fitted to predict the occurrence of fatal or nonfatal stroke. The model was then applied to all patients to compute their predicted untreated risk of stroke, R_c. They were then ranked according to this risk and subgrouped in three classes of increasing predicted risk of stroke. In each subgroup, the observed number of strokes among control and treated subjects allow to compute R_c and R_T, respectively. Figure 6 shows the absolute benefit ($R_c - R_T$) for stroke according to these three classes for males and females. Accounting for the large confidence intervals of the third risk class, the absolute benefit looks roughly proportional to R_c, with $R_T = aR_c$.

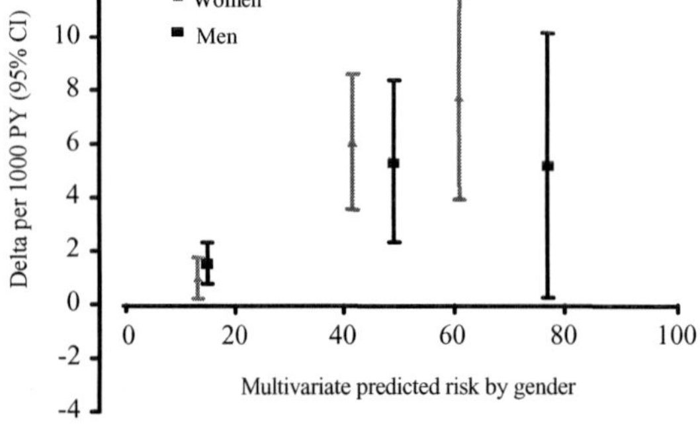

Figure 6 In patients in the control groups of 11 major randomized controlled trials of antihypertensive treatment INDANA database, a Cox model was fitted to predict the occurrence of fatal or nonfatal stroke. The model was thus applied to all patients, whatever the group they have allocated to, to compute their predicted untreated risk of stroke, R_C. They were then ranked according to this risk and subgrouped in three classes of increasing predicted risk of stroke. In each subgroup the observed number of stroke among control and treated subjects allow to compute R_C and R_T, respectively.

THE NUMBER OF SPARED EVENTS

By applying Equation (1) to all subjects of the population of interest P, each characterized by [R_c, X], one can compute a prediction of the number of spared events if T is given to all individuals. Among subjects characterized by [R_c, X], some benefit more from treatment T. More means that $R_c - R_T$ is greater than some threshold, s. There are several ways to set s.

1. By applying an external constraint to the effect model, e.g., the amount of money patients and/or the community accept to pay for a given expected gain
2. The burden of side-effects is balanced by the gain if $R_c - R_T > s$; the so-called natural threshold as illustrated by the antiarrhythmic and the streptokinase cases
3. An arbitrary value defining "responders."

Whatever is the definition, all the subjects for whom $R_c - R_T > s$ compose the Therapy Target Population (TTP). There are as many TTPs as ways of defining s.

Thus knowing the effect model, an external constraint or any other condition on the value of the acceptable individual or community benefit, one can find s and the perimeter of the therapy target population. Computation requires knowing the distribution of patients' characteristics in the population. From s one can derive the number of spared events in that population over a given time period. It should be kept in mind that all these values are time dependent. Note there are as many numbers of spared events as there are populations.

PREDICTING CLINICAL EFFICACY BY INTEGRATING THE FOUR CONCEPTS

We believe that predicting the effect model should be the major concern in designing new therapies. As we showed above, empirical as well as theoretical considerations support that relation. We have also shown that the effect model can be simulated from the numerical therapeutic model. Further, the effect model can be used to predict the number of events saved in a given population and in a given period of time, thanks to the treatment T. This number should become the thread by which innovation and development are driven. This explains why we said that the approach we are proposing is "top–down."

A FEW WORDS ON METHODOLOGICAL ISSUES

A core methodology is been devised for numerical modeling in systems physiopathology and drug discovery (17). At first sight, it seems to be an easy task: Just collect all the necessary pieces of information in the literature and put them in a computer to draw pictures. But everyone who tried this knows that many difficulties arise. First, we must find all the pieces and, if only one is missing, we may well not achieve our objective. Second, we must be sure that all the mathematical solutions represent the real world we are interested in properly. Third, we should solve all the methodological and technological hurdles that will be faced along the process by achieving a compromise between a too stringent reduction, which would make the model irrelevant to its goal and a too big amount of details that will make the objective impossible to reach. Fourth, it takes time to build the model and to make it run properly and in an efficient way.

In addition, beyond pure science, numerical modeling of diseases is science applied to improving care to patients. As a consequence, the researcher should pay attention to both quality and transparency. For ethical reasons the former is compulsory, whereas the later is required by regulators. Actually, regulations are already mandatory and one can bet they will be even more stringent in the future in order to protect consumers. Models and processes which they are part of should comply with the regulations.

Overall, modeling diseases develop along three axes: the phenomenon or subsystems axis, the time axis, and the integration axis (Fig. 1). Clearly, there is not a unique way to design the model that would help achieving the researcher's aims. All this makes modeling in physiopathology different. It explains why an appropriate strategy and relevant methodologies are needed.

The way the challenges are dealt with depends on the main objective of the research. However, there are a few general principles that are quite helpful in finding appropriate ways to solve the problems.

1. Modeling of biological processes can be achieved piecemeal, with each piece or submodel having clearly identified input and output and biomarkers that could be used to validate the global model (this is the "Lego®-like" principle).
2. Each piece can be numerically solved at a different level of complexity from the others.
3. At any time in the progress of the modeling process a submodel can be replaced by a more detailed one (the "plug-in" principle).

However, replacement by a more detailed submodel should be prepared (consistency in input and output) and may require new developments and recalibration of the whole model.

REFERENCES

1. Boissel JP, Gueyffier F, Cucherat M, et al. Pharmacogenetics and responders to a therapy: Theoretical background and practical problems. Clin Chem Lab Med 2003; 41:564–572.
2. Boissel JP, Cucherat M, Nony P, et al. New insights on the relation between untreated and treated outcomes for a given therapy. J Clin Epidemiol 2008; 61:301–307.
3. Nony P, Cucherat M, Boissel JP. Revisiting the effect compartment through timing errors in drug administration. Trends Pharmacol Sci 1998; 19:49–54.
4. Chabaud S, Girard P, Nony P, et al. Clinical trial simulation using therapeutic effect modelling: Application to ivabradine efficacy in patients with angina pectoris. J Pharmacokinet Pharmacodyn 2002; 29:339–363.
5. Hausheer FH, Kochat H, Parker AR, et al. New approaches to drug discovery and development: A mechanism-based approach to pharmaceutical research and its application to BNP7787, a novel chemoprotective agent. Cancer Chemother Pharmacol 2003; 52(suppl 1):S3–S15.
6. Kapetanovic IM. Computer-aided drug discovery and development (CADDD): In silico-chemico-biological approach. Chem Biol Interact 2008; 171(2):165–176.
7. Liu ET. Expression genomics and drug development: Towards predictive pharmacology. Brief Funct Genomic Proteomic 2005; 3:303–321.
8. Yamashita F, Hashida M. In silico approaches for predicting ADME properties of drugs. Drug Metab Pharmacokinet 2004; 19:327–338.
9. Ekins S, Mestres J, Testa B. In silico pharmacology for drug discovery: Applications to targets and beyond. Br J Pharmacol 2007; 152(1):21–37.
10. Rognan D. Chemogenomic approaches to rational drug design. Br J Pharmacol 2007; 152(1):38–52.
11. Noble D. Will genomics revolutionise pharmaceutical R&D? Trends Biotechnol 2003; 8:333–337.
12. Dronne MA, Grenier E, Dumont T, et al. Role of astrocytes in grey matter during stroke: A modelling approach. Brain Res 2007; 1138:231–242.
13. Grenier E, Dronne MA, Descombes S, et al. A numerical study of the blocking of migraine by Rolando sulcus. Prog Biophys Mol Biol 2008; 97(1):54–59.
14. Boissel JP, Collet JP, Lièvre M, et al. An effect model for the assessment of drug benefit: Example of antiarrhythmic drugs in post-myocardial infarction patients. J Cardiovasc Pharmacol 1993; 22:356–363.
15. Sharp SJ, Thompson SG, Altman DG. The relation between treatment benefit and underlying risk in meta-analysis. BMJ 1996; 313:735–738.
16. Arends LR, Hoes AW, Lubsen J, et al. Baseline risk as predictor of treatment benefit: Three clinical meta-re-analyses. Stat Med 2000; 19:3497–3518.
17. Boissel JP, Ribba B, Grenier E, et al. Modelling methodology in physiopathology. Prog Biophys Mol Biol 2008; 97(1):28–39.
18. Veng-Pedersen P, Modi NB. Pharmacodynamic system analysis of the biophase level predictor and the transduction function. J Pharm Sci 1992; 81:925–934.
19. Jusko WJ. Guidelines for collection and analysis of pharmacokinetic data. In: Evans WE, Schentag JJ, Jusko WJ, eds. Applied Pharmacokinetics. Spokane, WA: Applied Therapeutics Inc., 1987:9–54.
20. Hammerlein A, Derendorf H, Lowenthal DT. Pharmacokinetic and pharmacodynamic changes in the elderly: Clinical implications. Clin Pharmacokinet 1998; 35:49–64.
21. Kappel F, Peer RO. A mathematical model for fundamental regulation processes in the cardiovascular system. J Math Biol 1993; 31:611–631.
22. Duffull SB, Chabaud S, Nony P, et al. A Pharmacokinetic simulation model for ivabradine in healthy volunteers. Eur J Pharm Sci 2000; 10:285–294.
23. Duffull SB, Aarons L. Development of a sequential linked pharmacokinetic and pharmacodynamic simulation model for ivabradine in healthy volunteers. Eur J Pharm Sci 2000; 10:275–284.

24. Lubsen J, Tijssen JG. Large trials with simple protocol: Indications and contraindications. Control Clin Trials 1989; 10:S151–S160.
25. Glasziou PP, Irwig LM. An evidence based approach to individualising treatment. BMJ 1995; 311:1356–1359.
26. Boissel JP, Gueyffier F, Haugh MC. Response to Inclusion of women and minorities in clinical trials and the NIH revitalization act of 1993: The perspective of NIH clinical trialists. Control Clin Trials 1995; 16:286–288.
27. Stewart LA, Clarke MJ. Practical methodology of meta-analyses (overviews) using updated individual patient data. Cochrane Working Group. Stat Med 1995; 14:2057–2079.
28. Collet JP, Boissel JP; and the VALIDATA Group. Sick population-treated population: The need for a better definition. Eur J Clin Pharmacol 1991; 41:489–491.
29. Garg R, Yusuf S. Overview of randomized trials of angiotensin-converting enzyme inhibitors on mortality and morbidity in patients with heart failure. Collaborative Group on ACE Inhibitor Trials. JAMA 1995; 273:1450–1456.
30. McIntosh MW. The population risk as an explanatory variable in research synthesis of clinical trials. Stat Med 1996; 15:1713–1728.
31. Yusuf S, Peto R, Lewis JA, et al. Beta-blockade during and after myocardial infarction: An overview of the randomized trials. Prog Cardiovasc Dis 1985; 27:335–371.
32. The Beta-Blocker Pooling Project Research Group . The Beta-blocker Pooling Project (BBPP): Subgroup findings from randomized trials in post infarction patients. Eur Heart J 1988; 9:8–16.
33. Cucherat M, Boissel JP. A mathematical model for the determination of the optimum value of the treatment threshold for a continuous risk factor. Eur J Epidemiol 1998; 14:23–29.
34. Gueyffier F, Bulpitt C, Boissel JP, et al.; for the INDANA Group. Antihypertensive drugs in very old people: A subgroup meta-analysis of randomised controlled trials. Lancet 1999; 353:793–796.

Annex 1: List of Acronyms and Abbreviations

Anthony C. Cartwright
Pharmaceutical Regulatory Consultant, Harpenden, U.K.

LIST OF ACRONYMS AND ABBREVIATIONS

The following is a list of the technical and regulatory acronyms and abbreviations used in the text of this book. It is designed as a reference for readers.

5-HT	5-Hydroxytryptamine
AAALAC	Association for Assessment and Accreditation of Laboratory Animal Care International
ABPI	Association of the British Pharmaceutical Industry
ACDP	Advisory Committee on Dangerous Pathogens (United Kingdom)
ACE	Angiotensin-converting enzyme
ACS	American Chemical Society
ADME	Absorption, distribution, metabolism, and excretion
ADR	Adverse drug reaction
AED	Antiepilepsy drug
AF	Assessment (safety) factor
AFSSAPS	Agence Française de Sécurité Sanitaire des Produits de Santé (France)
ALARP	As low as reasonably practicable
ALT	Alanine transaminase
AMA	American Medical Association
AMI	Acute myocardial infarction
ANC	Absolute neutrophil count
ANCOVA	Analysis of covariance
ANDA	Abbreviated new drug application (FDA)
ANOVA	Analysis of variance
ANS	Autonomic nervous system
ANSI	American National Standards Institute
APD	Action potential duration
APEC	Asia-Pacific Economic Cooperation
APHIS	Animal and Plant Health Inspection Service (United States)
API	Active pharmaceutical ingredient
APT	Attached proton test (NMR)
AS	Active substance
ASA	Active systemic anaphylaxis
ASEAN	Association of Southeast Asian Nations
ASMF	Active substance master file (European Union)
ASSM	Active substance starting material
AST	Aspartate aminotransferase
AUC	Area under curve (usually of plasma concentration versus time)
AUC_t	Area under curve to time t
AUC_∞	Area under curve to infinity
AUEC	Area under effect curve
AV	Atrioventricular
BA	Bioavailability
BCF	Bioconcentration factor
BCOP	Bovine corneal opacity and permeability
BCS	Biopharmaceutical classification system
BDDCS	Biopharmaceutical drug disposition classification system
BE	Bioequivalence
BfArM	Bundesinstitut für Arzneimittel und Medizinprodukte (German agency)

BLA	Biologic license application (FDA)
BP	British pharmacopoeia
BPCA	Best Pharmaceuticals for Children Act (United States)
BPWP	Blood Products Working Party (CHMP)
BsAb	Bispecific monoclonal antibodies
BSE	Bovine spongiform encephalopathy
BW	Body weight
CA	Chemical abstracts
CA	Competent authority (European Union)
CAM	Complementary and alternative medicine
CANDA	Computer-aided new drug application
CAS	Chemical abstracts service
CBER	Center for Biologics Evaluation and Research (FDA)
CCAC	Canadian Council on Animal Care
CCDS	Company core data sheet
CCR5	Chemokine (C-C motif) receptor 5
CD	Cluster of differentiation
CD	Compact disc
CDER	Center for Drug Evaluation and Research (FDA)
cDNA	Complementary or cloned DNA
CDRH	Center for Devices and Radiological Health (FDA)
CD-ROM	Compact disc–read-only memory
CEP	Certificate of suitability to monograph of the European Pharmacopoeia
CEPIDEC	Center d'épidemiologie sur les causes de mortalité (WHO)
CFC	Chlorofluorocarbon
CFR	Code of Federal Regulations (United States)
CFSE	Carboxyfluoresceine succinimidyl ester
CFU	Colony-forming units
cGMP	Current good manufacturing practice
CHF	Congestive heart failure
CHL	Chinese hamster lung
CHMP	Committee for Medicinal Products for Human Use (European Union)
CHN	Carbon, hydrogen, and nitrogen
CHNS	Carbon, hydrogen, nitrogen, and sulfur
CHO	Chinese hamster ovary
CIE	Commission Internationale de l'Eclairage
CIOMS	Council for International Organizations of Medical Sciences
CJD	Creutzfeldt–Jakob disease
CL	Clearance
Clcr	Creatinine clearance (also CrCl)
Cl_{nr}	Nonrenal clearance
CL_r	Renal clearance
C_{max}	Maximum plasma concentration of a drug
CMC	Chemistry, manufacturing, and control
CMC	Carboxymethylcellulose
CMD(h)	Coordination Group for Mutual Recognition and Decentralized Procedures (human) (European Union)
CMH	Cochran-Mantel-Haenszel chi-Squared Test
C_{min}	Minimum plasma concentration of a drug
CMR	Carcinogenic, mutagenic, and reproductive toxicity
CNS	Central nervous system
C of A	Certificate of analysis
COLIPA	European Cosmetic, Toiletry, and Perfume Association
COMAR	Code of reference materials (database)
COMP	Committee for Orphan Medicinal Products (European Union)
COPE	Committee on Publication Ethics (United Kingdom)
CORT	Corticosterone
COSY	Correlation spectroscopy (NMR)
CP	Chinese pharmacopoeia
CPDB	Carcinogenic potency database

CPMP	Committee for Proprietary Medicinal Products (predecessor of CHMP, *qv*)
CQA	Critical quality attribute
CRA	Clinical research assistant/associate
CrCl	Creatinine clearance (also Clcr)
CRF	Case report form
CRM	Certified reference material
CRO	Contract Research Organization
CRS	Chemical reference substance
C_{ss}	Steady-state blood or plasma concentration after repeat dosing
$C_{ss\ max}$	Maximum blood or plasma concentration at steady state
$C_{ss\ min}$	Minimum blood or plasma concentration at steady state
CSTEE	Scientific Committee on Toxicity, Ecotoxicity, and the Environment
CT	Clinical trial
CTA	Cell transformation assay
CTD	Common technical document
c-TOC	Cumulative table of contents
CV	Curriculum vitae
CYPs	Cytochrome P_{450} enzymes
DAD	Diode array detector
DAMOS	Drug application methodology and optical storage
DCP	Decentralized procedure (European Union)
DDI	Drug–drug interaction
DEFRA	Department for Environment, Food, and Rural Affairs (United Kingdom)
DEHP	Di (2-ethylhexyl) phthalate
DEPT	Distortionless enhancement by polarization transfer (NMR)
DHPC	Dear health care professional communication (European Union)
DL	Detection limit
DLT	Digital linear tapes
DMB	Demineralized bone matrix
DMC	Data Monitoring Committee
DMF	Drug master file
DNA	Desoxyribonucleic acid
DPI	Dry powder inhaler
DPS	Description of Pharmacovigilance Systems (European Union)
DR	Delayed release
DRF	Dose-range finding
DSC	Differential scanning calorimetry
DSMB	Data and Safety Monitoring Board
DTD	Document-type definition
DTH	Delayed-type hypersensitivity
EA	Environmental assessment (United States)
EBV	Epstein–Barr virus
EC	Ethics Committee
EC	European Community
EC_{50}	Plasma concentration of drug associated with 50% of the maximal response
ECETOC	European Center for Ecotoxicity & Toxicology of Chemicals
e-CFR	Electronic Code of Federal Regulations (United States)
ECG	Electrocardiogram
eCRF	Electronic case report form
ECRIN	European clinical research infrastructures network
e-CTD	Electronic common technical document (ICH)
EDQM	European Directorate for Quality of Medicines and Healthcare
EEA	European economic area
EEC	European Economic Community (now termed European Union)
EEC	Expected environmental concentration (United States)
EEG	Electroencephalogram
EFGCP	European Forum for Good Clinical Practice
EFPIA	European Federation of Pharmaceutical Industry Associations
EGFr	Epidermal growth factor receptor
EIC	Expected introduction concentration (United States)

EIS	Environmental impact statement (United States)
EM	Extensive metabolizer
EM	Effect model
E_{max}	Model relating drug concentration to effects
EMEA	European Medicines Agency
EMRO	Eastern Mediterranean Regional Office (WHO)
EMS	Eosinophilia–myalgia syndrome
ENCePP	European Network of Centers for Pharmacoepidemiology and Pharmacovigilance
eNDA	Electronic new drug application (United States)
EPA	Environmental Protection Agency (United States)
EPAR	European Public Assessment Report (EMEA)
EPPV	Early postmarketing phase vigilance (Japan)
ERA	Environmental risk assessment (European Union)
ErbB	Epidermal growth factor receptor family of proteins
ERMS	European risk management strategy
ESTRI	Electronic Standards for Transmission of Regulatory Information (ICH)
EU	European Union
EU27	European Union of 27 Member States from January 2007
EU-RMP	European Risk Management Plan (European Union)
Eurostat	Statistical Office of the European Communities
EVCAM	European Committee on Validation of Alternative Methods
EVDAS	Eudravigilance Data Warehouse and Analysis System (European Union)
EWP	Efficacy Working Party (CHMP)
EWG	Expert Working Group (ICH)
F	Bioavailability
Fab	Fragment, antigen binding
FAS	Full analysis set
FC	Food consumption
FDA	Food and Drug Administration
FDAAA	Food and Drug Administration Amendments Act
FD&C Act	Food, Drug, and Cosmetics Act (United States)
FDAMA	Food and Drug Administration Modernization Act (United States)
FDC	Fixed-dose combination product (United States)
FISH	Fluorescence in situ hybridization
FOB	Functional observational battery
FONSI	Finding of no significant impact (United States)
FOPP	Follow-on protein product
Fpen	Market penetration factor
FR	Federal Register
fu	Unbound drug in plasma
GABA	Gammaaminobutyric acid
GAO	General Accounting Office (United States)
GC	Gas chromatography
GCC	Gulf Cooperation Countries
GCG	Global Cooperation Group (ICH)
GCP	Good clinical practice
GDNF	Glial cell-line–derived neurotrophic factor
GHTF	Global Harmonization Task Force
GI	Gastrointestinal
GIF	Graphics interchange format
GIP	Geriatric investigation plan
GLP	Good laboratory practice
GMC	General Medical Council (United Kingdom)
GMDN	Global medical device nomenclature
GMO	Genetically modified organism
GMP	Good manufacturing practice
GPMT	Guinea Pig Maximization Test
GPSP	Good Postmarketing Study Practice (Japan)
GRAS	Generally recognized as safe (FDA)
GVP	Good vigilance practice

HAS	Haute Autorité de Santé (France)
HC	Health Canada
HDPE	High-density polyethylene
HER	Human epidermal growth factor receptor
hERG	Human ether-a-go-go–related gene
HESI	Health and Environmental Science Institute
HETCOR	Heteronuclear correlation spectroscopy (NMR)
HFA	Hydrofluoroalkane
HIPAA	Health Insurance Portability and Accountability Act (United States)
HIV	Human immunodeficiency virus
HL7	Health Level Seven (United States)
HMA	Heads of Medicines Agencies (European Union)
HMBC	Heteronuclear multiple-bond spectroscopy (NMR)
HNSTD	Highest nonseverely toxic dose
HPLC	High-pressure liquid chromatography
HPRT	Hypoxanthine-guanine phosphororibyl transferase
HSA	Human serum albumin
HSQC	Heteronuclear single-quantum spectroscopy (NMR)
HT	Human toxicity
HVD	Highly variable drugs
IEC	Independent Ethics Committee
ICCVAM	Interagency Coordinating Committee on the Validation of Alternative Methods (United States)
ICE	Isolated chicken eye
ICH	International Conference on Harmonization of Technical Requirements for Registration of Pharmaceuticals for Human Use
IDMC	Independent Data Monitoring Committee
IEC	Independent Ethics Committee
IEC	International Electrotechnical Commission
IFPMA	International Federation of Pharmaceutical Manufacturers Associations
IgE	Immunoglobulin E
IgG	Immunoglobulin G
IKr	Rapid component of the delayed rectifier potassium current
IL	Interleukin
ILSI	International Life Sciences Institute
IM	Intramuscular
IMB	Irish Medicines Board
IND	Investigational New Drug (United States)
INDANA	Individual Data Analysis of Antihypertensive Interest Trials
INN	International nonproprietary name (WHO)
INSEE	Institut National de la Statistique et des Etudes Economiques (France)
IPC	In-process control
IQWiG	Institut für Qualität und Wirtschaftlichkeit im Gesundheitswesen (Germany)
IR	Infrared
IR	Immediate release
IRB	Institutional Review Board (FDA)
IRG	Isoresponse group
ISE	Integrated summary of efficacy (FDA)
ISO	International Organization for Standardization
ISS	Integrated summary of safety (FDA)
ITT	Intention to treat
IUPAC	International Union of Pure and Applied Chemistry
IV	Intravenous
IVD	In vitro diagnostic
IVIVC	In vitro–in vivo correlation
IVIVE	In vitro–in vivo extrapolation
IWGT	International Workshop on Genotoxicity Testing
JAN	Japanese accepted name
JMDN	Japanese medical device nomenclature
J-NDA	Japanese new drug application
JP	Japanese Pharmacopoeia

JPEG	Joint Photographic Experts Group
JPMA	Japan Pharmaceutical Manufacturers Association
J-PSUR	Japanese Periodic Safety Update Report
K_D	Distribution coefficient
KLH	Keyhole limpet hemocyanin
K_{ow}	Octanol–water partition coefficient
k_z	Terminal phase rate constant
LAL	Limulus amoebocyte lysate
LC	Liquid chromatography
LCR	Laryngeal cough reflex
LD_{50}	Median lethal dose
LDL	Low-density lipoprotein cholesterol
LDPE	Low-density polyethylene
LIF	Lakemedels-industriforeningen (Swedish Association of the Pharmaceutical Industry)
LLNA	Local lymph node assay
LOCF	Last observation carried forward
LSO	Local safety officer (European Union)
MA	Marketing authorization (European Union)
MAA	Marketing authorization application (European Union)
MAb/mAb	Monoclonal antibody
MABEL	Minimally anticipated biologically effect level
MAH	Marketing authorization holder (European Union)
MAP	Monophasic action potential
MDD	Maximum daily dose
MDD	Medical Device Directive (European Union)
MDEG	Medical Device Experts Group European Union)
MD5	Computing command line utility
MDI	Metered dose inhaler
MEDDEVs	Guidelines relating to the Medical Device Directives (European Union)
MedDRA	Medical Dictionary for Regulatory Activities
MEEC	Maximum expected environmental concentration (United States)
MEEKC	Microemulsion electrokinetic chromatography
MFD	Maximum feasible dose
MHLW	Ministry of Health, Labor and Welfare (Japan)
MHRA	Medicines and Healthcare products Regulatory Agency (United Kingdom)
MIR	Multiple internal reflection infrared
MKT	Mean kinetic temperature
MLA	Mouse lymphoma assay
MM	Molecular mass
MMS	Mammalian Mutagenicity Study Group
MN	Micronuclei
MNT	Micronucleus test
MOA	Mode of action
MOU	Memorandum of understanding
MP	Medicinal product
MPA	Medical Products Agency (Sweden)
MPP	*N*-methyl-4-phenyl-pyridinium cation
MPTP	*N*-methyl-4-phenyl-1,2,3,6-tetrahydropyridine
M_r	Relative molecular mass
MR	Mutual recognition (European Union)
MR	Modified release
MRI	Magnetic resonance imaging
MRP	Mutual recognition procedure (European Union)
MRT	Mean residence time
MS	Mass spectroscopy
MS	Member state (of the European Union)
MS	Master of Science (degree)
MSc	Master of Science (degree)
MSSO	Maintenance and Support Services Organization (for MedDRA)
MTD	Maximum tolerated dose

NBF	Neutral buffered formalin
NCE	New chemical entity
NCI	National Cancer Institute (United States)
NDA	New drug application
NEPA	National Environmental Policy Act (United States)
NF	National formulary (United States)
NfG	Note for guidance (European Union)
NHP	Nonhuman primate
NIBSC	National Institute for Biological Standards and Control (United Kingdom)
NICE	National Institute for Health and Clinical Excellence (United Kingdom)
NICEATM	National Toxicology Program Interagency Center for the Evaluation of Alternative Toxicology Methods (United States)
NIEHS	National Institute of Environmental Health Studies (United States)
NIH	National Institutes of Health (United States)
NIHS	National Institute of Health Sciences (Japan)
NK	Natural killer (cells)
NMP	Nanoparticle medicinal product
NMR	Nuclear magnetic resonance
NOAEL	No observed adverse effect level
NOEC	No observed effect concentration
NOEL	No observed effect level
Non-GMO	Non–genetically modified organism (European Union)
NONMEM	Software for nonlinear mixed effects modeling
NSAID	Nonsteroidal anti-inflammatory drug
NSE	Number of spared events (model)
NSRL	No significant risk level
NtA	Notice to applicants (European Union)
NTM	Numerical therapeutic model
NTR	Narrow therapeutic range
OECD	Organization for Economic Cooperation and Development
OEHHA	Office of Environmental Health Hazard Assessment (United States, California)
OHRP	Office for Human Research Protections (United States)
OIE	World Organization for Animal Health (formerly, the Office International des Epizooties)
OJ	Official Journal of the European Communities
OJL	Official Journal L series
OMP	Orphan medicinal product
OOS	Out of specification
OOT	Out of trend
OPHS	Office of Public Health and Science (United States)
ORI	Office for research integrity (United States)
OSCS	Oversulfated chondroitin sulfate
OSPAR	Convention for the Protection of the Marine Environment of the North-East Atlantic
OTC	Over the counter
OMB	Office of Management and Budget (United States)
PACE	Pressure-assisted capillary electrophoresis
PAL	Pharmaceutical Affairs Law (Japan)
PANDRH	Pan-American Network on Drug Regulatory Harmonization
PASG	Pharmaceutical Analytical Sciences Group (United Kingdom)
PASS	Postauthorization safety studies (European Union)
PBDK	Physiologically based pharmacokinetics
PBT	Persistence, bioaccumulation, toxicity assessment
PC	Personal computer
PCA	Passive cutaneous anaphylaxis
PD	Pharmacodynamics
PDE	Permitted daily exposure
PDE	Phosphorodiesterase
PDF	Portable document format (Adobe)
PDG	Pharmacopoeial Discussion Group
PDUFA	Prescription Drug User Fee Act (United States)
PE	Polyethylene
PE	Pharmacoeconomic

PEC	Predicted environmental concentration
PEG	Polyethylene glycol
PF	Pharmacopée Française
P-gp/PGP	*P*-glycoprotein
Ph. Eur.	European Pharmacopoeia
PhRMA	Pharmaceutical Research Manufacturers of America
PhVWP	Pharmacovigilance Working Party (CHMP)
PI	Package insert (United States)
PI	Propidium iodide
PIL	Package information leaflet (European Union)
PIP	Paediatric Investigation Plan (European Union)
PK	Pharmacokinetic
pK_a	Acid dissociation constant
PM	Poor metaboliser
PMA	Phorbol myristate acetate
PMDA	Pharmaceuticals and Medical Device Agency (Japan)
PMF	Plasma master file (European Union)
PMS	Postmarketing surveillance
PNEC	Predicted no-effect concentration
PNG	Portable network graphics
POC	Proof of concept
POTW	Publicly Owned Treatment Works (United States)
PRCA	Pure red cell aplasia
PREA	Pediatric Research Equity Act (United States)
PSUR	Periodic safety update report
PTO	Patent and Trademark Office (United States)
PTV	Programmable temperature vaporization
PUMA	Pediatric use marketing authorization (European Union)
PV	Pharmacovigilance
PVC	Polyvinyl chloride
PWR	Pediatric written request (FDA)
QA	Quality assurance
QbD	Quality by design
QC	Quality control
QL	Quantitation limit
QOS	Quality overall summary (CTD)
QP	Qualified person (European Union)
QPPV	Qualified Person for Pharmacovigilance (European Union)
QRD	Quality Review of Document (EMEA group)
QSAR	Quantitative Structure Activity Relationship
(Q)WBA	(Quantitative) Whole Body autoradiography
QWP	Quality Working Party (CHMP/CVMP)
R_{ac}	Accumulation ratio
R&D	Research and development
RBC	Red blood cells
rDNA	Recombinant DNA
REMS	Risk Evaluation and Minimization Strategies (FDA)
RH	Relative humidity
RHI	Regional harmonization initiative
RI	Replication index
RiskMAP	Risk Minimization Action Plan (FDA)
RMP	Risk Management Plan (European Union)
RLD	Reference-listed drug (FDA)
RM	Reference material
RMP	Risk Management Plan (European Union)
RN	Registry number
RPS	Regulated product submission
RS	Reference standard
RTF	Refusal to file (FDA)
RTG	Rainbow trout gonad (cell line)
SAD	Seasonal affective disorder

SADC	South African Development Community
SC	Subcutaneous
SCE	Sister chromatid exchange
SCENHIR	Scientific Committee on Emerging and Newly Identified Health Risks (European Union)
scFv	Single-chain variable fragments
SCGE	Single-cell gel electrophoresis
SCO	Serum cortisol level
SDO	Standards Developing Organization (United States)
SEDAMM	Soumission Électronique des Dossiers d'Autorisation de Mise sur le Marché (France)
SGML	Standard graphical markup language
SHE	Syrian hamster embryo cell transformation assay
SIT	Specific immunotherapy
SJP	Society of the Japanese Pharmacopoeia
SM	Starting material
SMF	System master file (European Union)
SmPC	Summary of product characteristics (European Union)
SOP	Standard operating procedure
SPA	Special protocol assessment (FDA)
SPC	Summary of Product Characteristics (European Union)
SPE	Solid-phase extraction
SPL	Structured product labeling (FDA)
SPME	Solid-phase microextraction
SRBC	Sheep red blood cells
STED	Summary technical document (medical devices)
STP	Sewage treatment plant
STP	Society for Toxicologic Pathology
STS	Standard toxicology studies
SUPAC	Scale-up and postapproval changes (FDA)
SVG	Scalable vector graphics
SWP	Safety Working Party (CHMP)
$t_{1/2}$	Half-life of elimination
TAMC	Total aerobic microbial count
TD_{50}	Dose rate in milligram per kilogram per day in animal species which administered chronically will halve the probability of remaining tumorless
TDAR	T-cell–dependent antibody response
TdP	Torsades de pointes
TER	Transcutaneous electrical resistance (test)
TFT	Trifluorothymidine
TG	Technical guidance
TGA	Therapeutic Goods Administration (Australia)
TGA	Thermogravimetric analysis
TGD	Technical guidance document (European Union)
TK	Toxicokinetics
TK	Thymidine kinase
TKTD	Toxicokinetic–toxicodynamic
TLC	Thin-layer chromatography
T_{max}	Time of maximum plasma concentration
TNF	Tumor necrosis factor
TNSS	Total nasal symptom score
TOC	Table of contents
TOCSY	Total correlation spectroscopy (NMR)
TPD	Therapeutic Products Directorate (Canada)
TRI	Triangulation, reverse-use dependency, and instability
TRP	Transient receptor potential
TRS	Technical report series (WHO)
TSE	Transmissible spongiform encephalopathy
TTC	Threshold of toxicological concern
TTP	Therapy target population
TYMC	Total combined yeasts/moulds count
UCSD	University of California, San Diego

UDS	Unscheduled DNA synthesis
ULN	Upper limit of normal
US	United States of America
USAN	U.S. approved name
USP	United States Pharmacopeia
UV	Ultraviolet
V	Volume of distribution
VAMF	Vaccine antigen master file (European Union)
VEGF	Vascular endothelial growth factor
V_{max}	Maximum velocity of the action potential upstroke
V_{ss}	Volume of distribution at steady state
VSD	Virtual safe dose
WBA	Whole-body autoradiography
WBC	White blood cell
WCI	World Class International Consulting Ltd.
WHO	World Health Organization
WMA	World Medical Association
WoE	Weight of evidence
XML	Extensible markup language
XRD	X-ray diffraction
XRPD	X-ray powder diffraction

ABBREVIATIONS FOR COUNTRY CODES FOR EUROPEAN UNION MEMBER STATES (EU27)

The European Union Publications Office Institutional Style Guide recommends abbreviations for country codes for the European Union Member States. These follow ISO 3166 except for Greece and the United Kingdom for which the abbreviations EL and United Kingdom are used.

Short name in English	Country code
Austria	AT
Belgium	BE
Bulgaria	BG
Cyprus	CY
Czech Republic	CZ
Germany	DE
Denmark	DK
Estonia	EE
Greece	EL
Spain	ES
Finland	FI
France	FR
Hungary	HU
Ireland	IE
Italy	IT
Lithuania	LT
Luxembourg	LU
Latvia	LV
Malta	MT
Netherlands	NL
Poland	PL
Portugal	PT
Romania	RO
Sweden	SE
Slovenia	SI
Slovakia	SK
United Kingdom	UK

Annex 2: List of Key Technical and Regulatory Information Sources

Anthony C. Cartwright
Pharmaceutical Regulatory Consultant, Harpenden, U.K.

Brian R. Matthews
Alcon Laboratories, Hemel Hempstead, U.K.

LIST OF KEY TECHNICAL AND REGULATORY INFORMATION SOURCES—WEB SITES AND OTHER PUBLISHED INFORMATION

The editors and the publishers are providing the following references to Web sites and publications as a service to readers and their inclusion does not mean that we are offering an opinion on their value.

Web Sites (All Accessed November 24, 2008)

1. International Conference on Harmonization (ICH): http://www.ich.org/cache/compo/276-254-1.html
   The ICH Web site gives access to adopted and draft ICH guidelines on Quality, Safety, Efficacy and Multidisciplinary topics; ICH conferences; the Global Cooperation Group and MedDRA. The complete guidelines on the Common Technical Document (CTD) and electronic CTD are also available here.
2. European Union
   European Commission, Directorate-General Enterprise, Pharmaceuticals: http://ec.europa.eu/enterprise/pharmaceuticals/index_en.htm
   Eudralex Rules Governing Medicinal Products in the European Union, Legislation, and Guidelines: http://ec.europa.eu/enterprise/pharmaceuticals/eudralex/eudralex_en.htm
   Community Register of Products Approved Through the Centralized Procedure: http://ec.europa.eu/enterprise/pharmaceuticals/register/register_en.htm
   Pharmaceuticals Unit of the European Commission: http://ec.europa.eu/enterprise/pharmaceuticals/commit/profile_en.htm
   News on Pharmaceuticals: http://ec.europa.eu/enterprise/pharmaceuticals/pharmacos/new_en.htm
   European Medicines Agency (EMEA): http://www.emea.europa.eu/
   Scientific Guidelines for Human Products: http://www.emea.europa.eu/htms/human/humanguidelines/background.htm
   EMEA Organogram: http://www.emea.europa.eu/htms/aboutus/organigramme.htm
   Committee for Human Medicinal Products (CHMP): http://www.emea.europa.eu/htms/general/contacts/CHMP/CHMP.html
   What's New: http://www.emea.europa.eu/whatsnewp.htm
   European Directorate for the Quality of Medicines and Healthcare (EDQM): http://www.edqm.eu/site/page_628.php
   Coordinating Group for Mutual Recognition and Decentralized Procedures (human) (CMD(h): http://www.hma.eu/cmdh.html
   Heads of Medicines Agencies (HMA): http://www.hma.eu/human_heads.html
   European Federation of Pharmaceutical Industry Associations (EFPIA): http://www.efpia.org/Content/Default.asp?
   National Agencies in the European Union Austria (AGES): http://www.ages.at/
   Belgium (FAMHP): https://portal.health.fgov.be/portal/page?_pageid=56,12590472&_dad=portal&_schema=PORTAL
   Bulgaria (BDA): http://www.bda.bg/

Cyprus: http://www.moh.gov.cy
Czech Republic (SUKL): http://www.sukl.cz/healthcare
Denmark (DMA): http://www.dkma.dk/1024/visUKLSForside.asp?artikelID=728
Estonia (SAM): http://www.sam.ee/
Finland (NAM): http://www.nam.fi/
France (AFSSAPS): http://afssaps.sante.fr/
Germany (BfArM): http://www.bfarm.de/cln_030/nn_424276/EN/Home/homepage_node.html__nnn=true
Germany (PEI): http://www.pei.de/EN/home/node-en.html?__nnn=true
Greece (EOF): http://eof1.eof.gr/eof_en/enhome.html
Hungary (OGYI): http://www.ogyi.hu/main_page/
Iceland (IMCA): http://www.imca.is/
Ireland (IMB): http://www.imb.ie/
Italy (AIFA): http://www.agenziafarmaco.it/section8983.html
Latvia (ZVA): http://www.zva.gov.lv/
Liechtenstein: http://www.llv.li/
Lithuania (VVKT): http://www.vvkt.lt/
Luxembourg (Ministère de la Santé): http://www.ms.etat.lu/
Malta (Medicines Authority): http://www.medicinesauthority.gov.mt/
Netherlands (MEB): http://www.cbg-meb.nl/cbg/en/default.htm
Norway (NoMA): http://www.legemiddelverket.no/
Poland (URPL): http://www.bip.urpl.gov.pl/
Portugal (INFARMED): http://www.infarmed.pt/portal/page/portal/INFARMED
Romania (ANM): http://www.anm.ro/home_anm.asp
Slovakia (SUKL): http://www.sukl.sk/sk
Slovenia (JAZMP): http://www.jazmp.si/
Spain (AEMPS): http://www.agemed.es/
Sweden (MPA): http://www.lakemedelsverket.se/Tpl/StartPage____395.aspx
United Kingdom (MHRA): http://www.mhra.gov.uk/index.htm

3. Japan
   Ministry of Health Labor and Welfare (MHLW): http://www.mhlw.go.jp/english/index.html#
   Pharmaceutical and Medical Devices Agency (PMDA): http://www.pmda.go.jp/english/index.html
   National Institute of Health Sciences (NIHS): http://www.nihs.go.jp/english/usefulinfo/drugs.html
   Information in English on Japanese Regulatory Affairs: http://www.nihs.go.jp/mhlw/jouhou/yakuji/yakuji-e0703.pdf
   Japanese Pharmaceutical Manufacturers Association (JPMA): http://www.jpma.or.jp/english/
4. United States
   Federal Register: http://www.gpoaccess.gov/fr/
   Food and Drugs Administration (FDA): http://www.fda.gov/default.htm
   Center for Drug Evaluation and Research (CDER): http://www.fda.gov/cder/
   CDER Regulatory and Scientific Guidances: http://www.fda.gov/cder/regulatory/default.htm#Regulatory%20&%20Scientific%20Guidances
   Center for Biologics Evaluation and Research (CBER): http://www.fda.gov/cber/index.html
   Center for Devices and Radiological Health (CDRH): http://www.fda.gov/cdrh/index.html
   Drugs@FDA (Drug Product Approvals Information and Assessment Reports): http://www.accessdata.fda.gov/scripts/cder/drugsatfda/
   Pharmaceutical Research Manufacturers of America (PhRMA): http://www.phrma.org/
5. Other Regulatory Agencies
   Health Canada, Drugs and Health Products: http://www.hc-sc.gc.ca/dhp-mps/index-eng.php
   Australia, Therapeutic Goods Administration (TGA): http://www.tga.gov.au/
6. World Health Organization: http://www.who.int/en/
   Pharmaceutical Products: http://www.who.int/topics/pharmaceutical_products/en/
   Norms and Standards: Quality, Safety, and Efficacy of medicines: http://www.who.int/medicines/areas/quality_safety/en/index.html

International Nonproprietary Names (INNs): http://www.who.int/medicines/services/inn/en/index.html
Expert Committee on Specifications for Pharmaceutical Preparations: http://www.who.int/medicines/services/expertcommittees/pharmprep/en/index.html

7. Pharmacopoeias
   European Pharmacopoeia (Ph. Eur.): http://www.edqm.eu/site/News_and_General_Information-43.html
   Legal Status and Background: http://www.edqm.eu/site/Legal_Status_Background-50.html
   British Pharmacopoeia (BP): http://www.pharmacopoeia.co.uk/
   Japanese Pharmacopoeia: http://jpdb.nihs.go.jp/jp14e/
   Pharmacopée Française (PF): http://agmed.sante.gouv.fr/htm/pharma/chap1.htm
   Deutsche Arzneibuch: http://www.bfarm.de/cln_030/nn_1199278/DE/Arzneimittel/2_zulassung/zulVerfahren/azBuch/arzneibuch-node.html__nnn=true
   United States Pharmacopeia (USP): http://www.usp.org/
   International Pharmacopoeia (WHO): http://www.who.int/medicines/publications/pharmacopoeia/en/index.html
8. Organization for Economic Cooperation and Development (OECD): http://www.oecd.org/home/0,2987,en_2649_201185_1_1_1_1_1,00.html
   Chemicals Testing Guidelines: http://www.oecd.org/document/40/0,3343,en_2649_34377_37051368_1_1_1_1,00.html
9. International Standards Organization (ISO): http://www.iso.org/iso/home.htm
10. Informa: http://www.informa.com/
    Books: http://www.informa.com/media_type/books
    CRC Press, Pharmaceutical Science Books: http://www.crcpress.co.uk/shopping_cart/categories/categories_products.asp?parent_id=101
11. International Excipients Council (IPEC)
    International Excipients Council (IPEC) Americas: http://www.ipecamericas.org/
    International Excipients Council (IPEC) Europe: http://www.ipec-europe.org/
    Japan Pharmaceutical Excipients Council (JPEC): http://www.jpec.gr.jp/english/
12. The Pharmaceutical Quality Group (forum for support of quality professionals within the pharmaceutical industry): http://www.pqg.org/index.php
13. International Pharmaceutical Federation (FIP): http://www.fip.org/www/
14. Clinical trials register (register of United States federally supported and privately supported clinical trials around the world, comprising over 65,000 trials in 159 countries): http://clinicaltrials.gov/
15. National Library of Medicine (United States): http://www.nlm.nih.gov/

Published Information

1. Regulatory Affairs Journal (RAJ) Pharma, 12 issues per year, ISSN 1740-1240, publisher: Informa Healthcare
2. Regulatory Affairs Journal (RAJ) Devices, 6 issues per year, ISSN 0969 4129, publisher: Informa Healthcare
3. SCRIP World Pharmaceutical News, daily email alerts and weekly newsletter, ISSN 0143 7690, publisher: Informa Healthcare
4. Regulatory Rapporteur, monthly journal from The Organization for Professionals in Regulatory Affairs (TOPRA), ISSN 1742-8955
5. Drug Information Journal, publication of the Drug Information Association, monthly journal, ISSN 0092-8615
6. Regulatory Focus, publication of the Regulatory Affairs Professionals Society, monthly journal
7. Pharmeuropa, publication of the European Pharmacopoeia, 4 issues per year, EDQM
8. Pharmeuropa Bio, publication of the European Pharmacopoeia, published periodically as a single issue, EDQM
9. Pharmeuropa Scientific Notes, scientific news linked to the work of the European Pharmacopoeia, published periodically as a single issue, EDQM
10. Pharmacopeia Forum, publication of the United States Pharmacopeia, bimonthly

Index